DISCOVERING STATISTICS USING
IBM® SPSS® STATISTICS

D0207106

CATISFIED CUSTOMERS

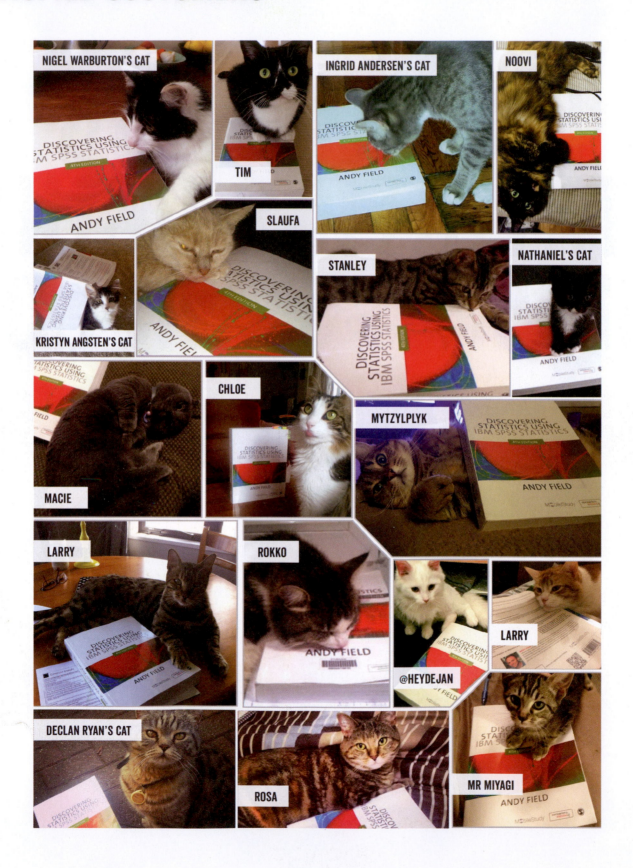

DISCOVERING STATISTICS USING IBM® SPSS® STATISTICS

NORTH AMERICAN EDITION

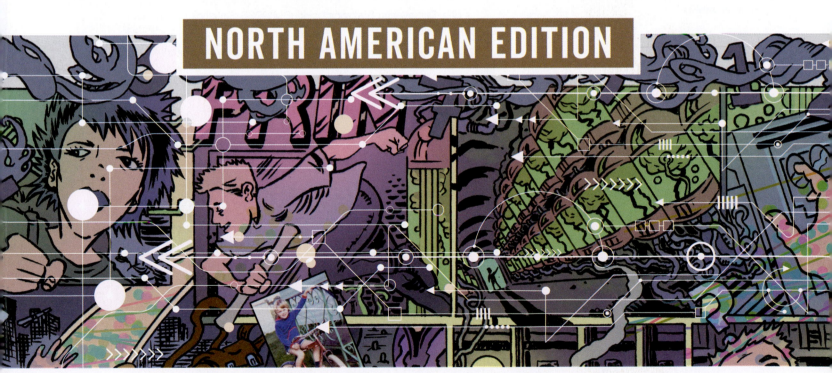

ANDY FIELD

Los Angeles | London | New Delhi
Singapore | Washington DC | Melbourne

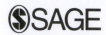

FOR INFORMATION:

SAGE Publications, Inc.
2455 Teller Road
Thousand Oaks, California 91320
E-mail: order@sagepub.com

SAGE Publications Ltd.
1 Oliver's Yard
55 City Road
London EC1Y 1SP
United Kingdom

SAGE Publications India Pvt. Ltd.
B 1/I 1 Mohan Cooperative Industrial Area
Mathura Road, New Delhi 110 044
India

SAGE Publications Asia-Pacific Pte. Ltd.
3 Church Street
#10-04 Samsung Hub
Singapore 049483

© Andy Field 2018

First edition published 2000
Second edition published 2005
Third edition published 2009. Reprinted 2009, 2010, 2011 (twice), 2012
Fourth edition published 2013. Reprinted 2014 (twice), 2015, 2016, 2017

All rights reserved. No part of this book may be reproduced or utilized in any form or by any means, electronic or mechanical, including photocopying, recording, or by any information storage and retrieval system, without permission in writing from the publisher.

Throughout the book, screenshots and images from IBM® SPSS® Statistics software ('SPSS') are reprinted courtesy of International Business Machines Corporation, © International Business Machines Corporation. SPSS Inc. was acquired by IBM in October 2009.

Printed in Canada

Library of Congress Cataloging-in-Publication Data

Name: Field, Andy, author

Title: Discovering Statistics Using IBM SPSS Statistics / Andy Field, University of Sussex

Description: Fifth edition | SAGE [2018] | Includes bibliographical references and index

Identifiers LCCN 2017954637 | ISBN 9781526440273; 9781526436566 (pbk)

Editor: Jai Seaman
Development editors: Sarah Turpie & Nina Smith
Assistant editor, digital: Chloe Statham
Production editor: Ian Antcliff
Copyeditor: Richard Leigh
Indexer: David Rudeforth
Marketing manager: Ben Griffin-Sherwood
Cover design: Wendy Scott
Typeset by: C&M Digitals (P) Ltd, Chennai, India
Printed in Canada by: TC Transcontinental Printing

Illustration: James Iles

CONTENTS

EXTENDED CONTENTS

PREFACE

Karma Police, arrest this man, he talks in maths, he buzzes like a fridge, he's like a detuned radio
Radiohead, 'Karma Police', *OK Computer* (1997)

Introduction

Many behavioral and social science students (and researchers for that matter) despise statistics. Most of us have a non-mathematical background, which makes understanding complex statistical equations very difficult. Nevertheless, the evil goat-warriors of Satan force our non-mathematical brains to apply themselves to what is the very complex task of becoming a statistics expert. The end result, as you might expect, can be quite messy. The one weapon that we have is the computer, which allows us to neatly circumvent the considerable disability of not understanding mathematics. Computer programs such as IBM SPSS Statistics, SAS, R, JASP and the like provide an opportunity to teach statistics at a conceptual level without getting too bogged down in equations. The computer to a goat-warrior of Satan is like catnip to a cat: it makes them rub their heads along the ground and purr and dribble ceaselessly. The only downside of the computer is that it makes it really easy to make a complete idiot of yourself if you don't understand what you're doing. Using a computer without any statistical knowledge at all can be a dangerous thing. Hence this book.

My first aim is to strike a balance between theory and practice: I want to use the computer as a tool for teaching statistical concepts in the hope that you will gain a better understanding of both theory and practice. If you want theory and you like equations then there are certainly more technical books. However, if you want a stats book that also discusses digital rectal stimulation, then you have just spent your money wisely.

Too many books create the impression that there is a 'right' and 'wrong' way to do statistics. Data analysis is more subjective than is often made out. Therefore, although I make recommendations, within the limits imposed by the senseless destruction of rainforests, I hope to give you enough background in theory to enable you to make your own decisions about how best to conduct your analysis.

A second (ridiculously ambitious) aim is to make this the only statistics book that you'll ever need to buy (sort of). It's a book that I hope will become your friend from your first year at university right through to your professorship. The start of the book is aimed at first-year undergraduates (Chapters 1–10), and then we move onto second-year undergraduate-level material (Chapters 6, 9 and 11–16) before a dramatic climax that should keep postgraduates tickled (Chapters 17–21). There should be something for everyone in each chapter, and to help you gauge the difficulty of material, I flag the level of each section within each chapter (more on that later).

My final and most important aim is to make the learning process fun. I have a sticky history with maths. This extract is from my school report at the age of 11:

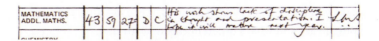

The '27=' in the report is to say that I came equal 27th with another student out of a class of 29. That's pretty much bottom of the class. The 43 is my exam mark as a percentage. Oh dear. Four years later (at 15), this was my school report:

NAME ..Andrew Field.............. FORM ..4Q.. SUBJECT ..Mathematics..

Andrew's progress in Mathematics has been remarkable. From being a weaker candidate who lacked confidence he has developed into a budding Mathematician. He should achieve a good grade.

Date ..27/6/88..   B.A. Create .....Subject Teacher

	EXAM	
	ATTAINMENT	
	EFFORT	

The catalyst of this remarkable change was a good teacher: my brother, Paul. I owe my life as an academic to Paul's ability to teach me stuff in an engaging way—something my maths teachers failed to do. Paul's a great teacher because he cares about bringing out the best in people, and he was able to make things interesting and relevant to me. Everyone should have a brother Paul to teach them stuff when they're throwing their maths book at their bedroom wall, and I will attempt to be yours.

I strongly believe that people appreciate the human touch, and so I inject a lot of my own personality and sense of humor (or lack of) into *Discovering Statistics Using ...* books. Many of the examples in this book, although inspired by some of the craziness that you find in the real world, are designed to reflect topics that play on the minds of the average student (i.e., sex, drugs, rock and roll, celebrity, people doing crazy stuff). There are also some examples that are there simply because they made me laugh. So, the examples are light-hearted (some have said 'smutty', but I prefer 'light-hearted') and by the end, for better or worse, I think you will have some idea of what goes on in my head on a daily basis. I apologize to those who think it's crass, hate it, or think that I'm undermining the seriousness of science, but, come on, what's not funny about a man putting an eel up his anus?

I never believe that I meet my aims, but previous editions have certainly been popular. I enjoy the rare luxury of having complete strangers emailing me to tell me how wonderful I am. (Admittedly, there are also emails accusing me of all sorts of unpleasant things, but I've usually got over them after a couple of months.) With every new edition, I fear that the changes I make will ruin my previous hard work. Let's see what you're going to get and what's different this time around.

What do you get for your money?

This book takes you on a journey (I try my best to make it a pleasant one) not just of statistics but also of the weird and wonderful contents of the world and my brain. It's full of daft examples, bad jokes, and smut. Aside from the smut, I have been forced, reluctantly, to include some academic content. It contains everything I know about statistics (actually, more than I know . . .). It also has these features:

- **Everything you'll ever need to know**: I want this book to be good value for money, so it guides you from complete ignorance (Chapter 1 tells you the basics of doing research) to being an expert in multilevel linear modeling (Chapter 21). Of course, no book can contain everything, but I think this one has a fair crack. It's pretty good for developing your biceps also.
- **Stupid faces**: You'll notice that the book is riddled with 'characters', some of them my own. You can find out more about the pedagogic function of these 'characters' in the next section.
- **Data sets**: There are about 132 data files associated with this book on the companion website. Not unusual in itself for a statistics book, but my data sets contain more sperm (not literally) than other books. I'll let you judge for yourself whether this is a good thing.
- **My life story**: Each chapter is book-ended by a chronological story from my life. Does this help you to learn about statistics? Probably not, but it might provide light relief between chapters.
- **SPSS tips**: SPSS does confusing things sometimes. In each chapter, there are boxes containing tips, hints and pitfalls related to SPSS.
- **Self-test questions**: Given how much students hate tests, I thought that the best way to commit commercial suicide was to liberally scatter tests throughout each chapter. These range from simple questions to test what you have just learned to going back to a technique that you read about several chapters before and applying it in a new context. All of these questions have answers so that you can check on your progress.
- **Online resources**: The website contains an insane amount of additional material, which no one reads, but it is described in the section about the online resources so that you know what you're ignoring.
- **Digital stimulation**: No, not the aforementioned type of digital stimulation, but brain stimulation. Many of the features on the website will be accessible from tablets and smartphones, so that when you're bored in the cinema you can read about the fascinating world of heteroscedasticity instead.
- **Reporting your analysis**: Every chapter has a guide to writing up your analysis. How one writes up an analysis varies a bit from one discipline to another, but my guides should get you heading in the right direction.
- **Glossary**: Writing the glossary was so horribly painful that it made me stick a vacuum cleaner into my ear to suck out my own brain. You can find my brain in the bottom of the vacuum cleaner in my house.
- **Real-world data**: Students like to have 'real data' to play with. The trouble is that real research can be quite boring. I trawled the world for examples of research on really fascinating topics (in my opinion). I then stalked the authors of the research until they gave me their data. Every chapter has a real research example.

What do you get that you didn't get last time?

I suppose if you have spent your hard-earned money on the previous edition, it's reasonable that you want a good reason to spend more of your hard-earned money on this edition. In some respects, it's hard to quantify all of the changes in a list: I'm a better writer than I was five years ago, so there is a lot of me rewriting things because I think I can do it better than before. I spent 6 months solidly on the updates, so, suffice it to say that a lot has changed; but anything you might have liked about the previous edition probably hasn't changed:

- **IBM SPSS compliance**: This edition was written using version 25 of IBM SPSS Statistics. IBM releases new editions of SPSS Statistics more often than I bring out new editions of this book, so, depending on when you buy the book, it may not reflect the latest version. This shouldn't worry you because the procedures covered in this book are unlikely to be affected (see Section 4.12).
- **New! Chapter**: In the past four years the open science movement has gained a lot of momentum. Chapter 3 is new and discusses issues relevant to this movement such as p-hacking, HARKing, researcher degrees of freedom, and pre-registration of research. It also has an introduction to Bayesian statistics.
- **New! Bayes**: Statistical times are a-changing, and it's more common than it was four years ago to encounter Bayesian methods in social science research. IBM SPSS Statistics doesn't really do Bayesian estimation, but you can implement Bayes factors. Several chapters now include sections that show how to obtain and interpret Bayes factors. Chapter 3 also explains what a Bayes factor is.
- **New! Robust methods**: Statistical times are a-changing . . . oh, hang on, I just said that. Although IBM SPSS Statistics does bootstrap (if you have the premium version), there are a bunch of statistics based on trimmed data that are available in R. I have included several sections on robust tests and syntax to do them (using R).
- **New! Pointless fiction**: Having got quite into writing a statistics textbook in the form of a fictional narrative (*An Adventure in Statistics*) I staved off boredom by fleshing out Brian and Jane's story (which goes with the diagrammatic summaries at the end of each chapter). Of course, it is utterly pointless, but maybe someone will enjoy the break from the stats.
- **New! Misconceptions**: Since the last edition my cat of 20 years died, so I needed to give him a more spiritual role. He has become the Correcting Cat, and he needed a foil, so I created the Misconception Mutt, who has a lot of common

misconceptions about statistics. So, the mutt (based on my cocker spaniel) gets stuff wrong and the cat appears from the spiritual ether to correct him. All of which is an overly elaborate way to point out some common misconceptions.
- **New-ish! The linear model theme**: In the past couple of editions of this book I've been keen to scaffold the content on the linear model to focus on the commonalities between models traditionally labeled as regression, ANOVA, ANCOVA, t-tests, etc. I've always been mindful of trying not to alienate teachers who are used to the historical labels, but I have again cranked up a level the general linear model theme.
- **New-ish! Characters**: I loved working with James Iles on *An Adventure in Statistics* so much that I worked with him to create new versions of the characters in the book (and other design features like their boxes). They look awesome.

Every chapter had a thorough edit/rewrite, I've redone all of the figures, and obviously updated the SPSS Statistics screenshots and output. Here is a chapter-by-chapter rundown of the more substantial changes:

- **Chapter 1** (**Doing research**): I changed the way I discuss hypotheses. I changed my suicide example to be about memes.
- **Chapter 2** (**Statistical theory**): I restructured this chapter around the acronym of SPINE,[1] so you'll notice that subheadings have changed and so on. The content is all there, just rewritten and reorganized into a better narrative. I've expanded my description of null hypothesis significance testing (NHST).
- **Chapter 3** (**Current thinking in statistics**): This chapter is completely new. It co-opts some of the critique of NHST that used to be in Chapter 2 but moves this into a discussion of open science, p-hacking, HARKing, researcher degrees of freedom, pre-registration, and ultimately Bayesian statistics (primarily Bayes factors).
- **Chapter 4** (**IBM SPSS Statistics**): Obviously reflects changes to SPSS Statistics since the previous edition. There's a new section on 'extending' SPSS Statistics that covers installing the PROCESS tool, the *Essentials for R* plugin and *WRS2* package (for robust tests).
- **Chapter 5** (**Graphs**): No substantial changes, I just tweaked a few examples.
- **Chapter 6** (**Assumptions**): The content is more or less as it was. I have a much stronger steer away from tests of normality and homogeneity (although I still cover them) because I now offer some robust alternatives to common tests.
- **Chapter 7** (**Nonparametric models**): No substantial changes to content.
- **Chapter 8** (**Correlation**): I completely rewrote the section on partial correlations.

---

1 Thanks to a colleague, Jennifer Mankin, for distracting me from the acronym that more immediately sprang to my childish mind.

- **Chapter 9** (**The linear model**): I restructured this chapter a bit and wrote new sections on robust regression and Bayesian regression.
- **Chapter 10** (*t*-tests): I did an overhaul of the theory section to tie it in more with the linear model theme. I wrote new sections on robust and Bayesian tests of two means.
- **Chapter 11** (**Mediation and moderation**): No substantial changes to content.
- **Chapters 12–13** (**GLM 1–2**): I changed the main example to be about puppy therapy. I thought that the Viagra example was a bit dated, and I needed an excuse to get some photos of my spaniel into the book. This was the perfect solution. I wrote new sections on robust and Bayesian (Chapter 12 only) variants of these models.
- **Chapter 14** (**GLM 3**): I tweaked the example—it's still about the beer-goggles effect, but I linked it to some real research so that the findings now reflect some actual science that's been

done. I added sections on robust and Bayesian variants of models for factorial designs.
- **Chapters 15–16** (**GLM 4–5**): I added some theory to Chapter 14 to link it more closely to the linear model (and to the content of Chapter 21). I give a clearer steer now to ignoring Mauchly's test and routinely applying a correction to *F* (although, if you happen to like Mauchly's test, I doubt that the change is dramatic enough to upset you). I added sections on robust variants of models for repeated-measures designs. I added some stuff on pivoting trays in tables. I tweaked the example in Chapter 16 a bit so that it doesn't compare males and females but instead links to some real research on dating strategies.
- For **Chapters 17** (**MANOVA**); **18** (**Factor analysis**); **19** (**Categorical data**); **20** (**Logistic regression**); and **21** (**Multilevel models**) there are no major changes, except to improve the structure in Chapter 19.

Goodbye

The first edition of this book was the result of two years (give or take a few weeks to write up my PhD) of trying to write a statistics book that I would enjoy reading. With each new edition I try not just to make superficial changes but also to rewrite and improve everything (one of the problems with getting older is that you look back at your past work and think you can do things better). This fifth edition is the culmination of about seven years of full-time work (on top of my actual job). This book has consumed the last 20 years or so of my life, and each time I get a nice email from someone who found it useful, I am reminded that it is the most useful thing I'll ever do with my academic life. It began and continues to be a labour of love. It still isn't perfect, and I still love to have feedback (good or bad) from the people who matter most: you.

Andy

 www.facebook.com/profandyfield

 @ProfAndyField

 www.youtube.com/user/ProfAndyField

 www.discoveringstatistics.com/category/blog/

HOW TO USE THIS BOOK

When the publishers asked me to write a section on 'How to use this book' it was tempting to write 'Buy a large bottle of Olay anti-wrinkle cream (which you'll need to fend off the effects of ageing while you read), find a comfy chair, sit down, fold back the front cover, begin reading and stop when you reach the back cover.' However, I think they wanted something more useful. 🙂

What background knowledge do I need?

In essence, I assume that you know nothing about statistics, but that you have a very basic grasp of computers (I won't be telling you how to switch them on, for example) and maths (although I have included a quick revision of some very basic concepts).

Do the chapters get more difficult as I go through the book?

Yes, more or less: Chapters 1–10 are first–year degree level, Chapters 9–16 move into second-year degree level, and Chapters 17–21 discuss more technical topics. However, my aim is to tell a statistical story rather than worry about what level a topic is at. Many books teach different tests in isolation and never really give you a grasp of the similarities between them; this, I think, creates an unnecessary mystery. Most of the tests in this book are the same thing expressed in slightly different ways. I want the book to tell this story, and I see it as consisting of seven parts:

- Part 1 (Doing research and introducing linear models): Chapters 1–4.
- Part 2: (Exploring data): Chapters 5–7.
- Part 3: (Linear models with continuous predictors): Chapters 8–9.
- Part 4: (Linear models with continuous or categorical predictors): Chapters 10–16.
- Part 5: (Linear models with multiple outcomes): Chapter 17–18.
- Part 6 (Linear models with categorical outcomes): Chapters 19–20.
- Part 7 (Linear models with hierarchical data structures): Chapter 21.

This structure might help you to see the method in my madness. If not, to help you on your journey, I've coded each section with an icon. These icons are designed to give you an idea of the difficulty of the section. It doesn't mean that you can skip the sections, but it will let you know whether a section is at about your level, or whether it's going to push you. It's based on a wonderful categorization system using the letter 'I':

Introductory, which I hope means that everyone will understand these sections. These are for people just starting their undergraduate courses.

Intermediate. Anyone with a bit of background in statistics should be able to get to grips with these sections. They are aimed at people who are perhaps in the second year of their degree, but they can still be quite challenging in places.

In at the deep end. These topics are difficult. I'd expect final-year undergraduates and recent postgraduate students to be able to tackle these sections.

Incinerate your brain. These are difficult topics. I would expect these sections to be challenging for undergraduates, but postgraduates with a reasonable background in research methods shouldn't find them too much of a problem.

Why do I keep seeing silly faces everywhere?

Brian Hemorrhage: Brian is a really nice guy, and he has a massive crush on Jane Superbrain. He's seen her around the university campus carrying her jars of brains (see below). Whenever he sees her, he gets a knot in his stomach and he imagines slipping a ring onto her finger on a beach in Hawaii, as their friends and family watch through gooey eyes. Jane never even notices him; this makes him very sad. His friends have told him that the only way she'll marry him is if he becomes a statistics genius (and changes his surname). Therefore, he's on a mission to learn statistics. It's his last hope of impressing Jane, settling down and living happily ever after. At the moment he knows nothing, but he's about to go on a journey that will take him from statistically challenged to a genius, in 808 pages. Along his journey he pops up and asks questions, and at the end of each chapter he flaunts his newly found knowledge to Jane in the hope that she'll go on a date with him.

Confusius: The great philosopher Confucius had a lesser-known brother called Confusius. Jealous of his brother's great wisdom and modesty, Confusius vowed to bring confusion to the world. To this end, he built the Confusion machine. He puts statistical terms into it, and out of it come different names for the same concept. When you see Confusius he will be alerting you to statistical terms that mean the same thing.

Correcting Cat: This cat lives in the ether and appears to taunt the Misconception Mutt by correcting his misconceptions. He also appears when I want to make a bad cat-related pun. He exists in loving memory of my own ginger cat who, after 20 years as the star of this book, sadly passed away, which he promised me he'd never do. You can't trust a cat.

Cramming Sam: Samantha thinks that statistics is a boring waste of time. She just wants to pass her exam and forget that she ever had to know anything about normal distributions. She appears and gives you a summary of the key points that you need to know. If, like Samantha, you're cramming for an exam, she will tell you the essential information to save you having to trawl through hundreds of pages of my drivel.

Jane Superbrain: Jane is the cleverest person in the universe. She has acquired a vast statistical knowledge, but no one knows how. She is an enigma, an outcast, and a mystery. Brian has a massive crush on her. Jane appears to tell you advanced things that are a bit tangential to the main text. Can Brian win his way into her heart? You'll have to read and find out.

Labcoat Leni: Leni is a budding young scientist and he's fascinated by real research. He says, 'Andy, I like an example about using an eel as a cure for constipation as much as the next guy, but all of your data are made up. We need some real examples, buddy!' Leni walked the globe, a lone data warrior in a thankless quest for real data. When you see Leni you know that you will get some real data, from a real research study, to analyze.

Misconception Mutt: Since the last edition, I acquired a spaniel called Ramsey. I needed some way to get him into the book, so here he is as the Misconception Mutt. He follows his owner to statistics lectures and finds himself learning about stats. Sometimes, he gets things wrong, though, and, when he does, something very strange happens. A ginger cat materializes out of nowhere and corrects him.

Oditi's Lantern: Oditi believes that the secret to life is hidden in numbers and that only by large-scale analysis of those numbers shall the secrets be found. He didn't have time to enter, analyze, and interpret all of the data in the world, so he established the cult of undiscovered numerical truths. Working on the principle that if you gave a million monkeys typewriters, one of them would recreate Shakespeare, members of the cult sit at their computers crunching numbers in the hope that one of them will unearth the hidden meaning of life. To help his cult Oditi has set up a visual vortex called 'Oditi's Lantern'. When Oditi appears, it is to implore you to stare into the lantern, which basically means that there is a video tutorial to guide you.

Oliver Twisted: With apologies to Charles Dickens, Oliver, like the more famous fictional London urchin, asks, 'Please, Sir, can I have some more?' Unlike Master Twist, though, Master Twisted always wants more statistics information. Who wouldn't? Let us not be the ones to disappoint a young, dirty, slightly smelly boy who dines on gruel. When Oliver appears, he's telling you that there is additional information on the companion website. (It took a long time to write, so, someone, please, actually read it.)

Satan's Personal Statistics Slave: Satan is a busy boy—he has all of the lost souls to torture in hell, then there are the fires to keep fuelled, not to mention organizing enough carnage on the planet's surface to keep Norwegian black metal bands inspired. Like many of us, this leaves little time for him to analyze data, and this makes him very sad. So, he has his own personal slave, who, also like some of us, spends all day dressed in a gimp mask and tight leather pants in front of IBM SPSS Statistics, analyzing Satan's data. Consequently, he knows a thing or two about SPSS, and when Satan's busy spanking a goat, he pops up in a box with SPSS tips.

Smart Alex: Alex was aptly named because she's, like, super smart. She likes teaching people, and her hobby is posing people questions so that she can explain the answers to them. Alex appears at the end of each chapter to pose you some questions. Her answers are on the companion website.

What online resources do you get with the book?

I've put a cornucopia of additional funk on that worldwide interweb thing. To enter my world of delights, go to https://edge.sagepub.com/field5e. The website contains resources for students and lecturers alike, with additional content from some of the characters from the book:

- **Testbank**: There is a (hopefully) comprehensive testbank of multiple-choice and numeracy-based/algorithmic questions for your instructors to use. It comes as a file that you can upload into your instution's online teaching system. Furthermore, there are additional testbanks of multiple-choice questions for your instructors.
- **Data files**: You need data files to work through the examples in the book, and they are on the website. We did this to force you to go there and, once you're there, SAGE will flash up subliminal messages that make you buy more of their books.
- **Resources for other subject areas**: I am a psychologist and, although I tend to base my examples around the weird and wonderful, I do have a nasty habit of resorting to psychology when I don't have any better ideas. My publishers have recruited some non-psychologists to provide data files and an instructor's testbank of multiple-choice questions for those studying or teaching in business and management, education, sport sciences and health sciences. You have no idea how happy I am that I didn't have to write those.
- **YouTube**: Whenever you see Oditi in the book it means that there is a screencast to accompany the chapter. These are hosted on my YouTube channel (www.youtube.com/user/ProfAndyField), which I have amusingly called μ-Tube (see what I did there?).
- **Self-assessment multiple-choice questions**: Organized by difficulty, or what you need to practice, these allow you to test whether wasting your life reading this book has paid off so that you can annoy your friends by walking with an air of confidence into the examination. If you fail said exam, please don't sue me.
- **Flashcard glossary**: As if a printed glossary wasn't enough, my publishers insisted that you'd like an electronic one too. Have fun here flipping through terms and definitions covered in the textbook; it's better than actually learning something.
- **Oliver Twisted's pot of gruel**: Oliver Twisted will draw your attention to the 300 pages or so of more information that we have put online so that (1) the planet suffers a little less, and (2) you won't die when the book falls from your bookshelf onto your head.
- **Labcoat Leni solutions**: There are full answers to the Labcoat Leni tasks.
- **Smart Alex answers**: Each chapter ends with a set of tasks for you to test your newly acquired expertise. The chapters are also littered with self-test questions. The companion website contains detailed answers. Will I ever stop writing?
- **PowerPoint slides**: I can't come and teach you all in person (although you can watch my lectures on YouTube). Instead, I rely on a crack team of highly skilled and super-intelligent pan-dimensional beings called 'lecturers'. I have personally grown each and every one of them in a greenhouse in my garden. To assist in their mission to spread the joy of statistics I have provided them with PowerPoint slides for each chapter. If you see something weird on their slides that upsets you, then remember that's probably my fault.
- **Links**: There are the obligatory links to other useful sites.
- **SAGE**: My publishers are giving you a ton of free material from their books, journals and digital products. If you want it.
- **SAGE Research Methods** is a digital platform full of research methods stuff. Some of it, including videos and a 'test yourself' maths diagnostic tool, is available for free on the companion website.
- **Cyberworms of knowledge**: I have used nanotechnology to create cyberworms that crawl down your broadband, wifi or 4G, pop out of a port on your computer, tablet, iPad or phone, and fly through space into your brain. They rearrange your neurons so that you understand statistics. You don't believe me? You'll never know for sure unless you visit the online resources . . .

Happy reading, and don't get distracted by social media.

THANK YOU

Colleagues: This book (in the SPSS, SAS, and R version) wouldn't have happened if not for Dan Wright's unwarranted faith in a then postgraduate to write the first SPSS edition. Numerous other people have contributed to previous editions of this book. I don't have room to list them all, but particular thanks to Dan (again), David Hitchin, Laura Murray, Gareth Williams, Lynne Slocombe, Kate Lester, Maria de Ridder, Thom Baguley, Michael Spezio, and my wife Zoë who have given me invaluable feedback during the life of this book. Special thanks to Jeremy Miles. Part of his 'help' involves ranting on at me about things I've written being, and I quote, 'bollocks'. Nevertheless, working on the SAS and R versions of this book with him has influenced me enormously. He's also been a very nice person to know over the past few years (apart from when he's ranting on at me about . . .). For this edition, J. W. Jacobs, Ann-Will Kruijt, Johannes Petzold, and E.-J. Wagenmakers provided particularly useful feedback.

Thanks to the following for allowing me to use their raw data—it's an honour for me to include their fascinating research in my book: Rebecca Ang, Philippe Bernard, Hakan Çetinkaya, Tomas Chamorro-Premuzic, Graham Davey, Mike Domjan, Gordon Gallup, Nicolas Guéguen, Sarah Johns, Eric Lacourse, Nate Lambert, Sarah Marzillier, Karlijn Massar, Geoffrey Miller, Peter Muris, Laura Nichols, Nick Perham, Achim Schüetzwohl, Mirjam Tuk, and Lara Zibarras.

I appreciate everyone who has taken time to write nice reviews of this book on the various Amazon (and other) websites around the world; the success of this book has been in no small part due to these people being so positive and constructive in their feedback. Thanks also to everyone who participates so enthusiastically in my Facebook and Twitter pages: I always hit motivational dark times when I'm writing, but feeling the positive vibes from readers always gets me back on track (especially the photos of cats, dogs, parrots, and lizards with my books 🙂). I continue to be amazed and bowled over by the nice things that people say about the book (and disproportionately upset by the less positive things).

Not all contributions are as tangible as those above. Very early in my career, Graham Hole made me realize that teaching research methods didn't have to be dull. My approach to teaching has been to steal his good ideas, and he has had the good grace not to ask for them back! He is a rarity in being brilliant, funny, and nice.

Software: This book wouldn't exist without the generous support of International Business Machines Corporation (IBM), who allow me to beta test IBM® SPSS® Statistics software ('SPSS'), kept me up to date with the software while I wrote this update, and kindly granted permission for me to include screenshots and images from SPSS. I wrote this edition on MacOS but used Windows for the screenshots. Mac and Mac OS are trademarks of Apple Inc., registered in the United States and other countries; Windows is a registered trademark of Microsoft Corporation in the United States and other countries. I don't get any incentives for saying this (perhaps I should, hint, hint . . .) but the following software packages are invaluable to me when writing: TechSmith's (www.techsmith.com) Camtasia (which I use to produce videos) and Snagit (which I use for screenshots) for Mac; the Omnigroup's (www.omnigroup.com) OmniGraffle, which I use to create most of the diagrams and flowcharts (it is awesome); and R (in particular, Hadley Wickham's *ggplot2* package) and R Studio, which I use for data visualizations.

Publishers: My publishers, SAGE, are rare in being a large, successful company that manages to maintain a family feel. For this edition I was particularly grateful for them trusting me enough to leave me alone to get on with things because my deadline was insane. Now that I have emerged from my attic, I'm fairly sure that I'm going to be grateful to Jai Seaman and Sarah Turpie for what they have been doing and will do to support the book. A long-overdue thank you to Richard Leigh, who has copyedited my books over many years and never gets thanked because his job begins after I've written the acknowledgements! My long-suffering production editor, Ian Antcliff, deserves special mention not only for the fantastic job he does but also for being the embodiment of calm when the pressure is on. I'm also grateful to Karen and Ziyad who don't work directly on my books but are such an important part of my fantastic relationship with SAGE.

James Iles redesigned the characters in this book and produced the artwork for the pedagogic boxes. I worked with James on another book where there was a lot more artwork (*An Adventure in Statistics*) and it was an incredible experience. I'm delighted that that experience didn't put him off working with me again. It's an honour to have his artwork in another of my books.

Music: I always write, listening to music. For this edition, I predominantly enjoyed (my neighbours less so): AC/DC, A Forest of Stars, Alice Cooper, Alter of Plagues, Anathema, Animals as Leaders, Anthrax, Billy Cobham, Blackfield, Deafheaven, Deathspell Omega, Deep Purple, Enslaved, Faith No More, Genesis (Peter Gabriel era), Ghost, Ghost Bath, Glenn Hughes, Gojira, Gorguts, Iced Earth, Ihsahn, The Infernal Sea, Iron Maiden, Judas Priest, Katatonia, Kiss, Marillion, Meshuggah, Metallica, MGLA, Motörhead, Primal Rock Rebellion, Opeth, Oranssi Pazuzu, Rebirth of Nefast, Royal Thunder, Satyricon, Skuggsja, Status Quo (R.I.P. Rick 🙁), Steven Wilson, Thin Lizzy, Wolves in the Throne Room.

Friends and family: All this book-writing nonsense requires many lonely hours of typing. Without some wonderful friends to drag me out of my dimly lit room from time to time I'd be even more of a gibbering cabbage than I already am. Across many editions, my eternal gratitude goes to Graham Davey, Ben Dyson, Kate Lester, Mark Franklin, and their lovely families for reminding me that there is more to life than work. I throw a robust set of horns to my brothers of metal, Rob Mepham, Nick Paddy, and Ben Anderson, for letting me deafen them with my drumming. Thanks to my parents and Paul and Julie for being my family. Special cute thanks to my niece and nephew, Oscar and Melody: I hope to teach you many things that will annoy your parents.

For someone who spends his life writing, I'm constantly surprised at how incapable I am of finding words to express how wonderful my wife Zoë is. She has a never-ending supply of patience, love, support, and optimism (even when her husband is a grumpy, sleep-deprived, withered, self-doubting husk). I never forget, not even for a nanosecond, how lucky I am. Finally, since the last edition, I made a trivial contribution to creating two humans: Zach and Arlo. I thank them for the realization of how utterly pointless work is and for the permanent feeling that my heart has expanded to bursting point from trying to contain my love for them.

Like the previous editions, this book is dedicated to my brother Paul and my cat Fuzzy (now in the spirit cat world), because one of them was an intellectual inspiration and the other woke me up in the morning by sitting on me and purring in my face until I gave him cat food: mornings were considerably more pleasant when my brother got over his love of cat food for breakfast. 🙂

SYMBOLS USED IN THIS BOOK

Mathematical operators

Σ This symbol (called sigma) means 'add everything up'. So, if you see something like $\sum x_i$ it means 'add up all of the scores you've collected'.

Π This symbol means 'multiply everything'. So, if you see something like $\prod x_i$ it means 'multiply all of the scores you've collected'.

$\sqrt{X}$ This means 'take the square root of x'.

Greek symbols

α Alpha, the probability of making a Type I error

β Beta, the probability of making a Type II error

β_i Beta, the standardized regression coefficient

ε Epsilon, usually stands for 'error', but is also used to denote sphericity

η^2 Eta squared, an effect size measure

μ Mu, the mean of a population of scores

ρ Rho, the correlation in the population; also used to denote Spearman's correlation coefficient

σ Sigma, the standard deviation in a population of data

σ^2 Sigma squared, the variance in a population of data

$\sigma_{\bar{x}}$ Another variation on sigma, which represents the standard error of the mean

τ Kendall's tau (non-parametric correlation coefficient)

ϕ Phi, a measure of association between two categorical variables, but also used to denote the dispersion parameter in logistic regression

χ^2 Chi-square, a test statistic that quantifies the association between two categorical variables

χ_F^2 Another use of the letter chi, but this time as the test statistic in Friedman's ANOVA, a non-parametric test of differences between related means

ω^2 Omega squared (an effect size measure). This symbol also means 'expel the contents of your intestine immediately into your trousers'; you will understand why in due course

Latin symbols

b_i	The regression coefficient (unstandardized); I tend to use it for any coefficient in a linear model	s	The standard deviation of a sample of data
df	Degrees of freedom	s^2	The variance of a sample of data
e_i	The error associated with the ith person	SS	The sum of squares, or sum of squared errors to give it its full title
F	F-statistic	SS_A	The sum of squares for variable A
H	Kruskal–Wallis test statistic	SS_M	The model sum of squares (i.e., the variability explained by the model fitted to the data)
k	The number of levels of a variable (i.e., the number of treatment conditions), or the number of predictors in a regression model	SS_R	The residual sum of squares (i.e., the variability that the model can't explain – the error in the model)
ln	Natural logarithm	SS_T	The total sum of squares (i.e., the total variability within the data)
MS	The mean squared error (mean square): the average variability in the data	t	Testicle statistic for a t-test. Yes, I did that deliberately to check whether you're paying attention
N, n, n_i	The sample size. N usually denotes the total sample size, whereas n usually denotes the size of a particular group	T	Test statistic for Wilcoxon's matched-pairs signed-rank test
P	Probability (the probability value, p-value or significance of a test are usually denoted by p)	U	Test statistic for the Mann–Whitney test
r	Pearson's correlation coefficient	W_s	Test statistic for Rilcoxon's wank-sum test. See what I did there? It doesn't matter because no one reads this page
r_s	Spearman's rank correlation coefficient		
r_b, r_{pb}	Biserial correlation coefficient and point-biserial correlation coefficient, respectively		
R	The multiple correlation coefficient	$\overline{X}$	The mean of a sample of scores
R^2	The coefficient of determination (i.e., the proportion of data explained by the model)	z	A data point expressed in standard deviation units

A MATHS REVIEW

There are good websites that can help you if any of the maths in this book confuses you. The pages at studymaths.co.uk, www.gcflearnfree.org/math, and www.mathsisfun.com look useful, but there are many others, so use a search engine to find something that suits you. Some resources are also available on the book's website so you can try there if you run out of inspiration. I will quickly remind you of three important things:

Two negatives make a positive: Although in life two wrongs don't make a right, in mathematics they do. When we multiply a negative number by another negative number, the result is a positive number. For example, $-2 \times -4 = 8$.

A negative number multiplied by a positive one make a negative number: If you multiply a positive number by a negative number then the result is another negative number. For example, $2 \times -4 = -8$, or $-2 \times 6 = -12$.

BODMAS and PEMDAS: These two acronyms are different ways of remembering the order in which mathematical operations are performed. BODMAS stands for Brackets, Order, Division, Multiplication, Addition, and Subtraction; whereas PEMDAS stems from Parentheses, Exponents, Multiplication, Division, Addition, and Subtraction. Having two widely used acronyms is confusing (especially because multiplication and division are the opposite way around), but they do mean the same thing:

- Brackets/Parentheses: When solving any expression or equation you deal with anything in brackets/parentheses first.
- Order/Exponents: Having dealt with anything in brackets, you next deal with any order terms/exponents. These refer to power terms such as squares. Four squared, or 4^2, used to be called four raised to the order of 2, hence the word 'order' in BODMAS. These days, the term 'exponents' is more common (so by all means use BEDMAS as your acronym if you find that easier).
- Division and Multiplication: The next things to evaluate are any division or multiplication terms. The order in which you handle them is from the left to the right of the expression/equation. That's why BODMAS and PEMDAS can list them the opposite way around, because they are considered at the same time (so, BOMDAS or PEDMAS would work as acronyms, too).
- Addition and Subtraction: Finally, deal with any addition or subtraction. Again, go from left to right, doing any addition or subtraction in the order that you meet the terms. (So, BODMSA would work as an acronym too, but it's hard to say.)

Let's look at an example of BODMAS/PEMDAS in action: what would be the result of $1 + 3 \times 5^2$? The answer is 76 (not 100 as some of you might have thought). There are no brackets, so the first thing is to deal with the order/exponent: 5^2 is 25, so the equation becomes $1 + 3 \times 25$. Moving from left to right, there is no division, so we do the multiplication: 3×25, which gives us 75. Again, going from left to right, we look for addition and subtraction terms—there are no subtractions, so the first thing we come across is the addition: $1 + 75$, which gives us 76 and the expression is solved. If I'd written the expression as $(1 + 3) \times 5^2$, then the answer would have been 100 because we deal with the brackets first: $(1 + 3) = 4$, so the expression becomes 4×5^2. We then deal with the order/exponent (5^2 is 25), which results in $4 \times 25 = 100$.

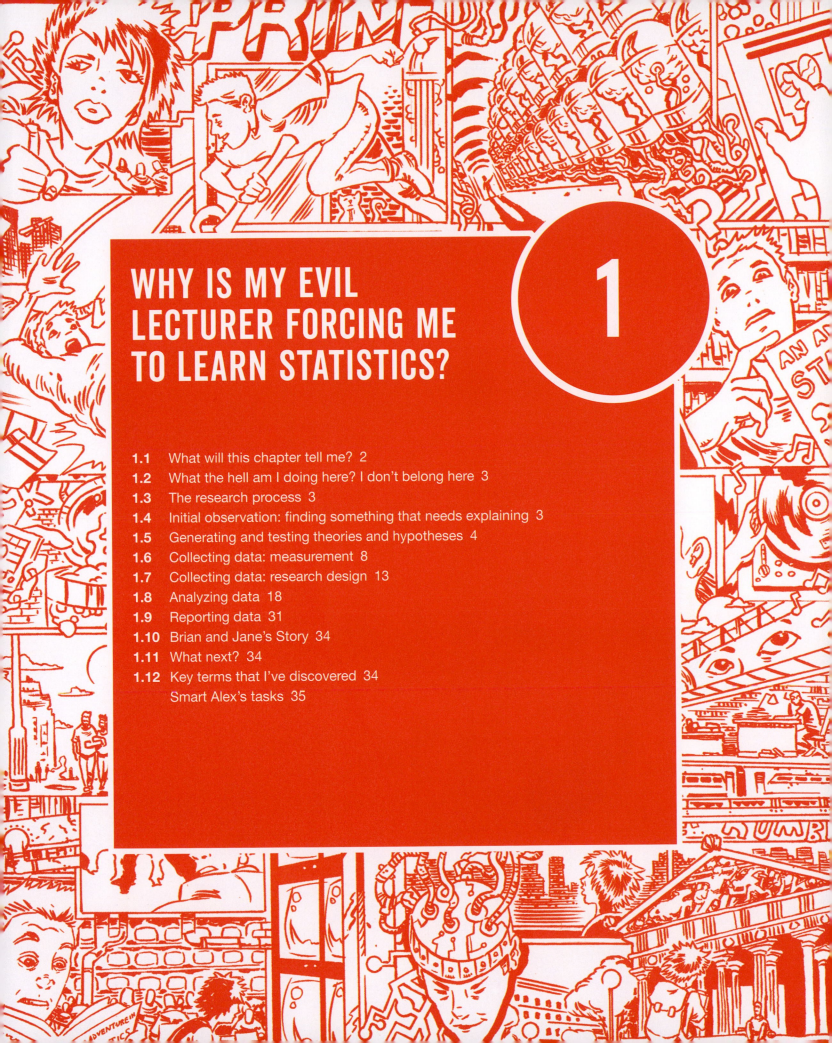

WHY IS MY EVIL LECTURER FORCING ME TO LEARN STATISTICS?

1

1.1 What will this chapter tell me?

I was born on 21 June 1973. Like most people, I don't remember anything about the first few years of life and, like most children, I went through a phase of driving my dad mad by asking 'Why?' every five seconds. With every question, the word 'dad' got longer and whinier: 'Dad, why is the sky blue?', 'Daaad, why don't worms have legs?' Daaaaaaaaad, where do babies come from?' Eventually, my dad could take no more and whacked me around the face with a golf club.[1]

My torrent of questions reflected the natural curiosity that children have: we all begin our voyage through life as inquisitive little scientists. At the age of 3, I was at my friend Obe's party (just before he left England to return to Nigeria, much to my distress). It was a hot day, and there was an electric fan blowing cold air around the room. My 'curious little scientist' brain was working through what seemed like a particularly pressing question: 'What happens when you stick your finger in a fan?' The answer, as it turned out, was that it hurts—a lot.[2] At the age of 3, we intuitively know that to answer questions you need to collect data, even if it causes us pain.

My curiosity to explain the world never went away, which is why I'm a scientist. The fact that you're reading this book means that the inquisitive 3-year-old in you is alive and well and wants to answer new and exciting questions, too. To answer these questions you need 'science' and science has a **pilot fish** called 'statistics' that hides under its belly eating ectoparasites. That's why your evil lecturer is forcing you to learn statistics. Statistics is a bit like sticking your finger into a revolving fan blade: sometimes it's very painful, but it does give you answers to interesting questions. I'm going to try to convince you in this chapter that statistics are an important part of doing research. We will overview the whole research process, from why we conduct research in the first place, through how theories are generated, to why we need data to test these theories. If that doesn't convince you to read on then maybe the fact that we discover whether Coca-Cola kills sperm will. Or perhaps not.

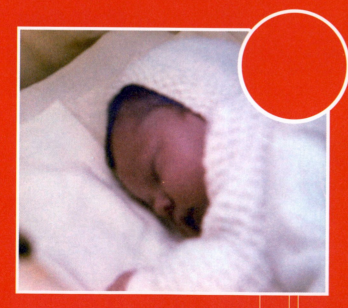

Figure 1.1   When I grow up, please don't let me be a statistics lecturer

1   He was practicing in the garden when I unexpectedly wandered behind him at the exact moment he took a back swing. It's rare that a parent enjoys the sound of their child crying, but, on this day, it filled my dad with joy because my wailing was tangible evidence that he hadn't killed me, which he thought he might have done. Had he hit me with the club end rather than the shaft he probably would have. Fortunately (for me, but not for you), I survived, although some might argue that this incident explains the way my brain functions.

2   In the 1970s, fans didn't have helpful protective cages around them to prevent idiotic 3-year-olds sticking their fingers into the blades.

1.2 What the hell am I doing here? I don't belong here ▮▮▮▮

You're probably wondering why you have bought this book. Maybe you liked the pictures, maybe you fancied doing some weight training (it *is* heavy) or perhaps you needed to reach something in a high place (it *is* thick). The chances are, though, that given the choice of spending your hard-earned cash on a statistics book or something more entertaining (a nice novel, a trip to the cinema, etc.) you'd choose the latter. So, why have you bought the book (or downloaded an illegal PDF of it from someone who has way too much time on their hands if they're scanning 900 pages for fun)? It's likely that you obtained it because you're doing a course on statistics or you're doing some research, and you need to know how to analyze data. It's possible that you didn't realize when you started your course or research that you'd have to know about statistics but now find yourself inexplicably wading, neck high, through the Victorian sewer that is data analysis. The reason why you're in the mess that you find yourself in is that you have a curious mind. You might have asked yourself questions like why people behave the way they do (psychology) or why behaviors differ across cultures (anthropology), how businesses maximize their profit (business), how the dinosaurs died (palaeontology), whether eating tomatoes protects you against cancer (medicine, biology), whether it is possible to build a quantum computer (physics, chemistry), whether the planet is hotter than it used to be and in what regions (geography, environmental studies). Whatever it is you're studying or

researching, the reason why you're studying it is probably that you're interested in answering questions. Scientists are curious people, and you probably are too. However, it might not have occurred to you that to answer interesting questions, you need data and explanations for those data.

The answer to 'What the hell are you doing here?' is simple: to answer interesting questions you need data. One of the reasons why your evil statistics lecturer is forcing you to learn about numbers is that they are a form of data and are vital to the research process. Of course, there are forms of data other than numbers that can be used to test and generate theories. When numbers are involved, the research involves **quantitative methods**, but you can also generate and test theories by analyzing language (such as conversations, magazine articles and media broadcasts). This involves **qualitative methods** and it is a topic for another book not written by me. People can get quite passionate about which of these methods is *best*, which is a bit silly because they are complementary, not competing, approaches and there are much more important issues in the world to get upset about. Having said that, all qualitative research is rubbish.[3]

1.3 The research process ▮▮▮▮

How do you go about answering an interesting question? The research process is broadly summarized in Figure 1.2. You begin with an observation that you want to

understand, and this observation could be anecdotal (you've noticed that your cat watches birds when they're on TV but not when jellyfish are on)[4] or could be based on some data (you've got several cat owners to keep diaries of their cat's TV habits and noticed that lots of them watch birds). From your initial observation you consult relevant theories and generate explanations (hypotheses) for those observations, from which you can make predictions. To test your predictions you need data. First you collect some relevant data (and to do that you need to identify things that can be measured) and then you analyze those data. The analysis of the data may support your hypothesis or generate a new one, which, in turn, might lead you to revise the theory. As such, the processes of data collection and analysis and generating theories are intrinsically linked: theories lead to data collection/ analysis and data collection/analysis informs theories. This chapter explains this research process in more detail.

1.4 Initial observation: finding something that needs explaining ▮▮▮▮

The first step in Figure 1.2 was to come up with a question that needs an answer. I spend rather more time than I should watching reality TV. Over many years, I used to swear that I wouldn't get hooked on reality TV, and yet year upon year I would find myself glued to the TV screen waiting for the next contestant's meltdown (I am a psychologist, so really this is just research). I used to wonder why there is so much arguing in these shows, and why so many contestants have really unpleasant personalities (my money is on narcissistic

---

3   This is a joke. Like many of my jokes, there are people who won't find it remotely funny. Passions run high between qualitative and quantitative researchers, so its inclusion will likely result in me being hunted down, locked in a room and forced to do discourse analysis by a horde of rabid qualitative researchers.

4   In his younger days my cat actually did climb up and stare at the TV when birds were being shown.

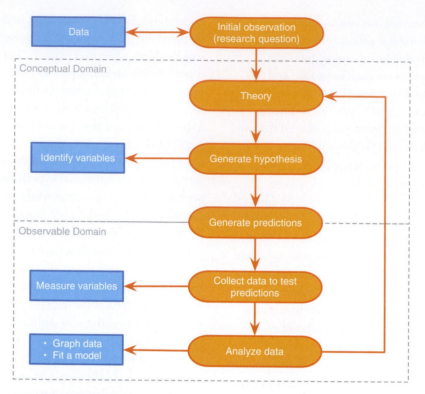

Figure 1.2 The research process

personality disorder).[5] A lot of scientific endeavor starts this way: not by watching reality TV, but by observing something in the world and wondering why it happens.

Having made a casual observation about the world (reality TV contestants on the whole have extreme personalities and argue a lot), I need to collect some data to see whether this observation is true (and not a biased observation). To do this, I need to define one or more **variables** to measure that quantify the thing I'm trying to measure. There's one variable in this example: the personality of the contestant. I could measure this variable by giving them one of the many well-established questionnaires that measure personality characteristics. Let's say that I did this and I found that 75% of contestants did have narcissistic personality disorder. These data support my observation: a lot of reality TV contestants have extreme personalities.

1.5 Generating and testing theories and hypotheses ▮▮▮▮

The next logical thing to do is to explain these data (Figure 1.2). The first step is to look for relevant theories. A **theory** is an explanation or set of principles that is well substantiated by repeated testing and explains a broad phenomenon. We might begin by looking at theories of narcissistic personality disorder, of which there are currently very few. One theory of personality disorders in general links them to early attachment (put simplistically, the bond formed between a child and their main caregiver). Broadly speaking, a child can form a secure (a good thing) or an insecure (not so good) attachment to their caregiver, and the theory goes that insecure attachment explains later personality disorders (Levy Johnson, Clouthier, Scala, &

Temes, 2015). This is a theory because it is a set of principles (early problems in forming interpersonal bonds) that explains a general broad phenomenon (disorders characterized by dysfunctional interpersonal relations). There is also a critical mass of evidence to support the idea. Theory also tells us that those with narcissistic personality disorder tend to engage in conflict with others despite craving their attention, which perhaps explains their difficulty in forming close bonds.

Given this theory, we might generate a **hypothesis** about our earlier observation (see Jane Superbrain Box 1.1). A hypothesis is a proposed explanation for a fairly narrow phenomenon or set of observations. It is not a guess, but an informed, theory-driven attempt to explain what has been observed. Both theories and hypotheses seek to explain the world, but a theory explains a wide set of phenomena with a small set of well-established principles, whereas a hypothesis typically seeks to explain a narrower phenomenon and is, as yet, untested. Both theories and hypotheses exist in the conceptual domain, and you cannot observe them directly.

To continue the example, having studied the attachment theory of personality disorders, we might decide that this theory implies that people with personality disorders seek out the attention that a TV appearance provides because they lack close interpersonal relationships. From this we can generate a hypothesis: people with narcissistic personality disorder use reality TV to satisfy their craving for attention from others. This is a conceptual statement that explains our original observation (that rates of narcissistic personality disorder are high on reality TV shows).

---

5 This disorder is characterized by (among other things) a grandiose sense of self-importance, arrogance, lack of empathy for others, envy of others and belief that others envy them, excessive fantasies of brilliance or beauty, the need for excessive admiration, and exploitation of others.

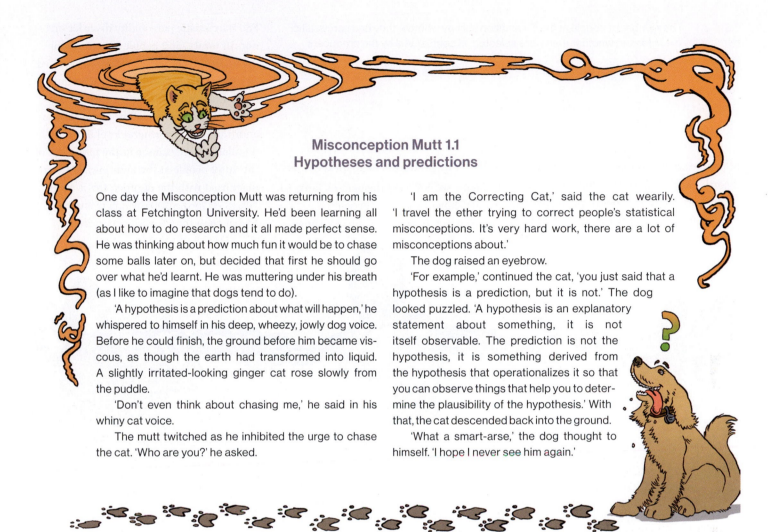

Misconception Mutt 1.1
Hypotheses and predictions

One day the Misconception Mutt was returning from his class at Fetchington University. He'd been learning all about how to do research and it all made perfect sense. He was thinking about how much fun it would be to chase some balls later on, but decided that first he should go over what he'd learnt. He was muttering under his breath (as I like to imagine that dogs tend to do).

'A hypothesis is a prediction about what will happen,' he whispered to himself in his deep, wheezy, jowly dog voice. Before he could finish, the ground before him became viscous, as though the earth had transformed into liquid. A slightly irritated-looking ginger cat rose slowly from the puddle.

'Don't even think about chasing me,' he said in his whiny cat voice.

The mutt twitched as he inhibited the urge to chase the cat. 'Who are you?' he asked.

'I am the Correcting Cat,' said the cat wearily. 'I travel the ether trying to correct people's statistical misconceptions. It's very hard work, there are a lot of misconceptions about.'

The dog raised an eyebrow.

'For example,' continued the cat, 'you just said that a hypothesis is a prediction, but it is not.' The dog looked puzzled. 'A hypothesis is an explanatory statement about something, it is not itself observable. The prediction is not the hypothesis, it is something derived from the hypothesis that operationalizes it so that you can observe things that help you to determine the plausibility of the hypothesis.' With that, the cat descended back into the ground.

'What a smart-arse,' the dog thought to himself. 'I hope I never see him again.'

To test this hypothesis, we need to move from the conceptual domain into the observable domain. That is, we need to operationalize our hypothesis in a way that enables us to collect and analyze data that have a bearing on the hypothesis (Figure 1.2). We do this using predictions. Predictions emerge from a hypothesis (Misconception Mutt 1.1), and transform it from something unobservable into something that is. If our hypothesis is that people with narcissistic personality disorder use reality TV to satisfy their craving for attention from others, then a prediction we could make based on this hypothesis is that people with narcissistic personality disorder are more likely to audition for reality TV than those without. In making this prediction we can move from the conceptual domain into the observable domain, where we can collect evidence.

In this example, our prediction is that people with narcissistic personality disorder are more likely to audition for reality TV than those without. We can measure this prediction by getting a team of clinical psychologists to interview each person at a reality TV audition and diagnose them as having narcissistic personality disorder or not. The population rates of narcissistic personality disorder are about 1%, so we'd be able to see whether the ratio of narcissistic personality disorder to not is higher at the audition than in the general population. If it is higher then our prediction is correct: a disproportionate number of people with narcissistic personality disorder turned up at the audition. Our prediction, in turn, tells us something about the hypothesis from which it derived.

This is tricky stuff, so let's look at another example. Imagine that, based on a different theory, we generated a different hypothesis. I mentioned earlier that people with narcissistic personality disorder tend to engage in conflict, so a different hypothesis is that producers of reality TV shows select people who have narcissistic personality disorder to be contestants because they believe that conflict makes good TV. As before, to test this hypothesis we need to bring it into the observable domain by generating a prediction from it. The prediction would be that (assuming no bias in the number

Table 1.1 The number of people at the TV audition split by whether they had narcissistic personality disorder and whether they were selected as contestants by the producers

	No Disorder	Disorder	Total
Selected	3	9	12
Rejected	6805	845	7650
Total	6808	854	7662

of people with narcissistic personality disorder applying for the show) a disproportionate number of people with narcissistic personality disorder will be selected by producers to go on the show. Imagine we collected the data in Table 1.1, which shows how many people auditioning to be on a reality TV show had narcissistic personality disorder or not. In total, 7662 people turned up for the audition. Our first prediction (derived from our first hypothesis) was that the percentage of people with narcissistic personality disorder will be higher at the audition than the general level in the population. We can see in the table that of the 7662 people at the audition, 854 were diagnosed with the disorder; this is about 11% (854/7662 × 100), which is much higher than the 1% we'd expect in the

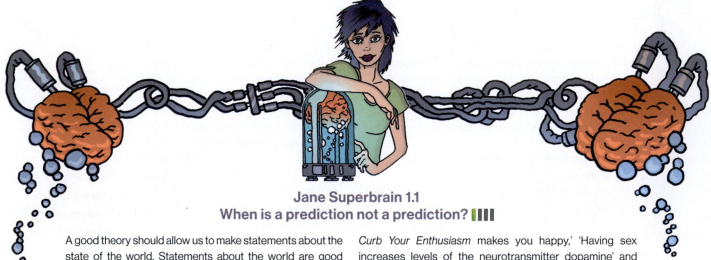

Jane Superbrain 1.1
When is a prediction not a prediction? ▌▌▌▌

A good theory should allow us to make statements about the state of the world. Statements about the world are good things: they allow us to make sense of our world, and to make decisions that affect our future. One current example is global warming. Being able to make a definitive statement that global warming is happening, and that it is caused by certain practices in society, allows us to change these practices and, hopefully, avert catastrophe. However, not all statements can be tested using science. Scientific statements are ones that can be verified with reference to empirical evidence, whereas non-scientific statements are ones that cannot be empirically tested. So, statements such as 'The Led Zeppelin reunion concert in London in 2007 was the best gig ever,'[6] 'Lindt chocolate is the best food' and 'This is the worst statistics book in the world' are all non-scientific; they cannot be proved or disproved. Scientific statements can be confirmed or disconfirmed empirically. 'Watching *Curb Your Enthusiasm* makes you happy,' 'Having sex increases levels of the neurotransmitter dopamine' and 'Velociraptors ate meat' are all things that can be tested empirically (provided you can quantify and measure the variables concerned). Non-scientific statements can sometimes be altered to become scientific statements, so 'The Beatles were the most influential band ever' is non-scientific (because it is probably impossible to quantify 'influence' in any meaningful way) but by changing the statement to 'The Beatles were the best-selling band ever,' it becomes testable (we can collect data about worldwide album sales and establish whether the Beatles have, in fact, sold more records than any other music artist). Karl Popper, the famous philosopher of science, believed that non-scientific statements were nonsense and had no place in science. Good theories and hypotheses should, therefore, produce predictions that are scientific statements.

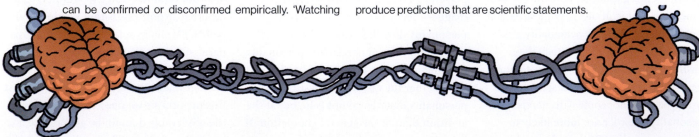

---

6 It was pretty awesome actually.

general population. Therefore, prediction 1 is correct, which in turn supports hypothesis 1. The second prediction was that the producers of reality TV have a bias towards choosing people with narcissistic personality disorder. If we look at the 12 contestants that they selected, 9 of them had the disorder (a massive 75%). If the producers did not have a bias we would have expected only 11% of the contestants to have the disorder (the same rate as was found when we considered everyone who turned up for the audition). The data are in line with prediction 2 which supports our second hypothesis. Therefore, my initial observation that contestants have personality disorders was verified by data, and then using theory I generated specific hypotheses that were operationalized by generating predictions that could be tested using data. Data are *very* important.

I would now be smugly sitting in my office with a contented grin on my face because my hypotheses were well supported by the data. Perhaps I would quit while I was ahead and retire. It's more likely, though, that having solved one great mystery, my excited mind would turn to another. I would lock myself in a room to watch more reality TV. I might wonder at why contestants with narcissistic personality disorder, despite their obvious character flaws, enter a situation that will put them under intense public scrutiny.[7] Days later, the door would open, and a stale odor would waft out like steam rising from the New York subway. Through this green cloud, my bearded face would emerge, my eyes squinting at the shards of light that cut into my pupils. Stumbling forwards, I would open my mouth to lay waste to my scientific rivals with my latest profound hypothesis: 'Contestants with narcissistic personality disorder believe that they will win'. I would croak before collapsing on the floor. The prediction from this hypothesis is that if I

ask the contestants if they think that they will win, the people with a personality disorder will say 'yes'.

Let's imagine I tested my hypothesis by measuring contestants' expectations of success in the show, by asking them, 'Do you think you will win?' Let's say that 7 of 9 contestants with narcissistic personality disorder said that they thought that they would win, which confirms my hypothesis. At this point I might start to try to bring my hypotheses together into a theory of reality TV contestants that revolves around the idea that people with narcissistic personalities are drawn towards this kind of show because it fulfils their need for approval and they have unrealistic expectations about their likely success because they don't realize how unpleasant their personalities are to other people. In parallel, producers tend to select contestants with narcissistic tendencies because they tend to generate interpersonal conflict.

One part of my theory is untested, which is the bit about contestants with narcissistic personalities not realizing how others perceive their personality. I could operationalize this hypothesis through a prediction that if I ask these contestants whether their personalities were different from those of other people they would say 'no'. As before, I would collect more data and ask the contestants with narcissistic personality disorder whether they believed that their personalities were different from the norm. Imagine that all 9 of them said that they thought their personalities *were* different from the norm. These data contradict my hypothesis. This is known as **falsification**, which is the act of disproving a hypothesis or theory.

It's unlikely that we would be the only people interested in why individuals who go on reality TV have extreme

personalities. Imagine that these other researchers discovered that: (1) people with narcissistic personality disorder think that they are more interesting than others; (2) they also think that they deserve success more than others; and (3) they also think that others like them because they have 'special' personalities.

This additional research is even worse news for my theory: if contestants didn't realize that they had a personality different from the norm, then you wouldn't expect them to think that they were more interesting than others, and you certainly wouldn't expect them to think that others will *like* their unusual personalities. In general, this means that this part of my theory sucks: it cannot explain all of the data, predictions from the theory are not supported by subsequent data, and it cannot explain other research findings. At this point I would start to feel intellectually inadequate and people would find me curled up on my desk in floods of tears, wailing and moaning about my failing career (no change there then).

At this point, a rival scientist, Fester Ingpant-Stain, appears on the scene adapting my theory to suggest that the problem is not that personality-disordered contestants don't realize that they have a personality disorder (or at least a personality that is unusual), but that they falsely believe that this special personality is perceived positively by other people. One prediction from this model is that if personality-disordered contestants are asked to evaluate what other people think of them, then they will overestimate other people's positive perceptions. You guessed it, Fester Ingpant-Stain collected yet more data. He asked each contestant to fill out a questionnaire evaluating all of the other contestants' personalities, and also to

---

7 One of the things I like about many reality TV shows in the UK is that the winners are very often nice people, and the odious people tend to get voted out quickly, which gives me faith that humanity favors the nice.

complete the questionnaire about themselves but answering from the perspective of each of their housemates. (So, for every contestant there is a measure of what they thought of every other contestant, and also a measure of what they believed every other contestant thought of them.) He found out that the contestants with personality disorders did overestimate their housemates' opinions of them; conversely, the contestants without personality disorders had relatively accurate impressions of what others thought of them. These data, irritating as it would be for me, support Fester Ingpant-Stain's theory more than mine: contestants with personality disorders do realize that they have unusual personalities but believe that these characteristics are ones that others would feel positive about. Fester Ingpant-Stain's theory is quite good: it explains the initial observations and brings together a range of research findings. The end result of this whole process (and my career) is that we should be able to make a general statement about the state of the world. In this case we could state 'Reality TV contestants who have personality disorders overestimate how much other people like their personality characteristics'.

1.6 Collecting data: measurement ▮▮▮

In looking at the process of generating theories and hypotheses, we have seen the importance of data in testing those hypotheses or deciding between competing theories. This section looks at data collection in more detail. First we'll look at measurement.

1.6.1 Independent and dependent variables ▮▮▮

To test hypotheses we need to measure variables. Variables are things that can change (or vary); they might vary between people (e.g., IQ, behavior) or locations (e.g., unemployment) or even time (e.g., mood, profit, number of cancerous cells). Most hypotheses can be expressed in terms of two variables: a proposed cause and a proposed outcome. For example, if we take the scientific statement, 'Coca-Cola is an effective spermicide'[8] then the proposed cause is 'Coca-Cola' and the proposed effect is dead sperm. Both the cause and the outcome are variables: for the cause we could vary the type of drink, and for the outcome, these drinks will kill different amounts of sperm. The key to testing scientific statements is to measure these two variables.

A variable that we think is a cause is known as an **independent variable** (because its value does not depend on any other variables). A variable that we think is an effect is called a **dependent variable** because the value of this variable depends on the cause (independent variable). These terms are very closely tied to experimental methods in which the cause is manipulated by the experimenter (as we will see in Section 1.7.2). However, researchers can't always manipulate variables (for example, if you wanted see whether smoking causes lung cancer you wouldn't lock a bunch of people in a room for 30 years and force them to smoke). Instead, they sometimes use correlational methods (Section 1.7), for which it doesn't make sense to talk of dependent and independent variables because all variables are essentially dependent variables. I prefer to use the terms **predictor variable** and **outcome variable** in place of dependent and independent variable. This is not a personal whimsy: in experimental work the cause (independent variable) is a predictor, and the effect (dependent variable) is an outcome, and in correlational work we can talk of one or more (predictor) variables predicting (statistically at least) one or more outcome variables.

1.6.2 Levels of measurement ▮▮▮

Variables can take on many different forms and levels of sophistication. The relationship between what is being measured and the numbers that represent what is being measured is known as the **level of measurement**. Broadly speaking, variables can be categorical or continuous, and can have different levels of measurement.

A **categorical variable** is made up of categories. A categorical variable that you should be familiar with already is your species (e.g., human, domestic cat, fruit bat, etc.). You are a human or a cat or a fruit bat: you cannot be a bit of a cat and a bit of a bat, and neither a batman nor (despite many fantasies to the contrary) a catwoman exist (not even one in a PVC suit). A categorical variable is one that names distinct entities. In its simplest

SELF TEST
Based on what you have read in this section, what qualities do you think a scientific theory should have?

---

8   Actually, there is a long-standing urban myth that a post-coital douche with the contents of a bottle of Coke is an effective contraceptive. Unbelievably, this hypothesis has been tested and Coke does affect sperm motility (movement), and some types of Coke are more effective than others—Diet Coke is best, apparently (Umpierre, Hill & Anderson, 1985). In case you decide to try this method out, I feel it worth mentioning that despite the effects on sperm motility a Coke douche is ineffective at preventing pregnancy.

Cramming Sam's Tips
Variables

 When doing and reading research you're likely to encounter these terms:

- *Independent variable*: A variable thought to be the cause of some effect. This term is usually used in experimental research to describe a variable that the experimenter has manipulated.
- *Dependent variable*: A variable thought to be affected by changes in an independent variable. You can think of this variable as an outcome.

- *Predictor variable*: A variable thought to predict an outcome variable. This term is basically another way of saying 'independent variable'. (Although some people won't like me saying that; I think life would be easier if we talked only about predictors and outcomes.)
- *Outcome variable*: A variable thought to change as a function of changes in a predictor variable. For the sake of an easy life this term could be synonymous with 'dependent variable'.

form it names just two distinct types of things, for example male or female. This is known as a **binary variable**. Other examples of binary variables are being alive or dead, pregnant or not, and responding 'yes' or 'no' to a question. In all cases there are just two categories and an entity can be placed into only one of the two categories. When two things that are equivalent in some sense are given the same name (or number), but there are more than two possibilities, the variable is said to be a **nominal variable**.

It should be obvious that if the variable is made up of names it is pointless to do arithmetic on them (if you multiply a human by a cat, you do not get a hat). However, sometimes numbers are used to denote categories. For example, the numbers worn by players in a sports team.

In rugby, the numbers on shirts denote specific field positions, so the number 10 is always worn by the fly-half [9] and the number 2 is always the hooker (the ugly-looking player at the front of the scrum). These numbers do not tell us anything other than what position the player plays. We could equally have shirts with FH and H instead of 10 and 2. A number 10 player is not necessarily better than a number 2 (most managers would not want their fly-half stuck in the front of the scrum!). It is equally daft to try to do arithmetic with nominal scales where the categories are denoted by numbers: the number 10 takes penalty kicks, and if the coach found that his number 10 was injured, he would not get his number 4 to give number 6 a piggy-back and then take the kick. The only way that nominal data

can be used is to consider frequencies. For example, we could look at how frequently number 10s score compared to number 4s.

So far, the categorical variables we have considered have been unordered (e.g., different brands of Coke with which you're trying to kill sperm), but they can be ordered too (e.g., increasing concentrations of Coke with which you're trying to skill sperm). When categories are ordered, the variable is known as an **ordinal variable**. Ordinal data tell us not only that things have occurred, but also the order in which they occurred. However, these data tell us nothing about the differences between values. In TV shows like *The X Factor*, *American Idol*, and *The Voice*, hopeful singers compete to win a recording contract. They are hugely popular shows, which could (if you take a depressing view)

---

9 Unlike, for example, NFL football where a quarterback could wear any number from 1 to 19.

Jane Superbrain 1.2
Self-report data ▌▌▌▌

A lot of self-report data are ordinal. Imagine two judges on *The X Factor* were asked to rate Billie's singing on a 10-point scale. We might be confident that a judge who gives a rating of 10 found Billie more talented than one who gave a rating of 2, but can we be certain that the first judge found her five times more talented than the second? What if both judges gave a rating of 8; could we be sure that they found her equally talented? Probably not: their ratings will depend on their subjective feelings about what constitutes talent (the quality of singing? showmanship? dancing?). For these reasons, in any situation in which we ask people to rate something subjective (e.g., their preference for a product, their confidence about an answer, how much they have understood some medical instructions) we should probably regard these data as ordinal, although many scientists do not.

reflect the fact that Western society values 'luck' more than hard work.[10] Imagine that the three winners of a particular *X Factor* series were Billie, Freema and Elizabeth. The names of the winners don't provide any information about where they came in the contest; however, labeling them according to their performance does—first, second and third. These categories are ordered. In using ordered categories we now know that the woman who won was better than the women who came second and third. We still know nothing about the differences between categories, though. We don't, for example, know how much better the winner was than the runners-up: Billie might have been an easy victor, getting many more votes than Freema and Elizabeth or it might have been a very close contest that she won by only a single vote. Ordinal data, therefore, tell us more than nominal data (they tell us the order in which things happened) but they still do not tell us about the differences between points on a scale.

The next level of measurement moves us away from categorical variables and into continuous variables. A **continuous variable** is one that gives us a score for each person and can take on any value on the measurement scale that we are using. The first type of continuous variable that you might encounter is an **interval variable**. Interval data are considerably more useful than ordinal data, and most of the statistical tests in this book rely on having data measured at this level at least. To say that data are interval, we must be certain that equal intervals on the scale represent equal differences in the property being measured. For example, on www.ratemyprofessors.com, students are encouraged to rate their lecturers on several dimensions (some of the lecturers' rebuttals of their negative evaluations are worth a look). Each dimension (helpfulness, clarity, etc.) is evaluated using a 5-point scale. For this scale to be interval it must be the case that the difference between helpfulness ratings of 1 and 2 is the same as the difference between (say)

---

10 I am in no way bitter about spending years learning musical instruments and trying to create original music, only to be beaten to musical fame and fortune by 15-year-olds who can sing, sort of.

3 and 4 or 4 and 5. Similarly, the difference in helpfulness between ratings of 1 and 3 should be identical to the difference between ratings of 3 and 5. Variables like this that look interval (and are treated as interval) are often ordinal—see Jane Superbrain Box 1.2.

Ratio variables go a step further than interval data by requiring that in addition to the measurement scale meeting the requirements of an interval variable, the ratios of values along the scale should be meaningful. For this to be true, the scale must have a true and meaningful zero point. In our lecturer ratings this would mean that a lecturer rated as 4 would be twice as helpful as a lecturer rated with a 2 (who would, in turn, be twice as helpful as a lecturer rated as 1). The time to respond to something is a good example of a ratio variable. When we measure a reaction time, not only is it true that, say, the difference between 300 and 350 ms (a difference of 50 ms) is the same as the difference between 210 and 260 ms or between 422 and 472 ms, but it is also true that distances along the scale are divisible: a reaction time of 200 ms is twice as long as a reaction time of 100 ms and half as long as a reaction time of 400 ms. Time also has a meaningful zero point: 0 ms does mean a complete absence of time.

Continuous variables can be, well, continuous (obviously) but also discrete. This is quite a tricky distinction (Jane Superbrain Box 1.3). A truly continuous variable can be measured to any level of precision, whereas a **discrete variable** can take on only certain values (usually whole numbers) on the scale. What does this actually mean? Well, our example of rating lecturers on a 5-point scale is an example of a discrete variable. The range of the scale is 1–5, but you can enter only values of 1, 2, 3, 4 or 5; you cannot enter a value of 4.32 or 2.18. Although a continuum exists underneath the scale (i.e., a rating of 3.24 makes sense), the actual values that the variable takes on are limited. A continuous variable would be something like age, which can be measured at an infinite level of precision (you could be 34 years, 7 months, 21 days, 10 hours, 55 minutes, 10 seconds, 100 milliseconds, 63 microseconds, 1 nanosecond old).

1.6.3 Measurement error ▮▮▮

It's one thing to measure variables, but it's another thing to measure them accurately. Ideally we want our measure to be calibrated such that values have the same meaning over time and across situations.

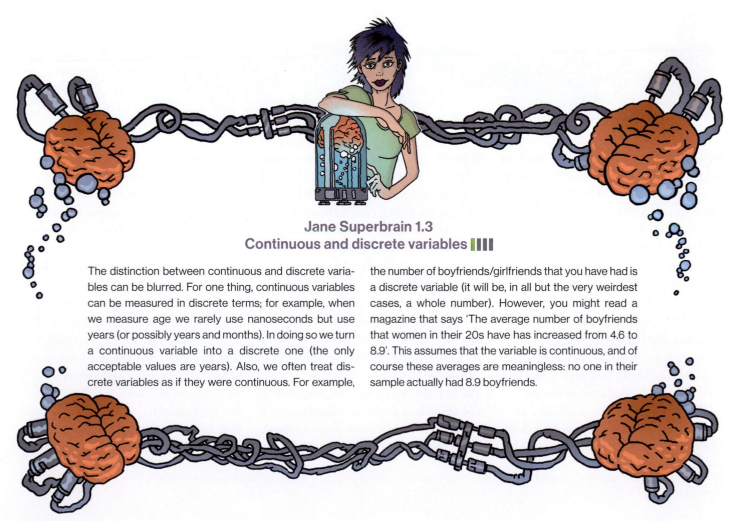

Jane Superbrain 1.3
Continuous and discrete variables ▮▮▮

The distinction between continuous and discrete variables can be blurred. For one thing, continuous variables can be measured in discrete terms; for example, when we measure age we rarely use nanoseconds but use years (or possibly years and months). In doing so we turn a continuous variable into a discrete one (the only acceptable values are years). Also, we often treat discrete variables as if they were continuous. For example, the number of boyfriends/girlfriends that you have had is a discrete variable (it will be, in all but the very weirdest cases, a whole number). However, you might read a magazine that says 'The average number of boyfriends that women in their 20s have has increased from 4.6 to 8.9'. This assumes that the variable is continuous, and of course these averages are meaningless: no one in their sample actually had 8.9 boyfriends.

Cramming Sam's Tips
Levels of measurement

- Variables can be split into categorical and continuous, and within these types there are different levels of measurement:
- Categorical (entities are divided into distinct categories):
  - Binary variable: There are only two categories (e.g., dead or alive).
  - Nominal variable: There are more than two categories (e.g., whether someone is an omnivore, vegetarian, vegan or fruitarian).
  - Ordinal variable: The same as a nominal variable but the categories have a logical order (e.g., whether people got a fail, a pass, a merit or a distinction in their exam).

- Continuous (entities get a distinct score):
  - Interval variable: Equal intervals on the variable represent equal differences in the property being measured (e.g., the difference between 6 and 8 is equivalent to the difference between 13 and 15).
  - Ratio variable: The same as an interval variable, but the ratios of scores on the scale must also make sense (e.g., a score of 16 on an anxiety scale means that the person is, in reality, twice as anxious as someone scoring 8). For this to be true, the scale must have a meaningful zero point.

Weight is one example: we would expect to weigh the same amount regardless of who weighs us or where we take the measurement (assuming it's on Earth and not in an anti-gravity chamber). Sometimes, variables can be measured directly (profit, weight, height) but in other cases we are forced to use indirect measures such as self-report, questionnaires, and computerized tasks (to name a few).

It's been a while since I mentioned sperm, so let's go back to our Coke as a spermicide example. Imagine we took some Coke and some water and added them to two test tubes of sperm. After several minutes, we measured the motility (movement) of the sperm in the two samples and discovered no

difference. A few years passed, as you might expect given that Coke and sperm rarely top scientists' research lists, before another scientist, Dr Jack Q. Late, replicated the study. Dr Late found that sperm motility was worse in the Coke sample. There are two measurement-related issues that could explain his success and our failure: (1) Dr Late might have used more Coke in the test tubes (sperm might need a critical mass of Coke before they are affected); (2) Dr Late measured the outcome (motility) differently than us.

The former point explains why chemists and physicists have devoted many hours to developing standard units of measurement. If you had reported that you'd used 100ml

of Coke and 5ml of sperm, then Dr Late could have ensured that he had used the same amount—because millilitres are a standard unit of measurement we would know that Dr Late used exactly the same amount of Coke that we used. Direct measurements such as the millilitre provide an objective standard: 100ml of a liquid is known to be twice as much as only 50ml.

The second reason for the difference in results between the studies could have been to do with how sperm motility was measured. Perhaps in our original study we measured motility using absorption spectrophotometry, whereas Dr Late used laser light-scattering techniques.[11] Perhaps his measure is more sensitive than ours.

---

11 In the course of writing this chapter I have discovered more than I think is healthy about the measurement of sperm motility.

There will often be a discrepancy between the numbers we use to represent the thing we're measuring and the actual value of the thing we're measuring (i.e., the value we would get if we could measure it directly). This discrepancy is known as **measurement error**. For example, imagine that you know as an absolute truth that you weigh 83kg. One day you step on the bathroom scales and they read 80kg. There is a difference of 3kg between your actual weight and the weight given by your measurement tool (the scales): this is a measurement error of 3kg. Although properly calibrated bathroom scales should produce only very small measurement errors (despite what we might want to believe when it says we have gained 3kg), self-report measures will produce larger measurement error because factors other than the one you're trying to measure will influence how people respond to our measures. For example, if you were completing a questionnaire that asked you whether you had stolen from a shop, would you admit it or might you be tempted to conceal this fact?

1.6.4 Validity and reliability

One way to try to ensure that measurement error is kept to a minimum is to determine properties of the measure that give us confidence that it is doing its job properly. The first property is **validity**, which is whether an instrument measures what it sets out to measure. The second is **reliability**, which is whether an instrument can be interpreted consistently across different situations.

Validity refers to whether an instrument measures what it was designed to measure (e.g., does your lecturer helpfulness rating scale actually measure lecturers' helpfulness?); a device for measuring sperm *motility* that actually measures sperm *count* is not valid. Things like reaction times and physiological measures are valid in the sense that a reaction time does, in fact, measure the

SELF TEST
What is the difference between reliability and validity?

time taken to react and skin conductance does measure the conductivity of your skin. However, if we're using these things to infer other things (e.g., using skin conductance to measure anxiety), then they will be valid only if there are no other factors other than the one we're interested in that can influence them.

Criterion validity is whether you can establish that an instrument measures what it claims to measure through comparison to objective criteria. In an ideal world, you assess this by relating scores on your measure to real-world observations. For example, we could take an objective measure of how helpful lecturers were and compare these observations to students' ratings of helpfulness on ratemyprofessor.com. When data are recorded simultaneously using the new instrument and existing criteria, then this is said to assess **concurrent validity**; when data from the new instrument are used to predict observations at a later point in time, this is said to assess **predictive validity**.

Assessing criterion validity (whether concurrently or predictively) is often impractical because objective criteria that can be measured easily may not exist. Also, with measuring attitudes, you might be interested in the person's perception of reality and not reality itself (you might not care whether a person *is* a psychopath but whether they *think* they are a psychopath). With self-report measures/questionnaires we can also assess the degree to which individual items represent the construct being measured, and cover the full range of the construct (**content validity**).

Validity is a necessary but not sufficient condition of a measure. A second consideration is reliability, which is the ability of the measure to produce the same results under the same conditions. To be valid the instrument must first be reliable. The easiest way to assess reliability is to test the same group of people twice: a reliable instrument will produce similar scores at both points in time (**test-retest reliability**). Sometimes, however, you will want to measure something that does vary over time (e.g., moods, blood-sugar levels, productivity). Statistical methods can also be used to determine reliability (we will discover these in Chapter 18).

1.7 Collecting data: research design

We've looked at the question of *what* to measure and discovered that to answer scientific questions we measure variables (which can be collections of numbers or words). We also saw that to get accurate answers we need accurate measures. We move on now to look at research design: *how* data are collected. If we simplify things quite a lot then there are two ways to test a hypothesis: either by observing what naturally happens or by manipulating some aspect of the environment and observing the effect it has on the variable that interests us. In **correlational** or **cross-sectional research** we observe what naturally goes on in the world without directly interfering with it, whereas in **experimental research** we manipulate one variable to see its effect on another.

1.7.1 Correlational research methods

In correlational research we observe natural events; we can do this by either taking a snapshot of many variables at a single point in time or by measuring variables repeatedly at different time points (known as **longitudinal research**). For example, we might measure pollution levels in a stream and the numbers of certain types of fish living there; lifestyle variables (smoking, exercise, food intake) and disease (cancer, diabetes); workers' job satisfaction under different managers; or children's school performance across regions with different demographics. Correlational research provides a very natural view of the question we're researching because we're not influencing what happens and the measures of the variables should not be biased by the researcher being there (this is an important aspect of **ecological validity**).

What's the difference between experimental and correlational research?

At the risk of sounding like I'm absolutely obsessed with using Coke as a contraceptive (I'm not, but my discovery that people in the 1950s and 1960s actually tried this has, I admit, intrigued me), let's return to that example. If we wanted to answer the question, 'Is Coke an effective contraceptive?' we could administer questionnaires about sexual practices (quantity of sexual activity, use of contraceptives, use of fizzy drinks as contraceptives, pregnancy, etc.). By looking at these variables, we could see which variables correlate with pregnancy and, in particular, whether those reliant on Coca-Cola as a form of contraceptive were more likely to end up pregnant than those using other contraceptives, and less likely than those using no contraceptives at all. This is the only way to answer a question like this because we cannot manipulate any of these variables particularly easily. Even if we could, it would be totally unethical to insist on

some people using Coke as a contraceptive (or indeed to do anything that would make a person likely to produce a child that they didn't intend to produce). However, there is a price to pay, which relates to causality: correlational research tells us nothing about the causal influence of variables.

1.7.2 Experimental research methods ▮▮▮▮

Most scientific questions imply a causal link between variables; we have seen already that dependent and independent variables are named such that a causal connection is implied (the dependent variable *depends* on the independent variable). Sometimes the causal link is very obvious in the research question, 'Does low self-esteem cause dating anxiety?' Sometimes the implication might be subtler; for example, in 'Is dating anxiety all in the mind?' the implication is that a person's mental outlook causes them to be anxious when dating. Even when the cause-effect relationship is not explicitly stated, most research questions can be broken down into a proposed cause (in this case, mental outlook) and a proposed outcome (dating anxiety). Both the cause and the outcome are variables: for the cause, some people will perceive themselves in a negative way (so it is something that varies); and, for the outcome, some people will get more anxious on dates than others (again, this is something that varies). The key to answering the research question is to uncover how the proposed cause and the proposed outcome relate to each other; are the people who have a low opinion of themselves the same people who are more anxious on dates?

David Hume, an influential philosopher, defined a cause as 'An object precedent and contiguous to another, and where all the objects resembling the former are placed in like relations of precedency and

contiguity to those objects that resemble the latter' (1739–40/1965).[12] This definition implies that (1) the cause needs to precede the effect, and (2) causality is equated to high degrees of correlation between contiguous events. In our dating example, to infer that low self-esteem caused dating anxiety, it would be sufficient to find that low self-esteem and feeling anxious when on a date co-occur, and that the low self-esteem emerged before the dating anxiety did.

In correlational research variables are often measured simultaneously. The first problem with doing this is that it provides no information about the contiguity between different variables: we might find from a questionnaire study that people with low self-esteem also have dating anxiety but we wouldn't know whether it was the low self-esteem or the dating anxiety that came first. Longitudinal research addresses this issue to some extent, but there is still a problem with Hume's idea that causality can be inferred from corroborating evidence, which is that it doesn't distinguish between what you might call an 'accidental' conjunction and a causal one. For example, it could be that both low self-esteem and dating anxiety are caused by a third variable (e.g., poor social skills which might make you feel generally worthless but also puts pressure on you in dating situations). Therefore, low self-esteem and dating anxiety do always co-occur (meeting Hume's definition of cause) but only because poor social skills causes them both.

This example illustrates an important limitation of correlational research: the **tertium quid** ('A third person or thing of indeterminate character'). For example, a correlation has been found between having breast implants and suicide (Koot, Peeters, Granath, Grobbee, & Nyren, 2003). However, it is unlikely that

---

12 As you might imagine, his view was a lot more complicated than this definition alone, but let's not get sucked down that particular wormhole.

having breast implants causes you to commit suicide—presumably, there is an external factor (or factors) that causes both; for example, low self-esteem might lead you to have breast implants and also attempt suicide. These extraneous factors are sometimes called **confounding variables** or confounds for short.

The shortcomings of Hume's definition led John Stuart Mill (1865) to suggest that, in addition to a correlation between events, all other explanations of the cause–effect relationship must be ruled out. To rule out confounding variables, Mill proposed that an effect should be present when the cause is present and that when the cause is absent, the effect should be absent also. In other words, the only way to infer causality is through comparing two controlled situations: one in which the cause is present and one in which the cause is absent. This is what *experimental methods* strive to do: to provide a comparison of situations (usually called *treatments* or *conditions*) in which the proposed cause is present or absent.

As a simple case, we might want to look at the effect of feedback style on learning about statistics. I might, therefore, randomly split[13] some students into three different groups, in which I change my style of feedback in the seminars on my course:

- **Group 1 (supportive feedback)**: During seminars I congratulate all students in this group on their hard work and success. Even when they get things wrong, I am supportive and say things like 'that was very nearly the right answer, you're coming along really well' and then give them a nice piece of chocolate.
- **Group 2 (harsh feedback)**: This group receives seminars in which I give relentless verbal abuse to all of the students even when they give the correct answer. I demean their contributions and am patronizing and dismissive of

everything they say. I tell students that they are stupid, worthless, and shouldn't be doing the course at all. In other words, this group receives normal university-style seminars. ☺
- **Group 3 (no feedback)**: Students are not praised or punished, instead I give them no feedback at all.

The thing that I have manipulated is the feedback style (supportive, harsh or none). As we have seen, this variable is known as the independent variable and, in this situation, it is said to have three levels, because it has been manipulated in three ways (i.e., the feedback style has been split into three types: supportive, harsh and none). The outcome in which I am interested is statistical ability, and I could measure this variable using a statistics exam after the last seminar. As we have seen, this outcome variable is the dependent variable because we assume that these scores will depend upon the type of teaching method used (the independent variable). The critical thing here is the inclusion of the 'no feedback' group because this is a group in which our proposed cause (feedback) is absent, and we can compare the outcome in this group against the two situations in which the proposed cause is present. If the statistics scores are different in each of the feedback groups (cause is present) compared to the group for which no feedback was given (cause is absent), then this difference can be attributed to the type of feedback used. In other words, the style of feedback used caused a difference in statistics scores (Jane Superbrain Box 1.4).

1.7.3 Two methods of data collection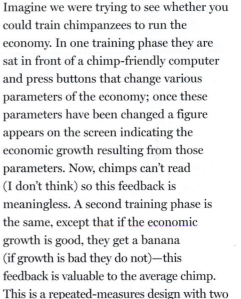

When we use an experiment to collect data, there are two ways to manipulate the independent variable. The first is to test different entities. This method is the one

described above, in which different groups of entities take part in each experimental condition (a **between-groups**, **between-subjects** or **independent design**). An alternative is to manipulate the independent variable using the same entities. In our motivation example, this means that we give a group of students supportive feedback for a few weeks and test their statistical abilities and then give this same group harsh feedback for a few weeks before testing them again and, then, finally, give them no feedback and test them for a third time (a **within-subject** or **repeated-measures design**). As you will discover, the way in which the data are collected determines the type of test that is used to analyze the data.

1.7.4 Two types of variation

Imagine we were trying to see whether you could train chimpanzees to run the economy. In one training phase they are sat in front of a chimp-friendly computer and press buttons that change various parameters of the economy; once these parameters have been changed a figure appears on the screen indicating the economic growth resulting from those parameters. Now, chimps can't read (I don't think) so this feedback is meaningless. A second training phase is the same, except that if the economic growth is good, they get a banana (if growth is bad they do not)—this feedback is valuable to the average chimp. This is a repeated-measures design with two conditions: the same chimps participate in condition 1 *and* in condition 2.

Let's take a step back and think what would happen if we did *not* introduce an experimental manipulation (i.e., there were no bananas in the second training phase, so condition 1 and condition 2 were identical). If there is no experimental manipulation

---

13 This random assignment of students is important, but we'll get to that later.

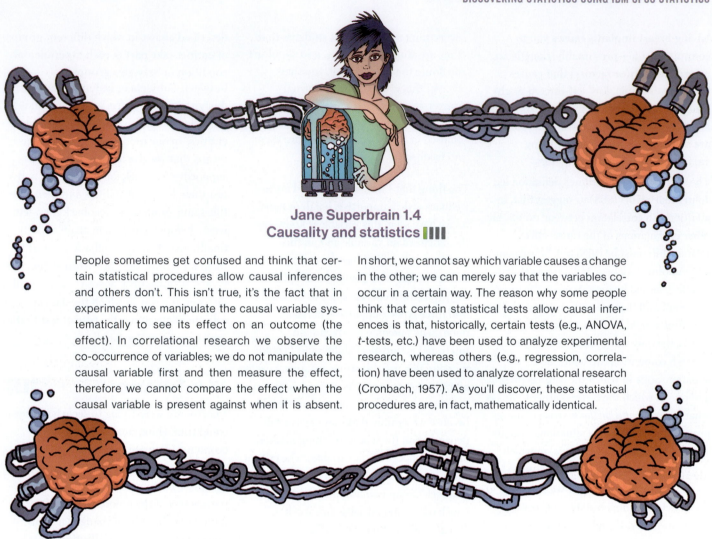

Jane Superbrain 1.4
Causality and statistics ▌▌▌▌

People sometimes get confused and think that certain statistical procedures allow causal inferences and others don't. This isn't true, it's the fact that in experiments we manipulate the causal variable systematically to see its effect on an outcome (the effect). In correlational research we observe the co-occurrence of variables; we do not manipulate the causal variable first and then measure the effect, therefore we cannot compare the effect when the causal variable is present against when it is absent.

In short, we cannot say which variable causes a change in the other; we can merely say that the variables co-occur in a certain way. The reason why some people think that certain statistical tests allow causal inferences is that, historically, certain tests (e.g., ANOVA, *t*-tests, etc.) have been used to analyze experimental research, whereas others (e.g., regression, correlation) have been used to analyze correlational research (Cronbach, 1957). As you'll discover, these statistical procedures are, in fact, mathematically identical.

then we expect a chimp's behavior to be similar in both conditions. We expect this because external factors such as age, sex, IQ, motivation and arousal will be the same for both conditions (a chimp's biological sex, etc. will not change from when they are tested in condition 1 to when they are tested in condition 2). If the performance measure (i.e., our test of how well they run the economy) is reliable, and the variable or characteristic that we are measuring (in this case ability to run an economy) remains stable over time, then a participant's performance in condition 1 should be very highly related to their performance in condition 2. So, chimps who score highly in

condition 1 will also score highly in condition 2, and those who have low scores for condition 1 will have low scores in condition 2. However, performance won't be *identical*, there will be small differences in performance created by unknown factors. This variation in performance is known as **unsystematic variation**.

If we introduce an experimental manipulation (i.e., provide bananas as feedback in one of the training sessions), then we do something different to participants in condition 1 than in condition 2. So, the *only* difference between conditions 1 and 2 is the manipulation that the experimenter has made (in this case

that the chimps get bananas as a positive reward in one condition but not in the other).[14] Therefore, any differences between the means of the two conditions are probably due to the experimental manipulation. So, if the chimps perform better in one training phase than in the other, this *has* to be due to the fact that bananas were used to provide feedback in one training phase but not in the other. Differences in performance created by a specific experimental manipulation are known as **systematic variation**.

Now let's think about what happens when we use different participants—an independent design. In this design, we still

---

14 Actually, this isn't the only difference because, by condition 2, they have had some practice (in condition 1) at running the economy; however, we will see shortly that these practice effects are easily eradicated.

have two conditions, but this time different participants participate in each condition. Going back to our example, one group of chimps receives training without feedback, whereas a second group of different chimps does receive feedback on their performance via bananas.[15] Imagine again that we didn't have an experimental manipulation. If we did nothing to the groups, then we would still find some variation in behavior between the groups because they contain different chimps who will vary in their ability, motivation, propensity to get distracted from running the economy by throwing their own feces, and other factors. In short, the factors that were held constant in the repeated-measures design are free to vary in the independent design. So, the unsystematic variation will be bigger than for a repeated-measures design. As before, if we introduce a manipulation (i.e., bananas), then we will see additional variation created by this manipulation. As such, in both the repeated-measures design and the independent design there are always two sources of variation:

- **Systematic variation**: This variation is due to the experimenter doing something in one condition but not in the other condition.
- **Unsystematic variation**: This variation results from random factors that exist between the experimental conditions (such as natural differences in ability, the time of day, etc.).

Statistical tests are often based on the idea of estimating how much variation there is in performance, and comparing how much of this is systematic to how much is unsystematic.

In a repeated-measures design, differences between two conditions can be caused by only two things: (1) the manipulation that was carried out on the participants or (2) any other factor that might affect the way in

which an entity performs from one time to the next. The latter factor is likely to be fairly minor compared to the influence of the experimental manipulation. In an independent design, differences between the two conditions can also be caused by one of two things: (1) the manipulation that was carried out on the participants or (2) differences between the characteristics of the entities allocated to each of the groups. The latter factor, in this instance, is likely to create considerable random variation both within each condition and between them. When we look at the effect of our experimental manipulation, it is always against a background of 'noise' created by random, uncontrollable differences between our conditions. In a repeated-measures design this 'noise' is kept to a minimum and so the effect of the experiment is more likely to show up. This means that, other things being equal, repeated-measures designs are more sensitive to detect effects than independent designs.

1.7.5 Randomization

In both repeated-measures and independent designs it is important to try to keep the unsystematic variation to a minimum. By keeping the unsystematic variation as small as possible we get a more sensitive measure of the experimental manipulation. Generally, scientists use the **randomization** of entities to treatment conditions to achieve this goal. Many statistical tests work by identifying the systematic and unsystematic sources of variation and then comparing them. This comparison allows us to see whether the experiment has generated considerably more variation than we would have got had we just tested participants without the experimental manipulation. Randomization is important because it eliminates most

other sources of systematic variation, which allows us to be sure that any systematic variation between experimental conditions is due to the manipulation of the independent variable. We can use randomization in two different ways depending on whether we have an independent or repeated-measures design. Let's look at a repeated-measures design first. I mentioned earlier (in a footnote) that when the same entities participate in more than one experimental condition they are naive during the first experimental condition but they come to the second experimental condition with prior experience of what is expected of them. At the very least they will be familiar with the dependent measure (e.g., the task they're performing). The two most important sources of systematic variation in this type of design are:

- **Practice effects**: Participants may perform differently in the second condition because of familiarity with the experimental situation and/or the measures being used.
- **Boredom effects**: Participants may perform differently in the second condition because they are tired or bored from having completed the first condition.

Although these effects are impossible to eliminate completely, we can ensure that they produce no systematic variation between our conditions by **counterbalancing** the order in which a person participates in a condition.

We can use randomization to determine in which order the conditions are completed. That is, we randomly determine whether a participant completes condition 1 before condition 2 or condition 2 before condition 1. Let's look at the teaching method example and imagine that there were just two conditions: no feedback and harsh

---

15 Obviously I mean that they receive a banana as a reward for their correct response and not that the bananas develop little banana mouths that sing them a little congratulatory song.

Why is randomization important?

feedback. If the same participants were used in all conditions, then we might find that statistical ability was higher after the harsh feedback. However, if every student experienced the harsh feedback after the no feedback seminars then they would enter the harsh condition already having a better knowledge of statistics than when they began the no feedback condition. So, the apparent improvement after harsh feedback would not be due to the experimental manipulation (i.e., it's not because harsh feedback works), but because participants had attended more statistics seminars by the end of the harsh feedback condition compared to the no feedback one. We can use randomization to ensure that the number of statistics seminars does not introduce a systematic bias by randomly assigning students to have the harsh feedback seminars first or the no feedback seminars first.

If we turn our attention to independent designs, a similar argument can be applied. We know that participants in different experimental conditions will differ in many respects (their IQ, attention span, etc.). Although we know that these confounding variables contribute to the variation between conditions, we need to make sure that these variables contribute to the unsystematic variation and *not* to the systematic variation. A good example is the effects of alcohol on behavior. You might give one group of people 5 pints of beer, and keep a second group sober, and then count how many times you can persuade them to do a fish impersonation. The effect that alcohol has varies because people differ in their tolerance: teetotal

people can become drunk on a small amount, while alcoholics need to consume vast quantities before the alcohol affects them. If you allocated a bunch of hardened drinkers to the condition that consumed alcohol, and teetotal people to the no alcohol condition, then you might find that alcohol doesn't increase the number of fish impersonations you get. However, this finding could be because (1) alcohol does not make people engage in frivolous activities or (2) the hardened drinkers were unaffected by the dose of alcohol. You have no way to dissociate these explanations because the groups varied not just on dose of alcohol but also on their tolerance of alcohol (the systematic variation created by their past experience with alcohol cannot be separated from the effect of the experimental manipulation). The best way to reduce this eventuality is to randomly allocate participants to conditions: by doing so you minimize the risk that groups differ on variables other than the one you want to manipulate.

1.8 Analyzing data ▮▮▮▮

The final stage of the research process is to analyze the data you have collected. When the data are quantitative this involves both looking at your data graphically (Chapter 5) to see what the general trends in the data are, and also fitting statistical models to the data (all other chapters). Given that the rest of the book is dedicated to this process, we'll begin here by looking at a few fairly basic ways to look at and summarize the data you have collected.

1.8.1 Frequency distributions ▮▮▮▮

What is frequency distribution and when is it normal?

Once you've collected some data a very useful thing to do is to plot a graph of how many times each score occurs. This is known as a **frequency distribution** or **histogram**, which is a graph plotting values of observations on the horizontal axis, with a bar showing how many times each value occurred in the data set. Frequency distributions can be very useful for assessing properties of the distribution of scores. We will find out how to create these types of charts in Chapter 5.

Frequency distributions come in many different shapes and sizes. It is quite important, therefore, to have some general descriptions for common types of distributions. In an ideal world our data would be distributed symmetrically around the center of all scores. As such, if we drew a vertical line through the center of the distribution then it should look the same on both sides. This is known as a **normal distribution** and is characterized by the bell-shaped curve with which you might already be familiar. This shape implies that the majority of scores lie around the center of the distribution (so the largest bars on the histogram are around the central value). Also, as we get further away from the center, the bars get smaller, implying that as scores start to deviate from the center their frequency is decreasing. As we move still further away from the center our scores become very infrequent (the bars are very short). Many naturally occurring things have this shape of distribution. For example, most men in the UK are around 175 cm tall;[16] some are a bit taller or shorter, but most cluster around this value. There will be very few men who are really

---

16 I am exactly 180 cm tall. In my home country this makes me smugly above average. However, I often visit the Netherlands, where the average male height is 185 cm (a little over 6ft, and a massive 10 cm higher than the UK), and where I feel like a bit of a dwarf.

tall (i.e., above 205 cm) or really short (i.e., under 145 cm). An example of a normal distribution is shown in Figure 1.3.

There are two main ways in which a distribution can deviate from normal: (1) lack of symmetry (called **skew**) and (2) pointyness (called **kurtosis**). Skewed distributions are not symmetrical and instead the most frequent scores (the tall bars on the graph) are clustered at one end of the scale. So, the typical pattern is a cluster of frequent scores at one end of the scale and the frequency of scores tailing off towards the other end of the scale. A skewed distribution can be either *positively skewed* (the frequent scores are clustered at the lower end and the tail points towards the higher or more positive scores) or *negatively skewed* (the frequent scores are clustered at the higher end and the tail points towards the lower or more negative scores). Figure 1.4 shows examples of these distributions.

Distributions also vary in their kurtosis. Kurtosis, despite sounding like some kind of exotic disease, refers to the degree to which scores cluster at the ends of the distribution (known as the *tails*) and this tends to express itself in how pointy a distribution is (but there are other factors that can affect how pointy the distribution looks—see Jane Superbrain Box 1.5). A distribution with *positive kurtosis* has many scores in the tails (a so-called heavy-tailed distribution) and is pointy. This is known as a **leptokurtic** distribution. In contrast, a distribution with *negative kurtosis* is relatively thin in the tails (has light tails) and tends to be flatter than normal. This distribution is called **platykurtic**. Ideally, we want our data to be normally distributed (i.e., not too skewed, and not too many or too few scores at the extremes). For everything there is to know about kurtosis, read DeCarlo (1997).

In a normal distribution the values of skew and kurtosis are 0 (i.e., the tails of the

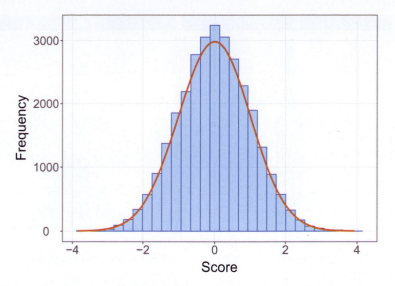

Figure 1.3   A 'normal' distribution (the curve shows the idealized shape)

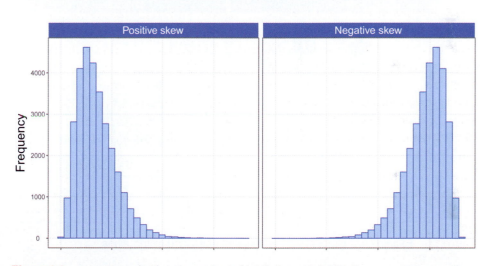

Figure 1.4   A positively (left) and negatively (right) skewed distribution

distribution are as they should be).[17] If a distribution has values of skew or kurtosis above or below 0 then this indicates a deviation from normal: Figure 1.5 shows distributions with kurtosis values of +2.6 (left panel) and −0.09 (right panel).

1.8.2   The mode ▌▐▐▐

We can calculate where the center of a frequency distribution lies (known as the **central tendency**) using three measures

commonly used: the mean, the mode and the median. Other methods exist, but these three are the ones you're most likely to come across.

The **mode** is the score that occurs most frequently in the data set. This is easy to spot in a frequency distribution because it will be the tallest bar. To calculate the mode, place the data in ascending order (to make life easier), count how many times each score occurs, and the score that

---

17 Sometimes no kurtosis is expressed as 3 rather than 0, but SPSS uses 0 to denote no excess kurtosis.

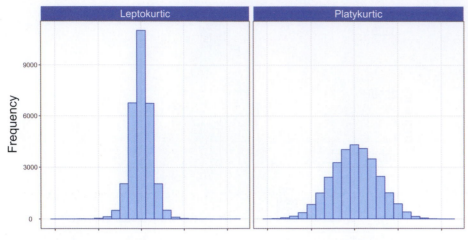

Figure 1.5 Distributions with positive kurtosis (leptokurtic, left) and negative kurtosis (platykurtic, right)

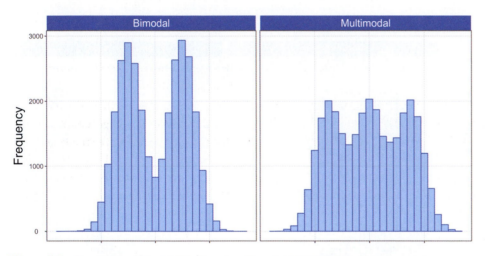

Figure 1.6 Examples of bimodal (left) and multimodal (right) distributions

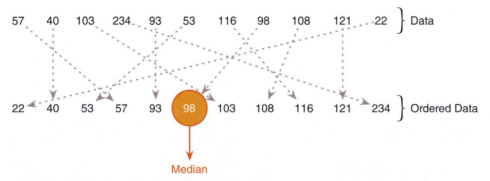

Median

Figure 1.7 The median is simply the middle score when you order the data

frequencies of certain scores are very similar, then the mode can be influenced by only a small number of cases.

1.8.3 The median

Another way to quantify the center of a distribution is to look for the middle score when scores are ranked in order of magnitude. This is called the **median**. Imagine we looked at the number of friends that 11 users of the social networking website Facebook had. Figure 1.7 shows the number of friends for each of the 11 users: 57, 40, 103, 234, 93, 53, 116, 98, 108, 121, 22.

To calculate the median, we first arrange these scores into ascending order: 22, 40, 53, 57, 93, 98, 103, 108, 116, 121, 234. Next, we find the position of the middle score by counting the number of scores we have collected (n), adding 1 to this value, and then dividing by 2. With 11 scores, this gives us $(n + 1)/2 = (11 + 1)/2 = 12/2 = 6$. Then, we find the score that is positioned at the location we have just calculated. So, in this example, we find the sixth score (see Figure 1.7).

This process works very nicely when we have an odd number of scores (as in this example), but when we have an even number of scores there won't be a middle value. Let's imagine that we decided that because the highest score was so big (almost twice as large as the next biggest number), we would ignore it. (For one thing, this person is far too popular and we hate them.) We have only 10 scores now. Figure 1.8 shows this situation. As before, we rank-order these scores: 22, 40, 53, 57, 93, 98, 103, 108, 116, 121. We then calculate the position of the middle score, but this time it is $(n + 1)/2 = 11/2 = 5.5$, which means that the median is halfway between the fifth and sixth scores. To get the median we add these two scores and

occurs the most is the mode. One problem with the mode is that it can take on several values. For example, Figure 1.6 shows an example of a distribution with two modes (there are two bars that are the highest), which is said to be **bimodal**, and three modes (data sets with more than two modes are **multimodal**). Also, if the

divide by 2. In this example, the fifth score in the ordered list was 93 and the sixth score was 98. We add these together (93 + 98 = 191) and then divide this value by 2 (191/2 = 95.5). The median number of friends was, therefore, 95.5.

The median is relatively unaffected by extreme scores at either end of the distribution: the median changed only from 98 to 95.5 when we removed the extreme score of 234. The median is also relatively unaffected by skewed distributions and can be used with ordinal, interval and ratio data (it cannot, however, be used with nominal data because these data have no numerical order).

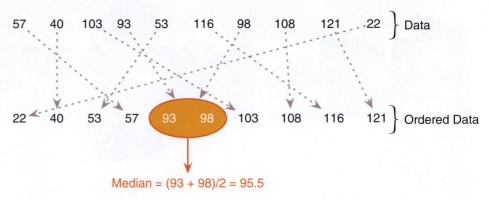

Median = (93 + 98)/2 = 95.5

Figure 1.8 When the data contain an even number of scores, the median is the average of the middle two values

1.8.4 The mean

The **mean** is the measure of central tendency that you are most likely to have heard of because it is the average score, and the media love an average score.[18] To calculate the mean we add up all of the scores and then divide by the total number of scores we have. We can write this in equation form as:

$$\bar{X} = \frac{\sum_{i=1}^{n} x_i}{n} \qquad (1.1)$$

This equation may look complicated, but the top half simply means 'add up all of the scores' (the x_i means 'the score of a particular person'; we could replace the letter i with each person's name instead), and the bottom bit means, 'divide this total by the number of scores you have got (n)'. Let's calculate the mean for the Facebook data. First, we add up all the scores:

$$\sum_{i=1}^{n} x_i = 22+40+53+57+93+98+103 \\ +108+116+121+234 = 1045 \qquad (1.2)$$

We then divide by the number of scores (in this case 11) as in equation (1.3):

$$\bar{X} = \frac{\sum_{i=1}^{n} x_i}{n} = \frac{1045}{11} = 95 \qquad (1.3)$$

The mean is 95 friends, which is not a value we observed in our actual data. In this sense the mean is a statistical model—more on this in the next chapter.

If you calculate the mean without our most popular person (i.e., excluding the value 234), the mean drops to 81.1 friends. This reduction illustrates one disadvantage of the mean: it can be influenced by extreme scores. In this case, the person with 234 friends on Facebook increased the mean by about 14 friends; compare this difference with that of the median. Remember that the median changed very little – from 98 to 95.5 – when we excluded the score of 234, which illustrates how the median is typically less affected by extreme scores than the mean. While we're being negative about the

SELF TEST

Compute the mean but excluding the score of 234.

mean, it is also affected by skewed distributions and can be used only with interval or ratio data.

If the mean is so lousy then why do we use it so often? One very important reason is that it uses every score (the mode and median ignore most of the scores in a data set). Also, the mean tends to be stable in different samples (more on that later too).

1.8.5 The dispersion in a distribution

It can also be interesting to quantify the spread or dispersion, of scores. The easiest way to look at dispersion is to take the largest score and subtract from it the smallest score. This is known as the **range** of scores. For our Facebook data we saw that if we order the scores, we get 22, 40, 53, 57, 93, 98, 103, 108, 116, 121, 234. The highest score is 234 and the lowest is 22;

SELF TEST

Compute the range but excluding the score of 234.

---

18 I wrote this on 15 February, and to prove my point, the BBC website ran a headline today about how PayPal estimates that Britons will spend an average of £71.25 each on Valentine's Day gifts. However, uSwitch.com said that the average spend would be only £22.69. Always remember that the media is full of lies and contradictions.

Cramming Sam's Tips
Central tendency

- The mean is the sum of all scores divided by the number of scores. The value of the mean can be influenced quite heavily by extreme scores.

- The median is the middle score when the scores are placed in ascending order. It is not as influenced by extreme scores as the mean.
- The mode is the score that occurs most frequently.

Interquartile Range

22   40   **53**   57   93   **98**   103   108   **116**   121   234   } Ordered Data

Lower Quartile   Median
(Second Quartile)   Upper Quartile

Figure 1.9 Calculating quartiles and the interquartile range

therefore, the range is 234–22 = 212. One problem with the range is that because it uses only the highest and lowest score, it is affected dramatically by extreme scores.

If you have done the self-test task you'll see that without the extreme score the range drops from 212 to 99—less than half the size.

One way around this problem is to calculate the range but excluding values at the extremes of the distribution. One convention is to cut off the top and bottom 25% of scores and calculate the range of the middle 50% of scores—known as the **interquartile range**. Let's do this with the Facebook data. First we

need to calculate what are called **quartiles**. Quartiles are the three values that split the sorted data into four equal parts. First, we calculate the median, which is also called the *second quartile*, which splits our data into two equal parts. We already know that the median for these data is 98. The **lower quartile** is the median of the lower half of the data and the **upper quartile** is the median of the upper half of the data. As a rule of thumb the median is not included in the two halves when they are split (this is convenient if you have an odd number of values), but you can include it (although

which half you put it in is another question). Figure 1.9 shows how we would calculate these values for the Facebook data. Like the median, if each half of the data had an even number of values in it, then the upper and lower quartiles would be the average of two values in the data set (therefore, the upper and lower quartile need not be values that actually appear in the data). Once we have worked out the values of the quartiles, we can calculate the interquartile range, which is the difference between the upper and lower quartile. For the Facebook data this value would be 116–53 = 63. The advantage of the interquartile range is that it isn't affected by extreme scores at either end of the distribution. However, the problem with it is that you lose a lot of data (half of it, in fact).

It's worth noting here that quartiles are special cases of things called **quantiles**. Quantiles are values that split a data set into equal portions. Quartiles are quantiles that split the data into four equal parts, but there are other quantiles such as **percentiles** (points that split the data into

100 equal parts), **noniles** (points that split the data into nine equal parts) and so on.

If we want to use all the data rather than half of it, we can calculate the spread of scores by looking at how different each score is from the center of the distribution. If we use the mean as a measure of the center of a distribution, then we can calculate the difference between each score and the mean, which is known as the **deviance** (Eq. 1.4):

$$deviance = x_i - \bar{x} \qquad (1.4)$$

If we want to know the total deviance then we could add up the deviances for each data point. In equation form, this would be:

$$total\ deviance = \sum_{i=1}^{n}(x_i - \bar{x}) \qquad (1.5)$$

The sigma symbol (Σ) means 'add up all of what comes after', and the 'what comes after' in this case is the deviances. So, this equation simply means 'add up all of the deviances'.

Let's try this with the Facebook data. Table 1.2 shows the number of friends for each person in the Facebook data, the mean, and the difference between the two. Note that because the mean is at the center of the distribution, some of the deviations are positive (scores greater than the mean) and some are negative (scores smaller than the mean). Consequently, when we add the scores up, the total is zero. Therefore, the 'total spread' is nothing. This conclusion is as silly as a tapeworm thinking they can have a coffee with the Queen of England if they don a bowler hat and pretend to be human. Everyone knows that the Queen drinks tea.

To overcome this problem, we could ignore the minus signs when we add the deviations up. There's nothing wrong with doing this, but people tend to square the deviations, which has a similar effect (because a negative number multiplied by another negative number becomes

SELF TEST
Twenty-one heavy smokers were put on a treadmill at the fastest setting. The time in seconds was measured until they fell off from exhaustion:

18, 16, 18, 24, 23, 22, 22, 23, 26, 29, 32, 34, 34, 36, 36, 43, 42, 49, 46, 46, 57

Compute the mode, median, mean, upper and lower quartiles, range and interquartile range.

positive). The final column of Table 1.2 shows these squared deviances. We can add these squared deviances up to get the **sum of squared errors, SS** (often just called the *sum of squares*); unless your scores are all exactly the same, the resulting value will be bigger than zero, indicating that there is some deviance from the mean. As an equation, we would write: equation (1.6) in which the sigma symbol means 'add up all of the things that follow' and what follows is the squared deviances (or *squared errors* as they're more commonly known):

$$sum\ of\ squared\ errors\ (SS) = \sum_{i=1}^{n}(x_i - \bar{x})^2 \qquad (1.6)$$

We can use the sum of squares as an indicator of the total dispersion or total deviance of scores from the mean. The

problem with using the total is that its size will depend on how many scores we have in the data. The sum of squares for the Facebook data is 32,246, but if we added another 11 scores that value would increase (other things being equal, it will more or less double in size). The total dispersion is a bit of a nuisance then because we can't compare it across samples that differ in size. Therefore, it can be useful to work not with the *total* dispersion, but the *average* dispersion, which is also known as the **variance**. We have seen that an average is the total of scores divided by the number of scores, therefore, the variance is simply the sum of squares divided by the number of observations (N). Actually, we normally divide the SS by the number of observations minus 1 as in equation (1.7) (the reason why is explained in the next chapter and Jane Superbrain Box 2.2):

Table 1.2 Table showing the deviations of each score from the mean

Number of Friends (x_i)	Mean ($\bar{x}$)	Deviance ($x_i - \bar{x}$)	Deviance squared ($x_i - \bar{x}$)^2
22	95	−73	5329
40	95	−55	3025
53	95	−42	1764
57	95	−38	1444
93	95	−2	4
98	95	3	9
103	95	8	64
108	95	13	169
116	95	21	441
121	95	26	676
234	95	139	19321
		$\sum_{i=1}^{n} x_i - \bar{x} = 0$	$\sum_{i=1}^{n}(x_i - \bar{x})^2 = 32246$

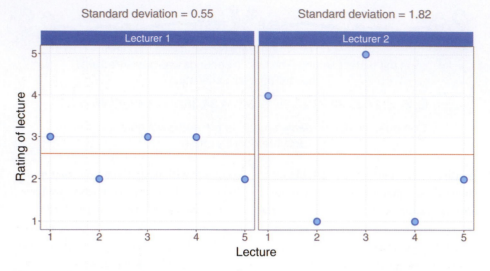

Figure 1.10 Graphs illustrating data that have the same mean but different standard deviations

$$\text{variance}\left(s^2\right) = \frac{\text{SS}}{N-1}$$

$$= \frac{\sum_{i=1}^{n}\left(x_i - \bar{x}\right)^2}{N-1} = \frac{32{,}246}{10} = 3224.6 \qquad (1.7)$$

As we have seen, the variance is the average error between the mean and the observations made. There is one problem with the variance as a measure: it gives us a measure in units squared (because we squared each error in the calculation). In our example we would have to say that the average error in our data was 3224.6 friends squared. It makes very little sense to talk about friends squared, so we often take the square root of the variance (which ensures that the measure of average error is in the same units as the original measure). This measure is known as the **standard deviation** and is the square root of the variance (Eq. 1.8).

$$s = \sqrt{\frac{\sum_{i=1}^{n}\left(x_i - \bar{x}\right)^2}{N-1}} \qquad (1.8)$$

$$= \sqrt{3224.6}$$

$$= 56.79$$

The sum of squares, variance and standard deviation are all measures of the dispersion or spread of data around the mean. A small standard deviation (relative to the value of the mean itself) indicates that the data points are close to the mean. A large standard deviation (relative to the mean) indicates that the data points are distant from the mean. A standard deviation of 0 would mean that all the scores were the same. Figure 1.10 shows the overall ratings (on a 5-point scale) of two lecturers after each of five different lectures. Both lecturers had an average rating of 2.6 out of 5 across the lectures. However, the first lecturer had a standard deviation of 0.55 (relatively small compared to the mean). It should be clear from the left-hand graph that ratings for this lecturer were consistently close to the mean rating. There was a small fluctuation, but generally her lectures did not vary in popularity. Put another way, the scores are not spread too widely around the mean. The second lecturer, however, had a standard deviation of 1.82 (relatively high compared to the mean). The ratings for this second lecturer are more spread from the mean than the first: for some lectures she received very high ratings, and for others her ratings were appalling.

1.8.6 Using a frequency distribution to go beyond the data ▌▌▌

Another way to think about frequency distributions is not in terms of how often scores actually occurred, but how likely it is that a score would occur (i.e., probability). The word 'probability' causes most people's brains to overheat (myself included) so it seems fitting that we use an example about throwing buckets of ice over our heads. Internet memes tend to follow the shape of a normal distribution, which we discussed a while back. A good example of this is the ice bucket challenge from 2014. You can check Wikipedia for the full story, but it all started (arguably) with golfer Chris Kennedy tipping a bucket of iced water on his head to raise awareness of the disease amyotrophic lateral sclerosis (ALS, also known as Lou Gehrig's disease).[19] The idea is that you are challenged and have 24 hours to post a video of you having a bucket of iced water poured over your head; in this video you also challenge at least three other people. If you fail to complete the challenge your forfeit is to donate to charity (in this case, ALS). In reality many people completed the challenge *and* made donations.

The ice bucket challenge is a good example of a meme: it ended up generating something like 2.4 million videos on Facebook and 2.3 million on YouTube. I mentioned that memes often follow a normal distribution, and Figure 1.12 shows this: the insert shows the 'interest' score from Google Trends for the phrase

---

19 Chris Kennedy did not invent the challenge, but he's believed to be the first to link it to ALS. There are earlier reports of people doing things with ice-cold water in the name of charity, but I'm focusing on the ALS challenge because it is the one that spread as a meme.

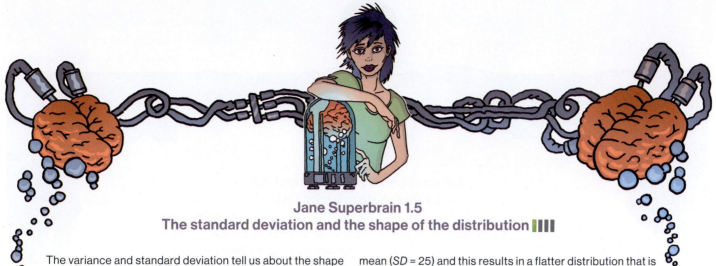

Jane Superbrain 1.5
The standard deviation and the shape of the distribution ▮▮▮▮

The variance and standard deviation tell us about the shape of the distribution of scores. If the mean represents the data well then most of the scores will cluster close to the mean and the resulting standard deviation is small relative to the mean. When the mean is a worse representation of the data, the scores cluster more widely around the mean and the standard deviation is larger. Figure 1.11 shows two distributions that have the same mean (50) but different standard deviations. One has a large standard deviation relative to the mean ($SD = 25$) and this results in a flatter distribution that is more spread out, whereas the other has a small standard deviation relative to the mean ($SD = 15$) resulting in a pointier distribution in which scores close to the mean are very frequent but scores further from the mean become increasingly infrequent. The message is that as the standard deviation gets larger, the distribution gets fatter. This can make distributions look platykurtic or leptokurtic when, in fact, they are not.

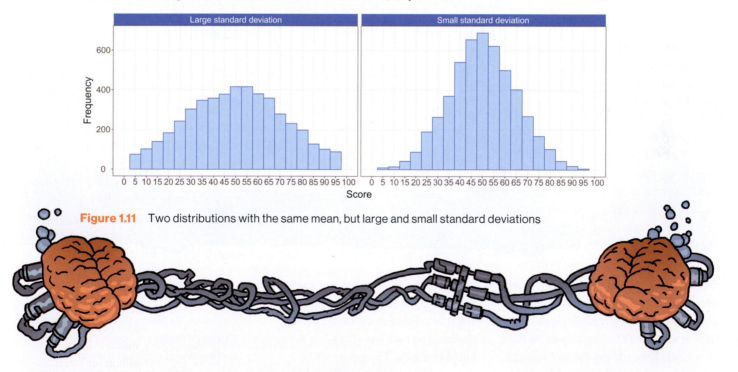

Figure 1.11   Two distributions with the same mean, but large and small standard deviations

'ice bucket challenge' from August to September 2014.[20] The 'interest' score that Google calculates is a bit hard to unpick but essentially reflects the relative number of times that the term 'ice bucket challenge' was searched for on Google. It's not the total number of searches, but the relative number. In a sense it shows the trend of the popularity of searching for 'ice bucket challenge'. Compare the line with the perfect normal distribution in Figure 1.3 – they look fairly similar, don't

20 You can generate the insert graph for yourself by going to Google Trends, entering the search term 'ice bucket challenge' and restricting the dates shown to August 2014 to September 2014.

Labcoat Leni's Real Research 1.1
Is Friday 13th unlucky? ▌▌▌▌

Scanlon, T. J., et al. (1993). *British Medical Journal*, *307*, 1584–1586.

Many of us are superstitious, and a common superstition is that Friday the 13th is unlucky. Most of us don't literally think that someone in a hockey mask is going to kill us, but some people are wary. Scanlon and colleagues, in a tongue-in-cheek study (Scanlon, Luben, Scanlon, & Singleton, 1993), looked at accident statistics at hospitals in the south-west Thames region of the UK. They took statistics both for Friday the 13th and Friday the 6th (the week before) in different months in 1989, 1990, 1991 and 1992. They looked at both emergency admissions of accidents and poisoning, and also transport accidents.

Calculate the mean, median, standard deviation and interquartile range for each type of accident and on each date. Answers are on the companion website.

	Accidents and Poisoning		Traffic Accidents	
Date	Friday 6th	Friday 13th	Friday 6th	Friday 13th
October 1989	4	7	9	13
July 1990	6	6	6	12
September 1991	1	5	11	14
December 1991	9	5	11	10
March 1992	9	7	3	4
November 1992	1	6	5	12

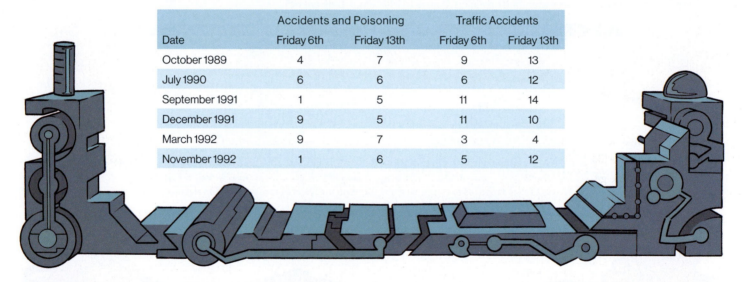

they? Once it got going (about 2–3 weeks after the first video) it went viral, and popularity increased rapidly, reaching a peak at around 21 August (about 36 days after Chris Kennedy got the ball rolling). After this peak, popularity rapidly declines as people tire of the meme.

The main histogram in Figure 1.12 shows the same pattern but reflects something a bit more tangible than 'interest scores'. It shows the number of videos posted on YouTube relating to the ice bucket challenge on each day after Chris Kennedy's initial challenge. There were 2323 thousand in total (2.32 million) during the period shown. In a sense it shows approximately how many people took up the challenge each day.[21] You can see that nothing much happened for 20 days, and early on relatively few people took up the challenge. By about 30 days after the initial challenge things are hotting up (well, cooling down, really) as the number of videos rapidly accelerated from 29,000 on day 30 to 196,000 on day 35. At day 36, the challenge hits its peak (204,000 videos posted), after which the decline sets in as it

---

21 Very very approximately indeed. I have converted the Google interest data into videos posted on YouTube by using the fact that I know that 2.33 million videos were posted during this period and by making the (not unreasonable) assumption that behavior on YouTube will have followed the same pattern over time as the Google interest score for the challenge.

Cramming Sam's Tips
Dispersion

- The deviance or error is the distance of each score from the mean.
- The sum of squared errors is the total amount of error in the mean. The errors/deviances are squared before adding them up.
- The variance is the average distance of scores from the mean. It is the sum of squares divided by the number of scores. It tells us about how widely dispersed scores are around the mean.
- The standard deviation is the *square root of the variance*. It is the variance converted back to the original units of

measurement of the scores used to compute it. Large standard deviations relative to the mean suggest data are widely spread around the mean, whereas small standard deviations suggest data are closely packed around the mean.
- The range is the distance between the highest and lowest score.
- The interquartile range is the range of the middle 50% of the scores.

becomes 'yesterday's news'. By day 50 it's only the type of people like me, and statistics lectures more generally, who don't check Facebook for 50 days, who suddenly become aware of the meme and want to get in on the action to prove how down with the kids we are. It's too late, though: people at that end of the curve are uncool, and the trendsetters who posted videos on day 25 call us lame and look at us dismissively. It's OK though, because we can plot sick histograms like the one in Figure 1.12; take that, hipster scum!

I digress. We can think of frequency distributions in terms of probability. To explain this, imagine that someone asked you 'How likely is it that a person posted an ice bucket video after 60 days?' What would your answer be? Remember that the height of the bars on the histogram reflects how

many videos were posted. Therefore, if you looked at the frequency distribution before answering the question you might respond 'not very likely' because the bars are very short after 60 days (i.e., relatively few videos were posted). What if someone asked you 'How likely is it that a video was posted 35 days after the challenge started?' Using the histogram, you might say 'It's relatively likely' because the bar is very high on day 35 (so quite a few videos were posted). Your inquisitive friend is on a roll and asks 'How likely is it that someone posted a video 35 to 40 days after the challenge started?' The bars representing these days are shaded orange in Figure 1.12. The question about the likelihood of a video being posted 35–40 days into the challenge is really asking 'How big is the orange area of Figure 1.12 compared to the total size of all bars?' We can find out the size of the dark blue region

by adding the values of the bars (196 + 204 + 196 + 174 + 164 + 141 = 1075); therefore, the orange area represents 1075 thousand videos. The total size of all bars is the total number of videos posted (i.e., 2323 thousand). If the orange area represents 1075 thousand videos, and the total area represents 2323 thousand videos, then if we compare the orange area to the total area we get 1075/2323 = 0.46. This proportion can be converted to a percentage by multiplying by 100, which gives us 46%. Therefore, our answer might be 'It's quite likely that someone posted a video 35–40 days into the challenge because 46% of all videos were posted during those 6 days'. A very important point here is that the size of the bars relates directly to the probability of an event occurring.

Hopefully these illustrations show that we can use the frequencies of different scores,

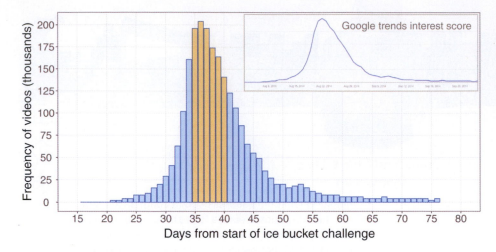

Figure 1.12 Frequency distribution showing the number of ice bucket challenge videos on YouTube by day since the first video (the insert shows the actual Google Trends data on which this example is based)

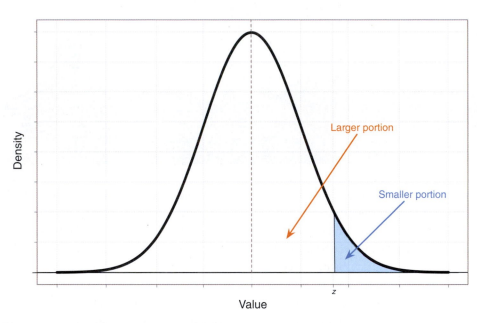

Figure 1.13 The normal probability distribution

meeting the deadline is 0 (not a chance in hell). If probabilities don't make sense to you then you're not alone; just ignore the decimal point and think of them as percentages instead (i.e., a 0.10 probability that something will happen is a 10% chance that something will happen) or read the chapter on probability in my other excellent textbook (Field, 2016).

I've talked in vague terms about how frequency distributions can be used to get a rough idea of the probability of a score occurring. However, we can be precise. For any distribution of scores we could, in theory, calculate the probability of obtaining a score of a certain size—it would be incredibly tedious and complex to do it, but we could. To spare our sanity, statisticians have identified several common distributions. For each one they have worked out mathematical formulae (known as **probability density functions, PDF**) that specify idealized versions of these distributions. We could draw such a function by plotting the value of the variable (x) against the probability of it occurring (y).[22] The resulting curve is known as a **probability distribution**; for a normal distribution (Section 1.8.1) it would look like Figure 1.13, which has the characteristic bell shape that we saw already in Figure 1.3.

A probability distribution is just like a histogram except that the lumps and bumps have been smoothed out so that we see a nice smooth curve. However, like a frequency distribution, the area under this curve tells us something about the probability of a value occurring. Just like we did in our ice bucket example we could use the area under the curve between two values to tell us how likely it is that a score fell within a particular range. For example, the blue shaded region in Figure 1.13 corresponds to the probability of a score being z or greater. The normal distribution

and the area of a frequency distribution, to estimate the probability that a particular score will occur. A probability value can range from 0 (there's no chance whatsoever of the event happening) to 1 (the event will definitely happen). So, for example, when I talk to my publishers I tell them there's a probability of 1 that I

will have completed the revisions to this book by July. However, when I talk to anyone else, I might, more realistically, tell them that there's a 0.10 probability of me finishing the revisions on time (or put another way, a 10% chance or 1 in 10 chance that I'll complete the book in time). In reality, the probability of my

---

22 Actually we usually plot something called the *density*, which is closely related to the probability.

is not the only distribution that has been precisely specified by people with enormous brains. There are many distributions that have characteristic shapes and have been specified with a probability density function. We'll encounter some of these other distributions throughout the book, for example the *t*-distribution, chi-square (χ^2) distribution, and *F*-distribution. For now, the important thing to remember is that all of these distributions have something in common: they are all defined by an equation that enables us to calculate precisely the probability of obtaining a given score.

As we have seen, distributions can have different means and standard deviations. This isn't a problem for the probability density function—it will still give us the probability of a given value occurring—but it is a problem for us because probability density functions are difficult enough to spell, let alone use to compute probabilities.

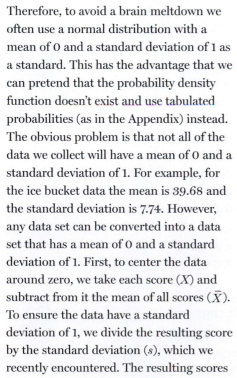

What is a z-score?

Therefore, to avoid a brain meltdown we often use a normal distribution with a mean of 0 and a standard deviation of 1 as a standard. This has the advantage that we can pretend that the probability density function doesn't exist and use tabulated probabilities (as in the Appendix) instead. The obvious problem is that not all of the data we collect will have a mean of 0 and a standard deviation of 1. For example, for the ice bucket data the mean is 39.68 and the standard deviation is 7.74. However, any data set can be converted into a data set that has a mean of 0 and a standard deviation of 1. First, to center the data around zero, we take each score (X) and subtract from it the mean of all scores ($\bar{X}$). To ensure the data have a standard deviation of 1, we divide the resulting score by the standard deviation (s), which we recently encountered. The resulting scores

A.1. Table of the standard normal distribution

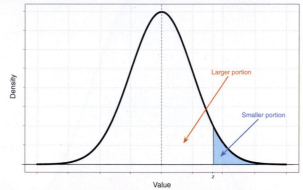

z	Larger Portion	Smaller Portion	y	z	Larger Portion	Smaller Portion	y
.00	.50000	.50000	.3989	.12	.54776	.45224	.3961
.01	.50399	.49601	.3989	.13	.55172	.44828	.3956
.02	.50798	.49202	.3989	.14	.55567	.44433	.3951
.03	.51197	.48803	.3988	.15	.55962	.44038	.3945
.04	.51595	.48405	.3986	.16	.56356	.43644	.3939
1.56	.94062	.05938	.1182	1.86	.96856	.03144	.0707
1.57	.94179	.05821	.1163	1.87	.96926	.03074	.0694
1.58	.94295	.05705	.1145	1.88	.96995	.03005	.0681
1.59	.94408	.05592	.1127	1.89	.97062	.02938	.0669
1.60	.94520	.05480	.1109	1.90	.97128	.02872	.0656
1.61	.94630	.05370	.1092	1.91	.97193	.02807	.0644
1.62	.94738	.05262	.1074	1.92	.97257	.02743	.0632
1.63	.94845	.05155	.1057	1.93	.97320	.02680	.0620
1.64	.94950	.05050	.1040	1.94	.97381	.02619	.0608
1.65	.95053	.04947	.1023	1.95	.97441	.02559	.0596
1.66	.95154	.04846	.1006	1.96	.97500	.02500	.0584
1.67	.95254	.04746	.0989	1.97	.97558	.02442	.0573
1.68	.95352	.04648	.0973	1.98	.97615	.02385	.0562
2.27	.98840	.01160	.0303	2.57	.99492	.00508	.0147
2.28	.98870	.01130	.0297	2.58	.99506	.00494	.0143
2.29	.98899	.01101	.0290	2.59	.99520	.00480	.0139
2.30	.98928	.01072	.0283	2.60	.99534	.00466	.0136
2.31	.98956	.01044	.0277	2.61	.99547	.00453	.0132
2.32	.98983	.01017	.0270	2.62	.99560	.00440	.0129
2.33	.99010	.00990	.0264	2.63	.99573	.00427	.0126

Figure 1.14 Using tabulated values of the standard normal distribution

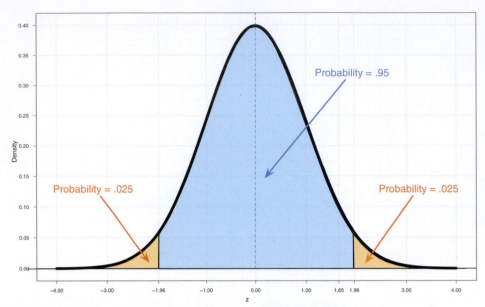

Figure 1.15   The probability density function of a normal distribution

are denoted by the letter z and are known as **Z-scores**. In equation form, the conversion that I've just described is:

$$z = \frac{X - \overline{X}}{s} \qquad (1.9)$$

The table of probability values that have been calculated for the standard normal distribution is shown in the Appendix. Why is this table important? Well, if we look at our ice bucket data, we can answer the question 'What's the probability that someone posted a video on day 60 or later?' First, we convert 60 into a z-score. We saw that the mean was 39.68 and the standard deviation was 7.74, so our score of 60 expressed as a z-score is 2.63 (Eq. 1.10):

$$z = \frac{60 - 39.68}{7.74} = 2.63 \qquad (1.10)$$

We can now use this value, rather than the original value of 60, to compute an answer to our question.

Figure 1.14 shows (an edited version of) the tabulated values of the standard normal distribution from the Appendix of this book. This table gives us a list of values of z, and the density (y) for each value of z, but, most

important, it splits the distribution at the value of z and tells us the size of the two areas under the curve that this division creates. For example, when z is 0, we are at the mean or center of the distribution so it splits the area under the curve exactly in half. Consequently, both areas have a size of 0.5 (or 50%). However, any value of z that is not zero will create different sized areas, and the table tells us the size of the larger and smaller portions. For example, if we look up our z-score of 2.63, we find that the smaller portion (i.e., the area above this value or the blue area in Figure 1.14) is 0.0043 or only 0.43%. I explained before that these areas relate to probabilities, so in this case we could say that there is only a 0.43% chance that a video was posted 60 days or more after the challenge started. By looking at the larger portion (the area below 2.63) we get 0.9957 or put another way, there's a 99.57% chance that an ice bucket video was posted on YouTube within 60 days of the challenge starting. Note that these two proportions

add up to 1 (or 100%), so the total area under the curve is 1.

Another useful thing we can do (you'll find out just how useful in due course) is to work out limits within which a certain percentage of scores fall. With our ice bucket example, we looked at how likely it was that a video was posted between 35 and 40 days after the challenge started; we could ask a similar question such as 'What is the range of days between which the middle 95% of videos were posted?' To answer this question we need to use the table the opposite way around. We know that the total area under the curve is 1 (or 100%), so to discover the limits within which 95% of scores fall we're asking 'What is the value of z that cuts off 5% of the scores?' It's not quite as simple as that because if we want the *middle* 95%, then we want to cut off scores from both ends. Given the distribution is symmetrical, if we want to cut off 5% of scores overall but we want to take some from both extremes of scores, then the percentage of scores we want to cut from each end will be 5%/2 = 2.5% (or 0.025 as a proportion). If we cut off 2.5% of scores from each end then in total we'll have cut off 5% scores, leaving us with the middle 95% (or 0.95 as a proportion)— see Figure 1.15. To find out what value of z cuts off the top area of 0.025, we look down the column 'smaller portion' until we reach 0.025, we then read off the corresponding value of z. This value is 1.96 (see Figure 1.14) and because the distribution is symmetrical around zero, the value that cuts off the bottom 0.025 will be the same but a minus value (−1.96). Therefore, the middle 95% of z-scores fall between −1.96 and 1.96. If we wanted to know the limits between which

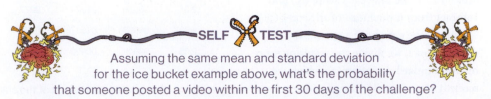

SELF TEST

Assuming the same mean and standard deviation
for the ice bucket example above, what's the probability
that someone posted a video within the first 30 days of the challenge?

Cramming Sam's Tips
Distributions and z-scores

- A frequency distribution can be either a table or a chart that shows each possible score on a scale of measurement along with the number of times that score occurred in the data.
- Scores are sometimes expressed in a standard form known as z-scores.

- To transform a score into a z-score you subtract from it the mean of all scores and divide the result by the standard deviation of all scores.
- The sign of the z-score tells us whether the original score was above or below the mean; the value of the z-score tells us how far the score was from the mean in standard deviation units.

the middle 99% of scores would fall, we could do the same: now we would want to cut off 1% of scores or 0.5% from each end. This equates to a proportion of 0.005. We look up 0.005 in the *smaller portion* part of the table and the nearest value we find is 0.00494, which equates to a z-score of 2.58 (see Figure 1.14). This tells us that 99% of z-scores lie between −2.58 and 2.58. Similarly (have a go), you can show that 99.9% of them lie between −3.29 and 3.29. Remember these values (1.96, 2.58 and 3.29) because they'll crop up time and time again.

1.8.7 Fitting statistical models to the data ▮▮▮▮

Having looked at your data (and there is a lot more information on different ways to do this in Chapter 5), the next step of the research process is to fit a statistical model to the data. That is to go where

eagles dare, and no one should fly where eagles dare; but to become scientists we have to, so the rest of this book attempts to guide you through the various models that you can fit to the data.

1.9 Reporting data ▮▮▮▮

1.9.1 Dissemination of research ▮▮▮▮

Having established a theory and collected and started to summarize data, you might want to tell other people what you have found. This sharing of information is a fundamental part of being a scientist. As discoverers of knowledge, we have a duty of care to the world to present what we find in a clear and unambiguous way, and with enough information that others can challenge our conclusions. It is good practice, for example, to make your data available to others and to be open with the

resources you used. Initiatives such as the Open Science Framework (https://osf.io) make this easy to do. Tempting as it may be to cover up the more unsavory aspects of our results, science is about truth, openness and willingness to debate your work.

Scientists tell the world about our findings by presenting them at conferences and in articles published in scientific **journals**. A scientific journal is a collection of articles written by scientists on a vaguely similar topic. A bit like a magazine, but more tedious. These articles can describe new research, review existing research or might put forward a new theory. Just like you have magazines such as *Modern Drummer*, which is about drumming or *Vogue*, which is about fashion (or Madonna, I can never remember which), you get journals such as *Journal of Anxiety Disorders*, which publishes articles about anxiety disorders, and *British Medical Journal*, which publishes articles about medicine (not specifically British medicine, I hasten to add). As a scientist, you submit your work to one of these journals and they will consider publishing it. Not everything a scientist writes will be published. Typically, your manuscript will be given to an 'editor' who will be a fairly eminent scientist working in that research area who has agreed, in return for their soul, to make decisions about whether or not to publish articles. This editor will send your manuscript out to review, which means they send it to other experts in your research area and ask those experts to assess the quality of the work. Often (but not always) the reviewer is blind to who wrote the manuscript. The reviewers' role is to provide a constructive and even-handed overview of the strengths and weaknesses of your article and the research contained within it. Once these reviews are complete the editor reads them and then assimilates the comments with his or her own views on the manuscript and decides whether to publish it (in reality, you'll be asked to make revisions at least once before a final acceptance).

The review process is an excellent way to get useful feedback on what you have done, and very often throws up things that you hadn't considered. The flip side is that when people scrutinize your work, they don't always say nice things. Early on in my career I found this process quite difficult: often you have put months of work into the article and it's only natural that you want your peers to receive it well. When you do get negative feedback, and even the most respected scientists do, it can be easy to feel like you're not good enough. At those times, it's worth remembering that if you're not affected by criticism, then you're probably not human; every scientist I know has moments when they doubt themselves.

1.9.2 Knowing how to report data ▮▮▮

An important part of publishing your research is how you present and report your data. You will typically do this through a combination of graphs (see Chapter 5) and written descriptions of the data. Throughout this book I will give you guidance about how to present data and write up results. The difficulty is that different disciplines have different conventions. In my area of science (psychology), we typically follow the publication guidelines of the American Psychological Association or APA (American Pyschological Association, 2010), but even within psychology different journals have their own idiosyncratic rules about how to report data. Therefore, my advice will be broadly based on the APA guidelines, with a bit of my own personal opinion thrown in when there isn't a specific APA 'rule'. However, when reporting data for assignments or for publication, it is always advisable to check the specific guidelines of your tutor or the journal.

Despite the fact that some people would have you believe that if you deviate from any of the 'rules' in even the most subtle of ways then you will unleash the four horsemen of the apocalypse onto the world to obliterate humankind, the 'rules' are no substitute for common sense. Although some people treat the APA style guide like a holy sacrament, its job is not to lay down intractable laws, but to offer a guide so that everyone is consistent in what they do. It does not tell you what to do in every situation, but does offer sensible guiding principles that you can extrapolate to most situations you'll encounter.

1.9.3 Some initial guiding principles ▮▮▮

When reporting data, your first decision is whether to use text, a graph or a table. You want to be succinct, so you shouldn't

present the same values in multiple ways: if you have a graph showing some results then don't also produce a table of the same results: it's a waste of space. The APA gives the following guidelines:

- Choose a mode of presentation that optimizes the understanding of the data.
- If you present three or fewer numbers then try using a sentence.
- If you need to present between 4 and 20 numbers consider a table.
- If you need to present more than 20 numbers then a graph is often more useful than a table.

Of these, I think the first is most important: I can think of countless situations where I would want to use a graph rather than a table to present 4–20 values because a graph will show up the pattern of data most clearly. Similarly, I can imagine some graphs presenting more than 20 numbers being an absolute mess. This takes me back to my point about rules being no substitute for common sense, and the most important thing is to present the data in a way that makes it easy for the reader to digest. We'll look at how to present graphs in Chapter 5 and we'll look at tabulating data in various chapters when we discuss how best to report the results of particular analyses. A second general issue is how many decimal places to use when reporting numbers. The guiding principle from the APA (which I think is sensible) is that the fewer decimal places the better, which means that you should round as much as possible but bear in mind the precision of the measure you're reporting. This principle again reflects making it easy for the reader to understand the data. Let's look at an example. Sometimes when a person doesn't respond to someone, they will ask 'What's wrong, has the cat got your tongue?' Actually, my cat had a large collection of carefully preserved human tongues that he kept in a box under the stairs. Periodically, he'd get one out, pop it in his mouth and wander around the neighborhood scaring people with his big tongue. If I measured the difference in length between his actual tongue and his fake human tongue, I might report this difference as 0.0425 meters, 4.25 centimeters or 42.5 millimeters. This example illustrates three points: (1) I needed a different number of decimal places (4, 2 and 1, respectively) to convey the same information in each case; (2) 4.25 cm is probably easier for someone to digest than 0.0425 m because it uses fewer decimal places; and (3) my cat was odd. The first point demonstrates that it's not the case that you should always use, say, two decimal places; you should use however many you need in a particular situation. The second point implies that if you have a very small measure it's worth considering whether you can use a different scale to make the numbers more palatable.

Finally, every set of guidelines will include advice on how to report specific analyses and statistics. For example, when describing data with a measure of central tendency, the APA suggests you use M (capital M in italics) to represent the mean but is fine with you using the mathematical notation ($\bar{X}$) too. However, you should be consistent: if you use M to represent the mean you should do so throughout your article. There is also a sensible principle that if you report a summary of the data such as the mean, you should also report the appropriate measure of the spread of scores. Then people know not just the central location of the data, but also how spread out they were. Therefore, whenever we report the mean, we typically report the standard deviation also. The standard deviation is usually denoted by SD, but it is also common to simply place it in parentheses as long as you indicate that you're doing so in the text. Here are some examples from this chapter:

✓ Andy has 2 friends on Facebook. On average, a sample of other users ($N = 11$), had considerably more, $M = 95$, $SD = 56.79$.

✓ The average number of days it took someone to post a video of the ice bucket challenge was $\bar{X} = 39.68$, $SD = 7.74$.

✓ By reading this chapter we discovered that (SD in parentheses), on average, people have 95 (56.79) friends on Facebook and on average it took people 39.68 (7.74) days to post a video of them throwing a bucket of iced water over themselves.

Note that in the first example, I used N to denote the size of the sample. This is a common abbreviation: a capital N represents the entire sample and a lower-case n represents a subsample (e.g., the number of cases within a particular group).

Similarly, when we report medians, there is a specific notation (the APA suggests Mdn) and we should report the range or interquartile range as well (the APA does not have an abbreviation for either of these terms, but IQR is commonly used for the interquartile range). Therefore, we could report:

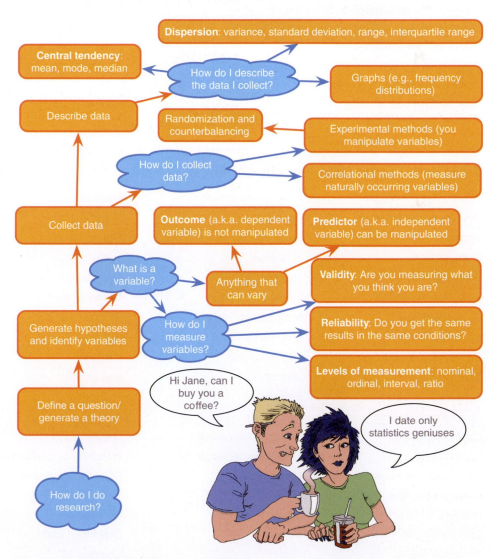

Figure 1.16 What Brian learnt from this chapter

✓ Andy has 2 friends on Facebook. A sample of other users ($N = 11$) typically had more, $Mdn = 98$, $IQR = 63$.

✓ Andy has 2 friends on Facebook. A sample of other users ($N = 11$) typically had more, $Mdn = 98$, $range = 212$.

1.10 Brian and Jane's Story ▊▊▊▊

Brian had a crush on Jane. He'd seen her around campus a lot, always rushing with a big bag and looking sheepish. People called her a weirdo, but her reputation for genius was well earned. She was mysterious, no one had ever spoken to her or knew why she scuttled around the campus with such purpose. Brian found her quirkiness sexy. He probably needed to reflect on that someday.

As she passed him on the library stairs, Brian caught her shoulder. She looked horrified. 'Sup,' he said with a smile.

Jane looked sheepishly at the bag she was carrying.

'Fancy a brew?' Brian asked.

Jane looked Brian up and down. He was handsome, but he looked like he might be an idiot . . . and Jane didn't trust people, especially guys. To her surprise, Brian tried to woo her with what he'd learnt in his statistics lecture that morning. Maybe she was wrong about his idiocy, maybe he was a statistics guy . . . that would make him more appealing, after all stats guys always told the best jokes.

Jane took his hand and led him to the Statistics section of the library. She pulled out a book called *An Adventure in Statistics* and handed it to him. Brian liked the cover. Jane turned and strolled away enigmatically.

1.11 What next? ▊▊▊▊

It is all very well discovering that if you stick your finger into a fan or get hit around the face with a golf club it hurts, but what if these are isolated incidents? It's better if we can somehow extrapolate from our data and draw more general conclusions. Even better, perhaps we can start to make predictions about the world: if we can predict when a golf club is going to appear out of nowhere then we can better move our faces. The next chapter looks at fitting models to the data and using these models to draw conclusions that go beyond the data we collected.

My early childhood wasn't all full of pain, on the contrary it was filled with a lot of fun: the nightly 'from how far away can I jump into bed' competition (which sometimes involved a bit of pain) and being carried by my brother and dad to bed as they hummed Chopin's *Marche Funèbre* before lowering me between two beds as though being buried in a grave. It was more fun than it sounds.

1.12 Key terms that I've discovered

Between-groups design	Experimental research	Multimodal	Randomization
Between-subjects design	Falsification	Negative skew	Range
Bimodal	Frequency distribution	Nominal variable	Ratio variable
Binary variable	Histogram	Nonile	Reliability
Boredom effect	Hypothesis	Normal distribution	Repeated-measures design
Categorical variable	Independent design	Ordinal variable	Second quartile
Central tendency	Independent variable	Outcome variable	Skew
Concurrent validity	Interquartile range	Percentile	Standard deviation
Confounding variable	Interval variable	Platykurtic	Sum of squared errors
Content validity	Journal	Positive skew	Systematic variation
Continuous variable	Kurtosis	Practice effect	*Tertium quid*
Correlational research	Leptokurtic	Predictive validity	Test–retest reliability
Counterbalancing	Level of measurement	Predictor variable	Theory
Criterion validity	Longitudinal research	Probability density function (PDF)	Unsystematic variance
Cross-sectional research	Lower quartile	Probability distribution	Upper quartile
Dependent variable	Mean	Qualitative methods	Validity
Deviance	Measurement error	Quantile	Variables
Discrete variable	Median	Quantitative methods	Variance
Ecological validity	Mode	Quartile	Within-subject design
			z-scores

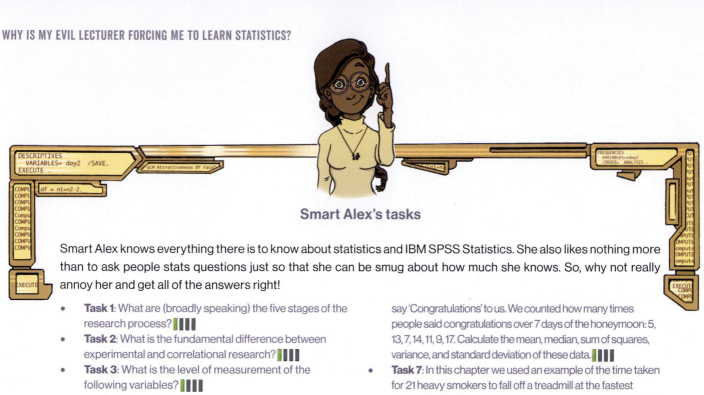

Smart Alex's tasks

Smart Alex knows everything there is to know about statistics and IBM SPSS Statistics. She also likes nothing more than to ask people stats questions just so that she can be smug about how much she knows. So, why not really annoy her and get all of the answers right!

- **Task 1**: What are (broadly speaking) the five stages of the research process? ▌▌▌▌
- **Task 2**: What is the fundamental difference between experimental and correlational research? ▌▌▌▌
- **Task 3**: What is the level of measurement of the following variables? ▌▌▌▌

  - The number of downloads of different bands' songs on iTunes
  - The names of the bands that were downloaded
  - Their positions in the download chart
  - The money earned by the bands from the downloads
  - The weight of drugs bought by the bands with their royalties
  - The type of drugs bought by the bands with their royalties
  - The phone numbers that the bands obtained because of their fame
  - The gender of the people giving the bands their phone numbers
  - The instruments played by the band members
  - The time they had spent learning to play their instruments

- **Task 4**: Say I own 857 CDs. My friend has written a computer program that uses a webcam to scan the shelves in my house where I keep my CDs and measure how many I have. His program says that I have 863 CDs. Define measurement error. What is the measurement error in my friend's CD-counting device? ▌▌▌▌
- **Task 5**: Sketch the shape of a normal distribution, a positively skewed distribution and a negatively skewed distribution. ▌▌▌▌
- **Task 6**: In 2011 I got married and we went to Disney World in Florida for our honeymoon. We bought some bride and groom Mickey Mouse hats and wore them around the parks. The staff at Disney are really nice and, upon seeing our hats, would say 'Congratulations' to us. We counted how many times people said congratulations over 7 days of the honeymoon: 5, 13, 7, 14, 11, 9, 17. Calculate the mean, median, sum of squares, variance, and standard deviation of these data. ▌▌▌▌
- **Task 7**: In this chapter we used an example of the time taken for 21 heavy smokers to fall off a treadmill at the fastest setting (18, 16, 18, 24, 23, 22, 22, 23, 26, 29, 32, 34, 34, 36, 36, 43, 42, 49, 46, 46, 57). Calculate the sum of squares, variance and standard deviation of these data. ▌▌▌▌
- **Task 8**: Sports scientists sometimes talk of a 'red zone', which is a period during which players in a team are more likely to pick up injuries because they are fatigued. When a player hits the red zone it is a good idea to rest them for a game or two. At a prominent London football club that I support, they measured how many consecutive games the 11 first-team players could manage before hitting the red zone: 10, 16, 8, 9, 6, 8, 9, 11, 12, 19, 5. Calculate the mean, standard deviation, median, range and interquartile range. ▌▌▌▌
- **Task 9**: Celebrities always seem to be getting divorced. The (approximate) lengths of some celebrity marriages in days are: 240 (J-Lo and Cris Judd), 144 (Charlie Sheen and Donna Peele), 143 (Pamela Anderson and Kid Rock), 72 (Kim Kardashian, if you can call her a celebrity), 30 (Drew Barrymore and Jeremy Thomas), 26 (W. Axl Rose and Erin Everly), 2 (Britney Spears and Jason Alexander), 150 (Drew Barrymore again, but this time with Tom Green), 14 (Eddie Murphy and Tracy Edmonds), 150 (Renée Zellweger and Kenny Chesney), 1657 (Jennifer Aniston and Brad Pitt). Compute the mean, median, standard deviation, range and interquartile range for these lengths of celebrity marriages. ▌▌▌▌
- **Task 10**: Repeat Task 9 but excluding Jennifer Anniston and Brad Pitt's marriage. How does this affect the mean, median, range, interquartile range, and standard deviation? What do the differences in values between Tasks 9 and 10 tell us about the influence of unusual scores on these measures? ▌▌▌▌

Answers & additional resources are available on the book's website at
https://edge.sagepub.com/field5e

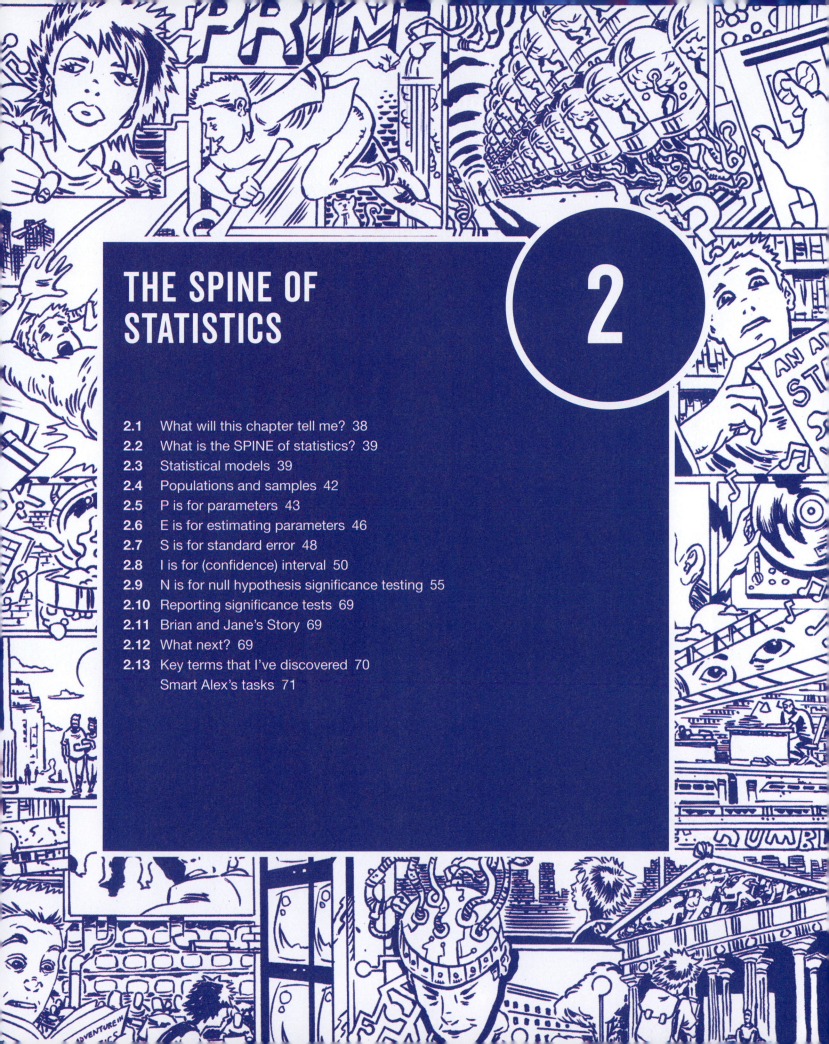

THE SPINE OF STATISTICS

2

2.1 What will this chapter tell me?

Although I had learnt a lot about golf clubs randomly appearing out of nowhere and hitting me around the face, I still felt that there was much about the world that I didn't understand. For one thing, could I learn to predict the presence of these golf clubs that seemed inexplicably drawn towards my apparently magnetic head? A child's survival depends upon being able to predict reliably what will happen in certain situations; consequently they develop a model of the world based on the data they have (previous experience) and they then test this model by collecting new data/experiences. Based on how well the new experiences fit with their original model, a child might revise their model of the world.

According to my parents (conveniently I have no memory of these events), while at nursery school one model of the world that I was enthusiastic to try out was 'If I get my penis out, it will be really funny'. To my considerable disappointment, this model turned out to be a poor predictor of positive outcomes. Thankfully for all concerned, I soon revised this model of the world to be 'If I get my penis out at nursery school the teachers and mummy and daddy will be quite annoyed'. This revised model may not have been as much fun but was certainly a better 'fit' of the observed data. Fitting models that accurately reflect the observed data is important to establish whether a hypothesis (and the theory from which it derives) is true.

You'll be relieved to know that this chapter is not about my penis but is about fitting statistical models. We edge sneakily away from the frying pan of research methods and trip accidentally into the fires of statistics hell. We will start to see how we can use the properties of data to go beyond our observations and to draw inferences about the world at large. This chapter and the next lay the foundation for the rest of the book.

Figure 2.1 The face of innocence … but what are the hands doing?

2.2 What is the SPINE of statistics?

To many students, statistics is a bewildering mass of different tests, each with their own set of equations. The focus is often on 'difference'. It feels like you need to learn a lot of different stuff. What I hope to do in this chapter is to focus your mind on some core concepts that many statistical models have in common. In doing so, I want to set the tone for you focusing on the *similarities* between statistical models rather than the differences. If your goal is to use statistics as a tool, rather than to bury yourself in the theory, then I think this approach makes your job a lot easier. In this chapter, I will first argue that most statistical models are variations on the very simple idea of predicting an outcome variable from one or more predictor variables. The mathematical form of the model changes, but it usually boils down to a representation of the relations between an outcome and one or more predictors. If you understand that, then there are five key concepts to get your head around. If you understand these, you've gone a long way towards understanding any statistical model that you might want to fit. They are the SPINE of statistics, which is a clever acronym for:

- **Standard error**
- **Parameters**
- **Interval estimates (confidence intervals)**
- **Null hypothesis significance testing**
- **Estimation**

I cover each of these topics, but not in this order because PESIN doesn't work nearly so well as an acronym.[1]

2.3 Statistical models ▮▯▯▯

We saw in the previous chapter that scientists are interested in discovering something about a phenomenon that we assume exists (a 'real-world' phenomenon). These real-world phenomena can be anything from the behavior of interest rates in the economy to the behavior of undergraduates at the end-of-exam party. Whatever the phenomenon, we collect data from the real world to test predictions from our hypotheses about that phenomenon. Testing these hypotheses involves building statistical models of the phenomenon of interest.

Let's begin with an analogy. Imagine an engineer wishes to build a bridge across a river. That engineer would be pretty daft if she built any old bridge, because it might fall down. Instead, she collects data from the real world: she looks at existing bridges and sees from what materials they are made, their structure, size and so on (she might even collect data about whether these bridges are still standing). She uses this information to construct an idea of what her new bridge will be (this is a 'model'). It's expensive and impractical for her to build a full-sized version of her bridge, so she builds a scaled-down version. The model may differ from reality in several ways—it will be smaller, for a start—but the engineer will try to build a model that best fits the situation of interest based on the data available. Once the model has been built, it can be used to predict things about the real world: for example, the engineer might test whether the bridge can withstand strong winds by placing her model in a wind tunnel. It is important that the model accurately represents the real world, otherwise any conclusions she extrapolates to the real-world bridge will be meaningless.

Why do we build statistical models?

Scientists do much the same: they build (statistical) models of real-world processes to predict how these processes operate under certain conditions (see Jane Superbrain Box 2.1). Unlike engineers, we don't have access to the real-world situation and so we can only *infer* things about psychological, societal, biological or economic processes based upon the models we build. However, like the engineer, our models need to be as accurate as possible so that the predictions we make about the real world are accurate too; the statistical model should represent the data collected (the *observed data*) as closely as possible. The degree to which a statistical model represents the data collected is known as the **fit** of the model.

Figure 2.2 shows three models that our engineer has built to represent her real-world bridge. The first is an excellent representation of the real-world situation and is said to be a *good fit*. If the engineer uses this model to make predictions about the real world then, because it so closely resembles reality, she can be confident that her predictions will be accurate. If the model collapses in a strong wind, then there is a good chance that the real bridge would collapse also. The second model has some similarities to the real world: the model includes some of the basic structural features, but there are some big differences too (e.g., the absence of one of the supporting towers). We might consider this model to have a *moderate fit* (i.e., there are some similarities to reality but also some important differences). If our engineer uses this model to make predictions about the real world then her predictions could be inaccurate or even catastrophic. For example, perhaps the model predicts that the bridge will collapse in a strong wind, so after the real bridge is built it gets closed every time a strong wind occurs, creating 100-mile tailbacks with everyone stranded

---

1 There is another, more entertaining, acronym that fits well with the anecdote at the start of the chapter, but I decided not to use it because in a séance with Freud he advised me that it could lead to pesin envy.

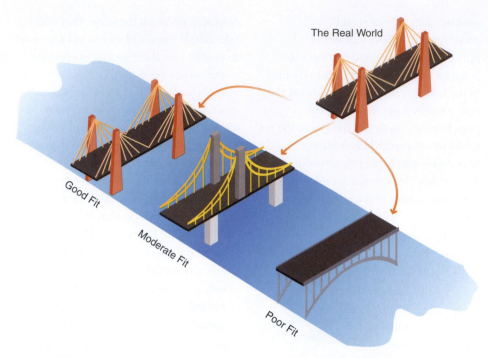

Figure 2.2 Fitting models to real-world data (see text for details)

in the snow, feasting on the crumbs of old sandwiches that they find under the seats of their cars. All of which turns out to be unnecessary because the real bridge was safe—the prediction from the model was wrong because it was a bad representation of reality. We can have a little confidence in predictions from this model, but not complete confidence. The final model is completely different than the real-world situation; it bears no structural similarities to the real bridge and is a *poor fit*. Any predictions based on this model are likely to be completely inaccurate. Extending this analogy to science, if our model is a poor fit to the observed data, then the predictions we make from it will be equally poor.

Although it's easy to visualize a model of a bridge, you might be struggling to understand what I mean by a 'statistical

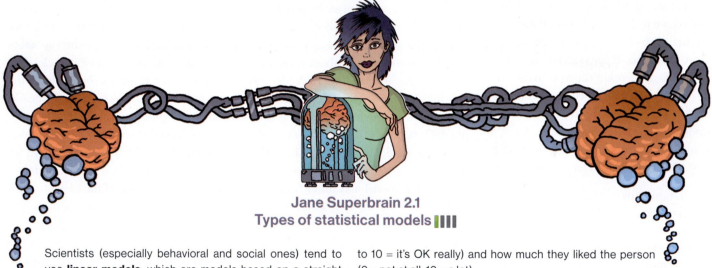

Jane Superbrain 2.1
Types of statistical models ▮▮▮▮

Scientists (especially behavioral and social ones) tend to use **linear models**, which are models based on a straight line. As you read scientific research papers, you'll see that they are riddled with 'analysis of variance (ANOVA)' and 'regression', which are identical statistical systems based on the linear model (Cohen, 1968). In fact, most of the chapters in this book explain this 'general linear model'.

Imagine we were interested in how people evaluated dishonest acts.[2] Participants evaluate the dishonesty of acts based on watching videos of people confessing to those acts. Imagine we took 100 people and showed them a random dishonest act described by the perpetrator. They then evaluated the honesty of the act (from 0 = appalling behavior

to 10 = it's OK really) and how much they liked the person (0 = not at all, 10 = a lot).

We can represent these hypothetical data on a scatterplot in which each dot represents an individual's rating on both variables (see Section 5.8). Figure 2.3 shows two versions of the same data. We can fit different models to the same data: on the left we have a linear (straight) model and on the right a non-linear (curved) one. Both models show that the more likeable a perpetrator is, the more positively people view their dishonest act. However, the curved line shows a subtler pattern: the trend to be more forgiving of likeable people kicks in when the likeableness rating rises above 4. Below 4 (where the perpetrator is not very likeable), all deeds are rated fairly low (the red line is quite flat),

2 This example came from a media story about the Honesty Lab set up by Stefan Fafinski and Emily Finch. However, this project no longer seems not to exist and I can't find the results published anywhere. I like the example though, so I kept it.

but as the perpetrator becomes likeable (above about 4) the slope of the line becomes steeper, suggesting that as likeableness rises above this value, people become increasingly forgiving of dishonest acts. Neither of the two models is necessarily correct, but one model will fit the data better than the other; this is why it is important for us to assess how well a statistical model fits the data.

Linear models tend to get fitted to data because they are less complex and because non-linear models are often not taught (despite 900 pages of statistics hell, I don't get into non-linear models in this book). This could have had interesting consequences for science: (1) many published statistical models might not be the ones that fit best (because the authors didn't try out non-linear models); and (2) findings might have been missed because a linear model was a poor fit and the scientists gave up rather than fitting non-linear models (which perhaps would have been a 'good enough' fit). It is useful to plot your data first: if your plot seems to suggest a non-linear model then don't apply a linear model because that's all you know, fit a non-linear model (after complaining to me about how I don't cover them in this book).

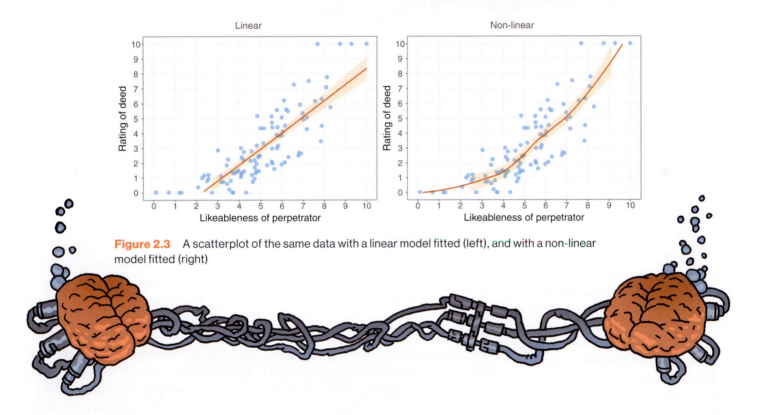

Figure 2.3   A scatterplot of the same data with a linear model fitted (left), and with a non-linear model fitted (right)

model'. Even a brief glance at some scientific articles will transport you into a terrifying jungle of different types of 'statistical model': you'll encounter tedious-looking names like *t*-test, ANOVA, regression, multilevel models, and structural equation modeling. It might make you yearn for a career in journalism, where the distinction between opinion and evidence need not trouble you. Fear not, though; I have a story that may help.

Many centuries ago there existed a cult of elite mathematicians. They spent 200 years trying to solve an equation that they believed would make them immortal.

However, one of them forgot that when you multiply two minus numbers you get a plus, and instead of achieving eternal life they unleashed Cthulhu from his underwater city. It's amazing how small computational mistakes in maths can have these sorts of consequences. Anyway, the only way they could agree to get Cthulhu to return to his entrapment was by promising to infect the minds of humanity with confusion. They set about this task with gusto. They took the simple and elegant idea of a statistical model and reinvented it in hundreds of seemingly different ways (Figures 2.4 and 2.5). They

described each model as though it were completely different from the rest. 'Ha!' they thought, 'that'll confuse students.' And confusion did indeed infect the minds of students. The statisticians kept their secret that all statistical models could be described in one simple, easy-to-understand equation locked away in a wooden box with Cthulhu's head burned into the lid. 'No one will open a box with a big squid head burnt into it', they reasoned. They were correct, until a Greek fisherman stumbled upon the box and, thinking it contained some vintage calamari, opened it. Disappointed with the

$$outcome_i = model + error_i$$

REGRESSION MODERATION
ANOVA
ANCOVA MULTILEVEL MODEL
T-TEST
CORRELATION MEDIATION

Figure 2.4 Thanks to the Confusion machine, a simple equation is made to seem like lots of unrelated tests

contents, he sold the script inside the box on eBay. I bought it for €3 plus postage. This was money well spent, because it means that I can now give you the key that will unlock the mystery of statistics forever. Everything in this book (and statistics generally) boils down to equation (2.1).

$$outcome_i = (model) + error_i \qquad (2.1)$$

This equation means that the data we observe can be predicted from the model we choose to fit plus some amount of error.[3] The 'model' in the equation will vary depending on the design of your study, the type of data you have and what it is you're trying to achieve with your model. Consequently, the model can also vary in its complexity. No matter how long the equation that describes your model might be, you can just close your eyes, reimagine it as the word 'model' (much less scary) and think of the equation above: we predict an outcome variable from some model (that may or may not be hideously complex) but we won't do so perfectly so there will be some error in there too. Next time you

encounter some sleep-inducing phrase like 'hierarchical growth model', just remember that in most cases it's just a fancy way of saying 'predicting an outcome from some variables'.

2.4 Populations and samples ▮▮▮▮

Before we get stuck into a specific form of statistical model, it's worth remembering that scientists are usually interested in finding results that apply to an entire **population** of entities. For example, psychologists want to discover processes that occur in all humans, biologists might be interested in processes that occur in all cells, economists want to build models that apply to all salaries, and so on. A population can be very general (all human beings) or very narrow (all male ginger cats called Bob). Usually, scientists strive to infer things about general populations rather than narrow ones. For example, it's not very interesting to conclude that psychology students with

brown hair who own a pet hamster named George recover more quickly from sports injuries if the injury is massaged (unless you happen to be a psychology student with brown hair who has a pet hamster named George, like René Koning).[4] It will have a much wider impact if we can conclude that *everyone's* (or most people's) sports injuries are aided by massage.

Remember that our bridge-building engineer could not make a full-sized model of the bridge she wanted to build and instead built a small-scale model and tested it under various conditions. From the results obtained from the small-scale model she inferred things about how the full-sized bridge would respond. The small-scale model may respond differently than a full-sized version of the bridge, but the larger the model, the more likely it is to behave in the same way as the full-sized bridge. This metaphor can be extended to scientists: we rarely, if ever, have access to every member of a population (the real-sized bridge). Psychologists cannot collect data from every human being, and ecologists cannot observe every male ginger cat called Bob. Therefore, we collect data from a smaller subset of the population known as a **sample** (the scaled-down bridge) and use these data to infer things about the population as a whole. The bigger the sample, the more likely it is to reflect the whole population. If we take several random samples from the population, each of these samples will give us slightly different results but, on average, the results from large samples should be similar.

---

3 The little i (e.g., outcome$_i$) refers to the *i*th score. Imagine, we had three scores collected from Andy, Zach and Zoë. We could replace the *i* with their name, so if we wanted to predict Zoë's score we could change the equation to: outcome$_{Zoë}$ = model + error$_{Zoë}$. The *i* reflects the fact that the value of the outcome and the error will be different for each person.

4 A brown-haired psychology student with a hamster called Sjors (Dutch for George, apparently), who emailed me to weaken my foolish belief that I'd generated an obscure combination of possibilities.

2.5 P is for parameters ▌▌▌▌

Remember that parameters are the 'P' in the SPINE of statistics. Statistical models are made up of variables and **parameters**. As we have seen, variables are measured constructs that vary across entities in the sample. In contrast, parameters are not measured and are (usually) constants believed to represent some fundamental truth about the relations between variables in the model. Some examples of parameters with which you might be familiar are: the mean and median (which estimate the center of the distribution) and the correlation and regression coefficients (which estimate the relationship between two variables).

Statisticans try to confuse you by giving estimates of different parameters different symbols and letters ($\overline{X}$ for the mean, r for the correlation, b for regression coefficients) but it's much less confusing if we just use the letter b. If we're interested only in summarizing the outcome, as we are when we compute a mean, then we won't have any variables in the model, only a parameter, so we could write our equation as:

$$\text{outcome}_i = \left(b_0\right) + \text{error}_i \qquad (2.2)$$

However, often we want to predict an outcome from a variable, and if we do this we expand the model to include this variable (predictor variables are usually denoted with the letter X). Our model becomes:

$$\text{outcome}_i = \left(b_0 + b_1 X_i\right) + \text{error}_i \qquad (2.3)$$

Now we're predicting the value of the outcome for a particular entity (i) not just from the value of the outcome when there are no predictors (b_0) but from the entity's score on the predictor variable (X_i). The predictor variable has a parameter (b_1) attached to it, which tells us something about the relationship between the predictor (X_i) and outcome.

If we want to predict an outcome from two predictors then we can add another predictor to the model too:

$$\text{outcome}_i = \left(b_0 + b_1 X_{1i} + b_2 X_{2i}\right) + \text{error}_i \qquad (2.4)$$

In this model, we're predicting the value of the outcome for a particular entity (i) from the value of the outcome when there are no predictors (b_0) and the entity's score on two predictor variables (X_{1i} and X_{2i}). Each predictor variable has a parameter (b_1, b_2) attached to it, which tells us something about the relationship between that predictor and the outcome. We could carry on expanding the model with more variables, but that will make our brains hurt, so let's not. In each of these equations I have kept brackets around the model, which aren't necessary, but I think it helps you to see which part of the equation is the model in each case.

Hopefully what you can take from this section is that this book boils down to a very simple idea: we can predict values of an outcome variable based on a model. The form of the model changes, but there will always be some error in prediction, and there will always be parameters that tell us about the shape or form of the model.

To work out what the model looks like, we estimate the parameters (i.e., the value(s) of b). You'll hear the phrase 'estimate the parameter' or 'parameter estimates' a lot in statistics, and you might wonder why we use the word 'estimate'. Surely statistics has evolved enough that we can compute exact values of things and not merely estimate them? As I mentioned before, we're interested in drawing conclusions about a population (to which we don't have access). In other words, we want to know what our model might look like in the whole population. Given that our model is defined by parameters, this amounts to saying that we're not interested in the parameter values in our sample, but we care about the parameter values in the population.

The problem is that we don't know what the parameter values are in the population because we didn't measure the population, we measured only a sample. However, we can use the sample data to *estimate* what the population parameter values are likely to be. That's why we use the word 'estimate', because when we calculate parameters based on sample data they are only estimates of what the true parameter value is in the population. Let's make these ideas a bit more concrete with a very simple model indeed: the mean.

2.5.1 The mean as a statistical model ▌▌▌▌

We encountered the mean in Section 1.8.4, where I briefly mentioned that it was a statistical model because it is a hypothetical value and not necessarily one that is observed in the data. For example, if we took five statistics lecturers and measured the number of friends that they had, we might find the following data: 1, 2, 3, 3 and 4. If we want to know the mean number of friends, this can be calculated by adding the values we obtained, and dividing by the number of values measured: $(1 + 2 + 3 + 3 + 4)/5 = 2.6$. It is impossible to have 2.6 friends (unless you chop someone up with a chainsaw and befriend their arm, which is probably not beyond your average statistics lecturer) so the mean value is a *hypothetical* value: it is a model created to summarize the data and there will be error in prediction. As in equation (2.2), the model is:

$$\text{outcome}_i = b_0 + \text{error}_i$$

in which the parameter, b_0, is the mean of the outcome. The important thing is that we can use the value of the mean (or any parameter) computed in our sample to estimate the value in the population (which is the value in which we're interested). We give estimates little hats to compensate them for the lack of

Figure 2.5 Thanks to the Confusion machine, there are lots of terms that basically refer to error

self-esteem they feel at not being true values. Who doesn't love a hat?

$$\text{outcome}_i = \left(\hat{b}_0\right) + \text{error}_i \qquad (2.5)$$

When you see equations where these little hats are used, try not to be confused, all the hats are doing is making explicit that the values underneath them are estimates. Imagine the parameter as wearing a little baseball cap with the word 'estimate' printed along the front. In the case of the mean, we estimate the population value by assuming that it is the same as the value in the sample (in this case 2.6).

2.5.2 Assessing the fit of a model: sums of squares and variance revisited ▌▌▌▌

It's important to assess the fit of any statistical model (to return to our bridge analogy, we need to know how representative the model bridge is of the bridge that we want to build). With most statistical models we can determine whether the model represents the data

well by looking at how different the scores we observed in the data are from the values that the model predicts. For example, let's look what happens when we use the model of the mean to predict how many friends the first lecturer has. The first lecture was called Andy; it's a small world. We observed that lecturer 1 had one friend and the model (i.e., the mean of all lecturers) predicted 2.6. By rearranging equation (2.1) we see that this is an error of −1.6:[5]

$$\begin{aligned} \text{outcome}_{\text{lecturer 1}} &= \hat{b}_0 + \varepsilon_{\text{lecturer 1}} \\ 1 &= 2.6 + \varepsilon_{\text{lecturer 1}} \\ \varepsilon_{\text{lecturer 1}} &= 1 - 2.6 \\ &= -1.6 \end{aligned} \qquad (2.6)$$

You might notice that all we have done here is calculate the deviance, which we encountered in Section 1.8.5. The *deviance* is another word for *error* (Figure 2.5). A more general way to think of the deviance or error is by rearranging equation (2.1) into:

$$\text{deviance} = \text{outcome}_i - \text{model}_i \qquad (2.7)$$

In other words, the error or deviance for a particular entity is the score predicted by

the model for that entity subtracted from the corresponding observed score. Figure 2.6 shows the number of friends that each statistics lecturer had, and the mean number that we calculated earlier on. The line representing the mean can be thought of as our model, and the dots are the observed data. The diagram also has a series of vertical lines that connect each observed value to the mean value. These lines represent the error or deviance of the model for each lecturer. The first lecturer, Andy, had only 1 friend (a glove puppet of a pink hippo called Professor Hippo) and we have already seen that the error for this lecturer is −1.6; the fact that it is a negative number shows that our model *overestimates* Andy's popularity: it predicts that he will have 2.6 friends but, in reality, he has only 1 (bless!).

We know the accuracy or 'fit' of the model for a particular lecturer, Andy, but we want to know the fit of the model *overall*. We saw in Section 1.8.5 that we can't add deviances because some errors are positive and others negative and so we'd get a total of zero:

$$\begin{aligned} \text{total error} &= \text{sum of errors} \\ &= \sum_{i=1}^{n} \left(\text{outcome}_i - \text{model}_i\right) \\ &= (-1.6) + (-0.6) + (0.4) + (0.4) + (1.4) = 0 \end{aligned} \qquad (2.8)$$

We also saw in Section 1.8.5 that one way around this problem is to square the errors. This would give us a value of 5.2:

$$\begin{aligned} \text{sum of squared errors (SS)} &= \sum_{i=1}^{n} \left(\text{outcome}_i - \text{model}_i\right)^2 \\ &= (-1.6)^2 + (-0.6)^2 + (0.4)^2 + (0.4)^2 + (1.4)^2 \\ &= 2.56 + 0.36 + 0.16 + 0.16 + 1.96 \\ &= 5.20 \end{aligned} \qquad (2.9)$$

Does this equation look familiar? It ought to, because it's the same as equation (1.6) for the sum of squares in Section 1.8.5—the only difference is that equation (1.6) was specific to when our model is the mean, so the

---

5 Remember that I'm using the symbol $\hat{b}_0$ to represent the mean. If this upsets you then replace it (in your mind) with the more traditionally used symbol, $\overline{X}$.

'model' was replaced with the symbol for the mean ($\bar{x}$), and the outcome was replaced by the letter x, which is commonly used to represent a score on a variable (Eq. 2.10).

$$\sum_{i=1}^{n}\left(\text{outcome}_i - \text{model}_i\right)^2 = \sum_{i=1}^{n}\left(x_i - \bar{x}\right)^2 \qquad (2.10)$$

However, when we're thinking about models more generally, this illustrates that we can think of the total error in terms of this general equation:

$$\text{total error} = \sum_{i=1}^{n}\left(\text{observed}_i - \text{model}_i\right)^2 \qquad (2.11)$$

This equation shows how something we have used before (the sum of squares) can be used to assess the total error in any model (not just the mean).

We saw in Section 1.8.5 that although the sum of squared errors (SS) is a good measure of the accuracy of our model, it depends upon the quantity of data that has been collected—the more data points, the higher the SS. We also saw that we can overcome this problem by using the average error,

rather than the total. To compute the average error we divide the sum of squares (i.e., the total error) by the number of values (N) that we used to compute that total. We again come back to the problem that we're usually interested in the error in the model in the population (not the sample). To estimate the

mean error in the population we need to divide not by the number of scores contributing to the total, but by the **degrees of freedom** (df), which is the number of scores used to compute the total adjusted for the fact that we're trying to estimate the population value (Jane Superbrain Box 2.2):

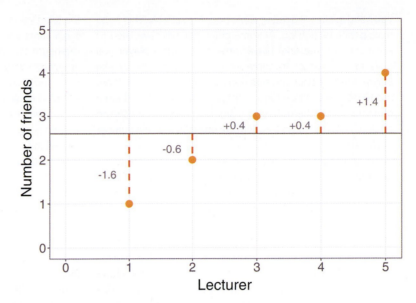

Figure 2.6   Graph showing the difference between the observed number of friends that each statistics lecturer had, and the mean number of friends

Jane Superbrain 2.2
Degrees of freedom ▮▮▮▮

The concept of degrees of freedom (df) is very difficult to explain. I'll begin with an analogy. Imagine you're the manager of a sports team (I'll try to keep it general so you can think of whatever sport you follow, but in my mind I'm thinking about soccer). On the morning of the game you have a team sheet with (in the case of soccer) 11 empty slots relating to the positions on the playing field. Different players have different positions on the field that determine their role (defense, attack, etc.) and to some extent their physical location (left,

right, forward, back). When the first player arrives, you have the choice of 11 positions in which to place this player. You place their name in one of the slots and allocate them to a position (e.g., striker) and, therefore, one position on the pitch is now occupied. When the next player arrives, you have 10 positions that are still free: you have 'a degree of freedom' to choose a position for this player (they could be put in defense, midfield, etc.). As more players arrive, your choices become increasingly limited: perhaps you have enough defenders so you need to start allocating some people to attack, where you have positions unfilled. At some point, you will have filled 10 positions and the final player arrives. With this player you have no 'degree of freedom' to choose where he or she plays—there is only one position left. In this scenario, there are 10 degrees of freedom: for 10 players you have a degree of choice over where they play, but for 1 player you have no choice. The degrees of freedom are one less than the number of players.

In statistical terms, the degrees of freedom relate to the number of observations that are free to vary. If we take a sample of four observations from a population, then these four scores are free to vary in any way (they can be any value). We use these four sampled values to estimate the mean of the population. Let's say that the mean of the sample was 10 and, therefore, we estimate that the population mean is also 10. The value of this parameter is now fixed: we have held one parameter constant. Imagine we now want to use this sample of four observations to estimate the mean squared error in the population. To do this, we need to use the value of the population mean, which we estimated to be

the fixed value of 10. With the mean fixed, are all four scores in our sample free to be sampled? The answer is no, because to ensure that the population mean is 10 only three values are free to vary. For example, if the values in the sample we collected were 8, 9, 11, 12 (mean = 10) then the first value sampled could be any value from the population, say 9. The second value can also be any value from the population, say 12. Like our football team, the third value sampled can also be any value from the population, say 8. We now have values of 8, 9 and 12 in the sample. The final value we sample, the final player to turn up to the soccer game, cannot be any value in the population, it *has to be* 11 because this is the value that makes the mean of the sample equal to 10 (the population parameter that we have held constant). Therefore, if we hold one parameter constant, then the degrees of freedom must be one fewer than the number of scores used to calculate that parameter. This fact explains why when we use a sample to estimate the mean squared error (or indeed the standard deviation) of a population, we divide the sums of squares by $N - 1$ rather than N alone. There is a lengthier explanation in one of my other books (Field, 2016).

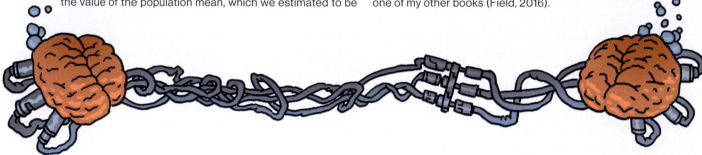

$$\text{mean squared error} = \frac{\text{SS}}{df}$$

$$= \frac{\sum_{i=1}^{n} \left(\text{outcome}_i - \text{model}_i\right)^2}{N-1} \qquad (2.12)$$

Does this equation look familiar? Again, it ought to, because it's a more general form of the equation for the variance (Eq. 1.7). Our model is the mean, so let's replace the 'model' with the mean ($\bar{x}$), and the 'outcome' with the letter x (to represent a score on the outcome). Lo and behold, the equation transforms into that of the variance:

$$\text{mean squared error} = \frac{\text{SS}}{df} = \frac{\sum_{i=1}^{n} \left(x_i - \bar{x}\right)^2}{N-1} = \frac{5.20}{4} = 1.30 \qquad (2.13)$$

To sum up, we can use the sum of squared errors and the mean squared error to assess the fit of a model. The mean squared error is also known as the variance. As such, the variance is a special case of a more general principle that we can apply to more complex models, which is that the fit of the model can be assessed with either the sum of squared errors or the mean squared error. Both measures give us an idea of how well a model fits the data: large values relative to the model indicate a lack of fit. Think back to Figure 1.10, which showed students' ratings of five lectures given by two lecturers. These lecturers differed in their mean squared error:[6] lecturer 1 had a smaller mean squared error than lecturer 2. Compare their graphs: the ratings for lecturer 1 were

consistently close to the mean rating, indicating that the mean is a good representation of the observed data—it is a good fit. The ratings for lecturer 2, however, were more spread out from the mean: for some lectures, she received very high ratings, and for others her ratings were terrible. Therefore, the mean is not such a good representation of the observed scores—it is a poor fit.

2.6 E is for estimating parameters ▮▮▮▮

We have seen that models are defined by parameters, and these parameters need to be estimated from the data that we collect. Estimation is the 'E' in the SPINE of statistics. We used an example of the

---

6 I reported the standard deviation, but this value is the square root of the variance (a.k.a. the mean square error).

mean because it was familiar, but it will also illustrate a general principle about how parameters are estimated. Let's imagine that one day we walked down the road and fell into a hole. Not just any old hole, though, but a hole created by a rupture in the space-time continuum. We slid down the hole, which turned out to be a sort of U-shaped tunnel under the road, and we emerged out of the other end to find that not only were we on the other side of the road, but we'd gone back in time a few hundred years. Consequently, statistics had not been invented and neither had the equation to compute the mean. Happier times, you might think. A slightly odorous and beardy vagrant accosts you, demanding to know the average number of friends that a lecturer has. If we didn't know the equation for computing the mean, how might we do it? We could guess and see how well our guess fits the data. Remember, we want the value of the parameter $\hat{b}_0$ in this equation:

$$\text{outcome}_i = \hat{b}_0 + \text{error}_i$$

We know already that we can rearrange this equation to give us the error for each person:

$$\text{error}_i = \text{outcome}_i - \hat{b}_0$$

If we add the error for each person, then we'll get the sum of squared errors, which we can use as a measure of 'fit'. Imagine we begin by guessing that the mean number of friends that a lecturer has is 2. We can compute the error for each lecturer by subtracting this value from the number of friends they actually had. We then square this value to get rid of any minus signs, and we add up these squared errors. Table 2.1 shows this process, and we find that by guessing a value of 2, we end up with a total squared error of 7. Now let's take another guess; this time we'll guess the value is 4. Again, we compute the sum of squared errors as a measure of 'fit'. This

model (i.e., guess) is worse than the last because the total squared error is larger than before: it is 15. We could carry on guessing and calculating the error for each guess. We *could*—if we were nerds with nothing better to do—but you're probably rad hipsters too busy doing whatever it is that rad hipsters do. I, however, am a badge-wearing nerd, so I have plotted the results in Figure 2.7, which shows the sum of squared errors that you would get for various values of the parameter $\hat{b}_0$. Note that, as we just calculated, when b is 2 we get an error of 7, and when it is 4 we get an error of 15. The shape of the line is interesting, though, because it curves to a minimum value—a value that produces

the lowest sum of squared errors. The value of b at the lowest point of the curve is 2.6, and it produces an error of 5.2. Do these values seem familiar? They should, because they are the values of the mean and sum of squared errors that we calculated earlier. This example illustrates that the equation for the mean is designed to estimate that parameter to minimize the error. In other words, it is the value that has the least error. This doesn't necessarily mean that the value is a *good* fit to the data, but it is a better fit than any other value you might have chosen.

Throughout this book, we will fit lots of different models, with parameters other than the mean that need to be estimated.

Table 2.1   Guessing the mean

Number of Friends (x_i)	b_1	Squared error $(x_i - b_1)^2$	b_2	Squared error $(x_i - b_2)^2$
1	2	1	4	9
2	2	0	4	4
3	2	1	4	1
3	2	1	4	1
4	2	4	4	0
		$\sum_{i=1}^{n}(x_i - b_1)^2 = 7$		$\sum_{i=1}^{n}(x_i - b_2)^2 = 15$

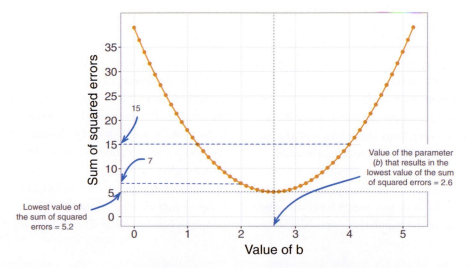

Figure 2.7   Graph showing the sum of squared errors for different 'guesses' of the mean

Although the equations for estimating these parameters will differ from that of the mean, they are based on this principle of minimizing error: they will give you the parameter that has the least error given the data you have. Again, it's worth reiterating that this is not the same thing as the parameter being accurate, unbiased or representative of the population: it could just be the best of a bad bunch. This section has focussed on the principle of minimizing the sum of squared errors, and this is known as the **method of least squares** or **ordinary least squares** (OLS). However, we'll also encounter other estimation methods later in the book.

2.7 S is for standard error ▊▊▊▊

We have looked at how we can fit a statistical model to a set of observations to summarize those data. It's one thing to summarize the data that you have actually collected, but in Chapter 1 we saw that good theories should say something about the wider world. It is one thing to be able to say that a sample of high-street stores in Brighton improved profits by placing cats in their store windows, but it's more useful to be able to say, based on our sample, that all high-street stores can increase profits by placing cats in their window displays. To do this, we need to go beyond the data, and to go beyond the data we need to begin to look at how representative our samples are of the population of interest. This idea brings us to the 'S' in the SPINE of statistics: the *standard error*.

In Chapter 1 we saw that the standard deviation tells us about how well the mean represents the sample data. However, if we're using the sample mean to estimate this parameter in the population, then we need to know how well it represents the value in the population, especially because

samples from a population differ. Imagine that we were interested in the student ratings of all lecturers (so lecturers in general are the population). We could take a sample from this population, and when we do we are taking one of many possible samples. If we were to take several samples from the same population, then each sample would have its own mean, and some of these sample means will be different. Figure 2.8 illustrates the process of taking samples from a population. Imagine for a fleeting second that we eat some magic beans that transport us to an astral plane where we can see for a few short, but beautiful, seconds the ratings of all lectures in the world. We're in this astral plane just long enough to compute the mean of these ratings (which, given the size of the population, implies we're there for quite some time). Thanks to our astral adventure we know, as an absolute fact, that the mean of all ratings is 3 (this is the *population mean*, μ, the parameter that we're trying to estimate).

Back in the real world, where we don't have magic beans, we also don't have access to the population, so we use a sample. In this sample we calculate the average rating, known as the *sample mean*, and discover it is 3; that is, lecturers were rated, on average, as 3. 'That was fun,' we think to ourselves, 'Let's do it again.' We take a second sample and find that lecturers were rated, on average, as only 2. In other words, the sample mean is different in the second sample than in the first. This difference illustrates **sampling variation**: that is, samples vary because they contain different members of the population; a sample that, by chance, includes some very good lecturers will have a higher average than a sample that, by chance, includes some awful lecturers. Imagine that we're so excited by this sampling malarkey that we take another

seven samples, so that we have nine in total (as in Figure 2.8). If we plotted the resulting sample means as a frequency distribution or histogram,[7] we would see that three samples had a mean of 3, means of 2 and 4 occurred in two samples each, and means of 1 and 5 occurred in only one sample each. The end result is a nice symmetrical distribution known as a **sampling distribution**. A sampling distribution is the frequency distribution of sample means (or whatever parameter you're trying to estimate) from the same population. You need to imagine that we're taking hundreds or thousands of samples to construct a sampling distribution—I'm using nine to keep the diagram simple. The sampling distribution is a bit like a unicorn: we can imagine what one looks like, we can appreciate its beauty, and we can wonder at its magical feats, but the sad truth is that you'll never see a real one. They both exist as ideas rather than physical things. You would never go out and actually collect thousands of samples and draw a frequency distribution of their means, instead very clever statisticians have worked out what these distributions look like and how they behave. Likewise, you'd be ill-advised to search for unicorns.

The sampling distribution of the mean tells us about the behavior of samples from the population, and you'll notice that it is centered at the same value as the mean of the population (i.e., 3). Therefore, if we took the average of all sample means we'd get the value of the population mean. We can use the sampling distribution to tell us how representative a sample is of the population. Think back to the standard deviation. We used the standard deviation as a measure of how representative the mean was of the observed data. A small standard deviation represented a scenario in which most data points were close to

---

7 This is a graph of possible values of the sample mean plotted against the number of samples that have a mean of that value—see Section 1.8.1 for more details.

the mean, whereas a large standard deviation represented a situation in which data points were widely spread from the mean. If our 'observed data' are *sample* means then the standard deviation of these sample means would similarly tell us how widely spread (i.e., how representative) sample means are around their average. Bearing in mind that the average of the sample means is the same as the population mean, the standard deviation of the sample means would therefore tell us how widely sample means are spread around the population mean: put another way, it tells us whether sample means are typically representative of the population mean.

The standard deviation of sample means is known as the **standard error of the mean** (**SE**) or **standard error** for short. In the land where unicorns exist, the standard error could be calculated by taking the difference between each sample mean and the overall mean, squaring these differences, adding them up, and then dividing by the number of samples. Finally, the square root of this value would need to be taken to get the standard deviation of sample means: the standard error. In the real world, it would be crazy to collect hundreds of samples, and so we compute the standard error from a mathematical approximation. Some exceptionally clever statisticians have demonstrated something called the **central limit theorem**, which tells us that as samples get large (usually defined as greater than 30), the sampling distribution has a normal distribution with a mean equal to the population mean, and a standard deviation shown in equation (2.14):

$$\sigma_{\bar{X}} = \frac{s}{\sqrt{N}} \qquad (2.14)$$

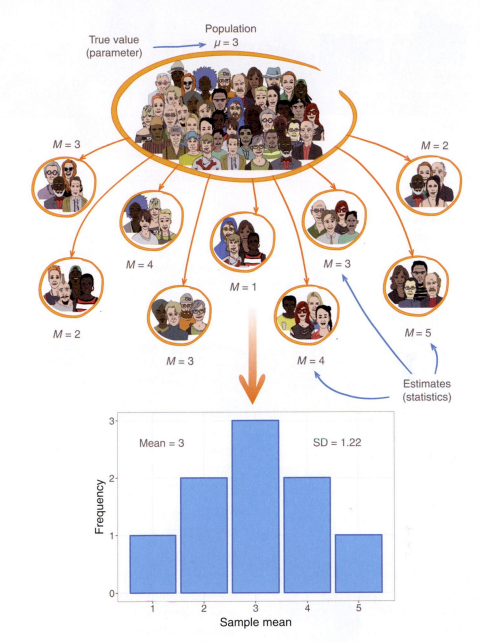

Figure 2.8   Illustration of the standard error (see text for details)

We will return to the central limit theorem in more detail in Chapter 6, but I've mentioned it here because it tells us that if our sample is large we can use equation (2.14) to approximate the standard error (because it is the standard deviation of the sampling distribution).[8] When the sample is relatively small (fewer than 30) the sampling distribution is not normal: it has a different shape, known as a *t*-distribution, which we'll come back to later. A final point is that our discussion here has been about the mean, but everything we have learnt about sampling distributions applies to other parameters too: any parameter that can be estimated in a sample has a hypothetical sampling distribution and standard error.

8 In fact, it should be the *population* standard deviation (σ) that is divided by the square root of the sample size; however, it's rare that we know the standard deviation of the population and, for large samples, this equation is a reasonable approximation.

Cramming Sam's Tips
The standard error

- The standard error of the mean is the standard deviation of sample means. As such, it is a measure of how representative of the population a sample mean is likely to be. A large standard error (relative to the sample mean) means that there is a lot of variability between the means

of different samples and so the sample mean we have might not be representative of the population mean. A small standard error indicates that most sample means are similar to the population mean (i.e., our sample mean is likely to accurately reflect the population mean).

2.8 I is for (confidence) interval ▮▮▮▮

The 'I' in the SPINE of statistics is for 'interval'; confidence interval, to be precise. As a brief recap, we usually use a sample value as an estimate of a parameter (e.g., the mean) in the population. We've just seen that the estimate of a parameter (e.g., the mean) will differ across samples, and we can use the standard error to get some idea of the extent to which these estimates differ across samples. We can also use this information to calculate boundaries within which we believe the population value will fall. Such boundaries are called **confidence intervals**. Although what I'm about to describe applies to any

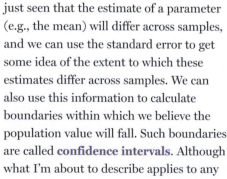

What is a confidence interval?

parameter, we'll stick with the mean to keep things consistent with what you have already learnt.

2.8.1 Calculating confidence intervals ▮▮▮▮

Domjan, Blesbois, & Williams (1998) examined the learnt release of sperm in Japanese quail. The basic idea is that if a quail is allowed to copulate with a female quail in a certain context (an experimental chamber), then this context will serve as a cue to a mating opportunity and this in turn will affect semen release (although during the test phase the poor quail were tricked into copulating with a terry cloth with an embalmed female quail head stuck on top).[9] Anyway, if we look at the mean amount of sperm released in the experimental chamber, there is a true mean (the mean in the population); let's

imagine it's 15 million sperm. Now, in our sample, we might find the mean amount of sperm released was 17 million. Because we don't know what the true value of the mean is (the population value), we don't know how good (or bad) our sample value of 17 million is as an estimate of it. So rather than fixating on a single value from the sample (the **point estimate**), we could use an **interval estimate** instead: we use our sample value as the midpoint, but set a lower and upper limit as well. So, we might say, we think the true value of the mean sperm release is somewhere between 12 million and 22 million sperm (note that 17 million falls exactly between these values). Of course, in this case, the true value (15 million) does falls within these limits. However, what if we'd set smaller limits—what if we'd said we think the true value falls between 16 and 18 million (again, note that 17 million is in

---

9 This may seem a bit sick, but the male quails didn't appear to mind too much, which probably tells us all we need to know about males.

the middle)? In this case the interval does not contain the population value of the mean.

Let's imagine that you were particularly fixated with Japanese quail sperm, and you repeated the experiment 100 times using different samples. Each time you did the experiment you constructed an interval around the sample mean as I've just described. Figure 2.9 shows this scenario: the dots represent the mean for each sample, with the lines sticking out of them representing the intervals for these means. The true value of the mean (the mean in the population) is 15 million and is shown by a vertical line. The first thing to note is that the sample means are different from the true mean (this is because of sampling variation as described earlier). Second, although most of the intervals do contain the true mean (they cross the vertical line, meaning that the value of 15 million sperm falls somewhere between the lower and upper boundaries), a few do not.

The crucial thing is to construct the intervals in such a way that they tell us something useful. For example, perhaps we might want to know how often, in the long run, an interval contains the true value of the parameter we're trying to estimate (in this case, the mean). This is what a confidence interval does. Typically, we look at 95% confidence intervals, and sometimes 99% confidence intervals, but they all have a similar interpretation: they are limits constructed such that, for a certain percentage of samples (be that 95% or 99%), the true value of the population parameter falls within the limits. So, when you see a 95% confidence interval for a mean, think of it like this: if we'd collected 100 samples, and for each sample calculated the mean and a confidence interval for it (a bit like in Figure 2.9), then for 95 of these samples, the confidence interval contains the value of the mean in the population, and in 5 of the samples the confidence interval does not contain the population mean.

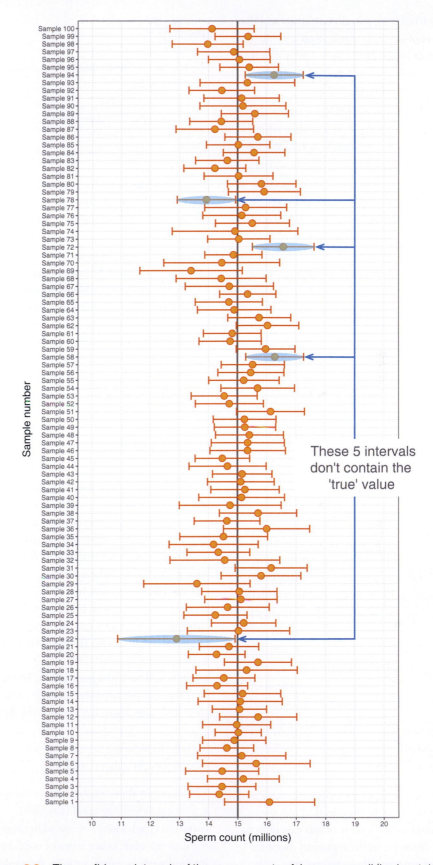

These 5 intervals don't contain the 'true' value

Figure 2.9 The confidence intervals of the sperm counts of Japanese quail (horizontal axis) for 100 different samples (vertical axis)

The trouble is, you do not know whether the confidence interval from a particular sample is one of the 95% that contain the true value or one of the 5% that do not (Misconception Mutt 2.1).

To calculate the confidence interval, we need to know the limits within which 95% of sample means will fall. We know (in large samples) that the sampling distribution of means will be normal, and the normal distribution has been precisely defined such that it has a mean of 0 and a standard deviation of 1. We can use this information to compute the probability of a score occurring or the limits between which a certain percentage of scores fall (see Section 1.8.6). It was no coincidence that when I explained all of this in Section 1.8.6 I used the example of how we would work out the limits between which 95% of scores fall; that is precisely what we need to know if we want to construct a 95% confidence interval. We discovered in Section 1.8.6 that 95% of z-scores fall between -1.96 and 1.96. This means that if our sample means were normally distributed with a mean of 0 and a standard error of 1, then the limits of our confidence interval would be -1.96 and $+1.96$. Luckily we know from the central limit theorem that in large samples (above about 30) the sampling distribution *will* be normally distributed (see Section 2.7). It's a pity then that our mean and standard deviation are unlikely to be 0 and 1— except it's not, because we can convert scores so that they do have a mean of 0 and standard deviation of 1 (z-scores) using equation (1.9):

$$z = \frac{X - \overline{X}}{s}$$

If we know that our limits are -1.96 and 1.96 as z-scores, then to find out the corresponding scores in our raw data we can replace z in the equation (because there are two values, we get two equations):

$$1.96 = \frac{X - \overline{X}}{s} \qquad -1.96 = \frac{X - \overline{X}}{s}$$

We rearrange these equations to discover the value of X:

$$1.96 \times s = X - \overline{X} \qquad -1.96 \times s = X - \overline{X}$$

$$(1.96 \times s) + \overline{X} = X \qquad (-1.96 \times s) + \overline{X} = X$$

Therefore, the confidence interval can easily be calculated once the standard deviation (s in the equation) and mean ($\overline{X}$ in the equation) are known. However, we use the standard error and not the standard deviation because we're interested in the variability of *sample* means, not the variability in observations within the sample. The lower boundary of the confidence interval is, therefore, the mean minus 1.96 times the standard error, and the upper boundary is the mean plus 1.96 standard errors:

$$\text{lower boundary of confidence interval} = \overline{X} - (1.96 \times SE)$$
$$\text{upper boundary of confidence interval} = \overline{X} + (1.96 \times SE)$$
$$(2.15)$$

As such, the mean is always in the center of the confidence interval. We know that 95% of confidence intervals contain the population mean, so we can assume this confidence interval contains the true mean; therefore, if the interval is small, the sample mean must be very close to the true mean. Conversely, if the confidence interval is very wide then the sample mean could be very different from the true mean, indicating that it is a bad representation of the population. You'll find that confidence intervals will come up time and time again throughout this book.

2.8.2 Calculating other confidence intervals ▮▮▮▮

The example above shows how to compute a 95% confidence interval (the most common type). However, we sometimes want to calculate other types of confidence interval such as a 99% or 90% interval. The 1.96 and -1.96 in equation (2.15) are

the limits within which 95% of z-scores occur. If we wanted to compute confidence intervals for a value other than 95% then we need to look up the value of z for the percentage that we want. For example, we saw in Section 1.8.6 that z-scores of -2.58 and 2.58 are the boundaries that cut off 99% of scores, so we could use these values to compute 99% confidence intervals. In general, we could say that confidence intervals are calculated as:

$$\text{lower boundary of confidence interval} = \overline{X} - \left(z_{\frac{1-p}{2}} \times SE \right)$$
$$\text{upper boundary of confidence interval} = \overline{X} + \left(z_{\frac{1-p}{2}} \times SE \right)$$
$$(2.16)$$

in which p is the probability value for the confidence interval. So, if you want a 95% confidence interval, then you want the value of z for $(1 - 0.95)/2 = 0.025$. Look this up in the 'smaller portion' column of the table of the standard normal distribution (look back at Figure 1.14) and you'll find that z is 1.96. For a 99% confidence interval we want z for $(1 - 0.99)/2 = 0.005$, which from the table is 2.58 (Figure 1.14). For a 90% confidence interval we want z for $(1 - 0.90)/2 = 0.05$, which from the table is 1.64 (Figure 1.14). These values of z are multiplied by the standard error (as above) to calculate the confidence interval. Using these general principles, we could work out a confidence interval for any level of probability that takes our fancy.

2.8.3 Calculating confidence intervals in small samples ▮▮▮▮

The procedure that I have just described is fine when samples, are large, because the central limit theorem tells us that the sampling distribution will be normal. However, for small samples, the sampling distribution is not normal—it has a t-distribution. The t-distribution is a family of probability distributions that change shape as the sample size gets bigger (when the sample is very big, it has

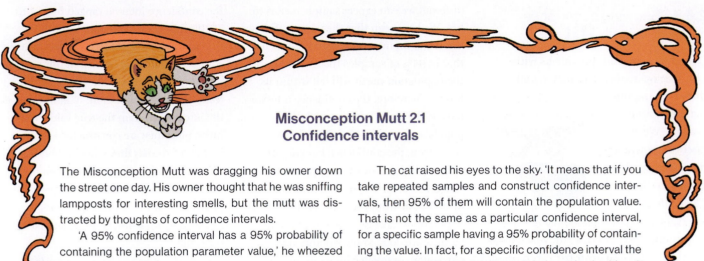

Misconception Mutt 2.1
Confidence intervals

The Misconception Mutt was dragging his owner down the street one day. His owner thought that he was sniffing lampposts for interesting smells, but the mutt was distracted by thoughts of confidence intervals.

'A 95% confidence interval has a 95% probability of containing the population parameter value,' he wheezed as he pulled on his lead.

A ginger cat emerged. The owner dismissed his perception that the cat had emerged from a solid brick wall. His dog pulled towards the cat in a stand-off. The owner started to check his text messages.

'You again?' the mutt growled.

The cat considered the dog's reins and paced around, smugly displaying his freedom. 'I'm afraid you will see very much more of me if you continue to voice your statistical misconceptions,' he said. 'They call me the Correcting Cat for a reason'.

The dog raised his eyebrows, inviting the feline to elaborate.

'You can't make probability statements about confidence intervals,' the cat announced.

'Huh?' said the mutt.

'You said that a 95% confidence interval has a 95% probability of containing the population parameter. It is a common mistake, but this is not true. The 95% reflects a *long-run* probability.'

'Huh?'

The cat raised his eyes to the sky. 'It means that if you take repeated samples and construct confidence intervals, then 95% of them will contain the population value. That is not the same as a particular confidence interval, for a specific sample having a 95% probability of containing the value. In fact, for a specific confidence interval the probability that it contains the population value is either 0 (it does not contain it) or 1 (it does contain it). You have no way of knowing which it is.' The cat looked pleased with himself.

'What's the point of that?' the dog asked.

The cat pondered the question. 'It is important if you want to control error,' he eventually answered. 'If you assume that the confidence interval contains the population value then you will be wrong only 5% of the time if you use a 95% confidence interval.'

The dog sensed an opportunity to annoy the cat. 'I'd rather know how likely it is that the interval contains the population value,' he said.

'In which case, you need to become a Bayesian,' the cat said, disappearing indignantly into the brick wall.

The mutt availed himself of the wall, hoping it might seal the cat in forever.

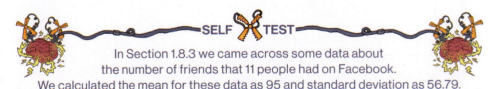

the shape of a normal distribution). To construct a confidence interval in a small sample we use the same principle as before, but instead of using the value for z we use the value for t:

and tells us which of the t-distributions to use. For a 95% confidence interval, we find the value of t for a two-tailed test with

probability of 0.05, for the appropriate degrees of freedom.

$$\text{lower boundary of confidence interval} = \bar{X} - (t_{n-1} \times SE)$$
$$\text{upper boundary of confidence interval} = \bar{X} + (t_{n-1} \times SE)$$

(2.17)

The $n-1$ in the equations is the degrees of freedom (see Jane Superbrain Box 2.2)

SELF TEST

In Section 1.8.3 we came across some data about the number of friends that 11 people had on Facebook. We calculated the mean for these data as 95 and standard deviation as 56.79.

- Calculate a 95% confidence interval for this mean.
- Recalculate the confidence interval assuming that the sample size was 56.

2.8.4 Showing confidence intervals visually ▮▮▮▮

Confidence intervals provide us with information about a parameter, and, therefore, you often see them displayed on graphs. (We will discover more about how to create these graphs in Chapter 5.) The confidence interval is usually displayed using something called an error bar, which looks like the letter 'I'. An error bar can represent the standard deviation or the standard error, but more often than not it shows the 95% confidence interval of the mean. So, often when you see a graph showing the mean, perhaps displayed as a bar or a symbol (Section 5.6), it is accompanied by this funny I-shaped bar.

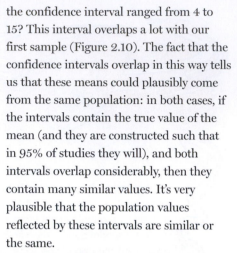

What is an error bar?

We have seen that any two samples can have slightly different means (and the standard error tells us a little about how different we can expect sample means to be). We have seen that the 95% confidence interval is an interval constructed such that in 95% of samples the true value of the population mean will fall within its limits. Therefore, the confidence interval tells us the limits within which the population mean is likely to fall. By comparing the confidence intervals of different means (or other parameters) we can get some idea about whether the means came from the same or different populations. (We can't be entirely sure because we don't know whether our particular confidence intervals are ones that contain the population value or not.)

Taking our previous example of quail sperm, imagine we had a sample of quail and the mean sperm release had been 9 million sperm with a confidence interval of 2 to 16. Therefore, if this is one of the 95% of intervals that contains the population value, then the population mean is between 2 and 16 million sperm. What if we now took a second sample of quail and found the confidence interval ranged from 4 to 15? This interval overlaps a lot with our first sample (Figure 2.10). The fact that the confidence intervals overlap in this way tells us that these means could plausibly come from the same population: in both cases, if the intervals contain the true value of the mean (and they are constructed such that in 95% of studies they will), and both intervals overlap considerably, then they contain many similar values. It's very plausible that the population values reflected by these intervals are similar or the same.

What if the confidence interval for our second sample ranged from 18 to 28? If we compared this to our first sample we'd get Figure 2.11. These confidence intervals don't overlap at all, so one confidence interval, which is likely to contain the population mean, tells us that the population mean is somewhere between 2 and 16 million, whereas the other confidence interval, which is also likely to contain the population mean, tells us that the population mean is somewhere between 18 and 28 million. This contradiction suggests two possibilities: (1) our confidence intervals both contain the population mean, but they come from different populations (and, therefore, so do our samples); or (2) both samples come from the same population but one (or both) of the confidence intervals doesn't contain the population mean (because, as you know, in 5% of cases they don't). If we've used 95% confidence intervals, then we know that the second possibility is unlikely (this happens only 5 times in 100 or 5% of the time), so the first explanation is more plausible.

I can hear you all thinking, 'So what if the samples come from a different population?' Well, it has a very important implication in experimental research. When we do an experiment, we introduce some form of manipulation between two or more conditions (see Section 1.7.2). If we have taken two random samples of people, and we have tested them on some

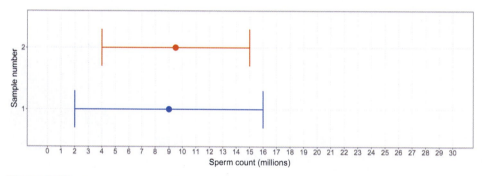

Figure 2.10 Two overlapping 95% confidence intervals

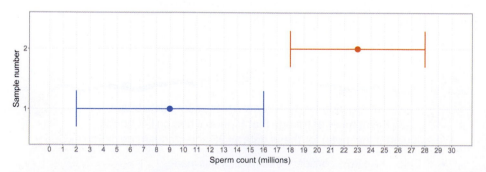

Figure 2.11 Two 95% confidence intervals that don't overlap

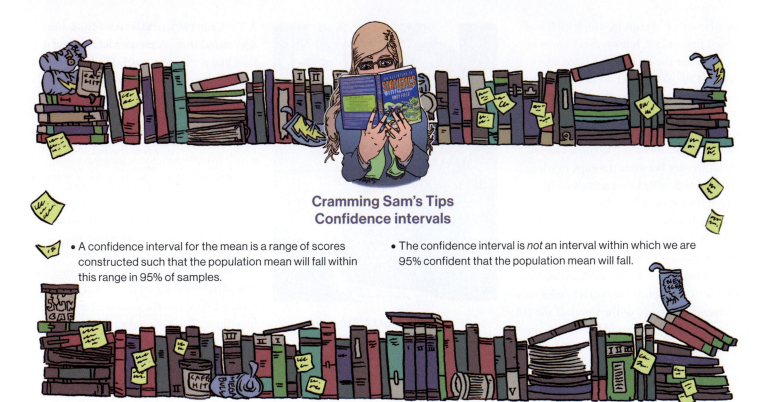

Cramming Sam's Tips
Confidence intervals

• A confidence interval for the mean is a range of scores constructed such that the population mean will fall within this range in 95% of samples.

• The confidence interval is *not* an interval within which we are 95% confident that the population mean will fall.

measure, then we expect these people to belong to the same population. We'd also, therefore, expect their confidence intervals to reflect the same population value for the mean. If their sample means and confidence intervals are so different as to suggest that they come from different populations, then this is likely to be because our experimental manipulation has induced a difference between the samples. Therefore, error bars showing 95% confidence intervals are useful, because if the bars of any two means do not overlap (or overlap by only a small amount) then we can infer that these means are from different populations— they are significantly different. We will return to this point in Section 2.9.9.

2.9 N is for null hypothesis significance testing ▊▊▊▊

In Chapter 1 we saw that research was a six-stage process (Figure 1.2). This chapter has looked at the final stage:

• Analyze the data: fit a statistical model to the data—this model will test your original predictions. Assess this model to see whether it supports your initial predictions.

I have shown that we can use a sample of data to estimate what's happening in a larger population to which we don't have access. We have also seen (using the mean as an example) that we can fit a statistical model to a sample of data and assess how well it fits. However, we have yet to see how fitting models like these can help us to test our research predictions. How do statistical models help us to test complex hypotheses such as 'Is there a relationship between the amount of gibberish that people speak and the amount of vodka jelly they've eaten?' or 'Does reading this chapter improve your knowledge of research methods?' This brings us to the 'N' in the SPINE of statistics: null hypothesis significance testing.

Null hypothesis significance testing (NHST) is a cumbersome name for an equally cumbersome process. NHST is the most commonly taught approach to testing research questions with statistical models. It arose out of two different approaches to the problem of how to use data to test theories: (1) Ronald Fisher's idea of computing probabilities to evaluate evidence, and (2) Jerzy Neyman and Egon Pearson's idea of competing hypotheses.

2.9.1 Fisher's *p*-value ▊▊▊▊

Fisher (1925/1991) (Figure 2.12) described an experiment designed to test a claim by a woman that she could determine, by tasting a cup of tea, whether the milk or the tea was added first to the cup. Fisher thought that he should give the woman some cups of tea, some of which had the milk added first and some of which had the milk added last, and see whether she could correctly identify them. The woman would know that there are an equal number of cups in which milk was added first or last, but wouldn't know in which order the cups were placed. If we take the simplest situation in which there are only two cups, then the woman has a 50% chance of guessing correctly. If she did guess correctly, we wouldn't be that confident in concluding that she can tell the difference between the cups in which the milk was added first and those in which it was added last, because even by guessing she would be correct half of the time. But what if we complicated things by having six cups? There are 20 orders in which these cups can be arranged and the woman would guess the correct order only 1 time in 20 (or 5% of the time). If she got the order correct we would be much more confident that she could genuinely tell the difference (and bow down in awe of her finely tuned palate). If you'd like to know more about Fisher and his tea-tasting antics, see David Salsburg's excellent book *The Lady Tasting Tea* (Salsburg, 2002). For our purposes the take-home point is that only when there was a very small probability that the woman could complete the tea task by guessing alone would we conclude that she had a genuine skill in detecting whether milk was poured into a cup before or after the tea.

It's no coincidence that I chose the example of six cups above (where the tea-taster had a 5% chance of getting the task right by guessing), because scientists tend to use 5% as a threshold for confidence: only when there is a 5% chance (or 0.05 probability) of getting the result we have (or one more extreme) if no effect exists are we confident enough to accept that the effect is genuine.[10] Fisher's basic point was that you should calculate the probability of an event and evaluate this probability within the research context. Although Fisher felt a $p = 0.01$ would be strong evidence to back up a hypothesis, and perhaps a $p = 0.20$ would be weak

Figure 2.12　Sir Ronald A. Fisher, the cleverest person ever ($p < 0.0001$)

evidence, he never said $p = 0.05$ was in any way a magic number. Fast forward 100 years or so, and everyone treats 0.05 as though it *is* a magic number.

2.9.2　Types of hypothesis

In contrast to Fisher, Neyman and Pearson believed that scientific statements should be split into testable hypotheses. The hypothesis or prediction from your theory would normally be that an effect will be present. This hypothesis is called the **alternative hypothesis** and is denoted by H_1. (It is sometimes also called the **experimental hypothesis**, but because this term relates to a specific type of methodology it's probably best to use 'alternative hypothesis'.) There is another type of hypothesis called the **null hypothesis**, which is denoted by H_0. This hypothesis is the opposite of the alternative hypothesis and so usually states that an effect is absent.

Often when I write, my thoughts are drawn towards chocolate. I believe that I would eat less of it if I could stop thinking about it. However, according to Morewedge, Huh,

& Vosgerau (2010), that's not true. In fact, they found that people ate less of a food if they had previously imagined eating it. Imagine we did a similar study. We might generate the following hypotheses:

- Alternative hypothesis: if you imagine eating chocolate you will eat less of it.
- Null hypothesis: if you imagine eating chocolate you will eat the same amount as normal.

The null hypothesis is useful because it gives us a baseline against which to evaluate how plausible our alternative hypothesis is. We can evaluate whether we think that the data we have collected are more likely, given the null or alternative hypothesis. A lot of books talk about accepting or rejecting these hypotheses, implying that you look at the data and either accept the null hypothesis (and therefore reject the alternative) or accept the alternative hypothesis (and reject the null). In fact, this isn't quite right because the way that scientists typically evaluate these hypotheses using p-values (which we'll come onto shortly) doesn't provide evidence for such black-and-white decisions. So, rather than talking about accepting or rejecting a hypothesis, we should talk about 'the chances of obtaining the result we have (or one more extreme), assuming that the null hypothesis is true'.

Imagine in our study that we took 100 people and measured how many pieces of chocolate they usually eat (day 1). On day 2, we got them to imagine eating chocolate and again measured how much chocolate they ate that day. Imagine that we found that 75% of people ate less chocolate on the second day than on the first. When we analyze our data, we are really asking, 'Assuming that imagining eating chocolate has no effect whatsoever, is it likely that 75% of people would eat less chocolate on the second day?' Intuitively, the answer is that the chances are very low: if the null hypothesis is true, then everyone should

---

10　Of course it might not be true—we're just prepared to believe that it is.

eat the same amount of chocolate on both days. Therefore, we are very unlikely to have got the data that we did if the null hypothesis were true.

What if we found that only 1 person (1%) ate less chocolate on the second day? If the null hypothesis is true and imagining eating chocolate has no effect whatsoever on consumption, then no people should eat less on the second day. The chances of getting these data if the null hypothesis is true are quite high. The null hypothesis is quite plausible given what we have observed.

SELF TEST

What are the null and alternative hypotheses for the following questions:

- 'Is there a relationship between the amount of gibberish that people speak and the amount of vodka jelly they've eaten?'
- 'Does reading this chapter improve your knowledge of research methods?'

When we collect data to test theories we work in these terms: we cannot talk about the null hypothesis being true or the experimental hypothesis being true, we can talk only in terms of the probability of obtaining a particular result or statistic if, hypothetically speaking, the null hypothesis were true. It's also worth remembering that our alternative hypothesis is likely to be one of many possible models that we could fit

Jane Superbrain 2.3
Who said statistics was dull? Part 1 ▮▮▮▮

Students often think that statistics is dull, but back in the early 1900s it was anything but dull, with prominent figures entering into feuds on a regular basis. Ronald Fisher and Jerzy Neyman had a particularly impressive feud. On 28 March 1935 Neyman delivered a talk to the Royal Statistical Society, at which Fisher was present, in which he criticized some of Fisher's most important work. Fisher directly attacked Neyman in his discussion of the paper at the same meeting: he more or less said that Neyman didn't know what he was talking about and didn't understand the background material on which his work was based. He said, 'I put it to you, sir, that you are a fool, an imbecile, a man so incapacitated by stupidity that in a battle of wits with a single-cell amoeba, the amoeba would fancy its chances.' He didn't really say that, but he opened the discussion without proposing a vote of thanks, which would have been equally as rude in those days.

Relations soured so much that while they both worked at University College London, Neyman openly attacked many of Fisher's ideas in lectures to his students. The two feuding groups even took afternoon tea (a common practice in the British academic community of the time and one which, frankly, we should reinstate) in the same room but at different times. The truth behind who fuelled these feuds is, perhaps, lost in the mists of time, but Zabell (1992) makes a sterling effort to unearth it. Basically, the founders of modern statistical methods, despite being super-humanly intelligent,[11] acted like a bunch of squabbling children.

---

11 Fisher, in particular, was a world leader in genetics, biology and medicine as well as possibly the most original mathematical thinker ever (Barnard, 1963; Field, 2005d; Savage, 1976).

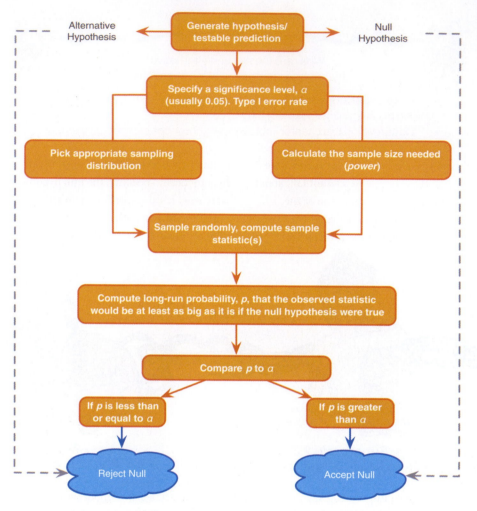

Figure 2.13   Flow chart of null hypothesis significance testing

to the data, so even if we believe it to be more likely than the null hypothesis, there may be other models of the data that we haven't considered that are a better fit, which again means that we cannot talk about the hypothesis as being definitively true or false, but we can talk about its plausibility relative to other hypotheses or models that we have considered.

Hypotheses can be directional or non-directional. A directional hypothesis states that an effect will occur, but it also states the direction of the effect. For example, 'If you imagine eating chocolate you will eat less of it' is a one-tailed hypothesis because it states the direction of the effect (people will eat less). A non-directional hypothesis states that an

effect will occur, but it doesn't state the direction of the effect. For example, 'Imagining eating chocolate affects the amount of chocolate you eat' does not tell us whether people will eat more or less.

2.9.3   The process of NHST ▌▌▌▌

NHST is a blend of Fisher's idea of using the probability value p as an index of the weight of evidence against a null hypothesis, and Jerzy Neyman and Egon Pearson's idea of testing a null hypothesis *against* an alternative hypothesis (Neyman & Pearson, 1933). There was no love lost between these competing statisticians (Jane Superbrain Box 2.3). NHST is a system designed to tell us whether the

alternative hypothesis is likely to be true—it helps us to decide whether to confirm or reject our predictions.

Figure 2.13 outlines the steps in NHST. As we have seen before, the process starts with a research hypothesis that generates a testable prediction. These predictions are decomposed into a null (there is no effect) and alternative hypothesis (there is an effect). At this point you decide upon the long-run error rate that you are prepared to accept, alpha (α). In other words, how often are you prepared to be wrong? This is the significance level, the probability of accepting an effect in our population as true, when no such effect exists (it is known as the Type I error rate, which we'll discuss in more detail in due course). It is important that we fix this error rate before we collect data, otherwise we are cheating (see Jane Superbrain Box 2.4). You should determine your error rate based on the nuances of your research area, and what it is you're trying to test. Put another way, it should be a meaningful decision. In reality, it is not: everyone uses 0.05 (a 5% error rate) with barely a thought for what it means or why they're using it. Go figure.

Having not given the thought you should have to your error rate, you choose a sampling distribution. This involves working out what statistical model to fit to the data that will test your hypothesis, looking at what parameters that model has, and then deciding on the shape of the sampling distribution attached to those parameters. Let's take my example of whether thinking about chocolate is related to consumption. You could measure how much people think about chocolate during the day and how much of it they eat in the same day. If the null hypothesis is true (there is no effect), then there should be no relationship at all between these variables. If it reduces consumption, then we'd expect a negative relationship between the two. One model we could fit that tests this hypothesis is the linear model that I described earlier, in which we predict consumption (the outcome) from thought about chocolate (the predictor). Our model is basically equation (2.3), but I'll replace the outcome and letter X with our variable names:

$$\text{consumption}_i = (b_0 + b\,\text{thought}_i) + \text{error}_i \qquad (2.18)$$

The parameter, b, attached to the variable **thought** tests our hypothesis: it quantifies the size and strength of relationship between thinking and consuming. If the null is true, b will be zero; otherwise it will be a value different from 0, the size and direction of which depends on what the relationship between the thought and consumption variables is. It turns out (see Chapter 9) that this parameter has a sampling distribution that has a t-distribution. So, that's what we'd use to test our hypothesis. We also need to establish how much data to collect to stand a reasonable chance of finding the effect we're looking for. This is called the power of the test, and I'll elaborate on this concept shortly.

Now the fun begins and you collect your data. You fit the statistical model that tests your hypothesis to the data. In the chocolate example, we'd estimate the parameter that represents the relationship between thought and consumption and its confidence interval. It's usually also possible to compute a test statistic that maps the parameter to a long-run

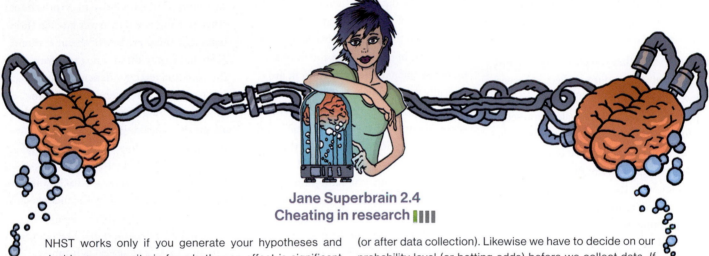

Jane Superbrain 2.4
Cheating in research ▮▮▮▮

NHST works only if you generate your hypotheses and decide on your criteria for whether an effect is significant before collecting the data. Imagine I wanted to place a bet on who would win the soccer World Cup. Being English, I might bet on England to win the tournament. To do this, I'd: (1) place my bet, choosing my team (England) and odds available at the betting shop (e.g., 6/4); (2) see which team wins the tournament; (3) collect my winnings (or more likely not).

To keep everyone happy, this process needs to be equitable: the betting shops set their odds such that they're not paying out too much money (which keeps them happy), but so that they do pay out sometimes (to keep the customers happy). The betting shop can offer any odds before the tournament has ended, but it can't change them once the tournament is over (or the last game has started). Similarly, I can choose any team before the tournament, but I can't then change my mind halfway through or after the final game.

The research process is similar: we can choose any hypothesis (soccer team) before the data are collected, but we can't change our minds halfway through data collection (or after data collection). Likewise we have to decide on our probability level (or betting odds) before we collect data. *If we do this, the process works.* However, researchers sometimes cheat. They don't formulate hypotheses before they conduct their experiments, they change them when the data are collected (like me changing my team after the World Cup is over) or, worse still, they decide on them after the data are collected (see Chapter 3). With the exception of procedures such as *post hoc* tests, this is cheating. Similarly, researchers can be guilty of choosing which significance level to use after the data are collected and analyzed, like a betting shop changing the odds after the tournament.

If you change your hypothesis or the details of your analysis, you increase the chances of finding a significant result, but you also make it more likely that you will publish results that other researchers can't reproduce (which is embarrassing). If, however, you follow the rules carefully and do your significance testing at the 5% level you at least know that in the long run at most only 1 result out of every 20 will risk this public humiliation. (Thanks to David Hitchin for this box, and apologies to him for introducing soccer into it.)

probability value (the p-value). In our chocolate example, we can compute a statistic known as t, which has a specific sampling distribution from which we can get a probability value (p). This probability value tells us how likely it would be to get a value of t at least as big as the one we have if the null hypothesis is true. As I keep mentioning, this p is a long-run probability: it is computed by working out how often you'd get specific values of the test statistic (in this case t) if you repeated your exact sampling process an infinite number of times.

It is important that you collect the amount of data that you set out to collect, otherwise the p-value you obtain will not be correct. It is possible to compute a p-value representing the long-run probability of getting a t-value at least as big as the one you have in repeated samples of, say, 80, but there is no way of knowing the probability of getting a t-value at least as big as the one you have in repeated samples of 74, where the intention was to collect 80 but term ended and you couldn't find any more participants. If you cut data collection short (or extend it) for this sort of arbitrary reason, then whatever p-value you end up with is certainly not the one you want. Again, it's cheating: you're changing your team after you have placed your bet, and you will likely end up with research egg on your face when no one can replicate your findings.

Having hopefully stuck to your original sampling frame and obtained the appropriate p-value, you compare it to your original alpha value (usually 0.05). If the p you obtain is less than or equal to the original α, scientists typically use this as grounds to reject the null hypothesis outright; if the p is greater than α, then they accept that the null hypothesis is plausibly true (they reject the alternative hypothesis). We can never be completely sure that either hypothesis is correct; all we can do is to calculate the probability that our model would fit at least as well as

it does if there were no effect in the population (i.e., the null hypothesis is true). As this probability decreases, we gain greater confidence that the alternative hypothesis is more plausible than the null hypothesis. This overview of NHST is a lot to take in, so we will revisit a lot of the key concepts in detail the next few sections.

2.9.4 Test statistics ▮▮▮▮

I mentioned that NHST relies on fitting a model to the data and then evaluating the probability of this model, given the assumption that no effect exists. I mentioned in passing that the fit of a model or the parameters within it, are typically mapped to a probability value through a test statistic. I was deliberately vague about what the 'test statistic' is, so let's lift the veil of secrecy. To do this we need to return to the concepts of systematic and unsystematic variation that we encountered in Section 1.7.4. Systematic variation is variation that can be explained by the model that we've fitted to the data (and, therefore, due to the hypothesis that we're testing). Unsystematic variation is variation that cannot be explained by the model that we've fitted. In other words, it is error or variation not attributable to the effect we're investigating. The simplest way, therefore, to test whether the model fits the data or whether our hypothesis is a good explanation of the data we have observed, is to compare the systematic variation against the unsystematic variation. In effect, we look at a signal-to-noise ratio: we compare how good the model/hypothesis is against how bad it is (the error):

$$\text{Test statistic} = \frac{\text{signal}}{\text{noise}}$$
$$= \frac{\text{variance explained by the model}}{\text{variance not explained by the model}}$$
$$= \frac{\text{effect}}{\text{error}} \qquad (2.19)$$

Likewise, the best way to test a parameter is to look at the size of the parameter

relative to the background noise (the sampling variation) that produced it. Again, it's a signal-to-noise ratio: the ratio of how big a parameter is to how much it can vary across samples:

$$\text{Test statistic} = \frac{\text{signal}}{\text{noise}}$$
$$= \frac{\text{size of parameter}}{\text{sampling variation in the parameter}}$$
$$= \frac{\text{effect}}{\text{error}} \qquad (2.20)$$

The ratio of effect relative to error is a **test statistic**, and you'll discover later in the book that there are lots of them: t, χ^2 and F, to name only three. The exact form of the equation changes depending on which test statistic you're calculating, but the important thing to remember is that they all, crudely speaking, represent the same thing: signal-to-noise or the amount of variance explained by the model we've fitted to the data compared to the variance that can't be explained by the model (see Chapters 9 and 10, in particular, for a more detailed explanation). The reason why this ratio is so useful is intuitive, really: if our model is good then we'd expect it to be able to explain more variance than it can't explain. In this case, the test statistic will be greater than 1 (but not necessarily significant). Similarly, larger parameters (bigger effects) that are likely to represent the population (smaller sampling variation) will produce larger test statistics.

A test statistic is a statistic for which we know how frequently different values occur. I mentioned the t-distribution, chi-square (χ^2) distribution and F-distribution in Section 1.8.6 and said that they are all defined by an equation that enables us to calculate precisely the probability of obtaining a given score. Therefore, if a test statistic comes from one of these distributions we can calculate the probability of obtaining a certain value (just as we could estimate the probability of getting a score of a certain size from a frequency distribution in Section 1.8.6).

This probability is the *p*-value that Fisher described, and in NHST it is used to estimate how likely (in the long run) it is that we would get a test statistic at least as big as the one we have *if there were no effect* (i.e., the null hypothesis were true).

Test statistics can be a bit scary, so let's imagine that they're cute kittens. Kittens are typically very small (about 100g at birth, on average), but every so often a cat will give birth to a big one (say, 150g). A 150g kitten is rare, so the probability of finding one is very small. Conversely, 100g kittens are very common, so the probability of finding one is quite high. Test statistics are the same as kittens in this respect: small ones are quite common and large ones are rare. So, if we do some research (i.e., give birth to a kitten) and calculate a test statistic (weigh the kitten), we can calculate the probability of obtaining a value/weight at least that large. The more variation our model explains compared to the variance it can't explain, the bigger, the test statistic will be (i.e., the more the kitten weighs), and the more unlikely it is to occur by chance (like our 150g kitten). Like kittens, as test statistics get bigger, the probability of them occurring becomes smaller. If this probability falls below a certain value ($p < 0.05$ if we blindly apply the conventional 5% error rate), we presume that the test statistic is as large as it is because our model explains a sufficient amount of variation to reflect a genuine effect in the real world (the population). The test statistic is said to be *statistically significant*. Given that the statistical model that we fit to the data reflects the hypothesis that we set out to test, then a significant test statistic tells us that the model would be unlikely to fit this well if the there was no effect in the population (i.e., the null hypothesis was true). Typically, this is taken as a reason to reject the null hypothesis and gain confidence that the alternative hypothesis is true. If, however, the probability of obtaining a test statistic at least as big as the one we have

(if the null hypothesis were true) is too large (typically $p > 0.05$), then the test statistic is said to be non-significant and is used as grounds to reject the alternative hypothesis (see Section 3.2.1 for a discussion of what 'statistically significant' means).

2.9.5 One- and two-tailed tests

We saw in Section 1.9.2 that hypotheses can be directional (e.g., 'The more someone reads this book, the more they want to kill its author') or non-directional (i.e., 'Reading more of this book could increase or decrease the reader's desire to kill its author'). A statistical model that tests a directional hypothesis is called a **one-tailed test**, whereas one testing a non-directional hypothesis is known as a **two-tailed test**.

Imagine we wanted to discover whether reading this book increased or decreased the desire to kill me. If we have no directional hypothesis then there are three possibilities. (1) People who read this book want to kill me more than those who don't, so the difference (the mean for those reading the book minus the mean for non-readers) is positive. Put another way, as the amount of time spent reading this book increases, so does the desire to kill me—a positive relationship. (2) People who read this book want to kill me less than those who don't, so the difference (the mean for those reading the book minus the mean for non-readers) is negative. Alternatively, as the amount of time spent reading this book increases, the desire to kill me decreases—a negative relationship. (3) There is no difference between readers and non-readers in their desire to kill me—the mean for readers minus the mean for non-readers is exactly zero. There is no relationship between reading this book and wanting to kill me. This final option is the null hypothesis.

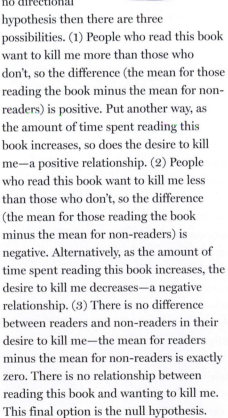

The direction of the test statistic (i.e., whether it is positive or negative) depends on whether the difference or direction of relationship, is positive or negative. Assuming that there is a positive difference or relationship (the more you read, the more you want to kill me), then to detect this difference we take account of the fact that the mean for readers is bigger than for non-readers (and so derive a positive test statistic). However, if we've predicted incorrectly and reading this book makes readers want to kill me *less*, then the test statistic will be negative instead.

What are the consequences of this? Well, if at the 0.05 level we needed to get a test statistic bigger than, say, 10 and the one we got was actually −12, then we would reject the hypothesis even though a difference does exist. To avoid this, we can look at both ends (or tails) of the distribution of possible test statistics. This means we will catch both positive and negative test statistics. However, doing this has a price because, to keep our criterion probability of 0.05, we split this probability across the two tails: we have 0.025 at the positive end of the distribution and 0.025 at the negative end. Figure 2.14 shows this situation—the orange tinted areas are the areas above the test statistic needed at a 0.025 level of significance. Combine the probabilities (i.e., add the two tinted areas together) at both ends and we get 0.05, our criterion value.

If we have made a prediction, then we put all our eggs in one basket and look only at one end of the distribution (either the positive or the negative end, depending on the direction of the prediction we make). In Figure 2.14, rather than having two small orange tinted areas at either end of the distribution that show the significant values, we have a bigger area (the blue tinted area) at only one end of the distribution that shows significant values. Note that this blue area contains within it one of the orange areas as well as an extra

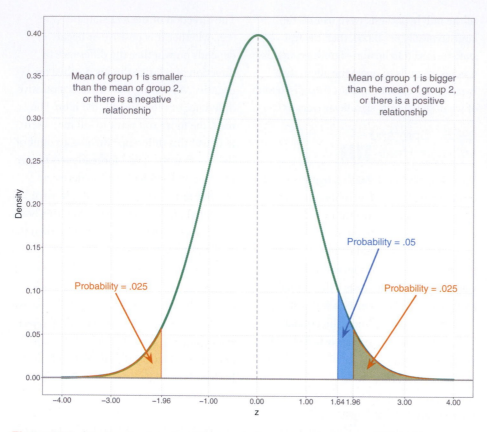

Figure 2.14 Diagram to show the difference between one- and two-tailed tests

bit of blue area. Consequently, we can just look for the value of the test statistic that would occur if the null hypothesis were true with a probability of 0.05. In Figure 2.14, the blue tinted area is the area above the positive test statistic needed at a 0.05 level of significance (1.64); this value is smaller than the value that begins the area for the 0.025 level of significance (1.96). This means that if we make a specific prediction then we need a smaller test statistic to find a significant result (because we are looking in only one tail of the distribution), but if our prediction happens to be in the wrong direction then we won't detect the effect that does exist. This final point is very important, so let me rephrase it: if you do a one-tailed test and the results turn out to be in the opposite direction to what you predicted, you must ignore them, resist all temptation to interpret them, and accept (no matter how much it pains you) the

null hypothesis. If you *don't* do this, then you have done a two-tailed test using a different level of significance from the one you set out to use (and Jane Superbrain Box 2.4 explains why that is a bad idea).

I have explained one- and two-tailed tests because people expect to find them explained in statistics textbooks. However, there are a few reasons why you should think long and hard about whether one-tailed tests are a good idea. Wainer (1972) quotes John Tukey (one of the great modern statisticians) as responding to the question 'Do you mean to say that one should *never* do a one-tailed test?' by saying 'Not at all. It depends upon to whom you are speaking. *Some people will believe anything*' (emphasis added). Why might Tukey have been so sceptical? As I have said already, if the result of a one-tailed test is in the opposite direction to what you expected, *you cannot and must not reject the null hypothesis.* In other words, you

must completely ignore that result even though it is poking you in the arm and saying 'look at me, I'm intriguing and unexpected'. The reality is that when scientists see interesting and unexpected findings their instinct is to want to explain them. Therefore, one-tailed tests are like a mermaid luring a lonely sailor to his death by being beguiling and interesting: they lure lonely scientists to their academic death by throwing up irresistible and unpredicted results.

One context in which a one-tailed test *could* be used, then, is if a result in the opposite direction to that expected would result in the same action as a non-significant result (Lombardi & Hurlbert, 2009; Ruxton & Neuhaeuser, 2010). There are some limited circumstances in which this might be the case. First, if a result in the opposite direction would be theoretically meaningless or impossible to explain even if you wanted to (Kimmel, 1957). Second, imagine you're testing a new drug to treat depression. You predict it will be better than existing drugs. If it is not better than existing drugs (non-significant *p*) you would not approve the drug; however, if it was significantly worse than existing drugs (significant *p* but in the opposite direction) you would also not approve the drug. In both situations, the drug is not approved. Finally, one-tailed tests encourage cheating. If you do a two-tailed test and find that your *p* is 0.06, then you would conclude that your results were not significant (because 0.06 is bigger than the critical value of 0.05). Had you done this test one-tailed, however, the *p* you would get would be half of the two-tailed value (0.03). This one-tailed value would be significant at the conventional level (because 0.03 is less than 0.05). Therefore, if we find a two-tailed *p* that is just non-significant, we might be tempted to pretend that we'd always intended to do a one-tailed test because our 'one-tailed' *p*-value is significant. But we can't change our rules after we have collected data (Jane Superbrain Box 2.4), so we must conclude that the effect is not significant. Although scientists hopefully don't do this sort of thing deliberately, people do get confused about what is and isn't permissible. Two recent surveys of practice in ecology journals concluded that 'all uses of one-tailed tests in the journals surveyed seemed invalid' (Lombardi & Hurlbert, 2009) and that only 1 in 17 papers using one-tailed tests were justified in doing so (Ruxton & Neuhaeuser, 2010).

One way around the temptation to cheat is to pre-register your study (which we'll discuss in detail in the following chapter). At a simple level pre-registration means that you commit publicly to your analysis strategy before collecting data. This could be a simple statement on your own website or, as we shall see, a formally submitted article outlining your research intentions. One benefit of pre-registering your research is that it then becomes transparent if you change your analysis plan (e.g., by switching from a two-tailed to a one-tailed test). It is much less tempting to halve your p-value to take it below 0.05 if the world will know you have done it!

2.9.6 Type I and Type II errors ▍▍▍▍

Neyman and Pearson identified two types of errors that we can make when we test hypotheses. When we use test statistics to tell us about the true state of the world, we're trying to see whether there is an effect in our population. There are two possibilities: there is, in reality, an effect in the population or there is, in reality, no effect in the population. We have no way of knowing which of these possibilities is true; however, we can look at test statistics and their associated probability to help us to decide which of the two is more likely. It is important that we're as accurate as possible. There are two mistakes we can make: a Type I and a Type II error. A **Type I error** occurs when we believe that there is a genuine effect in our population, when in fact there isn't. If we use the conventional criterion for alpha then the probability of this error is

0.05 (or 5%) when there is no effect in the population—this value is the **α-level** that we encountered in Figure 2.13. Assuming that there is no effect in our population, if we replicated our data collection 100 times, we could expect that on five occasions we would obtain a test statistic large enough to make us think that there was a genuine effect in the population even though there isn't. The opposite is a **Type II error**, which occurs when we believe that there is no effect in the population when, in reality, there is. This would occur when we obtain a small test statistic (perhaps because there is a lot of natural variation between our samples). In an ideal world, we want the probability of this error to be very small (if there is an effect in the population then it's important that we can detect it). Cohen (1992) suggests that the maximum acceptable probability of a Type II error would be 0.2 (or 20%)—this is called the **β-level**. That would mean that if we took 100 samples of data from a population in which an effect exists, we would fail to detect that effect in 20 of those samples (so we'd miss 1 in 5 genuine effects).

There is a trade-off between these two errors: if we lower the probability of accepting an effect as genuine (i.e., make α smaller) then we increase the probability that we'll reject an effect that does genuinely exist (because we've been so strict about the level at which we'll accept that an effect is genuine). The exact relationship between the Type I and Type II error is not straightforward because they are based on different assumptions: to make a Type I error there must be no effect in the population, whereas to make a Type II error the opposite is true (there must be an effect that we've missed). So, although we know that as the probability of making a Type I error decreases, the probability of making a Type II error increases, the exact nature of the relationship is usually left for the researcher to make an educated guess (Howell, 2012, gives a great explanation of the trade-off between errors).

2.9.7 Inflated error rates ▍▍▍▍

As we have seen, if a test uses a 0.05 level of significance then the chances of making a Type I error are only 5%. Logically, then, the probability of no Type I errors is 0.95 (95%) for each test. However, in science it's rarely the case that we can get a definitive answer to our research question using a single test on our data: we often need to conduct several tests. For example, imagine we wanted to look at factors that affect how viral a video becomes on YouTube. You might predict that the amount of humor and innovation in the video will be important factors. To test this, you might look at the relationship between the number of hits and measures of both the humor content and the innovation. However, you probably ought to also look at whether innovation and humor content are related too. Therefore, you would need to do three tests. If we assume that each test is independent (which in this case they won't be, but it enables us to multiply the probabilities), then the overall probability of no Type I errors will be $0.95^3 = 0.95 \times 0.95 \times 0.95 = 0.857$, because the probability of no Type I errors is 0.95 for each test and there are three tests. Given that the probability of no Type I errors is 0.857, then the probability of making at least one Type I error is this number subtracted from 1 (remember that the maximum probability of any event occurring is 1). So, the probability of at least one Type I error is $1 - 0.857 = 0.143$ or 14.3%. Therefore, across this group of tests, the probability of making a Type I error has increased from 5% to 14.3%, a value greater than the criterion that is typically used. This error rate across statistical tests conducted on the same data is known as the **familywise** or **experimentwise error rate**. Our scenario with three tests is relatively simple, and the effect of carrying out several tests is not too severe, but imagine that we increased the number of tests from three to ten. The familywise error rate can be calculated

Carlo Bonferroni

Olive Dunn

Figure 2.15 The king and queen of correction

using equation (2.21) (assuming you use a 0.05 level of significance):

$$familywise\ error = 1 - 0.95^n \qquad (2.21)$$

In this equation n is the number of tests carried out on the data. With ten tests carried out, the familywise error rate is $1 - 0.95^{10} = 0.40$, which means that there is a 40% chance of having made at least one Type I error.

To combat this build-up of errors, we can adjust the level of significance for individual tests such that the overall Type I error rate (α) across all comparisons remains at 0.05. There are several ways in which the familywise error rate can be controlled. The most popular (and easiest) way is to divide α by the number of comparisons, k, as in equation (2.22):

$$P_{Crit} = \frac{\alpha}{k} \qquad (2.22)$$

Therefore, if we conduct 10 tests, we use 0.005 as our criterion for significance. In doing so, we ensure that the cumulative Type I error remains below 0.05. This method is known as the **Bonferroni correction**, because it uses an inequality

described by Carlo Bonferroni, but despite the name its modern application to confidence intervals can be attributed to Olive Dunn (Figure 2.15). There is a trade-off for controlling the familywise error rate and that is a loss of statistical power, which is the next topic on our agenda.

2.9.8 Statistical power

We have seen that it is important to control the Type I error rate so that we don't too often mistakenly think that an effect is genuine when it is not. The opposite problem relates to the Type II error, which is how often we will miss an effect in the population that genuinely exists. If we set the Type II error rate high then we will be likely to miss a lot of genuine effects, but if we set it low we will be less likely to miss effects. The ability of a test to find an effect is known as its statistical **power** (not to be confused with statistical powder, which is an illegal substance that makes you better understand statistics). The power of a test is the probability that a given test will find an effect assuming that one exists in the

population. This is the opposite of the probability that a given test will *not* find an effect assuming that one exists in the population, which, as we have seen, is the β-level (i.e., Type II error rate). Therefore, the power of a test can be expressed as $1 - \beta$. Given that Cohen (1988, 1992) recommends a 0.2 probability of failing to detect a genuine effect (see above), the corresponding level of power would be $1 - 0.2$ or 0.8. Therefore, we typically aim to achieve a power of 0.8 or put another way, an 80% chance of detecting an effect if one genuinely exists. The power of a statistical test depends on:[12]

1 How big the effect is, because bigger effects will be easier to spot. This is known as the effect size and we'll discuss it in Section 3.5.

2 How strict we are about deciding that an effect is significant. The stricter we are, the harder it will be to 'find' an effect. This strictness is reflected in the α-level. This brings us back to our point in the previous section about correcting for multiple tests. If we use a more conservative Type I error rate for each test (such as a Bonferroni correction) then the probability of rejecting an effect that does exist is increased (we're more likely to make a Type II error). In other words, when we apply a Bonferroni correction, the tests will have less power to detect effects.

3 The sample size: We saw earlier in this chapter that larger samples are better approximations of the population; therefore, they have less sampling error. Remember that test statistics are basically a signal-to-noise ratio, so given that large samples have less 'noise', they make it easier to find the 'signal'.

Given that power ($1 - \beta$), the α-level, sample size, and the size of the effect are

---

12 It will also depend on whether the test is a one- or two-tailed test (see Section 2.9.5), but, as we have seen, you'd normally do a two-tailed test.

all linked, if we know three of these things, then we can find out the remaining one. There are two things that scientists do with this knowledge:

1 **Calculate the power of a test**: Given that we've conducted our experiment, we will have already selected a value of α, we can estimate the effect size based on our sample data, and we will know how many participants we used. Therefore, we can use these values to calculate $1 - \beta$, the power of our test. If this value turns out to be 0.8 or more, them we can be confident that we have achieved sufficient power to detect any effects that might have existed, but if the resulting value is less, then we might want to replicate the experiment using more participants to increase the power.

2 **Calculate the sample size necessary to achieve a given level of power**: We can set the value of α and $1 - \beta$ to be whatever we want (normally, 0.05 and 0.8, respectively). We can also estimate the likely effect size in the population

by using data from past research. Even if no one had previously done the exact experiment that we intend to do, we can still estimate the likely effect size based on similar experiments. Given this information, we can calculate how many participants we would need to detect that effect (based on the values of α and $1 - \beta$ that we've chosen).

The point of calculating the power of a test after the experiment has always been lost on me a bit: if you find a non-significant effect then you didn't have enough power, if you found a significant effect, then you did. Using power to calculate the necessary sample size is the more common and, in my opinion, more useful thing to do. The actual computations are very cumbersome, but there are computer programs available that will do them for you. *G*Power* is a free and powerful (excuse the pun) tool, there is a package *pwr* that can be used in the open source statistics package R, and various websites, including powerandsamplesize.com. There are also commercial software packages such as

nQuery Adviser (www.statsols.com/nquery-sample-size-calculator), *Power and Precision* (www.power-analysis.com) and *PASS* (www.ncss.com/software/pass). Also, Cohen (1988) provides extensive tables for calculating the number of participants for a given level of power (and vice versa).

2.9.9 Confidence intervals and statistical significance

I mentioned earlier (Section 2.8.4) that if 95% confidence intervals didn't overlap then we could conclude that the means come from different populations, and, therefore, that they are significantly different. I was getting ahead of myself a bit because this comment alluded to the fact that there is a relationship between statistical significance and confidence intervals. Cumming & Finch (2005) have three guidelines that are shown in Figure 2.16:

1 95% confidence intervals that just about touch end-to-end (as in the top left panel of Figure 2.16) represent a *p*-value

Oliver Twisted
Please, Sir, can I have some more . . . power?

'I've got the power!' sings Oliver as he pops a huge key up his nose and starts to wind the clockwork mechanism of his brain. If, like Oliver, you like to wind up your brain, the companion website contains links to the various packages for doing power analysis and sample-size estimation. If that doesn't quench your thirst for knowledge, then you're a grain of salt.

for testing the null hypothesis of no differences of approximately 0.01.

2 If there is a gap between the upper end of one 95% confidence interval and the lower end of another (as in the top right panel of Figure 2.16), then $p < 0.01$.

3 A p-value of 0.05 is represented by *moderate* overlap between the bars (the bottom panels of Figure 2.16).

These guidelines are poorly understood by many researchers. In one study (Belia, Fidler, Williams, & Cumming, 2005), 473 researchers from medicine, psychology and behavioral neuroscience were shown a graph of means and confidence intervals for two independent groups and asked to move one of the error bars up or down on the graph until they showed a 'just significant difference' (at $p < 0.05$). The

sample ranged from new researchers to very experienced ones but, surprisingly, this experience did not predict their responses. In fact, only a small percentage of researchers could position the confidence intervals correctly to show a just significant difference (15% of psychologists, 20% of behavioral neuroscientists and 16% of medics). The most frequent response was to position the confidence intervals more or less at the point where they stop overlapping (i.e., a p-value of approximately 0.01). Very few researchers (even experienced ones) realized that moderate overlap between confidence intervals equates to the standard p-value of 0.05 for accepting significance.

What do we mean by moderate overlap? Cumming (2012) defines it as half the

length of the average margin of error (MOE). The MOE is half the length of the confidence interval (assuming it is symmetric), so it's the length of the bar sticking out in one direction from the mean. In the bottom left of Figure 2.16 the confidence interval for sample 1 ranges from 4 to 14 so has a length of 10 and an MOE of half this value (i.e., 5). For sample 2, it ranges from 11.5 to 21.5 so again a distance of 10 and an MOE of 5. The average MOE is, therefore $(5 + 5)/2 = 5$. Moderate overlap would be half of this value (i.e., 2.5). This is the amount of overlap between the two confidence intervals in the bottom left of Figure 2.16. Basically, if the confidence intervals are the same length, then $p = 0.05$ is represented by an overlap of about a quarter of the confidence interval. In the more likely scenario of confidence intervals with different lengths, the interpretation of overlap is more difficult. In the bottom right of Figure 2.16 the confidence interval for sample 1 again ranges from 4 to 14 so has a length of 10 and an MOE of 5. For sample 2, it ranges from 12 to 18 and, so a distance of 6 and an MOE of half this value, 3. The average MOE is, therefore, $(5 + 3)/2 = 4$. Moderate overlap would be half of this value (i.e., 2) and this is what we get in the bottom right of Figure 2.16: the confidence intervals overlap by 2 points on the scale, which equates to a p of around 0.05.

2.9.10 Sample size and statistical significance

When we discussed power, we saw that it is intrinsically linked with the sample size. Given that power is the ability of a test to find an effect that genuinely exists, and we 'find' an effect by having a statistically significant result (i.e., $p < 0.05$), there is also a connection between the sample size and the p-value associated with a test statistic. We can demonstrate this connection with two examples. Apparently, male mice 'sing'

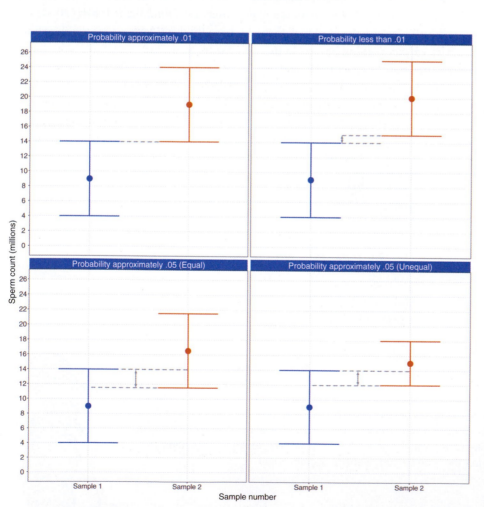

Figure 2.16 The relationship between confidence intervals and statistical significance

to female mice to try to attract them as mates (Hoffmann, Musolf, & Penn, 2012), I'm not sure what they sing, but I like to think it might be 'This mouse is on fire' by AC/DC or perhaps 'Mouses of the Holy' by Led Zeppelin or even 'The mouse Jack built' by Metallica. It's probably not 'Terror and hubris in the mouse of Frank Pollard' by Lamb of God. That would just be weird. Anyway, many a young man has spent time wondering how best to attract female mates, so to help them out, imagine we did a study in which we got two groups of 10 heterosexual young men and got them to go up to a woman that they found attractive and either engage them in conversation (group 1) or sing them a song (group 2). We measured how long it was before the woman ran away. Imagine we repeated this experiment but using 100 men in each group.

Figure 2.17 shows the results of these two experiments. The summary statistics from the data are identical: in both cases the singing group had a mean of 10 and a standard deviation of 3, and the conversation group had a mean of 12 and a standard deviation of 3. Remember that the only difference between the two experiments is that one collected 10 scores per sample, and the other 100 scores per sample.

Notice in Figure 2.17 that the means for each sample are the same in both graphs, but the confidence intervals are much narrower when the samples contain 100 scores than when they contain only 10 scores. You might think that this is odd given that I said that the standard deviations are all the same (i.e., 3). If you think back to how the confidence interval is computed it is the mean plus or minus 1.96 times the standard error. The standard error is the standard deviation divided by the square root of the sample size (see Eq. 2.14), therefore as the sample size gets larger, the standard error (and, therefore, confidence interval) will get smaller.

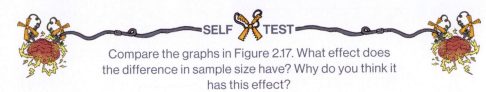

SELF TEST
Compare the graphs in Figure 2.17. What effect does the difference in sample size have? Why do you think it has this effect?

We saw in the previous section that if the confidence intervals of two samples are the same length then a p of around 0.05 is represented by an overlap of about a quarter of the confidence interval. Therefore, we can see that even though the means and standard deviations are identical in both graphs, the study that has only 10 scores per sample is not significant (the bars overlap quite a lot; in fact, $p = 0.15$), but the study that has 100 scores per sample shows a highly significant difference (the bars don't overlap at all; for these data $p < 0.001$). Remember, the means and standard deviations are *identical* in the two graphs, but the sample size affects the standard error and hence the significance.

Taking this relationship to the extreme, we can illustrate that with a big enough sample even a completely meaningless difference between two means can be deemed significant, with $p < 0.05$. Figure 2.18 shows such a situation. This time, the singing

group has a mean of 10.00 ($SD = 3$) and the conversation group has a mean of 10.01 ($SD = 3$): a difference of 0.01—a very small difference indeed. The main graph looks very odd: the means look identical and there are no confidence intervals. In fact, the confidence intervals are so narrow that they merge into a single line. The figure also shows a zoomed image of the confidence intervals (note that by zooming in the values on the vertical axis range from 9.98 to 10.02, so the entire range of values in the zoomed image is only 0.04). As you can see, the sample means are 10 and 10.01 as mentioned before,[13] but by zooming in we can see the confidence intervals. Note that the confidence intervals show an overlap of about a quarter, which equates to a significance value of about $p = 0.05$ (for these data the actual value of p is 0.044). How is it possible that we have two sample means that are almost identical (10 and 10.01), and have the same standard

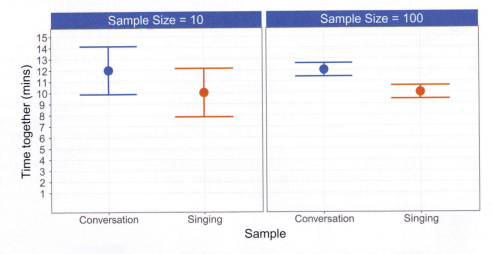

Figure 2.17 Graph showing two data sets with the same means and standard deviations but based on different-sized samples

---

13 The mean of the singing group looks bigger than 10, but this is only because we have zoomed in so much that its actual value of 10.00147 is noticeable.

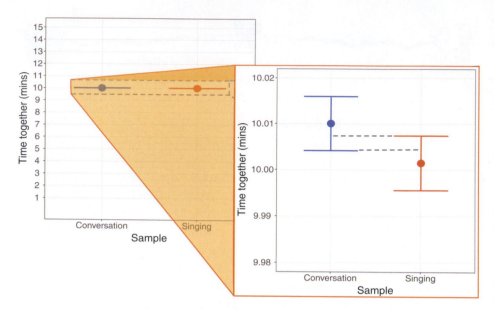

Figure 2.18 A very small difference between means based on an enormous sample size ($n = 1,000,000$ per group)

deviations, but are deemed significantly different? The answer is again the sample size: there are 1 million cases in each sample, so the standard errors are minuscule.

This section has made two important points. First, the sample size affects whether a difference between samples is deemed significant or not. *In large samples, small differences can be significant, and in small samples large differences can be non-significant.* This point relates to power: large samples have more power to detect effects. Second, even a difference of practically zero can be deemed 'significant' if the sample size is big enough. Remember that test statistics are effectively the ratio of signal to noise, and the standard error is our measure of 'sampling noise'. The

Cramming Sam's Tips
Null hypothesis significance testing

- NHST is a widespread method for assessing scientific theories. The basic idea is that we have two competing hypotheses: one says that an effect exists (the *alternative hypothesis*) and the other says that an effect doesn't exist (the *null hypothesis*). We compute a test statistic that represents the alternative hypothesis and calculate the probability that we would get a value as big as the one we have if the null hypothesis were true. If this probability is less than 0.05 we reject the idea that there is no effect, say that we have a *statistically significant* finding and throw a little party. If the probability is greater than 0.05 we do not reject the idea that there is no effect, we say that we have a *non-significant* finding and we look sad.

- We can make two types of error: we can believe that there is an effect when, in reality, there isn't (a *Type I error*); and we can believe that there is not an effect when, in reality, there is (a *Type II error*).
- The power of a statistical test is the probability that it will find an effect when one exists.
- The significance of a test statistic is directly linked to the sample size: the same effect will have different *p*-values in different-sized samples, small differences can be deemed 'significant' in large samples, and large effects might be deemed 'non-significant' in small samples.

standard error is estimated from the sample size, and the bigger the sample size, the smaller the standard error. Therefore, bigger samples have less 'noise', so even a tiny signal can be detected.

2.10 Reporting significance tests ▮▮▮▮

In Section 1.9 we looked at some general principles for reporting data. Now that we have learnt a bit about fitting statistical models, we can add to these guiding principles. We learnt in this chapter that we can construct confidence intervals (usually 95% ones) around a parameter such as the mean. A 95% confidence interval contains the population value in 95% of samples, so if your sample is one of those 95%, the confidence interval contains useful information about the population value. It is important to tell readers the type of confidence interval used (e.g., 95%), and in general we use the format [*lower boundary*, *upper boundary*] to present the values. So, if we had a mean of 30 and the confidence interval ranged from 20 to 40, we might write $M = 30$, 95% CI [20, 40]. If we were reporting lots of 95% confidence intervals it might be easier to state the level at the start of our results and just use the square brackets:

✓ 95% confidence intervals are reported in square brackets. Fear reactions were higher, $M = 9.86$ [7.41, 12.31] when Andy's cat Fuzzy wore a fake human tongue compared to when he did not, $M = 6.58$ [3.47, 9.69].

We also saw that when we fit a statistical model we calculate a test statistic and a p-value associated with it. Scientists typically conclude that an effect (our model) is significant if this p-value is less than 0.05. APA style is to remove the zero before the decimal place (so you'd report $p = .05$ rather than $p = 0.05$) but, because many other journals don't have this idiosyncratic rule, this is an APA rule that

I don't follow in this book. Historically, people would report p-values as being either less than or greater than 0.05. They would write things like:

✗ Fear reactions were significantly higher when Andy's cat Fuzzy wore a fake human tongue compared to when he did not, $p < 0.05$.

If an effect was very significant (e.g., if the p-value was less than 0.01 or even 0.001), they would also use these two criteria to indicate a 'very significant' finding:

✗ The number of cats intruding into the garden was significantly less when Fuzzy wore a fake human tongue compared to when he didn't, $p < 0.01$.

Similarly, non-significant effects would be reported in much the same way (note this time the p is reported as greater than 0.05):

✗ Fear reactions were not significantly different when Fuzzy wore a David Beckham mask compared to when he did not, $p > 0.05$.

In the days before computers, it made sense to use these standard benchmarks for reporting significance because it was a bit of a pain to compute exact significance values (Jane Superbrain Box 3.1). However, computers make computing p-values a piece of ps, so we have no excuse for using these conventions. We should report exact p-values because it gives the reader more information than simply knowing that the p-value was less or more than a random threshold like 0.05. The possible exception is the threshold of 0.001. If we find a p-value of 0.0000234 then for the sake of space and everyone's sanity it would be reasonable to report $p < 0.001$:

✓ Fear reactions were significantly higher when Andy's cat Fuzzy wore a fake human tongue compared to when he did not, $p = 0.023$.
✓ The number of cats intruding into the garden was significantly less when Fuzzy wore a fake human tongue compared to when he did not, $p = 0.007$.

2.11 Brian and Jane's Story ▮▮▮▮

Brian was feeling a little deflated after his encounter with Jane. She had barely said a word to him. Was he that awful to be around? He was enjoying the book she'd handed to him though. The whole thing had spurred him on to concentrate more during his statistics lectures. Perhaps that's what she wanted.

He'd seen Jane flitting about campus. She always seemed to have a massive bag with her. It seemed to contain something large and heavy, judging from her posture. He wondered what it was as he daydreamed across the campus square. His thoughts were broken by being knocked to the floor.

'Watch where you're going,' he said angrily.

'I'm so sorry …' the dark-haired woman replied. She looked flustered. She picked up her bag as Brian looked up to see it was Jane. Now *he* felt flustered.

Jane looked as though she wanted to expand on her apology but the words had escaped. Her eyes darted as though searching for them.

Brian didn't want her to go, but in the absence of anything to say, he recited his statistics lecture to her. It was weird. Weird enough that as he finished she shrugged and ran off.

2.12 What next? ▮▮▮▮

Nursery school was the beginning of an educational journey that I am a still on several decades later. As a child, your belief systems are very adaptable. One minute you believe that sharks can miniaturize themselves, swim up pipes from the sea to the swimming pool you're currently in before restoring their natural size and eating you; the next minute you don't, simply because your parents say it's not possible. At the age of 3 any hypothesis is plausible, every way of life is acceptable, and multiple incompatible perspectives can

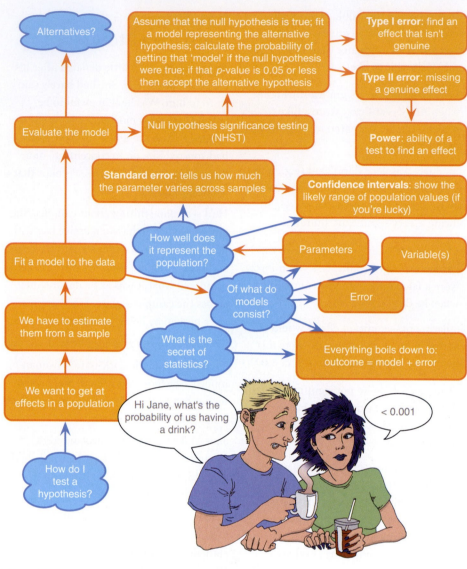

Figure 2.19   What Brian learnt from this chapter

be accommodated. Then a bunch of idiot adults come along and force you to think more rigidly. Suddenly, a cardboard box is not a high-tech excavator, it's a cardboard box, and there are 'right' and 'wrong' ways to live your life. As you get older, the danger is that—left unchecked—you plunge yourself into a swimming-pool-sized echo chamber of your own beliefs, leaving your cognitive flexibility in a puddle at the edge of the pool. If only sharks *could* compress themselves . . .

Before you know it, you're doggedly doing things and following rules that you can't remember the reason for. One of my early beliefs was that my older brother Paul (more on him later . . .) was 'the clever one'. Far be it from me to lay the blame at anyone's feet for this belief, but it probably didn't help that members of my immediate family used to say things like, 'Paul is the clever one, but at least you work hard'. Like I said, over time, if nothing challenges your view you can get very fixed in a way of doing things or a mode of thinking. If you spend your life thinking you're not 'the clever one', how can you ever change that? You need something unexpected and profound to create a paradigm shift. The next chapter is all about breaking ways of thinking that scientists have been invested in for a long time.

2.13 Key terms that I've discovered

α-level	Experimental hypothesis	One-tailed test	Sampling variation
Alternative hypothesis	Experimentwise error rate	Ordinary least squares	Standard error
β-level	Familywise error rate	Parameter	Standard error of the mean (SE)
Bonferroni correction	Fit	Point estimate	Test statistic
Central limit theorem	Interval estimate	Population	Two-tailed test
Confidence interval	Linear model	Power	Type I error
Degrees of freedom	Method of least squares	Sample	Type II error
Deviance	Null hypothesis	Sampling distribution	

Smart Alex's tasks

- **Task 1**: Why do we use samples?
- **Task 2**: What is the mean and how do we tell if it's representative of our data?
- **Task 3**: What's the difference between the standard deviation and the standard error?
- **Task 4**: In Chapter 1 we used an example of the time taken for 21 heavy smokers to fall off a treadmill at the fastest setting (18, 16, 18, 24, 23, 22, 22, 23, 26, 29, 32, 34, 34, 36, 36, 43, 42, 49, 46, 46, 57). Calculate the standard error and 95% confidence interval of these data.
- **Task 5**: What do the sum of squares, variance and standard deviation represent? How do they differ?
- **Task 6**: What is a test statistic and what does it tell us?
- **Task 7**: What are Type I and Type II errors?
- **Task 8**: What is statistical power?
- **Task 9**: Figure 2.17 shows two experiments that looked at the effect of singing versus conversation on how much time a woman would spend with a man. In both experiments the means were 10 (singing) and 12 (conversation), the standard deviations in all groups were 3, but the group sizes were 10 per group in the first experiment and 100 per group in the second. Compute the values of the confidence intervals displayed in the figure.
- **Task 10**: Figure 2.18 shows a similar study to the one above, but the means were 10 (singing) and 10.01

(conversation), the standard deviations in both groups were 3, and each group contained 1 million people. Compute the values of the confidence intervals displayed in the figure.

- **Task 11**: In Chapter 1 (Task 8), we looked at an example of how many games it took a sportsperson before they hit the 'red zone'. Calculate the standard error and confidence interval for those data.
- **Task 12**: At a rival club to the one I support, they similarly measured the number of consecutive games it took their players before they reached the red zone. The data are: 6, 17, 7, 3, 8, 9, 4, 13, 11, 14, 7. Calculate the mean, standard deviation, and confidence interval for these data.
- **Task 13**: In Chapter 1 (Task 9) we looked at the length in days of 11 celebrity marriages. Here are the lengths in days of eight marriages, one being mine and the other seven being those of some of my friends and family (in all but one case up to the day I'm writing this, which is 8 March 2012, but in the 91-day case it was the entire duration—this isn't my marriage, in case you're wondering): 210, 91, 3901, 1339, 662, 453, 16672, 21963, 222. Calculate the mean, standard deviation and confidence interval for these data.

Answers & additional resources are available on the book's website at
https://edge.sagepub.com/field5e

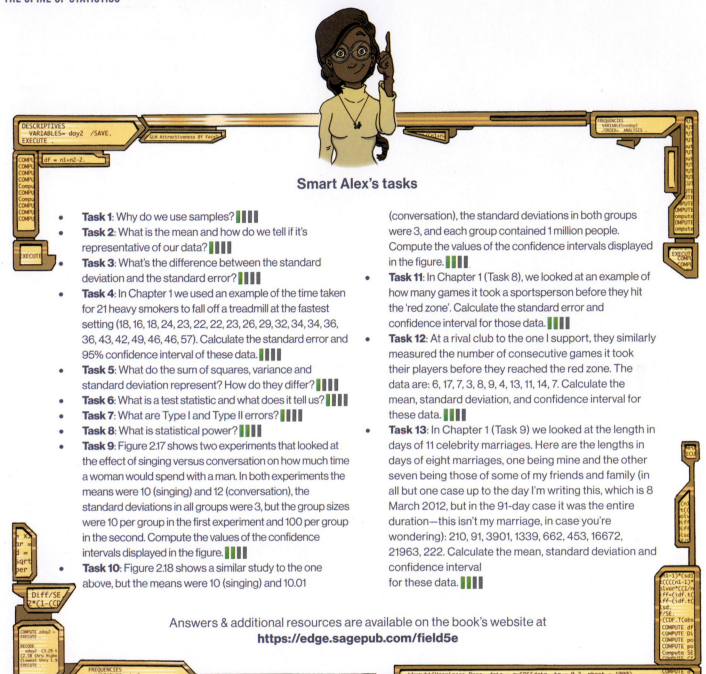

THE PHOENIX
OF STATISTICS

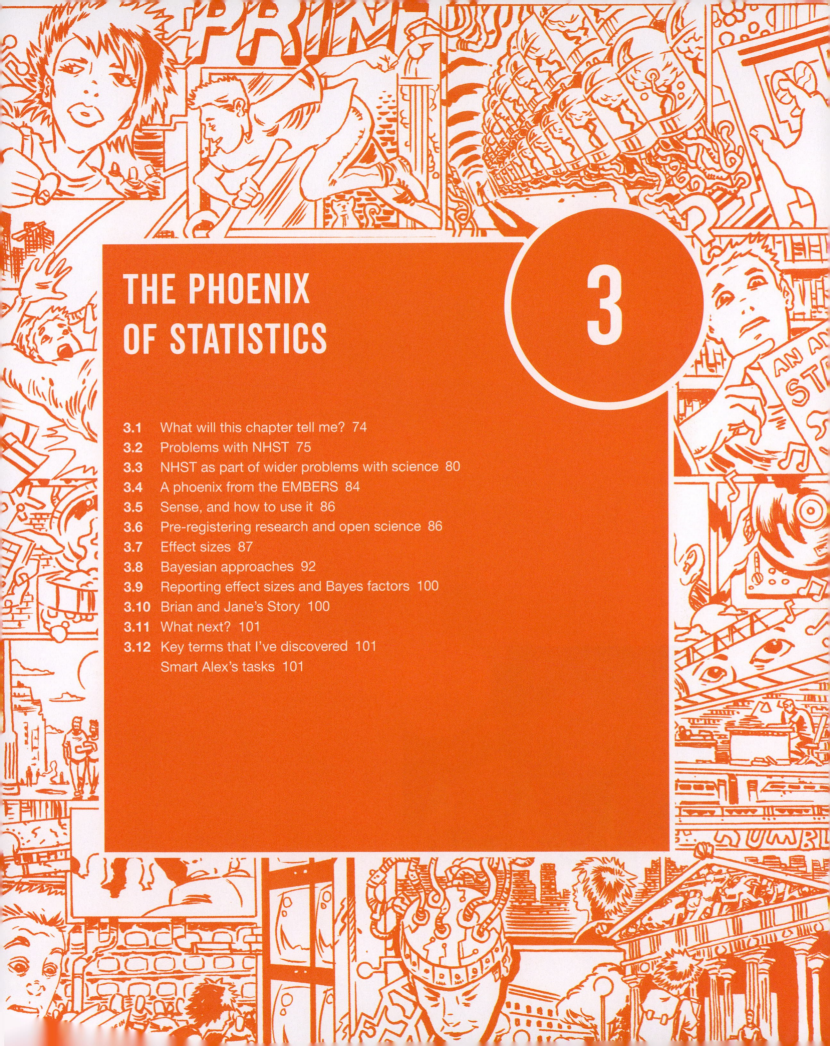

3

3.1 What will this chapter tell me?

At the end of the previous chapter I indulged in a self-pitying tale about my family thinking that my brother, Paul, was 'the clever one'. Perhaps you're anticipating an anecdote about how, on my fourth birthday, I presented my parents with a time machine that I had invented (tweaking Einstein's relativity theory in the process) and invited them to take a trip back a few years so that they could slap their earlier selves whenever they said, 'Paul is the clever one'. That didn't happen; apart from anything else, my parents meeting themselves would probably have caused a rift in space-time or something and we'd all be leaking into each other's parallel universes. Which, of course, might be fun if we ended up in a parallel universe where statistics hasn't been invented.

Being an Englishman, there was only one culturally appropriate way to deal with these early low family expectations: emotional repression. I silently internalized their low opinion of my intellect into a deep-seated determination to prove them wrong. If I was a hard-working idiot, I reasoned, then I could work hard at being less of an idiot? The plan started slowly: I distracted myself by playing football, guitar, and dribbling on myself. Then, at some point during my teens, when the resentment had festered like the contents of a can of Surströmming, I rose like a phoenix of statistics and worked myself into the ground for 20 years to get a degree, PhD, publish tons of research papers, get grants, and write textbooks. If you do enough of that stuff eventually you get to call yourself 'Professor', which sounds posh, like you might be clever. 'Who is the clever one now?' I thought to myself as I watched my brother enjoying a life in consultancy, free from stress-related psychological problems and earning a lot more money than me.

The point is that I'd got locked into a pattern of thinking in which my internal value as a human was connected to my academic achievements. As long as I 'achieved', I could justify taking up valuable space on this planet. I'd been invested in this 'habit' for close to 40 years. Like null hypothesis significance testing from the previous chapter, I'd found a system, a recipe to follow, that rewarded me. I thought I needed to achieve because that's how I'd been taught to think, I defined myself by academic success because I'd spent years defining myself by academic success. I never questioned the system, I never looked for its weaknesses, I never investigated an alternative.

Figure 3.1 'Hi! Can we interest you in a paradigm shift?'

Then Zach and Arlo came along[1] and changed everything. There's something about the overwhelming, unconditional love that these two little humans evoke that shows up a career in academia for the hollow, pointless waste of time that it has been. Has 20 years of my research made even a miniscule difference to anyone anywhere? No, of course it hasn't. Would the world be worse off if I never existed to write statistics books? No: there are a lot of people better at statistics than I am. Do my family now think I'm cleverer than my brother? Obviously not. I thought my 'system' was a way to the truth, but its logic was flawed. Gazing into Zach and Arlo's eyes, I see two hearts that would be irrevocably broken if I ceased to exist. If I work a 60-hour week 'achieving', the world will not care, but two little boys *will* care that they don't see their dad. If I write 100 more articles the world won't change, but if I read Zach and Arlo 100 more stories, their world *will* be better. Some people do science that changes the world, but I'm not one of them. I'll never be an amazing scientist, but I *can* be an amazing dad.

That was my personal paradigm shift, and this chapter is about a statistical one. We shift away from the dogmatic use of null hypothesis significance testing (NHST) by exploring its limitations and how they feed into wider issues in

science. We then discover some alternatives. It's possible that this chapter might come across as me trying to deter you from science by banging on about how awful it is. Science *is* awesome, and I believe that mostly it's done well by people who care about finding out the truth, but it's not perfect. Those of you who don't plan to be scientists need to know how to scrutinize scientific work, so that when politicians and the media try to convince you of things using science, you have the skills to read the science with a critical mind, aware of what may have influenced the results and conclusions. Many of you *will* want to become scientists, and for you this chapter is here because science needs you to make it better. Unhelpful scientific culture won't be overturned by the 'set-in-their-ways' professors, but it will be by the brilliant minds of the future. That's you, because you are better than some of the dodgy stuff that currently happens.

3.2 Problems with NHST ▮▮▮▮

We saw in the previous chapter that NHST is the dominant method for testing theories using statistics. It is compelling because it offers a rule-based framework for deciding whether to believe a hypothesis. It is also appealing to teach because even if your students don't understand the logic behind NHST, most of them can follow the rule

that a $p < 0.05$ is 'significant' and a $p > 0.05$ is not. No one likes to get things wrong, and NHST seems to provide an easy way to disentangle the 'correct' conclusion from the 'incorrect' one. Like timid bakers, we want a delicious fruitcake rather than egg on our face, so we diligently follow the steps in the NHST recipe. Unfortunately, when you bite into an NHST cake it tastes like one of those fermented sharks that Icelandic people bury in small holes for three months to maximize their rancidness.[2] Here are two quotes about NHST that sum up just what a putrefying shark it is:

The almost universal reliance on merely refuting the null hypothesis is a terrible mistake, is basically unsound, poor scientific strategy, and one of the worst things that ever happened in the history of psychology. (Meehl, 1978, p. 817)

NHST; I resisted the temptation to call it Statistical Hypothesis Inference Testing. (Cohen, 1994, p. 997)

This section rakes the stones, gravel and sand to reveal the decaying shark carcass that is NHST. We momentarily gag on its aroma of urine-infused cheese before … OK, I've gone too far with the shark thing, haven't I?

3.2.1 Misconceptions about statistical significance ▮▮▮▮

People have devoted entire books to the problems with NHST (for example, Ziliak & McCloskey, 2008), but we have space only to scratch the surface. We'll start with three common misconceptions (in no particular order) about what a statistically significant result

---

1 This turn of phrase makes it sound a bit like I opened my wardrobe one morning and two cheeky grins emanated out of the darkness. My wife certainly would have preferred that mode of delivery. Although their arrival was not in any way unexpected, their endless capacity to melt my heart was.

2 I probably shouldn't knock *hákarl* until I've tried it, but if I'm ever in Iceland I'm going to play my vegetarian trump card to the max to avoid doing so. Apologies to any Icelandic readers for being such a massive, non-Viking, wuss. I'm rubbish at growing beards and invading European countries too.

(typically defined as a *p*-value of less than 0.05) allows you to conclude. For a good paper on NHST misconceptions, see Greenland et al. (2016).

Misconception 1:
A significant result means
that the effect is important

Statistical significance is not the same thing as importance because the *p*-value from which we determine significance is affected by sample size (Section 2.9.10). Therefore, don't be fooled by that phrase 'statistically significant', because very small and unimportant effects will be statistically significant if sufficiently large amounts of data are collected (Figure 2.18), and very large and important effects will be missed if the sample size is too small.

Misconception 2:
A non-significant result means
that the null hypothesis is true

Actually, no. If the *p*-value is greater than 0.05 then you could decide to reject the alternative hypothesis,[3] but this is not the same as the null hypothesis being true. A non-significant result tells us only that the effect is not big enough to be found (given our sample size), it doesn't tell us that the effect size is zero. The null hypothesis is *a hypothetical* construct. We can assume an effect of zero in a population to calculate a probability distribution under the null hypothesis, but it isn't reasonable to think that a real-world effect is zero to an infinite number of decimal places. Even if a population effect had a size of 0.000000001, that is not the same as zero, and given a big enough sample that effect could be detected and deemed 'statistically significant' (Cohen, 1990, 1994). Therefore, a non-significant result should never be interpreted as 'no difference between means' or 'no relationship between variables'.

Misconception 3:
A significant result means
that the null hypothesis is false

Wrong again. A significant test statistic is based on probabilistic reasoning, which limits what we can conclude. Cohen (1994) points out that formal reasoning relies on an initial statement of fact, followed by a statement about the current state of affairs, and an inferred conclusion. This syllogism illustrates what he meant:

- If a *person plays the flute*, then the person is not *a member of the band Iron Maiden*.
  - This person is a member of Iron Maiden.
  - Therefore, this person does not play the flute.

The syllogism starts with a statement of fact that allows the end conclusion to be reached because you can deny that the person plays the flute (the antecedent) by denying that 'not a member of Iron Maiden' (the consequent) is true. A comparable version of the null hypothesis is:

- If the *null hypothesis* is correct, then *this test statistic value* cannot occur.
  - This test statistic value has occurred.
  - Therefore, the null hypothesis is not correct.

This is all very nice, except that the null hypothesis is not characterized by a statement of fact such as, 'If the null hypothesis is correct, then this test statistic cannot occur'; instead it reflects a statement of probability such as, 'If the null hypothesis is correct, then this test statistic is highly unlikely'. Not starting out with a statement of fact messes up the consequent logic. The syllogism becomes:

- If the *null hypothesis* is correct, then it is highly unlikely *to get this test statistic value*.
  - This test statistic value has occurred.
  - Therefore, the null hypothesis is highly unlikely.

If, like me, logic makes your brain pulsate in an unpleasant way then it might not be obvious why the syllogism no longer makes sense, so let's convert it to something more tangible by replacing the phrase 'null hypothesis' with 'person plays guitar', and the phrase 'to get this test statistic value' with 'the person is a member of the band Iron Maiden'. Let's see what we get:

- If *person plays guitar* is correct, then it is highly unlikely that *the person is a member of the band Iron Maiden*.
  - This person is a member of Iron Maiden.
  - Therefore, *person plays* guitar is highly unlikely.

Let's break this syllogism down. The first statement is true. On Earth, there are (very approximately) 50 million people who play guitar and only 3 of them are in Iron Maiden. Therefore, the probability of someone being in Iron Maiden, given, that they play guitar, is about 3/50 million or 6×10^{-8}. In other words, it *is* very unlikely. The consequent logical statements assert that given someone *in* a member of Iron Maiden it is unlikely that they play guitar, which is demonstrably false. Iron Maiden have six band members and three of them play guitar (I'm not including bass guitar in any of this), so, given that a person is in Iron Maiden, the probability that they play guitar is 3/6 or 0.5 (50%), which is quite likely (not unlikely). This example illustrates a common fallacy in hypothesis testing.

To sum up, although NHST is the result of trying to find a system that can test which of two competing hypotheses (the null or the alternative) is likely to be correct, it fails because the significance of the test provides no evidence about either hypothesis.

3.2.2 All-or-nothing thinking ▌▌▌▌

Perhaps the biggest practical problem created by NHST is that it encourages all-or-nothing thinking: if $p < 0.05$, then an effect

---

3 You shouldn't because your study might be underpowered, but people do.

is significant, but if $p > 0.05$, it is not. One scenario that illustrates the ridiculousness of this thinking is that you have two effects, based on the same sample sizes, and one has $p = 0.0499$, and the other $p = 0.0501$. If you apply the NHST recipe book, then the first effect is significant and the second is not. You'd reach completely opposite conclusions based on p-values that differ by only 0.0002. Other things being equal, these p-values would reflect basically the same-sized effect, but the 'rules' of NHST encourage people to treat them as completely opposite.

There is nothing magic about the criterion of $p < 0.05$, it is merely a convenient rule of thumb that has become popular for arbitrary reasons (see Jane Superbrain Box 3.1). When I outlined the NHST process (Section 2.9.3) I said that you should set an idiosyncratic alpha that is meaningful for your research question, but that most people choose 0.05 without thinking about why they are doing so.

Let's look at how the recipe-book nature of NHST encourages us to think in these black-and-white terms, and how misleading that can be. Students are often very scared of statistics. Imagine that a scientist claimed to have found a cure for statistics anxiety: a potion containing badger sweat, a tear from a newborn child, a teaspoon of Guinness, some cat saliva and sherbet. She called it *antiSTATic*. Treat yourself by entertaining the fiction that 10 researchers all did a study in which they compared anxiety levels in students who had taken *antiSTATic* to those who had taken a placebo potion (water). If *antiSTATic* doesn't work, then there should be a difference of zero between these group means (the null hypothesis), but if it does work then those who took *antiSTATic* should be less anxious than those who took the placebo (which will show up in a positive difference between the groups). The results of the 10 studies are shown in Figure 3.2, along with the p-value within each study.

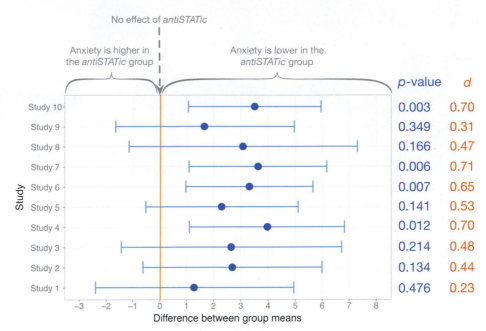

Figure 3.2 Results of 10 different studies looking at the difference between two interventions. The circles show the mean difference between groups

SELF TEST

Based on what you have learnt so far, which of the following statements best reflects your view of *antiSTATic*?

a The evidence is equivocal; we need more research.
b All the mean differences show a positive effect of *antiSTATic*; therefore, we have consistent evidence that *antiSTATic* works.
c Four of the studies show a significant result ($p < 0.05$), but the other six do not. Therefore, the studies are inconclusive: some suggest that *antiSTATic* is better than placebo, but others suggest there's no difference. The fact that more than half of the studies showed no significant effect means that *antiSTATic* is not (on balance) more successful in reducing anxiety than the control.
d I want to go for C, but I have a feeling it's a trick question.

Based on what I have told you about NHST, you should have answered C: only 4 of the 10 studies have a 'significant' result, which isn't very compelling evidence for *antiSTATic*. Now pretend you know nothing about NHST, look at the confidence intervals, and think about what we know about overlapping confidence intervals.

I would hope that some of you have changed your mind to option B. If you're still sticking with option C, then let me try to convince you otherwise. First, 10 out of 10 studies show a positive effect of *antiSTATic* (none of the mean differences are below zero), and even though sometimes this positive effect is not always 'significant', it is consistently positive. The

SELF TEST

Now you've looked at the confidence intervals, which of the earlier statements best reflects your view of *antiSTATic*?

Jane Superbrain 3.1
Why do we use 0.05? ▮▮▮

Given that the criterion of 95% confidence or a 0.05 probability, is so ubiquitous in NHST, you'd expect a very solid justification for it, wouldn't you? Think again. The mystery of how the 0.05 criterion came to be is complicated. Fisher believed that you calculate the probability of an event and evaluate this probability within the research context. Although Fisher felt that $p = 0.01$ would be strong evidence to back up a hypothesis, and perhaps $p = 0.20$ would be weak evidence, he objected to Neyman's use of an alternative hypothesis (among other things). Conversely, Neyman objected to Fisher's exact probability approach (Berger, 2003; Lehmann, 1993). The confusion created by the schism between Fisher and Neyman was like a lightning bolt bringing life to NHST: a bastard child of both approaches. I use the word 'bastard' advisedly.

During the decades in which Fisher and Neyman's ideas were stitched together into a horrific Frankenstein, the probability level of 0.05 rose to prominence. The likely reason is that in the days before computers, scientists had to compare their test statistics against published tables of 'critical values' (they did not have software to calculate exact probabilities for them). These critical values had to be calculated by

exceptionally clever people like Fisher, who produced tables of these values in his influential textbook *Statistical methods for research workers* (Fisher, 1925).[4] To save space, Fisher tabulated critical values for specific probability values (0.05, 0.02 and 0.01). This book had a monumental impact (to get some idea of its influence 25 years after publication, see Mather, 1951; Yates, 1951) and the tables of critical values were widely used—even Fisher's rivals, Neyman and Pearson, admitted that the tables influenced them (Lehmann, 1993). This disastrous combination of researchers confused about the Fisher and Neyman–Pearson approaches and the availability of critical values for only certain levels of probability created a trend to report test statistics as being significant at the now infamous $p < 0.05$ and $p < 0.01$ (because critical values were readily available at these probabilities). Despite this, Fisher believed that the dogmatic use of a fixed level of significance was silly: 'No scientific worker has a fixed level of significance at which from year to year, and in all circumstances, he rejects hypotheses; he rather gives his mind to each particular case in the light of his evidence and his ideas' (Fisher, 1956).

confidence intervals overlap with each other substantially in all studies, suggesting consistency across the studies: they all throw up (potential) population effects of a similar size. Remember that the confidence interval will contain the actual population value in 95% of samples. Look at how much of each confidence

interval falls above zero across the 10 studies: even in studies for which the confidence interval includes zero (implying that the population effect might be zero) most of the interval is greater than zero. Again, this suggests very consistent evidence that the population value could be greater than zero

(i.e., *antiSTATic* works). Therefore, looking at the confidence intervals rather than focusing on significance allows us to see the consistency in the data and not a bunch of apparently conflicting results (based on NHST): in all the studies the effect of *antiSTATic* was positive, and, taking all 10 studies into account, there's

---

4 You can read this online at http://psychclassics.yorku.ca/Fisher/Methods/

good reason to think that the population effect is plausibly greater than zero.

3.2.3 NHST is influenced by the intentions of the scientist ▮▮▮▮

Another problem is that the conclusions from NHST depend on what the researcher intended to do before collecting data. You might reasonably wonder how a statistical procedure can be affected by the intentional states of the researcher. I wondered about that too, and it took me some considerable effort to get my head around the reason why. Let's begin by reminding ourselves that, assuming you chose an alpha of 0.05, NHST works on the principle that you will make a Type I error in 5% of an infinite number of repeated, identical, experiments. The 0.05 value of alpha is a long-run probability, an **empirical probability**.

An empirical probability is the proportion of events that have the outcome in which you're interested in an indefinitely large collective of events (Dienes, 2011). For example, if you define the collective as everyone who has ever eaten fermented Icelandic shark, then the empirical probability of gagging will be the proportion of people (who have ever eaten fermented Icelandic shark) who gagged. The crucial point is that the probability applies to the collective and not to the individual events. You can talk about there being a 0.1 probability of gagging when eating putrefying shark, but the individuals who ate the shark either gagged or didn't, so their *individual* probability of gagging was either 0 (they didn't gag) or 1 (they gagged).

NHST is based on long-run probabilities too. The alpha, typically set to be a 0.05 probability of a Type I error, is a long-run probability and means that *across repeated identical experiments* the probability of making a Type I error is 0.05. That probability does not apply to an individual study, in which you either have ($p = 1$) or have not ($p = 0$) made a Type I error. The

probability of a Type II error (β) is also a long-run probability. If you set it at 0.2 then over an infinite number of identical experiments you can expect to miss an effect that genuinely exists in 20% of experiments. However, in an individual study the probability is not 0.2, it is either 1 (you have missed the effect) or 0 (you have not missed the effect).

Imagine the null hypothesis is true and there is no effect to be detected. Let's also imagine you went completely mad and carried out 1 million identical replications of an experiment designed to detect this effect. In each replication you compute a test statistic t_{null} (Kruschke, 2013) that arose from the null hypothesis. What you have is 1 million values of t_{null}. That does not satiate your need for replication, though, so you do another replication and compute another test statistic, t. You want a p-value for this new test statistic so you use your 1 million previous values of t_{null} to calculate one. The resulting p-value is a long-run probability: it is the relative frequency of the most recent observed t compared to the previous 1 million values of t_{null}.

This point is important, so let's write it again: the p-value is the probability of getting a test statistic at least as large as the one observed relative to all possible values of t_{null} *from an infinite number of identical replications of the experiment*. Like determining the likelihood of gagging after eating putrefying shark by looking at the proportion of people who gag across a collective of all people who've eaten putrefying shark, the p-value is the frequency of the observed test statistic relative to all possible values that could be observed *in the collective of identical experiments*. This is effectively what happens when you compute a p-value, except (thankfully) you don't have to conduct 1 million previous experiments, instead you use a computer.

The typical decision rule is that if this long-run probability (the p-value) is less

than 0.05 we are inclined to believe that the null hypothesis is not true. Again, this decision rule is a long-run probability: it will control the Type I error rate to 5% across an indefinite set of identical replications of the experiment. Similarly, if the Type II error rate is set at 0.2, then it controls this error rate to be 20% across an indefinite set of replications of the experiment. These two probabilities are used prior to data collection to determine the sample size necessary to detect the effect of interest (Section 2.9.3). The scientist collects data until they reach this predetermined number of observations; in doing so, the p-value represents the relative frequency of the observed test statistic relative to all the t_{null}s that could be observed in the collective of identical experiments with the *exact same sampling procedure* (Kruschke, 2010a, 2013).

Imagine that you aim, before the experiment, to collect data from 100 people, but once you start collecting data you find that you can only find 93 people willing to participate. Your decision rule before the experiment is based on collecting data from 100 people, so the p-value that you want to compute should also be based on collecting data from 100 people. That is, you want the relative frequency of the observed test statistic relative to all the test statistics that could be observed in the collective of identical experiments that aimed to get 100 observations. So, you *should* be computing a p-value based on degrees of freedom for 100 participants. However, you only have 93, so you end up computing a p-value based on degrees of freedom for 93 participants. This is the wrong p-value: you end up computing the relative frequency of your test statistic compared to all possible t_{null}s from experiments of size 93, but what you set out to do was to compare your test statistic to all possible t_{null}s from experiments of size 100. The space of possible t_{null}s has been influenced by an arbitrary variable such as the *availability of participants* rather than sticking to the original sampling scheme (Kruschke, 2013). In short, your decision rule has changed because of collecting the data—this is like changing your prediction after the bet is placed (think back to Jane Superbrain Box 2.4). The *p*-value that you need in this scenario should be based on relative frequency of the observed test statistic compared to all possible t_{null}s from the collective of experiments where the intention was to collect 100 participants but (for the same reasons as in your experiment) only 93 participants were available. This *p*-value is too idiosyncratic to compute.

Researchers sometimes use data collection rules other than collecting up to a predetermined sample size. For example, they might be interested in collecting data for a fixed amount of time rather than a fixed sample size (Kruschke, 2010b). Imagine you collect data for 1 week. If you repeated this experiment many times you would get different-sized samples of data because it would be unlikely that you would get the same amount of data in different week-long periods. As such, your *p*-value needs to be the long-run probability of your observed test statistic relative to all possible t_{null}s from identical experiments in which data were collected for 1 week. However, the *p*-value that is chugged out of your favored software package will not be this relative frequency: it will be the relative frequency of your observed test statistic to all possible t_{null}s from identical experiments with the same sample size as yours and *not* the same *duration* of data collection (Kruschke, 2013).

These scenarios illustrate two different intentions for collecting data: whether you wanted to collect a certain number of data points *or* collect data for a specific period. The resulting *p*-values for those two intentions would be different, because one should be based on t_{null}s from replications using the same sample size whereas the

other should be based on t_{null}s from replications using the same duration of data collection. Thus, the *p*-value is affected by the intention of the researcher.

3.3 NHST as part of wider problems with science ▐▐▌▌

There are direct consequences attached to the problems we have just looked at. For example, the consequences of the misconceptions of NHST are that scientists overestimate the importance of their effects (misconception 1); ignore effects that they falsely believe don't exist because of 'accepting the null' (misconception 2); and pursue effects that they falsely believe exist because of 'rejecting the null' (misconception 3).

Given that a lot of science is directed at informing policy and practice, the practical implications could be things like developing treatments that, in reality, have trivial efficacy or not developing ones that have potential. NHST also plays a role in some wider issues in science.

To understand why, we need a trip to the seedy underbelly of science. Science should be objective, and it should be driven, above all else, by a desire to find out truths about the world. It should not be self-serving, at least not if that gets in the way of the truth. Unfortunately, scientists compete for scarce resources to do their work: research funding, jobs, lab space, participant time and so on. It is easier to get these scarce resources if you are 'successful', and being 'successful' is tied up with NHST (as we shall see). Also, scientists are not emotionless robots (seriously!) but are people who have, for the most part, spent their whole lives succeeding at academic pursuits (school grades, college grades, etc.). They also work in places where they are surrounded

by incredibly intelligent and 'successful' people and they probably don't want to be known as 'Professor Thicko whose experiments never work'. Before you get out the world's tiniest violin to play me a tune, I'm not looking for sympathy, but it does get wearing feeling inferior to people with more grants, better papers and more ingenious research programs, trust me. What does this have to do with NHST? Let's find out.

3.3.1 Incentive structures and publication bias ▐▐▌▌

Imagine that two scientists, Beth and Danielle, are both interested in the psychology of people with extreme views. If you've spent any time on social media, you're likely to have encountered this sort of scenario: someone expresses an opinion, someone else responds with some unpleasant insult, followed by the directly opposing view, and there ensues a pointless and vile exchange where neither person manages to convince the other of their position. Beth and Danielle want to know whether people with extreme views literally can't see the grey areas. They set up a test in which participants see words displayed in various shades of grey and for each word they try to match its color by clicking along a gradient of greys from near white to near black. Danielle finds that politically moderate participants are significantly more accurate (the shades of grey they chose more closely matched the color of the word) than participants with extreme political views to the left or right. Beth found no significant differences in accuracy between groups with different political views. This is a real study (Labcoat Leni's Real Research 3.1).

What are the consequences for our two scientists? Danielle has an interesting, surprising and media-friendly result, she writes it up and submits it to a journal for publication. Beth does not have a sexy result, but being a positive kind of person, and trusting the rigor of her study, she decides to send it to a journal anyway. The chances are that Danielle's article will be published but Beth's will not, because significant findings are about seven times more likely to be published than non-significant ones (Coursol & Wagner, 1986). This phenomenon is called **publication bias**. In my own discipline, Psychology, over 90% of journal articles report significant results (Fanelli, 2010b, 2012). This bias is driven partly by reviewers and editors rejecting articles with non-significant results (Hedges, 1984), and partly by scientists not submitting articles with

non-significant results because they are aware of this editorial bias (Dickersin, Min, & Meinert, 1992; Greenwald, 1975).

To return to Danielle and Beth, assuming that they are equal in all other respects, Danielle now has a better-looking track record than Beth: she will be a stronger candidate for jobs, research funding, and internal promotion. Danielle's study was no different than Beth's (except for the results) but doors are now a little more open for her. The effect of a single research paper may not be dramatic, but over time, and over research papers, it can be the difference between a 'successful' career and an 'unsuccessful' one.

The current incentive structures in science are individualistic rather than collective. Individuals are rewarded for 'successful' studies that can be published and can therefore form the basis for funding applications or tenure; 'success' is, therefore, defined largely by results being significant. If a person's career as a scientist depends on significant results, then they might feel pressure to get significant results. In the USA, scientists from institutions that publish a large amount are more likely to publish results supporting their hypothesis (Fanelli, 2010a). Given that a good proportion of hypotheses ought to be wrong, and these wrong hypotheses ought to be distributed

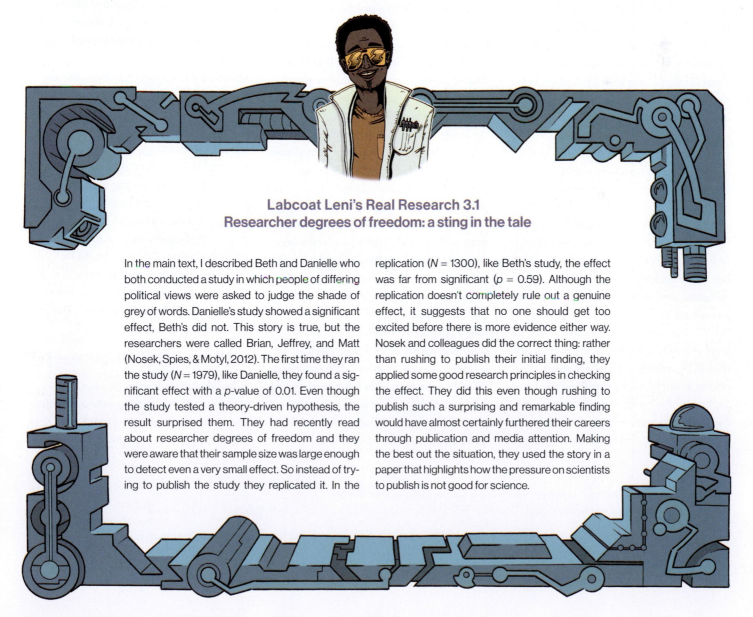

Labcoat Leni's Real Research 3.1
Researcher degrees of freedom: a sting in the tale

In the main text, I described Beth and Danielle who both conducted a study in which people of differing political views were asked to judge the shade of grey of words. Danielle's study showed a significant effect, Beth's did not. This story is true, but the researchers were called Brian, Jeffrey, and Matt (Nosek, Spies, & Motyl, 2012). The first time they ran the study ($N = 1979$), like Danielle, they found a significant effect with a p-value of 0.01. Even though the study tested a theory-driven hypothesis, the result surprised them. They had recently read about researcher degrees of freedom and they were aware that their sample size was large enough to detect even a very small effect. So instead of trying to publish the study they replicated it. In the

replication ($N = 1300$), like Beth's study, the effect was far from significant ($p = 0.59$). Although the replication doesn't completely rule out a genuine effect, it suggests that no one should get too excited before there is more evidence either way. Nosek and colleagues did the correct thing: rather than rushing to publish their initial finding, they applied some good research principles in checking the effect. They did this even though rushing to publish such a surprising and remarkable finding would have almost certainly furthered their careers through publication and media attention. Making the best out the situation, they used the story in a paper that highlights how the pressure on scientists to publish is not good for science.

across researchers and institutions, Fanelli's work implies that those working in high-stress 'publish or perish' environments less often have wrong hypotheses. One explanation is that high-stress environments attract better scientists who derive better-quality hypotheses, but an alternative is that the incentive structures in academia encourage people in high-pressure environments to cheat more. Scientists wouldn't do that, though.

3.3.2 Researcher degrees of freedom ▌▌▌▌

As well as reviewers and editors tending to reject articles reporting non-significant results, and scientists not submitting them, there are other ways that scientists contribute to publication bias. The first is by selectively reporting their results to focus on significant findings and exclude non-significant ones. At the extreme, this could entail not including details of other experiments that had results that contradict the significant finding. The second is that researchers might capitalize on **researcher degrees of freedom** (Simmons, Nelson & Simonsohn, 2011) to show their results in the most favorable light possible. 'Researcher degrees of freedom' refers to the fact that a scientist has many decisions to make when designing and analyzing a study. We have already seen some decisions related to NHST that apply here: the alpha level, the level of power, and how many participants should be collected. There are many others, though: which statistical model to fit, how to deal with extreme scores, which control variables to consider, which measures to use, and so on. Focusing on

analysis alone, when 29 different research teams were asked to answer the same research question (are soccer referees more likely to give red cards to players with dark skin than to those with light skin?) using the same data set, 20 teams found a significant effect, whereas nine did not. There was also a wide variety in the analytical models that the teams used to address the question (Silberzahn & Uhlmann, 2015; Silberzahn et al., 2015). These researcher degrees of freedom could be misused to, for example, exclude cases to make the result significant.

Fanelli (2009) assimilated data[5] from 18 published studies containing surveys about questionable research practices and found that for scientists reporting on their own behavior, across all studies, on average 1.97% admitted fabricating or falsifying data or altering results to improve the outcome. That's a small amount, but other questionable practices were admitted across studies by 9.54%, and there were higher rates for things like dropping observations based on a gut feeling (15.3%), using inappropriate research designs (13.5%), not publishing key findings (12.1%), allowing industry funders to either write the first draft of the report (29.6%) and influence when the study is terminated (33.7%). The last two practices are important because companies that fund research often have a conflict of interest in the results and so might spin them or terminate data collection earlier or later than planned, to suit their agenda. Plans for data collection should be determined before the study, not adjusted during it. The percentages reported by Fanelli will almost certainly be underestimates because scientists were

reporting on their own behavior, and, therefore, admitting to activities that harm their credibility. Fanelli also assimilated studies in which scientists reported on *other scientists'* behavior. He found that, on average, 14.12% had observed fabricating or falsifying data or altering results to improve the outcome, and on average a disturbingly high 28.53% reported other questionable practices. Fanelli details response rates to specific research practices mentioned in the studies he assimilated.[6] It's noteworthy that there were high rates of scientists responding that they were aware of others failing to report contrary data in an article (69.3%), choosing a statistical technique that provided a more favorable outcome (45.8%), reporting only significant results (58.8%), and excluding data based on a gut feeling (20.3%).

Questionable research practices are not necessarily the fault of NHST, but NHST nurtures these temptations by fostering black-and-white thinking, in which significant results garner much greater personal rewards than non-significant ones. For example, having spent months of hard work planning a project and collecting data, it's easy to imagine that if your analysis spews out a difficult-to-publish p-value of 0.08, it might be tempting to change your analytic decisions to see whether the p-value drops below the easier-to-publish 0.05 threshold. Doing so creates scientific noise.

3.3.3 *p*-hacking and HARKing ▌▌▌▌

p-hacking (Simonsohn, Nelson, & Simmons, 2014)[7] and hypothesizing after the results are known or **HARKing** (Kerr, 1998) are researcher degrees of freedom

---

5 Using something called meta-analysis, which you'll discover later in this chapter.

6 See Table S3 in the supplementary materials of the article (http://journals.plos.org/plosone/article?id=10.1371/journal.pone.0005738).

7 Simonsohn et al. coined the phrase '*p*-hacking', but the problem of selective reporting and trying multiple analyses to obtain significance is an old problem that has variously been termed *fishing, cherry-picking, data snooping, data mining, data diving, selective reporting*, and *significance chasing* (De Groot, 1956/2014; Peirce, 1878). Nevertheless, I enjoy the imagery of a maniacal scientist driven so mad in the pursuit of a *p*-value less than 0.05 that they hack at their computer with a machete.

that relate closely to NHST. *p*-hacking refers to researcher degrees of freedoms that lead to the selective reporting of significant *p*-values. It is a broad term that encompasses some of the practices we have already discussed such as trying multiple analyses and reporting only the one that yields significant results, deciding to stop collecting data at a point other than when the predetermined (prior to data collection) sample size is reached, and including (or not) data based on the effect they have on the *p*-value. The term also encompasses practices such as including (or excluding) variables in an analysis based on how those variables affect the *p*-value, measuring multiple *outcomes* or *predictor* variables but reporting only those for which the effects are significant, merging groups of variables or scores to yield significant results, and transforming or otherwise manipulating, scores to yield significant *p*-values. HARKing refers to the practice in research articles of presenting a hypothesis that was made *after* data collection as though it were made *before* data collection.

Let's return to Danielle, our researcher interested in whether perceptions of the shade of grey differed across people of different political extremes. Imagine that she had recorded not only the political views of her participants, but also a bunch of other variables about their lifestyle and personality such as what music they like, their openness to new experiences, how tidy they are, various mental health questionnaires measuring different things, their biological sex, gender, sexual orientation, age, and so on. You get the picture. At the end of data collection, Danielle has her measure of how accurately participants perceive the colors of the grey words and, let's say, 20 other variables. She does 20 different analyses to see whether each of these 20 variables predicts perception of the color grey. The only one that significantly predicts it is 'political group', so she

reports this effect and doesn't mention the other 19 analyses. This is *p*-hacking. She then tries to explain this finding to herself by looking at the literature on embodiment (the link between the mind and body) and decides that the results might be because the perception of grey is an 'embodiment' of political extremism. When she writes up the report she pretends that the motivation for the study was to test a hypothesis about embodiment of political views. This is HARKing.

The example above illustrates that HARKing and *p-hacking* are not mutually exclusive: a person might *p*-hack to find a significant result that they then HARK. If you were so bad at comedy that you had to write statistics textbooks for a living, you might call this *p*-harking.

Both *p*-hacking and HARKing are cheating the system of NHST (explained in Section 2.9.3). In both cases, it means that you're not controlling the Type I error rate (because you're deviating from the process that ensures that it is controlled) and therefore you have no idea how many Type I errors you will make in the long run (although it will certainly be more than 5%). You are making it less likely that your findings will replicate. More important, it is morally dubious to put spurious results or scientific noise, into the public domain. It wastes a lot of people's time and money.

Scientists have looked at whether there is general evidence for *p*-hacking in science. There have been different approaches to the problem but, broadly speaking, they have focussed on looking at the distribution of *p*-values you would expect to get if *p*-hacking doesn't happen with the distributions you might expect if it does. To keep things simple, one example is to extract the reported *p*-values across a range of studies on a topic or within a discipline, and plot their frequency. Figure 3.3 reproduces some of the data from Masicampo and Lalande (2012), who extracted the reported *p*-values from

12 issues of three prominent psychology journals. The line shows a **p-curve,** which is the number of *p*-values you would expect to get for each value of *p*. In this paper the *p*-curve was derived from the data; it was a statistical summary of *p*-values that they extracted from the articles, and shows that smaller *p*-values are more frequently reported (left of the figure) than larger, non-significant ones (right of the figure). You'd expect this because of the bias towards publishing significant results. You can also compute *p*-curves based on statistical theory, whereby the curve is affected by the effect size and the sample size but has a characteristic shape (Simonsohn et al., 2014). When there is no effect (effect size = 0) the curve is flat—all *p*-values are equally likley. For effect sizes greater than 1, the curve has an exponential shape, which looks similar to curve in Figure 3.3, because smaller *p*s (reflecting more significant results) occur more often than larger *p*s (reflecting less significant results).

The dots in Figure 3.3 are the number of *p*-values reported for each value of *p* that Masicampo and Lalande (2012) extracted from a journal called *Psychological Science*. Note that the dots are generally close to the line: that is, the frequency of reported *p*-values matches what the model (the line) predicts. The interesting part of this graph (yes, there is one) is the shaded box, which shows *p*-values just under the 0.05 threshold. The dot in this part of the graph is much higher than the line. This shows that there are a lot more *p*-values just under the 0.05 threshold reported in *Psychological Science* than you would expect based on the curve. Masicampo and Lalande argue that this shows *p*-hacking: researchers are engaging in practices to nudge their *p*-values below the threshold of significance and that's why these values are over-represented in the published literature. Others have used similar analyses to replicate this finding in psychology (Leggett, Thomas, Loetscher, &

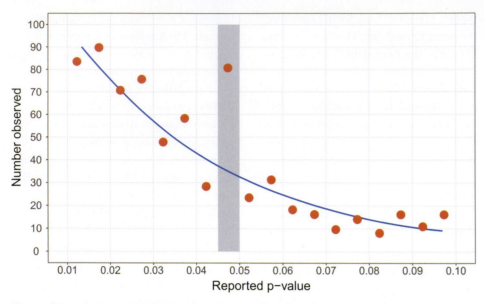

Figure 3.3   Reproduction of the data in Masicampo and Lalande (2012, Figure 2)

Nicholls, 2013) and science more generally (Head, Holman, Lanfear, Kahn, & Jennions, 2015). The analysis of *p*-curves looks for *p*-hacking by examining any *p*-value reported within and across the articles studied. Often, it is an automated process using a computer to search journal articles and extract anything that looks like a *p*-value.

Another approach is to focus on *p*-values in individual studies that report multiple experiments. The logic here is that if a scientist reports, say, four studies that examine the same effect, then you can work out, based on the size of effect being measured and sample size of the studies, what the probability is that you would get a significant result in all four studies. A low probability means that it is highly unlikely that the researcher would get these results: they are 'too good to be true' (Francis, 2013), the implication being that *p*-hacking or other dubious behavior may have occurred. Research using these **tests of excess success** also finds evidence for *p*-hacking in psychology, epigenetics and science generally (Francis, 2014a, 2014b; Francis, Tanzman, & Matthews, 2014).

All of this paints a pretty bleak picture of science, but it also illustrates the importance of statistical literacy and a critical mind. Science is only as good as the people who do it, and what you will learn in this book (and others) more than anything else are the skills to evaluate scientific results and data for yourself.

Having thoroughly depressed you, it's worth ending this section on a cheery note. I was tempted to tell you a tale of a small army of puppies found bouncing on their owner's bed to the sound of 'The chase is better than the cat' by Motörhead. While you process that image, I'll mention that some of the work that seems to suggest rampant *p*-hacking is not without its problems. For example, Lakens (2015) has shown that the *p*-curves used by Masicampo and Lalande probably don't resemble what you would actually see if *p*-hacking were taking place; when you use *p*-curves that better resemble *p*-hacking behavior, the evidence vanishes. Hartgerink, van Aert, Nuijten, Wicherts, & van Assen (2016) noted that *p*-values are often misreported in articles, and that studies of *p*-hacking typically look for a 'bump' just below the 0.05 threshold (as shown Figure 3.3) rather than more subtle forms of excess at the 0.05 threshold. Looking at 258,050 tests from 30,701 psychology research articles, Hartgerink et al. found evidence for a 'bump' in only

three of the eight journals that they looked at. They found widespread rounding errors in the *p*-values reported, and when they corrected these errors the 'bump' remained in only one of the eight journals examined. Perhaps their most important conclusion was that it was very difficult to extract enough information (even in this huge study) to model accurately what *p*-hacking might look like, and therefore ascertain whether it had happened. This sentiment is echoed by other researchers, who have pointed out that the shape of the *p*-curve may indicate very little about whether *p*-hacking has occurred (Bishop & Thompson, 2016; Bruns & Ioannidis, 2016). There is also an argument that TES aren't useful because they don't test what they set out to test (e.g., Morey, 2013; Vandekerckhove, Guan, & Styrcula, 2013).

You might reasonably wonder why I've wasted your time convincing you that scientists are all duplicitous, self-interested, nefarious hacks only to then say 'Actually, they might not be'. If you spend enough time in science you'll realize that this is what it's like: as soon as you believe one thing, someone will come along and change your mind. That's OK, though, and remaining open-minded is good because it helps you not to get sucked into dubious research behaviors. Given that you will be the future of science, what matters is not what has happened until now, but what you go out into the world and do in the name of science. To that end, I hope to have convinced you not to *p*-hack or HARK or fudge your data and so on. You are better than that, and science will be better for the fact that you are.

3.4 A phoenix from the EMBERS ▌▌▌▌

The pitfalls of NHST and the way that scientists use it have led to a shift in the pervasive view of how to evaluate hypotheses. It's not quite the paradigm

Cramming Sam's Tips
Problems with NHST

- A lot of scientists misunderstand NHST. A few examples of poorly understood things related to significance testing are:
  - A significant effect is not necessarily an important one.
  - A non-significant result does not mean that the null hypothesis is true.
  - A significant result does not mean that the null hypothesis is false.
- NHST encourages all-or-nothing thinking whereby an effect with a p-value just below 0.05 is perceived as important whereas one with a p-value just above 0.05 is perceived as unimportant.
- NHST is biased by researchers deviating from their initial sampling frame (e.g., by stopping data collection earlier than planned).
- There are lots of ways that scientists can influence the p-value. These are known as *researcher degrees of freedom* and include selective exclusion of data, fitting different statistical models but reporting only the one with the most favorable results, stopping data collection at a point other than that decided at the study's conception, and including only control variables that influence the p-value in a favorable way.
- Incentive structures in science that reward publication of significant results also reward the use of researcher degrees of freedom.
- p-hacking refers to practices that lead to the selective reporting of significant p-values, most commonly trying multiple analyses and reporting only the one that yields significant results.
- Hypothesizing after the results are known (HARKing) occurs when scientists present a hypothesis that was made after data analysis as though it were made at the study's conception.

shift to which I alluded at the start of this chapter, but certainly the tide is turning. The ubiquity of NHST is strange, given that its problems have been acknowledged for decades (e.g., Rozeboom, 1960). In a post on a discussion board of the American Statistical Association (ASA), George Cobb noted a circularity: the reason why so many colleges and graduate schools still teach $p = 0.05$ is that it's what the scientific community uses, and the reason why the scientific community still (predominantly) uses $p = 0.05$ is that it's what they were taught in college or graduate school (Wasserstein & Lazar, 2016). This cycle of habit probably

extends to teaching too, in that statistics teachers will tend to teach what they know, which will be what they were taught, which is typically NHST. Therefore, their students learn NHST too, and those that go on to teach statistics will pass NHST down to their students. It doesn't help that because most scientists are taught NHST and that's all they know, they use it in their research papers, which means that lecturers probably *must* teach it so that their students can understand research papers. NHST is a hard habit to break because it requires a lot of people making the effort to broaden their statistical horizons.

Things are changing, though. In my discipline (Psychology), an American Psychological Association (APA) task force produced guidelines for the reporting of data in its journals. This report acknowledged the limitations of NHST while appreciating that a change in practice would not happen quickly; consequently, it didn't recommend against NHST but suggested instead that scientists report things like confidence intervals and effect sizes to help them (and readers) evaluate the research findings without dogmatic reliance on p-values (Wilkinson, 1999). At the extreme, in 2015 the journal *Basic and Applied Social*

Psychology banned *p*-values from its articles, but that was silly because it throws the baby out with the bathwater. The ASA has published some guidelines about *p*-values (Wasserstein & American Statistical Association, 2016) which we'll get to shortly. In the following sections, we'll explore some of the ways you might address the problems with NHST. It will be like a statistical phoenix rising from the EMBERS of significant testing (but not in that order):

• **E**ffect sizes
• **M**eta-analysis
• **B**ayesian **E**stimation
• **R**egistration
• **S**ense

3.5 Sense, and how to use it ▮▮▮▮

It is easy to blame NHST for the evils in the scientific world, but some of the problem is not the process so much as how people misuse and misunderstand it. As we have seen (Jane Superbrain Box 3.1), Fisher never encouraged people to brainlessly use a 0.05 level of significance and throw a party for effects with *p*-values below that threshold while hiding away those with *p*s above the threshold in a dungeon of shame. Nor did anyone suggest that we evolve a set of incentive structures in science that reward dishonesty and encourage the use of researcher degrees of freedom. You *can* blame NHST for being so difficult to understand, but you can also blame scientists for not working harder to understand it. The first way to combat the problems of NHST is, therefore, to apply sense when you use it. A statement by the ASA on *p*-values (Wasserstein & American Statistical Association, 2016) offers a set of six principles for scientists using NHST.

1 The ASA points out that *p*-values *can* indicate how incompatible the data are with a specified statistical model. In the case of what we've been discussing this means we can use the value of *p* (its exact value, not whether it is above or below an arbitrary threshold) to indicate how incompatible the data are with the null hypothesis. Smaller *p*s indicate more incompatibility with the null. You are at liberty to use the degree of incompatibility to inform your own beliefs about the relative plausibility of the null and alternative hypotheses, as long as . . .

2 . . . you don't interpret *p*-values as a measure of the probability that the hypothesis in question is true. They are also not the probability that the data were produced by random chance alone.

3 Scientific conclusions and policy decisions *should not* be based only on whether a *p*-value passes a specific threshold. Basically, resist the black-and-white thinking that *p*-values encourage.

4 Don't *p*-hack. The ASA says that '*p*-values and related analyses should not be reported selectively. Conducting multiple analyses of the data and reporting only those with certain *p*-values (typically those passing a significance threshold) renders the reported *p*-values essentially uninterpretable.' Be fully transparent about the number of hypotheses explored during the study, and all data collection decisions and statistical analyses.

5 Don't confuse statistical significance with practical importance. A *p*-value does not measure the size of an effect and is influenced by the sample size, so you should never interpret a *p*-value in any way that implies that it quantifies the size or importance of an effect.

6 Finally, the ASA notes that 'By itself, a *p*-value does not provide a good measure of evidence regarding a model or hypothesis.' In other words, even if your small *p*-value suggests that your data are compatible with the alternative hypothesis or your large *p*-value suggests that your data are compatible with the null, there may be many other hypotheses (untested and perhaps not conceived of) that are more compatible with the data than the hypotheses you have tested.

If you have these principles in mind when you use NHST then you will avoid some of the common pitfalls when interpreting and communicating your research.

3.6 Pre-registering research and open science ▮▮▮▮

The ASA's fourth principle is a call for transparency in what is reported. This notion of transparency is gaining considerable momentum because it is a simple and effective way to protect against some of the pitfalls of NHST. There is a need for transparency in scientific intentions both before the study is conducted and after it is complete (whether that is through sharing your data or stating clearly any deviations from your intended protocol). These goals are encapsulated by the term **open science**, which refers to a movement to make the process, data and outcomes of research freely available to everyone. Part of this movement is about providing free access to scientific journals, which have traditionally been available only to individuals and institutions who pay to subscribe. The part most relevant to our discussion, though, is **pre-registration** of research, which refers to the practice of making all aspects of your research process (rationale, hypotheses, design, data processing strategy, data analysis strategy) publicly available before data collection begins. This can be done in a **registered report** in an academic journal (e.g., Chambers, Dienes, McIntosh, Rotshtein, & Willmes, 2015; Nosek & Lakens, 2014) or more informally (e.g., on a public website such as the Open Science Framework). A formal registered report is a submission to an academic journal that outlines an intended research protocol (rationale, hypotheses, design, data processing strategy, data analysis strategy) before data are collected. The submission is reviewed by relevant experts just as a submission of a completed study would be (Section 1.9). If the protocol is deemed to be rigorous enough and the research question novel enough, the protocol is accepted by the journal typically with a guarantee to publish the findings no matter what they are.

Other initiatives that facilitate open science principles include the **Peer Reviewers' Openness Initiative** (Morey et al., 2016), which asks scientists to commit to the principles of open science when they act as expert reviewers for journals. Signing up is a pledge to review submissions only if the data, stimuli, materials, analysis scripts and so on are made publicly available (unless there is a good reason not to, such as a legal requirement). Another, the Transparency and Openness Promotion (TOP) guidelines, is a set of standards for open science principles that can be applied to journals. The eight standards cover citations (1), pre-registration of study and analysis protocols (2 and 3), replication transparency with data, analysis scripts, design and analysis plans, research materials (4–7), and replication (8). For each standard, there are levels defined from 0 to 3. For example, in data transparency, Level 0 is defined as the journal merely encouraging data-sharing or saying nothing, and Level 3 is data posted on a trusted repository and results reproduced independently before publication (Nosek et al., 2015). Using these standards, a journal's level of commitment to open science can be 'badged'. As junior scientists, you can contribute by aiming to prioritize 'open' materials (questionnaires, stimuli, etc.) over equivalent proprietary ones, and you can make your own materials available when you publish your own work.

Open science practices help to combat many of the wider problems into which NHST feeds. For example, pre-registering research encourages adherence to an agreed protocol, thus discouraging the misuse of researcher degrees of freedom. Better still, if the protocol has been reviewed by experts, then the scientist gets useful feedback *before* collecting data (rather than after, by which time it is too late to change things). This feedback can be on the methods and design, but also on the analytic plan, which ought to improve the study and in turn science. Also, by guaranteeing publication of the results—no matter what they are— registered reports should reduce publication bias and discourage questionable research practices aimed at nudging p-values below the 0.05 threshold. Of course, by having a public record of the planned analysis strategy, deviations from it will be transparent. In other words, p-hacking and HARKing will be discouraged. Data being in the public domain also makes it possible to check for researcher degrees of freedom that may have influenced the results.

None of these developments change the flaws inherent in NHST but they tighten up the way that it is used, and the degree to which scientists can conceal misuses of it. Finally, by promoting methodological rigor and theoretical importance above the results themselves (Nosek & Lakens, 2014), open science initiatives push incentive structures towards quality not quantity, documenting research rather than 'storytelling' and collaboration rather than competition. Individualists need not fear either because there is evidence that practicing open science principles is associated with higher citation rates of your papers, more media exposure and better-quality feedback on your work from experts (McKiernan et al., 2016). It's a win–win.

3.7 Effect sizes

One of the problems we identified with NHST was that significance does not tell us about the importance of an effect. The solution to this criticism is to measure the size of the effect that we're testing in a standardized way. When we measure the size of an effect (be that an experimental manipulation or the strength of a relationship between variables) it is known as an **effect size**. An effect size is an objective and (usually) standardized measure of the magnitude of observed effect. The fact that the measure is 'standardized' means that we can compare effect sizes across different studies that have measured different variables or have used different scales of measurement (so an effect size based on reaction time in milliseconds could be compared to an effect size based on heart rates). Effect sizes add information that you don't get from a p-value, so reporting them is a habit well worth getting into.

Many measures of effect size have been proposed, the most common of which are Cohen's d, Pearson's correlation coefficient r (Chapter 8) and the odds ratio (Chapters 19 and 20). There are others, but these three are the simplest to understand. Let's look at Cohen's d first.

3.7.1 Cohen's d

Think back to our example of singing as a mating strategy (Section 2.9.10). Remember we had some men singing to women, and others starting conversations with them. The outcome was how long it took before the woman ran away.[8] If we wanted to quantify the effect between the singing and conversation groups, how might we do it? A

---

8 Although the fictional studies I use to explain effect sizes focus on males trying to attract females by singing, this is only because the idea came from replicating, in humans, research showing that male mice 'sing' to female mice to attract them (Hoffmann et al., 2012). Please use your creativity to change the examples to whatever sex-pairings best reflect your life.

straightforward thing to do would be to take the difference between means. The conversation group had a mean of 12 minutes (before the recipient ran away), and the singing group a mean of 10 minutes. So, the effect of singing compared to conversation is 10 – 12 = –2 minutes. This is an effect size. Singing had a detrimental effect on how long the recipient stayed, by –2 minutes. That's simple enough to compute and understand, but it has two small inconveniences. First, the difference in means will be expressed in the units of measurement for the outcome variable. In this example, this inconvenience isn't an inconvenience at all because minutes mean something to us: we can all imagine what an extra 2 minutes of time with someone would be like. We can also have an idea of what 2 minutes with someone is relative to the amount of time we usually spend talking to random people. However, if we'd measured what the recipient thought of the singer rather than how much time they spent with them, then interpretation is trickier: 2 units of 'thought' or 'positivity' or whatever is less tangible to us than two minutes of time. The second inconvenience is that although the difference between means gives us an indication of the 'signal', it does not tell us about the 'noise' in the measure. Is 2 minutes of time a lot or a little relative to the 'normal' amount of time spent talking to strangers?

We can remedy these problems in the same way. We saw in Chapter 2 that the standard deviation is a measure of 'error' or 'noise' in the data, and we saw in Section 1.8.6 that if we divide by the standard deviation then the result is a score expressed in standard deviation units (i.e., a z-score). Therefore, if we divide the difference between means by the standard

deviation we get a signal-to-noise ratio, but we also get a value that is expressed in standard deviation units (and can, therefore, be compared in different studies that used different outcome measures). What I have just described is **Cohen's d**, and we can express it formally as:

$$\hat{d} = \frac{\bar{X}_1 - \bar{X}_2}{s} \tag{3.1}$$

I have put a hat on the d to remind us that we're interested in the effect size in the population, but because we can't measure that directly, we estimate it from the sample.[9] We've seen these hats before; they mean 'estimate of'. Therefore, d is the difference between means divided by the standard deviation. However, we had two means and, therefore, two standard deviations, so which one should we use? Sometimes we assume that group variances (and therefore standard deviations) are equal (see Chapter 6), and, if they are, we can pick a standard deviation from either group because it won't matter much. In our singing for a date example, the standard deviations were identical in the two groups (SD = 3) so whichever one we pick, we get:

$$\frac{\bar{X}_{\text{Singing}} - \bar{X}_{\text{Conversation}}}{s} = \frac{10-12}{3} = -0.667 \tag{3.2}$$

This effect size means that if a person sang rather than having a normal conversation, the time the recipient stayed was reduced by 0.667 standard deviations. That's quite a bit.

Cohen (1988, 1992) made some widely used suggestions about what constitutes a large or small effect: d = 0.2 (small), 0.5 (medium) and 0.8 (large). For our singing data this would mean we have a medium to large effect size. However, as Cohen acknowledged, these benchmarks

encourage the kind of lazy thinking that we were trying to avoid and ignore the context of the effect such as the measurement instruments and norms within the research area. Lenth put it nicely when he said that when we interpret effect sizes, we're not trying to sell T-shirts: 'I'll have the Metallica tour effect size in a medium, please' (Baguley, 2004; Lenth, 2001).

When groups do not have equal standard deviations, there are two common options. First, use the standard deviation of the control group or baseline. This option makes sense because any intervention or experimental manipulation might be expected to change not just the mean but also the spread of scores. Therefore, the control group/baseline standard deviation will be a more accurate estimate of the 'natural' standard deviation for the measure you're using. In our singing study, we would use the conversation group standard deviation because you wouldn't normally go up to someone and start singing. Therefore, d would represent the amount of time less that a person spent with someone who sang at them compared to someone who talked to them relative to the normal variation in time that people spend with strangers who talk to them.

The second option is to pool the standard deviations of the two groups using (if your groups are independent) this equation:

$$s_p = \sqrt{\frac{(N_1-1)s_1^2 + (N_2-1)s_2^2}{N_1+N_2-2}} \tag{3.3}$$

in which N is the sample size of each group and s is the standard deviation. For the singing data, because the standard deviations and sample sizes are the same

---

9 The value for the population is expressed as:

$$d = \frac{\bar{\mu}_1 - \bar{\mu}_2}{\sigma}$$

It's the same equation, but because we're dealing with population values rather than sample values, the hat over the d is gone, the means are expressed with μ and the standard deviation with σ.

in the two groups, this pooled estimate will be 3, the same as the standard deviation:

$$s_p = \sqrt{\frac{(10-1)3^2 + (10-1)3^2}{10+10-2}} = \sqrt{\frac{81+81}{18}} = \sqrt{9} = 3$$

(3.4)

When the group standard deviations are different, this pooled estimate can be useful; however, it changes the meaning of d because we're now comparing the difference between means against all the background 'noise' in the measure, not just the noise that you would expect to find in normal circumstances.

If you did the self-test you should have got the same result as before: −0.667. That's because the difference in sample size did not affect the means or standard deviations, and therefore will not affect the effect size. Other things being equal, effect sizes are not affected by sample size, unlike p-values. Therefore, by using effect sizes, we overcome one of the major problems with NHST. The situation is more complex because, like any parameter, you will get better estimates of the population value in large samples than small ones. So, although the sample size doesn't affect the computation of your effect size in the sample, it does affect how closely the sample effect size matches that of the population (the *precision*).

If you did the self-test then you will have found that the effect size for our larger study was $d = -0.003$. In other words, very small indeed. Remember that when we looked at p-values, this very small effect was deemed statistically significant.

When we looked at the data sets in Figures 2.17 and 2.18 and their corresponding p-values, we concluded the following:

- Figure 2.17: Two experiments with identical means and standard deviations yield completely opposite conclusions when using a p-value to interpret them (the study based on 10 scores per group was not significant but the study based on 100 scores per group was).
- Figure 2.18: Two virtually identical means are deemed to be significantly different based on a p-value.

SELF TEST
Compute Cohen's d for the effect of singing when a sample size of 100 was used (right-hand graph in Figure 2.17).

SELF TEST
Compute Cohen's d for the effect in Figure 2.18. The exact mean of the singing group was 10, and for the conversation group was 10.01. In both groups the standard deviation was 3.

SELF TEST
Look at Figures 2.17 and 2.18. Compare what we concluded about these three data sets based on p-values with what we conclude using effect sizes.

If we use effect sizes to guide our interpretations, we would conclude the following:

- Figure 2.17: Two experiments with identical means and standard deviations yield identical conclusions when using an effect size to interpret them (both studies had $d = -0.667$).
- Figure 2.18: Two virtually identical means are deemed to be not very different at all based on an effect size ($d = -0.003$, which is tiny).

With these examples, I hope to have convinced you that effect sizes offer us something potentially less misleading than NHST.

3.7.2 Pearson's r

Let's move on to Pearson's correlation coefficient, r, which is a measure of the strength of relationship between two variables. We'll cover this statistic in detail in Chapter 8. For now all you need to know is that it is a measure of the strength of a relationship between two continuous variables or between one continuous variable and a categorical variable containing two categories. It can vary from −1 (a perfect negative relationship) through 0 (no relationship) to +1 (a perfect positive relationship).

Imagine we continued our interest in whether singing was an effective dating strategy. This time, however, we focussed not on whether the person sang or not, but instead on whether the duration of the performance mattered. Rather than having two groups (singing versus conversation) all the participants in our study sang a song, but for different durations ranging from a 1-minute edit to a 10-minute extended remix. Once the performance had stopped, we timed how long the lucky recipient hung around for a chat. Figure 3.4 shows five different outcomes of this study. The top left and middle panels show that the longer the song was, the less time the recipient stayed to chat. This is called a negative relationship: as one variable increases, the other decreases. A perfect negative relationship (top left) has $r = -1$ and means that as you increase the length of the song, you decrease the subsequent conversation by a proportionate amount. A slightly smaller negative relationship (top middle, $r = -0.5$) means that an increase in the duration of the song also decreases the subsequent conversation, but by a smaller amount. A positive correlation coefficient (bottom row) shows the opposite trend: as the song increases in length so does the subsequent conversation. In general terms, as one variable increases so does the other. If the positive relationship is perfect ($r = 1$, bottom left), then the increases are by a proportionate amount whereas values less than 1 reflect increases in one variable that are not equivalent to the increase in the other. A correlation coefficient of 0 (top right) shows a situation where there is no relationship at all: as the length of the song increases there is no fluctuation whatsoever in the duration of the subsequent conversation.

Another point to note from Figure 3.4 is that the strength of the correlation reflects how tightly packed the observations are around the model (straight line) that summarizes the relationship between variables. With a perfect relationship ($r = -1$ or 1) the observed data basically fall exactly on the line (the model is a perfect fit to the data), but for a weaker relationship ($r = -0.5$ or 0.5) the observed data are scattered more widely around the line.

Although the correlation coefficient is typically known as a measure of the relationship between two continuous variables (as we have just described), it can also be used to quantify the difference in means between two groups. Remember that r quantifies the relationship between two variables; it turns out that if one of those variables is a categorical variable that represents two groups where one group is coded with a 0 and the other is coded with a 1, then what you get is a standardized measure of the difference between two means (a bit like Cohen's d). I'll explain this in Chapter 8, but for now trust me. Like with d, Cohen (1988, 1992) suggested some 'T-shirt sizes' for r:

- $r = 0.10$ (**small effect**): In this case the effect explains 1% of the total variance. (You can convert r to the proportion of variance by squaring it—see Section 8.4.2.)
- $r = 0.30$ (**medium effect**): The effect accounts for 9% of the total variance.
- $r = 0.50$ (**large effect**): The effect accounts for 25% of the variance.

It's worth bearing in mind that r is not measured on a linear scale, so an effect with $r = 0.6$ isn't twice as big as one with $r = 0.3$. Also, as with d, although it's tempting to wheel out these 'canned' effect sizes when you can't be bothered to think properly about your data, you ought to evaluate an effect size within the context of your specific research question.

There are many reasons to like r as an effect size measure, one of them being that it is constrained to lie between 0 (no effect) and 1 (a perfect effect).[10] However, there are situations in which d may be favored; for example, when group sizes are very discrepant, r can be quite biased compared to d (McGrath & Meyer, 2006).

3.7.3 The odds ratio ▮▮▮▮

The final effect size we'll look at is the **odds ratio**, which is a popular effect size for counts. Imagine a final scenario for our dating research in which we had groups of people who either sang a song or started up a conversation (like our Cohen's d example). However, this time, the outcome was not the length of time before the recipient ran away, instead at the end of the song or conversation the recipient was asked 'Would you go on a date with me' and their response ('yes' or 'no') was recorded.

Here we have two categorical variables (singing versus conversation and date versus no date) and the outcome is a count (the number of recipients in each combination of those categories). Table 3.1 summarizes the data in a 2×2 **contingency table**, which is a table representing the cross-classification of two or more *categorical variables*. The levels of each variable are arranged in a grid, and the number of observations falling into each category is contained in the cells of the table. In this example, we see that the categorical variable of whether someone sang or started a conversation is represented by rows of the table, and the categorical variable for the response to the question

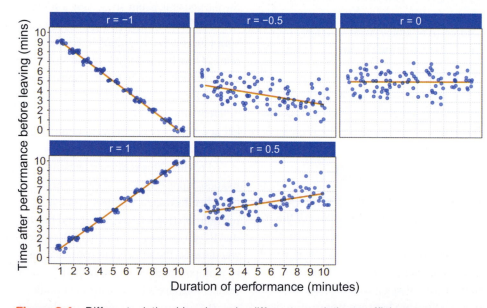

Figure 3.4 Different relationships shown by different correlation coefficients

---

10 The correlation coefficient can also be negative (but not below –1), which is useful because the sign of r tells us about the direction of the relationship, but when quantifying group differences the sign of r merely reflects the way in which the groups were coded (see Chapter 10).

about a date is represented by columns. This creates four cells representing the combinations of those two variables (singing–yes, singing–no, conversation–yes, conversation–no) and the numbers in these cells are the frequencies of responses in each combination of categories. From the table we can see that 12 recipients said 'yes' to a date after they were sung to, compared to 26 after a conversation. There were 88 recipients who said 'no' after singing, compared to 74 after a conversation. Looking at the column and row totals, we can see that there were 200 date-pairs in the study; in 100 of them the strategy was to sing, whereas in the remainder it was a conversation. In total, 38 recipients agreed to a date and 162 did not.

To understand these counts, we might ask how much more likely a person was to say 'yes' to a singer than to someone who starts a conversation. To quantify this effect, we need to calculate some odds. The **odds** of an event occurring are defined as the probability of an event occurring divided by the probability of that event not occurring:

$$odds = \frac{P(\text{event})}{P(\text{no event})} \qquad (3.5)$$

To begin with, we want to know the odds of a 'yes' response to singing, which will be the probability of a 'yes' response to a singer divided by the probability of a 'no' response to a singer. The probability of a 'yes' response for singers is the number of yes responses divided by the total number of singers and the probability of a 'no' response for singers is the number of no responses divided by the total number of singers:

$$Odds_{\text{yes to a singer}} = \frac{P(\text{yes to a singer})}{P(\text{no to a singer})}$$
$$= \frac{\text{Yes responses to singers}/\text{number of singers}}{\text{No responses to singers}/\text{number of singers}}$$
$$= \frac{12/100}{88/100} \qquad (3.6)$$

To get both probabilities you divide by the total number of singers (in this case 100),

Table 3.1 A contingency table of some fabricated data about singing and dates

		Response to date question		
		Yes	No	Total
Dating behavior	Singing	12	88	100
	Conversation	26	74	100
	Total	**38**	**162**	**200**

so the result of equation (3.6) is the same as dividing the number of 'yes' responses to a singer by the corresponding number of 'no' responses, which gives us 0.14:

$$Odds_{\text{yes to a singer}} = \frac{\text{Number of yes responses to a singer}}{\text{Number of no responses to a singer}}$$
$$= \frac{12}{88} = 0.14 \qquad (3.7)$$

Next, we want to know the odds of a 'yes' response if the person started a conversation ('talkers'). We compute this in the same way: it is, the *probability* of 'yes' response divided by the *probability* of a 'no' response for talkers, which is the same as dividing the *number* of 'yes' responses to talkers by the corresponding *number* of 'no' responses:

$$Odds_{\text{yes to a talker}} = \frac{\text{Number of yes responses to a talker}}{\text{Number of no responses to a talker}}$$
$$= \frac{26}{74} = 0.35 \qquad (3.8)$$

Finally, the odds ratio is the odds of a 'yes' to a singer divided by the odds of a 'yes' to a talker:

$$Odds\ ratio = \frac{Odds_{\text{yes to a singer}}}{Odds_{\text{yes to a talker}}}$$
$$= \frac{0.14}{0.35} = 0.4 \qquad (3.9)$$

This ratio tells us that the odds of a 'yes' response were 0.4 times as large to a singer as to someone who started a conversation. If the odds ratio is 1, then it means that the odds of one outcome are the same as the odds of the other; because it is less than 1, we know that the odds of a 'yes' response after singing are *worse* than the odds after a conversation. We can flip

this odds ratio on its head to see the relative odds of a 'yes' after a conversation compared to after singing by dividing 1 by the odds ratio. Doing so gives us $1/0.4 = 2.5$. We can state the same relationship, then, by saying that the odds of a 'yes' response to a person who started a conversation were 2.5 times as large as to one who sang.

Once you understand what odds are, the odds ratio becomes a very intuitive way to quantify the effect. In this case, if you were looking for love, and someone told you that the odds of a date after a conversation were 2.5 times those after singing, then you probably wouldn't need to finish this book to know it would be wise to keep your Justin Bieber impersonation, no matter how good it is, to yourself. It's always good to keep your Justin Bieber impersonation to yourself.

3.7.4 Effect sizes compared to NHST

Effect sizes overcome many of the problems associated with NHST:

- They encourage interpreting effects on a continuum and not applying a categorical decision rule such as 'significant' or 'not significant'. This is especially true if you ignore the 'canned effect size' benchmarks.
- Effect sizes are affected by sample size (larger samples yield better estimates of the population effect size), but, unlike *p*-values, there is no decision rule attached to effect sizes (see above), so the interpretation of effect sizes is not confounded by sample size (although it is

important in contextualizing the degree to which an effect size might represent the population). Because of this, effect sizes are less affected than *p*-values by things like early or late termination of data collection or sampling over a time period rather than until a set sample size is reached.

- Of course, there are still some researcher degrees of freedom (not related to sample size) that researchers could use to maximize (or minimize) effect sizes, but there is less incentive to do so because effect sizes are not tied to a decision rule in which effects either side of a certain threshold have qualitatively opposite interpretations.

3.7.5 Meta-analysis ▮▮▮

I have alluded many times to how scientists often test similar theories and hypotheses. An important part of science is replicating results, and it is rare that a single study gives a definitive answer to a scientific question. In Section 3.2.2 we looked at an example of 10 experiments that had all explored whether a potion called *antiSTATic* reduces statistics anxiety compared to a placebo (water). The summary of these studies was shown in Figure 3.2. Earlier we saw that, based on *p*-values, we would conclude that there were inconsistent results: four studies show a significant effect of the potion and six do not. However, based on the confidence intervals, we would conclude the opposite: that the findings across the studies were quite consistent and that it was likely that the effect in the population was positive. Also shown in this figure, although you wouldn't have known what they were at the time, are the values of Cohen's *d* for each study.

The 10 studies summarized in Figure 3.2 have *d*s ranging from 0.23 (other things being equal, smallish) to 0.71 (other things being equal, fairly large). The effect sizes are all positive: no studies showed worse anxiety after taking *antiSTATic*. Therefore,

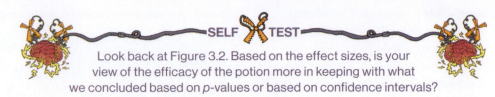

SELF TEST

Look back at Figure 3.2. Based on the effect sizes, is your view of the efficacy of the potion more in keeping with what we concluded based on *p*-values or based on confidence intervals?

the effect sizes are very consistent: all studies show positive effects and *antiSTATic*, at worst, had an effect of about a quarter of a standard deviation, and, at best, an effect of almost three-quarters of a standard deviation. Our conclusions are remarkably like what we concluded when we looked at the confidence intervals, that is, there is consistent evidence of a positive effect in the population. Wouldn't it be nice if we could use these studies to get a definitive estimate of the effect in the population? Well, we can, and this process is known as **meta-analysis**. It sounds hard, doesn't it? What wouldn't be hard would be to summarize these 10 studies by taking an average of the effect sizes:

$$\bar{d} = \frac{\sum_{i=1}^{k} d_i}{n}$$

$$= \frac{0.23+0.44+0.48+0.70+0.53+0.65+0.71+0.47+0.31+0.70}{10}$$

$$= 0.52 \qquad (3.10)$$

Congratulations, you have done your first meta-analysis—well, sort of. It wasn't that hard, was it? There's more to it than that, but at a very basic level a meta-analysis involves computing effect sizes for a series of studies that investigated the same research question, and taking an average of those effect sizes. At a less simple level, we don't use a conventional average, we use what's known as a weighted average: in a meta-analysis each effect size is weighted by its precision (i.e., how good an estimate of the population it is) before the average is computed. By doing this, large studies, which will yield effect sizes that are more likely to closely approximate the population, are given more 'weight' than smaller studies, which should have

yielded imprecise effect size estimates. Because the aim of meta-analysis is not to look at *p*-values and assess 'significance', it overcomes the same problems of NHST that we discussed for effect sizes.

Of course, it has its own problems, but let's not get into those because meta-analysis is not easily done in IBM SPSS Statistics. If you're interested, I have written some fairly accessible tutorials on doing a meta-analysis using SPSS (Field & Gillett, 2010) and also using a free software package called R (Field, 2012). There are also numerous good books and articles on meta-analysis that will get you started (e.g., Cooper, 2010; Field, 2001, 2003, 2005b, 2005c; Hedges, 1992; Hunter & Schmidt, 2004; Lakens, Hilgard, & Staaks, 2016).

3.8 Bayesian approaches ▮▮▮

The final alternative to NHST is based on a different philosophy on analyzing data called **Bayesian statistics**. Bayesian statistics is a topic for an entire textbook (I particularly recommend these ones: Kruschke, 2014; McElreath, 2016), so we'll explore only the key concepts. Bayesian statistics is about using the data you collect to update your beliefs about a model parameter or a hypothesis. In some senses, NHST is also about updating your beliefs, but Bayesian approaches model the process explicitly.

To illustrate this idea, imagine you've got a crush on someone in your class. This is in no way autobiographical, but imagine

Cramming Sam's Tips
Effect sizes and meta-analysis

- An effect size is a way of measuring the size of an observed effect, usually relative to the background error.
- Cohen's *d* is the difference between two means divided by the standard deviation of the mean of the control group or by a pooled estimate based on the standard deviations of both groups.
- Pearson's correlation coefficient, *r*, is a versatile effect size measure that can quantify the strength (and direction) of relationship between two continuous variables, and can also quantify the difference between groups along a continuous variable. It ranges from −1 (a perfect negative relationship) through 0 (no relationship at all) to +1 (a perfect positive relationship).
- The odds ratio is the ratio of the *odds* of an event occurring in one category compared to another. An odds ratio of 1 indicates that the *odds* of a particular outcome are equal in both categories.
- Estimating the size of an effect in the population by combining effect sizes from different studies that test the same hypothesis is called meta-analysis.

you're super nerdy and love stats and heavy metal music, and this other person seems way cooler than that. They hang around with the 'in crowd' and you haven't noticed any tendencies for them to wear Iron Maiden T-shirts, not even in a 'fashion statement' kind of way. This person never notices you. What is your belief that this person has a romantic interest in you? It's probably low, maybe a 10% chance.

A couple of days later you're in a class and out of the corner of your eye you notice your crush looking at you. Naturally, you avoid eye contact, but in your peripheral vision you keep spotting them looking at you. Towards the end of class, your curiosity gets the better of you and, certain that they're looking somewhere else, you turn to them. To your horror, they're looking directly back at you and your eyes meet.

They smile at you. It's a sweet, friendly smile that melts your heart a little bit.

You have some new data about your situation—how will it affect your original belief? The data contained lots of positive signs, so maybe you now think there's a 30% chance that they like you. Your belief after inspecting the data is different from your belief before you looked at it: you have updated your beliefs based on new information. That's a sensible way to live your life and is the essence of Bayesian statistics.

3.8.1 Bayesian statistics and NHST

There are important differences between Bayesian statistics and the classical methods that this book (and SPSS) focuses on. NHST evaluates the probability of getting a test statistic at least as large as the one you have, given the null hypothesis. In doing so, you quantify the probability of the data obtained given a hypothesis is true:[11] $p(\text{data}|\text{hypothesis})$. Specifically, you ask what is the probability of my test statistic (data) or one bigger, given that the *null* hypothesis is true, $p(\text{test statistic}|\text{null hypothesis})$. This isn't a test of the null hypothesis, it's a test of the data given the null hypothesis is true. To test the null hypothesis, you need to answer the question 'What is the probability of the hypothesis given the data we have collected, $p(\text{hypothesis}|\text{data})$?' In the case of the null, we want the probability of the null hypothesis given the observed test statistic, $p(\text{null hypothesis}|\text{test statistic})$.

---

11 I'm simplifying the situation because, in reality, NHST asks about the data *and more extreme data*.

Table 3.2 Contingency table of whether people are humans or green lizard aliens based on whether their DNA matches a sample of alien DNA that the government has

		DNA sample matches alien DNA		
		Match	No match	Total
Accused	Green lizard alien	1	0	1
	Human	99	1900	1999
	Total	**100**	**1900**	**2000**

A simple example will illustrate that p(data|hypothesis) is not the same as p(null hypothesis|data). The probability that you are a professional actor or actress given that you have appeared in a blockbuster movie, p(actor or actress|been in a blockbuster), is very high because I suspect nearly everyone appearing in a blockbuster movie is a professional actor or actress. However, the inverse probability that you have been in a blockbuster movie given that you are a professional actor or actress, p(been in a blockbuster|actor or actress), is very small because the vast majority of actors and actresses do not end up in a blockbuster movie. Let's look at this distinction in more detail.

As you're probably aware, a significant proportion of people are scaly green lizard aliens disguised as humans. They live happily and peacefully among us, causing no harm to anyone and contributing lots to society by educating humans through things like statistics textbooks. I've said too much. The way the world is going right now, it's only a matter of time before people start becoming intolerant of the helpful lizard aliens and try to eject them from their countries. Imagine you have been accused of being a green lizard alien. The government has a hypothesis that you are an alien. They sample your DNA and compare it to an alien DNA sample that they have. It turns out that your DNA matches the alien DNA.

Table 3.2 illustrates your predicament. To keep things simple, imagine that you live on a small island of 2000 inhabitants. One of those people (but not you) *is* an alien, and his or her DNA will match that of an alien. It is not possible to be an alien and for your DNA *not* to match, so that cell of the table contains a zero. Now, the remaining 1999 people are not aliens and the majority (1900) have DNA that does not match that of the alien sample that the government has; however, a small number do (99), including you.

We'll start by look at the probability of the data given the hypothesis. This is the probability that the DNA matches, given that the person is, in fact, an alien, p(match|alien). The conditional probability of a DNA match given the person is an alien would be the probability of being an alien *and* having a DNA match, p(alien ∩ match), divided by the probability of being an alien, p(alien). There was one person from the population of 2000 who was an alien and had a DNA match, and there was also one person out of 2000 who was an alien, so the conditional probability is 1:

$$p(A \mid B) = \frac{p(B \cap A)}{p(B)}$$

$$p(\text{match} \mid \text{alien}) = \frac{p(\text{alien} \cap \text{match})}{p(\text{alien})}$$

$$= \frac{1/2000}{1/2000} = \frac{0.0005}{0.0005} = 1 \qquad (3.11)$$

Given the person is an alien, their DNA *must* match the alien DNA sample. Asking what the probability of the data (DNA match) is given the hypothesis is true (they are an alien) is a daft question: if you already know the person is an alien, then it is a certainty that their DNA will match the alien sample. Also, if you

already knew they were an alien then you would not need to bother collecting data (DNA). Calculating the probability of the data, given that the hypothesis is true, does not tell you anything useful about that hypothesis because it is conditional on the assumption that the hypothesis is true, which it might not be. Draw your own conclusions about NHST, which is based on this logic.

If you were a government employee tasked with detecting green lizards pretending to be humans, the question that matters is: given the data (the fact that your DNA matches), what is the probability of the theory that you *are* an alien? The conditional probability of a person being an alien given that their DNA matches would be the probability of being an alien and having a DNA match, p(alien ∩ match), divided by the probability of a DNA sample matching the alien sample in general, p(match). There is one person in the population of 2000 who is an alien and has a DNA match, and there were 100 out of 2000 DNA samples that match that of an alien:

$$p(\text{alien} \mid \text{match}) = \frac{p(\text{match} \cap \text{alien})}{p(\text{match})}$$

$$= \frac{1/2000}{100/2000} = \frac{0.0005}{0.05} = 0.01 \qquad (3.12)$$

This illustrates the importance of asking the right question. If the government agent followed the logic of NHST, he or she would believe you were certainly a green lizard alien because the probability of you having a DNA match given you are an alien is 1. However, by turning the question around, they would realize that there is only a 0.01 or 1% probability that you are an alien given that your DNA matches. Bayesian statistics addresses a more useful question than NHST: What is the probability of your hypothesis being true, given the data collected?

3.8.2 Bayes' theorem

The conditional probabilities that we just discussed can be obtained using Bayes' theorem, which states that the conditional probability of two events can be obtained from their individual probabilities and the inverse conditional probability (Eq. 3.13). We can replace the letter A with our model or hypothesis, and replace letter B with the data collected to get a sense of how Bayes' theorem might be useful in testing hypotheses:

$$p(A|B) = \frac{p(B|A) \times p(A)}{p(B)}$$

$$p(\text{Model}|\text{Data}) = \frac{p(\text{Data}|\text{Model}) \times p(\text{Model})}{p(\text{Data})}$$

$$\text{Posterior probability} = \frac{\text{likelihood} \times \text{prior probability}}{\text{marginal likelihood}}$$

$$(3.13)$$

The terms in this equation have special names that we'll explore in detail now.

The **posterior probability** is our belief in a hypothesis (or parameter, but more on that later) after we have considered the data (hence it is *posterior* to considering the data). In the alien example, it is our belief that a person is an alien given that their DNA matches alien DNA, $p(\text{alien}|\text{match})$. This is the value that we are interested in finding out: the probability of our hypothesis given the data.

The **prior probability** is our belief in a hypothesis (or parameter) before considering the data. In our alien example, it is the government's belief in your guilt before they consider whether your DNA matches or not. This would be the base rate for aliens, $p(\text{aliens})$, which in our example is 1 in 2000 or 0.0005.

The **marginal likelihood** or *evidence*, is the probability of the observed data, which in this example is the probability of matching DNA, $p(\text{match})$. The data show that there were 100 matches in 2000 cases, so this value is 100/2000 or 0.05. The **likelihood** is the probability that the observed data could be produced given the hypothesis or model being considered. In the alien example, it is, therefore, the probability that you would find that the DNA matched given that someone was in

fact an alien, $p(\text{match}|\text{alien})$, which we saw before is 1.

If we put this together, our belief in you being a lizard alien given that your DNA matches that of a lizard alien is a function of how likely a match is if you *were* an alien ($p = 1$), our prior belief in you being an alien ($p = 0.0005$) and the probability of getting a DNA match (0.05). Our belief, having considered the data, is 0.01: there is a 1% chance that you are an alien:

$$p(\text{alien}|\text{match}) = \frac{p(\text{match}|\text{alien}) \; p(\text{alien})}{p(\text{match})}$$

$$= \frac{1 \cdot 0.0005}{0.05} = 0.01 \qquad (3.14)$$

This is the same value that we calculated in equation (3.12), all of which shows that Bayes' theorem is another route to obtaining the posterior probability.

Note that our prior belief in you being an alien (before we knew the DNA results) was 1 in 2000 or a probability of 0.0005. Having examined the data, that your DNA matched an alien, our belief increases to 1 in 100 or 0.01. After knowing that your DNA matched an alien, we are more convinced that you are an alien, than before we knew, which is as it should be; however, this is a far cry from being *certain* that you are an alien because your DNA matched.

3.8.3 Priors on parameters ▮▮▮▯

In the alien example our belief was a hypothesis (you are an alien). We can also use Bayesian logic to update beliefs about parameter values (e.g., we can produce a Bayesian estimate of a b-value in a linear model by updating our prior beliefs in the value of that b using the data we collect).

Let's return to the example of you having a crush on someone in your class. One day, before you notice them smiling at you, you're lamenting this unrequited love to a friend and, being a stats nerd too, they ask you to estimate how much your crush likes you from 0 (hates you) to 10 (wants to marry you). You reply, '1.' Self-esteem was

never your strong point. They then ask, 'Is it possible that it's a 5?' You don't think it is. What about a 4? No chance, you think. You turn your nose up at the possibility of a 3, but concede that there's an outside chance that it could be a 2, and you're sure it won't be a 0 because your crush doesn't seem like the hateful type. Through this process, you've established that you think that your crush will like you somewhere between 0 and 2 on the scale, but you're most confident of a score of 1. As in the alien example, you can represent this belief *prior* to examining more data, but as a distribution, not a single probability value.

Immediately after the class in which you notice your crush smiling at you, you rush to your friend's room to report on this exciting development. Your friend asks you again to estimate where your crush's feelings lie on the 0–10 scale. You reply, '2.' You're less confident than before that it is a 1, and you're prepared to entertain the possibility that it could be 3. Your beliefs after you have examine the data are represented by a posterior probability *distribution*.

Unlike a belief in a hypothesis where the prior probability is a single value, when estimating a parameter, the prior probability is a *distribution* of possibilities. Figure 3.5 shows two examples of a **prior distribution**. The top one mirrors our example and is said to be an **informative prior distribution**. We know from what we have learnt about probability distributions that they represent the plausibility of values: the peak is the value with the highest probability (the most plausible value) and the tails represent values with low probability. The top distribution is centered on 1, the value that you were most confident about when your friend first asked you to estimate your crush's feelings. This is the value you think is the most probable. However, the fact that there is a curve around this point shows that you are prepared to entertain other values with varying degrees of certainty. For example, you believe that values above 1 (the peak) are decreasingly likely until by the time the value is 2 your belief is close to 0: you think that it is impossible that your crush would like you *more than* 2 on the scale. The same is true if we look at values below 1: you think that values below the peak are decreasingly likely until we reach a rating of 0 (they hate you), at which point the probability of this belief is 0. Basically, you think that it is impossible that they hate you. Your self-esteem is low, but not *that* low. To sum up, you feel strongest that your crush likes you 1 on the scale, but you're prepared to accept that it could be any value between 0 and 2; as you approach those extremes your beliefs become weaker and weaker. This distribution is informative because it narrows down your beliefs: we know, for example, that you are not prepared to believe values above 2. We also know from the shape of the distribution that you are most confident of a value between about 0.5 and 1.5.

Figure 3.5 shows another type of prior distribution known as an **uninformative prior distribution**. It is a flat line, which means that you're prepared to believe all possible outcomes with equal probability. In our example, this means you're equally prepared to believe that your crush hates you (0) as you are that they love you (10) or sort of like you (5) or any value in

between. Basically, you don't have a clue and you're prepared to believe anything. It's kind of a nice place to be when it comes to unrequited love. This prior distribution is called uninformative because the distribution doesn't tell us anything useful about your beliefs: it doesn't narrow them down.

Next you examine some data. In our example, you observe the person on whom you have a crush. These data are then mashed together (using Bayes' theorem) with your prior distribution to create a new distribution: the **posterior distribution**. Basically, a toad called the reverend Toadmas Lillipad-Bayes sticks out his big sticky tongue and grabs your prior distribution and pops it in his mouth, he then lashes the tongue out again and grabs the observed data. He swirls the numbers around in his mouth, does that funny expanding throat thing that toads sometimes do, before belching out the posterior distribution. I think that's how it works.

Figure 3.5 shows the resulting posterior. It differs from the prior distribution. How it differs depends on both the original prior and the data. So, in Figure 3.5, if the data were the same you'd get different posteriors from the two prior distributions (because we're putting different initial beliefs into the toad's mouth). In general, if the prior is uninformative then the posterior will be

heavily influenced by the data. This is sensible because if your prior belief is very open-ended then the data have a lot of scope to shape your posterior beliefs. If you start off not knowing what to believe, then your posterior beliefs should reflect the data. If the prior is strongly informative (i.e., you have very constrained beliefs), then the data will influence the posterior less than for an uninformative prior. The data are working against the prior. If the data are consistent with your already narrow beliefs, then the effect would be to narrow your belief further.

By comparing the prior and posterior distributions in Figure 3.5 we can speculate about what the data might have looked like. For the informative prior, the data were probably quite a bit more positive than your initial beliefs (just like our story in which you observed positive signs from your crush) because your posterior belief is more positive than the prior: after inspecting the data, you're most willing to believe that your crush would rate you as a 2 (rather than your prior belief of 1). Because your prior belief was narrow (your beliefs were very constrained) the data haven't managed to pull your belief too far away from its initial state. The data have, however, made your range of beliefs wider than it was: you're now prepared to believe values in a range of about 0.5 to 3.5 (the distribution has got wider). In the case of the uninformative prior, the data that would yield the posterior in Figure 3.5 would be data distributed like the posterior: in other words, the posterior looks like the data. The posterior is more influenced by the data than the uninformative prior.

We can use the posterior distribution to quantify the plausible values of a parameter (whether that be a value such as your belief in what your crush thinks about you or a *b*-value in a linear model). If we want a point estimate (a single value) of the parameter then we can use the value where the probability distribution peaks. In other words, use the value that we think is most

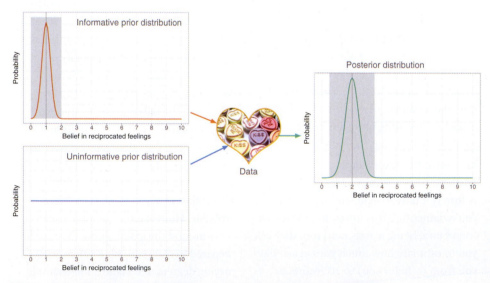

Figure 3.5   The process of updating your beliefs with Bayesian statistics

probable. If we want an interval estimate then we can use the values of the estimate that enclose a percentage of the posterior distribution. The interval is known as a **credible interval**, which is the limits between which (usually) 95% of values fall in the posterior distribution fall. Unlike a confidence interval, which for a given sample may or may not contain the true value, a credible interval can be turned into a probability statement such as 95% probability that the parameter of interest falls within the limits of the interval.

3.8.4 Bayes factors

We'll now look at how we can use Bayes' theorem to compare two competing hypotheses. We'll return to the alien example (Section 3.8.1). In NHST terms,

SELF TEST
Use Table 3.2 and Bayes' theorem to calculate $p(human | match)$.

the alternative hypothesis equates to something happening. In this case, it would be that you are an alien. We know that the probability of you being an alien is 0.01. In NHST, we also have a null hypothesis that reflects no effect or that nothing happened. The equivalent here is the hypothesis that you are human, given the DNA match, $p(human | match)$. What would this probability be?'

First, we need to know the probability of a DNA match, given the person was human, $p(match | human)$. There were 1999 humans and 99 had a DNA match, so this

probability would be 99/1999 or 0.0495. Next we need to know the probability of being human, $p(human)$. There were 1999 out of 2000 people who are human (Table 3.2), so the probability would be 1999/2000 or 0.9995. Finally, we need the probability of a DNA match, $p(match)$, which we worked out before as 0.05. Putting these values into Bayes' theorem, we get a probability of 0.99:

$$p(human | \text{match}) = \frac{p(\text{match} | \text{human})\ p(\text{human})}{p(\text{match})}$$

$$= \frac{0.0495\ 0.9995}{0.05} = 0.99 \quad (3.15)$$

Cramming Sam's Tips
Summary of the Bayesian process

1 Define a prior that represents your subjective beliefs about a hypothesis (the prior is a single value) or a parameter (the prior is a distribution of possibilities). The prior can range from completely uninformative, which means that you are prepared to believe pretty much anything, to strongly informative, which means that your initial beliefs are quite narrow and specific.

2 Inspect the relevant data. In our frivolous example, this was observing the behavior of your crush. In science, the process would be a bit more formal than that.

3 Bayes' theorem is used to update the prior distribution with the data. The result is a posterior probability, which can be a single value representing your new belief in a hypothesis or a distribution that represents your beliefs in plausible values of a parameter, after seeing the data.

4 A posterior distribution can be used to obtain a point estimate (perhaps the peak of the distribution) or an interval estimate (a boundary containing a certain percentage, for example 95%, of the posterior distribution) of the parameter in which you were originally interested.

We know, then, that the probability of you being an alien given the data is 0.01 or 1%, but the probability of you being human given the data is 0.99 or 99%. We can evaluate the evidence of whether you are alien or human (given the data we have) using these values. In this case, you are far more likely to be human than you are to be an alien.

We know the posterior probability of being human given a DNA match is 0.99, and we previously worked out that the posterior probability of being alien given a DNA match is 0.01. We can compare these values by computing their ratio, which is known as the **posterior odds.** In this example, the value is 99, which means that you are 99 times more likely to be human than alien despite your DNA matching the alien sample:

$$\text{posterior odds} = \frac{p(\text{hypothesis 1}|\text{data})}{p(\text{hypothesis 2}|\text{data})}$$
$$= \frac{p(\text{human}|\text{match})}{p(\text{alien}|\text{match})} = \frac{0.99}{0.01} = 99 \quad (3.16)$$

This example of the posterior odds shows that we can use Bayes' theorem to compare two hypotheses. NHST compares a hypothesis about there being an effect (the alternative hypothesis) with one that the effect is zero (the null hypothesis). We can use Bayes' theorem to compute the probability of both hypotheses given the data, which means we can compute a posterior odds of the alternative hypothesis relative to the null. This value quantifies how much more likely the alternative hypothesis is given the data relative to null (given the data).

Equation (3.17) shows how this can be done:

$$\frac{p(\text{alternative}|\text{data})}{p(\text{null}|\text{data})}$$
$$= \frac{p(\text{data}|\text{alternative}) \times p(\text{alternative})/p(\text{data})}{p(\text{data}|\text{null}) \times p(\text{null})/p(\text{data})}$$
$$= \frac{p(\text{data}|\text{alternative})}{p(\text{data}|\text{null})} \times \frac{p(\text{alternative})}{p(\text{null})}$$
$$\text{posterior odds} = \text{Bayes factor} \times \text{prior odds} \quad (3.17)$$

The first two lines show Bayes' theorem for the alternative hypothesis divided by Bayes' theorem for the null. Because both incorporate the probability of the data, $p(\text{data})$, which is the marginal likelihood for both the alternative and null hypotheses, these terms drop out in the third line. The final line shows some special names that are given to the three components of the equation. On the left, we have the posterior odds, which we have just explained. On the right, we have the **prior odds**, which compare the probability of the alternative hypothesis to the null *before* you look at the data.

We have seen that priors reflect subjective beliefs before looking at the data, and the priors odds are no different. A value of 1 would reflect a belief that the null and alternative hypotheses are equally likely. This value might be appropriate if you were testing a completely novel hypothesis. Usually, though, our hypothesis would be based on past research and we would have a stronger belief in the alternative hypothesis than the null, in which case you might want your prior odds to be greater than 1.

The final term in equation (3.17) is the **Bayes factor**. It represents the degree to which our beliefs change because of looking at the data (i.e., the change in our prior odds). A Bayes factor less than 1 supports the null hypothesis because it suggests that our beliefs in the alternative (relative to the null) have weakened. Looking at what the Bayes factor is (Eq. 3.17), it should be clear that a value less than 1 also means that the probability of the data given the null is higher than the probability of the data given the alternative hypothesis. Therefore, it makes sense that your beliefs in the alternative (relative to the null) should weaken. For the opposite reason, a Bayes factor greater than 1 suggests that the observed data are more likely given the alternative hypothesis than the null and your belief in the alternative (relative to the null) strengthens. A value of exactly 1 means that the data are equally likely under the alternative and null hypotheses and your prior beliefs are unchanged by the data.

For example, a Bayes factor of 10 means that the observed data are ten times more likely under the alternative hypothesis than the null hypothesis. Like effect sizes, some benchmarks have been suggested, but need to be used with extreme caution: values between 1 and 3 are considered evidence for the alternative hypothesis that is 'barely worth mentioning', values between 3 and 10 are considered 'evidence that has substance' for the alternative hypothesis, and values greater than 10 are strong evidence for the alternative hypothesis (Jeffreys, 1961).

3.8.5 Benefits of Bayesian approaches

We've seen already that Bayesian hypothesis testing asks a more sensible question than NHST, but it's worth reflecting on how it overcomes the problems associated with NHST.

First of all, we looked at various misconceptions around what conclusions you can draw from NHST about the null and alternative hypothesis. Bayesian approaches specifically evaluate the evidence for the null hypothesis so, unlike NHST, you *can* draw conclusions about the likelihood that the null hypothesis is true.

Second, p-values are confounded by the sample size and the stopping rules applied to data collection. Sample size is not an issue for Bayesian analysis. Theoretically you can update your beliefs based on one new data point. That single data point might not have much influence over a strong prior belief, but it could do. Imagine that you think my lizard alien example is ridiculous. You are utterly convinced that no green lizard alien people live among us. You have extensive data from every day of your life on which your belief is based. One day during a statistics lecture, you notice your professor licking her eyeball with a strangely

serpentine tongue. After the lecture, you follow her, and when she thinks no one can see she peels her face off to reveal scaly green cheeks. This one new observation might substantially change your beliefs. However, perhaps you can explain this away as a prank. You turn to run across campus to tell your friends, and as you do, you face a crowd of hundreds of students all peeling away their faces to reveal their inner space lizard. After soiling your pants, my guess is that your prior belief would change substantially. My point is that the more data you observe, the stronger the effect it can have on updating your beliefs, but that's the only role that sample size plays in Bayesian statistics: bigger samples provide more information.

This point relates to the intentions of the researcher too. Your prior belief can be updated based on any amount of new information, therefore you do not need to determine how much data to collect before the analysis. It does not matter when you stop collecting data or what your rule is for collecting data, because any new data relevant to your hypothesis can be used to update your prior beliefs in the null and alternative hypothesis. Finally, Bayesian analysis is focussed on estimating parameter values (which quantify effects) or evaluating relative evidence for the alternative hypothesis (Bayes factors), and so there is no black-and-white thinking involved, only estimation and interpretation. Thus, behavior such as p-hacking is circumvented. Pretty cool, eh? I think Chicago put it best in that cheesy song, 'I'm addicted to you Bayes, you're a hard habit to break'.

There are downsides too. The major objection to Bayesian statistics is the reliance on a prior, which is a subjective decision and open to researcher degrees of freedom. You can imagine a scenario where someone tweaks the prior to get a bigger Bayes factor or a parameter value that they like the look of. Personally, I see priors as a strength because you can't report Bayesian models without explaining your decisions about the prior. In doing so readers are fully informed of what researcher degrees of freedom you might have employed.

Cramming Sam's Tips
Bayes factors

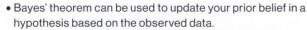

- Bayes' theorem can be used to update your prior belief in a hypothesis based on the observed data.
- The probability of the alternative hypothesis given the data relative to the probability of the null hypothesis given the data is quantified by the *posterior odds*.
- A *Bayes factor* is the ratio of the probability of the data given the alternative hypothesis to that for the null hypothesis. A

Bayes factor greater than 1 suggests that the observed data are more likely given the alternative hypothesis than given the null. Values less than 1 suggest the opposite. Values between 1 and 3 reflect evidence for the alternative hypothesis that is 'barely worth mentioning', values between 1 and 3 is evidence that 'has substance', and values between 3 and 10 are 'strong' evidence (Jeffreys, 1961).

3.9 Reporting effect sizes and Bayes factors

If you're going to use NHST, then, because *p*-values depend on things like the sample size, it is highly advisable to report effect sizes, which quantify the size of the observed effect, as well as *p*-values. Using the examples from the previous chapter (Section 2.10), we could report significance tests like this (note the presence of exact *p*-values and effect sizes):[12]

✓ Fear reactions were significantly higher when Andy's cat Fuzzy wore a fake human tongue compared to when he did not, $p = 0.023$, $d = 0.54$.
✓ The number of cats intruding into the garden was significantly less when Fuzzy wore a fake human tongue compared to when he did not, $p = 0.007$, $d = 0.76$.
✓ Fear reactions were not significantly different when Fuzzy wore a David Beckham mask compared to when he did not, $p = 0.18$, $d = 0.22$.

You could also report Bayes factors, which are denoted by BF_{10} in the form that I have described them (i.e., when a value greater than 1 is evidence for the alternative hypothesis). Alternatively, the Bayes factor can be expressed as the reciprocal of the value I have described, which reverses the interpretation (i.e., a value greater than 1 is evidence for the null hypothesis) and this form is denoted as BF_{01}. Currently, because most journals expect to find NHST, it is not that common to see papers that report purely Bayesian statistics (although the number is increasing), and instead scientists augment their *p*-values with a Bayes factor. As you will see, Bayes factors are computed with an estimate of error (a percentage) so report this value as well.

✓ The number of cats intruding into the garden was significantly less when Fuzzy wore a fake human tongue compared to

when he did not, $p = 0.007$, $d = 0.76$, $BF_{10} = 5.67 \pm 0.02\%$.

3.10 Brian and Jane's Story

Brian had been ruminating about his last encounter with Jane. It hadn't gone well. He had a plan, though. His lecturer had casually mentioned some problems with significance testing but glossed over them

the way lecturers do when they don't understand something. Brian decided to go to the library, where he read up on this thing called Bayesian statistics. 'If my lecturer doesn't understand the limitations of NHST, let alone Bayesian estimation,' Brian thought, 'Jane has got to be impressed with what I've been reading'. Brian noted everything down. A lot of it was too complex, but if he could get across some buzzwords, then maybe, just maybe, Jane would say more than a few words to him.

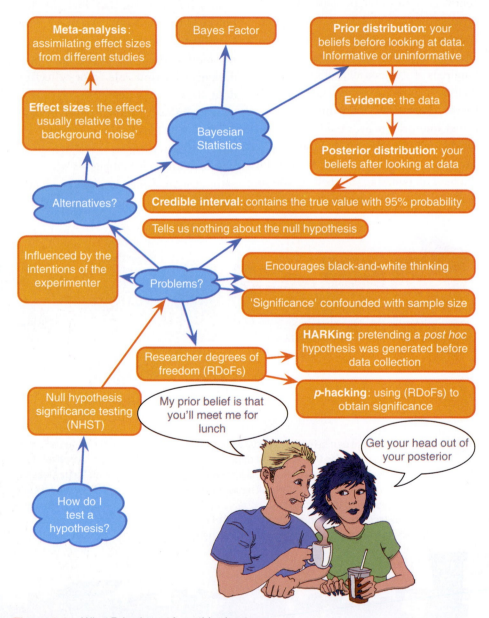

Figure 3.6 What Brian learnt from this chapter

12 If you use APA format, you need to drop the zero before the decimal point, so report $p = .023$, not $p = 0.023$.

100

Brian waited on the library steps to see whether he could spot Jane in the campus square. Sure enough, he saw her figure in the distance. He took one final look at his notes, and as she walked past him he said, 'Hey, Jane. 'Sup?'

Jane stopped. She looked down at the guy she'd bumped into last week. Why wouldn't he leave her alone? All she ever wanted was for people to just leave her alone. She smiled a fake smile.

'Everyone say's you're really smart,' Brian continued. 'Maybe you can help me with some stuff I've been reading?'

Notes shaking in his hand, Brian told Jane everything he could about significance testing, effect sizes, and Bayesian statistics.

3.11 What next? ▮▮▮▮

We began this chapter by seeing how an early belief that I couldn't live up to my brother's vast intellect panned out. We skipped forward in my life story, so that I could crowbar a photo of my children into the book. It was a pleasant diversion, but now we must return to my own childhood. At age 3, I wasn't that concerned about the crippling weight of low expectation from my family. I had more pressing issues, namely talk of me leaving nursery (note: not 'being thrown out'). Despite the 'incidents', nursery was a safe and nurturing place to be. I can't honestly remember how I felt about leaving nursery, but given how massively neurotic I am, it's hard to believe that I was anything other than anxious. I had friends at nursery, and whatever place I was going to was, no doubt, full of other children. Children I didn't know, scary children, children who would want to steal my toys, children who were best avoided because they were new and unfamiliar. At some point in our lives, though, we must leave the safety of a familiar place and try out new things: toddlers must find new pastures in which to wave their penis. The new pasture into which I headed was primary school (or 'elementary school' as I believe it's called in the USA). This was a scary new environment, a bit like SPSS might be for you. So, we'll hold hands and face it together.

3.12 Key terms that I've discovered

Bayes factor
Bayesian statistics
Cohen's d
Contingency table
Credible interval
Effect size
Empirical probability
HARKing

Informative prior distribution
Likelihood
Marginal likelihood
Meta-analysis
Odds
Odds ratio
Open science
p-curve

p-hacking
Peer Reviewers' Openness Initiative
Posterior distribution
Posterior odds
Posterior probability
Pre-registration
Prior distribution

Prior odds
Prior probability
Publication bias
Registered reports
Researcher degrees of freedom
Tests of excess success
Uninformative prior distribution

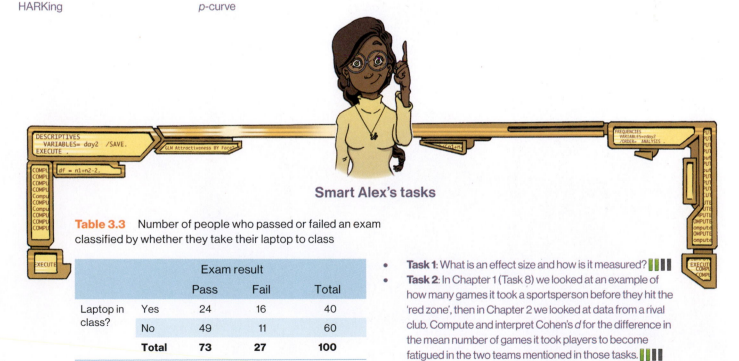

Smart Alex's tasks

Table 3.3 Number of people who passed or failed an exam classified by whether they take their laptop to class

		Exam result		
		Pass	Fail	Total
Laptop in class?	Yes	24	16	40
	No	49	11	60
	Total	**73**	**27**	**100**

- **Task 1**: What is an effect size and how is it measured? ▮▮▮
- **Task 2**: In Chapter 1 (Task 8) we looked at an example of how many games it took a sportsperson before they hit the 'red zone', then in Chapter 2 we looked at data from a rival club. Compute and interpret Cohen's d for the difference in the mean number of games it took players to become fatigued in the two teams mentioned in those tasks. ▮▮▮▮

- **Task 3**: Calculate and interpret Cohen's *d* for the difference in the mean duration of the celebrity marriages in Chapter 1 (Task 9) and mine and my friends' marriages in Chapter 2 (Task 13). ▌▌▌

- **Task 4**: What are the problems with null hypothesis significance testing? ▌▌▌

- **Task 5**: What is the difference between a confidence interval and a credible interval? ▌▌▌

- **Task 6**: What is meta-analysis? ▌▌▌

- **Task 7**: Describe what you understand by the term 'Bayes factor'? ▌▌▌

- **Task 8**: Various studies have shown that students who use laptops in class often do worse on their modules (Payne-Carter, Greenberg, & Waller, 2016; Sana, Weston, & Cepeda, 2013). Table 3.3 shows some fabricated data that mimic what has been found. What is the odds ratio for passing the exam if the student uses a laptop in class compared to if they don't? ▌▌▌

- **Task 9**: From the data in Table 3.3, what is the conditional probability that someone used a laptop in class, given that that they passed the exam, *p*(laptop|pass). What is the conditional probability that someone didn't use a laptop in class, given that they passed the exam, *p*(no laptop |pass)? ▌▌▌

- **Task 10**: Using the data in Table 3.3, what are the posterior odds of someone using a laptop in class (compared to not using one), given that they passed the exam? ▌▌▌

Answers & additional resources are available on the book's website at

https://edge.sagepub.com/field5e

THE IBM SPSS
STATISTICS ENVIRONMENT

4

4.1 What will this chapter tell me?

At about 5 years old I moved from nursery to primary school. Even though my older brother (you know, Paul, 'the clever one') was already there, I was really apprehensive on my first day. My nursery school friends were all going to different schools and I was terrified about meeting new children. I arrived in my classroom, and as I'd feared, it was full of scary children. In a fairly transparent ploy to make me think that I'd be spending the next 6 years building sand castles, the teacher told me to play in the sandpit. While I was nervously trying to discover whether I could build a pile of sand high enough to bury my head in it, a boy came to join me. His name was Jonathan Land, and he was really nice. Within an hour, he was my new best friend (5-year-olds are fickle . . .) and I loved school. We remained close friends all through primary school. Sometimes, new environments seem scarier than they really are. This chapter introduces you to what might seem like a scary new environment: IBM SPSS Statistics. I won't lie, the SPSS environment is a more unpleasant environment in which to spend time than a sandpit, but try getting a plastic digger to do a least squares regression for you. For the purpose of this chapter, I intend to be a 5-year-old called Jonathan. Thinking like a 5-year-old comes quite naturally to me, so it should be fine. I will hold your hand, and show you how to use the diggers, excavators, grabbers, cranes, front loaders, telescopic handlers, and tractors[1] in the sandpit of IBM SPSS Statistics. In short, we're going to learn the tools of IBM SPSS Statistics, which will enable us, over subsequent chapters, to build a magical sand palace of statistics. Or thrust our faces into our computer monitor. Time will tell.

1 Yes, I have been spending a lot of time with a vehicle-obsessed 2-year-old boy recently.

Figure 4.1 All I want for Christmas is ... some tasteful wallpaper

4.2 Versions of IBM SPSS Statistics

This book is based primarily on version 25 of IBM SPSS Statistics (I generally call it SPSS for short). IBM regularly improves and updates SPSS, but this book covers only a small proportion of the functionality of SPSS, and focuses on tools that have been in the software a long time and work well. Consequently, improvements made in new versions of SPSS Statistics are unlikely to impact the contents of this book. With a bit of common sense, you can get by with a book that doesn't explicitly cover the latest version (or the version you're using). So, although this edition was written using version 25, it will happily cater for earlier versions (certainly back to version 18), and most likely for versions 26 onwards (unless IBM does a major overhaul just to keep me on my toes).

IBM SPSS Statistics comes in four flavors:[2]

- **Base**: Most of the functionality covered in this book is in the base package. The exceptions are exact tests and bootstrapping, which are available only in the *premium* edition.
- **Standard**: This has everything in the base package but also covers generalized linear models (which we don't get into in this book).
- **Professional**: This has everything in the standard edition, but with missing value imputation and decision trees and forecasting (again, not covered in this text).
- **Premium**: This has everything in the professional package but also exact tests and bootstrapping (which we cover in this book), and structural equation modeling and complex sampling (which we don't cover).

There is also a subscription model where you can buy monthly access to a base package (as described above but also including bootstrapping) and, for an extra fee, add-ons for:

- **Custom tables and advanced statistics users**: is similar to the standard package above in that it adds generalized linear models. It also includes logistic regression, survival analysis, Bayesian analysis and more customization of tables.
- **Complex sampling and testing users**: adds functionality for missing data and complex sampling as well as categorical principal components analysis, multidimensional scaling, and correspondence analysis.
- **Forecasting and decision trees users**: as the name suggests, this adds functionality for forecasting and decision trees as well as neural network predictive models.

If you are subscribing, then most of the contents of this book appear in the base subscription package, with a few things (e.g., Bayesian statistics and logistic regression) requiring the advanced statistics add on.

4.3 Windows, Mac OS and Linux

SPSS Statistics works on and Windows, Mac OS and Linux (and Unix-based operating systems such as IBM AIX, HP-UX, and Solaris). SPSS Statistics is built on a program called Java, which means that the Windows, Mac OS and Linux versions differ very little (if at all). They look a bit different, but only in the way that, say, Mac OS looks different from Windows anyway.[3]

Figure 4.2   The start-up window of IBM SPSS

---

2 You can look at a detailed comparison here: https://www.ibm.com/marketplace/spss-statistics/purchase
3 You can get the Mac OS version to display itself like the Windows version, but I have no idea why you'd want to do that.

I have taken the screenshots from Windows because that's the operating system that most readers will use, but you can use this book if you have a Mac (or Linux). In fact, I wrote this book using a Mac.

4.4 Getting started ▌▌▌▌

SPSS mainly uses two windows: the **data editor** (this is where you input your data and carry out statistical functions) and the **viewer** (this is where the results of any analysis appear). You can also activate the **syntax editor** window (see Section 4.10), which is for entering text commands (rather than using dialog boxes). Most beginners ignore the syntax window and click merrily away with their mouse, but using syntax does open up additional functions and can save time in the long run. Strange people who enjoy statistics can find numerous uses for syntax and dribble excitedly when discussing it. At times I'll force you to use syntax, but only because I wish to drown in my own saliva. When SPSS loads, the start-up window in Figure 4.2 appears. At the top left is a box

labeled *New Files*, where you can select to open an empty data editor window or begin a database query (something not covered in this book). Underneath, in the box labeled *Recent Files*, there will appear a list of any SPSS data files (on the current computer) on which you've recently worked. If you want to open an existing file, select it from the list and then click 〔 Open 〕. If you want to open a file that isn't in the list, select 🗁 Open another file... and click 〔 Open 〕 to open a window for browsing to the file you want (see Section 4.12). The dialog box also has an overview of what's new in this release and contains links to tutorials and support, and a link to the online developer community. If you don't want this dialog to appear when SPSS starts up, then select ☑ Don't show this dialog in the future.

4.5 The data editor ▌▌▌▌

Unsurprisingly, the data editor window is where you enter and view data (Figure 4.3). At the top of this window (or the top of the screen on a Mac) is a menu

bar like ones you've probably seen in other programs. As I am sure you're aware, you can navigate menus by using your mouse/trackpad to move the on-screen arrow to the menu you want and pressing (*clicking*) the left mouse button once. The click will reveal a list of menu items in a list, which again you can click using the mouse. In SPSS if a menu item is followed by a ▶ then clicking on it will reveal another list of options (a *submenu*) to the right of that menu item; if it doesn't then clicking on it will activate a window known as a *dialog box*. Any window in which you have to provide information or a response (i.e., 'have a dialog' with the computer) is a dialog box. When referring to selecting items in a menu, I will use the menu item names connected by arrows to indicate moving down items or through submenus. For example, if I were to say that you should select the *Save As* ... option in the *File* menu, you will see *File* ▶ *Save As* ...

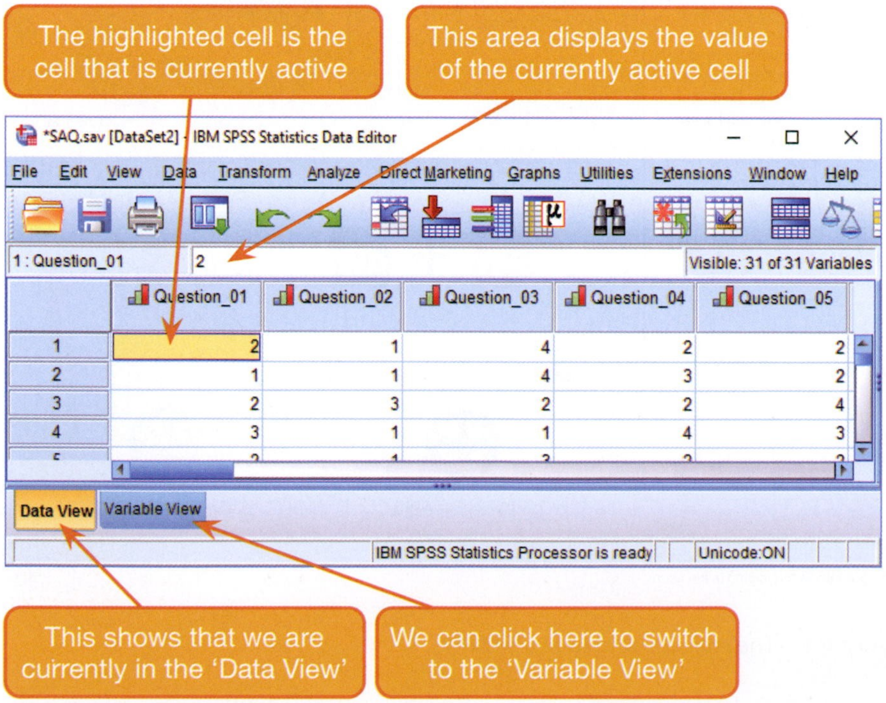

The data editor has a **data view** and a **variable view**. The data view is for entering data, and the variable view is for defining characteristics of the variables within the data editor. To switch between the views, select one of the tabs at the bottom of the data editor (〔 Data View 〕〔 Variable View 〕); the highlighted tab indicates which view you're in (although it's obvious). Let's look at some features of the data editor that are consistent in both views. First, the menus.

Some letters are underlined within menu items in Windows, which tells you the *keyboard shortcut* for accessing that item. With practice these shortcuts are faster than using the mouse. In Windows, menu items can be activated by simultaneously pressing *Alt* on the keyboard and the underlined letter. So, to access the *File* ▶ *Save As* ... menu item you would simultaneously press *Alt* and F on the keyboard to activate the *File* menu, then, keeping your finger on the *Alt* key, press A. In Mac OS, keyboard shortcuts are listed in the menus, for example, you can save a file by simultaneously pressing ⌘ and S (I denote these shortcuts as ⌘ + S). Below is a brief reference guide to each of the menus:

The highlighted cell is the cell that is currently active

This area displays the value of the currently active cell

This shows that we are currently in the 'Data View'

We can click here to switch to the 'Variable View'

Figure 4.3 The SPSS Data Editor

SPSS Tip 4.1
Save time and avoid RSI ▌▌▌▌

By default, when you go to open a file, SPSS looks in the directory in which it is stored, which is usually not where you store your data and output. So, you waste time navigating your computer trying to find your data. If you use SPSS as much as I do then this has two consequences: (1) all those seconds have added up to weeks navigating my computer when I could have been doing something useful like playing my drum kit; (2) I have increased my chances of getting RSI in my wrists, and if I'm going to get RSI in my wrists I can think of more enjoyable ways to achieve it than navigating my computer (drumming again, obviously). Luckily, we can avoid wrist death by using *Edit* ▶ 🗊 Optio*n*s… to

open the *Options* dialog box (Figure 4.4) and selecting the 'File Locations' tab.

In this dialog box we can select the folder in which SPSS will initially look for data files and other files. For example, I keep my data files in a single folder called, rather unimaginatively, 'Data'. In the dialog box in Figure 4.4 I have clicked on Browse… and then navigated to my data folder. SPSS will now use this as the default location when I open files, and my wrists are spared the indignity of RSI. You can also select the option for SPSS to use the *Last folder used*, in which case SPSS remembers where you were last time it was loaded and uses that folder as the default location when you open or save files.

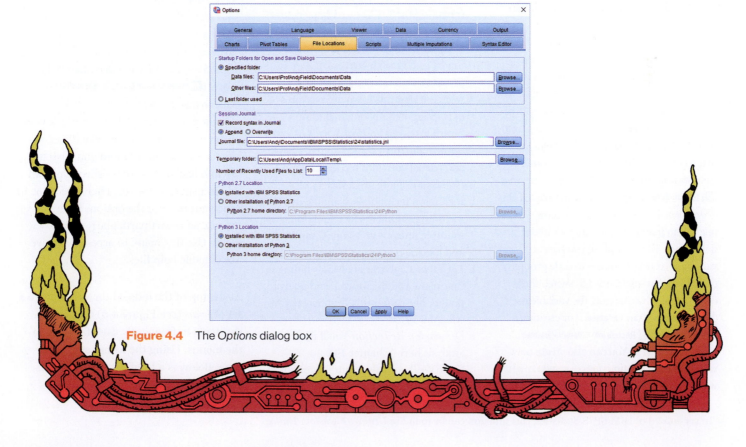

Figure 4.4   The *Options* dialog box

- *File* This menu contains all the options that you expect to find in *File* menus: you can save data, graphs or output, open previously saved files and print graphs, data or output.

- *Edit* This menu contains edit functions for the data editor. For example, it is possible to *cut* and *paste* blocks of numbers from one part of the data editor to another (which is handy when you realize that you've entered lots of numbers in the wrong place). You can insert a new variable into the data editor (i.e., add a column) using ⊞ Insert Variable, and add a new row of data between two existing rows using ⊞ Insert Cases. Other useful options for large data sets are the ability to skip to a particular row (⊞ Go to Case...) or column (⊞ Go to Variable...) in the data editor. Finally, although for most people the default preferences are fine, you can change them by selecting ⊞ Options....

- *View* This menu deals with system specifications such as whether you have grid lines on the data editor or whether you display value labels (exactly what value labels are will become clear later).

- *Data* This menu is all about manipulating the data in the data editor. Some of the functions we'll use are the ability to split the file (⊞ Split File...) by a grouping variable (see Section 6.10.4), to run analyses on only a selected sample of cases (⊞ Select Cases...), to weight cases by a variable (⊞ Weight Cases...) which is useful for frequency data (Chapter 19), and to convert the data from wide format to long or vice versa (⊞ Restructure...) which we'll use in Chapter 12.

- *Transform* This menu contains items relating to manipulating variables in the data editor. For example, if you have a variable that uses numbers to code groups of cases then you might want to switch these codes around by changing the variable itself (⊞ Recode into Same Variables...) or creating a new variable (⊞ Recode into Different Variables...); see SPSS Tip 11.2. You can also create new variables from existing ones (e.g., you might want a variable that is the sum of 10 existing

variables) using the *compute* function (⊞ Compute Variable...); see Section 6.12.6.

- *Analyze* The fun begins here, because the statistical procedures lurk in this menu. Below is a rundown of the bits of the statistics menu that we'll use in this book:

  - *Descriptive Statistics* We'll use this for conducting descriptive statistics (mean, mode, median, etc.), frequencies and general data exploration. We'll use *Crosstabs...* for exploring frequency data and performing tests such as chi-square, Fisher's exact test and Cohen's kappa (Chapter 19).

  - *Compare Means* We'll use this menu for *t*-tests (related and unrelated—Chapter 10) and one-way independent ANOVA (Chapter 12).

  - *General Linear Model* This menu is for linear models involving categorical predictors, typically experimental designs in which you have manipulated a predictor variable using different cases (independent design), the same cases (repeated measures deign) or a combination of these (mixed designs). It also caters for multiple outcome variables, such as in multivariate analysis of variance (MANOVA)—see Chapters 13–17.

  - *Mixed Models* We'll use this menu in Chapter 21 to fit a multilevel linear model and growth curve.

  - *Correlate* It doesn't take a genius to work out that this is where measures of correlation hang out, including bivariate correlations such as Pearson's *r*, Spearman's rho (ρ) and Kendall's tau (τ) and partial correlations (see Chapter 8).

  - *Regression* There are a variety of regression techniques available in SPSS, including simple linear regression, multiple linear regression (Chapter 9), and logistic regression (Chapter 20).

  - *Loglinear* Loglinear analysis is hiding in this menu, waiting for you, and ready to pounce like a tarantula from its burrow (Chapter 19).

  - *Dimension Reduction* You'll find factor analysis here (Chapter 19).

  - *Scale* We'll use this menu for reliability analysis in Chapter 18.

  - *Nonparametric Tests* Although, in general, I'm not a fan of these tests, in Chapter 7 I prostitute my principles to cover the Mann–Whitney test, the Kruskal–Wallis test, Wilcoxon's test and Friedman's ANOVA.

- *Graphs* This menu is used to access the Chart Builder (discussed in Chapter 5), which is your gateway to, among others, bar charts, histograms, scatterplots, box–whisker plots, pie charts and error bar graphs.

- *Utilities* There's plenty of useful stuff here, but we don't get into it. I will mention that ⊞ Data File Comments... is useful for writing notes about the data file to remind yourself of important details that you might forget (where the data come from, the date they were collected and so on).

- *Extensions (formerly Add-ons)* Use this menu to access other IBM software that augments SPSS Statistics. For example, *IBM SPSS Sample Power* computes the sample size required for studies and power statistics (see Section 2.9.7), and if you have the premium version you'll find IBM SPSS AMOS listed here, which is software for structural equation modeling. Because most people won't have these add-ons (including me) I'm not going to discuss them in the book. We'll also use the *Utilities* submenu to install custom dialog boxes (⊞ Install Custom Dialog (Compatibility mode)...) later in this chapter.[4]

- *Window* This menu allows you to switch from window to window. So, if you're looking at the output and you wish to switch back to your data sheet, you can do so using this menu. There are icons to shortcut most of the options in this menu, so it isn't particularly useful.

- *Help* Use this menu to access extensive searchable help files.

At the top of the data editor window are a set of *icons* (see Figure 4.3) that are shortcuts to frequently used facilities in the menus. Using the icons saves you time. Below is a brief list of these icons and their functions.

---

4 In version 23 of IBM SPSS Statistics, this function can be found in *Utilities* ▶ *Custom Dialogs* . . . .

Use this icon to open a previously saved file (if you are in the data editor, SPSS assumes you want to open a data file; if you are in the output viewer, it will offer to open a viewer file).

 Use this icon to save files. It will save the file you are currently working on (be it data, output or syntax). If the file hasn't already been saved it will produce the *Save Data As* dialog box.

 Use this icon for printing whatever you are currently working on (either the data editor or the output). The exact print options will depend on your printer. By default, SPSS prints everything in the output window, so a useful way to save trees is to print only a selection of the output (see SPSS Tip 4.5).

 Clicking on this icon activates a list of the last 12 dialog boxes that were used; select any box from the list to reactivate the dialog box. This icon is a useful shortcut if you need to repeat parts of an analysis.

 The big arrow on this icon implies to me that clicking it activates a miniaturizing ray that shrinks you before sucking you into a cell in the data editor, where you will spend the rest of ofq your days cage-fighting decimal points. It turns out my intuition is wrong, though, and this icon opens the 'Case' tab of the *Go To* dialog box, which enables you to go to a specific case (row) in the data editor. This shortcut is useful for large data files. For example, if we were analyzing a survey with 3000 respondents, and wanted to look at participant 2407's responses, rather than tediously scrolling down the data editor to find row 2407 we could click this icon, enter 2407 in the response box and click Go (Figure 4.5, left).

Figure 4.5 The *Go To* dialog boxes for a case (left) and a variable (right)

 As well as data files with huge numbers of cases, you sometimes have ones with huge numbers of variables. Like the previous icon, clicking this one opens the *Go To* dialog box but in the 'Variable' tab, which enables you to go to a specific variable (column) in the data editor. For example, the data file we use in Chapter 18 (**SAQ.sav**) contains 23 variables and each variable represents a question on a questionnaire and is named accordingly. If we wanted to go to Question 15, rather than getting wrist cramp by scrolling across the data editor to find the column containing the data for Question 15, we could click this icon, scroll down the variable list to Question 15 and click Go (Figure 4.5, right).

 Clicking on this icon opens a dialog box that shows you the variables in the data editor on the left and summary information about the selected variable on the right. Figure 4.6 shows the dialog box for the same data file that we discussed for the previous icon. I have selected the first variable in the list on the left, and on the right we see the variable name (Question_01), the label (Statistics makes me cry), the measurement level (ordinal), and the value labels (e.g., the number 1 represents the response of 'strongly agree').

 If you select a variable (column) in the data editor by clicking on the name of the variable (at the top of the column) so that the column is highlighted, then clicking this icon will produce a table of descriptive statistics for that variable in the viewer window. To get descriptive statistics for multiple variables hold down *Ctrl* as you click at the top of the columns you want to summarize to highlight them, then click the icon.

Figure 4.6 Dialog box for the *Variables* icon

 I initially thought that this icon would allow me to spy on my neighbors, but this shining diamond of excitement was snatched cruelly from me as I discovered that it enables me to search for words or numbers in the data editor or viewer. In the data editor, clicking this icon initiates a search within the variable (column) that is currently active. This shortcut is useful if you realize from plotting the data that you have made an error, for example typed 20.02 instead of 2.02 (see Section 5.4), and you need to find the error—in this case by searching for 20.02 within the relevant variable and replacing it with 2.02 (Figure 4.7).

 Clicking on this icon inserts a new case in the data editor (it creates a blank row at the point that is currently highlighted in the data editor).

 Clicking on this icon creates a new variable to the left of the variable that is currently active (to activate a variable click the name at the top of the column).

 Clicking on this icon is a shortcut to the *Data* ▸ 🔲 Split File… dialog box (see Section 6.10.4). In SPSS, we differentiate groups of cases by using a coding variable (see Section 4.6.5), and this function runs any analyses separately for groups coded with such a variable. For example, imagine we test males and females on their statistical ability. We would code each participant with a number that represents their sex (e.g., 1 = female, 0 = male). If we then want to know the mean statistical ability for males and females separately we ask SPSS to split the file by the variable **Sex** and then run descriptive statistics.

Figure 4.7 The *Find and Replace* dialog box

 This icon shortcuts to the *Data* ▸ 🔲 Weight Cases… dialog box. As we shall see, you sometimes need to use the *weight cases* function when you analyze frequency data (see Section 19.7.2). It is also useful for some advanced issues in survey sampling.

 This icon is a shortcut to the *Data* ▸ 🔲 Select Cases… dialog box, which can be used if you want to analyze only a portion of your data. This function allows you to specify what cases you want to include in the analysis.

Clicking on this icon either displays or hides the value labels of any coding variables in the data editor. We use a coding variable to input information about category or group membership. We discuss this in Section 4.6.5. Briefly, if we wanted to record participant sex, we could create a variable called **Sex** and assign 1 as female and 0 as male. We do this by assigning *value labels* describing the category (e.g. 'female') to the number assigned to the category (e.g. 1). In the data editor, we'd enter a number 1 for any females and 0 for any males. Clicking this icon toggles between the numbers you entered (you'd see a column of 0s and 1s) and the value labels you assigned to those numbers (you'd see a column displaying the word 'male' or 'female' in each cell).

4.6 Entering data into IBM SPSS Statistics ▌▌▌▌

4.6.1 Data formats ▌▌▌▌

There are two common data entry formats, which are sometimes referred to as **wide format data** and **long format data**. Most of the time, we enter data into SPSS in wide format, although you can switch between wide and long formats using the *Data* ▶ 🔲 Restructure... menu. In the wide format *each row represents data from one entity and each column represents a variable*. There is no discrimination between predictor (independent) and outcome (dependent) variables: both appear in a separate column. The key point is that each row represents one entity's data (be that entity a human, mouse, tulip, business or water sample) and any information about that entity should be entered across the data editor. Contrast this with long format, in which scores on an outcome variable appear in a single column and rows represent a combination of the

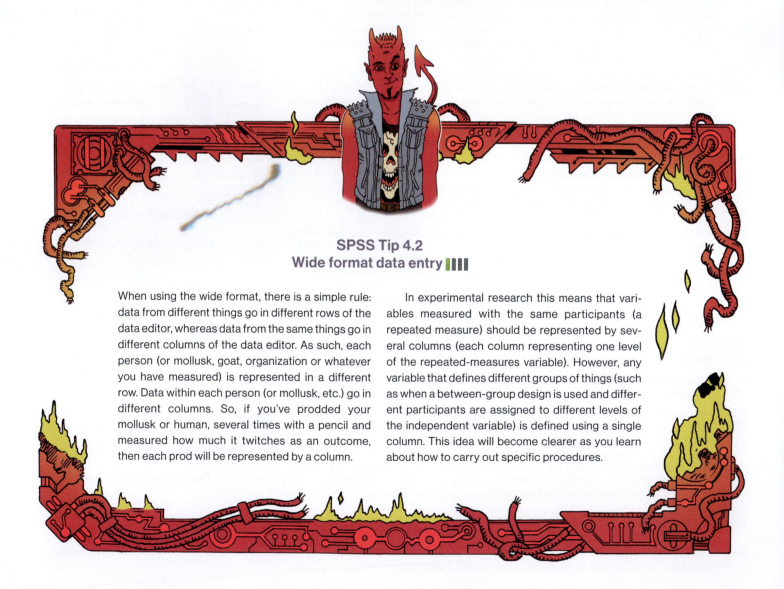

SPSS Tip 4.2
Wide format data entry ▌▌▌▌

When using the wide format, there is a simple rule: data from different things go in different rows of the data editor, whereas data from the same things go in different columns of the data editor. As such, each person (or mollusk, goat, organization or whatever you have measured) is represented in a different row. Data within each person (or mollusk, etc.) go in different columns. So, if you've prodded your mollusk or human, several times with a pencil and measured how much it twitches as an outcome, then each prod will be represented by a column.

In experimental research this means that variables measured with the same participants (a repeated measure) should be represented by several columns (each column representing one level of the repeated-measures variable). However, any variable that defines different groups of things (such as when a between-group design is used and different participants are assigned to different levels of the independent variable) is defined using a single column. This idea will become clearer as you learn about how to carry out specific procedures.

attributes of those scores. In long format data, scores from a single entity can appear over multiple rows, where each row represents a combination of the attributes of the score (the entity from which the score came, to which level of an independent variable the score belongs, the time point at which the score was recorded, etc.).

We use the long format in Chapter 21, but for everything else in this book we use the wide format, so let's look at an example of how to enter data in this way. Imagine you were interested in how perceptions of pain created by hot and cold stimuli were influenced by whether or not you swore while in contact with the stimulus (Stephens, Atkins, & Kingston, 2009). You could place some people's hands in a bucket of very cold water for a minute and ask them to rate how painful they thought the experience was on a scale of 1 to 10. You could then ask them to hold a hot potato and again measure their perception of pain. Half the participants are encouraged to shout profanities during the experiences. Imagine I was a participant in the swearing group. You would have a single row representing my data, so there would be a different column for my name, the group I was in, my pain perception for cold water and my pain perception for a hot potato: Andy, Swearing Group, 7, 10.

The column with the information about the group to which I was assigned is a grouping variable: I can belong to either the group that could swear or the group that was forbidden, but not both. This variable is a between-group or independent measure (different people

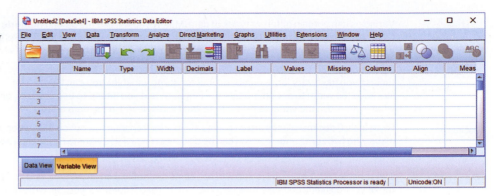

Figure 4.8   The variable view of the SPSS Data Editor

belong to different groups). In SPSS we typically represent group membership with numbers, not words, but assign labels to those numbers. As such, group membership is represented by a single column in which the group to which the person belonged is defined using a number (see Section 4.6.5). For example, we might decide that if a person was in the swearing group we assign them the number 1, and if they were in the non-swearing group we assign them a 0. We then assign a *value label* to each number, which is text that describes what the number represents. To enter group membership, we would input the numbers we have decided to use into the data editor, but the value labels remind us which groups those numbers represent (see Section 6.10.4).

The two pain scores make up a repeated measure because all of the participants produced a score after contact with a hot and cold stimulus. Levels of this variable (see SPSS Tip 4.2) are entered in separate columns (one for pain from a hot stimulus and one for pain from a cold stimulus).

The data editor is made up of lots of *cells*, which are boxes in which data values can be placed. When a cell is active, it becomes

highlighted in orange (as in Figure 4.3). You can move around the data editor, from cell to cell, using the arrow keys ← ↑ ↓ → (on the right of the keyboard) or by clicking the mouse on the cell that you wish to activate. To enter a number into the data editor, move to the cell in which you want to place the data value, type the value, then press the appropriate arrow button for the direction in which you wish to move. So, to enter a row of data, move to the far left of the row, type the first value and then press → (this process inputs the value and moves you into the next cell on the right).

4.6.2 The variable view ▮▮▮▮

Before we input data into the data editor, we need to create the variables using the variable view. To access this view click the 'Variable View' tab at the bottom of the data editor (Data View Variable View); the contents of the window will change (see Figure 4.8).

Every row of the variable view represents a variable, and you set characteristics of each variable by entering information into the following labeled columns (play around, you'll get the hang of it):

| Name | Enter a name in this column for each variable. This name will appear at the top of the corresponding column in the data view, and helps you to identify variables. You can more or less write what you like, but there are certain symbols you can't use (mainly those that have other uses in SPSS such as +, −, &), and you can't use spaces. (I find 'hard spaces' useful, where you use an underscore instead of a space; for example Andy_Field instead of Andy Field.) If you use a character that SPSS doesn't like, you'll get an error message saying that the variable name is invalid when you try to move out of the cell. |

Type	Variables can contain different types of data. Mostly you will use the default **numeric variables** (the variable contains numbers). Other types are **string variables** (strings of letters), **currency variables** (i.e., £s, $s, €s) and **date variables** (e.g., 21-06-1973). So, for example, if you want to type in a person's name, you would need to change the variable type from numeric to string.
Width	By default, SPSS sets up new variables as *numeric* and storing 8 digits/characters. For numeric variables 8 digits is usually fine (unless you have very large numbers), but for string variables you can't write a lot in only 8 characters so you might want to increase the width to accommodate the length of text you intend to use. Width differs from Columns (see below) in that it affects what is stored in the variable rather than what is displayed in the data editor.
Decimals	By default, 2 decimal places are displayed. (If you don't change this option then when you enter whole numbers SPSS adds a decimal place with two zeros after it, which can be disconcerting.) If you want to change the number of decimal places for a given variable, replace the 2 with a new value or increase or decrease the values using ⬍.
Label	The variable Name (see above) has some restrictions on characters, and you also wouldn't want to use huge long names at the top of your columns because they become hard to read. Therefore, you can write a longer variable description in this column. This may seem pointless, but is one of the best habits you can get into (see SPSS Tip 4.3).
Values	This column is for assigning numbers to represent groups of people (see Section 4.6.5).
Missing	This column is for assigning numbers to missing data (see Section 4.6.7).
Columns	Enter a number into this column to determine the width of the column (i.e., how many characters are displayed in the column). The information entered under Columns differs from that for Width (see above) which determines the width of the variable itself. So, you could have a variable with a width of 10 characters but by setting the *column* width to 8 you would see only 8 of the 10 characters of the variable in the data editor. It can be useful to increase the column width if you have a string variable (Section 4.6.3) or a variable with value labels (Section 4.6.5) that exceed 8 characters.
Align	You can use this column to select the alignment of the data in the corresponding column of the data editor. You can choose to align the data to the Left, Right or Center.
Measure	Use this column to define the scale of measurement for the variable as Nominal, Ordinal or Scale (Section 1.6.2).
Role	SPSS Statistics has some procedures that attempt to run analyses automatically without you needing to think about what you're doing (one example is *Analyze* ▶ *Regression* ▶ Automatic Linear Modeling...). To think on your behalf, SPSS needs to know whether a variable is a predictor (Input), an outcome (Target), both (Both, although I'm not sure how that works out in practice), a variable that splits the analysis by different groups (Split), a variable that selects out part of the data (Partition) or a variable that has no predefined role (None). It's rarely a good idea to let a computer do your thinking for you, so I'm not a fan of the sorts of automated procedures that take advantage of these role assignments (they have their place, but that place is not in this book). Therefore, I'm not going to mention roles again.

Let's use the variable view to create some variables. Imagine we were interested in looking at the differences between lecturers and students. We took a random sample of five psychology lecturers from the University of Sussex and five psychology students and then measured how many friends they had, their weekly alcohol consumption (in units), their yearly income and how neurotic they were (higher score is more neurotic). These data are in Table 4.1.

Table 4.1 Some data with which to play

Name	Birth Date	Job	No. of Friends	Alcohol (Units)	Income (p.a.)	Neuroticism
Ben	03-Jul-1977	Lecturer	5	10	20,000	10
Martin	24-May-1969	Lecturer	2	15	40,000	17
Andy	21-Jun-1973	Lecturer	0	20	35,000	14
Paul	16-Jul-1970	Lecturer	4	5	22,000	13
Graham	10-Oct-1949	Lecturer	1	30	50,000	21
Carina	05-Nov-1983	Student	10	25	5,000	7
Karina	08-Oct-1987	Student	12	20	100	13
Doug	16-Sep-1989	Student	15	16	3000	9
Mark	20-May-1973	Student	12	17	10,000	14
Zoë	12-Nov-1984	Student	17	18	10	13

4.6.3 Creating a string variable ▮▮▮

The first variable in Table 4.1 is the name of the lecturer/student. This variable is a *string variable* because it consists of names (which are strings of letters). To create this variable in the variable view:

1 Click in the first white cell in the column labeled *Name*.
2 Type the word 'Name'.
3 Move from this cell using the arrow keys on the keyboard (you can also just click in a different cell, but this is a very slow way of doing it).

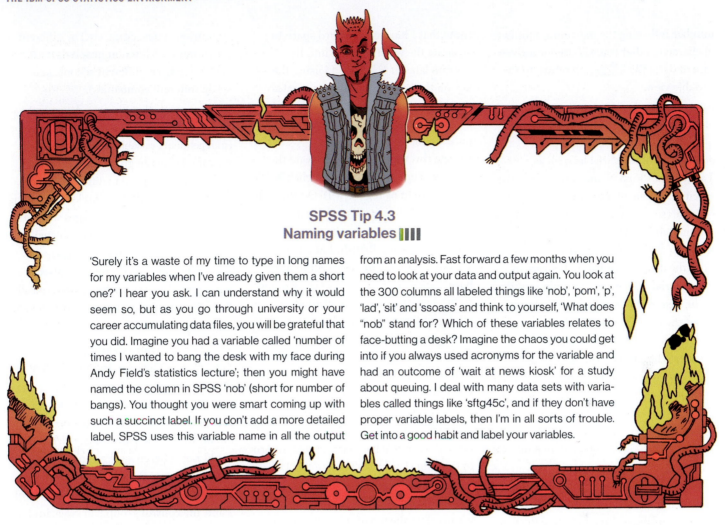

SPSS Tip 4.3
Naming variables ▌▌▌▌

'Surely it's a waste of my time to type in long names for my variables when I've already given them a short one?' I hear you ask. I can understand why it would seem so, but as you go through university or your career accumulating data files, you will be grateful that you did. Imagine you had a variable called 'number of times I wanted to bang the desk with my face during Andy Field's statistics lecture'; then you might have named the column in SPSS 'nob' (short for number of bangs). You thought you were smart coming up with such a succinct label. If you don't add a more detailed label, SPSS uses this variable name in all the output from an analysis. Fast forward a few months when you need to look at your data and output again. You look at the 300 columns all labeled things like 'nob', 'pom', 'p', 'lad', 'sit' and 'ssoass' and think to yourself, 'What does "nob" stand for? Which of these variables relates to face-butting a desk? Imagine the chaos you could get into if you always used acronyms for the variable and had an outcome of 'wait at news kiosk' for a study about queuing. I deal with many data sets with variables called things like 'sftg45c', and if they don't have proper variable labels, then I'm in all sorts of trouble. Get into a good habit and label your variables.

Well done, you've just created your first variable. Notice that once you've typed a name, SPSS creates default settings for the variable (such as assuming it's numeric and assigning 2 decimal places). However, we don't want a numeric variable (i.e., numbers), we want to enter people's names, so we need a *string* variable, so we have to change the variable type. Move into the column labeled Type using the arrow keys on the keyboard. The cell will now look like this Numeric ... . Click ... to activate the *Variable Type* dialog box. By default, the numeric variable type is selected (◉ Numeric)—see the top of Figure 4.9. To change the variable to a string variable, click ◉ String (bottom left of Figure 4.9). Next, if you need to enter text of more than 8 characters (the default width), then change this default value to a

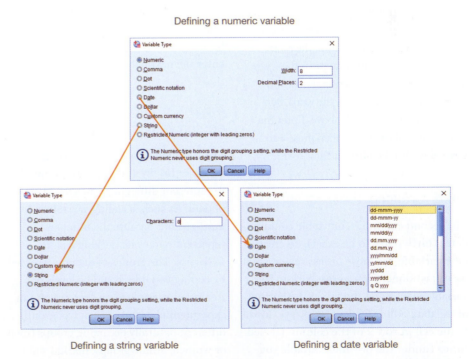

Defining a numeric variable

Defining a string variable

Defining a date variable

Figure 4.9 Defining numeric, string and date variables

number reflecting the maximum number of characters that you will use for a given case of data. Click OK to return to the variable view.

Next, because I want you to get into good habits, move to the cell in the Label column and type a description of the variable, such as 'Participant's First Name'. Finally, we can specify the scale of measurement for the variable (see Section 1.6.2) by going to the column labeled *Measure* and selecting Nominal, Ordinal or Scale from the drop-down list. In the case of a string variable, it represents a description of the case and provides no information about the order of cases or the magnitude of one case compared to another. Therefore, select Nominal.

Once the variable has been created, return to the data view by clicking on the 'Data View' tab at the bottom of the data editor (Data View Variable View). The contents of the window will change, and notice that the first column now has the label *Name*. We can enter the data for this variable in the column underneath. Click the white cell at the top of the column labeled *Name* and type the first name, 'Ben'. To register this value in this cell, move to a different cell and because we are entering data down a column, the most sensible way to do this is to press the ↓ key on the keyboard. This action moves you down to the next cell, and the word 'Ben' should appear in the cell above. Enter the next name, 'Martin', and then press ↓ to move down to the next cell, and so on.

4.6.4 Creating a date variable ▮▮▮

The second column in our table contains dates (birth dates to be exact). To create a date variable, we more or less repeat what we've just done. First, move back to the variable view using the tab at the bottom of the data editor (Data View Variable View). Move to the cell in row 2 of the column labeled *Name* (under the previous variable you created). Type the word 'Birth_Date'

(note that I have used a hard space to separate the words). Move into the column labeled Type using the → key on the keyboard (doing so creates default settings in the other columns). As before, the cell you have moved into will indicate the default of Numeric ..., and to change this we click ... to activate the *Variable Type* dialog box, and click ◉ Date (bottom right of Figure 4.9). On the right of the dialog box is a list of date formats, from which you can choose your preference; being British, I am used to the day coming before the month and have chosen dd-mmm-yyyy (i.e., 21-Jun-1973), but Americans, for example, more often put the month before the date and so might select mm/dd/yyyy (06/21/1973). When you have selected a date format, click OK to return to the variable view. Finally, move to the cell in the column labeled *Label* and type 'Date of Birth'.

Once the variable has been created, return to the data view by clicking on the 'Data View' tab (Data View Variable View). The second column now has the label *Birth_Date*; click the white cell at the top of this column and type the first value, 03-Jul-1977. To register this value in this cell, move down to the next cell by pressing the ↓ key. Now enter the next date, and so on.

4.6.5 Creating coding variables ▮▮▮

I've mentioned coding or grouping variables briefly already; they use numbers to represent different groups or categories of data. As such, a coding variable is *numeric*, but because the numbers represent names its variable type is Nominal. The groups of data represented by coding variables could be levels of a treatment variable in an experiment (an experimental group or a control group), different naturally occurring groups (men or women, ethnic groups, marital status, etc.), different geographic locations

(countries, states, cities, etc.) or different organizations (different hospitals within a healthcare trust, different schools in a study, different companies).

In experiments that use an independent design, coding variables represent predictor (independent) variables that have been measured between groups (i.e., different entities were assigned to different groups). We do not, generally, use this kind of coding variable for experimental designs where the independent variable was manipulated using repeated measures i.e., participants take part in all experimental conditions). For repeated-measures designs we typically use different columns to represent different experimental conditions.

Think back to our swearing and pain experiment. This *was* an independent design because we had two groups representing the two levels of our independent variable: one group could swear during the pain tasks, the other could not. Therefore, we can use a coding variable. We might assign the experimental group (swearing) a code of 1 and the control group (no swearing) a code of 0. To input these data you would create a variable (which you might call **group**) and type the value 1 for any participants in the experimental group, and 0 for any participant in the control group. These codes tell SPSS that the cases that have been assigned the value 1 should be treated as belonging to the same group, and likewise for the cases assigned the value 0. The codes you use are arbitrary because the numbers themselves won't be analyzed, so although people typically use 0, 1, 2, 3, etc., if you're a particularly arbitrary person feel free to code one group as 616 and another as 11 and so on.

We have a coding variable in our data that describes whether a person was a lecturer or student. To create this coding variable, we follow the same steps as before, but we will also have to record which numeric

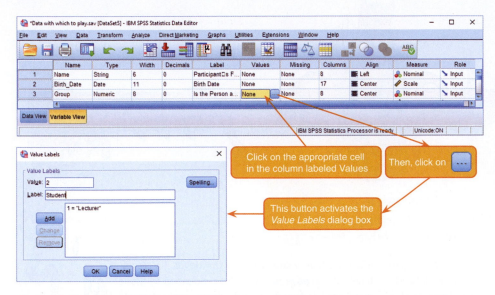

Figure 4.10   Defining coding variables and their values

click in the white space below, next to where it says *Label* (or press *Tab* or *Alt* and L at the same time) and type in an appropriate label for that group. In Figure 4.10 I have already defined a code of 1 for the lecturer group, and then I have typed in 2 as a code and given this a label of *Student*. To add this code to the list click [Add]. When you have defined all your coding values you might want to check for spelling mistakes in the value labels by clicking [Spelling...]. To finish, click [OK]; if you do this before you have clicked [Add] to register your most recent code in the list, SPSS displays a warning that any 'pending changes will be lost'. This message is telling you to go back and click [Add] before continuing. Finally, coding variables represent categories and so the scale of measurement is nominal (or ordinal if the categories have a meaningful order). To specify this level of measurement, go to the column labeled *Measure* and select [Nominal] (or [Ordinal] if the groups have a meaningful order) from the drop-down list.

Having defined your codes, switch to the data view and for each participant type

codes are assigned to which groups. First, return to the variable view ([Data View] [Variable View]) if you're not already in it and move to the cell in the third row under the column labeled *Name*. Type a name (let's call it **Group**). I'm still trying to instil good habits, so move along the third row to the column called *Label* and give the variable a full description such as, 'Is the person a lecturer or a student?' To

define the group codes, move along the row to the column labeled [Values]. The cell will indicate the default of [None] [...]. Click [...] to access the *Value Labels* dialog box (see Figure 4.10).

The *Value Labels* dialog box is used to specify group codes. First, click in the white space next to where it says *Value* (or press *Alt* and U at the same time) and type in a code (e.g., 1). The second step is to

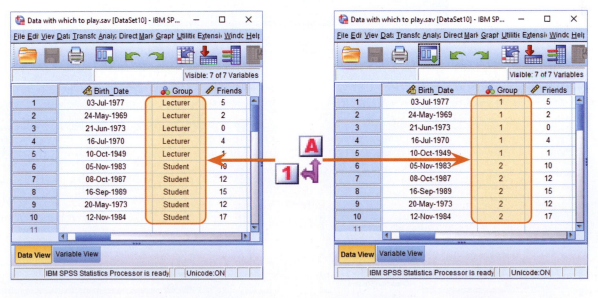

Value labels on   Value labels off

Figure 4.11   Coding values in the data editor with the value labels switched off and on

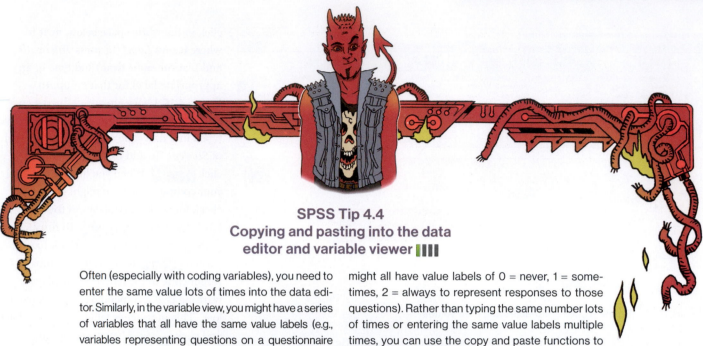

SPSS Tip 4.4
Copying and pasting into the data editor and variable viewer ▌▌▌▌

Often (especially with coding variables), you need to enter the same value lots of times into the data editor. Similarly, in the variable view, you might have a series of variables that all have the same value labels (e.g., variables representing questions on a questionnaire might all have value labels of 0 = never, 1 = sometimes, 2 = always to represent responses to those questions). Rather than typing the same number lots of times or entering the same value labels multiple times, you can use the copy and paste functions to

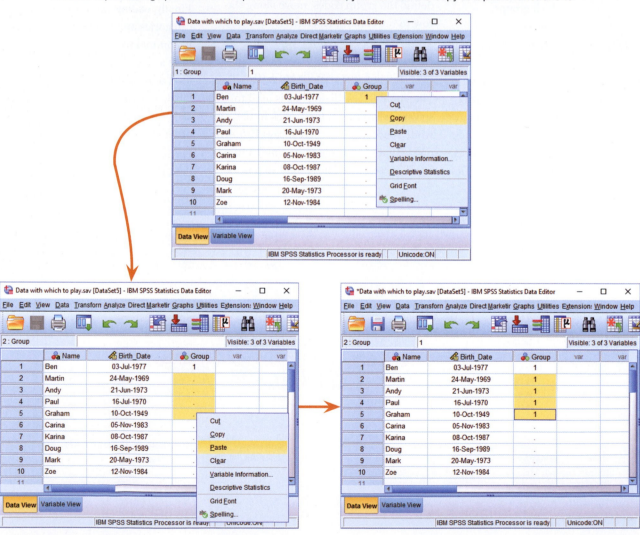

Figure 4.12 Copying and pasting into empty cells

speed things up. All you need to do is to select the cell containing the information that you want to copy (whether that is a number or text in the data view or a set of value labels or another characteristic within the variable view) and click with the right mouse button to activate a menu within which you can click (with the left mouse button) on *Copy* (top of Figure 4.12). Next, highlight any cells into which you want to place what you have copied by dragging the mouse over them while holding down the left mouse button. These cells will be highlighted in orange. While the pointer is over the highlighted cells, click with the right mouse button to activate a menu from which you should click *Paste* (bottom left of Figure 4.12). The highlighted cells will be filled with the value that you copied (bottom right of Figure 4.12). Figure 4.12 shows the process of copying the value '1' and pasting it into four blank cells in the same column.

the numeric value that represents their group membership into the column labeled *Group*. In our example, if a person was a lecturer, type '1', but if they were a student then type '2' (see SPSS Tip 4.4). SPSS can display either the numeric codes or the value labels that you assigned to them, and you can toggle between the two states by clicking (see Figure 4.11). Figure 4.11 shows how the data should be arranged: remember that each row of the data editor represents data from one entity: the first five participants were lecturers, whereas participants 6–10 were students.

4.6.6 Creating a numeric variable

Our next variable is **Friends**, which is numeric. Numeric variables are the easiest ones to create because they are the default format in SPSS. Move back to the variable view using the tab at the bottom of the

data editor (Data View Variable View). Go to the cell in row 4 of the column labeled *Name* (under the previous variable you created). Type the word 'Friends'. Move into the column labeled Type using the → key on the keyboard. As with the previous variables we have created, SPSS has assumed that our new variable is Numeric , and because our variable *is* numeric we don't need to change this setting.

The scores for the number of friends have no decimal places (unless you are a very strange person indeed, you can't have 0.23 of a friend). Move to the Decimals column and type '0' (or decrease the value from 2 to 0 using) to tell SPSS that you don't want to display decimal places.

Let's continue our good habit of naming variables and move to the cell in the column labeled *Label* and type 'Number of Friends'. Finally, number of friends is measured on the ratio scale of

measurement (see Section 1.6.2) and we can specify this by going to the column labeled *Measure* and selecting Scale from the drop-down list (this will have been done automatically, but it's worth checking).

Once the variable has been created, you can return to the data view by clicking on the 'Data View' tab at the bottom of the data editor (Data View Variable View). The contents of the window will change, and you'll notice that the fourth column now has the label *Friends*. To enter the data, click the white cell at the top of the column labeled *Friends* and type the first value, 5. Because we're entering scores down the column the most sensible way to record this value in this cell is to press the ↓ key on the keyboard. This action moves you down to the next cell, and the number 5 is stored in the cell above. Enter the next number, 2, and then press ↓ to move down to the next cell, and so on.

4.6.7 Missing values

Although we strive to collect complete sets of data, often scores are missing. Missing data can occur for a variety of reasons: in long questionnaires participants accidentally (or, depending on how paranoid you're feeling, deliberately to irritate you) miss out questions; in experimental procedures

SELF TEST

Why is the 'Number of Friends' variable a 'scale' variable?

SELF TEST

Having created the first four variables with a bit of guidance, try to enter the rest of the variables in Table 4.1 yourself.

Labcoat Leni's Real Research 4.1
Gonna be a rock 'n' roll singer ▌▌▌▌

Oxoby, R. J. (2008). *Economic Enquiry, 47*(3), 598–602.

AC/DC are one one of the best-selling hard rock bands in history, with around 100 million certified sales, and an estimated 200 million actual sales. In 1980 their original singer Bon Scott died of alcohol poisoning and choking on his own vomit. He was replaced by Brian Johnson, who has been their singer ever since.[5] Debate rages with unerring frequency within the rock music press over who is the better frontman. The conventional wisdom is that Bon Scott was better, although personally, and I seem to be somewhat in the minority here, I prefer Brian Johnson. Anyway, Robert Oxoby, in a playful paper, decided to put this argument to bed once and for all (Oxoby, 2008).

Using a task from experimental economics called the ultimatum game, individuals are assigned the role of either proposer or responder and paired randomly. Proposers are allocated $10 from which they have to make a financial offer to the responder (i.e., $2). The responder can accept or reject this offer. If the offer is rejected neither party gets any money, but if the offer is accepted the responder keeps the offered amount (e.g., $2), and the proposer keeps the original amount minus what they offered (e.g., $8). For half of the participants the song 'It's a long way to the top'

sung by Bon Scott was playing in the background, for the remainder 'Shoot to thrill' sung by Brian Johnson was playing. Oxoby measured the offers made by proposers, and the minimum offers that responders accepted (called the minimum acceptable offer). He reasoned that people would accept lower offers and propose higher offers when listening to something they like (because of the 'feel-good factor' the music creates). Therefore, by comparing the value of offers made and the minimum acceptable offers in the two groups, he could see whether people have more of a feel-good factor when listening to Bon or Brian. The offers made (in $) are[6] as follows (there were 18 people per group):

- Bon Scott group: 1, 2, 2, 2, 2, 3, 3, 3, 3, 3, 4, 4, 4, 4, 4, 5, 5, 5
- Brian Johnson group: 2, 3, 3, 3, 3, 3, 4, 4, 4, 4, 4, 5, 5, 5, 5, 5, 5, 5

Enter these data into the SPSS Data Editor, remembering to include value labels, to set the *measure* property, to give each variable a proper label, and to set the appropriate number of decimal places. Answers are on the companion website, and my version of how this file should look can be found in *Oxoby (2008) Offers.sav*.

---

mechanical faults can lead to a score not being recorded; and in research on delicate topics (e.g., sexual behavior) participants may exert their right not to answer a question. However, just because we have missed out on some data for a participant, that doesn't mean that we have to ignore the data we do have (although it creates statistical difficulties). The simplest way to record a missing score is to leave the cell in the data editor empty, but it can be helpful to tell SPSS explicitly that a score is missing. We do this, much like a coding variable, by choosing a number to

---

5 Well, until all that weird stuff with W. Axl Rose in 2016, which I'm trying to pretend didn't happen.
6 These data are estimated from Figures 1 and 2 in the paper because I couldn't get hold of the author to get the original data files.

represent the missing data point. You then tell SPSS to treat that number as missing. For obvious reasons, it is important to choose a code that cannot also be a naturally occurring data value. For example, if we use the value 9 to code missing values and several participants genuinely scored 9, then SPSS will wrongly treat those scores as missing. You need an 'impossible' value, so people usually pick a score greater than the maximum possible score on the measure. For example, in an experiment in which attitudes are measured on a 100-point scale (so scores vary from 1 to 100) a good code for missing values might be something like 101, 999 or, my personal favorite, 666 (because missing values *are* the devil's work).

To specify missing values click in the column labeled Missing in the variable view (Data View Variable View) and then click

--- to activate the *Missing Values* dialog box in Figure 4.13. By default, SPSS assumes that no missing values exist, but you can define them in one of two ways. The first is to select discrete values (by clicking on the radio button next to where it says *Discrete missing values*), which are single values that represent missing data. SPSS allows you to specify up to three values to represent missing data. The reason why you might choose to have several numbers to represent missing values is that you can assign a different meaning to each discrete value. For example, you could have the number 8 representing a response of 'not applicable', a code of 9 representing a 'don't know' response, and a code of 99 meaning that the participant failed to give any response. SPSS treats these values in the same way (it ignores them), but different codes can be helpful to

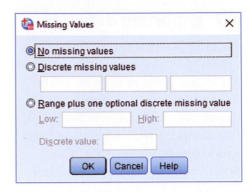

Figure 4.13 Defining missing values

remind you of why a particular score is missing. The second option is to select a range of values to represent missing data and this is useful in situations in which it is necessary to exclude data falling between two points. So, we could exclude all scores between 5 and 10. With this last option you can also (but don't have to) specify one discrete value.

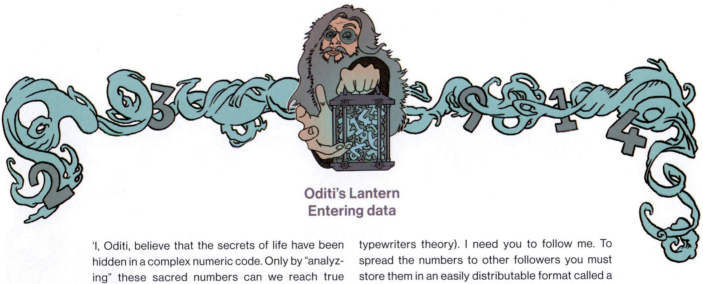

Oditi's Lantern
Entering data

'I, Oditi, believe that the secrets of life have been hidden in a complex numeric code. Only by "analyzing" these sacred numbers can we reach true enlightenment. To crack the code I must assemble thousands of followers to analyze and interpret these numbers (it's a bit like the chimps and typewriters theory). I need you to follow me. To spread the numbers to other followers you must store them in an easily distributable format called a "data file". You, my follower, are loyal and loved, and to assist you my lantern displays a tutorial on how to do this.'

4.7 Importing data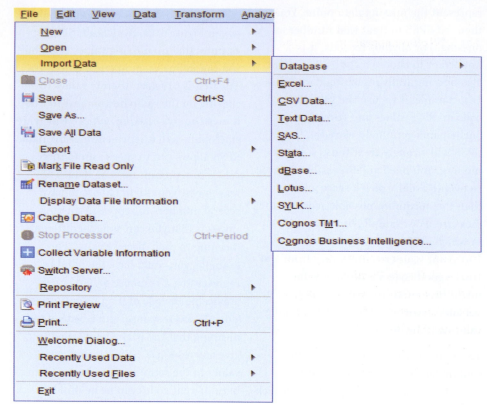

We can import data into SPSS from other software packages such as Microsoft Excel, R, SAS, and Systat by using the *File* ▶ *Import Data* menu and selecting the corresponding software from the list (Figure 4.14). If you want to import from a package that isn't listed (e.g., R or Systat), then export the data from these packages as tab-delimited text data (*.txt* or *.dat*) or comma-separated values (*.csv*) and select the *Text Data* or *CSV Data* options in the menu.

Figure 4.14   The *Import Data* menu

Oditi's Lantern
Importing data into SPSS

'I, Oditi, have become aware that some of the sacred numbers that hide the secrets of life are contained within files other than those of my own design. We cannot afford to miss vital clues that lurk among these rogue files. Like all good cults, we must convert all to our cause, even data files. Should you encounter one of these files, you must convert it to the SPSS format. My lantern shows you how.'

4.8 The SPSS viewer

The *SPSS viewer* appears in a different window than the data editor and displays the output of any procedures in SPSS: tables of results, graphs, error messages and pretty much everything you could want, except for photos of your cat. Although the *SPSS viewer* is all-singing and all-dancing, my prediction in previous editions of this book that it will one day include a tea-making facility have not come to fruition (IBM, take note ☺). Figure 4.15 shows the viewer. On the right there is a large space in which all output is displayed. Graphs (Section 5.9) and tables displayed here can be edited by double-clicking on them. On the left, is a tree diagram of the output. This tree diagram provides an easy way to access parts of the output, which is useful when you have conducted tons of analyses. The tree structure is self-explanatory: every time you do something in SPSS (such as

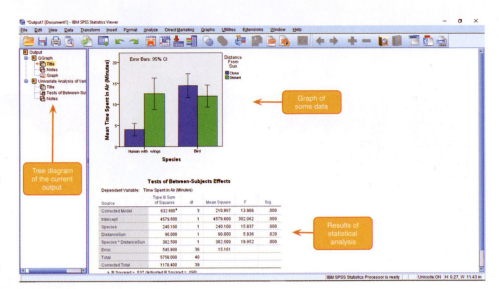

Figure 4.15 The SPSS viewer

drawing a graph or running a statistical procedure), it lists this procedure as a main heading.

In Figure 4.15, I ran a graphing procedure followed by a univariate analysis of variance (ANOVA), and these names appear as main headings in the tree diagram. For each procedure there are subheadings that represent different parts of the analysis. For example, in the ANOVA procedure, which you'll learn more about later in the book, there are

Oditi's Lantern
Editing tables

'I, Oditi, impart to you, my loyal posse, the knowledge that SPSS will conceal the secrets of life within tables of output. Like the author of this book's personality, these tables appear flat and lifeless; however, if you give them a poke they have hidden depths. Often you will need to seek out the hidden codes within the tables. To do this, double-click on them. This will reveal the "layers" of the table. Stare into my lantern and find out how.'

SPSS Tip 4.5
Printing and saving the planet

Rather than printing all of your SPSS output, you can help the planet by printing only a selection. Do this by using the tree diagram in the SPSS viewer to select parts of the output for printing. For example, if you decided that you wanted to print a particular graph, click on the word *Graph* in the tree structure to highlight the graph in the output. Then, in the *Print* menu you can print just the selected part of the output (Figure 4.16). Note that if you click a main heading (such as *Univariate Analysis of Variance*) SPSS will highlight all the subheadings under that heading, which is useful for printing all the output from a single statistical procedure.

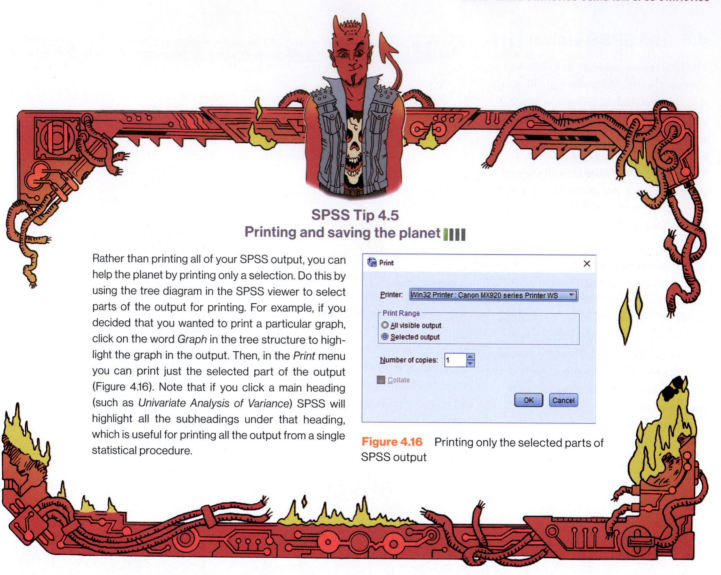

Figure 4.16 Printing only the selected parts of SPSS output

sections such as *Tests of Between-Subjects Effects* (this is the table containing the main results). You can skip to any one of these sub-components by clicking on the appropriate branch of the tree diagram.

So, if you wanted to skip to the between-groups effects, you would move the on-screen arrow to the left-hand portion of the window and click where it says *Tests of Between-Subjects Effects*. This action will highlight this part of the output in the main part of the viewer (see SPSS Tip 4.5).

Some of the icons in the viewer are the same as those for the data editor (so refer back to our earlier list), but others are unique.

 In the viewer, this icon activates a dialog box for printing SPSS output (see SPSS Tip 4.5).

 This icon returns you to the data editor. I'm not sure what the big red star is all about.

 This icon takes you to the last output in the viewer (it returns you to the last procedure you conducted).

 This icon *promotes* the currently active part of the tree structure to a higher branch of the tree. For example, in Figure 4.15 the *Tests of Between-Subjects Effects* are a sub-component under the heading of *Univariate Analysis of Variance*. If we wanted to promote this part of the output to a higher level (i.e., to make it a main heading) then we'd click this icon.

 This icon is the opposite of the above: it *demotes* parts of the tree structure. For example, in Figure 4.15, if we didn't want the *Univariate Analysis of Variance* to be a unique section we could select this heading and click this icon to demote it so that it becomes part of the previous heading (the *Graph* heading). This icon is useful for combining parts of the output relating to a specific research question.

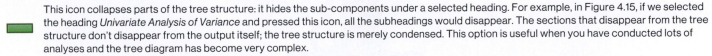

This icon collapses parts of the tree structure: it hides the sub-components under a selected heading. For example, in Figure 4.15, if we selected the heading *Univariate Analysis of Variance* and pressed this icon, all the subheadings would disappear. The sections that disappear from the tree structure don't disappear from the output itself; the tree structure is merely condensed. This option is useful when you have conducted lots of analyses and the tree diagram has become very complex.

This icon expands any collapsed sections. By default, main headings are displayed in the tree diagram in their expanded form. If you have opted to collapse part of the tree diagram (using the icon above) then use this icon to undo your dirty work.

This icon and the following one show or hide parts of the output. If you select part of the output in the tree diagram and click this icon, that part of the output disappears. Some procedures produce copious amounts of output, so this icon can be useful to hide the less relevant parts so you can differentiate the wood from the trees. The output isn't erased, just hidden from view. This icon is like the *collapse* icon above, except that it affects the output rather than the tree structure.

This icon undoes the work of the previous one: if you have hidden part of the output from view, then clicking this icon makes it reappear. By default, this icon is not active because none of the output is hidden, but it becomes active if you use the icon above to hide stuff.

Given this icon looks like it has a slot in which to insert a CD, I had high hopes for it activating some blistering thrash metal, which, as everyone knows, is the best music to have on when doing stats. I was disappointed to discover that rather than blasting out Anthrax at high volume it instead inserts a new heading into the tree diagram. For example, if you had several statistical tests that related to one of many research questions, you could insert a main heading and demote the headings of the relevant analyses to fall under this new heading.

Assuming you had done the above, you can use this icon to provide your new heading with a title. The title you type appears in the output. So, you might have a heading like 'Hypothesis 1: thrash metal music helps you to understand stats' which tells you that the analyses under this heading relate to your first hypothesis.

This final icon is used to place a text box in the output window. You can type anything into this box. In the context of the previous two icons, you might use a text box to explain your hypothesis (e.g., 'Hypothesis 1 is that thrash metal music helps you to understand stats. This hypothesis stems from research showing that a 10-minute blast of classical music can enhance learning and memory (Rauscher, Shaw, & Ky, 1993). Thrash metal music has similar rhythmic and melodic complexity to classical music, so why not?').

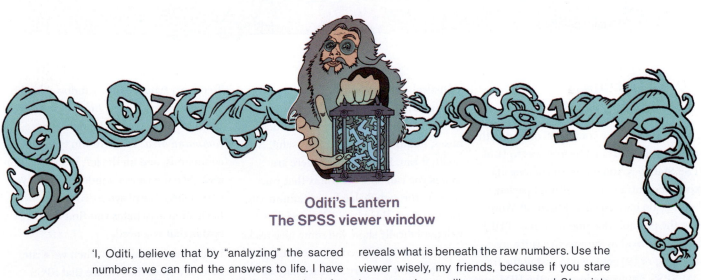

Oditi's Lantern
The SPSS viewer window

'I, Oditi, believe that by "analyzing" the sacred numbers we can find the answers to life. I have given you the tools to spread these numbers far and wide, but to interpret these numbers we need "the viewer". The viewer is like an X-ray that reveals what is beneath the raw numbers. Use the viewer wisely, my friends, because if you stare long enough you will see your very soul. Stare into my lantern and see a tutorial on the viewer.'

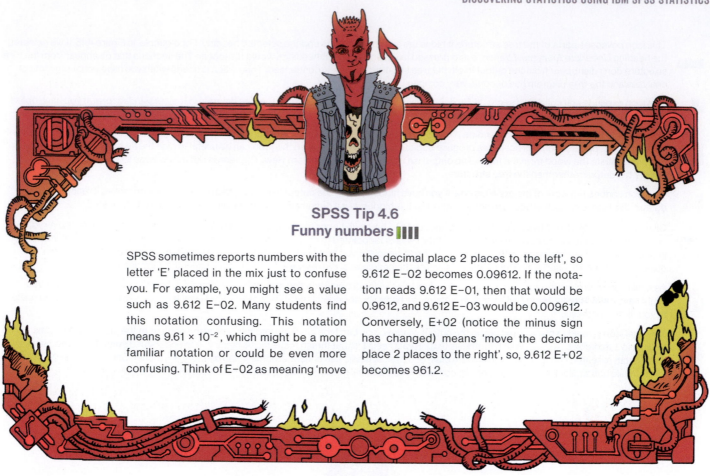

SPSS Tip 4.6
Funny numbers ▮▮▮

SPSS sometimes reports numbers with the letter 'E' placed in the mix just to confuse you. For example, you might see a value such as 9.612 E−02. Many students find this notation confusing. This notation means 9.61×10^{-2}, which might be a more familiar notation or could be even more confusing. Think of E−02 as meaning 'move the decimal place 2 places to the left', so 9.612 E−02 becomes 0.09612. If the notation reads 9.612 E−01, then that would be 0.9612, and 9.612 E−03 would be 0.009612. Conversely, E+02 (notice the minus sign has changed) means 'move the decimal place 2 places to the right', so, 9.612 E+02 becomes 961.2.

4.9 Exporting SPSS output ▮▮▮

If you want to share your SPSS output with other people who don't have access to IBM SPSS Statistics, you have two choices: (1) export the output into a software package that they do have (such as Microsoft Word) or in the portable document format (PDF) that can be read by various free software packages; or (2) get them to install the free IBM SPSS **Smartreader** from the IBM SPSS website. The SPSS Smartreader is basically a free version of the viewer so you can view output but not run new analyses.

4.10 The syntax editor ▮▮▮

I mentioned earlier that sometimes it's useful to use SPSS syntax. Syntax is a language of commands for carrying out statistical analyses and data manipulations. Most people prefer to do the things they need to do using dialog boxes, but SPSS syntax can be useful. No, really, it can. For one thing, there are things you can do with syntax that you can't do through dialog boxes (admittedly, most of these things are advanced, but I will periodically show you some nice tricks using syntax). The second benefit to syntax is if you carry out very similar analyses on data sets. In these situations, it is often quicker to do the analysis and save the syntax as you go along. Then you can adapt it to new data sets (which is frequently quicker than going through dialog boxes). Finally, using syntax creates a record of your analysis, and makes it reproducible, which is an important part of engaging in open science practices (Section 3.6).

To open a syntax editor window, like the one in Figure 4.17, use *File* ▸ *New* ▸ ⊕ *Syntax*. The area on the right (the *command area*) is where you type syntax commands, and on the left is a navigation area (like the viewer window). When you have a large file of syntax commands the navigation area helps you find the bit of syntax that you need.

Like grammatical rules when we write, there are rules that ensure that SPSS 'understands' the syntax. For example, each line must end with a full stop. If you make a syntax error (i.e., break one of the rules), SPSS produces an error message in the viewer window. The messages can be indecipherable until you gain experience of translating them, but they helpfully identify the line in the syntax window in which the error occurred. Each line in the syntax window is numbered so you can

Oditi's Lantern
Exporting SPSS output

'That I, the almighty Oditi, can discover the secrets within the numbers, they must spread around the world. But non-believers do not have SPSS, so we must send them a link to the IBM SPSS Smartreader.

I have also given to you, my subservient brethren, a tutorial on how to export SPSS output into Word. These are the tools you need to spread the numbers. Go forth and stare into my lantern.'

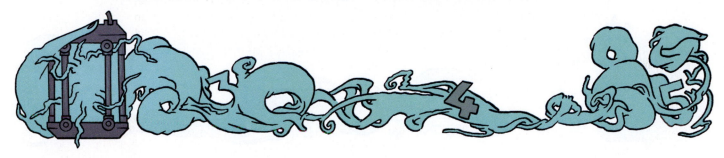

easily find the line in which the error occurred, even if you don't understand what the error is! Learning SPSS syntax is time-consuming, so in the beginning the easiest way to generate syntax is to use dialog boxes to specify the analysis you want to do and then click Paste (many dialog boxes have this button). Doing so pastes the syntax to do the analysis you specified in the dialog box. Using dialog boxes in this way is a good way to get a feel for syntax.

Once you've typed in your syntax you run it using the *Run* menu. *Run* ▶ 🌐 All will run all the syntax in the window or you can highlight a selection of syntax using the mouse and select *Run* ▶ ▶ Selection (or click ▶ in the syntax window) to process the selected syntax. You can also run the syntax a command at a time from either the current command (*Run* ▶ *Step Through* ▶ *From Current*) or the beginning (*Run* ▶ *Step*

Through ▶ *From Start*). You can also process the syntax from the cursor to the end of the syntax window by selecting *Run* ▶ →I To End.

A final note. You can have multiple data files open in SPSS simultaneously. Rather than having a syntax window for each data file, which could get confusing, you can use one syntax window, but select the data file that you want to run the syntax commands on before you run them using the drop-down list DataSet1 ▼.

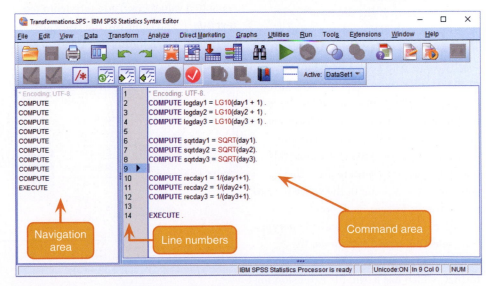

Figure 4.17   A syntax window with some syntax in it

Oditi's Lantern
Sin-tax

'I, Oditi, leader of the cult of undiscovered numerical truths, require my brethren to focus only on the discovery of those truths. To focus their minds I shall impose a tax on sinful acts. Sinful acts (such as dichotomizing a continuous variable) can distract from the pursuit of truth. To implement this tax, followers will need to use the sin-tax window. Stare into my lantern to see a tutorial on how to use it.'

4.11 Saving files

Most of you should be familiar with how to save files. Like most software, SPSS has a save icon and you can use *File* ▶ *Save* or *File* ▶ *Save as …* or *Ctrl* + S (⌘ + S on Mac OS). If the file hasn't been saved previously then initiating a save will open the *Save As* dialog box (see Figure 4.18). SPSS will save whatever is in the window that was active when you initiated the save; for example, if you are in the data editor when you initiate the save, then SPSS will save the data file (not the output or syntax). You use this dialog box as you would in any other software: type a name in the space next to where it says *File name*. If you have sensitive data, you can password encrypt it by selecting ☑ Encrypt file with password. By default, the file will be saved in an SPSS format, which has a *.sav* file extension for data files, *.spv* for viewer documents, and *.sps* for syntax files. Once a file has

previously been saved, it can be saved again (updated) by clicking on 💾.

You can save data in formats other than SPSS. Three of the most useful are Microsoft Excel files (*.xls*, *.xlsx*), comma-separated values (*.csv*) and tab-delimited text (*.dat*). The latter two file types are plain text, which means that they can be

opened by virtually any spreadsheet software you can think of (including Excel, OpenOffice, Numbers, R, SAS, and Systat). To save your data file in of these formats (and others), click SPSS Statistics (*.sav) and select a format from the drop-down list (Figure 4.18). If you select a

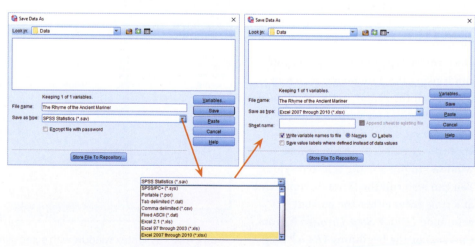

Figure 4.18   The *Save Data As* dialog box

format other than SPSS, the

☐ S̲ave value labels where defined instead of data values option becomes active. If you leave this option unchecked, coding variables (Section 4.6.5) will be exported as numeric values in the data editor; if you select it then coding variables will be exported as string variables containing the value labels. You can also choose to include the variable names in the exported file (usually a good idea) as either the *Names* at the top of the data editor columns or the full *Labels* that you gave to the variables.

4.12 Opening files ▮▮▯▯

This book relies on you working with data files that you can download from the companion website. You probably don't need me to tell you how to open these file, but just in case … To load a file into SPSS use the 📁 icon or select *File ▸ Open ▸* and then 🔲 D̲ata to open a data file, 📄 O̲utput to open a viewer file or 📄 S̲yntax to open a syntax file. This process opens a dialog box (Figure 4.19), with which I'm sure you're familiar. Navigate to wherever you saved the file that you need. SPSS will list the files of the type you asked to open (so, data

files if you selected 🔲 D̲ata). Open the file you want by either selecting it and clicking on [Open] or double-clicking on the icon next to the file you want (e.g., double-clicking on 🔲). If you want to open data in a format other than SPSS (*.sav*), then click [SPSS Statistics (*.sav, *.zsav) ▼] to display a list of alternative file formats. Click the appropriate file type—(Microsoft Excel file (*.xls*), text file (*.dat, *.txt*), etc.)—to list files of that type in the dialog box.

4.13 Extending IBM SPSS Statistics ▮▮▯▯

IBM SPSS Statistics has some powerful tools for users to build their own functionality. For example, you can create your own dialog boxes and menus to run syntax that you may have written. SPSS Statistics also interfaces with a powerful open source statistical computing language called R (R Core Team, 2016). There are two extensions to SPSS that we use in this book. One is a tool called *PROCESS* and the other is the *Essentials for R for Statistics* plugin, which will give us access to R so that we can implement robust models using the

WRS2 package (Mair, Schoenbrodt, & Wilcox, 2015).

4.13.1 The *PROCESS* tool ▮▮▮▯

The *PROCESS* tool (Hayes, 2018) wraps up a range of functions written by Andrew Hayes and Kristopher Preacher (e.g., Hayes & Matthes, 2009; Preacher & Hayes, 2004, 2008a) to do moderation and mediation analyses, which we look at in Chapter 11. While using these tools, spare a thought of gratitude to Hayes and Preacher for using their spare time to do cool stuff like this that makes it possible for you to analyze your data without having a nervous breakdown. Even if you think you are having a nervous breakdown, trust me, it's not as big as the one you'd be having if *PROCESS* didn't exist. The *PROCESS* tool is what's known as a custom dialog box and it can be installed in three steps below (Mac OS users ignore step 2):

1 *Download the install file*. Download the file **process.spd** from Andrew Hayes's website: http://www.processmacro.org/download.html. Save this file onto your computer.

2 *Start up IBM SPSS Statistics as an administrator*. To install the tool in Windows, you need to start IBM SPSS Statistics as an administrator. To do this, make sure that SPSS isn't already running, and click the *Start* menu (⊞). Locate the icon for SPSS ([🔲 IBM SPSS Statistics 24]), which, if it's not in your most used list, will be listed under 'I' for IBM SPSS Statistics. The text next to the icon will refer to the version of SPSS Statistics that you have installed (if you have a subscription it will say 'Subscription' rather than a version number). Click on this icon with the *right mouse button* to activate the menu in Figure 4.20. Within this menu select (you're back to using the left mouse button now) [Run as administrator]. This action opens SPSS Statistics but

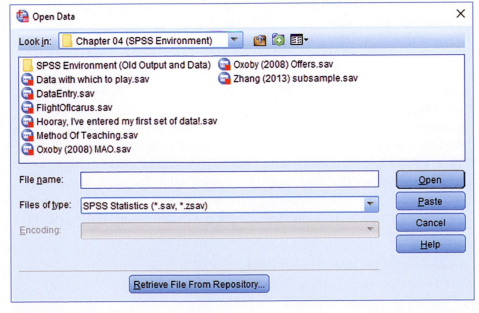

Figure 4.19 Dialog box to open a file

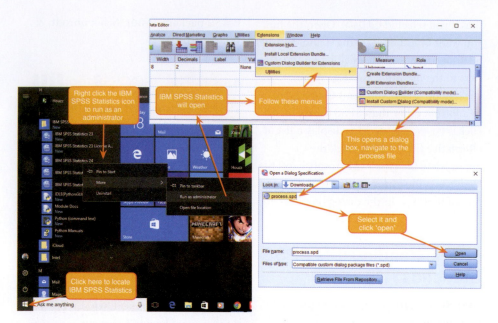

Figure 4.20 Installing the *PROCESS* menu

allows it to make changes to your computer. A dialog box will appear that asks you whether you want to let SPSS make changes to your computer and you should reply 'yes'.

3 Once SPSS has loaded select *Extensions* ▶ *Utilities* ▶ 📋 Install Custom Dialog (Compatibility mode)..., which activates a dialog box for opening files (Figure 4.20).[7] Locate the file **process. spd**, select it, and click 🔲 Open 🔲. This installs the *PROCESS* menu and dialog boxes into SPSS. If you get an error message, the most likely explanation is that you haven't opened SPSS as an administrator (see step 2).

4.13.2 Essentials for *R* ▮▮▮▮

At various points in the book we're going to use robust tests that use R. To get SPSS Statistics to interface with R, we need to install: (1) the version of R

that is compatible with our version of SPSS Statistics; and (2) the *Essentials for R for Statistics* plugin from IBM. At the time of writing, the R plugin isn't available for SPSS Statistics version 25, but by the time the book is published it may well be. These instructions are for SPSS Statistics version 24 but you can hopefully extrapolate to other versions. First, let's get the plugin and installation documentation from IBM:

1 Create an account on IBM.com (www-01.ibm.com).

2 Go to https://www-01.ibm.com/ marketing/iwm/iwm/web/preLogin. do?source=swg-tspssp

3 There will be a long list of stuff you can download. Select *IBM SPSS Statistics Version 24—Essentials for R* (or whatever version of SPSS Statistics you're using) and click continue.

4 Complete the privacy information, and read and agree (or not) to IBM's terms and conditions.

5 Download the version of *IBM SPSS Statistics Version 24—Essentials for R* for your operating system (Windows, Mac OS, Linux, etc.) and the corresponding installation instructions (labeled *Installation Documentation 24.0 Multilingual for xxx*, where *xxx* is the operating system you use). By default the website uses an app called the *Download Director* to manage the download. This app never works for me (on a Mac) and if you have the same problem, switch the tab at the top of the list of downloads to 'Download using http' (Download using Download Director Download using http) and download the files directly through your browser.

6 Open the installation documentation (it should be a PDF file) and check which version of R you need to install.[8]

Having got the *Essentials for R* plugin, *don't install it yet*. You need to check which version of R you need, and download it. SPSS Statistics typically uses an old version of R (because IBM needs to check that the *Essentials for R* plugin is stable before releasing it and by the time they have done that R has updated). Finding old versions of R is tediously overcomplicated; I've tried to illustrate the process in Figure 4.21.

7 Go to https://www.r-project.org/

8 Click the link labeled *CRAN* (under the *Download* heading) to go to a page to select a CRAN mirror. A CRAN mirror is a location from which to download R. It doesn't matter which you choose; because I'm based in the UK, I picked one of the UK links in Figure 4.21.

---

7 If you're using a version of SPSS earlier than 24, you need to select *Utilities* ▶ *Custom Dialogs* ▶.

8 At the time of writing, the installation documentation for SPSS Statistics 24 links to a PDF file for version 23, which says that you need R 3.1. This is true for version 23 of SPSS Statistics, but version 24 requires R 3.2 onwards.

9 On the next page, click the link for the operating system you use (Windows, Mac or Linux).

10 You will already know what version of R you're looking for because I told you to check before getting to this point (e.g., SPSS Statistics version 24 uses R version 3.2).[9] What happens next differs for Windows and Mac OS:

- Windows: If you selected the link to the Windows version you'll be directed to a page for R for Windows. Click the link labeled *Install R for the first time* to go a page to download R for Windows. *Do not* click the link at the top of the page, but scroll down to the section labeled *Other builds*, and click the link to *Previous releases*. The resulting page lists previous versions of R. Select the version you want (for SPSS Statistics 24, select R 3.2.5, for other versions of SPSS consult the documentation).

- Mac OS: If you selected the link to the OS X version you'll be directed to a page for R for Mac OS X. On this page click the link to the *old* directory. This takes you to a directory listing. You need to scroll down a bit until you find the *.pkg* files. Click the link to the *.pkg* file of the version of R that you want (for SPSS Statistics 24, click R 3.2.4, for other versions consult the documentation).

You should now have the install files for R and for the *Essentials for R* plugin in your download folder. Find them. First, install R by double-clicking the install file and going through the usual install process for your operating system. Having installed R, install the *Essentials for R* plugin by double-clicking the install file to initiate a

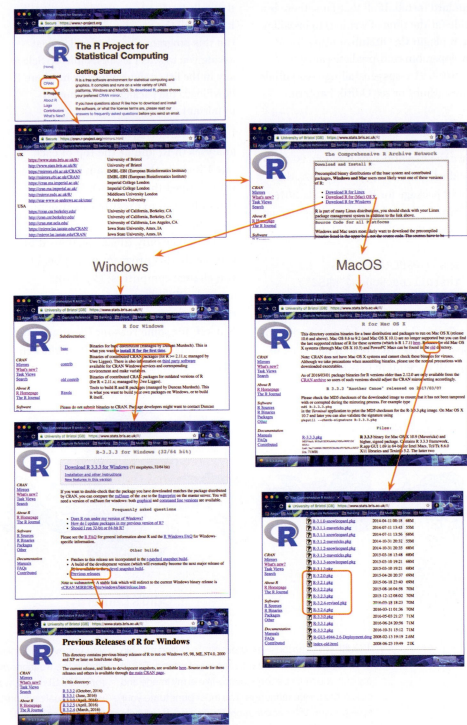

Windows

MacOS

Figure 4.21 Finding an old version of R is overly complicated ...

---

9 There will be several versions of R 3.2 which are denoted as 3.2.x, where x is a minor update. It shouldn't matter whether you install version 3.2.1 or 3.2.5, but you may as well go for the last of the releases. In the case of R 3.2, the last update before release 3.3 was 3.2.5.

standard install. If all that fails, there is a guide (at the time of writing) to installing the R plugin via GitHub at https://developer.ibm.com/predictiveanalytics/2016/03/21/r-spss-installing-r-essentials-from-github/ or see Oditi's Lantern.

4.13.3 The WRS2 package

Once the *Essentials for R* plugin is installed (see above) we can access the WRS2 package for R (Mair et al., 2015) by opening a syntax window and typing and executing the following syntax:

```
BEGIN PROGRAM R.
install.packages("WRS2")
END PROGRAM.
```

The first and last lines (remember the full stops) tell SPSS to talk to R and then to stop. All the stuff in between is language that tells R what to do. In this case it tells R to install the package *WRS2*. When you run this program a window will appear asking you to select a CRAN mirror. Select any in the list (it determines from where R downloads the package, so it's not an important decision).

I supply various syntax files for robust analyses in R, and at the top of each one I include this program (for those who skipped this section). However, you only need to execute this program once, not every time you run an analysis. The only times you'd need to re-execute this program would be: (1) if you change computers; (2) if you upgrade SPSS Statistics or need to reinstall the *Essentials for R* plugin or R itself, for some reason; (3) something goes wrong and you think it might help to reinstall *WRS2*.

4.13.4 Accessing the extensions

Once the *PROCESS* tool has been added to SPSS Statistics it appears in the *Analyze* ▶ *Regression* menu. If you can't see it then the install hasn't worked and you'll need to work through this section again. At the time of writing *WRS2* can be accessed only using syntax.

4.14 Brian and Jane's Story

Brian had been stung by Jane's comment. He was many things, but he didn't think he had his head up his own backside. He retreated from Jane to get on with his single life. He listened to music, met his friends, and played *Uncharted 4*.

Oditi's Lantern
SPSS extensions

'I, Oditi, am bearded like a great pirate sailing my ship of idiocy across the vacant seas of your mind. To join my cult you must become pirate-like in my image and speak the pirate language. You must punctuate your speech with the exclamation 'Rrrrrrrrrr'. It will help you uncover the unknown numerical truths embedded in the treasure maps of data. The Rrrrrrr plugin for SPSS Statistics will help, and my lantern is primed with a visual cannon-ball of an installation guide that will blow your mind.'

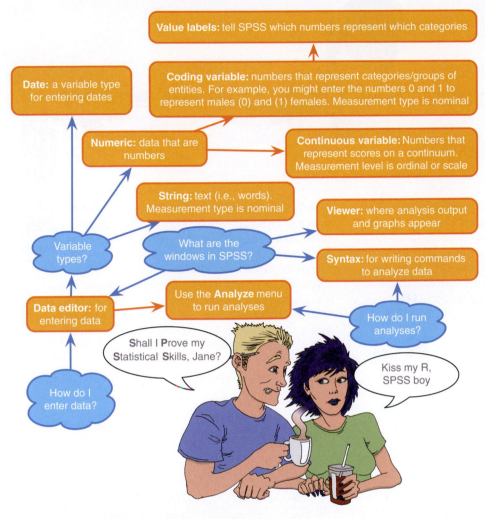

Figure 4.22   What Brian learnt from this chapter

Truthfully, he mainly played *Uncharted 4*. The more he played, the more he thought of Jane, and the more he thought of Jane, the more convinced he became that she'd be the sort of person who was into video games. When he next saw her he tried to start a conversation about games, but it went nowhere. She said computers were good only for analyzing data. The seed was sown, and Brian went about researching statistics packages. There were a lot of them. Too many. After hours on Google, he decided that one called SPSS looked the easiest to learn. He would learn it, and it would give him something to talk about with Jane. Over the following week he read books, blogs, watched tutorials on YouTube, bugged his lecturers, and practiced his new skills. He was ready to chew the statistical software fat with Jane.

He searched around campus for her: the library, numerous cafés, the quadrangle—she was nowhere. Finally, he found her in the obvious place: one of the computer rooms at the back of campus called the Euphoria cluster. Jane was studying numbers on the screen, but it didn't look like SPSS. 'What the hell . . . ,' Brian thought to himself as he sat next to her and asked . . .

4.15 What next? ▮▮▮▮

At the start of this chapter we discovered that I feared my new environment of primary school. My fear wasn't as irrational as you might think, because, during the time I was growing up in England, some idiot politician had decided that all school children had to drink a small bottle of milk at the start of the day. The government supplied the milk, I think, for free, but most free things come at some kind of price. The price of free milk turned out to be lifelong trauma. The milk was usually delivered early in the morning and then left in the hottest place someone could find until we innocent children hopped and skipped into the playground oblivious to the gastric hell that awaited. We were greeted with one of these bottles of warm milk and a very small straw. We were then forced to drink it through grimacing faces. The straw was a blessing because it filtered out the lumps formed in the gently curdling milk. Politicians take note: if you want children to enjoy school, don't force-feed them warm, lumpy milk.

But despite gagging on warm milk every morning, primary school was a very happy time for me. With the help of Jonathan Land, my confidence grew. With this new confidence, I began to feel comfortable not just at school but in the world more generally. It was time to explore.

4.16 Key terms that I've discovered

Currency variable	Date variable	Smartreader	Variable view
Data editor	Long format data	String variable	Viewer
Data view	Numeric variable	Syntax editor	Wide format data

Smart Alex's tasks

- **Task 1**: Smart Alex's first task for this chapter is to save the data that you've entered in this chapter. Save it somewhere on the hard drive of your computer (or a USB stick if you're not working on your own computer). Give it a sensible title and save it somewhere easy to find (perhaps create a folder called 'My Data Files' where you can save all of your files when working through this book). ▌▌▌▌

- **Task 2**: What are the following icons shortcuts to? ▌▌▌▌

- **Task 3**: The data below show the score (out of 20) for 20 different students, some of whom are male and some female, and some of whom were taught using positive reinforcement (being nice) and others who were taught using punishment (electric shock). Enter these data into SPSS and save the file as Method Of Teaching.sav. (Hint: the data should not be entered in the same way that they are laid out below.) ▌▌▌▌

Male		Female	
Electric Shock	Being Nice	Electric Shock	Being Nice
15	10	6	12
14	9	7	10
20	8	5	7
13	8	4	8
13	7	8	13

- **Task 4**: Thinking back to Labcoat Leni's Real Research 4.1, Oxoby also measured the minimum acceptable offer; these MAOs (in dollars) are below (again, they are approximations based on the graphs in the paper). Enter these data into the SPSS Data Editor and save this file as **Oxoby (2008) MAO.sav**. ▌▌▌▌
  Bon Scott group: 2, 3, 3, 3, 3, 4, 4, 4, 4, 4, 4, 4, 4, 5, 5, 5, 5, 5
  Brian Johnson group: 0, 1, 2, 2, 3, 3, 3, 3, 3, 4, 4, 4, 4, 4, 4, 4, 4, 1

- **Task 5**: According to some highly unscientific research done by a UK department store chain and reported in *Marie Claire* magazine (http://ow.ly/9Dxvy), shopping is good for you. They found that the average woman spends 150 minutes and walks 2.6 miles when she shops, burning off around 385 calories. In contrast, men spend only about 50 minutes shopping, covering 1.5 miles. This was based on strapping a pedometer on a mere 10 participants. Although I don't have the actual data, some simulated data based on these means are below. Enter these data into SPSS and save them as **Shopping Exercise.sav**. ▌▌▌▌

Male		Female	
Distance	Time	Distance	Time
0.16	15	1.40	22
0.40	30	1.81	140
1.36	37	1.96	160
1.99	65	3.02	183
3.61	103	4.82	245

- **Task 6**: This task was inspired by two news stories that I enjoyed. The first was about a Sudanese man who was forced to marry a goat after being caught having sex with it (http://ow.ly/9DyyP). I'm not sure whether he treated the goat to a nice dinner in a posh restaurant beforehand but, either way, you have to feel sorry for the goat. I'd barely had time to recover from that story when another appeared about an Indian man forced to marry a dog to atone for stoning two dogs and stringing them up in a tree 15 years earlier (http://ow.ly/9DyFn). Why anyone would think it's a good idea to enter a dog into matrimony with a man with a history of violent behavior towards dogs is beyond me. Still, I wondered whether a goat or dog made a better spouse. I found some other people who had been forced to marry goats and dogs and measured their life satisfaction and how much they like animals. Enter these data into SPSS and save as **Goat or Dog.sav**. ▌▌▌▌

Goat		Dog	
Animal Liking	Life Satisfaction	Animal Liking	Life Satisfaction
69	47	16	52
25	6	65	66
31	47	39	65
29	33	35	61
12	13	19	60
49	56	53	68
25	42	27	37
35	51	44	72
51	42		
40	46		
23	27		
37	48		

Cups of Tea	Cognitive Functioning
3	40
3	54
4	34
1	46

- **Task 7**: One of my favorite activities, especially when trying to do brain-melting things like writing statistics books, is drinking tea. I am English, after all. Fortunately, tea improves your cognitive function—well, it does in old Chinese people, at any rate (Feng, Gwee, Kua, & Ng, 2010). I may not be Chinese and I'm not *that* old, but I nevertheless, enjoy the idea that tea might help me think. Here are some data based on Feng et al.'s study that measured the number of cups of tea drunk and cognitive functioning in 15 people. Enter these data into SPSS and save the file as **Tea Makes You Brainy 15.sav**. ▮▮▮

Cups of Tea	Cognitive Functioning
2	60
4	47
3	31
4	62
2	44
3	41
5	49
5	56
2	45
5	56
1	57

- **Task 8**: Statistics and maths anxiety are common and affect people's performance on maths and stats assignments; women, in particular, can lack confidence in mathematics (Field, 2010). Zhang, Schmader, & Hall, (2013) did an intriguing study, in which students completed a maths test in which some put their own name on the test booklet, whereas others were given a booklet that already had either a male or female name on it. Participants in the latter two conditions were told that they would use this other person's name for the purpose of the test. Women who completed the test using a different name performed better than those who completed the test using their own name. (There were no such effects for men.) The data below are a random subsample of Zhang et al.'s data. Enter them into SPSS and save the file as **Zhang (2013) subsample.sav** ▮▮▮

Male			Female		
Female Fake Name	Male Fake Name	Own Name	Female Fake Name	Male Fake Name	Own Name
33	69	75	53	31	70
22	60	33	47	63	57
46	82	83	87	34	33
53	78	42	41	40	83
14	38	10	62	22	86
27	63	44	67	17	65
64	46	27	57	60	64
62	27			47	37
75	61			57	80
50	29				

- **Task 9**: What is a coding variable? ▮▮▮
- **Task 10**: What is the difference between wide and long format data? ▮▮▮

Answers & additional resources are available on the book's website at
https://edge.sagepub.com/field5e

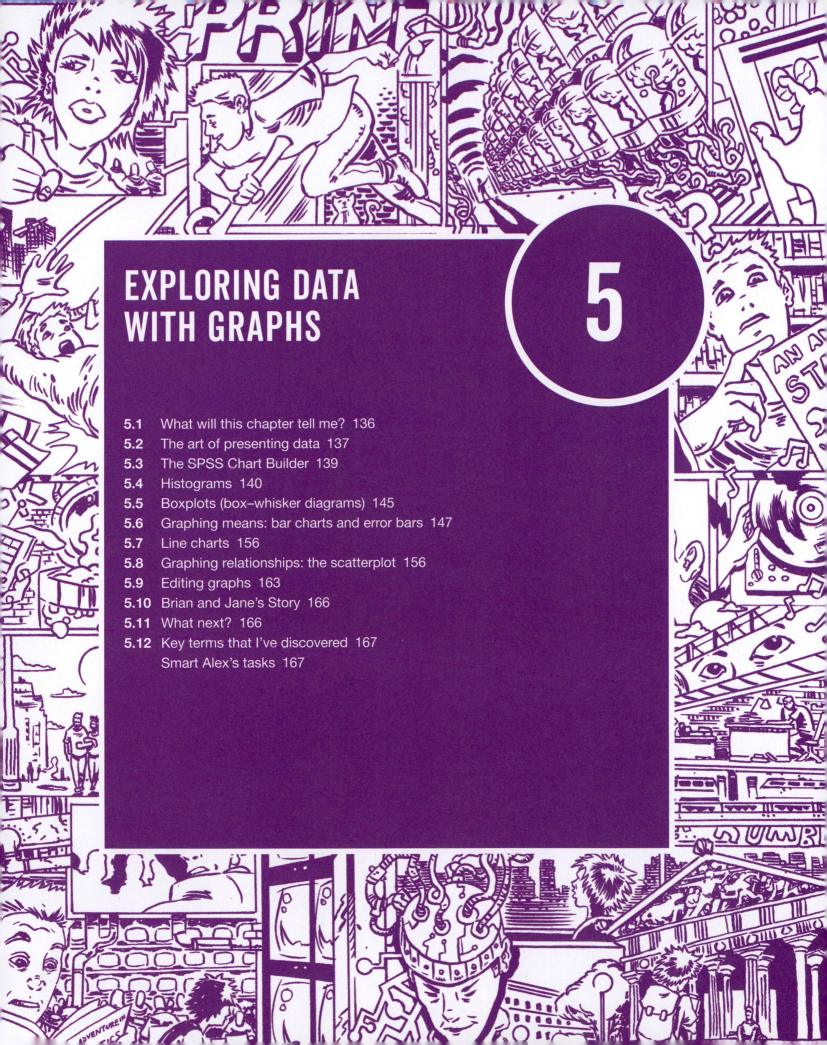

EXPLORING DATA WITH GRAPHS

5

5.1 What will this chapter tell me?

As I got older I found the joy of exploring. Many a happy holiday was spent clambering over the rocky Cornish coastline with my dad. At school they taught us about maps and the importance of knowing where you are going and what you are doing. I had a more relaxed view of exploration and there is a little bit of lifelong theme of me wandering off to whatever looks most exciting at the time and assuming I know where I am going.[1] I got lost at a holiday camp once when I was about 3 or 4. I remember nothing about it, but my parents tell a tale of frantically running around trying to find me while I was happily entertaining myself (probably by throwing myself out of a tree or something). My older brother, 'the clever one', apparently wasn't 'the observant one' and got a bit of flak for neglecting his duty of watching me. In his defense he was probably mentally deriving equations to bend time and space at the time. He did that a lot when he was 7. The careless explorer in me hasn't really gone away: in new cities I tend to just wander off and hope for the best, usually get lost and so far haven't managed to die (although I tested my luck once by unknowingly wandering through part of New Orleans where tourists get attacked a lot— it seemed fine to me). To explore data in the way that the 6-year-old me used to explore the world is to spin around 8000 times while drunk and then run along the edge of a cliff. With data you can't be as careless as I am in new cities. To negotiate your way around your data you need a map, maps of data are called graphs, and it is into this tranquil and tropical ocean that we now dive (with a compass and ample supply of oxygen, obviously).

1   It terrifies me that my sons might have inherited this characteristic.

Figure 5.1   Explorer Field borrows a bike and gets ready to ride it recklessly around a caravan site

5.2 The art of presenting data ▌▌▌▌

Wright (2003) adopts Rosenthal's view that researchers should 'make friends with their data'. Although it is true that statisticians need all the friends they can get, Rosenthal didn't mean that: he was urging researchers not to rush data analysis. Wright uses the analogy of a fine wine: you should savor the bouquet and delicate flavors to truly enjoy the experience. He's considerably overstating the joys of data analysis, but rushing your analysis is, I suppose, a bit like downing a bottle of wine: the consequences are messy and incoherent. So, how do we make friends with data? The first thing is to look at a graph; for data this is the equivalent of a profile picture. Although it is definitely wrong to judge people based on their appearance, with data the opposite is true.

5.2.1 What makes a good graph? ▌▌▌

What makes a good profile picture? Lots of people seem to think that it's best to jazz it up: have some impressive background location, strike a stylish pose, mislead people by inserting some status symbols that you've borrowed, adorn yourself with eye-catching accessories, wear your best clothes to conceal the fact you usually wear a onesie, look like you're having the most fun ever so that people think your life is perfect. This is OK for a person's profile picture, but it is not OK for data: you should avoid impressive backgrounds, eye-catching accessories or symbols that distract the eye, don't fit models that are unrepresentative of reality, and definitely don't mislead anyone into thinking the data matches your predictions perfectly.

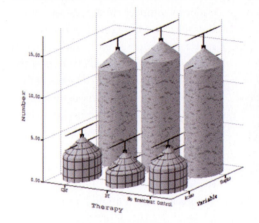

Error Bars show 95.0 % CI of Mean

Bars show Means

Figure 5.2 A cringingly bad example of a graph from the first edition of this book (left) and Florence Nightingale (right) who would have mocked my efforts

Unfortunately, all stats software (including IBM SPSS Statistics) enables you to do all of things that I just told you not to do (see Section 5.9). You may find yourself losing consciousness at the excitement of coloring your graph bright pink (really, it's amazing how excited my students get at the prospect of bright pink graphs—personally, I'm not a fan). Much as pink graphs might send a twinge of delight down your spine, remember why you're drawing the graph—it's not to make yourself (or others) purr with delight at the pinkness, it's to present information (dull, but true).

Tufte (2001) points out that graphs should do the following, among other things:

✓ Show the data.
✓ Induce the reader to think about the data being presented (rather than some other aspect of the graph, like how pink it is).
✓ Avoid distorting the data.
✓ Present many numbers with minimum ink.
✓ Make large data sets (assuming you have one) coherent.
✓ Encourage the reader to compare different pieces of data.
✓ Reveal the underlying message of the data.

Graphs often don't do these things (see Wainer, 1984, for some examples), and there is a great example of how not to draw a graph in the first edition of this book (Field, 2000). Overexcited by SPSS's ability to add distracting fluff to graphs (like 3-D effects, fill effects and so on—Tufte calls these **chartjunk**), I went into some weird orgasmic state and produced the absolute abomination in Figure 5.2. I literally don't know what I was thinking. Data visualization pioneer, Florence Nightingale, also wouldn't have known what I was thinking.[2] The only positive is that it's not bloody pink! What do you think is wrong with this graph?

✗ The bars have a 3-D effect: Never use 3-D plots for a graph plotting two variables because it obscures the data.[3] In particular, 3-D effects make it hard to see the values of the bars: in Figure 5.2, for example, the 3-D effect makes the error bars almost impossible to read.
✗ Patterns: The bars also have patterns, which, although very pretty, distract the eye from what matters (namely the data). These are completely unnecessary.

---

2 You may be more familiar with Florence Nightingale as a pioneer of modern nursing, but she also pioneered the visualization of data, not least by inventing the pie chart. Pretty amazing woman.

3 If you do 3-D plots when you're plotting only two variables, then a bearded statistician will come to your house, lock you in a room and make you write Ι μυστ νοτ δο 3–Δ γραπησ 75,172 times on the blackboard. Seriously.

✗ Cylindrical bars: Were my data so sewage-like that I wanted to put them in silos? The cylinder effect muddies the data and distracts the eye from what is important.

✗ Badly labeled *y*-axis: 'Number' of what? Delusions? Fish? Cabbage-eating sea lizards from the eighth dimension? Idiots who don't know how to draw graphs?

Now take a look at the alternative version of this graph (Figure 5.3). Can you see what improvements have been made?

✓ A 2-D plot: The completely unnecessary third dimension is gone, making it much easier to compare the values across therapies and thoughts/behaviors.

✓ I have superimposed the summary statistics (means and confidence intervals) over the raw data so readers get a full sense of the data (without it being overwhelming).

✓ The *y*-axis has a more informative label: We now know that it was the number of obsessions per day that was measured. I've also added a legend to inform readers that obsessive thoughts and actions are differentiated by color.

✓ Distractions: There are fewer distractions like patterns, cylindrical bars and the like.

✓ Minimum ink: I've got rid of superfluous ink by getting rid of the axis lines and by using subtle grid lines to make it easy to read values from the *y*-axis. Tufte would be pleased.

5.2.2 Lies, damned lies, and . . . erm . . . graphs

Governments lie with statistics, but scientists shouldn't. How you present your data makes a huge difference to the message conveyed to the audience. As a big fan of cheese, I'm often curious about whether the urban myth that it gives you nightmares is true. Shee (1964) reported the case of a man who had nightmares about his workmates: 'He dreamt of one, terribly mutilated, hanging from a meat-hook.[4] Another he dreamt of falling into a bottomless abyss. When cheese was withdrawn from his diet the nightmares ceased.' This would not be good news if you were the minister for cheese in your country.

Figure 5.4 shows two graphs that, believe it or not, display the same data: the number of nightmares had after eating cheese. The graph on the left shows how the graph should probably be scaled. The *y*-axis reflects the maximum of the scale, and this creates the correct impression: that people have more nightmares about colleagues hanging from meat-hooks if they eat cheese before bed. However, as minister for cheese, you want people to think the opposite; all you do is rescale the graph (by extending the *y*-axis way beyond the average number of nightmares) and suddenly the difference in nightmares has diminished considerably. Tempting as it is, don't do this (unless, of course, you plan to be a politician at some point in your life).

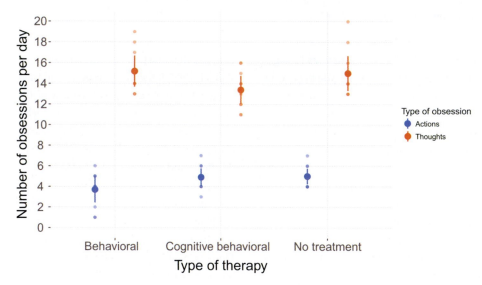

Figure 5.3 Figure 5.2 drawn properly

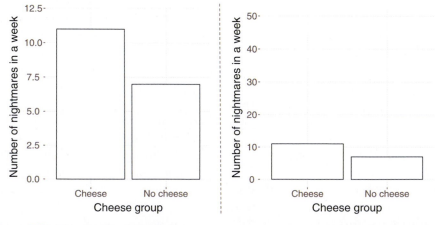

Figure 5.4 Two graphs about cheese

---

4 I have similar dreams, but that has more to do with some of my workmates than cheese.

Cramming Sam's Tips
Graphs

- The vertical axis of a graph is known as the *y*-axis (or ordinate).
- The horizontal axis of a graph is known as the *x*-axis (or abscissa).

If you want to draw a good graph follow the cult of Tufte:

- Don't create false impressions of what the data show (likewise, don't hide effects) by scaling the *y*-axis in some weird way.

- Avoid chartjunk: Don't use patterns, 3-D effects, shadows, pictures of spleens, photos of your Uncle Fred, pink cats or anything else.
- Avoid excess ink: This is a bit radical, but if you don't need the axes, then get rid of them.

5.3 The SPSS Chart Builder ▌▌▌▌

You are probably drooling like a rabid dog to get into the statistics and to discover the answer to your fascinating research question, so surely graphs are a waste of your precious time? Data analysis is a bit like Internet dating (it's not, but bear with me). You can scan through the vital statistics and find a perfect match (good IQ, tall, physically fit, likes arty French films, etc.) and you'll think you have found the perfect answer to your question. However, if you haven't looked at a picture, then you don't really know how to interpret this information—your perfect match might turn out to be Rimibald the Poisonous, King of the Colorado River Toads, who has genetically combined himself with a human to further his plan to start up a lucrative rodent farm (they like to eat small rodents).[5] Data analysis is much the same: inspect your data with a picture, see how it looks and only then can you interpret the more vital statistics.

Although SPSS's graphing facilities are quite versatile (you can edit most things— see Section 5.9), they are still quite limited for repeated-measures data.[6] To draw graphs in SPSS we use the all-singing and all-dancing **Chart Builder**.[7]

Figure 5.5 shows the basic *Chart Builder* dialog box, which is accessed through the *Graphs* ▶ 📊 Chart Builder... menu.

---

5 On the plus side, he would have a long sticky tongue and if you smoke his venom (which, incidentally, can kill a dog), you'll hallucinate (if you're lucky, you'll hallucinate that you weren't on a date with a Colorado river toad–human hybrid).

6 For this reason, some of the graphs in this book were created using a package called *ggplot2* for the software R, in case you're wondering why you can't replicate them in SPSS.

7 Unfortunately it's dancing like an academic at a conference disco and singing 'I will always love you' in the wrong key after 34 pints of beer.

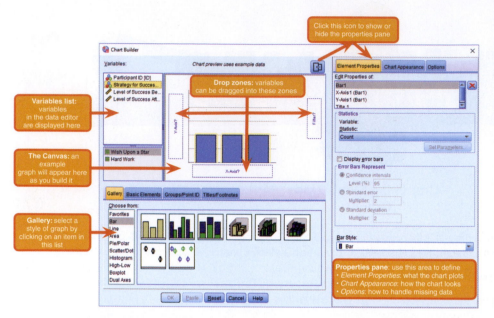

Figure 5.5 The SPSS Chart Builder

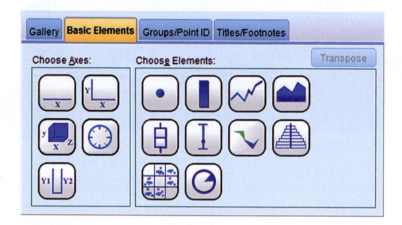

Figure 5.6 Building a graph from basic elements

There are some important parts of this dialog box:

- *Gallery*: For each type of graph, a gallery of possible variants is shown. Double-click an icon to select a particular type of graph.
- *Variable list*: The variables in the data editor are listed here. These can be dragged into *drop zones* to specify what is displayed on the graph.
- *The canvas*: This is the main area in the dialog box and is where the graph is previewed as you build it.

- *Drop zones*: You can drag variables from the variable list into zones designated with blue dotted lines, called drop zones.
- *Properties panel*: the right-hand panel is where you determine what the graph displays, its appearance, and how to handle missing values.

There are two ways to build a graph: the first is by using the gallery of predefined graphs; and the second is by building a graph on an element-by-element basis. The gallery is the default option and this

tab (Gallery Basic Elements Groups/Point ID Titles/Footnotes) is automatically selected; however, if you want to build your graph from basic elements then click the 'Basic Elements' tab (Gallery Basic Elements Groups/Point ID Titles/Footnotes) to change the bottom of the dialog box in Figure 5.5 to look like Figure 5.6.

We will have a look at building various graphs throughout this chapter rather than trying to explain everything in this introductory section (see also SPSS Tip 5.1). Most graphs that you are likely to need can be obtained using the gallery view, so I will tend to stick with this method.

5.4 Histograms ▌▌▌▌

We encountered histograms (frequency distributions) in Chapter 1; they're a useful way to look at the shape of your data and spot problems (more on that in the next chapter). We will now learn how to create one in SPSS. My wife and I spent our honeymoon at Disney in Orlando.[8] It was two of the best weeks of my life. Although some people find the Disney experience a bit nauseating, in my view there is absolutely nothing wrong with spending time around people who are constantly nice and congratulate you on your marriage. The world could do with more 'nice' in it. The one blip in my tolerance of Disney was their obsession with dreams coming true and wishing upon a star. Don't misunderstand me, I love the idea of having dreams (I haven't yet given up on the idea that one day Steve Harris from Iron Maiden might call requiring my drumming services for their next world tour, nor have I stopped thinking, despite all the physical evidence to the contrary, that I could step in and help my favorite soccer team at their time of need). Dreams are good, but a completely blinkered view that they'll come true without any work on your part

---

8 Although not necessarily representative of our Disney experience, I have put a video of a bat fellating itself at the Animal Kingdom on my YouTube channel. It won't help you to learn statistics.

SPSS Tip 5.1
Strange dialog boxes ▌▌▌▌

When you first load the Chart Builder a dialog box appears that seems to signal an impending apocalypse (Figure 5.7). In fact, SPSS is helpfully reminding you that for the Chart Builder to work, you need to have set the level of measurement correctly for each variable. That is, when you defined each variable you must have set them correctly to be scale, ordinal or nominal (see Section 4.6.2). This is because SPSS needs to know whether variables are categorical (nominal) or continuous (scale) when it creates the graphs. If you have been diligent and set these properties when you entered the data then click OK to make the dialog disappear. If you forgot to set the level of measurement for any variables then click Define Variable Properties... to go to a new dialog box in which you can change the properties of the variables in the data editor.

Figure 5.7   Initial dialog box when the Chart Builder is opened

---

is not. My chances of playing drums for Iron Maiden will be greatly enhanced by me practicing, forging some kind of name for myself as a professional drummer, and incapacitating their current drummer (sorry, Nicko). I think it highly unlikely that merely 'wishing upon a star' will make my dream come true. I wonder if the seismic increase in youth internalizing disorders (Twenge, 2000) is, in part, caused by millions of Disney children reaching the rather depressing realization that 'wishing upon a star' didn't work.

Sorry, I started that paragraph in the happy glow of honeymoon memories, but somewhere in the middle it took a turn for the negative. Anyway, I collected some data from 250 people on their level of success

using a composite measure involving their salary, quality of life and how closely their life matches their aspirations. This gave me a score from 0 (complete failure) to 100 (complete success). I then implemented an intervention: I told people that, for the next 5 years, they should either wish upon a star for their dreams to come true or work as hard as they could to make their dreams come true. I measured their success again 5 years later. People were randomly allocated to these two instructions. The data are in **Jiminy Cricket.sav**. The variables are **Strategy** (hard work or wishing upon a star), **Success_Pre** (their baseline level of success) and **Success_Post** (their level of success 5 years later).

First, access the Chart Builder as in Figure 5.5 and select *Histogram* in the list labeled *Choose from* to bring up the gallery shown in Figure 5.8. This gallery has four icons, representing different types of histogram. Select the appropriate one either by double-clicking on it or by dragging it onto the canvas:

SELF TEST

What does a histogram show?

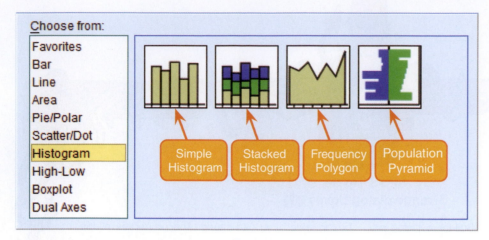

Figure 5.8 The histogram gallery

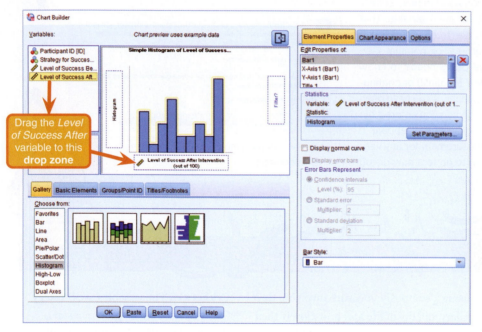

Figure 5.9 Defining a histogram in the Chart Builder

frequencies for each population on the horizontal: the populations appear back to back on the graph. This option is useful for comparing distributions across groups.

Let's do a simple histogram. Double-click the icon for a simple histogram (Figure 5.8). The *Chart Builder* dialog box will show a preview of the graph on the canvas. At the moment it's not very exciting (top of Figure 5.9) because we haven't told SPSS which variables to plot. The variables in the data editor are listed on the left-hand side of the Chart Builder, and any of these variables can be dragged into any of the spaces surrounded by blue dotted lines (the *drop zones*).

A histogram plots a single variable (x-axis) against the frequency of scores (y-axis), so all we need to do is select a variable from the list and drag it into X-Axis?. Let's do this for the post-intervention success scores. Click this variable (**Level of Success After**) in the list and drag it to X-Axis? as shown in Figure 5.9; you will now find the histogram previewed on the canvas. (It's not a preview of the actual data—it displays the general form of the graph, not what your specific data will look like.) You can edit what the histogram displays using the properties panel (SPSS Tip 5.2). To draw the histogram click OK.

The resulting histogram is shown in Figure 5.11. The distribution is quite lumpy: although there is a peak of scores around 50 (the midpoint of the scale), there are quite a few scores at the high end, and fewer at the low end. This creates the impression of negative skew, but it's not quite as simple as that. To help us to dig a bit deeper it might be helpful to plot the histogram separately for those who wished upon a star and those who worked hard: after all, if the intervention was a success then their distributions should be from different populations.

To compare frequency distributions of several groups simultaneously we can use

- *Simple histogram*: Use this option to visualize frequencies of scores for a single variable.
- *Stacked histogram*: If you have a grouping variable (e.g., whether people worked hard or wished upon a star) you can produce a histogram in which each bar is split by group. In this example, each bar would have two colors, one representing people who worked hard and the other those who wished upon a star. This option is a good way to compare the relative frequency of scores

across groups (e.g., were those who worked hard more successful than those who wished upon a star?).

- *Frequency polygon*: This option displays the same data as the simple histogram, except that it uses a line instead of bars to show the frequency, and the area below the line is shaded.
- *Population pyramid*: Like a stacked histogram, this shows the relative frequency of scores in two populations. It plots the variable (e.g., success after 5 years) on the vertical axis and the

SPSS Tip 5.2
The properties pane ▊▊▊▊

You can edit the histogram using the properties pane (Figure 5.10), which you can show or hide by clicking ▣. First, you can change the statistic displayed using the *Element Properties* tab: the default is *Histogram* but if you want to express values as a percentage rather than a frequency, select *Histogram Percent*. You can also decide manually how you want to divide up your data by clicking Set Parameters… . In the resulting dialog box you can determine properties of the 'bins' used to make the histogram. Think of a bin as, well, a rubbish bin (this is a pleasing analogy, as you will see): on each rubbish bin you write a score (e.g., 3) or a range of scores (e.g., 1–3), then you go through each score in your data set and throw it into the rubbish bin with the appropriate label on it (so a score of 2 gets thrown into the bin labeled 1–3). When you have finished throwing your data into these rubbish bins, you count how many scores are in each bin. A histogram is created in much the same way; either SPSS decides how the bins are labeled (the default) or you decide. Our success scores range from 0 to 100, therefore we might decide that our bins should begin with 0 and we could set the ⊙ Custom value for anchor: property to 0. We might also decide that we want each bin to contain scores between whole numbers (i.e., 0–1, 1–2, 2–3, etc.), in which case we could set the ⊙ Interval width: to be 1. This is what I've done in Figure 5.10, but for the time being leave the default settings (i.e., everything set to ⊙ Automatic).

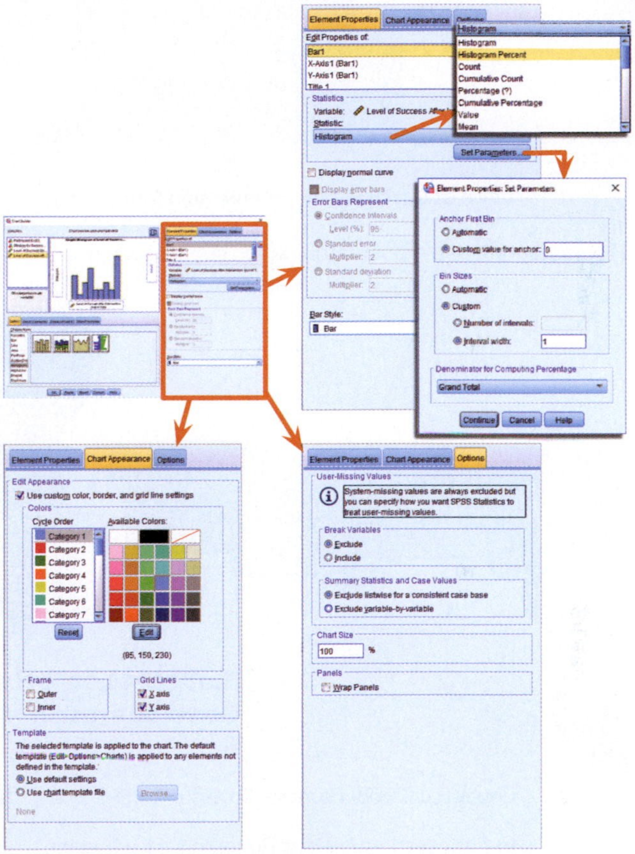

In the *Chart Appearance* tab we can change the default color scheme (the bars will be colored blue, but we can change this by selecting *Category 1* and choosing a different color). We can also choose to have an inner or outer frame (I wouldn't bother) and switch off the grid lines (I'd leave them). It's also possible to apply templates so you can create color schemes, save them as templates and apply them to subsequent charts.

Finally, the *Options* tab enables us to determine how we deal with user-defined missing values (the defaults are fine for our purposes) and whether to wrap panels on charts where we have panels representing different categories (this is useful when you have a variable containing a lot of categories).

Figure 5.10   The properties pane

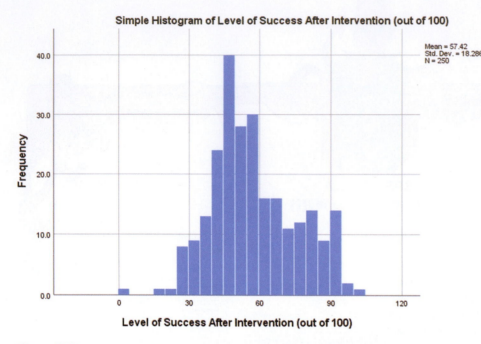

Figure 5.11 Histogram of the post-intervention success scores

a population pyramid. Click the population pyramid icon (Figure 5.8) to display the template for this graph on the canvas. Then, from the variable list, select the variable representing the success scores after the intervention and drag it into `Distribution Variable?` to set it as the variable that you want to plot. Then select the variable **Strategy** and drag it to `Split Variable?` to set it as the variable for which you want to plot different distributions. The dialog should now look like Figure 5.12 (note I have hidden the properties pane)—the variable names are displayed in the drop zones, and the canvas now displays a preview of our graph (e.g., there are two histograms representing each strategy for success). Click `OK` to produce the graph.

The resulting population pyramid is show, in Figure 5.13. It shows that for those who wished upon a star there is a fairly normal

Labcoat Leni's Real Research 5.1
Gonna be a rock 'n' roll singer (again) ▌▌▌▌

Oxoby, R. J. (2008). *Economic Enquiry, 47*(3), 598–602.

In Labcoat Leni's Real Research 4.1 we came across a study that compared economic behavior while different music by AC/DC played in the background. Specifically, Oxoby manipulated whether the background song was sung by AC/DC's original singer (Bonn Scott) or his replacement (Brian Johnson). He measured how many offers participants accepted (**Oxoby (2008) Offers.sav**) and the minimum offer they would accept (**Oxoby (2008) MOA.sav**). See Labcoat Leni's Real Research 4.1 for more detail on the study. We entered the data for this study in the previous chapter; now let's graph it. Produce separate population pyramids for the number of offers and the minimum acceptable offer, and in both cases split the data by which singer was singing in the background music. Compare these plots with Figures 1 and 2 in the original article.

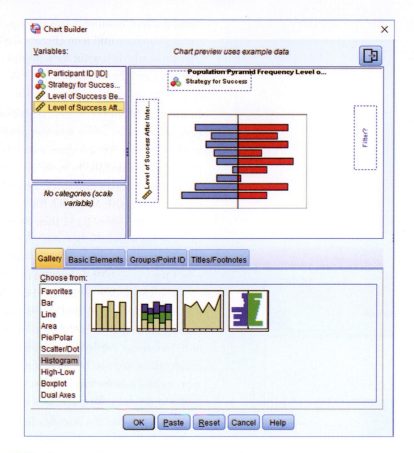

Figure 5.12 Defining a population pyramid in the Chart Builder

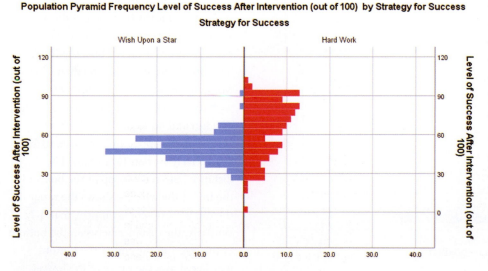

Figure 5.13 Population pyramid of success scores (5 years after different strategies were implemented)

SELF TEST

Produce a histogram and population pyramid for the success scores *before* the intervention.

distribution centered at about the midpoint of the success scale (50%). A small minority manage to become successful just by wishing, but most just end up sort of averagely successful. Those who work hard show a skewed distribution, where a large proportion of people (relative to those wishing) become very successful, and fewer people are around or below the midpoint of the success scale. Hopefully, this example shows how a population pyramid can be a very good way to visualize differences in distributions in different groups (or populations).

5.5 Boxplots (box–whisker diagrams) ▌▌▌▌

A **boxplot** or **box–whisker diagram** is one of the best ways to display your data. At the center of the plot is the *median*. This statistic is surrounded by a box the top and bottom of which are the limits within which the middle 50% of observations fall (the interquartile range, IQR). Sticking out of the top and bottom of the box are two whiskers which show the top and bottom 25% of scores (approximately). First, we will plot some boxplots using the Chart Builder, and then we'll look at what they tell us in more detail.

In the Chart Builder (Figure 5.5) select *Boxplot* in the list labeled *Choose from* to bring up the gallery in Figure 5.14. Boxplots display a summary for a single outcome variable; for example, we might choose level of success after 5 years for our 'wishing upon a star' example. There are three types of boxplot you can produce:

- *1-D Boxplot*: This option produces a single boxplot of all scores for the chosen outcome (e.g., level of success after 5 years).
- *Simple boxplot*: This option produces multiple boxplots for the chosen outcome by splitting the data by a categorical variable. For example, we had two groups:

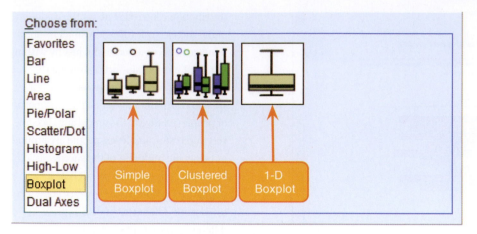

Figure 5.14   The boxplot gallery

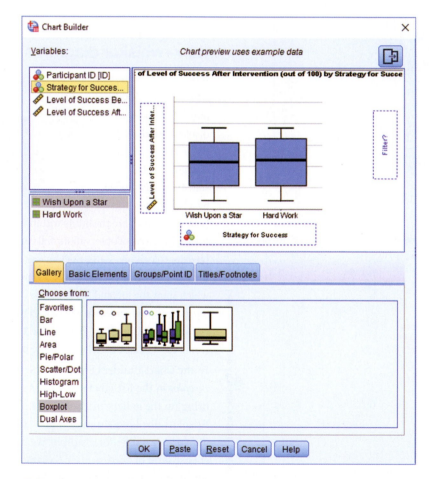

Figure 5.15   Completed dialog box for a simple boxplot

and workers, but within these groups we could also have different-colored boxplots for those who believe in the power of wishing and those who do not.

In the data file of the success scores we have information about whether people worked hard or wished upon a star. Let's plot this information. To make a boxplot of the post-intervention success scores for our two groups, double-click the *simple boxplot* icon (Figure 5.14), then from the variable list select the **Success_Post** variable and drag it into $\boxed{\text{Y-Axis?}}$ and select the variable **Strategy** and drag it to $\boxed{\text{X-Axis?}}$. The dialog box should now look like Figure 5.15—note that the variable names are displayed in the drop zones, and the canvas displays a preview of our graph (there are two boxplots: one for wishers and one for hard workers). Click $\boxed{\text{OK}}$ to produce the graph.

Figure 5.16 shows the boxplots for the success data. The blue box represents the IQR (i.e., the middle 50% of scores). The box is much longer in the hard-work group than for those who wished upon a star, which means that the middle 50% of scores are more spread out in the hard-work group. Within the box, the thick horizontal line shows the median. The workers had a higher median than the wishers, indicating greater success overall. The top and bottom of the blue box represent the upper and lower quartile, respectively (see Section 1.8.5). The distance between the top of the box and the top of the whisker shows the range of the top 25% of scores (approximately); similarly, the distance between the bottom of the box and the end of the bottom whisker shows the range of the lowest 25% of scores (approximately). I say 'approximately' because SPSS looks for unusual cases before creating the whiskers: any score greater than the upper quartile plus 1.5 times the IQR is deemed to be an 'outlier' (more on those in Chapter 5), and any case greater than the

wishers and workers. It would be useful to use this option to display different boxplots (on the same graph) for these groups (unlike the 1-D boxplot, which lumps the data from these groups together).

• *Clustered boxplot*: This option is the same as the simple boxplot, except that it splits the data by a second categorical variable. Boxplots for this second variable are produced in different colors. For example, imagine we had also measured whether our participants *believed* in the power of wishing. We could produce boxplots not just for the wishers

upper quartile plus 3 times the IQR is an 'extreme case'. The same rules are applied to cases below the lower quartile. When there are no unusual cases, the whiskers show the top and bottom 25% of scores exactly, but when there are unusual cases, they show the top and bottom 25% of scores only approximately because the unusual cases are excluded. The whiskers also tell us about the range of scores because the top and bottom of the whiskers show the lowest and highest scores *excluding unusual cases*.

In terms of the success scores, the range of scores was much wider for the workers than the wishers, but the wishers contained an outlier (which SPSS shows as a circle) and an extreme score (which SPSS shows as an asterisk). SPSS labels these cases with the row number from the data editor (in this case, rows 204 and 229), so that we can identify these scores in the data, check that they were entered correctly or look for reasons why they might have been unusual. Like histograms, boxplots also tell us whether the distribution is symmetrical or skewed. If the whiskers are the same length then the distribution is symmetrical (the range of the top and bottom 25% of scores is the same); however, if the top or bottom whisker is much longer than the opposite whisker then the distribution is asymmetrical (the range of the top and bottom 25% of scores is different). The scores from those wishing upon a star look symmetrical because the two whiskers are of similar lengths, but the hard-work group show signs of skew because the lower whisker is longer than the upper one.

5.6 Graphing means: bar charts and error bars ▮▮▮▮

Bar charts are the usual way for people to display means, although they are not ideal because they use a lot of ink to display only one piece of information. How you create bar graphs in SPSS depends on

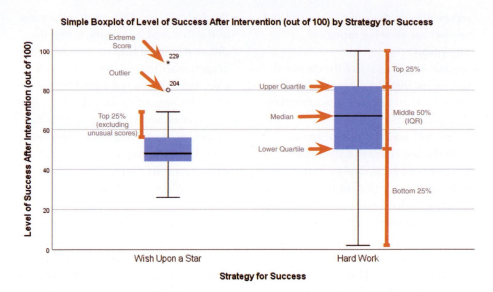

Figure 5.16 Boxplot of success scores, 5 years after implementing a strategy of working hard or wishing upon a star

Produce boxplots for the success scores *before* the intervention.

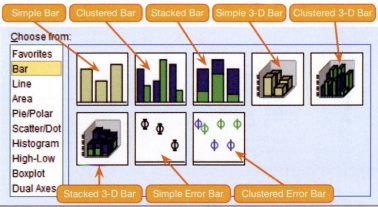

Figure 5.17 The bar chart gallery

whether the means come from independent cases and so are independent or come from the same cases and so are related. We'll look at both situations. Our starting point is always the Chart Builder (Figure 5.5). In this dialog box select *Bar* in the list labeled *Choose from* to bring up the gallery shown in Figure 5.17. This gallery has eight icons, representing different types of bar chart that you can select by double-clicking one or by dragging it onto the canvas.

- *Simple bar*: Use this option to display the means of scores across different groups or categories of cases. For example, you might want to plot the mean ratings of two films.
- *Clustered bar*: If you have a second grouping variable you can produce a simple bar chart (as above) but with different colored bars to represent levels of a second grouping variable. For example, you could have ratings of the two films, but for each film have a bar

representing ratings of 'excitement' and another bar showing ratings of 'enjoyment'.

- *Stacked bar*: This is like the clustered bar, except that the different-colored bars are stacked on top of each other rather than placed side by side.
- *Simple 3-D bar*: This is also like the clustered bar, except that the second grouping variable is displayed not by different-colored bars, but by an additional axis. Given what I said in Section 5.2 about 3-D effects obscuring the data, my advice is to stick to a clustered bar chart and not use this option.
- *Clustered 3-D bar*: This is like the clustered bar chart above, except that you can add a third categorical variable on an extra axis. The means will almost certainly be impossible for anyone to read on this type of graph, so don't use it.

- *Stacked 3-D bar*: This graph is the same as the clustered 3-D graph, except the different-colored bars are stacked on top of each other instead of standing side by side. Again, this is not a good type of graph for presenting data clearly.
- *Simple error bar*: This is the same as the simple bar chart, except that, instead of bars, the mean is represented by a dot, and a line represents the precision of the estimate of the mean (usually, the 95% confidence interval is plotted, but you can plot the standard deviation or standard error of the mean instead). You can add these error bars to a bar chart anyway, so really the choice between this type of graph and a bar chart with error bars is largely down to personal preference. (Including the bar adds a lot of superfluous ink, so if you want to be Tuftian about it you'd probably use this option over a bar chart.)

- *Clustered error bar*: This is the same as the clustered bar chart, except that the mean is displayed as a dot with an error bar around it. These error bars can also be added to a clustered bar chart.

5.6.1 Simple bar charts for independent means

To begin with, imagine that a film company director was interested in whether there was really such a thing as a 'chick flick' (a film that has the stereotype of appealing to women more than to men). He took 20 men and 20 women and showed half of each sample a film that was supposed to be a 'chick flick' (*The Notebook*). The other half watched a documentary about notebooks as a control. In all cases the company director measured participants' arousal[9] as an indicator of how much they enjoyed the film. Load the data in a file called **Notebook.sav** from the companion website.

Let's plot the mean rating of the two films. To do this, double-click the icon for a simple bar chart in the Chart Builder (Figure 5.17). On the canvas you will see a graph and two drop zones: one for the *y*-axis and one for the *x*-axis. The *y*-axis needs to be the outcome variable, the thing you've measured or more simply the thing for which you want to display the mean. In this case it would be **arousal**, so select arousal from the variable list and drag it into the *y*-axis drop zone (Y-Axis?). The *x*-axis should be the variable by which we want to split the arousal data. To plot the means for the two films, select the variable **film** from the variable list and drag it into the drop zone for the *x*-axis (X-Axis?).

Figure 5.18 shows some other useful options in the *Element Properties* tab (if this isn't visible click [icon]). There are

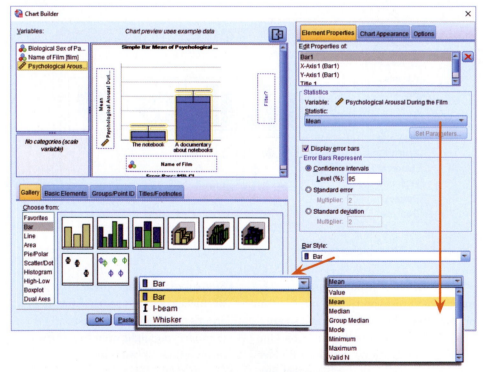

Figure 5.18   Dialog boxes for a simple bar chart with error bar

---

9 I had an email from someone expressing her 'disgust' at measuring arousal while watching a film. This reaction surprised me because to a psychologist (like me) 'arousal' means a heightened emotional response—the sort of heightened emotional response you might get from watching a film you like. Apparently if you're the sort of person who complains about the contents of textbooks then 'arousal' means something different. I can't think what.

three important features of this tab. The first is that, by default, the bars will display the mean value. This is fine, but note that you

can plot other summary statistics such as the median or mode. Second, just because you've selected a simple bar chart, that doesn't mean that you *have* to have a bar chart. You can select to show an I-beam (the bar is reduced to a line with horizontal bars at the top and bottom) or just a whisker (the bar is reduced to a vertical line). The I-beam and whisker options might be useful when you're not planning on adding error bars, but because we are going to show error bars we should stick with a bar. Finally, you can add error bars to your chart to create an **error bar chart** by selecting ☑ Display error bars. You have a choice of what your error bars represent. Normally, error bars show the 95% confidence interval (see Section 2.8), and I have selected this option (◉ Confidence intervals).[10] Note that you can change the width of the confidence interval displayed by changing the '95' to a different value. You can also change the options so that, instead of the confidence interval, the error bars display the standard error (by default 2 standard errors, but you can change this value to 1) or standard deviation (again, the default is 2, but this value can be changed). The completed dialog box is in Figure 5.18. Click OK to produce the graph.

Figure 5.19 shows the resulting bar chart. This graph displays the means (and the confidence intervals of those means) and shows us that, on average, people were more aroused by *The Notebook* than by a documentary about

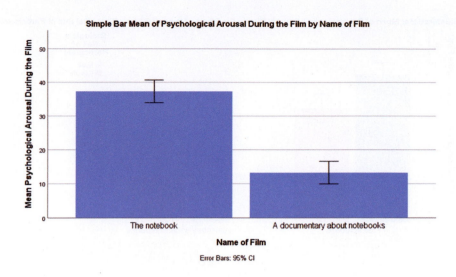

Figure 5.19   Bar chart of the mean arousal for each of the two films

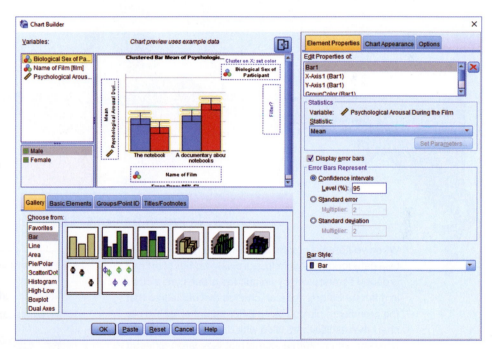

Figure 5.20   Dialog boxes for a clustered bar chart with error bar

notebooks. However, we originally wanted to look for differences between the sexes, so this graph isn't telling us what we need to know. We need a *clustered graph*.[11]

5.6.2 Clustered bar charts for independent means ▌▌▌▌

To do a clustered bar chart for means that are independent (i.e., have come from different groups) we need to

---

10 It's also worth mentioning at this point that because confidence intervals are constructed assuming a normal distribution, you should plot them only when this is a reasonable assumption (see Section 2.8).

11 You can also use a drop-line graph, which is described in Section 5.8.6.

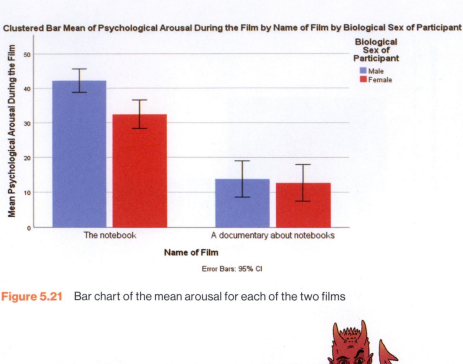

Clustered Bar Mean of Psychological Arousal During the Film by Name of Film by Biological Sex of Participant

Figure 5.21 Bar chart of the mean arousal for each of the two films

double-click the clustered bar chart icon in the Chart Builder (Figure 5.17). On the canvas you will see a graph similar to the simple bar chart, but with an extra drop zone: Cluster on X: set color. All we need to do is to drag our second grouping variable into this drop zone. As with the previous example, select **arousal** from the variable list and drag it into Y-Axis?, then select **film** from the variable list and drag it into X-Axis?. Dragging the variable **sex** into Cluster on X: set color will result in different-colored bars representing males and females (but see SPSS Tip 5.3). As in the previous section, select error bars in the properties dialog box. Figure 5.20 shows the completed Chart Builder. Click OK to produce the graph.

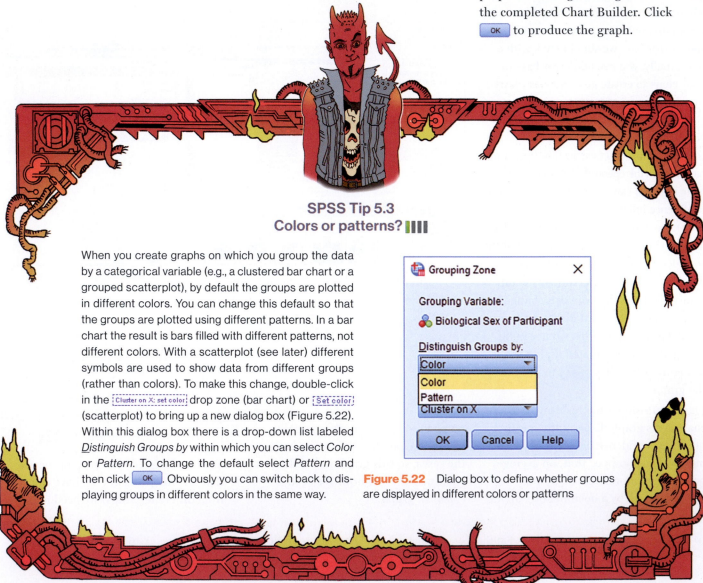

SPSS Tip 5.3
Colors or patterns? ▮▮▮▮

When you create graphs on which you group the data by a categorical variable (e.g., a clustered bar chart or a grouped scatterplot), by default the groups are plotted in different colors. You can change this default so that the groups are plotted using different patterns. In a bar chart the result is bars filled with different patterns, not different colors. With a scatterplot (see later) different symbols are used to show data from different groups (rather than colors). To make this change, double-click in the Cluster on X: set color drop zone (bar chart) or Set color (scatterplot) to bring up a new dialog box (Figure 5.22). Within this dialog box there is a drop-down list labeled *Distinguish Groups by* within which you can select *Color* or *Pattern*. To change the default select *Pattern* and then click OK . Obviously you can switch back to displaying groups in different colors in the same way.

Figure 5.22 Dialog box to define whether groups are displayed in different colors or patterns

Figure 5.21 shows the resulting bar chart. Like the simple bar chart, this graph tells us that arousal was overall higher for *The Notebook* than for the documentary about notebooks, but it splits this information by biological sex. The mean arousal for *The Notebook* shows that males were actually more aroused during this film than females. This indicates that they enjoyed the film more than the women did. Contrast this with the documentary, for which arousal levels are comparable in males and females. On the face of it, this contradicts the idea of a 'chick flick': it seems that men enjoy these films more than women (deep down we're all romantics at heart . . .).

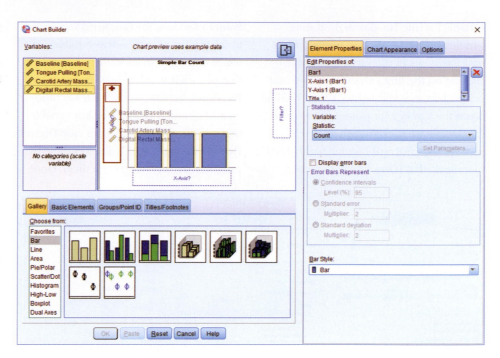

5.6.3 Simple bar charts for related means ▮▮▮

How do I plot a bar graph of repeated-measures data?

Graphing means from the same entities is trickier, but, as they say, if you're going to die, die with your boots on. So, let's put our boots on and hopefully not die. Hiccups can be a serious problem: Charles Osborne apparently got a case of hiccups while slaughtering a hog (well, who wouldn't?) that lasted 67 years. People have many methods for stopping hiccups (a surprise, holding your breath), and medical science has put its collective mind to the task too. The official treatment methods include tongue-pulling maneuvers, massage of the carotid artery, and, believe it or not, digital rectal massage (Fesmire, 1988). I don't know the details of what the digital rectal massage involved, but I can imagine. Let's say we wanted to put digital rectal massage to the test (erm, as a cure for hiccups). We took 15 hiccup sufferers, and during a bout of hiccups administered each of the three procedures (in random order and at intervals of 5 minutes) after taking a baseline of how many hiccups

Figure 5.23 Specifying a simple bar chart for repeated-measures data

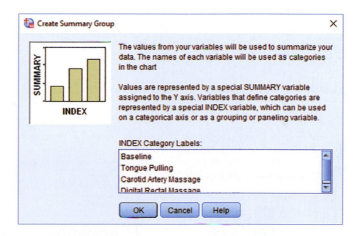

Figure 5.24 The *Create Summary Group* dialog box

they had per minute. We counted the number of hiccups in the minute after each procedure. Load the file **Hiccups. sav**. Note that these data are laid out in different columns; there is no grouping variable that specifies the interventions, because each patient experienced all interventions. In the previous two examples, we used grouping variables to specify aspects of the graph (e.g., we used the grouping variable **film** to specify the x-axis). For repeated-measures data we don't have these grouping variables, and

so the process of building a graph is a little more complicated (but only a little).

To plot the mean number of hiccups, go to the Chart Builder and double-click the icon for a simple bar chart (Figure 5.17). As before, you will see a graph on the canvas with drop zones for the *x*- and *y*-axis. Previously, we specified the column in our data that contained data from our outcome measure on the *y*-axis, but for these data our outcome variable (number of hiccups) is spread over four columns. We need to drag all four of these variables from the variable list into the *y-axis* drop zone simultaneously. To do this, we first select multiple items in the variable list by clicking on the first variable that we want (which becomes highlighted), then holding down the *Ctrl* key (or *Cmd* if you're on a Mac) while we click any others. Each variable you click will become highlighted to indicate that it has been selected.

Sometimes (as is the case here) you want to select a list of consecutive variables, in which case you can click the first variable that you want to select (in this case **Baseline**), hold down the *Shift* key (also on a Mac) and then click the last variable that you want to select (in this case **Digital Rectal Massage**); this will select these two variables and any in between them. Once

you have selected the four variables, click any one of them (while still pressing *Cmd* or *Shift* if you're on a Mac) and drag to the ⬚ Y-Axis? drop zone. This action transfers all selected variables to that drop zone (see Figure 5.23).

Once you have dragged the four variables onto the *y-axis* drop zone a new dialog box appears (Figure 5.24). This box tells us that SPSS is creating two temporary variables. One is called **Summary**, which is going to be the outcome variable (i.e., what we measured—in this case the number of hiccups per minute). The other is called **Index**, which will represent our independent variable (i.e., what we manipulated—in this case the type of intervention). SPSS uses these temporary names because it doesn't know what our variables represent, but we will change them to be something more helpful. First, click OK to get rid of this dialog box.

To edit the names of the **Summary** and **Index** variables, we use the *Element Properties* tab, which we have used before; if you can't see it then click ⬚ Figure 5.25 shows the options that need to be set. In the left panel, note that I have selected to display error bars (see the previous two sections for more information). The middle panel is accessed by clicking on *X-Axis1 (Bar1)* in the list labeled *Edit Properties of*, which allows us to edit properties of the horizontal axis. First, we'll give the axis a sensible title. I have typed *Intervention* in the space labeled *Axis Label*, which will now be the *x*-axis label on the graph. We can also change the order of our variables by selecting a variable in the list labeled *Order* and moving it up or down using ⬆ and ⬇. This is useful if the levels of our predictor variable have a meaningful order that is not reflected in the order of variables in the data editor. If we change our mind about displaying one of our variables then we can remove it from the list by selecting it and clicking on ⊠.

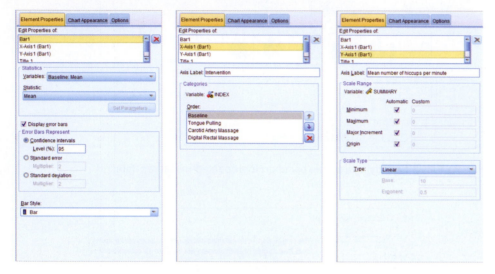

Figure 5.25 Setting *Element Properties* for a repeated-measures graph

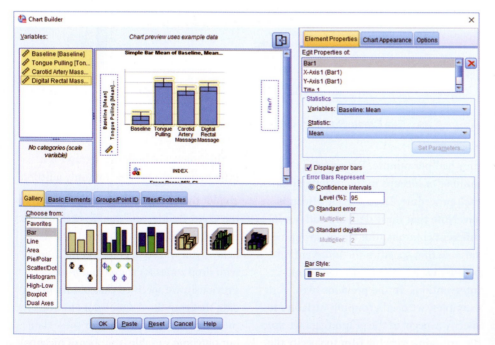

Figure 5.26 Completed Chart Builder for a repeated-measures graph

The right panel of Figure 5.25 is accessed by clicking on *Y-Axis1 (Bar1)* in the list labeled *Edit Properties of*, which allows us to edit properties of the vertical axis. The main change that I have made here is to give the axis a label so that the final graph has a useful description on the axis (by default it will display 'Mean', which is too vague). I have typed 'Mean number of hiccups per minute' in the box labeled *Axis Label*. Also note that you can use this dialog box to set the scale of the vertical axis (the minimum value, maximum value and the major increment, which is how often a mark is made on the axis). Mostly you can let SPSS construct the scale automatically—if it doesn't do it sensibly

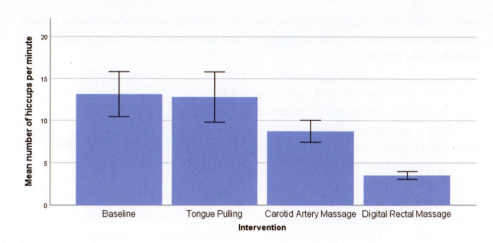

Figure 5.27 Bar chart of the mean number of hiccups at baseline and after various interventions

Labcoat Leni's Real Research 5.2
Seeing red ▌▌▌▌

Johns, S. E. et al. (2012). *PLoS One, 7*(4), e34669.

It is believed that males have a biological predisposition towards the color red because it is sexually salient. The theory suggests that women use the color red as a proxy signal for genital color to indicate ovulation and sexual proceptivity. If this hypothesis is true, then using the color red in this way would have to attract men (otherwise, it's a pointless strategy). In a novel study, Johns, Hargrave, & Newton-Fisher (2012) tested this idea by manipulating the color of four pictures of female genitalia to make them increasing shades of red (pale pink, light pink, dark pink, red). Heterosexual males rated the resulting 16 pictures from 0 (unattractive) to 100 (attractive). The data are in the file **Johns et al. (2012).sav**. Draw an error bar graph of the mean ratings for the four different colors. Do you think men preferred red genitals (remember, if the theory is correct, then red should be rated highest). Answers are on the companion website. (We analyze these data at the end of Chapter 16.)

you can edit it later. Figure 5.26 shows the completed Chart Builder. Click on OK to produce the graph.

The resulting bar chart in Figure 5.27 displays the mean number of hiccups (and associated confidence interval)[12] at baseline and after the three interventions. Note that the axis labels that I typed in have appeared on the graph. We can conclude that the amount of hiccups after tongue pulling was about the same as at baseline; however, carotid artery massage reduced hiccups, but not by as much as digital rectal massage. The moral? If you have hiccups, go amuse yourself with something digital for a few minutes. Lock the door first.

5.6.4 Clustered bar charts for related means

Now we have seen how to plot means that are related (i.e., display scores from the same group of cases in different conditions), you might well wonder what you do if you have a second independent variable that had been measured in the same sample. You'd do a clustered bar chart, right? Wrong? The SPSS Chart Builder doesn't appear to be able to cope with this situation at all—at least not that I can work out from playing about with it. (Cue a deluge of emails along the general theme of 'Dear Professor Field, I was recently looking through my FEI Titan 80-300 scanning transmission electron microscope and I think I may have found your brain. I have enclosed it for you— good luck finding it in the envelope. May I suggest that you take better care next time there is a slight gust of wind or else, I fear, it might blow out of your head again. Yours, Professor Enormobrain. PS Doing clustered charts for related means in SPSS

is simple for anyone whose mental acumen can raise itself above that of a louse.')

5.6.5 Clustered bar charts for 'mixed' designs

The Chart Builder *can* produce graphs of a mixed design (see Chapter 16). A mixed design has one or more independent variables measured using different groups, and one or more independent variables measured using the same entities. The Chart Builder can produce a graph, provided you have only one repeated-measure variable.

My students like to message on their phones during my lectures (I assume they text the person next to them things like, 'This loser is so boring—need to poke my eyes out to relieve tedium. LOL' or tweeting 'In lecture of @profandyfield #WillThePainNeverEnd).' With all this typing on phones, though, what will become of humanity? Maybe we'll evolve miniature thumbs or lose the ability to write correct English. Imagine we conducted an experiment in which a group of 25 people were encouraged to message their friends and post on social media using their mobiles over a six-month period. A second group of 25 people were banned

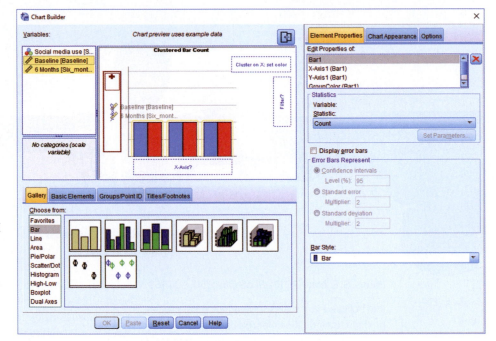

Figure 5.28 Selecting the repeated-measures variable in the Chart Builder

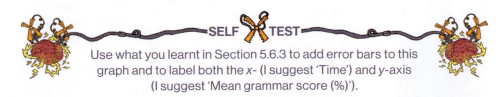

SELF TEST

Use what you learnt in Section 5.6.3 to add error bars to this graph and to label both the *x*- (I suggest 'Time') and *y*-axis (I suggest 'Mean grammar score (%)').

---

12 The error bars on graphs of repeated-measures designs should be adjusted as we will see in Chapter 10; so if you're graphing your own data have a look at Section 10.6.2 before you do.

from messaging and social media for the same period by being given armbands that administered painful shocks in the presence of microwaves (like those emitted from phones).[13] The outcome was a percentage score on a grammatical test that was administered both before and after the intervention. The first independent variable was, therefore, social media use (encouraged or banned) and the second was the time at which grammatical ability was assessed (baseline or after 6 months). The data are in the file **Social Media.sav**.

To graph these data we begin as though we are creating a clustered bar chart (Section 5.6.2). However, because one of our independent variables was a repeated measure, we specify the outcome variable as we did for a bar chart of related means (Section 5.6.3). Our repeated-measures variable is time (whether grammatical ability was measured at baseline or after 6 months) and is represented in the data file by two columns, one for the baseline data and the other for the follow-up data. In the Chart Builder, select these two variables simultaneously by clicking on one and then holding down the *Ctrl* key (*Cmd* on a Mac) and clicking on the other. When they are both highlighted click either one (keep *Cmd* pressed on a Mac) and drag it into Y-Axis? as shown in Figure 5.28. The second variable (whether people were encouraged to use social media or were banned) was measured using different participants and is represented in the data file by a grouping variable (**Social media use**). Drag this variable from the variable list into Cluster on X: set color. The two groups will be displayed as different-colored bars. The finished Chart Builder is in Figure 5.29. Click OK to produce the graph.

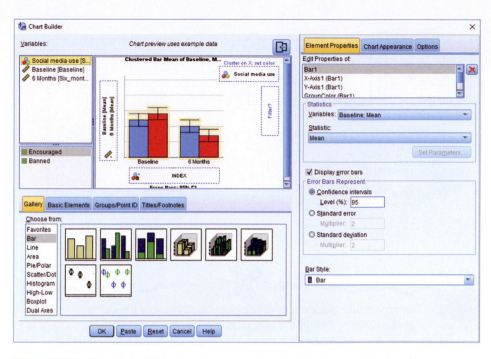

Figure 5.29   Completed dialog box for an error bar graph of a mixed design

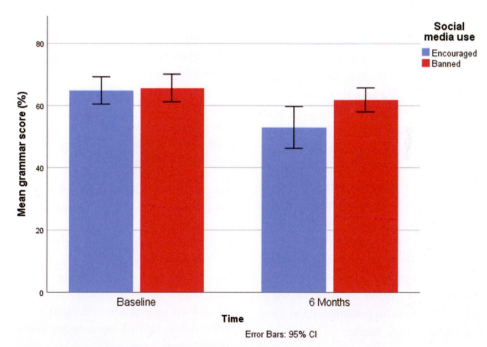

Figure 5.30   Error bar graph of the mean grammar score over 6 months in people who were encouraged to use social media compared to those who were banned

13 This turned out to be a very bad idea because other people's phones also emit microwaves. Let's just say there are now 25 people battling chronic learned helplessness.

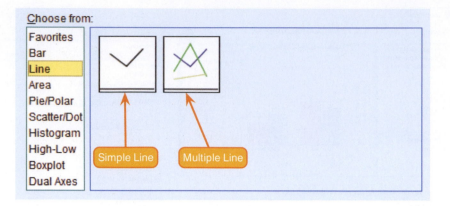

Figure 5.31 The line chart gallery

SELF TEST

The procedure for producing line graphs is basically the same as for bar charts. Follow the previous sections for bar charts but selecting a simple line chart instead of a simple bar chart, and a multiple line chart instead of a clustered bar chart. Produce line chart equivalents of each of the bar charts in the previous section. If you get stuck, the self-test answers on the companion website provide help.

Figure 5.30 shows the resulting bar chart. It shows that at baseline (before the intervention) the grammar scores were comparable in our two groups; however, after the intervention, the grammar scores were lower in those encouraged to use social media than in those banned from using it. If you compare the two blue bars you can see that social media users' grammar scores have fallen over the 6 months; compare this to the controls (green bars) whose grammar scores are similar over time. We might, therefore, conclude that social media use has a detrimental effect on people's understanding of English grammar. Consequently, civilization will crumble and Abaddon will rise cackling from his bottomless pit to claim our wretched souls. Maybe.

5.7 Line charts ▮▮▮▮

Line charts are bar charts but with lines instead of bars. Therefore, everything we have just done with bar charts we can do with line charts instead. As ever, our starting point is the Chart Builder (Figure 5.5). In this dialog box select *Line* in the list labeled *Choose from* to bring up the gallery shown in Figure 5.31. This gallery has two icons,

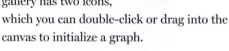

which you can double-click or drag into the canvas to initialize a graph.

- *Simple line*: Use this option to display the means of scores across different groups of cases.
- *Multiple line*: This option is equivalent to the clustered bar chart: it will plot means of an outcome variable for different categories/groups of a predictor variable and also produce different-colored lines for each category/group of a second predictor variable.

5.8 Graphing relationships: the scatterplot ▮▮▮▮

Sometimes we need to look at the relationships between variables (rather than their means or frequencies).

A **scatterplot** is a graph that plots each person's score on one variable against their score on another. It visualizes the relationship between the variables, but also helps us to identify unusual cases that might bias that relationship. In fact, we encountered a scatterplot when we discussed effect sizes (see Section 3.7.2). Producing a scatterplot using SPSS is dead easy. As ever, open the Chart Builder dialog box (Figure 5.5). Select *Scatter/Dot* in the list labeled *Choose from* to bring up the gallery shown in Figure 5.32. This gallery has eight icons, representing different types of scatterplot. Select one by double-clicking on it or by dragging it to the canvas.

- *Simple scatter*: Use this option to plot values of one continuous variable against another.
- *Grouped scatter*: This is like a simple scatterplot, except that you can display points belonging to different groups in different colors (or symbols).
- *Simple 3-D scatter*: Use this option to plot values of one continuous variable against values of two others.
- *Grouped 3-D scatter*: Use this option to plot values of one continuous variable against two others, but differentiating groups of cases with different-colored dots.
- *Summary point plot*: This graph is the same as a bar chart (see Section 5.6), except that a dot is used instead of a bar.
- *Simple dot plot*: Otherwise known as a **density plot**, this graph is like a histogram (see Section 5.4), except that, rather than having a summary bar representing the frequency of scores, individual scores are displayed as dots. Like histograms, they are useful for looking at the shape of the distribution of scores.
- *Scatterplot matrix*: This option produces a grid of scatterplots showing the relationships between multiple pairs of variables in each cell of the grid.
- *Drop-line:* This option produces a plot similar to a clustered bar chart (see, for example, Section 5.6.2) but with a dot representing a summary statistic (e.g., the mean) instead of a bar, and

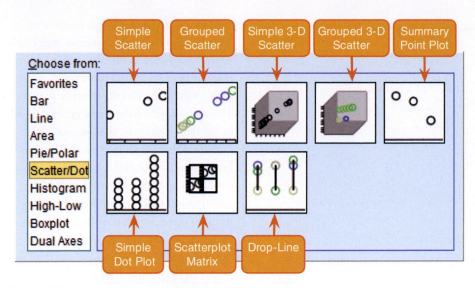

Figure 5.32　The scatter/dot gallery

drop zones: one for the *y*-axis and one for the *x*-axis. Typically the *y*-axis displays the outcome variable and the *x*-axis the predictor.[14] In this case the outcome is exam performance (**Exam Performance (%)**), so select it from the variable list and drag it into `Y-Axis?`, and the predictor is exam anxiety (**Exam Anxiety**), so drag it into `X-Axis?`. Figure 5.33 shows the completed Chart Builder. Click `OK` to produce the graph.

Figure 5.34 shows the resulting scatterplot; yours won't have a funky line on it yet, but don't get too depressed about it because I'm going to show you how to add one very soon. The scatterplot tells us that the majority of

with a line connecting the 'summary' (e.g., mean) of each group. These graphs are useful for comparing statistics, such as the mean, across groups or categories.

5.8.1 Simple scatterplot ▮▮▮

This type of scatterplot is for looking at just two variables. For example, a psychologist was interested in the effects of exam stress on exam performance. She devised and validated a questionnaire to assess state anxiety relating to exams (called the Exam Anxiety Questionnaire or EAQ). This scale produced a measure of anxiety scored out of 100. Anxiety was measured before an exam, and the percentage mark of each student on the exam was used to assess the exam performance. The first thing that the psychologist should do is draw a scatterplot of the two variables (her data are in the file **ExamAnxiety.sav**, so load this file into SPSS).

In the Chart Builder double-click the icon for a simple scatterplot (Figure 5.33). On the canvas you will see a graph and two

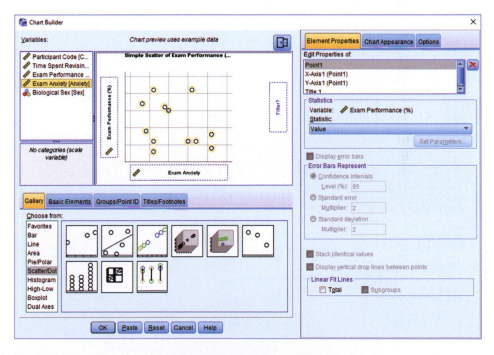

Figure 5.33　Completed *Chart Builder* dialog box for a simple scatterplot

---

14　This makes sense in experimental research because changes in the independent variable (the variable that the experimenter has manipulated) cause changes in the dependent variable (outcome). In English we read from left to right, so by having the causal variable on the horizontal, we naturally scan across the changes in the 'cause' and see the effect on the vertical plane as we do. In correlational research, variables are measured simultaneously and so no cause-and-effect relationship can be established, so although we can still talk about *predictors* and *outcomes*, these terms do not imply causal relationships.

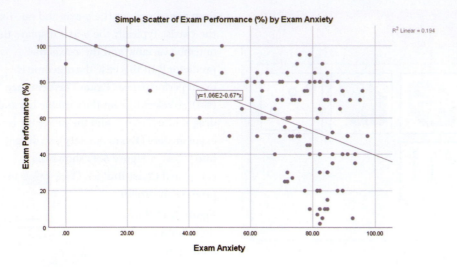

Figure 5.34 Scatterplot of exam anxiety and exam performance

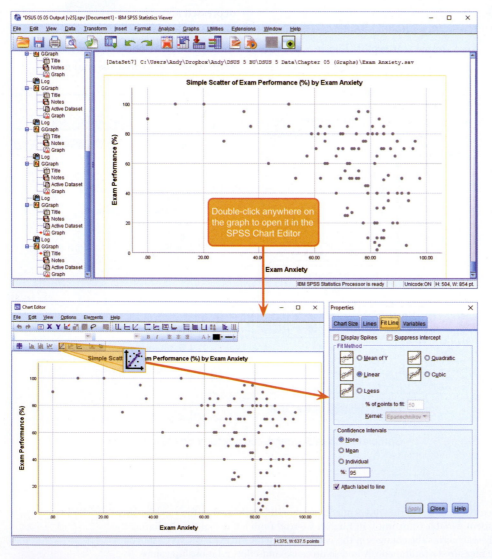

Figure 5.35 Opening the Chart Editor and *Properties* dialog box for a simple scatterplot

students suffered from high levels of anxiety (there are very few cases that had anxiety levels below 60). Also, there are no obvious outliers in that most points seem to fall within the vicinity of other points. There also seems to be some general trend in the data, shown by the line, such that higher levels of anxiety are associated with lower exam scores and low levels of anxiety are almost always associated with high examination marks. Another noticeable trend in these data is that there were no cases having low anxiety and low exam performance—in fact, most of the data are clustered in the upper region of the anxiety scale.

Often it is useful to plot a line that summarizes the relationship between variables on a scatterplot (this is called a **regression line**, and we will discover more about them in Chapter 9). Figure 5.35 shows the process of adding a regression line to the scatterplot. First, any graph in the *SPSS viewer* can be edited by double-clicking on it to open the SPSS **Chart Editor** (we explore this window in detail in Section 5.9). Once in the Chart Editor, click 📈 to open the *Properties* dialog box. Using this dialog box, we can add a line to the graph that represents the overall mean of all data, a linear (straight line) model, a quadratic model, a cubic model and so on (these trends are described in Section 12.4.5). We'll add a linear regression line, so select ⦿ Linear. By default, SPSS attaches a label to the line (☑ Attach label to line) containing the equation for the line (more on this in Chapter 9). Often, this label obscures the data, so I tend to switch this option off. Click Apply to register any property changes to the scatterplot. To exit the Chart Editor simply close the window (❎). The scatterplot should now look like Figure 5.34. A variation on the scatterplot is the catterplot, which is useful for plotting unpredictable data that views you as its human servant (Jane Superbrain Box 5.1).

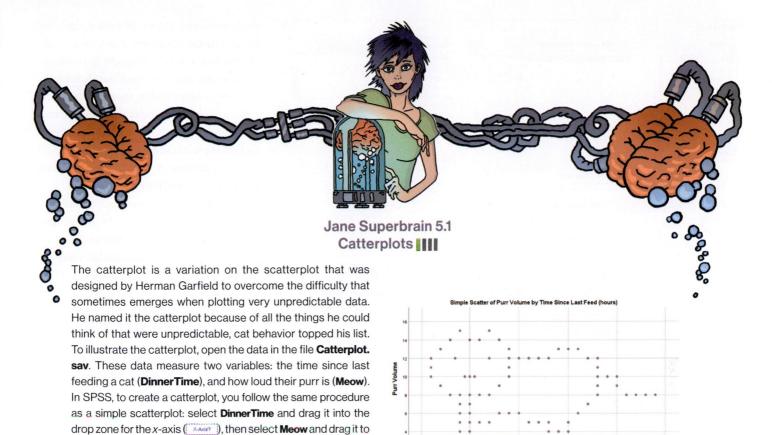

Jane Superbrain 5.1
Catterplots ▮▮▮▮

The catterplot is a variation on the scatterplot that was designed by Herman Garfield to overcome the difficulty that sometimes emerges when plotting very unpredictable data. He named it the catterplot because of all the things he could think of that were unpredictable, cat behavior topped his list. To illustrate the catterplot, open the data in the file **Catterplot. sav**. These data measure two variables: the time since last feeding a cat (**DinnerTime**), and how loud their purr is (**Meow**). In SPSS, to create a catterplot, you follow the same procedure as a simple scatterplot: select **DinnerTime** and drag it into the drop zone for the *x*-axis (X-Axis?), then select **Meow** and drag it to the *y*-axis drop zone (Y-Axis?). Click OK to produce the graph.

The catterplot is shown in Figure 5.36. You might expect that there is a positive relationship between the variables: the longer the time since being fed, the more vocal the cat becomes. However, the graph shows something quite different: there doesn't seem to be a consistent relationship.[15]

Figure 5.36   A catterplot

5.8.2 Grouped scatterplot ▮▮▮▮

Imagine that we want to see whether male and female students had different reactions to exam anxiety. We can visualize this with a grouped scatterplot, which displays scores on two continuous variables, but colors the data points by a third categorical variable. To create this plot for the exam anxiety data, double-click the grouped scatter icon in the Chart Builder (Figure 5.32). As in the previous example, drag **Exam Performance** (%) from the variable list into Y-Axis? , and drag **Exam Anxiety** into X-Axis? . There is an additional drop zone (Set color!) into which we can drop any categorical variable. If we want to visualize the relationship between exam anxiety and performance separately for male and female students then we can drag **Biological Sex** into Set color!. (If you want to display the different sexes using different symbols rather than colors then

---

15 I'm hugely grateful to Lea Raemaekers for sending me these data.

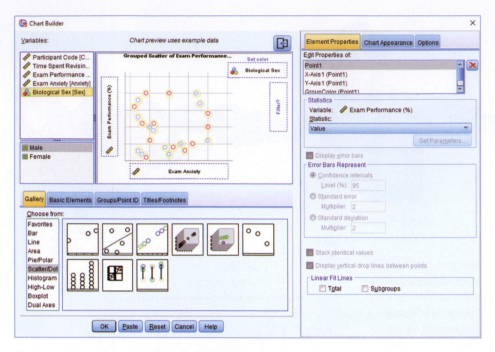

Figure 5.37 Completed *Chart Builder* dialog box for a grouped scatterplot

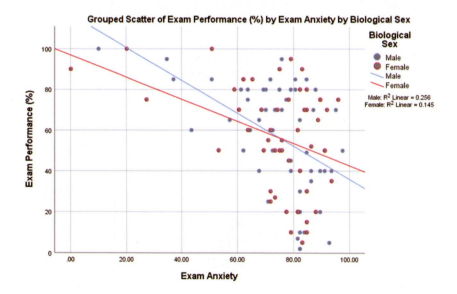

Figure 5.38 Scatterplot of exam anxiety and exam performance split by sex

read SPSS Tip 5.3.) Figure 5.37 shows the completed Chart Builder. Click OK to produce the graph.

Figure 5.38 shows the resulting scatterplot; as before, I have added regression lines, but this time I have added different lines for each group. We saw in the previous section

that graphs can be edited using the Chart Editor (Figure 5.35), and that we could fit a regression line that summarized the whole data set by clicking on. We could do this again, but having split the data by sex it might be more useful to fit separate lines for our two groups. To do this, double-click

the plot to open the Chart Editor, then click to open the *Properties* dialog box (Figure 5.35), and again select Linear. Click Apply, and close the Chart Editor window to return to the viewer. Note that SPSS has plotted separate lines for the men and women (Figure 5.38). These lines tell us that the relationship between exam anxiety and exam performance was slightly stronger in males (the line is steeper), indicating that men's exam performance was more adversely affected by anxiety than women's exam anxiety. (Whether this difference is significant is another issue— see Section 8.6.1.)

5.8.3 Simple and grouped 3-D scatterplots

One of the few times you can use a 3-D graph without a statistician locking you up in a room and whipping you with his beard is a scatterplot. A 3-D scatterplot displays the relationship between three variables, and the reason why it's sometimes OK to use a 3-D graph in this context is that the third dimension tells us something useful (it isn't there to look pretty). As an example, imagine our researcher decided that exam anxiety might not be the only factor contributing to exam performance. So, she also asked participants to keep a revision[16] diary from which she calculated the number of hours spent for the exam. She might want to look at the relationships between these variables simultaneously, and she could do this using a 3-D scatterplot. Personally, I don't think a 3-D scatterplot is a clear way to present data—a matrix scatterplot is better—but if you want to do one, see Oliver Twisted.

5.8.4 Matrix scatterplot

Instead of plotting several variables on the same axes on a 3-D scatterplot (which can be difficult to interpret), I think it's better to plot a matrix of 2-D scatterplots. This type of plot allows you to see the

16 In the UK, the term 'revision' is synonymous with study.

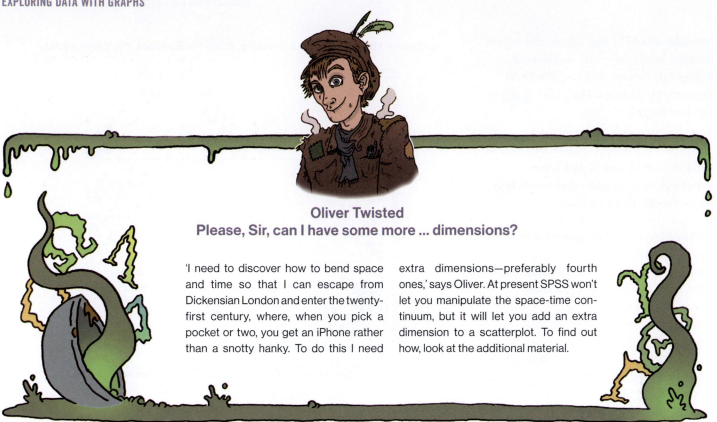

Oliver Twisted
Please, Sir, can I have some more ... dimensions?

'I need to discover how to bend space and time so that I can escape from Dickensian London and enter the twenty-first century, where, when you pick a pocket or two, you get an iPhone rather than a snotty hanky. To do this I need extra dimensions—preferably fourth ones,' says Oliver. At present SPSS won't let you manipulate the space-time continuum, but it will let you add an extra dimension to a scatterplot. To find out how, look at the additional material.

relationship between all combinations of many different pairs of variables. Let's continue with the example of the relationships between exam performance, exam anxiety and time spent revising. First, access the Chart Builder and double-click the icon for a scatterplot matrix (Figure 5.32). A different type of graph than you've seen before will appear on the canvas, and it has only one drop zone (Scattermatrix?). We need to drag all the variables that we want to see plotted against each other into this drop zone. We have dragged multiple variables into a drop zone in previous sections, but, to recap, we first select multiple items in the variable list. To do this, select the first variable (**Time Spent Revising**) by clicking on it with the mouse. The variable will be highlighted. Now, hold down the *Ctrl* key (*Cmd* on a Mac) and click the other two variables (**Exam Performance (%)** and **Exam Anxiety**).[17] The three

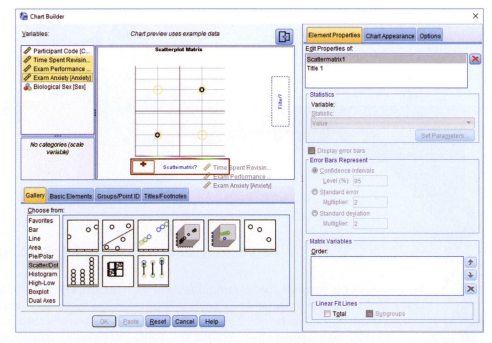

Figure 5.39 *Chart Builder* dialog box for a matrix scatterplot

---

17 We could also have clicked on **Time Spent Revising**, then held down the *Shift* key and clicked on **Exam Anxiety**.

variables should be highlighted and can be dragged into 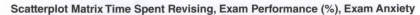 as shown in Figure 5.39 (you need to keep *Cmd* held down as you drag on a Mac). Click OK to produce the graph.

The six scatterplots in Figure 5.40 represent the various combinations of each variable plotted against each other variable. Using the grid references to help locate specific plots, we have:

- **A2**: revision time (Y) against exam performance (X)
- **A3**: revision time (Y) against anxiety (X)
- **B1**: exam performance (Y) against revision time (X)
- **B3**: exam performance (Y) against anxiety (X)
- **C1**: anxiety (Y) against revision time (X)
- **C2**: anxiety (Y) against exam performance (X)

Notice that the three scatterplots below the diagonal of the matrix are the same

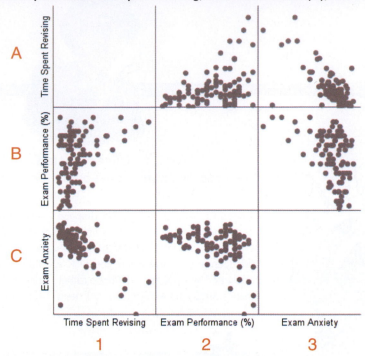

Scatterplot Matrix Time Spent Revising, Exam Performance (%), Exam Anxiety

Figure 5.40 Matrix scatterplot of exam performance, exam anxiety and revision time. Grid references have been added for clarity

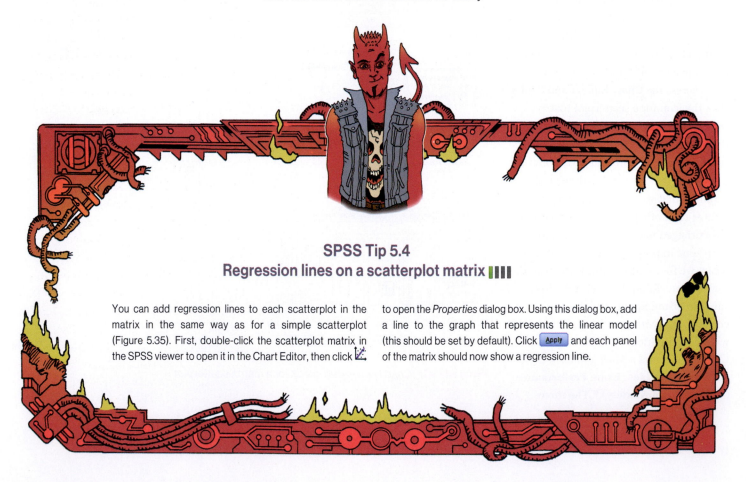

SPSS Tip 5.4
Regression lines on a scatterplot matrix ▎▎▎▎

You can add regression lines to each scatterplot in the matrix in the same way as for a simple scatterplot (Figure 5.35). First, double-click the scatterplot matrix in the SPSS viewer to open it in the Chart Editor, then click ⬚ to open the *Properties* dialog box. Using this dialog box, add a line to the graph that represents the linear model (this should be set by default). Click Apply and each panel of the matrix should now show a regression line.

plots as the ones above the diagonal, but with the axes reversed. From this matrix we can see that revision time and anxiety are inversely related (i.e., the more time spent revising, the less exam anxiety the person experienced). Also, in the scatterplot of revision time against anxiety (grids A3 and C1) it looks as though there is one possible unusual case—a single participant who spent very little time revising yet suffered very little anxiety about the exam. Because all of the participants who had low anxiety scored highly on the exam (grid C2), we can deduce that this person probably also did well on the exam (it was probably Smart Alex). We could examine this case more closely if we believed that their behavior was caused by some external factor (such as a dose of *antiSTATic*).[18] Matrix scatterplots are very convenient for examining pairs of relationships between variables (see SPSS Tip 5.4).

However, they can become very confusing indeed if you plot them for more than about three or four variables.

5.8.5 Simple dot plot or density plot ▌▌▌

I mentioned earlier that the simple dot plot or density plot as it is also known, is a histogram except that each data point is plotted (rather than using a single summary bar to show each frequency). Like a histogram, the data are still placed into bins (SPSS Tip 5.2), but a dot is used to represent each data point. You should be able to follow the instructions for a histogram to draw one.

5.8.6 Drop-line graph ▌▌▌

I also mentioned earlier that the drop-line plot is fairly similar to a clustered bar chart (or line chart), except that each mean is represented by a dot (rather than a bar), and within groups these dots are

linked by a line (contrast this with a line graph, where dots are joined across groups, rather than within groups). The best way to see the difference is to plot one, and to do this you can apply what you were told about clustered line graphs (Section 5.6.2) to this new situation.

5.9 Editing graphs ▌▌▌

We have already seen how to add regression lines to scatterplots using the Chart Editor (Section 5.8.1). We'll now look in more detail at the Chart Editor window. Remember that to open this window you double-click the graph you want to edit in the viewer window (Figure 5.35). You can edit almost every aspect of the graph: in the Chart Editor you can click virtually anything that you want to change and change it. Once in the Chart Editor (Figure 5.41) there are several icons that you can click to change aspects of the graph. Whether a particular icon is active depends on the type of chart that you are editing (e.g., the icon to fit a regression line will not be active for a bar chart). The figure tells you what most of the icons do, but most of them are fairly self-explanatory. Play around and you'll find icons for that to add elements to the graph (such as grid lines, regression lines, data labels).

You can also edit parts of the graph by selecting them and changing their properties using the *Properties* dialog box. To select part of the graph, double-click it; it will become highlighted in orange and a new dialog box will appear (Figure 5.42). This *Properties* dialog box enables you to change virtually anything about the item that you have selected. You can change the bar colors, the axis titles, the scale of each axis, and so on. You can also do things like make the bars three-dimensional and pink, but we know better than to do things like that. There are both written (see

SELF TEST

Doing a simple dot plot in the Chart Builder is quite similar to drawing a histogram. Reload the **Jiminy Cricket.sav** data and see if you can produce a simple dot plot of the success scores after the intervention. Compare the resulting graph to the earlier histogram of the same data (Figure 5.11). Remember that your starting point is to double-click the icon for a simple dot plot in the Chart Builder (Figure 5.32), then use the instructions for plotting a histogram (Section 5.4)—there is guidance on the companion website.

SELF TEST

Doing a drop-line plot in the Chart Builder is quite similar to drawing a clustered bar chart. Reload the **Notebook.sav** data and see if you can produce a drop-line plot of the arousal scores. Compare the resulting graph to the earlier clustered bar chart of the same data (Figure 5.21). The instructions in Section 5.6.2 should help. Now see if you can produce a drop-line plot of the **Social Media.sav** data from earlier in this chapter. Compare the resulting graph to the earlier clustered bar chart of the same data (Figure 5.30). The instructions in Section 5.6.5 should help.

Remember that your starting point for both tasks is to double-click on the icon for a drop-line plot in the Chart Builder (Figure 5.32). There is full guidance for both examples in the additional material on the companion website.

---

18 If this joke isn't funny it's your own fault for skipping or not paying attention to, Chapter 3. Or it might just not be funny.

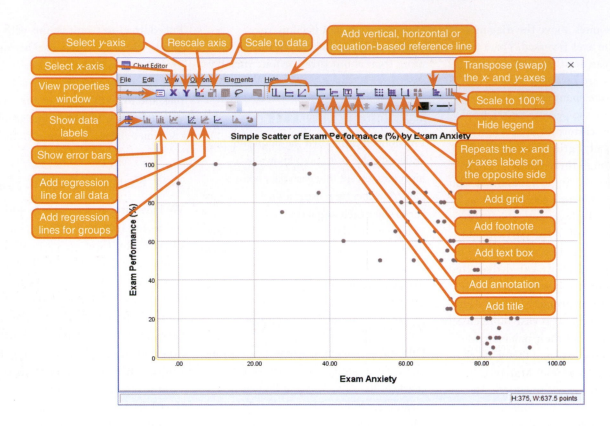

Figure 5.41   The Chart Editor

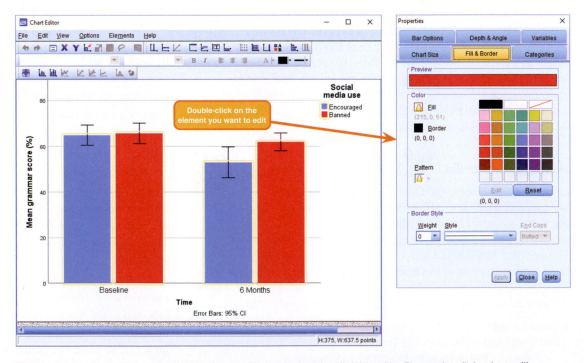

Figure 5.42   To select an element in the graph simply double-click it and its *Properties* dialog box will appear

Oliver Twisted
Please, Sir, can I have some more . . . graphs?

'Blue and green should never be seen!' shrieks Oliver with so much force that his throat starts to hurt. 'This graph offends my delicate artistic sensibilities. It *must* be changed immediately!' Never fear, Oliver. Using the editing functions in SPSS, it's possible to create some very tasteful graphs. These facilities are so extensive that I could probably write a whole book on them. In the interests of saving trees, I have prepared a tutorial that can be downloaded from the companion website. We look at an example of how to edit an error bar chart to make it conform to some of the guidelines that I talked about at the beginning of this chapter. In doing so we will look at how to edit the axes, add grid lines, change the bar colors, change the background and borders. It's a very extensive tutorial.

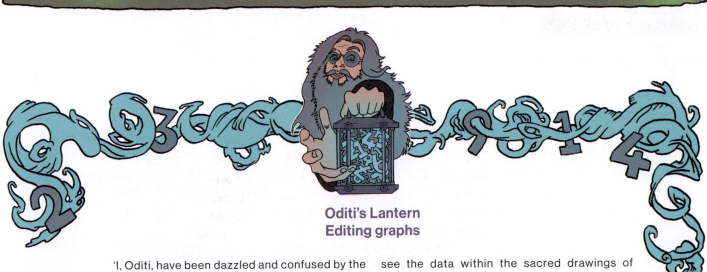

Oditi's Lantern
Editing graphs

'I, Oditi, have been dazzled and confused by the pinkness of many a graph. Those who seek to prevent our worthy mission do bedazzle us with their pink and lime green monstrosities. These colors burn our retinas until we can no longer see the data within the sacred drawings of truth. To complete our mission to find the secret of life we must make the sacred drawings palatable to the human eye. Stare into my latern to find out how.'

5.10 Brian and Jane's Story ▌▌▌▌

During their brief encounter in the Euphoria cluster on campus, Brian had noticed some pictures on Jane's screen. He knew that these pictures were graphs—he wasn't that stupid—but he didn't understand what they showed or how to create them. He began to wonder whether he could create one for Jane, sort of like painting her a picture with numbers. She had seemed entranced by the minimalist, stylish images on her screen. 'If she was enthralled by those plain images,' Brian considered, 'imagine how impressed she'd be by some 3-D effects.' He made a mental note that she'd probably love it if he colored the bars pink.

Across town Jane wandered the long corridors of a labyrinthine basement at the university. The walls were made from hundreds of jars. Each contained a brain suspended in lightly glowing green fluid. It was like an ossuary of minds. There was no natural light, but the glow of the jars created enough ambient light to see. Jane loved it down here. The jars were beautiful, elegant designs with brass bases and lids etched with green lines that made an elaborate external circuit board peppered with glistening lights. Jane wondered whether anyone had noticed that she had cracked the electronic lock or that she spent so many evenings down here? Her parents and teachers had never appreciated how clever she was. At school they'd mocked her 'human flatworm' theory of learning. She would prove them wrong. She stopped and turned to a jar labeled 'Florence'. She took a small electronic device from her pocket and pressed it

Figure 5.43 What Brian learnt from this chapter

against the jar. With a quiet hum, the jar moved forward from the wall, and the lid slowly opened. Jane took a knife and fork from her pocket.

Brian admired his poster. It was the most magnificent pink 3-D graph he had ever seen. He turned it to show the dude behind the counter at the printers. The guy smiled at Brian; it was a patronizing smile. Brian rushed to campus, poster tube under his arm. He had a spring in his step, and a plan in his mind.

5.11 What next? ▌▌▌▌

We have discovered that when it comes to graphs, minimal is best: no pink, no 3-D effects, no pictures of Errol your pet ferret superimposed on the data—oh,

and did I mention no pink? Graphs are a useful way to visualize life. At the age of about 5 I was trying to visualize my future and, like many boys, my favored career choices were going into the army (goodness only knows why, but a possible explanation is being too young to comprehend morality and death) and becoming a famous sports person. On balance, I seemed to favor the latter, and like many a UK-born child my sport of choice was football (or soccer as people outside of the UK sometimes like to call it to avoid confusion with a game in which a ball is predominantly passed through the hands, and not the feet, but is bizarrely also called football). In the next chapter we learn what became of my football career.

5.12 Key terms that I've discovered

Bar chart

Boxplot (box–whisker diagram)

Chart Builder

Chart Editor

Chartjunk

Density plot

Error bar chart

Line chart

Regression line

Scatterplot

Smart Alex's tasks

- **Task 1**: Using the data from Chapter 3 (which you should have saved, but if you didn't, re-enter it from Table 4.1), plot and interpret an error bar chart showing the mean number of friends for students and lecturers. ▮▮▮

- **Task 2**: Using the same data, plot and interpret an error bar chart showing the mean alcohol consumption for students and lecturers. ▮▮▮

- **Task 3**: Using the same data, plot and interpret an error line chart showing the mean income for students and lecturers. ▮▮▮

- **Task 4**: Using the same data, plot and interpret error a line chart showing the mean neuroticism for students and lecturers. ▮▮▮

- **Task 5**: Using the same data, plot and interpret a scatterplot with regression lines of alcohol consumption and neuroticism grouped by lecturer/student. ▮▮▮

- **Task 6**: Using the same data, plot and interpret a scatterplot matrix with regression lines of alcohol consumption, neuroticism and number of friends. ▮▮▮

- **Task 7**: Using the **Zhang (2013) subsample.sav** data from Chapter 4 (Task 8), plot a clustered error bar chart of the mean test accuracy as a function of the type of name participants completed the test under (x-axis) and

whether they were male or female (different colored bars). ▮▮▮

- **Task 8**: Using the **Method Of Teaching.sav** data from Chapter 4 (Task 3), plot a clustered error line chart of the mean score when electric shocks were used compared to being nice, and plot males and females as different colored lines. ▮▮▮

- **Task 9**: Using the **Shopping Exercise.sav** data from Chapter 4 (Task 5), plot two error bar graphs comparing men and women (x-axis): one for the distance walked, and the other for the time spent shopping. ▮▮▮

- **Task 10**: Using the **Goat or Dog.sav** data from Chapter 4 (Task 6), plot two error bar graphs comparing scores when married to a goat or a dog (x-axis): one for the animal liking variable, and the other for the life satisfaction. ▮▮▮

- **Task 11**: Using the same data as above, plot a scatterplot of animal liking scores against life satisfaction (plot scores for those married to dogs or goats in different colors). ▮▮▮

- **Task 12**: Using the **Tea Makes You Brainy 15.sav** data from Chapter 4 (Task 7), plot a scatterplot showing the number of cups of tea drunk (x-axis) against cognitive functioning (y-axis). ▮▮▮

Answers & additional resources are available on the book's website at
https://edge.sagepub.com/field5e

THE BEAST OF BIAS

6

6.1 What will this chapter tell me?

Like many young boys in the UK my first career choice was to become a soccer star. My granddad (Harry) had been something of a local soccer hero in his day, and I wanted nothing more than to emulate him. Harry had a huge influence on me: he had been a goalkeeper, and consequently I became a goalkeeper too. This decision, as it turned out, wasn't a great one because I was quite short for my age, which meant that I got overlooked to play in goal for my school in favor of a taller boy. Admittedly I am biased, but I think I was the better goalkeeper technically, though I did have an Achilles heel that was quite fatal to my goalkeeping career: the opposition could lob the ball over my head. Instead of goal, I typically got played at left back ('left back in the changing room', as the joke used to go) because, despite being right-footed, I could kick with my left one too. The trouble was, having spent years learning my granddad's goalkeeping skills, I didn't have a clue what a left back was supposed to do.[1] Consequently, I didn't exactly shine in the role, and for many years that put an end to my belief that I could play soccer. This example shows that a highly influential thing (like your granddad) can bias the conclusions you come to and that this can lead to quite dramatic consequences. The same thing happens in data analysis: sources of influence and bias lurk within the data, and unless we identify and correct for them we'll end up becoming goalkeepers despite being a short-arse. Or something like that.

1 In the 1970s at primary school, 'teaching' soccer involved shoving 11 boys onto a pitch and watching them chase the ball. It didn't occur to teachers to develop your technique, tactical acumen or even tell you the rules.

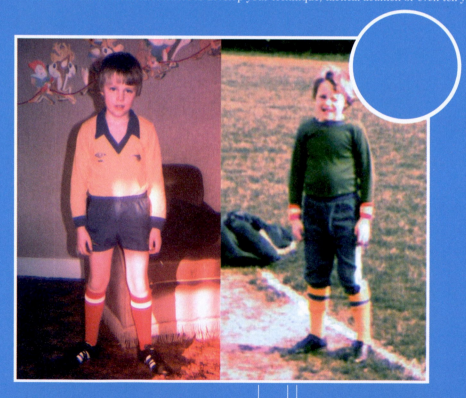

Figure 6.1   My first failed career choice was a soccer star

6.2 What is bias? ||||

If you support a sports team then at some point in your life you've probably accused a referee of being 'biased' (or worse). If not, perhaps you've watched a TV show like *The Voice* and felt that one of the judges was 'biased' towards the singers that they mentored. In these contexts, bias means that the summary information from the person ('Jasmin's singing was note perfect throughout') is at odds with the objective truth (pitch analysis shows that 33% of Jasmin's notes were sharp or flat). Similarly, in statistics the summary statistics that we estimate can be at odds with the true values. A 'unbiased estimator' is one that yields an expected value that is the same as the thing it is trying to estimate.[2]

T o review: we saw in Chapter 2 that, having collected data, we fit a model representing the hypothesis that we want to test. A common model is the linear model, which takes the form of equation (2.4). To remind you, it looks like this:

$$\text{outcome}_i = \left(b_0 + b_1 X_{1i} + b_2 X_{2i}\right) + \text{error}_i \qquad (6.1)$$

In short, we predict an outcome variable from a model described by one or more predictor variables (the Xs in the equation) and parameters (the bs in the equation) that tell us about the relationship between the predictor and the outcome variable. The model will not predict the outcome perfectly, so for each observation there is some amount of error.

We often obtain values for the parameters in the model using the method of least squares (Section 2.6). These parameter values (in our sample) estimate the parameter values in the population (because we want to draw conclusions that extend beyond our sample). For each parameter in the model we also compute

an estimate of how well it represents the population such as a standard error (Section 2.7) or confidence interval (Section 2.8). The parameters can be used to test hypotheses by converting them to a test statistic with an associated probability (p-value, Section 2.9.1). Statistical bias enters the process I've just summarized in (broadly) three ways:

1 things that bias the parameter estimates (including effect sizes);
2 things that bias standard errors and confidence intervals;
3 things that bias test statistics and p-values.

The last two are linked by the standard error: confidence intervals and test statistics are computed using the standard error, so if the standard error is biased then the corresponding confidence interval and test statistic (and associated p-value) will be biased too. Needless to say, if the statistical information we use to infer things about the world is biased, then our inferences will be too.

Sources of bias come in the form of a two-headed, fire-breathing, green-scaled beast that jumps out from behind a mound of blood-soaked moss to try to eat us alive. One of its heads goes by the name of unusual scores or 'outliers', whereas the other is called 'violations of assumptions'. These are probably names that led to it being teased at school, but, it could breathe fire from both heads, so it could handle it. Onward into battle . . .

6.3 Outliers ||||

Before we get to assumptions, we'll look at the first head of the beast of bias: outliers. An **outlier** is a score very different from the rest of the data. Let's look at an example. When I published my first book (the first edition of this book), I was very

excited and I wanted everyone in the world to love my new creation and me. Consequently, I obsessively checked the book's ratings on Amazon.co.uk. Customer ratings can range from 1 to 5 stars, where 5 is the best. Back in around 2002, my first book had seven ratings (in the order given) of 2, 5, 4, 5, 5, 5, and 5. All but one of these ratings are similar (mainly 5 and 4) but the first rating was quite different from the rest—it was a rating of 2 (a mean and horrible rating). Figure 6.2 plots the seven reviewers on the horizontal axis and their ratings on the vertical axis. The blue horizontal line shows the mean rating (4.43, as it happens). All of the scores except one lie close to this line. The rating of 2 lies way below the mean and is an example of an outlier—a weird and unusual person (I mean, score) that deviates from the rest of humanity (I mean, data set). The orange horizontal line shows the mean excluding the outlier (4.83). This line is higher than the original mean, indicating that by ignoring this score the mean increases (by 0.4). This example shows how a single score, from some mean-spirited badger turd, can bias a parameter such as the mean: the first rating of 2 drags the average down. Based on this biased estimate, new customers might erroneously conclude that my book is worse than the population actually thinks it is. I am consumed with bitterness about this whole affair, but it has given me a great example of an outlier.

Outliers bias parameter estimates, but they have an even greater impact on the error associated with that estimate. Back in Section 2.5.1 we looked at an example showing the number of friends that five statistics lecturers had. The data were 1, 3, 4, 3, 2, the mean was 2.6 and the sum of squared error was 5.2. Let's replace one of the scores with an outlier by changing the 4 to a 10. The data are now: 1, 3, 10, 3, and 2.

---

2 You might recall that when estimating the population variance we divide by $N-1$ instead of N (see Section 2.5.2). This has the effect of turning the estimate from a biased one (using N) to an unbiased one (using $N-1$).

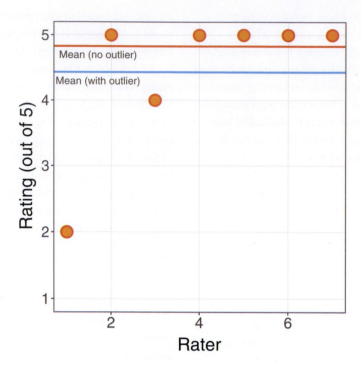

Figure 6.2   The first seven customer ratings of the first edition of this book on www.amazon.co.uk (in about 2002). The first score biases the mean

SELF TEST
Compute the mean and sum of squared error for the new data set.

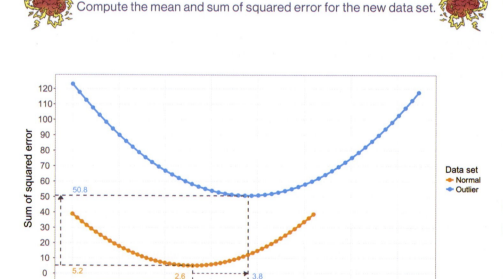

Figure 6.3   The effect of an outlier on a parameter estimate (the mean) and its associated estimate of error (the sum of squared errors)

If you did the self-test, you should find that the mean of the data with the outlier is 3.8 and the sum of squared error is 50.8. Figure 6.3 shows these values. Like Figure 2.7, it shows the sum of squared error (*y*-axis) associated with different potential values of the mean (the parameter we're estimating, *b*). For both the original data set and the one with the outlier the estimate for the mean is the optimal estimate: it is the one with the least error, which you can tell because the curve converges on the values of the mean (2.6 and 3.8). The presence of the outlier, however, pushes the curve to the right (it makes the mean higher) and upwards (it makes the sum of squared error larger). By comparing the horizontal to vertical shift in the curve you should see that the outlier affects the sum of squared error more dramatically than the parameter estimate itself. This is because we use squared errors, so any bias created by the outlier is magnified by the fact that deviations are squared.[3]

The dramatic effect of outliers on the sum of squared errors is important because it is used to compute the standard deviation, which in turn is used to estimate the standard error, which itself is used to calculate confidence intervals around the parameter estimate and test statistics. If the sum of squared errors is biased, the associated standard error, confidence interval and test statistic will be too.

6.4 Overview of assumptions ||||

The second head of the beast of bias is called 'violation of assumptions'. An assumption is a condition that ensures that what you're attempting to do works. For example, when we assess a model using a test statistic, we have usually made some

---

3 In this example, the difference between the outlier and the mean (the deviance) is $10 - 3.8 = 6.2$. The deviance squared is $6.2^2 = 38.44$. Therefore, of the 50.8 units of error that we have, a whopping 38.44 are attributable to the outlier.

assumptions, and if these assumptions are true then we know that we can take the test statistic (and associated *p*-value) at face value and interpret it accordingly. Conversely, if any of the assumptions are not true (usually referred to as a violation) then the test statistic and *p*-value will be inaccurate and could lead us to the wrong conclusion.

What are assumptions?

The statistical procedures common in the social sciences are often presented as unique tests with idiosyncratic assumptions, which can be confusing. However, because most of these procedures are variations of the linear model (see Section 2.3) they share a common set of assumptions. These assumptions relate to the quality of the model itself, and the test statistics used to assess it (which are usually **parametric tests** based on the normal distribution). The main assumptions that we'll look at are:

- additivity and linearity;
- normality of something or other;
- homoscedasticity/homogeneity of variance;
- independence.

6.5 Additivity and linearity ▮▮▮▮

The first assumption we'll look at is additivity and linearity. The vast majority of statistical models in this book are based on the linear model, which we reviewed a few pages back. The assumption of additivity and linearity means that the relationship between the outcome variable and predictors is accurately described by equation (2.4). It means that scores on the outcome variable are, in reality, linearly related to any predictors, and that if you have several predictors then their combined effect is best described by adding their effects together.

This assumption is the most important because if it is not true then, even if all other assumptions are met, your model is invalid because your description of the process you want to model is wrong. If the relationship between variables is curvilinear, then describing it with a linear model is wrong (think back to Jane Superbrain Box 2.1). It's a bit like calling your pet cat a dog: you can try to get it to go into a kennel or fetch a ball or sit when you tell it to, but don't be surprised when it coughs up a hairball because no matter how often you describe it as a dog, it is in fact a cat. Similarly, if you describe your statistical model inaccurately it won't behave itself and there's no point in interpreting its parameter estimates or worrying about significance tests of confidence intervals: the model is wrong.

6.6 Normally distributed something or other ▮▮▮▮

The second assumption relates to the normal distribution, which we encountered in Chapter 1. Many people wrongly take the 'assumption of normality' to mean that the data need to be normally distributed (Misconception Mutt 6.1). In fact, it relates in different ways to things we want to do when fitting models and assessing them:

- **Parameter estimates**: The mean is a parameter, and we saw in Section 6.3 (the Amazon ratings) that extreme scores can bias it. This illustrates that estimates of parameters are affected by non-normal distributions (such as those with outliers). Parameter estimates differ in how much they are biased in a non-normal distribution: the median, for example, is less biased by skewed distributions than the mean. We've also seen that any model we fit will include some error: it won't predict the outcome variable perfectly for every case. Therefore, for each case there is an error term (the *deviance* or *residual*). If these residuals are normally distributed in the

population then using the method of least squares to estimate the parameters (the *b*s in equation (2.4)) will produce better estimates than other methods.

   ○ For the estimates of the parameters that define a model (the *b*s in equation (2.4)) to be optimal (to have the least possible error given the data) the residuals (the error$_i$ in equation (2.4)) in the population must be normally distributed. This is true mainly if we use the method of least squares (Section 2.6), which we often do.

- **Confidence intervals**: We use values of the standard normal distribution to compute the confidence interval (Section 2.8.1) around a parameter estimate (e.g., the mean or a *b* in equation (2.4)). Using values of the standard normal distribution makes sense only if the parameter estimates comes from one.

   ○ For confidence intervals around a parameter estimate (e.g., the mean or a *b* in equation (2.4)) to be accurate, that estimate must have a normal sampling distribution.

- **Null hypothesis significance testing**: If we want to test a hypothesis about a model (and, therefore, the parameter estimates within it) using the framework described in Section 2.9 then we assume that the parameter estimates have a normal distribution. We assume this because the test statistics that we use (which we will learn about in due course) have distributions related to the normal distribution (such as the *t*-, *F*- and chi-square distributions), so if our parameter estimate is normally distributed then these test statistics and *p*-values will be accurate (see Jane Superbrain Box 6.1 for some more information).

   ○ For significance tests of models (and the parameter estimates that define them) to be accurate the *sampling distribution* of what's being tested must be normal. For example, if testing whether two

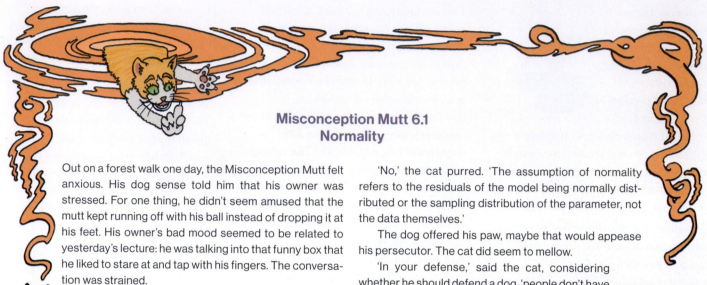

Misconception Mutt 6.1
Normality

Out on a forest walk one day, the Misconception Mutt felt anxious. His dog sense told him that his owner was stressed. For one thing, he didn't seem amused that the mutt kept running off with his ball instead of dropping it at his feet. His owner's bad mood seemed to be related to yesterday's lecture: he was talking into that funny box that he liked to stare at and tap with his fingers. The conversation was strained.

'I don't get it though,' his owner said into the box. 'What is the assumption of normality?'

The mutt wanted to help. He had enjoyed his owner taking him along to statistics lectures: he got a lot of strokes, but seemed to be learning statistics too.

'It means that your data need to be normally distributed,' the mutt said. His owner stopped shouting into his phone briefly to wonder why his dog was suddenly whining.

A nearby puddle started to ripple. The mutt turned in time to see some ginger ears appearing from the water. He sighed a depressed sigh.

Having emerged from the puddle, the Correcting Cat sauntered over. He quietly biffed the dog's nose.

'No,' the cat purred. 'The assumption of normality refers to the residuals of the model being normally distributed or the sampling distribution of the parameter, not the data themselves.'

The dog offered his paw, maybe that would appease his persecutor. The cat did seem to mellow.

'In your defense,' said the cat, considering whether he should defend a dog, 'people don't have direct access to the sampling distribution, so they have to make an educated guess about its shape. One way to do that is to look at the data because if the data are normally distributed then it's reasonable to assume that the errors in the model and the sampling distribution are also.'

The dog smiled. His tongue flopped out and he lurched to lick the cat's forehead.

The cat looked disgusted with himself and turned to return to his puddle. As his form liquidized into the ground, he turned and said, 'It doesn't change the fact that you were wrong!'

means are different, the data do not need to be normally distributed, but the sampling distribution of means (or differences between means) does. Similarly, if looking at relationships between variables, the significance tests of the parameter estimates that define those relationships (the *b*s in equation (2.4)) will be accurate only when the sampling distribution of the estimate is normal.

6.6.1 The central limit theorem revisited ▮▮▮▯

To understand when and if we need to worry about the assumption of normality, we need to revisit the central limit theorem,[4] which we encountered in Section 2.7. Imagine we have a population of scores that is not normally distributed. Figure 6.5 shows such a population, containing scores of how many friends

statistics lecturers have. It is very skewed, with most lecturers having no friends, and the frequencies declining as the number of friends increases to the maximum score of 7 friends. I'm not tricking you, this population is as far removed from the bell-shaped normal curve as it looks. Imagine that I take samples of five scores from this population and in each sample I estimate a parameter (let's say I compute the mean) and then replace the scores.

---

4 The 'central' in the name refers to the theorem being important and far-reaching and has nothing to do with centers of distributions.

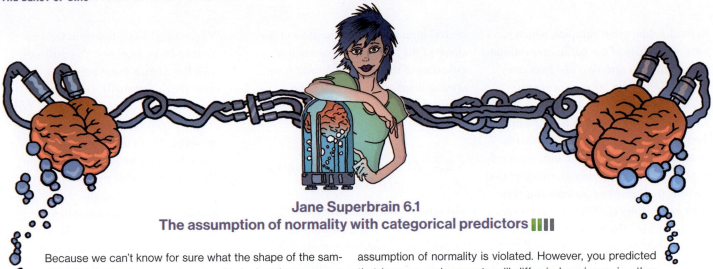

Jane Superbrain 6.1
The assumption of normality with categorical predictors ▮▮▮▮

Because we can't know for sure what the shape of the sampling distribution is, researchers tend to look at the scores on the outcome variable (or the residuals) when assessing normality. When you have a categorical predictor variable you wouldn't expect the overall distribution of the outcome (or residuals) to be normal. For example, if you have seen the movie *The Muppets*, you will know that muppets live among us. Imagine you predicted that muppets are happier than humans (on TV they seem to be). You collect happiness scores in some muppets and some humans and plot the frequency distribution. You get the graph on the left of Figure 6.4 and decide that because the data are not normal it is likely that the

assumption of normality is violated. However, you predicted that humans and muppets will differ in happiness; in other words, you predict that they come from different populations. If we plot separate frequency distributions for humans and muppets (right of Figure 6.4) you'll notice that within each group the distribution of scores is very normal. The data are as you predicted: muppets are happier than humans, and so the center of their distribution is higher than that of humans. When you combine the scores this creates a bimodal distribution (i.e., two humps). This example illustrates that it is not the normality of the outcome (or residuals) overall that matters, but normality at each unique level of the predictor variable.

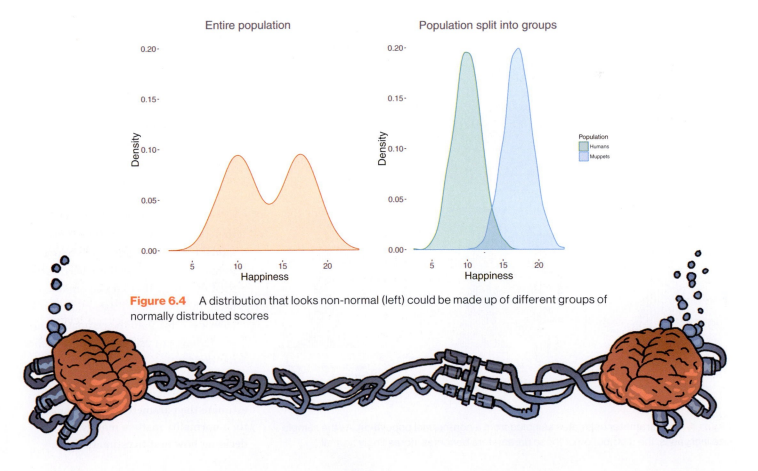

Figure 6.4 A distribution that looks non-normal (left) could be made up of different groups of normally distributed scores

In total, I take 5000 samples, which gives me 5000 values of the parameter estimate (one from each sample). The frequency distribution of the 5000 parameter estimates from the 5000 samples is on the far left of Figure 6.5. This is the sampling distribution of the parameter estimate. Note that it is a bit skewed, but not nearly as skewed as the population. Imagine that I repeat this sampling process, but this time my samples each contain 30 scores instead of five. The resulting distribution of the 5000 parameter estimates is in the center of Figure 6.5. The skew is gone and the distribution looks normal. Finally, I repeat the whole process but this time take samples of 100 scores rather than 30. Again, the resulting distribution is basically normal (right of Figure 6.5). As the sample sizes get bigger the sampling distributions become more normal, until a point at which the sample is big enough that the sampling distribution *is* normal—even though the population of scores is very non-normal indeed. This is the central limit theorem: regardless of the shape of the population, parameters estimates of that population will have a normal distribution provided the samples are 'big enough' (see Jane Superbrain Box 6.2).

6.6.2 When does the assumption of normality matter? ▮▮▮▮

The central limit theorem means that *there are a variety of situations in which we can assume normality regardless of the shape of our sample data* (Lumley et al., 2002). Let's think back to the things affected by normality:

1 For confidence intervals around a parameter estimate (e.g., the mean or a *b* in equation (2.4)) to be accurate, that estimate must come from a normal sampling distribution.

The central limit theorem tells us that in large samples, the estimate will have come from a normal distribution regardless of what the sample or population data look like. Therefore, if we are interested in computing confidence intervals then we don't need to worry about the assumption of normality if our sample is large enough.

2 For significance tests of models to be accurate the sampling distribution of what's being tested must be normal. Again, the central limit theorem tells us that in large samples this will be true no matter what the shape of the population. Therefore, the shape of our data shouldn't affect significance tests *provided our sample is large enough*. However, the extent to which test statistics perform as they should do in large samples varies across different test statistics, and we will deal with these idiosyncratic issues in the appropriate chapter.

3 For the estimates of model parameters (the *b*s in equation (2.4)) to be optimal (using the method of least squares) the residuals in the population must be normally distributed. The method of least squares will always give you an estimate of the model parameters that minimizes error, so in that sense you don't need to assume normality of anything to fit a linear model and estimate the parameters that define it (Gelman & Hill, 2007). However, there are other methods for estimating model parameters, and if you happen to have normally distributed errors then the estimates that you obtained using the method of least squares will have less error than the estimates you would have got using any of these other methods.

To sum up, then, if all you want to do is estimate the parameters of your model then normality matters mainly in deciding how best to estimate them.

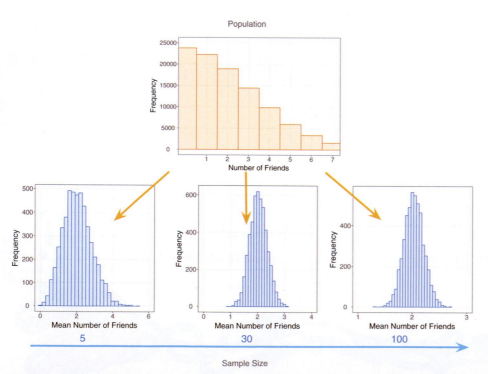

Figure 6.5 Parameter estimates sampled from a non-normal population. As the sample size increases, the distribution of those parameters becomes increasingly normal

Oditi's Lantern
The central limit theorem

'I, Oditi, believe that the central limit theorem is key to unlocking the hidden truths that the cult strives to find. The true wonder of the CLT cannot be understood by a static diagram and the ramblings of a damaged mind. Only by staring into my lantern can you see the CLT at work in all its wonder. Go forth and look into the abyss.'

Jane Superbrain 6.2
Size really does matter

How big is 'big enough' for the central limit theorem to kick in? The widely accepted value is a sample size of 30, and we saw in Figure 6.4 that with samples of this size we got a sampling distribution that approximated normal. We also saw that samples of 100 yielded a better approximation of normal. There isn't a simple answer to how big is 'big enough': it depends on the population distribution. In light-tailed distributions (where outliers are rare) an N as sm[...] be 'big enough', but in heavy-tailed distributi[...] ers are common) then up to 100 or even [...] sary. If the distribution has a lot [...] might need a very l[...] theorem to wo[...] you're trying to e[...]

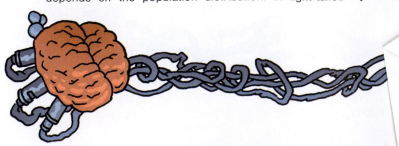

Jane Superbrain 6.3
Stealth outliers ▐▐▐▐

We tend to think of outliers as one or two very extreme scores, but sometimes they soak themselves in radar-absorbent paint and contort themselves into strange shapes to avoid detection. These 'stealth outliers' (that's my name for them, no one else calls them that) hide undetected in data sets, radically affecting analyses. Imagine you collected happiness scores, and when you plotted the frequency distribution it looked like Figure 6.6 (left). You might decide that this distribution is normal, because it has the characteristic bell-shaped curve. However, it is not: it is a **mixed normal distribution** or **contaminated normal distribution** (Tukey, 1960). The happiness scores on the left of Figure 6.6 are made up of two distinct populations: 90% of scores are from humans, but 10% are from muppets (we saw in Jane Superbrain Box 6.1 that they live among us). Figure 6.6 (right) reproduces this overall distribution (the blue one), but also shows the unique distributions for the humans (red) and muppets (Kermit-colored green) that contribute to it.

The human distribution is a perfect normal distribution, but the curve for the muppets is flatter and heavier in the tails, showing that muppets are more likely than humans to be extremely happy (like Kermit) or extremely miserable (like Statler and Waldorf). When these populations combine, the muppets contaminate the perfectly normal distribution of humans: the combined distribution (blue) has slightly more scores in the extremes than a perfect normal distribution (orange). The muppet scores have affected the overall distribution even though (1) they make up only 10% of the scores; and (2) their scores are more frequent at the extremes of 'normal' and not radically different like you might expect an outlier to be. These extreme scores inflate estimates of the population variance (think back to Jane Superbrain Box 1.5). Mixed normal distributions are very common and they reduce the power of significance tests–see Wilcox (2010) for a thorough account of the problems associated with these distributions.

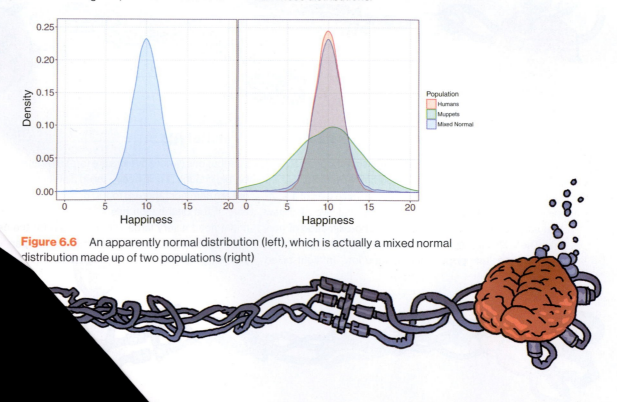

Figure 6.6　An apparently normal distribution (left), which is actually a mixed normal distribution made up of two populations (right)

If you want to construct confidence intervals around those parameters or compute significance tests relating to those parameters, then the assumption of normality matters in small samples, but because of the central limit theorem we don't really need to worry about this assumption in larger samples (but see Jane Superbrain Box 6.2). In practical terms, provided your sample is large, outliers are a more pressing concern than normality. Although we tend to think of outliers as isolated very extreme cases, you can have outliers that are less extreme but are not isolated cases. These outliers can dramatically reduce the power of significance tests (Jane Superbrain Box 6.3).

6.7 Homoscedasticity/ homogeneity of variance ▌▌▌▌

The second assumption relates to variance (Section 1.8.5) and is called homoscedasticity (also known as homogeneity of variance). It impacts two things:

- **Parameters**: Using the method of least squares (Section 2.6) to estimate the parameters in the model, we get optimal estimates if the variance of the outcome variable is equal across different values of the predictor variable.
- **Null hypothesis significance testing**: Test statistics often assume that the variance of the outcome variable is equal across different values of the predictor variable. If this is not the case then these test statistics will be inaccurate.

6.7.1 What is homoscedasticity/ homogeneity of variance? ▌▌▌▌

In designs in which you test groups of cases this assumption means that these groups come from populations with the same variance. In correlational designs,

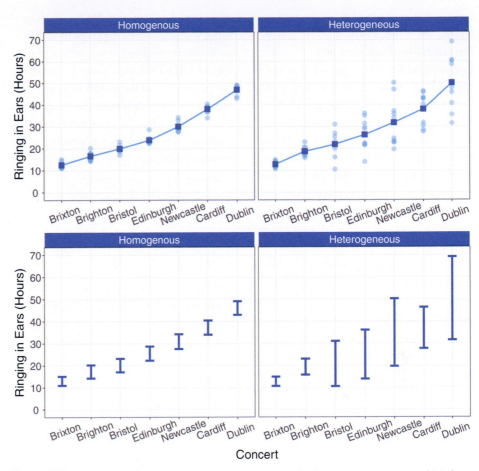

Figure 6.7 Graphs illustrating data with homogeneous (left) and heterogeneous (right) variances

this assumption means that the variance of the outcome variable should be stable at all levels of the predictor variable. In other words, as you go through levels of the predictor variable, the variance of the outcome variable should not change. Let's illustrate this idea with an example. An audiologist was interested in the effects of loud concerts on people's hearing. She sent 10 people to concerts of the loudest band in history, Manowar,[5] in Brixton (London), Brighton, Bristol, Edinburgh, Newcastle, Cardiff and Dublin and measured for how many hours after the concert they had ringing in their ears.

The top of Figure 6.7 shows the number of hours that each person (represented by a circle) had ringing in his or her ears after each concert. The squares show the average number of hours of ringing and the line connecting these means shows a cumulative effect of the concerts on ringing in the ears (the means increase). The graphs on the left and right show similar means but different *spreads* of the scores (circles) around their mean. To make this difference clearer, the bottom of Figure 6.7 removes the data and replaces them with a bar that shows the range of the scores displayed in the top figure. In the left-hand graphs these bars are of similar lengths, indicating that the

5 Before they realized that it's a bad idea to encourage bands to be loud, the *Guinness Book of World Records* cited a 1984 Manowar concert as the loudest. Before that Deep Purple held the honor for a 1972 concert of such fearsome volume that it rendered three members of the audience unconscious.

spread of scores around the mean was roughly the same at each concert. This is **homogeneity of variance** or **homoscedasticity**:[6] the spread of scores for hearing damage is the same at each level of the concert variable (i.e., the spread of scores is the same at Brixton, Brighton, Bristol, Edinburgh, Newcastle, Cardiff and Dublin). This is not the case on the right-hand side of Figure 6.7: the spread of scores is different at each concert. For example, the spread of scores after the Brixton concert is small (the vertical distance from the lowest score to the highest score is small), but the scores for the Dublin show are very spread out around the mean (the vertical distance from the lowest score to the highest score is large). The uneven spread of scores is easiest to see if we look at the bars in the lower right-hand graph. This scenario illustrates **heterogeneity of variance** or **heteroscedasticity**: at some levels of the concert variable the variance of scores is different than that at other levels (graphically, the vertical distance from the lowest to highest score is different after different concerts).

6.7.2 When does homoscedasticity/homogeneity of variance matter? ▮▮▮▮

If we assume equality of variance then the parameter estimates for a linear model are optimal using the method of least squares. The method of least squares will produce 'unbiased' estimates of parameters even when homogeneity of variance can't be assumed, but they won't be optimal. That just means that better estimates can be achieved using a method

other than least squares, for example, by using **weighted least squares** in which each case is weighted by a function of its variance. If all you care about is estimating the parameters of the model in your sample then you don't need to worry about homogeneity of variance in most cases: the method of least squares will produce unbiased estimates (Hayes & Cai, 2007). However, unequal variances/heteroscedasticity creates a bias and inconsistency in the estimate of the standard error associated with the parameter estimates in your model (Hayes & Cai, 2007). As such, confidence intervals, significance tests (and, therefore, p-values) for the parameter estimates will be biased, because they are computed using the standard error. Confidence intervals can be 'extremely inaccurate' when homogeneity of variance/homoscedasticity cannot be assumed (Wilcox, 2010). Therefore, if you want to look at the confidence intervals around your model parameter estimates or to test the significance of the model or its parameter estimates then homogeneity of variance matters. Some test statistics are designed to be accurate even when this assumption is violated, and we'll discuss these in the appropriate chapters.

6.8 Independence ▮▮▮▮

This assumption means that the errors in your model (the error$_i$ in equation (2.4)) are not related to each other. Imagine Paul and Julie were participants in an experiment where they had to indicate whether they remembered having seen particular photos. If Paul and Julie were to confer about whether they'd seen certain photos then their answers would *not* be independent: Julie's response to a given question would depend on Paul's answer.

We know already that if we estimate a model to predict their responses, there will be error in those predictions and, because Paul and Julie's scores are not independent, the errors associated with these predicted values will also not be independent. If Paul and Julie were unable to confer (if they were locked in different rooms) then the error terms should be independent (unless they're telepathic): the error in predicting Paul's response should not be influenced by the error in predicting Julie's response.

The equation that we use to estimate the standard error (equation (2.14)) is valid only if observations are independent. Remember that we use the standard error to compute confidence intervals and significance tests, so if we violate the assumption of independence then our confidence intervals and significance tests will be invalid. If we use the method of least squares, then model parameter estimates will still be valid but not optimal (we could get better estimates using a different method). In general, if this assumption is violated, we should apply the techniques covered in Chapter 21, so it is important to identify whether the assumption is violated.

6.9 Spotting outliers ▮▮▮▮

When they are isolated, extreme cases and outliers are fairly easy to spot using graphs such as histograms and boxplots; it is considerably trickier when outliers are more subtle (using z-scores may be useful—Jane Superbrain Box 6.4). Let's look at an example. A biologist was worried about the potential health effects of music festivals. She went to the Download Music Festival[7] (those of you outside the UK can pretend it is Roskilde Festival, Ozzfest, Lollapalooza, Wacken or something) and measured the

---

6 My explanation is simplified because usually we're making the assumption about the errors in the model and not the data themselves, but the two things are related.

7 www.downloadfestival.co.uk

hygiene of 810 concert-goers over the three days of the festival. She tried to measure every person on every day but, because it was difficult to track people down, there were missing data on days 2 and 3. Hygiene was measured using a standardized technique (don't worry, it *wasn't* licking the person's armpit) that results in a score ranging between 0 (you smell like a corpse that's been left to rot up a skunk's arse) and 4 (you smell of sweet roses on a fresh spring day). I know from bitter experience that sanitation is not always great at these places (the Reading Festival seems particularly bad) and so the biologist predicted that personal hygiene would go down dramatically over the three days of the festival. The data can be found in **DownloadFestival.sav**.

The resulting histogram is shown in Figure 6.8 (left). The first thing that should leap out at you is that there is one case that is very different from the others. All the scores are squashed up at one end of the distribution because they are less than 5 (yielding a very pointy distribution) except for one, which has a value of 20. This score is an obvious outlier and is particularly odd because a value of 20 exceeds the top of our scale (our hygiene scale ranged from 0 to 4). It must be a mistake. However, with 810 cases, how on earth do we find out which case it was? You could just look through the data, but that would certainly give you a headache, and so instead we can use a boxplot (see Section 5.5) which is another very useful way to spot outliers.

The outlier that we detected in the histogram shows up as an extreme score (*) on the boxplot (Figure 6.8, right). IBM SPSS Statistics helpfully tells us the row number (611) of this outlier. If we go to the data editor (data view), we can skip straight to this case by clicking on [image] and typing '611' in the resulting dialog box. Looking at row 611 reveals a score of 20.02, which is probably a mistyping

Using what you learnt in Section 5.4, plot a histogram of the hygiene scores on day 1 of the festival.

Using what you learnt in Section 5.5, plot a boxplot of the hygiene scores on day 1 of the festival.

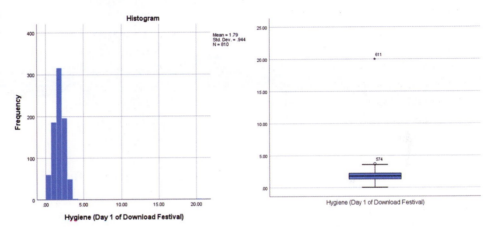

Figure 6.8　Histogram (left) and boxplot (right) of hygiene scores on day 1 of the Download Festival

Now we have removed the outlier in the data, re-plot the histogram and boxplot.

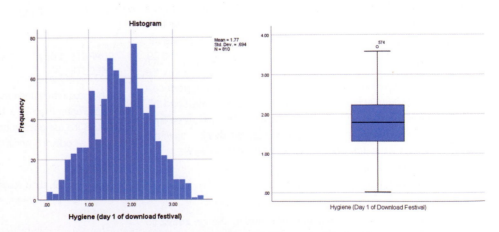

Figure 6.9　Histogram (left) and boxplot (right) of hygiene scores on day 1 of the Download Festival after removing the extreme score

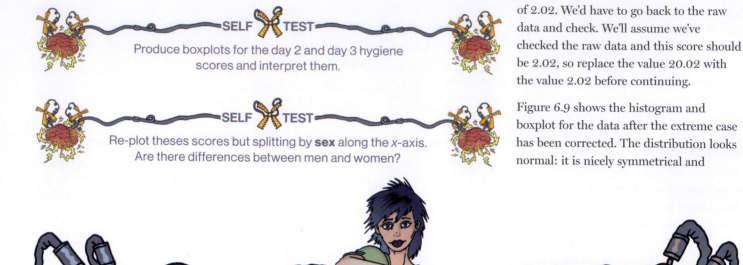

SELF TEST
Produce boxplots for the day 2 and day 3 hygiene scores and interpret them.

SELF TEST
Re-plot theses scores but splitting by **sex** along the *x*-axis. Are there differences between men and women?

of 2.02. We'd have to go back to the raw data and check. We'll assume we've checked the raw data and this score should be 2.02, so replace the value 20.02 with the value 2.02 before continuing.

Figure 6.9 shows the histogram and boxplot for the data after the extreme case has been corrected. The distribution looks normal: it is nicely symmetrical and

Jane Superbrain 6.4
Using *z*-scores to find outliers

We saw in Section 1.8.6 that *z*-scores express scores in terms of a distribution with a mean of 0 and a standard deviation of 1. By converting our data to *z*-scores we can use benchmarks that we can applyto any data set to search for outliers. Activate the *Analyze* ▶ *Descriptive Statistics* ▶ *Descriptives*...dialog box, select the variable(s) to convert (such as day 2 of the hygiene data) and tick ☑ Save standardized values as variables (Figure 6.10). SPSS will create a new variable in the data editor (with the same name as the selected variable, but prefixed with *z*).

To look for outliers we can count how many *z*-scores fall within certain important limits. If we ignore whether the *z*-score is positive or negative (called the 'absolute value'), then in a normal distribution we'd expect about 5% to be greater than 1.96 (we often use 2 for convenience), 1% to have absolute values greater than 2.58, and none to be greater than about 3.29. To get SPSS to do the counting for you, use the syntax file **Outliers.sps** (on the companion website), which will produce a table for day 2 of the Download Festival hygiene data. Load this file and run the syntax (see Section 4.10). The first three lines use the *descriptives* function on the variable **day2** to save the *z*-scores in the data editor (as a variable called **zday2**).

```
DESCRIPTIVES
VARIABLES= day2/SAVE.
EXECUTE.
```

Next, we use the compute command to change **zday2** so that it contains the absolute values (i.e. converts all minus values to plus values).

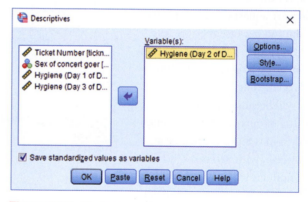

Figure 6.10   Saving *z*-scores

```
COMPUTE zday2= abs(zday2).
EXECUTE.
```

The next commands recode the variable **zday2** so that if a value is greater than 3.29 it's assigned a code of 1, if it's greater than 2.58 it's assigned a code of 2, if it's greater than 1.96 it's assigned a code of 3, and if it's less than 1.95 it gets a code of 4.

```
RECODE
zday2 (3.29 thru highest = 1)(2.58 thru highest = 2)
(1.96 thru highest = 3)(Lowest thru 1.95 = 4).
EXECUTE.
```

We then use the *value labels* command to assign helpful labels to the codes we defined above.

```
VALUE LABELS zday2
4 'Normal range' 3 'Potential Outliers (z
> 1.96)' 2 'Probable Outliers (z > 2.58)' 1
'Extreme (z-score > 3.29)'.
```

Finally, we use the *frequencies* command to produce a table (Output 6.1) telling us the percentage of 1s, 2s, 3s and 4s found in the variable **zday2**.

```
FREQUENCIES
VARIABLES= zday2
/ORDER=ANALYSIS.
```

Thinking about what we know about the absolute values of z-scores, we would expect to see only 5% (or less) with values greater than 1.96, 1% (or less) with values greater than 2.58, and very few cases above 3.29. The column labeled *Cumulative Percent* tells us the corresponding percentages for the hygiene scores on day 2: 0.8% of cases were above 3.29 (extreme cases), 2.3% (compared to the 1% we'd expect) had values greater than 2.58, and 6.8% (compared to the 5% we would expect) had values greater than 1.96. The remaining cases (which, if you look at the *Valid Percent*, constitute 93.2%) were in the normal range. These percentages are broadly consistent with what we'd expect in a normal distribution (around 95% were in the normal range).

Zscore: Hygiene (Day 2 of Download Festival)

		Frequency	Percent	Valid Percent	Cumulative Percent
Valid	Extreme (z-score> 3.29)	2	.2	.8	.8
	Probable Outliers (z > 2.58)	4	.5	1.5	2.3
	Potential Outliers (z > 1.96)	12	1.5	4.5	6.8
	Normal range	246	30.4	93.2	100.0
	Total	264	32.6	100.0	
Missing	System	546	67.4		
Total		810	100.0		

Output 6.1

doesn't seem too pointy or flat. Neither plot indicates any particularly extreme scores: the boxplot suggests that case 574 is a mild outlier, but the histogram doesn't seem to show any cases as being particularly out of the ordinary.

6.10 Spotting normality ▮▮▮▮

6.10.1 Using graphs to spot normality ▮▮▮▮

Frequency distributions are not only good for spotting outliers, they are the natural choice for looking at the shape of the distribution, as we can see for the day 1 scores in Figure 6.9. An alternative is the **P-P plot** (probability–probability plot), which plots the cumulative probability of a variable against the cumulative probability of a particular distribution (in this case we would specify a normal distribution). The data are ranked and sorted, then for each rank the corresponding z-score is calculated to create an 'expected value' that the score should have in a normal distribution. Next, the score itself is converted to a z-score (see Section 1.8.6). The actual z-score is plotted against the expected z-score. If the data are normally distributed then the actual z-score will be the same as the expected z-score and you'll get a lovely straight diagonal line. This ideal scenario is helpfully plotted on the graph and your job is to compare the data points to this line. If values fall on the diagonal of the plot then the variable is normally distributed; however, when the data sag consistently above or below the diagonal then this shows that the kurtosis differs from a normal distribution, and

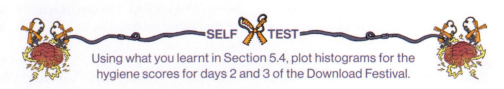

SELF TEST

Using what you learnt in Section 5.4, plot histograms for the hygiene scores for days 2 and 3 of the Download Festival.

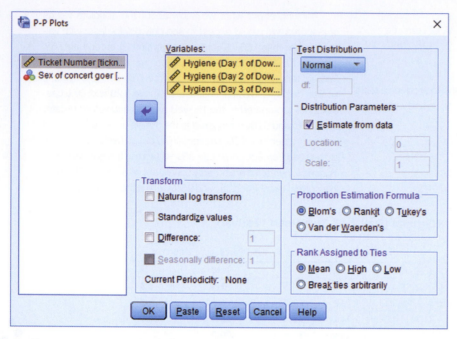

Figure 6.11 Dialog box for obtaining P-P plots

when the data points are S-shaped, the problem is skewness.

To get a P-P plot use *Analyze* ▸ *Descriptive Statistics* ▸ 🖼 *P-P Plots...* to access the dialog box in Figure 6.11.[8] There's not a lot to say about this dialog box because the default options produce plots that compare the selected variables to a normal distribution, which is what we want (although there is a drop-down list of other distributions against which you can compare your variables). Select the three hygiene score variables in the variable list (click the day 1 variable, then hold down *Shift* and select the day 3 variable), transfer them to the box labeled *Variables* by dragging or clicking on 🔽, and click OK.

Figure 6.12 shows the histograms (from the self-test tasks) and the corresponding P-P plots. We looked at the day 1 scores in the previous section and concluded that they looked quite normal. The P-P plot echoes this view because the data points fall very close to the 'ideal' diagonal line. However, the distributions for days 2 and

3 look positively skewed. This can be seen in the P-P plots by the data points deviating away from the diagonal. These plots suggest that relative to day 1, hygiene scores on days 2 and 3 were more clustered around the low end of the scale (more people were less hygienic); so people became smellier as the festival progressed. The skew on days 2 and 3 occurs because a minority insisted on upholding their levels of hygiene over the course of the festival (baby wet-wipes are indispensable, I find).

6.10.2 Using numbers to spot normality ▐▐▐

Graphs are particularly useful for looking at normality in big samples; however, in smaller samples it can be useful to explore the distribution of the variables using the *frequencies* command (*Analyze* ▸ *Descriptive Statistics* ▸ 🔢 *Frequencies...*). The main dialog box is shown in Figure 6.13. The variables in the data editor are listed on the

left-hand side, and they can be transferred to the box labeled *Variable(s)* by clicking on a variable (or highlighting several with the mouse) and dragging or clicking 🔽. If a variable listed in the *Variable(s)* box is selected, it can be transferred back to the variable list by clicking on the arrow button (which should now be pointing in the opposite direction). By default, SPSS produces a tabulated frequency distribution of all scores. There are two other dialog boxes that we'll look at: the *Statistics* dialog box is accessed by clicking Statistics..., and the *Charts* dialog box is accessed by clicking Charts....

The *Statistics* dialog box allows you to select ways to describe a distribution, such as measures of central tendency (mean, mode, median), measures of variability (range, standard deviation, variance, quartile splits), and measures of shape (kurtosis and skewness). Select the mean, mode, median, standard deviation, variance and range. To check that a distribution of scores is normal, we can look at the values of kurtosis and skewness (see Section 1.8.1). The *Charts* dialog box is a simple way to plot the frequency distribution of scores (as a bar chart, a pie chart or a histogram). We've already plotted histograms of our data, so we don't need to select these options, but you could use them in future analyses. When you have selected the appropriate options, return to the main dialog box by clicking Continue, and click OK to run the analysis.

Output 6.2 shows the table of descriptive statistics for the three variables in this example. On average, hygiene scores were 1.77 (out of 5) on day 1 of the festival, but went down to 0.96 and 0.98 on days 2 and 3, respectively. The other important measures for our purposes are the

8 You'll notice in the same menu something called a Q-Q plot, which is very similar and which we'll discuss later.

skewness and kurtosis (see Section 1.8.1), both of which have an associated standard error. There are different ways to calculate skewness and kurtosis, but SPSS uses methods that give values of zero for a normal distribution. Positive values of skewness indicate a pile-up of scores on the left of the distribution, whereas negative values indicate a pile-up on the right. Positive values of kurtosis indicate a heavy-tailed distribution, whereas negative values indicate a light-tailed distribution. The further the value is from zero, the more likely it is that the data are not normally distributed. For day 1 the skew value is very close to zero (which is good) and kurtosis is a little negative. For days 2 and 3, though, there is a skewness of around 1 (positive skew) and larger kurtosis.

We can convert these values to a test of whether the values are significantly different from 0 (i.e. normal) using z-scores (Section 1.8.6). Remember that a z-score is the distance of a score from the mean of its distribution standardized by dividing by an estimate of how much scores vary (the standard deviation). We want our z-score to represent the distance of our score for skew/kurtosis from the mean of the sampling distribution for skew/kurtosis values from a normal distribution. The mean of this sampling distribution will be zero (on average, samples from a normally distributed population will have skew/kurtosis of 0). We then standardize this distance using an estimate of the variation in sample values of skew/kurtosis, which would be the standard deviation of the sampling distribution, which we know is called the standard error. Therefore, we end up dividing the estimates of skew and kurtosis by their standard errors:

$$z_{skewness} = \frac{S-0}{SE_{skewness}} \qquad z_{kurtosis} = \frac{K-0}{SE_{kurtosis}}$$

(6.2)

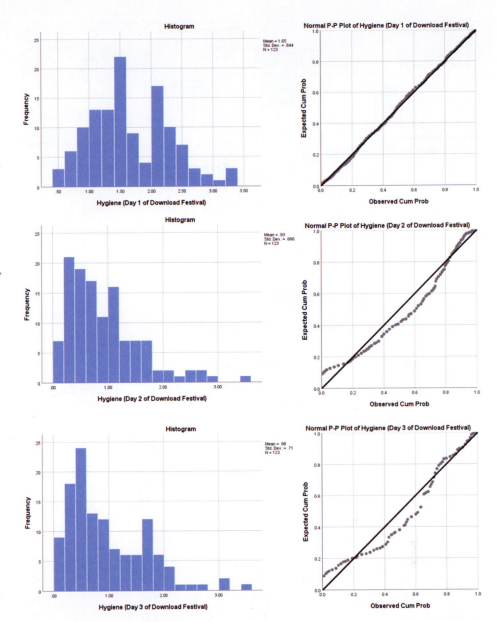

Figure 6.12 Histograms (left) and P-P plots (right) of the hygiene scores over the three days of the Download Festival

The values of S (skewness) and K (kurtosis) and their respective standard errors are produced by SPSS (Output 6.2). The resulting z-scores can be compared against values that you would expect to get if skew and kurtosis were not different from 0 (see Section 1.8.6). So, an absolute value greater than 1.96 is significant at $p < 0.05$, above 2.58 is significant at $p < 0.01$ and above 3.29 is significant at $p < 0.001$.

For the hygiene scores, the z-score of skewness is $-0.004/0.086 = 0.047$ on day 1, $1.095/0.150 = 7.300$ on day 2 and $1.033/0.218 = 4.739$ on day 3. It is clear, then, that although on day 1 scores are not at all skewed, on days 2 and 3 there is a very significant positive skew (as was evident from the histogram). The kurtosis z-scores are: $-0.410/0.172 = -2.38$ on day 1, $0.822/0.299 = 2.75$ on day 2 and $0.732/0.433 = 1.69$ on day 3. These values indicate significant problems with skew, kurtosis or both (at $p < 0.05$) for all three days; however, because of the large sample,

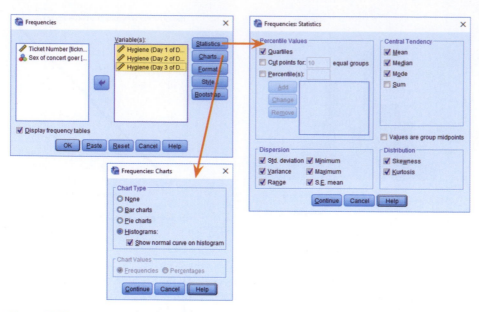

shape of the distribution visually, interpret the value of the skewness and kurtosis statistics, and possibly don't even worry about normality at all.

The **Kolmogorov–Smirnov test** and **Shapiro–Wilk test** compare the scores in the sample to a normally distributed set of scores with the same mean and standard deviation. If the test is non-significant ($p > 0.05$) it tells us that the distribution of the sample is not significantly different from a normal distribution (i.e., it is probably normal). If, however, the test is significant ($p < 0.05$) then the distribution in question is significantly different from a normal distribution (i.e., it is non-normal). These tests are tempting: they lure you with an easy way to decide whether scores are normally distributed (nice!). However, Jane Superbrain Box 6.5 explains some really good reasons not to use them. If you insist on using them, bear Jane's advice in mind and always plot your data as well and try to make an

Figure 6.13   Dialog boxes for the *frequencies* command

Statistics

		Hygiene (Day 1 of Download Festival)	Hygiene (Day 2 of Download Festival)	Hygiene (Day 3 of Download Festival)
N	Valid	810	264	123
	Missing	0	546	687
Mean		1.7711	.9609	.9765
Std. Error of Mean		.02437	.04436	.06404
Median		1.7900	.7900	.7600
Mode		2.00	.23	.44[a]
Std. Deviation		.69354	.72078	.71028
Variance		.481	.520	.504
Skewness		-.004	1.095	1.033
Std. Error of Skewness		.086	.150	.218
Kurtosis		-.410	.822	.732
Std. Error of Kurtosis		.172	.299	.433
Range		3.67	3.44	3.39
Minimum		.02	.00	.02
Maximum		3.69	3.44	3.41
Percentiles	25	1.3050	.4100	.4400
	50	1.7900	.7900	.7600
	75	2.2300	1.3500	1.5500

a. Multiple modes exist. The smallest value is shown

Output 6.2

Figure 6.14   Andrey Kolmogorov, wishing he had a Smirnov

this isn't surprising and so we can take comfort from the central limit theorem. Although I felt obliged to explain the *z*-score conversion, there is a very strong case for never using significance tests to assess assumptions (see Jane Superbrain Box 6.5). In larger samples you should certainly not do them; instead, look at the

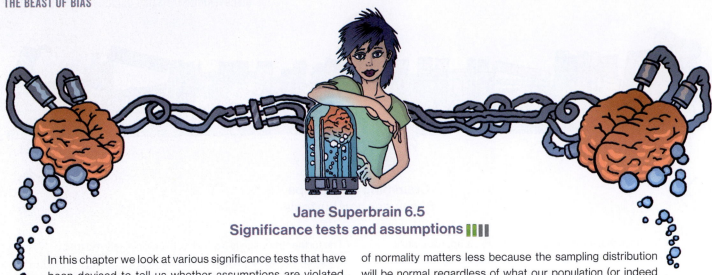

Jane Superbrain 6.5
Significance tests and assumptions ▍▍▍▍

In this chapter we look at various significance tests that have been devised to tell us whether assumptions are violated. These include tests of whether a distribution is normal (the Kolmogorov–Smirnov and Shapiro–Wilk tests), tests of homogeneity of variances (Levene's test), and tests of significance for skew and kurtosis. I cover these tests mainly because people expect to see these sorts of things in introductory statistics books, and not because they are a good idea. All these tests are based on null hypothesis significance testing, and this means that (1) in large samples they can be significant even for small and unimportant effects, and (2) in small samples they will lack power to detect violations of assumptions (Section 2.9.10).

We have also seen in this chapter that the central limit theorem means that as sample sizes get larger, the assumption of normality matters less because the sampling distribution will be normal regardless of what our population (or indeed sample) data look like. So, in large samples, where normality matters less (or not at all), a test of normality is more likely to be significant and make us worry about and correct for something that doesn't need to be corrected for or worried about. Conversely, in small samples, where we should worry about normality, a significance test won't have the power to detect non-normality and so is likely to encourage us not to worry about something that we probably ought to.

The best advice is that if your sample is large then don't use significance tests of normality, in fact don't worry too much about normality at all. In small samples pay attention if your significance tests are significant but resist being lulled into a false sense of security if they are not.

informed decision about the extent of non-normality based on converging evidence.

The Kolmogorov–Smirnov (K-S; Figure 6.14) test is accessed through the *explore* command (*Analyze* ▶ *Descriptive Statistics* ▶ 🔍 *Explore…*). Figure 6.15 shows the dialog boxes for this command. First, enter any variables of interest in the box labeled *Dependent List* by highlighting them on the left-hand side and transferring them by clicking ➡. For this example, select the hygiene scores for the three days. If you click ⬚Statistics… a dialog box appears, but the default option is fine (it will produce

means, standard deviations and so on). The more interesting option for our current purposes is accessed by clicking ⬚Plots… . In this dialog box select the option ☑ Normality plots with tests, and this will produce both the K-S test and some *normal quantile–quantile (Q-Q) plots*. A **Q-Q plot** is like the P-P plot that we encountered in Section 6.10, except that it plots the quantiles (Section 1.8.5) of the data instead of every individual score. The expected quantiles

are a straight diagonal line, whereas the observed quantiles are plotted as individual points. The Q-Q plot can be interpreted in the same way as a P-P plot: kurtosis is shown up by the dots sagging above or below the line, whereas skew is shown up by the dots snaking around the line in an 'S' shape. If you have a lot of scores Q-Q plots can be easier to interpret than P-P plots because they display fewer values.

By default, SPSS will produce boxplots (split according to group if a factor has been specified) and stem-and-leaf diagrams as well. We also need to click ⬚Options… to tell SPSS how to deal with

Cramming Sam's Tips
Skewness and kurtosis

- To check that the distribution of scores is approximately normal, look at the values of skewness and kurtosis in the output.
- Positive values of skewness indicate too many low scores in the distribution, whereas negative values indicate a build-up of high scores.
- Positive values of kurtosis indicate a heavy-tailed distribution, whereas negative values indicate a light-tailed distribution.

- The further the value is from zero, the more likely it is that the data are not normally distributed.
- You can convert these scores to z-scores by dividing by their standard error. If the resulting score (when you ignore the minus sign) is greater than 1.96 then it is significant ($p < 0.05$).
- Significance tests of skew and kurtosis should not be used in large samples (because they are likely to be significant even when skew and kurtosis are not too different from normal).

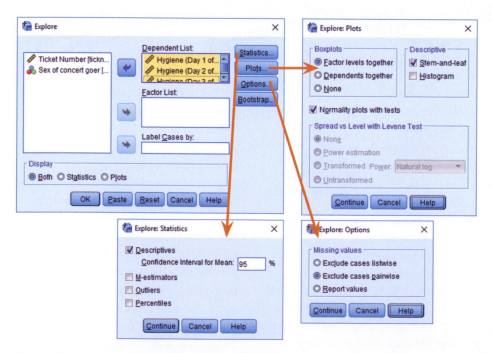

Figure 6.15 Dialog boxes for the *explore* command

missing values. This is important because although we start off with 810 scores on day 1, by day 2 we have only 264 and by day 3 only 123. By default SPSS will use only cases for which there are valid scores on all selected variables. This would mean that for day 1, even though we have 810 scores, SPSS will use only the 123 cases for which there are scores on all three days. This is known as excluding cases *listwise*. However, we want it to use all of the scores it has on a given day, which is known as *pairwise* (SPSS Tip 6.1). Once you have clicked Options... select *Exclude cases pairwise*, then click Continue to return to the main dialog box and click OK to run the analysis.

SPSS produces a table of descriptive statistics (mean, etc.) that should have the same values as the tables obtained using

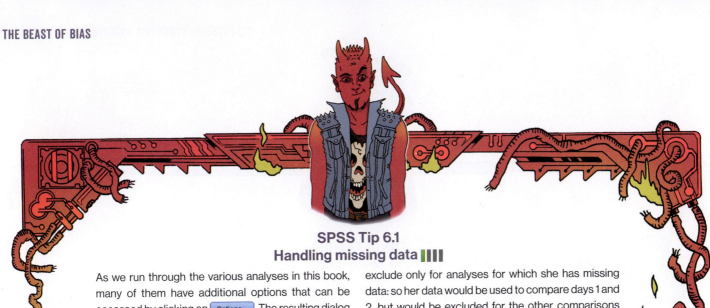

SPSS Tip 6.1
Handling missing data ▐▐▐▐

As we run through the various analyses in this book, many of them have additional options that can be accessed by clicking on Options... The resulting dialog box will offer some selection of the following possibilities: exclude cases 'pairwise', 'analysis by analysis' or 'listwise', and sometimes 'replace with mean'. Let's imagine we wanted to use our hygiene scores to compare mean scores on days 1 and 2, days 1 and 3, and days 2 and 3. First, we can exclude cases listwise, which means that if a case has a missing value for any variable, then the case is excluded from the whole analysis. So, for example, if we had the hygiene score for a person (let's call her Melody) at the festival on days 1 and 2, but not day 3, then Melody's data will be excluded for all of the comparisons mentioned above. Even though we have her data for days 1 and 2, we won't use them for that comparison—*they would be completely excluded from the analysis*. Another option is to exclude cases on a *pairwise* (a.k.a. *analysis-by-analysis* or *test-by-test*) basis, which means that Melody's data will be

exclude only for analyses for which she has missing data: so her data would be used to compare days 1 and 2, but would be excluded for the other comparisons (because we don't have her score on day 3).

Sometimes SPSS will offer to replace the missing score with the average score for this variable and then include that case in the analysis. The problem is that this will likely suppress the true value of the standard deviation (and, more importantly, the standard error). The standard deviation will be suppressed because for any replaced case there will be no difference between the mean and the score, whereas if data had been collected for that missing case there would, almost certainly, have been some difference between the mean and the score. If the sample is large and the number of missing values small then this may not be a serious consideration. However, if there are many missing values this choice is potentially dangerous because smaller standard errors are more likely to lead to significant results that are a product of the data replacement rather than a genuine effect.

the *frequencies* procedure. The table for the K-S test (Output 6.3) includes the test statistic itself, the degrees of freedom (which should equal the sample size)[9] and the significance value of this test. Remember that a significant value (*Sig.* less than 0.05) indicates a deviation from normality. For day 1 the K-S test is just about not significant ($p = 0.097$), albeit surprisingly close to significant given how normal the day 1 scores looked in the histogram (Figure 6.12). This has occurred because the sample size on day 1 is very

large ($N = 810$) so the test is highly powered: it shows how in large samples even small and unimportant deviations from normality might be deemed significant by this test (Jane Superbrain

Box 6.5). For days 2 and 3 the test is highly significant, indicating that these distributions are not normal, which is likely to reflect the skew seen in the histograms for these data (Figure 6.12).

Tests of Normality

	Kolmogorov-Smirnov[a]			Shapiro-Wilk		
	Statistic	df	Sig.	Statistic	df	Sig.
Hygiene (Day 1 of Download Festival)	.029	810	.097	.996	810	.032
Hygiene (Day 2 of Download Festival)	.121	264	.000	.908	264	.000
Hygiene (Day 3 of Download Festival)	.140	123	.000	.908	123	.000

a. Lilliefors Significance Correction

Output 6.3

---

9 It is not $N - 1$ because the test compares the sample to an idealized normal, so the sample mean isn't used as an estimate of the population mean, which means that all scores are free to vary.

Oliver Twisted
Please, Sir, can I have some more ... normality tests?

'There is another test reported in the table (the Shapiro–Wilk test)', whispers Oliver as he creeps up behind you, knife in hand, 'and a footnote saying that the 'Lilliefors significance correction' has been applied. What the hell is going on?' Well, Oliver, all

will be revealed in the additional material for this chapter on the companion website: you can find out more about the K-S test, and information about the Lilliefors correction and Shapiro–Wilk test. What are you waiting for?

Cramming Sam's Tips
Normality tests

- The K-S test can be used (but shouldn't be) to see if a distribution of scores significantly differs from a normal distribution.
- If the K-S test is significant (*Sig.* in the SPSS table is less than 0.05) then the scores are significantly different from a normal distribution.
- Otherwise, scores are approximately normally distributed.

- The Shapiro–Wilk test does much the same thing, but it has more power to detect differences from normality (so this test might be significant when the K-S test is not).
- *Warning*: In large samples these tests can be significant even when the scores are only slightly different from a normal distribution. Therefore, I don't particularly recommend them and they should always be interpreted in conjunction with histograms, P-P or Q-Q plots, and the values of skew and kurtosis.

6.10.3 Reporting the K-S test

If you must use the K-S test, its statistic is denoted by D and you should report the degrees of freedom (df) in brackets after the D. The results in Output 6.3 could be reported as:

✓ The hygiene scores on day 1, $D(810) = 0.029$, $p = 0.097$, did not deviate significantly from normal; however, day 2, $D(264) = 0.121$, $p < 0.001$, and day 3, $D(123) = 0.140$, $p < 0.001$, scores were both significantly non-normal.

6.10.4 Normality within groups and the *split file* command

When predictor variables are formed of categories, if you decide that you need to check the assumption of normality then you need to do it within each group separately (Jane Superbrain Box 6.1). For example, for the hygiene scores we have data for males and females (in the variable **sex**). If we made some prediction about there being differences in hygiene between males and females at a music festival then we should look at normality within males and females separately. There are several ways to produce basic descriptive statistics for separate groups. First, I will introduce you to the *split file* command, in which you specify a coding variable that SPSS uses to carry out separate analyses on each category of cases.

If we want to obtain separate descriptive statistics for males and females in our festival hygiene scores, we can split the file, and then use the *frequencies* command described in Section 6.10.2. To split the file, select *Data* ▶ 🗔 Split File... or click on 🗔. In the resulting dialog box (Figure 6.16) select the option *Organize output by groups*. Once this option is selected, the *Groups*

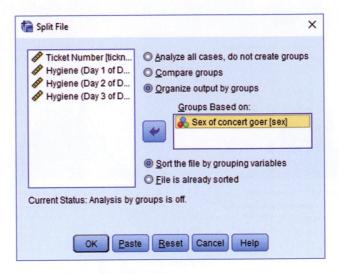

Figure 6.16 Dialog box for the *split file* command

Male

Statistics[a]

		Hygiene (Day 1 of Download Festival)	Hygiene (Day 2 of Download Festival)	Hygiene (Day 3 of Download Festival)
N	Valid	315	104	56
	Missing	0	211	259
Mean		1.6021	.7733	.8291
Std. Error of Mean		.03620	.05847	.07210
Median		1.5800	.6700	.7300
Mode		2.00	.23	.44
Std. Deviation		.64241	.59630	.53954
Variance		.413	.356	.291
Skewness		.200	1.476	.719
Std. Error of Skewness		.137	.237	.319
Kurtosis		-.101	3.134	-.268
Std. Error of Kurtosis		.274	.469	.628
Range		3.47	3.35	2.09
Minimum		.11	.00	.02
Maximum		3.58	3.35	2.11
Percentiles	25	1.1400	.2975	.4400
	50	1.5800	.6700	.7300
	75	2.0000	1.0725	1.1950

a. Sex of Concert Goer = Male

Female

Statistics[a]

		Hygiene (Day 1 of Download Festival)	Hygiene (Day 2 of Download Festival)	Hygiene (Day 3 of Download Festival)
N	Valid	495	160	67
	Missing	0	335	428
Mean		1.8787	1.0829	1.0997
Std. Error of Mean		.03164	.06078	.09896
Median		1.9400	.8900	.8500
Mode		2.02	.85	.38
Std. Deviation		.70396	.76876	.81001
Variance		.496	.591	.656
Skewness		-.176	.870	.869
Std. Error of Skewness		.110	.192	.293
Kurtosis		-.397	.089	.069
Std. Error of Kurtosis		.219	.381	.578
Range		3.67	3.38	3.39
Minimum		.02	.06	.02
Maximum		3.69	3.44	3.41
Percentiles	25	1.4100	.4700	.4400
	50	1.9400	.8900	.8500
	75	2.3500	1.5475	1.7000

a. Sex of Concert Goer = Female

Output 6.4

Can I analyze groups of data?

Tests of Normality

	Sex of concert goer	Kolmogorov-Smirnov[a]			Shapiro-Wilk		
		Statistic	df	Sig.	Statistic	df	Sig.
Hygiene (Day 1 of Download Festival)	Male	.035	315	.200[*]	.993	315	.119
	Female	.053	495	.002	.993	495	.029

*. This is a lower bound of the true significance.

a. Lilliefors Significance Correction

Output 6.5

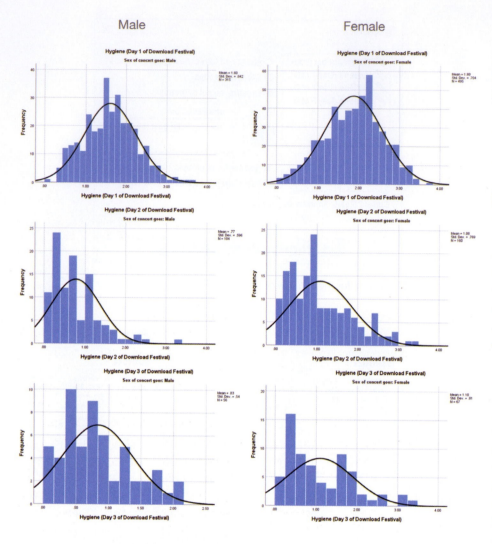

Figure 6.17 Distributions of hygiene scores for males (left) and females (right) over three days (top to bottom) of the music festival

If you're determined to ignore my advice, you can do K-S tests within the different groups by repeating the analysis we did earlier (Figure 6.15); because the *split file* command is switched on, we'd get the K-S test performed on males and females separately. An alternative method is to split the analysis by group from within the *explore* command itself. First, switch *split file* off by clicking *Data* ▶ Split File... (or clicking 🔳) to activate the dialog box in Figure 6.16. Select *Analyze all cases, do not create groups* and click OK. The *split file* function is now off and analyses will be conducted on the data as whole. Next, activate the *explore* command just as we did before: *Analyze* ▶ *Descriptive Statistics* ▶ Explore.... We can ask for separate tests for males and females by placing **sex** in the box labeled *Factor List* as in Figure 6.20 and selecting the same options as described earlier. Let's do this for the day 1 hygiene scores. You should see the table in Output 6.5, which shows that the distribution of hygiene scores was normal for males (the value of *Sig.* is greater than 0.05) but not for females (the value of *Sig.* is smaller than 0.05).

SPSS also produces a normal Q-Q plot (see Figure 6.18). Despite the K-S having completely different outcomes for males and females, the Q-Q plots are remarkably similar: there is no sign of a major problem with kurtosis (the dots do not particularly sag above or below the line) and there is some slight skew (the female graph in particular has a slight S-shape). However, both graphs show that the quantiles fall very close to the diagonal line, which, let's not forget, represents a perfect normal distribution. For the females the graph is at odds with the significant K-S test, and this illustrates my earlier point that if you have a large sample then tests such as K-S will lead you to conclude that even very minor deviations from normality are 'significant'.

Based on box will activate. Select the variable containing the group codes by which you wish to repeat the analysis (in this example select **sex**), and drag it to the box or click on ➡. By default, SPSS will sort the file by these groups (i.e., it will list one category followed by the other in the data editor). Once you have split the file, use the *frequencies* command as above. Let's request statistics for all three days as in Figure 6.13.

Output 6.4 shows the results, which have been split into two tables: the results for males and the results for females. Males scored lower than females on all three days

of the festival (i.e., they were smellier). Figure 6.17 shows the histograms of hygiene scores split according to the sex of the festival-goer. Male and female scores have similar distributions. On day 1 they are fairly normal (although females perhaps show a very slight negative skew, which indicates a higher proportion of them were at the higher end of hygiene scores than males). On days 2 and 3 both males and females show the positive skew that we saw in the sample overall. It looks as though proportionally more females are in the skewed end of the distribution (i.e., up the hygienic end).

6.11 Spotting linearity and heteroscedasticity/heterogeneity of variance ▮▮▮▮

6.11.1 Using graphs to spot problems with linearity or homoscedasticity ▮▮▮

The reason for looking at the assumption of linearity and homoscedasticity together is that we can check both with a single graph. Both assumptions relate to the errors (a.k.a. residuals) in the model and we can plot the values of these residuals against the corresponding values of the outcome predicted by our model in a scatterplot. The resulting plot shows whether there is a systematic relationship between what comes out of the model (the predicted values) and the errors in the model. Normally we convert the predicted values and errors to z-scores,[10] so this plot is sometimes referred to as *zpred vs. zresid*. If linearity and homoscedasticity hold true then there should be no systematic relationship between the errors in the model and what the model predicts. If this graph funnels out, then the chances are that there is heteroscedasticity in the data. If there is any sort of curve in this graph then the assumption of linearity is likely to be suspect.

Figure 6.19 shows examples of the plot of standardized residuals against standardized predicted values. The top left-hand graph shows a situation in which the assumptions of linearity and homoscedasticity have been met. The top right-hand graph shows a similar plot for a data set that violates the assumption of homoscedasticity. Note that the points form a funnel: they become more spread out across the graph. This funnel shape is typical of heteroscedasticity and indicates

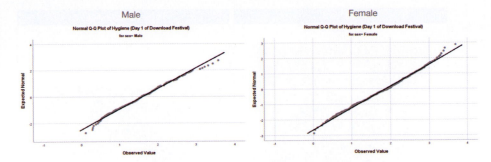

Figure 6.18 Normal Q-Q plots of hygiene scores for day 1 of the music festival

SELF TEST

Compute and interpret a K-S test and Q-Q plots for males and females for days 2 and 3 of the music festival.

increasing variance across the residuals. The bottom left-hand graph shows a plot of some data in which there is a non-linear relationship between the outcome and the predictor: there is a clear curve in the residuals. Finally, the bottom right-hand graph panel illustrates data that not only have a non-linear relationship but also show heteroscedasticity: there is a curved trend in the residuals *and* at one end of the plot the spread of residuals is very small, whereas at the other end the residuals are widely dispersed. When these assumptions have been violated you won't see these exact patterns, but hopefully these plots will help you to understand the general anomalies to look out for. We'll look at an example of how to use this graph in Chapter 9.

6.11.2 Spotting heteroscedasticity/heterogeneity of variance using numbers ▮▮▮▮

Remember that homoscedasticity/homogeneity of variance means that as you go through levels of one variable, the variance of the other should not change. If you've collected groups of data then this

means that the variance of your outcome variable or variables should be the same in each group. SPSS produces something called **Levene's test** (Levene, 1960), which tests the null hypothesis that the variances in different groups are equal. It works by doing a one-way ANOVA (see Chapter 12) on the deviation scores; that is, the absolute difference between each score and the mean of the group from which it came (see Glass, 1966, for a very readable explanation).[11] For now, all you need to know is that if Levene's test is significant at $p \leq 0.05$ then people tend to conclude that the null hypothesis is incorrect and that the variances are significantly different—therefore, the assumption of homogeneity of variances has been violated. If Levene's test is non-significant (i.e., $p > 0.05$) people take this to mean that the variances are roughly equal and the assumption is tenable (but please read Jane Superbrain Box 6.6).

Although Levene's test can be selected as an option in many of the statistical tests that require it, if you insist on using it then look at it when you're exploring data because it informs the model you

---

10 Theses standardized errors are called standardized residuals, which we'll discuss in Chapter 9.

11 We haven't covered ANOVA yet, so this explanation won't make much sense to you now, but in Chapter 12 we will look in more detail at how Levene's test works.

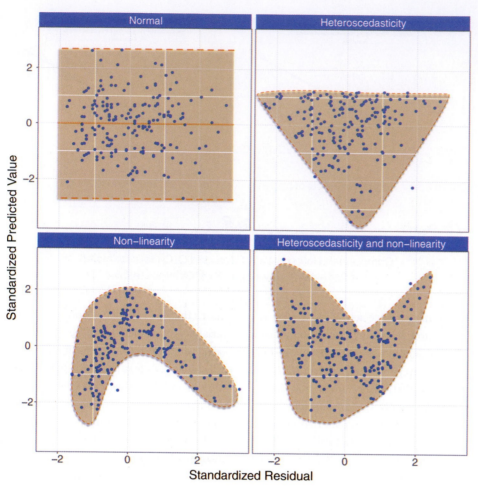

Figure 6.19   Plots of standardized residuals against predicted (fitted) values

subsequently fit. For the same reason as you probably shouldn't use the K-S test (Jane Superbrain Box 6.5), you probably shouldn't use Levene's test (Jane Superbrain Box 6.6): in large samples trivial differences in group variances can produce a Levene's test that is significant, and in small samples the test will only pick up on big differences.

Some people also look at **Hartley's F_{max}** also known as the **variance ratio** (Pearson & Hartley, 1954). This is the ratio of the variances between the group with the biggest variance and the group with the smallest. This ratio was compared to critical values in a table published by Hartley. The critical values depend on the number of cases per group and the number of variances being compared. For example,

with sample sizes (n) of 10 per group, an F_{max} of less than 10 is more or less always going to be non-significant, with 15–20 per group the ratio needs to be less than about 5, and with samples of 30–60 the ratio should be below about 2 or 3. This ratio isn't used very often, and because it is a significance test it has the same problems as Levene's test. Nevertheless, if you want the critical values (for a 0.05 level of significance) see Oliver Twisted.

6.11.3 If you still decide to do Levene's test

After everything I've said, you're not going to do Levene's test, are you? Oh, you are. OK then. Sticking with the hygiene scores, we'll compare the

variances of males and females on day 1 of the festival. Use *Analyze* ▶ *Descriptive Statistics* ▶ 🔍 *Explore…* to open the dialog box in Figure 6.20. Transfer the **day1** variable from the list on the left-hand side to the box labeled *Dependent List* by clicking ➡ next to this box; because we want to split the output by the grouping variable to compare the variances, select the variable **sex** and transfer it to the box labeled *Factor List* by clicking ➡ (or dragging). Click ▭Plots…▭ to open the other dialog box in Figure 6.20. To get Levene's test we select one of the options where it says *Spread vs Level with Levene Test*. If you select ◉ Untransformed then Levene's test is carried out on the raw data (a good place to start). Click ▭Continue▭ to return to the main *Explore* dialog box and ▭OK▭ to run the analysis.

Output 6.6 shows Levene's test, which can be based on differences between scores and the mean or between scores and the median. The median is slightly preferable (because it is less biased by outliers). When using both the mean ($p = 0.030$) and the median ($p = 0.037$) the significance values are less than 0.05, indicating a significant difference between the male and female variances. To calculate the variance ratio, we need to divide the largest variance by the smallest. You should find the variances in your output, but if not we obtained these values in Output 6.4. The male variance was 0.413 and the female one 0.496, the variance ratio is, therefore, $0.496/0.413 = 1.2$. The variances are practically equal. So, why does Levene's test tell us they are significantly different? The answer is because the sample sizes are so large: we had 315 males and 495 females, so even this very small difference in variances is shown up as significant by Levene's test (Jane Superbrain Box 6.5). Hopefully this example will convince you to treat this test cautiously.

Jane Superbrain 6.6
Is Levene's test worth the effort? ▮▮▮▮

Statisticians used to recommend testing for homogeneity of variance using Levene's test and, if the assumption was violated, using an adjustment to correct for it. People have stopped using this approach for two reasons. First, violating this assumption matters only if you have unequal group sizes; if group sizes are equal this assumption is pretty much irrelevant and can be ignored. Second, tests of homogeneity of variance work best when you have equal group sizes and large samples (when it doesn't matter if you have violated the assumption) and are less effective with unequal group sizes and smaller samples—which is exactly when the assumption matters. Plus, there are adjustments to correct for violations of this assumption that can be applied (as we shall see): typically, a correction is applied to offset whatever degree of heterogeneity is in the data (no heterogeneity = no correction). The take-home point is that you might as well always apply the correction and forget about the assumption. If you're really interested in this issue, I like the article by Zimmerman (2004).

Oliver Twisted
Please, Sir, can I have some more ... Hartley's F_{max}?

'What kind of fool uses the variance ratio, let alone worries about its significance?' I ask.

'Me, me, me!' cackles Oliver, threatening me with his gruel-covered wooden spoon. 'Give me more significance!' he demands.

Well, there's no fool like a Dickensian bubo of a fool, so to protect my head from the wooden spoon the full table of critical values is on the companion website.

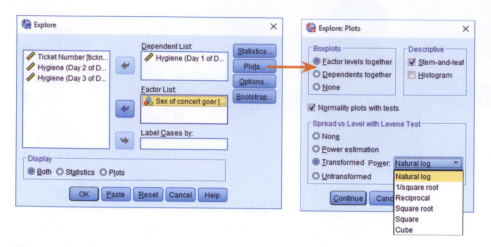

6.11.4 Reporting Levene's test

Using the labels from Output 6.6, Levene's test can be reported in this general form: $F(df_1, df_2)$ = test statistic, p = p-value. For Output 6.6 we would write (note I've used the value based on the median):

- For the hygiene scores on day 1 of the festival, the variances for males and females were significantly different, $F(1, 808) = 4.35$, $p = 0.037$.

Figure 6.20 Exploring groups of data and obtaining Levene's test

Test of Homogeneity of Variance

		Levene Statistic	df1	df2	Sig.
Hygiene (Day 1 of Download Festival)	Based on Mean	4.736	1	808	.030
	Based on Median	4.354	1	808	.037
	Based on Median and with adjusted df	4.354	1	805.066	.037
	Based on trimmed mean	4.700	1	808	.030

Output 6.6

6.12 Reducing bias

Having looked at potential sources of bias, the next issue is how to reduce the impact of bias. We'll look at four approaches for correcting problems with the data, which can be remembered with the handy acronym of TWAT (or WATT if you prefer):

Cramming Sam's Tips
Homogeneity of variance

- Homogeneity of variance/homoscedasticity is the assumption that the spread of outcome scores is roughly equal at different points on the predictor variable.
- The assumption can be evaluated by looking at a plot of the standardized predicted values from your model against the standardized residuals (*zpred vs. zresid*).
- When comparing groups, this assumption can be tested with Levene's test and the variance ratio (Hartley's F_{max}).
  - If Levene's test is significant (*Sig.* in the SPSS table is less than 0.05) then the variances are significantly different in different groups.
  - Otherwise, homogeneity of variance can be assumed.
  - The variance ratio is the largest group variance divided by the smallest. This value needs to be smaller than the critical values in the additional material.
- *Warning*: There are good reasons not to use Levene's test or the variance ratio. In large samples they can be significant when group variances are similar, and in small samples they can be non-significant when group variances are very different.

- **Trim the data**: Delete a certain quantity of scores from the extremes.
- **Winsorizing**: Substitute outliers with the highest value that isn't an outlier.
- **Apply a robust estimation method**: A common approach is to use bootstrapping.
- **Transform the data**: Apply a mathematical function to scores to correct problems.

Probably the best of these choices is to use **robust tests**, which is a term applied to a family of procedures to estimate statistics that are unbiased even when the normal assumptions of the statistic are not met (Section 6.12.3). Let's look at each technique in more detail.

6.12.1 Trimming the data

Trimming the data means deleting some scores from the extremes. In its simplest form it could be deleting the data from the person who contributed the outlier. However, this should be done only if you have good reason to believe that this case is not from the population that you intended to sample. Imagine you were investigating factors that affected how much cats purr and one cat didn't purr at all; this would likely be an outlier (all cats purr). Upon inspection, if you discovered that this cat was a dog wearing a cat costume (hence why it didn't purr), then you'd have grounds to exclude this case because it comes from a different population (dogs who like to dress as cats) than your target population (cats).

More often, trimming involves removing extreme scores using one of two rules: (1) a percentage based rule; and (2) a standard deviation based rule. A percentage based rule would be, for example, deleting the 10% of highest and lowest scores. Let's look at an example. Meston and Frohlich (2003) report a

SELF TEST
Compute the mean and variance of the attractiveness ratings. Now compute them for the 5%, 10% and 20% trimmed data.

study showing that heterosexual people rate a picture of someone of the opposite sex as more attractive after riding a roller-coaster compared to before. Imagine we took 20 people as they came off the Rockit roller-coaster at Universal studios in Orlando[12] and asked them to rate the attractiveness of someone in a photograph on a scale of 0 (looks like Jabba the Hut) to 10 (my eyes can't cope with such beauty and have exploded). Figure 6.21 shows these scores. Most people gave ratings above the midpoint of the scale: they were pretty positive in their ratings. However, there were two people who gave zeros. If we were to trim 5% of the data from either end, this would mean deleting one score at each extreme (there are 20 scores and 5% of 20 is 1). Figure 6.21 shows that this involves deleting a 0 and a 10. We could compute a 5% trimmed mean by working out the mean for this trimmed data set. Similarly, Figure 6.21 shows that with 20 scores, a 10% trim would mean deleting two scores from each extreme, and a 20% trim would entail deleting four scores from each extreme. If you take trimming to its extreme then you get the median, which is the value left when you have trimmed all but the middle score. If we calculate the mean in a sample that has been trimmed in this way, it is called

(unsurprisingly) a **trimmed mean**. A similar robust measure of location is an **M-estimator**, which differs from a trimmed mean in that the amount of trimming is determined empirically. In other words, rather than the researcher deciding before the analysis how much of the data to trim, an M-estimator determines the optimal amount of trimming necessary to give a robust estimate of, say, the mean. This has the obvious advantage that you never over- or under-trim your data; however, the disadvantage is that it is not always possible to reach a solution.

If you do the self-test you should find that the mean rating was 6. The 5% trimmed mean is 6.11, and the 10% and 20% trimmed means are both 6.25. The means get higher in this case because the extreme scores were both low (two people who gave ratings of 0) and trimming reduces their impact (which would have been to lower the mean). For the overall sample the variance was 8, for the 5%, 10%, and 20% trimmed data you get 5.87, 3.13 and 1.48, respectively. The variances get smaller (and more stable) because, again, scores at the extreme have no impact (because they are trimmed). We saw earlier that the accuracy of the mean and variance depends on a symmetrical

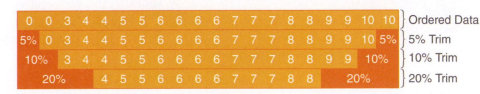

Figure 6.21   Illustration of trimming data

12 I have a video of my wife and me on this rollercoaster during our honeymoon. I swear quite a lot on it, but I might stick it on my YouTube channel so you can laugh at what a sissy I am.

distribution, but a trimmed mean (and variance) will be relatively accurate even when the distribution is not symmetrical, because by trimming the ends of the distribution we remove outliers and skew that bias the mean. Some robust methods work by taking advantage of the properties of the trimmed mean.

Standard deviation based trimming involves calculating the mean and standard deviation of a set of scores, and then removing values that are a certain number of standard deviations greater than the mean. A good example is reaction time data (which are notoriously messy), where it is very common to remove reaction times greater than (or below) 2.5 standard deviations from the mean (Ratcliff, 1993). For the roller-coaster data the variance was 8 and the standard deviation is the square root of this value, 2.83. If we wanted to trim 2.5 times the standard deviation, then we'd use 2.5 × 2.83 = 7.08. The mean was 6, therefore, we would delete scores greater than 6 + 7.08 = 13.08, of which there were none (it was only a 10-point scale); we would also delete scores less than 6 − 7.08 = −1.08, which again means deleting no scores (because the lowest score was zero). In short, applying this rule wouldn't affect the mean or standard deviation, which is odd, isn't it? The example illustrates the fundamental problem with standard deviation based trimming, which is that the mean and standard deviation are both highly influenced by outliers (see Section 6.3); therefore, the outliers in the data bias the criterion that you use to reduce their impact. In this case, the inflated standard deviation also inflates the trimming rule to beyond the limits of the data.

There isn't a simple way to implement these methods in SPSS. You can calculate a 5% trimmed mean using the *explore* command (Figure 6.15), but it won't remove the cases from the data. To do tests on a trimmed sample you need to use the *Essentials for R* plugin (I'll elaborate in Section 6.12.3).

6.12.2 Winsorizing

Winsorizing the data involves replacing outliers with the next highest score that is *not* an outlier. It's perfectly natural to feel uncomfortable at the idea of changing the scores you collected to different values. It feels a bit like cheating. Bear in mind though that if the score you're changing is very unrepresentative of the sample and biases your statistical model then it's better than reporting and interpreting a biased model.[13] What would be cheating is not dealing with extreme cases because they bias the results in favor of your hypothesis or changing scores in a systematic way other than to reduce bias (again, perhaps to support your hypothesis).

There are some variations on winsorizing, such as replacing extreme scores with a score 3 standard deviations from the mean. A *z*-score of 3.29 constitutes an outlier (see Section 6.9), so we can calculate what score would give rise to a *z*-score of 3.29 (or perhaps 3) by rearranging the *z*-score equation: $X = (z \times s) + \overline{X}$. All we're doing is calculating the mean ($\overline{X}$) and standard deviation (*s*) of the data and, knowing that *z* is 3 (or 3.29 if you want to be exact), adding three times the standard deviation to the mean and replacing our outliers with that score. This is something you would need to do manually in SPSS or using the *select cases* command (see Oditi's Lantern).

6.12.3 Robust methods

By far the best option if you have irksome data (other than sticking a big samurai sword through your head) is to estimate parameters and their standard errors with methods that are robust to violations of assumptions and outliers. In other words, use methods that are relatively unaffected by irksome data. The first set of tests are ones that do not rely on the assumption of normally distributed data (see Chapter 7).[14] These non-parametric tests have been developed for only a limited range of situations; happy days if you want to compare two means, but sad and lonely days listening to Joy Division if you have a complex experimental design. Despite having a chapter dedicated to them, there are better methods these days.

These better methods fall under the banner of 'robust methods' (see Field & Wilcox, in press, for a gentle introduction). They have developed as computers have become more sophisticated (applying these methods without a computer would be only marginally less painful than

Figure 6.22   Illustration of winsorizing data

---

13 It is worth making the point that having outliers is interesting in itself, and if you don't think they represent the population then you need to ask yourself why they are different. The answer to the question might be a fruitful topic of more research.

14 For convenience a lot of textbooks refer to these tests as *non-parametric tests* or *assumption-free* tests and stick them in a separate chapter. Neither of these terms is particularly accurate (none of these tests are assumption-free), but in keeping with tradition I've banished them to their own chapter (Chapter 7) and labeled it 'Non-parametric models'.

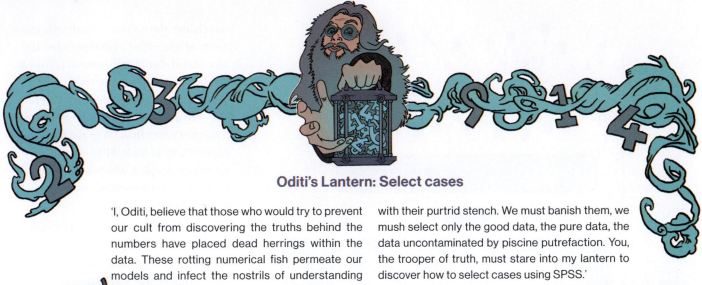

Oditi's Lantern: Select cases

'I, Oditi, believe that those who would try to prevent our cult from discovering the truths behind the numbers have placed dead herrings within the data. These rotting numerical fish permeate our models and infect the nostrils of understanding with their purtrid stench. We must banish them, we mush select only the good data, the pure data, the data uncontaminated by piscine putrefaction. You, the trooper of truth, must stare into my lantern to discover how to select cases using SPSS.'

ripping off your skin and diving into a bath of salt). How robust methods work is the topic of a book in its own right (I recommend Wilcox, 2017), but two simple concepts will give you the general idea. The first we have already looked at: parameter estimates based on trimmed data such as the trimmed mean and M-estimators. The second is the **bootstrap** (Efron & Tibshirani, 1993). The problem that we have is that we don't know the shape of the sampling distribution, but normality in our data allows us to infer that the sampling distribution is normal. Lack of normality prevents us from knowing the shape of the sampling distribution unless we have big samples. Bootstrapping gets around this problem by estimating the properties of the sampling distribution from the sample data. Figure 6.23 illustrates the process: in effect, the sample data are treated as a population from which smaller samples (called bootstrap samples) are taken

(putting each score back before a new one is drawn from the sample). The parameter of interest (e.g., the mean) is calculated in each bootstrap sample. This process is repeated perhaps 2000 times. The result is 2000 parameter estimates, one from each bootstrap sample. There are two things we can do with these estimates: the first is to order them and work out the limits within which 95% of them fall. For example, in Figure 6.23, 95% of bootstrap sample means fall between 5.15 and 6.80. We can use these values as estimate of the limits of the 95% confidence interval of the parameter. The result is known as a percentile bootstrap confidence interval (because it is based on the values between which 95% of bootstrap sample estimates fall). The second thing we can do is to calculate the

standard deviation of the parameter estimates from the bootstrap samples and use it as the standard error of parameter estimates. When we use bootstrapping, we're effectively getting the computer to use our sample data to mimic the sampling process described in Section 2.7. An important point to remember is that because bootstrapping is based on taking random samples from the data you've collected, the estimates you get will be slightly different every time. This is nothing to worry about. For a fairly gentle introduction to the bootstrap, see Wright, London & Field (2011).

Some procedures in SPSS have a bootstrap option, which can be accessed by clicking Bootstrap… to activate the dialog box in Figure 6.24 (see Oditi's Lantern).[15] Select ☑ Perform bootstrapping to activate bootstrapping for the procedure you're currently doing. In terms of the options, SPSS will compute a 95% percentile confidence interval (◉ Percentile), but you

---

15 This button is active in the base subscription version and the premium stand-alone version of IBM SPSS Statistics.

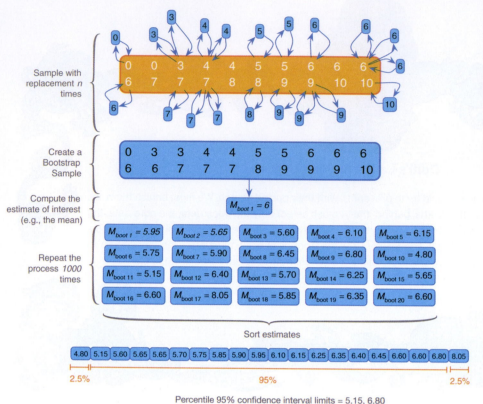

Figure 6.23   Illustration of the percentile bootstrap

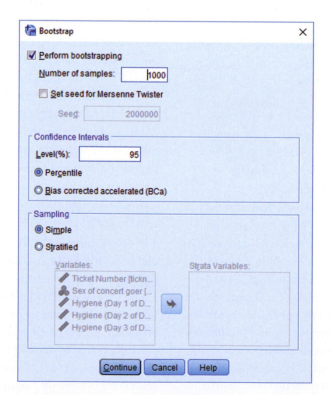

Figure 6.24   The standard *Bootstrap* dialog box

can change the method to a slightly more accurate one called a bias corrected and accelerated confidence interval (Efron & Tibshirani, 1993) by selecting

◉ Bias corrected accelerated (BCa). You can also change the confidence level by typing a number other than 95 in the box labeled *Level(%)*. By default, SPSS uses 1000 bootstrap samples, which is a reasonable number, and you certainly wouldn't need to use more than 2000.

There are versions of common procedures such as ANOVA, ANCOVA, correlation and multiple regression based on trimmed means that enable you to ignore everything we have discussed about bias in this chapter. That's a happy story, but one with a tragic ending because you can't implement them directly in SPSS. The definitive guide to these tests is Wilcox's (2017) outstanding book. Thanks to Wilcox, these tests can be implemented using a free statistics program called R (www.r-project.org). You can access these tests in SPSS Statistics using the R plugin and WRS2 package (Section 4.13.3) and I'll describe some of these tests as we go along. If you want more detail and fancy getting into R (shameless plug alert) try my R version of this textbook (Field, Miles, & Field, 2012).

6.12.4 Transforming data ▌▌▌▌

The final way to combat problems with normality and linearity is to transform the data. The idea behind **transformations** is that you do something to every score to correct for distributional problems, outliers, lack of linearity or unequal variances. Some students (understandably) think that transforming data sounds dodgy (the phrase 'fudging your results' springs into some people's minds!), but it isn't because you do the same thing to all of your scores.

Think back to our roller-coaster example (where we had 20 ratings of attractiveness from people coming off a

**Oditi's Lantern
Bootstrapping**

'I, Oditi, believe that R is so-called because it makes you shout "Arrghhh!!?" You, my followers, are precious to me and I would not want you to place your sensitive body parts into that guillotine. Instead, stare into my lantern to see how we can use bootstrapping in SPSS.'

roller-coaster). Imagine we also recorded their fear during the ride (on a scale from 0 to 10). Figure 6.25 plots the attractiveness scores against the fear scores, summarizing their relationship (top left) and showing the means of the two variables (bottom left). Let's take the square root of the attractiveness ratings (but not the fear scores). The form of the relationship between attractiveness and fear (top middle) is changed (the slope is less steep) but the relative position of scores is unchanged (the relationship is still positive; the line still slopes up). If we look at the means (lower panels of Figure 6.25) the transformation creates a difference between means (middle) that didn't exist before the transformation (left). If we transform *both* variables (right-hand panels) then the relationship remains intact but the

similarity between means is restored also.

Therefore, if you are looking at relationships between variables you can transform only the problematic variable, but if you are looking at differences between variables (e.g., changes in a variable over time) you must transform all the relevant variables. Our festival hygiene data were not normal on days 2 and 3 of the festival, so we might want to transform them. However, if we want to look at how hygiene levels changed across the three days (i.e., compare the mean on day 1 to the means on days 2 and 3 to see if people got smellier) we must also transform the day 1 data (even though scores were not

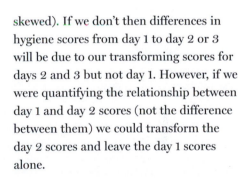

What is a data transformation?

skewed). If we don't then differences in hygiene scores from day 1 to day 2 or 3 will be due to our transforming scores for days 2 and 3 but not day 1. However, if we were quantifying the relationship between day 1 and day 2 scores (not the difference between them) we could transform the day 2 scores and leave the day 1 scores alone.

6.12.5 Choosing a transformation ▌▌▌▌

Various transformations exist that correct the problems we've discussed; the most common are summarized in Table 6.1.[16] You basically try one out, see if it helps and if it doesn't then try a different one. Whether these transformations are necessary or useful is a complex issue (see Jane Superbrain Box 6.7).

---

16 You'll notice in this section that I write X_i. We saw in Chapter 1 that this refers to the observed score for the ith person (so, think of the i as the person's name, for example, for Oscar, $X_i = X_{Oscar}$ = Oscar's score, and for Melody, $X_i = X_{Melody}$ = melody's score).

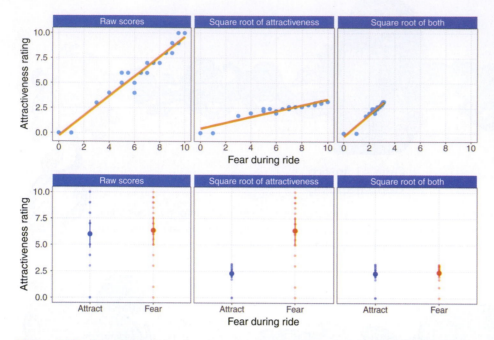

Figure 6.25 The effect of transforming attractiveness scores on its relationship to fear (top) and its mean relative to fear (bottom)

Trying out different transformations is time-consuming, but if heterogeneity of variance is the issue there is a shortcut to seeing if they solve the problem. In Section 6.11.3 we used the *explore* function to get Levene's test for the raw scores (⦿ Untransformed). If the variances turn out to be unequal, as they did in our example, then select ⦿ Transformed in the same dialog box (Figure 6.20). A drop-down menu becomes active that lists transformations including the ones that I have just described. Select a transformation from this list (*Natural log* perhaps or *Square root*) and SPSS will compute Levene's test on the transformed scores; you decide whether variances are still unequal by looking at the output of the test.

Table 6.1 Data transformations and their uses

Data transformation	Can correct for
Log transformation (log(X_i)): Taking the logarithm of a set of numbers squashes the right tail of the distribution, which reduces positive skew. This transformation can also sometimes make a curvilinear relationship linear. Because you can't get a log value of zero or negative numbers, you may need to add a constant to all scores before taking the log: if you have scores of zero then do log(X_i + 1); if you have negative numbers add whatever value makes the smallest score positive.	Positive skew, positive kurtosis, unequal variances, lack of linearity
Square root transformation ($\sqrt{X_i}$): Like the log transformation, taking the square root of scores has a greater impact on large scores than small ones. Consequently, taking the square root scores brings large scores closer to the center, which will reduce positive skew. Although zeros are fine, negative numbers don't have a square root so you may need to add a constant before transforming.	Positive skew, positive kurtosis, unequal variances, lack of linearity
Reciprocal transformation (1/X_i): Dividing 1 by each score also reduces the impact of large scores. The transformed variable will have a lower limit of 0 (very large numbers will become close to 0). This transformation reverses the scores: large scores become small (close to zero) after the transformation, and small scores become large. For example, scores of 1 and 100 become 1/1 = 1, and 1/10 = 0.01 after transforming: their relative size swaps. To avoid this, reverse the scores before the transformation by converting each score to the highest score for the variable minus the score you're looking at. So, rather than using 1/X_i as the transformation, use 1/($X_{Highest}$−X_i). You can't take the reciprocal of 0 (because 1/0 = infinity), so if you have zeros in the data add a constant to all scores before doing the transformation.	Positive skew, positive kurtosis, unequal variances
Reverse score transformations: Any one of the above transformations can be used to correct negatively skewed data if you first reverse the scores. To do this, subtract each score from the highest score on the variable or the highest score + 1 (depending on whether you want your lowest score to be 0 or 1). Don't forget to reverse the scores back afterwards or that the interpretation of the variable is reversed: big scores have become small and small scores have become big.	Negative skew

Jane Superbrain 6.7
To transform or not to transform, that is the question ▮▮▮▮

Not everyone thinks that transforming data is a good idea. Glass, Peckham, & Sanders (1972, p. 241) commented that 'the payoff of normalizing transformations in terms of more valid probability statements is low, and they are seldom considered to be worth the effort'. The issue is complicated, but the core question is whether a statistical model performs better when applied to transformed data or data that violate the assumption that the transformation corrects. The answer will depend on which 'model' you're applying and how robust it is (see Section 6.12).

For example, the *F*-test (see Chapter 12) is often claimed to be robust (Glass et al., 1972). Early findings suggested that *F* performed as it should in skewed distributions and that transforming the data helped as often as it hindered the accuracy of *F* *(Games & Lucas, 1966)*. However, in a lively but informative exchange Levine and Dunlap (1982) showed that transformations of skew did improve the performance of *F*; Games (1983) argued that this conclusion was incorrect; and Levine and Dunlap (1983) disagreed in a response to the response. In a response to the response to the response, Games (1984) raised several important issues:

1 The central limit theorem (Section 6.6.1) tells us that in samples larger than about 30 the sampling distribution will be normal regardless. This is theoretically true, but recent work has shown that with heavy-tailed distributions much larger samples are necessary to invoke the central limit theorem (Wilcox, 2017). Transformations might be useful for such distributions.

2 Transforming the data changes the hypothesis being tested. For example, when comparing means, converting from raw scores to log scores means that you're now comparing geometric means rather than arithmetic means. Transformation also means that you're addressing a different construct than the one originally measured, and this has obvious implications for interpreting the data (Grayson, 2004).

3 It is tricky to determine normality one way or another in small samples (see Jane Superbrain Box 6.5).

4 The consequences for the statistical model of applying the 'wrong' transformation could be worse than the consequences of analyzing the untransformed scores.

Given these issues, unless you're correcting for a lack of linearity I would use robust procedures, where possible, in preference to transforming the data.

6.12.6 The *compute* function ▮▮▮

If you do decide to transform scores, use the *compute* command, which enables you to create new variables. To access the *Compute Variable* dialog box, select *Transform* ▸ 🖩 Compute Variable.... Figure 6.26 shows the main dialog box; it has a list of functions on the right-hand side, a calculator-like keyboard in the center and a blank space that I've labeled the command area. You type a name for a new variable in the area labeled *Target Variable* and then you use the command area to tell SPSS how to create this new variable. You can:

- *Create new variables from existing variables*: For example, you could use it like a calculator to add variables (i.e., add two columns in the data editor to make a third) or to apply a function to an existing variable (e.g., take the square root).

- *Create new variables from functions*: There are hundreds of built-in functions that SPSS has grouped together. In the dialog box these groups are listed in the area labeled *Function group*. Upon selecting a function group, a list of available functions within that group will appear in the box labeled *Functions and Special Variables*. When you select a function, a description of it appears in the white box indicated in Figure 6.26.

You can enter variable names into the command area by selecting the variable required from the variables list and then clicking ⬈. Likewise, you can select a function from the list of available functions and enter it into the command area by clicking ⬆.

First type a variable name in the box labeled *Target Variable*, then click Type & Label... and another dialog box appears, where you can give the variable a descriptive label and specify whether it is a numeric or string variable (see Section 4.6.2). When you have written your command for SPSS to execute, click OK to run the command and create the new variable. If you type in a variable name that already exists, SPSS will tell you so and ask you whether you want to replace this existing variable. If you respond with *Yes* then SPSS will replace the data in the existing column with the result of the *compute* command; if you respond with *No* then nothing will happen and you will need to rename the target variable. If you're computing a lot of new variables it can be quicker to use syntax (see SPSS Tip 6.2).

Some useful functions are listed in Table 6.2, which shows the standard form of the function, the name of the function, an example of how the function can be used and what the resulting variable would contain if that command were executed. If you want to know more, the SPSS help files have details of all the functions available through the *Compute Variable* dialog box (click Help when you're in the dialog box).

6.12.7 The log transformation using SPSS Statistics ▮▮▮

Let's use *compute* to transform our data. Open the *Compute* dialog box by selecting *Transform* ▸ 🖩 Compute Variable.... Enter the name **logday1** into the box labeled *Target Variable*, click Type & Label... and give the variable a more descriptive name such as *Log transformed hygiene scores for day 1 of Download festival*. In the box labeled *Function group* select *Arithmetic* and then in the box labeled *Functions and Special Variables* select *Lg10* (this is the log transformation to base 10; *Ln* is the natural log) and transfer it to the command area by clicking ⬆. The command will appear as 'LG10(?)' and the question mark needs to be replaced with a variable name; replace it with the variable **day1** by selecting the variable in the list and dragging it across, clicking ⬈ or just by typing 'day1' where the question mark is.

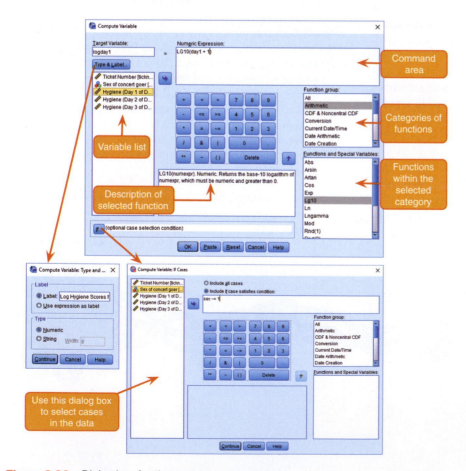

Figure 6.26 Dialog box for the *compute* command

Let's look at some of the simple functions:

Addition: This button places a plus sign in the command area. For example, with our hygiene data, 'day1 + day2' creates a column in which each row contains the hygiene score from the column labeled *day1* added to the score from the column labeled *day2* (e.g., for participant 1: 2.64 + 1.35 = 3.99).

Subtraction: This button places a minus sign in the command area. For example, if we wanted to calculate the change in hygiene from day 1 to day 2 we could type 'day2 – day1'. This command creates a column in which each row contains the score from the column labeled *day1* subtracted from the score from the column labeled *day2* (e.g., for participant 1: 2.64 – 1.35 = 1.29).

Multiply: This button places a multiplication sign in the command area. For example, 'day1*day2' creates a column that contains the score from the column labeled *day1* multiplied by the score from the column labeled *day2* (e.g., for participant 1: 2.64 × 1.35 = 3.56).

Divide: This button places a division sign in the command area. For example, 'day1/day2' creates a column that contains the score from the column labeled *day1* divided by the score from the column labeled *day2* (e.g., for participant 1: 2.64/1.35 = 1.96).

Exponentiation: This button raises the preceding term by the power of the succeeding term. So, 'day1**2' creates a column that contains the scores in the *day1* column raised to the power of 2 (i.e., the square of each number in the *day1* column: for participant 1, $2.64^2 = 6.97$). Likewise, 'day1**3' creates a column with values of **day1** cubed.

Less than: This operation is helpful for 'include case' functions. If you click on the ⏸ button, a dialog box appears that allows you to select certain cases on which to carry out the operation. So, if you typed 'day1 < 1', then SPSS would carry out the *compute* function only for those participants whose hygiene score on day 1 of the festival was less than 1 (i.e., if *day1* was 0.99 or less). So, we might use this if we wanted to look only at the people who were already smelly on the first day of the festival.

Less than or equal to: This operation is the same as above, except that in the example above, cases that are exactly 1 would be included as well.

More than: This operation is used to include cases above a certain value. So, if you clicked on ⏸ and typed 'day1 > 1' then SPSS would carry out any analysis only on cases for which hygiene scores on day 1 of the festival were greater than 1 (i.e., 1.01 and above). This could be used to exclude people who were already smelly at the start of the festival. We might want to exclude them because these people will contaminate the data (not to mention our nostrils) because they reek of putrefaction to begin with so the festival cannot further affect their hygiene.

More than or equal to: This operation is the same as above but will include cases that are exactly 1 as well.

Equal to: You can use this operation to include cases for which participants have a specific value. So, if you clicked on ⏸ and typed 'day1 = 1' then only cases that have a value of exactly 1 for the **day1** variable are included. This is most useful when you have a coding variable and you want to apply a function to only one of the groups. For example, if we wanted to compute values only for females at the festival we could type 'sex = 1', then the compute function will be applied on only females (who are coded as 1 in the data).

Not equal to: This operation will include all cases except those with a specific value. So, 'sex ~= 1' (as in Figure 6.26) will execute the *compute* command only on males and exclude females (because they have a 1 in the sex column).

Table 6.2 Some useful *compute* functions

Function	Name	Example Input	Output
MEAN(?,?,...)	Mean	Mean(day1, day2, day3)	For each row, SPSS calculates the average hygiene score across the three days of the festival
SD(?,?,...)	Standard deviation	SD(day1, day2, day3)	Across each row, SPSS calculates the standard deviation of the values in the columns labeled *day1*, *day2* and *day3*
SUM(?,?,...)	Sum	SUM(day1, day2)	For each row, SPSS adds the values in the columns labeled *day1* and *day2*
SQRT(?)	Square root	SQRT(day2)	Produces a column containing the square root of each value in the column labeled *day2*
ABS(?)	Absolute value	ABS(day1)	Produces a variable that contains the absolute value of the values in the column labeled *day1* (absolute values are ones where the signs are ignored: so –5 becomes +5 and +5 stays as +5)
LG10(?)	Base 10 logarithm	LG10(day1)	Produces a variable that contains the logarithmic (to base 10) values of the variable *day1*
RV.NORMAL (mean, stddev)	Normal random numbers	Normal(20, 5)	Produces a variable of pseudo-random numbers from a normal distribution with a mean of 20 and a standard deviation of 5

For the day 2 hygiene scores there is a value of 0 in the original data, and there is no logarithm of the value 0. To overcome this problem we add a constant to our original scores before we take the log. Any constant will do (although sometimes it can matter), provided that it makes all of the scores greater than 0. In this case our lowest score is 0 so adding 1 will do the job. Although this problem affects the day 2 scores, we must be consistent and apply the same constant to the day 1 scores. To do this, make sure the cursor is still inside the brackets and click ⊞ and then ① (or just type '+1'). The expression should read LG10(day1 + 1) as in Figure 6.26. Click ⊞OK⊞ to create a new variable **logday1** that contains the log of the day 1 scores after 1 was added to them.

6.12.8 The square root transformation using SPSS Statistics ▊▊▊▊

Use the same process to apply a square root transformation. Enter a name such

as **sqrtday1** in the box labeled *Target Variable* (and click ⊞Type & Label...⊞ to give the variable a descriptive name). In the list box labeled *Function group* select *Arithmetic*, select *Sqrt* in the box labeled *Functions and Special Variables* and drag it to the command area (or click ⊞↑⊞). The command appears as SQRT(?); replace the question mark with the variable **day1** by selecting the variable in the list and dragging it, clicking ⊞→⊞ or typing 'day1' where the question mark is. The final expression will read SQRT(day1). Click ⊞OK⊞ to create the variable.

6.12.9 The reciprocal transformation using SPSS Statistics ▊▊▊▊

To do a reciprocal transformation on the data from day 1, we could use a name such as **recday1** in the box labeled *Target Variable*. Then we click ① followed by ⊞/⊞. Ordinarily you would select the variable name that you want to transform from the list and drag it

across, click ⊞→⊞ or type its name. However, because the day 2 data contain a zero value and you can't divide by 0 we add a constant to our variable like we did for the log transformation. As before, 1 is a convenient number for these data. So, instead of selecting the variable that we want to transform, click ⊞()⊞ to place a pair of brackets into the command area; make sure the cursor is between these two brackets, select the variable you want to transform from the list and transfer it by dragging, clicking ⊞→⊞ or typing its name. Now click ⊞+⊞ and then ① (or type '+ 1'). The box labeled *Numeric Expression* should now contain the text 1/(day1 + 1). Click ⊞OK⊞ to create a new variable containing the transformed values.

6.12.10 The effect of transformations ▊▊▊▊

Figure 6.27 shows the distributions for days 1 and 2 of the festival after the three different transformations. Compare these to the untransformed distributions in Figure 6.12. All three transformations have cleaned up the hygiene scores for day 2: the positive skew is reduced (the square root transformation is especially useful). However, because our hygiene scores on day 1 were more or less symmetrical, they have become slightly negatively skewed for the log and square root transformation, and positively skewed for the reciprocal transformation.[17] If we're using scores from day 2 alone or looking at the relationship between day 1 and day 2, then we could use the transformed scores; however, if we wanted to look at the *change* in scores then we'd have to weigh up whether the benefits of the transformation for the day 2 scores outweigh the problems it creates in the day

SELF TEST
Have a go at creating similar variables **logday2** and **logday3** for the day 2 and day 3 data. Plot histograms of the transformed scores for all three days.

SELF TEST
Repeat this process for **day2** and **day3** to create variables called **sqrtday2** and **sqrtday3**. Plot histograms of the transformed scores for all three days.

SELF TEST
Repeat this process for **day2** and **day3**. Plot histograms of the transformed scores for all three days.

---

17 The reversal of the skew for the reciprocal transformation is because, as I mentioned earlier, the reciprocal reverses the order of scores.

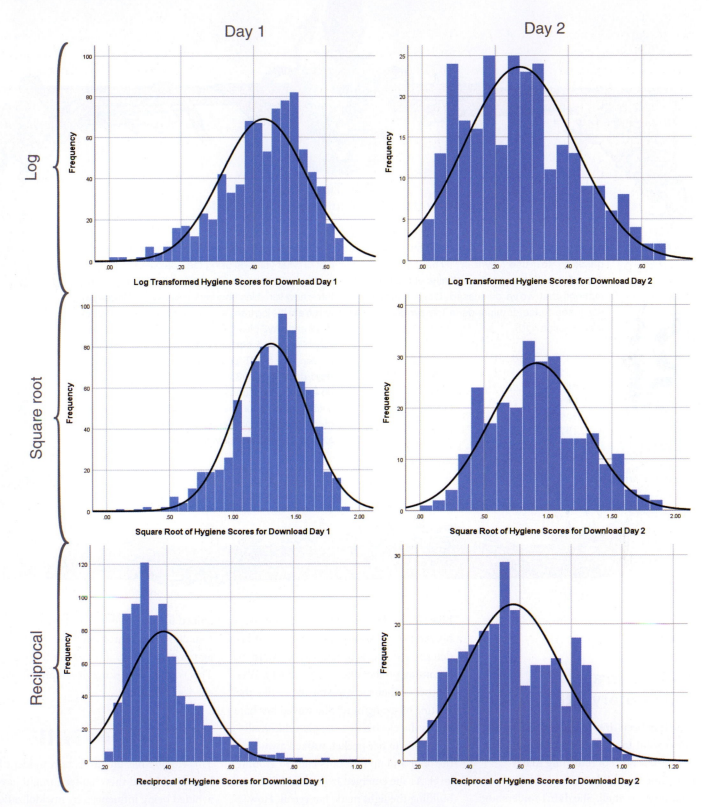

Figure 6.27 Distributions of the hygiene data on day 1 and day 2 after various transformations

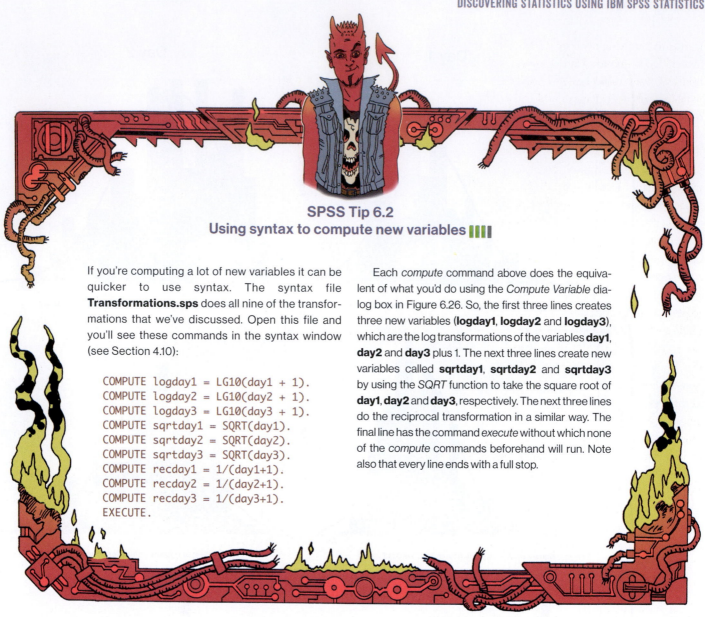

SPSS Tip 6.2
Using syntax to compute new variables ▮▮▮▮

If you're computing a lot of new variables it can be quicker to use syntax. The syntax file **Transformations.sps** does all nine of the transformations that we've discussed. Open this file and you'll see these commands in the syntax window (see Section 4.10):

```
COMPUTE logday1 = LG10(day1 + 1).
COMPUTE logday2 = LG10(day2 + 1).
COMPUTE logday3 = LG10(day3 + 1).
COMPUTE sqrtday1 = SQRT(day1).
COMPUTE sqrtday2 = SQRT(day2).
COMPUTE sqrtday3 = SQRT(day3).
COMPUTE recday1 = 1/(day1+1).
COMPUTE recday2 = 1/(day2+1).
COMPUTE recday3 = 1/(day3+1).
EXECUTE.
```

Each *compute* command above does the equivalent of what you'd do using the *Compute Variable* dialog box in Figure 6.26. So, the first three lines creates three new variables (**logday1**, **logday2** and **logday3**), which are the log transformations of the variables **day1**, **day2** and **day3** plus 1. The next three lines create new variables called **sqrtday1**, **sqrtday2** and **sqrtday3** by using the *SQRT* function to take the square root of **day1**, **day2** and **day3**, respectively. The next three lines do the reciprocal transformation in a similar way. The final line has the command *execute* without which none of the *compute* commands beforehand will run. Note also that every line ends with a full stop.

1 scores—data analysis is invariably frustrating. ☺

6.13 Brian and Jane's Story ▮▮▮▮

Jane had been thinking about Brian. She didn't want to—it was a distraction from work—but he kept returning to her thoughts. Guys didn't interest her or relationships at all. She didn't understand other people, they were so ... unpredictable. Jane liked certainty, it made her feel safe.

Like all people, Brian freaked her out with his random invasions of her space, but his determined efforts to impress her were becoming predictable, in a good way. Was it possible that she was starting to like the routine of seeing him? She waved her hand across the electronic lock, placed a small metal box in her pocket, walked away from the door and up the stairs back to ground level. As she emerged from the Pleiades building the light made her recoil. How long had she been down there? As her eyes adjusted, a fuzzy Brian in the distance

gained focus. He smiled and waved at her. Usually seeing him skewed her, made her agitated, but today it made her feel safe. This was an interesting development. Was she transforming?

6.14 What next? ▮▮▮▮

This chapter has taught us how to identify bias. Had I read this chapter I might have avoided being influenced by my idolization of my granddad[18] and realized that I could be a useful midfield player instead of

---

18 Despite worshipping the ground my granddad walked on, I ended up supporting the local rivals of the north London team that he supported.

fruitlessly throwing my miniature body around a goal-mouth. Had I played midfield, a successful career in soccer would undoubtedly have unfolded in front of me. Or, as anyone who has seen me play will realize, perhaps not. I sort of had the last laugh on the goalkeeping front. At the end of my time at primary school we had a five-a-side tournament between local schools so that kids from different schools could get to know each other before going to secondary school together. My goalkeeping nemesis was, of course, chosen to play and I was the substitute. In the first game he had a shocker, and I was called up to play in the second game during which I made a series of dramatic and acrobatic saves (at least they are in my memory). I did likewise in the next game, and my nemesis had to sit out the whole of the rest of the tournament. Of course, five-a-side goals are shorter than normal goals, so I didn't have my usual handicap. When I arrived at secondary school I didn't really know a position other than goalkeeper, and I was still short, so I gave up football. Years later when I started playing again, I regretted the years I didn't spend honing ball skills (my teammates regret it too). During my non-sporting years I read books and immersed myself in music. Unlike 'the clever one' who was reading Albert Einstein's papers (well, Isaac Asimov) as an embryo, my literary preferences were more in keeping with my intellect . . .

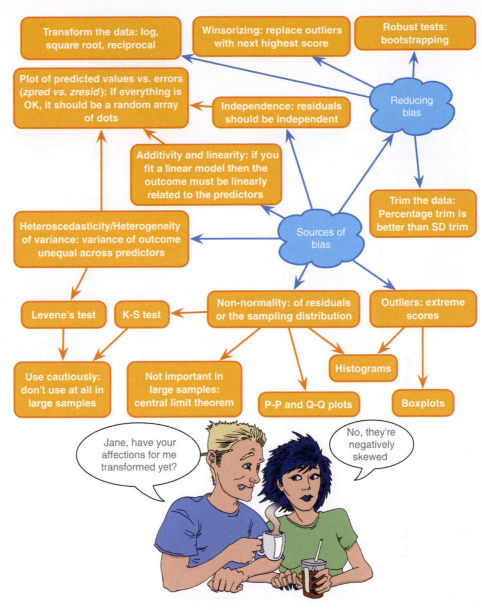

Figure 6.28   What Brian learnt from this chapter

6.15 Key terms that I've discovered

Bootstrap	Homoscedasticity	Outlier	Transformation
Contaminated normal distribution	Independence	P-P plot	Trimmed mean
Hartley's F_{max}	Kolmogorov–Smirnov test	Parametric test	Variance ratio
Heterogeneity of variance	Levene's test	Q-Q plot	Weighted least squares
Heteroscedasticity	M-estimator	Robust test	
Homogeneity of variance	Mixed normal distribution	Shapiro–Wilk test	

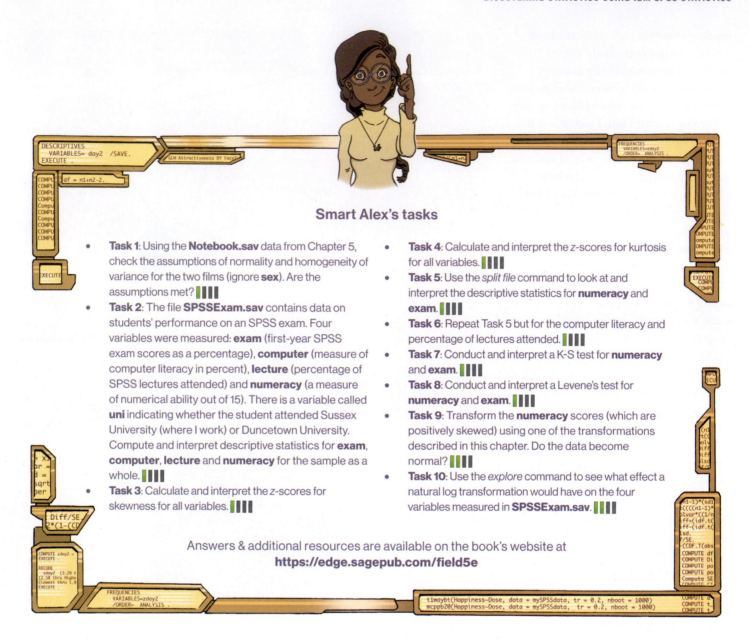

Smart Alex's tasks

- **Task 1**: Using the **Notebook.sav** data from Chapter 5, check the assumptions of normality and homogeneity of variance for the two films (ignore **sex**). Are the assumptions met? ▌▌▌

- **Task 2**: The file **SPSSExam.sav** contains data on students' performance on an SPSS exam. Four variables were measured: **exam** (first-year SPSS exam scores as a percentage), **computer** (measure of computer literacy in percent), **lecture** (percentage of SPSS lectures attended) and **numeracy** (a measure of numerical ability out of 15). There is a variable called **uni** indicating whether the student attended Sussex University (where I work) or Duncetown University. Compute and interpret descriptive statistics for **exam**, **computer**, **lecture** and **numeracy** for the sample as a whole. ▌▌▌

- **Task 3**: Calculate and interpret the *z*-scores for skewness for all variables. ▌▌▌

- **Task 4**: Calculate and interpret the *z*-scores for kurtosis for all variables. ▌▌▌

- **Task 5**: Use the *split file* command to look at and interpret the descriptive statistics for **numeracy** and **exam**. ▌▌▌

- **Task 6**: Repeat Task 5 but for the computer literacy and percentage of lectures attended. ▌▌▌

- **Task 7**: Conduct and interpret a K-S test for **numeracy** and **exam**. ▌▌▌

- **Task 8**: Conduct and interpret a Levene's test for **numeracy** and **exam**. ▌▌▌

- **Task 9**: Transform the **numeracy** scores (which are positively skewed) using one of the transformations described in this chapter. Do the data become normal? ▌▌▌

- **Task 10**: Use the *explore* command to see what effect a natural log transformation would have on the four variables measured in **SPSSExam.sav**. ▌▌▌

Answers & additional resources are available on the book's website at
https://edge.sagepub.com/field5e

NON-PARAMETRIC MODELS

7

7.1 What will this chapter tell me?

When we were learning to read at primary school, we used to read versions of stories by the famous storyteller Hans Christian Andersen. One of my favorites was the story of the ugly duckling. This duckling was a big ugly grey bird, so ugly that even a dog would not bite him. The poor duckling was ridiculed, ostracized and pecked by the other ducks. Eventually, it became too much for him and he flew to the swans, the royal birds, hoping that they would end his misery by killing him because he was so ugly. Still, life sometimes throws up surprises, and as he stared into the water, he saw not an ugly grey bird but a beautiful swan. Data are much the same. Sometimes they're just big, grey and ugly and don't do any of the things that they're supposed to do. When we get data like these, we swear at them, curse them, peck them and hope that they'll fly away and be killed by the swans. Alternatively, we can try to force our data into becoming beautiful swans. That's what this chapter is all about: trying to make an ugly duckling of a data set turn into a swan. Be careful what you wish your data to be, though: a swan can break your arm.[1]

---

1 Although it is theoretically possible, apparently you'd have to be weak boned, and swans are nice and wouldn't do that sort of thing.

Figure 7.1   I came first in the competition for who has the smallest brain

7.2 When to use non-parametric tests ▌▌▌▌

In the previous chapter we looked at several ways to reduce bias. Sometimes, however, no matter how hard you try, you will find that you can't correct the problems in your data. This is especially irksome if you have a small sample and can't rely on the central limit theorem to get you out of trouble. The historical solution is a small family of models called **non-parametric tests** or 'assumption-free tests' that make fewer assumptions than the linear model that we looked at in the previous chapter.[2] Robust methods have superseded non-parametric tests, but we'll look at them anyway because (1) the range of robust methods in IBM SPSS Statistics is limited; and (2) non-parametric tests act as a gentle introduction to using a statistical test to evaluate a hypothesis. Some people believe that non-parametric tests have less power than their parametric counterparts, but this is not always true (Jane Superbrain Box 7.1).

In this chapter, we'll explore four of the most common non-parametric procedures: the Mann–Whitney test, the Wilcoxon signed-rank test, Friedman's test and the Kruskal–Wallis test. All four tests overcome distributional problems by **ranking** the data: that is, finding the lowest score and giving it a rank of 1, then finding the next highest score and giving it a rank of 2, and so on. This process results in high scores being represented by

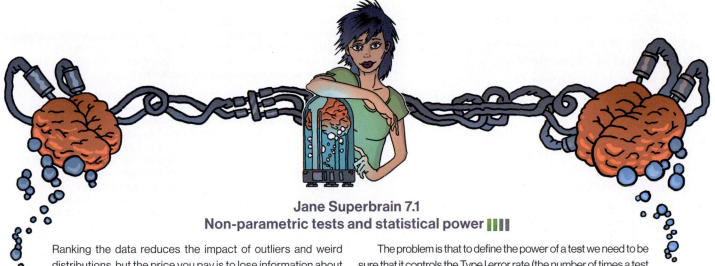

Jane Superbrain 7.1
Non-parametric tests and statistical power ▌▌▌▌

Ranking the data reduces the impact of outliers and weird distributions, but the price you pay is to lose information about the magnitude of differences between scores. Consequently, non-parametric tests can be less powerful than their parametric counterparts. Remember that statistical power (Section 2.9.7) is the ability of a test to find an effect that genuinely exists, so we're saying that if there is a genuine effect then a non-parametric test is less likely to detect it than a parametric one. This statement is true only *if the assumptions described in Chapter 6 are met*. If we use a parametric test and a non-parametric test on the same data, and those data meet the appropriate assumptions, then the parametric test will have greater power to detect the effect than the non-parametric test.

The problem is that to define the power of a test we need to be sure that it controls the Type I error rate (the number of times a test will find a significant effect when there is no effect to find—see Section 2.9.5). We saw in Chapter 2 that this error rate is normally set at 5%. When the sampling distribution is normally distributed then the Type I error rate of tests based on this distribution is indeed 5%, and so we can work out the power. However, when the sampling distribution is not normal the Type I error rate of tests based on this distribution won't be 5% (in fact we don't know what it is because it will depend on the shape of the distribution) and so we have no way of calculating power (because it is linked to the Type I error rate—see Section 2.9.7). So, if someone tells you that non-parametric tests have less power than parametric tests, tell them that this is true only if the sampling distribution is normally distributed.

---

2 Some people describe non-parametric tests as 'distribution-free tests' and claim that they make *no* distributional assumptions. In fact they do, they just don't assume a *normal* distribution: the ones in this chapter, for example, all assume a continuous distribution.

large ranks, and low scores being represented by small ranks. The model is then fitted to the ranks and not the raw scores. By using ranks we eliminate the effect of outliers. Imagine you have 20 data points and the two highest scores are 30 and 60 (a difference of 30); these scores will become ranks of 19 and 20 (a difference of 1). In much the same way, ranking irons out problems with skew.

7.3 General procedure of non-parametric tests using SPSS Statistics ▎▎▎▎

The tests in this chapter use a common set of dialog boxes, which I'll describe here before we look at the specific tests. If you're comparing groups containing different entities, select *Analyze* ▸ *Nonparametric Tests* ▸

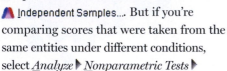

▲ Independent Samples.... But if you're comparing scores that were taken from the same entities under different conditions, select *Analyze* ▸ *Nonparametric Tests* ▸ ▲ Related Samples.... Both menus take you to a similar dialog box that has three tabs:

As Figure 7.2 shows, regardless of whether you have scores from the same entities or different ones, this tab offers the choice to compare scores automatically (i.e., SPSS selects a test for you, which I don't recommend because it's not a good idea to let a computer think on your behalf) or to select the analysis yourself (◉ Customize analysis).

Selecting this tab activates a screen in which you select the variables that you want to analyze. Within this screen, if you set roles for your variables when you entered the data (Section 4.6.2) then SPSS uses these roles to guess what analysis you want to do (◉ Use predefined roles). If you have not set roles or if you don't think it's wise to let SPSS guess what you want to do, then you can specify the variables within the analysis yourself (◉ Use custom field assignments). This tab changes depending on whether you have independent or related samples, but in both cases your variables are listed in the left-hand box labeled *Fields* (e.g., Figure 7.7). By default, all variables are listed (All), but you can filter this list to show only nominal/categorical variables (🔴) or only scale variables (📏). This is helpful in big data sets to help you find your outcome (which will typically be scale, 📏) and predictor variable (which is likely to be nominal, 🔴). You can toggle between showing the variable name or the variable label in the list by clicking 🔖. The right-hand box labeled *Test Fields* is where you place outcome variables within an analysis, and for tests of independent samples there will be a box labeled *Groups* to place categorical predictors. We'll look at the exact configuration of this tab within each analysis. For now, just note that it's similar regardless of which non-parametric test you do.

Selecting this tab activates the options for what test to perform. You can let SPSS pick a test for you (◉ Automatically choose the tests based on the data), but I'd advise you to make the decisions yourself (◉ Customize tests). Regardless of the test you're doing you can set the significance level (the default is 0.05), the confidence interval level (the default is 95%) and whether to exclude cases listwise or test by test (see SPSS Tip 6.1) by clicking on *Test Options* (see Figure 7.3). Similarly, if you have categorical variables and missing values you can choose to exclude or include these missing values by selecting *User-Missing Values* and checking the appropriate option (see Figure 7.3). The default option is to exclude them, which makes sense a lot of the time.

Scores from different entities

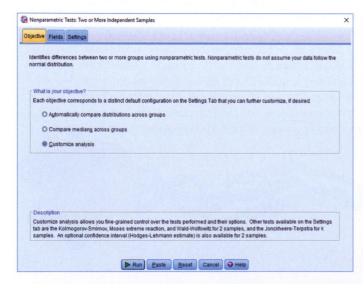

Scores from the same entities

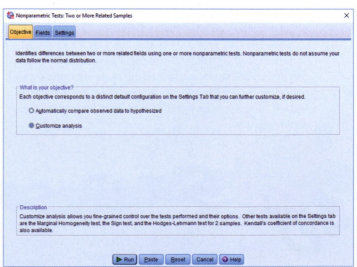

Figure 7.2   Dialog boxes for the 'Objective' tab of the *Nonparametric Tests* menu

The general process for any non-parametric analysis, then, is:

1 Choose ⊙ <u>C</u>ustomize analysis in the `Objective` `Fields` `Settings` tab (Figure 7.2), because I don't think you should trust a computer to analyze your data.

2 In the `Objective` `Fields` `Settings` tab, if SPSS fails to correctly guess what analysis you want to do (⊙ <u>U</u>se predefined roles) then select ⊙ Use <u>c</u>ustom field assignments and specify your predictor and outcome variables.

3 In the `Objective` `Fields` `Settings` tab you can let SPSS pick a test for you

(⊙ <u>A</u>utomatically choose the tests based on the data), but you have more options if you select ⊙ <u>C</u>ustomize tests. I recommend the latter option. Change test or missing values options if necessary, although the defaults are fine (Figure 7.3).

Test options

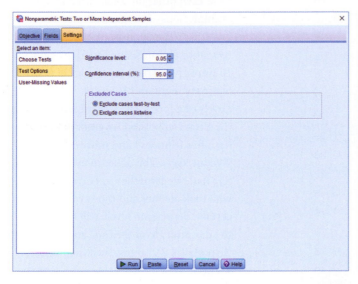

User-missing values

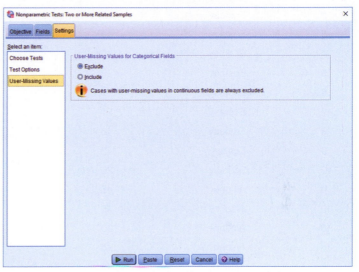

Figure 7.3   Dialog box for the 'Settings' tab when choosing *Test Options* and *User-Missing Values*

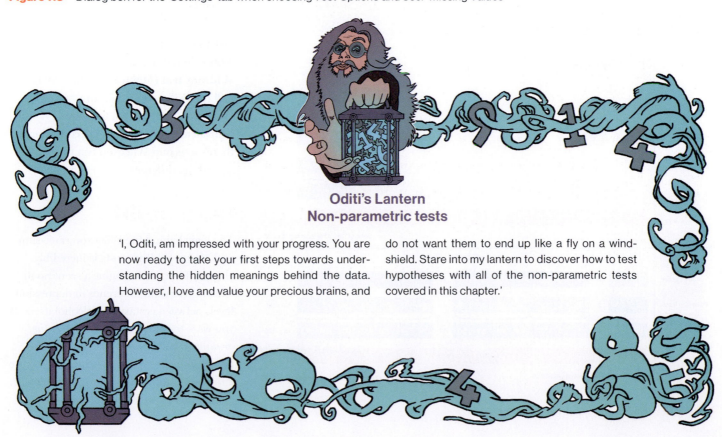

Oditi's Lantern
Non-parametric tests

'I, Oditi, am impressed with your progress. You are now ready to take your first steps towards understanding the hidden meanings behind the data. However, I love and value your precious brains, and do not want them to end up like a fly on a windshield. Stare into my lantern to discover how to test hypotheses with all of the non-parametric tests covered in this chapter.'

SELF TEST

What are the null hypotheses for these hypotheses?

7.4 Comparing two independent conditions: the Wilcoxon rank-sum test and Mann–Whitney test ▮▮▮▮

Imagine that you have a hypothesis that two groups of different entities will differ from each other on some variable. For example, a psychologist collects data to investigate the depressant effects of two recreational drugs. She tested 20 clubbers: 10 were given an ecstasy tablet to take on a Saturday night and 10 drank alcohol. Levels of depression were measured using the Beck Depression Inventory (BDI) the day after (Sunday) and midweek (Wednesday). The data are in Table 7.1. She had two hypotheses: between those who took alcohol and those who took ecstasy, depression levels will be different the day after (hypothesis 1) and mid-week (hypothesis 2). To test these hypotheses, we need to fit a model that compares the distribution in the alcohol group to that in the ecstasy group.

Table 7.1 Data for drug experiment

Participant	Drug	BDI (Sunday)	BDI (Wednesday)
1	Ecstasy	15	28
2	Ecstasy	35	35
3	Ecstasy	16	35
4	Ecstasy	18	24
5	Ecstasy	19	39
6	Ecstasy	17	32
7	Ecstasy	27	27
8	Ecstasy	16	29
9	Ecstasy	13	36
10	Ecstasy	20	35
11	Alcohol	16	5
12	Alcohol	15	6
13	Alcohol	20	30
14	Alcohol	15	8
15	Alcohol	16	9
16	Alcohol	13	7
17	Alcohol	14	6
18	Alcohol	19	17
19	Alcohol	18	3
20	Alcohol	18	10

There are two choices to compare the distributions in two conditions containing scores from different entities: the **Mann–Whitney test** (Mann & Whitney, 1947) and the **Wilcoxon rank-sum test** (Wilcoxon, 1945). Both tests are equivalent, and to add to the confusion there's a second Wilcoxon test that does something different.

7.4.1 Theory ▮▮▮▮

The logic behind the Wilcoxon rank-sum and Mann–Whitney tests is incredibly elegant. First, let's imagine a scenario in which there is no difference in depression levels between ecstasy and alcohol users. If you were to rank the data *ignoring the group to which a person belonged* from lowest to highest (i.e., give the lowest score a rank of 1 and the next lowest a rank of 2, etc.), if there's no difference between the

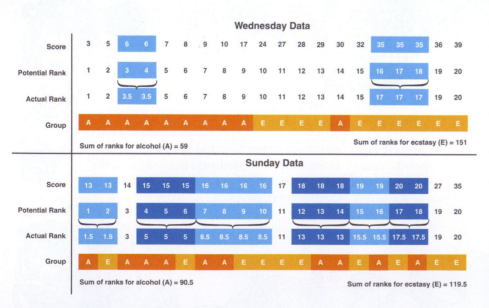

Figure 7.4 Ranking the depression scores

groups then you should find a similar number of high and low ranks in each group; specifically, if you added up the ranks, then you'd expect the summed total of ranks in each group to be about the same. Now let's imagine that the ecstasy group is more depressed than the alcohol group. What do you think would happen to the ranks? If you rank the scores as before, then you would expect more higher ranks to be in the ecstasy group and more lower ranks to be in the alcohol group. Again, if we summed the ranks in each group, we'd expect the sum of ranks to be higher in the ecstasy group than in the alcohol group. The Mann–Whitney and Wilcoxon rank-sum tests use this principle. In fact, when the groups have unequal numbers of participants in them, the test statistic (W_s) for the Wilcoxon rank-sum test is simply the sum of ranks in the group that contains the fewer people; when the group sizes are equal it's the value of the smaller summed rank.

Let's look at how ranking works. Figure 7.4 shows the ranking process for both the Wednesday and Sunday data. To begin with, let's focus on Wednesday, because the ranking is more straightforward. First, we arrange the scores in ascending order and attach a label to remind us from which group each score came (I've used A for alcohol and E for ecstasy). Starting at the lowest score, we assign potential ranks starting with 1 and going up to the number of scores we have. I've called these 'potential ranks' because sometimes the same score occurs more than once in a data set (e.g., in these data a score of 6 occurs twice, and a score of 35 occurs three times). These are called *tied ranks*, and we rank them with the value of the average potential rank for

How do I rank scores?

Figure 7.5   Frank Wilcoxon

those scores. For example, our two scores of 6 would've been ranked as 3 and 4, so we assign a rank of 3.5, the average of these values. Likewise, the three scores of 35 have potential ranks of 16, 17 and 18, so we assign a rank of 17, the average of these three values ((16 + 17 + 18)/3 = 17). Once we've ranked the data, we add the ranks for the two groups. First we add the ranks of the scores from the alcohol group (you should find the sum is 59) and then add the ranks of the scores from the ecstasy group (this value is 151). Our test statistic is the lower of these sums, which for these data is the sum for the Wednesday data, $W_s = 59$.

Having done the self-test, you should find that the sum of ranks is 90.5 for the alcohol group and 119.5 for the ecstasy group. The test statistic is the lower of these sums, which is the sum for the alcohol group, $W_s = 90.5$.

How do we determine whether this test statistic is significant? It turns out that the mean ($\overline{W}_s$) and standard error ($SE_{\overline{W}_s}$) of this test statistic can be calculated from the sample sizes of each group (n_1 is the

sample size of group 1 and n_2 is the sample size of group 2):

$$\overline{W}_s = \frac{n_1\left(n_1 + n_2 + 1\right)}{2} \qquad (7.1)$$

$$SE_{\overline{W}_s} = \sqrt{\frac{n_1 n_2\left(n_1 + n_2 + 1\right)}{12}} \qquad (7.2)$$

We have equal-sized groups with 10 people in each, so n_1 and n_2 are both 10. Therefore, the mean and standard deviation are:

$$\overline{W}_s = \frac{10\left(10 + 10 + 1\right)}{2} = 105 \qquad (7.3)$$

$$SE_{\overline{W}_s} = \sqrt{\frac{\left(10 \times 10\right)\left(10 + 10 + 1\right)}{12}} = 13.23 \qquad (7.4)$$

If we know the test statistic, the mean test statistic and the standard error, then we can convert the test statistic to a z-score using the equation that we came across in Chapter 1 (equation (1.9)):

$$z = \frac{X - \overline{X}}{s} = \frac{W_s - \overline{W}_s}{SE_{\overline{W}_s}} \qquad (7.5)$$

We also know that we can use Table A.1 in the Appendix to ascertain a p-value for a z-score (and more generally that a z greater than 1.96 or smaller than −1.96 is significant at $p < 0.05$). The z-scores for the Sunday and Wednesday depression scores are:

$$z_{\text{Sunday}} = \frac{W_s - \overline{W}_s}{SE_{\overline{W}_s}} = \frac{90.5 - 105}{13.23} = -1.10 \qquad (7.6)$$

$$z_{\text{Wednesday}} = \frac{W_s - \overline{W}_s}{SE_{\overline{W}_s}} = \frac{59 - 105}{13.23} = -3.48 \qquad (7.7)$$

So, there is a significant difference between the groups on Wednesday, but not on Sunday.

The procedure I've described is the Wilcoxon rank-sum test. The

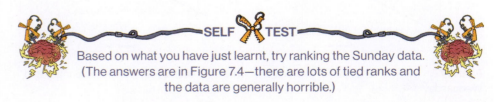

SELF TEST

Based on what you have just learnt, try ranking the Sunday data. (The answers are in Figure 7.4—there are lots of tied ranks and the data are generally horrible.)

DISCOVERING STATISTICS USING IBM SPSS STATISTICS

See whether you can use what you have learnt about data entry to enter the data in Table 7.1 into SPSS.

Mann–Whitney test is basically the same, but uses a test statistic U, which has a direct relationship with the Wilcoxon test statistic. SPSS produces both statistics in the output. If you're interested, U is calculated using an equation in which n_1 and n_2 are the sample sizes of groups 1 and 2 respectively, and R_1 is the sum of ranks for group 1:

$$U = n_1 n_2 + \frac{n_1(n_1 + 1)}{2} - R_1 \qquad (7.8)$$

For our data we'd get the following (remember we have 10 people in each group and the sum of ranks for group 1, the ecstasy group, was 119.5 for the Sunday data and 151 for the Wednesday data):

$$U_{\text{Sunday}} = (10 \times 10) + \frac{10(11)}{2} - 119.50 = 35.50 \qquad (7.9)$$

$$U_{\text{Wednesday}} = (10 \times 10) + \frac{10(11)}{2} - 151.00 = 4.00 \qquad (7.10)$$

7.4.2 Inputting data and provisional analysis

The data editor will have three columns. The first column is a coding variable (called something like **Drug**), which will have two codes (for convenience I suggest 1 = ecstasy group and 2 = alcohol group). When you enter this variable into SPSS,

remember to enter value labels to set the codes as discussed in Section 4.6.5. The second column will have values for the dependent variable (BDI) measured the day after (call this variable **Sunday_BDI**) and the third will have the midweek scores on the same questionnaire (call this variable **Wednesday_BDI**). You can, if you like, add a fourth column that is a variable to identify the participant (with a code or number). Save the file as **Drug.sav**.

Your first step should always be graphs and exploratory analysis. Given that we have a small sample (10 per group), there's probably some worth in tests of normality and homogeneity of variance (but see Jane Superbrain Box 6.5). For normality, because we're looking for group differences, we need to run the analyses separately for each group.

The results of our exploratory analysis are shown in Output 7.1 and Figure 7.6. The normal Q-Q plots show quite clear deviations from normality for ecstasy on Sunday and alcohol on Wednesday because the dots deviate from the diagonal line. The tables in Output 7.1 confirm these observations: for the Sunday data the distribution for ecstasy, $D(10) = 0.28$, $p = 0.03$, appears to be non-normal, whereas the alcohol data, $D(10) = 0.17$, $p = 0.20$, were normal; conversely, for the Wednesday data, although the data for ecstasy were normal, $D(10) = 0.24$, $p = 0.13$, the data for alcohol were significantly non-normal, $D(10) = 0.31$, $p = 0.009$. Remember that we can tell this by whether the significance of the Kolmogorov–Smirnov and Shapiro–Wilk tests are less than 0.05 (and, therefore, significant) or greater than 0.05 (and, therefore, non-significant, *ns*). These findings signal that the sampling distribution might be non-normal for the Sunday and Wednesday data and, given that our sample is small, a non-parametric test would be appropriate. To whatever extent you see the point in Levene's test, it shows that the variances are not

Use SPSS to test for normality and homogeneity of variance in these data (see Sections 6.10 and 6.11).

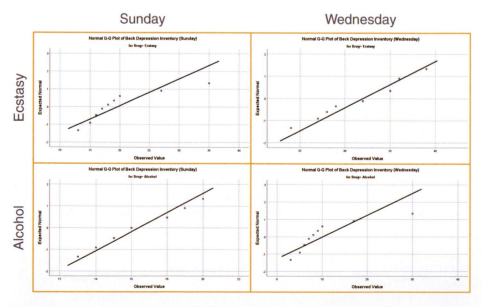

Figure 7.6 Normal Q-Q plots of depression scores after ecstasy and alcohol on Sunday and Wednesday

218

Tests of Normality

	Type of Drug	Kolmogorov–Smirnov[a]			Shapiro–Wilk		
		Statistic	df	Sig.	Statistic	df	Sig.
Beck Depression Inventory (Sunday)	Ecstasy	.276	10	.030	.811	10	.020
	Alcohol	.170	10	.200*	.959	10	.780
Beck Depression Inventory (Wednesday)	Ecstasy	.235	10	.126	.941	10	.566
	Alcohol	.305	10	.009	.753	10	.004

*. This is a lower bound of the true significance.

a. Lilliefors Significance Correction

Test of Homogeneity of Variance

		Levene Statistic	df1	df2	Sig.
Beck Depression Inventory (Sunday)	Based on Mean	3.644	1	18	.072
	Based on Median	1.880	1	18	.187
	Based on Median and with adjusted df	1.880	1	10.076	.200
	Based on trimmed mean	2.845	1	18	.109
Beck Depression Inventory (Wednesday)	Based on Mean	.508	1	18	.485
	Based on Median	.091	1	18	.766

Output 7.1

significantly different between the drug groups on Sunday, $F(1, 18) = 3.64$, $p = 0.072$, and Wednesday, $F(1, 18) = 0.51$, $p = 0.485$ (Output 7.1, bottom table), suggesting that the assumption of homogeneity has been met.

7.4.3 The Mann–Whitney test using SPSS Statistics ▐▐▐▐

To run a Mann–Whitney test follow the general procedure outlined in Section 7.3 by first selecting *Analyze* ▶ *Nonparametric Tests* ▶ 🔺 *Independent Samples*…. When you reach the `Objective` `Fields` `Settings` tab, if you need to assign variables then select

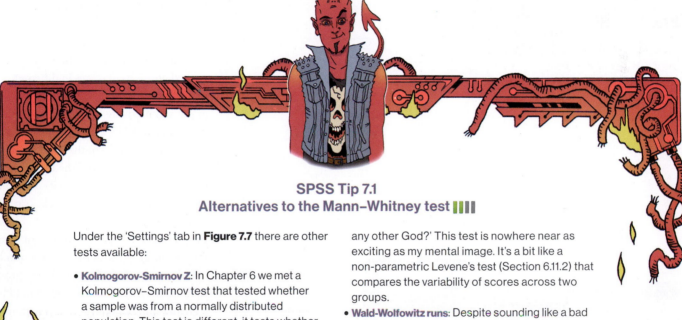

SPSS Tip 7.1
Alternatives to the Mann–Whitney test ▐▐▐▐

Under the 'Settings' tab in **Figure 7.7** there are other tests available:

- **Kolmogorov–Smirnov Z**: In Chapter 6 we met a Kolmogorov–Smirnov test that tested whether a sample was from a normally distributed population. This test is different: it tests whether two groups have been drawn from the same population (regardless of what that population may be). In effect, this test does much the same as the Mann–Whitney test, but it tends to have more power when sample sizes are less than about 25 per group, and so is worth selecting if that's the case.

- **Moses Extreme Reactions**: This test makes me think of a bearded man standing on Mount Sinai reading a stone tablet and then suddenly bursting into a wild rage, smashing the tablet and screaming 'What do you mean, do not worship

any other God?' This test is nowhere near as exciting as my mental image. It's a bit like a non-parametric Levene's test (Section 6.11.2) that compares the variability of scores across two groups.

- **Wald–Wolfowitz runs**: Despite sounding like a bad case of diarrhea, this test is another variant on the Mann–Whitney test. In this test the scores are rank-ordered as in the Mann–Whitney test, but rather than analyzing the ranks, this test looks for 'runs' of scores from the same group within the ranked order. If there's no difference between groups, then ranks from the two groups should be randomly interspersed. However, if the groups are different then you should see more ranks from one group at the lower end and more ranks from the other group at the higher end. By looking for clusters of scores in this way the test can determine if the groups differ.

◉ Use <u>c</u>ustom field assignments and specify the model (Figure 7.7, top) by dragging **Beck Depression Inventory (Sunday)** and **Beck Depression Inventory (Wednesday)** into the box labeled *Test Fields* (or select them in the box labeled *Fields* and click). Next, transfer **Type of Drug** to the box labeled *Groups*. Activate the | Objective | Fields | Settings | tab, select ◉ <u>C</u>ustomize tests and check ☑ Mann-W<u>h</u>itney U (2 samples) (Figure 7.7, bottom). The dialog box lists tests other than the Mann–Whitney test, which are explained in SPSS Tip 7.1. Click ▶ Run to run the analysis.

7.4.4 Output from the Mann–Whitney test ▮▮▮▮

With all non-parametric tests, the output contains a summary table that you need to double-click to open the *model viewer* window (see Figure 7.8). The model viewer is divided into two panels: the left-hand panel shows the summary table of any analyses that you have done, and the right-hand panel shows the details of the analysis. In this example, we analyzed group differences for both Sunday and Wednesday, so the summary table has two rows: one for

Sunday and one for Wednesday. To see the results of the Sunday analysis in the right-hand panel, click the row of the table for Sunday in the left-hand panel. Once selected, the row in the left-hand panel becomes shaded (as shown in Figure 7.8). To see the results for the Wednesday data we would click somewhere on the second row of the table in the left-hand panel. This row would become shaded within the table and the output in the right-hand panel would change to show the details for Wednesday's data.

I explained earlier that the Mann–Whitney test works by looking at differences in the ranked positions of scores in different groups. The first part of the output is a graph summarizing the data after they have been ranked. SPSS shows us the distribution of ranks in the two groups (alcohol and ecstasy) and the mean rank in each condition (see Output 7.2). Remember that the Mann–Whitney test relies on scores being ranked from lowest to highest; therefore, the group with the lowest mean rank is the group with the greatest number of lower scores in it. Conversely, the group that has the highest mean rank should have a greater number of high scores within it. Therefore, this graph can be used to ascertain which group had the highest scores, which is useful for interpreting a significant result. For example, we can see for the Sunday data that the distributions in the two groups are almost identical (the ecstasy has a couple of higher ranks but otherwise the bars look the same) and the mean ranks are similar (9.05 and 11.95); on Wednesday, however, the distribution of ranks is shifted upwards in the ecstasy group compared to the alcohol group, which is reflected in a much larger mean rank (15.10 compared to 5.90).

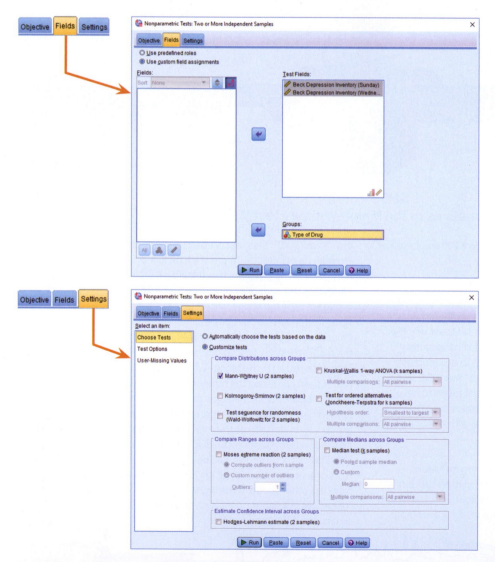

Figure 7.7 Dialog boxes for the Mann–Whitney test

Underneath the graph a table shows the test statistics for the Mann–Whitney test, the Wilcoxon procedure and the corresponding z-score. Note that the values of U, W_s and the z-score are the same as we calculated in Section 7.4.1 (phew!). The rows labeled *Asymptotic Sig.* and *Exact Sig.* tell us the probability that a test statistic of at least that magnitude would occur if there were no difference between groups. The two *p*-values are interpreted in the same way but are computed differently: our sample is fairly small so we'll use the exact method (see Jane Superbrain Box 7.2). For these data, the Mann–Whitney test is non-significant for the depression scores taken on the Sunday because the *p*-value of 0.280 is greater than the critical value of 0.05. This finding indicates that ecstasy is no more of a depressant, the day after taking it, than alcohol: both groups report comparable levels of depression. This confirms what we concluded from the mean ranks and distribution of ranks. For the midweek measures the results are highly significant because the exact *p*-value of

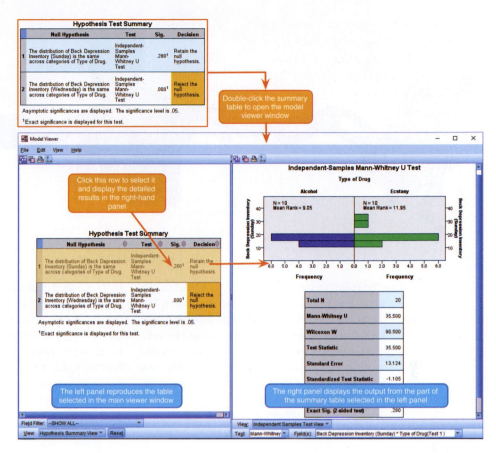

Figure 7.8 With non-parametric tests you must double-click the summary table within the viewer window to open the model viewer window

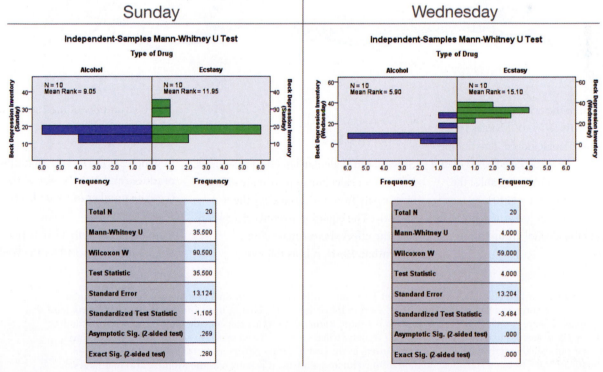

Output 7.2

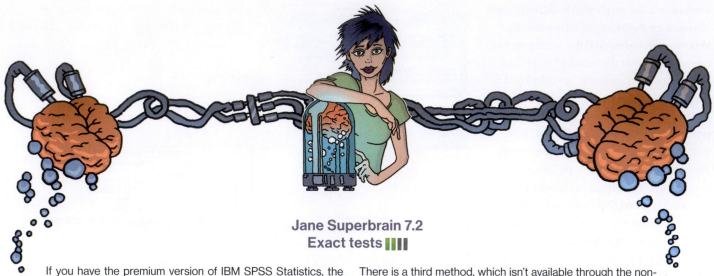

Jane Superbrain 7.2
Exact tests ▎▎▎

If you have the premium version of IBM SPSS Statistics, the *p*-value for non-parametric tests is computed in two ways. The *asymptotic method* gives you a sort of approximation that in large samples will be a perfectly serviceable answer. However, when samples are small or the data are particularly poorly distributed, it doesn't give you a good answer. The *exact method* is more computationally difficult (but we don't care because our computer is doing the computations for us) and gives us an exact significance value. You should use this exact significance in small samples (by which I mean anything under 50 really).

There is a third method, which isn't available through the non-parametric menus that we're using, but is available for some other tests, so we might as well learn about it now. The **Monte Carlo method**[3] is a slightly less labor-intensive method than computing an exact *p*-value. This method is like bootstrapping (Section 6.12.3) and involves creating a distribution similar to that found in the sample and then taking several samples (the default is 10,000) from this distribution. From those samples the mean significance value and the confidence interval around it can be created.

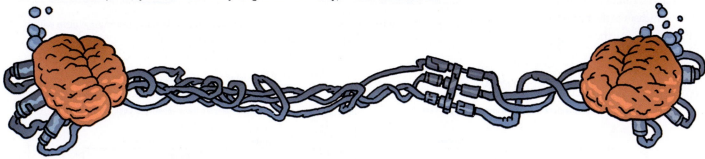

0.000 is less than the critical value of 0.05. In this case we write $p < 0.001$ because the observed p is very small indeed. This finding also confirms what we suspected based on the distribution of ranks and mean ranks: the ecstasy group (mean rank = 15.10) had significantly higher levels of depression midweek than the alcohol group (mean rank = 5.90).

7.4.5 Calculating an effect size ▎▎▎▎

SPSS doesn't calculate an effect size for us, but we can calculate approximate effect sizes easily from the *z*-score for the test-statistic. The equation to convert a *z*-score into the effect size estimate (from Rosenthal, 1991), *r*, is as follows:

$$r = \frac{z}{\sqrt{N}} \qquad (7.11)$$

in which z is the z-score that SPSS produces and N is the size of the study (i.e., the number of total observations) on which z is based.

Output 7.2 tells us that z is −1.11 for the Sunday data and −3.48 for the Wednesday

---

3 It's called the Monte Carlo method because back in the late nineteenth century when Karl Pearson was trying to simulate data he didn't have a computer to do it for him. So he used to toss coins. A lot. That is, until a friend suggested that roulette wheels, if unbiased, were excellent random number generators. Rather than trying to persuade the Royal Society to fund trips to Monte Carlo casinos to collect data from their roulette wheels, he purchased copies of *Le Monaco*, a weekly Paris periodical that published exactly the data that he required, at the cost of 1 franc (Pearson, 1894; Plackett, 1983). When simulated data are used to test a statistical method or to estimate a statistic, it is known as the Monte Carlo method even though we use computers now and not roulette wheels.

data. In both cases the total number of observations was 20 (10 ecstasy users and 10 alcohol users). The effect sizes are therefore:

$$r_{Sunday} = \frac{-1.11}{\sqrt{20}} = -0.25 \qquad (7.12)$$

$$r_{Wednesday} = \frac{-3.48}{\sqrt{20}} = -0.78 \qquad (7.13)$$

These values represents a small to medium effect for the Sunday data (it is below the 0.3 criterion for a medium effect size) and a huge effect for the Wednesday data (the effect size is well above the 0.5 threshold for a large effect). The Sunday data show how a substantial effect size can be non-significant in a small sample (see Section 2.9.10).

7.4.6 Writing the results ▌▌▌▌

For the Mann–Whitney test, report only the test statistic (denoted by U) and its significance. In keeping with good practice (Section 3.8), include the effect size and report exact values of p (rather than summary values such as $p < 0.05$). We could report something like:

✓ Depression levels in ecstasy users ($Mdn = 17.50$) did not differ significantly from alcohol users ($Mdn = 16.00$) the day after the drugs were taken, $U = 35.50$, $z = -1.11$, $p = 0.280$, $r = -0.25$. However, by Wednesday, ecstasy users ($Mdn = 33.50$) were significantly more depressed than alcohol users ($Mdn = 7.50$), $U = 4.00$, $z = -3.48$, $p < 0.001$, $r = -0.78$.

I've reported the median for each condition because this statistic is more appropriate than the mean for non-parametric tests. You can get these values by running descriptive statistics (Section 6.10.2) or you could report the mean ranks instead of the median.

If you want to report Wilcoxon's test rather than Mann–Whitney's U statistic you could write:

✓ Depression levels in ecstasy users ($Mdn = 17.50$) did not significantly differ from alcohol users ($Mdn = 16.00$) the day after the drugs were taken, $W_s = 90.50$, $z = -1.11$, $p = 0.280$, $r = -0.25$. However, by Wednesday, ecstasy users ($Mdn = 33.50$) were significantly more depressed than alcohol users ($Mdn = 7.50$), $W_s = 59.00$, $z = -3.48$, $p < 0.001$, $r = -0.78$.

7.5 Comparing two related conditions: the Wilcoxon signed-rank test ▌▌▌▌

The **Wilcoxon signed-rank test** (Wilcoxon, 1945), not to be confused with the rank-sum test in the previous section,

Cramming Sam's Tips
Mann–Whitney test

- The Mann–Whitney test and Wilcoxon rank-sum test compare two conditions when different participants take part in each condition and the resulting data have unusual cases or violate any assumption in Chapter 6.
- Look at the row labeled *Asymptotic Sig.* or *Exact Sig.* (if your sample is small). If the value is less than 0.05 then the two groups are significantly different.

- The values of the mean ranks tell you how the groups differ (the group with the highest scores will have the highest mean rank).
- Report the U-statistic (or W_s if you prefer), the corresponding z and the significance value. Also report the medians and their corresponding ranges (or draw a boxplot).
- Calculate the effect size and report this too.

Table 7.2 Ranking data in the Wilcoxon signed-rank test

BDI Sunday	BDI Wednesday	Difference	Sign	Rank	Positive Ranks	Negative Ranks
Ecstasy						
15	28	13	+	2.5	2.5	
35	35	0	Exclude			
16	35	19	+	6	6	
18	24	6	+	1	1	
19	39	20	+	7	7	
17	32	15	+	4.5	4.5	
27	27	0	Exclude			
16	29	13	+	2.5	2.5	
13	36	23	+	8	8	
20	35	15	+	4.5	4.5	
Total =					**36**	**0**
Alcohol						
16	5	−11	−	9		9
15	6	−9	−	7		7
20	30	10	+	8	8	
15	8	−7	−	3.5		3.5
16	9	−7	−	3.5		3.5
13	7	−6	−	2		2
14	6	−8	−	5.5		5.5
19	17	−2	−	1		1
18	3	−15	−	10		10
18	10	−8	−	5.5		5.5
Total =					**8**	**47**

SELF TEST

Split the file by **Drug** (see Section 6.10.4).

is used in situations where you want to compare two sets of scores that are related in some way (e.g., they come from the same entities). Imagine the psychologist in the previous section was interested in the *change* in depression levels, within people, for each of the two drugs. She now wants to compare the BDI scores on Sunday to those on Wednesday. Remember that the distributions of scores for both drugs were non-normal on one of the two days, implying (because the sample is small) that the sampling distribution will be non-normal too (see Output 7.1), so the

psychologist would have to use a non-parametric test.

7.5.1 Theory of the Wilcoxon signed-rank test ▮▮▮▮

The Wilcoxon signed-rank test is based on ranking the differences between scores in the two conditions you're comparing. Once these differences have been ranked (just like in Section 7.4.1), the sign of the difference (positive or negative) is assigned to the rank.

Table 7.2 shows the ranking for comparing depression scores on Sunday to those on Wednesday for the two drugs separately. First, we calculate the difference between scores on Sunday and Wednesday (that's just Sunday's score subtracted from Wednesday's). If the difference is zero (i.e., the scores are the same on Sunday and Wednesday) we exclude this score from the ranking. We make a note of the sign of the difference (positive or negative) and then rank the differences (starting with the smallest), ignoring whether they are positive or negative. The ranking process is the same as in Section 7.4.1, and we deal with tied scores in exactly the same way. Finally, we collect together the ranks that came from a positive difference between the conditions, and add them up to get the sum of positive ranks (T_+). We also add up the ranks that came from negative differences between the conditions to get the sum of negative ranks (T_-). For ecstasy, $T_+ = 36$ and $T_- = 0$ (in fact there were no negative ranks), and for alcohol, $T_+ = 8$ and $T_- = 47$. The test statistic is T_+, and so it is 36 for ecstasy and 8 for alcohol.

To calculate the significance of the test statistic (T), we again look at the mean ($\overline{T}$) and standard error ($SE_{\overline{T}}$), which, like the Mann–Whitney and rank-sum test in the previous section, are functions of the sample size, n (because we used the same participants, there is only one sample size):

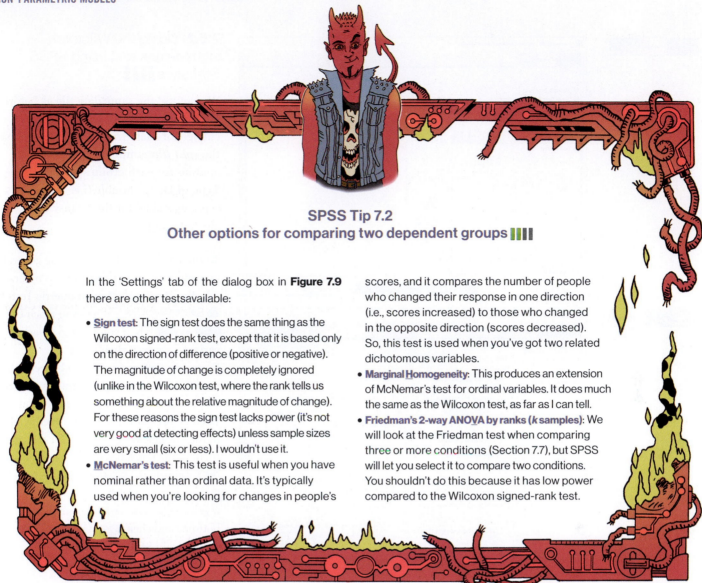

SPSS Tip 7.2
Other options for comparing two dependent groups ▮▮▮

In the 'Settings' tab of the dialog box in **Figure 7.9** there are other tests available:

- **Sign test**: The sign test does the same thing as the Wilcoxon signed-rank test, except that it is based only on the direction of difference (positive or negative). The magnitude of change is completely ignored (unlike in the Wilcoxon test, where the rank tells us something about the relative magnitude of change). For these reasons the sign test lacks power (it's not very good at detecting effects) unless sample sizes are very small (six or less). I wouldn't use it.

- **McNemar's test**: This test is useful when you have nominal rather than ordinal data. It's typically used when you're looking for changes in people's

scores, and it compares the number of people who changed their response in one direction (i.e., scores increased) to those who changed in the opposite direction (scores decreased). So, this test is used when you've got two related dichotomous variables.

- **Marginal Homogeneity**: This produces an extension of McNemar's test for ordinal variables. It does much the same as the Wilcoxon test, as far as I can tell.

- **Friedman's 2-way ANOVA by ranks (*k* samples)**: We will look at the Friedman test when comparing three or more conditions (Section 7.7), but SPSS will let you select it to compare two conditions. You shouldn't do this because it has low power compared to the Wilcoxon signed-rank test.

$$\bar{T} = \frac{n(n+1)}{4} \tag{7.14}$$

$$SE_{\bar{T}} = \sqrt{\frac{n(n+1)(2n+1)}{24}} \tag{7.15}$$

In both groups, *n* is 10 (because that's how many participants were used). However, remember that for our ecstasy group we excluded two people because they had differences of zero, therefore the sample size we use is 8, not 10, which gives us:

$$\bar{T}_{\text{Ecstasy}} = \frac{8(8+1)}{4} = 18 \tag{7.16}$$

$$SE_{\bar{T}_{\text{Ecstasy}}} = \sqrt{\frac{8(8+1)(16+1)}{24}} = 7.14 \tag{7.17}$$

For the alcohol group there were no exclusions and we get:

$$\bar{T}_{\text{Alcohol}} = \frac{10(10+1)}{4} = 27.50 \tag{7.18}$$

$$SE_{\bar{T}_{\text{Alcohol}}} = \sqrt{\frac{10(10+1)(20+1)}{24}} = 9.81 \tag{7.19}$$

As with the Mann–Whitney test, if we know the test statistic, the mean of test statistics and the standard error, then we can convert the test statistic to a *z*-score using the standard equation that we revisited in the previous section:

$$z = \frac{X - \bar{X}}{s} = \frac{T - \bar{T}}{SE_{\bar{T}}} \tag{7.20}$$

If we calculate the *z* for the ecstasy and alcohol depression scores we get:

$$z_{\text{Ecstasy}} = \frac{T - \bar{T}}{SE_{\bar{T}}} = \frac{36 - 18}{7.14} = 2.52 \tag{7.21}$$

$$z_{\text{Alcohol}} = \frac{T - \bar{T}}{SE_{\bar{T}}} = \frac{8 - 27.5}{9.81} = -1.99 \tag{7.22}$$

If these values are greater than 1.96 (ignoring the minus sign) then the test is significant at $p < 0.05$. So, it looks as though there is a significant difference between depression scores on Wednesday and Sunday for both ecstasy and alcohol.

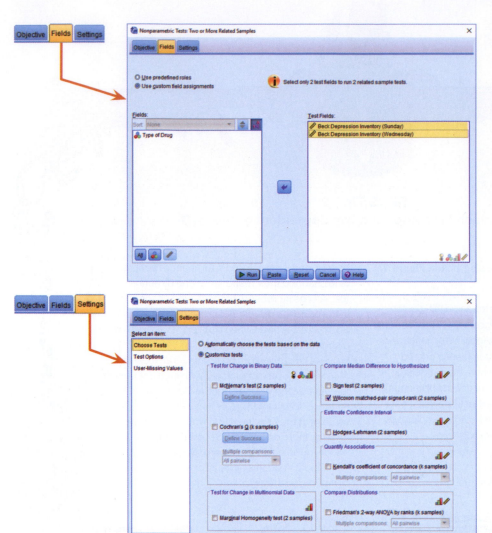

7.5.2 Doing the Wilcoxon signed-rank test using SPSS Statistics ▌▌▌▌

We can use the same data file as before, but because we want to look at the change for each drug *separately*, we use the *split file* command to repeat the analysis for each group specified in the **Type of Drug** variable (i.e. we'll get a separate model for the ecstasy and alcohol groups).

To run a Wilcoxon test follow the general procedure outlined in Section 7.3 by first selecting Analyze ▶ Nonparametric Tests ▶ 🔺 Related Samples…. When you reach the Objective Fields Settings tab, if you need to assign variables yourself then select ◉ Use custom field assignments and specify the model (Figure 7.9, top) by dragging **Beck Depression Inventory (Sunday)** and **Beck Depression Inventory (Wednesday)** to the box labeled *Test Fields* (or select them in the box labeled *Fields* and click ⬇). Activate the Objective Fields Settings tab, select ◉ Customize tests, check ☑ Wilcoxon matched-pair signed-rank (2 samples) (Figure 7.9, bottom) and click ▶ Run . Other options are explained in SPSS Tip 7.2.

Figure 7.9   Dialog boxes for the Wilcoxon signed-rank test

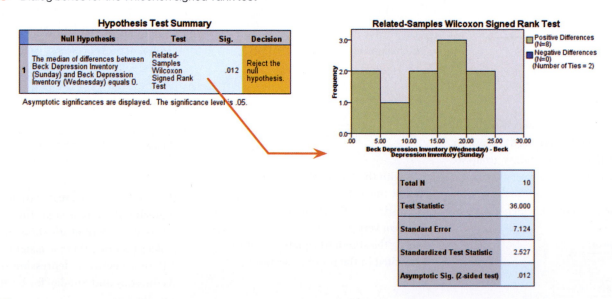

Output 7.3

7.5.3 Output for the ecstasy group

If you have split the file, then the first set of results obtained will be for the ecstasy group (Output 7.3). The summary table tells you that the significance of the test was 0.012 and helpfully suggests that you reject the null hypothesis. Let's not be bossed around by SPSS, though. If you double-click this table to enter the model viewer you will see a histogram of the distribution of differences. These differences are the Sunday scores subtracted from the Wednesday scores (which we're told underneath the histogram) and correspond to the values in the *Difference* column in Table 7.2. A positive difference means more depression on Wednesday than Sunday, a negative difference means more depression on Sunday than Wednesday, and a difference of zero means that depression levels were identical on Sunday and Wednesday. The histogram is color-coded based on whether ranks are positive or negative: positive ranks appear as brown bars, and negative ranks as blue bars. You might notice that there are no blue bars, which tells us that there were no negative ranks. Therefore, the histogram is a very quick indication of the ratio of positive to negative ranks: in this case all ranks are positive (or tied) and none are negative. We are told the same in the legend to the histogram: there were 8 positive differences, 0 negative differences and 2 ties.

In Section 7.5.1 I explained that the test statistic, T, is the sum of positive ranks, so our test value here is 36. I also showed how this value can be converted to a z-score, and in doing so we can compute exact significance values based on the normal distribution. Underneath the histogram in Output 7.3 is a table that tells us the test statistic (36), its standard error (7.12) and the z-score (2.53) which all correspond (more or less) to the values we computed by hand in Section 7.5.1. This z-score has a significance value of $p = 0.012$. This value is less than the standard critical value of 0.05, so we conclude that there is a significant change in depression scores from Sunday to Wednesday (i.e., we reject the null hypothesis). From the histogram we know that this test statistic is based on there being many more positive differences (i.e., scores being higher on Wednesday than Sunday), so we can conclude that when taking ecstasy there was a significant increase in depression (as measured by the BDI) from the morning after to midweek.

7.5.4 Output for the alcohol group

For the alcohol group (Output 7.4) the summary table tells us that the significance of the test was 0.047 and again suggests that we reject the null hypothesis. As before, double-click this table to enter the model viewer. Notice that for the alcohol group (unlike the ecstasy group) we have different-colored bars: the brown bars represent positive differences and the blue bars negative differences. For the ecstasy group we saw only brown bars, but for the alcohol group we see the opposite: the bars are predominantly blue. This indicates that on the whole differences between Wednesday and Sunday were negative. In other words, scores were generally higher on Sunday than they were on Wednesday. Again, these differences are the same as those in the *Difference* column in Table 7.2. The legend of the graph confirms that there was only 1 positive difference, 9 negative differences and 0 ties. The table below the histogram tells us the test statistic (8), its standard error (9.80),

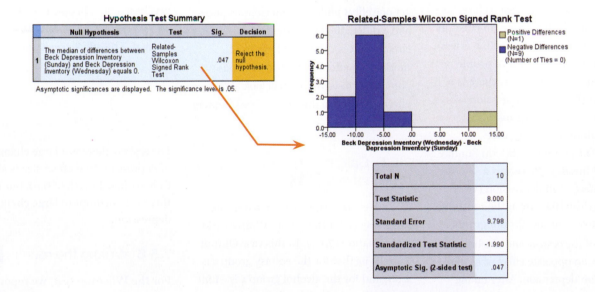

Output 7.4

Cramming Sam's Tips

- The Wilcoxon signed-rank test compares two conditions when the scores are related (e.g., scores come from the same participants) and the resulting data have unusual cases or violate any assumption in Chapter 6.
- Look at the row labeled *Asymptotic Sig. (2-sided test)*. If the value is less than 0.05 then the two conditions are significantly different.

- Look at the histogram and numbers of positive or negative differences to tell you how the groups differ (the greater number of differences in a particular direction tells you the direction of the result).
- Report the *T*-statistic, the corresponding *z*, the exact significance value and an effect size. Also report the medians and their corresponding ranges (or draw a boxplot).

and the corresponding *z*-score (−1.99). These are the values we calculated in Section 7.5.1. The *p*-value associated with the *z*-score is 0.047, which means that there's a probability of 0.047 that we would get a value of *z* at least as large as the one we have if there were no effect in the population; because this value is less than the critical value of 0.05, people would typically conclude that there is a significant difference in depression scores. We know that because the histogram showed predominantly negative differences (i.e., scores higher on Sunday than on Wednesday) there was a significant *decline* in depression (as measured by the BDI) from the morning after to midweek in the alcohol group.

The results of the ecstasy and alcohol groups show an opposite effect of alcohol and ecstasy on depression. After taking alcohol, depression is higher the morning

after than midweek, whereas after taking ecstasy, depression increases from the morning after to the middle of the week. A different effect across different groups or conditions is known as *moderation* (i.e., you get one effect under certain circumstances and a different effect under other circumstances). You can't look at moderation effects directly using non-parametric tests, but we will look at these effects in detail in due course (see Chapters 11 and 14).

7.5.5 Calculating an effect size

The effect size can be calculated in the same way as for the Mann–Whitney test (see equation (7.11)). In this case Output 7.4 tells us that for the ecstasy group *z* is 2.53, and for the alcohol group *z* is −1.99. For alcohol, we had 20 observations

(we tested only 10 people but each contributed 2 scores, and it is the total number of observations, not the number of people, that is important). For ecstasy we ended up excluding two cases, so the *z* was based on 8 people contributing 2 scores each, which means 16 observations in total. The effect size is therefore:

$$r_{\text{Ecstasy}} = \frac{2.53}{\sqrt{16}} = 0.63 \qquad (7.23)$$

$$r_{\text{Alcohol}} = \frac{-1.99}{\sqrt{20}} = -0.44 \qquad (7.24)$$

For ecstasy there is a large change in levels of depression (the effect size is above Cohen's benchmark of 0.5), but for alcohol there is a medium to large change in depression.

7.5.6 Writing the results

For the Wilcoxon test, we report the test statistic (denoted by the letter *T*), its

Labcoat Leni's Real Research 7.1
Having a quail of a time? ▌▌▌▌

Matthews, R. C. et al. (2007). *Psychological Science*, *18*(9), 758–762.

We encountered some research in Chapter 2 in which we discovered that you can influence aspects of male quail's sperm production through 'conditioning'. The basic idea is that the male is granted access to a female for copulation in a certain chamber (e.g., one that is colored green) but gains no access to a female in a different context (e.g., a chamber with a tilted floor). The male, therefore, learns that when he is in the green chamber his luck is in, but if the floor is tilted then frustration awaits. For other males the chambers will be reversed. The human equivalent (well, sort of) would be if you always managed to pull at Coalition but never at Digital.[4] During the test phase, males get to mate in both chambers. The question is: after the males have learnt that they will get a mating opportunity in a certain context, do they produce more sperm or better-quality sperm when mating in that context compared to the control context? (i.e., are you more of a stud at Coalition? OK, I'm going to stop this analogy now.)

Mike Domjan and his colleagues predicted that if conditioning evolved because it increases reproductive fitness then males who mated in the context that had previously signalled a mating opportunity would fertilize a significantly greater number of eggs than quails that mated in their control context (Matthews, Domjan, Ramsey, & Crews, 2007). They put this hypothesis to the test in an experiment that is utter genius. After training, they allowed 14 females to copulate with two males (counterbalanced): one male copulated with the female in the chamber that had previously signalled a reproductive opportunity (**Signalled**), whereas the second male copulated with the same female but in the chamber that had not previously signalled a mating opportunity (**Control**). Eggs were collected from the females for 10 days after the mating and a genetic analysis was used to determine the father of any fertilized eggs.

The data from this study are in the file **Matthews et al. (2007).sav**. Labcoat Leni wants you to carry out a Wilcoxon signed-rank test to see whether more eggs were fertilized by males mating in their signalled context compared to males in their control context.

Answers are on the companion website (or look at page 760 in the original article).

---

4 These are both clubs in Brighton that I don't go to because I'm too old for that sort of thing, but even when I was younger I was far too socially inept to cope with nightclubs.

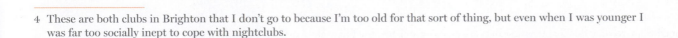

exact significance and an effect size (see Section 3.8). So, we could report something like:

✓ For ecstasy users, depression levels were significantly higher on Wednesday (Mdn = 33.50) than on Sunday (Mdn = 17.50), T = 36, p = 0.012, r = 0.63. However, for alcohol users the opposite was true: depression levels were significantly lower on Wednesday (Mdn = 7.50) than on Sunday (Mdn = 16.0), T = 8, p = 0.047, r = −0.44.

You can get the median values by running descriptive statistics (Section 6.10.2). Alternatively, we could report the values of z:

✓ For ecstasy users, depression levels were significantly higher on Wednesday (Mdn = 33.50) than on Sunday (Mdn = 17.50), z = 2.53, p = 0.012, r = 0.63. However, for alcohol users the opposite was true: depression levels were significantly lower on Wednesday (Mdn = 7.50) than on Sunday (Mdn = 16.0), z = −1.99, p = 0.047, r = −0.44.

7.6 Differences between several independent groups: the Kruskal–Wallis test

Having looked at models that compare two groups or conditions, we now move onto models that can compare more than two conditions: the Kruskal-Wallis test compares groups or conditions containing independent scores, whereas the Friedman test is used when scores are related. Let's look at the Kruskal–Wallis test first

Figure 7.10   William Kruskal

(Kruskal & Wallis, 1952), which assesses the hypothesis that multiple independent groups come from different populations. If you'd like to know a bit more about William Kruskal (Figure 7.10) there is a lovely biography by Fienberg, Stigler, & Tanur (2007).

I read a story in a newspaper (yes, back when they existed) claiming that the chemical genistein, which is naturally occurring in soya, was linked to lowered sperm counts in Western males. When you read the actual study, it had been conducted on rats, and it found no link to lowered sperm counts, but there was evidence of abnormal sexual development in male rats (probably because genistein acts like estrogen). As journalists tend to do, a study showing no link between soya and sperm counts was used as the

scientific basis for an article about soya being the cause of declining sperm counts in Western males (never trust what you read). Imagine the rat study was enough for us to want to test this idea in humans. We recruit 80 males and split them into four groups that vary in the number of soya 'meals' they ate per week over a year-long period (a 'meal' would be a dinner containing 75 g of soya[5]). The first group was a control and ate no soya meals (i.e., none in the whole year); the second group had one soya meal per week (that's 52 over the year); the third group had four soya meals per week (208 over the year); and the final group had seven soya meals a week (364 over the year). At the end of the year, the participants were sent away to produce some sperm that I could count (when I say 'I', I mean someone else in a laboratory as far away from me as humanly possible).[6]

7.6.1 Theory of the Kruskal–Wallis test

Like the other tests we've seen, the Kruskal–Wallis test is used with ranked data. To begin with, scores are ordered from lowest to highest, ignoring the group to which the score belongs. The lowest score is assigned a rank of 1, the next highest a rank of 2 and so on (see Section 7.4.1 for more detail). Once ranked, the scores are collected back into their groups and their ranks are added within each group. The sum of ranks within each group is denoted by R_i (where i denotes the group). Table 7.3 shows the raw data for this example along with the ranks.

Once the sum of ranks has been calculated within each group, the test statistic, H, is calculated as follows:

$$H = \frac{12}{N(N+1)}\sum_{i=1}^{k}\frac{R_i^2}{n_i} - 3(N+1) \tag{7.25}$$

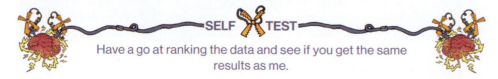

SELF TEST
Have a go at ranking the data and see if you get the same results as me.

5 Or 'soy', as you will know it.

6 In case any medics are reading this chapter, these data are made up and, because I have absolutely no idea what a typical sperm count is, they're probably ridiculous. I apologize, and you can laugh at my ignorance.

In this equation R_i is the sum of ranks for each group, N is the total sample size (in this case 80) and n_i is the sample size of a particular group (in this example group sample sizes are all 20). The middle part of the equation means that for each group we square the sum of ranks, divide this value by the sample size for that group, then add up these values. The rest of the equation involves calculating values based on the total sample size. For these data we get:

$$
\begin{aligned}
H &= \frac{12}{80(81)}\left(\frac{927^2}{20}+\frac{883^2}{20}+\frac{883^2}{20}+\frac{547^2}{20}\right)-3(81) \\
&= \frac{12}{6480}(42966.45+38984.45+38984.45+14960.45)-243 \\
&= 0.0019(135895.8)-243 \\
&= 251.659-243 \\
&= 8.659
\end{aligned}
\tag{7.26}
$$

This test statistic has a distribution from the family of chi-square distributions (see Chapter 19). Whereas the standard normal distribution is defined by a mean of 0 and a standard deviation of 1, the chi-square distribution is defined by a single value: the degrees of freedom, which is one less than the number of groups (i.e., $k-1$), in this case 3.

7.6.2 Follow-up analysis ▌▌▌▌

The Kruskal–Wallis test tells us that, overall, groups come from different populations. However, it doesn't tell us which groups differ. Are all of the groups different or just some of them? The simplest way to break down the overall effect is to compare all pairs of groups (known as **pairwise comparisons**). In our current example, this would entail six tests: none vs. 1 meal; none vs. 4 meals; none vs. 7 meals; 1 vs. 4 meals; 1 vs. 7 meals; and 4 vs. 7 meals. A very simple approach would be to perform 6 Mann–Whitney tests, one for each of these comparisons. However, we saw in

Section 2.9.7 that when we do lots of tests on the same data we inflate the familywise error rate: there will be a more than 5% chance that we'll make at least one Type I error. Ideally, we want a 5% chance of making a Type I error over *all the tests* we do, and we've seen that one method to achieve this is to use a lower probability as the threshold for significance. Therefore, we *could* perform six tests (one to compare each pair of groups) if we adjust the p-value so that overall, across all the tests, the Type I error rate remains at 5%. This is what a pairwise comparison does.

By being stricter about the p-value you deem to be significant you reduce the power of the tests—you might throw the baby out with the bathwater (Section 2.9.8). An alternative is to use a stepped procedure. The one SPSS uses begins by ordering the groups based on the sum of ranks from smallest to largest (if there are ties, the median decides the order rather than the sum).[7] For our data the rank sums were 7 meals (rank sum = 547,

Table 7.3 Data for the soya example with ranks

No Soya		1 Soya Meal		4 Soya Meals		7 Soya Meals	
Sperm (million/ml)	Rank	Sperm (million/ml)	Rank	Sperm (million/ml)	Rank	Sperm (million/ml)	Rank
3.51	4	3.26	3	4.03	6	3.10	1
5.76	9	3.64	5	5.98	10	3.20	2
8.84	17	6.29	11	9.59	19	5.60	7
9.23	18	6.36	12	12.03	21	5.70	8
12.17	22	7.66	14	13.13	24	7.09	13
15.10	30	15.33	32	13.54	27	8.09	15
15.17	31	16.22	34	16.81	35	8.71	16
15.74	33	17.06	36	18.28	37	11.80	20
24.29	41	19.40	38	20.98	40	12.50	23
27.90	46	24.80	42	29.27	48	13.25	25
34.01	55	27.10	44	29.59	49	13.40	26
45.15	59	41.16	57	29.95	50	14.90	28
47.20	60	56.51	61	30.87	52	15.02	29
69.05	65	67.60	64	33.64	54	20.90	39
75.78	68	70.79	66	43.37	58	27.00	43
77.77	69	72.64	67	58.07	62	27.48	45
96.19	72	79.15	70	59.38	63	28.30	47
100.48	73	80.44	71	101.58	74	30.67	51
103.23	75	120.95	77	109.83	76	32.78	53
210.80	80	184.70	79	182.10	78	41.10	56
Total (R_i)	**927**		**883**		**883**		**547**
Average ($\overline{R}_i$)	*46.35*		*44.15*		*44.15*		*27.35*

---

7 Each group has 20 scores, so the median will be the average of the 10th and 11th scores when the scores are in ascending order. The data in Table 7.3 are presented in ascending order in each group, so we can see that the medians are: $(27.90 + 34.01)/2 = 30.96$ (0 meals); $(24.80 + 27.10)/2 = 25.95$ (1 meal); $(29.27 + 29.59)/2 = 29.43$ (4 meals); $(13.25 + 13.40)/2 = 13.33$ (7 meals).

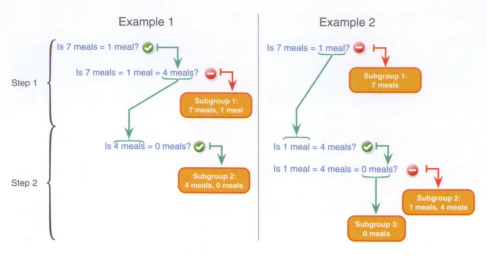

Figure 7.11 The non-parametric step-down procedure

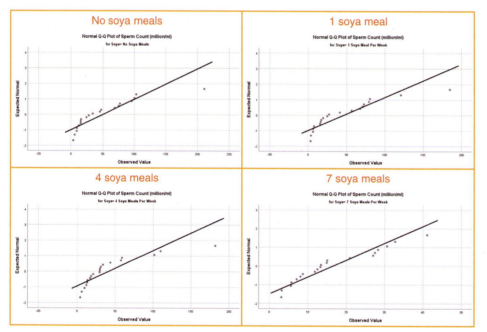

Figure 7.12 Normal Q-Q plots of sperm counts after different doses of soya meals per week

In Example 1 in Figure 7.11, we start with the first two groups in our ordered list (7 meals and 1 meal). They are equivalent (not significantly different) so we add in the third ordered group (4 meals). Doing so makes the groups significantly different (they are not equivalent), so we carry 4 meals into the second step, and conclude that 7 meals and 1 meal are equivalent (they are said to be *homogeneous groups*). In step 2, we compare 4 meals to the one remaining group (0 meals). These groups are equivalent (not significantly different), so we put them together in a different subgroup of 'homogeneous groups' and stop the process. In Example 2, we start with the first two groups in our ordered list (7 meals and 1 meal). They are significantly different (i.e. not equivalent) so we carry 1 meal into the second step, and conclude that 7 meals is a group on its own. In step 2, we compare 1 meal to 4 meals. They are equivalent (not significantly different) so we try to add in 0 meals, but doing so makes the groups significantly different (not equivalent), so we conclude that 4 meals and 1 meal are a homogeneous group, and (because there are no other groups to compare it with) remove 0 meals into a subgroup on its own. These follow-up procedures are quite complicated, so don't worry if you don't fully understand them—we will discuss these types of test in more detail later in the book.

7.6.3 Inputting data and provisional analysis

As a minimum, the data editor will have two columns. One column should be a coding variable (called something like **Soya**) containing four numeric codes that define the groups (for convenience I suggest 1 = no soya, 2 = one soya meal per week, 3 = four soya meals per week and 4 = seven soya meals per week). Remember to define values so that you know which group is represented by

median = 13.33), 4 meals (rank sum = 883, median = 29.43), 1 meal (rank sum = 883, median = 25.95), no meals (rank sum = 927, median = 30.96), resulting in the order 7 meals, 1 meal, 4 meals, and no meals. Figure 7.11 shows how the step-down process works. Step 1 is to see whether the first ordered group is the same as the second (i.e., is there a non-significant difference?). If they are equivalent you then put in the third ordered group and see if all three are equivalent. If they are, you put in the fourth group and see if all four are equivalent. If at any point you find a significant difference (i.e., the groups are not equivalent) then you stop, carry the group that you included last into the next step, and consider the groups you don't carry forward as a subset (i.e., they are equivalent). In step 2 you repeat the same process.

SELF TEST

See whether you can enter the data in Table 7.3 into SPSS (you don't need to enter the ranks). Then conduct some exploratory analyses on the data (see Sections 6.10 and 6.11).

Tests of Normality

	Number of Soya Meals Per Week	Kolmogorov–Smirnov[a]			Shapiro–Wilk		
		Statistic	df	Sig.	Statistic	df	Sig.
Sperm Count (Millions)	No Soya Meals	.181	20	.085	.805	20	.001
	1 Soya Meal Per Week	.207	20	.024	.826	20	.002
	4 Soya Meals Per Week	.267	20	.001	.743	20	.000
	7 Soya Meals Per Week	.204	20	.028	.912	20	.071

a. Lilliefors Significance Correction

Test of Homogeneity of Variance

		Levene Statistic	df1	df2	Sig.
Sperm Count (Millions)	Based on Mean	5.117	3	76	.003
	Based on Median	2.860	3	76	.042
	Based on Median and with adjusted df	2.860	3	58.107	.045
	Based on trimmed mean	4.070	3	76	.010

Output 7.5

which code (see Section 4.6.5). Another column should contain the sperm counts measured at the end of the year (call this variable **Sperm**). The data can be found in the file **Soya.sav**.

The results of our exploratory analysis are shown Figure 7.12 and Output 7.4. The normal Q-Q plots show quite clear deviations from normality for all four groups because the dots deviate from the diagonal line. We don't really need to do anything more than look at these graphs—the evidence of non-normality is plain to see, and formal tests can be problematic (see Jane Superbrain Box 6.5). However, given that within each group the sample is quite small ($n = 20$), if tests of normality are significant then this can be informative (because if the test has detected a deviation in such a small sample, then it's probably a

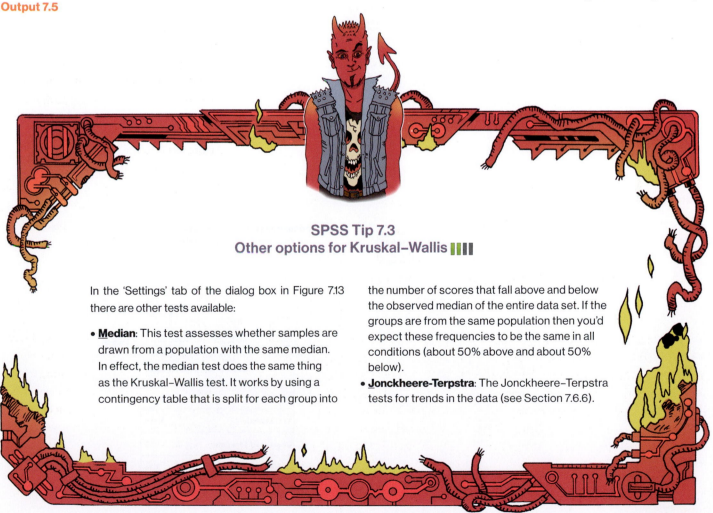

SPSS Tip 7.3
Other options for Kruskal–Wallis

In the 'Settings' tab of the dialog box in Figure 7.13 there are other tests available:

- **Median**: This test assesses whether samples are drawn from a population with the same median. In effect, the median test does the same thing as the Kruskal–Wallis test. It works by using a contingency table that is split for each group into the number of scores that fall above and below the observed median of the entire data set. If the groups are from the same population then you'd expect these frequencies to be the same in all conditions (about 50% above and about 50% below).

- **Jonckheere-Terpstra**: The Jonckheere–Terpstra tests for trends in the data (see Section 7.6.6).

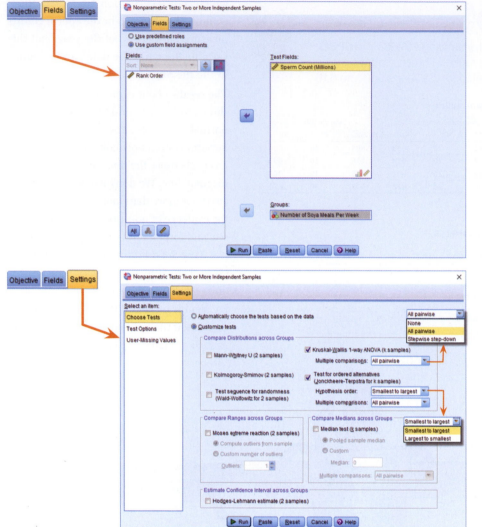

fairly substantial deviation). If you do these tests (Output 7.4) you'll find that the more accurate Shapiro–Wilk test is significant ($p < 0.05$) for all groups except the 7 meals (but even for that group it is close to being significant). Notwithstanding the pointlessness of Levene's test, its output would favor the conclusion that homogeneity of variance cannot be assumed, $F(3, 76) = 5.12$, $p = 0.003$, because the p-value is less than 0.05. As such, the information converges on a sad story: that data are probably not normally distributed, and the groups probably have heterogeneous variances.

7.6.4 Doing the Kruskal–Wallis test using SPSS Statistics

To run a Kruskal–Wallis test, follow the general procedure outlined in Section 7.3 by first selecting *Analyze* ▶ *Nonparametric Tests* ▶ ⚠ Independent Samples…. When you reach the Objective Fields Settings tab, if you need to assign variables then select

◉ Use custom field assignments and specify the model by dragging **Sperm Count** (**Millions**) into the box labeled *Test Fields* (or select it in the box labeled

Figure 7.13 Dialog boxes for the Kruskal–Wallis test

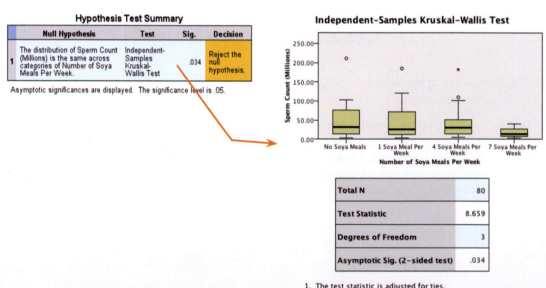

Hypothesis Test Summary

	Null Hypothesis	Test	Sig.	Decision
1	The distribution of Sperm Count (Millions) is the same across categories of Number of Soya Meals Per Week.	Independent-Samples Kruskal-Wallis Test	.034	Reject the null hypothesis.

Asymptotic significances are displayed. The significance level is .05.

Independent-Samples Kruskal-Wallis Test

Total N	80
Test Statistic	8.659
Degrees of Freedom	3
Asymptotic Sig. (2-sided test)	.034

1. The test statistic is adjusted for ties.

Output 7.6

Fields and click). Next, drag **Soya** to the box labeled *Groups* (Figure 7.13, top). Activate the Objective Fields Settings tab, select ⊙ Customize tests (SPSS Tip 7.3) and check ☑ Kruskal-Wallis 1-way ANOVA (k samples) (Figure 7.13, bottom). Next to this option there is a drop-down list labeled *Multiple comparisons*. Within this list are two options, which we discussed earlier: to compare every group against every other group (*All pairwise*) or to use a step-down method (*Stepwise step-down*). You can also ask for the Jonckheere–Terpstra trend test, which is useful to see whether the medians of the groups increase or decrease in a linear way. For the time being don't select this option, but we will look at this test in due course. To run the analysis click ▶ Run .

7.6.5 Output from the Kruskal–Wallis test ▮▮▮▮

Output 7.6 shows the summary table, which tells us the *p*-value of the test (0.034) and gives us a little message of advice telling us to reject the null hypothesis. Double-click this summary table to open up the model viewer, which contains the same summary table in the left pane, and a detailed output in the right pane. The detailed output shows a boxplot of the data, and a table containing the Kruskal–Wallis test statistic, *H* (8.659, the same value that we calculated earlier), its associated degrees of freedom (we had 4 groups, so the degrees of freedom are 4, 1 or 3) and the significance. The significance value is 0.034; because this value is less than 0.05, people would typically conclude that the amount of soya meals eaten per week significantly affects sperm counts.

As we discussed earlier, the overall effect tells us that sperm counts were different in some of the groups, but we don't know specifically which groups differed. The boxplots of the data (Output 7.6) can help here. The first thing to note is that there

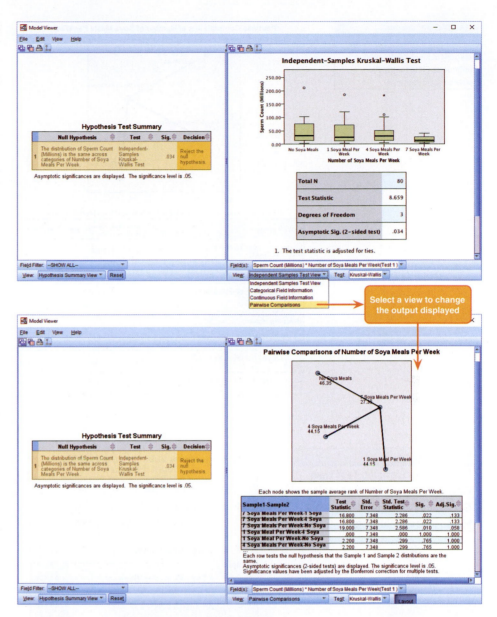

Figure 7.14   Changing the main output view to the pairwise comparisons view

are a few men who produced a particularly rampant quantity of sperm (the circles and asterisks that lie above the top whiskers). These are weirdos—I mean, outliers. Using the control as our baseline, the medians of the first three groups seem quite similar; however, the median of the 7 soya meals per week group seems a little lower; perhaps this is where the difference lies. We can find out using follow-up analyses like those we discussed in Section 7.6.2.

The output you see will depend on whether you selected *All pairwise* or

Stepwise step-down in the drop-down list labeled *Multiple comparisons* when you ran the analysis (Figure 7.13). In both cases, the output of these tests won't be immediately visible in the model viewer. The right-hand pane of the model viewer shows the main output by default (labeled the *Independent Samples Test View*), but we can change what is visible using the drop-down list labeled *View*. Clicking on this drop-down list reveals options including *Pairwise Comparisons* (if you selected *All pairwise* when you ran

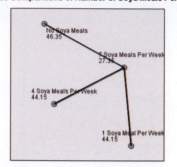

Pairwise Comparisons
- Independent Samples Test View
- Categorical Field Information
- Continuous Field Information
- **Pairwise Comparisons**

Pairwise Comparisons of Number of Soya Meals Per Week

No Soya Meals
46.35

7 Soya Meals Per Week
27.35

4 Soya Meals Per Week
44.15

1 Soya Meal Per Week
44.15

Each node shows the sample average rank of Number of Soya Meals Per Week.

Sample1-Sample2	Test Statistic	Std. Error	Std. Test Statistic	Sig.	Adj.Sig.
7 Soya Meals Per Week-1 Soya Meal Per Week	16.800	7.348	2.286	.022	.133
7 Soya Meals Per Week-4 Soya Meals Per Week	16.800	7.348	2.286	.022	.133
7 Soya Meals Per Week-No Soya Meals	19.000	7.348	2.586	.010	.058
1 Soya Meal Per Week-4 Soya Meals Per Week	.000	7.348	.000	1.000	1.000
1 Soya Meal Per Week-No Soya Meals	2.200	7.348	.299	.765	1.000
4 Soya Meals Per Week-No Soya Meals	2.200	7.348	.299	.765	1.000

Each row tests the null hypothesis that the Sample 1 and Sample 2 distributions are the same.
Asymptotic significances (2-sided tests) are displayed. The significance level is .05.

Output 7.7

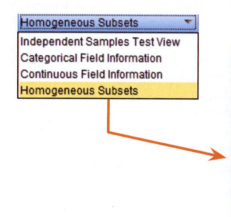

Homogeneous Subsets
- Independent Samples Test View
- Categorical Field Information
- Continuous Field Information
- **Homogeneous Subsets**

Homogeneous Subsets based on Sperm Count (Millions)

		Subset	
		1	2
Sample[1]	7 Soya Meals Per Week	27.350	
	1 Soya Meal Per Week		44.150
	4 Soya Meals Per Week		44.150
	No Soya Meals		46.350
Test Statistic		[2]	.118
Sig. (2-sided test)			.943
Adjusted Sig. (2-sided test)			.943

Homogeneous subsets are based on asymptotic significances. The significance level is .05.

[1]Each cell shows the sample average rank of Sperm Count (Millions).

[2]Unable to compute because the subset contains only one sample.

Output 7.8

the analysis) or *Homogenous Subsets* (if you selected *Stepwise step-down*). Selecting one of these will display its output in the right-hand pane of the

model viewer. To switch back to the main output use the same drop-down list to select *Independent Samples Test View* (Figure 7.14).

Let's look at the pairwise comparisons (Output 7.7). The diagram at the top shows the average rank within each group: so, for example, the average rank in the 7 meal group was 27.35, and for the no soya meals group was 46.35. This diagram highlights differences between groups using different-colored lines to connect them (in the current example, there are no significant differences between groups, so all the connecting lines are black). The table underneath shows the comparisons between all possible pairs of groups. In each case the test statistic is the difference between the mean ranks of those groups. For 7 vs. 1 soya meal, this will be $44.15 - 27.35 = 16.80$, for 0 vs. 4 soya meals it will be $46.35 - 44.15 = 2.20$, and so on. These test statistics are converted into z-scores (the column labeled *Std. Test Statistic*) by dividing by their standard errors, from which a p-value can be ascertained. For example, the 7 vs. 1 soya meal comparison has a z-score of 2.286 and the exact p-value for this z is 0.022. As I mentioned in Section 7.6.2, to control the Type I error rate we have to adjust the p-value for the number of tests we have done. The column labeled *Adj.Sig.* contains these adjusted p-values, and we interpret the values in this column (no matter how tempted we are to interpret the column labeled *Sig.*). Looking at the *Adj.Sig.* column, none of the values fall below the criterion of 0.05 (although the comparison between 7 soya meals and no soya meals comes fairly close with $p = 0.058$, and this reminds us that significance testing encourages black-and-white thinking and effect sizes might be useful).

To sum up, despite the significant overall effect, none of the specific comparisons between groups indicates a significant difference in sperm counts due to differing amounts of soya consumption. The effect we got seems to mainly reflect the fact that eating soya seven times per week lowers (I know this from the mean ranks) sperm counts compared to eating no soya, although even this comparison was just non-significant.

If you choose the *Stepwise step-down* procedure to follow up the Kruskal–Wallis test then you'll see the output in Output 7.8 (to see this output remember to select *Homogeneous Subsets* in the *View* drop-down menu, which will be listed only if you chose *Stepwise step-down* in Figure 7.13, top). The step-down method doesn't compare every group with every other group, which means that the *p*-values are not adjusted so strictly (because we're not doing as many significance tests on the same data). The output of the step-down procedure is a table that clusters equivalent (homogeneous) groups in the same columns (and color-codes them to make the differences clear). From column 1, we can see that the group that ate 7 soya meals a week clusters on its own. In other words, comparing it with the next highest-ranking group (the 1 soya meal group) produced a significant difference. Consequently, the 1 soya meal group was moved into a different subset in column 2 and was then compared to the next highest-ranking group (4 soya meals), which did not lead to a significant difference, so they were both compared to the no soya meals group, which also

produced no significant difference (think back to Figure 7.11). The fact that these three groups (1, 4 and no soya meals) are clustered within the same column (and have the same background color) tells us that they are equivalent (i.e., homogeneous). The *Adjusted Sig.* tells us that the *p*-value associated with comparing the 1, 4 and no soya meals groups was 0.943, which means not at all significant. In short, having 7 soya meals per week seemed to lower sperm counts significantly compared to all other groups, but all other doses of soya had no significant effect on sperm counts.

7.6.6 Testing for trends: The Jonckheere–Terpstra test

Back in Section 7.6.4 I mentioned the **Jonckheere–Terpstra test**,

☑ Test for ordered alternatives (Jonckheere-Terpstra for k samples) (Jonckheere, 1954; Terpstra, 1952), which tests for an ordered pattern to the medians of the groups you're comparing. It does the same thing as the Kruskal–Wallis test (i.e., test for a difference between the medians of the groups) but it incorporates information about whether the order of the groups is meaningful. As such, you should use this test when you

expect the groups you're comparing to produce a meaningful order of medians. In the current example we expect that the more soya a person eats, the lower their sperm count. The control group should have the highest sperm count, those having one soya meal per week should have a lower sperm count, the sperm count in the four meals per week group should be lower still, and the seven meals per week group should have the lowest sperm count. Therefore, there is an order to our medians: they should decrease across the groups.

Conversely, there might be situations where you expect your medians to increase. For example, there's a phenomenon in psychology known as the 'mere exposure effect': the more you're exposed to something, the more you'll like it. Back in the days when people paid money for music, record companies put this effect to use by making sure songs were played on radio for about 2 months prior to their release, so that by the day of release, everyone loved the song and rushed out to buy it.[8] Anyway, if you took three groups of people, exposed them to a song 10 times, 20 times and 30 times respectively and measured how much they liked the song, you'd expect the medians to

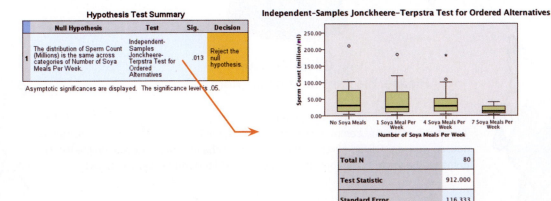

Output 7.9

---

8 For most chart music it had the opposite effect on me.

increase. Those who heard it 10 times would like it less than those who heard it 20 times who in turn would like it less than those who heard it 30 times. The Jonckheere–Terpstra test was designed for these situations, and there are two options (Figure 7.13):

- *Smallest to largest*: This option tests whether the first group differs from the second group, which in turn differs from the third group, which in turn differs from the fourth and so on until the last group.
- *Largest to smallest*: This option tests whether the last group differs from the group before, which in turn differs from the group before and so on until the first group.

In both cases the test looks at differences across ordered groups; it does not distinguish whether medians increase or decrease. The test determines whether the medians of the groups *ascend or descend* in the order specified by the coding variable; therefore, the coding variable must code groups *in the order that you expect the medians to change* (to reiterate, it doesn't matter whether you expect them to increase or decrease). For our soya example, we coded our groups as 1 = no soya, 2 = one soya meal per week, 3 = four soya meals per week and 4 = seven soya meals per week, so the Jonckheere–Terpstra test would assess whether the median sperm count increases or decreases across the groups when they're ordered in that way. If we wanted to test a different order then we'd have to specify a different coding variable that reflected this different order. Figure 7.13 shows how to specify the test, so rerun the analysis (as in Section 7.6.4) but select ☑ Test for ordered alternatives (Jonckheere-Terpstra for k samples) (*Smallest to largest*) instead of a Kruskal–Wallis test. Output 7.9 shows the output from the Jonckheere–Terpstra test for the soya data. Like the Kruskal–Wallis test, the viewer window will display only a summary table, which tells us the *p*-value of the test (0.013) and advises us to reject the null hypothesis. Double-click this table to show detailed results in the model viewer. The output tells us the value of test statistic, J, which is 912. In large samples (more than about eight per group) this test statistic can be converted to a *z*-score, which for these data is −2.476. As with any *z*-score, we can ascertain a *p*-value, which in this case is 0.013 and indicates a significant trend in the medians because it is lower than the typical critical value of 0.05. The sign of the *z*-value tells us the direction of the trend: a positive value indicates ascending medians (i.e., the medians get bigger as the values of the coding variable get bigger), whereas a negative value (as we have here) indicates descending medians (the medians get smaller as the value of the coding variable gets bigger). In this example, because we set the test option to be *Smallest to largest* (Figure 7.13) and we coded the variables as 1 = no soya, 2 = one soya meal per week, 3 = four soya meals per week and 4 = seven soya meals per week, the negative value of *z* means that the medians get smaller as we go from no soya, to one soya meal, to four soya meals and onto seven soya meals.[9]

Oliver Twisted
Please, Sir, can I have some more ... Jonck?

'I want to know how the Jonckheere–Terpstra Test works,' complains Oliver. Of course you do, Oliver, sleep is hard to come by these days. I am only too happy to oblige, my little syphilitic friend. The material for this chapter on the companion website explains how to do the test by hand. I bet you're glad you asked.

---

9   If you're bored, rerun the test but specify *Largest to smallest*. The results will be identical, except that the *z* will be 2.476 rather than −2.476. This positive value shows an ascending trend, rather than a descending one. This will happen because by selecting *Largest to smallest* we would be looking at the medians in the opposite direction (i.e., from 7 to 4 to 1 to 0 meals) compared to *Smallest to largest* (i.e., 0 to 1 to 4 to 7 meals).

7.6.7 Calculating an effect size

There isn't an easy way to convert a Kruskal–Wallis test statistic that has more than 1 degree of freedom to an effect size, r. You could use the significance value of the Kruskal–Wallis test statistic to find an associated value of z from a table of probability values for the normal distribution (like Table A.1 in the Appendix). From this you could use the conversion to r that we used in Section 7.4.5 (see equation (7.11)). However, this kind of effect size is rarely useful (because it's summarizing a general effect). In most cases it's more interesting to know the effect size for a focussed comparison (such as when comparing two things). For this reason, I'd suggest calculating

Table 7.4 Calculating effect sizes for pairwise comparisons

Comparison	z	$\sqrt{N}$	r
7 vs. 1 meal	2.286	6.32	0.362
7 vs. 4 meals	2.286	6.32	0.362
7 vs. no meals	2.586	6.32	0.409
1 vs. 4 meals	0.000	6.32	0.000
1 vs. no meals	0.299	6.32	0.047
4 vs. no meals	0.299	6.32	0.047

effect sizes for the pairwise tests we used to follow up the main analysis. Table 7.4 shows how you would do this for these data. For each comparison the z-score comes from the column labeled *Std. Test Statistic* in Output 7.7. Each comparison compared two groups of 20 people, so the total N for a given comparison is 40.

We use the square root of this value ($\sqrt{40} = 6.32$) to compute r, which is $z/\sqrt{N}$). We can see from the table that the effect sizes were medium to large for 7 meals compared to all other groups; despite the significance tests for these comparisons being non-significant, there seems to be something going on. All other

Cramming Sam's Tips

- The Kruskal–Wallis test compares several conditions when different participants take part in each condition and the resulting data have unusual cases or violate any assumption in Chapter 6.
- Look at the row labeled *Asymptotic Sig.* A value less than 0.05 is typically taken to mean that the groups are significantly different.
- Pairwise comparisons compare all possible pairs of groups with a *p*-value that is corrected so that the error rate across all tests remains at 5%.

- If you predict that the medians will increase or decrease across your groups in a specific order then test this with the Jonckheere–Terpstra test.
- Report the *H*-statistic, the degrees of freedom and the significance value for the main analysis. For any follow-up tests, report an effect size, the corresponding *z* and the significance value. Also report the medians and their corresponding ranges (or draw a boxplot).

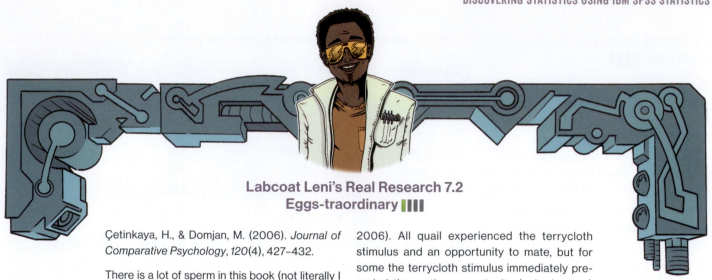

Labcoat Leni's Real Research 7.2
Eggs-traordinary ▮▮▮▮

Çetinkaya, H., & Domjan, M. (2006). *Journal of Comparative Psychology*, *120*(4), 427–432.

There is a lot of sperm in this book (not literally I hope, stats is not *that* exciting). We saw in Labcoat Leni's Real Research 7.1 that male quail fertilized more eggs if they had been trained to be able to predict when a mating opportunity would arise. Some quail develop fetishes. Really. In studies where a terrycloth object acts as a sign that a mate will shortly become available, some quail start to direct their sexual behavior towards the terrycloth object. (I will regret this analogy, but in human terms if every time you were going to have sex with your boyfriend you gave him a green towel a few moments before seducing him, then after enough seductions he would start getting really excited by green towels. If you're planning to dump your boyfriend a towel fetish could be an entertaining parting gift.)[10] In evolutionary terms, this fetishistic behavior seems counterproductive because sexual behavior becomes directed towards something that cannot provide reproductive success. However, perhaps this behavior serves to prepare the organism for the 'real' mating behavior.

Hakan Çetinkaya and Mike Domjan sexually conditioned male quail (Çetinkaya & Domjan, 2006). All quail experienced the terrycloth stimulus and an opportunity to mate, but for some the terrycloth stimulus immediately preceded the mating opportunity (paired group) whereas others experienced a 2-hour delay (this acted as a control group because the terrycloth stimulus did not predict a mating opportunity). In the paired group, quail were classified as fetishistic or not depending on whether they engaged in sexual behavior with the terrycloth object.

During a test trial the quail mated with a female and the researchers measured the percentage of eggs fertilized, the time spent near the terrycloth object, the latency to initiate copulation, and copulatory efficiency. If this fetishistic behavior provides an evolutionary advantage then we would expect the fetishistic quail to fertilize more eggs, initiate copulation faster and be more efficient in their copulations.

The data from this study are in the file **Çetinkaya & Domjan (2006).sav.** Labcoat Leni wants you to carry out a Kruskal–Wallis test to see whether fetishist quail produced a higher percentage of fertilized eggs and initiated sex more quickly.

Answers are on the companion website (or look at pages 429–430 in the original article).

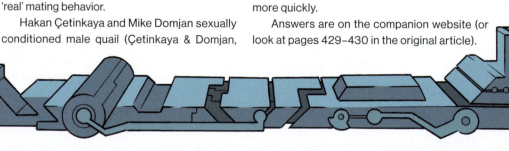

comparisons yield tiny effect sizes (less than $r = 0.1$).

We can calculate an effect size for the Jonckheere–Terpstra test using the same equation. Using the values of z (-2.476) and N (80) from Output 7.9, we get a medium effect size:

$$r_{J\text{-}T} = \frac{-2.476}{\sqrt{80}} = -0.28 \tag{7.27}$$

7.6.8 Writing and interpreting the results ▮▮▮▮

For the Kruskal–Wallis test, report the test statistic (which we saw earlier is denoted by *H*), its degrees of freedom and its

---

10 Green towels are just the beginning of where you could take this … go fill your boots. Or get him to.

significance. So, we could report something like:

✓ Sperm counts were significantly affected by eating soya meals, $H(3)= 8.66$, $p = 0.034$.

However, we need to report the follow-up tests as well (including their effect sizes):

✓ Sperm counts were significantly affected by eating soya meals, $H(3)= 8.66$, $p = 0.034$. Pairwise comparisons with adjusted p-values showed that there were no significant differences between sperm counts when people ate 7 soya meals per week compared to 4 meals ($p = 0.133$, $r = 0.36$), 1 meal ($p = 0.133$, $r = 0.36$) or no meals ($p = 0.058$, $r = 0.41$). There were also no significant differences in sperm counts between those eating 4 soya meals per week and those eating 1 meal ($p = 1.00$, $r = 0.00$) and no meals ($p = 1.00$, $r = 0.05$). Finally, there were no significant differences in sperm counts between those eating 1 soya meal per week and those eating none ($p = 1.00$, $r = 0.05$).

✓ Sperm counts were significantly affected by eating soya meals, $H(3)= 8.66$, $p = 0.034$. Step-down follow-up analysis showed that if soya is eaten every day it significantly reduces sperm counts compared to eating none; however, eating soya less than every day has no significant effect on sperm counts, $p = 0.943$ ('Phew!' says the vegetarian book author).

Alternatively, we might want to report our trend:

✓ Sperm counts were significantly affected by eating soya meals, $H(3)= 8.66$, $p = 0.034$. A Jonckheere–Terpstra test revealed that as more soya was eaten, the median sperm count significantly decreased, $\mathcal{J} = 912$, $z = -2.48$, $p = 0.013$, $r = -0.28$.

7.7 Differences between several related groups: Friedman's ANOVA ▌▌▌

The final test we'll look at is **Friedman's ANOVA** (Friedman, 1937), which tests

differences between three or more conditions when the scores across conditions are related (usually because the same entities have provided scores in all conditions). As with all the tests in this chapter, Friedman's ANOVA is used to counteract the presence of unusual cases or when one of the assumptions from Chapter 6 has been violated.

In the Western world we are brainwashed by the media into believing that stick-thin emaciated celebrity corpses are attractive. We all end up terribly depressed that we're not perfect because we don't have someone to photoshop out our imperfections, our lips are not slug-like enough, our teeth are not white enough, and we have jobs so we can't spend 8 hours a day at the gym working on a sixpack (not bitter at all . . .). Parasites exploit our vulnerability to make loads of money on diets and exercise regimes that will help us attain the body beautiful, which we think will fill the emotional void in our lives. Not wishing

to miss this great opportunity, I developed the Andikins diet.[11] The principle is to follow my exemplary (ahem) lifestyle: eat no meat, drink lots of Darjeeling tea, fill up on truckloads of lovely European cheese, fresh crusty bread, pasta, and chocolate at every available opportunity (especially when writing books), enjoy an occasional beer, play soccer and drums as much as humanly possible (preferably not simultaneously). To test the efficacy of my wonderful new diet, I took 10 people who thought that they needed to lose weight and put them on this diet for two months. Their weight was measured in kilograms at the start of the diet and then after one month and two months.

7.7.1 Theory of Friedman's ANOVA ▌▌▌

As with all of the tests in this chapter, Friedman's ANOVA works on ranked data. To begin with, we'll place the data for

SELF TEST

Have a go at ranking the data and see if you get the same results as in Table 7.5.

Table 7.5   Data for the diet example with ranks

	Weight			Weight		
	Start	Month 1	Month 2	Start (Ranks)	Month 1 (Ranks)	Month 2 (Ranks)
Person 1	63.75	65.38	81.34	1	2	3
Person 2	62.98	66.24	69.31	1	2	3
Person 3	65.98	67.70	77.89	1	2	3
Person 4	107.27	102.72	91.33	3	2	1
Person 5	66.58	69.45	72.87	1	2	3
Person 6	120.46	119.96	114.26	3	2	1
Person 7	62.01	66.09	68.01	1	2	3
Person 8	71.87	73.62	55.43	2	3	1
Person 9	83.01	75.81	71.63	3	2	1
Person 10	76.62	67.66	68.60	3	1	2
R_i				19	20	21

11 Not to be confused with the Atkins diet, obviously. ☺

different conditions into different columns (in this case there were three conditions, so we have three columns)—see Table 7.5. Each row represents the weight of a different person, each column represents their weight at a different point in time. Next, we rank the data *for each person*. So, we start with person 1, we look at their scores (in this case person 1 weighed 63.75 kg at the start, 65.38 kg after one month on the diet, and 81.34 kg after two months on the diet), and then we give the lowest one a rank of 1, the next highest a rank of 2 and so on (see Section 7.4.1 for more detail). When you've ranked the data for the first person, you move onto the next person, and starting at 1, rank their lowest score, then rank the next highest as 2, and so on. You do this for every row, and then add up the ranks for each condition (R_i, where i denotes the particular condition).

Once the sum of ranks has been calculated for each group, the test statistic, F_r, is calculated as follows:

$$F_r = \left[\frac{12}{Nk(k+1)} \sum_{i=1}^{k} R_i^2 \right] - 3N(k+1) \qquad (7.28)$$

In this equation, R_i is the sum of ranks for each group, N is the total sample size (in this case 10) and k is the number of conditions (in this case 3). The equation is similar to the one for the Kruskal–Wallis test (equation (7.25)). The middle part of the equation tells us to square the sum of ranks for each condition and add up these values. The rest of the equation calculates constants based on the total sample size and the number of conditions. For these data we get:

$$\begin{aligned} F_r &= \left[\frac{12}{(10 \times 3)(3+1)} \left(19^2 + 20^2 + 21^2 \right) \right] - (3 \times 10)(3+1) \\ &= \frac{12}{120}(361 + 400 + 441) - 120 \\ &= 0.1(1202) - 120 \\ &= 120.2 - 120 \\ &= 0.2 \end{aligned} \qquad (7.29)$$

When the number of people tested is greater than about 10, this test statistic, like the Kruskal–Wallis test in the previous section, has a chi-square distribution (see Chapter 19) with degrees of freedom that are one less than the number of groups (i.e., $k - 1$), in this case 2.

7.7.2 Inputting data and provisional analysis ▮▮▮▮

When the data are collected using the same participants in each condition, scores are entered in different columns. The data editor will have at least three columns of data. One column is for the data from the start of the diet (called something like **Start**), another will have values for the weights after one month (called **Month1**) and the final column will have the weights at the end of the diet (called **Month2**). The data can be found in the file **Diet.sav**.

Exploratory analysis is shown Figure 7.15 and in Output 7.10. The normal Q-Q plots show quite clear deviations from normality at all three time points because the dots deviate from the diagonal line. These graphs are evidence enough that our data are not normal, and because our sample size is small we can't rely on the central limit theorem to get us out of trouble. If you're keen on normality tests, then p-values less than 0.05 (or whatever threshold you choose) in these tests would support the belief of a lack of normality because the small sample size would mean that these tests would only have power to detect severe deviations from normal. (It's worth reminding you that non-significance in this context tells us nothing useful because our sample size is so small.) If you do these tests (Output 7.10), you'll find that the more accurate Shapiro–Wilk test is significant at the start of the diet ($p = 0.009$), after 1 month ($p = 0.001$), but not at the end of the diet ($p = 0.121$). The tests and Q-Q plots converge on a belief in non-normal data or unusual cases at all time points.

7.7.3 Doing Friedman's ANOVA using SPSS Statistics ▮▮▮▮

Follow the general procedure outlined in Section 7.3 by first selecting *Analyze* ▶ *Nonparametric Tests* ▶ 🔺 Related Samples...

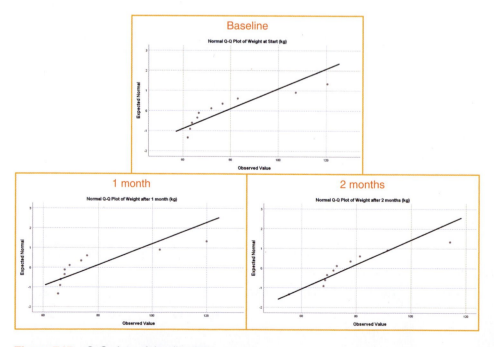

Figure 7.15   Q-Q plots of the diet data

When you reach the [Objective][Fields][Settings] tab, if you need to assign variables then select ⦿ Use custom field assignments and specify the model (Figure 7.16, top) by dragging **Start, Month1** and **Month2** into the box labeled *Test Fields* (or select them in the box labeled *Fields* and click [→]). Activate the [Objective][Fields][Settings] tab, select ⦿ Customize tests (see SPSS Tip 7.4) and check ☑ Friedman's 2-way ANOVA by ranks (k samples). Next to this option is a drop-down list labeled *Multiple comparisons* (Figure 7.16, bottom), just as there was for the Kruskal–Wallis test. Within this list are the two options we discussed earlier: compare every group against every other group (*All pairwise*) or use a step-down method (*Stepwise step-down*). To run the analysis click [▶ Run].

━SELF✂TEST━

Using what you know about inputting data, enter these data into SPSS and run exploratory analyses (see Chapter 6).

7.7.4 Output from Friedman's ANOVA ▮▮▮▮

The summary table (Output 7.11) tells us the *p*-value of the test (0.905) and advises us to retain the null hypothesis. Double-click this table to display more details in the model viewer window. As well as the summary table, we now see some histograms and a table containing the test statistic, F_r, for the Friedman test (0.2, which we calculated earlier), its degrees of freedom (in this case we had 3 groups so the they are $3-1$ or 2) and the associated *p*-value (significance). The significance value is 0.905, which is well above 0.05, and would typically lead to a conclusion that the weights didn't significantly change over the course of the diet.

The histograms in the output show the distribution of ranks across the three groups. It's clear that the mean rank changes very little over time: it is 1.90 (baseline), 2.00 (1 month) and 2.10 (2 months). This explains the lack of significance of the test statistic.

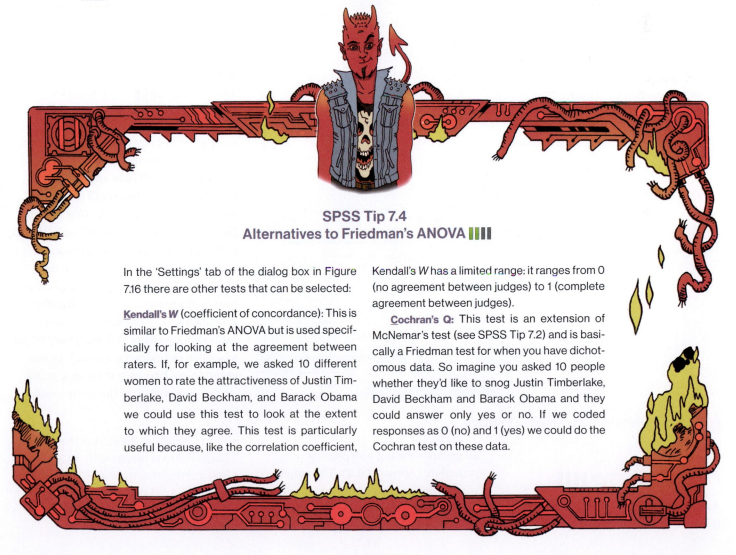

SPSS Tip 7.4
Alternatives to Friedman's ANOVA ▮▮▮▮

In the 'Settings' tab of the dialog box in Figure 7.16 there are other tests that can be selected:

Kendall's W (coefficient of concordance): This is similar to Friedman's ANOVA but is used specifically for looking at the agreement between raters. If, for example, we asked 10 different women to rate the attractiveness of Justin Timberlake, David Beckham, and Barack Obama we could use this test to look at the extent to which they agree. This test is particularly useful because, like the correlation coefficient, Kendall's *W* has a limited range: it ranges from 0 (no agreement between judges) to 1 (complete agreement between judges).

Cochran's Q: This test is an extension of McNemar's test (see SPSS Tip 7.2) and is basically a Friedman test for when you have dichotomous data. So imagine you asked 10 people whether they'd like to snog Justin Timberlake, David Beckham and Barack Obama and they could answer only yes or no. If we coded responses as 0 (no) and 1 (yes) we could do the Cochran test on these data.

Tests of Normality

	Kolmogorov–Smirnov[a]			Shapiro–Wilk		
	Statistic	df	Sig.	Statistic	df	Sig.
Weight at Start (kg)	.228	10	.149	.784	10	.009
Weight after 1 month (kg)	.335	10	.002	.685	10	.001
Weight after 2 months (kg)	.203	10	.200[*]	.877	10	.121

*. This is a lower bound of the true significance.

a. Lilliefors Significance Correction

Output 7.10

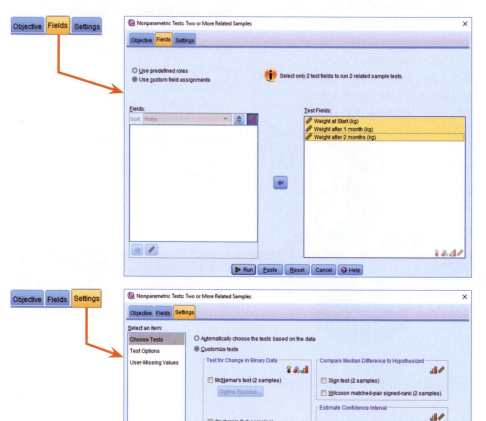

Figure 7.16   Dialog boxes for Friedman's ANOVA

7.7.5 Following up Friedman's ANOVA ▮▮▮▮

As with the Kruskal–Wallis test, we can follow up a Friedman test by comparing all groups or using a step-down procedure (Section 7.6.2). The output you see depends on whether you selected *All pairwise* or *Stepwise step-down* in the drop-down list labeled *Multiple comparisons* when you ran the analysis (Figure 7.16). As with the Kruskal–Wallis test, to see the output of the follow-up tests we use the drop-down menu labeled *View*. This drop-down list will include *Pairwise Comparisons* (if you selected *All pairwise* when you ran the analysis) or *Homogeneous Subsets* (if you selected *Stepwise step-down*). For the current data you won't see anything in the drop-down list because SPSS produces these tests only if the overall analysis is significant; because our overall analysis wasn't significant, we have no follow-up tests. This decision is sensible because it is logically dubious to want to unpick an effect that isn't significant in the first place. If you have data that yield a significant overall effect you would examine follow-up analyses in the same way as we did for the Kruskal–Wallis test.

7.7.6 Calculating an effect size ▮▮▮▮

It's most sensible (in my opinion) to calculate effect sizes for any comparisons you've done after the main test. In this example we didn't have any follow-up analyses because the overall effect was non-significant. However, effect sizes for these comparisons might still be useful so that people can see the magnitude of group differences. This is a dilemma because SPSS doesn't compute follow-up tests in the presence of a non-significant Friedman test. What we'd have to do instead is a series of Wilcoxon tests (from which we can extract a z-score). In this example, we have only three groups, so we can compare all of the groups with three tests:

- Test 1: Weight at the start of the diet compared to at one month.
- Test 2: Weight at the start of the diet compared to at two months.
- Test 3: Weight at one month compared to at two months.

Output 7.12 shows the three Wilcoxon signed-rank tests. As we saw in Section 7.5.5, it's straightforward to get an effect size r from the Wilcoxon signed-rank test. For the first comparison (start weight vs. 1 month), z is −0.051 (Output 7.12), and because this is based on comparing two conditions each containing 10 observations, we had 20 observations in total (remember it isn't important that the observations come from the same people). The effect size is tiny:

$$r_{\text{Start–1 Month}} = \frac{-0.051}{\sqrt{20}} = -0.01 \qquad (7.30)$$

For the second comparison (start weight vs. 2 months), z is −0.255 (Output 7.12) based on 20 observations, again yielding a tiny effect size:

$$r_{\text{Start–2 Months}} = \frac{-0.255}{\sqrt{20}} = -0.06 \qquad (7.31)$$

The final comparison (1 month vs. 2 months) had a z of −0.153 (Output 7.12) based on 20 observations. The effect size is again miniscule:

$$r_{\text{1 Month–2 Months}} = \frac{-0.153}{\sqrt{20}} = -0.03 \qquad (7.32)$$

Unsurprisingly, given the lack of significance of the Friedman test, these effect sizes are all very close to zero, indicating virtually non-existent effects.

7.7.7 Writing and interpreting the results

For Friedman's ANOVA we report the test statistic, denoted by χ^2_F, its degrees of freedom and its significance.[12] So, we could report something like:

✓ The weight of participants did not significantly change over the two months of the diet, $\chi^2(2) = 0.20$, $p = 0.91$.

Although with no significant initial analysis we wouldn't report follow-up tests for these data, in case you need to, you should write something like this:

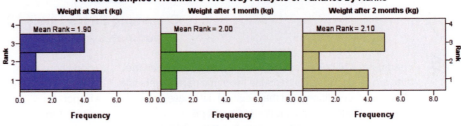

Hypothesis Test Summary

	Null Hypothesis	Test	Sig.	Decision
1	The distributions of Weight at Start (kg), Weight after 1 month (kg) and Weight after 2 months (kg) are the same.	Related-Samples Friedman's Two-Way Analysis of Variance by Ranks	.905	Retain the null hypothesis.

Asymptotic significances are displayed. The significance level is .05.

Related-Samples Friedman's Two-Way Analysis of Variance by Ranks

Total N	10
Test Statistic	.200
Degrees of Freedom	2
Asymptotic Sig. (2-sided test)	.905

1. Multiple comparisons are not performed because the overall test retained the null hypothesis of no differences.

Output 7.11

SELF TEST
Carry out the three Wilcoxon tests suggested above (see Figure 7.9).

Baseline - 1 Month	
Total N	10
Test Statistic	27.000
Standard Error	9.811
Standardized Test Statistic	-.051
Asymptotic Sig. (2-sided test)	.959

Baseline - 2 Months	
Total N	10
Test Statistic	25.000
Standard Error	9.811
Standardized Test Statistic	-.255
Asymptotic Sig. (2-sided test)	.799

1 Month - 2 Months	
Total N	10
Test Statistic	26.000
Standard Error	9.811
Standardized Test Statistic	-.153
Asymptotic Sig. (2-sided test)	.878

Output 7.12

✓ The weight of participants did not significantly change over the two months of the diet, $\chi^2(2) = 0.20$, $p = 0.91$. Wilcoxon tests were used to follow up this finding. It appeared that weight didn't significantly change from the start of the diet to one month, $T = 27$, $r = -0.01$, from the start of the diet to two months, $T = 25$, $r = -0.06$ or from one month to two months, $T = 26$, $r = -0.03$. We can conclude that the Andikins diet, like its creator, is a complete failure.

12 The test statistic is sometimes denoted without the F as χ^2.

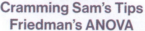

Cramming Sam's Tips
Friedman's ANOVA

- Friedman's ANOVA compares several conditions when the data are related (usually because the same participants take part in each condition) and the resulting data have unusual cases or violate any assumption in Chapter 6.

- Look at the row labeled *Asymptotic Sig.* If the value is less than 0.05 then typically people conclude that the conditions are significantly different.

- You can follow up the main analysis with pairwise comparisons. These tests compare all possible pairs of conditions using a *p*-value that is adjusted such that the overall Type I error rate remains at 5%.

- Report the χ^2 statistic, the degrees of freedom and the significance value for the main analysis. For any follow-up tests, report an effect size, the corresponding *z* and the significance value.

- Report the medians and their ranges (or draw a boxplot).

7.8 Brian and Jane's Story ▌▌▌▌

'Jane is an anomaly,' Brian thought, 'an awkward, mysterious, complicated set of observations that don't conform to my conventional assumptions.' Maybe that was why he couldn't shake her from his mind. She had a brilliant mind that compelled her to him. He'd never felt that comfortable around girls. He'd had a few relationships, but they all ended the same way: he'd get dumped for some more exciting guy. His dad said he was too nice, and maybe he was—he'd certainly had his fair share of being a doormat. He'd never lost hope—he dreamed of a Disney fairy tale, where the last person he expected turned out to be his soul mate. Maybe he was compensating for growing up without a

mum. He probably needed to get real. Maybe with Jane the writing was on the wall. He should forget about her and knuckle down to his studies. The trouble was, he got excited every time he understood something new about statistics, and Jane was the person with whom he wanted to share his excitement.

'The campus guy is an anomaly,' Jane thought, 'an awkward, mysterious, complicated set of observations that don't conform to my conventional assumptions.' Maybe that was why she couldn't shake him from her mind. She was sure that he had the mind of an idiot, but his efforts to impress her were sweet. He certainly didn't fit into her mental model of men. She'd got into the habit of being curt to men; it was what they deserved. She'd liked a guy, Josh, at school. She was crazy about him, but she was a nerd and he was one of the popular

kids. She was not on his radar. Not until he asked her to the cinema. She could hardly breathe when he asked. She bought clothes, spent hours getting ready, and became the stereotype that she'd fought against being. She arrived and waited. Most of her class were in the foyer. She hadn't expected that and it made her feel conspicuous and awkward. He was late and every minute increased her embarrassment and stress. She wanted to run. She should have run: it would have spared her. When he paraded confidently through the doors like a film star with Eliza Hamilton on his arm she knew instantly that she'd been set up. They turned to Jane, placed their fingers in an L on their foreheads, and mouthed 'loser', causing the foyer to erupt in laughter. Jane felt a rage that she hadn't known she was capable of, but she didn't run. She put her headphones in, calmly walked past Josh and smiled. It was a smile that wiped his own one from his face.

The funny thing was that Josh wasn't popular for long. First a bunch of emails leaked in which he tried to sell information about the school football team, then he posted highly personal information about Eliza all over social media. He became an outcast. He lost his place in the team, his grades plummeted and no one spoke to him. He became a total loser. He protested his innocence, but only a genius could have hacked his online life so comprehensively. Jane smiled at the memory. Campus guy seemed different. He was sweet . . . but not enough to let her guard down.

7.9 What next? ▮▮▮

'You promised us swans,' I hear you cry, 'and all we got was Kruskal this and Wilcoxon that. Where were the bloody swans?!' Well, the Queen owns them all so I wasn't allowed to have them. Nevertheless, this chapter did negotiate Dante's eighth circle of hell (Malebolge), where data of deliberate and knowing evil dwell. That is, data don't always behave themselves. Unlike the data in this chapter, my formative years at school were spent being very well behaved and uninteresting. However, a mischievous and rebellious streak was growing inside. Perhaps the earliest signs were my taste in music. Even from about the age of 3 music was my real passion: one of my earliest

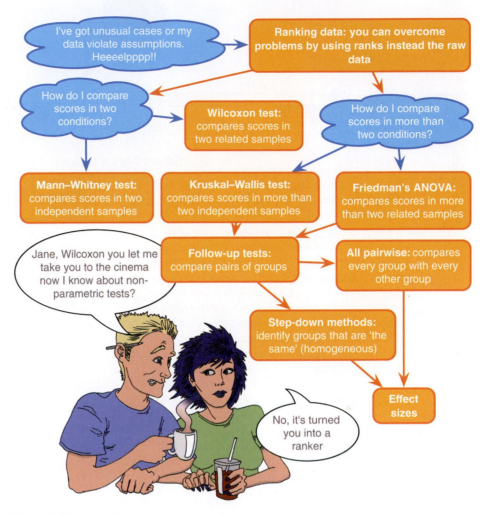

Figure 7.17   What Brian learnt from this chapter

memories is of listening to my dad's rock and soul records (back in the days of vinyl) while waiting for my older brother to come home from school. I still have a nostalgic obsession with vinyl. The first record I asked my parents to buy me was 'Take on the world' by Judas Priest, which I'd heard on *Top of the Pops* (a now defunct UK TV show) and liked. Watching the Priest on *Top of the Pops* is a very vivid memory—it had a huge impact. This record came out in 1978 when I was 5. Some people think that this sort of music corrupts young minds. Let's see if it did . . .

7.10 Key terms that I've discovered

Cochran's Q	Kruskal–Wallis test	Moses extreme reactions	Wald–Wolfowitz runs
Friedman's ANOVA	Mann–Whitney test	Non-parametric tests	Wilcoxon rank-sum test
Jonckheere–Terpstra test	McNemar's test	Pairwise comparisons	Wilcoxon signed-rank test
Kendall's W	Median test	Ranking	
Kolmogorov–Smirnov Z	Monte Carlo method	Sign test	

Smart Alex's tasks

- **Task 1**: A psychologist was interested in the cross-species differences between men and dogs. She observed a group of dogs and a group of men in a naturalistic setting (20 of each). She classified several behaviors as being dog-like (urinating against trees and lampposts, attempts to copulate with anything that moved, and attempts to lick their own genitals). For each man and dog she counted the number of dog-like behaviors displayed in a 24-hour period. It was hypothesized that dogs would display more dog-like behaviors than men. Analyze the data in **MenLikeDogs.sav** with a Mann–Whitney test.

- **Task 2**: Both Ozzy Osbourne and Judas Priest have been accused of putting backward masked messages on their albums that subliminally influence poor unsuspecting teenagers into doing things like blowing their heads off with shotguns. A psychologist was interested in whether backward masked messages could have an effect. He created a version of Britney Spears's 'Baby one more time' that contained the masked message 'deliver your soul to the dark lord' repeated in the chorus. He took this version, and the original, and played one version (randomly) to a group of 32 people. Six months later he played them whatever version they hadn't heard the time before. So each person heard both the original and the version with the masked message, but at different points in time. The psychologist measured the number of goats that were sacrificed in the week after listening to each version. Test the hypothesis that the backward message would lead to more goats being sacrificed using a Wilcoxon signed-rank test (**DarkLord.sav**).

- **Task 3**: A media researcher was interested in the effect of television programs on domestic life. She hypothesized that through 'learning by watching', certain programs encourage people to behave like the characters within them. She exposed 54 couples to three popular TV shows, after which the couple were left alone in the room for an hour. The experimenter measured the number of times the couple argued. Each couple viewed all TV shows but at different points in time (a week apart) and in a counterbalanced order. The TV shows were *EastEnders* (which portrays the lives of extremely miserable, argumentative, London folk who spend their lives assaulting each other, lying and cheating), *Friends* (which portrays unrealistically considerate and nice people who love each other oh so very much—but I love it anyway), and a

National Geographic program about whales (this was a control). Test the hypothesis with Friedman's ANOVA (**Eastenders.sav**).

- **Task 4**: A researcher was interested in preventing coulrophobia (fear of clowns) in children. She did an experiment in which different groups of children (15 in each) were exposed to positive information about clowns. The first group watched adverts in which Ronald McDonald is seen cavorting with children and singing about how they should love their mums. A second group was told a story about a clown who helped some children when they got lost in a forest (what a clown was doing in a forest remains a mystery). A third group was entertained by a real clown, who made balloon animals for the children. A final, control, group had nothing done to them at all. Children rated how much they liked clowns from 0 (not scared of clowns at all) to 5 (very scared of clowns). Use a Kruskal–Wallis test to see whether the interventions were successful (**coulrophobia.sav**).

- **Task 5**: Thinking back to Labcoat Leni's Real Research 4.1, test whether the number of offers was significantly different in people listening to Bon Scott compared to those listening to Brian Johnson (**Oxoby (2008) Offers.sav**). Compare your results to those reported by Oxoby (2008).

- **Task 6**: Repeat the analysis above, but using the minimum acceptable offer (**Oxoby (2008) MAO.sav**)—see Chapter 4, Task 3.

- **Task 7**: Using the data in **Shopping Exercise.sav** (Chapter 4, Task 4), test whether men and women spent significantly different amounts of time shopping.

- **Task 8**: Using the same data, test whether men and women walked significantly different distances while shopping.

- **Task 9**: Using the data in **Goat or Dog.sav** (Chapter 4, Task 5), test whether people married to goats and dogs differed significantly in their life satisfaction.

- **Task 10**: Use the **SPSSExam.sav** (Chapter 6, Task 2) data to test whether students at the Universities of Sussex and Duncetown differed significantly in their SPSS exam scores, their numeracy, their computer literacy, and the number of lectures attended.

- **Task 11**: Use the **DownloadFestival.sav** data from Chapter 6 to test whether hygiene levels changed significantly over the three days of the festival.

Answers & additional resources are available on the book's website at
https://edge.sagepub.com/field5e

CORRELATION

8

8.1 What will this chapter tell me?

When I was 8 years old, my parents bought me a guitar for Christmas. Even then, I'd desperately wanted to play the guitar for years. I could not contain my excitement at getting this gift (had it been an *electric* guitar I think I would actually have exploded with excitement). The guitar came with a 'learn to play' book, and after some time trying to play what was on page 1 of this book, I readied myself to unleash a riff of universe-crushing power on the world (well, 'Skip to my Lou'). But, I couldn't do it. I burst into tears and ran upstairs to hide.[1] My dad sat with me and said something like 'Don't worry, Andy, everything is hard to begin with, but the more you practice the easier it gets.' With his comforting words, my dad was inadvertently teaching me about the relationship or correlation, between two variables. These two variables could be related in three ways: (1) *positively related*, meaning that the more I practiced my guitar, the better a guitar player I would become (i.e., my dad was telling me the truth); (2) *not related* at all, meaning that as I practiced the guitar my playing ability would remain completely constant (i.e., my dad had fathered a cretin); or (3) *negatively related*, which would mean that the more I practiced the guitar the worse a guitar player I would become (i.e., my dad had fathered an indescribably strange child). This chapter looks first at how we can express the relationships between variables statistically by looking at two measures: *covariance* and the *correlation coefficient*. We discover how to carry out and interpret correlations in SPSS Statistics. The chapter ends by looking at more complex measures of relationships; in doing so, it acts as a precursor to the chapter on the linear model.

1 This is not a dissimilar reaction to the one I have when publishers ask me for new editions of statistics textbooks.

Figure 8.1 I don't have a photo from Christmas 1981, but this was taken about that time at my grandparents' house. I'm trying to play an E by the looks of it, no doubt because it's in 'Take on the world'.

8.2 Modeling relationships ▍▍▍▍

In Chapter 5 I stressed the importance of looking at your data graphically before fitting models to them. Our starting point with a correlation analysis is, therefore, to look at scatterplots of the variables we have measured. I am not going to repeat how to produce these graphs, but I urge you (if you haven't done so already) to read Section 5.8 before embarking on the rest of this chapter.

What is a correlation?

Way back in Chapter 2 we started talking about fitting models to your data that represent the hypothesis you're trying to test. In the previous chapter we looked at this process using a specific set of models that are useful when the data contain unusual cases or fail to meet the assumptions we discussed in Chapter 6. However, when these assumptions are met we can use the general linear model, which is an incredibly versatile and simple model that we've already encountered. I mentioned in Section 2.3 that everything in statistics boils down to one simple idea (expressed in equation (2.1)):

$$\text{outcome}_i = (\text{model}) + \text{error}_i \quad (8.1)$$

To recap, this equation means that the data we observe can be predicted from the model we choose to fit to the data plus some error in prediction. The 'model' in the equation varies depending on the design of your study, the type of data you have and what you're trying to achieve. If we want to model a relationship between variables then we're trying to predict an outcome variable from a predictor variable. Therefore, we need to include the predictor variable in the model. As we saw in equation (2.3),

we denote predictor variables with the letter X, so our model will be:

$$\text{outcome}_i = (b_1 X_i) + \text{error}_i \quad (8.2)$$

This equation means 'the outcome for an entity is predicted from their score on the predictor variable plus some error'. The model is described by a parameter, b_1, which in this context represents the relationship between the predictor variable (X) and the outcome. If we work with the raw data then to make predictions we need to know where the outcome is centered: in other words, what the value of the outcome is when the predictor is absent from the model (i.e., it is zero). This gives us a starting point for our prediction (i.e., if there were no predictor variable, what value of the outcome would we expect?). We add this value into the model as a constant, b_0, known as the intercept (we discuss this in detail in the next chapter). If we work with standardized scores (i.e., z-scores) then both the predictor and outcome have a mean of 0, so we already know the average value of the outcome when the predictor isn't in the model: it's 0. In other words, the intercept drops out of the model, leaving us with b_1:

$$z(\text{outcome})_i = b_1 z(X_i) + \text{error}_i \quad (8.3)$$

This equation means that the outcome expressed as z-scores can be predicted from the predictor variable (also expressed as z-scores) multiplied by b_1. When working with standardized scores like this, b_1 is called the Pearson product-moment correlation coefficient, and when we're not formally expressing the model as in equation (8.3), it is denoted by the letter r. Remember that we use the sample to estimate a value for b_1 (i.e., r) in the population, and its value quantifies the strength and direction of relationship between the predictor and outcome. How do we estimate this parameter? Like a quest for fire, we could search across the land . . . or we could use maths.

8.2.1 A detour into the murky world of covariance ▍▍▍▍

The simplest way to look at whether two variables are associated is to look at whether they *covary*. To understand what covariance is, we first need to think back to the concept of variance that we met in Chapter 1. Remember that the variance of a single variable represents the average amount that the data vary from the mean. Numerically, it is described by:

$$\text{variance}(s^2) = \frac{\sum_{i=1}^{n}(x_i - \bar{x})^2}{N-1} = \frac{\sum_{i=1}^{n}(x_i - \bar{x})(x_i - \bar{x})}{N-1} \quad (8.4)$$

The mean of the sample is represented by $\bar{x}$, x_i is the data point in question and N is the number of observations. If two variables are related, then changes in one variable should be met with similar changes in the other variable. Therefore, when one variable deviates from its mean we would expect the other variable to deviate from its mean in a similar way.

To illustrate what I mean, imagine we took five people and subjected them to a certain number of advertisements promoting toffee sweets, and then measured how many packets of those sweets each person bought during the next week. The data are in Table 8.1, as well as the mean and standard deviation (s) of each variable.

If there were a relationship between these two variables, then as one variable deviates from its mean, the other variable should deviate from its mean in the same or the directly opposite way. Figure 8.2 shows the data for each participant (orange circles represent the number of packets bought and blue circles represent the number of adverts watched); the orange line is the average number of packets bought and the blue line is the average number of adverts watched. The vertical lines represent the differences (remember that these differences are called *deviations* or

Table 8.1 Some data about toffees and adverts

Participant:	1	2	3	4	5	Mean	s
Adverts Watched	5	4	4	6	8	5.4	1.673
Packets Bought	8	9	10	13	15	11.0	2.915

residuals) between the observed values and the mean of the relevant variable. The first thing to notice about Figure 8.2 is that there is a similar pattern of deviations for both variables. For the first three participants the observed values are below the mean for both variables, for the last two people the observed values are above the mean for both variables. This pattern is indicative of a potential relationship between the two variables (because it seems that if a person's score is below the mean for one variable then their score for the other will also be below the mean).

So, how do we calculate the exact similarity between the patterns of differences of the two variables displayed in Figure 8.2? One possibility is to calculate the total amount of deviation, but we would have the same problem as in

the single-variable case: the positive and negative deviations would cancel out (see Section 1.8.5). Also, by adding the deviations, we would gain little insight into the *relationship* between the variables. In the single-variable case, we squared the deviations to eliminate the problem of positive and negative deviations cancelling each other out. When there are two variables, rather than squaring each deviation, we can multiply the deviation for one variable by the corresponding deviation for the second variable. If both deviations are positive or negative then this will give us a positive value (indicative of the deviations being in the same direction), but if one deviation is positive and one negative then the resulting product will be negative (indicative of the deviations being opposite

in direction). When we multiply the deviations of one variable by the corresponding deviations of a second variable, we get the **cross-product deviations**. As with the variance, if we want an average value of the combined deviations for the two variables, we divide by the number of observations (we actually divide by $N-1$ for reasons explained in Jane Superbrain Box 2.2). This averaged sum of combined deviations is known as the **covariance**. We can write the covariance in equation form:

$$\text{covariance}(x,y) = \frac{\sum_{i=1}^{n}(x_i-\bar{x})(y_i-\bar{y})}{N-1} \quad (8.5)$$

Notice that the equation is the same as the equation for variance (equation (1.7)), except that instead of squaring the deviances, we multiply them by the corresponding deviance of the second variable.

For the data in Table 8.1 and Figure 8.2 we get a value of 4.25:

$$
\begin{aligned}
\text{covariance}(x,y) &= \frac{\sum_{i=1}^{n}(x_i-\bar{x})(y_i-\bar{y})}{N-1}\\
&= \frac{\begin{array}{c}(-0.4)(-3)+(-1.4)(-2)+(-1.4)\\(-1)+(0.6)(2)+(2.6)(4)\end{array}}{N-1} \quad (8.6)\\
&= \frac{1.2+2.8+1.4+1.2+10.4}{4}\\
&= \frac{17}{4} = 4.25
\end{aligned}
$$

A positive covariance indicates that as one variable deviates from the mean, the other variable deviates in the same direction. On the other hand, a negative covariance indicates that as one variable deviates from the mean (e.g., increases), the other deviates from the mean in the opposite direction (e.g., decreases). However, the covariance depends upon the scales of measurement used: it is not a standardized measure. For example, if we use the data above and assume that they represented two variables measured in miles then the covariance is 4.25 square

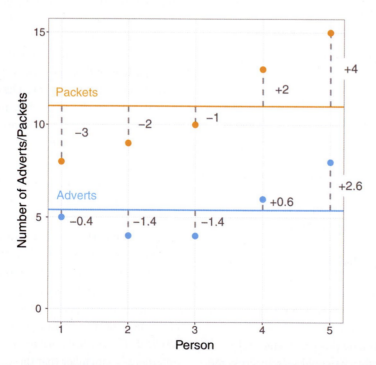

Figure 8.2 Graphical display of the differences between the observed data and the means of two variables

miles (as calculated above). If we convert these data into kilometers (by multiplying all values by 1.609) and calculate the covariance then we find that it increases to 11 square kilometers. This dependence on the scale of measurement is a problem because we cannot compare covariances in an objective way—we cannot say whether a covariance is particularly large or small relative to another data set unless both data sets were measured in the same units.

8.2.2 Standardization and the correlation coefficient

To overcome the problem of dependence on the measurement scale, we need to convert the covariance into a standard set of units. This process is known as **standardization**. We need a unit of measurement into which any variable can be converted, and typically we use the *standard deviation*. We came across this measure in Section 1.8.5 and saw that, like the variance, it is a measure of the average deviation from the mean. If we divide any distance from the mean by the standard deviation, it gives us that distance in standard deviation units. For example, for the data in Table 8.1, the standard deviation for the number of packets bought is approximately 3.0 (the exact value is 2.92). In Figure 8.2 we can see that the observed value for participant 1 was 3 packets less than the mean (so there was an error of −3 packets of sweets). If we divide this deviation, −3, by the standard deviation, which is approximately 3, then we get a value of −1. This tells us that the difference between participant 1's score and the mean was −1 standard deviation. In this way we can express the deviation from the mean for a participant in standard units by dividing the observed deviation by the standard deviation.

It follows from this logic that if we want to express the covariance in a standard unit of measurement we can divide by the standard deviation. However, there are two variables and, hence, two standard deviations. When we calculate the covariance we calculate two deviations (one for each variable) and multiply them. We do the same for the standard deviations: we multiply them and divide the covariance by the product of this multiplication. The standardized covariance is known as a *correlation coefficient* and is defined as follows:

$$r = \frac{\text{cov}_{xy}}{s_x s_y} = \frac{\sum_{i=1}^{n}(x_i - \bar{x})(y_i - \bar{y})}{(N-1)s_x s_y} \quad (8.7)$$

in which s_x is the standard deviation of the first variable and s_y is the standard deviation of the second variable (all other letters are the same as in the equation defining covariance). This coefficient, the *Pearson product-moment correlation coefficient* or **Pearson's correlation coefficient**, r, was invented by Karl Pearson with Florence Nightingale David[2]

doing a lot of the hard maths to derive distributions for it (see Figure 8.3 and Jane Superbrain Box 8.1).[3] If we look back at Table 8.1 we see that the standard deviation for the number of adverts watched (s_x) was 1.673, and for the number of packets of crisps bought (s_y) was 2.915. If we multiply these together we get $1.673 \times 2.915 = 4.877$. Now all we need to do is take the covariance, which we calculated a few pages ago as being 4.250, and divide by these multiplied standard deviations. This gives us $r = 4.250/4.877 = 0.871$.

By standardizing the covariance we end up with a value that has to lie between −1 and +1 (if you find a correlation coefficient less than −1 or more than +1 you can be sure that something has gone hideously wrong). We saw in Section 3.7.2 that a coefficient of +1 indicates that the two variables are perfectly positively correlated: as one variable increases, the other increases by a proportionate amount. This does not mean that the change in one variable *causes* the other to change, only that their changes coincide (Misconception Mutt 8.1). Conversely, a

Figure 8.3   Karl Pearson and Florence Nightingale David

---

2 Not to be confused with the Florence Nightingale in Chapter 5 who she was named after.

3 Pearson's product-moment correlation coefficient is denoted by r but, just to confuse us, when we square r (as in Section 8.4.2.2) an upper-case R is typically used.

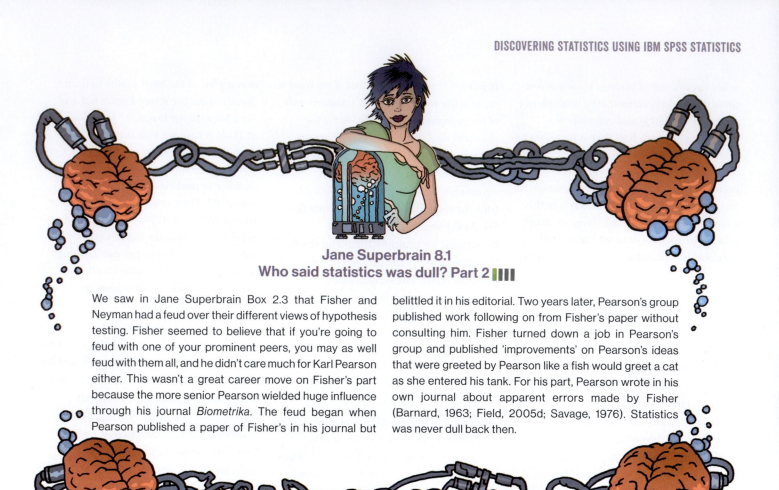

Jane Superbrain 8.1
Who said statistics was dull? Part 2 ▮▮▮▮

We saw in Jane Superbrain Box 2.3 that Fisher and Neyman had a feud over their different views of hypothesis testing. Fisher seemed to believe that if you're going to feud with one of your prominent peers, you may as well feud with them all, and he didn't care much for Karl Pearson either. This wasn't a great career move on Fisher's part because the more senior Pearson wielded huge influence through his journal *Biometrika*. The feud began when Pearson published a paper of Fisher's in his journal but

belittled it in his editorial. Two years later, Pearson's group published work following on from Fisher's paper without consulting him. Fisher turned down a job in Pearson's group and published 'improvements' on Pearson's ideas that were greeted by Pearson like a fish would greet a cat as she entered his tank. For his part, Pearson wrote in his own journal about apparent errors made by Fisher (Barnard, 1963; Field, 2005d; Savage, 1976). Statistics was never dull back then.

coefficient of −1 indicates a perfect negative relationship: if one variable increases, the other decreases by a proportionate amount. A coefficient of zero indicates no linear relationship at all and so as one variable changes, the other stays the same. We also saw that because the correlation coefficient is a standardized measure of an observed effect, it is a commonly used effect size measure and that values of ±0.1 represent a small effect, ±0.3 is a medium effect and ±0.5 is a large effect (although we should interpret the effect size within the context of the research literature and not use these canned effect sizes).

We have just described a **bivariate correlation**, which is a correlation between two variables. Later in the chapter

we'll look at variations on the correlation that adjust for one or more additional variables.

8.2.3 The significance of the correlation coefficient ▮▮▮▮

Although we can interpret the size of a correlation coefficient directly (Section 3.7.2), we have seen in Chapter 2 that scientists like to test hypotheses using probabilities. In the case of a correlation coefficient we can test the hypothesis that the correlation is different from zero (i.e., different from 'no relationship'). If we find that our observed coefficient was very unlikely to be (at least) as big as it would be if there were no effect in the population then we might gain confidence that the

relationship that we have observed is statistically meaningful.

There are two ways that we can go about testing this hypothesis. The first is to use the trusty z-scores that keep cropping up in this book. As we have seen, z-scores are useful because we know the probability of a given value of z occurring, if the distribution from which it comes is normal. There is one problem with Pearson's r, which is that it is known to have a sampling distribution that is not normally distributed. This would be a bit of a nuisance, except that thanks to our friend Fisher (1921) we can adjust r so that its sampling distribution *is* normal:

$$z_r = \frac{1}{2}\log_e\left(\frac{1+r}{1-r}\right)$$

(8.8)

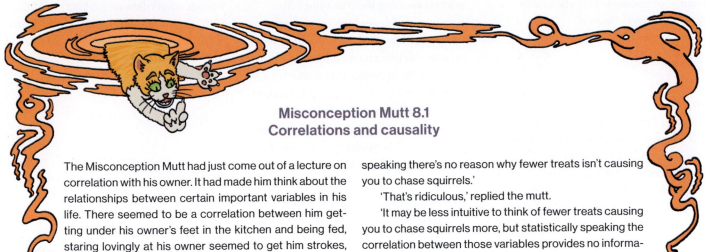

Misconception Mutt 8.1
Correlations and causality

The Misconception Mutt had just come out of a lecture on correlation with his owner. It had made him think about the relationships between certain important variables in his life. There seemed to be a correlation between him getting under his owner's feet in the kitchen and being fed, staring lovingly at his owner seemed to get him strokes, and walking nicely and not chasing squirrels got him treats.

As his mind wandered, he thought about how much he liked the campus. It was full of trees, grass and squirrels that he really wanted to chase if only it didn't stop the supply of treats. 'I really like treats,' he thought, 'and I really like chasing squirrels. But chasing squirrels is negatively correlated with treats, so if I chase the squirrel it will cause the number of treats to go down.'

He was hungry, so he started walking extra nicely. A squirrel darted up to him and jumped on his head as if trying to goad him into a chase. It started to grow, becoming heavier and more ginger until the grinning Correcting Cat stared down into the mutt's eyes from on top of his head.

'Correlation coefficients give no indication of the direction of causality,' he said to the dog. 'You might think that chasing squirrels causes fewer treats, but statistically speaking there's no reason why fewer treats isn't causing you to chase squirrels.'

'That's ridiculous,' replied the mutt.

'It may be less intuitive to think of fewer treats causing you to chase squirrels more, but statistically speaking the correlation between those variables provides no information about cause: it is purely a measure of the degree to which variables covary. Think back to the lecture: the correlation measures whether differences between scores on one variable and its mean correspond to differences between scores on a second variable and its mean. Causality does not feature in the computation.'

'Another issue,' the cat continued, 'is that there could be other measured or unmeasured variables affecting the two things that correlate. This is known as the third-variable problem or the *tertium quid* (Section 1.7.2). Perhaps the time of day affects both how many treats you get and how many squirrels you chase.'

Annoyingly, the cat had a point, so the mutt shook him from his head and, after watching his feline form shrink back to that of a squirrel, chased him across campus.

The resulting z_r has a standard error given by:

$$SE_{z_r} = \frac{1}{\sqrt{N-3}} \qquad (8.9)$$

For our advert example, our $r = 0.871$ becomes 1.337 with a standard error of 0.707.

We can then transform this adjusted r into a z-score in the usual way. If we want a z-score that represents the size of the correlation relative to a particular value, then we compute a z-score by subtracting the value that we want to test against and then dividing by the standard error. Normally we want to see whether the correlation is different from 0, in which case we subtract 0 from the observed value of z_r and divide by the standard error, which is the same as simply dividing z_r by its standard error:

$$z = \frac{z_r}{SE_{z_r}} \qquad (8.10)$$

For our advert data this gives us 1.337/0.707 = 1.891. We can look up this value of z (1.89) in the table for the normal distribution in the Appendix and get the one-tailed probability from the column labeled 'Smaller Portion' (think back to Section 1.8.6). In this case the value is 0.02938. To get the two-tailed probability we multiply this value by 2, which gives us 0.05876. As such the correlation is not significant, because $p > 0.05$.

In fact, the hypothesis that the correlation coefficient is different from 0 is usually (SPSS, for example, does this) tested not using a z-score, but using a different test statistic called a t-statistic with $N - 2$

degrees of freedom. This statistic can be obtained directly from r:

$$t_r = \frac{r\sqrt{N-2}}{\sqrt{1-r^2}} \qquad (8.11)$$

So you might wonder why I told you about z-scores. Partly it was to keep the discussion framed in concepts with which you are already familiar (we don't encounter the t-statistic properly for a few chapters), but also it is useful background information for the next section.

8.2.4 Confidence intervals for r

We saw in Chapter 2 that 95% confidence intervals tell us about the likely value (in this case of the correlation) in the population (assuming that your sample is one of the 95% for which the confidence

interval contains the true value). To compute confidence intervals for r, we take advantage of what we learnt in the previous section about converting r to z_r (to make the sampling distribution normal), and use the associated standard errors. By using z_r we can construct a confidence interval in the usual way. For example, a 95% confidence interval is calculated (see Eq. 2.15) as:

$$\text{lower boundary of confidence interval} = \bar{X} - (1.96 \times SE)$$
$$\text{upper boundary of confidence interval} = \bar{X} + (1.96 \times SE)$$
$$(8.12)$$

In the case of our transformed correlation coefficients, these equations become:

$$\text{lower boundary of confidence interval} = z_r - \left(1.96 \times SE_{z_r}\right)$$
$$\text{upper boundary of confidence interval} = z_r + \left(1.96 \times SE_{z_r}\right)$$
$$(8.13)$$

For our advert data we get $1.337 - (1.96 \times 0.707) = -0.049$, and $1.337 + (1.96 \times 0.707) = 2.723$. Remember that these values are in the z_r metric, but we can convert back to a correlation coefficient using:

$$r = \frac{e^{2z_r}-1}{e^{2z_r}+1} \qquad (8.14)$$

This gives us an upper bound of $r = 0.991$ and a lower bound of -0.049 (because this value is so close to zero the transformation to z has no impact).

I was moaning earlier on about how SPSS doesn't make tea for you. Another thing that it doesn't do is compute these confidence intervals for you, although there is a good macro available by Weaver & Koopman (2014). However, SPSS Statistics does something even better (than computing confidence intervals, not than making tea): it computes a bootstrap

Cramming Sam's Tips
Correlation

- A crude measure of the relationship between variables is the covariance.
- If we standardize this value we get Pearson's correlation coefficient, r.
- The correlation coefficient has to lie between −1 and +1.
- A coefficient of +1 indicates a perfect positive relationship, a coefficient of −1 indicates a perfect

negative relationship, and a coefficient of 0 indicates no linear relationship.
- The correlation coefficient is a commonly used measure of the size of an effect: values of ±0.1 represent a small effect, ±0.3 is a medium effect and ±0.5 is a large effect. However, interpret the size of correlation within the context of the research you've done rather than blindly following these benchmarks.

confidence interval. We learnt about the percentile bootstrap confidence interval in Section 6.12.3: it is a confidence interval that is derived from the actual data and, therefore, we know it will be accurate even when the sampling distribution of *r* is not normal. This is very good news indeed.

8.3 Data entry for correlation analysis ▎▎▎▎

When looking at relationships between variables, each variable is entered in a separate column in the data editor. So, for each variable you have measured, create a variable with an appropriate name, and enter a participant's scores across one row. If you have categorical variables (such as the participant's sex) these can also be entered in a column (remember to define appropriate value labels). For example, if we wanted to calculate the correlation between the two variables in Table 8.1 we would enter these data as in Figure 8.4; each variable is entered in a separate column, and each row represents a single individual's data (so the first consumer saw 5 adverts and bought 8 packets).

8.4 Bivariate correlation ▎▎▎▎

Figure 8.5 shows a general procedure to follow when computing a bivariate correlation coefficient. First, check for sources of bias as outlined in Chapter 6. The two most important ones in this context are linearity and normality. Remember that we're fitting a linear model to the data, so if the relationship between variables is not linear then this model is invalid (a transformation might help to make the relationship linear). To meet this requirement, the outcome variable needs to be measured at the interval or ratio level (see Section 1.6.2), as does the predictor variable (one exception is that a predictor variable can be a categorical variable with only two

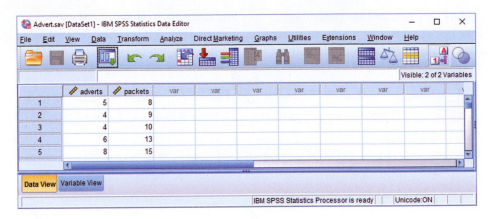

Figure 8.4   Data entry for correlation

Enter the advert data and use the chart editor to produce a scatterplot (number of packets bought on the *y*-axis, and adverts watched on the *x*-axis) of the data.

categories—we'll get onto this in Section 8.4.5). As far as normality is concerned, we care about this only if we want confidence intervals or significance tests and if the sample size is small (Section 6.6.1).

If the data have outliers, are not normal (and the sample is small) or your variables are measured at the ordinal level then you can use Spearman's rho (Section 8.4.3) or Kendall's tau (Section 8.4.4), which are versions of the correlation coefficient applied to ranked data (just like the tests in the previous chapter). Ranking the data reduces the impact of outliers. Furthermore, given that normality matters

only for inferring significance and computing confidence intervals, we could use a bootstrap to compute the confidence interval, then we don't need to worry about the distribution.

In Chapter 5 we looked at an example relating to exam anxiety: a psychologist was interested in the effects of exam stress and revision on exam performance. She had devised and validated a questionnaire to assess state anxiety relating to exams (called the Exam Anxiety Questionnaire or EAQ). This scale produced a measure of anxiety scored out of 100. Anxiety was measured

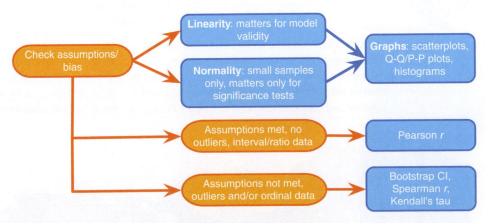

Figure 8.5   The general process for conducting correlation analysis

before an exam, and the percentage mark of each student on the exam was used to assess the exam performance.

She also measured the number of hours spent revising. These data are in **Exam Anxiety.sav**. We have already created scatterplots for these data (Section 5.8) so we don't need to do that again; however, we will look at the distributions of the three main variables.

It's clear from the P-P plots in Figure 8.6 that exam performance is the most normally distributed of the variables (the dots hover close to the line), but for

SELF TEST
Create P-P plots of the variables **Revise**, **Exam**, and **Anxiety**.

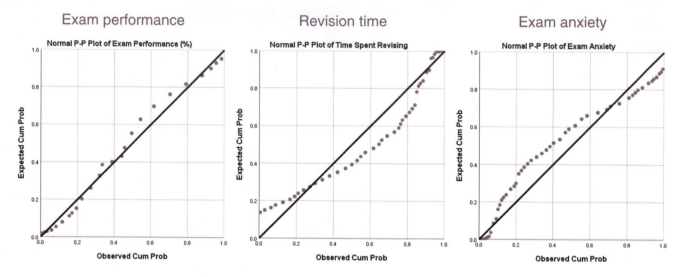

Figure 8.6 P-P plots for the exam anxiety variables

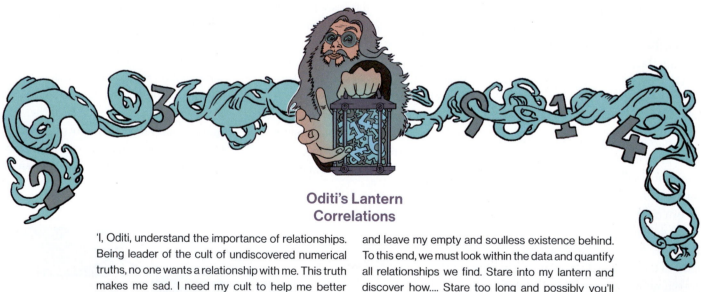

Oditi's Lantern
Correlations

'I, Oditi, understand the importance of relationships. Being leader of the cult of undiscovered numerical truths, no one wants a relationship with me. This truth makes me sad. I need my cult to help me better understand relationships so that I might have one and leave my empty and soulless existence behind. To this end, we must look within the data and quantify all relationships we find. Stare into my lantern and discover how.... Stare too long and possibly you'll never have another relationship.'

exam revision and exam anxiety there is evidence of skew (the dots snake around the diagonal line). This skew is a problem if we want to do significance tests or look at confidence intervals. The sample contains 103 observations, which is reasonably large, and possibly large enough for the central limit theorem to relieve us of concerns about normality. However, it would be advisable to use a bootstrap to get robust confidence intervals. We might also consider using a rank-based method to compute the correlation coefficient itself.

8.4.1 General procedure for correlations using SPSS Statistics ▮▮▮▮

You conduct bivariate correlation using the dialog box accessed through *Analyze* ▸ *Correlate* ▸ 🔲 *Bivariate...*. In this dialog box (Figure 8.7) the variables in the data editor are listed on the left-hand side and there is an empty box labeled *Variables* on the right-hand side. You can select any variables from the list using the mouse and transfer them to the *Variables* box by dragging them there or clicking 🔲. SPSS creates a table (called a *correlation matrix*) of correlation coefficients for all combinations of specified variables. For our current example, select the variables **Exam performance**, **Exam anxiety** and **Time spent revising** and transfer them to the *Variables* box. Having selected the variables of interest, you can choose between three correlation coefficients: the default is Pearson's product-moment correlation coefficient (☑ Pearson), but you can also select Spearman's rho (☑ Spearman) and Kendall's tau (☑ Kendall's tau-b)—we'll explore the differences in due course. You can choose more than one of these correlation coefficients if you like.

You can also specify whether the test is one- or two-tailed. In Section 2.9.5 I advised against one-tailed tests, so I would leave the default of ⦿ Two-tailed, but if you

don't like my advice then you might select ⦿ One-tailed if your hypothesis is directional (e.g., 'The more anxious someone is about an exam, the worse their mark will be') and ⦿ Two-tailed if it is non-directional (i.e., 'I'm not sure whether exam anxiety will improve or reduce exam marks').

Clicking [Style...] opens the dialog box in Figure 8.8, which allows you to apply formatting to the table of correlations in the output. The drop-down menu in the column called *Value* is for selecting types of cells in the output: you could select the cells containing the correlation coefficients (as I have in the figure), the cells containing the sample sizes (*N*), the means, the significance values or all the cells in the table. Clicking in the column labeled *Condition* opens a dialog box for setting a condition for formatting. In the figure I have set a condition of formatting cells that contain an absolute value greater than or equal to 0.5. Doing so will apply the formatting only to cells

containing correlation coefficients greater than 0.5 or smaller than −0.5. Clicking in the column labeled *Format* opens a dialog box for defining what formatting you wish to apply. In the figure I have chosen to change the background color of the cells to yellow, but you can do other things such as change the text color or make it bold or italic and so on. By using the settings in Figure 8.8, I have (1) chosen to format only the cells containing the correlation coefficients; and (2) asked that cells containing a value greater than 0.5 or smaller than −0.5 have a yellow background. The effect will be that these cells are highlighted in the output: I will very quickly be able to see which variables were strongly correlated because their cells will be highlighted. Not that I recommend this (because it's the sort of thing that will encourage you to blindly apply the *p* < 0.05 criterion), but I imagine that some people use this facility to highlight cells with significance

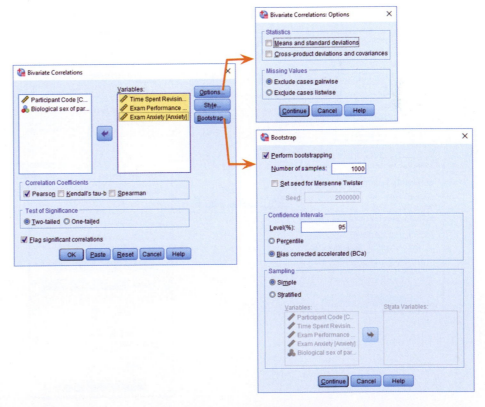

Figure 8.7 Dialog box for conducting a bivariate correlation

values less than 0.05. You know better than to do that though.

You can set several formatting rules by clicking [Add] to create a new rule (it will appear as a new row in the dialog box) and then editing the rule. For example, we could set up four rules to use different background colors for cells containing correlations based on whether they are tiny (absolute values between 0 and 0.1), small to medium (absolute values between 0.1 and 0.3), medium to large (absolute values between 0.3 and 0.5) or large (absolute values greater than 0.5). These four rules would be quite similar so once we have set the first rule we could save time by selecting it and clicking [Duplicate] to create a new rule that duplicates its settings (which we could then tweak). You can delete a rule by selecting the row representing the rule and clicking [Delete].

Going back to the main dialog box (Figure 8.7), clicking [Options...] opens a dialog

box with two *Statistics* options and two options for missing values.

The *Statistics* options are enabled only when Pearson's correlation is selected (otherwise they appear 'greyed out').

These two options are meaningful only for interval data, which the Pearson correlation requires, so it makes sense that these options are deactivated if the Pearson correlation has not been selected. Selecting the tick-box labeled *Means and standard deviations* produces the mean and standard deviation of the variables selected for analysis. Selecting the tick-box labeled *Cross-product deviations and covariances* produces the values of these statistics for the variables in the analysis. The cross-product deviations are the values of the numerator (top half) of equation (8.5). The covariances between variables are what you would get from applying equation (8.5) to your variables. In other words, the covariance values are the cross-product deviations divided by

$N - 1$ and represent the unstandardized correlation coefficient. In most instances you won't need these options, but they occasionally come in handy (see Oliver Twisted). At this point we need to decide how to deal with missing values (look back to SPSS Tip 6.1).

Finally, to get bootstrapped confidence intervals for the correlation coefficient click [Bootstrap...]. We discussed this dialog box in Section 6.12.3; to recap, select ☑ Perform bootstrapping to activate bootstrapping for the correlation coefficient, and to get a 95% confidence interval click ⦿ Percentile or ⦿ Bias corrected accelerated (BCa). For this analysis, let's ask for a bias corrected (BCa) confidence interval.

8.4.2 Pearson's correlation coefficient using SPSS Statistics

To obtain Pearson's correlation coefficient, run through the general procedure that we've just looked at (Section 8.4.1), selecting ☑ Pearson (the default) and then clicking [OK] (Figure 8.7). In the viewer we'll see a matrix of results (Output 8.1), which is not as bewildering as it looks. For one thing, the information in the top part of the table is the same as in the bottom half (which I have faded out), so we can ignore half of the table. The first row tells us about time spent revising. This row is subdivided into other rows, the first of which contains the correlation coefficients with the exam performance ($r = 0.397$) and exam anxiety ($r = -0.709$). The second major row in the table tells us about exam performance, and from this part of the table we get the correlation coefficient for its relationship with exam anxiety, $r = -0.441$. Directly underneath each correlation coefficient we're told the significance value of the correlation and the sample size (N) on which it is based. The significance values are all less than 0.001 (as indicated by the double asterisk after the coefficient).

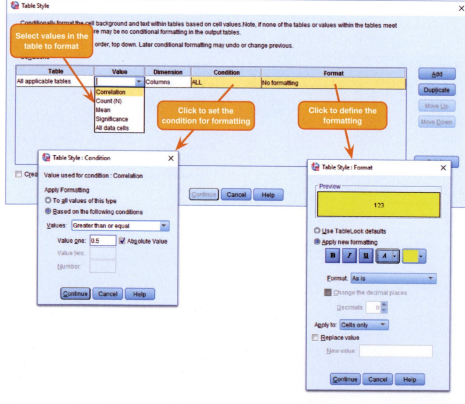

Figure 8.8 The *Table Style* dialog box

Oliver Twisted: Please, Sir, can I have some more ... options?

Oliver is so excited to get onto analyzing his data that he doesn't want me to spend pages waffling on about options that you will probably never use. 'Stop writing, you waffling fool,' he says. 'I want to analyze my data.' Well, he's got a point. If you want to find out more about what the  do in correlation, then look at the companion website.

This significance value tells us that the probability of getting a correlation coefficient at least this big in a sample of 103 people if the null hypothesis were true (there was no relationship between these variables) is very low (close to zero in fact). All the significance values are below the standard criterion of 0.05, indicating a 'statistically significant' relationship.

Given the lack of normality in some of the variables, we should be more concerned with the bootstrapped confidence intervals than the significance *per se*: this is because the bootstrap confidence intervals will be unaffected by the distribution of scores, whereas the significance value might be. These confidence intervals are labeled *BCa 95% Confidence Interval*, and you're given two values: the upper boundary and the lower boundary. For the relationship between revision time and exam performance the interval is 0.245 to 0.524, for revision time and exam anxiety it is −0.863 to −0.492, and for exam anxiety and exam performance it is −0.564 to −0.301.

There are two important points here. First, because the confidence intervals are derived empirically using a random sampling

procedure (i.e., bootstrapping) the results will be slightly different each time you run the analysis. The confidence intervals you get won't be the same as in Output 8.1, and that's normal and nothing to worry about. Second, think about what a correlation of zero represents: it is no effect whatsoever. A confidence interval is the boundary between

		Correlation coefficient, r
		Significance, *p*-value
		Confidence intervals

Correlations

			Time Spent Revising	Exam Performance (%)	Exam Anxiety
Time Spent Revising	Pearson Correlation		1	.397**	−.709**
	Sig. (2–tailed)			.000	.000
	N		103	103	103
	Bootstrap[c]	Bias	0	−.002	−.004
		Std. Error	0	.070	.112
	BCa 95% Confidence Interval	Lower	.	.245	−.863
		Upper	.	.524	−.492
Exam Performance (%)	Pearson Correlation		.397**	1	−.441**
	Sig. (2–tailed)		.000		.000
	N		103	103	103
	Bootstrap[c]	Bias	−.002	0	.004
		Std. Error	.070	0	.065
	BCa 95% Confidence Interval	Lower	.245	.	−.564
		Upper	.524	.	−.301
Exam Anxiety	Pearson Correlation		−.709**	−.441**	1
	Sig. (2–tailed)		.000	.000	
	N		103	103	103
	Bootstrap[c]	Bias	−.004	.004	0
		Std. Error	.112	.065	0
	BCa 95% Confidence Interval	Lower	−.863	−.564	.
		Upper	−.492	−.301	.

**. Correlation is significant at the 0.01 level (2–tailed).

c. Unless otherwise noted, bootstrap results are based on 1000 bootstrap samples

Duplicate information

Output 8.1

which the population value falls in 95% of samples. If we assume that our sample is one of the 95% of samples that produces a confidence interval containing the true value of the correlation (and be aware that this assumption might be wrong), then if the interval crosses zero it means that (1) the population value could be zero (i.e., no effect at all), and (2) we can't be sure if the true relationship is positive or negative because the population value could plausibly be a negative or positive value. None of our confidence intervals cross zero, therefore we might take this information to mean that there is a genuine effect in the population. In psychological terms, this means that there is a complex interrelationship between the three variables: (1) as anxiety about an exam increases, the percentage mark obtained in that exam significantly decreases; (2) as the amount of time revising increases, the percentage obtained in the exam significantly increases; and (3) as revision time increases, the student's anxiety about the exam significantly decreases.

What is the coefficient of determination?

Although we cannot make direct conclusions about causality from a correlation coefficient (Misconception Mutt 8.1), we can take it a step further by squaring it. The correlation coefficient squared (known as the **coefficient of determination**, R^2) is a measure of the amount of variability in one variable that is shared by the other. Exam performance in our data varied, reflecting the fact that people's scores were not identical; they will have varied for any number of reasons (different ability, different levels of preparation and so on). If we add up this individual variability we get an estimate of how much variability in exam performances there was in total (this is computed with equation (1.7) from Section 1.8.5). It turns out that the variance is about 672 for the exam performance scores. Imagine that this value is the surface area of a biscuit. The size of the biscuit represents 100% of the variance in exam scores. Imagine that exam anxiety comes along and takes a bite out of the biscuit. R^2 tells us how much of this exam performance 'biscuit' has been eaten by exam anxiety. In other words, it tells us the proportion of variance (what proportion of the 672 units of variation) overlaps with exam anxiety. These two variables had a correlation of −0.4410 and so the value of R^2 is $−0.4410^2 = 0.194$, which means that 0.194 of the variability in exam performance is shared by exam anxiety. It's a bit easier to think of this value as a percentage rather than a proportion, which we can obtain by multiplying by 100. In this example, then, exam anxiety shares 19.4% of the variability in exam performance. To put this value into perspective, this leaves 80.6% of the variability unexplained.

You'll often see people write things about R^2 that imply causality: they might write 'the variance in *y accounted for* by *x*' or 'the variation in one variable *explained* by the other'. Although R^2 is a useful measure of the substantive importance of an effect, it cannot be used to infer causal relationships. Exam anxiety might well share 19.4% of the variation in exam scores, but it does not necessarily cause this variation.

8.4.3 Spearman's correlation coefficient ▊▊▊▊

Spearman's correlation coefficient, denoted by r_s (Figure 8.9), is a non-parametric statistic that is useful to minimize the effects

What if my data violate the assumptions of the linear model?

of extreme scores or the effects of violations of the assumptions discussed in Chapter 6. You'll sometimes hear the test referred to as Spearman's rho (pronounced 'row', as in 'row your boat gently down the stream').

Figure 8.9 Charles Spearman, ranking furiously

Spearman's test works by first ranking the data (see Section 7.4.1), and then applying Pearson's equation (equation (8.7)) to those ranks (Spearman, 1910).

I was born in England, which has some bizarre traditions. One such oddity is the World's Biggest Liar competition held annually at the Santon Bridge Inn in Wasdale (in the Lake District). The contest honors a local publican, 'Auld Will Ritson', who in the nineteenth century was famous in the area for his far-fetched stories (one such tale being that Wasdale turnips were big enough to be hollowed out and used as garden sheds). Each year locals are encouraged to attempt to tell the biggest lie in the world (lawyers and politicians are apparently banned from the competition). Over the years there have been tales of mermaid farms, giant moles, and farting sheep blowing holes in the ozone layer. (I am thinking of entering next year and reading out some sections of this book.)

Imagine I wanted to test a theory that more creative people will be able to create taller tales. I gathered together 68 past contestants from this competition and noted where they were placed in the competition (first, second, third, etc.); I also gave them a creativity questionnaire (maximum score 60). The position in the competition is an ordinal variable (see Section 1.6.2) because the places are categories but have a meaningful order (first place is better than second place and so on). Therefore, Spearman's correlation coefficient should be used (Pearson's *r* requires interval or ratio data). The data for this study are in the file **The Biggest Liar.sav**. The data are in two columns: one labeled **Creativity** and one labeled **Position** (there's a third variable in there, but we will ignore it for the time being). For the **Position** variable, each of the categories described above has been coded with a numerical value. First place has been coded with the value 1, with other positions being labeled 2, 3 and so on. Note that for each numeric code I have provided a value label (just like we did for coding variables). I have also set the *Measure* property of this variable to Ordinal.

The procedure for doing a Spearman correlation is the same as for a Pearson correlation, except that in the *Bivariate Correlations* dialog box (Figure 8.7) we need to select ☑ Spearman and deselect the option for a Pearson correlation. As with the Pearson correlation, we should use the Bootstrap... option to get some robust confidence intervals.

The output for a Spearman correlation (Output 8.2) is like that of the Pearson correlation, giving the correlation coefficient between the two variables (−0.373), the significance value of this coefficient (0.002) and the sample size (68).[4] We also have the BCa 95% confidence interval, ranging from −0.602 to −0.122.[5] The fact that the confidence interval does not cross zero (and the significance is less than 0.05) tells us that

Correlations

				Creativity	Position in Biggest Liar Competition
Spearman's rho	Creativity	Correlation Coefficient		1.000	−.373**
		Sig. (2−tailed)		.	.002
		N		68	68
		Bootstrap[c]	Bias	.000	.001
			Std. Error	.000	.125
			BCa 95% Confidence Interval Lower	.	−.602
			Upper	.	−.122
	Position in Biggest Liar Competition	Correlation Coefficient		−.373**	1.000
		Sig. (2−tailed)		.002	.
		N		68	68
		Bootstrap[c]	Bias	.001	.000
			Std. Error	.125	.000
			BCa 95% Confidence Interval Lower	−.602	.
			Upper	−.122	.

**. Correlation is significant at the 0.01 level (2−tailed).

c. Unless otherwise noted, bootstrap results are based on 1000 bootstrap samples

Output 8.2

SELF TEST
Did creativity cause success in the World's Biggest Liar competition?

there is a significant negative relationship between creativity scores and how well someone did in the World's Biggest Liar competition: as creativity increased, position decreased. This might seem contrary to what we predicted until you remember that a low number means that you did well in the competition (a low number such as 1 means you came first, and a high number like 4 means you came fourth). Therefore, our hypothesis is supported: as creativity increased, so did success in the competition.

8.4.4 Kendall's tau (non-parametric) ▌▌▌▌

Kendall's tau, denoted by τ, is another non-parametric correlation and it should be used rather than Spearman's coefficient when you have a small data set with a large number of tied ranks. This means that if you rank the scores and many

scores have the same rank, then Kendall's tau should be used. Although Spearman's statistic is the more popular of the two coefficients, there is much to suggest that Kendall's statistic is a better estimate of the correlation in the population (see Howell, 2012). As such, we can draw more accurate generalizations from Kendall's statistic than from Spearman's. To carry out Kendall's correlation on the World's Biggest Liar data follow the same steps as for Pearson and Spearman correlations but select ☑ Kendall's tau-b and deselect the Pearson and Spearman options.

The output is much the same as for Spearman's correlation (Output 8.3). Notice that the value of the correlation coefficient is closer to zero than the Spearman correlation (it has changed from −0.373 to −0.300). Despite the difference in the correlation coefficients, we can still interpret this result as being a

---

4   It is good to check that the value of *N* corresponds to the number of observations that were made. If it doesn't then data may have been excluded for some reason.

5   Remember that these confidence intervals are based on a random sampling procedure so the values you get will differ slightly from mine, and will change if you rerun the analysis.

Correlations

				Creativity	Position in Biggest Liar Competition
Kendall's tau_b	Creativity	Correlation Coefficient		1.000	-.300**
		Sig. (2-tailed)		.	.001
		N		68	68
		Bootstrap^c	Bias	.000	.001
			Std. Error	.000	.094
			BCa 95% Confidence Interval	Lower	.	-.473
				Upper	.	-.117
	Position in Biggest Liar Competition	Correlation Coefficient		-.300**	1.000
		Sig. (2-tailed)		.001	.
		N		68	68
		Bootstrap^c	Bias	.001	.000
			Std. Error	.094	.000
			BCa 95% Confidence Interval	Lower	-.473	.
				Upper	-.117	.

**. Correlation is significant at the 0.01 level (2-tailed).

c. Unless otherwise noted, bootstrap results are based on 1000 bootstrap samples

Output 8.3

SELF TEST

Conduct a Pearson correlation analysis of the advert data from the beginning of the chapter.

highly significant relationship because the significance value of 0.001 is less than 0.05 and the robust confidence interval does not cross zero (−0.473 to −0.117). However, Kendall's value is likely a more accurate gauge of what the correlation in the population would be. As with any correlation, we cannot assume that creativity caused success in the World's Biggest Liar competition.

8.4.5 Biserial and point-biserial correlations ▐▐▐▐

Often it is necessary to investigate relationships between two variables when one of the variables is dichotomous (i.e., it is categorical with only two categories). An example of a dichotomous variable is being pregnant, because a woman can be either pregnant or not (she cannot be 'a bit pregnant'). The biserial and point-biserial correlation coefficients should be used in these situations. These correlations are distinguished only by a

conceptual difference, but their statistical calculation is quite different. The difference between the use of biserial and point-biserial correlations depends on whether the dichotomous variable is discrete or continuous. This difference is very subtle. A discrete or true, dichotomy is one for which there is no underlying continuum between the categories. An example of this is whether someone is dead or alive: a person can be only dead or alive, they can't be 'a bit dead'. Although you might describe a person as being 'half-dead'—especially after a heavy drinking session—there is a definition of clinically dead and if you don't meet that definition then you are alive. As such, there is no continuum between the two categories. However, it is possible to have a dichotomy for which a continuum does exist. An example is passing or failing a statistics test: some people will only just fail while others will fail by a large margin; likewise some people will scrape a pass while others will excel. Although participants fall into only two categories

(pass or fail) there is an underlying continuum along which people lie.

The **point-biserial correlation** coefficient (r_{pb}) is used when one variable is a discrete dichotomy (e.g., pregnancy), whereas the **biserial correlation** coefficient (r_b) is used when one variable is a continuous dichotomy (e.g., passing or failing an exam). The biserial correlation coefficient cannot be calculated directly in SPSS Statistics; first you must calculate the point-biserial correlation coefficient and then adjust the value. Let's look at an example.

Imagine that I was interested in the relationship between the sex of a cat and how much time it spent away from home (I love cats, so these things interest me). I had heard that male cats disappeared for substantial amounts of time on long-distance roams around the neighborhod (something about hormones driving them to find mates) whereas female cats tended to be more homebound. I used this as a purr-fect (sorry!) excuse to go and visit lots of my friends and their cats. I took a note of the sex of the cat (**Sex**) and then asked the owners to note down the number of hours that their cat was absent from home over a week (**Time**). The time spent away from home is measured at a ratio level—and let's assume it meets the other assumptions of parametric data—while the sex of the cat is a discrete dichotomy. The data are in the file **Roaming Cats.sav**.

We want to calculate a point-biserial correlation, and this is simplicity itself: it is a Pearson correlation when the dichotomous variable is coded with 0 for one category and 1 for the other (in practice, you can use any values because SPSS changes the lower one to 0 and the higher one to 1 when it does the calculations). In the saved data I coded the **Sex** variable 1 for male and 0 for female. The **Time** variable contains the roaming time over the week in hours.

SELF TEST

Using the **Roaming Cats.sav** file, compute a Pearson correlation between **Sex** and **Time**.

Correlations

			Time away from home (hours)	Sex of cat
Time away from home (hours)	Pearson Correlation		1	.378**
	Sig. (2–tailed)			.003
	N		60	60
	Bootstrap[c]	Bias	0	-.005
		Std. Error	0	.113
		BCa 95% Confidence Interval	Lower	.	.153
			Upper	.	.588
Sex of cat	Pearson Correlation		.378**	1
	Sig. (2–tailed)		.003	
	N		60	60
	Bootstrap[c]	Bias	-.005	0
		Std. Error	.113	0
		BCa 95% Confidence Interval	Lower	.153	.
			Upper	.588	.

**. Correlation is significant at the 0.01 level (2–tailed).

c. Unless otherwise noted, bootstrap results are based on 1000 bootstrap samples

Output 8.4

Congratulations: if you did the self-test task then you have conducted your first point-biserial correlation. Despite the horrible name, it's really quite easy to do. You should find that you have the same output as Output 8.4, which shows the correlation matrix of **Time** and **Sex**. The point-biserial correlation coefficient is $r_{pb} = 0.378$, which has a significance value of 0.003. The significance test for this correlation is the same as performing an independent-samples t-test on the data (see Chapter 10). The sign of the correlation (i.e., whether the relationship was positive or negative) depends entirely on which way round the dichotomous variable was coded. To prove this point, the data file has an extra variable called **Recode** which is the same as the variable **Sex** except that the coding is reversed (1 = female, 0 = male). If you repeat the Pearson correlation using **Recode** instead of **Sex** you will find that the correlation coefficient becomes −0.378. The sign of the coefficient is completely dependent on

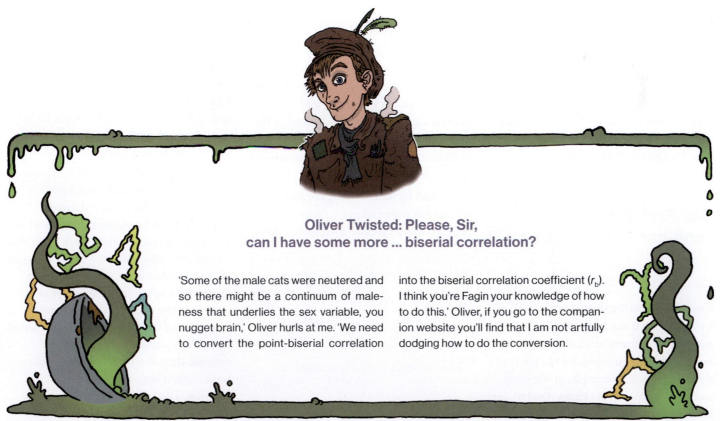

Oliver Twisted: Please, Sir, can I have some more ... biserial correlation?

'Some of the male cats were neutered and so there might be a continuum of male-ness that underlies the sex variable, you nugget brain,' Oliver hurls at me. 'We need to convert the point-biserial correlation into the biserial correlation coefficient (r_b). I think you're Fagin your knowledge of how to do this.' Oliver, if you go to the companion website you'll find that I am not artfully dodging how to do the conversion.

Cramming Sam's Tips
Correlations

- Spearman's correlation coefficient, r_s, is a non-parametric statistic and requires only ordinal data for both variables.
- Kendall's correlation coefficient, τ, is like Spearman's r_s but probably better for small samples.
- The point-biserial correlation coefficient, r_{pb}, quantifies the relationship between a continuous variable and a variable that is a discrete dichotomy (e.g., there is no

continuum underlying the two categories, such as dead or alive).

- The biserial correlation coefficient, r_b, quantifies the relationship between a continuous variable and a variable that is a continuous dichotomy (e.g., there is a continuum underlying the two categories, such as passing or failing an exam).

which category you assign to which code, and so we must ignore all information about the direction of the relationship. However, we can still interpret R^2 as before. In this example, $R^2 = 0.378^2 = 0.143$, which equates to the sex of the cat sharing 14.3% of the variability in the time spent away from home.

8.5 Partial and semi-partial correlation ▮▮▮▯

8.5.1 Semi-partial (or part) correlation ▮▮▮▯

I mentioned earlier that there is a type of correlation that can be done that allows you to look at the relationship between two variables, accounting for the effect of a third variable. For example, in the exam anxiety data (in the file **Exam Anxiety.sav**) exam performance (EP) was negatively

related to exam anxiety (EA), but positively related to revision time (RT), and revision time itself was negatively related to exam anxiety. If revision time is related to both exam anxiety and exam performance, then to get a measure of the unique relationship between exam anxiety and exam performance we need to account for revision time.

Let's begin by transforming the correlation coefficients from Output 8.1 into proportions of variance by squaring them:

$$\text{Proportion of variance shared by EP and EA} = r^2_{\text{EP–EA}}$$
$$= -0.441^2$$
$$= 0.194$$
$$\text{Proportion of variance shared by EP and RT} = r^2_{\text{EP–RT}}$$
$$= 0.397^2$$
$$= 0.158$$
$$\text{Proportion of variance shared by EA and RT} = r^2_{\text{EA–RT}}$$
$$= -0.709^2$$
$$= 0.503$$

(8.15)

If we multiply the resulting proportions by 100 to turn them into percentages (which people typically find easier to understand) we see that exam performance shares 19.4% of its variance with exam anxiety, and 15.8% with revision time. Revision time shares about half of its variance (50.3%) with exam anxiety.

Figure 8.10 depicts the variance of each variable as a shape: exam performance and anxiety are blue and red squares respectively, and revision time is a sort of weird yellow spacecraft or spark plug. The area of each shape is 100% of the variance of the variable it represents. The figure shows the overlap in the variances of these variables. The depicted overlaps are proportional (approximately) to the actual data: if two variables share 10% of variance then about 10% of the surface areas of their shapes overlap in the diagram. The left-hand side shows only the overlap of shapes/variances and the right duplicates the image but with specific areas marked out and labeled.

To see what the overlaps mean, let's look at exam performance and exam anxiety. Notice that the bottom of the square for exam performance overlaps with the top of the square for exam anxiety. For both variables (squares) this overlap is about 19.4% of their surface area, which corresponds to the amount of variance they share (equation (8.15)). This overlap is (I hope) clear on the left-hand side, but on the right I have marked out these areas as A and C. What distinguishes area A from area C is that area C also overlaps with revision time, whereas area A does not. Together (A + C) these areas are the overlap between exam performance and exam anxiety (19.4%), but this overlap can be decomposed into variance in exam performance that is unique to exam anxiety (A) and variance in exam performance that is not unique to exam anxiety because it is also shared with revision time (C).

Similarly, look at the overlap between exam performance and revision time by focusing on the area where the yellow shape overlaps with the blue square. This overlap is represented by a sort of L-shape that in total represents 15.8% of the surface area of both shapes (i.e., 15.8% of shared variance). I have marked out the overlap between the blue square and yellow shape on the right with areas labeled B and C. Like before, these areas combine (B + C) to represent the overlap between exam performance and revision time (15.8%), but this overlap can be decomposed into variance in exam performance that is unique to revision time (B) and variance in exam performance that is not unique to revision time because it is also shared with exam anxiety (C).

We know from equation (8.15) that revision time and exam anxiety share 50.3% of their variance. This is represented in Figure 8.10 by the overlap between the yellow shape and red square, which is represented by areas C and D. Like before, these areas combine (C + D) to represent the total overlap between exam anxiety and revision time (50.3%), but this shared variance can be decomposed into variance in exam anxiety that is unique to revision time (D) and variance in exam anxiety that is not unique to revision time because it is also shared with exam performance (C).

Figure 8.10 contains the size of the areas A, B, C and D as well as some of their sums. It also explains what areas represent in isolation and in combination; for example, area B is the variance in exam performance that is *uniquely shared* with revision time (1.5%), but in combination with area C it is the *total* variance in exam performance that is shared with revision time (15.8%). Think about that. The correlation between exam performance and revision time tells us that they share 15.8% of the variance, but in reality, only 1.5% is unique to revision time whereas the remaining 14.3% (area C) is also shared with exam anxiety. This issue gets to the heart of **semi-partial correlation** (also referred to as a **part correlation**), which we'll return to in the next chapter. The areas of unique variance in Figure 8.10 represent the semi-partial correlation. For example, area A is the unique variance in exam performance shared with exam anxiety expressed as a proportion of the variance in exam performance. It is 5.1%. This means that 5.1% of the variance in exam performance is shared uniquely with exam anxiety (and no other variable that we're currently using

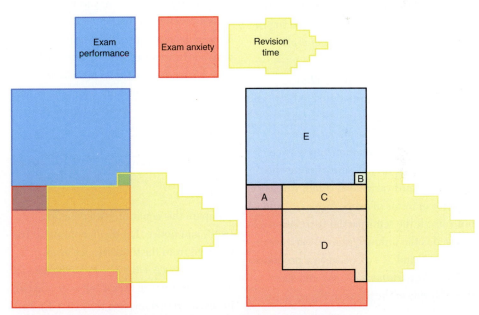

A = variance in exam performance uniquely shared with exam anxiety (5.1%)

B = variance in exam performance uniquely shared with revision time (1.5%)

C = variance in exam performance shared by both exam anxiety and revision time (14.3%)

D = variance shared by exam anxiety and revision time but not exam performance (36%)

E = variance in exam performance not shared by any measured variable (79.1%)

A + C = variance shared by exam performance and exam anxiety (19.4%)

C + B = variance shared by exam performance and revision time (15.8%)

C + D = variance shared by revision time and exam anxiety (50.3%)

A + B + C = variance in exam performance accounted for by revision time and exam anxiety (20.9%)

Figure 8.10   Diagram showing the principle of partial correlation

to predict exam performance). As a proportion this value is 0.051, and remember that to get these proportions we square the correlation coefficient, so to get back to r we would take the square root of 0.051, which is 0.226 (because r can be negative the value could also be –0.226). This is the semi-partial correlation between exam performance and exam anxiety. It is correlation between exam performance and exam anxiety ignoring the portion of that relationship that is also shared by revision time. Likewise, we can look at area B which represents the unique variance in exam performance shared with revision time (1.5%). As a proportion this value is 0.015, and to get this proportion back to a correlation coefficient, we take the square root, which gives us 0.122 (or –0.122). This value is the semi-partial correlation between exam performance and revision time. It is the correlation between exam performance and revision time ignoring the portion of that relationship that is also shared by exam anxiety.

As such, the semi-partial correlation expresses the unique relationship between two variables as a function of their total variance. In general terms, imagine we want to look at the relationship between two variables X and Y, adjusting for the effect of a third variable Z. Again it's easier to think in terms of proportions (r^2) and not r itself; the semi-partial correlation squared is the uniquely shared variance between X and Y, expressed as a propotion of the *total variance in Y*. We can show this in terms of the areas in Figure 8.10:

$$sr^2_{EP-EA} = \frac{A}{A+B+C+E} = \frac{5.1}{100} = 0.051$$
$$sr_{EP-EA} = \sqrt{sr^2_{EP-EA}} = \sqrt{0.051} = \pm0.226$$

(8.16)

The semi-partial correlation squared (sr^2) between exam performance and exam anxiety is the uniquely overlapping area (A) expressed as a function of the total area of exam performance (A + B + C + E). The values of each area are underneath Figure

8.10 if you want to plug the numbers into equation (8.16) as I have. As such a semi-partial correlation is the relationship between X and Y accounting for the overlap in X and Z, but not the overlap in Y and Z. In our specific example, the semi-partial correlation for exam performance and exam anxiety (area A) quantifies their relationship accounting for the overlap between exam anxiety and revision time (area C) *but not* the overlap between exam performance and revision time (area B).

8.5.2 Partial correlation ▌▌▌▌

Another way to express the unique relationship between two variables (i.e., the relationship accounting for other variables) is the **partial correlation**. Remember that the semi-partial correlation expresses the unique relationship between two variables, X and Y, as a function of the *total variance in Y*. We can instead express this unique variance in terms of the *variance in Y left over when other variables have been considered*. The following equation shows this in terms of the areas in Figure 8.10:

$$pr^2_{EP-EA} = \frac{A}{A+E} = \frac{5.1}{5.1+79.1} = 0.061$$
$$pr_{EP-EA} = \sqrt{sr^2_{EP-EA}} = \sqrt{0.066} = \pm0.247$$

(8.17)

Compare this equation with equation (8.16) and notice that the denominator has changed from the total area for exam performance (A + B + C + E) to the area for exam performance that is left over once we have considered revision time (A + E). In other words, the areas in exam performance that overlap with revision time (areas B and C) are removed from the denominator. Hopefully this makes clear the point that the partial correlation expresses the unique relationship between X and Y as a function of the *variance in Y left over when other variables have been considered*. Returning to our exam example, partial correlation squared (pr^2)

between exam performance and exam anxiety is the uniquely overlapping area (A) expressed as a function of the area of exam performance that does *not* overlap with revision time (A + E). Again you can use the values in Figure 8.10 to follow equation (8.17).

By ignoring the variance in Y that overlaps with Z, a partial correlation adjusts for both the overlap that X and Z have *and* the overlap in Y and Z, whereas a semi-partial correlation adjusts only for the overlap that X and Z have. In our specific example, the partial correlation for exam performance and exam anxiety (area A) adjusts for both the overlap in exam anxiety and revision time (area C) *and* the overlap in exam performance and revision time (area B).

8.5.3 Partial correlation using SPSS Statistics ▌▌▌▌

Reload the **Exam Anxiety.sav** file so that we can conduct a partial correlation between exam anxiety and exam performance while adjusting for the effect of revision time. Access the *Partial Correlations* dialog box (Figure 8.11) using the *Analyze* ▶ *Correlate* ▶ 🔢 Pa*r*tial... menu. Your variables will be listed in the left-hand box, and on the right the box labeled *Variables* is for specifying the variables that you want to correlate and the box labeled *Controlling for* is for declaring the variables for which you want to adjust. If we want to look at the unique effect of exam anxiety on exam performance we'd correlate the variables **exam** and **anxiety** while adjusting for **revise** as shown in Figure 8.11. In this instance we are going to adjust for one variable, which is known as a *first-order partial correlation*. It is possible to control for the effects of two variables (a *second-order partial correlation*), three variables (a *third-order partial correlation*) and so on by dragging more variables to the *Controlling for* box.

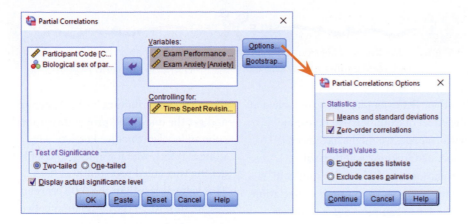

Figure 8.11 Main dialog box for conducting a partial correlation

Clicking [Options...] accesses options such as those in bivariate correlation. Within this dialog box you can select *Zero-order correlations*, which are the Pearson correlation coefficients without adjusting for other variables. If we select this option SPSS will produce a correlation matrix of **anxiety**, **exam** and **revise**, which might be useful if you haven't already looked at the raw (or zero-order) correlations between variables, but we have (Output 8.1), so don't tick this option for this example. As we have done throughout, use the [Bootstrap...] option to get some robust confidence intervals.

Output 8.5 shows the partial correlation of exam anxiety and exam performance, controlling for revision time. Note that the top and bottom of the table contain identical values, so we can ignore one half of the table. The partial correlation between exam performance and exam anxiety is −0.247, which is the same value as we computed in equation (8.17). This value is considerably less than when we didn't adjust for the effect of revision time ($r = -0.441$). Although this correlation is still statistically significant (its *p*-value is still below 0.05) and the confidence

Correlations

Control Variables				Exam Performance (%)	Exam Anxiety
Time Spent Revising	Exam Performance (%)	Correlation		1.000	−.247
		Significance (2–tailed)		.	.012
		df		0	100
		Bootstrap[a]	Bias	.000	.007
			Std. Error	.000	.100
			BCa 95% Confidence Interval	Lower	.	−.430
				Upper	.	−.030
	Exam Anxiety	Correlation		−.247	1.000
		Significance (2–tailed)		.012	.
		df		100	0
		Bootstrap[a]	Bias	.007	.000
			Std. Error	.100	.000
			BCa 95% Confidence Interval	Lower	−.430	.
				Upper	−.030	.

a. Unless otherwise noted, bootstrap results are based on 1000 bootstrap samples

Output 8.5 Output from a partial correlation

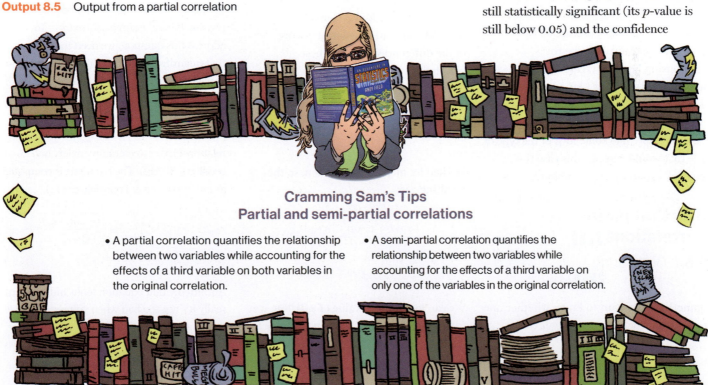

Cramming Sam's Tips
Partial and semi-partial correlations

- A partial correlation quantifies the relationship between two variables while accounting for the effects of a third variable on both variables in the original correlation.

- A semi-partial correlation quantifies the relationship between two variables while accounting for the effects of a third variable on only one of the variables in the original correlation.

interval [–0.430, –0.030] still doesn't contain zero, the relationship is diminished. In terms of variance, the value of R^2 for the partial correlation is 0.061, which means that exam anxiety shares only 6.1% of the variance in exam performance that is left over by revision time (compared to 19.4% when revision time was not factored in). Running this analysis has shown us that exam anxiety alone does explain some of the variation in exam scores, but there is a complex relationship between anxiety, revision and exam performance that might otherwise have been ignored. Although causality is still not certain, because relevant variables are being included, the third-variable problem is, at least, being addressed to some degree.

Partial correlations can be done when variables are dichotomous (including the 'third' variable). For example, we could look at the relationship between bladder relaxation (did the person wet themselves or not?) and the number of large tarantulas crawling up the person's leg, adjusting for fear of spiders (the first variable is dichotomous, but the second variable and 'control' variable are both continuous). Similarly, to use an earlier example, we could examine the relationship between creativity and success in the World's Biggest Liar competition, adjusting for whether someone had previous experience in the competition (and therefore had some idea of the type of tale that would win). In this case the 'control' variable is dichotomous.[6]

8.6 Comparing correlations ▮▮▮▮

8.6.1 Comparing independent *rs* ▮▮▮▮

Sometimes we want to know whether one correlation coefficient is bigger than

another. For example, when we looked at the effect of exam anxiety on exam performance, we might have been interested to know whether this correlation was different in men and women. We can compute the correlation in these two samples, but how do we assess whether the difference is meaningful?

If you do the self-test you'll find that the correlations are $r_{Male} = -0.506$ and $r_{Female} = -0.381$. These two samples are independent (they contain different entities). To compare these correlations we can use what we discovered in Section 8.2.3 to convert these coefficients to z_r (because it makes the sampling distribution normal). Do the conversion and you should obtain z_r (males) = –0.557 and z_r (females) = –0.401. We can calculate a z-score of the differences between these correlations using:

Can I compare the sizes of two correlations?

$$z_{Difference} = \frac{z_{r_1} - z_{r_2}}{\sqrt{\frac{1}{N_1-3} + \frac{1}{N_2-3}}}$$ (8.18)

We had 52 men and 51 women, so the resulting z is:

$$z_{Difference} = \frac{-0.557-(-0.401)}{\sqrt{\frac{1}{49}+\frac{1}{48}}} = \frac{-0.156}{0.203} = -0.768$$ (8.19)

We can look up this value of z (0.768, we can ignore the minus sign) in the table for

the normal distribution in the Appendix and get the one-tailed probability from the column labeled 'Smaller Portion'. The value is 0.221, which we need to double to get the two-tailed probability of 0.442. This value is less than the 0.05 criterion z-score of 1.96, so the correlation between exam anxiety and exam performance is not significantly different in men and women.

8.6.2 Comparing dependent *rs* ▮▮▮▮

If you want to compare correlation coefficients that come from the same entities then things are a little more complicated. You can use a *t*-statistic to test whether a difference between two dependent correlations is significant. For example, in our exam anxiety data we might want to see whether the relationship between exam anxiety (x) and exam performance (y) is stronger than the relationship between revision (z) and exam performance. To address this question, we need the three *rs* that quantify the relationships between these variables: r_{xy}, the relationship between exam anxiety and exam performance (–0.441); r_{zy}, the relationship between revision and exam performance (0.397); and r_{xz}, the relationship between exam anxiety and revision (–0.709). The *t*-statistic is computed as follows (Chen & Popovich, 2002):

$$t_{Difference} = (r_{xy} - r_{zy})\sqrt{\frac{(n-3)(1+r_{xz})}{2\left(1-r_{xy}^2-r_{xz}^2-r_{zy}^2+2r_{xy}r_{xz}r_{zy}\right)}}$$ (8.20)

Admittedly that equation looks hideous, but really it's not too bad when you remember that it just uses the three correlation

6 Both these examples are, in fact, simple cases of hierarchical regression (see Chapter 9), and the first example is also an example of analysis of covariance. This may not mean much yet, but it illustrates what I have repeatedly said about all statistical models being variations of the same linear model.

coefficients and the sample size N. Place the numbers from the exam anxiety example in (N was 103) and you should end up with:

$$t_{\text{Difference}} = (-0.838)\sqrt{\frac{29.1}{2(1-0.194-0.503-0.158+0.248)}}$$
$$= -5.09$$

(8.21)

This value can be checked against the appropriate critical value for t with $N-3$ degrees of freedom (in this case 100). The critical values in the table (see the Appendix) are 1.98 ($p < 0.05$) and 2.63 ($p < 0.01$), two-tailed. As such we can say that the correlation between exam anxiety and exam performance was significantly higher than the correlation between revision time and exam performance (this isn't a massive surprise, given that these relationships went in the opposite directions from each other).

8.6.3 Comparing rs using SPSS Statistics ▌▌▌▌

You can't compare correlations through the dialog boxes in SPSS, but there is a macro available by Weaver and Wuensch (2013).

8.7 Calculating the effect size ▌▌▌▌

Calculating effect sizes for correlation coefficients couldn't be easier because, as we saw earlier in the book, correlation coefficients *are* effect sizes. So, no calculations (other than those you have already done) are necessary. However, although the Spearman and Kendall correlations are comparable to Pearson's r in many respects (their power, for example, is similar under parametric conditions), there are important differences (Strahan, 1982).

First, we can square the value of Pearson's r to get the proportion of shared variance, R^2. For Spearman's r_s we can do this too

Can I use r^2 for non-parametric correlations?

because it uses the same equation as Pearson's r. However, the resulting R_s^2 is the proportion of variance in the *ranks* that two variables share. Having said this, R_s^2 is usually a good approximation of R^2 (especially in conditions of near-normal distributions). Kendall's τ is not numerically similar to either r or r_s, and so τ^2 does not tell us about the proportion of variance shared by two variables (or the ranks of those two variables).

The second difference relates to a more general point that when using correlations as effect sizes (both when reporting your own analysis and when interpreting others) be mindful that the choice of correlation coefficient can make a substantial difference to the apparent size of the effect. For example, Kendall's τ is 66–75% smaller than both Spearman's r_s and Pearson's r, but r and r_s are generally similar in size (Strahan, 1982). Therefore, if τ is used as an effect size it is not comparable to r and r_s. The point-biserial and biserial correlations differ in size too, so you should think very carefully about whether your dichotomous variable has an underlying continuum or whether it is a truly discrete variable.

8.8 How to report correlation coefficents ▌▌▌▌

Reporting correlation coefficients is pretty easy: you report how big they are, their confidence intervals and significance value (the significance value is probably the least important because the correlation coefficient is an effect size). Some general points (see Sections 1.9.3 and 3.8) are as follows: (1) if you use APA style there should be no zero before the decimal point for the correlation coefficient or the probability value (because neither can exceed 1); (2) coefficients are usually reported to 2 or 3 decimal places because this is a reasonable level of precision; (3) generally people report 95% confidence intervals; (4) each correlation coefficient is represented by a different letter (and some of them are Greek); and (5) report exact p-values. Let's take a few examples from this chapter:

✓ There was not a significant relationship between the number of adverts watched and the number of packets of sweets purchased, $r = 0.87$, $p = 0.054$.
✓ Bias corrected and accelerated bootstrap 95% CIs are reported in square brackets. Exam performance was significantly correlated with exam anxiety, $r = -0.44$ [−0.56, −0.30], and time spent revising, $r = 0.40$ [0.25, 0.52]; the time spent revising was also correlated with exam anxiety, $r = -0.71$ [−0.86, −0.49] (all ps < 0.001).
✓ Creativity was significantly related to how well people did in the World's Biggest Liar competition, $r_s = -0.37$, 95% BCa CI [−0.60, −0.12], $p = 0.002$.
✓ Creativity was significantly related to a person's placing in the World's Biggest Liar competition, $\tau = -0.30$, 95% BCa CI [−0.47, −0.12], $p = 0.001$. (Note that I've quoted Kendall's τ.)

Table 8.2 An example of reporting a table of correlations

	Exam Performance	Exam Anxiety	Revision Time
Exam Performance	1	−0.44*** [−0.56,−0.30]	0.40*** [0.25, 0.52]
Exam Anxiety	103	1	−0.71*** [−0.86,−0.49]
Revision Time	103	103	1

ns = not significant ($p > 0.05$), * $p < 0.05$, ** $p < 0.01$, *** $p < 0.001$. BCa bootstrap 95% CIs reported in brackets.

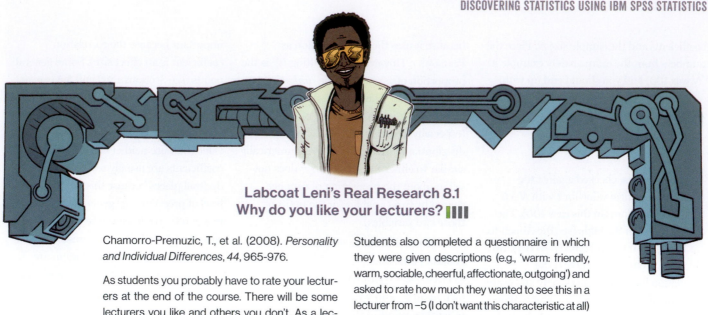

Labcoat Leni's Real Research 8.1
Why do you like your lecturers? ▮▮▮▮

Chamorro-Premuzic, T., et al. (2008). *Personality and Individual Differences, 44*, 965-976.

As students you probably have to rate your lecturers at the end of the course. There will be some lecturers you like and others you don't. As a lecturer I find this process horribly depressing (although this has a lot to do with the fact that I tend to focus on negative feedback and ignore the good stuff). There is some evidence that students tend to pick courses of lecturers they perceive to be enthusiastic and good communicators. In a fascinating study, Tomas Chamorro-Premuzic and his colleagues (Chamorro-Premuzic, Furnham, Christopher, Garwood, & Martin, 2008) tested the hypothesis that students tend to like lecturers who are like themselves. (This hypothesis will have the students on my course who like my lectures screaming in horror.)

The authors measured students' own personalities using a very well-established measure (the NEO-FFI) which measures five fundamental personality traits: neuroticism, extroversion, openness to experience, agreeableness and conscientiousness.

Students also completed a questionnaire in which they were given descriptions (e.g., 'warm: friendly, warm, sociable, cheerful, affectionate, outgoing') and asked to rate how much they wanted to see this in a lecturer from −5 (I don't want this characteristic at all) through 0 (the characteristic is not important) to +5 (I really want this characteristic in my lecturer). The characteristics were the same as those measured by the NEO-FFI.

As such, the authors had a measure of how much a student had each of the five core personality characteristics, but also a measure of how much they wanted to see those same characteristics in their lecturer. Tomas and his colleagues could then test whether, for instance, extroverted students want extroverted lecturers. The data from this study are in the file **Chamorro-Premuzic.sav**. Run Pearson correlations on these variables to see if students with certain personality characteristics want to see those characteristics in their lecturers. What conclusions can you draw? Answers are on the companion website (or look at Table 3 in the original article, which shows you how to report a large number of correlations).

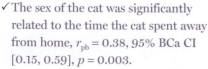

✓ The sex of the cat was significantly related to the time the cat spent away from home, $r_{pb} = 0.38$, 95% BCa CI [0.15, 0.59], $p = 0.003$.

✓ The sex of the cat was significantly related to the time the cat spent away from home, $r_b = 0.48$, $p = 0.003$.

A table is a good way to report lots of correlations. Our exam anxiety correlations could be reported as in

Table 8.2. Note that above the diagonal I have reported the correlation coefficients and used symbols to represent different levels of significance. The confidence intervals are reported underneath. Under the table there is a legend to tell readers what symbols represent. (None of the correlations were non-significant or had p bigger than 0.001 so most of the table footer is there to give you a template—you would normally include

only symbols that you had actually used in the table.) Finally, in the lower part of the table I have reported the sample sizes. These are all the same (103), but when you have missing data it is useful to report the sample sizes in this way because different values of the correlation will be based on different sample sizes. You could alternatively use the bottom part of the table to report exact p-values.

8.9 Brian and Jane's Story ▮▮▮▮

Brian wandered out of his lecture about correlations. An hour learning about relationships had made him feel flat. He knew it was stupid: the lecturer wasn't talking about those kinds of relationships, but the word kept buzzing around his head, distracting him. Jane was complex. On the one hand she always gave him the brush-off, on the other he sensed a playful humor in her responses to him. Was he kidding himself or was that her idea of flirting? Should he be trying to get involved with someone who thought *that* was how to flirt? It was confusing. He wanted to get in her head. Did she think they were developing a positive relationship? Was it negative? Non-existent? He decided to be direct.

Jane's morning had been strange. She'd worked late into the night, but woke early in a mental haze. She was acting on autopilot. She ate breakfast, spent more time on her make-up than she normally would, and put on some of her favorite clothes. She didn't remember doing any of it. It didn't register as odd to her. Neither did it seem unusual to her that she passed by the entrance of the Leviathan lecture hall at exactly 10:50, the time that Brian's statistics lecture kicked out. Seeing Brian dispersed the fog in her mind. She panicked as she realized that her unconscious mind had brought her to this place at this exact time. She felt overdressed and awkward. He seemed sad, so she asked him if he was OK. He told her about the lecture he had just had, and the familiarity of correlations calmed her.

8.10 What next? ▮▮▮▮

At the age of 8 my dad taught me a valuable lesson, which is that if you really

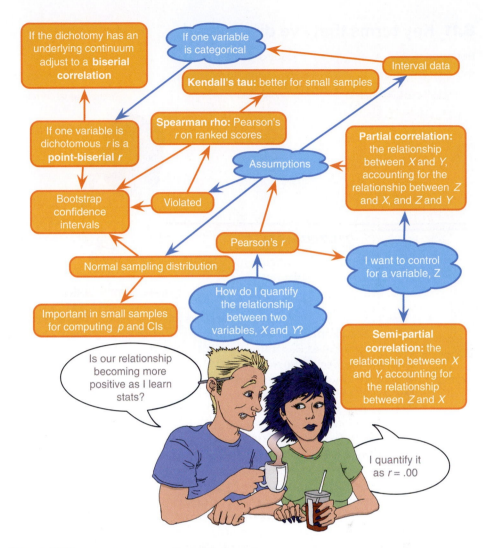

Figure 8.12 What Brian learnt from this chapter

want something then you need to work at it, and the harder you work at it the more likely you are to get what you want. I did practice my guitar and before long the tears gave way to a competent version of 'Skip to my Lou'. My dad had also aspired to be a musician when he was young and encouraged my new passion.[7] He found me a guitar teacher and found the money for lessons. These lessons illustrate how being a good student often depends on finding the right teacher. Ken Steers, despite his best efforts, was on a completely different wavelength than me. I wanted to learn some crushing metal

riffs, and he wanted me to work through Bert Weedon's *Play in a Day* and learn trad jazz classics. As an adult, I wish I had paid more attention to Ken because I'd have been a better guitar player than I am. I was a terrible student and adopted a strategy of selective practice: I'd practice if I wanted to do something but not if I thought it was 'boring'. Perhaps this is why I am so obsessed with trying not to be a boring teacher. Nevertheless, my dad and Ken did get me going and soon enough, like my favorite record of the time, I was ready to 'Take on the world'. Well, Wales at any rate . . .

---

7 My dad, like me, never made it in his band, but, unlike me, did sing on the UK TV show *Stars in Their Eyes*, which made us all pretty proud.

8.11 Key terms that I've discovered

Biserial correlation

Bivariate correlation

Coefficient of determination

Covariance

Cross-product deviations

Kendall's tau

Partial correlation

Pearson correlation coefficient

Point-biserial correlation

Semi-partial correlation

Spearman's correlation coefficient

Standardization

Smart Alex's tasks

- **Task 1**: A student was interested in whether there was a positive relationship between the time spent doing an essay and the mark received. He got 45 of his friends and timed how long they spent writing an essay (**hours**) and the percentage they got in the essay (**essay**). He also translated these grades into their degree classifications (**grade**): in the UK, a student can get a first-class mark (the best), an upper-second-class mark, a lower second, a third, a pass or a fail (the worst). Using the data in the file **EssayMarks.sav**, find out what the relationship was between the time spent doing an essay and the eventual mark in terms of percentage and degree class (draw a scatterplot too). ▮▮▮▮

- **Task 2**: Using the **Notebook.sav** data from Chapter 3, find out the size of relationship between the participant's sex and arousal. ▮▮▮▮

- **Task 3**: Using the notebook data again, quantify the relationship between the film watched and arousal. ▮▮▮▮

- **Task 4**: As a statistics lecturer I am interested in the factors that determine whether a student will do well on a statistics course. Imagine I took 25 students and looked at their grades for my statistics course at the end of their first year at university: first, upper second, lower second and third class (see Task 1). I also asked these students what grade they got in their high school maths exams. In the UK, GCSEs are school exams taken at age 16 that are graded A, B, C, D, E or F (an A grade is the best). The data for this study are in the file **grades.sav**. To what degree does GCSE maths grade correlate with first-year statistics grade? ▮▮▮▮

- **Task 5**: In Figure 2.3 we saw some data relating to people's ratings of dishonest acts and the likeableness of the perpetrator (for a full description see Jane Superbrain Box 2.1). Compute the Spearman correlation between ratings of dishonesty and likeableness of the perpetrator. The data are in **HonestyLab.sav**.

- **Task 6**: In Chapter 4 (Task 6) we looked at data from people who had been forced to marry goats and dogs and measured their life satisfaction and, also, how much they like animals (**Goat or Dog.sav**). Is there a significant correlation between life satisfaction and the type of animal to which a person was married? ▮▮▮▮

- **Task 7**: Repeat the analysis above, taking account of animal liking when computing the correlation between life satisfaction and the animal to which a person was married. ▮▮▮▮

- **Task 8**: In Chapter 4 (Task 7) we looked at data based on findings that the number of cups of tea drunk was related to cognitive functioning (Feng et al., 2010). The data are in the file **Tea Makes You Brainy 15.sav**. What is the correlation between tea drinking and cognitive functioning? Is there a significant effect? ▮▮▮▮

- **Task 9**: The research in the previous task was replicated but in a larger sample ($N = 716$), which is the same as the sample size in Feng et al.'s research (**Tea Makes You Brainy 716.sav**). Conduct a correlation between tea drinking and cognitive functioning. Compare the correlation coefficient and significance in this large sample with the previous task. What statistical point do the results illustrate? ▮▮▮▮

- **Task 10**: In Chapter 6 we looked at hygiene scores over three days of a rock music festival (**Download Festival.sav**). Using Spearman's correlation, were hygiene scores on day 1 of the festival significantly correlated with those on day 3? ▮▮▮▮

- **Task 11**: Using the data in **Shopping Exercise.sav** (Chapter 4, Task 5), find out if there is a significant relationship between the time spent shopping and the distance covered. ▮▮▮▮

- **Task 12**: What effect does accounting for the participant's sex have on the relationship between the time spent shopping and the distance covered? ▮▮▮▮

Answers & additional resources are available on the book's website at
https://edge.sagepub.com/field5e

THE LINEAR MODEL (REGRESSION)

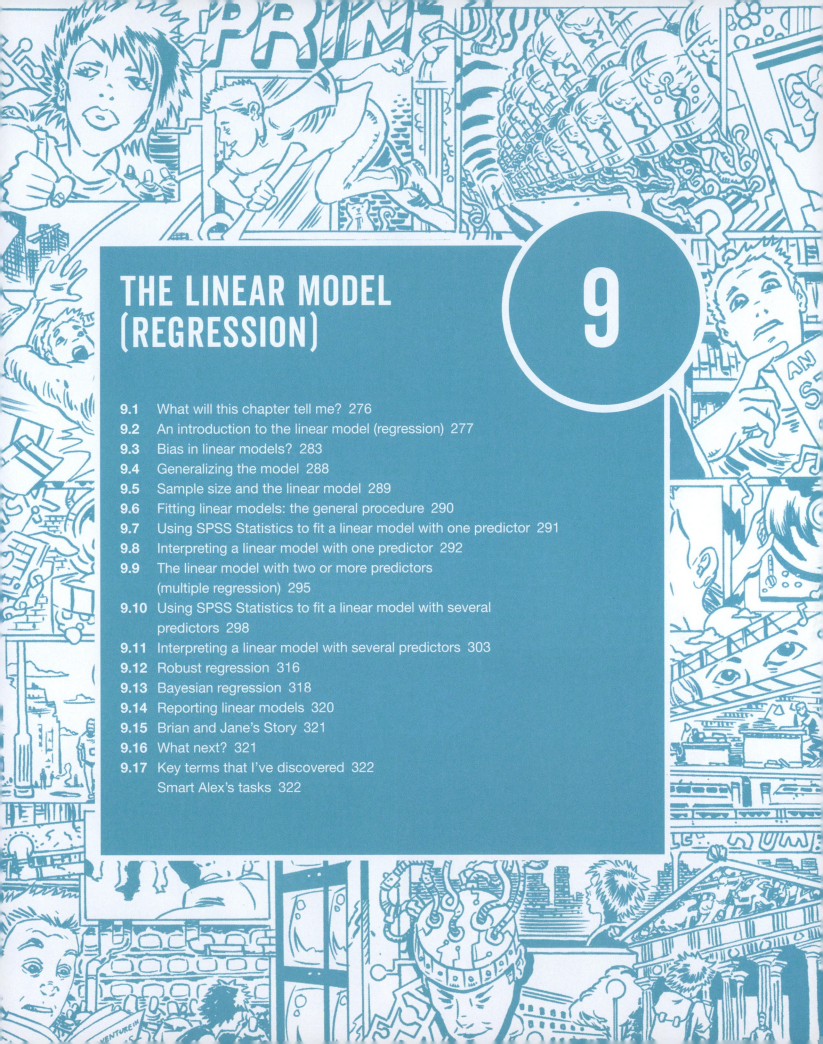

9

9.1 What will this chapter tell me?

Although none of us can know the future, predicting it is so important that organisms are hard-wired to learn about predictable events in their environment. We saw in the previous chapter that I received a guitar for Christmas when I was eight. My first foray into public performance was a weekly talent show at a holiday camp called 'Holimarine' in Wales (it doesn't exist any more because I am old and this was 1981). I sang a Chuck Berry song called 'My ding-a-ling'[1] and to my absolute amazement I won the competition.[2] Suddenly other 8-year-olds across the land (well, a ballroom in Wales) worshipped me (I made lots of friends after the competition). I had tasted success, it tasted like praline chocolate, and so I wanted to enter the competition in the second week of our holiday. To ensure success, I needed to know why I had won in the first week. One way to do this would have been to collect data and to use these data to predict people's evaluations of children's performances in the contest from certain variables: the age of the performer, what type of performance they gave (singing, telling a joke, magic tricks), and perhaps how cute they looked. Obviously actual talent wouldn't be a factor. A linear model (regression) fitted to these data would enable us to predict the future (success in next week's competition) based on values of the variables we'd measured. If, for example, singing was an important factor in getting a good audience evaluation, I could sing again the following week; but if jokers tended to do better then I might switch to a comedy routine. When I was eight I wasn't the pathetic nerd that I am today, so I didn't know about linear models (nor did I wish to); however, my dad thought that success was due to the winning combination of a cherub-looking 8-year-old singing songs that can be interpreted in a filthy way. He wrote a song for me to sing about the keyboard player in the Holimarine Band 'messing about with his organ'. He said 'take this song, son, and steal the show' … and that's what I did: I came first again. There's no accounting for taste.

1 It appears that even then I had a passion for lowering the tone.

2 I have a very grainy video of this performance recorded by my dad's friend on a video camera the size of a medium-sized dog that had to be accompanied at all times by a 'battery pack' the size and weight of a tank (see Oditi's Lantern).

Figure 9.1 Me playing with my ding-a-ling in the Holimarine Talent Show. Note the groupies queuing up at the front

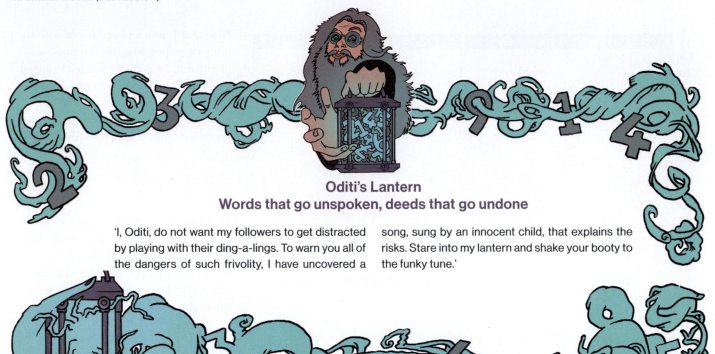

Oditi's Lantern
Words that go unspoken, deeds that go undone

'I, Oditi, do not want my followers to get distracted by playing with their ding-a-lings. To warn you all of the dangers of such frivolity, I have uncovered a song, sung by an innocent child, that explains the risks. Stare into my lantern and shake your booty to the funky tune.'

9.2 An introduction to the linear model (regression) ▮▮▮▮

9.2.1 The linear model with one predictor ▮▮▮▮

In the previous chapter we started getting down to the nitty-gritty of the linear model that we've been discussing since way back in Chapter 2. We saw that if we wanted to look at the relationship between two variables we could use the model in equation (2.3):

$$\text{outcome}_i = (b_1 X_i) + \text{error}_i \qquad (9.1)$$

I mentioned then that if we work with raw scores we must add information about where the outcome variable is centered. I wrote that we add a constant, b_0, known as the intercept to the model that represents the value of the outcome when the predictor is absent (i.e., it is zero). The resulting model is:

$$\text{outcome}_i = (b_0 + b_1 X_i) + \text{error}_i$$
$$Y_i = (b_0 + b_1 X_i) + \varepsilon_i \qquad (9.2)$$

This equation keeps the fundamental idea that an outcome for a person can be predicted from a model (the stuff in parentheses) and some error associated with that prediction (ε_i). We still predict an outcome variable (Y_i) from a predictor variable (X_i) and a parameter, b_1, associated with the predictor variable that quantifies the relationship it has with the outcome variable. This model differs from that of a correlation only in that it uses an *unstandardized* measure of the relationship (b_1) and consequently we include a parameter, b_0, that tells us the value of the outcome when the predictor is zero.

As a quick diversion, let's imagine that instead of b_0 we use the letter c, and instead of b_1 we use the letter m. Let's also ignore the error term. We could predict our outcome as follows:

$$\text{outcome}_i = mx + c$$

Or if you're American, Canadian or Australian let's use the letter b instead of c:

$$\text{outcome}_i = mx + b$$

Perhaps you're French, Dutch or Brazilian, in which case let's use a instead of m:

$$\text{outcome}_i = ax + b$$

Do any of these equations look familiar? If not, there are two explanations: (1) you didn't pay enough attention at school; or (2) you're Latvian, Greek, Italian, Swedish, Romanian, Finnish, Russian or from some other country that has a different variant of the equation of a straight line. The different forms of the equation illustrate how the symbols or letters in an equation can be somewhat arbitrary choices.[3] Whether we write $mx + c$ or $b_1 X + b_0$ doesn't really matter; what matters is what the symbols represent. So, what do the symbols represent?

---

3   For example, you'll sometimes see equation (9.2) written as $Y_i = (\beta_0 + \beta_1 X_i) + \varepsilon_i$. The only difference is that this equation has got βs in it instead of bs. Both versions are the same thing, they just use different letters to represent the coefficients.

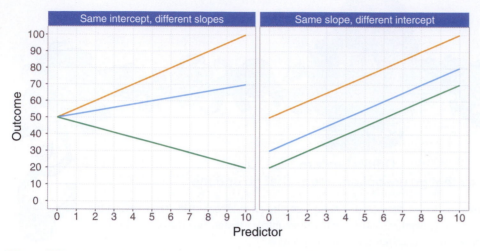

Figure 9.2   Lines that share the same intercept but have different gradients, and lines with the same gradients but different intercepts

I have talked throughout this book about fitting 'linear models', and linear simply means 'straight line'. All the equations above are forms of the equation of a straight line. Any straight line can be defined by two things: (1) the slope (or gradient) of the line (usually denoted by b_1); and (2) the point at which the line crosses the vertical axis of the graph (known as the *intercept* of the line, b_0). These parameters b_1 and b_0 are known as the regression coefficients and will crop up throughout this book, where you see them referred to generally as b (without any subscript) or b_i (meaning the b associated with variable i). Figure 9.2 (left) shows a set of lines that have the same intercept but different gradients. For these three models, b_0 is the same in each but b_1 is different for each line. Figure 9.2 (right) shows models that have the same gradients (b_1 is the same in each model) but different intercepts (b_0 is different in each model).

In Chapter 8 we saw how relationships can be either positive or negative (and I don't mean whether you and your partner argue all the time). A model with a positive b_1 describes a positive relationship, whereas a line with a negative b_1 describes a negative relationship. Looking at Figure 9.2 (left),

the orange line describes a positive relationship whereas the green line describes a negative relationship. As such, we can use a linear model (i.e., a straight line) to summarize the relationship between two variables: the gradient (b_1) tells us what the model looks like (its shape) and the intercept (b_0) locates the model in geometric space.

Let's look at an example. Imagine that I was interested in predicting physical and downloaded album sales (outcome) from the amount of money spent advertising that album (predictor). We could adapt the linear model (equation (9.2)) by replacing the predictor and outcome with our variable names:

$$Y_i = b_0 + b_1 X_i + \varepsilon_i$$
$$\text{album sales}_i = b_0 + b_1 \text{advertising budget}_i + \varepsilon_i$$

(9.3)

Once we have estimated the values of the bs we would be able to make a prediction about album sales by replacing 'advertising' with a number representing how much we wanted to spend advertising an album. For example, imagine that b_0 turned out to be 50 and b_1 turned out to be 100. Our model would be:

$$\text{album sales}_i = 50 + (100 \times \text{advertising budget}_i) + \varepsilon_i$$

(9.4)

Note that I have replaced the bs with their numeric values. Now, we can make a prediction. Imagine we wanted to spend £5 on advertising. We can replace the variable 'advertising budget' with this value and solve the equation to discover how many album sales we will get:

$$\text{album sales}_i = 50 + (100 \times 5) + \varepsilon_i$$
$$= 550 + \varepsilon_i$$

(9.5)

So, based on our model we can predict that if we spend £5 on advertising, we'll sell 550 albums. I've left the error term in there to remind you that this prediction will probably not be perfectly accurate. This value of 550 album sales is known as a **predicted value**.

9.2.2 The linear model with several predictors ▌▐▌

Life is usually complicated and there will be numerous variables that might be related to the outcome that you want to predict. To take our album sales example, variables other than advertising are likely to affect sales. For example, how much someone hears songs from the album on the radio or the 'look' of the band. One of the beautiful things about the linear model is that it expands to include as many predictors as you like. We hinted at this in Chapter 2 (equation (2.4)). An additional predictor can be placed in the model and given a b to estimate its relationship to the outcome:

$$Y_i = (b_0 + b_1 X_{1i} + b_2 X_{2i}) + \varepsilon_i$$

(9.6)

All that has changed is the addition of a second predictor (X_2) and an associated parameter (b_2). To make things more concrete, if we add the number of plays of the band on the radio per week (airplay) to the model in equation (9.3), we get:

$$\text{album sales}_i = b_0 + b_1 \text{advertising budget}_i + b_2 \text{airplay}_i + \varepsilon_i$$

$$(9.7)$$

The new model includes a *b*-value for both predictors (and, of course, the constant, b_0). By estimating the *b*-values, we can make predictions about album sales based not only on the amount spent on advertising but also on airplay.

The resulting model is visualized in Figure 9.3. The tinted trapezium (the regression *plane*) is described by equation (9.7) and the dots represent the observed data points. Like a regression line, a regression plane aims to give the best prediction for the observed data. However, there are invariably differences between the model and the real-life data (this fact is evident because most of the dots do not lie exactly on the plane). The vertical distances between the plane and each data point are the errors or *residuals* in the model. The *b*-value for advertising describes the slope of the left and right sides of the plane, whereas the *b*-value for airplay describes the slope of the top and bottom of the plane. Just like with one predictor, these two slopes describe the shape of the model (what it looks like) and the intercept locates the model in space.

It is easy enough to visualize a linear model with two predictors, because it is possible to plot the plane using a 3-D scatterplot. However, with three, four or even more predictors you can't immediately visualize what the model looks like or what the *b*-values represent, but you can apply the principles of these basic models to more complex scenarios. For example, in general, we can add as many predictors as we like, provided we give them a *b*, and the linear model expands accordingly:

$$Y_i = \left(b_0 + b_1 X_{1i} + b_2 X_{2i} + \cdots + b_n X_{ni}\right) + \varepsilon_i$$

$$(9.8)$$

Y is the outcome variable, b_1 is the coefficient of the first predictor (X_1), b_2 is the coefficient of the second predictor (X_2), b_n is the coefficient of the *n*th

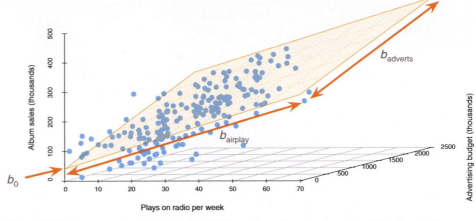

Figure 9.3 Scatterplot of the relationship between album sales, advertising budget and radio play

predictor (X_{ni}), and ε_i is the error for the *i*th entity. (The parentheses aren't necessary, they're there to make the connection to equation (9.2).) This equation illustrates that we can add predictors to the model until we reach the final one (X_n), and each time we add one, we assign it a regression coefficient (*b*).

To sum up, regression analysis is a term for fitting a linear model to data and using it to predict values of an **outcome variable** (a.k.a. dependent variable) from one or more **predictor variables** (a.k.a. independent variables). With one predictor variable, the technique is sometimes referred to as **simple regression**, but with several predictors it is called **multiple regression**. Both are merely terms for the linear model.

9.2.3 Estimating the model ▌▌▌▌

We have seen that the linear model is a versatile model for summarizing the relationship between one or more predictor variables and an outcome variable. No matter how many predictors we have, the model can be described entirely by a constant (b_0) and by parameters associated with each predictor (*b*s). You might wonder how we estimate these parameters, and the quick answer is that we typically use the method of least squares that was described in Section 2.6. We saw then that we could assess the fit of a model (the example we used was the mean) by looking at the deviations between the model and the data collected. These deviations were the vertical distances between what the model predicted and each data point that was observed. We can do the same to assess the fit of a regression line (or plane).

Figure 9.4 shows some data about advertising budget and album sales. A model has been fitted to these data (the straight line). The blue circles are the observed data. The line is the model. The orange dots on the line are the predicted values. We saw earlier that predicted values are the values of the outcome variable calculated from the model. In other words, if we estimated the values of *b* that define the model and put these values into the linear model (as we did in equation (9.4)), then insert different values for advertising budget, the predicted values are the resulting estimates of album sales. If we insert the observed values of advertising budget into the model to get these predicted values, then we can gauge how well the model fits

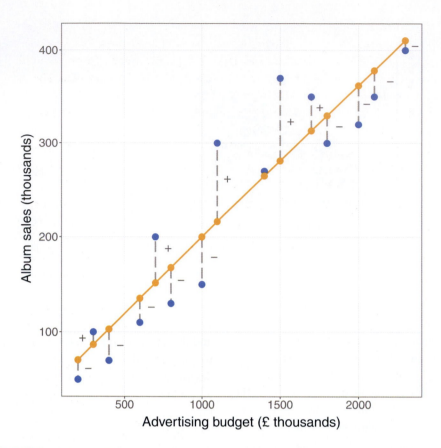

Figure 9.4 A scatterplot of some data with a line representing the general trend. The vertical lines (dotted) represent the differences (or residuals) between the line and the observed data

(i.e., makes accurate predictions). If the model is a perfect fit to the data then for a given value of the predictor(s) the model will predict the same value of the outcome as was observed. In terms of Figure 9.4 this would mean that the orange and blue dots fall in the same locations. They don't, because the model is not perfect (and it never will be): sometimes it overestimates the observed value of the outcome and sometimes it underestimates it. With the linear model the differences between what the model predicts and the observed data are usually called **residuals** (they are the same as *deviations* when we looked at the mean); they are the vertical dashed lines in Figure 9.4.

We saw in Chapter 2, equation (2.11), that to calculate the total error in a model we square the differences between the observed values of the outcome, and the predicted values that come from the model:

$$\text{total error} = \sum_{i=1}^{n} \left(\text{observed}_i - \text{model}_i \right)^2 \qquad (9.9)$$

Sometimes the predicted value of the outcome is less than the actual value and sometimes it is greater. Consequently, some residuals are positive but others are negative, and if we summed them they would cancel out. The solution is to square them before we add them up (this idea should be familiar from Section 2.5.2). Therefore, to assess the error in a linear model, just like when we assessed the fit of the mean using the variance, we use a sum of squared errors, and because we call these errors residuals, this total is called the *sum of squared residuals* or **residual sum of**

squares (SS_R). The residual sum of squares is a gauge of how well a linear model fits the data: if the squared differences are large, the model is not representative of the data (there is a lot of error in prediction); if the squared differences are small, the line is representative.

Let's get back to how we estimate the *b*-values. If you were particularly bored, you could draw every possible straight line (linear model) through your data and calculate the residual sum of squares for each one. You could then compare these 'goodness-of-fit' measures and keep the line with the smallest SS_R because it would be the best-fitting model. We have better things to do, so like when we estimate the mean, we use the method of least squares to estimate the parameters (*b*) that define the regression model for which the sum of squared errors is the minimum it can be (given the data). This method is known as **ordinary least squares** (**OLS**) regression. How exactly the method of least squares does this is beyond me: it uses a mathematical technique for finding maxima and minima to find the *b*-values that describe the model that minimizes the sum of squared differences.

I don't know much more about it than that, to be honest, so with one predictor I tend to think of the process as a little bearded wizard called Nephwick the Line Finder who just magically finds lines of best fit. Yes, he lives inside your computer. For more complex models, Nephwick invites his brother Clungglewad the Beta Seeker for tea and cake and together they stare into the tea leaves in their cups until the optimal beta-values are revealed to them. Then they compare beard growth since their last meeting. I'm pretty sure that's how the method of least squares works.

9.2.4 Assessing the goodness of fit, sums of squares, *R* and R^2 ▐▐▐▐

Once Nephwick and Clungglewad have found the values of *b* that define the model of best fit we assess how well this model fits the observed data (i.e., the **goodness of fit**). We do this because even though the model is the best one available, it can still be a lousy fit (the best of a bad bunch). We saw above that the residual sum of squares measures how much error there is in the model: it quantifies the error in prediction, but it doesn't tell us whether using the model is better than nothing. We need to compare the model against a baseline to see whether it 'improves' how well we can predict the outcome. So, we fit a baseline model and use equation (9.9) to calculate the fit of this model. Then we fit the best model, and calculate the error, SS_R, within it using equation (9.9). If the best model is any good, it should have significantly less error within it than the baseline model.

What would be a good baseline model? Let's go back to our example of predicting album sales (Y) from the amount of money spent advertising that album (X). In my fictional world where I am a statistician employed by a record company or my favorite football team, my boss one day bursts into my office. He says, 'Andy, I know you wanted to be a rock star but have ended up working as my stats-monkey, but how many albums will we sell if we spend £100,000 on advertising?' If I didn't have an accurate model of the relationship between album sales and advertising, what would my best guess be? Probably the best answer I could give would be the mean number of album sales (say, 200,000) because—on average—that's how many albums we expect to sell. This response might satisfy a brainless record company executive (who didn't offer my band a record contract). The next

day he bursts in again and demands to know how many albums we will sell if we spend £1 on advertising. In the absence of any better information, I'm again going to have to say the average number of sales (200,000). This is getting embarrassing for me: whatever amount of money is spent on advertising, I predict the same levels of sales. My boss will think I'm an idiot.

The mean of the outcome is a model of 'no relationship' between the variables: as one variable changes the prediction for the other remains constant (see Section 3.7.2). I hope this illustrates that the mean of the outcome is a good baseline of 'no relationship'. Using the mean of the outcome as a baseline model, we can calculate the difference between the observed values and the values predicted by the mean (equation (9.9)). We saw in Section 2.5.1 that we square these differences to give us the sum of squared differences. This sum of squared differences is known as the **total sum of squares** (denoted by SS_T) and it represents how good the mean is as a model of the observed outcome scores (Figure 9.5, top left).

We then fit a more sophisticated model to the data, such as a linear model, and again work out the differences between what this new model predicts and the observed data (again using equation (9.9)). This value is the residual sum of squares (SS_R) discussed in the previous section. It represents the degree of inaccuracy when the best model is fitted to the data (Figure 9.5, top right).

We can use the values of SS_T and SS_R to calculate how much better the linear model is than the baseline model of 'no relationship'. The improvement in prediction resulting from using the linear model rather than the mean is calculated as the difference between SS_T and SS_R

(Figure 9.5, bottom). This difference shows us the reduction in the inaccuracy of the model resulting from fitting the regression model to the data. This improvement is the **model sum of squares** (SS_M). Figure 9.5 shows each sum of squares graphically where the model is a line (i.e., one predictor) but the same principles apply with more than one predictor.

If the value of SS_M is large, the linear model is very different from using the mean to predict the outcome variable. This implies that the linear model has made a big improvement to predicting the outcome variable. If SS_M is small then using the linear model is little better than using the mean (i.e., the best model is no better than predicting from 'no relationship'). A useful measure arising from these sums of squares is the proportion of improvement due to the model. This is calculated by dividing the sum of squares for the model by the total sum of squares to give a quantity called R^2:

$$R^2 = \frac{SS_M}{SS_T} \qquad (9.10)$$

To express this value as a percentage multiply it by 100. R^2 represents the amount of variance in the outcome explained by the model (SS_M) relative to how much variation there was to explain in the first place (SS_T); it is the same as the R^2 we met in Section 8.4.2 and it is interpreted in the same way: it represents the proportion of the variation in the outcome that can be predicted from the model. We can take the square root of this value to obtain Pearson's correlation coefficient for the relationship between the values of the outcome predicted by the model and the observed values of the outcome.[4] As such, the correlation coefficient provides us with a good estimate of the overall fit of the regression

How can I tell whether my model is good?

---

4 This is the correlation between the orange and blue dots in Figure 9.4. With only one predictor in the model this value will be the same as the Pearson correlation coefficient between the predictor and outcome variable.

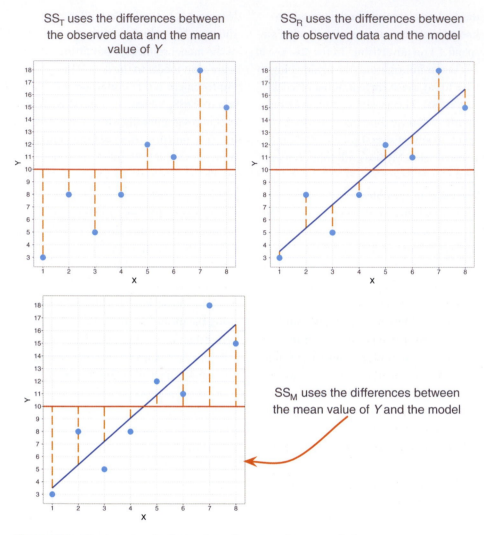

SS$_T$ uses the differences between the observed data and the mean value of Y

SS$_R$ uses the differences between the observed data and the model

SS$_M$ uses the differences between the mean value of Y and the model

Figure 9.5   Diagram showing from where the sums of squares derive

the associated degrees of freedom (this is comparable to calculating the variance from the sums of squares—see Section 2.5.2). For SS$_M$ the degrees of freedom are the number of predictors in the model (k), and for SS$_R$ they are the number of observations (N) minus the number of parameters being estimated (i.e., the number of b coefficients including the constant). We estimate a b for each predictor and the intercept (b_0), so the total number of bs estimated will be $k + 1$, giving us degrees of freedom of $N - (k + 1)$ or, more simply, $N–k–1$. Thus

$$MS_M = \frac{SS_M}{k} \qquad MS_R = \frac{SS_R}{N - k - 1} \qquad (9.11)$$

There is more on mean squares in Chapter 12. The **F-statistic** computed from these mean squares,

$$F = \frac{MS_M}{MS_R} \qquad (9.12)$$

is a measure of how much the model has improved the prediction of the outcome compared to the level of inaccuracy of the model. If a model is good, then the improvement in prediction from using the model should be large (MS$_M$ will be large) and the difference between the model and the observed data should be small (MS$_R$ will be small). In short, for a good model the numerator in equation (9.12) will be bigger than the denominator, resulting in a large F-statistic (greater than 1 at least). This F has an associated probability distribution from which a p-value can be derived to tell us the probability of getting an F at least as big as the one we have if the null hypothesis were true. The null hypothesis in this case is a flat model (predicted values of the outcome are the same regardless of the value of the

model (i.e., the correspondence between predicted values of the outcome and the actual values), and R^2 provides us with a gauge of the substantive size of the model fit.[5]

A second use of the sums of squares in assessing the model is the F-test. I mentioned way back in Chapter 2 that test statistics (like F) are usually the amount of systematic variance divided by the amount of unsystematic variance or, put another

way, the model compared to the error in the model. This is true here: F is based upon the ratio of the improvement due to the model (SS$_M$) and the error in the model (SS$_R$). I say 'based upon' because the sums of squares depend on the number of differences that were added up, and so the average sums of squares (referred to as the **mean squares** or MS) are used to compute F. The mean sum of squares is the sum of squares divided by

---

5   When the model contains more than one predictor, people sometimes refer to R^2 as multiple R^2. This is another example of how people attempt to make statistics more confusing than it needs to be by referring to the same thing in different ways. The meaning and interpretation of R^2 are the same regardless of how many predictors you have in the model or whether you choose to call it multiple R^2: it is the squared correlation between values of the outcome predicted by the model and the values observed in the data.

predictors). If you want to go old school, you can compare the F-statistic against critical values for the corresponding degrees of freedom (as in the Appendix). The F-statistic is also used to calculate the significance of R^2 using the following equation:

$$F = \frac{(N-k-1)R^2}{k(1-R^2)} \qquad (9.13)$$

in which N is the number of cases or participants, and k is the number of predictors in the model. This F tests the null hypothesis that R^2 is zero (i.e., there is no improvement in the sum of squared error due to fitting the model).

9.2.5 Assessing individual predictors ▋▋▋

We've seen that any predictor in a linear model has a coefficient (b_1). The value of b represents the change in the outcome resulting from a unit change in a predictor. If a predictor was useless at predicting the outcome, then what might we expect the change in the outcome to be as values of the predictor change? If a predictor had 'no relationship' with the outcome then the change would be zero. Think back to Figure 9.5. In the panel representing SS_T we saw that the line representing 'no relationship' or 'mean of the outcome' is flat: as the predictor variable changes, the predicted value of the outcome does *not* change (it is a constant value). A 'flat' model, a model in which the same predicted value arises from all values of the predictor variables, will have b-values of 0 for the predictors.

A regression coefficient of 0 means: (1) a unit change in the predictor variable results in no change in the predicted value of the outcome (the predicted value of the outcome is constant); and (2) the linear model is 'flat' (the line or plane doesn't deviate from the horizontal). Therefore, logically, if a variable significantly predicts

an outcome, it should have a b-value that is *different* from zero. This hypothesis is tested using a **t-statistic** that tests the null hypothesis that the value of b is 0. If the test is significant, we might interpret this information as supporting a hypothesis that the b-value is significantly different from 0 and that the predictor variable contributes significantly to our ability to estimate values of the outcome.

Like F, the t-statistic is based on the ratio of explained variance against unexplained variance or error. What we're interested in here is not so much variance but whether the b we have is big compared to the amount of error in that estimate. Remember that the standard error for b tells us something about how different b-values would be across different samples (think back to Section 2.7). If the standard error is very small, then most samples are likely to have a b-value similar to the one in our sample (because there is little variation across samples). Therefore, the standard error is a good estimate of how much error there is likely to be in our b. The following equation shows how the t-test is calculated:

$$t = \frac{b_{observed} - b_{expected}}{SE_b} = \frac{b_{observed}}{SE_b} \qquad (9.14)$$

You'll find a general version of this equation in Section 10.5.1 (equation (10.5)). The $b_{expected}$ is the value of b that we would expect to obtain if the null hypothesis were true. The null hypothesis is that b is 0, and so this value is replaced by 0 and drops out of the equation. The resulting t is the observed value of b divided by the standard error with which it is associated. The t, therefore, tells us whether the observed b is different from 0 relative to the variation in bs across samples. When the standard error is small even a small deviation from zero can reflect a significant difference because b is representative of the majority of possible samples.

The statistic t has a probability distribution that differs according to the

degrees of freedom for the test. In this context, the degrees of freedom are $N - k - 1$, where N is the total sample size and k is the number of predictors. With only one predictor, this reduces to $N - 2$. Using the appropriate t-distribution, it's possible to calculate a p-value that indicates the probability of getting a t at least as large as the one we observed if the null hypothesis were true (i.e., if b was in fact 0 in the population). If this observed p-value is less than 0.05, then scientists tend to assume that b is significantly different from 0; put another way, the predictor makes a significant contribution to predicting the outcome. However, remember the potential pitfalls of blindly applying this 0.05 rule. If you want to pretend it's 1935 then instead of computing an exact p, you can compare your observed t against critical values in a table (in the Appendix).

9.3 Bias in linear models? ▋▋▋

In Chapter 6 we saw that statistical models can be biased by unusual cases or by failing to meet certain assumptions. Therefore, the next

How can I tell whether my model is biased?

questions to ask are whether the model: (1) is influenced by a small number of cases; and (2) generalizes to other samples. These questions are, in some sense, hierarchical because we wouldn't want to generalize a bad model. However, it is a mistake to think that because a model fits the observed data well we can draw conclusions beyond our sample. **Generalization** (Section 9.4) is a critical additional step, and if we find that our model is not generalizable, then we must restrict any conclusions to the sample used. First, let's look at bias. To answer the question of whether the model is influenced by a small number of cases, we can look for outliers and influential cases (the difference is explained in Jane Superbrain Box 9.1).

9.3.1 Outliers

An outlier is a case that differs substantially from the main trend in the data (see Section 6.3). Outliers can affect the estimates of the regression coefficients. For example, Figure 9.6 uses the same data as Figure 9.4 except that the score of one album has been changed

to be an outlier (in this case an album that sold relatively few copies despite a very large advertising budget). The blue line shows the original model, and the orange line shows the model with the outlier included. The outlier makes the line flatter (i.e., b_1 gets smaller) and increases the intercept (b_0 gets larger). If outliers affect the estimates of the bs that define the model then it is important to detect them. But how?

An outlier, by its nature, is very different from the other scores. In which case, do you think that the model will predict an

outlier's score very accurately? Probably not: in Figure 9.6 it's evident that even though the outlier has dragged the model towards it, the model still predicts it very badly (the line is a long way from the outlier). Therefore, if we compute the residuals (the differences between the observed values of the outcome and the values predicted by the model), outliers could be spotted because they'd have large values. In other words, we'd look for cases that the model predicts inaccurately.

Remember that residuals represent the error present in the model. If a model fits the sample data well then all residuals will be small (if the model was a perfect fit of the sample data—all data points fall on the regression line—then all residuals would be zero). If a model is a poor fit to the sample data then the residuals will be large. Up to now we have discussed *normal* or **unstandardized residuals**. These are the raw differences between predicted and observed values of the outcome variable. They are measured in the same units as the outcome variable, which makes it difficult to apply general rules (because what constitutes 'large' depends on the outcome variable). All we can do is to look for residuals that stand out as being particularly large.

To overcome this problem, we can use **standardized residuals**, which are the residuals converted to z-scores (see Section 1.8.6) and so are expressed in standard deviation units. Regardless of the variables in your model, standardized residuals (like any z-scores) are distributed around a mean of 0 with a standard deviation of 1. Therefore, we can compare standardized residuals from different models and use what we know about z-scores to apply universal guidelines for what is expected. For example, in a normally distributed sample, 95% of z-scores should lie between −1.96 and +1.96, 99% should lie between −2.58 and +2.58, and 99.9% (i.e., nearly all of them) should lie between

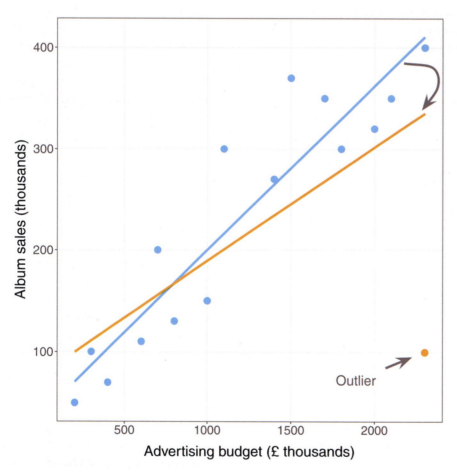

Figure 9.6 Graph demonstrating the effect of an outlier. The blue line represents the original regression line for these data, whereas the orange line represents the regression line when an outlier is present

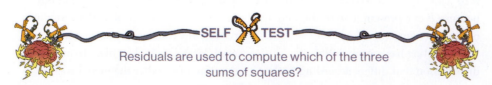

SELF TEST

Residuals are used to compute which of the three sums of squares?

−3.29 and +3.29 (see Chapter 1). Based on this: (1) standardized residuals with an absolute value greater than 3.29 (we can use 3 as an approximation) are cause for concern because in an average sample a value this high is unlikely to occur; (2) if more than 1% of our sample cases have standardized residuals with an absolute value greater than 2.58 (2.5 will do) there is evidence that the level of error within our model may be unacceptable; and (3) if more than 5% of cases have standardized residuals with an absolute value greater than 1.96 (2 for convenience) then the model may be a poor representation of the data.

A third form of residual is the **studentized residual**, which is the unstandardized residual divided by an estimate of its standard deviation that varies point by point. These residuals have the same properties as the standardized residuals but usually provide a more precise estimate of the error variance of a specific case.

9.3.2 Influential cases

It is also possible to look at whether certain cases exert undue influence over the parameters of the model. In other words, if we were to delete a certain case, how different would the regression coefficients be? This analysis helps to determine whether the model is stable across the sample or whether it is biased by a few influential cases. This process can also unveil outliers.

There are several statistics used to assess the influence of a case. The **adjusted predicted value** for a case is the predicted value of the outcome for that case from a model in which the case is excluded. In effect, you estimate the model parameters excluding a particular case and use this new model to predict the outcome for the case that was excluded. If a case does not exert a large influence over the model then the adjusted predicted value should be similar to the predicted value when the case is included. Put simply, if the model is stable then the predicted value of a case should be the same regardless of whether that case was used to estimate the model.

We can also look at the **deleted residual**, which is the difference between the adjusted predicted value and the original observed value. The deleted residual can be divided by the standard error to give a standardized value known as the **studentized deleted residual**. This residual can be compared across different regression analyses because it is measured in standard units.

The deleted residuals are very useful to assess the influence of a case on the ability of the model to predict that case. However, they do not provide any information about how a case influences the model as a whole (i.e., the impact that a case has on the model's ability to predict *all* cases). **Cook's distance** is a measure of the overall influence of a case on the model, and Cook and Weisberg (1982) have suggested that values greater than 1 may be cause for concern.

The **leverage** (sometimes called **hat values**) gauges the influence of the observed value of the outcome variable over the predicted values. The average leverage value is defined as $(k + 1)/n$, in which k is the number of predictors in the model and n is the number of cases.[6] The maximum value for leverage is $(N−1)/N$; however, IBM SPSS Statistics calculates a version of the leverage that has a maximum value of 1 (indicating that the case has complete influence over prediction).

- If no cases exert undue influence over the model then all leverage values should be close to the average value $((k + 1)/n)$.
- We should investigate cases with values greater than twice the average, $2(k + 1)/n$ (Hoaglin & Welsch, 1978) or three times the average, $3(k + 1)/n$ (Stevens, 2002).

We will see how to use these cut-off points later. However, cases with large leverage values will not necessarily have a large influence on the regression coefficients because they are measured on the outcome variables, not the predictors.

Related to the leverage values are the **Mahalanobis distances**, which measure the distance of cases from the mean(s) of the predictor variable(s). Look for the cases with the highest values. These distances have a chi-square distribution, with degrees of freedom equal to the number of predictors (Tabachnick & Fidell, 2012). One way to establish a cut-off point is to find the critical value of chi-square for the desired alpha level (values for $p = 0.05$ and 0.01 are in the Appendix). For example, with three predictors, a distance greater than 7.81 ($p = 0.05$) or 11.34 ($p = 0.01$) would be cause for concern. As general context, based on Barnett and Lewis (1978), with large samples ($N = 500$) and five predictors, values above 25 are cause for concern. In smaller samples ($N = 100$) and fewer predictors (namely, three), values greater than 15 are problematic. In very small samples ($N = 30$) with only two predictors, values greater than 11 should be examined.

Another approach is to look at how the estimates of b in a model change as a result of excluding a case (i.e., compare the values of b estimated from the full data to those estimated from the data excluding the particular case). The change in bs tells us how much influence a case has on the

---

6 You may come across the average leverage denoted as p/n, in which p is the number of parameters being estimated. In regression, we estimate parameters for each predictor and also for a constant, and so p is equivalent to the number of predictors plus one $(k + 1)$.

Figure 9.7 Prasanta Chandra Mahalanobis staring into his distances

Table 9.1 The difference in the parameters of the regression model when one case is excluded

Parameter (*b*)	Case 30 Included	Case 30 Excluded	Difference
Constant (intercept)	29.00	31.00	−2.00
Predictor (gradient)	−0.90	−1.00	0.10
Model (regression line):	$Y = -0.9X + 29$	$Y = -1X + 31$	
Predicted Y	28.10	30.00	−1.90

SELF TEST

Once you have read Section 9.7, fit a linear model first with all the cases included and then with case 30 deleted.

parameters of the model. To take a hypothetical example, imagine two variables that have a perfect negative relationship except for a single case (case 30). These data are in the file **DFBeta.sav.** The results of these two models are summarized in Table 9.1, which shows: (1) the parameters for the regression model when the extreme case is included or excluded; (2) the resulting regression equations; and (3) the value of Y predicted from participant 30's score on the X variable (which is obtained by replacing the X in the regression equation with participant 30's score for X, which was 1). When case 30 is excluded, these data have a perfect negative relationship; hence the coefficient for the predictor (b_1) is −1, and the coefficient for the constant (the intercept, b_0) is 31. However, when case 30 is included, both parameters are reduced[7] and the difference between the parameters is also displayed. The difference between a parameter estimated using all cases and estimated when one case is excluded is known as the **DFBeta.** DFBeta is calculated for every case and for each of the parameters in the model. So, in our hypothetical example, the DFBeta for the constant is −2, and the DFBeta for the predictor variable is 0.1. The values of DFBeta help us to identify cases that have a large influence on the parameters of the

model. The units of measurement used will affect these values, and so you can use **standardized DFBeta** to apply universal cut-off points. Standardized DFBetas with absolute values above 1 indicate cases that substantially influence the model parameters (although Stevens, 2002, suggests looking at cases with absolute values greater than 2).

A related statistic is the **DFFit**, which is the difference between the predicted values for a case when the model is estimated including or excluding that case: in this example the value is −1.90 (see Table 9.1). If a case has no influence then its DFFit should be zero—hence, we expect non-influential cases to have small DFFit values. As with DFBeta, this statistic depends on the units of measurement of the outcome, and so a DFFit of 0.5 will be very small if the outcome ranges from 1 to 100, but very large if the outcome varies from 0 to 1. To overcome this problem we can look at standardized versions of the DFFit values (**standardized DFFit**) which are expressed in standard deviation units. A final measure is the **covariance ratio** (**CVR**), which quantifies the degree to which a case influences the variance of the regression parameters. A description of the computation of this statistic leaves me dazed and confused, so suffice to say that when this ratio is close to 1 the case is

having very little influence on the variances of the model parameters. Belsey, Kuh, & Welsch (1980) recommend the following:

- If $\text{CVR}_i > 1 + [3(k + 1)/n]$ then deleting the ith case will damage the precision of some of the model's parameters.
- If $\text{CVR}_i < 1 - [3(k + 1)/n]$ then deleting the ith case will improve the precision of some of the model's parameters.

In both inequalities, k is the number of predictors, CVR_i is the covariance ratio for the ith participant, and n is the sample size.

9.3.3 A final comment on diagnostic statistics ▌▌▌

I'll conclude this section with a point made by Belsey et al. (1980): diagnostics are tools to see how well your model fits the sampled data and *not* a way of justifying the removal of data points to effect some desirable change in the regression parameters (e.g., deleting a case that changes a non-significant *b*-value into a significant one). Stevens (2002) similarly notes that if a case is a significant outlier but is not having an influence (e.g., Cook's distance is less than 1, DFBetas and DFFit are small) there is no real need to worry about that point because it's not having a large impact on the model parameters. Nevertheless, you should still be interested in *why* the case didn't fit the model.

---

7 The value of b_1 is reduced because the variables no longer have a perfect linear relationship and so there is now variance that the predictor cannot explain.

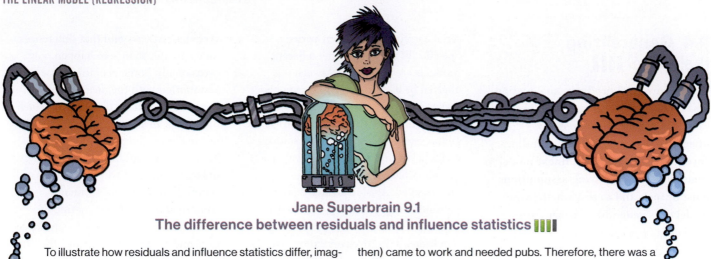

Jane Superbrain 9.1
The difference between residuals and influence statistics ▊▊▊▊

To illustrate how residuals and influence statistics differ, imagine that the Mayor of London in 1900 was interested in how drinking affected mortality. London is divided up into different regions called boroughs, and so he measured the number of pubs and the number of deaths over a period of time in eight of his boroughs. The data are in a file called **pubs.sav**.

The scatterplot of these data (Figure 9.8) reveals that without the last case there is a perfect linear relationship (the orange line). However, the presence of the last case (case 8) changes the line of best fit dramatically (although this line is still a significant fit to the data—fit the model and see for yourself).

The residuals and influence statistics are interesting (Output 9.1). The standardized residual for case 8 is the second *smallest*: it produces a very small residual (most of the non-outliers have larger residuals) because it sits very close to the line that has been fitted to the data. According to the residual it is not an outlier, but how is that possible when it is so different from the rest of the data? The answer lies in the influence statistics, which are all massive for case 8: it exerts a huge influence over the model—so huge that the model predicts that case very well.

When you see a statistical oddity like this, ask what's happening in the real world. District 8 is the City of London, a tiny area of only 1 square mile in the center of London where very few people lived but where thousands of commuters (even then) came to work and needed pubs. Therefore, there was a massive number of pubs. (I'm very grateful to David Hitchin for this example, and he in turn got it from Dr Richard Roberts.)

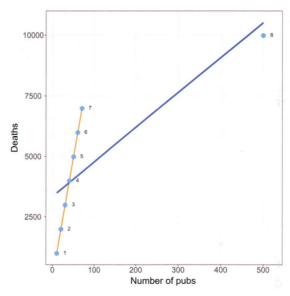

Figure 9.8   Relationship between the number of pubs and the number of deaths in 8 London districts

Case Summaries[a]

	Standardized Residual	Mahalanobis Distance	Cook's Distance	Centered Leverage Value	DFFIT	DFBETA Intercept	DFBETA pubs
1	−1.33839	.28515	.21328	.04074	−495.72692	−509.65184	1.39249
2	−.87895	.22370	.08530	.03196	−305.09716	−321.12768	.80153
3	−.41950	.16969	.01814	.02424	−137.20167	−147.10661	.33016
4	.03995	.12314	.00015	.01759	12.38769	13.45081	−.02658
5	.49940	.08403	.02294	.01200	147.81622	161.44976	−.27267
6	.95885	.05237	.08092	.00748	273.00807	297.67748	−.41116
7	1.41830	.02817	.17107	.00402	391.72124	422.81664	−.44422
8	−.27966	6.03375	227.14286	.86196	−39478.585	3351.95531	−85.66108
Total N	8	8	8	8	8	8	8

a. Limited to first 100 cases.

Output 9.1

9.4 Generalizing the model ▌▌▌▌

The linear model produces an equation that is correct for the sample of observed values. However, we are usually interested in generalizing our findings beyond the sample. For a linear model to generalize the underlying assumptions must be met, and to test whether the model does generalize we can cross-validate it.

9.4.1 Assumptions of the linear model ▌▌▌▌

We have already looked at the main assumptions of the linear model and how to assess them in Chapter 6. The main ones in order of importance (Field & Wilcox, 2017; Gelman & Hill, 2007) are:

- *Additivity and linearity*: The outcome variable should, in reality, be linearly related to any predictors, and, with several predictors, their combined effect is best described by adding their effects together. In other words, the process we're trying to model can be described by the linear model. If this assumption isn't met then the model is invalid. You can sometimes transform variables to make their relationships linear (see Chapter 6).
- **Independent errors**: For any two observations the residual terms should be uncorrelated (i.e., independent). This eventuality is sometimes described as a lack of **autocorrelation**. If we violate this assumption then the model standard errors will be invalid, as will the confidence intervals and significance tests based upon them. In terms of the model parameters themselves, the estimates from the method of least squares will be valid but not optimal (see Section 6.8). This assumption can be tested with the **Durbin–Watson test**, which tests for

serial correlations between errors. Specifically, it tests whether adjacent residuals are correlated. As such it is affected by the order of cases and only makes sense when your cases have a meaningful order (which they don't in the album sales example). The test statistic varies between 0 and 4, with a value of 2 meaning that the residuals are uncorrelated. A value greater than 2 indicates a negative correlation between adjacent residuals, whereas a value below 2 indicates a positive correlation. The size of the Durbin–Watson statistic depends upon the number of predictors in the model and the number of observations. If this test is relevant to you, look up the critical values in Durbin and Watson (1951). As a very conservative rule of thumb, values less than 1 or greater than 3 are cause for concern.

- *Homoscedasticity* (see Section 6.7): At each level of the predictor variable(s), the variance of the residual terms should be constant. This assumption means that the residuals at each level of the predictor(s) should have the same variance (**homoscedasticity**); when the variances are very unequal there is said to be **heteroscedasticity**. Violating this assumption invalidates confidence intervals and significance tests; estimates of the model parameters (b) using the method of least squares are valid but not optimal. This problem is overcome using weighted least squares regression, in which each case is weighted by a function of its variance or using robust regression.
- *Normally distributed errors* (see Section 6.6): It can be helpful if the residuals in the model are random, normally distributed variables with a mean of 0. This assumption means that the differences between the predicted and observed data are most frequently zero or

very close to zero, and that differences much greater than zero happen only occasionally. Some people confuse this assumption with the idea that predictors have to be normally distributed, which they don't. In small samples a lack of normality invalidates confidence intervals and significance tests, whereas in large samples it will not because of the central limit theorem. If you are concerned only with estimating the model parameters (and not significance tests and confidence intervals) then this assumption barely matters. If you bootstrap confidence intervals then you can ignore this assumption.

There are other considerations that we haven't touched on (see Berry, 1993):

- *Predictors are uncorrelated with 'external variables'*: *External variables* are variables that haven't been included in the model and that influence the outcome variable.[8] These variables are like the 'third variable' that we discussed in the correlation chapter. This assumption means that there should be no external variables that correlate with any of the variables included in the regression model. Obviously, if external variables do correlate with the predictors, then the conclusions we draw from the model become unreliable (because other variables exist that can predict the outcome just as well).
- *Variable types*: All predictor variables must be quantitative or categorical (with two categories), and the outcome variable must be quantitative, continuous and unbounded. By 'quantitative' I mean that they should be measured at the interval level and by 'unbounded' I mean that there should be no constraints on the variability of the outcome. If the outcome is a measure ranging from 1 to 10 yet the data collected vary between 3 and 7, then these data are constrained.

---

8 Some authors refer to these external variables as part of an error term that includes any random factor in the way in which the outcome varies. However, to avoid confusion with the residual terms in the regression equations I have chosen the label 'external variables'. Although this term implicitly washes over any random factors, I acknowledge their presence.

- *No perfect* **multicollinearity**: If your model has more than one predictor then there should be no perfect linear relationship between two or more of the predictors. So, the predictor variables should not correlate too highly (see Section 9.9.3).
- *Non-zero variance*: The predictors should have some variation in value (i.e., they should not have variances of 0). This is self-evident really.

As we saw in Chapter 6, violating these assumptions has implications mainly for significance tests and confidence intervals; the estimates of *b*s are not dependent on these assumptions (although least squares methods will be optimal when the assumptions are met). However, the 95% confidence interval for a *b* tells us the boundaries within which the population values of that *b* are likely to fall.[9] Therefore, if confidence intervals are inaccurate (as they are when these assumptions are broken) we cannot accurately estimate the likely population value. In other words, we can't generalize our model to the population. When the assumptions are met then *on average* the regression model from the sample is the same as the population model. However, you should be clear that even when the assumptions are met, it is possible that a model obtained from a sample is not the same as the population model—but the likelihood of them being the same is increased.

9.4.2 Cross-validation of the model ▮▮▮▮

Even if we can't be confident that the model derived from our sample accurately represents the population, we can assess how well our model might predict the outcome in a different sample. Assessing the accuracy of a model across different samples is known as **cross-validation**. If a model can be generalized, then it must be capable of accurately predicting the same outcome variable from the same set of predictors in a different group of people. If the model is applied to a different sample and there is a severe drop in its predictive power, then the model does *not* generalize. First, we should collect enough data to obtain a reliable model (see the next section). Once we have a estimated the model there are two main methods of cross-validation:

- *Adjusted R^2*: Whereas R^2 tells us how much of the variance in Y overlaps with predicted values from the model in our sample, **adjusted R^2** tells us how much variance in Y would be accounted for if the model had been derived from the population from which the sample was taken. Therefore, the adjusted value indicates the loss of predictive power or **shrinkage**. SPSS derives the adjusted R^2 using Wherry's equation. This equation has been criticized because it tells us nothing about how well the model would predict scores of a different sample of data from the same population. Stein's formula,

$$\text{adjusted } R^2 = 1 - \left[\left(\frac{n-1}{n-k-1} \right) \left(\frac{n-2}{n-k-2} \right) \left(\frac{n+1}{n} \right) \right] (1 - R^2)$$

(9.15)

does tell us how well the model cross-validates (see Stevens, 2002), and the more mathematically minded of you might want to try using it instead of what SPSS chugs out. In Stein's formula, R^2 is the unadjusted value, n is the number of cases and k is the number of predictors in the model.

- *Data splitting*: This approach involves randomly splitting your sample data, estimating the model in both halves of the data and comparing the resulting models. When using stepwise methods (see Section 9.9.1), cross-validation is particularly important; you should run the stepwise regression on a random selection of about 80% of your cases.

Then force this model on the remaining 20% of the data. By comparing values of R^2 and *b* in the two samples you can tell how well the original model generalizes (see Tabachnick & Fidell, 2012).

9.5 Sample size and the linear model ▮▮▮▮

In the previous section I said that it's important to collect enough data to obtain a reliable regression model. Also, larger samples enable us to assume that our *b*s  have normal sampling distributions because of the central limit theorem (Section 6.6.1). Well, how much is enough?

You'll find a lot of rules of thumb floating about, the two most common being that you should have 10 cases of data for each predictor in the model or 15 cases of data per predictor. These rules are very pervasive but they oversimplify the issue to the point of being useless. The sample size required depends on the size of effect that we're trying to detect (i.e., how strong the relationship is that we're trying to measure) and how much power we want to detect these effects. The simplest rule of thumb is that the bigger the sample size, the better: the estimate of R that we get from regression is dependent on the number of predictors, k, and the sample size, N. In fact, the expected R for random data is $k/(N-1)$ and so with small sample sizes random data can appear to show a strong effect: for example, with six predictors and 21 cases of data, $R = 6/(21-1) = 0.3$ (a medium effect size by Cohen's criteria described in Section 3.7.2). Obviously for random data we'd want the expected R to be 0 (no effect), and for this to be true we need large samples (to take

_____

9 Assuming your sample is one of the 95% that generates a confidence interval containing the population value. Yes, I do have to keep making this point—it's important.

289

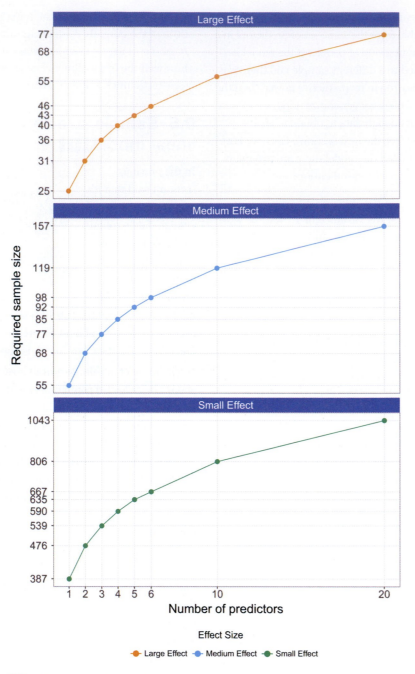

Effect Size

• Large Effect • Medium Effect • Small Effect

Figure 9.9 The sample size required to test the overall regression model depending on the number of predictors and the size of expected effect, R2 = 0.02 (small), 0.13 (medium) and 0.26 (large)

speaking, if your aim is to test the overall fit of the model: (1) if you expect to find a large effect then a sample size of 77 will always suffice (with up to 20 predictors) and if there are fewer predictors then you can afford to have a smaller sample; (2) if you're expecting a medium effect, then a sample size of 160 will always suffice (with up to 20 predictors), you should always have a sample size above 55, and with six or fewer predictors you'll be fine with a sample of 100; and (3) if you're expecting a small effect size then just don't bother unless you have the time and resources to collect hundreds of cases of data. Miles and Shevlin (2001) produce more detailed graphs that are worth a look, but the take-home message is that if you're looking for medium to large effects sample sizes don't need to be massive, regardless of how many predictors you have.

9.6 Fitting linear models: the general procedure ▌▌▌▌

Figure 9.10 shows the general process of fitting linear models. First, we should produce scatterplots to get some idea of whether the assumption of linearity is met, and to look for outliers or obvious unusual cases. At this stage we might transform the data to correct problems. Having done this initial screen for problems, we fit a model and save the various diagnostic statistics that we discussed in Section 9.3. If we want to generalize our model beyond the sample or we are interested in interpreting significance tests and confidence intervals, then we examine these residuals to check for homoscedasticity, normality, independence and linearity (although this will likely be fine, given our earlier screening). If we find problems then we take corrective action and re-estimate the model. This process might seem complex, but it's not

the previous example, if we had 100 cases rather than 21, then the expected R would be a more acceptable 0.06).

Figure 9.9 shows the sample size required[10] to achieve a high level of power (I've taken Cohen's, 1988, benchmark of 0.8) to test

that the model is significant overall (i.e., R^2 is not equal to zero). I've varied the number of predictors and the size of expected effect: I used $R^2 = 0.02$ (small), 0.13 (medium) and 0.26 (large), which correspond to benchmarks in Cohen (1988). Broadly

10 I used the program G*power, mentioned in Section 2.9.8, to compute these values.

as bad as it seems. Also, it's probably wise to use bootstrapped confidence intervals when we first estimate the model because then we can basically forget about things like normality.

9.7 Using SPSS Statistics to fit a linear model with one predictor ▌▌▌▌

Earlier on I asked you to imagine that I worked for a record company and that my boss was interested in predicting album sales from advertising. There are data for this example in the file **Album Sales.sav**. This data file has 200 rows, each one representing a different album. There are also several columns, one of which contains the sales (in thousands) of each album in the week after release (**Sales**) and one containing the amount (in thousands of pounds) spent promoting the album before release (**Adverts**). The other columns represent how many times songs from the album were played on a prominent national radio station in the week before release (**Airplay**), and how attractive people found the band's image out of 10 (**Image**). Ignore these last two variables for now; we'll use them later. Note how the data are laid out (Figure 9.11): each variable is in a column and each row represents a different album. So, the first album had £10,260 spent advertising it, sold 330,000 copies, received 43 plays on radio the week before release, and was made by a band with a pretty sick image.

Figure 9.12 shows that a positive relationship exists: the more money spent advertising the album, the more it sells. Of course there are some albums that sell well regardless of advertising (top left of scatterplot), but there are none that sell badly when advertising levels are high (bottom right of scatterplot). The scatterplot shows the line of best fit for these data: bearing in mind that the mean would be represented by a flat line at

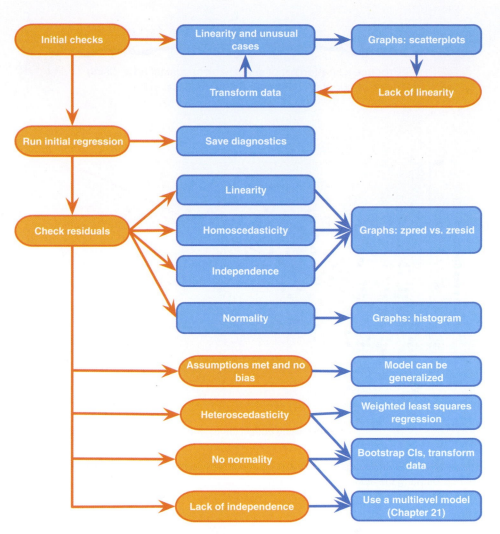

Figure 9.10   The process of fitting a regression model

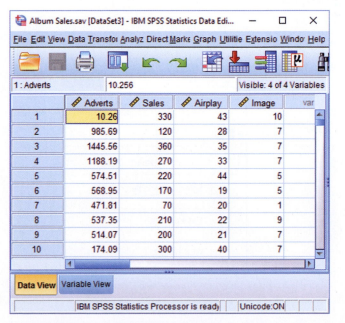

Figure 9.11   The data editor for fitting a linear model

291

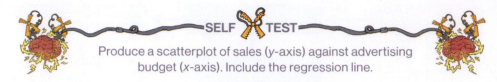

Produce a scatterplot of sales (*y*-axis) against advertising budget (*x*-axis). Include the regression line.

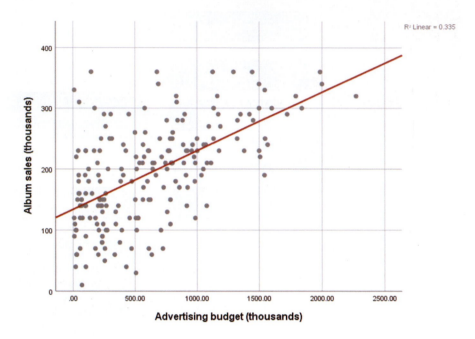

Figure 9.12   Scatterplot showing the relationship between album sales and the amount spent promoting the album

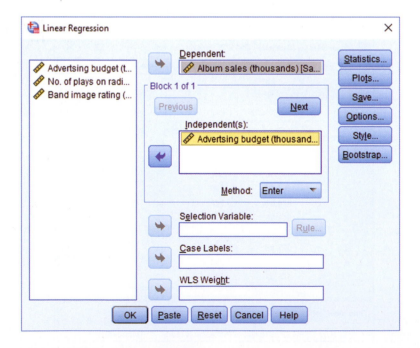

Figure 9.13   Main dialog box for regression

around the 200,000 sales mark, the regression line is noticeably different.

To fit the model, access the main dialog box by selecting *Analyze* ▶ *Regression* ▶ Linear... (Figure 9.13). First, we define the outcome variable (in this example **Sales**). Select **Sales** from the list on the left-hand side, and transfer it to the space labeled *Dependent* by dragging it or clicking ⟶. In this model we're going to enter only one predictor (**Adverts**) so select it from the list and click ⟶ (or drag it) to transfer it to the box labeled *Independent(s)*. There are a ton of options available, but we'll explore these when we build up the model in due course. For now, request bootstrapped confidence intervals for the regression coefficients by clicking Bootstrap... (see Section 6.12.3). Select ☑ Perform bootstrapping to activate bootstrapping, and to get a 95% confidence interval select ◉ Bias corrected accelerated (BCa). Click OK in the main dialog box to fit the model.

9.8 Interpreting a linear model with one predictor ▌▌▌▌

9.8.1 Overall fit of the model ▌▌▌▌

The first table is a summary of the model (Output 9.2). This summary table provides the value of R and R^2 for the model. For these data, R has a value of 0.578 and because there is only one predictor, this value is the correlation between advertising and album sales (you can confirm this by running a correlation using what you learnt in Chapter 8). The value of R^2 is 0.335, which tells us that advertising expenditure can account for 33.5% of the variation in album sales. This means that 66.5% of the variation in album sales remains unaccounted for: there might be other variables that have an influence also.

The next part of the output (Output 9.3) reports the various sums of squares described in Figure 9.5, the degrees of freedom associated with each and the resulting mean squares (equation (9.11)). The most important part of the table is the F-statistic (equation (9.12)) of 99.59 and its associated significance value of $p < 0.001$ (expressed this way because the value in the column labeled *Sig.* is less than 0.001). This p-value tells us that there is less than a 0.1% chance that an F-statistic at least this large would happen if the null hypothesis were true. Therefore, we could conclude that our model results in significantly better prediction of album sales than if we used the mean value of album sales. In short, the linear model overall predicts album sales significantly.

9.8.2 Model parameters

Output 9.4 provides estimates of the model parameters (the beta values) and the significance of these values. We saw in equation (9.2) that b_0 was the Y intercept, and this value is 134.14 (B for the constant in Output 9.4). This value can be interpreted as meaning that when no money is spent on advertising (when $X = 0$), the model predicts that 134,140 albums will be sold (remember that our unit of measurement is thousands of albums). We can also read off the value of b_1 from the table, which is 0.096. Although this value is the slope of the line for the model, it is more useful to think of this value as representing *the change in the outcome associated with a unit change in the predictor*. In other words, if our predictor variable is increased by one unit (if the advertising budget is increased by 1), then our model predicts that 0.096 extra albums will be sold. Our units of measurement were thousands of pounds and thousands of albums sold, so we can say that for an increase in advertising of

How do I interpret *b*-values?

Model Summary

Model	R	R Square	Adjusted R Square	Std. Error of the Estimate
1	.578[a]	.335	.331	65.991

a. Predictors: (Constant), Advertsing budget (thousands)

Output 9.2

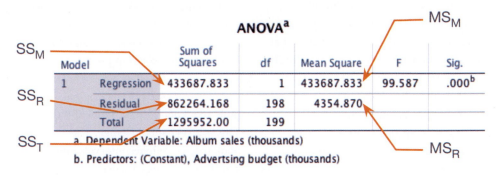

ANOVA[a]

Model		Sum of Squares	df	Mean Square	F	Sig.
1	Regression	433687.833	1	433687.833	99.587	.000[b]
	Residual	862264.168	198	4354.870		
	Total	1295952.00	199			

a. Dependent Variable: Album sales (thousands)
b. Predictors: (Constant), Advertsing budget (thousands)

Output 9.3

Coefficients[a]

Model		Unstandardized Coefficients		Standardized Coefficients	t	Sig.
		B	Std. Error	Beta		
1	(Constant)	134.140	7.537		17.799	.000
	Advertising budget (thousands)	.096	.010	.578	9.979	.000

a. Dependent Variable: Album sales (thousands)

Bootstrap for Coefficients

Model		Bootstrap[a]					
		B	Bias	Std. Error	Sig. (2–tailed)	BCa 95% Confidence Interval	
						Lower	Upper
1	(Constant)	134.140	-.049	8.087	.001	118.699	150.746
	Advertising budget (thousands)	.096	.000	.008	.001	.079	.113

a. Unless otherwise noted, bootstrap results are based on 1000 bootstrap samples

Output 9.4

£1000 the model predicts 96 (0.096 × 1000 = 96) extra album sales. This investment is pretty useless for the record company: it invests £1000 and gets only 96 extra sales! Fortunately, as we already know, advertising accounts for only one-third of album sales.

We saw earlier that if a predictor is having a significant impact on our ability to predict the outcome then its b should be different from 0 (and large relative to its standard error). We also saw that the t-test

and associated p-value tell us whether the b-value is significantly different from 0. The column *Sig.* contains the exact probability that a value of t at least as big as the one in the table would occur if the value of b in the population were zero. If this probability is less than 0.05, then people interpret that as the predictor being a 'significant' predictor of the outcome (see Chapter 2). For both ts, the probabilities are given as 0.000 (zero to 3 decimal places), and so we can say that the

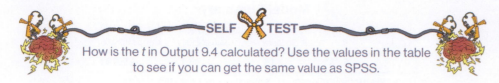

How is the *t* in Output 9.4 calculated? Use the values in the table to see if you can get the same value as SPSS.

probability of these *t* values (or larger) occurring if the values of *b* in the population were zero is less than 0.001. In other words, the *b*s are significantly different from 0. In the case of the *b* for advertising budget this result means that the advertising budget makes a significant contribution ($p < 0.001$) to predicting album sales.

If our sample is one of the 95% producing confidence intervals that contain the population value then the bootstrap confidence interval tells us that the population value of *b* for advertising budget is likely to fall between 0.079 and 0.113, and because this interval doesn't include zero we might conclude that there is a genuine positive relationship between advertising budget and album sales in the population. Also, the significance associated with this confidence interval is $p = 0.001$, which is highly significant. Note that the bootstrap process involves re-estimating the standard error (it changes from 0.01 in the original table to a bootstrap estimate of 0.008). This is a very small change. The bootstrap confidence intervals and significance values are useful to report and interpret because they do not rely on assumptions of normality or homoscedasticity.

9.8.3 Using the model ▮▮▮▮

We have discovered that we have a model that significantly improves our ability to predict album sales. The next stage is often to use that model to make predictions about the future. First we specify the model by replacing the *b*-values in equation (9.2) with the values from Output 9.4. We can also replace the *X* and *Y* with the variable names:

$$\text{album sales}_i = b_0 + b_1\text{advertising budget}_i$$
$$= 134.14 + \left(0.096 \times \text{advertising budget}_i\right)$$

(9.16)

We can make a prediction about album sales by replacing the advertising budget with a value of interest. For example, if we spend £100,000 on advertising a new album, remembering that our units are already in thousands of pounds, we simply replace the advertising budget with 100.

Cramming Sam's Tips
Linear models

- A linear model (regression) is a way of predicting values of one variable from another based on a model that describes a straight line.
- This line is the line that best summarizes the pattern of the data.
- To assess how well the model fits the data use:

  ○ R^2, which tells us how much variance is explained by the model compared to how much variance there is to explain in the first place. It is the proportion of variance in the outcome variable that is shared by the predictor variable.

  ○ *F*, which tells us how much variability the model can explain relative to how much it can't explain (i.e., it's the ratio of how good the model is compared to how bad it is).

  ○ the *b*-value, which tells us the gradient of the regression line and the strength of the relationship between a predictor and the outcome variable. If it is significant (*Sig.* < 0.05 in the SPSS output) then the predictor variable significantly predicts the outcome variable.

SELF TEST

How many albums would be sold if we spent £666,000 on advertising the latest album by Deafheaven?

We discover that album sales should be around 144,000 for the first week of sales:

$$\begin{aligned} \text{album sales}_i &= 134.14 + \left(0.096 \times \text{advertising budget}_i\right) \\ &= 134.14 + \left(0.096 \times 100\right) \\ &= 143.74 \end{aligned} \qquad (9.17)$$

9.9 The linear model with two or more predictors (multiple regression) ▌▌▌

Imagine that the record company executive wanted to extend the model of albums sales to incorporate other predictors. Before an album is released, the executive notes the amount spent on advertising, the number of times songs from the album are played on a prominent radio station the week before release (**Airplay**), and ratings of the band's image (**Image**). He or she does this for 200 albums (each by a different band). The credibility of the band's image was rated by a random sample of the target audience on a scale from 0 (dad dancing at a disco) to 10 (sicker than a dog that's eaten a bag of onions). The mode rating was used because the executive was interested in what most people thought, not the average opinion.

When we build a model with several predictors, everything we have discussed so far applies. However, there are some additional things to think about. The first is what variables to enter into the model. A great deal of care should be taken in selecting predictors for a model because the estimates of the regression coefficients depend upon the variables in the model (and the order in which they are entered). *Do not enter hundreds of predictors, just because you've measured them, and expect the resulting model to make sense.* SPSS Statistics will happily generate output based on any garbage you decide to feed it—it will not judge you, but others will. Select predictors based on a sound theoretical rationale or well-conducted past research that has demonstrated their importance.[11] In our example, it seems logical that the band's image and radio play ought to affect sales, so these are sensible predictors. It would not be sensible to measure how much the album cost to make because this won't affect sales directly: you would just add noise to the model. If predictors are being added that have never been looked at before (in your research context) then select these variables based on their substantive *theoretical* importance. The key point is that the most important thing when building a model is to use your brain—which is slightly worrying if your brain is as small as mine.

9.9.1 Methods of entering predictors into the model ▌▌▌

Having chosen predictors, you must decide the order to enter them into the model. When predictors are completely uncorrelated the order of variable entry has very little effect on the parameters estimated; however, we rarely have uncorrelated predictors, and so the method of variable entry has consequences and is, therefore, important.

Other things being equal, use **hierarchical regression**, in which you select predictors based on past work and decide in which order to enter them into the model. Generally speaking, you should enter known predictors (from other research) into the model first in order of their importance in predicting the outcome. After having entered known predictors, you can add new predictors into the model simultaneously, in a stepwise manner or hierarchically (entering the new predictor suspected to be the most important first).

An alternative is forced entry (or *Enter* as it is known in SPSS), in which you force all predictors into the model simultaneously. Like hierarchical, this method relies on good theoretical reasons for including the chosen predictors, but unlike hierarchical, you make no decision about the order in which variables are entered. Some researchers believe that this method is the only appropriate method for theory testing (Studenmund & Cassidy, 1987), because stepwise techniques are influenced by random variation in the data and so seldom give replicable results if the model is retested.

The final option, **stepwise regression**, is generally frowned upon by statisticians. Nevertheless, SPSS Statistics makes it easy to do and actively encourages it in the *Automatic Linear Modeling* process (probably because this function is aimed at people who don't know better)—see Oditi's Lantern. I'm assuming that you wouldn't wade through 900 pages of my drivel unless you wanted to know better, so we'll give stepwise a wide berth. However, you probably ought to know what it does so you can understand why to avoid it. The stepwise method bases decisions about the order in which predictors enter the model on a purely mathematical criterion. In the *forward* method, an initial model is defined that contains only the constant (b_0). The computer then searches for the predictor (out of the ones available) that best predicts the outcome variable—it does this by selecting the predictor that has the highest simple correlation with the outcome. If this predictor significantly improves the model's ability to predict the outcome then it is

---

11 Preferably past research that is methodologically and statistically rigorous and yielded reliable, generalizable models.

retained and the computer looks to add a second predictor from the available pool of variables. The next predictor the computer tries will be the one that has the largest semi-partial correlation with the outcome. Remember that the semi-partial correlation quantifies the unique overlap between two variables X and Y: it 'partials out' or accounts for the relationship that X has with other predictors. Therefore, the computer looks for the variable that has the largest *unique* overlap with the outcome. This variable is retained if it significantly improves the fit of the model, otherwise it is rejected and the process stops. If it is retained and there are still potential predictors left out of the model then these are reviewed and the one with the largest semi-partial correlation with the outcome is entered, evaluated and retained if it significantly improves the fit, and so on until there are no more potential predictors or none of the potential predictors significantly improves the model if it is entered.

Let's make this process a bit more concrete. In Section 8.5, we used an example of the relationships between exam performance, exam anxiety and revision time. Imagine our goal is to predict exam performance from the other two variables. Think back to Figure 8.10. If we build the model stepwise, the first step is to see which of exam anxiety and revision time overlaps more with exam performance. The area of overlap between exam performance and exam anxiety is area A + C (19.4%), whereas for revision time it is area B + C (15.8%). Therefore, exam anxiety will enter the model first and is retained only if it significantly improves the model's predictive power. If not, no predictor variables are entered.

In our exam performance example, there is only one other potential predictor (revision time), so this will be entered next. Remember that its unique overlap with exam performance is area B in Figure 8.10: we ignore the part of its overlap with exam performance that is shared with exam anxiety

(area C), because that variable is already in the model. If area B is big enough to improve the fit of the model significantly then revision time will be retained. If not the final model will contain only exam anxiety.

In the case where we had another potential predictor (let's say we measured the difficulty of the exam) and exam anxiety had been entered first, then the unique overlap for revision time and exam performance (area B) would be compared to the equivalent area for exam difficulty. The variable with the bigger area would be entered next, evaluated and retained only if its inclusion improved the fit of the model.

The *stepwise* method in SPSS Statistics is the same as the forward method, except that each time a predictor is added to the equation, a removal test is made of the least useful predictor. As such, the regression equation is constantly reassessed to see whether redundant predictors can be removed. The *backward* method is the opposite of the forward method in that the

Oditi's Lantern
Automatic Linear Modeling

'I, Oditi, come with a warning. Your desparation to bring me answers to numerical truths so as to gain a privileged place within my heart may lead you into the temptation that is SPSS's *Automatic Linear Modeling*. This feature promises answers without thought, and like a cat who is promised a fresh salmon, you will drool and purr in anticipation. If you want to find out more then stare into my lantern, but be warned, sometimes what looks like a juicy salmon is a rotting pilchard in disguise.'

model initially contains all predictors and the contribution of each is evaluated with the p-value of its t-test. This significance value is compared against a removal criterion (which can be either an absolute value of the test statistic or a p-value). If a predictor meets the removal criterion (i.e., it is not making a statistically significant contribution to the model) it is removed and the model is re-estimated for the remaining predictors. The contribution of the remaining predictors is then reassessed.

Which of these methods should you use? The short answer is 'not stepwise', because variables are selected based upon mathematical criteria. The issue is that these criteria (e.g., the semi-partial correlation) are at the mercy of sampling variation. That is, a particular variable might have a large semi-partial correlation in your sample but a small one in a different sample. Therefore, models built using stepwise methods are less likely to generalize across samples because the selection of variables in the model is affected by the sampling process. Also, because the criterion for retaining variables is based on statistical significance, your sample size affects the model you get: in large samples significance tests are highly powered, resulting in predictors being retained that make trivial contributions to predicting the outcome, and in small samples where power is low, predictors that make a large contribution may get overlooked. Consequently, there is the danger of overfitting (having too many variables in the model that essentially make little contribution to predicting the outcome) and underfitting (leaving out important predictors) the model. Stepwise methods

Which method of regression should I use?

also take important methodological decisions out of the hands of the researcher.

The main problem with stepwise methods is that they assess the fit of a variable based on the other variables in the model. Jeremy Miles (who has worked with me on other books) illustrates this problem by imagining getting dressed using a stepwise method. You wake up one morning and on your dressing table (or floor, if you're me) you have underwear, some jeans, a T-shirt and jacket. Imagine these items are predictor variables. It's a cold day and you're trying to keep warm. A stepwise method will put your trousers on first because they fit your goal best. It then looks around and tries the other clothes (variables). It tries to put your underwear on you but it won't fit over your jeans. It decides they are 'a poor fit' and discards them. It tries a jacket—that fits, but your T-shirt doesn't go over the top and is discarded. You end up leaving the house in jeans and a jacket with nothing underneath. You are very cold. Later in the day during a university seminar you stand up and your trousers fall down (because your body has shrunk from the cold), exposing you to your year group. It's a mess. The problem is that the underwear was a poor fit only because when you tried to put it on you were already wearing jeans. In stepwise methods, variables might be considered bad predictors only because of what has already been put in the model. For these reasons, stepwise methods are best avoided except for exploratory model building. If you do decide to use a stepwise method then let the statistical blood be on your hands, not mine. Use the backward method rather than the forward method to minimize **suppressor effects**, which occur when a predictor has a significant effect only when another variable is held constant.

Forward selection is more likely than backward elimination to exclude predictors involved in suppressor effects. As such, the forward method runs a higher risk of making a Type II error (i.e., missing a predictor that does in fact predict the outcome). It is also advisable to cross-validate your model by splitting the data (see Section 9.4.2).

9.9.2 Comparing models

Hierarchical and (although obviously you'd never use them) stepwise methods involve adding predictors to the model in stages, and it is useful to assess the improvement to the model at each stage. Given that larger values of R^2 indicate better fit, a simple way to quantify the improvement when predictors are added is to compare the R^2 for the new model to that for the old model. We can assess the significance of the change in R^2 using equation (9.13), but because we're looking at the change in models we use the change in R^2 (R^2_{change}) and the change in the number of predictors (k_{change}), as well as the R^2 (R^2_{new}) and number of predictors (k_{new}) in the new model:

$$F_{\text{change}} = \frac{\left(N - k_{\text{new}} - 1\right)R^2_{\text{change}}}{k_{\text{change}}\left(1 - R^2_{\text{new}}\right)} \qquad (9.18)$$

We can compare models using this F-statistic. The problem with R^2 is that when you add more variables to the model, it always goes up. So, if you are deciding which of two models fits the data better, the model with more predictor variables in will always fit better. The **Akaike information criterion** (AIC)[12] is a measure of fit that penalizes the model for having more variables. If the AIC is bigger, the fit is worse; if the AIC is smaller, the fit is better. If you use the *Automatic Linear Model* function in SPSS Statistics then you can use the AIC to

---

12 Hirotsugu Akaike (pronounced 'A-ka-ee-kay') was a Japanese statistician who gave his name to the AIC, which is used in a huge range of different places.

select models rather than the change in R^2. The AIC doesn't mean anything on its own: you cannot say that an AIC value of 10 is small or that a value of 1000 is large. The only thing you do with the AIC is compare it to other models with the same outcome variable: if it's getting smaller, the fit of your model is improving.

9.9.3 Multicollinearity ▮▮▮▮

A final consideration for models with more than one predictor is multicollinearity, which exists when there is a strong correlation between two or more predictors. **Perfect collinearity** exists when at least one predictor is a perfect linear combination of the others (the simplest example being two predictors that are perfectly correlated—they have a correlation coefficient of 1). If there is perfect collinearity between predictors it becomes impossible to obtain unique estimates of the regression coefficients because there are an infinite number of combinations of coefficients that would work equally well. Put simply, if we have two predictors that are perfectly correlated, then the values of b for each variable are interchangeable. The good news is that perfect collinearity is rare in real-life data. The bad news is that less than perfect collinearity is virtually unavoidable. Low levels of collinearity pose little threat to the model estimates, but as collinearity increases there are three problems that arise:

- **Untrustworthy bs**: As collinearity increases, so do the standard errors of the b coefficients. Big standard errors for b coefficients mean more variability in these bs across samples, and a greater chance of (1) predictor equations that are unstable across samples too; and (2) b coefficients in the sample that are unrepresentative of those in the population. Crudely put, multicollinearity leads to untrustworthy b-values. Don't lend them money and

don't let them go for dinner with your boy- or girlfriend.
- **It limits the size of R**: Remember that R is a measure of the correlation between the predicted values of the outcome and the observed values and that R^2 indicates the variance in the outcome for which the model accounts. Imagine a situation in which a single variable predicts the outcome variable with $R = 0.80$ and a second predictor variable is added to the model. This second variable might account for a lot of the variance in the outcome (which is why it is included in the model), but the variance it accounts for is the same variance accounted for by the first variable (the second variable accounts for very little *unique* variance). Hence, the overall variance in the outcome accounted for by the two predictors is little more than when only one predictor is used (R might increase from 0.80 to 0.82). If, however, the two predictors are completely uncorrelated, then the second predictor is likely to account for *different* variance in the outcome than that accounted for by the first predictor. The second predictor might account for only a little of the variance in the outcome, but the variance it does account for is different from that of the other predictor (and so when both predictors are included, R is substantially larger, say 0.95).
- **Importance of predictors**: Multicollinearity between predictors makes it difficult to assess the individual importance of a predictor. If the predictors are highly correlated, and each accounts for similar variance in the outcome, then how can we know which of the two variables is important? We can't—the model could include either one, interchangeably.

A 'ball park' method of identifying multicollinearity (that will miss subtler forms) is to scan the correlation matrix for predictor variables that correlate very highly (values of r above 0.80 or 0.90). SPSS Statistics can compute the **variance**

inflation factor (VIF), which indicates whether a predictor has a strong linear relationship with the other predictor(s), and the **tolerance** statistic, which is its reciprocal (1/VIF). Some general guidelines have been suggested for interpreting the VIF:

- If the largest VIF is greater than 10 (or the tolerance is below 0.1) then this indicates a serious problem (Bowerman & O'Connell, 1990; Myers, 1990).
- If the average VIF is substantially greater than 1 then the regression may be biased (Bowerman & O'Connell, 1990).
- Tolerance below 0.2 indicates a potential problem (Menard, 1995).

Other measures that are useful in discovering whether predictors are dependent are the *eigenvalues of the scaled, uncentered cross-products matrix*, the *condition indexes* and the *variance proportions*. These statistics will be covered as part of the interpretation of SPSS output (see Section 9.11.5). If none of this made sense, Hutcheson and Sofroniou (1999) explain multicollinearity very clearly.

9.10 Using SPSS Statistics to fit a linear model with several predictors ▮▮▮▮

Remember the general procedure in Figure 9.10. First, we could look at scatterplots of the relationships between the outcome variable and the predictors. Figure 9.14 shows a matrix of scatterplots for our album sales data, but I have shaded all the scatterplots except the three related to the outcome, album sales. Although the data are messy, the three predictors have reasonably linear relationships with the album sales and there are no obvious outliers (except maybe in the bottom left of the scatterplot with band image).

9.10.1 Main options ▌▌▌▌

Past research shows that advertising budget is a significant predictor of album sales, and so we should include this variable in the model first, entering the new variables (**Airplay** and **Image**) afterwards. This method is hierarchical (we decide the order that variables are entered based on past research). To do a hierarchical regression we enter the predictors in blocks, with each block representing one step in the hierarchy. Access the main *Linear Regression* dialog box by selecting *Analyze* ▶ *Regression* ▶ Linear…. We encountered this dialog box when we looked at a model with only one predictor (Figure 9.13). To set up the first block we do what we did before: drag **Sales** to the box labeled *Dependent* (or click ➡). We also need to specify the predictor variable for the first block, which we decided should be advertising budget. Drag this variable from the left-hand list to the box labeled *Independent(s)* (or click ➡). Underneath the *Independent(s)* box is a drop-down menu for specifying the *Method* of variable entry (see Section 9.9.1). You can select a different method for each block by clicking on Enter ▼. The default option is forced entry, and this is the option we want, but if you were carrying out more exploratory work, you might use a different method.

Having specified the first block, we can specify a second by clicking Next. This process clears the *Independent(s)* box so that you can enter the new predictors (note that it now reads *Block 2 of 2* above this box to indicate that you are in the second block of the two that you have so far specified). We decided that the second block would contain both of the new predictors, so select **Airplay** and **Image** and drag them to the *Independent(s)* box (or click ➡). The dialog box should look like Figure 9.15. To move between blocks use the Previous and Next buttons (e.g., to move back to block 1, click Previous).

It is possible to select different methods of variable entry for different blocks. For

SELF TEST

Produce a matrix scatterplot of **Sales**, **Adverts**, **Airplay** and **Image** including the regression line.

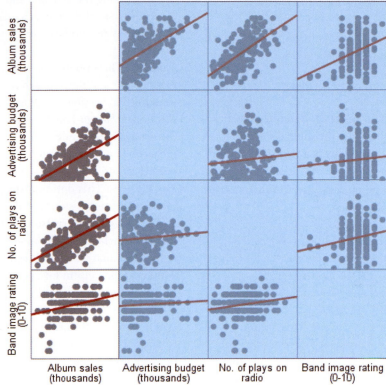

Figure 9.14   Matrix scatterplot of the relationships between advertising budget, airplay, image rating and album sales

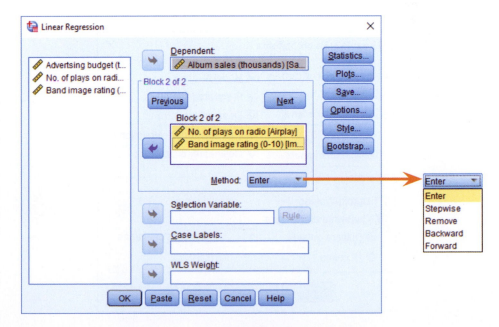

Figure 9.15   Main dialog box for block 2 of the multiple regression

example, having specified forced entry for the first block, we could now specify a stepwise method for the second. Given that we have no previous research regarding the effects of image and airplay on album sales, we might be justified in doing this. However, because of the problems with stepwise methods, I am going to stick with forced entry for both blocks.

9.10.2 Statistics ▌▌▌

In the main *Regression* dialog box click Statistics... to open the dialog box in Figure 9.16. Below is a run-down of the options available. Select the options you require and click Continue to return to the main dialog box.

- *Estimates*: This option is selected by default because it gives us the estimated *b*-values for the model as well as the associated *t*-test and *p*-value (see Section 9.2.5).
- *Confidence intervals*: This option produces confidence intervals for each *b*-value in the model. Remember that if the model assumptions are not met these confidence intervals will be inaccurate and bootstrap confidence intervals should be used instead.

- *Covariance matrix*: This option produces a matrix of the covariances, correlation coefficients and variances between the *b*-values for each variable in the model. A variance–covariance matrix displays variances along the diagonal and covariances as off-diagonal elements. Correlations are produced in a separate matrix.
- *Model fit*: This option produces the *F*-test, *R*, R^2 and the adjusted R^2 (described in Sections 9.2.4. and 9.4.2).
- *R squared change*: This option displays the change in R^2 resulting from the inclusion of a new predictor (or block of predictors)—see Section 9.9.2.
- *Descriptives*: This option displays a table of the mean, standard deviation and number of observations of the variables included in the model. A correlation matrix is produced too, which can be helpful for spotting multicollinearity.
- *Part and partial correlations*: This option produces the zero-order correlation (the Pearson correlation) between each predictor and the outcome variable. It also produces the semi-partial (part) and partial correlation between each predictor and the outcome (see Sections 8.5 and 9.9.1).

- *Collinearity diagnostics*: This option produces collinearity statistics such as the VIF, tolerance, eigenvalues of the scaled, uncentered cross-products matrix, condition indexes and variance proportions (see Section 9.9.3).
- *Durbin-Watson*: This option produces the Durbin–Watson test statistic, which tests the assumption of independent errors when cases have some meaningful sequence. In this case it isn't useful because our cases do not have a meaningful order.
- *Casewise diagnostics*: This option produces a table that lists the observed value of the outcome, the predicted value of the outcome, the difference between these values (the residual) and this difference standardized. You can choose to have this information for all cases, but that will result in a big table in large samples. The alternative option is to list only cases for which the standardized residual is greater than 3 (when the ± sign is ignored). I usually change this to 2 (so that I don't miss cases with standardized residuals not quite reaching the threshold of 3) A summary table of residual statistics indicating the minimum, maximum, mean and standard deviation of both the values predicted by the model and the residuals is also produced (see Section 9.10.4).

9.10.3 Regression plots ▌▌▌

Once you are back in the main dialog box, click Plots... to activate the dialog box in Figure 9.17, which we can use to test some assumptions of the model. Most of these plots involve various *residual* values, which were described in Section 9.3. The left-hand side lists several variables:

- **DEPENDNT**: the outcome variable.
- ***ZPRED**: the standardized predicted values of the outcome based on the model. These values are standardized forms of the values predicted by the model.
- ***ZRESID**: the standardized residuals or errors. These values are the standardized

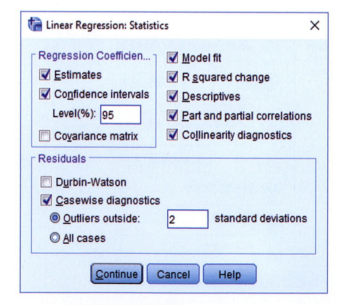

Figure 9.16 *Statistics* dialog box for regression analysis

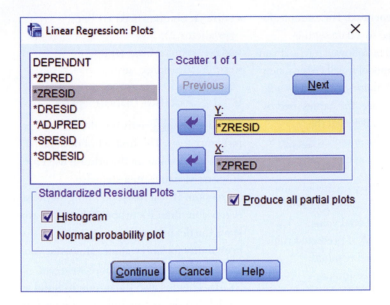

Figure 9.17 The *Plots* dialog box

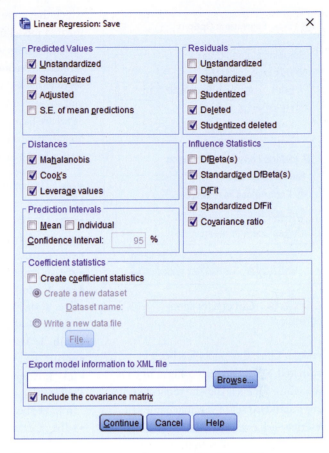

Figure 9.18 Dialog box for regression diagnostics

differences between the observed values of the outcome and those predicted by the model.

- ***DRESID**: the deleted residuals described in Section 9.3.2.
- ***ADJPRED**: the adjusted predicted values described in Section 9.3.2.
- ***SRESID**: the studentized residual described in Section 9.3.1.
- ***SDRESID**: the studentized deleted residual described in Section 9.3.2.

In Section 6.11.1 we saw that a plot of ***ZRESID** (*y*-axis) against ***ZPRED** (*x*-axis) is useful for testing the assumptions of independent errors, homoscedasticity, and linearity. A plot of ***SRESID** (*y*-axis) against ***ZPRED** (*x*-axis) will show up heteroscedasticity also. Although often these two plots are virtually identical, the latter is more sensitive on a case-by-case basis. To create these plots drag a variable from the list to the space labeled either *X* or *Y* (which refer to the axes) or select the variable and click ↩. When you have selected two variables for the first plot (as in Figure 9.17) you can specify a new plot (up to nine different plots) by clicking on Next. This process clears the dialog box and you can specify a second plot. Click Next or Previous to move between plots you have specified.

Ticking the box labeled *Produce all partial plots* will produce scatterplots of the residuals of the outcome variable and each of the predictors when both variables are regressed separately on the remaining predictors. Regardless of whether the previous sentence made any sense to you, these plots have important characteristics that make them worth inspecting. First, the gradient of the regression line between the two residual variables is equivalent to the coefficient of the predictor in the regression equation. As such, any obvious outliers on a partial plot represent cases that might have undue influence on a predictor's coefficient, *b*. Second, non-linear relationships between a predictor and the outcome variable are much more evident on these plots. Finally, they are useful for spotting collinearity.

There are two other tick-boxes labeled *Standardized Residual Plots*. One produces a histogram of the standardized residuals and the other produces a normal probability plot, both of which are useful for checking for normality of errors). Click Continue to return to the main dialog box.

9.10.4 Saving regression diagnostics ▌▌▌▌

Section 9.3 described numerous variables that we can use to diagnose outliers and influential cases. We can save these diagnostic variables for our model in the data editor (SPSS calculates them and places the values in new columns in the data editor) by clicking Save... to access the dialog box in Figure 9.18. Most of the available options were explained in Section 9.3, and Figure 9.18 shows what I consider to be a reasonable set of diagnostic statistics. Standardized (and Studentized) versions of these diagnostics are generally easier to interpret, and so I tend to select them in preference to the unstandardized

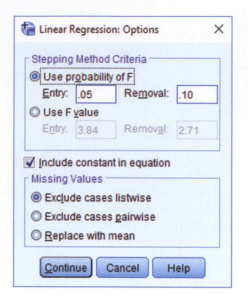

Figure 9.19 Options for linear regression

which they were generated. For example, for the first model fitted to a data set the variable names will be followed by a 1, if you estimate a second model it will create a new set of variables with names followed by a 2, and so on. For reference, the names used by SPSS are listed below. Selected the diagnostics you require and click **Continue** to return to the main dialog box.

- **pre_1**: unstandardized predicted value;
- **zpr_1**: standardized predicted value;
- **adj_1**: adjusted predicted value;
- **sep_1**: standard error of predicted value;
- **res_1**: unstandardized residual;
- **zre_1**: standardized residual;
- **sre_1**: Studentized residual;
- **dre_1**: deleted residual;
- **sdr_1**: Studentized deleted residual;
- **mah_1**: Mahalanobis distance;
- **coo_1**: Cook's distance;
- **lev_1**: centered leverage value;
- **sdb0_1**: standardized DFBeta (intercept);
- **sdb1_1**: standardized DFBeta (predictor 1);

- **sdb2_1**: standardized DFBeta (predictor 2);
- **sdf_1**: standardized DFFit;
- **cov_1**: covariance ratio.

9.10.5 Further options

Clicking **Options...** activates the dialog box in Figure 9.19. The first set of options allows you to change the criteria used for entering variables in a stepwise regression. If you insist on doing stepwise regression, then it's probably best that you leave the default criterion of 0.05 probability for entry alone. However, you can make this criterion more stringent (0.01). There is also the option to build a model that doesn't include a constant (i.e., has no Y intercept). This option should also be left alone. Finally, you can select a method for dealing with missing data points (see SPSS Tip 6.1 for a description). Just a hint, but leave the default of listwise alone because using pairwise can lead to absurdities such as R^2 that is negative or greater than 1.0.

versions. Once the model has been estimated, SPSS creates a column in your data editor for each statistic requested; it uses a standard set of variable names to describe each one. After the name, there will be a number that refers to the model from

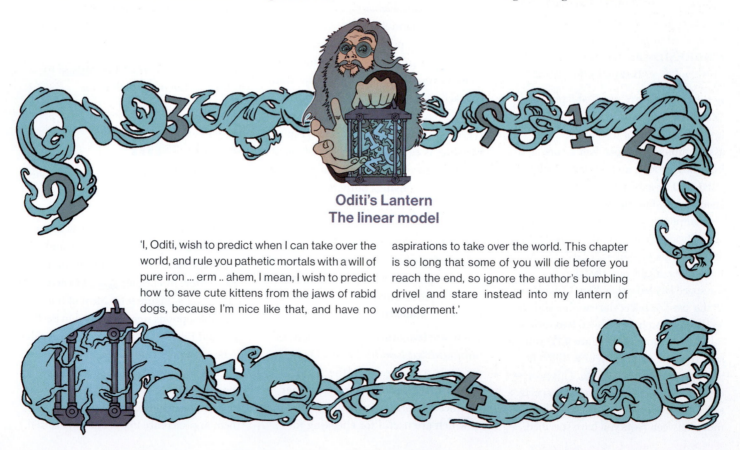

**Oditi's Lantern
The linear model**

'I, Oditi, wish to predict when I can take over the world, and rule you pathetic mortals with a will of pure iron ... erm .. ahem, I mean, I wish to predict how to save cute kittens from the jaws of rabid dogs, because I'm nice like that, and have no aspirations to take over the world. This chapter is so long that some of you will die before you reach the end, so ignore the author's bumbling drivel and stare instead into my lantern of wonderment.'

9.11 Interpreting a linear model with several predictors ▮▮▮▮

Having selected the relevant options and returned to the main dialog box, click `OK` and watch in awe as SPSS Statistics spews forth quite terrifying amounts of output in the viewer window.

9.11.1 Descriptives ▮▮▮▮

The output described in this section is produced using the options in Figure 9.16. If you selected the *Descriptives* option, you'll get Output 9.5, which tells us the mean and standard deviation of each variable in our model. This table is a useful summary of the variables. You'll also see a correlation matrix containing the Pearson correlation coefficient between every pair of variables, the one-tailed significance of each correlation, and the number of cases contributing to each correlation. Along the diagonal of the matrix the values for the correlation coefficients are all 1.00 (a perfect positive

correlation) because they are the correlation of each variable with itself. We can use the correlation matrix to get a sense of the relationships between predictors and the outcome, and for a preliminary look for multicollinearity. If there is no multicollinearity in the data then there should be no substantial correlations ($r > 0.9$) between predictors. If we look only at the predictors (ignore album sales) then the highest correlation is between the ratings of the band's image and the amount of airplay, which is significant at the 0.01 level ($r = 0.182$, $p = 0.005$). Despite the significance, the coefficient itself is small and so there is no collinearity to worry about. If we look at the outcome variable, then it's apparent that of the predictors airplay correlates best with the outcome ($r = 0.599$, $p < 0.001$).

9.11.2 Summary of the model ▮▮▮▮

Output 9.6 describes the overall fit of the model. There are two models in the table because we chose a hierarchical method

with two blocks and the summary statistics are repeated for each model/block. Model 1 refers to the first stage in the hierarchy when only advertising budget is used as a predictor. Model 2 refers to when all three predictors are used. We can tell this from the footnotes under the table. If you selected the *R squared change* and *Durbin-Watson* options, then these values are included also (we didn't select Durbin–Watson so it is missing from Output 9.6).

The column labeled *R* contains the multiple correlation coefficient between the predictors and the outcome. When only advertising budget is used as a predictor, this is the simple correlation between advertising and album sales (0.578). In fact, all of the statistics for model 1 are the same as the simple regression model earlier (see Section 9.8). The next column gives us a value of R^2, which we already know is a measure of how much of the variability in the outcome is accounted for by the predictors. For the first model its value is 0.335, which means that advertising budget accounts for 33.5% of the variation

Cramming Sam's Tips
Descriptive statistics

• Use the descriptive statistics to check the correlation matrix for multicollinearity; that is, predictors that correlate too highly with each other, $r > 0.9$.

Descriptive Statistics

	Mean	Std. Deviation	N
Album sales (thousands)	193.20	80.699	200
Advertising budget (thousands)	614.41	485.65521	200
No. of plays on radio	27.50	12.270	200
Band image rating (0–10)	6.77	1.395	200

Correlations

		Album sales (thousands)	Advertising budget (thousands)	No. of plays on radio	Band image rating (0–10)
Pearson Correlation	Album sales (thousands)	1.000	.578	.599	.326
	Advertising budget (thousands)	.578	1.000	.102	.081
	No. of plays on radio	.599	.102	1.000	.182
	Band image rating (0–10)	.326	.081	.182	1.000
Sig. (1–tailed)	Album sales (thousands)	.	.000	.000	.000
	Advertising budget (thousands)	.000	.	.076	.128
	No. of plays on radio	.000	.076	.	.005
	Band image rating (0–10)	.000	.128	.005	.
N	Album sales (thousands)	200	200	200	200
	Advertising budget (thousands)	200	200	200	200
	No. of plays on radio	200	200	200	200
	Band image rating (0–10)	200	200	200	200

Output 9.5

Model Summary^c

Model	R	R Square	Adjusted R Square	Std. Error of the Estimate	Change Statistics				
					R Square Change	F Change	df1	df2	Sig. F Change
1	.578^a	.335	.331	65.991	.335	99.587	1	198	.000
2	.815^b	.665	.660	47.087	.330	96.447	2	196	.000

a. Predictors: (Constant), Advertising budget (thousands)

b. Predictors: (Constant), Advertising budget (thousands), Band image rating (0–10), No. of plays on radio

c. Dependent Variable: Album sales (thousands)

Output 9.6

in album sales. However, when the other two predictors are included as well (model 2), this value increases to 0.665 or 66.5% of the variance in album sales. If advertising accounts for 33.5%, then image and airplay must account for an additional 33%.[13]

The adjusted R^2 gives us some idea of how well our model generalizes, and ideally we'd like its value to be the same as or very close to, the value of R^2. In this example the difference for the final model is small (it is 0.665 − 0.660 = 0.005 or about 0.5%). This shrinkage means that if the model were derived from the population rather than a sample it would account for

approximately 0.5% less variance in the outcome. If you apply Stein's formula (equation (9.15)) you'll get an adjusted value of 0.653 (Jane Superbrain Box 9.2), which is very close to the observed value of R^2 (0.665), indicating that the cross-validity of this model is very good.

The change statistics are provided only if requested and they tell us whether the change in R^2 is significant (i.e., how much does the model fit improve as predictors are added?). The change is reported for each block of the hierarchy: for model 1, R^2 changes from 0 to 0.335 and gives rise to an F-statistic of 99.59, which is significant with a probability less than

0.001. In model 2, in which image and airplay have been added as predictors, R^2 increases by 0.330, making the R^2 of the new model 0.665 with a significant ($p < 0.001$) F-statistic of 96.44 (Jane Superbrain Box 9.2).

Output 9.7 shows the F-test of whether the model is significantly better at predicting the outcome than using the mean outcome (i.e., no predictors). The F-statistic represents the ratio of the improvement in prediction that results from fitting the model, relative to the inaccuracy that still exists in the model (see Section 9.2.4). This table again reports the information for each model separately. The table contains the sum of squares for the model (the value of SS_M from Section 9.2.4), the residual sum of squares (the value of SS_R from Section 9.2.4) and their respective degrees of freedom. For SS_M the df are the number of predictors (1 for the first model and 3 for the second). For SS_R the df are the number of observations (200) minus the number of coefficients in the regression model. The first model has two coefficients (one each for the predictor and constant) whereas the second has four (the constant plus one for each of the three predictors). Therefore, model 1 has 198 residual degrees of freedom whereas model 2 has 196. Remember, the mean sum of squares (MS) is the SS divided by the df and the F-statistic is average improvement in prediction by the model (MS_M) divided by the average error in prediction (MS_R). The p-value tells us the probability of getting an F at least as large as the one we have if the null hypothesis were true (if we used the outcome mean to predict album sales). The F-statistic is 99.59, $p < 0.001$ for the initial model and 129.498, $p < 0.001$ for the second. We can interpret these results as meaning that both models significantly improved our ability to predict the outcome variable compared to not fitting the model.

---

13 That is, 33% = 66.5% − 33.5% (this value is the *R Square Change* in the table).

Jane Superbrain 9.2
Maths frenzy ▮▮▮▮

We can look at how some of the values in the output are computed by thinking back to the theory part of the chapter. For example, looking at the change in R^2 for the first model, we have only one predictor (so $k = 1$) and 200 cases ($N = 200$), so the F comes from equation (9.13):[14]

$$F_{Model1} = \frac{(200-1-1)0.334648}{1(1-0.334648)} = 99.59$$

In model 2 in Output 9.6, two predictors have been added (image and airplay), so the new model has 3 predictors (k_{new}) and the previous model had only 1, which is a change of 2 (k_{change}). The addition of these two predictors increases R^2 by 0.330 (R^2_{change}), making the R^2 of the new model 0.665 (R^2_{new}).[15] The F-statistic for this change comes from equation (9.18):

$$F_{change} = \frac{(N-3-1)0.33}{2(1-0.664668)} = 96.44$$

We can apply Stein's formula (equation (9.15)) to R^2 to get an idea of its likely value in different samples. We replace n with the sample size (200) and k with the number of predictors (3):

$$\text{adjusted } R^2 = 1 - \left[\left(\frac{n-1}{n-k-1}\right)\left(\frac{n-2}{n-k-2}\right)\left(\frac{n+1}{n}\right)\right](1-R^2)$$

$$= 1 - \left[(1.015)(1.015)(1.005)\right](0.335)$$

$$= 1 - 0.347$$

$$= 0.653$$

9.11.3 Model parameters ▮▮▮▮

Output 9.8 shows the model parameters for both steps in the hierarchy. The first step in our hierarchy was to include advertising budget, and so the parameters for this first model are identical to those we obtained earlier in this chapter in Output 9.4. Therefore, we will focus on the parameters for the final model (in which all predictors were included). The

ANOVA[a]

Model		Sum of Squares	df	Mean Square	F	Sig.
1	Regression	433687.833	1	433687.833	99.587	.000[b]
	Residual	862264.168	198	4354.870		
	Total	1295952.00	199			
2	Regression	861377.418	3	287125.806	129.498	.000[c]
	Residual	434574.582	196	2217.217		
	Total	1295952.00	199			

a. Dependent Variable: Album sales (thousands)

b. Predictors: (Constant), Advertising budget (thousands)

c. Predictors: (Constant), Advertising budget (thousands), Band image rating (0–10), No. of plays on radio

Output 9.7

---

14 To get the same values as SPSS we have to use the exact value of R^2, which is 0.3346480676231 (if you don't believe me, double-click in the table in the SPSS output that reports this value, then double-click the cell of the table containing the value of R^2 and you'll see that what was 0.335 now has a lot more decimal places).

15 The more precise value is 0.664668.

Cramming Sam's Tips
The model summary

- The fit of the linear model can be assessed using the *Model Summary* and *ANOVA* tables from SPSS.
- R^2 tells you the proportion of variance explained by the model.
- If you have done a hierarchical regression, assess the improvement of the model at each stage by looking at the change in R^2 and whether it is significant (values less than 0.05 in the column labeled *Sig. F Change*).
- The *F*-test tells us whether the model is a significant fit to the data overall (look for values less than 0.05 in the column labeled *Sig.*).

format of the table of coefficients depends on the options selected in Figure 9.16; for example, the confidence intervals *b*, collinearity diagnostics and the part and partial correlations will be present only if you checked those options.

Earlier in the chapter we saw that a linear model with several predictors takes the form of equation (9.8), which contains several unknown parameters (the *b*-values). The first column in Output 9.8 contains estimates for these *b*-values, which indicate the individual contribution of each predictor to the model. By replacing the *Xs* in equation (9.8) with variables names and taking the *b*-values from Output 9.8 we can define our specific model as:

$$
\begin{aligned}
\text{sales}_i &= b_0 + b_1\text{advertising}_i + b_2\text{airplay}_i + b_3\text{image}_i \\
&= -26.61 + \left(0.085\ \text{advertising}_i\right) \\
&\quad + \left(3.367\ \text{airplay}_i\right) + \left(11.086\ \text{image}_i\right)
\end{aligned}
$$

(9.19)

The *b*-values quantify the relationship between album sales and each predictor.

The *direction* of the coefficient—positive or negative—corresponds to whether the relationship with the outcome is positive or negative. All three predictors have positive *b*-values, indicating positive relationships. So, as advertising budget, plays on the radio, and image rating increase so do album sales. The *size* of the *b* indicates the degree to which each predictor affects the outcome *if the effects of all other predictors are held constant*:

- **Advertising budget**: $b = 0.085$ indicates that as advertising budget increases by one unit, album sales increase by 0.085 units. Both variables were measured in thousands; therefore, for every £1000 more spent on advertising, an extra 0.085 thousand albums (85 albums) are sold. This interpretation is true only if the effects of band image and airplay are held constant.
- **Airplay**: $b = 3.367$ indicates that as the number of plays on radio in the week before release increases by one, album sales increase by 3.367 units. Every additional play of a song on radio (in the week before release) is associated with an extra 3.367 thousand albums (3367 albums) being sold. This interpretation is true only if the effects of the band's image and advertising budget are held constant.
- **Image**: $b = 11.086$ indicates that if a band can increase their image rating by 1 unit they can expect additional album sales of 11.086 units. Every unit increase in the band's image rating is associated with an extra 11.086 thousand albums (11,086 albums) being sold. This interpretation is true only if the effects of airplay and advertising are held constant.

Each of the beta values has an associated standard error indicating to what extent these values vary across different samples. The standard errors are also used to compute a *t*-statistic that tests whether the *b*-value is significantly different from 0 (Section 9.2.5). Remember that if a predictor's *b* is zero then its relationship to the outcome is zero also. By testing whether an observed b is significantly different from zero, we're testing whether the relationship between the predictor and outcome is different from zero. The p-value associated with a b's t-statistic (in the column *Sig.*) is the probability of getting a t at least as big as the one we have if the population value of b was zero (i.e., if there was no relationship between that predictor and the outcome).

For this model, the advertising budget, $t(196) = 12.26$, $p < 0.001$, the amount of radio play prior to release, $t(196) = 12.12$, $p < 0.001$ and band image, $t(196) = 4.55$, $p < 0.001$, are all significant predictors of album sales.[17] Remember that these significance tests are accurate only if the assumptions discussed in Chapter 6 are met. From the magnitude of the t-statistics we can see that the advertising budget and radio play had a similar impact, whereas the band's image had less impact.

The standardized versions of the b-values are sometimes easier to interpret (because they are not dependent on the units of measurement of the variables). The standardized beta values (in the column labeled *Beta*, β_i) tell us the number of standard deviations that the outcome changes when the predictor changes by one standard deviation. Because the standardized beta values are measured in standard deviation units they are directly comparable: the values for airplay and advertising budget are virtually identical (0.512 and 0.511, respectively), suggesting that both variables have a comparably large effect, whereas image (standardized beta of 0.192) has a relatively smaller effect (this concurs with what the magnitude of the t-statistics told us). To interpret these values literally, we need to know the

Coefficients[a]

Model		Unstandardized Coefficients		Standardized Coefficients			95.0% Confidence Interval for B	
		B	Std. Error	Beta	t	Sig.	Lower Bound	Upper Bound
1	(Constant)	134.140	7.537		17.799	.000	119.278	149.002
	Advertising budget (thousands)	.096	.010	.578	9.979	.000	.077	.115
2	(Constant)	-26.613	17.350		-1.534	.127	-60.830	7.604
	Advertising budget (thousands)	.085	.007	.511	12.261	.000	.071	.099
	No. of plays on radio	3.367	.278	.512	12.123	.000	2.820	3.915
	Band image rating (0–10)	11.086	2.438	.192	4.548	.000	6.279	15.894

a. Dependent Variable: Album sales (thousands)

Coefficients[a]

Model		Correlations			Collinearity Statistics	
		Zero-order	Partial	Part	Tolerance	VIF
1	Advertising budget (thousands)	.578	.578	.578	1.000	1.000
2	Advertising budget (thousands)	.578	.659	.507	.986	1.015
	No. of plays on radio	.599	.655	.501	.959	1.043
	Band image rating (0–10)	.326	.309	.188	.963	1.038

a. Dependent Variable: Album sales (thousands)

Output 9.8[16]

standard deviations of the variables, and these values can be found in Output 9.5.

- **Advertising budget**: Standardized $\beta = 0.511$ indicates that as advertising budget increases by one standard deviation (£485,655), album sales increase by 0.511 standard deviations. The standard deviation for album sales is 80,699, so this constitutes a change of 41,240 sales (0.511 × 80,699). Therefore, for every £485,655 more spent on advertising, an extra 41,240 albums are sold. This interpretation is true only if the effects of the band's image and airplay are held constant.
- **Airplay**: Standardized $\beta = 0.512$ indicates that as the number of plays on radio in the week before release increases by 1 standard deviation (12.27), album sales increase by 0.512 standard deviations. The standard deviation for

album sales is 80,699, so this is a change of 41,320 sales (0.512 × 80,699). Basically, if the station plays the song an extra 12.27 times in the week before release, 41,320 extra album sales can be expected. This interpretation is true only if the effects of the band's image and advertising are held constant.
- **Image**: Standardized $\beta = 0.192$ indicates that a band rated one standard deviation (1.40 units) higher on the image scale can expect additional album sales of 0.192 standard deviations units. This is a change of 15,490 sales (0.192 × 80,699). A band with an image rating 1.40 higher than another band can expect 15,490 additional sales. This interpretation is true only if the effects of airplay and advertising are held constant.

SELF TEST

Think back to what the confidence interval of the mean represented (Section 2.8). Can you work out what the confidence intervals for b represent?

---

16 To spare your eyesight I have split this part of the output into two tables; however, it should appear as one long table.

17 For all of these predictors I wrote $t(196)$. The number in brackets is the degrees of freedom. We saw in Section 9.2.5 that the degrees of freedom are $N - k - 1$, where N is the total sample size (in this case 200) and k is the number of predictors (in this case 3). For these data we get $200 - 3 - 1 = 196$.

Output 9.8 also contains the confidence intervals for the *b*s (again these are accurate only if the assumptions discussed in Chapter 6 are met). A bit of revision. Imagine that we collected 100 samples of data measuring the same variables as our current model. For each sample we estimate the same model that we have in this chapter, including confidence intervals for the unstandardized beta values. These boundaries are constructed such that in 95% of samples they contain the population value of *b* (see Section 2.8). Therefore, 95 of our 100 samples will yield confidence intervals for *b* that contain the population value. The trouble is that we don't know if our sample is one of the 95% with confidence intervals containing the population values or one of the 5% that misses.

The typical pragmatic solution to this problem is to assume that your sample is one of the 95% that hits the population value. If you assume this, then you can reasonably interpret the confidence interval as providing information about the population value of *b*. A narrow confidence interval suggests that all samples would yield estimates of *b* that are fairly close to the population value, whereas wide intervals suggest a lot of uncertainty about what the population value of *b* might be. If the interval contains zero then it suggests that the population value of *b* might be zero—in other words, no relationship between that predictor and the outcome—and could be positive but might be negative. All of these statements are reasonable if you're prepared to believe that your sample is one of the 95% for which the intervals contain the population value. Your belief will be wrong 5% of the time, though.

In our model of album sales, the two best predictors (advertising and airplay) have very tight confidence intervals, indicating that the estimates for the current model are likely to be representative of the true population values. The interval for the band's image is wider (but still does not cross zero), indicating that the parameter for this variable is less representative, but nevertheless significant.

If you asked for part and partial correlations, then they appear in separate columns of the table. The zero-order correlations are the Pearson's correlation coefficients and correspond to the values in Output 9.5. Semi-partial (part) and partial correlations were described in Section 8.5; in effect, the part correlations quantify the unique relationship that each predictor has with the outcome. If you opted to do a stepwise regression, you would find that variable entry is based initially on the variable with the largest zero-order correlation and then on the part correlations of the remaining variables. Therefore, airplay would be entered first (because it has the largest zero-order correlation), then advertising budget (because its part correlation is bigger than that of image rating) and then finally the band's image rating—try running a forward stepwise regression on these data to see if I'm right. Finally, Output 9.8 contains collinearity statistics, but we'll discuss these in Section 9.11.5.

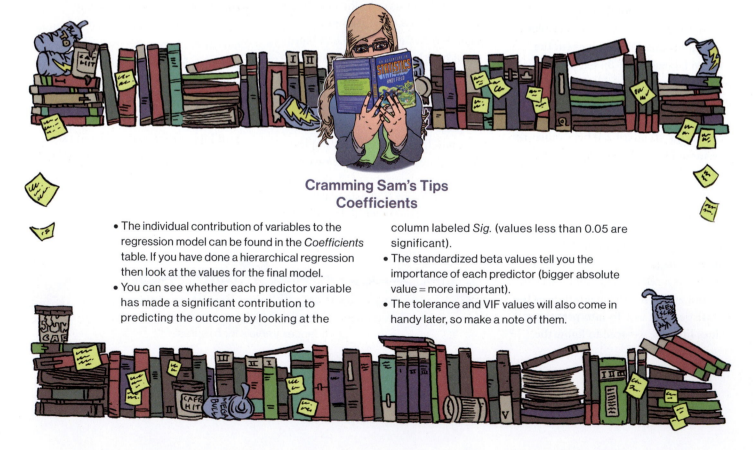

Cramming Sam's Tips
Coefficients

- The individual contribution of variables to the regression model can be found in the *Coefficients* table. If you have done a hierarchical regression then look at the values for the final model.
- You can see whether each predictor variable has made a significant contribution to predicting the outcome by looking at the column labeled *Sig.* (values less than 0.05 are significant).
- The standardized beta values tell you the importance of each predictor (bigger absolute value = more important).
- The tolerance and VIF values will also come in handy later, so make a note of them.

9.11.4 Excluded variables

At each stage of fitting a linear model a summary is provided of predictors that are not yet in the model.

We had a two-block hierarchy with one predictor entered (and two excluded) in block 1, and three predictors entered (and none excluded) in block 2. Output 9.9 details excluded variables only for the first block of our hierarchy, because in the second block no predictors were excluded. The table includes an estimate of the *b*-value and associated *t*-statistic for each predictor *if* it entered the model at this point. Using a stepwise method, the predictor with the highest *t*-statistic will enter the model next, and predictors will continue to be entered until there are none left with *t*-statistics that have significance values less than 0.05. The partial correlation also indicates what contribution (if any) an excluded predictor would make if it entered the model.

9.11.5 Assessing multicollinearity

I promised to come back to the measures of collinearity in Output 9.8, so here we

Excluded Variables[a]

Model		Beta In	t	Sig.	Partial Correlation	Collinearity Statistics		Minimum Tolerance
						Tolerance	VIF	
1	No. of plays on radio	.546[b]	12.513	.000	.665	.990	1.010	.990
	Band image rating (0–10)	.281[b]	5.136	.000	.344	.993	1.007	.993

a. Dependent Variable: Album sales (thousands)
b. Predictors in the Model: (Constant), Advertising budget (thousands)

Output 9.9

Collinearity Diagnostics[a]

Model	Dimension	Eigenvalue	Condition Index	Variance Proportions			
				(Constant)	Advertising budget (thousands)	No. of plays on radio	Band image rating (0–10)
1	1	1.785	1.000	.11	.11		
	2	.215	2.883	.89	.89		
2	1	3.562	1.000	.00	.02	.01	.00
	2	.308	3.401	.01	.96	.05	.01
	3	.109	5.704	.05	.02	.93	.07
	4	.020	13.219	.94	.00	.00	.92

a. Dependent Variable: Album sales (thousands)

Output 9.10

go. The output contains the VIF and tolerance statistics (with tolerance being 1 divided by the VIF), and we need to apply the guidelines from Section 9.9.3. The VIF values are well below 10 and the tolerance statistics are well above 0.2. The average VIF, obtained by adding the VIF values for each predictor and dividing by the number of predictors (*k*), is also very close to 1:

$$\overline{\text{VIF}} = \frac{\sum_{i=1}^{k} \text{VIF}_i}{k} = \frac{1.015 + 1.043 + 1.038}{3} = 1.032$$

(9.20)

Cramming Sam's Tips
Multicollinearity

- To check for multicollinearity, use the VIF values from the table labeled *Coefficients*.
- If these values are less than 10 then that indicates there probably isn't cause for concern.
- If you take the average of VIF values, and it is not substantially greater than 1, then there's also no cause for concern.

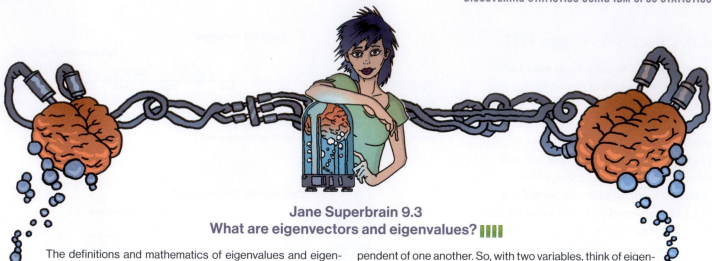

Jane Superbrain 9.3
What are eigenvectors and eigenvalues? ▍▍▍▍

The definitions and mathematics of eigenvalues and eigenvectors are complicated and most of us need not worry about them (although they do crop up again in Chapters 17 and 18). Although the mathematics is hard, we can get a sense of what they represent visually. Imagine we have two variables: the age of a zombie (how long it has been a zombie), and how many blows to the head it takes to kill it.[18] These two variables are normally distributed and can be considered together as a bivariate normal distribution. If these variables are correlated their scatterplot forms an ellipse: if we draw a dashed line around the outer values of the scatterplot we get an oval shape (Figure 9.20). Imagine two lines to measure the length and height of this ellipse: these represent the *eigenvectors* of the correlation matrix for these two variables (a vector is a set of numbers that tells us the location of a line in geometric space). Note that the two straight lines in Figure 9.20 are at 90 degrees to each other, which means that they are inde-

pendent of one another. So, with two variables, think of eigenvectors as lines measuring the length and height of the ellipse that surrounds the scatterplot of data for those variables. If we add a third variable (e.g., force of blow) our scatterplot gets a third dimension (depth), the ellipse turns into something shaped like a rugby ball (or American football), and we get an extra eigenvector to measure the extra dimension. If we add a fourth variable, a similar logic applies (although it's harder to visualize).

Each eigenvector has an *eigenvalue* that tells us its length (i.e., the distance from one end of the eigenvector to the other). By looking at the eigenvalues for a data set, we know the dimensions of the ellipse (length and height) or rugby ball (length, height, depth); more generally, we know the dimensions of the data. Therefore, the eigenvalues quantify how evenly (or otherwise) the variances of the matrix are distributed.

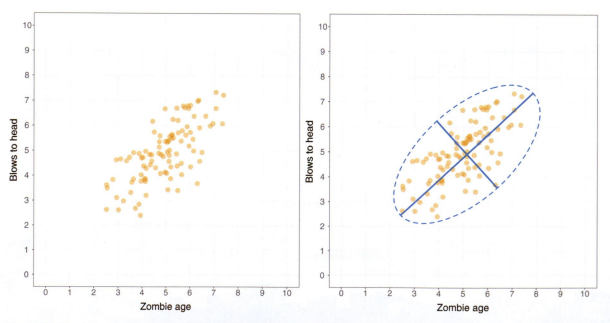

Figure 9.20   A scatterplot of two correlated variables forms an ellipse

---

18 Assuming you can ever kill a zombie, that is.

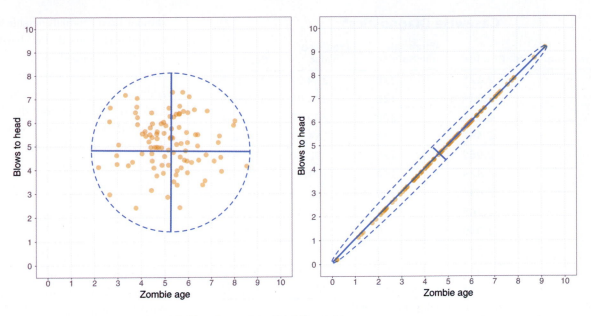

Figure 9.21   Perfectly uncorrelated (left) and correlated (right) variables

In the case of two variables, the *condition* of the data is related to the ratio of the larger eigenvalue to the smaller. Figure 9.21 shows the two extremes: when there is no relationship at all between variables (left), and when there is a perfect relationship (right). When there is no relationship, the data cloud will be contained roughly within a circle (or a sphere if we had three variables). If we draw lines that measure the height and width of this circle they'll be the same length, which means they'll have the same eigenvalues.

Consequently, when we divide the largest eigenvalue by the smallest we'll get a value of 1. When the variables are perfectly correlated (i.e., there is perfect collinearity) the data cloud (and ellipse around it) will collapse to a straight line. The height of the ellipse will be very small indeed (it will approach zero). Therefore, the largest eigenvalue divided by the smallest will tend to infinity (because the smallest eigenvalue is close to zero). An infinite condition index is a sign of deep trouble.

It seems unlikely, therefore, that we need to worry about collinearity among predictors.

The other information we get about collinearity is a table of eigenvalues of the scaled, uncentered cross-products matrix, condition indexes and variance proportions. I discuss collinearity and variance proportions at length in Section 20.8.2, so here I'll just give you the headline: look for large variance proportions on the same *small* eigenvalues (Jane Superbrain Box 9.3). Therefore, in Output 9.10 inspect the bottom few rows of the table (these are the small eigenvalues) and look for variables that *both* have high variance proportions for that eigenvalue. The variance proportions vary between 0 and 1, and you'd like to see each predictor having a high proportion on a different eigenvalue to other predictors (in other words, the large proportions are distributed across different eigenvalues).

For our model, each predictor has most of its variance loading onto a different dimension than other predictors (advertising has 96% of variance on dimension 2, airplay has 93% of variance on dimension 3 and image rating has 92% of variance on dimension 4). These data represent no multicollinearity. For an example of when collinearity exists in the data and some suggestions about what can be done, see Chapters 20 (Section 20.8.2) and 18 (Section 18.3.3).

Casewise Diagnostics[a]

Case Number	Std. Residual	Album sales (thousands)	Predicted Value	Residual
1	2.125	330	229.92	100.080
2	−2.314	120	228.95	−108.949
10	2.114	300	200.47	99.534
47	−2.442	40	154.97	−114.970
52	2.069	190	92.60	97.403
55	−2.424	190	304.12	−114.123
61	2.098	300	201.19	98.810
68	−2.345	70	180.42	−110.416
100	2.066	250	152.71	97.287
164	−2.577	120	241.32	−121.324
169	3.061	360	215.87	144.132
200	−2.064	110	207.21	−97.206

a. Dependent Variable: Album sales (thousands)

Output 9.11

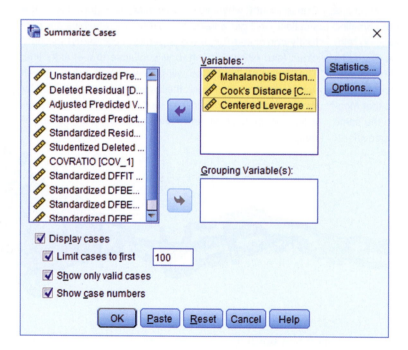

Figure 9.22 The *Summarize Cases* dialog box

9.11.6 Bias in the model: casewise diagnostics

The final stage of the general procedure outlined in Figure 9.10 is to check the residuals for evidence of bias. The first step is to examine the casewise diagnostics. Output 9.11 shows any cases that have a standardized residual less than −2 or greater than 2 (remember that we changed the default criterion from 3 to 2 in Figure 9.16). In an ordinary sample we would expect 95% of cases to have standardized residuals within about ±2 (Jane Superbrain Box 6.4). We have a sample of 200, therefore it is reasonable to expect about 10 cases (5%) to have

standardized residuals outside these limits. Output 9.11 shows that we have 12 cases (6%) that are outside of the limits: pretty close to what we would expect. In addition, 99% of cases should lie within ±2.5 and only 1% of cases should lie outside these limits. We have two cases that lie outside of the limits (cases 164 and 169), which is 1% and what we would expect. These diagnostics give us no cause for concern, except that case 169 has a standardized residual greater than 3, which is probably large enough for us to investigate this case further.

In Section 9.10.4 we opted to save various diagnostic statistics. You should find that the data editor contains columns for these variables. You can check these values in the data editor or list values in the viewer window. To create a table of values in the viewer, select *Analyze* ▶ *Reports* ▶ **Case Summaries**... to access the dialog box in Figure 9.22. Select and drag the variables that you want to list into the box labeled *Variables* (or click ▶). By default, the output is limited to the first 100 cases, but if you want to list all cases deselect this option (also see SPSS Tip 9.1). It is also useful to select *Show case numbers* to enable you to identify the case numbers of any problematic cases.

To save space, Output 9.12 shows the influence statistics for 12 cases that I selected. None of them has a Cook's distance greater than 1 (even case 169 is well below this criterion), and so no case appears to have an undue influence on the model. The average leverage can be calculated as $(k+1)/n = 4/200 = 0.02$, and we should look for values either twice (0.04) or three times (0.06) this value (see Section 9.3.2). All cases are within the boundary of three times the average, and only case 1 is close to two times the average. For the Mahalanobis distances we saw earlier in the chapter that with a sample of 100 and three predictors, values greater than 15 are problematic. Also, with 3 predictors values greater than 7.81 are

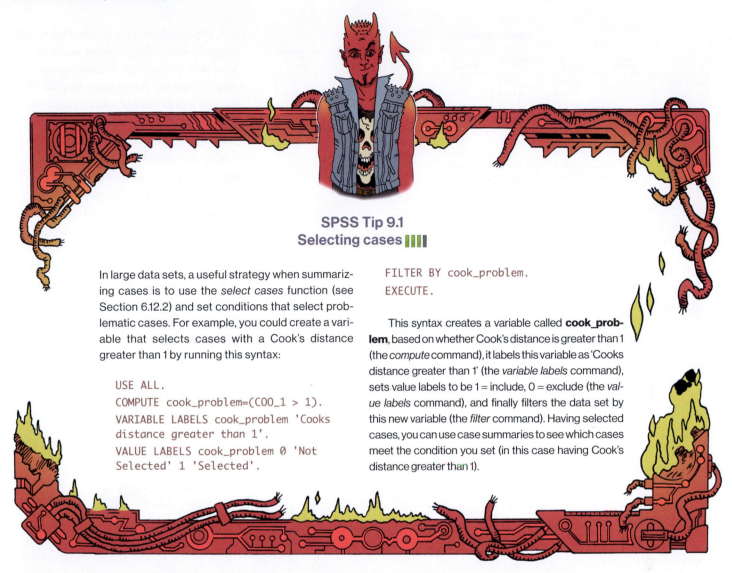

SPSS Tip 9.1
Selecting cases ▐▐▐

In large data sets, a useful strategy when summarizing cases is to use the *select cases* function (see Section 6.12.2) and set conditions that select problematic cases. For example, you could create a variable that selects cases with a Cook's distance greater than 1 by running this syntax:

```
USE ALL.
COMPUTE cook_problem=(COO_1 > 1).
VARIABLE LABELS cook_problem 'Cooks
distance greater than 1'.
VALUE LABELS cook_problem 0 'Not
Selected' 1 'Selected'.
```

```
FILTER BY cook_problem.
EXECUTE.
```

This syntax creates a variable called **cook_problem**, based on whether Cook's distance is greater than 1 (the *compute* command), it labels this variable as 'Cooks distance greater than 1' (the *variable labels* command), sets value labels to be 1 = include, 0 = exclude (the *value labels* command), and finally filters the data set by this new variable (the *filter* command). Having selected cases, you can use case summaries to see which cases meet the condition you set (in this case having Cook's distance greater than 1).

significant ($p < 0.05$). None of our cases comes close to exceeding the criterion of 15 although case 1 would be deemed 'significant'.

The DFBeta statistics tell us how much influence each case has on the model parameters. An absolute value greater than 1 is a problem, and all cases in Output 9.12 have values within ±1, which is good news.

For the covariance ratio we need to use the following criteria (Section 9.3.2):

- $CVR_i > 1 + [3(k + 1)/n] = 1 + [3(3 + 1)/200] = 1.06$
- $CVR_i < 1 − [3(k + 1)/n] = 1 − [3(3 + 1)/200] = 0.94$.

Therefore, we are looking for any cases that deviate substantially from these boundaries. Most of our 12 potential outliers have CVR values within or just outside these boundaries. The only case that causes concern is case 169 (again), whose CVR is some way below the bottom limit. However, given the Cook's distance for this case, there is probably little cause for alarm. You will have requested other diagnostic statistics and you can apply what we learnt earlier in the chapter when glancing over them.

From this minimal set of diagnostics there's nothing to suggest that there are influential cases (although we'd need to look at all 200 cases to confirm this conclusion); we appear to have a fairly reliable model that has not been unduly influenced by any subset of cases.

9.11.7 Bias in the model: assumptions ▐▐▐

The general procedure outlined in Figure 9.10 suggests that having fitted a model, we need to look for evidence of bias, and the second stage of this process is to check the assumptions described in Chapter 6. We saw in Section 6.11.1 that we can look for heteroscedasticity and non-linearity using a plot of standardized predicted values against standardized residuals. We asked for this plot in Section 9.10.3. If everything is OK then this graph should

Case Summaries[a]

	Case Number	COVRATIO	Standardized DFFIT	Standardized DFBETA Intercept	Standardized DFBETA Adverts	Standardized DFBETA Airplay	Standardized DFBETA Image
1	1	.97127	.48929	−.31554	−.24235	.15774	.35329
2	2	.92018	−.21110	.01259	−.12637	.00942	−.01868
3	10	.94392	.26896	−.01256	−.15612	.16772	.00672
4	47	.91458	−.31469	.06645	.19602	.04829	−.17857
5	52	.95995	.36742	.35291	−.02881	−.13667	−.26965
6	55	.92486	−.40736	.17427	−.32649	−.02307	−.12435
7	61	.93654	.15562	.00082	−.01539	.02793	.02054
8	68	.92370	−.30216	−.00281	.21146	−.14766	−.01760
9	100	.95888	.35732	.06113	.14523	−.29984	.06766
10	164	.92037	−.54029	.17983	.28988	−.40088	−.11706
11	169	.85325	.46132	−.16819	−.25765	.25739	.16968
12	200	.95435	−.31985	.16633	−.04639	.14213	−.25907
Total N	12	12	12	12	12	12	12

a. Limited to first 100 cases.

	Mahalanobis Distance	Cook's Distance	Centered Leverage Value
1	8.39591	.05870	.04219
2	.59830	.01089	.00301
3	2.07154	.01776	.01041
4	2.12475	.02412	.01068
5	4.81841	.03316	.02421
6	4.19960	.04042	.02110
7	.06880	.00595	.00035
8	2.13106	.02229	.01071
9	4.53310	.03136	.02278
10	6.83538	.07077	.03435
11	3.14841	.05087	.01582

Output 9.12

look like a random array of dots. Figure 9.23 (top left) shows the graph for our model. Note how the points are randomly and evenly dispersed throughout the plot. This pattern is indicative of a situation in which the assumptions of linearity and homoscedasticity have been met; compare it with the examples in Figure 6.19.

Figure 9.23 also shows the partial plots, which are scatterplots of the residuals of the outcome variable and each of the predictors when both variables are regressed separately on the remaining predictors. Obvious outliers on a partial plot represent cases that might have undue influence on a predictor's *b* coefficient. Non-linear relationships and heteroscedasticity can be detected using these plots as well. For image (Figure 9.23, top right) the partial plot shows the strong positive relationship to album sales. There are no obvious outliers and the cloud of dots is evenly spaced out around the line,

Cramming Sam's Tips
Residuals

- Look for cases that might be influencing the model.
- Look at standardized residuals and check that no more than 5% of cases have absolute values above 2, and that no more than about 1% have absolute values above 2.5. Any case with a value above about 3 could be an outlier.
- Look in the data editor for the values of Cook's distance: any value above 1 indicates a case that might be influencing the model.
- Calculate the average leverage and look for values greater than twice or three times this average value.

- For Mahalanobis distance, a crude check is to look for values above 25 in large samples (500) and values above 15 in smaller samples (100). However, Barnett and Lewis (1978) should be consulted for more refined guidelines.
- Look for absolute values of DFBeta greater than 1.
- Calculate the upper and lower limit of acceptable values for the covariance ratio, CVR. Cases that have a CVR that fall outside these limits may be problematic.

indicating homoscedasticity. The plot for airplay (Figure 9.23, bottom left) also shows a strong positive relationship to album sales, there are no obvious outliers, and the cloud of dots is evenly spaced around the line, again indicating homoscedasticity. For image (Figure 9.23, bottom right) the plot again shows a positive relationship to album sales, but the dots show funnelling, indicating greater spread for bands with a high image rating. There are no obvious outliers on this plot, but the funnel-shaped cloud indicates a violation of the assumption of homoscedasticity.

To test the normality of residuals, we look at the histogram and normal probability plot selected in Figure 9.17 and shown in Figure 9.24. Compare these plots to examples of non-normality in Section 6.10.1. For the album sales data, the distribution is very normal: the histogram is symmetrical and approximately bell-shaped. In the P-P plot the dots lie almost exactly along the diagonal, which we know indicates a normal distribution (see Section 6.10.1); hence this plot also suggests that the residuals are normally distributed.

9.12 Robust regression ▮▮▮▮

Our model appears, in most senses, to be both accurate for the sample and generalizable to the population. The only slight glitch is some concern over whether image ratings violated the assumption of homoscedasticity. Therefore, we could conclude that in our sample advertising budget and airplay are equally important in predicting album sales. The image of the band is a significant predictor of album sales, but is less important than the other predictors (and probably needs verification because of possible heteroscedasticity). The assumptions

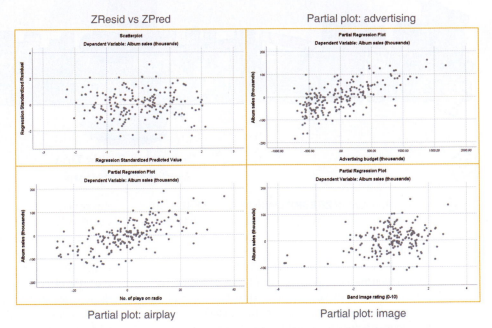

ZResid vs ZPred   Partial plot: advertising

Partial plot: airplay   Partial plot: image

Figure 9.23 Plot of standardized predicted values against standardized residuals (top left), and partial plots of album sales against advertising (top right), airplay (bottom left) and image of the band (bottom right)

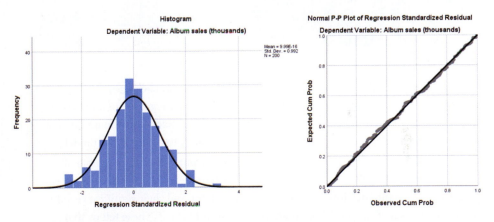

Figure 9.24 Histogram and normal P-P plot for the residuals from our model

seem to have been met, and so we can probably assume that this model would generalize to any album being released. You won't always (ever?) have such nice data: there will be times when you uncover problems that cast a dark shadow of evil over your model. It will invalidate significance tests, confidence intervals and generalization of the model (use Chapter 6 to remind yourself of the

implications of violating model assumptions).

Luckily, a lot of the problems can be overcome. If confidence intervals and significance tests of the model parameters are in doubt then use bootstrapping to generate confidence intervals and p-values. If homogeneity of variance is the issue then estimate the model with standard errors designed for

Cramming Sam's Tips
Model assumptions

- Look at the graph of **ZRESID*** plotted against **ZPRED***. If it looks like a random array of dots then this is good. If the dots get more or less spread out over the graph (look like a funnel) then the assumption of homogeneity of variance is probably unrealistic. If the dots have a pattern to them (i.e., a curved shape) then the assumption of linearity is probably not true. If the dots seem to have a pattern and are more spread out at some points on the plot than others then this could reflect violations of both homogeneity of variance *and* linearity.

- Any of these scenarios puts the validity of your model into question. Repeat the above for all partial plots too.
- Look at the histogram and P-P plot. If the histogram looks like a normal distribution (and the P-P plot looks like a diagonal line), then all is well. If the histogram looks non-normal and the P-P plot looks like a wiggly snake curving around a diagonal line then things are less good. Be warned, though: distributions can look very non-normal in small samples even when they are normal.

heteroscedastic residuals (Hayes & Cai, 2007)—you can do this using the PROCESS tool described in Chapter 11. Finally, if the model parameters themselves are in doubt, estimate them using robust regression.

To get robust confidence intervals and significance tests of the model parameters re-estimate your model, selecting the same options as before but clicking Bootstrap... in the main dialog box (Figure 9.13) to access the dialog box explained in Section 6.12.3. To recap, select ☑ Perform bootstrapping to activate bootstrapping, and to get a 95% confidence interval click ◉ Percentile or ◉ Bias corrected accelerated (BCa). For this analysis, let's ask for a bias corrected (BCa) confidence interval. Bootstrapping won't work if you have set options to save

diagnostics, so click Save... to open the dialog box in Figure 9.18 and *deselect everything*. Back in the main dialog box click OK to estimate the model.

The output will contain a table of bootstrap confidence intervals for each predictor and their significance value (Output 9.13).[19] These tell us that advertising, $b = 0.09$ [0.07, 0.10], $p = 0.001$, airplay, $b = 3.37$ [2.77, 3.97], $p = 0.001$, and the band's image, $b = 11.09$ [6.26, 15.28], $p = 0.001$, all significantly predict album sales. These bootstrap confidence intervals and significance values do not rely on assumptions of normality or homoscedasticity, so they give us an accurate estimate of the population value of b for each predictor (assuming our sample is one of the 95% with confidence

intervals that contain the population value).

To estimate the bs themselves using a robust method we can use the R plugin. If you have installed this plugin (Section 4.13.2) then you can access a dialog box (Figure 9.25) to run robust regression using R by selecting *Analyze* ▶ *Regression* ▶ 🔧 Robust Regression . If you haven't installed the plugin then this menu won't be there! Drag the outcome (album sales) to the box labeled *Dependent* and any predictors in the final model (in this case advertising budget, airplay and image rating) to the box labeled *Independent Variables*. Click OK to estimate the model.

Output 9.14 shows the resulting robust b-values, their robust standard errors and t-statistics. Compare these with the

---

19 Remember that because of how bootstrapping works the values in your output will be different than mine, and different if you rerun the analysis.

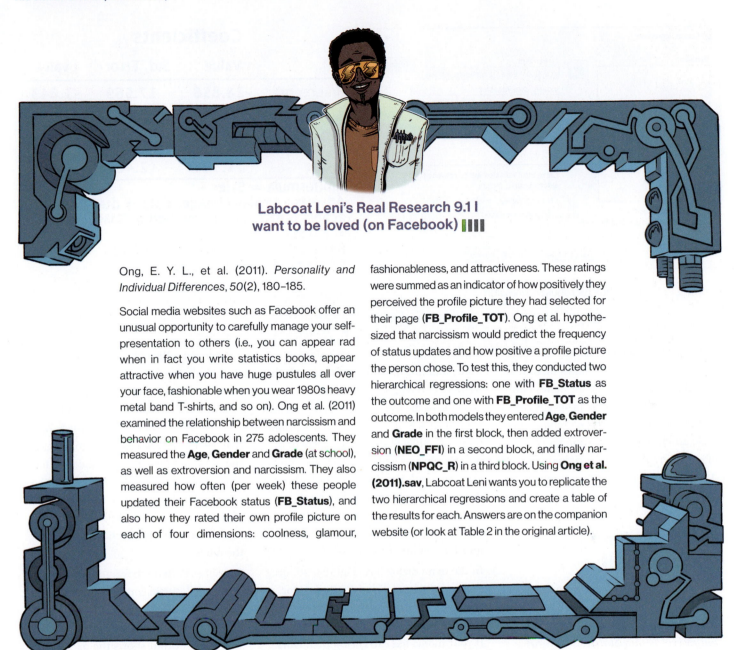

Labcoat Leni's Real Research 9.1 I want to be loved (on Facebook) ▮▮▮▮

Ong, E. Y. L., et al. (2011). *Personality and Individual Differences*, *50*(2), 180–185.

Social media websites such as Facebook offer an unusual opportunity to carefully manage your self-presentation to others (i.e., you can appear rad when in fact you write statistics books, appear attractive when you have huge pustules all over your face, fashionable when you wear 1980s heavy metal band T-shirts, and so on). Ong et al. (2011) examined the relationship between narcissism and behavior on Facebook in 275 adolescents. They measured the **Age**, **Gender** and **Grade** (at school), as well as extroversion and narcissism. They also measured how often (per week) these people updated their Facebook status (**FB_Status**), and also how they rated their own profile picture on each of four dimensions: coolness, glamour,

fashionableness, and attractiveness. These ratings were summed as an indicator of how positively they perceived the profile picture they had selected for their page (**FB_Profile_TOT**). Ong et al. hypothe-sized that narcissism would predict the frequency of status updates and how positive a profile picture the person chose. To test this, they conducted two hierarchical regressions: one with **FB_Status** as the outcome and one with **FB_Profile_TOT** as the outcome. In both models they entered **Age**, **Gender** and **Grade** in the first block, then added extrover-sion (**NEO_FFI**) in a second block, and finally nar-cissism (**NPQC_R**) in a third block. Using **Ong et al. (2011).sav**, Labcoat Leni wants you to replicate the two hierarchical regressions and create a table of the results for each. Answers are on the companion website (or look at Table 2 in the original article).

non-robust versions in Output 9.8. The values are not much different (mainly because our original model didn't seem to violate its assumptions); for example, the *b* for image rating has changed from 11.09 (Output 9.8.) to 11.39 (Output 9.14), the associated standard error was 2.44 and the robust version is 2.47, and the associated *t*-statistic has changed from 4.55 to 4.62. Essentially our interpretation of the model won't have changed, but this is still a useful sensitivity analysis in that if robust

Bootstrap for Coefficients

Model		B	Bias	Std. Error	Sig. (2–tailed)	Lower	Upper
1	(Constant)	134.14	.156	7.613	.001	119.470	150.048
	Advertising budget (thousands)	.096	.000	.008	.001	.081	.111
2	(Constant)	−26.61	−.028	15.733	.077	−54.589	2.715
	Advertising budget (thousands)	.085	.000	.007	.001	.071	.098
	No. of plays on radio	3.367	.005	.305	.001	2.773	3.972
	Band image rating (0–10)	11.086	−.019	2.223	.001	6.264	15.283

a. Unless otherwise noted, bootstrap results are based on 1000 bootstrap samples

Output 9.13

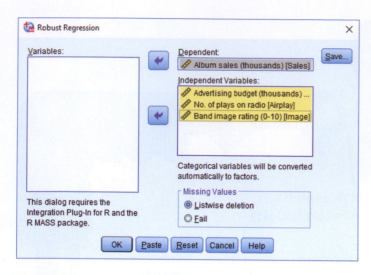

Figure 9.25 Dialog box for robust regression

Coefficients

	Value	Std. Error	t value
(Intercept)	–28.858	17.569	–1.643
Adverts	.086	.007	12.219
Airplay	3.371	.281	11.984
Image	11.394	2.469	4.615

rlm(formula = Sales ~ Adverts+Airplay+Image, data = dta, na. action = na.exclude, method = "MM", model = FALSE)
Residual standard error: 45.37396
Degrees of freedom: 196

Output 9.14

estimates are giving us basically the same results as non-robust estimates then we know that the non-robust estimates have not been unduly biased by properties of the data. So, this is always a useful double check, and if the robust estimates are hugely different from the original estimates then you can use and report the robust versions.

9.13 Bayesian regression ▮▮▮▮

In Section 3.8.4 we looked at Bayesian approaches. To access a dialog box (Figure 9.26) to fit a Bayesian linear model select *Analyze* ▸ *Bayesian Statistics* ▸ *Linear Regression*. You can fit the model either using default priors (so called ◉ Reference priors), which set distributions that represent very diffuse prior beliefs, or conjugate priors, which allow you to specify more specific priors. One of the key strengths of Bayesian statistics (in my opinion) is that you can set evidence-based priors that you update with the data that you collect. However, this is not a trivial

undertaking, and it requires a deeper understanding of the models being fit than we have covered. So, to help you to dip your toe in the water of Bayesian statistics we will stick to using the reference priors built into SPSS Statistics. The benefit of reference priors is that they enable you to get going with Bayesian models without drowning in a lot of quite technical material, but the cost is that you are building only very uninformative prior information into your models.[20]

In the main dialog box (Figure 9.26) drag **Sales** to the box labeled *Dependent* (or click ➡) and drag **Adverts**, **Airplay** and **Image** to the *Covariate(s)* box (or click ➡). If your model has categorical predictors (which we'll look at in the following chapter) drag them to the *Factor(s)* box. If you want to both compute Bayes factors and estimate the model parameters then select ◉ Use *Both* Methods.

If you want a credible interval other than 95% then click Criteria and change the 95 to the value you want. Click Priors... to set your priors, although we'll stick with ◉ Reference priors. Click Bayes Factor... if you

want to get a Bayes factor for your model. By default, the full model will be compared to the null model and there are four methods to compute them. I have selected ◉ JZS method (Jeffreys, 1961; Zellner & Siow, 1980). Click Plots... to inspect the prior and posterior distributions for each predictor. Drag all predictors to the box labeled *Plot covariate(s)* (or click ➡) and select ☑ Bayesian predicted distribution. In the main dialog box click OK to fit the model.

Output 9.15 (left) shows the Bayes factor for the full model compared to the null model, which I assume is the model including only the intercept. The right side of the output shows the parameter estimates based on Bayesian estimation. The Bayes factor is 1.066×10^{43} (that's what the E+43 means). In other words it is massive. In short, the probability of the data given the model including all three predictors is 1.07×10^{43} greater than the probability of the data given the model with only the intercept. We should shift our belief in the model (relative to the null model) by a factor of

---

20 Another downside of this convenience is that I find it hard to know what these priors actually represent (especially in the case of regression).

1.07×10^{43}! This is very strong evidence for the model.

The Bayesian estimate of b can be found in the columns labeled *Posterior Mode* and *Posterior Mean*. In fact the columns contain identical values, but they won't always. The reason for the two columns is that we use the peak of the posterior distribution as our estimate and that peak can be defined by either the mode of the posterior or its mean. The values are 0.085 for advertising budget, 3.367 for airplay and 11.086 for image, compared to the values of 0.085, 3.37 and 11.09 (Output 9.8). from the non-Bayesian model. They are basically the same, which is not all that surprising because we started off with very diffuse priors (and so these priors will have had very little influence over the estimates – think back to Section 3.8). We can see this fact in Output 9.16, which shows the prior distribution for the b for advertising budget as a red line (in your output you will see similar plots for the other two predictors): the line is completely flat, representing a completely open and diffuse belief about the model parameters. The green line is the posterior distribution, which is quantified in Output 9.15 (right).

Perhaps the most useful parts of Output 9.15 are the 95% credible intervals for the model parameters. Unlike confidence intervals, credible intervals contain the population value with a probability of 0.95 (95%). For advertising budget, therefore, there is a 95% probability that the population value of b lies between 0.071 and 0.099, for airplay the population value is plausibly between 2.820 and 3.915, and for image it plausibly lies between 6.279 and 15.894. These intervals are constructed assuming that an effect exists, so you cannot use them to test hypotheses, only to establish plausible population values of the bs in the model.

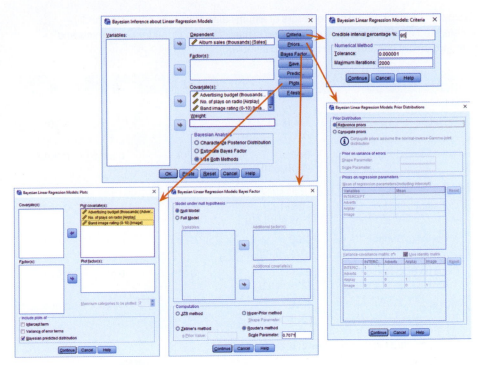

Figure 9.26 Dialog box for Bayesian regression

Bayes Factor Model Summary[a,b]

Bayes Factor[c]	R	R Square	Adjusted R Square	Std. Error of the Estimate
1.066E+43	.815	.665	.660	47.09

a. Method: JZS

b. Model: (Intercept), Advertising budget (thousands), No. of plays on radio, Band image rating (0–10)

c. Bayes factor: Testing model versus null model (Intercept).

Bayesian Estimates of Coefficients[a,b,c]

Parameter	Posterior			95% Credible Interval	
	Mode	Mean	Variance	Lower Bound	Upper Bound
(Intercept)	−26.61	−26.61	304.13	−60.830	7.604
Advertising budget (thousands)	.085	.085	.000	.071	.099
No. of plays on radio	3.367	3.367	.078	2.820	3.915
Band image rating (0–10)	11.086	11.086	6.004	6.279	15.894

a. Dependent Variable: Album sales (thousands)

b. Model: (Intercept), Advertising budget (thousands), No. of plays on radio, Band image rating (0–10)

c. Assume standard reference priors.

Output 9.15

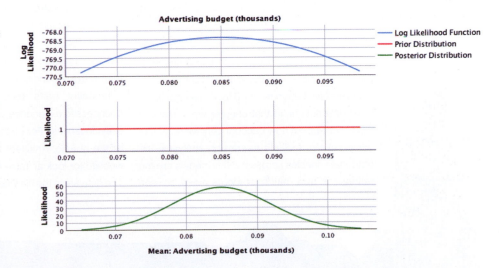

Output 9.16

Table 9.2 Linear model of predictors of album sales. 95% bias corrected and accelerated confidence intervals reported in parentheses. Confidence intervals and standard errors based on 1000 bootstrap samples

	b	SE B	β	p
Step 1				
Constant	134.14 (120.11, 148.79)	7.95		0.001
Advertising Budget	0.10 (0.08, 0.11)	0.01	0.58	0.001
Step 2				
Constant	−26.61 (−55.40, 8.60)	16.30		0.097
Advertising Budget	0.09 (0.07, 0.10)	0.01	0.51	0.001
Plays on BBC Radio 1	3.37 (2.74, 4.02)	0.32	0.51	0.001
Image	11.09 (6.46, 15.01)	2.22	0.19	0.001

Note. $R^2 = 0.34$ for Step 1; $\Delta R^2 = 0.33$ for Step 2 (all $ps < 0.001$).

9.14 Reporting linear models

If your model has several predictors than you can't really beat a summary table as a concise way to report your model. As a bare minimum report the betas along with their standard errors and confidence interval (or credible interval if you've gone Bayesian). If you haven't gone Bayesian, report the significance value and perhaps the standardized beta. Include some general fit statistics about the model such as R^2 or the Bayes factor. Personally, I like to see the constant as well because then readers of your work can construct the full regression model if they need to. For hierarchical regression you should report these values at each

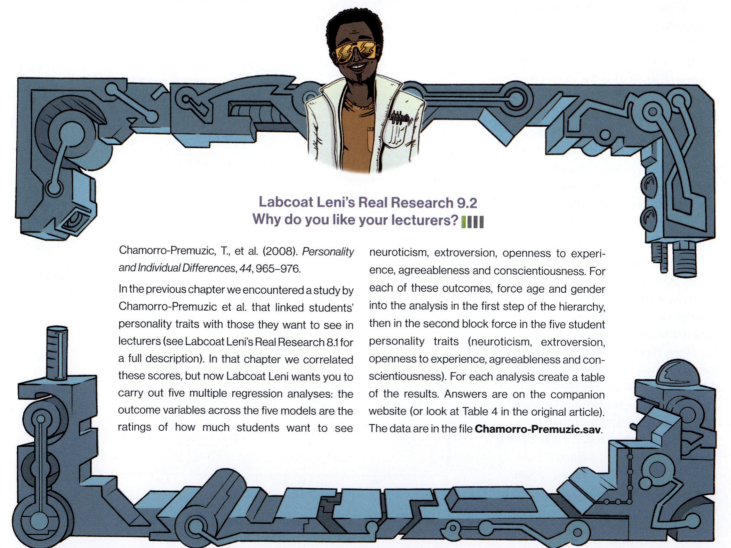

Labcoat Leni's Real Research 9.2
Why do you like your lecturers?

Chamorro-Premuzic, T., et al. (2008). *Personality and Individual Differences, 44*, 965–976.

In the previous chapter we encountered a study by Chamorro-Premuzic et al. that linked students' personality traits with those they want to see in lecturers (see Labcoat Leni's Real Research 8.1 for a full description). In that chapter we correlated these scores, but now Labcoat Leni wants you to carry out five multiple regression analyses: the outcome variables across the five models are the ratings of how much students want to see

neuroticism, extroversion, openness to experience, agreeableness and conscientiousness. For each of these outcomes, force age and gender into the analysis in the first step of the hierarchy, then in the second block force in the five student personality traits (neuroticism, extroversion, openness to experience, agreeableness and conscientiousness). For each analysis create a table of the results. Answers are on the companion website (or look at Table 4 in the original article). The data are in the file **Chamorro-Premuzic.sav**.

stage of the hierarchy. For the example in this chapter we might produce a table like that in Table 9.2.

Things to note are as follows: (1) I've rounded off to 2 decimal places throughout because this is a reasonable level of precision given the variables measured; (2) if you are following APA format (which I'm not), do not put zeros before the decimal point for the standardized betas, R^2 and p-values (because these values shouldn't exceed 1); (3) I've reported exact p-values, which is good practice; (4) the R^2 for the initial model and the change in R^2 (denoted by ΔR^2) for each subsequent step of the model are reported below the table; and (5) in the title I have mentioned that confidence intervals and standard errors in the table are based on bootstrapping, which is important for readers to know.

9.15 Brian and Jane's Story ▮▮▮▮

Jane put the fork down next to the jar and suppressed her reflex to gag. When she'd started at this university she'd had utter conviction in testing her flatworm theory. She would be her own single-case design. She knew that experimenting on herself would confound everything, but she wanted some evidence to firm up her beliefs. If it didn't work on her then she could move on, but if she found evidence for some effect then that was starting point for better research. She felt conflicted, though. Was it the experiments making her mind so unfocussed or was it the interest from campus guy? She hadn't come here looking for a relationship, she hadn't expected it, and it wasn't in the plan. Usually she was so good at ignoring other people, but his kindness was slowly corroding her shell. As she got up and replaced the jar on the shelf she told herself that the nonsense with campus guy had to stop. She needed to draw a line.

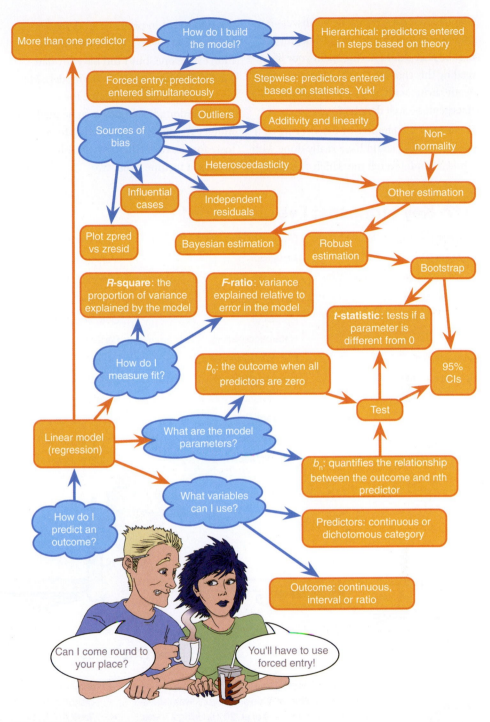

Figure 9.27   What Brian learnt from this chapter

9.16 What next? ▮▮▮▮

This chapter is possibly the longest book chapter ever written, and if you feel like you aged several years while reading it then, well, you probably have (look around, there are cobwebs in the room, you have a long beard, and when you go outside you'll discover a second ice age has been and gone, leaving only you and a few woolly mammoths to populate the planet). However, on the plus side, you now know more or less everything you'll ever need to know about statistics. Seriously—you'll discover in the coming chapters that everything else we discuss is a variation of this chapter. So, although you may be near death, having spent your life reading this chapter (and I'm certainly near death having written it), you are officially a stats genius—well done!

We started the chapter by discovering that at 8 years old I could have really done with a linear model to tell me which variables are important in predicting talent competition success. Unfortunately I didn't have one, but I did have my dad (and he's better than a linear model). He correctly predicted the recipe for superstardom, but in doing so he made me hungry for more. I was starting to get a taste for the rock-idol lifestyle: I had friends, a fortune (well, two fake-gold-plated winner's medals), fast cars (a bike) and dodgy-looking 8-year-olds were giving me suitcases full of lemon sherbet to lick off mirrors. The only things needed to complete the job were a platinum-selling album and a heroin addiction. However, before I could get those my parents and teachers were about to impress reality upon my young mind . . .

9.17 Key terms that I've discovered

Adjusted predicted value	Durbin–Watson test	Model sum of squares	Standardized DFBeta
Adjusted R^2	F-statistic	Multicollinearity	Standardized DFFit
Akaike information criterion (AIC)	Generalization	Multiple regression	Standardized residuals
Autocorrelation	Goodness of fit	Ordinary least squares (OLS)	Stepwise regression
b_i	Hat values	Outcome variable	Studentized deleted residuals
β_i	Heteroscedasticity	Perfect collinearity	Studentized residuals
Cook's distance	Hierarchical regression	Predicted Value	Suppressor effects
Covariance ratio (CVR)	Homoscedasticity	Predictor variable	t-statistic
Cross-validation	Independent errors	Residual	Tolerance
Deleted residual	Leverage	Residual sum of squares	Total sum of squares
DFBeta	Mahalanobis distance	Shrinkage	Unstandardized residuals
DFFit	Mean squares	Simple regression	Variance inflation factor (VIF)

Smart Alex's tasks

- **Task 1**: In Chapter 4 (Task 7) we looked at data based on findings that the number of cups of tea drunk was related to cognitive functioning (Feng et al., 2010). Using a linear model that predicts cognitive functioning from tea drinking, what would cognitive functioning be if someone drank 10 cups of tea? Is there a significant effect? (see Chapter 8, Task 9) (**Tea Makes You Brainy 716.sav**) ▮▮▮
- **Task 2**: Estimate a linear model for the **pubs.sav** data in Jane Superbrain Box 9.1 predicting **mortality** from the number of **pubs**. Try repeating the analysis but bootstrapping the confidence intervals. ▮▮▮
- **Task 3**: In Jane Superbrain Box 2.1 we encountered data (**HonestyLab.sav**) relating to people's ratings of dishonest acts and the likeableness of the perpetrator.

Run a linear model with bootstrapping to predict ratings of dishonesty from the likeableness of the perpetrator. ▮▮▮
- **Task 4**: A fashion student was interested in factors that predicted the salaries of catwalk models. She collected data from 231 models (**Supermodel.sav**). For each model she asked them their salary per day (**salary**), their age (**age**), their length of experience as a model (**years**), and their industry status as a model as their percentile position rated by a panel of experts (**beauty**). Use a linear model to see which variables predict a model's salary. How valid is the model? ▮▮▮
- **Task 5**: A study was carried out to explore the relationship between **Aggression** and several potential predicting factors in 666 children who had an older sibling. Variables

measured were **Parenting_Style** (high score = bad parenting practices), **Computer_Games** (high score = more time spent playing computer games), **Television** (high score = more time spent watching television), **Diet** (high score = the child has a good diet low in harmful additives), and **Sibling_Aggression** (high score = more aggression seen in their older sibling). Past research indicated that parenting style and sibling aggression were good predictors of the level of aggression in the younger child. All other variables were treated in an exploratory fashion. Analyze them with a linear model (**Child Aggression.sav**). ▮▮▮▮

- **Task 6**: Repeat the analysis in Labcoat Leni's Real Research 9.1 using bootstrapping for the confidence intervals. What are the confidence intervals for the regression parameters? ▮▮▮▮

- **Task 7**: Coldwell, Pike, & Dunn (2006) investigated whether household chaos predicted children's problem behavior over and above parenting. From 118 families they recorded the age and gender of the youngest child (**child_age** and **child_gender**). They measured dimensions of the child's perceived relationship with their mum: (1) warmth/enjoyment (**child_warmth**), and (2) anger/hostility (**child_anger**). Higher scores indicate more warmth/enjoyment and anger/hostility respectively. They measured the mum's perceived relationship with her child, resulting in dimensions of positivity (**mum_pos**) and negativity (**mum_neg**). Household chaos (**chaos**) was assessed. The outcome variable was the child's adjustment (**sdq**): the higher the score, the more problem behavior the child was reported to display. Conduct a hierarchical linear model in three steps: (1) enter child age and gender; (2) add the variables measuring parent–child positivity, parent–child negativity, parent–child warmth, parent–child anger; (3) add chaos. Is household chaos predictive of children's problem behavior over and above parenting? (**Coldwell et al. (2006).sav**). ▮▮▮▮

Answers & additional resources are available on the book's website at
https://edge.sagepub.com/field5e

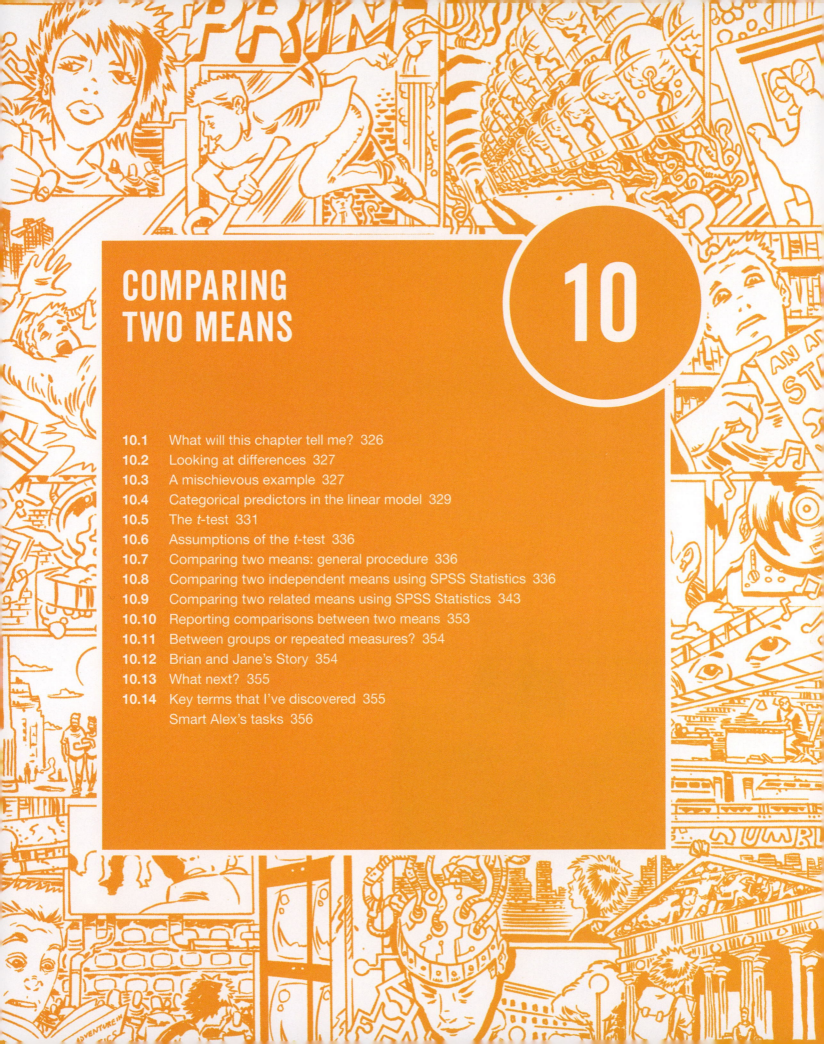

COMPARING TWO MEANS

10

10.1 What will this chapter tell me?

We saw in the previous chapter that I had successfully conquered the holiday camps of Wales with my singing and guitar playing (and the Welsh know a thing or two about good singing). I had jumped on a snowboard called oblivion and thrown myself down the black run known as world domination. About 10 meters after starting this slippery descent I hit the lumpy patch of ice called 'adults'. I was 9, life was fun, but every adult I encountered seemed obsessed with my future. 'What do you want to be when you grow up?' they would ask. Would I be a surgeon, a lawyer, a teacher? I was 9 and 'grown up' was a lifetime away. All I knew was that I was going to marry Clair Sparks (more on her in the next chapter) and be a rock legend who didn't need to worry about such adult matters as having a job. It was a difficult question, but adults require answers and I wasn't going to let them know that I didn't care about 'grown-up' matters. Like all good scientists I drew upon past data: I hadn't tried conducting brain surgery, neither did I have experience of sentencing psychopaths to prison for eating their husbands, nor had I taught anyone. I had, however, had a go at singing and playing guitar; therefore, I predicted I would be a rock star. Even at this early age I realized that not all adults would appreciate the raw talent that would see me parading across the lighted stage in front of tens of thousands of people. Some of them might not think that rock stardom was good career prospect. I needed to convince them. Adults tend to think money is important, so I decided I should demonstrate that rock stars earn more money than, say, a 'respectable' profession such as being a teacher. I could gather some teachers and rock stars, find out what their salaries were and compare them. Effectively I'd be 'predicting' salary from two categories: rock star or teacher. I could have done all that, but I didn't know about statistics when I was 9. Happy days.

Figure 10.1 Practicing for my career as a rock star by slaying the baying throng of Grove Primary School at the age of 10. (Note the girl with her hands covering her ears.)

10.2 Looking at differences ▮▮▮▮

The last two chapters have focussed on relationships between continuous variables, but sometimes researchers want to look at differences between groups of people or between people in different treatment conditions. Experimental research, for example, takes advantage of the fact that if we systematically manipulate what happens to people we can make causal inferences about the effects of those manipulations. The simplest form of experiment is one in which we split the sample into an experimental group and a control group that is identical to the experimental group in all respects except the one expected to have an impact on the outcome (see Field & Hole, 2003). For example, we might also want to compare statistics knowledge before and after a lecture. One group, the experimental group, has to sit through the lecture, while the other group, the control group, gets to miss it and stay in bed. Here are some other examples of scenarios where we'd compare two conditions:

- Is the movie *Scream 2* scarier than the original, *Scream*? We could measure heart rates (which indicate anxiety) during both films and compare them.
- Is your work better when you do it listening to Andy's favorite music? You could get some people to write an essay (or book) listening to my favorite music (as listed in the Acknowledgements), and then write a different one in silence (this is a control) and compare the grades.
- Do diet pills work? We take two groups of people and randomly assign one group to a program of diet pills and the other group to a program of sugar pills (which

they believe will help them lose weight). If the people who take the diet pills lose more weight than those on the sugar pills we can infer that the diet pills caused the weight loss.

Systematic manipulation of the independent (predictor) variable is a powerful tool because it goes one step beyond merely observing variables.[1] This chapter is the first of many that look at these research scenarios. We start with the simplest scenario: when we have two groups or, to be more specific, when we want to compare two means. We discovered in Chapter 1 that we can expose different entities to different experimental manipulations (a *between-groups* or *independent* design) or take a single group of entities and expose them to different experimental manipulations at different points in time (a *repeated-measures* or *within-subject* design). Researchers sometimes get tempted to compare artificially created groups by, for example, dividing people into groups based on a median score; avoid this temptation (Jane Superbrain Box 10.1).

10.3 A mischievous example ▮▮▮▮

Two news stories caught my eye related to physics (Di Falco, Ploschner, & Krauss, 2010). In the first headline (November 2010) the *Daily Mirror* (a UK newspaper) reported 'Scientists make Harry Potter's invisible cloak'. I'm not really a Harry Potter aficionado,[2] so it wasn't his mention that caught my attention, but the idea of being able to don a cloak that would render me invisible and able to get up to mischief. That idea was very exciting

indeed; where could I buy one? By February 2011 the same newspaper was reporting on a different piece of research (Chen et al., 2011) with a similarly exaggerated headline: 'Harry Potter-style "invisibility cloak" built by scientists'. Needless to say, scientists hadn't actually made Harry Potter's cloak of invisibility or anything close to it, but never let that get in the way of a headline. What Chen et al. had made wasn't so much a 'cloak' of invisibility as a 'calcite lump' of invisibility. This lump could hide small objects (centimeters and millimeters in scale): you could conceal my brain but little else. Nevertheless, with a suitably large piece of calcite in tow, I could theoretically hide my whole body (although people might get suspicious of the apparently autonomous block of calcite manoeuvring itself around the room on a trolley). Di Falco et al. had created a flexible material (Metaflex) with optical properties that meant that if you layered it up you might be able to create something around which light would bend. Not exactly a cloak in the clothing sense of the word, but easier to wear than, say, a slab of calcite.

Although the newspapers overstated the case a little, these are exciting pieces of research that bring the possibility of a cloak of invisibility closer to a reality. I imagine a future in which we have some cloaks of invisibility to test out. Given my slightly mischievous streak, the future me is interested in the effect that wearing a cloak of invisibility has on the tendency for mischief. I take 24 participants and place them in an enclosed community. The community is riddled with hidden cameras so that we can record mischievous acts. Half of the participants are given cloaks of invisibility; they are told not to tell anyone else about their cloak and that they can

---

1 People sometimes get confused and think that certain statistical procedures allow causal inferences and others don't (see Jane Superbrain Box 1.4).

2 Perhaps I should be, given that a UK newspaper once tagged me 'the Harry Potter of the social sciences' (http://www .discoveringstatistics.com/docs/thes_170909.pdf). I wasn't sure whether this made me a heroic wizard battling against the evil forces of statistics or an adult with a mental age of 11.

Jane Superbrain 10.1
Are median splits the devil's work? ▌▌▌▌

Sometimes scientists use a 'median split' of a predictor variable. For example, there is a stereotype that science fiction fans are recluses with no social skills. If you wanted to test this you might measure social skills and knowledge of the film *Star Wars*. You might then take the median score on *Star Wars* knowledge and classify anyone with a score above the median as a '*Star Wars* fan', and those below as a 'non-fan'. In doing this you 'dichotomize' a continuous variable. This practice is quite common, but there are several problems with median splits (MacCallum, Zhang, Preacher, & Rucker 2002):

1 Imagine there are four people: Peter, Birgit, Jip and Kiki. We measure how much they know about *Star Wars* as a percentage and get Jip (100%), Kiki (60%), Peter (40%) and Birgit (0%). If we split these four people at the median (50%) then we're saying that Jip and Kiki are the same (they get a score of 1 = fanatic) and Peter and Birgit are the same (they both get a score of 0 = not a fanatic). Median splits change the original information quite dramatically: Peter and Kiki are originally very similar but

become opposite after the split, whereas Jip and Kiki are relatively dissimilar originally but become identical after the split.

2 Effect sizes get smaller. If you correlate two continuous variables then the effect size will be larger than if you correlate the same variables after one of them has been dichotomized. Effect sizes also get smaller in linear models.

3 There is an increased chance of finding spurious effects.

So, if your supervisor has just told you to do a median split, have a good think about whether it is the right thing to do and read up on the topic (I recommend DeCoster, Gallucci, & Iselin, 2011; DeCoster, Iselin, & Gallucci, 2009; MacCallum et al., 2002). According to MacCallum et al., one of the rare situations in which dichotomizing a continuous variable is justified is when there is a clear theoretical rationale for distinct categories of people based on a meaningful break point (i.e., not the median); for example, phobic versus not phobic based on diagnosis by a trained clinician might be a legitimate dichotomization of anxiety.

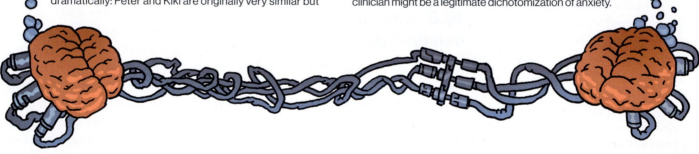

wear it whenever they like. I measure how many mischievous acts they perform in a week. These data are in Table 10.1.

The file **Invisibility.sav** shows how you should have entered the data: the variable **Cloak** records whether a person was given a cloak (**Cloak** = 1) or not (**Cloak** = 0), and **Mischief** is how many mischievous acts were performed.

Output 10.1 (your table will have more stuff in it—I edited mine down to save

SELF TEST
Enter these data into SPSS.

SELF TEST
Produce some descriptive statistics for these data (using *Explore*).

Table 10.1 Data from **Invisibility.sav**

Participant	Cloak	Mischief
1	0	3
2	0	1
3	0	5
4	0	4
5	0	6
6	0	4
7	0	6
8	0	2
9	0	0
10	0	5
11	0	4
12	0	5
13	1	4
14	1	3
15	1	6
16	1	6
17	1	8
18	1	5
19	1	5
20	1	4
21	1	2
22	1	5
23	1	7
24	1	5

space) shows that people with invisibility cloaks performed more mischievous acts, $M = 5$, 95% CI [3.95, 6.05], $SD = 1.65$, than those with no cloak, $M = 3.75$, 95% CI [2.53, 4.97], $SD = 1.91$. Both groups' scores are normally distributed according to the K-S tests because the value of *Sig.* is greater than 0.05, but these tests will be highly underpowered because they are based on Ns of 12 (Jane Superbrain Box 6.5).

10.4 Categorical predictors in the linear model ▮▮▮▮

If we want to compare differences between the means of two groups, all we are doing is predicting an outcome based on membership of two groups.

For our invisibility example, we're predicting the number of mischievous acts from whether someone had a cloak of invisibility. This is a linear model with one dichotomous predictor. The *b* for the model reflects the differences between the mean levels of mischief in the two groups, and the resulting *t*-test will, therefore, tell us whether the difference between means is different from zero (because, remember, the *t*-test tests whether *b* = 0).

You might be thinking '*b*s show relationships, not differences between means—what is this fool going on about?'. You might be starting to mistrust me or are stuffing the book in a box to post it back for a refund. I wouldn't blame you, because I used to think this too. To tame a land like the complex, thorny, weed-infested and Andy-eating tarantula-inhabited world of

statistics you need an epiphany. Mine was a paper by Cohen (1968) that showed me that when we compare means we are using a special case of the linear model. This revelation turned my statistical world into a beautiful meadow filled with bleating little lambs all jumping for joy at the wonder of life.

Recall from Chapter 2 that all statistical models are versions of the simple idea:

$$outcome_i = (model) + error_i \qquad (10.1)$$

When using a linear model this general equation becomes equation (9.2) in which the model is defined by parameters: b_0 tells us the value of the outcome when the predictor is zero, and b_1 quantifies the relationship between the predictor (X_i) and outcome (Y_i). We've seen this equation lots of times, but let's

Descriptives

Cloak of invisibility				Statistic	Std. Error
Mischievous Acts	No Cloak	Mean		3.75	.552
		95% Confidence Interval for Mean	Lower Bound	2.53	
			Upper Bound	4.97	
		5% Trimmed Mean		3.83	
		Median		4.00	
		Variance		3.659	
		Std. Deviation		1.913	
		Skewness		-.789	.637
		Kurtosis		-.229	1.232
	Cloak	Mean		5.00	.477
		95% Confidence Interval for Mean	Lower Bound	3.95	
			Upper Bound	6.05	
		5% Trimmed Mean		5.00	
		Median		5.00	
		Variance		2.727	
		Std. Deviation		1.651	
		Skewness		.000	.637
		Kurtosis		.161	1.232

Tests of Normality

	Cloak of invisibility	Kolmogorov–Smirnov[a]			Shapiro–Wilk		
		Statistic	df	Sig.	Statistic	df	Sig.
Mischievous Acts	No Cloak	.219	12	.118	.913	12	.231
	Cloak	.167	12	.200[*]	.973	12	.936

*. This is a lower bound of the true significance.

a. Lilliefors Significance Correction

Output 10.1

tailor it to our example. The equation predicting the variable **Mischief** from the group to which a person belongs (the variable **Cloak**) is:

$$Y_i = (b_0 + b_1 X_{1i}) + \varepsilon_i$$
$$\text{Mischief}_i = (b_0 + b_1 \text{Cloak}_i) + \varepsilon_i \qquad (10.2)$$

Cloak is a nominal variable: people had a 'cloak' or 'no cloak'. We can't put words into a statistical model because it will burn a hole in the ozone layer. Instead, we convert this variable into numbers, just like we do when we enter nominal variables into IBM SPSS Statistics (see Section 4.6.5). When we enter nominal variables into SPSS it doesn't matter what numbers we choose, because SPSS converts them into sensible values behind the scenes. But the numbers we choose to represent our categories in a mathematical model are important: they change the meaning of the resulting b-values. There are different 'standard' ways to code variables (which we won't get into here), one of which is to use **dummy variables**. We'll look at these in Section 11.5.1, but the summary is that we code a baseline category with a 0, and other categories with a 1. In this example there are two categories. Our baseline category is no cloak (the control condition), and we assign these participants a 0 for the variable **Cloak**. The 'experimental' group contains those who were given a cloak, and we assign these participants a 1. This is the coding I've used in the SPSS file. Let's plug these numbers into the model and see what happens.

First, imagine someone is in the no cloak condition. Knowing that they are in that group, the best prediction we could make of the number of mischievous acts would be the group mean because this value is the summary statistic with the least squared error. So, the value of Y in the equation will be the group mean $\bar{X}_{\text{No Cloak}}$ (which is 3.75 in Output 10.1) and the value of the **Cloak** variable will be 0. If we ignore the error term, equation (10.2) becomes:

$$\text{Mischief}_i = b_0 + b_1 \text{Cloak}_i$$
$$\bar{X}_{\text{No Cloak}} = b_0 + (b_1 \times 0)$$
$$b_0 = \bar{X}_{\text{No Cloak}} \qquad (10.3)$$
$$b_0 = 3.75$$

Note that b_0 (the intercept) is equal to the mean of the group coded as 0 (i.e., the no cloak group).

Now let's use the model to predict mischief in people who had an invisibility cloak. As before, the predicted value of the outcome would be the mean of the group to which the person belonged, because this is the summary statistic with the least squared error. The predicted value for someone in the cloak group is, therefore, the mean of the cloak group $\bar{X}_{\text{Cloak}}$, which is 5 in Output 10.1. The value of the **Cloak** variable is 1 (because this is the value we used to code group membership). Remember that b_0 is equal to the mean of the no cloak group ($\bar{X}_{\text{No Cloak}}$). If we place all of the values into equation (10.2) and rearrange it a bit we get:

$$\text{Mischief}_i = b_0 + b_1 \text{Cloak}_i$$
$$\bar{X}_{\text{Cloak}} = b_0 + (b_1 \times 1)$$
$$\bar{X}_{\text{Cloak}} = b_0 + b_1 \qquad (10.4)$$
$$\bar{X}_{\text{Cloak}} = \bar{X}_{\text{No Cloak}} + b_1$$
$$b_1 = \bar{X}_{\text{Cloak}} - \bar{X}_{\text{No Cloak}}$$

which shows that b_1 represents the difference between the group means (in this case $5 - 3.75 = 1.25$).

The take-home message is that we use the same linear model that we have used throughout the book to compare group means. In a model with a categorical predictor with two predictors, b_1 represents the difference between group means, and b_0 is equal to the mean of the group coded as 0. We have seen in the previous chapter that a t-statistic is used to ascertain whether a model parameter (b_1) is equal to 0; in this context, therefore, it would test whether the difference between group means is equal to 0.

If you do the self-test you should obtain the table in Output 10.2. First, notice that the value of the constant (b_0) is 3.75, the same as the mean of the base category (the no cloak group). Second, notice that the value of the regression coefficient b_1 is 1.25, which is the difference between the two group means ($5 - 3.75 = 1.25$). Finally, the t-statistic, which tests whether b_1 is significantly different from zero, is not significant because the significance value is greater than 0.05, which means that the difference between means (1.25) is not significantly different from 0. This section demonstrates that differences between

---

SELF TEST

To prove that I'm not making it up as I go along, fit a linear model to the data in **Invisibility.sav** with **Cloak** as the predictor and **Mischief** as the outcome using what you learnt in the previous chapter. **Cloak** is coded using zeros and ones as described above.

Coefficients[a]

Model		Unstandardized Coefficients B	Unstandardized Coefficients Std. Error	Standardized Coefficients Beta	t	Sig.
1	(Constant)	3.750	.516		7.270	.000
	Cloak of invisibility	1.250	.730	.343	1.713	.101

Output 10.2

means can be represented by linear models, which is a theme to which we'll return many times over the coming chapters.

10.5 The *t*-test ▌▌▌▌

We have looked at how we can include a categorical predictor in a linear model to test for differences between two means. This approach is useful in showing you the joy that is the linear model, and to keep the thread of linear models running through the book. Historically, people think about comparing two means as a separate test, and SPSS keeps this historical convention with its menu structure. This is not as bonkers as it may seem, because the linear model framework gets complicated when we want to tackle repeated-measures designs. Therefore, when testing the difference between two means, researchers tend to apply the *t*-statistic but masquerading as something called the *t*-test (Student, 1908). In this section we'll look at the theoretical underpinnings of the test. There are two variants of this test:

- **Independent *t*-test**: This test is used when you want to compare two means that come from conditions consisting of different entities (this is sometimes called the *independent-measures* or *independent-means t*-test).
- **Paired-samples *t*-test**: This test, also known as the **dependent *t*-test**, is used when you want to compare two means that come from conditions consisting of the same or related entities (Figure 10.2).

10.5.1 Rationale for the *t*-test ▌▌▌▌

Both *t*-tests have a similar rationale, which is based on what we learnt in Chapter 2 about hypothesis testing:

Figure 10.2 Thanks to the Confusion machine there are lots of terms for the paired-samples *t*-test

- Two samples of data are collected and the sample means calculated. These means might differ by either a little or a lot.
- If the samples come from the same population, then we expect their means to be roughly equal (see Section 2.7). Although it is possible for their means to differ because of sampling variation, we would expect large differences between sample means to occur very infrequently. Under the null hypothesis we assume that the experimental manipulation has no effect on the participant's behavior: therefore, we expect means from two random samples to be very similar.
- We compare the difference between the sample means that we collected to the difference between the sample means that we would expect to obtain (in the long run) if there were no effect (i.e., if the null hypothesis were true). We use the standard error (see Section 2.7) as a gauge of the variability between sample means. If the standard error is small, then we expect most samples to have very similar means. When the standard error is large, large differences in sample means are more likely. If the difference between the samples we have collected is larger than we would expect based on the standard error then one of two things has happened:

  o There is no effect but sample means from our population fluctuate a lot and we happen to have collected two samples that produce very different means.

  o The two samples come from different populations, which is why they have different means, and this difference is, therefore, indicative of a genuine difference between the samples. In other words, the null hypothesis is unlikely.

- The larger the observed difference between the sample means (relative to the standard error), the more likely it is that the second explanation is correct: that is, that the two sample means differ because of the different testing conditions imposed on each sample.

Most test statistics are a signal-to-noise ratio: the 'variance explained by the model' divided by the 'variance that the model can't explain' (reread Section 2.9.4). In other words, effect divided by error. When comparing two means the 'model' that we fit (the effect) is the difference between the two group means. Means vary from sample to sample (sampling variation), and we can use the standard error as a measure of how much means fluctuate (in other words, the error in the estimate of the mean)—see Chapter 2. Therefore, we can use the standard error of the differences between the two means as an estimate of the error in our model (or the error in the difference between means). Therefore, the *t*-statistic can be expressed as:

$$t = \frac{\substack{\text{observed difference} \\ \text{between sample means}} - \substack{\text{expected difference} \\ \text{between population means} \\ \text{(if null hypothesis is true)}}}{\substack{\text{estimate of the standard error of the} \\ \text{difference between two sample means}}}$$

(10.5)

The top half of the equation is the 'model', which is that the difference between means is bigger than the expected difference under the null hypothesis, which in most cases will be 0. The bottom half is the 'error'. So, we're basically getting the test statistic by dividing the model (or effect) by the error in the model. The exact form that this equation takes depends on whether scores are independent (e.g., come from different entities) or related to each other (come from the same or related entities).

10.5.2 The paired-samples *t*-test equation explained ▌▌▌

We'll start with the simpler scenario of when scores in the two conditions that you want to compare are related; for example, the same entities have been tested in the different conditions of your experiment or perhaps you have data on a task from twins (you'd expect each person's score to be more similar to that of their twin than that of a stranger). If you choose not to think in terms of a linear model, then you can calculate the *t*-statistic using a numerical version of equation (10.5):

$$t = \frac{\bar{D} - \mu_D}{\sigma_{\bar{D}}} = \frac{\bar{D}}{\sigma_{\bar{D}}}$$

(10.6)

This equation compares the mean difference between our samples ($\bar{D}$) to the difference that we would expect to find between population means (μ_D), relative to the standard error of the differences ($\sigma_{\bar{D}}$). If the null hypothesis is true, then we expect no difference between the population means and $\mu_D = 0$ and it drops out of the equation.

Let's explore the logic of equation (10.6). Imagine you take a pair of samples from a population, calculate their means and then take the difference between them. We know from sampling theory (Section 2.7) that, on average, sample means will be very similar to the population mean; therefore, on average, most samples should have very similar means. Our pair of random samples should, therefore, have similar means, meaning that the difference between their means is zero or close to zero. Imagine we repeated this process lots of times. We should find that most pairs of samples have differences between means that are close to zero, but sometimes one or both of the samples will have a mean very deviant from the population mean and we'd obtain a large difference between sample means. In short, sampling variation means that it's possible to get a difference between two sample means that is quite large, but it will happen relatively infrequently. If we plotted the frequency distribution of differences between means of pairs of samples we'd get the sampling distribution of differences between means. We might expect this distribution to be normal around zero, indicating that most pairs of samples have differences close to zero, and only very infrequently do we get large differences between sample means. The standard deviation of this sampling distribution is called the **standard error of differences**. Like any standard error (refresh your memory of Section 2.7 if you need to), a small standard error suggests that the difference between

means of most pairs of samples will be very close to the population mean (in this case 0 if the null is true) and that substantial differences are very rare. A large standard error tells us that the difference between means of most pairs of samples can be quite variable: although the difference between means of most pairs of samples will still be centered around zero, substantial differences from zero are more common (than when the standard error is small). As such, the standard error is a good indicator of the size of the difference between sample means that we can expect from sampling variation. In other words, it's a good baseline for what could reasonably happen if the conditions under which scores are collected are stable.

The conditions under which scores are collected are not stable, though. In experiments, we systematically manipulate the conditions under which scores are collected. For example, to test whether looking like a human affects trust of robots, participants might have two interactions with a robot: in one the robot is concealed under clothes and realistic flesh, whereas in the other their natural titanium exoskeleton is visible. Each person's trust score in the first interaction could be different from the second; the question is whether this difference is the product of how the robot looked or just what you'd get if you test the same person twice. The standard error helps us to gauge this by giving us a scale of likely variability between samples. If the standard error is small then we know that even a modest difference between scores in the two conditions would be unlikely from two random samples. If the standard error is large then a modest difference between scores is plausible from two random samples. As such, the standard error of differences provides a scale of measurement for how plausible it is that an observed difference between sample means could be the product of taking two random samples from the same population. That's what the bottom of equation (10.6) represents: it places the observed difference between sample means in the context of what's plausible for random samples.

The top half of equation (10.6) represents the size of the observed effect. $\bar{D}$ is the average difference between people's scores in two conditions. For each person, if we took their score in one condition and subtracted it from their score in the other, this would give us a difference score for each person; $\bar{D}$ is the mean of these difference scores. Going back to our robot example, if the appearance of the robot had no effect on people's trust, scores would be similar in the two conditions and we'd get an average difference of 0 (or close to it). If the robot's appearance matters, we'd expect scores to differ in the two conditions, and the resulting average difference would be different from 0.

So, $\bar{D}$ represents the effect size, and as I said earlier, we place this effect size within the context of what's plausible for random samples by dividing by the standard error of differences. We know that the standard error can be estimated from the standard deviation divided by the square root of the sample size (equation (2.14) in Section 2.7). The standard error of differences ($\sigma_{\bar{D}}$) is likewise estimated from the standard deviation of differences within the sample (s_D) divided by the square root of sample size (N). Replacing this term in equation (10.6) gives:

$$t = \frac{\bar{D}}{\sigma_{\bar{D}}} = \frac{\bar{D}}{s_D / \sqrt{N}} \tag{10.7}$$

Therefore, t is a signal-to-noise ratio or the systematic variance compared to the unsystematic variance. The top half of equation (10.7) is the signal or effect, whereas the bottom places that effect within the context of the natural variation between samples (the noise or unsystematic variation). If the experimental manipulation creates difference between

conditions, then we would expect the effect (the signal) to be greater than the unsystematic variation (the noise) and, at the very least, t will be greater than 1. We can compare the obtained value of t against the maximum value we would expect to get if the null hypothesis were true in a t-distribution with the same degrees of freedom (these values can be found in the Appendix). If the observed t exceeds the critical value for the predetermined alpha (usually 0.05), scientists tend to assume that this reflects an effect of their independent variable. We *can* compare the observed t to tabulated critical values, but we can also urinate in a bucket and throw it from our window into the street. It doesn't mean that we should, and seeing as it's not 1908 we won't do either of these things; instead we'll urinate in a computer and let a toilet compute an exact p-value for t. I think it's that way around. If the exact p-value for t is below the predetermined alpha value (usually 0.05), scientists take this to support the conclusion that the differences between scores are not due to sampling variation and that their manipulation (e.g., dressing up a robot as a human) has had a significant effect.

10.5.3 The independent *t*-test equation explained ▌▌▌▌

When we want to compare scores that are independent (e.g., different entities have been tested in the different conditions of your experiment) we are in the same logical territory as when scores are related. The main difference is how we arrive at the values of interest. The equation for t based on independent scores is still a numerical version of equation (10.5). The main difference is that we're not dealing with difference scores because there's no connection between scores in the two conditions that we want to compare.

When scores in two groups come from different participants, pairs of scores will differ not only because of the experimental manipulation reflected by those

conditions, but also because of other sources of variance (individual differences between participants' motivation, IQ, etc.). These individual differences are eliminated when we use the same participants across conditions. Because the scores in the two conditions have no logical connection, we compare means on a *per condition* basis. We compute differences between the two sample means ($\bar{X}_1 - \bar{X}_2$) and not between individual pairs of scores. The difference between sample means is compared to the difference we would expect to get between the means of the two populations from which the samples come ($\mu_1 - \mu_2$):

$$t = \frac{\left(\bar{X}_1 - \bar{X}_2\right) - \left(\mu_1 - \mu_2\right)}{\text{estimate of the standard error}} \tag{10.8}$$

If the null hypothesis is true then the samples have been drawn from populations that have the same mean. Therefore, under the null hypothesis $\mu_1 = \mu_2$, which means that $\mu_1 - \mu_2 = 0$, and so $\mu_1 - \mu_2$ drops out of the equation leaving us with:

$$t = \frac{\bar{X}_1 - \bar{X}_2}{\text{estimate of the standard error}} \tag{10.9}$$

Now, imagine we took several pairs of samples—each pair containing one sample from the two different populations—and compared the means of these samples. From what we have learnt about sampling distributions, we know that many samples from a population will have similar means. If the populations from which we're drawing have the same mean (which under the null hypothesis they do) then pairs of samples should also have the same means, and the difference between sample means should be zero or close to it. We're now in very familiar territory to the paired-samples t-test because the sampling distribution of the differences between pairs of sample means would be normal with a mean equal to the difference between population means ($\mu_1 - \mu_2$), which under the null is zero. The sampling distribution would tell us by how much we can expect the means of two (or more) samples to differ (if the null were true). The standard deviation of this sampling distribution (the standard error) tells us how plausible differences between sample means are (under the null hypothesis). If the standard error is large then large differences between sample means can be expected; if it is small then only small differences between sample means are typical. As with related scores, it makes sense then to use the standard error to place the difference between sample means into the context of what's plausible given the null hypothesis. As such, equation (10.9) is conceptually the same as equation (10.6); all that differs is how we arrive at the effect (top half) and the standard error (the bottom half). The standard error, in particular, is derived quite differently for independent samples.

I've already reminded you that the standard error can be estimated from the

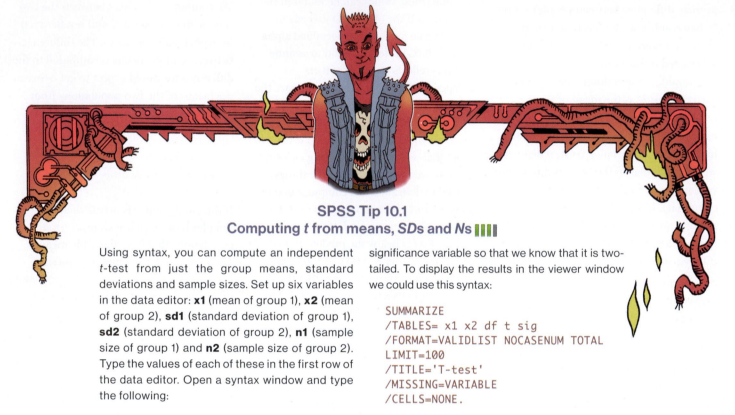

SPSS Tip 10.1
Computing t from means, SDs and Ns ▮▮▮▮

Using syntax, you can compute an independent t-test from just the group means, standard deviations and sample sizes. Set up six variables in the data editor: **x1** (mean of group 1), **x2** (mean of group 2), **sd1** (standard deviation of group 1), **sd2** (standard deviation of group 2), **n1** (sample size of group 1) and **n2** (sample size of group 2). Type the values of each of these in the first row of the data editor. Open a syntax window and type the following:

```
COMPUTE df = n1+n2-2.
COMPUTE poolvar =
(((n1-1)*(sd1**2))+((n2-1)*(sd2**2)))/df.
COMPUTE t = (x1-x2)/sqrt(poolvar*
((1/n1)+(1/n2))).
COMPUTE sig = 2*(1-(CDF.T(abs(t),df))).
Variable labels sig 'Significance
(2-tailed)'.
EXECUTE.
```

The first line computes the degrees of freedom, the second the pooled variance, s_p^2, the third t and the fourth its two-tailed significance. All these values will be created in new columns in the data editor. The line beginning 'Variable labels' labels the significance variable so that we know that it is two-tailed. To display the results in the viewer window we could use this syntax:

```
SUMMARIZE
/TABLES= x1 x2 df t sig
/FORMAT=VALIDLIST NOCASENUM TOTAL
LIMIT=100
/TITLE='T-test'
/MISSING=VARIABLE
/CELLS=NONE.
```

These commands produce a table of the variables **x1**, **x2**, **df**, **t** and **sig**, so you'll see the means of the two groups, the degrees of freedom, the value of t and its two-tailed significance.

You can run lots of t-tests at the same time by putting different values for the means, SDs and sample sizes in different rows. If you do this, though, I suggest having a string variable called **Outcome** in the file in which you type what was being measured (or some other information so that you can identify to what the t-test relates). I've used these commands in a syntax file called **Independent t from means.sps**. My file is a bit more complicated because it also calculates Cohen's d. For an example of how to use it see Labcoat Leni's Real Research 10.1.

Labcoat Leni's Real Research 10.1
You don't have to be mad here, but it helps ||||

Board, B. J., & Fritzon, K. (2005). *Psychology, Crime & Law, 11*, 17–32.

In the UK you often see the 'humorous' slogan 'You don't have to be mad to work here, but it helps' stuck up in work places. Board and Fritzon (2005) took this slogan a step further by measuring whether 39 senior business managers and chief executives from leading UK companies were mad (well, had personality disorders, PDs). They gave them the Minnesota Multiphasic Personality Inventory Scales for DSM III Personality Disorders (MMPI-PD), which measures 11 personality disorders: histrionic, narcissistic, antisocial, borderline, dependent, compulsive, passive-aggressive, paranoid, schizotypal, schizoid and avoidant. As a comparison group they chose 317 legally classified psychopaths from a high-security psychiatric hospital.

The authors report the means and standard deviations for these two groups in Table 2 of their paper. Run the syntax file **Independent t from means.sps** on the data in **Board and Fritzon 2005.sav** to see whether managers score significantly higher on personality disorder questionnaires than legally classified psychopaths. Report these results. What do you conclude? Answers are on the companion website (or see Table 2 in the original article).

standard deviation and the sample size. It is, therefore, straightforward to estimate the standard error for the sampling distribution of each population by using the standard deviation (s) and size (N) for each sample:

$$SE \text{ of sampling distribution of population 1} = \frac{s_1}{\sqrt{n_1}}$$

$$SE \text{ of sampling distribution of population 2} = \frac{s_2}{\sqrt{n_2}}$$

$$(10.10)$$

These values don't tell us about the standard error for the sampling distribution of *differences* between means, though. To estimate that, we need to first convert these standard errors to variances by squaring them:[3]

$$\begin{aligned} \text{variance of sampling} \\ \text{distribution of population 1} \end{aligned} = \left(\frac{s_1}{\sqrt{n_1}}\right)^2 = \frac{s_1^2}{n_1}$$

$$\begin{aligned} \text{variance of sampling} \\ \text{distribution of population 2} \end{aligned} = \left(\frac{s_2}{\sqrt{n_2}}\right)^2 = \frac{s_2^2}{n_2}$$

$$(10.11)$$

Having converted to variances, we can take advantage of the **variance sum law,** which states that the variance of a difference between two independent variables is equal to the sum of their variances (see, for example, Howell, 2012). Put simply, the variance of the sampling distribution of differences between two sample means will be equal to the sum of the variances of the two populations from which the samples were taken. This law means that we can estimate the variance of the sampling distribution of differences by adding together the variances of the sampling distributions of the two populations:

$$\begin{aligned} \text{variance of sampling} \\ \text{distribution of differences} \end{aligned} = \frac{s_1^2}{n_1} + \frac{s_2^2}{n_2}$$

$$(10.12)$$

---

3   Remember that a standard error is a standard deviation (it's just called a standard error because we're dealing with a sampling distribution), and the standard deviation is the square root of the variance.

We convert this variance back to a standard error by taking the square root:

$$SE \text{ of sampling distribution of differences} = \sqrt{\frac{s_1^2}{n_1} + \frac{s_2^2}{n_2}} \qquad (10.13)$$

If we pop this equation for the standard error of differences into equation (10.9) we get:

$$t = \frac{\bar{X}_1 - \bar{X}_2}{\sqrt{\frac{s_1^2}{n_1} + \frac{s_2^2}{n_2}}} \qquad (10.14)$$

Equation (10.14) is true only when the sample sizes are equal, which in naturalistic studies may not be possible. To compare two groups that contain different numbers of participants we use a pooled variance estimate instead, which takes account of the difference in sample size by *weighting* the variance of each sample by a function of the size of sample on which it's based:

$$s_p^2 = \frac{(n_1 - 1)s_1^2 + (n_2 - 1)s_2^2}{n_1 + n_2 - 2} \qquad (10.15)$$

This weighting makes sense because (as we saw in Chapter 1) large samples more closely approximate the population than small ones; therefore, they should carry more weight. In fact, rather than weighting by the sample size, we weight by the sample size minus 1 (the degrees of freedom).

The pooled variance estimate in equation (10.15) is a weighted average: each variance is multiplied (weighted) by its degrees of freedom, and then we divide by the sum of weights (or sum of the two degrees of freedom) to get an average. The resulting weighted average variance is plopped into the equation for *t*:

$$t = \frac{\bar{X}_1 - \bar{X}_2}{\sqrt{\frac{s_p^2}{n_1} + \frac{s_p^2}{n_2}}} \qquad (10.16)$$

One thing that might be apparent from equation (10.16) is that you don't actually need any raw data to compute *t*; you just need the group means, standard deviations and sample sizes (see SPSS Tip 10.1).

As with the paired-samples *t*, the Edwardians among you can compare the observed *t* to critical values in a table, but the rest of us flush a toilet and an exact *p*-value comes out that we use as information to decide whether the observed effect is indicative of something more theoretically interesting than sampling variation.

10.6 Assumptions of the *t*-test ▌▌▌▌

Both the independent *t*-test and the paired-samples *t*-test are *parametric tests* and as such are prone to the sources of bias discussed in Chapter 6. For the paired-samples *t*-test the assumption of normality

relates to the sampling distribution of the *differences* between scores, not the scores themselves (see Section 1.9.2). There are variants on these tests that overcome all of the potential problems, though.

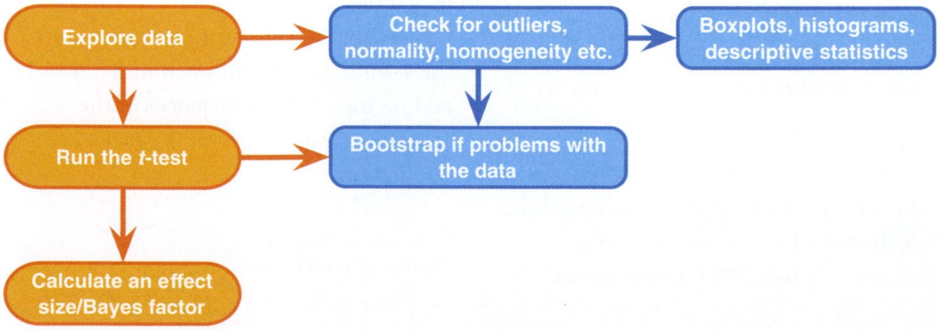

10.7 Comparing two means: general procedure ▌▌▌▌

I have probably bored most of you to the point of wanting to eat your own legs by now. Equations are tedious, and that is why computers were invented to help us minimize our contact with them. It's time to move on and do stuff. Figure 10.3 shows the general process for performing a *t*-test. As with fitting any model, we start by looking for the sources of bias identified in Chapter 6. Having satisfied ourselves that assumptions are met and outliers dealt with, we run the test. We can also consider using bootstrapping if any of the test assumptions were not met. Finally, we compute an effect size and Bayes factor.

10.8 Comparing two independent means using SPSS Statistics ▌▌▌▌

Thinking back to our invisibility data (**Invisibility.sav**), we have 12 people who were given an invisibility cloak and 12 who were not (the groups are coded using the variable **Cloak**). Remember that the number of mischievous acts they performed was measured (**Mischief**). I have already described how the data are arranged (see Section 10.3), so we can move onto doing the test itself.

10.8.1 Exploring data and testing assumptions ▌▌▌▌

We obtained descriptive statistics and looked at distributional assumptions in Section 10.3. We found evidence of normality in each group, and the mean

Figure 10.3 The general process for performing a *t*-test

number of mischievous acts was higher for those with a cloak ($M = 5$) than for those without ($M = 3.75$). To look at homogeneity of variance (Section 6.11) SPSS will produce Levene's test when we run the *t*-test.

SELF TEST

Produce an error bar chart of the **Invisibility.sav** data (**Cloak** will be on the *x*-axis and **Mischief** on the *y*-axis).

10.8.2 The independent *t*-test using SPSS Statistics ▮▮▮▮

To run an independent *t*-test, we access the main dialog box by selecting *Analyze* ▸ *Compare Means* ▸ Independent-Samples T Test... (see Figure 10.4). Once the dialog box is activated, select the outcome variable (**Mischief**) and drag it to the box labeled *Test Variable(s)* (or click). If you want to carry out *t*-tests on several outcome variables at once then you can select several variables and transfer them to the *Test Variable(s)* box. However, remember that by doing lots of tests you inflate the Type I error rate (see Section 2.9.7).

Next, we need to specify a predictor variable (the grouping variable). In this case, transfer **Cloak** to the box labeled *Grouping Variable*. The Define Groups... button will become active, so click it to activate the *Define Groups* dialog box. SPSS needs to know what numeric codes you assigned to your two groups, and there is a space for you to type the codes. In this example, we coded the no cloak group as 0 and our cloak group as 1, and so these are the values that we type into the boxes (as in Figure 10.4). Alternatively, you can specify a *Cut point* value whereby cases greater than or equal to that value are assigned to one group and values below the cut point to a second group. You can use this option to, for example, compare groups of participants based on something like a median split (see Jane Superbrain Box 10.1)—you would type the median value in the box labeled *Cut point*. When you have defined the groups, click Continue to return to the main dialog box.

Clicking Options... activates a dialog box in which you can change the width of the

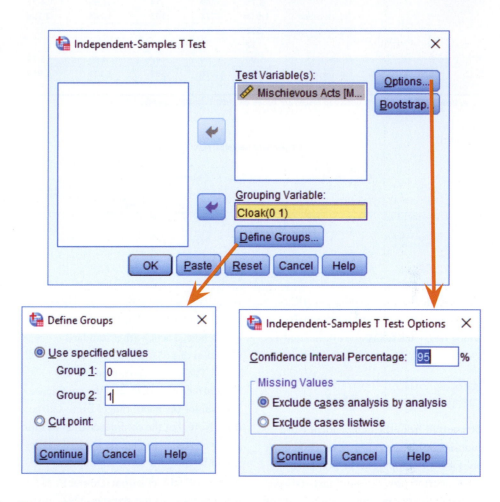

Figure 10.4 Dialog boxes for the independent-samples *t*-test

confidence interval in the output. The default setting of a 95% confidence interval is fine, but there may be times where you want to be stricter (and specify something like a 99% confidence interval) even though you run a higher risk of failing to detect a genuine effect (a Type II error) or more lenient (e.g., a 90% confidence interval) which of course increases the chance of falsely accepting a hypothesis (a Type I error). You can also select how to deal with missing values (see SPSS Tip 6.1).

If you are worried about the assumption of normality or simply want confidence intervals that don't rely on this assumption, then you can use bootstrapping (Section 6.12.3). Select this option by clicking Bootstrap... in the main dialog box to access the *Bootstrap* dialog box. We discussed this dialog box in Section 6.12.3; to recap, select ☑ Perform bootstrapping to activate bootstrapping, and to get a 95% confidence interval click ◉ Percentile or ◉ Bias corrected accelerated (BCa). For this

Group Statistics

	Cloak of invisibility		Statistic	Bootstrap[a]			
				Bias	Std. Error	BCa 95% Confidence Interval Lower	BCa 95% Confidence Interval Upper
Mischievous Acts	No Cloak	N	12				
		Mean	3.75	.00	.54	2.69	4.71
		Std. Deviation	1.913	−.124	.367	1.206	2.258
		Std. Error Mean	.552				
	Cloak	N	12				
		Mean	5.00	.00	.47	4.20	5.79
		Std. Deviation	1.651	−.110	.329	1.131	1.958
		Std. Error Mean	.477				

a. Unless otherwise noted, bootstrap results are based on 1000 bootstrap samples

Output 10.3

Independent Samples Test

		Levene's Test for Equality of Variances		t-test for Equality of Means					95% Confidence Interval of the Difference	
		F	Sig.	t	df	Sig. (2-tailed)	Mean Difference	Std. Error Difference	Lower	Upper
Mischievous Acts	Equal variances assumed	.545	.468	−1.713	22	.101	−1.250	.730	−2.763	.263
	Equal variances not assumed			−1.713	21.541	.101	−1.250	.730	−2.765	.265

Output 10.4

analysis, choose a bias corrected (BCa) confidence interval. Back in the main dialog box, click OK to run the analysis.

10.8.3 Output from the independent *t*-test ▮▮▮▮

The output from the independent *t*-test contains only three tables (two if you don't opt for bootstrapping). The first table (Output 10.3) provides summary statistics for the two experimental conditions (if you don't ask for bootstrapping this table will be a bit more straightforward). From this table, we can see that both groups had 12 participants (rows labelled *N*). The group that had no cloak performed, on average, 3.75 mischievous acts with a standard deviation of 1.913. What's more, the standard error of that group is 0.552. The bootstrap estimate of this standard error is a little smaller at 0.54, and the bootstrapped confidence interval for the

mean ranges from 2.69 to 4.71. Those who were given an invisibility cloak performed, on average, 5 acts, with a standard deviation of 1.651, and a standard error of 0.477. The bootstrap estimate of this standard error is 0.47, and the confidence interval for the mean ranges from 4.20 to 5.79. Note that the confidence intervals for the two groups overlap, implying that they might be from the same population.

The second table of output (Output 10.4) contains the main test statistics. There are two rows containing values for the test statistics: one is labelled *Equal variances assumed*, while the other is labelled *Equal variances not assumed*. In Chapter 6, we saw that parametric tests assume that the variances in experimental groups are roughly equal. We also saw in Jane Superbrain Box 6.6 that there are adjustments that can be made in situations in which the variances are not equal. The rows of the table relate to whether these adjustments have been applied.

We saw in Section 6.11 that Levene's test tests whether variances are different in different groups. However, because the results of this test will depend on the sample size you have, and we can adjust the degrees of freedom of the *t*-test to compensate for the degree to which variances are unequal, there is a good argument for ignoring Levene's test and always reading results from the row labelled *Equal variances not assumed* (see Jane Superbrain Box 6.6).

We are told the mean difference ($\bar{X}_{\text{No Cloak}} - \bar{X}_{\text{Cloak}} = 3.75 - 5 = -1.25$) and the standard error of the sampling distribution of differences, which is calculated using the lower half of equation (10.14). The *t*-statistic is the mean difference divided by this standard error ($t = -1.25/0.730 = -1.71$). A *p*-value associated with this *t* is calculated based on a *t*-distribution with particular degrees of freedom. For the independent *t*-test, the degrees of freedom are calculated by adding the two sample sizes and then subtracting the number of samples ($df = N_1 + N_2 - 2 = 12 + 12 - 2 = 22$). This value is then reduced to compensate for any imbalance in group variances (in this case it becomes 21.54). The resulting (two-tailed) *p*-value is 0.107, which represents the probability of getting a *t* of −1.71 or smaller if the null hypothesis were true. Assuming our alpha is 0.05, we'd conclude that there was no significant difference between the means of these two samples because the observed *p* of 0.101 is greater than the criterion of 0.05. In terms of the experiment, we can infer that having a cloak of invisibility did not significantly affect the amount of mischief a person got up to. Note that the value of *t* and the significance value are the same as when we ran the same test as a regression (see Output 10.2).[4]

---

4 The value of the *t*-statistic is the same but has a positive sign rather than negative. You'll remember from the discussion of the point-biserial correlation in Section 8.4.5 that when you correlate a dichotomous variable the direction of the correlation coefficient depends entirely upon which cases are assigned to which groups. Therefore, the direction of the *t*-statistic here is similarly influenced by which group we select to be the base category (the category coded as 0).

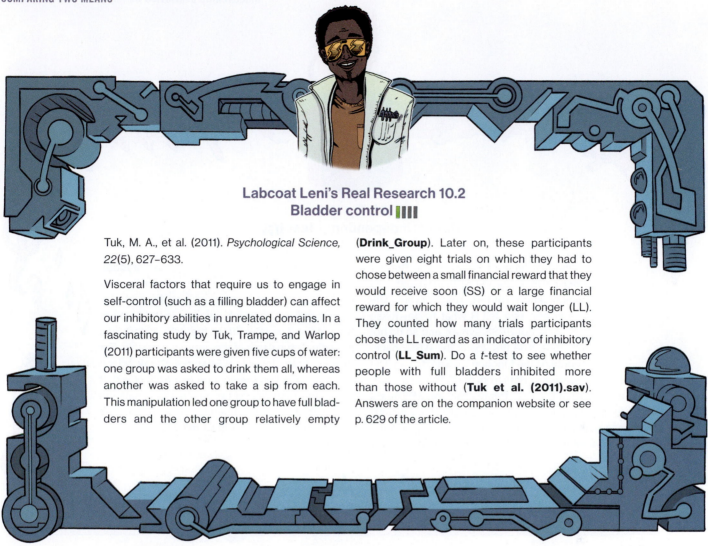

Labcoat Leni's Real Research 10.2
Bladder control ▌▌▌▌

Tuk, M. A., et al. (2011). *Psychological Science, 22*(5), 627–633.

Visceral factors that require us to engage in self-control (such as a filling bladder) can affect our inhibitory abilities in unrelated domains. In a fascinating study by Tuk, Trampe, and Warlop (2011) participants were given five cups of water: one group was asked to drink them all, whereas another was asked to take a sip from each. This manipulation led one group to have full bladders and the other group relatively empty (**Drink_Group**). Later on, these participants were given eight trials on which they had to chose between a small financial reward that they would receive soon (SS) or a large financial reward for which they would wait longer (LL). They counted how many trials participants chose the LL reward as an indicator of inhibitory control (**LL_Sum**). Do a *t*-test to see whether people with full bladders inhibited more than those without (**Tuk et al. (2011).sav**). Answers are on the companion website or see p. 629 of the article.

10.8.4 Robust tests of two independent means ▌▌▌▌

Assuming you had reason to doubt the assumption of normality, you could interpret the bootstrap confidence intervals (Output 10.5). It's also a legitimate approach to always use bootstrapped confidence intervals and not think about normality one way or another. The table shows a robust re-estimate of the standard error of the mean difference (0.703 rather than 0.730, the value in Output 10.4).[5] The difference between means was –1.25, and the bootstrap confidence interval ranges from –2.653 to 0.111, which implies that the difference

Bootstrap for Independent Samples Test

			Bootstrap[a]			
		Mean Difference	Bias	Std. Error	BCa 95% Confidence Interval	
					Lower	Upper
Mischievous Acts	Equal variances assumed	–1.250	.001	.703	–2.653	.111
	Equal variances not assumed	–1.250	.001	.703	–2.653	.111

Output 10.5

between means in the population could be negative, positive or even zero (because the interval ranges from a negative value to a positive one). In other words, if we assume that this is one of the 95% of intervals that will capture the population value, it's possible that the true difference between means is zero—no difference at all. The

bootstrap confidence interval confirms our conclusion that having a cloak of invisibility seems not to affect acts of mischief.

It's also possible to run a robust version of the *t*-test itself using R. To do this you need the Essentials for R plugin installed (Section 4.13.2) and you need to have installed the *WRS2* package (Section 4.13.3).

---

5 Remember that the values for the standard error and confidence interval you get could differ from mine because of the way bootstrapping works.

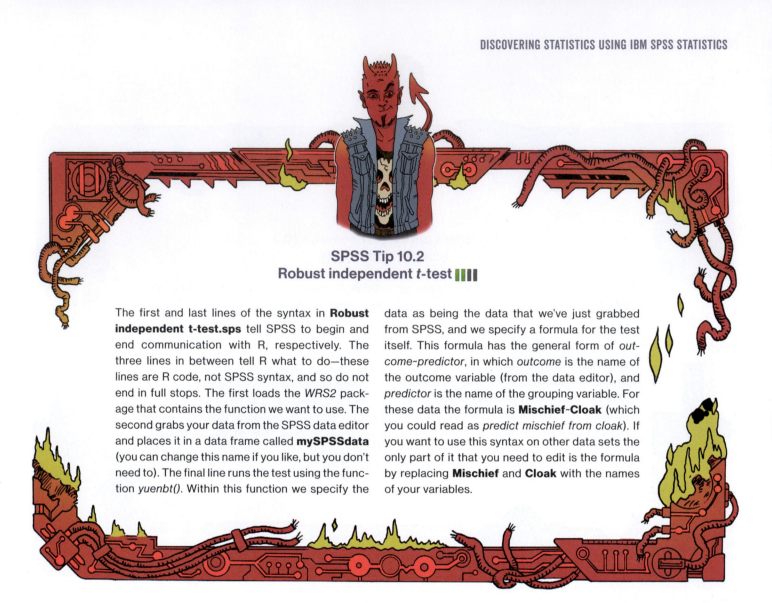

SPSS Tip 10.2
Robust independent *t*-test ▮▮▮▮

The first and last lines of the syntax in **Robust independent t-test.sps** tell SPSS to begin and end communication with R, respectively. The three lines in between tell R what to do—these lines are R code, not SPSS syntax, and so do not end in full stops. The first loads the *WRS2* package that contains the function we want to use. The second grabs your data from the SPSS data editor and places it in a data frame called **mySPSSdata** (you can change this name if you like, but you don't need to). The final line runs the test using the function *yuenbt()*. Within this function we specify the

data as being the data that we've just grabbed from SPSS, and we specify a formula for the test itself. This formula has the general form of *outcome~predictor*, in which *outcome* is the name of the outcome variable (from the data editor), and *predictor* is the name of the grouping variable. For these data the formula is **Mischief~Cloak** (which you could read as *predict mischief from cloak*). If you want to use this syntax on other data sets the only part of it that you need to edit is the formula by replacing **Mischief** and **Cloak** with the names of your variables.

The companion website contains a syntax file (**Robust independent t-test.sps**) for running a robust variant of the *t*-test based on Yuen (1974) which uses trimming and a bootstrap. The syntax in the file is as follows:

```
BEGIN PROGRAM R.
library(WRS2)
mySPSSdata = spssdata.GetData
FromSPSS(factorMode = "labels")
yuenbt(Mischief~Cloak, data =
mySPSSdata)
END PROGRAM.
```

```
Test statistic: -1.3607 (df=NA), p-value=0.16694
Trimmed mean difference: -1
95 percent confidence interval:
-2.5161 0.5161
```

Output 10.6

To obtain the robust test select and run these five lines of syntax, which are explained in SPSS Tip 10.2.

Having run the syntax, you'll find some rather uninspiring text output in the viewer (Output 10.6). This tells us that there is not a significant difference (because the confidence interval crosses zero and the *p*-value is greater than 0.05) in mischief scores across the two cloak groups, $Y_t = -1.36$ (−2.52, 0.52), $p = 0.167$.

10.8.5 Bayesian test of two independent means ▮▮▮▮

To access the dialog box to compute a Bayes factor and credible interval for independent means select *Analyze* ▸ *Bayesian Statistics* ▸ *Independent Samples Normal* (Figure 10.5). Select the outcome variable (**Mischief**) and drag it into the box labeled *Test Variable(s)* (or click ⬇), and select the group variable (**Cloak**) and drag it to the box labeled *Grouping Variable* (or click ⬇). As with the regular *t*-test use the Define Groups… button to tell SPSS that the first group (no cloak) was coded with 0 and group 2 (cloak) was coded with 1. To compute Bayes factors and estimate the model parameters select ⦿ Use *B*oth Methods.

Imagine that prior to collecting data we believed that people without an invisibility cloak might be likely to perform anywhere between 0 and 6 mischievous acts. Obviously, we can't believe in a value less than 0 (you can't have a negative number of mischievous acts), but we're not prepared to believe in a value over 6. With a cloak of invisibility, we think the number of mischievous acts might fall anywhere between 0 and about 8. We could model these prior beliefs with a normal distribution. The graphs with red lines in Output 10.7 show these priors. The left graph is the prior for the no cloak group. As you can see it is a normal distribution centered at 3 and ranging from about 0 to 6. This equates to us believing that without an invisibility cloak, 3 acts of mischief are most probable, but we're prepared to accept values as low as 0 or as high as 6 (although we think these values are unlikely). The right graph is the prior for the cloak group, which is also a normal distribution but centered at 4 and ranging from about 0 to 8. This equates to us believing that with an invisibility cloak, 4 acts of mischief are most probable, but we're prepared to accept (with low probability) values as low as 0 or as high as 8.

To set up these priors click [Priors...] and first specify the variances of the two groups (remember that group 1 is the no cloak and group 2 is the cloak group). I've used the variances from the data, which I obtained by squaring the standard deviations in Output 10.3. Using these variances simplifies things (we don't need to set a prior for the variance) but as you get more experience you might want to estimate the variance too. At the bottom select ⦿ Normal. For group 1 I have specified a distribution with mean (location) of 3 and standard deviation (scale) of 1. For group 2 I have specified a distribution with a mean of 4 and standard deviation of 1.75. To get an appropriate scale parameter, start by taking the range of plausible scores and divide by 6, then tweak it until the distribution looks correct. For example,

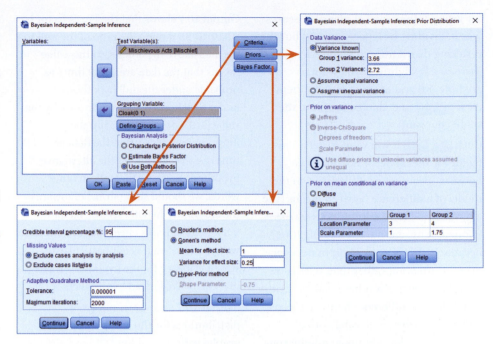

Figure 10.5 Dialog box for a Bayesian test of two independent means

Bayes Factor Independent Sample Test (Method = Gonen)[a]

	Mean Difference	Pooled Std. Error Difference	Bayes Factor[b]	t	df	Sig.(2–tailed)
Mischievous Acts	1.25	.730	.419	1.713	22	.101

a. Assumes unequal variance between groups.
b. Bayes factor: Null versus alternative hypothesis.

Posterior Distribution Characterization for Independent Sample Mean[a]

		Posterior		95% Credible Interval	
	Mode	Mean	Variance	Lower Bound	Upper Bound
Mischievous Acts	1.31	1.31	.434	.02	2.60

a. Prior for Variance: Diffuse. Prior for Mean: Normal.

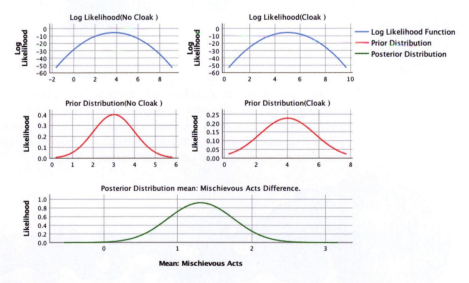

Output 10.7

for group 2 the range of plausible scores was 8, dividing by 6 gave me 1.33 but the resulting distribution was too narrow so I increased it slightly. If you want a credible interval other than 95% then click Criteria... and change the 95 to the value you want.

Click Bayes Factor... if you want to get a Bayes factor for the model of no difference between means (the null) relative to the model of a difference between means (the alternative). I have selected ● Gonen's method (Gönen, Johnson, Lu, & Westfall, 2005), because it allows us to incorporate some prior information rather than relying on defaults. Specifically, I have defined a prior belief that the difference between means will be 1 (having an invisibility cloak would lead to one additional mischievous act compared to not having the cloak), and I've set this effect to have a variance of 0.25. Back in the main dialog box click OK to fit the model.

Output 10.7 shows that the Bayes factor is 0.419. SPSS reports the ratio of the null to the alternative hypothesis, so this value means that the data are 0.419 times as probable under the null hypothesis as under the alternative. If we divide 1 by the value (1/0.419 = 2.39) we flip the interpretation to say that the data are 2.39 times as probable under the alternative hypothesis as under the null. In other words, we should shift our belief towards the alternative hypothesis by a factor of 2.39. Although this effect is in the direction we hypothesised, it is not particularly strong evidence for the hypothesis that invisibility cloaks lead to mischief.

Output 10.7 also shows our prior distributions for the group means (the graphs with the red lines). I have already described these. The Bayesian estimates of the difference between means (i.e., the b-value for group as a predictor of mischief) are in the columns labeled

Posterior Mode and Posterior Mean. The 95% credible interval for this estimate ranged from 0.02 to 2.60. In other words, *assuming that the effect exists,* the population value of the effect will be between 0.02 to 2.60 with 95% probability. This tells us nothing about the null hypothesis (because it assumes the effect exists) but helps us to ascertain the likely population value if we're prepared to accept that the effect exists. So, we can say with 95% probability that having a cloak of invisibility will increase mischievous acts by anything as low as 0.02 (i.e., not at all) to 2.60 (although, of course, you can't have 0.6 of a mischievous act!).

If you want to see what happens when you use reference priors, then repeat this process but select *Assume unequal variance* when you click on Priors... and ● Rouder's method when you click Bayes Factor.... Doing so will fit a model with diffuse, uninformative priors.

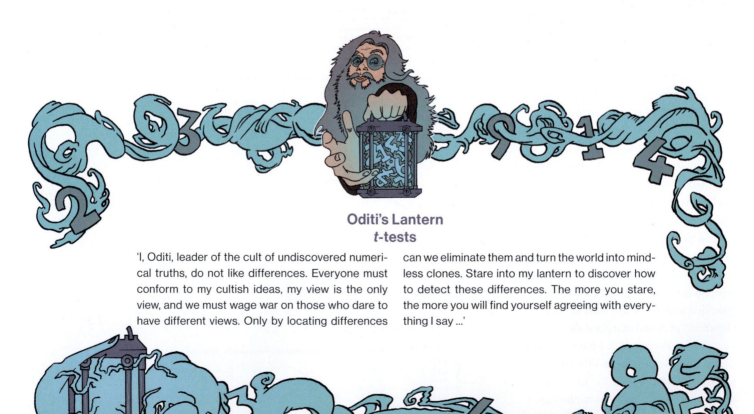

Oditi's Lantern
t-tests

'I, Oditi, leader of the cult of undiscovered numerical truths, do not like differences. Everyone must conform to my cultish ideas, my view is the only view, and we must wage war on those who dare to have different views. Only by locating differences can we eliminate them and turn the world into mindless clones. Stare into my lantern to discover how to detect these differences. The more you stare, the more you will find yourself agreeing with everything I say ...'

Cramming Sam's Tips
The independent *t*-test

- The independent *t*-test compares two means, when those means have come from different groups of entities.
- You should probably ignore the column labeled *Levene's Test for Equality of Variance* and always look at the row in the table labeled *Equal variances not assumed*.
- Look at the column labeled *Sig*. If the value is less than 0.05 then the means of the two groups are significantly different.

- Look at the table labeled *Bootstrap for Independent Samples Test* to get a robust confidence interval for the difference between means.
- Look at the values of the means to see how the groups differ.
- A robust version of the test can be computed using syntax.
- A Bayes factor can be computed that quantifies the ratio of how probable the data are under the alternative hypothesis compared to the null.
- Calculate and report the effect size. Go on, you can do it!☺

10.8.6 Effect sizes for two independent means ▮▮▮▮

The fact that the *t*-statistic is not statistically significant doesn't mean that our effect is unimportant, so it's worth quantifying it with an effect size (see Section 3.5). We can convert a *t*-value into an *r*-value using the following equation (e.g. Rosenthal, 1991; Rosnow & Rosenthal, 2005):

$$r = \sqrt{\frac{t^2}{t^2 + df}} \qquad (10.17)$$

Using the value of *t* and the *df* from Output 10.4, we get:

$$r = \sqrt{\frac{(-1.713)^2}{(-1.713)^2 + 22}} = \sqrt{\frac{2.93}{24.93}} = 0.34 \qquad (10.18)$$

Thinking back to our benchmarks for effect sizes, this represents a medium effect. Therefore, even though the effect was non-significant, it represents a fairly substantial effect.

We could instead compute Cohen's *d* (Section 3.7.1), using the two means (5 and 3.75) and the standard deviation of the control group (no cloak):

$$\hat{d} = \frac{\bar{X}_{\text{Cloak}} - \bar{X}_{\text{No Cloak}}}{s_{\text{No Cloak}}} = \frac{5 - 3.75}{1.91} = 0.65 \qquad (10.19)$$

This means that there is 0.65 standard deviations difference between the two groups in terms of their mischief making, which again is fairly substantial.

10.9 Comparing two related means using SPSS Statistics ▮▮▮▮

10.9.1 Entering data ▮▮▮▮

Let's imagine that we had collected the cloak of invisibility data using a repeated-measures design: we might have recorded everyone's natural level of mischievous acts in a week, then given them an invisibility

SELF TEST
Enter the data in Table 10.1 into the data editor as though a repeated-measures design was used.

cloak and counted the number of mischievous acts in the following week.[6] The data will be identical to the previous example, not because I am too lazy to generate different scores, but because it allows me to illustrate various things.

The data would now be arranged differently in the data editor. Instead of having a coding variable, and a single column with mischief scores in, we would arrange the data in two columns (one representing the **Cloak** condition and one representing the **No_Cloak** condition). The data are in **Invisibility RM.sav** if you had difficulty entering them yourself.

10.9.2 Exploring repeated-measures data ▌▌▌▌

We talked about the assumption of normality in Chapter 6. With the paired-samples *t*-test we're interested in the sampling distribution of the difference scores (not the raw scores). Therefore, if

you want to test for normality before a paired-samples *t*-test then you should compute the differences between scores, and then check if this new variable is normally distributed as a proxy for the sampling distribution (or use a big sample or robust test and not worry about normality☺). It is possible to have two measures that are highly non-normal and produce beautifully distributed differences.

We saw in Chapter 5 that you can visualize group differences using error bars. However, there's a problem when we graph error bars from repeated-measures designs. In one of the earlier self-tests I asked you to produce an error bar graph for the data when we treated it as an independent design. Compare that graph to the one you have just produced. Figure 10.6 shows both graphs; can you spot the difference? Hopefully you can't, because the graphs are identical (apart from axis labels). This is odd, isn't it? We discovered in Chapter 1 that repeated-measures designs eliminate some extraneous variables (such as age, IQ

and so on), so why doesn't the graph of the repeated-measures design reflect the increased sensitivity of the design? It's because SPSS Statistics treats the data as though scores are independent, and consequently the error bars do not reflect the 'true' error around the means for repeated-measures designs. Let's correct these error bars.

First, we need to calculate the average mischief for each participant. Select *Transform* ▶ 🖩 Compute Variable… to access the *Compute Variable* dialog box (see Section 6.12.6). Enter the name **Mean** into the box labeled *Target Variable*. In the list labeled *Function group* select *Statistical*, and then in the list labeled *Functions and Special Variables* select *Mean*. Click 🔼 to transfer this command to the command area. The command will appear as *MEAN(?,?)*, and we need to replace the question marks with variable names by typing them or transferring them from the variables list. Replace the first question mark with the variable **No_Cloak** and the second one with the variable **Cloak**. The completed dialog box should look like Figure 10.7. Clicking OK will create this new variable in the data editor.

The **grand mean** is the mean of all scores, and for the current data this value will be the mean of all 24 scores. The means we have just calculated are the average score for each participant; if we take the average of those mean scores, we will have the grand mean—phew, there were a lot of means in that sentence. We can use the *Descriptives* command (you could also use the *Explore* or *Frequencies* commands that we came across in Chapter 6, but as I've already covered those we'll try something different). Access the *Descriptives* dialog box (Figure 10.8) by

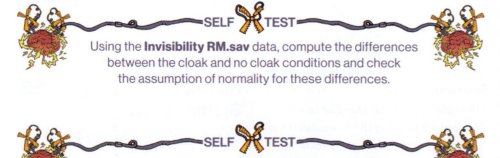

SELF TEST
Using the **Invisibility RM.sav** data, compute the differences between the cloak and no cloak conditions and check the assumption of normality for these differences.

SELF TEST
Produce an error bar chart of the **Invisibility RM.sav** data (**Cloak** on the *x*-axis and **Mischief** on the *y*-axis).

---

6 In theory we'd counterbalance the weeks so that some people had the cloak and then it was taken away, whereas others had no cloak but were then given one. However, given that the research scenario relied on participants not knowing about the cloaks of invisibility it might be best just to have a baseline phase and then give everyone their cloak at the same time (unaware that others were getting cloaks too).

selecting *Analyze* ▶ *Descriptive Statistics* ▶ 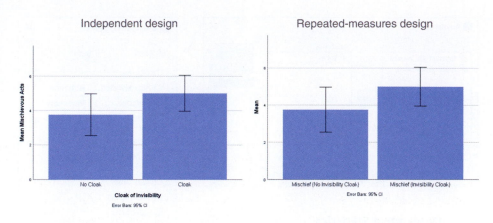 *Descriptives*.... Select the variable **Mean** from the list and drag it to the box labeled *Variable(s)* (or click ➡). Clicking Options... activates a second dialog box in which we want to check only the option for the mean (that's all we are interested in). The resulting output gives us the mean of the variable that we called **Mean**, which, remember, contains the average score for each participant across the cloak and no cloak conditions. This value (4.375) is the grand mean.

If you look at the variable labeled **Mean**, you'll notice that the values for each participant are different, which tells us that some people were, on average, more mischievous than others across the conditions. The fact that participants' mean mischief scores differ represents individual differences (it shows that some participants are generally more mischievous than others). According to Loftus and Masson (1994), these individual differences contaminate the error bars and need to be removed. In effect we want to equalize the means between participants (i.e., adjust the scores in each condition such that when we take the mean score across conditions, it is the same for all participants). To do this, we use the *compute* function to calculate an adjustment factor by subtracting each participant's mean score from the grand mean. Activate the *Compute Variable* dialog box, give the target variable a name (I suggest **Adjustment**) and use the command '4.375 – Mean'. This command will take the grand mean (4.375) and subtract from it each participant's average mischief level (see Figure 10.9).

This process creates a new variable in the data editor called **Adjustment** which contains the difference between each participant's mean mischief levels and the mean mischief level across all participants. Some of the values are positive

Figure 10.6 Two error bar graphs of the invisibility data

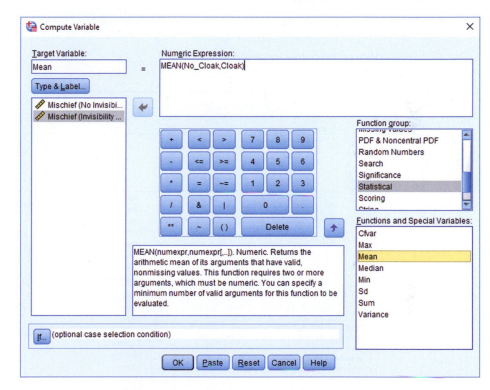

Figure 10.7 Using the *compute* function to calculate the mean of two columns

(participants who were less mischievous than average) and others negative (participants who were more mischievous than average). We can use these adjustment values to eliminate the between-subject differences in mischief.

First we'll adjust the scores in the **No_Cloak** condition again using the *Compute* command. Activate the *Compute* dialog box, and title the new variable **No_Cloak_Adjusted** (you can click Type & Label... to give this variable a label such as 'No Cloak Condition: Adjusted Values'). All we need to do is add each participant's score in the **No_Cloak** condition to their adjustment value. Select the variable **No_Cloak** and drag it to the command area (or click ➡), then click ➕ and select the variable **Adjustment** and transfer it to the

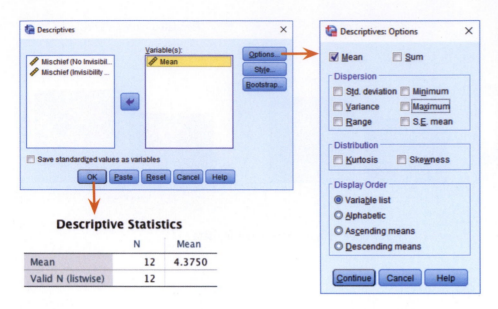

Descriptive Statistics

	N	Mean
Mean	12	4.3750
Valid N (listwise)	12	

Figure 10.8 Dialog boxes and output for descriptive statistics

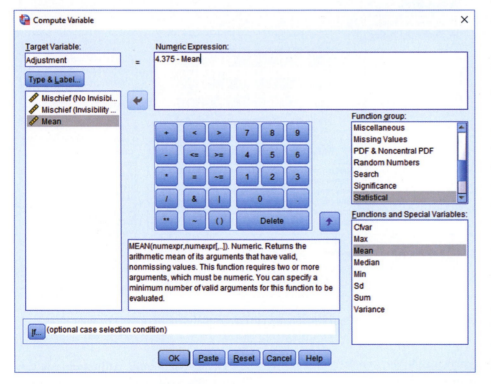

Figure 10.9 Calculating the adjustment factor

SELF TEST

Create an error bar chart of the mean of the adjusted values that you have just made (**Cloak_Adjusted** and **No_Cloak_Adjusted**).

command area. The completed dialog box is in Figure 10.10. Do the same thing for the variable **Cloak**: create a variable called **Cloak_Adjusted** that contains the values of **Cloak** added to the value in the **Adjustment** column.

The variables **Cloak_Adjusted** and **No_Cloak_Adjusted** represent the mischief experienced in each condition, adjusted to eliminate between-subject differences. If you don't believe me, use the *Compute* command to create a variable **Mean2** that is the average of **Cloak_Adjusted** and **No_Cloak_Adjusted**. You should find that the value in this column is the same for every participant (it will be 4.375, the grand mean), demonstrating that the between-subject variability in means is gone.

Compare the resulting error bar graph (Figure 10.11) to the graphs in Figure 10.6—what difference do you notice? The first point to make is that the means in the two conditions have not changed. However, the error bars have got smaller. Also, whereas in Figure 10.6 the error bars overlap, in this new graph they do not. Therefore, when we plot the proper error bars for the repeated-measures data it shows the extra sensitivity that this design has: the differences between conditions appear to be significant (the error bars don't overlap), whereas when different participants are used, there does not appear to be a significant difference (error bars overlap a lot). Remember that the means in both situations are identical, but the sampling error is smaller in the repeated-measures design—I expand upon this point in Section 10.10.

10.9.3 The paired-samples *t*-test using SPSS Statistics ▮▮▮▮

To conduct a paired-samples *t*-test, access the dialog box in Figure 10.12 by selecting *Analyze* ▶ *Compare Means* ▶

Paired-Samples T Test... We need to select pairs of variables to be analyzed. In this case we have only one pair (**Cloak** vs. **No_Cloak**). To select a pair click the first variable that you want to select (in this case **No_Cloak**), then hold down the *Ctrl* key (*Cmd* on a Mac) and select the second (in this case **Cloak**). Transfer these variables to the box labeled *Paired Variables* by clicking ➡. (You can select and transfer each variable individually, but selecting both variables as just described is quicker.) If you want to carry out several *t*-tests then you can select another pair of variables in the same way. Clicking Options... activates another dialog box that gives you the same options as for the independent *t*-test. Similarly, click Bootstrap... to access the bootstrap function (Section 6.12.3). As with the independent *t*-test, select ☑ Perform bootstrapping and ◉ Bias corrected accelerated (BCa). Back in the main dialog box, click OK to run the analysis.

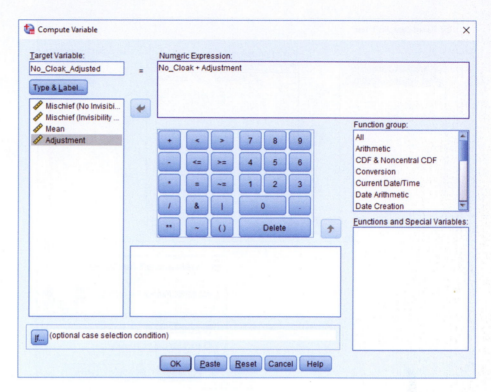

Figure 10.10   Adjusting the values of **No_Cloak**

10.9.4 Output from the paired-samples *t*-test ▌▌▌

The resulting output produces four tables (three if you don't select bootstrapping). Output 10.8 shows a table of summary statistics for the two experimental conditions (if you don't ask for bootstrapping this table will be a bit more straightforward). For each condition we are told the mean, the number of participants (*N*), the standard deviation and standard error. These values are the same as when we treated the data as an independent design and were described in Section 1.8.3.

Output 10.8 also shows the Pearson correlation between the two conditions. When repeated measures are used, scores in the experimental conditions will correlate to some degree (because the data in each condition come from the same entities you'd expect some constancy in their responses). The output contains the value of Pearson's *r* and the two-tailed significance value (see Chapter 8). For these data the experimental conditions yield a very large, highly significant, correlation coefficient, $r = 0.806$, $p = 0.002$, with a bootstrap confidence interval that doesn't include zero, BCa 95% CI [0.417, 0.943].

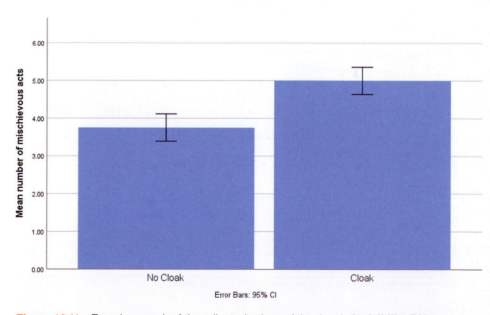

Figure 10.11   Error bar graph of the adjusted values of the data in **Invisibility RM.sav**

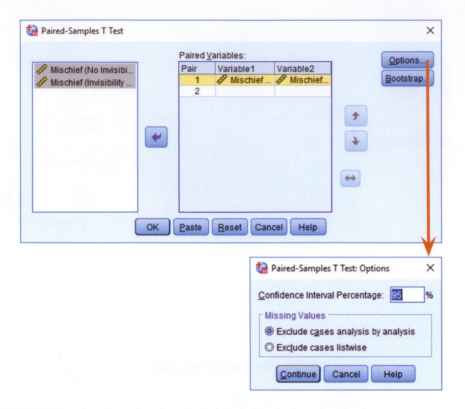

Figure 10.12   Main dialog box for paired-samples *t*-test

Paired Samples Statistics

			Statistic	Bootstrap[a] Bias	Std. Error	BCa 95% Confidence Interval Lower	Upper
Pair 1	Mischief (No Invisibility Cloak)	Mean	3.75	.02	.54	2.50	4.92
		N	12				
		Std. Deviation	1.913	-.129	.352	1.357	2.193
		Std. Error Mean	.552				
	Mischief (Invisibility Cloak)	Mean	5.00	.02	.47	3.83	6.17
		N	12				
		Std. Deviation	1.651	-.103	.310	1.115	1.946
		Std. Error Mean	.477				

a. Unless otherwise noted, bootstrap results are based on 1000 bootstrap samples

Paired Samples Correlations

		N	Correlation	Sig.	Bootstrap for Correlation[a] Bias	Std. Error	BCa 95% Confidence Interval Lower	Upper
Pair 1	Mischief (No Invisibility Cloak) & Mischief (Invisibility Cloak)	12	.806	.002	-.029	.160	.417	.943

a. Unless otherwise noted, bootstrap results are based on 1000 bootstrap samples

Output 10.8

Paired Samples Test

		Paired Differences Mean	Std. Deviation	Std. Error Mean	95% Confidence Interval of the Difference Lower	Upper	t	df	Sig. (2-tailed)
Pair 1	Mischief (No Invisibility Cloak) – Mischief (Invisibility Cloak)	-1.250	1.138	.329	-1.973	-.527	-3.804	11	.003

Output 10.9

Output 10.9 tabulates the main *t*-test results. The table contains the average difference score ($\overline{D}$ in equation (10.6)), which is $3.75 - 5 = -1.25$, the standard deviation of the difference scores (1.138), and the standard error of difference scores (0.329, the lower half of equation (10.6)). The test statistic, *t*, is calculated using equation (10.6) by dividing the mean of differences by the standard error of differences ($t = -1.25/0.329 = -3.804$). For a repeated-measures design the degrees of freedom are the sample size minus 1 ($df = N - 1 = 11$). The *p*-value in the column labeled *Sig.* is the long-run probability that a value of *t* at least the size of the one obtained could occur if the population mean of difference scores was 0. This two-tailed probability is very low ($p = 0.003$); it tells us that there is only a 0.3% chance that a value of *t* of −3.804 or smaller could occur if the null hypothesis were true. Assuming we'd decided on an alpha of 0.05 before the experiment, we'd conclude that there was a significant difference between the means of these two samples, because 0.003 is less than 0.05. In terms of the experiment, we might conclude that having a cloak of invisibility significantly affected the amount of mischief a person got up to, $t(11) = -3.80$, $p = 0.003$. This result was predicted by the error bar chart in Figure 10.11.

10.9.5   Robust tests of two dependent means

As you've probably noticed, I'm keen on looking at the robust bootstrap confidence intervals (especially in a small sample like we have here). Output 10.10 shows these confidence intervals. Remember that confidence intervals are constructed such that in 95% of samples the intervals contain the true value of the mean difference. So, assuming that this sample's confidence interval is one of the 95 out of 100 that contain the

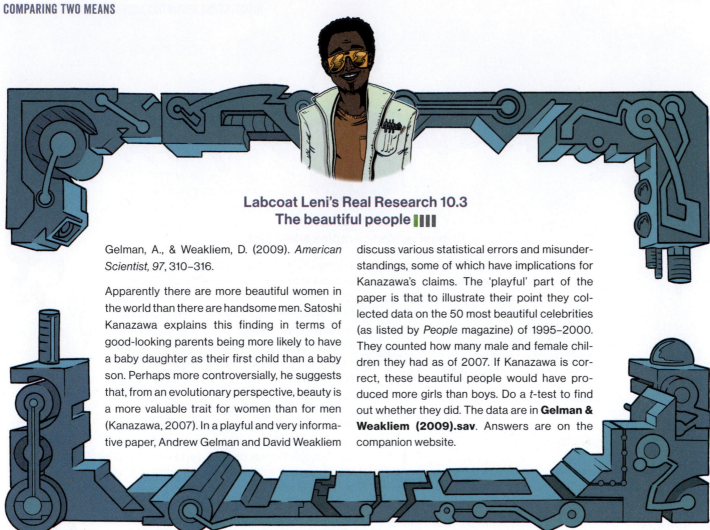

Labcoat Leni's Real Research 10.3
The beautiful people ▮▮▮▮

Gelman, A., & Weakliem, D. (2009). *American Scientist, 97*, 310–316.

Apparently there are more beautiful women in the world than there are handsome men. Satoshi Kanazawa explains this finding in terms of good-looking parents being more likely to have a baby daughter as their first child than a baby son. Perhaps more controversially, he suggests that, from an evolutionary perspective, beauty is a more valuable trait for women than for men (Kanazawa, 2007). In a playful and very informative paper, Andrew Gelman and David Weakliem

discuss various statistical errors and misunderstandings, some of which have implications for Kanazawa's claims. The 'playful' part of the paper is that to illustrate their point they collected data on the 50 most beautiful celebrities (as listed by *People* magazine) of 1995–2000. They counted how many male and female children they had as of 2007. If Kanazawa is correct, these beautiful people would have produced more girls than boys. Do a *t*-test to find out whether they did. The data are in **Gelman & Weakliem (2009).sav**. Answers are on the companion website.

---

population value, we can say that the true mean difference lies between −1.83 and −0.67. The importance of this interval is that it does not contain zero (both limits are negative), which tells us that the true value of the mean difference is unlikely to be zero. In other words, there is an effect in the population reflecting more mischievous acts performed when someone is given an invisibility cloak.

Bootstrap for Paired Samples Test

		Bootstrap[a]				BCa 95% Confidence Interval	
		Mean	Bias	Std. Error	Sig. (2-tailed)	Lower	Upper
Pair 1	Mischief (No Invisibility Cloak) – Mischief (Invisibility Cloak)	−1.250	−.005	.311	.008	−1.833	−.667

a. Unless otherwise noted, bootstrap results are based on 1000 bootstrap samples

Output 10.10

It's also possible to run a robust version of the paired-samples *t*-test using R. As was the case for the robust independent *t*-test, we need the *Essentials for R* plugin and *WRS2* package (Section 4.13). The companion website contains a syntax file (**Robust paired-samples t-test.sps**) for running a robust variant of the paired-samples *t*-test based on Yuen (1974) and

described by Wilcox (2012). The syntax in the file is as follows:

```
BEGIN PROGRAM R.
library(WRS2)
mySPSSdata = spssdata.
GetDataFromSPSS()
yuend(mySPSSdata$No_Cloak,
mySPSSdata$Cloak, tr = 0.2)
END PROGRAM.
```

Select and run these five lines of syntax (see SPSS Tip 10.3) and you'll find text output in the viewer (Output 10.11) which says that there was a significant difference (because the confidence interval does not cross zero and the *p*-value is less than 0.05) in mischief scores across the two cloak groups, Y_t (7) = −2.70 [−1.87, −0.13], *p* = 0.031.

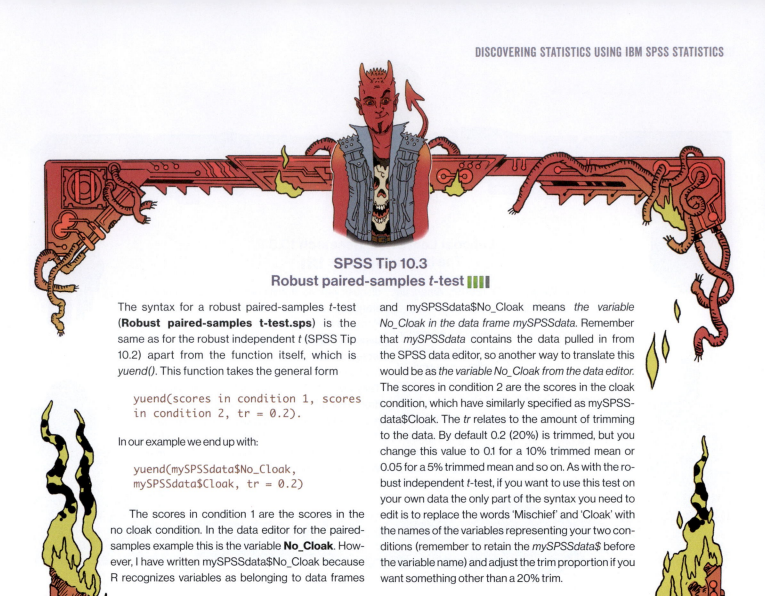

SPSS Tip 10.3
Robust paired-samples *t*-test

The syntax for a robust paired-samples *t*-test (**Robust paired-samples t-test.sps**) is the same as for the robust independent *t* (SPSS Tip 10.2) apart from the function itself, which is *yuend()*. This function takes the general form

```
yuend(scores in condition 1, scores
in condition 2, tr = 0.2).
```

In our example we end up with:

```
yuend(mySPSSdata$No_Cloak,
mySPSSdata$Cloak, tr = 0.2)
```

The scores in condition 1 are the scores in the no cloak condition. In the data editor for the paired-samples example this is the variable **No_Cloak**. However, I have written mySPSSdata$No_Cloak because R recognizes variables as belonging to data frames

and mySPSSdata$No_Cloak means *the variable No_Cloak in the data frame mySPSSdata*. Remember that *mySPSSdata* contains the data pulled in from the SPSS data editor, so another way to translate this would be as *the variable No_Cloak from the data editor*. The scores in condition 2 are the scores in the cloak condition, which have similarly specified as mySPSSdata$Cloak. The *tr* relates to the amount of trimming to the data. By default 0.2 (20%) is trimmed, but you change this value to 0.1 for a 10% trimmed mean or 0.05 for a 5% trimmed mean and so on. As with the robust independent *t*-test, if you want to use this test on your own data the only part of the syntax you need to edit is to replace the words 'Mischief' and 'Cloak' with the names of the variables representing your two conditions (remember to retain the *mySPSSdata$* before the variable name) and adjust the trim proportion if you want something other than a 20% trim.

```
Test statistic: -2.7027 (df = 7), p-value = 0.03052
Trimmed mean difference: -1
95 percent confidence interval:
-1.8749 -0.1251

Explanatory measure of effect size: 0.4
```

Output 10.11

10.9.6 Bayesian test of two means from paired samples

To access the dialog box to compute a Bayes factor and credible interval for dependent means select *Analyze ▶ Bayesian Statistics ▶ Related Samples*

Normal (Figure 10.13). Just like with the regular dependent *t*-test we start by selecting a pair of variables and transferring them to the *Paired Variables* list. We're expecting more mischievous acts in the cloak group than the no cloak

group, therefore, if we specify our pair as **Cloak** (*Variable 1*) and **No_Cloak** (*Variable 2*) we'd predict a positive mean difference (when we subtract scores for the no cloak condition from those for the cloak condition the results should be, on average, positive if our prediction is correct). Conversely, if we specify our pair as **No_Cloak** (*Variable 1*) and **Cloak** (*Variable 2*) we'd predict a negative mean difference. In terms of setting up the prior distribution it's probably easier to think about a positive value for the mean

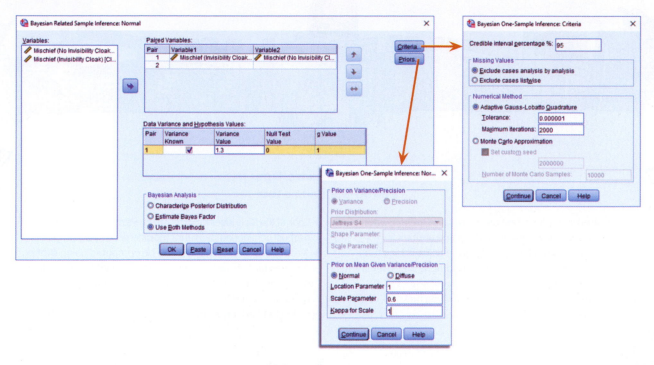

Figure 10.13 Dialog box for a Bayesian test of two related means

difference, so let's specify **Cloak** as *Variable 1* and **No_Cloak** as *Variable 2*. To do this, select **Cloak** from the *Variables* list and click ➡, then select **No_Cloak** and click ➡. Select the *Variance Known* option and type 1.3 in the cell labeled *Variance Value*. The value of 1.3 is obtained by squaring the standard deviation of difference scores in Output 10.9 to convert it to the variance. Using the variance for the data simplifies things (we don't need to set a prior for the variance) but as you get more experience you might want to estimate the variance too. To compute Bayes factors and estimate the model parameters select ⦿ Use Both Methods.

Imagine that prior to collecting data we believed that the difference between the number of mischievous acts committed with and without an invisibility cloak would be 1 but you were prepared to believe that it could range from about –1 to 3. In other words, you believe the most probable outcome is that having an invisibility cloak would lead to 1 additional mischievous act compared to not having a

Bayes Factor for Related–Sample T Test

	N	Mean Difference	Std. Deviation	Std. Error Mean	Bayes Factor	t	df	Sig.(2–tailed)
Mischief (Invisibility Cloak) – Mischief (No Invisibility Cloak)	12	1.25	1.138	.329	.005	3.804	11	.003

Bayes factor: Null versus alternative hypothesis.

Posterior Distribution Characterization for Related–Sample Mean Difference

			Posterior		95% Credible Interval	
	N	Mode	Mean	Variance	Lower Bound	Upper Bound
Mischief (Invisibility Cloak) – Mischief (No Invisibility Cloak)	12	1.21	1.21	.092	.62	1.81

Prior on Variance: Diffuse. Prior on Mean: Normal.

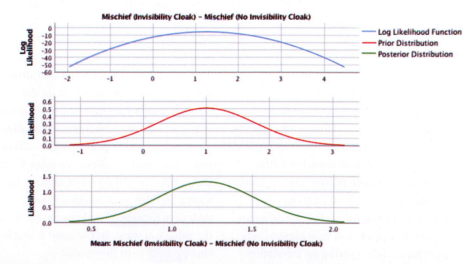

Output 10.12

Cramming Sam's Tips
Paired-samples *t*-test

- The paired-samples *t*-test compares two means, when those means have come from the same entities.
- Look at the column labeled *Sig*. If the value is less than 0.05 then the means of the two conditions are significantly different.
- Look at the values of the means to tell you how the conditions differ.

- Look at the table labeled *Bootstrap for Paired Samples Test* to get a robust confidence interval for the difference between means.
- A robust version of the test can be computed using syntax.
- A Bayes factor can be computed that quantifies the ratio of how probable the data are under the alternative hypothesis compared to the null.
- Calculate and report the effect size too.

cloak. However, you were prepared to accept that, at most, it might lead to 3 additional acts (but you think this possibility has a low probability – it is highly unlikely). Conversely, you're prepared to accept that you might be wrong and that having an invisibility cloak creates a sense of responsibility in the wearer that leads to fewer mischievous acts. At most, you be prepared to believe that a cloak could lead to 1 fewer act, but again you think this outcome is highly unlikely (you assign it low probability). The graph with the red line in Output 10.12 shows this prior distribution. It is a normal distribution centered at 1 (the most likely outcome in your prior opinion) and ranging from about –1 to 3. This equates to a prior belief that the difference between the number of mischievous acts committed when wearing an invisibility

cloak compared to when not is most likely 1, but could be as low as –1 or as high as 3 (although we think these more extreme values are unlikely).

To set up these priors click Priors... and select ⦿ Normal. I have specified a distribution with mean (location) of 1 and standard deviation (scale) of 0.6. As before, to get this scale parameter I took the range of beliefs, 3–(–1) = 4, and divided by 6, 4/6 = 0.67. I then rounded down (although this won't have had an impact on the results). Kappa I set to be the SPSS default of 1. If you'd rather use a noninformative reference prior (i.e., prior to data collection you're prepared to believe any value for the mean difference) then select Diffuse. If you want a credible interval other than 95% then click Criteria... and change the 95 to the value you want. In the main dialog box click OK to fit the model.

Output 10.12 shows that the Bayes factor is 0.005. SPSS reports the ratio of the null to the alternative hypothesis, so this value means that the data are 0.005 times as probable under the null hypothesis as under the alternative. We can flip the interpretation by dividing 1 by the value (1/0.005 = 200). This value tells us that the data are 200 times as probable under the alternative hypothesis as under the null. In other words, we should shift our belief towards the alternative hypothesis by a factor of 200. This is strong evidence for the hypothesis that invisibility cloaks lead to more mischief.

Output 10.12 also shows our prior distribution for the difference between group means (the graph with the red line). I have already described this graph. The Bayesian estimates of the difference between means are in the columns labeled

Posterior Mode and *Posterior Mean*. The 95% credible interval for this estimate ranged from 0.62 to 1.81. In other words, *assuming that the effect exists,* the population value of the effect will be between 0.62 to 1.81 with 95% probability. This tells us nothing about the null hypothesis (because it assumes the effect exists) but helps us to ascertain the likely population value if we're prepared to accept that the effect exists. So, we can say with 95% probability that having a cloak of invisibility will increase mischievous acts by anything as low as 0.62 to 1.81.

10.9.7 Effect sizes for two related means ▐▐▐

According to Rosenthal (1991), we can compute the effect size direct from the value of *t*, just as we did for the independent *t*-test. Using equation (10.17) and the values from Output 10.9, we get:

$$r = \sqrt{\frac{(-3.804)^2}{(-3.804)^2 + 11}} = \sqrt{\frac{14.47}{25.47}} = 0.75 \qquad (10.20)$$

Therefore, as well as being statistically significant, this effect is very large. Notice that the effect is a lot bigger than when we treated the data as though they were from an independent design (*r* = 0.34), which is odd, given that we used exactly the same scores. This difference reflects the finding that using a *t* from a paired-samples *t*-test leads to an overestimation of the population effect size (Dunlap, Cortina, Vaslow, & Burke, 1996).

You could instead compute Cohen's *d* (Section 3.7.1) as we did in Section 10.8.6. Note the change in design does not affect the calculation: the effect size is still 0.65:

$$\hat{d} = \frac{\bar{X}_{\text{Cloak}} - \bar{X}_{\text{No Cloak}}}{s_{\text{No Cloak}}} = \frac{5 - 3.75}{1.91} = 0.65 \qquad (10.21)$$

This consistency is a good thing because both studies did, in fact, show the same

difference between means (because the scores used in the examples are identical). Some argue that you need to factor in the between scores in treatment conditions by dividing the estimate of *d* by the square root of 1 minus the correlation between the scores (which you can find in Output 10.8, *r* = 0.806). The corrected *d* is 1.48:

$$\hat{d}_D = \frac{\hat{d}}{\sqrt{1-r}} = \frac{0.65}{\sqrt{1-0.806}} = \frac{0.65}{0.44} = 1.48 \qquad (10.22)$$

which is more than double the original size! My issue with this 'correction' is precisely that the effect size now expresses information not just about the observed difference between means but the study design used to measure it. However, I include it in case you disagree with me.

10.10 Reporting comparisons between two means ▐▐▐

As we have seen before, you usually state the finding to which the test relates and then report the test statistic, its degrees of freedom and its probability value. This applies whether we use a t-test or a robust test. Ideally report an estimate of the effect size and the Bayes factor too. If you used a robust test you should cite R (R Core Team, 2016) and the WRS2 package (Mair, Schoenbrodt, & Wilcox, 2017), because that's what was used to compute them.

10.10.1 Reporting *t*-tests ▐▐▐

For the data based on independent samples we could report this (to see from where the values come look at Output 10.4, Output 10.5, and Output 10.7):

✓On average, participants given a cloak of invisibility engaged in more acts of

mischief (*M* = 5, *SE* = 0.48), than those not given a cloak (*M* = 3.75, *SE* = 0.55). This difference, −1.25, BCa 95% CI [−2.65, 0.11], was not significant, *t*(21.54) = −1.71, *p* = 0.101; however, it represented an effect of *d* = 0.65.

For the data based on paired samples you might report (see Outputs 10.9, 10.10 and 10.12):

✓On average, participants given a cloak of invisibility engaged in more acts of mischief (*M* = 5, *SE* = 0.48), than those not given a cloak (*M* = 3.75, *SE* = 0.55). This difference, −1.25, BCa 95% CI [−1.83, −0.67], was significant, *t*(11) = −3.80, *p* = 0.003, and represented an effect of *d* = 0.65.

10.10.2 Reporting robust variants for the *t*-test ▐▐▐

If you'd done the robust test, you could start with a general statement such as:

✓R (R Core Team, 2016) was used to compute a robust variant of the *t*-test based on Yuen (1974) using the *WRS2* package (Mair et al., 2017).

Then for the data based on independent samples report this (see Output 10.6):

✓On average, participants given a cloak of invisibility engaged in more acts of mischief (*M* = 5, *SE* = 0.48), than those not given a cloak (*M* = 3.75, *SE* = 0.55). This difference was not significant, $Y_t = -1.36$, 95% CI [−2.52, 0.52], *p* = 0.167.

Then for the data based on paired samples report this (see Output 10.11):

✓On average, participants given a cloak of invisibility engaged in more acts of mischief (*M* = 5, *SE* = 0.48), than those not given a cloak (*M* = 3.75, *SE* = 0.55). This difference was significant, $Y_t(7) = -2.70$, 95% CI [−1.87, −0.13], *p* = 0.031.

10.10.3 Reporting Bayesian comparisons of means ▮▮▮

For the data based on independent samples, report (see Output 10.7):

✓ On average, participants given a cloak of invisibility engaged in more acts of mischief ($M = 5$, $SE = 0.48$), than those not given a cloak ($M = 3.75$, $SE = 0.55$). The prior distributions for the group means were set as a normal distributions with a mean of 3 and standard deviation of 1 for the no cloak group and a mean of 4 and standard deviation of 1.75 for the cloak group. The Bayes factor was estimated using Gönen's method (Gönen et al., 2005) with a prior difference between means of 1 with a variance of 0.25. The Bayesian estimate of the true difference between means was 1.31, 95% credible interval [0.02, 2.60]. The associated Bayes factor, $BF_{10} = 2.39$, suggested that the data were weakly more probable under the alternative hypothesis than the null.

For the data based on paired samples (see Output 10.12):

✓ On average, participants given a cloak of invisibility engaged in more acts of mischief ($M = 5$, $SE = 0.48$), than those not given a cloak ($M = 3.75$, $SE = 0.55$). The prior distribution for the difference between means was set to be a normal distribution with mean of 1 and standard deviation of 0.6. Kappa was set at 1. The Bayesian estimate of the true difference between means was 1.21, 95% credible interval [0.62, 1.81]. The associated Bayes factor, $BF_{10} = 200$, suggested that the data were 200 times more probable under the alternative hypothesis than under the null.

10.11 Between groups or repeated measures? ▮▮▮

The two examples in this chapter illustrate the difference between data collected using the same participants and data collected using different participants. The two examples use the same scores, but when analyzed as though the data came from the same participants the result was a significant difference between means and large Bayes factor, and when analyzed as though the data came from different participants there was no significant difference and a small Bayes factor. This finding may puzzle you—after all, the numbers being analyzed were identical, as reflected in the effect size (d) being the same. The explanation is that repeated-measures designs have relatively more power. When the same entities are used across conditions the unsystematic variance (error variance) is reduced dramatically, making it easier to detect the systematic variance. It is often assumed that the way in which you collect data is irrelevant, and in terms of the effect size it sort of is, but if you're interested in significance then it matters. Researchers have carried out studies using the same participants in experimental conditions, then replicated the study using different participants, and have found that the method of data collection interacts significantly with the results found (see Erlebacher, 1977).

10.12 Brian and Jane's Story ▮▮▮

Brian wished he had a cloak of invisibility. Term was nearly over and Jane was still giving him the brush-off. The nicer he was to her, the more caustic she became. What was it with her? He felt embarrassed by his behavior—it was a bit desperate. He wanted to be invisible for a while or at least avoid Jane, and so he'd been hiding out in the library.

A week ago, he'd been struggling through his statistics assignment. He didn't get the whole Bayesian thing. Weeks of learning about p-values, and suddenly his lecturer was throwing this whole new approach into the mix. Why don't lecturers see how confusing that is? Mind you, with his big spacey eyes his lecturer looked like his brain might have crashed out on mushrooms in the '70s. Surely only a drugs casualty has a name like Oditi? Anyway, Brian was stuck, frustrated, and about to give up when he recognized Alex and Sam from his lectures.

He went over and asked for help. They welcomed him. Alex was smart, that much was obvious, but her generosity and empathy in mentoring him and Sam was quite overwhelming. Alex was nervy around Brian and got a little tetchy whenever he mentioned Jane.

During his week of hiding out in the library he chatted to Alex more and more. She was big-hearted, funny and seemed to enjoy his company.

Today he was alone, though, and reflecting on what a good week it had been. He'd started to feel more human—like a regular guy with normal friends. He felt calm and relaxed for the first time since the term started. That was until a finger tapped his shoulder and he turned to face the knowing smile of Jane.

'Hi, stranger,' she said.

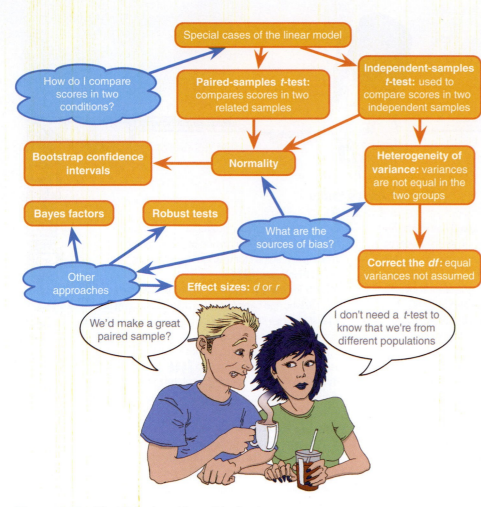

Figure 10.14 What Brian learnt from this chapter

10.13 What next? ▮▮▮▮

I'd announced to my parents that my career of choice was a rock star. Obviously I hadn't presented them with a robust *t*-test showing how much more money I would earn compared to a university professor, but even if I had, I'm not sure it would have mattered. My parents were quite happy for me to live this fantasy if I entertained the possibility that it might not work out and had a plan B. Preferably a plan B that was a little bit more sensible than being a rock star. At the age of 10, I think my plan B was probably to be a soccer star. One way or another I wanted my career to involve being a star, so if it wasn't rock, then soccer would do. However, we've seen already that I was at a genetic disadvantage when it came to soccer, but not so much when it came to rock stardom: my dad, after all, was quite musical. All I had to do was make it happen. The first step, I reasoned, was to build a fan base, and the best place to start a fan base is among your friends. With that in mind, I put on my little denim jacket with Iron Maiden patches sewn onto it, threw my guitar over my back and headed off down the rocky road of stardom. The first stop was my school.

10.14 Key terms that I've discovered

Dependent *t*-test

Dummy variables

Grand mean

Independent *t*-test

Paired-samples *t*-test

Standard error of differences

Variance sum law

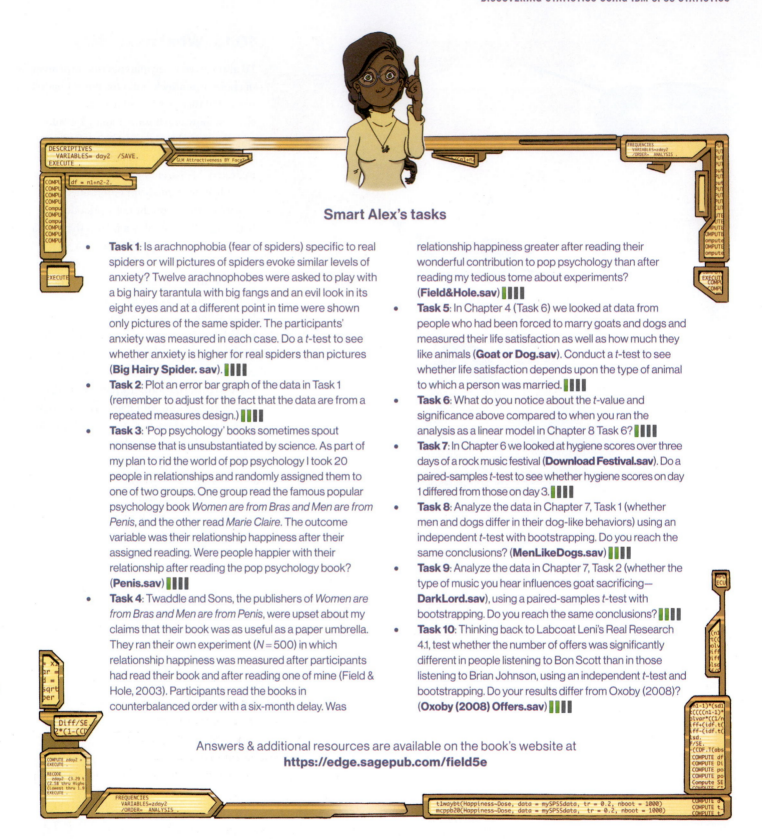

Smart Alex's tasks

- **Task 1**: Is arachnophobia (fear of spiders) specific to real spiders or will pictures of spiders evoke similar levels of anxiety? Twelve arachnophobes were asked to play with a big hairy tarantula with big fangs and an evil look in its eight eyes and at a different point in time were shown only pictures of the same spider. The participants' anxiety was measured in each case. Do a *t*-test to see whether anxiety is higher for real spiders than pictures (**Big Hairy Spider. sav**).

- **Task 2**: Plot an error bar graph of the data in Task 1 (remember to adjust for the fact that the data are from a repeated measures design.)

- **Task 3**: 'Pop psychology' books sometimes spout nonsense that is unsubstantiated by science. As part of my plan to rid the world of pop psychology I took 20 people in relationships and randomly assigned them to one of two groups. One group read the famous popular psychology book *Women are from Bras and Men are from Penis*, and the other read *Marie Claire*. The outcome variable was their relationship happiness after their assigned reading. Were people happier with their relationship after reading the pop psychology book? (**Penis.sav**)

- **Task 4**: Twaddle and Sons, the publishers of *Women are from Bras and Men are from Penis*, were upset about my claims that their book was as useful as a paper umbrella. They ran their own experiment (*N* = 500) in which relationship happiness was measured after participants had read their book and after reading one of mine (Field & Hole, 2003). Participants read the books in counterbalanced order with a six-month delay. Was relationship happiness greater after reading their wonderful contribution to pop psychology than after reading my tedious tome about experiments? (**Field&Hole.sav**)

- **Task 5**: In Chapter 4 (Task 6) we looked at data from people who had been forced to marry goats and dogs and measured their life satisfaction as well as how much they like animals (**Goat or Dog.sav**). Conduct a *t*-test to see whether life satisfaction depends upon the type of animal to which a person was married.

- **Task 6**: What do you notice about the *t*-value and significance above compared to when you ran the analysis as a linear model in Chapter 8 Task 6?

- **Task 7**: In Chapter 6 we looked at hygiene scores over three days of a rock music festival (**Download Festival.sav**). Do a paired-samples *t*-test to see whether hygiene scores on day 1 differed from those on day 3.

- **Task 8**: Analyze the data in Chapter 7, Task 1 (whether men and dogs differ in their dog-like behaviors) using an independent *t*-test with bootstrapping. Do you reach the same conclusions? (**MenLikeDogs.sav**)

- **Task 9**: Analyze the data in Chapter 7, Task 2 (whether the type of music you hear influences goat sacrificing— **DarkLord.sav**), using a paired-samples *t*-test with bootstrapping. Do you reach the same conclusions?

- **Task 10**: Thinking back to Labcoat Leni's Real Research 4.1, test whether the number of offers was significantly different in people listening to Bon Scott than in those listening to Brian Johnson, using an independent *t*-test and bootstrapping. Do your results differ from Oxoby (2008)? (**Oxoby (2008) Offers.sav**)

Answers & additional resources are available on the book's website at
https://edge.sagepub.com/field5e

MODERATION, MEDIATION AND MULTICATEGORY PREDICTORS

11

11.1 What will this chapter tell me?

Having successfully slain audiences at holiday camps around Britain, my next step towards global domination was my primary school. I had learnt another Chuck Berry song ('Johnny B. Goode'), but broadened my repertoire to include songs by other artists (I have a feeling 'Over the edge' by Status Quo was one of them).[1] When the opportunity came to play at a school assembly I jumped at it. The headmaster tried to ban me,[2] but the show went on. It was a huge success (10-year-olds are very easily impressed). My classmates carried me around the playground on their shoulders. I was a hero. Around this time I had a childhood sweetheart called Clair Sparks. Actually, we had been sweethearts since before my newfound rock legend status. I don't think the guitar playing and singing impressed her much, but she rode a motorbike (really, a little child's one) which impressed *me* quite a lot; I was utterly convinced that we would one day get married and live happily ever after. I was utterly convinced, that is, until she ran off with Simon Hudson. Being 10, she probably literally did run off with him—across the playground. I remember telling my parents and them asking me how I felt about it. I told them I was being philosophical about it. I probably didn't know what philosophical meant at the age of 10, but I knew that it was the sort of thing you said if you were pretending not to be bothered about being dumped.

If I hadn't been philosophical, I might have wanted to look at what had lowered Clair's relationship satisfaction. We've seen in previous chapters that we could predict things like relationship satisfaction using a linear model. Perhaps it's predicted from your partner's love of rock bands like Status Quo (I don't recall Clair liking that sort of thing). However, life is usually more complicated than this; for example, your partner's love of rock music probably depends on your own love of rock music.

1 This would have been about 1982, so just before they abandoned hard rock in favor of a long series of increasingly cringeworthy publicity stunts. Nevertheless, up to the age of 10 they were my favorite band.
2 Seriously! I grew up in an age when a headmaster would try to ban a 10-year-old from playing guitar in assembly. The guy used to play hymns on an acoustic guitar, and I can assume only that he lost all perspective on the situation and decided that a 10-year-old blasting out some Quo in a squeaky little voice with an electric guitar was subversive or something.

Figure 11.1 My 10th birthday. From left to right: my brother Paul (who still hides behind cakes rather than have his photo taken), Paul Spreckley, Alan Palsey, Clair Sparks and me

For example, if you both like rock music then your love of the same music might have an additive effect giving you huge relationship satisfaction (*moderation*) or perhaps the relationship between your partner's love of rock and your own relationship satisfaction can be explained by your own music tastes (*mediation*). In the previous chapter we also saw that a dichotomous variable (e.g., rock fan or not) can be a predictor in a linear model, but what if you wanted to categorize musical taste into several categories (rock, hip-hop, R&B, etc.)? Surely you can't use multiple categories as a predictor variable? This chapter extends what we know about the linear model to these more complicated scenarios. First we look at two common linear models—moderation and mediation—before expanding what we already know about categorical predictors.

11.2 The *PROCESS* tool ▐▐▐▐

The best way to tackle moderation and mediation is with the *PROCESS* tool, which needs to be installed using the instructions in Section 4.13.1.

11.3 Moderation: interactions in the linear model ▐▐▐▐

11.3.1 The conceptual model ▐▐▐▐

So far we have looked at individual predictors in the linear model. It is possible for a statistical model to include the combined effect of two or more predictor variables on an outcome, which is known conceptually as **moderation**,

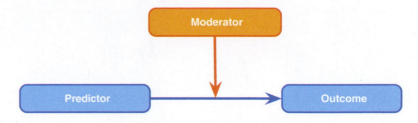

Figure 11.2   Diagram of the *conceptual* moderation model

and in statistical terms as an **interaction effect**. We'll start with the conceptual, using an example of whether violent video games make people antisocial. Video games are among the favorite online activities for young people: two-thirds of 5–16-year-olds have their own video games console, with 88% of boys aged 8–15 owning at least one games console (Ofcom, 2008). Although playing violent video games can enhance visuospatial acuity, visual memory, probabilistic inference, and mental rotation (Feng et al., 2007; Green & Bavelier, 2007; Green et al., 2010; Mishra et al., 2011), compared to games such as Tetris, these games have also been linked to increased aggression in youths (Anderson & Bushman, 2001). Another predictor of aggression and conduct problems is callous-unemotional traits such as lack of guilt, lack of empathy, and callous use of others for personal gain (Rowe et al., 2010). Imagine that a scientist explored the relationship between playing violent video games (such as Grand Theft Auto, MadWorld and Manhunt) and aggression. She measured aggressive behavior (**Aggress**), callous-unemotional traits (**CaUnTs**), and the number of hours per week they play video games (**Vid_Game**) in 442 youths (**Video Games.sav**).

Let's assume we're interested in the relationship between the hours spent

playing these games (predictor) and aggression (outcome). The conceptual model of moderation in Figure 11.2 shows that a **moderator** variable is one that affects the relationship between two others. If callous-unemotional traits were a moderator then we're saying that the strength or direction of the relationship between game playing and aggression is affected by callous-unemotional traits.

Suppose that we could classify people as having callous-unemotional traits or not. Our moderator variable would be categorical (callous or not callous). Figure 11.3 shows an example of moderation in this case: for people who are not callous there is no relationship between video games and aggression (the line is completely flat), but for people who are callous there is a positive relationship: as the time spent playing these games increases, so do aggression levels (the line slopes upwards). Therefore, callous-unemotional traits moderate the relationship between video games and aggression: there is a positive relationship for those with callous-unemotional traits but not for those without. This is a simple way to think about moderation, but it is not necessary that there is an effect in one group but not in the other, just that there is a difference in the relationship between video games and aggression in the two callousness groups. It could be that the effect is weakened or changes direction.

If we measure the moderator variable along a continuum it becomes trickier to

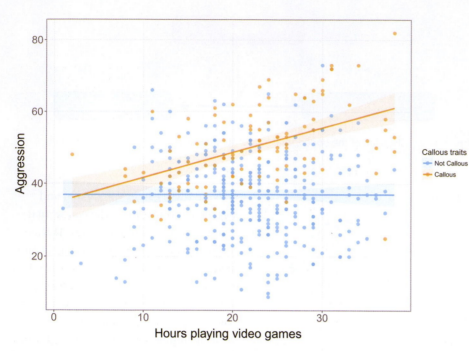

Figure 11.3 A categorical moderator (callous traits)

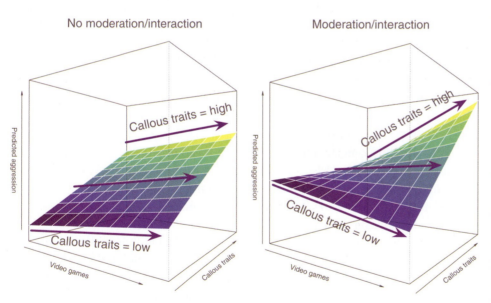

Figure 11.4 A continuous moderator (callous traits)

games and aggression. In Figure 11.4 (left) you can see that at the low end of the callous-unemotional traits scale, there is a slight positive relationship between playing video games and aggression (as time playing games increases so does aggression). At the high end of the callous-unemotional traits scale, we see a very similar relationship between video games and aggression (the top and bottom of the regression plane slope at the same angle). The same is also true at the middle of the callous-unemotional traits scale. This is a case of no interaction or no moderation.

Figure 11.4 (right) shows moderation: at low values of callous-unemotional traits the plane slopes downwards, indicating a slightly negative relationship between playing video games and aggression, but at the high end of callous-unemotional traits the plane slopes upwards, indicating a strong positive relationship between gaming and aggression. At the midpoint of the callous-unemotional traits scale, the relationship between video games and aggression is relatively flat. So, as we move along the callous-unemotional traits variable, the relationship between gaming and aggression changes from slightly negative to neutral to strongly positive. We can say that the relationship between violent video games and aggression is moderated by callous-unemotional traits.

11.3.2 The statistical model

Now we know what moderation is conceptually, let's look at how we test for moderation within a statistical model. Figure 11.5 shows the statistical model: we predict the outcome from the predictor variable, the proposed moderator, and the interaction of the two. It is the interaction effect that tells us whether moderation has

visualize, but the basic interpretation stays the same. Figure 11.4 shows two graphs that display the relationships between the time spent playing video games, aggression and callous-unemotional traits (measured along a continuum rather than as two groups). We're still interested in how the

relationship between video games and aggression changes as a function of callous-unemotional traits. We can explore this by comparing the slope of the regression plane for time spent gaming at low and high values of callous traits. To help you I have added arrows that show the relationship between video

occurred,[3] but *we must include the predictor and moderator for the interaction term to be valid*. This point is very important. In our example, then, we'd be looking at fitting a linear model predicting aggression (the outcome) from video game playing, callous-unemotional traits and their interaction.

We've encountered the general form of the linear model many times already. The following equation is to refresh your memory (but if it is completely new information to you then read Chapter 9 before continuing):

$$\text{outcome}_i = (\text{model}) + \text{error}_i$$
$$Y_i = (b_0 + b_1 X_{1i} + b_2 X_{2i} + \cdots + b_n X_{ni}) + \varepsilon_i$$
(11.1)

By replacing the Xs with the names of our predictor variables, and the Y with our outcome variable, the linear model becomes:

$$\text{Aggression}_i = (b_0 + b_1 \text{Gaming}_i + b_2 \text{Callous}_i) + \varepsilon_i$$
(11.2)

To test for moderation we need to consider the interaction between gaming and callous-unemotional traits. We have seen before that to add variables to a linear model we literally just add them in and assign them a parameter (b). Therefore, if we had two predictors labeled A and B, a model that tests for moderation would be expressed as:

$$Y_i = (b_0 + b_1 A_i + b_2 B_i + b_3 AB_i) + \varepsilon_i$$
(11.3)

If we replace the A and B with the names of the variables for this specific example we can express the model as:

$$\text{Aggression}_i = \begin{pmatrix} b_0 + b_1 \text{Gaming}_i \\ + b_2 \text{Callous}_i + b_3 \text{Interaction}_i \end{pmatrix} + \varepsilon_i$$
(11.4)

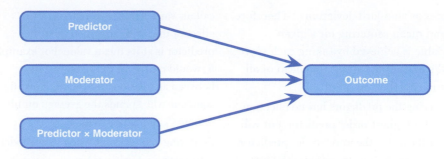

Figure 11.5 Diagram of the *statistical* moderation model

11.3.3 Centering variables ▮▮▮▮

When an interaction term is included in the model the b parameters have a specific meaning: for the individual predictors they represent the regression of the outcome on that predictor when the other predictor is zero. So, in equation (11.4), b_1 represents the relationship between aggression and gaming when callous traits are zero, and b_2 represents the relationship between aggression and callous traits when someone spends zero hours gaming per week. This interpretation isn't problematic because zero is a meaningful score for both predictors: it's plausible both that a child spends no hours playing video games, and that a child gets a score of 0 on the continuum of callous-unemotional traits. However, there are often situations where it makes no sense for a predictor to have a score of zero. Imagine that rather than measuring how much a child played violent video games we'd measured their heart rate while playing the games as an indicator of their physiological reactivity to them:

$$\text{Aggression}_i = \begin{pmatrix} b_0 + b_1 \text{Heart Rate}_i \\ + b_2 \text{Callous}_i + b_3 \text{Interaction}_i \end{pmatrix} + \varepsilon_i$$
(11.5)

In this model b_2 is the regression of the aggression on callous traits when

someone has a heart rate of zero while playing the games. This b makes no sense unless we're interested in knowing something about the relationship between callous traits and aggression in youths who die (and therefore have a heart rate of zero) while playing these games. It's fair to say that if video games killed the players, we'd have more to worry about than them developing aggression. The point is that the presence of the interaction term makes the bs for the main predictors uninterpretable in many situations.

For this reason, it is common to transform the predictors using **grand mean centering**. 'Centering' refers to the process of transforming a variable into deviations around a fixed point. This fixed point can be any value that you choose, but typically it's the grand mean. When we calculated z-scores in Chapter 1 we used grand mean centering because the first step was to take each score and subtract from it the mean of all scores. This is grand mean centering. Like z-scores, the subsequent scores are centered on zero, but unlike z-scores, we don't care about expressing the centered

What is centering and do I need to do it?

---

scores as standard deviations.[4] Therefore, grand mean centering for a given variable is achieved by taking each score and subtracting from it the mean of all scores (for that variable).

Centering the predictors has no effect on the b for highest-order predictor, but will affect the bs for the lower-order predictors. *Order* refers to how many variables are involved: the gaming × callous traits interaction is a higher-order effect than gaming alone because it involves two variables rather than one. In our model (equation (11.4)), whether we center the predictors has no effect on b_3 (the parameter for the interaction) but it changes the values of b_1 and b_2 (the parameters for gaming and callous traits). If we don't center the gaming and callous variables, the bs represent the effect of the predictor when the other predictor is zero. However, if we center the gaming and

callous variables, the bs represent the effect of the predictor when the other predictor is at its mean value. For example, b_2 would represent the relationship between aggression and callous traits for someone who spends the average number of hours gaming per week.

Centering is important when your model contains an interaction term because it makes the bs for lower-order effects interpretable. There are good reasons for not caring about the lower-order effects when the higher-order interaction involving those effects is significant; for example, if the gaming × callous traits interaction is significant, then it's not clear why we would be interested in the individual effects of gaming and callous traits. However, when the interaction is not significant, centering makes interpreting the main effects easier. With centered variables the bs for individual

predictors have two interpretations: (1) they are the effect of that predictor at the mean value of the sample; and (2) they are the average effect of the predictor across the range of scores for the other predictors. To explain the second interpretation, imagine we took everyone who spent no hours gaming, estimated the linear model between aggression and callous traits and noted the b. We then took everyone who played games for 1 hour and did the same, and then took everyone who gamed for 2 hours per week and did the same. We continued doing this until we had estimated linear models for every different value of the hours spent gaming. We'd have a lot of bs, each one representing the relationship between callous traits and aggression for different amounts of gaming. If we took an average of these bs then we'd get the same value as the b for callous traits (centered) when we

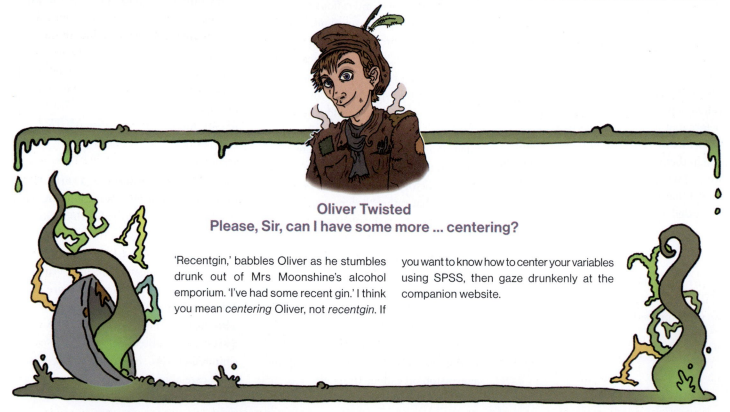

Oliver Twisted
Please, Sir, can I have some more ... centering?

'Recentgin,' babbles Oliver as he stumbles drunk out of Mrs Moonshine's alcohol emporium. 'I've had some recent gin.' I think you mean *centering* Oliver, not *recentgin*. If

you want to know how to center your variables using SPSS, then gaze drunkenly at the companion website.

---

4 Remember that with z-scores we go a step further and divide the centered scores by the standard deviation of the original variable, which changes the units of measurements to standard deviations. You can do this if you want to compare bs across predictors within a model, but remember that b would represent the change in the outcome associated with a *1 standard deviation* change in the predictor.

use it as a predictor with gaming (centered) and their interaction.

The *PROCESS* tool does the centering for us so we don't need to worry about how it's done, but because centering is useful in other analyses Oliver Twisted has some additional material that shows you how to do it manually for this example.

SELF TEST
Follow Oliver Twisted's instructions to create the centered variables **CUT_Centered** and **Vid_Centered**. Then use the *compute* command to create a new variable called **Interaction** in the **Video Games.sav** file, which is **CUT_Centered** multiplied by **Vid_Centered**.

11.3.4 Creating interaction variables ▮▮▮

Equation (11.4) contains a variable called 'Interaction', but the data file does not. You might ask how you enter a variable into the model that doesn't exist in the data. You create it, and it's easier than you might think. Mathematically speaking, when we look at the combined effect of two variables (an interaction) we are literally looking at the effect of the two variables multiplied together. The interaction variable in our example would be the scores on the time spent gaming multiplied by the scores for callous-unemotional traits. That's why interactions are denoted as *variable 1 × variable 2*. The *PROCESS* tool creates the interaction variable for you, but the self-help task gives you practice at doing it manually (for future reference).

11.3.5 Following up an interaction effect ▮▮▮

If the moderation effect is significant, we need to interpret it. In our example, we're predicting that the moderator (callous traits) influences the relationship between playing violent video games and aggression. If the interaction of callous traits and time spent gaming is a significant predictor of aggression then we know that we have a significant moderation effect, but we don't know the nature of the effect. It could be that the time spent gaming always has a positive relationship with aggression, but that the relationship gets stronger the more a person has callous traits. Alternatively, perhaps the time spent gaming *reduces* aggression in people low on callous traits but *increases* aggression in those high on callous traits (i.e., the relationship reverses). To find out what is going on we need to do a **simple slopes analysis** (Aiken & West, 1991; Rogosa, 1981).

The idea behind simple slopes analysis is no different from what was illustrated in Figure 11.4. When describing that figure I talked about comparing the relationship between the predictor (time spent gaming) and outcome (aggression) at low and high levels of the moderator (callous traits). For example, in Figure 11.4 (right), we saw that time spent gaming and aggression had a slightly negative relationship at low levels of callous traits, but a positive relationship at high levels of callous traits. This is the essence of simple slopes analysis: we work out the model equations for the predictor at outcome at low, high and average levels of the moderator. The 'high' and 'low' levels can be anything you like, but *PROCESS* uses 1 standard deviation above and below the mean value of the moderator. Therefore, in our example, we would get the linear model for aggression predicted from hours spent gaming for the average value of callous traits, for 1 standard deviation above the mean value of callous traits and for 1 standard deviation below the mean value of callous traits. We compare these slopes in terms of both their significance and the value and direction of the *b* to see whether the relationship between hours spent gaming and aggression changes at different levels of callous traits.

A related approach is to look at how the relationship between the predictor and outcome changes at lots of different values of the moderator (not just at high, low and mean values). *PROCESS* implements such an approach based on Johnson and Neyman (1936). Essentially, it estimates the model including only the predictor and outcome at lots of different values of the moderator. For each model it computes the significance of the *b* for the predictor so you can see for which values of the moderator the relationship between the predictor and outcome is significant. It returns a 'zone of significance',[5] which consists of two values of the moderator. Typically, between these two values of the moderator the predictor does not significantly predict the outcome, whereas below the lower value and above the upper value of the moderator the predictor significantly predicts the outcome.

11.3.6 Moderation analysis using IBM SPSS Statistics ▮▮▮

Given that moderation is demonstrated through a significant interaction between the predictor and moderator in a linear model, we could follow the general procedure in Chapter 9 (Figure 9.10). We would first center the predictor and moderator, then create the interaction term as discussed already, then run a

---

5 I must be careful not to confuse this with my wife, who is the Zoë of significance.

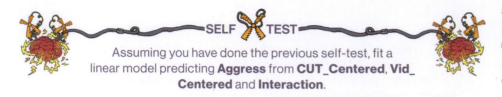

Assuming you have done the previous self-test, fit a linear model predicting **Aggress** from **CUT_Centered**, **Vid_Centered** and **Interaction**.

analysis. To access the dialog boxes in Figure 11.6 select *Analyze* ▶ *Regression* ▶ PROCESS, by Andrew F. Hayes (http://www.afhayes.com).[6] Drag (or click ⬛) the outcome variable (**Aggress**) from the box labeled *Data File Variables* to the box labeled *Outcome Variable (Y)*. Similarly, drag the predictor variable (**Vid_Game**) to the box labeled *Independent Variable (X)*. Finally, drag the moderator variable (**CaUnTs**) to the box labeled *M Variable(s)*, which is where you specify moderators (you can have more than one).

PROCESS can test 74 different types of model, and these models are listed in the drop-down box labeled *Model Number*.[7] The default model is 4 (mediation, which we'll look at next), so activate the drop-down list and select ⬛, which is a simple moderation model. The rest of the options in this dialog box are for models other than simple moderation, so we'll ignore them.

Clicking ⬛ Options... activates a dialog box containing four useful options for moderation. Selecting (1) *Mean center for products* centers the predictor and moderator for you; (2) *Heteroscedasticity-consistent SEs* removes the need to worry about the heteroscedasticity assumption; (3) *OLS/ML confidence intervals* produces confidence intervals for the model, and I've emphasized the importance of these many times; and (4) *Generate data for plotting* is helpful for interpreting and visualizing the simple slopes analysis. Talking of simple slopes, click ⬛ Conditioning for options related to this analysis. You can change whether you want simple slopes at ±1 standard deviation of the mean of the moderator (the default, which is fine) or at percentile points (*PROCESS* uses the 10th, 25th, 50th, 75th and 90th percentiles). Select the *Johnson-Neyman* method to get a zone of significance for the moderator.

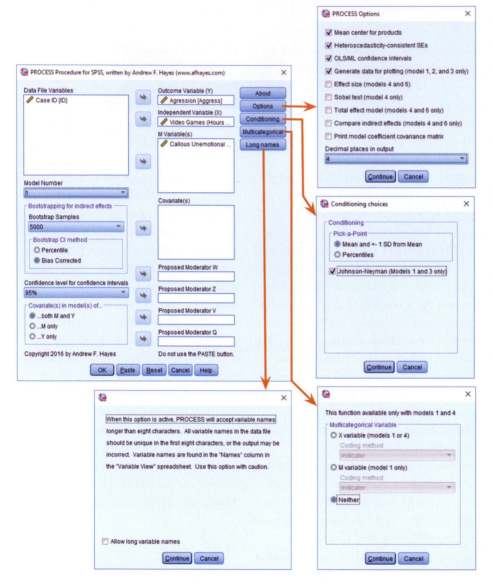

Figure 11.6 The dialog boxes for running moderation analysis

forced entry regression with the centered predictor, centered moderator and the interaction of the two centered variables as predictors. The advantage of this approach is that we can inspect sources of bias in the model.

Using the *PROCESS* tool has several advantages over using the normal regression menu in SPSS Statistics: (1) it centers predictors for us; (2) it computes the interaction term automatically; and (3) it produces simple slopes

---

6 If the menu isn't there, see Section 4.13.1.
7 Details of the models are in the **templates.pdf** file that downloads with the *PROCESS* tool and in Hayes (2018).

Clicking [Multicategorical] opens a dialog box where you can specify contrasts for the predictor (X) or moderator (M) if that variable is categorical with more than two categories. There is standard indicator or dummy coding (Section 11.5.1) as well as some others (including Helmert coding, which we look at in Section 12.4.4). Clicking [Long names] opens a dialog box with a check box for asking *PROCESS* to truncate your variable names. This option is here because *PROCESS* works only with variable names of 8 characters or less, so if you have variable names longer than this it won't work. One solution is to use this option to truncate the existing variable names, but you do need to be careful that you don't end up with multiple variables truncated to the same name (SPSS Tip 11.1). To be safe, I'd simply use variable names of

SELF TEST
Assuming you did the previous self-test, compare the table of coefficients that you got with those in Output 11.1.

8 characters or less and leave this option unselected. Back in the main dialog box, click [OK] to run the analysis.

11.3.7 Output from moderation analysis

The output appears as text rather than nicely formatted in tables. Try not to let this formatting disturb you. If your output looks odd or contains warnings or has a lot of zeros in it, it might be worth checking the variables that you input into *PROCESS* (SPSS Tip 11.1). Assuming everything has gone smoothly, you should see Output 11.1, which is the main moderation analysis. This output is pretty much the same as the table of coefficients that we saw in Chapter 9. We're told the b-value for each predictor, the associated standard errors (which have been adjusted for heteroscedasticity because we asked for them to be). Each b is compared to zero using a t-test, which is computed from the beta divided by its standard error. The confidence interval for the b is also produced (because we asked for it). Moderation is shown up by a significant

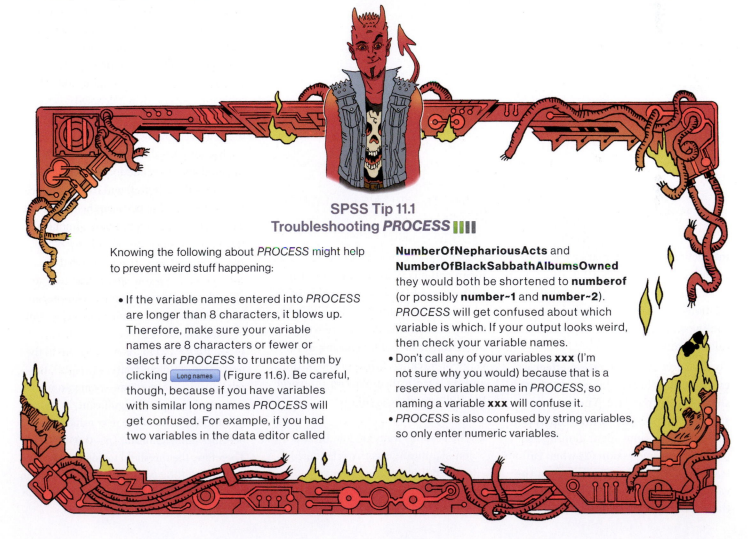

SPSS Tip 11.1
Troubleshooting *PROCESS*

Knowing the following about *PROCESS* might help to prevent weird stuff happening:

- If the variable names entered into *PROCESS* are longer than 8 characters, it blows up. Therefore, make sure your variable names are 8 characters or fewer or select for *PROCESS* to truncate them by clicking [Long names] (Figure 11.6). Be careful, though, because if you have variables with similar long names *PROCESS* will get confused. For example, if you had two variables in the data editor called

NumberOfNephariousActs and **NumberOfBlackSabbathAlbumsOwned** they would both be shortened to **numberof** (or possibly **number~1** and **number~2**). *PROCESS* will get confused about which variable is which. If your output looks weird, then check your variable names.

- Don't call any of your variables **xxx** (I'm not sure why you would) because that is a reserved variable name in *PROCESS*, so naming a variable **xxx** will confuse it.

- *PROCESS* is also confused by string variables, so only enter numeric variables.

```
* * * * * * * * * * * * * * * * * * * * * * * * * * * * * * * * * * * * * * * * * * * * * * * * * * * * * *
Model  =   1
    Y  =  Aggress
    X  =  Vid_Game
    M  =  CaUnTs

Sample size
       442

* * * * * * * * * * * * * * * * * * * * * * * * * * * * * * * * * * * * * * * * * * * * * * * * * * * * * *

Outcome: Aggress

Model Summary
        R       R-sq        MSE          F        df1        df2          p
     .6142      .3773    99.5266    90.5311     3.0000   438.0000      .0000

Model
                  coeff         se          t          p       LLCI       ULCI
constant        39.9671      .4750    84.1365      .0000    39.0335    40.9007
CaUnTs            .7601      .0466    16.3042      .0000      .6685      .8517
Vid_Game          .1696      .0759     2.2343      .0260      .0204      .3188
int_1             .0271      .0073     3.7051      .0002      .0127      .0414


Interactions:

    int_1   Vid_Game    X    CaUnTs
```

Output 11.1

```
Conditional effect of X on Y at values of the moderator(s):

   CaUnTs    Effect         se          t          p       LLCI       ULCI
   -9.6177    -.0907      .1058     -.8568      .3920     -.2986      .1173
     .0000     .1696      .0759     2.2343      .0260      .0204      .3188
    9.6177     .4299      .1010     4.2562      .0000      .2314      .6284
```

Values for quantitative moderators are the mean and plus/minus one SD from mean.

Values for dichotomous moderators are the two values of the moderator.

Output 11.2

interaction effect, and that's what we've got here, $b = 0.027$, 95% CI [0.013, 0.041], $t = 3.71$, $p = 0.0002$, indicating that the relationship between the time spent gaming and aggression is moderated by callous traits.

To interpret the moderation effect we examine the simple slopes, which are shown in Output 11.2. The output shows us the results of three models: the model for time spent gaming as a predictor of aggression (1) when callous traits are low (to be precise, when the value of callous traits is –9.6177); (2) at the mean value of callous traits

(because we centered callous traits its mean value is 0, as indicated in the output); and (3) when the value of callous traits is 9.6177 (i.e., high). We interpret these models as we would any other linear model by looking at the value of b (called *Effect* in the output), and its significance. We can interpret the three models as follows:

1 When callous traits are low, there is a non-significant negative relationship between time spent gaming and aggression, $b = -0.091$, 95% CI [–0.299, 0.117], $t = -0.86$, $p = 0.392$.

2 At the mean value of callous traits, there is a significant positive relationship between time spent gaming and aggression, $b = 0.170$, 95% CI [0.020, 0.319], $t = 2.23$, $p = 0.026$.

3 When callous traits are high, there is a significant positive relationship between time spent gaming and aggression, $b = 0.430$, 95% CI [0.231, 0.628], $t = 4.26$, $p < 0.001$.

These results tell us that the relationship between time spent playing violent video games and aggression only really emerges in people with average or greater levels of callous-unemotional traits.

Output 11.3 shows the results of the Johnson–Neyman method. First we're told that the boundaries of the zone of significance are –17.1002 and –0.7232. These are the values of the centered version of the callous-unemotional traits variable, and define regions within which the relationship between the time spent gaming and aggression is significant. The table underneath gives a detailed breakdown of these regions. Essentially it's doing a finer-grained simple slopes analysis: it takes different values of callous-unemotional traits and for each one computes the b (*Effect*) and its significance for the relationship between the time spent gaming and aggression. I have annotated the output to show the boundaries of the zone of significance. If you look at the column labeled p you can see that we start off with a significant negative relationship between time spent gaming and aggression, $b = -0.334$, 95% CI [–0.645, –0.022], $t = -2.10$, $p = 0.036$. As we move up to the next value of callous traits (–17.1002), the relationship between time spent gaming and aggression is still significant ($p = 0.0500$), but at the next value it becomes non-significant ($p = 0.058$). Therefore, the threshold for significance ends at –17.1002 (which we were told at the top of the output). As we increase the value of callous-unemotional traits the relationship between time spent gaming and aggression remains non-significant until the value of callous-unemotional traits is −0.723, at which point it crosses the threshold for significance again. For all subsequent values of callous-unemotional traits the relationship between time spent gaming and aggression is significant. Looking at the *b*-values themselves (in the column labeled *Effect*) we can also see that as callous-unemotional traits increases, the strength of relationship between time spent gaming and aggression goes from a small negative effect (*b* = −0.334) to a strong positive one (*b* = 0.830).

The final way to break down the interaction is to plot it. In Figure 11.6 we asked *PROCESS* to generate data for plotting, and these data are at the bottom of the output (see Figure 11.7). We're given values of the variable **Vid_Game** (−6.9622, 0, 6.9622) and of **CaUnTs** (−9.6177, 0, 9.6177). These values are not important in themselves, but they correspond to low, mean and high values of the variable. The final column contains the predicted values of the outcome (aggression) for these combinations of the predictors. For example, when **Vid_Game** and **CaUnTs** are both low (−6.9622 and −9.6177, respectively) the predicted value of aggression is 33.2879, when both variables are at their mean (0 and 0), the predicted value of aggression is 39.9671, and so on. To create the plot, put these values into the data editor. In a blank data editor window create coding variables that represent low, mean and high (use any codes you like). Then enter all combinations of these codes. For example, in Figure 11.7 I've created variables called **Games** and **CaUnTs** both of which are coding variables (1 = low, 2 = mean, 3 = high), then entered the combinations of these codes that correspond to the *PROCESS* output (low–low, mean–low, high–low, etc.), and then typed in the corresponding predicted values from the *PROCESS* output. Hopefully you can see from Figure 11.7 how the output from *PROCESS* corresponds to the new data file. If you can't work out how

```
********************* JOHNSON-NEYMAN TECHNIQUE *************************

Moderator value(s) defining Johnson-Neyman significance region(s):
      Value     % below     % above
   -17.1002     1.3575     98.6425
    -.7232     48.8688     51.1312

Conditional effect of X on Y at values of the moderator (M)
```

CaUnTs	Effect	se	t	p	LLCI	ULCI	
-18.5950	-.3336	.1587	-2.1027	.0361	-.6454	-.0218	Significant
-17.1002	-.2931	.1492	-1.9654	.0500	-.5863	.0000	
-16.4450	-.2754	.1451	-1.8987	.0583	-.5605	.0097	
-14.2950	-.2172	.1319	-1.6467	.1003	-.4765	.0420	
-12.1450	-.1590	.1194	-1.3319	.1836	-.3937	.0756	
-9.9950	-.1009	.1077	-.9361	.3497	-.3126	.1109	Not significant
-7.8450	-.0427	.0972	-.4390	.6609	-.2338	.1484	
-5.6950	.0155	.0882	.1757	.8606	-.1579	.1889	
-3.5450	.0737	.0813	.9059	.3655	-.0862	.2336	
-1.3950	.1319	.0771	1.7111	.0878	-.0196	.2833	
-.7232	.1501	.0763	1.9654	.0500	.0000	.3001	
.7550	.1901	.0759	2.5053	.0126	.0410	.3392	
2.9050	.2482	.0779	3.1878	.0015	.0952	.4013	
5.0550	.3064	.0829	3.6980	.0002	.1436	.4693	
7.2050	.3646	.0903	4.0360	.0001	.1871	.5422	
9.3550	.4228	.0997	4.2386	.0000	.2267	.6188	Significant
11.5050	.4810	.1106	4.3490	.0000	.2636	.6983	
13.6550	.5392	.1225	4.4013	.0000	.2984	.7799	
15.8050	.5973	.1352	4.4188	.0000	.3317	.8630	
17.9550	.6555	.1484	4.4160	.0000	.3638	.9473	
20.1050	.7137	.1621	4.4017	.0000	.3950	1.0324	
22.2550	.7719	.1762	4.3814	.0000	.4256	1.1181	
24.4050	.8301	.1905	4.3580	.0000	.4557	1.2044	

```
***************************************************************************
```

Output 11.3

Draw a multiple line graph of **Aggress** (*y*-axis) against **Games** (*x*-axis) with different colored lines for different values of **CaUnTs**.

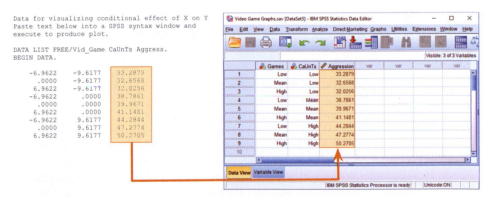

Figure 11.7   Entering data for graphing simple slopes

Now draw a multiple line graph of **Aggress** (*y*-axis) against **CaUnTs** (*x*-axis) with different colored lines for different values of **Games**.

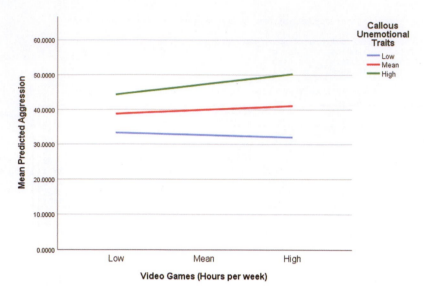

Figure 11.8 Simple slopes equations of the regression of aggression on video games at three levels of callous traits

to create the file yourself, use **Video Game Graph.sav**. Having transferred the output to a data file, we can draw a line graph using what we learnt in Chapter 5.

The graph from the self-test (Figure 11.8) shows what we found from the simple slopes analysis. When callous traits are low (blue line) there is a non-significant negative relationship between time spent gaming and aggression; at the mean value of callous traits (red line) there is small positive relationship between time spent gaming and aggression; and this relationship gets even stronger at high levels of callous traits (green line).

Cramming Sam's Tips
Moderation

- Moderation occurs when the relationship between two variables changes as a function of a third variable. For example, the relationship between watching horror films and feeling scared at bedtime might increase as a function of how vivid an imagination a person has.

- Moderation is tested using a linear model in which the outcome (fear at bedtime) is predicted from a predictor (how many horror films are watched), the moderator (imagination) and the interaction of the predictor variables.

- Predictors should be centered before the analysis.
- The interaction of two variables is their scores multiplied together.
- If the interaction is significant then the moderation effect is also significant.
- If moderation is found, follow up the analysis with simple slopes analysis, which looks at the relationship between the predictor and outcome at low, mean and high levels of the moderator.

11.3.8 Reporting Moderation Analysis ▮▮▮▮

Moderation can be reported in the same way as described in Section 9.13. My personal preference would be to produce a table like Table 11.1.

11.4 Mediation ▮▮▮▮

11.4.1 The conceptual model ▮▮▮▮

Whereas moderation alludes to the combined effect of two variables on an outcome, **mediation** refers to a situation when the relationship between a predictor variable and an outcome variable can be explained by their relationship to a third variable (the **mediator**). The top of Figure 11.9 shows a relationship between a predictor and an outcome (denoted by c). The bottom of the figure shows that these variables are also related to a third variable in specific ways: (1) the predictor also predicts the mediator through the path denoted by a; (2) the mediator predicts the outcome through the path denoted by b. The relationship between the predictor and outcome will probably be different when the mediator is also included in the model and so is denoted c'. The letters denoting each path (a, b, c and c') represent the unstandardized regression coefficient between the variables connected by the arrow; therefore, they symbolize the strength of relationship between variables. Mediation is said to have occurred if the strength of the relationship between the predictor and outcome is reduced by including the mediator (i.e., the b-value for c' is smaller than for c). Perfect mediation occurs when c' is zero: the relationship between the predictor and outcome is completely wiped out by including the mediator in the model.

Table 11.1 Linear model of predictors of aggression

	b	SE B	t	p
Constant	39.97 [39.03, 40.90]	0.475	84.13	$p < 0.001$
Callous Traits (centered)	0.76 [0.67, 0.85]	0.047	16.30	$p < 0.001$
Gaming (centered)	0.17 [0.02, 0.32]	0.076	2.23	$p = 0.026$
Callous Traits × Gaming	0.027 [0.01, 0.04]	0.007	3.71	$p < 0.001$

Note. $R^2 = 0.38$.

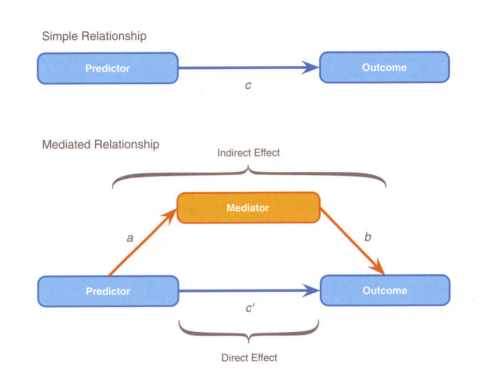

Figure 11.9 Diagram of a basic mediation model

This description is a bit abstract, so let's use an example. My wife and I often wonder what the important factors are in making a relationship last. For my part, I don't really understand why she'd want to be with a balding heavy rock fan with an oversized collection of vinyl and musical instruments and an unhealthy love of *Doctor Who* and numbers. It is important I gather as much information as possible about keeping her happy because the odds are stacked against me. For her part, I have no idea why she wonders: her very existence makes me happy. Perhaps if you are in a relationship you have wondered how to make it last too. During our cyber-travels, Mrs Field and I have discovered that physical attractiveness (McNulty, Neff, & Karney, 2008), conscientiousness and neuroticism (good for us) predict marital satisfaction (Claxton, O'Rourke, Smith, & Delongis, 2012). Pornography use probably doesn't: it is related to infidelity (Lambert, Negash, Stillman, Olmstead, & Fincham, 2012). Mediation is really all about the variables that explain relationships like these: it's

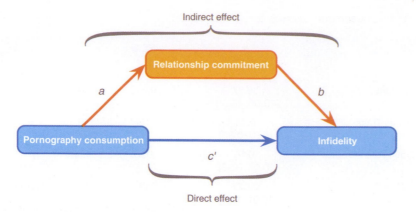

Figure 11.10   Diagram of a mediation model from Lambert et al. (2012)

unlikely that everyone who catches a glimpse of some porn suddenly rushes out of their house to have an affair – presumably it leads to some kind of emotional or cognitive change that undermines the love glue that holds us and our partners together. Lambert et al. tested this hypothesis. Figure 11.10 shows their mediator model: the initial relationship is that between pornography consumption (the predictor) and infidelity (the outcome), and they hypothesized that this relationship is mediated by commitment (the mediator). This model suggests that the relationship between pornography consumption and infidelity isn't a direct effect but operates though a reduction in relationship commitment. For this hypothesis to be true: (1) pornography consumption must predict infidelity in the first place (path c'); (2) pornography consumption must predict relationship commitment (path a); (3) relationship commitment must predict infidelity (path b); and (4) the relationship between pornography consumption and infidelity should be smaller when relationship commitment is included in the model than when it isn't. We can distinguish between the **direct effect** of pornography consumption on infidelity, which is the relationship between them controlling for relationship commitment, and the **indirect effect**, which is the effect of pornography consumption on infidelity through relationship commitment (Figure 11.10).

11.4.2 The statistical model ▍▍▍▍

Unlike moderation, the statistical model for mediation is basically the same as the conceptual model: it is characterized in Figure 11.9. Historically, this model was tested through a series of linear models that reflect the four conditions necessary to demonstrate mediation (Baron & Kenny, 1986). I have mentioned already that the letters denoting the paths in Figure 11.9 represent the unstandardized b-values for the relationships between variables denoted by the path. Therefore, to estimate any one of these paths, we want to know the unstandardized b for the two variables involved. For example, Baron and Kenny suggested that mediation is tested through three linear models (see also Judd & Kenny, 1981):

1 A linear model predicting the outcome from the predictor variable. The b-value coefficient for the predictor gives us the value of c in Figure 11.9.

2 A linear model predicting the mediator from the predictor variable. The b-value for the predictor gives us the value of a in Figure 11.9.

3 A linear model predicting the outcome from both the predictor variable and the mediator. The b-value for the predictor gives us the value of c' in Figure 11.9, and the b-value for the mediator gives us the value of b.

These models test the four conditions of mediation: (1) the predictor variable must significantly predict the outcome variable in model 1; (2) the predictor variable must significantly predict the mediator in model 2; (3) the mediator must significantly predict the outcome variable in model 3; and (4) the predictor variable must predict the outcome variable less strongly in model 3 than in model 1.

In Lambert et al.'s (2012) study, all participants had been in a relationship for at least a year. The researchers measured pornography consumption on a scale from 0 (low) to 8 (high), but this variable, as you might expect, was skewed (most people had low scores) so they analyzed log-transformed values (**LnPorn**). They also measured commitment to their current relationship (**Commit**) on a scale from 1 (low) to 5 (high). Infidelity was measured with questions asking whether the person had committed a physical act (**Phys_Inf**) that they or their partner would consider to be unfaithful (0 = no, 1 = one of them would consider the act unfaithful, 2 = both of them would consider it unfaithful),[8] and also using the number of people they had 'hooked up' with in the previous year (**Hook_Ups**), which would mean during a time period in which they were in their current relationship.[9] The actual data from Lambert et al.'s study are in the file **Lambert et al. (2012).sav**.

---

8 I've coded this variable differently from the original data to make interpretation of it more intuitive, but it doesn't affect the results.

9 A 'hook-up' was defined to participants as 'when two people get together for a physical encounter and don't necessarily expect anything further (e.g., no plan or intention to do it again)'.

Many people still use this approach to test mediation. I think it is very useful for illustrating the principles of mediation and for understanding what mediation means, but it has limitations. The main one is the fourth criterion: *the predictor variable must predict the outcome variable less strongly in model 3 than in model 1.* Although perfect mediation is shown when the relationship between the predictor and outcome is reduced to zero in model 3, in practice it rarely happens. Instead, you see a reduction in the relationship between the predictor and outcome, but it is not wiped out altogether. The question, therefore, arises of how much 'reduction' is sufficient to infer mediation.

Although Baron and Kenny advocated looking at the sizes of the *b*-values, in practice people tend to look for a change in significance; so, mediation would occur if the relationship between the predictor and outcome was significant ($p < 0.05$) when looked at in isolation (model 1) but not significant ($p > 0.05$) when the mediator is included too (model 3). This approach leads to all sorts of silliness because of the 'all-or-nothing' thinking that *p*-values encourage. You could have a situation in which the *b*-value for the relationship between the predictor and outcome changes very little in models with and without the mediator, but the *p*-values shift from one side of the threshold to the other (e.g., from $p = 0.049$ when the mediator isn't included to $p = 0.051$ when it is). Even though the *p*-values have changed from significant to not significant, the change is very small, and the size of the relationship between the predictor and outcome will not have changed very much at all. Conversely, you could have a situation where the *b* for the relationship between the predictor and the outcome reduces a lot when the mediator is included, but remains significant in both cases. For example,

SELF TEST

Run the three models necessary to test mediation for Lambert et al.'s data: (1) a linear model predicting **Phys_Inf** from **LnPorn**; (2) a linear model predicting **Commit** from **LnPorn**; and (3) a linear model predicting **Phys_Inf** from both **LnPorn** and **Commit**. Is there mediation?

perhaps when looked at in isolation the relationship between the predictor and outcome is $b = 0.46$, $p < 0.001$, but when the mediator is included as a predictor as well it reduces to $b = 0.18$, $p = 0.042$. You'd conclude (based on significance) that mediation hadn't occurred despite the relationship between the predictor and outcome reducing to more than half its original value.

An alternative is to estimate the indirect effect and its significance. The indirect effect is illustrated in Figures 11.9 and 11.10; it is the combined effects of paths *a* and *b*, and the significance of this test can be assessed using the **Sobel test** (Sobel, 1982). If the Sobel test is significant it means that the predictor significantly affects the outcome variable via the mediator. In other words, there is significant mediation. This test works well in large samples, but you're better off computing confidence intervals for the indirect effect using bootstrap methods (Section 6.12.3). Now that computers make it easy for us to estimate the indirect effect (i.e., the effect of mediation) and its confidence interval, this practice is becoming increasingly common and is preferable to Baron and Kenny's regressions and the Sobel test because it's harder to get sucked into the black-and-white thinking of significance testing (Section 3.2.2). People tend to apply Baron and Kenny's method in a way that is intrinsically bound to looking for 'significant' relationships, whereas estimating the indirect effect and its confidence interval allows us to simply report the degree of mediation observed in the data.

11.4.3 Effect sizes of mediation ▌▌▌

If we're going to look at the size of the indirect effect to judge the degree of mediation, then it's useful to have effect size measures to help us (see Section 3.5). Many effect size measures have been proposed and are discussed in detail elsewhere (MacKinnon, 2008; Preacher & Kelley, 2011). The simplest is to look at the *b*-value for the indirect effect and its confidence interval. Figure 11.9 shows us that the indirect effect is the combined effect of paths *a* and *b*. We have also seen that *a* and *b* are unstandardized model coefficients for the relationships between variables denoted by the path. To find the combined effect of these paths, we multiply these *b*-values:

$$\text{indirect effect} = ab \qquad (11.6)$$

The resulting value is an unstandardized regression coefficient like any other, and consequently is expressed in the original units of measurement. As we have seen, it is sometimes useful to look at standardized *b*-values, because these can be compared across different studies using different outcome measures (see Chapter 9). MacKinnon (2008) suggested standardizing this measure by dividing by the standard deviation of the outcome variable:

$$\text{indirect effect (partially standardized)} = \frac{ab}{s_{\text{Outcome}}} \qquad (11.7)$$

This standardizes the indirect effect with respect to the outcome variable, but not the predictor or mediator. As such, it is sometimes referred to as the partially standardized indirect effect. To fully standardize the indirect effect we would need to multiply the partially standardized measures by the standard deviation of the predictor variable (Preacher & Hayes, 2008b):

$$\text{indirect effect (standardized)} = \frac{ab}{s_{\text{Outcome}}} \times s_{\text{predictor}}$$

$$(11.8)$$

This measure is sometimes called the **index of mediation**. It is useful in that it can be compared across different mediation models that use different measures of the predictor, outcome and mediator. Reporting this measure would be particularly helpful if anyone decides to include your research in a meta-analysis.

A different approach to estimating the size of the indirect effect is to look at the size of the indirect effect relative to either the total effect of the predictor or the direct effect of the predictor. For example, if we wanted the ratio of the indirect effect (ab) to the total effect (c) we could use the b-values from the various linear models displayed in Figure 11.9:

$$P_M = \frac{ab}{c}$$

$$(11.9)$$

Similarly, if we wanted to express the indirect effect as a ratio of the direct effect (c'), the models give us the values we need:

$$R_M = \frac{ab}{c'}$$

$$(11.10)$$

These ratio-based measures only really re-describe the original indirect effect. Both are very unstable in small samples, and MacKinnon (2008) advises against using P_M and R_M in samples smaller than 500 and 5000, respectively. Also, although it is tempting to think of P_M as a proportion (because it is the ratio of the indirect effect to the *total* effect) it is not:

it can exceed 1 and even take negative values (Preacher & Kelley, 2011).

We used R^2 to assess the fit of a linear model. We can compute a form of R^2 for the indirect effect, which tells us the proportion of variance explained by the indirect effect. MacKinnon (2008) proposes several versions, but *PROCESS* computes the following:

$$R_M^2 = R_{Y,M}^2 - \left(R_{Y,MX}^2 - R_{Y,X}^2 \right)$$

$$(11.11)$$

This equation uses the proportion of variance in the outcome variables explained by the predictor($R_{Y,X}^2$), the mediator ($R_{Y,M}^2$), and both ($R_{Y,MX}^2$). It can be interpreted as the variance in the outcome that is shared by the mediator and the predictor, but that cannot be attributed to either in isolation. Again, this measure is not bounded to fall between 0 and 1, and it's possible to get negative values (which usually indicate suppression effects rather than mediation). Finally, Preacher and Kelley (2011) proposed a measure called kappa-squared (κ^2) that was available in versions of *PROCESS* prior to 2.16. Unfortunately, the maths behind the measure has subsequently been shown to be incorrect, leading to paradoxical (and undesirable) effects such as κ^2 *decreasing* as mediation effect *increases* (Wen & Fan, 2015). If you're using an older version of *PROCESS*, ignore this measure.

Probably the most useful of these measures are the unstandardized and standardized indirect effect. All the measures have accompanying confidence intervals and are unaffected by sample sizes (although note my earlier comments about the variability of P_M and R_M in small samples). However, P_M, R_M and R_M^2 cannot be interpreted easily because they aspire to being proportions but are not, and all of the measures are unbounded, which again makes interpretation tricky (Preacher & Kelley, 2011). For these

reasons Wen and Fan (2015) argue against effect size measures for mediation, although for simple mediation models they concede that P_M can be useful provided that it is accompanied by the total effect (as important context for the size of the indirect effect).

11.4.4 Mediation using SPSS Statistics

We can test Lambert's mediation model (Figure 11.10) using the *PROCESS* tool. Access the dialog boxes in Figure 11.11 by selecting *Analyze ▶ Regression ▶*
PROCESS, by Andrew F. Hayes (http://www.afhayes.com). Drag (or click ▶) the outcome variable (**Phys_ Inf**) from the box labeled *Data File Variables* to the box labeled *Outcome Variable (Y)*, then drag the predictor variable (**LnPorn**) to the box labeled *Independent Variable (X)*. Finally, drag the mediator variable (**Commit**) to the box labeled *M Variable(s)*, where you specify any mediators (you can have more than one).

Simple mediation is represented by model 4 (the default), therefore, make sure that [4 ▼] is selected in the drop-down list under *Model Number*. Unlike for moderation, there are other options in this dialog box that are useful. For example, to test the indirect effects we will use bootstrapping to generate a confidence interval around the indirect effect. By default *PROCESS* uses 5000 bootstrap samples, and will compute bias corrected and accelerated confidence intervals. These default options are fine, but be aware that you can ask for percentile bootstrap confidence intervals instead (see Section 6.12.3).

Click [Options...] to open another dialog box. Selecting (1) *Effect size* produces the estimates of the size of the indirect effect discussed in Section 11.4.3;[10] (2) *Sobel test* produces a significance test of the indirect effect devised by Sobel; (3) *Total effect model* produces the direct effect of the

---

10 R_M^2 and κ^2 are produced only for models with a single mediator.

predictor on the outcome (in this case the linear model of infidelity predicted from pornography consumption); and (4) *Compare indirect effects* will, when you have more than one mediator in the model, estimate the effect and confidence interval for the difference between the indirect effects resulting from these mediators. This final option is useful when you have more than one mediator to compare their relative importance in explaining the relationship between the predictor and outcome. However, we have only a single mediator, so we don't need to select this option (you can select it if you like, but it won't change the output produced). None of the options activated by clicking Conditioning apply to simple mediation models, so we can ignore this button. If our predictor variable (*X*) was categorical with more than two categories (which it's not), we could click Multicategorical to get *PROCESS* to automatically dummy-code (Section 11.5.1) it for us. Back in the main dialog box, click OK to run the analysis.

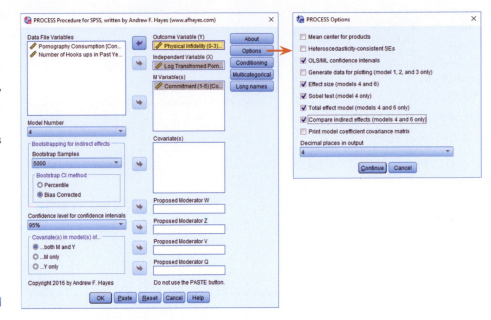

Figure 11.11   The dialog boxes for running mediation analysis

11.4.5 Output from mediation analysis

As with moderation, the output appears as text. Output 11.4 shows the first part of the output, which initially tells us the name of the outcome (*Y*), the predictor (*X*) and the mediator (*M*) variables (SPSS Tip 11.1). This is useful for double-checking we have entered the variables in the correct place: the outcome is infidelity,

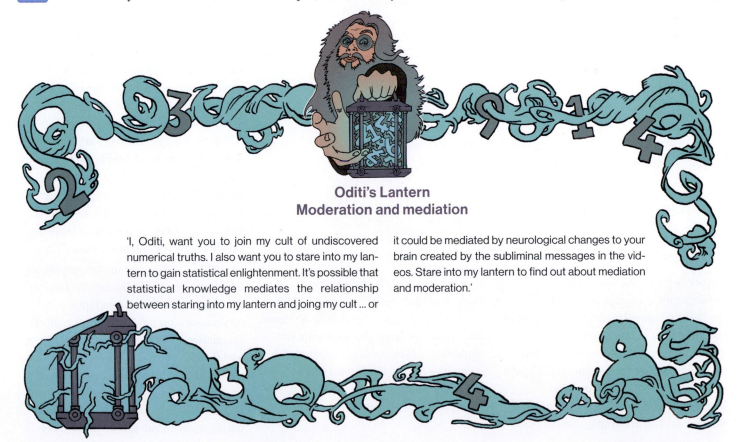

Oditi's Lantern
Moderation and mediation

'I, Oditi, want you to join my cult of undiscovered numerical truths. I also want you to stare into my lantern to gain statistical enlightenment. It's possible that statistical knowledge mediates the relationship between staring into my lantern and joing my cult ... or it could be mediated by neurological changes to your brain created by the subliminal messages in the videos. Stare into my lantern to find out about mediation and moderation.'

```
*******************************************************************
Model = 4
    Y = Phys_Inf
    X = LnPorn
    M = Commit
Sample size
      239
*******************************************************************
Outcome: Commit

Model Summary
       R      R-sq     MSE        F        df1        df2         p
    .1418    .0201   .5354   4.8633   1.0000   237.0000     .0284

Model
             coeff       se        t          p       LLCI      ULCI
constant    4.2027    .0545  77.1777     .0000    4.0954    4.3100
  LnPorn    -.4697    .2130  -2.2053     .0284    -.8892    -.0501
```

Output 11.4

```
*******************************************************************
Outcome: Phys_Inf

Model Summary
       R      R-sq     MSE        F        df1        df2         p
    .3383    .1144   .4379  15.2453   2.0000   236.0000     .0000

Model
             coeff       se        t          p       LLCI      ULCI
constant    1.3704    .2518   5.4433     .0000     .8744    1.8663
  Commit    -.2710    .0587  -4.6128     .0000    -.3867    -.1552
  LnPorn     .4573    .1946   2.3505     .0196     .0740     .8407
```

Output 11.5

```
******************** TOTAL EFFECT MODEL ********************
Outcome: Phys_Inf

Model Summary
       R      R-sq     MSE        F        df1        df2         p
    .1859    .0346   .4754   8.4866   1.0000   237.0000     .0039

Model
             coeff       se        t          p       LLCI      ULCI
constant     .2315    .0513   4.5123     .0000     .1304     .3326
  LnPorn     .5846    .2007   2.9132     .0039     .1893     .9800
```

Output 11.6

Output 11.5 shows the results of the regression of infidelity predicted from both pornography consumption (i.e., path c' in Figure 11.10) and commitment (i.e., path b in Figure 11.10). We can see that pornography consumption significantly predicts infidelity even with relationship commitment in the model, $b = 0.46$, 95% CI [−0.07, 0.84], $t = 2.35$, $p = 0.02$; relationship commitment also significantly predicts infidelity, $b = −0.27$, 95% CI [−0.39, −0.16], $t = −4.61$, $p < 0.001$. The R^2 value tells us that the model explains 11.4% of the variance in infidelity. The negative b for commitment tells us that as commitment increases, infidelity declines (and vice versa), but the positive b for consumptions indicates that as pornography consumption increases, infidelity increases also. These relationships are in the predicted direction.

Output 11.6 shows the total effect of pornography consumption on infidelity (outcome). You will get this bit of the output only if you selected *Total effect model* in Figure 11.11. The total effect is the effect of the predictor on the outcome when the mediator is not present in the model—in other words, path c in Figure 11.9. When relationship commitment is not in the model, pornography consumption significantly predicts infidelity, $b = 0.58$, 95% CI [0.19, 0.98], $t = 2.91$, $p = 0.004$. The R^2 value tells us that the model explains 3.46% of the variance in infidelity. As is the case when we include relationship commitment in the model, pornography consumption has a positive relationship with infidelity (as shown by the positive b-value).

Output 11.7 is the most important part of the output because it displays the results for the indirect effect of pornography consumption on infidelity (i.e., the effect via relationship commitment). First, we're told the effect of pornography consumption on infidelity in isolation (the total effect), and these values replicate the model in Output 11.6. Next, we're told the effect of pornography consumption on infidelity

the predictor is (log-transformed) pornography consumption, and the mediator is commitment. The next part of the output shows us the results of the linear model of commitment predicted from pornography consumption (i.e., path a in Figure 11.10). This output is interpreted just as we would interpret any linear model: pornography consumption

significantly predicts relationship commitment, $b = −0.47$, 95% CI [−0.89, −0.05], $t = −2.21$, $p = 0.028$. The R^2 values tells us that pornography consumption explains 2% of the variance in relationship commitment, and the fact that the b is negative tells us that as consumption increases, commitment declines (and vice versa).

when relationship commitment is included as a predictor as well (the direct effect). These values replicate those in Output 11.5. The first bit of new information is the *Indirect effect of X on Y*, which in this case is the indirect effect of pornography consumption on infidelity. We're given an estimate of this effect ($b = 0.127$) as well as a bootstrapped standard error and confidence interval. As we have seen many times before, 95% confidence intervals contain the true value of a parameter in 95% of samples. People tend to assume that their sample isn't one of the 5% that does not contain the true value and use them to infer the population value of an effect. In this case, assuming our sample is one of the 95% that 'hits' the true value, we know that the true b-value for the indirect effect falls between 0.017 and 0.297.[11] This range does not include zero, and remember that $b = 0$ would mean 'no effect whatsoever'; therefore, the fact that the confidence interval does not contain zero supports the idea that relationship commitment really does mediate the relationship between pornography consumption and infidelity.

The rest of Output 11.7 you will see only if you selected *Effect size* in Figure 11.11; it contains the effect size measures from Section 11.4.3. Rather than interpret them all, I'll note that for each one you get an estimate along with a confidence interval based on a bootstrapped standard error. As with the unstandardized indirect effect, if the confidence intervals don't contain zero then people assume that the true effect size is different from 'no effect'. In other words, there is mediation. All of the effect size measures have confidence intervals that don't include zero, so whatever one we look at we can assume that the indirect effect is probably greater than 'no effect'. Focusing on the most useful of these effect sizes, the

```
************** TOTAL, DIRECT, AND INDIRECT EFFECTS *************
Total effect of X on Y
     Effect         SE          t           p         LLCI        ULCI
     .5846        .2007      2.9132      .0039       .1893       .9800
Direct effect of X on Y
     Effect         SE          t           p         LLCI        ULCI
     .4573        .1946      2.3505      .0196       .0740       .8407
Indirect effect of X on Y
               Effect      Boot SE      BootLLCI     BootULCI
    Commit      .1273       .0708        .0170        .2972
Partially standardized indirect effect of X on Y
               Effect      Boot SE      BootLLCI     BootULCI
    Commit      .1818       .0997        .0215        .4156
Completely standardized indirect effect of X on Y
               Effect      Boot SE      BootLLCI     BootULCI
    Commit      .0405       .0220        .0052        .0922
Ratio of indirect to total effect of X on Y
               Effect      Boot SE      BootLLCI     BootULCI
    Commit      .2177      8.6658        .0082       1.2609
Ratio of indirect to direct effect of X on Y
               Effect      Boot SE      BootLLCI     BootULCI
    Commit      .2783     13.7610       -.0392       4.1073
R-squared mediation effect size (R-sq_med)
               Effect      Boot SE      BootLLCI     BootULCI
    Commit      .0138       .0104        .0009        .0462
```

Output 11.7

standardized b for the indirect effect, its value is $b = 0.041$, 95% BCa CI [0.005, 0.092].

The final part of the output (Output 11.8) shows the results of the Sobel test. As I have mentioned before, it is better to interpret the bootstrap confidence intervals than formal tests of significance; however, if you selected *Sobel test* in Figure 11.11, this is what you will see. Again, we're given the size of the indirect effect ($b = 0.127$), the standard error, associated z-score ($z = 1.95$) and p-value ($p = 0.051$).[12] The p-value isn't quite under the not-at-all magic 0.05 threshold, so technically we'd conclude that there isn't a significant indirect effect, but this shows you how misleading these kind of tests can be: every single effect size had a confidence

```
Normal theory tests for
indirect effect
    Effect       se         Z           p
    .1273      .0652    1.9526      .0509
```

Output 11.8

interval not containing zero, so there is compelling information that there is a small but meaningful mediation effect.

11.4.6 Reporting mediation analysis ▮▮▮

Some people report only the indirect effect in mediation analysis, and possibly the Sobel test. However, I have repeatedly favored using bootstrap confidence intervals, so you should report these:

---

11 Remember that because of the nature of bootstrapping you will get slightly different values in your output.

12 You might remember that in the linear model we calculate a test statistic (t) by dividing the regression coefficient by its standard error (as in equation (9.14)). We do the same here, except we get a z instead of a t: $z = 0.1273/0.0652 = 1.9525$.

Labcoat Leni's Real Research 11.1
I heard that Jane has a boil and kissed a tramp ▌▌▌

Massar, K., et al. (2012). *Personality and Individual Differences*, 52, 106–109.

Everyone likes a good gossip from time to time, but apparently it has an evolutionary function. One school of thought is that gossip is used as a way to derogate sexual competitors—especially by questioning their appearance and sexual behavior. For example, if you've got your eyes on a guy, but he has his eyes on Jane, then a good strategy is to spread gossip that Jane has a massive pus-oozing boil on her stomach and that she kissed a smelly vagrant called Aqualung. Apparently men rate gossiped-about women as less attractive, and they are more influenced by the gossip if it came from a woman with a high mate value (i.e., attractive and sexually desirable). Karlijn Massar and

her colleagues hypothesized that if this theory is true then (1) younger women will gossip more because there is more mate competition at younger ages; and (2) this relationship will be mediated by the mate value of the person (because for those with high mate value gossiping for the purpose of sexual competition will be more effective). Eighty-three women aged from 20 to 50 (**Age**) completed questionnaire measures of their tendency to gossip (**Gossip**) and their sexual desirability (**Mate_Value**). Test Massar et al.'s mediation model using Baron and Kenny's method (as they did) but also using *PROCESS* to estimate the indirect effect (**Massaret al. (2011).sav**). Answers are on the companion website (or look at Figure 1 in the original article, which shows the parameters for the various regressions).

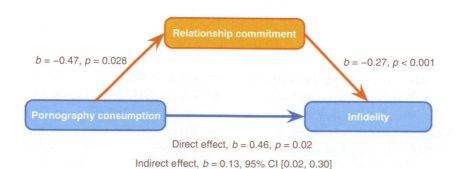

✓ There was a significant indirect effect of pornography consumption on infidelity through relationship commitment, $b = 0.127$, 95% BCa CI [0.017, 0.297].

This is fine, but it can be quite useful to present a diagram of the mediation model, and indicate on it the regression coefficients, the indirect effect and its bootstrapped confidence intervals. For the current example, we might produce something like Figure 11.12.

Figure 11.12   Model of pornography consumption as a predictor of infidelity, mediated by relationship commitment. The confidence interval for the indirect effect is a BCa bootstrapped CI based on 5000 samples

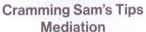

Cramming Sam's Tips
Mediation

- Mediation is when the strength of the relationship between a predictor variable and outcome variable is reduced by including another variable as a predictor. Essentially, mediation equates to the relationship between two variables being 'explained' by a third. For example, the relationship between watching horror films and feeling scared at bedtime might be explained by scary images appearing in your head.

- Mediation is tested by assessing the size of the *indirect effect* and its confidence interval. If the confidence interval contains zero then we tend to assume that a genuine mediation effect doesn't exist. If the confidence interval doesn't contain zero, then we tend to conclude that mediation has occurred.

11.5 Categorical predictors in regression ▮▮▮▮

We saw in the previous chapter that categorical predictors with two categories can be included in a linear model: we simply code the categories with 0 and 1.[13] However, often you'll collect data about groups of people in which there are more than two categories (e.g., ethnic group, gender, socio-economic status, diagnostic category). Given that we can include categorical predictors with two categories into a linear model (Section 10.4), and we can have several such predictors in a model, it follows that we can include a predictor with more than two categories by converting it to several variables each of

which has two categories. There are several different ways to achieve this, one of the most common is dummy coding (a.k.a. indicator coding).[14]

11.5.1 Dummy coding ▮▮▮▮

Imagine a survey had a question about religiosity that resulted in many categories such as Muslim, Jewish, Hindu, Catholic, Buddhist, Protestant, Jedi,[15] other. These groups cannot be distinguished using a single variable coded with zeros and ones. If we wanted to include this variable in a linear model we would need to create **dummy variables**, which is a way of representing groups of people using only zeros and ones. To do it, we create several variables; in fact, the number of variables we need is one less than the number of

groups we're recoding. There are eight basic steps:

1 Count the number of groups you want to recode and subtract 1.

2 Create as many new variables as the value you calculated in step 1. These are your dummy variables.

3 Choose one of your groups as a baseline against which all other groups will be compared. Normally you'd pick a group that might be considered a control or, if you don't have a specific hypothesis, the group that represents the majority of people (because it might be interesting to compare other groups against the majority).

4 Having chosen a baseline group, assign that group values of 0 for all dummy variables.

---

13 We saw in Section 10.2.2 why we use 0 and 1, and I elaborate on this issue in Section 12.2.1.

14 For some more detailed reading, see Hardy (1993).

15 Approximately 390,000 (almost 0.8%) people in England and Wales stated Jedi as their religion on the 2001 Census form, making it the fourth most popular religion. In my hometown of Brighton, 2.6% of the city claimed to be Jedi.

5 For your first dummy variable, assign the value 1 to the first group that you want to compare against the baseline group. Assign all other groups 0 for this variable.

6 For the second dummy variable assign the value 1 to the second group that you want to compare against the baseline group. Assign all other groups 0 for this variable.

7 Repeat this process until you run out of dummy variables.

8 Place all the dummy variables into the linear model in the same block.

Let's try this out using an example. In Chapter 6 we encountered a biologist who was worried about the potential health effects of music festivals. She originally collected data at a heavy metal festival (Download Festival), but was worried that her findings might not generalize beyond metal fans. Perhaps it's only metal fans who get smellier at festivals (at this point,

as a metal fan, I would sacrifice the biologist to Odin for her prejudices). To find out whether the type of music a person likes predicts whether hygiene decreases over the festival the biologist measured hygiene over the three days of the Glastonbury Music Festival, which has an eclectic clientele. Her hygiene measure ranged between 0 (you smell like you've bathed in sewage) and 4 (you smell like you've bathed in freshly baked bread). The data file (**GlastonburyFestival.sav**) contains the hygiene scores for each day of the festival and a variable called **change**, which is the change in hygiene from day 1 to day 3 of the festival. Not everyone could be followed up on day 3, so only a subset of the original sample has a change score. The biologist coded the festival-goer's musical affiliations into the categories 'indie kid' (people who mainly like alternative music), 'metaller' (people who like heavy metal), and 'crusty' (people who like hippy/folky/ambient stuff). Anyone

not falling into these categories was labeled 'no musical affiliation'. These groups were coded 1, 2, 3 and 4 respectively in the variable **music**. With four groups we need three dummy variables (one less than the number of groups). The first step is to choose a baseline group. We're interested in comparing those who have different musical affiliations against those who don't, so our baseline category will be 'no musical affiliation'. We code this group with 0 for all dummy variables. For our first dummy variable, we could look at the 'crusty' group by assigning anyone who is a crusty a code of 1, and everyone else a code of 0. For our second dummy variable, we could look at the 'metaller' group by assigning a 1 to anyone who is a metaller, and a 0 to everyone else. Our final dummy variable codes the final category 'indie kid' by assigning 1 to anyone who is an indie kid, and a 0 to everyone else. Table 11.2 illustrates the resulting coding scheme. Note that each group has a code of 1 on only one of the dummy variables (except the base category, which is always coded as 0).

11.5.2 The *recode* function

To create these dummy variables select *Transform* ▶ Recode into Different Variables... to access the dialog box in Figure 11.13. Select the variable you want to recode (in this case **music**) and drag it (or click →) to the box labeled *Numeric Variable* → *Output Variable*. To create a new variable we first type a name in the box labeled *Name* (let's call this first dummy variable **Crusty**). Give this variable a more descriptive name by typing something in the box labeled *Label* (I've labeled it 'No Affiliation vs. Crusty' which reflects what it represents). Click Change to transfer this new variable to the box labeled *Numeric Variable* → *Output Variable* (this box should now say *music* → *Crusty*).

The variable **change** has missing values because the biologist couldn't get follow-up

Table 11.2 Dummy coding for the Glastonbury Festival data

	Dummy Variable 1	Dummy Variable 2	Dummy Variable 3
No Affiliation	0	0	0
Indie Kid	0	0	1
Metaller	0	1	0
Crusty	1	0	0

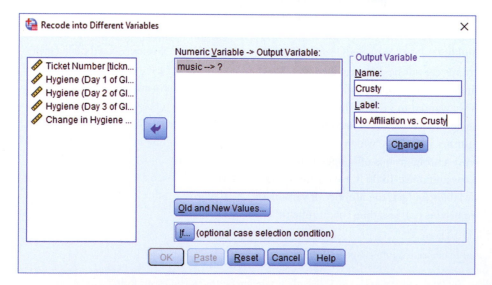

Figure 11.13 The *Recode* dialog box

measures for everyone on day 3. If we recode the **music** variable, we will include all cases (including those for which we have missing values on the variable **change**). You may not care about this, but if you do you can set a 'do if' condition along the lines of 'if there is a value for the variable **change** then recode the variable **music**', by clicking If... to access the dialog box in Figure 11.14. To set a condition that excludes cases for which the variable **change** has a missing value, select ⦿ Include if case satisfies condition:. Click 1 and then · (or type '1–' in the command area). In the box labeled *Function group* select *Missing Values* and in *Functions and Special Variables* select *Missing*, and click ⬆ to transfer the command into the command area. The command will appear as MISSING(?). Drag the variable **change** across to replace the question mark (or delete it and type *change*). The completed dialog box should look like Figure 11.14.

MISSING() returns 'true' (i.e., the value 1) for a case that has a system-missing or user-defined missing value for the specified variable; it returns 'false' (i.e., the value 0) if a case has a value. Hence, *MISSING(change)* returns a value of 1 for cases that have a missing value for the variable **change** and 0 for cases that do have values. By specifying *1–MISSING(change)* we reverse the command so that it returns 1 (true) for cases that have a value for the variable **change** and 0 (false) for system-missing or user-defined missing values. Therefore, this command says 'Do the following *recode* commands if the case has a value for the variable **change**'. If you don't have missing values you can skip this step. Click Continue to return to the main dialog box.

Now it's time to specify *how* to recode the values of the variable **music** into the values that we want for the new variable, **Crusty**. Click Old and New Values... to access the dialog box in Figure 11.15, which we'll use to recode values of the original variable into different values for the new variable.

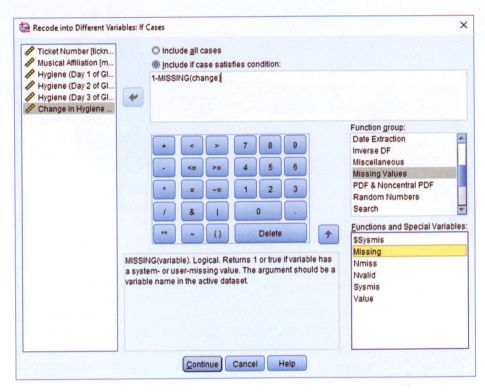

Figure 11.14 Setting an *if* command to include cases if the variable **change** is not a missing value

Try creating the remaining two dummy variables (call them **Metaller** and **Indie_Kid**) using the same principles.

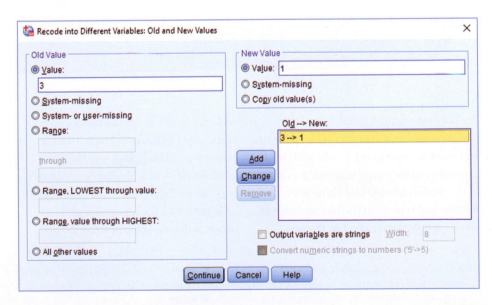

Figure 11.15 Dialog boxes for the *Recode* function

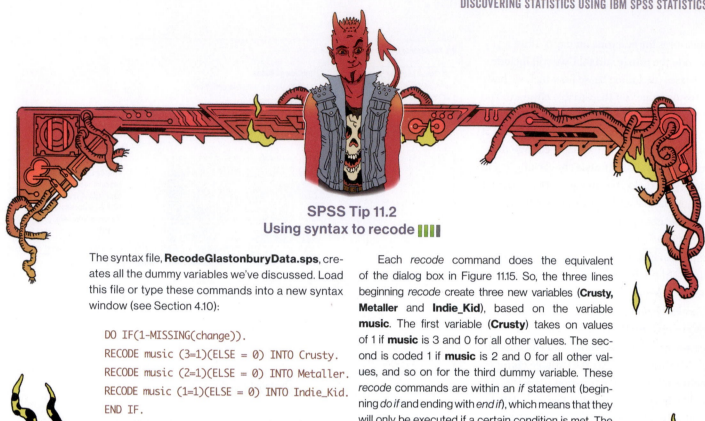

SPSS Tip 11.2
Using syntax to recode ▮▮▮▮

The syntax file, **RecodeGlastonburyData.sps**, creates all the dummy variables we've discussed. Load this file or type these commands into a new syntax window (see Section 4.10):

```
DO IF(1-MISSING(change)).
RECODE music (3=1)(ELSE = 0) INTO Crusty.
RECODE music (2=1)(ELSE = 0) INTO Metaller.
RECODE music (1=1)(ELSE = 0) INTO Indie_Kid.
END IF.
VARIABLE LABELS Crusty 'No Affiliation vs.
Crusty'.
VARIABLE LABELS Metaller 'No Affiliation
vs. Metaller'.
VARIABLE LABELS Indie_Kid 'No Affiliation
vs. Indie Kid'.
VARIABLE LEVEL Crusty Metaller Indie_Kid
(Nominal).
FORMATS Crusty Metaller Indie_Kid (F1.0).
EXECUTE.
```

Each *recode* command does the equivalent of the dialog box in Figure 11.15. So, the three lines beginning *recode* create three new variables (**Crusty, Metaller** and **Indie_Kid**), based on the variable **music**. The first variable (**Crusty**) takes on values of 1 if **music** is 3 and 0 for all other values. The second is coded 1 if **music** is 2 and 0 for all other values, and so on for the third dummy variable. These *recode* commands are within an *if* statement (beginning *do if* and ending with *end if*), which means that they will only be executed if a certain condition is met. The condition we have set is *1 – MISSING(change)*, which is the same as in Figure 11.14 (see main text).

The *variable labels* command assigns the text in the quotations as labels for the variables **Crusty, Metaller**, and **Indie_Kid** respectively. The *variable level* command then sets these three variables to be 'nominal', and the *formats* command changes the variables to have a width of 1 and 0 decimal places (hence the 1.0). The *execute* command executes the commands above (without it nothing works). Note that every line ends with a full stop.

For our first dummy variable, we want anyone who was a crusty to get a code of 1 and everyone else to get a code of 0. In the original variable crusty was coded as 3, so in the section labeled *Old Value* select ◉ Value: and type '3' in the box underneath. We want this value to be 1 in the new variable, so in the section labeled *New Value* select ◉ Value: and type '1' in the empty box. Click [Add] and the box labeled *Old → New* should now include *3 → 1* as in Figure 11.15. The next thing to do is to recode the remaining groups in **music** to have a value of 0 for the first dummy variable. To do this select ◉ All other values,[16] and in the section labeled *New Value* select ◉ Value: and type '0' in the empty box type and click [Add]. The box labeled *Old → New* should now include *ELSE → 0* in the list. Click [Continue] to return to the main dialog box, and click

---

16 Be careful about using ◉ All other values when you have missing values—remember that we set a 'do if', which means that we can use this option safe in the knowledge that missing values won't be recoded. An alternative method is to skip the 'do if' step and recode missing values specifically using the ◉ Range: option. It is a good idea to use the *frequencies* or *crosstabs* commands after a recode to check that you have caught all the missing values.

OK to create the first dummy variable. This variable appears as a new column in the data editor, and it will have a value of 1 for anyone originally classified as a crusty and a value of 0 for everyone else. It can be quicker to recode variables using syntax (see SPSS Tip 11.2).

11.5.3 Output for dummy variables

Let's assume you've created the three dummy coding variables (if you're stuck there is a data file called **GlastonburyDummy.sav** (the 'Dummy' refers to the fact it has dummy variables in it—I'm not implying that if you need to use this file you're a dummy☺). To put these dummy variables into a linear model you must enter them all into the model simultaneously (i.e. in the same block).

Output 11.9 shows the model statistics: by entering the three dummy variables we can explain 7.6% of the variance in the change in hygiene scores. In other words, the musical affiliation of the person explains 7.6% of the variance in the change in hygiene. The F associated with the R^2 change and the model fit assess the same thing when the model has only one block, so these tells us that the model is significantly better at predicting the change in hygiene scores than having no model (put another way, the 7.6% of explained variance is a significant amount).

Output 11.10 shows the *Coefficients* table for the dummy variables. Each dummy variable has a useful label (such as *No Affiliation vs. Crusty*) because I thought ahead and suggested typing useful labels when we created the variables (Figure 11.13); if we hadn't added labels the table would contain the less helpful variable names of *Crusty, Metaller* and *Indie_Kid*. The labels remind us of what each dummy variable represents.

Remember that a *b*-value tells us the change in the outcome due to a unit

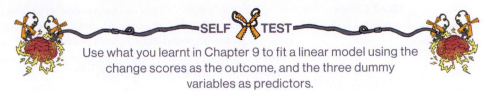

Use what you learnt in Chapter 9 to fit a linear model using the change scores as the outcome, and the three dummy variables as predictors.

Model Summary

Model	R	R Square	Adjusted R Square	Std. Error of the Estimate	R Square Change	F Change	df1	df2	Sig. F Change
					Change Statistics				
1	.276[a]	.076	.053	.68818	.076	3.270	3	119	.024

a. Predictors: (Constant), No Affiliation vs. Indie Kid, No Affiliation vs. Crusty, No Affiliation vs. Metaller

ANOVA[a]

Model		Sum of Squares	df	Mean Square	F	Sig.
1	Regression	4.646	3	1.549	3.270	.024[b]
	Residual	56.358	119	.474		
	Total	61.004	122			

a. Dependent Variable: Change in Hygiene Over The Festival

b. Predictors: (Constant), No Affiliation vs. Indie Kid, No Affiliation vs. Crusty, No Affiliation vs. Metaller

Output 11.9

Coefficients[a]

Model		Unstandardized Coefficients B	Std. Error	Standardized Coefficients Beta	t	Sig.	95.0% Confidence Interval for B Lower Bound	Upper Bound
1	(Constant)	-.554	.090		-6.134	.000	-.733	-.375
	No Affiliation vs. Crusty	-.412	.167	-.232	-2.464	.015	-.742	-.081
	No Affiliation vs. Metaller	.028	.160	.017	.177	.860	-.289	.346
	No Affiliation vs. Indie Kid	-.410	.205	-.185	-2.001	.048	-.816	-.004

a. Dependent Variable: Change in Hygiene Over The Festival

Bootstrap for Coefficients

Model		B	Bootstrap[a] Bias	Std. Error	Sig. (2–tailed)	BCa 95% Confidence Interval Lower	Upper
1	(Constant)	-.554	.005	.097	.001	-.736	-.349
	No Affiliation vs. Crusty	-.412	-.011	.179	.030	-.733	-.101
	No Affiliation vs. Metaller	.028	-.006	.149	.847	-.262	.293
	No Affiliation vs. Indie Kid	-.410	-.010	.201	.049	-.813	-.043

a. Unless otherwise noted, bootstrap results are based on 1000 bootstrap samples

Output 11.10

change in the predictor. For our dummy variables, a unit change in the predictor is the change from 0 to 1. By including all three dummy variables in the model, zero represents the baseline category of no affiliation. For the first dummy variable 1 represents 'crusty' and so the change from 0 to 1 is the change from no affiliation to crusty. As such, the first dummy variable represents the difference in the change in hygiene scores for a crusty relative to someone with no musical affiliation. This

difference is the difference between the two group means (see Section 10.4).

I've produced a table (Output 11.11) of the group means for each of the four groups and also the difference between the mean of each group and the mean of the no affiliation group. For example, the difference in the means of the no affiliation group and the crusty group is (−0.966) − (−0.554) = −0.412. The decrease in hygiene scores of the crusty group (−0.966) is larger in magnitude than for the no

OLAP Cubes

Change in Hygiene Over The Festival

Musical Affiliation	Mean
Indie Kid	−.9643
Metaller	−.5259
Crusty	−.9658
No Musical Affiliation	−.5543
Crusty – No Musical Affiliation	−.4115
Metaller – No Musical Affiliation	.0284
Indie Kid – No Musical Affiliation	−.4100
Total	−.6750

Output 11.11

Table 11.3 Linear model of predictors of the change in hygiene scores. 95% bias corrected and accelerated confidence intervals reported in parentheses. Confidence intervals and standard errors based on 1000 bootstrap samples

	b	$SE\ B$	β	p
Constant	−0.55 (−0.74, −0.35)	0.10		$p = 0.001$
No Affiliation vs. Crusty	−0.41 (−0.73, −0.10)	0.18	−0.23	$p = 0.030$
No Affiliation vs. Metaller	0.03 (−0.26, 0.29)	0.15	0.02	$p = 0.847$
No Affiliation vs. Indie Kid	−0.41 (−0.81, −0.04)	0.20	−0.19	$p = 0.049$

Note. $R^2 = 0.08$ ($p = 0.024$).

affiliation group (−0.554), showing that crusties' hygiene decreases more over the festival than that of those with no musical affiliation. The difference between these two group means (−0.412) is the *unstandardized* beta in Output 11.10. This example demonstrates that b-values for dummy variables tell us the difference in the mean of a particular group and the group that we chose as a baseline category.

As with any linear model, the b-value has an associated t-statistic and p-value that tests whether it is significantly different from 0. For these dummy variables it is, therefore, testing whether the difference between group means is significantly different from 0. For our first dummy variable, the t-test is significant and the beta value is negative, so the change in hygiene scores decreases as a person changes from having no affiliation to being a crusty. In other words, hygiene decreased significantly more in crusties compared to those with no musical affiliation.

Our next dummy variable compares metallers to those who have no musical affiliation. The b-value (0.028 in Output 11.10) is the difference in the group means for the no affiliation group and the metaller group: (−0.526) − (−0.554) = 0.028. The t-test is not

significant, which we could take to mean that the change in hygiene scores across the festival is similar in metallers to those with no affiliation.

The final dummy variable compares indie kids to those who have no musical affiliation. The b-value (−0.410 in Output 11.10) is the difference in the group means for the no affiliation group and the indie kid group: (−0.964) − (−0.554) = −0.410. The t-test is significant and the beta value has a negative value, so, as with the first dummy variable, we could say that the change in hygiene scores goes down as a person changes from having no affiliation to being an indie kid. In other words, hygiene decreased significantly more in indie kids than in those with no musical affiliation. We could report the results as in Table 11.3 (note that I've included the bootstrap confidence intervals). Overall the model shows that, compared to having no musical affiliation, crusties and indie kids get significantly smellier across the three days of the festival, but metallers don't.

11.6 Brian and Jane's Story ▮▮▮

Jane didn't like the library: it was full of people, and people made her uneasy. Why had she gone there to find campus guy? It

made no sense to her, but as days passed without seeing him, she missed his nervy recitals of his statistics lectures. She'd watched from afar as he left his lectures and scuttled to the library. Was he avoiding her? The thought that he might be made her want to see him. She wanted to make amends for her barbed comments and yet when she found him, and he spat statistical theory at her like one of Pavlov's dogs, she cut him down again. Why did she do that? Why wouldn't she give him a millimeter? Had she spent too much time alone in the basement of the Pleiades building? Was she so broken that she needed to torture this guy with an elaborate rite of passage to make herself feel better about her past? Jane had realized as she'd looked at his deflated face in the library that he would reach a limit with her quips. She'd wanted to keep him at arm's length, but it was only now that she realized how long her emotional arms could be. If she wanted their chats to continue she needed to give him some hope.

11.7 What next? ▮▮▮

We started this chapter by looking at my relative failures as a human being compared to Simon Hudson. I then bleated on excitedly about moderation and mediation, which could explain why Clair Sparks chose Simon Hudson all those years ago. Perhaps she could see the writing on the wall. I was true to my word to my parents though: I *was* philosophical about it. I set my sights on a girl called Zoë during the obligatory lunchtime game of kiss chase (not the same Zoë whom I ended up marrying). I don't think she was all that keen, which was just as well because I was about to be dragged out of her life forever. Not that I believe in these things, but if I did I would have believed that the scaly, warty, green, long-nailed hand of fate (I have always assumed that anything named as ominously as the hand of fate would need to look monstrous) had decided that I was far too young to be getting distracted by girls. Waggling its finger at me, it plucked me out of primary school and cast me into the flaming pit of hell, otherwise known as all-boys' grammar school. It's fair to say that my lunchtime primary school kiss chase games were the last I would see of girls for quite some time . . .

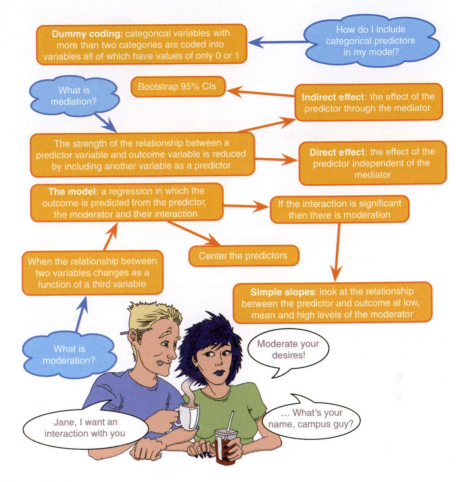

Figure 11.16   What Brian learnt from this chapter

11.8 Key terms that I've discovered

Direct effect	Index of mediation	Mediation	Moderator
Dummy variables	Indirect effect	Mediator	Simple slopes analysis
Grand mean centering	Interaction effect	Moderation	Sobel test

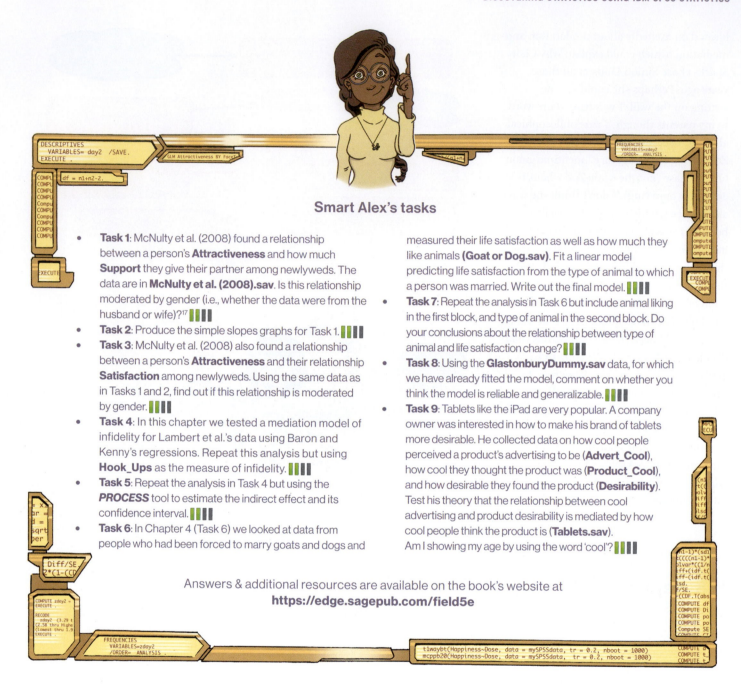

Smart Alex's tasks

- **Task 1**: McNulty et al. (2008) found a relationship between a person's **Attractiveness** and how much **Support** they give their partner among newlyweds. The data are in **McNulty et al. (2008).sav**. Is this relationship moderated by gender (i.e., whether the data were from the husband or wife)?[17] ▮▮▮

- **Task 2**: Produce the simple slopes graphs for Task 1. ▮▮▮

- **Task 3**: McNulty et al. (2008) also found a relationship between a person's **Attractiveness** and their relationship **Satisfaction** among newlyweds. Using the same data as in Tasks 1 and 2, find out if this relationship is moderated by gender. ▮▮▮

- **Task 4**: In this chapter we tested a mediation model of infidelity for Lambert et al.'s data using Baron and Kenny's regressions. Repeat this analysis but using **Hook_Ups** as the measure of infidelity. ▮▮▮

- **Task 5**: Repeat the analysis in Task 4 but using the **PROCESS** tool to estimate the indirect effect and its confidence interval. ▮▮▮

- **Task 6**: In Chapter 4 (Task 6) we looked at data from people who had been forced to marry goats and dogs and measured their life satisfaction as well as how much they like animals **(Goat or Dog.sav)**. Fit a linear model predicting life satisfaction from the type of animal to which a person was married. Write out the final model. ▮▮▮

- **Task 7**: Repeat the analysis in Task 6 but include animal liking in the first block, and type of animal in the second block. Do your conclusions about the relationship between type of animal and life satisfaction change? ▮▮▮

- **Task 8**: Using the **GlastonburyDummy.sav** data, for which we have already fitted the model, comment on whether you think the model is reliable and generalizable. ▮▮▮

- **Task 9**: Tablets like the iPad are very popular. A company owner was interested in how to make his brand of tablets more desirable. He collected data on how cool people perceived a product's advertising to be **(Advert_Cool)**, how cool they thought the product was **(Product_Cool)**, and how desirable they found the product **(Desirability)**. Test his theory that the relationship between cool advertising and product desirability is mediated by how cool people think the product is **(Tablets.sav)**. Am I showing my age by using the word 'cool'? ▮▮▮

Answers & additional resources are available on the book's website at
https://edge.sagepub.com/field5e

---

17 These are not the actual data from the study, but are simulated to mimic the findings in Table 1 of the original paper.

GLM 1: COMPARING SEVERAL INDEPENDENT MEANS

12

12.1 What will this chapter tell me?

There are pivotal moments in everyone's life, and one of mine was at the age of 11. Where I grew up in England there were three choices when leaving primary school and moving on to secondary school: (1) state school (where most people go); (2) grammar school (where clever people who pass an exam called the Eleven Plus go); and (3) private school (where rich people go). My parents were not rich and I am not clever, and consequently I failed my Eleven Plus, so private school and grammar school (where my older brother, 'the clever one', had gone) were out. There was no choice but for me to join my friends at the local state school. I could not have been happier: at age 11 I hadn't yet become a stats hermit and liked having friends. Imagine everyone's shock when my parents received a letter saying that some extra spaces had become available at the grammar school; although the local authority could scarcely believe it and had checked the Eleven Plus papers several million times to confirm their findings, I was next on their list. I could not have been unhappier. So, I waved goodbye to all my friends and trundled off to join my brother at Ilford County High School for Boys (a school at which the head teacher hit students with a cane if they were particularly bad and that, for some considerable time and with good reason, had 'H.M. Prison' painted in huge white letters on its roof). It was goodbye to normality, and hello to seven years of learning how not to function in society. I often wonder how my life would have turned out had I not gone to this school. In the parallel universes where the letter didn't arrive and the parallel Andy went to state school or where my parents were rich and he went to private school, what became of him? If we wanted to compare these three situations we couldn't use the methods in Chapter 10 because there are more than two conditions.[1] However, this chapter tells us all about the statistical models that we use to analyze situations in which we want to compare more than two independent means.

The model is typically called analysis of variance (or ANOVA to its friends) but, as we shall see, it is just a variant on the linear model. So, in effect, we're going to learn stuff that we already know from previous chapters. Hopefully that's reassuring.

Figure 12.1 My brother Paul (left) and I (right) in our very fetching school uniforms (note the fear in my face)

---

1 This is the least of our problems—there's also the small issue of reinventing physics to access the parallel universes.

12.2 Using a linear model to compare several means ▌▌▌

We saw in Chapter 10 that if we include a predictor variable containing two categories into the linear model then the resulting *b* for that predictor compares the difference between the mean score for the two categories. We also saw in Chapter 11 that if we want to include a categorical predictor that contains more than two categories, this can be achieved by recoding that variable into several categorical predictors each of which has only two categories (dummy coding).

When we do, the *b*s for predictors represent differences between means. Therefore, if we're interested in comparing more than two means we can use the linear model to do this. Remembering from Chapter 9 that we test the overall fit of a linear model with an *F*-statistic, we can do the same here: we first use an *F* to test whether we significantly predict the outcome variable by using group means (which tells us whether, overall, the group means are significantly different) and then use the specific model parameters (the *b*s) to tell us which means differ from which. It's not uncommon for researchers to think, and for people to be taught, that

you compare means with 'ANOVA' and that this is somehow different from 'regression' (i.e. the linear model), which you apply to look for relationships between variables. This artificial division is unhelpful (in my view) and exists largely for weird historical reasons (Misconception Mutt 12.1). The 'ANOVA' to which some people allude is simply the *F*-statistic that we encountered as a test of the fit of a linear model, it's just that the linear model consists of group means. This chapter will develop what we discovered in Chapters 10 and 11 about using dummy variables in the linear model to compare means.

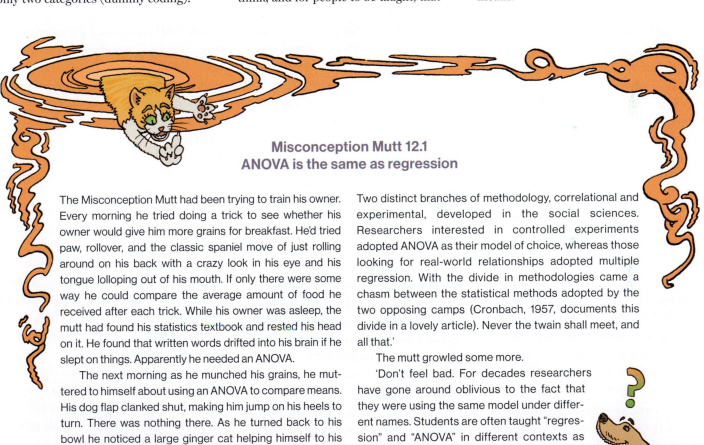

Misconception Mutt 12.1
ANOVA is the same as regression

The Misconception Mutt had been trying to train his owner. Every morning he tried doing a trick to see whether his owner would give him more grains for breakfast. He'd tried paw, rollover, and the classic spaniel move of just rolling around on his back with a crazy look in his eye and his tongue lolloping out of his mouth. If only there were some way he could compare the average amount of food he received after each trick. While his owner was asleep, the mutt had found his statistics textbook and rested his head on it. He found that written words drifted into his brain if he slept on things. Apparently he needed an ANOVA.

The next morning as he munched his grains, he muttered to himself about using an ANOVA to compare means. His dog flap clanked shut, making him jump on his heels to turn. There was nothing there. As he turned back to his bowl he noticed a large ginger cat helping himself to his food. The mutt growled.

'You know,' said the cat, 'ANOVA is the same thing as the linear model or regression.'

The growling continued.

'The reason why some people think of ANOVA and regression as separate statistical models is historical.

Two distinct branches of methodology, correlational and experimental, developed in the social sciences. Researchers interested in controlled experiments adopted ANOVA as their model of choice, whereas those looking for real-world relationships adopted multiple regression. With the divide in methodologies came a chasm between the statistical methods adopted by the two opposing camps (Cronbach, 1957, documents this divide in a lovely article). Never the twain shall meet, and all that.'

The mutt growled some more.

'Don't feel bad. For decades researchers have gone around oblivious to the fact that they were using the same model under different names. Students are often taught "regression" and "ANOVA" in different contexts as though they are different. But now, my canine friend, you know better. Nice grains, by the way.'

With that, the cat finished the last of the dog's food and darted out of the dog flap, leaving a ginger vapor trail.

Figure 12.2 Some puppy therapy for you in the form of my dog, Ramsey

Table 12.1 Data in **Puppies.sav**

	Control	15 minutes	30 minutes
	3	5	7
	2	2	4
	1	4	5
	1	2	3
	4	3	6
$\bar{x}$	**2.20**	**3.20**	**5.00**
s	**1.30**	**1.30**	**1.58**
s^2	**1.70**	**1.70**	**2.50**

Grand mean = **3.467** Grand SD = **1.767**

Grand variance = **3.124**

As a quick side note, there is a different way of teaching the use of the F-statistic to compare means known as the variance-ratio method. This approach is fine for simple designs, but becomes impossibly cumbersome in more complex situations such as analysis of covariance or when you have unequal sample sizes.[2] The linear model framework has various advantages. First, we're building on material that you have already learnt (this chapter is a natural progression from the bits of the book that you've hopefully already read). Second, the linear model extends very logically to the more complex situations (e.g., multiple predictors, unequal group sizes) without the need to get bogged down in mathematics. And third, IBM SPSS Statistics uses the linear model framework (known as the **general linear model**) for comparing means (on the whole).

Let's start with an example. You're about halfway through the book now, and there are a lot of equations in this chapter, so we probably need some puppy therapy. Puppy therapy is a form of animal-assisted therapy, in which puppy contact is introduced into the therapeutic process. Puppy rooms have been set up to de-stress students and staff at my own university (Sussex) in the UK along with universities in Bristol, Nottingham, Aberdeen and Lancaster. I've heard of similar things at Dalhousie and Simon Fraser in Canada and Tufts and Caldwell in the USA. My own contribution is to sometimes bring my adorable spaniel, Ramsey (Figure 12.2), into work to sit in my office and look cute at any students who break down in tears about some stats issue they have. He may pop up at strategic points in this chapter to help your mental state.

Despite this increase in puppies on campuses (which can only be a good thing) to reduce stress, the evidence base is pretty mixed. A review of animal-assisted therapy in childhood mental health found that of 24 studies, 8 found positive effects of animal-assisted therapy, 10 showed mixed findings, and 6 concluded that there was no effect (Hoagwood, Acri, Morrissey, & Peth-Pierce, 2017). Imagine we wanted to contribute to this literature by running a study in which we randomized people into three groups: (1) a control group (this could be a treatment as usual, a no treatment or ideally some kind of placebo group—for example, if our hypothesis was specifically about puppies we could give people in this group a cat disguised as a dog); (2) 15 minutes of puppy therapy (a low-dose group); and (3) 30 minutes of puppy contact (a high-dose group). The dependent variable was a measure of happiness ranging from 0 (as unhappy as I can possibly imagine being) to 10 (as happy as I can possibly imagine being). The design of this study mimics a very simple randomized controlled trial (as used in pharmacological, medical and psychological intervention trials) because people are randomized into a control group or groups containing the active intervention (in this case puppies, but in other cases a drug or a surgical procedure). We'd predict that any form of puppy therapy should be better than the control (i.e. higher happiness scores) but also formulate a dose-response hypothesis that as exposure time increases (from 0 minutes to 15 and 30) happiness will increase too. The data are in Table 12.1 and the file **Puppies.sav**.

If we want to predict happiness from group membership we can use the general equation that keeps popping up:

$$outcome_i = (model) + error_i \qquad (12.1)$$

---

2 Having said this, it is well worth the effort of trying to obtain equal sample sizes in your different conditions because unbalanced designs do cause statistical complications (see Section 12.3).

We've seen that with two groups we can use a linear model, by replacing the 'model' in equation (12.1) with one dummy variable that codes two groups (0 for one group and 1 for the other) and an associated *b*-value that would represent the difference between the group means (Section 10.4). We have three groups here, but we've also seen that this situation is easily incorporated into the linear model by including two dummy variables (each assigned a *b*-value), and that any number of groups can be included by extending the number of dummy variables to one less than the number of groups (Section 11.5).

We've also learnt already that when we use dummy variables we assign one group as the baseline category and assign that group a zero code on all dummy variables (remember in Section 11.5 that we chose the 'no musical affiliation' condition as a baseline). The baseline category should be the condition against which you intend to compare the other groups. In most well-designed experiments there will be a group of participants who act as a control for the other groups and, other things being equal, this will be your baseline category—although the group you choose will depend upon the particular hypotheses you want to test. In designs in which the group sizes are unequal it is important that the baseline category contains a large number of cases to ensure that the estimates of the *b*-values are reliable. In the puppy therapy example, we can take the control group (who received no puppy therapy) as the baseline category because we are interested in comparing both the 15- and 30-minute groups to this group. If the control group is the baseline category then the two dummy variables need to represent the other two conditions: so, let's call one of them **Long** (30-minute dose) and the other **Short** (15 minutes) to reflect the length of the dose of

Table 12.2 Dummy coding for the three-group experimental design

Group	Dummy variable 1 (Long)	Dummy variable 2 (Short)
Control	0	0
15 minutes of puppy therapy	0	1
30 minutes of puppy therapy	1	0

puppies. Putting these dummy variables into the model as predictors gives:

$$\text{Happiness}_i = b_0 + b_1 \text{Long}_i + b_2 \text{Short}_i + \varepsilon_i \quad (12.2)$$

in which a person's happiness is predicted from knowing their group code (i.e., the numeric code for the **Long** and **Short** dummy variables) and the intercept (b_0) of the model. The dummy variables can be coded in several ways, but the simplest way is to use dummy coding (Section 11.5). The baseline category is coded as 0 for all dummy variables. If a participant received 30 minutes of puppy therapy then they are coded with a 1 for the **Long** dummy variable and 0 for **Short**. If a participant received 15 minutes of puppy therapy then they are coded with the value 1 for the **Short** dummy variable and coded with 0 for **Long**. Using this coding scheme, each group is uniquely expressed by the combined values for the two dummy variables (see Table 12.2). When we are predicting an outcome from group membership, predicted values from the model (the value of happiness in equation (12.2)) are the group means. This is illustrated in Figure 12.3, which splits the data by group membership. If we are trying to predict the happiness of a new person and we know to which group they have been assigned (but we don't have their score yet) then our best guess will be the group mean because, on average, we'll be correct. For example, if we know that someone is going to receive 30 minutes of puppy therapy and we want to predict their happiness, our best guess will be 5 because we know that, on average, people who have 30 minutes with

a puppy rate their happiness as 5. If the group means are meaningfully different, then using the group means should be an effective way to predict scores (because we can successfully differentiate people's predicted happiness based on how much puppy time they received). We'll return to this point in the next section.

Let's first examine the model for the *control group*. Both the **Long** and **Short** dummy variables are coded 0 for people in the control group. Therefore, if we ignore the error term (ε_i), the model becomes:

$$\begin{aligned} \text{Happiness}_i &= b_0 + (b_1 \times 0) + (b_2 \times 0) \\ &= b_0 \\ \overline{X}_{\text{Control}} &= b_0 \end{aligned} \quad (12.3)$$

The 15- and 30-minute groups have dropped out of the model (because they are coded 0) and we're left with b_0. As we have just discovered, the predicted value of happiness will be the mean of the control group ($\overline{X}_{\text{control}}$), so we can replace **Happiness** with this value. This leaves us with the epiphany that b_0 in the model is always the mean of the baseline category. For someone in the *30-minute group*, the value of the dummy variable **Long** will be 1 and the value for **Short** will be 0. By replacing these values in equation (12.2) the model becomes:

$$\begin{aligned} \text{Happiness}_i &= b_0 + (b_1 \times 1) + (b_2 \times 0) \\ &= b_0 + b_1 \end{aligned} \quad (12.4)$$

which tells us that predicted happiness for someone in the 30-minute group is the sum of b_0 and the *b* for the dummy variable **Long** (b_1). We know already that b_0 is the mean of the control group ($\overline{X}_{\text{control}}$) and

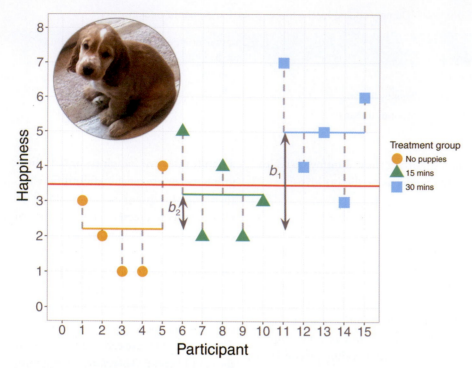

Figure 12.3 The puppy therapy data. The colored horizontal lines represent the mean happiness of each group. The shapes represent the happiness of individual participants (different shapes indicate different experimental groups). The red horizontal line is the average happiness of all participants. The puppy is my dog Ramsey, as a pup. You should never needlessly apply puppies to your graphs ... unless they're Ramsey, in which case it's fine

SELF TEST

To illustrate what is going on I have created a file called **Puppies Dummy.sav** that contains the puppy therapy data along with the two dummy variables (**dummy1** and **dummy2**) we've just discussed (Table 10.2). Fit a linear model predicting happiness from **dummy1** and **dummy2**. If you're stuck, read Chapter 9 again.

that the predicted value of **Happiness** for someone in the 30-minute group is the mean of that group ($\bar{X}_{30\,mins}$). Therefore, we can replace b_0 with $\bar{X}_{control}$ and **Happiness** with $\bar{X}_{30\,mins}$. The result is:

$$\begin{aligned} Happiness_i &= b_0 + b_1 \\ \bar{X}_{30\,mins} &= \bar{X}_{Control} + b_1 \\ b_1 &= \bar{X}_{30\,mins} - \bar{X}_{Control} \end{aligned} \quad (12.5)$$

which shows that the *b*-value for the dummy variable representing the 30-minute group is the difference between the means of that group and the control.

Finally, for someone assigned to the *15-minute group*, the dummy variable **Short** has a value of 1 and the dummy variable **Long** is 0. By replacing these values in equation (12.2) the model becomes:

$$\begin{aligned} Happiness_i &= b_0 + (b_1 \times 0) + (b_2 \times 1) \\ &= b_0 + b_2 \end{aligned} \quad (12.6)$$

which tells us that predicted happiness for someone in the 15-minute group is the sum of b_0 and the *b* for the dummy variable **Short** (b_2). Again, we can replace

b_0 with $\bar{X}_{control}$. The predicted value of **Happiness** for someone in the 15-minute group is the mean of that group so we can replace Happiness with $\bar{X}_{15\,mins}$. The result is:

$$\begin{aligned} Happiness_i &= b_0 + b_2 \\ \bar{X}_{15\,mins} &= \bar{X}_{Control} + b_2 \\ b_2 &= \bar{X}_{15\,mins} - \bar{X}_{Control} \end{aligned} \quad (12.7)$$

which shows that the *b*-value for the dummy variable representing the 15-minute (**Short**) group is the difference between means for the 15-minute group and the control.

Using dummy coding is only one of many ways to code dummy variables. We shall see later in this chapter (Section 12.4.2) that an alternative is contrast coding, in which you code the dummy variables in such a way that the *b*-values represent differences between groups that you specifically hypothesized before collecting data.

Output 12.1 shows the model from the self-test. Quickly remind yourself of the group means from Table 12.1. The overall fit of the model has been tested with an *F*-statistic (i.e., ANOVA), which is significant, $F(2, 12) = 5.12$, $p = 0.025$. Given that our model represents the group means, this *F* tells us that using group means to predict happiness scores is significantly better than using the mean of all scores: in other words, the group means are significantly different.

The *F*-test is an overall test that doesn't identify differences between specific means. However, the model parameters (the *b*-values) do. As we just discovered, the constant (b_0) is equal to the mean of the base category (the control group), 2.2. The *b*-value for the first dummy variable (b_1) is equal to the difference between the means of the 30-minute group and the control group ($5.0 - 2.2 = 2.8$). Finally, the *b*-value for the second dummy variable (b_2) is equal to the difference between the means of the 15-minute group and the control group (3.2 − 2.2 = 1). This demonstrates what we saw in equations (12.3), (12.5) and (12.7). We can see from the significance values of the associated t-tests that the difference between the 30-minute group and the control group (b_1) is significant because $p = 0.008$, which is less than 0.05; however, the difference between the 15-minute and the control group is not ($p = 0.282$).

We can extend this three-group scenario to four groups (see Section 11.5 for an example). As before, we specify one category as a base category (a control group) and assign this category a code of 0 for all dummy variables. The remaining three conditions will have a code of 1 for the dummy variable that describes that condition and a code of 0 for the other dummy variables (Table 12.3).

ANOVA[a]

Model		Sum of Squares	df	Mean Square	F	Sig.
1	Regression	20.133	2	10.067	5.119	.025[b]
	Residual	23.600	12	1.967		
	Total	43.733	14			

a. Dependent Variable: Happiness (0–10)

b. Predictors: (Constant), Dummy 2: 15 mins vs. Control, Dummy 1: 30 mins vs. Control

Coefficients[a]

Model		Unstandardized Coefficients B	Unstandardized Coefficients Std. Error	Standardized Coefficients Beta	t	Sig.
1	(Constant)	2.200	.627		3.508	.004
	Dummy 1: 30 mins vs. Control	2.800	.887	.773	3.157	.008
	Dummy 2: 15 mins vs. Control	1.000	.887	.276	1.127	.282

a. Dependent Variable: Happiness (0–10)

Output 12.1

Table 12.3 Dummy coding for the four-group experimental design

	Dummy variable 1	Dummy variable 2	Dummy variable 3
Group 1	1	0	0
Group 2	0	1	0
Group 3	0	0	1
Group 4 (base)	0	0	0

12.2.1 Logic of the F-statistic ▌▌▌▌

We learnt in Chapter 9 that the F-statistic (or F-ratio as it's also known) tests the overall fit of a linear model to a set of observed data. F is the ratio of how good the model is compared to how bad it is (its error). When the model is based on group means, our predictions from the model are those means. If the group means are the same then our ability to predict the observed data will be poor (F will be small), but if the means differ we will be able to better discriminate between cases from different groups (F will be large). So, in this context F basically tells us whether the group means are significantly different. Let me elaborate.

Figure 12.3 shows the puppy therapy data including the group means, the overall mean and the difference between each case and the group mean. We want to test the hypothesis that the means of three groups are different (so the null hypothesis is that the group means are the same). If the group means were all the same, then

we would not expect the control group to differ from the 15- or 30-minute groups, and we would not expect the 15-minute group to differ from the 30-minute group. In this situation, the three colored horizontal lines representing the group means in Figure 12.3 would be in the same vertical position (the exact position would be the grand mean—the red horizontal line in the figure). This is not the case in the figure: the colored lines are in different vertical positions, showing that the group means are different. We have just found out that in the model, b_1 represents the difference between the control and 30-minute group means, and b_2 represents the difference between the 15-minute and control group means. These two distances are represented in Figure 12.3 by the vertical arrows. If the null hypothesis is true and all the groups

have the same mean, then these b coefficients should be zero (because if the group means are equal then the difference between them will be zero). We can apply the same logic as for any linear model:

- The model that represents 'no effect' or 'no relationship between the predictor variable and the outcome' is one where the predicted value of the outcome is always the grand mean (the mean of the outcome variable).
- We can fit a different model to the data that represents our alternative hypotheses. We compare the fit of this model to the fit of the null model (i.e., using the grand mean).
- The intercept and one or more parameters (b) describe the model.
- The parameters determine the shape of the model that we have fitted; therefore, the bigger the coefficients, the

greater the deviation between the model and the null model (grand mean).

- In experimental research the parameters (*b*) represent the differences between group means. The bigger the differences between group means, the greater the difference between the model and the null model (grand mean).
- If the differences between group means are large enough, then the resulting model will be a better fit to the data than the null model (grand mean).
- If this is the case we can infer that our model (i.e., predicting scores from the group means) is better than not using a model (i.e., predicting scores from the grand mean). Put another way, our group means are significantly different from the null (that all means are the same).

We use the *F*-statistic to compare the improvement in fit due to using the model (rather than the null or grand mean, model) to the error that still remains. In other words, the *F*-statistic is the ratio of the explained to the unexplained variation. We calculate this variation using sums of squares (look back at Section 9.2.4 to refresh your memory), which might sound complicated, but isn't as bad as you think (see Jane Superbrain Box 12.1).

12.2.2 Total sum of squares (SS_T) ▮▮▮

To find the total amount of variation within our data we calculate the difference between each observed data point and the grand mean. We square these differences

and add them to give us the total sum of squares (SS_T):

$$SS_T = \sum_{i=1}^{N} \left(x_i - \bar{x}_{grand} \right)^2 \qquad (12.9)$$

The variance and the sums of squares are related such that variance, $s^2 = SS/(N-1)$, where N is the number of observations (Section 2.5.1). Therefore, we can calculate the total sum of squares from the variance of all observations (the **grand variance**) by rearranging the relationship ($SS = s^2(N-1)$). The grand variance is the variation between all scores, regardless of the group from which the scores come. Figure 12.4 shows the different sums of squares graphically (note the similarity to

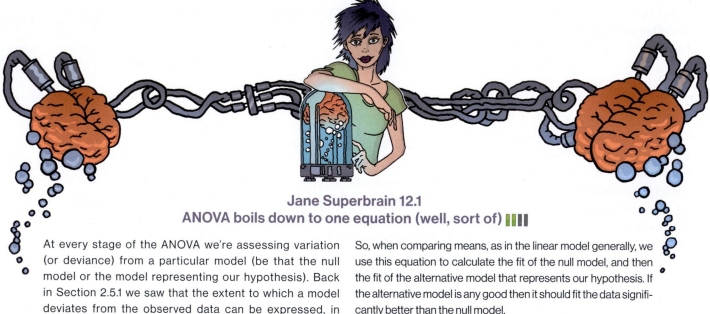

Jane Superbrain 12.1
ANOVA boils down to one equation (well, sort of) ▮▮▮

At every stage of the ANOVA we're assessing variation (or deviance) from a particular model (be that the null model or the model representing our hypothesis). Back in Section 2.5.1 we saw that the extent to which a model deviates from the observed data can be expressed, in general, in the form of equation (2.11), which I repeat here as equation (12.8):

$$\text{Total error} = \sum_{i=1}^{n} \left(\text{observed}_i - \text{model}_i \right)^2 \qquad (12.8)$$

So, when comparing means, as in the linear model generally, we use this equation to calculate the fit of the null model, and then the fit of the alternative model that represents our hypothesis. If the alternative model is any good then it should fit the data significantly better than the null model.

All of the sums of squares we look at in this chapter are variations on equation (12.8): all that changes is what we use as the model and observed data. As you read about the sums of squares, think back to equation (12.8) to remind yourself that the equations are just variants of looking at the difference between observed values and values predicted by a model.

Figure 9.5 which we looked at when we learnt about the linear model). The top left panel shows the total sum of squares: it is the sum of the squared distances between each point and the solid horizontal line (which represents the mean of all scores).

The grand variance for the puppy therapy data is given in Table 12.1, and there were 15 scores in all. Therefore, SS_T is 43.74:

$$
\begin{aligned}
SS_T &= s_{grand}^2 (N-1) \\
&= 3.124(15-1) \\
&= 3.124 \times 14 \\
&= 43.74
\end{aligned}
\tag{12.10}
$$

Before we move on, have a look back at Jane Superbrain Box 2.2 to refresh your memory on degrees of freedom. When we estimate population values, the degrees of freedom are typically one less than the number of scores used to calculate the estimate. This is because to get the estimates we hold something constant in the population (e.g., to get the variance we hold the mean constant), which leaves all but one of the scores free to vary. For SS_T, we used the entire sample (i.e., 15 scores) to calculate the sums of squares and so the total degrees of freedom (df_T) are one less than the total sample size ($N-1$). For the puppy therapy data, this value is 14.

12.2.3 Model sum of squares (SS_M) ▌▌▌▌

So far, we know that the total amount of variation within the outcome is 43.74 units. We now need to know how much of this variation the model can explain. Because our model predicts the outcome from the means of our treatment (puppy therapy) groups, the model sums of squares tell us how much of the total variation in the outcome can be explained by the fact that different scores come from entities in different treatment conditions.

The model sum of squares is calculated by taking the difference between the values predicted by the model and the grand mean (see Section 9.2.4, Figure 9.5). When making predictions from group

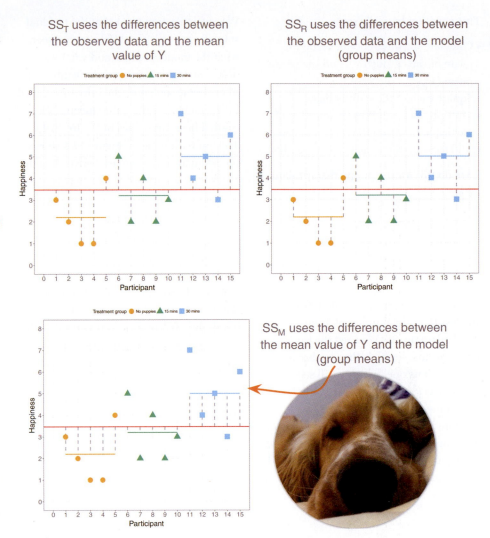

membership, the values predicted by the model are the group means (the colored horizontal lines in Figure 12.4). The bottom panel in Figure 12.4 shows the model sum of squared error: it is the sum of the squared distances between what the model predicts for each data point—the mean of the group to which the score belongs represented by the orange (control), green (15 mins) and blue (30 mins) horizontal lines—and the overall mean of the outcome (the red horizontal line). For example, the predicted value for the five participants in the control group (circles) is 2.2, for the participants (triangles) in the 15-minute

condition is 3.2, and for the participants (squares) in the 30-minute condition is 5. The model sum of squares requires us to calculate the differences between each participant's predicted value and the grand mean. These differences are squared and added together (for reasons I've explained before). Given that the predicted value for participants in a group is the same value (the group mean), the easiest way to calculate SS_M is by using:

$$
SS_M = \sum_{g=1}^{k} n_g \left(\bar{x}_g - \bar{x}_{grand} \right)^2
\tag{12.11}
$$

Figure 12.4 Graphical representation of the different sums of squares when comparing several means using a linear model. Also a picture of Ramsey as a puppy. Tufte would call him chartjunk, but I call him my adorable, crazy, spaniel

This equation basically says:

- Calculate the difference between the mean of each group ($\bar{x}_g$) and the grand mean ($\bar{x}_{grand}$).
- Square each of these differences.
- Multiply each result by the number of participants within that group (n_g).
- Add the values for each group together.

If we do this for the puppy therapy data, we get:

$$\begin{aligned} SS_M &= 5(2.200-3.467)^2 + 5(3.200-3.467)^2 + 5(5.000-3.467)^2 \\ &= 5(-1.267)^2 + 5(-0.267)^2 + 5(1.533)^2 \\ &= 8.025 + 0.355 + 11.755 \\ &= 20.135 \end{aligned} \qquad (12.12)$$

For SS_M, the degrees of freedom (df_M) are one less than the number of 'things' used to calculate the SS. We used the three group means, so df_M is the number of groups minus one (which you'll see denoted as $k-1$). So, in this example, we have three groups and the degrees of freedom are 2 (because the calculation of the sums of squares is based on the group means, two of which will be free to vary in the population if the third is held constant).

12.2.4 Residual sum of squares (SS_R) ▮▮▮▮

We now know that there are 43.74 units of variation to be explained in our outcome variable, and that our model explains 20.14 of them (nearly half). The residual sum of squares (SS_R) tells us how much of the variation *cannot* be explained by the model. This value is the amount of variation created by things that we haven't measured such as measurement error and individual differences in things that might affect happiness. The simplest way to calculate SS_R is to subtract SS_M from SS_T ($SS_R = SS_T - SS_M$), but this provides little insight into what SS_R represents and, of course, if you've messed up the calculations of either SS_M or SS_T (or both!) then SS_R will be incorrect also.

We saw in Section 9.2.4 that the residual sum of squares is the difference between what the model predicts and what was observed. When using group membership to predict an outcome the values predicted by the model are the group means (the colored horizontal lines in Figure 12.4). The top right panel of Figure 12.4 shows the residual sum of squared error: it is the sum of the squared distances between each point and the horizontal line for the group to which the score belongs.

We already know that for a given participant, the model predicts the mean of the group to which that person belongs. Therefore, SS_R is calculated by looking at the difference between the score obtained by a person and the mean of the group to which the person belongs. In graphical terms the dashed vertical lines in Figure 12.3 represent this sum of squares. These distances between each data point and the group mean are squared and added together to give the residual sum of squares, SS_R:

$$SS_R = \sum_{g=1}^{k} \sum_{i=1}^{n} (x_{ig} - \bar{x}_g)^2 \qquad (12.13)$$

Equation (12.13) says that the sum of squares for each group is the squared difference between each participant's score in a group (x_{ig}) and the group mean ($\bar{x}_g$), and the two sigma signs mean that we repeat this calculation for the first participant ($i = 1$) through to the last (n), in the first group ($g = 1$) through to the last (k). As such, we can also express SS_R as $SS_R = SS_{group\,1} + SS_{group\,2} + SS_{group\,3} + \ldots + SS_{group\,k}$. We know that the variance is the sums of squares divided by $n - 1$, and we can express the residual sum of squares in terms of the variance like we did for the total sum of squares. The result is:

$$SS_R = \sum_{g=1}^{k} s_g^2 (n_g - 1) \qquad (12.14)$$

which translates as 'multiply the variance for each group (s_g^2) by one less than the

number of people in that group ($n_g - 1$), then add the results for each group together'. For the puppy therapy data, we get:

$$\begin{aligned} SS_R &= s_{group1}^2 (n_1 - 1) + s_{group2}^2 (n_2 - 1) + s_{group3}^2 (n_3 - 1) \\ &= 1.70(5-1) + 1.70(5-1) + 2.50(5-1) \\ &= (1.70 \times 4) + (1.70 \times 4) + (2.50 \times 4) \\ &= 6.8 + 6.8 + 10 \\ &= 23.60 \end{aligned} \qquad (12.15)$$

The degrees of freedom for SS_R (df_R) are the total degrees of freedom minus the degrees of freedom for the model ($df_R = df_T - df_M = 14 - 2 = 12$). Put another way, this is $N - k$: the total sample size, N, minus the number of groups, k.

12.2.5 Mean squares ▮▮▮▮

SS_M tells us the *total* variation that the model (in this case the experimental manipulation) explains and SS_R tells us the *total* variation that is due to unmeasured factors. Because both values are sums, their size depends on the number of scores; for example, SS_M used the sum of three values (the group means), whereas SS_R and SS_T used the sum of 15 values. To eliminate this bias we calculate an average sum of squares (known as the *mean squares*, MS). Rather than dividing by the number of scores used for each SS, we divide by the degrees of freedom because we are trying to extrapolate to a population and so some parameters within that population will be held constant (remember we did this when calculating the variance—see Jane Superbrain Box 2.2). For the puppy therapy data we get a model mean squares (MS_M) of 10.067 and a residual mean squares (MS_R) of 1.96:

$$\begin{aligned} MS_M &= \frac{SS_M}{df_M} = \frac{20.135}{2} = 10.067 \\ MS_R &= \frac{SS_R}{df_R} = \frac{23.60}{12} = 1.96 \end{aligned} \qquad (12.16)$$

MS_M represents the average amount of variation explained by the model (e.g., the systematic variation), whereas MS_R is a gauge of the average amount of variation explained by unmeasured variables (the unsystematic variation).

12.2.6 The F-statistic ▮▮▮

The F-statistic (a.k.a. the F-ratio) is a measure of the ratio of the variation explained by the model and the variation attributable to unsystematic factors. In other words, it is the ratio of how good the model is to how bad it is (how much error there is). It is calculated by dividing the model mean squares by the residual mean squares:

$$F = \frac{MS_M}{MS_R} \qquad (12.17)$$

As with other test statistics that we have looked at (e.g., t) the F-statistic is a signal-to-noise ratio. In experimental research, it is the ratio of the experimental effect to the individual differences in performance. Because F is the ratio of systematic to unsystematic variance, if it is less than 1 it means that MS_R is greater than MS_M and that there is more unsystematic than systematic variance. In experimental research this means that the effect of natural variation is greater than differences brought about by the experiment. In this scenario, we can, therefore, be sure that our experimental manipulation has been unsuccessful (because it has bought about less change than if we left our participants alone) and F will be non-significant. For the puppy therapy data, the F-statistic is:

$$F = \frac{MS_M}{MS_R} = \frac{10.067}{1.967} = 5.12 \qquad (12.18)$$

meaning that the systematic variation is 5 times larger than the unsystematic variation; basically the experimental manipulation (puppy therapy groups) had some effect above and beyond the effect of individual differences in performance. Typically researchers are interested in

whether this ratio is significant: in other words, what would the probability be of getting an F at least this big if the experimental manipulation, in reality, had no effect at all on happiness (i.e. the null hypothesis). When the time comes that society collapses and we revert to a simple bartering system and all technology has been jettisoned into space, we will compare the obtained value of F against an F-distribution with the same degrees of freedom (see the Appendix). If the observed value exceeds the critical value we will probably conclude that our independent variable has had a genuine effect (because an F at least as big as the one we have observed would be very unlikely if there were no effect in the population). With 2 and 12 degrees of freedom the critical values are 3.89 ($p = 0.05$) and 6.93 ($p = 0.01$). The observed value, 5.12, is, therefore, significant at the 0.05 level of significance but not significant at the 0.01 level. Until the dawn of this anti-technological future, we can get the exact p from a computer: if it is less than the alpha level that we set before the experiment (e.g., 0.05) then scientists typically conclude that the variable that they manipulated had a genuine effect (more generally, predicting the outcome from group membership improves prediction).

12.2.7 Interpreting F ▮▮▮

I've already mentioned that F assesses the overall fit of the model to the data. When the model is one that predicts an outcome from group means, F evaluates whether 'overall' there are differences between means; it does not provide specific information about which groups were affected. It is an *omnibus* test. In our puppy therapy example in which there

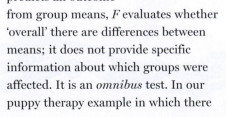

What does an ANOVA tell me?

are three groups, a significant F tells us that the means of these three samples are not equal (i.e., that $\overline{X}_1 = \overline{X}_2 = \overline{X}_3$ is *not* true). There are several ways in which the means can differ: (1) all three sample means could be significantly different ($\overline{X}_1 \neq \overline{X}_2 \neq \overline{X}_3$); (2) the means of groups 1 and 2 are similar to each other but different from that of group 3 ($\overline{X}_1 = \overline{X}_2 \neq \overline{X}_3$); (3) the means of groups 2 and 3 are similar to each other but different from that of group 1 ($\overline{X}_1 \neq \overline{X}_2 = \overline{X}_3$); or (4) the means of groups 1 and 3 are similar to each other but different from that of group 2 ($\overline{X}_1 = \overline{X}_3 \neq \overline{X}_2$).

You might feel that F is a bit pointless because, given that you've gone to the trouble of running an experiment, you probably had more specific predictions than 'there's a difference, somewhere or other'. You might wonder why you don't fit lots of models each of which compares only two means at a time; after all, this would tell you specifically whether pairs of group means differ. The reason why we don't do this was explained in Section 2.9.7: every time you run a test on the same data you inflate the Type I error rate. We'll return to this point in Section 12.5 when we discover how to establish where the group differences lie. For now, though, the reason why the F-test is useful is that as a single test (albeit of a non-specific hypothesis) it controls the Type I error rate. Having established that overall group means differ (i.e., the outcome can be significantly predicted using the group means) we can use the parameters of the model (the b-values) to tell us where the differences lie.

12.3 Assumptions when comparing means ▮▮▮

To compare means we use a linear model, so all of the potential sources of bias discussed in Chapter 6 apply. Normality is tested on scores *within groups*, not across the entire sample (see Jane Superbrain Box 6.1).

12.3.1 Homogeneity of variance ▮▮▮▮

As with any linear model, we assume that the variance of the outcome is steady as the predictor changes (in this context it means that variances in the groups are equal). When group sizes are unequal, violations of the assumption of homogeneity of variance can have quite serious consequences. This assumption can be tested using Levene's test (see Section 6.11.2). A conventional approach to the assumption is that if Levene's test is significant (i.e., the p-value is less than 0.05) then we conclude that the variances are significantly different and try to rectify the situation. However, the F-statistic can be adjusted to correct for the degree of heterogeneity and so you may as well just use the corrected F because small deviations from homogeneity will result in very small corrections (see Jane Superbrain Box 6.6). Two such corrections are the **Brown–Forsythe F** (Brown & Forsythe, 1974), and **Welch's F** (Welch, 1951). If you're really bored, these two

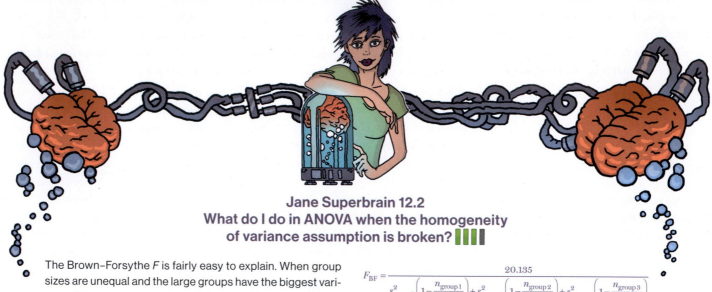

Jane Superbrain 12.2
What do I do in ANOVA when the homogeneity of variance assumption is broken? ▮▮▮▮

The Brown–Forsythe F is fairly easy to explain. When group sizes are unequal and the large groups have the biggest variance, F is conservative. If you think back to equation (12.14), this makes sense because to calculate SS_R variances are multiplied by their sample size (minus 1), so you get a large sample size cross-multiplied with a large variance, which will inflate the value of SS_R. F is proportionate to SS_M/SS_R, so if SS_R is big, then the F-statistic gets smaller (which is why it becomes conservative: its value is overly reduced). Brown and Forsythe get around this problem by weighting the group variances not by their sample size, but by the inverse of their sample sizes (they use n/N, which is the sample size as a proportion of the total sample size). This adjustment reduces the impact of large sample sizes with large variances:

$$F_{BF} = \frac{SS_M}{SS_{R_{BF}}} = \frac{SS_M}{\sum s_k^2 \left(1 - \frac{n_k}{N}\right)}$$

For the puppy therapy data, SS_M is the same as before (20.135), so F becomes:

$$F_{BF} = \frac{20.135}{s_{group1}^2\left(1 - \frac{n_{group1}}{N}\right) + s_{group2}^2\left(1 - \frac{n_{group2}}{N}\right) + s_{group3}^2\left(1 - \frac{n_{group3}}{N}\right)}$$

$$= \frac{20.135}{1.7\left(1 - \frac{5}{15}\right) + 1.7\left(1 - \frac{5}{15}\right) + 2.5\left(1 - \frac{5}{15}\right)}$$

$$= \frac{20.135}{3.933}$$

$$= 5.119$$

This statistic is evaluated using degrees of freedom for the model and error terms. For the model, df_M is the same as before (i.e., $k - 1 = 2$), but an adjustment is made to the residual degrees of freedom, df_R. Welch's (1951) F is an alternative adjustment that is more involved to explain—if you're interested, see Oliver Twisted. Both adjustments control the Type I error rate well (i.e., when there's no effect in the population you do get a non-significant F), but Welch's F has more power (i.e., is better at detecting an effect that exists) except when there is an extreme mean that has a large variance (Tomarken & Serlin, 1986).

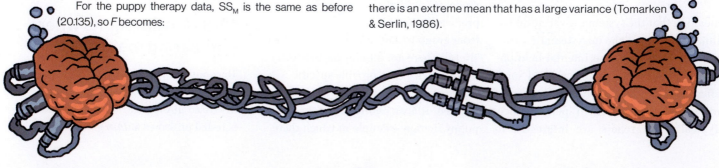

Oliver Twisted: Please, Sir, Can I have Some More . . . Welch's *F*?

'You're only telling us about the Brown–Forsythe *F* because you don't understand Welch's *F*,' taunts Oliver. 'Andy, Andy, brains all sandy ….' Whatever, Oliver. Like the Brown–Forsythe *F*, Welch's *F* adjusts *F* and the residual degrees of freedom to combat problems arising from violations of the homogeneity of variance assumption. There is a lengthy explanation about Welch's *F* in the additional material available on the companion website. Oh, and Oliver, microchips are made of sand.

statistics are discussed in Jane Superbrain Box 12.2. You can also use a robust version of *F* that does not assume homogeneity (Section 12.8).

12.3.2 Is ANOVA robust?

Is the *F*-statistic robust?

People often say 'ANOVA is a robust test', which means that it doesn't matter much if we break the assumptions, *F* will still be accurate. Remember from Chapter 6 that we mainly care about normality if we want to assess significance or construct confidence intervals. There are two issues to consider around the significance of *F*. First, does the *F* control the Type I error rate or is it significant even when there are no differences between means? Second, does *F* have enough power (i.e., is it able to detect differences when they are there)?

The myth that 'ANOVA is robust' is quite pervasive. It probably stems from research done a long time ago that investigated a limited range of situations. Examples of this work suggested that *when group sizes are equal* the *F*-statistic can be quite robust to violations of normality (Lunney, 1970), notably skew (Donaldson, 1968), and that when variances are proportional to the means the power of *F* is unaffected by heterogeneity of variance (Budescu, 1982; Budescu & Appelbaum, 1981).

The trouble is that this research is over 35 years old, so let's update things (for a more detailed review see Field & Wilcox, 2017). Recent simulations show that differences in skewness, non-normality and heteroscedasticity interact in complicated ways that impact power (Wilcox, 2017). For example, in the absence of normality, violations of homoscedasticity will affect *F* even when group sizes are equal (Wilcox, 2010, 2012, 2016), and when means are equal the error rate (which should be 5%) can be as high as 18%. Wilcox (2016) suggests that *F* can be considered robust only if the group distributions are identical; for example, groups are skewed to an identical degree, which in practice is probably unlikely. Heavy-tailed distributions are particularly problematic: if you set up a situation with power of 0.9 to detect an effect in a normal distribution and contaminate that distribution with 10% of scores from a normal distribution with a bigger variance (so you get heavier tails), power drops to 0.28 (despite the fact that only 10% of scores have changed). Similarly, Cohen's *d* drops from 1 when distributions are normal to 0.28 (Wilcox, Carlson, Azen, & Clark, 2013). Also, heavy-tailed samples have implications for the central limit theorem, which says that in samples of 30 or more the sampling distribution should be normal (Section 6.6.1); for heavy-tailed distributions samples need to be much larger, up to 160 in some cases (Wilcox, 2010). To sum up, *F* is not robust, despite what your supervisor might tell you.

Violations of the assumption of independence are very serious indeed. Scariano and Davenport (1987) showed that if scores are made to correlate moderately (say, with a Pearson coefficient of 0.5), then when comparing three groups of 10 observations per group the Type I error rate is 0.74 (remember that we'd expect it to be 0.05). In other words, you think you'll make a Type I error 5% of the time but in fact you'll make one 74% of the time!

12.3.3 What to do when assumptions are violated

Violations of assumptions are nowhere near the headache they used to be. In Chapter 6 we discussed methods for correcting problems (e.g., the bias reduction methods in Section 6.12), but these can mostly be avoided. If you routinely interpret Welch's F then you need never even think about homogeneity of variance, and you can bootstrap parameter estimates, which won't affect F itself, but at least you know that the model parameters are robust. There are also robust tests that use 20% trimmed means and a bootstrap, which we can implement in SPSS using R (see Section 4.13). There is also a school of thought that you should apply this robust test in all situations and don't bother thinking about assumptions. Finally, you can use the Kruskal–Wallis test from Chapter 7 (although personally I'd use the robust test). If you do apply the usual F-statistic then, at the very least, conduct a sensitivity analysis (i.e., apply a robust test to check that your conclusion doesn't change).

12.4 Planned contrasts (contrast coding)

I've already alluded to the need to follow up a significant F by looking at the model parameters, which tell us about specific differences between means. In fact, in Section 12.2 we saw that if we dummy-code the groups that we want to compare then the b for each dummy variable compares the mean of the baseline group to the group coded with a 1 in the particular dummy variable. In the puppy therapy example we ended up with two bs, one comparing the mean of the 15-minute group to the control group and the other comparing the 30-minute group to the control (Figure 12.3). Each of these bs is tested with a t-statistic that tells us whether the b is significantly different from 0 (i.e. the means differ).

The trouble is that with two dummy variables we end up with two t-tests, which inflates the familywise error rate (see Section 2.9.7). The other problem is that the dummy variables might not make all the comparisons that we want to make (e.g., the 15- and 30-minute groups are never compared).

There are a couple of solutions to these problems. The first is to use contrast coding rather than dummy coding. Contrast coding is a way of assigning weights to groups in dummy variables to carry out **planned contrasts** (also known

as planned comparisons). Weights are assigned in such a way that the contrasts are independent, which means that the overall Type I error rate is controlled. A second option is to compare every group mean to all others (i.e., to conduct several overlapping tests using a t-statistic each time) but using a stricter acceptance criterion that keeps the familywise error rate at 0.05. These are known as using ***post hoc* tests** (see Section 12.5). Typically planned contrasts are done to test specific hypotheses, whereas *post hoc* tests are used when there were no specific hypotheses. Let's first look at contrast coding and planned contrasts.

12.4.1 Choosing which contrasts to do

In the puppy therapy example the primary hypothesis would be that any dose of puppy therapy should change happiness compared to the control group. A second hypothesis might be that a 30-minute session should increase happiness more than a 15-minute one. To do planned contrasts, these hypotheses must be derived *before* the data are collected. The F-statistic is based upon splitting the total variation into two component parts: the variation due to the model or experimental manipulation (SS_M) and the variation due to unsystematic factors (SS_R) (see Figure 12.5).

Planned contrasts extend this logic by breaking down the variation due to the model/experiment into component parts (see Figure 12.6). The exact contrasts will depend upon the hypotheses you want to test. Figure 12.6 shows a situation in which the model variance is broken down based on the two hypotheses that we already discussed. Contrast 1 looks at how much variation in happiness is created by the two puppy conditions compared to the no puppy control condition. Next, the variation explained by puppy therapy (in general) is broken down to see what proportion is accounted for by a

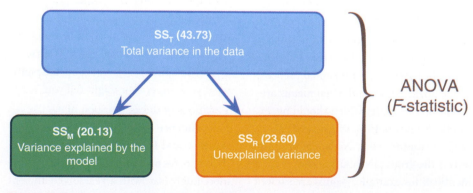

Figure 12.5 Partitioning variance for ANOVA

30-minute session relative to a 15-minute one (contrast 2).

Typically, students struggle with the notion of designing planned contrasts, but there are three rules that can help you to work out what to do.

1 If you have a control group, this is usually because you want to compare it against any other groups.
2 Each contrast must compare only two 'chunks' of variation.
3 Once a group has been singled out in a contrast it can't be used in another contrast.

Let's look at these rules in reverse order. First, if a group is singled out in one contrast, then it should not reappear in another contrast. The important thing is that we are breaking down one chunk of variation into smaller independent chunks. This independence matters for controlling the Type I error rate. It's like slicing up a cake: you begin with a cake (the total sum of squares) and then cut it into two pieces (SS_M and SS_R), then you take the piece of cake that represents SS_M and slice it again into smaller pieces. Once you have cut off a piece of cake you cannot stick that piece back onto the original slice, and you cannot stick it onto other pieces of cake, but you can divide it into smaller pieces of cake. Likewise, once a slice of variance has been split from a larger chunk, it cannot be attached to any other pieces of variance, it can only be subdivided into smaller chunks of variance. All this talk of cake is making me hungry, but hopefully it illustrates a point. So, in Figure 12.6 contrast 1 compares the control group to the experimental groups and because the control group is singled out, it is not incorporated into contrast 2.

Second, each contrast must compare only two chunks of variance. This rule is so that we can interpret the contrast. The original F tells us that some of our means differ, but not which ones, and if we were to perform a contrast on more than two

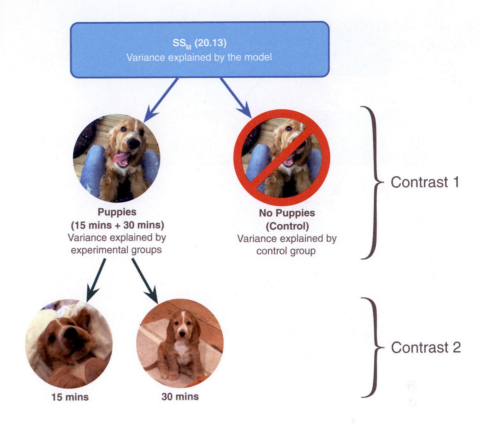

chunks of variance we would be no better off. By comparing only two chunks of variance we know that the result represents a significant difference (or not) between these two portions of variation. If you follow the independence of contrasts rule (the cake slicing), and always compare only two pieces of variance, then you should end up with $k - 1$ contrasts (where k is the number of conditions you're comparing); in other words, one fewer contrasts than the number of conditions you have in your design.

The first rule in the list reminds us that we often use at least one control condition, and it (or they) usually exists because we predict that the experimental conditions will differ from it (or them). As such, when planning contrasts the chances are that your first contrast will be one that compares all the experimental groups with the control group (or groups). Once you have done this first comparison,

Figure 12.6 Partitioning of the model/experimental variance into component contrasts

any remaining contrasts will depend upon which groups you predict will differ (based on the theory you're testing).

To further illustrate these principles Figures 12.7 and 12.8 show potential sets of contrasts for two different four-group experiments. In both examples there are three possible contrasts (one less than the number of groups) and every contrast compares only two chunks of variance. The first contrast is the same in both cases: the experimental groups are compared against the control group or groups. In Figure 12.7 there is only one control condition and this portion of variance is used only in the first contrast (because it cannot be broken down any further). In Figure 12.8 there were two control groups, and so the portion of variance due to the control conditions (contrast 1) can be broken down further to see whether the scores in the control groups differ from each other (contrast 3).

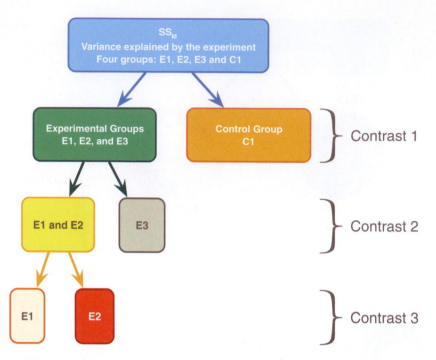

Figure 12.7   Partitioning variance for planned contrasts in a four-group experiment using one control group

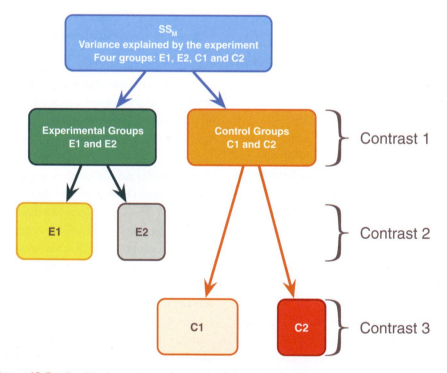

Figure 12.8   Partitioning variance for planned contrasts in a four-group experiment using two control groups

In Figure 12.7, the first contrast contains a chunk of variance that is due to the three experimental groups, and this chunk of variance is broken down by first looking at whether groups E1 and E2 differ from E3 (contrast 2). It is equally valid to use contrast 2 to compare groups E1 and E3 to E2 or to compare groups E2 and E3 to

E1. The exact contrast that you choose depends upon your hypotheses. For contrast 2 in Figure 12.7 to be valid we need to have a good theoretical reason to expect group E3 to be different from the other two groups. The third contrast in Figure 12.7 depends on the comparison chosen for contrast 2. Contrast 2 necessarily had to involve comparing two experimental groups against a third, and the experimental groups chosen to be combined must be separated in the final contrast. As a final point about Figures 12.7 and 12.8, notice that once a group has been singled out in a contrast, it is never used in any subsequent one.

When we carry out a planned contrast, we compare 'chunks' of variance and these chunks often consist of several groups. When you design a contrast that compares several groups to one other group, you are comparing the means of the groups in one chunk with the mean of the group in the other chunk. As an example, for the puppy therapy data I suggested that an appropriate first contrast would be to compare the two dose groups with the control group. The means of the groups are 2.20 (control), 3.20 (15 minutes) and 5.00 (30 minutes), and so the first comparison, which compared the two experimental groups to the control, is comparing 2.20 (the mean of the control group) to the average of the other two groups ((3.20 + 5.00)/2 = 4.10). If this first contrast turns out to be significant, then we can conclude that 4.10 is significantly greater than 2.20, which in terms of the experiment tells us that the average of the experimental groups is significantly different from the average of the controls. You can probably see that logically this means that, if the standard errors are the same, the experimental group with the highest mean (the 30-minute group) will be significantly different from

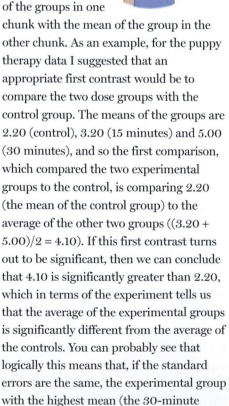

What does a planned contrast tell me?

the mean of the control group. However, the experimental group with the lower mean (the 15-minute group) might not necessarily differ from the control group; we have to use the final contrast to make sense of the experimental conditions. For the puppy data the final contrast looked at whether the two experimental groups differ (i.e., is the mean of the 30-minute group significantly different from the mean of the 15-minute group?). If this contrast turns out to be significant then we can conclude that having 30 minutes of puppy therapy significantly affected happiness compared to having 15 minutes. If the contrast is non-significant then we conclude that the dosage of puppy therapy made no significant difference to happiness. In this latter scenario it is likely that both doses affect happiness more than control, whereas the former case implies that having 15 minutes may be no different than having a control. However, the word *implies* is important here: it is possible that the 15-minute group might not differ from the control. To be completely sure you would need *post hoc* tests.

12.4.2 Defining contrasts using weights

Hopefully by now you have got some idea of how to plan which contrasts to do (i.e., if your brain hasn't exploded yet). The next issue is how to do them. To carry out contrasts we need to code our dummy variables in a way that results in *b*s that compare the 'chunks' that we set out in our contrasts. Remember that when we used dummy coding, we used values of 0 and 1 for the dummy variables and ended up with *b*-values that compared each group to a baseline group. We need to work out what values (instead of 0 and 1) to assign to each group to give us (in this example) two *b*-values, the first of which compares all puppy therapy scores to the control group, and the second of which compares 15 minutes of puppy therapy

with 30 minutes (and ignores the control group). The values assigned to the dummy variables are known as **weights**.

This procedure is horribly confusing, but there are a few basic rules for assigning values to the dummy variables to obtain the contrasts you want. I will explain these simple rules before showing how the process works. Remember the previous section when you read through these rules, and remind yourself of what I mean by a 'chunk' of variation.

- **Rule 1**: Choose sensible contrasts. Remember that you want to compare only two chunks of variation and that if a group is singled out in one contrast, that group should be excluded from any subsequent contrasts.
- **Rule 2**: Groups coded with positive weights will be compared against groups coded with negative weights. So, assign one chunk of variation positive weights and the opposite chunk negative weights.
- **Rule 3**: If you add up the weights for a given contrast the result should be zero.
- **Rule 4**: If a group is not involved in a contrast, automatically assign it a weight of zero, which will eliminate it from the contrast.
- **Rule 5**: For a given contrast, the weights assigned to the group(s) in one chunk of variation should be equal to the number of groups in the opposite chunk of variation.

OK, let's follow these rules to derive the weights for the puppy therapy data. The

first contrast we chose was to compare the two experimental groups against the control (Figure 12.9). The first chunk of variation contains the two experimental groups, and the second chunk contains only the control group. Rule 2 states that we should assign one chunk positive weights, and the other negative. It doesn't matter which way round we do this, but for convenience let's assign chunk 1 positive weights and chunk 2 negative weights, as in Figure 12.9. Using rule 5, the weight we assign to the groups in chunk 1 should be equivalent to the number of groups in chunk 2. There is only one group in chunk 2, and so we assign each group in chunk 1 a weight of 1. Likewise, we assign a weight to the group in chunk 2 that is equal to the number of groups in chunk 1. There are two groups in chunk 1 so we give the control group a weight of 2. Then we combine the sign of the weights with the magnitude to give us weights of −2 (control), +1 (15 minutes) and +1 (30 minutes), as in Figure 12.9. Rule 3 states that for a given contrast, the weights should add up to zero, and by following rules 2 and 5 this should be true (if you haven't followed the rules properly then it will become clear when you add the weights). Let's check by adding the weights: sum of weights = 1 + 1 − 2 = 0. Happy days.

The second contrast was to compare the two experimental groups, and so we want to ignore the control group. Rule 4 tells us that we should automatically assign this group a weight of 0 (to eliminate it). We are left with two chunks of variation:

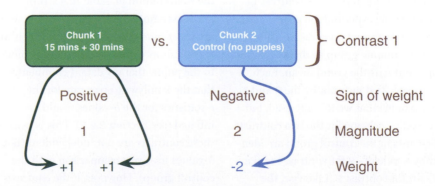

Figure 12.9   Assigning weights for contrast 1

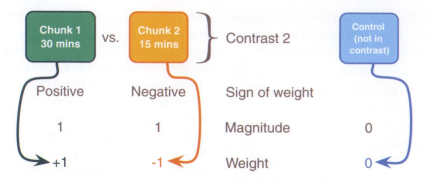

Figure 12.10 Assigning weights for contrast 2

Table 12.4 Orthogonal contrasts for the puppy therapy data

Group	Dummy variable 1 (Contrast$_1$)	Dummy variable 2 (Contrast$_2$)	Product Contrast$_1$ × Contrast$_2$
Control	−2	0	0
15 minutes	1	−1	−1
30 minutes	1	1	1
Total	0	0	0

chunk 1 contains the 15-minute group and chunk 2 contains the 30-minute group. By following rules 2 and 5 it should be obvious that one group is assigned a weight of +1 while the other is assigned a weight of −1 (Figure 12.10). If we add the weights for contrast 2 we find that they again add up to zero: sum of weights = 1 − 1 + 0 = 0. Even happier days.

The weights for each contrast are placed in the two dummy variables in the following equation:

$$\text{Happiness}_i = b_0 + b_1 \text{Contrast}_{1i} + b_2 \text{Contrast}_{2i}$$

$$(12.19)$$

Hence, the weights are used in a linear model in which b_1 represents contrast 1 (comparing the experimental groups to the control), b_2 represents contrast 2 (comparing the 30-minute group to the 15-minute group), and b_0 is the grand mean. Each group is specified now not by the 0 and 1 coding scheme that we initially used, but by the coding scheme for the two contrasts. Participants in the control group are identified by a code of −2 for contrast 1 and a code of 0 for contrast 2. Likewise, the 30-minute group is identified by a code of

1 for both variables, and the 15-minute group has a code of 1 for one contrast and a code of −1 for the other (Table 12.4).

It is important that the weights for a contrast sum to zero because it ensures that you are comparing two unique chunks of variation. Therefore, a t-statistic can be used (remember it assumes independence). A more important consideration is that when you multiply the weights for a particular group, these products should also add up to zero (see the final column of Table 12.4). If the products sum to zero then the contrasts are *independent* or **orthogonal**. When we used dummy coding and fit a linear model to the puppy therapy data, I commented that the familywise error rate for the t-statistics for the b-values would be inflated (see Section 2.9.7). This is because these contrasts are not independent (both b-values involve a comparison with the control group). However, if the contrasts *are* independent then the t-statistics for

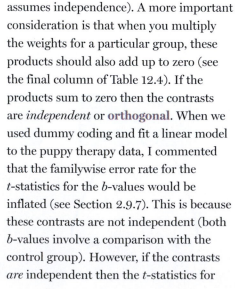

What are orthogonal contrasts?

the b-values are also independent and the resulting p-values are uncorrelated. You might think that it is very difficult to ensure that the weights you choose for your contrasts conform to the requirements for independence but, provided you follow the rules I have laid out, you should always derive a set of *orthogonal* contrasts. You should double-check by looking at the sum of the multiplied weights, and if this total is not zero then go back to the rules and see where you have gone wrong.

Earlier on, I mentioned that when you contrast-code dummy variables in a linear model the b-values represent the differences between the means that the contrasts were designed to test. Let's have a look at how this works (this next part is not for the faint-hearted). When we do planned contrasts, the intercept b_0 is equal to the grand mean (i.e., the value predicted by the model when group membership is not known), which in this example (and because group sizes are equal) is:

$$b_0 = \text{grand mean} = \frac{\bar{X}_{30\,\text{mins}} + \bar{X}_{15\,\text{mins}} + \bar{X}_{\text{Control}}}{3}$$

$$(12.20)$$

For a participant in the *control group*, we define their group membership using the values for the control group in Table 12.4 and the predicted value of happiness is the mean of the control group. The model can be expressed as:

$$\text{Happiness}_i = b_0 + b_1 \text{Contrast}_{1i} + b_2 \text{Contrast}_{2i}$$
$$\bar{X}_{\text{Control}} = \left(\frac{\bar{X}_{30\,\text{mins}} + \bar{X}_{15\,\text{mins}} + \bar{X}_{\text{Control}}}{3} \right)$$
$$+ \left(-2b_1 \right) + \left(b_2 \times 0 \right)$$

$$(12.21)$$

If we rearrange this equation and multiply everything by 3 (to get rid of the fraction) we get:

$$2b_1 = \left(\frac{\bar{X}_{30\,\text{mins}} + \bar{X}_{15\,\text{mins}} + \bar{X}_{\text{Control}}}{3} \right) - \bar{X}_{\text{Control}}$$
$$6b_1 = \bar{X}_{30\,\text{mins}} + \bar{X}_{15\,\text{mins}} + \bar{X}_{\text{Control}} - 3\bar{X}_{\text{Control}}$$
$$= \bar{X}_{30\,\text{mins}} + \bar{X}_{15\,\text{mins}} - 2\bar{X}_{\text{Control}}$$

$$(12.22)$$

We then divide everything by 2 to reduce the equation to its simplest form:

$$3b_1 = \left(\frac{\bar{X}_{30\,mins} + \bar{X}_{15\,mins}}{2} \right) - \bar{X}_{Control}$$

$$b_1 = \frac{1}{3}\left[\left(\frac{\bar{X}_{30\,mins} + \bar{X}_{15\,mins}}{2} \right) - \bar{X}_{Control} \right] \qquad (12.23)$$

We planned contrast 1 to look at the difference between the average of the two puppy groups and the control, and a final rearrangement of the equation shows how b_1 represents this difference:

$$3b_1 = \left(\frac{\bar{X}_{30\,mins} + \bar{X}_{15\,mins}}{2} \right) - \bar{X}_{Control}$$

$$= \frac{5 + 3.2}{2} - 2.2 \qquad (12.24)$$

$$= 1.9$$

Rather than being the true value of the difference between experimental and control groups, b_1 is actually a third of this difference ($b_1 = 1.9/3 = 0.633$)—it is divided by the number of groups in the contrast. Nevertheless, it is proportionate to the difference it set out to test.

For someone in the *30-minute group*, the predicted value of happiness is the mean for the 30-minute group, and their group membership is coded using the values for the 30-minute group in Table 12.4. The resulting model is in equation (12.25).

$$Happiness_i = b_0 + b_1 Contrast_{1i} + b_2 Contrast_{2i}$$

$$\bar{X}_{30\,mins} = b_0 + (b_1 \times 1) + (b_2 \times 1)$$

$$b_2 = \bar{X}_{30\,mins} - b_1 - b_0 \qquad (12.25)$$

We know already what b_1 and b_0 represent, so we place these values into the equation and then multiply by 3 to get rid of some of the fractions:

$$b_2 = \bar{X}_{30\,mins} - b_1 - b_0$$

$$= \bar{X}_{30\,mins} - \frac{1}{3}\left[\left(\frac{\bar{X}_{30\,mins} + \bar{X}_{15\,mins}}{2} \right) - \bar{X}_{Control} \right]$$

$$- \frac{\bar{X}_{30\,mins} + \bar{X}_{15\,mins} + \bar{X}_{Control}}{3}$$

$$3b_2 = 3\bar{X}_{30\,mins} - \left[\left(\frac{\bar{X}_{30\,mins} + \bar{X}_{15\,mins}}{2} \right) - \bar{X}_{Control} \right]$$

$$- \bar{X}_{30\,mins} + \bar{X}_{15\,mins} + \bar{X}_{Control} \qquad (12.26)$$

If we multiply everything by 2 to get rid of the fraction, expand the brackets and collect terms we get:

$$6b_2 = 6\bar{X}_{30\,mins} - \left(\bar{X}_{30\,mins} + \bar{X}_{15\,mins} - 2\bar{X}_{Control} \right)$$

$$- 2\left(\bar{X}_{30\,mins} + \bar{X}_{15\,mins} + \bar{X}_{Control} \right)$$

$$= 6\bar{X}_{30\,mins} - \bar{X}_{30\,mins} - \bar{X}_{15\,mins} + 2\bar{X}_{Control}$$

$$- 2\bar{X}_{30\,mins} - 2\bar{X}_{15\,mins} - 2\bar{X}_{Control}$$

$$= 3\bar{X}_{30\,mins} - 3\bar{X}_{15\,mins} \qquad (12.27)$$

Finally, let's divide the equation by 6 to find out what b_2 represents (remember that $3/6 = 1/2$):

$$b_2 = \frac{\bar{X}_{30\,mins} - \bar{X}_{15\,mins}}{2} \qquad (12.28)$$

We planned contrast 2 to look at the difference between the experimental groups:

$$\bar{X}_{30\,mins} - \bar{X}_{15\,mins} = 5 - 3.2 = 1.8 \qquad (12.29)$$

and b_2 represents this difference (equation (12.28)). Again, rather than being the absolute value of the difference between the experimental groups, b_2 is this difference divided by the number of groups in the contrast (1.8/2 = 0.9), but the key thing is what it represents and it is proportional to the difference between experimental group means.

Output 12.2 shows the result of the model from the self-test. The main ANOVA for the model is the same as when dummy coding was used (compare it to Output 12.1), showing that the model fit is the same (it should be because the model represents the group means and these have not changed); however, the b-values have changed because the values of our

dummy variables have changed. The first thing to notice is that the intercept is the grand mean, 3.467 (see, I wasn't telling lies). Second, the b for contrast 1 is one-third of the difference between the average of the experimental conditions and the control condition. Finally, the b for contrast 2 is half of the difference between the experimental groups (see above). The significance values of the t-statistics tell us that our puppy groups were significantly different from the control ($p = 0.029$) but that the 15- and 30-minutes of puppy therapy were not significantly different ($p = 0.065$).

12.4.3 Non-orthogonal contrasts ▮▮▮

Contrasts don't have to be orthogonal: non-orthogonal contrasts are contrasts that are related. The best way to get them is to disobey rule 1 in the previous section. Using my cake analogy again, non-orthogonal contrasts are where you slice up your cake and then try to stick slices of cake together again. Standard dummy coding (Section 12.2) is an example of non-orthogonal contrasts because the baseline group is used in each contrast. For the puppy therapy data another set of non-orthogonal contrasts might be to have the same initial contrast (comparing experimental groups against the control), but then to compare the 30-minute group to the control. This disobeys rule 1 because the control group is singled out in the first

Are non-orthogonal contrasts legit?

SELF TEST

To illustrate these principles, I have created a file called **Puppies Contrast.sav** in which the puppy therapy data are coded using the contrast coding scheme used in this section. Fit a linear model using happiness as the outcome and **dummy1** and **dummy2** as the predictor variables (leave all default options).

Coefficients[a]

Model		Unstandardized Coefficients		Standardized Coefficients		
		B	Std. Error	Beta	t	Sig.
1	(Constant)	3.467	.362		9.574	.000
	Dummy 1: Dose vs. Control	.633	.256	.525	2.474	.029
	Dummy 2: 15 mins vs 30 mins	.900	.443	.430	2.029	.065

a. Dependent Variable: Happiness (0–10)

Output 12.2

contrast but used again in the second contrast. The coding for this set of contrasts is shown in Table 12.5, and the last column makes clear that when you multiply and add the codings from the two contrasts the sum is not zero. This tells us that the contrasts are not orthogonal.

There is nothing intrinsically wrong with non-orthogonal contrasts, but you must be careful about how you interpret them because the contrasts are related and so the resulting test statistics and p-values will be correlated to some degree. Basically, the Type I error rate isn't controlled, so you should use a more conservative probability level to accept a given contrast as statistically significant (see Section 12.5).

12.4.4 Built-in contrasts

Although under most circumstances you will design your own contrasts, there are also 'off-the-shelf' contrasts that you can choose. Table 12.6 shows the built-in contrasts in SPSS Statistics for procedures such as logistic regression (see Section 20.5.7), factorial designs and repeated-measures designs (see Chapters 14 and 15). The exact codings are not provided in Table 12.6, but I give examples of the contrasts done in a three- and four-group situation (where the groups are labeled 1, 2, 3 and 1, 2, 3, 4, respectively). When you code categorical variables in the data editor, SPSS Statistics treats the lowest-value code as group 1, the next highest code as group 2, and so on. Therefore,

Cramming Sam's Tips
Planned contrasts

- If the *F* for the overall model is significant you need to find out which groups differ.
- When you have generated specific hypotheses before the experiment, use *planned contrasts*.
- Each contrast compares two 'chunks' of variance. (A chunk can contain one or more groups.)
- The first contrast will usually be experimental groups against control groups.
- The next contrast will be to take one of the chunks that contained more than one group (if there were any) and divide it in to two chunks.
- You repeat this process: if there are any chunks in previous contrasts that contained more than one group that haven't already been broken down into smaller chunks, then create new contrasts that breaks them down into smaller chunks.
- Carry on creating contrasts until each group has appeared in a chunk on its own in one of your contrasts.
- The number of contrasts you end up with should be one less than the number of experimental conditions. If not, you've done it wrong.
- In each contrast assign a 'weight' to each group that is the value of the number of groups in the opposite chunk in that contrast.
- For a given contrast, randomly select one chunk, and for the groups in that chunk change their weights to be negative numbers.
- Breathe a sigh of relief.

depending on which contrasts you want, you should code your grouping variable appropriately (and then use Table 12.6 as a guide to which contrasts you'll get). Some contrasts in Table 12.6 are orthogonal (i.e., Helmert and difference contrasts) while others are non-orthogonal (deviation, simple and repeated). You might also notice that simple contrasts are the same as those given by using the dummy variable coding described in Table 12.2.

Table 12.5 Non-orthogonal contrasts for the puppy therapy data

Group	Dummy variable 1 (Contrast$_1$)	Dummy variable 2 (Contrast$_2$)	Product Contrast1 × Contrast$_2$
Control	−2	−1	2
15 minutes	1	0	0
30 minutes	1	1	1
Total	0	0	3

12.4.5 Polynomial contrasts: trend analysis ▮▮▮▮

One type of contrast deliberately omitted from Table 12.6 is the **polynomial contrast**. This contrast tests for trends in the data, and in its most basic form it looks for a linear trend (i.e., that the group means increase proportionately). However, there are other trends such as quadratic, cubic and quartic trends that can be examined. Figure 12.11 shows examples of the types of trend that can exist in data sets. The *linear* trend should be familiar to you and represents a simple proportionate change in the value of the dependent variable across ordered categories (the diagram shows a positive linear trend, but of course it could be negative). A **quadratic trend** is where there is a curve in the line (the curve can be subtler than in the figure and in the opposite direction). An example of this is a situation in which a drug enhances performance on a task at first but then as the dose increases the performance tails off or drops. To find a quadratic trend you need at least three groups because with two groups the means of the dependent variable can't be connected by anything other than a straight line. A **cubic trend** is where there are two changes in the direction of the trend. So, for example, the mean of the dependent variable at first goes up across categories of the independent variable, then across the succeeding categories the means go down,

Table 12.6 Standard contrasts available in SPSS

Name	Definition	Contrast	Three Groups	Four Groups
Deviation (first)	Compares the effect of each category (except the first) to the overall experimental effect	1	2 vs. (1, 2, 3)	2 vs. (1, 2, 3, 4)
		2	3 vs. (1, 2, 3)	3 vs. (1, 2, 3, 4)
		3		4 vs. (1, 2, 3, 4)
Deviation (last)	Compares the effect of each category (except the last) to the overall experimental effect	1	1 vs. (1, 2, 3)	1 vs. (1, 2, 3, 4)
		2	2 vs. (1, 2, 3)	2 vs. (1, 2, 3, 4)
		3		3 vs. (1, 2, 3, 4)
Simple (first)	Each category is compared to the first category	1	1 vs. 2	1 vs. 2
		2	1 vs. 3	1 vs. 3
		3		1 vs. 4
Simple (last)	Each category is compared to the last category	1	1 vs. 3	1 vs. 4
		2	2 vs. 3	2 vs. 4
		3		3 vs. 4
Repeated	Each category (except the first) is compared to the previous category	1	1 vs. 2	1 vs. 2
		2	2 vs. 3	2 vs. 3
		3		3 vs. 4
Helmert	Each category (except the last) is compared to the mean effect of all subsequent categories	1	1 vs. (2, 3)	1 vs. (2, 3, 4)
		2	2 vs. 3	2 vs. (3, 4)
		3		3 vs. 4
Difference (reverse **Helmert**)	Each category (except the first) is compared to the mean effect of all previous categories	1	3 vs. (2, 1)	4 vs. (3, 2, 1)
		2	2 vs. 1	3 vs. (2, 1)
		3		2 vs. 1

but then across the last few categories the means rise again. To have two changes in the direction of the mean you must have at least four categories of the independent variable. The final trend that you are likely to come across is the **quartic trend**, and this trend has three changes of direction (so you need at least five categories of the independent variable).

Polynomial trends should be examined in data sets in which it makes sense to order the categories of the independent variable (so, for example, if you have administered five doses of a drug it makes sense to examine the five doses in order of magnitude). For the puppy therapy data there are three groups, and so we can expect to find only a linear or quadratic trend (it would be pointless to test for any higher-order trends).

Each of these trends has a set of codes for the dummy variables in the model, so we

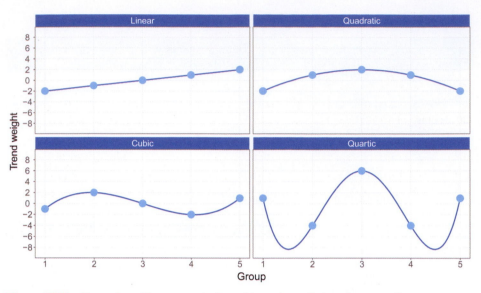

Figure 12.11   Examples of linear, quadratic, cubic and quartic trends across five groups

are doing the same thing that we did for planned contrasts except that the codings have already been devised to represent the type of trend of interest. In fact, the graphs in Figure 12.11 have been constructed by plotting the code values for the five groups. If you add the codes for a given trend the sum will equal zero, and if you multiply the codes you will find that the sum of the products also equals zero. Hence, these contrasts are orthogonal.

12.5 Post hoc procedures ▮▮▮

Often people have no specific *a priori* predictions about the data they have collected and instead they rummage around the data looking for any differences between means that they can find. It's a bit like *p*-hacking (Section 3.3), except that you adjust the *p* to make it harder to find significant differences. *Post hoc* tests consist of **pairwise comparisons** that are designed to compare all different combinations of the treatment groups. So, it is taking every pair of groups and performing a separate test on each. Now, this might seem like a particularly stupid thing to do in light of

what I have already told you about the problems of inflated familywise error rates (Section 2.9.7). However, pairwise comparisons control the familywise error by correcting the level of significance for each test such that the overall Type I error rate (α) across all comparisons remains at 0.05. There are several ways in which the familywise error rate can be controlled, and we have already discussed one of the most popular ones: the Bonferroni correction (Section 2.9.7).

There are other methods too (SPSS Statistics does about 18 different ones). Although I would love to go into the tedious detail of all 18 procedures, there really is little point. One reason is that there are excellent texts already available for those who wish to know (Klockars & Sax, 1986; Toothaker, 1993), but the main one is that to explain them I would have to learn about them first. I may be a nerd, but even I draw the line at reading up on 18 different *post hoc* tests. However, it *is* important that you know which *post hoc* tests perform best according to three important criteria. First, does the test control the Type I error rate? Second, does the test control the Type II error rate (i.e., does the test

have good statistical power)? Third, is the test robust?

12.5.1 Type I and Type II error rates for *post hoc* tests ▮▮▮

The Type I error rate and the statistical power of a test are linked. Therefore, there is always a trade-off: if a test is conservative (the probability of a Type I error is small) then it is likely to lack statistical power (the probability of a Type II error will be high). Therefore, it is important that multiple comparison procedures control the Type I error rate but without a substantial loss in power. If a test is too conservative then we are likely to reject differences between means that are, in reality, meaningful.

The *least-significant difference* (LSD) pairwise comparison makes no attempt to control the Type I error and is equivalent to performing multiple tests on the data. The only difference is that the LSD requires the overall ANOVA to be significant. The *Studentized Newman–Keuls* (SNK) procedure is also a very liberal test and lacks control over the familywise error rate. *Bonferroni's* and *Tukey's* tests both control the Type I error rate very well but are conservative (they lack statistical power). Of the two, Bonferroni has more power when the number of comparisons is small, whereas Tukey is more powerful when testing large numbers of means. Tukey generally has greater power than *Dunn* and *Scheffé*. The *Ryan, Einot, Gabriel and Welsch Q* procedure (REGWQ) has good power and tight control of the Type I error rate. In fact, when you want to test all pairs of means this procedure is probably the best. However, when group sizes are different this procedure should not be used.

12.5.2 Are *post hoc* procedures robust? ▮▮▮

Most research on *post hoc* tests has looked at whether the test performs well when the group sizes are different (an unbalanced design), when the population variances are very different, and when data are not normally distributed. The good news is that most multiple comparison procedures perform relatively well under small deviations from normality. The bad news is that they perform badly when group sizes are unequal and when population variances are different.

Hochberg's GT2 and *Gabriel's* pairwise test procedure were designed to cope with situations in which sample sizes are different. Gabriel's procedure is generally more powerful but can become too liberal when the sample sizes are very different. Also, Hochberg's GT2 is very unreliable when the population variances are different, and so should be used only when you are sure that this is not the case. There are several multiple comparison procedures

that have been specially designed for situations in which population variances differ. SPSS provides four options for this situation: *Tamhane's T2*, *Dunnett's T3*, *Games–Howell* and *Dunnett's C*. Tamhane's T2 is conservative and Dunnett's T3 and C keep very tight Type I error control. The Games–Howell procedure is the most powerful but can be liberal when sample sizes are small. However, Games–Howell is also accurate when sample sizes are unequal.

12.5.3 Summary of *post hoc* procedures

The choice of comparison procedure will depend on the exact situation you have and whether it is more important for you to keep strict control over the familywise error rate or to have greater statistical power. However, some general guidelines

can be drawn (Toothaker, 1993). When you have equal sample sizes and you are confident that your population variances are similar then use REGWQ or Tukey as both have good power and tight control over the Type I error rate. Bonferroni is generally conservative, but if you want guaranteed control over the Type I error rate then this is the test to use. If sample sizes are slightly different then use Gabriel's procedure because it has greater power, but if sample sizes are very different use Hochberg's GT2. If there is any doubt that the population variances are equal then use the Games–Howell procedure because this generally seems to offer the best performance. I recommend running the Games–Howell procedure in addition to any other tests you might select because of the uncertainty of knowing whether the population variances are equivalent.

Cramming Sam's Tips
Post hoc tests

- When you have no specific hypotheses before the experiment, follow up the model with *post hoc tests*.
- When you have equal sample sizes and group variances are similar use REGWQ or Tukey.
- If you want guaranteed control over the Type I error rate then use Bonferroni.

- If sample sizes are slightly different then use Gabriel's, but if sample sizes are very different use Hochberg's GT2.
- If there is any doubt that group variances are equal then use the Games–Howell procedure.

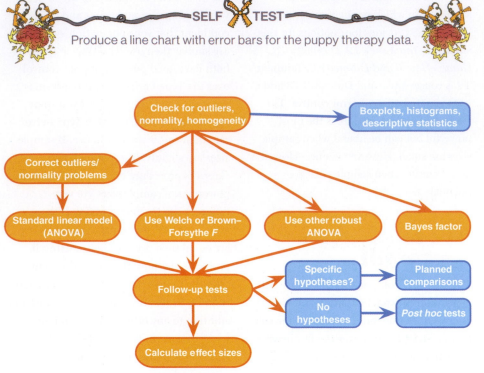

SELF TEST

Produce a line chart with error bars for the puppy therapy data.

Figure 12.12 Overview of the general procedure for one-way ANOVA

Although these general guidelines provide a convention to follow, be aware of the other procedures available and when they might be useful to use (e.g., Dunnett's test is the only multiple comparison that allows you to test means against a control mean).

12.6 Comparing several means using SPSS Statistics ▐▐▐

Because to compare means we simply fit a variant of the linear model, we could simply set up contrast dummy variables and run the analysis using the linear regression menus from Chapter 9. However, SPSS Statistics has a designated menu for situations where you're predicting an outcome from several group means (historically known as a one-way independent analysis of variance, ANOVA). Using this designated

menu has the advantage that you don't have to create dummy variables manually, as we shall see.

12.6.1 General procedure of one-way ANOVA ▐▐▐

The data are in the file **Puppies.sav** (although it's good practice to enter them yourself). We enter the data into two columns (three if you want to include a participant ID). One column (**Dose**) specifies how much puppy therapy the participant was given, and is a variable that codes the group to which the individual belonged (I have coded 1 = control, 2 = 15 minutes and 3 = 30 minutes). The other column (**Happiness**) contains the person's happiness score.

Because one-way **independent ANOVA** is a linear model with a different label attached, look back at the general procedure for linear models in Chapter 9. Figure 12.12 highlights the steps that are

specific to this version of the linear model. As with any analysis, begin by graphing the data and looking for and correcting sources of bias.

Noting my earlier comments about homogeneity of variance and normality (Section 12.3), there is a case for always proceeding with a robust test or at the very least always using Welch's F.

To fit the model select *Analyze* ▶ *Compare Means* ▶ 🔣 *One-Way ANOVA…* to access the main dialog in Figure 12.13. There is an empty box in which to place one or more dependent variables and another to specify a categorical predictor or *factor*. 'Factor' is another term for independent variable or categorical predictor and not to be confused with a very different type of factor that we learn about in Chapter 18. For the puppy therapy data drag **Happiness** from the variables list into the box labeled *Dependent List* (or click 🔽). Then drag the grouping variable **Dose** to the box labeled *Factor* (or click 🔽).

One thing that I dislike about SPSS Statistics is that in various procedures, such as this one, you are encouraged to carry out multiple tests (in this case by allowing you to specify several outcome variables at once). This is not a good thing because you lose control of the Type I error rate. If you had measured several dependent variables (e.g., you had measured not just happiness but other indicators of stress such as cortisol levels, non-verbal behavior, and heart rate) it would be preferable to analyze these data using MANOVA (Chapter 17) rather than treating each outcome measure separately.

12.6.2 Planned contrasts using SPSS ▐▐▐

Click ⬚Contrasts… to access a dialog box for specifying the contrast described in Section 12.4. The dialog box (Figure 12.14) has two sections. The first section is for specifying trend analyses. If you want to test for trends in the data then select

☑ <u>Polynomial</u>. Once this option is active you can select the degree of polynomial you would like from the drop-down list (Linear ▾). The puppy therapy data have only three groups and so the highest degree of trend there can be is quadratic (see Section 12.4.3). It is important from the point of view of trend analysis that we have coded the grouping variable in a meaningful order. We expect happiness to be smallest in the control group, to increase in the 15-minute group and then to increase again in the 30-minute group, so to detect a meaningful trend, these groups must be coded in ascending order. We have done this by coding the control group with the lowest value 1, the 15-minute group with the middle value 2 and the 30-minute group with the highest value of 3. If we coded the groups differently, this would influence both whether a trend is detected and, if one is detected, whether it has a meaningful interpretation. So, for the puppy therapy data select ☑ <u>Polynomial</u> and Quadratic ▾ . SPSS Statistics will test for the trend requested and all lower-order trends, so with *Quadratic* selected we'll get tests both for a linear and a quadratic trend.

The lower part of the dialog box in Figure 12.14 is for specifying weights for the planned contrasts that you have decided to do. We went through the process of generating weights for the contrasts we want in Section 12.4.2. The weights for contrast 1 were –2 (control group), +1 (15-minute group) and +1 (30-minute group). We will specify this contrast first. It is important to make sure that you enter the correct weight for each group; the first weight that you enter is the weight for the *first* group (i.e., the group coded with the lowest value in the data editor). For the puppy therapy data, the group coded with the lowest value was the control group (which had a code of 1) so we enter the weighting for this group first. Click in the box labeled *Coefficients*, type '–2' and click Add . Next, we input the weight for the second group, which for the puppy

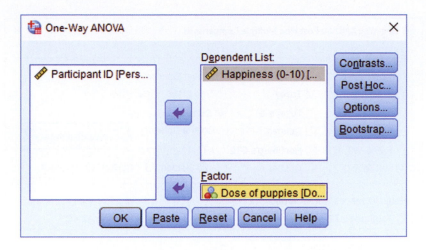

Figure 12.13   Main dialog box for one-way ANOVA

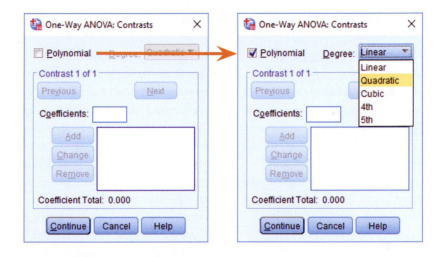

Figure 12.14   Dialog box for conducting planned contrasts

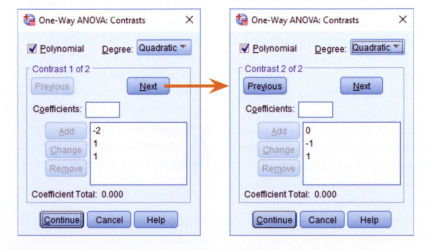

Figure 12.15   *Contrasts* dialog box completed for the two contrasts of the puppy therapy data

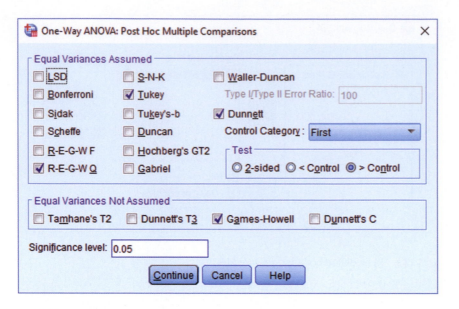

Figure 12.16   Dialog box for specifying *post hoc* tests

therapy data is the 15-minute group (because this group was coded in the data editor with the second-highest value). Click in the box labeled C*oefficients*, type '1' and click _Add_. Finally, we input the weight for the last group, which for the puppy therapy data is the 30-minute group (because this group was coded with the highest value in the data editor). Click in the box labeled *Coefficients*, type '1' and click _Add_. The dialog box should now look like Figure 12.15 (left).

Once the weights are assigned you can change or remove any one of them by selecting the weight that you want to change. The weight will appear in the box labeled *Coefficients* where you can type a new weight and then click _Change_. Alternatively, you can click a weight and remove it completely by selecting _Remove_. Underneath the weights the sum of weights is displayed, which, as we saw in Section 12.4.2, should equal zero. If the *Coefficient Total* is anything other than zero you should go back and check that the contrasts you have planned make sense and that you have followed the appropriate rules for assigning weights.

Once you have specified the first contrast, click _Next_. The weights that you have just entered will disappear and the dialog box will now read *Contrast 2 of 2*. We know from Section 12.4.2 that the weights for contrast 2 were: 0 (control group), –1 (15-minute group) and +1 (30-minute group). We specify this contrast as before. Remembering that the first weight we enter will be for the control group, we enter the value 0 as the first weight. Click in the box labeled *Coefficients*, then type '0' and click _Add_. Next, we input the weight for the 15-minute group by clicking in the box labeled *Coefficients*, typing '–1' and clicking _Add_. Finally, we input the weight for the 30-minute group by clicking in the box labeled *Coefficients*, typing '+1' and clicking _Add_. The dialog box should now look like Figure 12.15 (right). Notice that the weights add up to zero as they did for contrast 1. You must remember to input zero weights for any groups that are not in the contrast. When the contrasts have been specified, click _Continue_ to return to the main dialog box.

12.6.3 *Post hoc* tests in SPSS ▌▌▌▌

In theory, if we have done planned contrasts we shouldn't need to do *post hoc* tests (because we have already tested the hypotheses of interest). Likewise, if we

choose to conduct *post hoc* tests then we should not need to do planned contrasts (because we have no hypotheses to test). However, for the sake of space we will conduct some *post hoc* tests on the puppy therapy data as well as the contrasts we have just specified. Click _Post Hoc..._ in the main dialog box to access the dialog box in Figure 12.16.

In Section 12.5.3, I recommended various *post hoc* procedures for particular situations. For the puppy data there are equal sample sizes and so we need not use Gabriel's test. We should use Tukey's test and REGWQ and check the findings with the Games–Howell procedure. We have a specific hypothesis that both the 30- and 15-minute groups should differ from the control group and so we could use Dunnett's test to examine these hypotheses. Once you have selected ☑ Dunnett, change the control category from Last ▾ to First ▾ so that the no puppies category (which, remember, we coded with the lowest value and so is the first group) is used as the comparison group. You can choose whether to conduct a two-tailed (◉ 2-sided) or a one-tailed test. If you choose a one-tailed test (which I advised against in Section 2.9.5) then you need to predict whether you believe that the mean of the first group (i.e., no puppies) will be less than each experimental group (◉ < Control) or greater (◉ > Control). These are all the *post hoc* tests that we need to specify (see Figure 12.16). Click _Continue_ to return to the main dialog box.

12.6.4 Options ▌▌▌▌

Click _Options..._ to access the dialog box in Figure 12.17. You can ask for descriptive statistics, which will produce a table of the means, standard deviations, standard errors, ranges and confidence intervals within each group. This information will help us to interpret the results. Select ☑ Homogeneity of variance test if you want Levene's test (Section 6.11.2), although it's more important to select ☑ Brown-Forsythe or

☑ <u>Welch</u> so that you can interpret these if you're concerned about having unequal variances or, better still, use them by default. There is also an option to have a *Means plot* which produces a line graph of the group means. The resulting graph is a leprous tramp compared to what you can create using the Chart Builder and it's best to graph your data *before* the analysis, not during it. Finally, the options let us specify whether we want to exclude cases on a listwise basis or on a per analysis basis (see SPSS Tip 6.1 for an explanation). This option is useful only if you are fitting models to several outcome variables simultaneously (which hopefully you're not—ever).

12.6.5 Bootstrapping ▮▮▮

Also in the main dialog box is the alluring Bootstrap... button. We know that bootstrapping is a good way to overcome bias, and this button glistens and tempts us with the promise of untold riches, like a diamond in a bull's rectum. However, if you use bootstrapping it'll be as disappointing as if you reached for that diamond only to discover that it's a piece of glass. You might, not unreasonably, think that if you select bootstrapping you'd get a nice bootstrap of the *F*-statistic. You won't. You will get bootstrap confidence intervals around the means (if you ask for descriptive statistics), contrasts and *post hoc* tests. All of which are useful, but the main test won't be bootstrapped. For this example, we have a very small data set so bootstrapping is going to go haywire, so we won't select it. Click OK in the main dialog box to run the analysis.

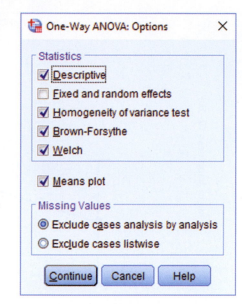

Figure 12.17 Options for one-way ANOVA

**Oditi's Lantern
One-way ANOVA**

'I, Oditi, have made great progress in unearthing the hidden truth behind the numbers. This morning, one of my loyal followers reported to me that, based on an ANOVA he'd done, all dogs are controlled by cats who hide small remote controls up their rectums and manipulate them with their tongues. Everytime you see a cat 'cleaning' itself, there will be a dog nearby chasing its tail. Listen carefully and you can hear the cat laughing to itself. Be warned, cats are merely piloting the technology, and soon they will control us too, turning us into heated chairs and food vendors. We must find out more. Stare into my lantern so that you too can use ANOVA.'

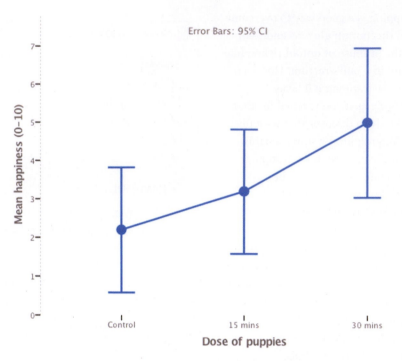

Figure 12.18 Error bar (95% CI) chart of the puppy therapy data

Descriptives

Happiness (0–10)

	N	Mean	Std. Deviation	Std. Error	95% Confidence Interval for Mean		Minimum	Maximum
					Lower Bound	Upper Bound		
Control	5	2.20	1.304	.583	.58	3.82	1	4
15 mins	5	3.20	1.304	.583	1.58	4.82	2	5
30 mins	5	5.00	1.581	.707	3.04	6.96	3	7
Total	15	3.47	1.767	.456	2.49	4.45	1	7

Output 12.3

Test of Homogeneity of Variances

Happiness (0–10)

Levene Statistic	df1	df2	Sig.
.092	2	12	.913

Output 12.4

12.7 Output from one-way independent ANOVA ▮▮▮

You should find that the output looks the same as what follows. If not, we should panic because one of us has done it wrong—hopefully not me or a lot of trees have died for nothing. Figure 12.18 shows a line chart with error bars from the self-test earlier in the chapter (I have edited my graph; see if you can make yours look like mine). All the

error bars overlap a fair bit, indicating that, at face value, there might not be between-group differences (see Section 2.9.9). The line that joins the means seems to indicate a linear trend in that, as the dose of puppy therapy increases, so does the mean level of happiness.

12.7.1 Output for the main analysis ▮▮▮

Output 12.3 shows the table of descriptive statistics for the puppy data. The means and standard deviations correspond to those shown in Table 12.1, which is reassuring. We are also given the standard error and confidence intervals for the mean. If this sample is one of the 95%

that have confidence intervals containing the true value then the true value of the mean is between 0.58 and 3.82 for the control group. We will refer back to this table as we wade through the output.

Output 12.4 shows Levene's test (see Section 6.11.2) for those of you hell-bent on using it. For these data the variances are very similar (hence Levene's test is very non-significant, with p close to 1); in fact, if you look at Output 12.3 you'll see that the variances of the control and 15-minute groups are identical.

Output 12.5 shows the main ANOVA summary table. The table is divided into between-group effects (effects due to the model—the experimental effect) and within-group effects (this is the unsystematic variation in the data). The between-group effect is further broken down into a linear and quadratic component as requested in Figure 12.14. The between-group effect labeled *Combined* is the overall experimental effect or, put another way, the improvement in the prediction of happiness scores resulting from using the group means. We are told the model sums of squares ($SS_M = 20.13$), which corresponds to the value calculated in Section 12.2.3. The degrees of freedom are 2 and the mean squares for the model corresponds to the value calculated in Section 12.2.5 (10.067).

The row labeled *Within Groups* gives details of the unsystematic variation within the data (the variation due to natural individual differences in happiness and different reactions to puppy therapy). The table tells us that the residual sum of squares (SS_R) is 23.60, as calculated in Section 12.2.4. The average amount of unsystematic variation, the mean squares (MS_R), is 1.967, as calculated in Section 12.2.5.

The test of whether the group means are the same is represented by the F-statistic for the combined between-group effect. This tells us whether predicting happiness from group means significantly improves the fit of the model. The value of F is 5.12, which we calculated in Section 12.2.6. The final column labeled *Sig.* tells us the probability of getting an F at least this big if there wasn't a difference between means in the population (see also SPSS Tip 12.1). In this case, there is a probability of 0.025 that an F-statistic of at least this size would occur if in reality the effect was zero. Assuming we set a cut-off point of 0.05 as a criterion for statistical significance before collecting data, most scientists would take the fact that 0.025 is less than the criterion of 0.05 as support for a significant effect of puppy therapy. At this stage we still do not know exactly what the effect of puppy therapy was (we don't know which groups differed). One interesting point is that we obtained a significant experimental effect yet the error bar plot suggested that no significant difference would be found. This contradiction illustrates how the error bar

ANOVA

Happiness (0–10)

			Sum of Squares	df	Mean Square	F	Sig.
Between Groups	(Combined)		20.133	2	10.067	5.119	.025
	Linear Term	Contrast	19.600	1	19.600	9.966	.008
		Deviation	.533	1	.533	.271	.612
	Quadratic Term	Contrast	.533	1	.533	.271	.612
Within Groups			23.600	12	1.967		
Total			43.733	14			

Output 12.5

chart can act only as a rough guide to the data.

Knowing that the overall effect of puppy therapy was significant, we can look at the trend analysis. First, let's look at the linear component. This contrast tests whether the means increase across groups in a linear way. For the linear trend the F-statistic is 9.97 and this value is significant at $p = 0.008$. Therefore, we can say that as the dose of puppy therapy increased from nothing to 15 minutes to 30 minutes, happiness increased proportionately. The quadratic trend tests whether the pattern of means is curvilinear (i.e., is represented by a curve that has one bend). The error bar graph of the data suggests that the means cannot be represented by a curve and the results for the quadratic trend bear this out. The F-statistic for the quadratic trend is non-significant (in fact, the value of F is less than 1, which immediately indicates that this contrast will not be significant).

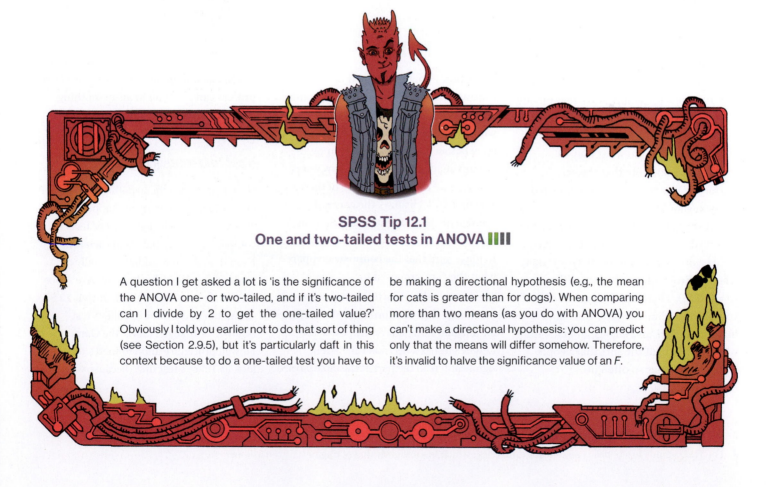

SPSS Tip 12.1
One and two-tailed tests in ANOVA ||||

A question I get asked a lot is 'is the significance of the ANOVA one- or two-tailed, and if it's two-tailed can I divide by 2 to get the one-tailed value?' Obviously I told you earlier not to do that sort of thing (see Section 2.9.5), but it's particularly daft in this context because to do a one-tailed test you have to be making a directional hypothesis (e.g., the mean for cats is greater than for dogs). When comparing more than two means (as you do with ANOVA) you can't make a directional hypothesis: you can predict only that the means will differ somehow. Therefore, it's invalid to halve the significance value of an F.

Robust Tests of Equality of Means

Happiness (0–10)

	Statistic[a]	df1	df2	Sig.
Welch	4.320	2	7.943	.054
Brown–Forsythe	5.119	2	11.574	.026

a. Asymptotically F distributed.

Output 12.6

Contrast Coefficients

	Dose of puppies		
Contrast	Control	15 mins	30 mins
1	-2	1	1
2	0	-1	1

Contrast Tests

		Contrast	Value of Contrast	Std. Error	t	df	Sig. (2–tailed)
Happiness (0–10)	Assume equal variances	1	3.80	1.536	2.474	12	.029
		2	1.80	.887	2.029	12	.065
	Does not assume equal variances	1	3.80	1.483	2.562	8.740	.031
		2	1.80	.917	1.964	7.720	.086

Output 12.7

Output 12.6 shows the Welch and Brown–Forsythe F-statistics. If you're interested in how these values are calculated then look at Jane Superbrain Box 12.2, but to be honest it's just bloody confusing; you're much better off just looking at the values in Output 12.6 and trusting that they do what they're supposed to do (note that the error degrees of freedom have been adjusted and you should remember this when you report the values). Based on whether the observed p is less than 0.05, the Welch F yields a non-significant result (p = 0.054) whereas Brown–Forsythe is significant (p = 0.026). This is confusing, but only if you like to imbue the 0.05 threshold with magical powers and engage in black-and-white thinking of the sort that people who use p-values so often do.

12.7.2 Output for planned contrasts ▮▮▮

In Section 12.6.2 we defined two planned contrasts: one to test whether the control group was different from the two groups which received puppy therapy, and one to see whether the two doses of puppy therapy made a difference to happiness. Output 12.7 shows the results of these. The first table displays the contrast coefficients that we entered in Section 12.6.2, and it is worth checking this table to make sure that the contrasts compare what they are supposed to.

The second table gives the statistics for each contrast in their raw form, but also corrected for unequal variances. Some people use Levene's test as a decision rule: if it is significant, read the part of the table labeled

Does not assume equal variances, and if it is not, use the part of the table labeled *Assume equal variances*. However, given the issues with Levene's test, it's probably more sensible to just routinely look at the corrected values. The table tells us the value of the contrast itself, which is the weighted sum of the group means. This value is obtained by taking each group mean, multiplying it by the weight for the contrast of interest, and then adding these values together.[3] The table also gives the standard error of each contrast and a t-statistic. The t-statistic is derived by dividing the contrast value by the standard error ($t = 3.8/1.5362 = 2.47$) and then, in the bottom two rows of the table, adjusted for the degree of heterogeneity. The significance value of the contrast is given in the final column, and this value is two-tailed. For contrast 1, we can say that taking puppy therapy significantly increased happiness compared to the control group ($p = 0.031$), but contrast 2 tells us that 30 minutes of puppy therapy did not significantly affect happiness compared to 15 minutes ($p = 0.086$). Contrast 2 is almost significant, which again demonstrates how the NHST process can lead you to all-or-nothing thinking (Section 3.2.2).

12.7.3 Output for *post hoc* tests ▮▮▮

If we had no specific hypotheses about the effect of puppy therapy on happiness then we would have selected *post hoc* tests to compare all group means to each other. Even though we wouldn't normally do contrasts and *post hoc* tests, to save space we did (Section 12.6.3) and Output 12.8 shows tables containing Tukey's test (known as Tukey's HSD),[4] the Games–Howell procedure and Dunnett's test. If we look at Tukey's test first (because we have

---

3 For the first contrast this value is $\sum \bar{X}W = (2.2 \times -2) + (3.2 \times 1) + (5 \times 1) = 3.8$.

4 The HSD stands for 'honestly significant difference', which has a slightly dodgy ring to it, if you ask me.

no reason to doubt that the population variances are unequal) we see that for each pair of groups the difference between group means is displayed, the standard error of that difference, the significance level of that difference and a 95% confidence interval. The first row of Output 12.8 compares the control group to the 15-minute group and reveals a non-significant difference (*Sig.* of 0.516 is greater than 0.05), and the second row compares the control group to the 30-minute group where there is a significant difference (*Sig.* of 0.021 is less than 0.05). It might seem odd that the planned contrast showed that any dose of puppy therapy produced a significant increase in happiness, yet the *post hoc* tests indicate that a 15-minute does not.

In Section 12.4.2, I explained that the first planned contrast would compare the experimental groups to the control group. Specifically, it would compare the average of the two group means of the experimental groups ((3.2 + 5.0)/2 = 4.1) to the mean of the control group (2.2). So, it was assessing whether the difference between these values (4.1 − 2.2 = 1.9) was significant. In the *post hoc* tests, when the 15-minute group is compared to the control it is testing whether the difference between the means of these two groups is significant. The difference in this case is only 1, compared to a difference of 1.9 for the planned contrast. This explanation illustrates how it is possible to have apparently contradictory results from planned contrasts and *post hoc* comparisons. More important, it illustrates the need to think carefully about what our planned contrasts test.

The third and fourth rows of Output 12.8 compare the 15-minute group to both the control group and the 30-minute group. The test involving the 15-minute and 30-minute groups shows that these group means did not differ (because the *p* of 0.147 is greater than our alpha of 0.05). Rows 5 and 6 repeat comparisons already discussed.

SELF TEST
Can you explain the contradiction between the planned contrasts and *post hoc* tests?

Multiple Comparisons

Dependent Variable: Happiness (0–10)

	(I) Dose of puppies	(J) Dose of puppies	Mean Difference (I–J)	Std. Error	Sig.	95% Confidence Interval	
						Lower Bound	Upper Bound
Tukey HSD	Control	15 mins	−1.000	.887	.516	−3.37	1.37
		30 mins	−2.800*	.887	.021	−5.17	−.43
	15 mins	Control	1.000	.887	.516	−1.37	3.37
		30 mins	−1.800	.887	.147	−4.17	.57
	30 mins	Control	2.800*	.887	.021	.43	5.17
		15 mins	1.800	.887	.147	−.57	4.17
Games-Howell	Control	15 mins	−1.000	.825	.479	−3.36	1.36
		30 mins	−2.800*	.917	.039	−5.44	−.16
	15 mins	Control	1.000	.825	.479	−1.36	3.36
		30 mins	−1.800	.917	.185	−4.44	.84
	30 mins	Control	2.800*	.917	.039	.16	5.44
		15 mins	1.800	.917	.185	−.84	4.44
Dunnett t (>control)[b]	15 mins	Control	1.000	.887	.227	−.87	
	30 mins	Control	2.800*	.887	.008	.93	

*. The mean difference is significant at the 0.05 level.

b. Dunnett t-tests treat one group as a control, and compare all other groups against it.

Output 12.8

Happiness (0–10)

	Dose of puppies	N	Subset for alpha = 0.05	
			1	2
Tukey HSD[a]	Control	5	2.20	
	15 mins	5	3.20	3.20
	30 mins	5		5.00
	Sig.		.516	.147
Ryan–Einot–Gabriel–Welsch Range	Control	5	2.20	
	15 mins	5	3.20	3.20
	30 mins	5		5.00
	Sig.		.282	.065

Means for groups in homogeneous subsets are displayed.

a. Uses Harmonic Mean Sample Size = 5.000.

Output 12.9

The second block of the table describes the Games–Howell test, and a quick inspection reveals the same pattern of results: the only groups that differed significantly were the 30-minute and control groups. These results give us confidence in our conclusions from Tukey's test because even if the population variances are not equal (which seems unlikely given that the sample variances are very similar), then the profile of results holds true. Finally, Dunnett's test is described, and you'll hopefully remember that we asked the computer to compare both experimental groups against the control using a one-tailed hypothesis that the mean of the control group would be smaller than both experimental groups. Even as a one-tailed hypothesis, levels of happiness in the 15-minute group are equivalent to the control group. However, the 30-minute group has a significantly higher happiness than the control group.

Output 12.9 shows the results of Tukey's test and the REGWQ test. These tests display subsets of groups that have the same means with associated *p*-values for each subset. Tukey's test has created two subsets of groups with statistically similar means. The first subset contains the control and 15-minute groups (indicating that these two groups have similar means, *p* = 0.516), whereas the second subset contains the 30- and 15-minute groups (which also have similar means, *p* = 0.147). The REGWQ test agrees with the first subset, suggesting that the control and 15-minute groups have similar means (*p* = 0.282) and the second that the 15- and 30-minute groups have similar means (*p* = 0.065). Tukey's LSD uses the **harmonic mean** sample size, which is a weighted version of the mean that takes account of the relationship between variance and sample size. Although you don't need to know the intricacies of the harmonic mean, it is useful that the harmonic sample size is used because it reduces bias that might be introduced through having unequal sample sizes. However, as we have seen, these tests are still biased when sample sizes are unequal.

Labcoat Leni's Real Research 12.1
Scraping the barrel? ▮▮▮▮

Gallup, G. G. J., et al. (2003). *Evolution and Human Behavior*, 24, 277–289.

Evolution has endowed us with many beautiful things (cats, dolphins, the Great Barrier Reef, etc.) all selected to fit their ecological niche. Given evolution's seemingly limitless capacity to produce beauty, it's something of a wonder how it managed to produce such a monstrosity as the human penis. One theory is sperm competition: the human penis has unusually large glans (the 'bell-end') compared to other primates, and this may have evolved so that the penis can displace seminal fluid from other males by 'scooping it out' during intercourse. Armed with various female masturbatory devices from Hollywood Exotic Novelties, an artificial vagina from California Exotic Novelties, and some water and cornstarch to make fake sperm, Gallup et al. (2003) put this theory to the test. They loaded the artificial vagina with 2.6 ml of fake sperm and inserted one of three female sex toys into it before withdrawing it: a control phallus that had no coronal ridge (i.e., no bell-end), a phallus with a minimal coronal ridge (small bell-end) and a phallus with a coronal ridge.

They measured sperm displacement as a percentage using the following expression (included here because it is more interesting than all of the other equations in this book):

$$\frac{\text{weight of vagina with semen} - \text{weight of vagina following insertion and removal of phallus}}{\text{weight of vagina with semen} - \text{weight of empty vagina}} \times 100$$

100% means that all the sperm was displaced, and 0% means that none of the sperm was displaced. If the human penis evolved as a sperm displacement device then Gallup et al. predicted: (1) that having a bell-end would displace more sperm than not; and (2) that the phallus with the larger coronal ridge would displace more sperm than the phallus with the minimal coronal ridge. The conditions are ordered (no ridge, minimal ridge, normal ridge), so we might also predict a linear trend. The data are in the file **Gallup et al.sav**. Draw an error bar graph of the means of the three conditions. Fit a model with planned contrasts to test the two hypotheses above. What did Gallup et al. find? Answers are on the companion website (or look at pages 280–281 in the original article).

12.8 Robust comparisons of several means ▌▌▌▌

It's possible to run a robust test of several means using R. We need the *Essentials for R* plugin and *WRS2* package installed (Section 4.13). The companion website contains a syntax file (**t1waybt.sps**) for running a robust variant of one-way independent ANOVA (*t1waybt*) with *post hoc* tests (*mcppb20*) described by Wilcox (2017). These tests assume neither normality nor homogeneity of variance, so you can ignore these assumptions and plough ahead. The syntax in the file is as follows:

```
BEGIN PROGRAM R.
library(WRS2)
mySPSSdata = spssdata.
GetDataFromSPSS(factorMode =
"labels")
t1waybt(Happiness~Dose, data =
mySPSSdata, tr = 0.2, nboot = 1000)
```

```
Effective number of bootstrap samples was 671.

Test statistic: 3
p-value: 0.08942
Variance explained 0.623
Effect size 0.789

mcppb20(formula = Happiness ~ Dose, data = mySPSSdata, tr = 0.2,
    nboot = 1000)

                          psihat  ci.lower  ci.upper  p-value
Control vs.  15  mins     -1     -3.33333   1.33333   0.381
Control vs.  30  mins     -3     -5.00000  -0.33333   0.010
15 mins vs.  30  mins     -2     -4.00000   0.66667   0.081
```

Output 12.10

```
mcppb20(Happiness~Dose, data =
mySPSSdata, tr = 0.2, nboot = 1000)
END PROGRAM.
```

Select and run these six lines of syntax (see SPSS Tip 10.3) and you'll find text output in the viewer (Output 12.10) that tells us that there was not a significant difference (because the *p*-value is greater than 0.05)

in happiness scores across the puppy therapy groups, $F_t = 3$, $p = 0.089$. The *post hoc* tests (which technically we should ignore because the overall test wasn't significant) show no significant difference between the control and 15-minute groups ($p = 0.381$) or between the 15- and 30-minute groups ($p = 0.081$), but there is a significant difference (that we should

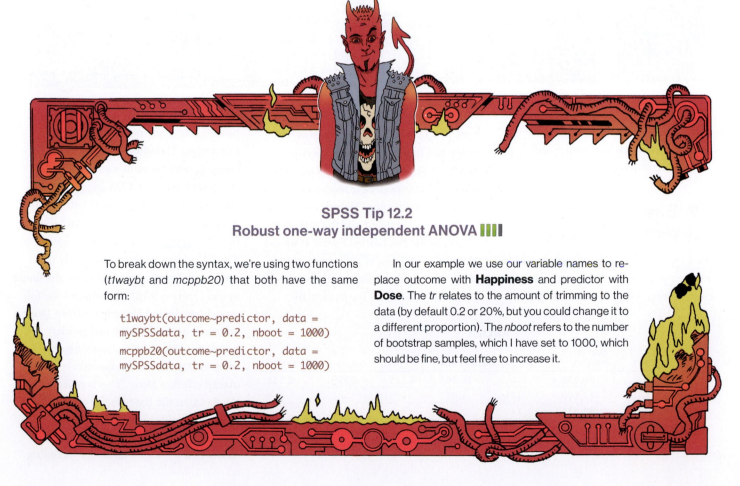

SPSS Tip 12.2
Robust one-way independent ANOVA ▌▌▌▌

To break down the syntax, we're using two functions (*t1waybt* and *mcppb20*) that both have the same form:

```
t1waybt(outcome~predictor, data =
mySPSSdata, tr = 0.2, nboot = 1000)
mcppb20(outcome~predictor, data =
mySPSSdata, tr = 0.2, nboot = 1000)
```

In our example we use our variable names to replace outcome with **Happiness** and predictor with **Dose**. The *tr* relates to the amount of trimming to the data (by default 0.2 or 20%, but you could change it to a different proportion). The *nboot* refers to the number of bootstrap samples, which I have set to 1000, which should be fine, but feel free to increase it.

Cramming Sam's Tips
One-way independent ANOVA

- One-way independent ANOVA compares several means, when those means have come from different groups of people; for example, if you have several experimental conditions and have used different participants in each condition. It is a special case of the linear model.
- When you have generated specific hypotheses before the experiment use *planned contrasts*, but if you don't have specific hypotheses use *post hoc* tests.
- There are lots of different *post hoc* tests: when you have equal sample sizes and homogeneity of variance is met, use REGWQ or Tukey's HSD. If sample sizes are slightly different then use Gabriel's procedure, but if sample sizes are very

- different use Hochberg's GT2. If there is any doubt about homogeneity of variance use the Games–Howell procedure.
- You can test for homogeneity of variance using Levene's test, but consider using a *robust* test in all situations (the Welch or Browne–Forsythe *F*) or Wilcox's *t1way()* function.
- Locate the *p*-value (usually in a column labeled *Sig.*). If the value is less than 0.05 then scientists typically interpret this as the group means being significantly different.
- For contrasts and *post hoc* tests, again look to the columns labeled *Sig.* to discover if your comparisons are significant (they will be if the significance value is less than 0.05).

ignore because the overall test wasn't significant) between the control and 30-minute groups ($p = 0.010$).

12.9 Bayesian comparison of several means ▋▋▋

Because we've been using a linear model to compare means, we can compute a Bayes factor (Section 3.8) just as we did in

Section 9.13. You can instead access the dialog box in Figure 12.19 by selecting *Analyze* ▸ *Bayesian Statistics* ▸ *One-way ANOVA*. First, specify the model by dragging the outcome (**Happiness**) to the box labeled *Dependent* and the predictor (**Dose**) to the box labeled *Factor*. If we use estimation on a model with a categorical predictor we will get Bayesian estimates for the group means (*not* the *b*s from the

model), which are not particularly interesting. Therefore, I have selected to compute only Bayes factors (◉ Estimate Bayes Factor). Click Bayes Factor... and select the default ◉ JZS method (Jeffreys, 1961; Zellner & Siow, 1980). In the main dialog box click OK to fit the model.

The Bayes factor here is essentially the same as the one we computed for the linear model (Section 9.13): it compares the full model (predicting **Happiness** from **Dose** and the intercept) to the null model (predicting **Happiness** from only the intercept). It, therefore, quantifies the overall effect of **Dose**. This is reflected by the fact that the Bayes factor appears as part of the table displaying the overall fit of the model (the *F*-statistic). Output 12.11 shows that the Bayes factor for including

ANOVA

Happiness (0–10)	Sum of Squares	df	Mean Square	F	Sig.	Bayes Factor[a]
Between Groups	20.133	2	10.067	5.119	.025	2.158
Within Groups	23.600	12	1.967			
Total	43.733	14				

a. Bayes factor: JZS

Output 12.11

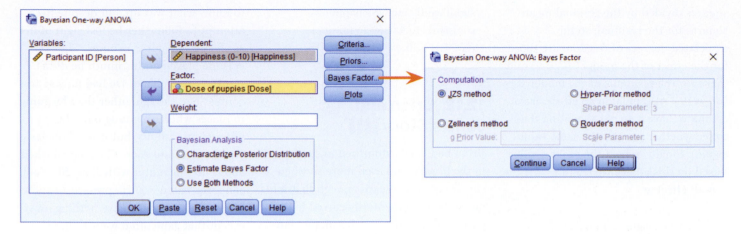

Figure 12.19   Dialog boxes for Bayesian ANOVA

Dose as a predictor (compared to not including it) is 2.158. This means that the data are 2.158 times more likely under the alternative hypothesis (dose of puppy therapy has an effect) than under the null (dose of puppy therapy has no effect). This value is not strong evidence, but nevertheless suggests we should shift our belief about puppy therapy towards it being effective by a factor of about 2.

12.10 Calculating the effect size ▌▌▌▌

SPSS Statistics doesn't provide an effect size, but we saw in equation (9.10) that we can compute R^2 for a linear model:

$$R^2 = \frac{SS_M}{SS_T} \qquad (12.30)$$

We know these values from the output, so we can calculate R^2 using the between-group effect (SS_M), and the total amount of variance in the data (SS_T)—although for some bizarre reason it's usually called **eta squared**, η^2. It is then a simple matter to take the square root of this value to give us the effect size r of 0.68:

$$r^2 = \eta^2 = \frac{SS_M}{SS_T} = \frac{20.13}{43.73} = 0.46 \qquad (12.31)$$
$$r = \sqrt{0.46} = 0.68$$

Therefore, the effect of puppy therapy on happiness is a substantive finding.

However, this measure of effect size is slightly biased because it is based purely on sums of squares from the sample and no adjustment is made for the fact that we're trying to estimate the effect size in the population. Therefore, we often use a slightly more complex measure called **omega squared** (ω^2). This effect size estimate is still based on the sums of squares that we've met in this chapter: it uses the variance explained by the model, and the average error variance:

$$\omega^2 = \frac{SS_M - (df_M)MS_R}{SS_T + MS_R} \qquad (12.32)$$

The df_M in the equation is the degrees of freedom for the effect, which you can get from the output. In this example we'd get:

$$\omega^2 = \frac{20.13 - (2)1.97}{43.73 + 1.97} = \frac{16.19}{45.70} = 0.35 \qquad (12.33)$$
$$\omega = 0.60$$

This adjustment has led to a slightly lower estimate to using r, and in general ω is a more accurate measure. Think of ω as you would r (because it's basically an unbiased estimate of r). People normally report ω^2, and it has been suggested that values of 0.01, 0.06 and 0.14 represent small, medium and large effects, respectively (Kirk, 1996). Remember, though, that these are rough guidelines and that effect sizes need to be interpreted within the context of the research literature.

Most of the time it isn't that interesting to

have effect sizes for the overall ANOVA because it's testing a general hypothesis. Instead, we really want effect sizes for the contrasts (because these compare only two things, so the effect size is easier to interpret). Planned contrasts are tested with the t-statistic and, therefore, we can use the same equation as in Section 10.9.5:

$$r_{Contrast} = \sqrt{\frac{t^2}{t^2 + df}} \qquad (12.34)$$

We know the value of t and the df from Output 10.7, and so we can compute r as follows:

$$r_{Contrast1} = \sqrt{\frac{2.474^2}{2.474^2 + 12}} = \sqrt{\frac{6.12}{18.12}} = 0.58 \qquad (12.35)$$

As well as being statistically significant, this effect represents a substantive finding. For contrast 2 we get:

$$r_{Contrast2} = \sqrt{\frac{2.029^2}{2.029^2 + 12}} = \sqrt{\frac{4.12}{16.12}} = 0.51 \qquad (12.36)$$

This too is a substantive finding.

12.11 Reporting results from one-way independent ANOVA ▌▌▌▌

We report the F-statistic and the degrees of freedom associated with it. Remember that F is the model mean squares divided by the residual mean squares for the residual, so the associated degrees of freedom are those for the effect of the model ($df_M = 2$) and those for the residuals of the model ($df_R = 12$). Also include an effect size estimate (have a go at calculating these to see whether you get the same values as me). Finally, report the exact p-value. Based on this advice we could report the overall effect as:

✓ There was a significant effect of puppy therapy on levels of happiness, $F(2, 12) = 5.12$, $p = 0.025$, $\omega = 0.60$.

Notice that the degrees of freedom are in parentheses. I've advised you to always report Browne–Forsythe or Welch's F, so let's do that instead (note the adjusted residual degrees of freedom, and change in F and p):

✓ There was not a significant effect of puppy therapy on levels of happiness, $F(2, 7.94) = 4.32$, $p = 0.054$, $\omega = 0.60$.
✓ There was a significant effect of puppy therapy on levels of happiness, $F(2, 11.57) = 5.12$, $p = 0.026$, $\omega = 0.60$.

The linear contrast can be reported in much the same way:

✓ There was a significant linear trend, $F(1, 12) = 9.97$, $p = 0.008$, $\omega = 0.62$, indicating that as the dose of puppy therapy increased, happiness increased proportionately.

The degrees of freedom have changed to reflect how F was calculated. We can do something similar for the planned contrasts:

✓ Planned contrasts revealed that having any dose of puppy therapy significantly increased happiness compared to having a control, $t(8.74) = 2.56$, $p = 0.031$, $r = 0.58$, but having 30 minutes did not

significantly increase happiness compared to having 15 minutes, $t(7.72) = 1.96$, $p = 0.086$, $r = 0.51$.

12.12 Brian and Jane's Story ▮▮▮

Brian had been astonished to see Jane in the library, and even more so when she dropped her guard enough to ask his name. He'd momentarily cursed his parents for calling him Brian, but Jane didn't seem to care. They'd even continued a bit of an awkward conversation about music before Jane quickly excused herself and ran off. She'd looked pale, like she'd remembered something horrific. Brian wasn't used to seeing her like that: she was always so self-assured. He wanted

to know what was up, but despite his attempt to follow her he lost sight of her as she darted across campus. The episode played on his mind: he didn't want her to feel bad. He had no way to contact her, though, other than by going to places where she hung out. He loitered on campus, but it was hopeless. He bumped into Alex. She seemed irked about his encounter with Jane. She said Jane was bad news. He pressed Alex for information, but she gave nothing away except that Jane often went to Blow Your Speakers, the record store in town, at the weekends. Brian knew it well: they had vinyl and coffee mornings on a Saturday. He'd always avoided it, fearing it would be full of pretentious hipsters. This Saturday he made an exception and sat among people in skinny trousers and strange hats, who sipped coffee,

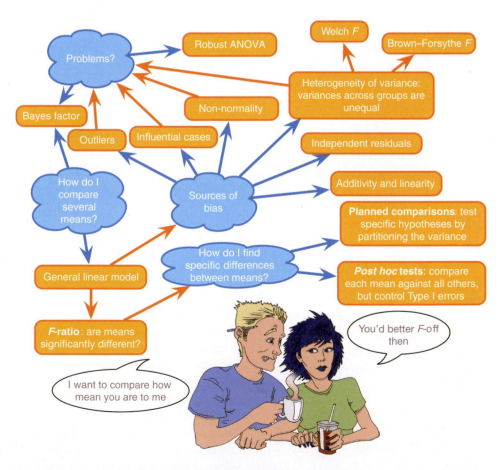

Figure 12.20 What Brian learnt from this chapter

twiddled their moustaches and tried to choose more obscure cuts to play than the last person. He'd had enough and got up to leave as Jane walked in. Jane was shocked to see Brian, but resisted the urge to flee. Brian smiled, rushed over and placed a reassuring hand on her shoulder. Jane flinched. Brian asked if she was alright and explained that he was worried he'd upset her. Jane was silent, unsure what to say, so Brian filled the silence with his statistics lecture. He needed to work on his social skills.

12.13 What next?

My life was changed by a letter popping through the letterbox one day saying that I could go to the local grammar school. When my parents told me, they were downbeat and not as celebratory as you might expect, but they knew how much I wanted to be with my friends. I had got used to my failure, and my initial reaction was to say that I wanted to go to the local school. I was unwavering in this view. Unwavering, that is, until my brother convinced me that being at the same

school as him would be really cool. It's hard to measure how much I looked up to him, and still do, but the fact that I willingly subjected myself to a lifetime of social dysfunction just to be with him is a measure of sorts. As it turned out, being at school with him was not always cool—he was bullied for being a boffin (that's what they called nerds in the UK in the 1980s), and being the younger brother of a boffin made me a target. Luckily, unlike my brother, I was stupid and played football, which seemed to be good enough reasons for them to leave me alone. Most of the time.

12.14 Key terms that I've discovered

Analysis of variance (ANOVA)	Eta squared, η^2	Omega squared	Quadratic trend
Brown–Forsythe F	General linear model	Orthogonal	Quartic trend
Cubic trend	Grand variance	Pairwise comparisons	Repeated contrast
Deviation contrast	Harmonic mean	Planned contrasts	Simple contrast
Difference contrast (reverse Helmert contrast)	Helmert contrast	Polynomial contrast	Weights
	Independent ANOVA	*Post hoc* tests	Welch's F

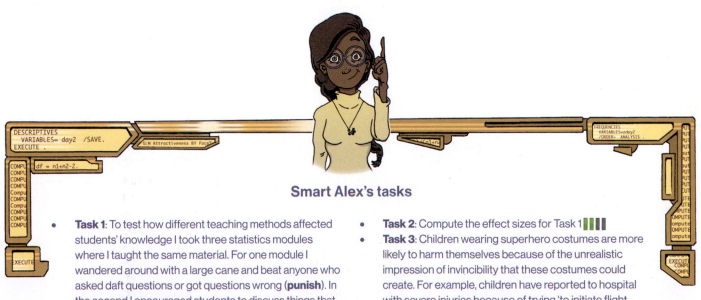

Smart Alex's tasks

- **Task 1**: To test how different teaching methods affected students' knowledge I took three statistics modules where I taught the same material. For one module I wandered around with a large cane and beat anyone who asked daft questions or got questions wrong (**punish**). In the second I encouraged students to discuss things that they found difficult and gave anyone working hard a nice sweet (**reward**). In the final course I neither punished nor rewarded students' efforts (**indifferent**). I measured the students' exam marks (**percentage**). The data are in the file **Teach.sav**. Fit a model with planned contrasts to test the hypotheses that: (1) reward results in better exam results than either punishment or indifference; and (2) indifference will lead to significantly better exam results than punishment.

- **Task 2**: Compute the effect sizes for Task 1
- **Task 3**: Children wearing superhero costumes are more likely to harm themselves because of the unrealistic impression of invincibility that these costumes could create. For example, children have reported to hospital with severe injuries because of trying 'to initiate flight without having planned for landing strategies' (Davies, Surridge, Hole, & Munro-Davies, 2007). I can relate to the imagined power that a costume bestows upon you; indeed, I have been known to dress up as Fisher by donning a beard and glasses and trailing a goat around on a lead in the hope that it might make me more knowledgeable about statistics. Imagine we had data (**Superhero.sav**) about the severity of **injury** (on a scale from 0, no injury, to 100, death) for children reporting to

the accident and emergency department at hospitals, and information on which superhero costume they were wearing (**hero**): Spiderman, Superman, the Hulk or a teenage mutant ninja turtle. Fit a model with planned contrasts to test the hypothesis that different costumes give rise to more severe injuries. ▌▌▌▌

- **Task 4**: In Chapter 7 (Section 7.6) there are some data looking at whether eating soya meals reduces your sperm count. Analyze these data with a linear model (ANOVA). What's the difference between what you find and what was found in Section 7.6.5? Why do you think this difference has arisen? ▌▌▌▌

- **Task 5**: Mobile phones emit microwaves, and so holding one next to your brain for large parts of the day is a bit like sticking your brain in a microwave oven and pushing the 'cook until well done' button. If we wanted to test this experimentally, we could get six groups of people and strap a mobile phone on their heads, then by remote control turn the phones on for a certain amount of time each day. After six months, we measure the size of any tumor (in mm^3) close to the site of the phone antenna (just behind the ear). The six groups experienced 0, 1, 2, 3, 4 or 5 hours per day of phone microwaves for six months. Do

tumors significantly increase with greater daily exposure? The data are in **Tumor.sav**. ▌▌▌▌

- **Task 6**: Using the Glastonbury data from Chapter 6 (**GlastonburyFestival.sav**), fit a model to see if the change in hygiene (**change**) is significant across people with different musical tastes (**music**). Do a simple contrast to compare each group against the no affiliation group. Compare the results to those described in Section 11.5. ▌▌▌▌

- **Task 7**: Labcoat Leni's Real Research 7.2 describes an experiment (Çetinkaya & Domjan, 2006) on quails with fetishes for terrycloth objects. There were two outcome variables (time spent near the terrycloth object and copulatory efficiency) that we didn't analyze. Read Labcoat Leni's Real Research 7.2 to get the full story, then fit a model with Bonferroni *post hoc* tests on the time spent near the terrycloth object. ▌▌▌▌

- **Task 8**: Repeat the analysis in Task 7 but using copulatory efficiency as the outcome. ▌▌▌▌

- **Task 9**: A sociologist wanted to compare murder rates (**Murder**) each month in a year at three high-profile locations in London (**Street**). Fit a model with bootstrapping on the *post hoc* tests to see in which streets the most murders happened. The data are in **Murder.sav**. ▌▌▌▌

Answers & additional resources are available on the book's website at
https://edge.sagepub.com/field5e

GLM 2: COMPARING MEANS ADJUSTED FOR OTHER PREDICTORS (ANALYSIS OF COVARIANCE)

13

13.1 What will this chapter tell me?

My road to rock stardom had taken a bit of a knock with my unexpected entry to an all-boys' grammar school (rock bands and grammar schools really didn't go together). I needed to be inspired and I turned to the masters: Iron Maiden. I first heard Iron Maiden at the age of 11 when a friend lent me *Piece of Mind* on a cassette and told me to listen to 'The Trooper'. It was, to put it mildly, an epiphany. I became their smallest (I was 11) biggest fan and obsessed about them in the unhealthiest of ways. I bombarded the man who ran their fan club (a guy called Keith) with letters, and, bless him, he replied to them all. Eventually my stalking paid off and Keith arranged for me to go backstage when they played what was then (and to me always will be) the Hammersmith Odeon in London on 5 November 1986 (*Somewhere on Tour*, in case you're interested). Not only was it the first time I had seen them live, but I got to meet them too. It is difficult to convey how exciting and anxiety-provoking that night was. It was all quite overwhelming. I was so utterly awe-struck that I managed to say precisely nothing to any of the band (but I do have some good photos where my speechlessness is tangible; see Figure 13.1). Soon to become a theme in my life, a social situation had provoked me to make an utter fool of myself.[1] When it was over I was in no doubt that this was the best day of my life. In fact, I thought, I should just kill myself there and then because nothing would ever be as good.[2] This may be true, but I have subsequently had other very nice experiences, so who is to say that they were not better? I could compare experiences to see which one is the best, but there is an important confound: my age. At the age of 13, meeting Iron Maiden was bowel-weakeningly exciting, but adulthood (sadly) dulls your capacity for this kind of unqualified excitement. To really see which experience was best, I would have to take account of the variance in enjoyment that is attributable to my age at the time. Doing so will give me a purer measure of how much variance in my enjoyment is attributable to the event itself. This chapter extends the previous one to look at situations in which you want to compare groups means, but also adjust those means for another variable (or variables) that you expect to affect the outcome. This involves a linear model in which an outcome is predicted from dummy

Figure 13.1 Dave Murray (guitarist from Iron Maiden) and me backstage in London in 1986 (my grimace reflects the utter terror I was feeling at meeting my hero)

---

1 In my teens I met many bands I liked, and Iron Maiden were by far the nicest.
2 Apart from my wedding day, as it turned out.

variables representing group membership but one or more other predictors (usually continuous variables) are included. These additional predictors are sometimes labeled covariates, and this configuration of the linear model is sometimes known as *analysis of covariance*.

13.2 What is ANCOVA? IIII

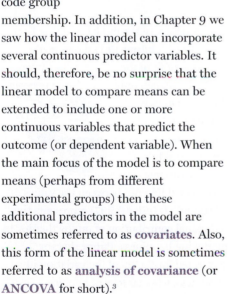

What's a covariate?

In the previous chapter we saw how we can compare multiple group means with the linear model by using dummy variables to code group membership. In addition, in Chapter 9 we saw how the linear model can incorporate several continuous predictor variables. It should, therefore, be no surprise that the linear model to compare means can be extended to include one or more continuous variables that predict the outcome (or dependent variable). When the main focus of the model is to compare means (perhaps from different experimental groups) then these additional predictors in the model are sometimes referred to as **covariates**. Also, this form of the linear model is sometimes referred to as **analysis of covariance** (or **ANCOVA** for short).[3]

In the previous chapter, we used an example about the effects of puppy therapy on happiness. Let's think about things other than puppy therapy that might influence happiness. Well, the

obvious one is how much you like dogs (a dog phobic is going to be about as happy after puppy therapy as I would be after tarantula therapy), but there are other things too such as individual differences in temperament. If these variables (the covariates) are measured, then it is possible to adjust for the influence they have on the outcome variable by including them in the linear model. From what we know of hierarchical regression (see Chapter 9) it should be clear that if we enter the covariate into the model first, and then enter the dummy variables representing the group means (e.g., the experimental manipulation), we can see what effect a predictor variable has, *adjusting for* the effect of the covariate. In essence, rather than predicting the outcome from group means, we predict it from group means that have been adjusted for the effect of covariate(s). There are two main reasons to include covariates in ANOVA:

- **To reduce within-group error variance**: When we predict an outcome from group means (e.g., when these represent the effect of an experiment), we compute an *F*-statistic by comparing the amount of variability in the outcome that the experiment can explain against the variability that it cannot explain. If we can attribute some of this 'unexplained' variance (SS_R) to other measured variables (covariates), then we

reduce the error variance, allowing us to assess more sensitively the difference between group means (SS_M).
- **Elimination of confounds**: In any experiment, there may be unmeasured variables that confound the results (i.e., variables other than the experimental manipulation that affect the outcome variable). If any variables are known to influence the outcome variable being measured, then including them as covariates can remove these variables as potential explanations for the effect of interest.

13.3 ANCOVA and the general linear model IIII

The researchers who conducted the puppy therapy study in the previous chapter suddenly realized that a participant's love of dogs would affect whether puppy therapy would affect happiness. Therefore, they repeated the study on different participants, but included a self-report measure of love of puppies from 0 (I am a weird person who hates puppies, please be deeply suspicious of me) to 7 (puppies are the best thing ever, one day I might marry one). The data are in Table 13.1 and in the file **Puppy Love.sav**, which contains the variables **Dose** (1 = control, 2 = 15 minutes, 3 = 30 minutes), **Happiness** (the person's happiness on a scale from 0 to 10), and **Puppy_love** (love of puppies from 0 to 7).

SELF TEST

Use IBM SPSS Statistics to find the means and standard deviations of both happiness and love of puppies across all participants and within the three groups. (Answers are in Table 13.2.)

---

3 As we've discussed before, these labels for special cases of the linear model (such as one-way independent ANOVA in the previous chapter, and ANCOVA here) reflect historical divisions in methods (see Misconception Mutt 12.1). They are unhelpful because they create the impression that we're using distinct statistical models when we're not. I want you to focus on the general linear model that underpins these special cases, but I can't really avoid using the ANOVA/ANCOVA labels now and again so that when your supervisor tells you to do ANOVA/ANCOVA you can find the relevant part of the book!

Table 13.1   Data from Puppy Love.sav

Dose	Participant's happiness	Love of puppies
Control	3	4
	2	1
	5	5
	2	1
	2	2
	2	2
	7	7
	2	4
	4	5
15 minutes	7	5
	5	3
	3	1
	4	2
	4	2
	7	6
	5	4
	4	2
30 minutes	9	1
	2	3
	6	5
	3	4
	4	3
	4	3
	4	2
	6	0
	4	1
	6	3
	2	0
	8	1
	5	0

Table 13.2   Means (and standard deviations) from **Puppy Love.sav**

Dose	Participant's happiness	Love of puppies
Control	3.22 (1.79)	3.44 (2.07)
15 minutes	4.88 (1.46)	3.12 (1.73)
30 minutes	4.85 (2.12)	2.00 (1.63)
Total	**4.37 (1.96)**	**2.73 (1.86)**

SELF TEST

Add two dummy variables to the file **Puppy Love.sav** that compare the 15-minute group to the control (**Dummy 1**) and the 30-minute group to the control (**Dummy 2**)—see Section 12.2 for help. If you get stuck use **Puppy Love Dummy.sav.**

In the previous chapter, we characterized this experimental scenario as equation (12.2), and knowing what we know about the linear model (Chapter 9) you can hopefully see that that equation can be extended to include the covariate as follows:

$$\text{Happiness}_i = b_0 + b_1\text{Long}_i + b_2\text{Short}_i + b_3\text{Covariate}_i + \varepsilon_i$$
$$\text{Happiness}_i = b_0 + b_1\text{Long}_i + b_2\text{Short}_i + b_3\text{Puppy\_love}_i + \varepsilon_i$$

$$(13.1)$$

We can compare the means of different groups using a linear model (see Section 12.2) in which groups are coded as the dummy variables **Long** and **Short**: **Long** takes the value of 1 only for the 30-minute group, **Short** takes a value of 1 only for the 15-minute group, and in all other situations they have a value of 0. We can add a covariate as a predictor to the model to test the difference between group means *adjusted for the covariate*. Let's look at a practical example.

The summary of the model resulting from the self-test (Output 13.1) shows us the goodness of fit of the model first when only the covariate is used in the model, and second when both the covariate and the dummy variables are used. The difference between the values of R^2 (0.288 − 0.061 = 0.227) represents the individual contribution of puppy therapy to predicting happiness. Puppy therapy accounted for 22.7% of the variation in happiness, whereas love of puppies accounted for only 6.1%. This additional information provides some insight into the substantive importance of puppy therapy. The next table is the ANOVA table, which is also divided into two sections. The top half represents the effect of the covariate alone, whereas the bottom half represents the whole model (i.e., covariate and puppy therapy included). Notice at the bottom of the ANOVA table (the bit for model 2) that the entire model (love of puppies and the dummy variables) accounts for 31.92 units

Can I run ANCOVA through the regression menu?

of variance (SS_M), there are 110.97 units in total (SS_T) and the unexplained variance (SS_R) is 79.05.

The interesting bit is the table of model coefficients (Output 13.2). The top half shows the effect when only the covariate is in the model, and the bottom half contains the whole model. The *b*-values for the dummy variables represent the difference between the means of the 15-minute group and the control group (**Dummy 1**) and the 30-minute group and the control group (**Dummy 2**)—see Section 12.2 for an explanation of why. The means of the 15- and 30-minute groups were 4.88 and 4.85 respectively, and the mean of the control group was 3.22. Therefore, the *b*-values for the two dummy variables should be roughly the same (4.88 − 3.22 = 1.66 for **Dummy 1** and 4.85 − 3.22 = 1.63 for **Dummy 2**). The astute among you might notice that the *b*-values in Output 13.2 are not only very different from each other (which shouldn't be the case because the 15- and 30-minute groups means are virtually the same), but also different from the values I've just calculated. Does this mean I've been lying to you for the past 50 pages about what the beta values represent? I'm evil, but I'm not *that* evil. The reason for this apparent anomaly is that with a covariate present, the *b*-values represent the differences between the means of each group and the control *adjusted for the covariate(s)*. In this case, they represent the difference in the means of the puppy therapy groups adjusted for the love of puppies.

These **adjusted means** come directly from the model. If we replace the *b*-values in equation (13.1) with the values in Output 13.2, our model becomes:

$$\text{Happiness}_i = 1.789 + 2.225\text{Long}_i \quad (13.2)$$
$$+ 1.786\text{Short}_i + 0.416\text{Puppy\_love}_i$$

Remember that **Long** and **Short** are dummy variables such that **Long** takes

SELF TEST
Fit a hierarchical regression with **Happiness** as the outcome. In the first block enter love of puppies (**Puppy_love**) as a predictor, and then in a second block enter both dummy variables (forced entry) – see Section 9.10 for help.

Model Summary

Model	R	R Square	Adjusted R Square	Std. Error of the Estimate
1	.246[a]	.061	.027	1.929
2	.536[b]	.288	.205	1.744

a. Predictors: (Constant), Love of puppies (0–7)

b. Predictors: (Constant), Love of puppies (0–7), Dummy 1: (control vs. 15 mins), Dummy 2: (control vs. 30 mins)

ANOVA[a]

Model		Sum of Squares	df	Mean Square	F	Sig.
1	Regression	6.734	1	6.734	1.809	.189[b]
	Residual	104.232	28	3.723		
	Total	110.967	29			
2	Regression	31.920	3	10.640	3.500	.030[c]
	Residual	79.047	26	3.040		
	Total	110.967	29			

a. Dependent Variable: Happiness (0–10)

b. Predictors: (Constant), Love of puppies (0–7)

c. Predictors: (Constant), Love of puppies (0–7), Dummy 1: (control vs. 15 mins), Dummy 2: (control vs. 30 mins)

Output 13.1

the value of 1 only for the 30-minute group, and **Short** takes a value of 1 only for the 15-minute group; in all other situations they have a value of 0. To get the adjusted means, we use this equation, but rather than replacing the covariate with an individual's score, we replace it with the mean value of the covariate from Table 13.2 (2.73) because we're interested in the predicted value for each group at the mean level of the covariate. For the control group, the dummy variables are both coded as 0, so we replace **Long** and **Short** in the model with 0. The adjusted mean will, therefore, be 2.925:

$$\text{Happiness}_{\text{Control}} = 1.789 + (2.225 \times 0) + (1.786 \times 0)$$
$$+ (0.416 \times \bar{X}_{\text{Puppy\_love}})$$
$$= 1.789 + (0.416 \times 2.73)$$
$$= 2.925 \quad (13.3)$$

For the 15-minute group, the dummy variable **Short** is 1 and **Long** is 0, so the adjusted mean is 4.71:

$$\text{Happiness}_{15\text{ mins}} = 1.789 + (2.225 \times 0) + (1.786 \times 1)$$
$$+ (0.416 \times \bar{X}_{\text{Puppy\_love}})$$
$$= 1.789 + 1.786 + (0.416 \times 2.73)$$
$$= 4.71 \quad (13.4)$$

Coefficients[a]

Model		Unstandardized Coefficients B	Unstandardized Coefficients Std. Error	Standardized Coefficients Beta	t	Sig.
1	(Constant)	3.657	.634		5.764	.000
	Love of puppies (0-7)	.260	.193	.246	1.345	.189
2	(Constant)	1.789	.867		2.063	.049
	Love of puppies (0-7)	.416	.187	.395	2.227	.035
	Dummy 1: (control vs. 15 mins)	1.786	.849	.411	2.102	.045
	Dummy 2: (control vs. 30 mins)	2.225	.803	.573	2.771	.010

a. Dependent Variable: Happiness (0-10)

Output 13.2

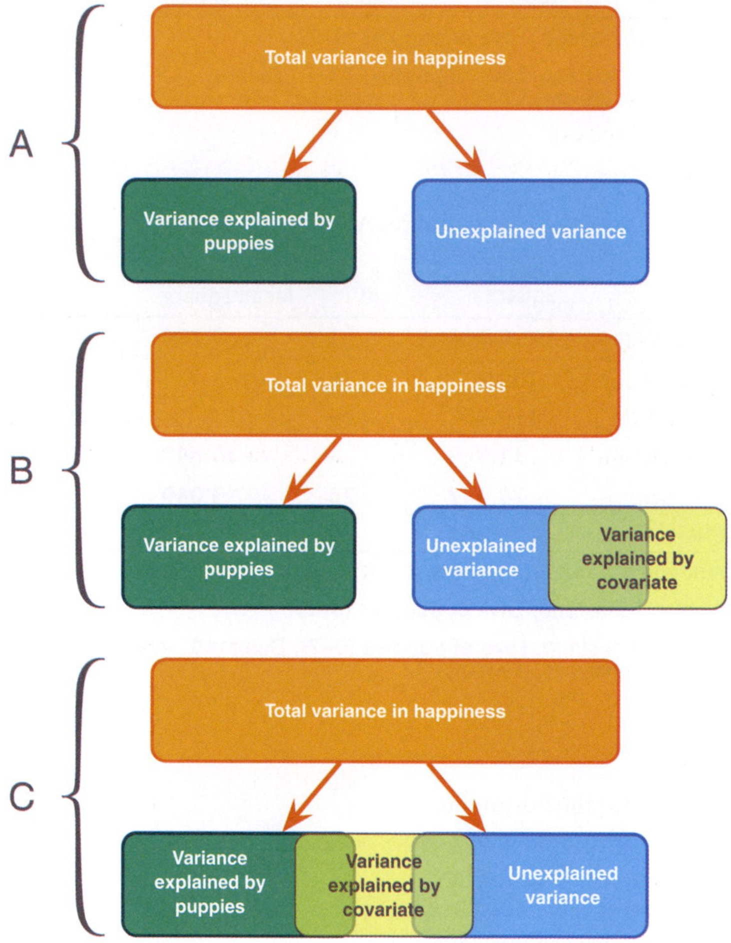

Figure 13.2 The role of the covariate in ANCOVA (see text for details)

For the 30-minute group, the dummy variable **Short** is 0 and **Long** is 1, so the adjusted mean is 5.15:

$$\begin{aligned}\overline{\text{Happiness}}_{30\text{ mins}} &= 1.789+(2.225\times 1)+(1.786\times 0)\\ &\quad +\left(0.416\times \bar{X}_{\text{Puppy_love}}\right)\\ &= 1.789+2.225+(0.416\times 2.73)\\ &= 5.15\end{aligned} \qquad \text{Eq. 13.5}$$

We can now see that the b-values for the two dummy variables represent the differences between these *adjusted* means ($4.71 - 2.93 = 1.78$ for **Dummy 1** and $5.15 - 2.93 = 2.22$ for **Dummy 2**). These adjusted means are the average amount of happiness for each group at the mean level of love of puppies. Some people think of this kind of model (i.e., ANCOVA) as 'controlling' for the covariate, because it compares the predicted group means at the average value of the covariate, so the groups are being compared at a level of the covariate that is the same for each group. However, as we shall see, the 'controlling for the covariate' analogy is not a good one.

13.4 Assumptions and issues in ANCOVA

Including covariates doesn't change the fact we're using the general linear model, so all the sources of potential bias (and counteractive measures) discussed in Chapter 6 apply. There are two additional considerations: (1) independence of the covariate and treatment effect; and (2) homogeneity of regression slopes.

13.4.1 Independence of the covariate and treatment effect

I said in the previous section that covariates can be used to reduce within-group error variance if the covariate explains some of this error variance, which will be the case if the covariate is independent of the experimental effect (group means). Figure 13.2 shows three different scenarios. Part A shows a basic model that compares group means (it is like Figure 12.5). The variance in the outcome (in our example happiness) can be partitioned into two parts that represent the experimental or treatment effect (in this case the administration of puppy therapy) and the error or unexplained variance (i.e., factors that affect happiness that we haven't measured). Part B shows the ideal scenario when including a covariate, which is that the covariate shares its variance only with the bit of happiness that is currently unexplained. In other words, it is completely independent of the treatment effect (it does not overlap with the effect of puppy therapy at all). Some argue that this scenario is the only one in which ANCOVA is appropriate (Wildt & Ahtola, 1978). Part C shows a situation in which the effect of the covariate overlaps with the experimental effect. In other words, the experimental effect is confounded with the effect of the covariate. In this situation, the covariate will reduce (statistically speaking) the experimental effect because it explains some of the variance that would otherwise be attributable to the experiment. When the covariate and the experimental effect (independent variable) are not independent, the treatment effect is obscured, spurious treatment effects can arise, and at the very least the interpretation of the ANCOVA is seriously compromised (Wildt & Ahtola, 1978).

The problem of the covariate and treatment sharing variance is common and is ignored or misunderstood by many people (Miller & Chapman, 2001). Miller and Chapman are not the only people to point this out, but their paper is very readable and they cite many examples of people misapplying ANCOVA. Their main point is that when treatment groups differ on the covariate, putting the covariate into the analysis will not 'control for' or 'balance out' those differences (Lord, 1967, 1969). This situation arises mostly when participants are not randomly assigned to experimental treatment conditions. For example, anxiety and depression are closely correlated (anxious people tend to be depressed), so if you wanted to compare an anxious group of people against a non-anxious group on some task, the chances are that the anxious group would also be more depressed than the non-anxious group. You might think that by adding depression as a covariate into the analysis you can look at the 'pure' effect of anxiety, but you can't. This situation matches part C of Figure 13.2 because the effect of the covariate (depression) would contain some of the variance from the effect of

anxiety. Statistically speaking, all that we know is that anxiety and depression share variance; we cannot separate this shared variance into 'anxiety variance' and 'depression variance', it will always be 'shared'. Another common example is if you happen to find that your experimental groups differ in their ages. Placing age into the analysis as a covariate will not solve this problem—it is still confounded with the experimental manipulation. The use of covariates cannot solve this problem (see Jane Superbrain Box 13.1).

This problem can be avoided by randomizing participants to experimental groups or by matching experimental groups on the covariate (in our anxiety example, you could try to find participants for the low-anxiety group who score high on depression). We can see whether this problem is likely to be an issue by checking whether experimental groups differ on the covariate before fitting the model. To use our anxiety example again, we could test whether our high- and low-anxiety groups

differ on levels of depression. If the groups do not significantly differ then we might consider it reasonable to use depression as a covariate.

13.4.2 Homogeneity of regression slopes

When a covariate is used we look at its overall relationship with the outcome variable: we ignore the group to which a person belongs. We assume that this relationship between covariate and outcome variable holds true for all groups of participants, which is known as the assumption of **homogeneity of regression slopes**. Think of the assumption like this: imagine a scatterplot for each group of participants with the covariate on one axis, the outcome on the other, and a regression line summarizing their relationship. If the assumption is met then the regression lines should look similar (i.e., the values of b in each group should be equal).

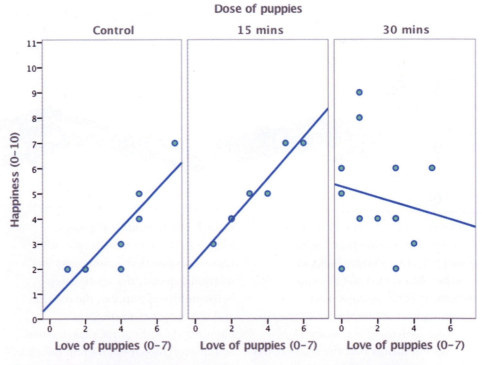

Figure 13.3 Scatterplot and regression lines of happiness against love of puppies for each of the experimental conditions

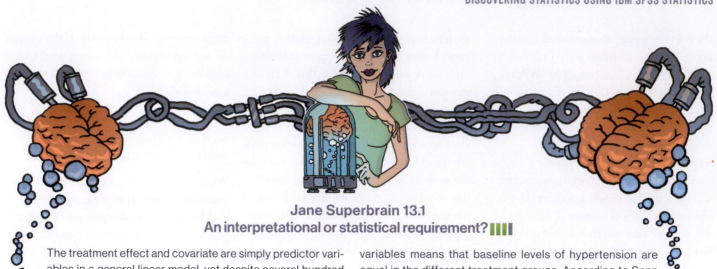

Jane Superbrain 13.1
An interpretational or statistical requirement? ▮▮▮▮

The treatment effect and covariate are simply predictor variables in a general linear model, yet despite several hundred pages discussing linear models, I haven't before mentioned that predictors should be completely independent. I've said that they shouldn't overlap too much (e.g., collinearity) but that's quite different from saying that they shouldn't overlap at all. If, in general, we don't care about predictors being independent in linear models, why should we care now? The short answer is we don't—there is no *statistical* requirement for the treatment variable and covariate to be independent.

However, there are situations in which ANCOVA can be biased when the covariate is not independent of the treatment variable. One situation, common in medical research, has been discussed a lot: an outcome (e.g., hypertension) is measured at baseline, and after a treatment intervention (with participants assigned to a treatment or control group). This design can be analyzed using an ANCOVA in which treatment effects on post-intervention hypertension are analyzed while covarying baseline levels of hypertension. In this scenario the independence of treatment and covariate

variables means that baseline levels of hypertension are equal in the different treatment groups. According to Senn (2006), the idea that ANCOVA is biased unless treatment groups are equal on the covariate applies only when there is *temporal additivity*. To use our hypertension example, temporal additivity is the assumption that both treatment groups would experience the same change in hypertension over time if the treatment had no effect. In other words, had we left the two groups alone, their hypertension would change by exactly the same amount. Given that the groups have different overall levels of hypertension to begin with, this assumption might not be reasonable, which undermines the argument for requiring group equality in baseline measures.

To sum up, the independence of the covariate and treatment makes interpretation more straightforward but is not a statistical requirement. ANCOVA can be unbiased when groups differ on levels of the covariate, but as Miller and Chapman point out, it creates an interpretational problem that ANCOVA cannot magic away.

Let's make this concept a bit more concrete. Remember that the main example in this chapter looks at whether different doses of puppy therapy affect happiness when including love of puppies as a covariate. The *homogeneity of regression slopes* assumption means that the relationship between the outcome (dependent variable) and the covariate is the same

in each of our treatment groups. Figure 13.3 shows a scatterplot with regression line that summarizes this relationship (i.e., the relationship between love of puppies, the covariate, and the outcome, participant's happiness) for the three experimental conditions (shown in different panels). There is a positive relationship (the regression line slopes upwards from left

to right) between love of puppies and participant's happiness in both the control (left panel) and 15-minute conditions (middle panel). In fact, the slopes of the lines for these two groups are very similar, showing that the relationship between happiness and love of puppies is very similar in these two groups. This situation is an example of *homogeneity* of regression

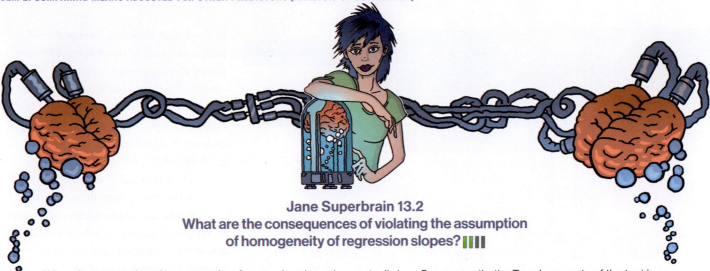

Jane Superbrain 13.2
What are the consequences of violating the assumption
of homogeneity of regression slopes? ▮▮▮

When the assumption of homogeneity of regression slopes is met the resulting *F*-statistic can be assumed to have the corresponding *F*-distribution; however, when the assumption is not met it can't, meaning that the resulting *F*-statistic is being evaluated against a distribution different than the one that it actually has. Consequently, the Type I error rate of the test is inflated and the power to detect effects is not maximized (Hollingsworth, 1980). This is especially true when group sizes are unequal (Hamilton, 1977) and when the standardized regression slopes differ by more than 0.4 (Wu, 1984).

slopes. However, in the 30-minute condition (right panel) there is a slightly negative relationship between happiness and love of puppies. The slope of this line differs from the slopes in the other two groups, suggesting *heterogeneity of regression slopes* (because the relationship between happiness and love of puppies is different in the 30-minute group compared to the other two groups).

Although in a traditional ANCOVA heterogeneity of regression slopes is a bad thing (Jane Superbrain Box 13.2), there are situations where you might expect regression slopes to differ across groups and that variability may be interesting. For example, when research is conducted across different locations, you might expect the effects to vary across those locations. Imagine you had a new treatment for backache, and you recruit

several physiotherapists to try it out in different hospitals. The effect of the treatment is likely to differ across these hospitals (because therapists will differ in expertise, the patients they see will have different problems and so on). As such, heterogeneity of regression slopes is not a bad thing *per se*. If you have violated the assumption of homogeneity of regression slopes or if the variability in regression slopes is an interesting hypothesis in itself, then you can explicitly model this variation using multilevel linear models (see Chapter 21).

13.4.3 What to do when assumptions are violated ▮▮▮

A bootstrap for the model parameters and *post hoc* tests can be used so that these, at least, are robust (see Chapter 6). The bootstrap won't help for the *F*-tests

though. There is a robust variant of ANCOVA that can be implemented using R, and we'll discuss this in Section 13.8.

13.5 Conducting ANCOVA using SPSS Statistics ▮▮▮

13.5.1 General procedure ▮▮▮

The general procedure is much the same as for any linear model, so remind yourself of the steps for fitting a linear model (Chapter 9). Figure 13.4 shows a simpler overview of the process that highlights some of the specific issues for ANCOVA-style models. As with any analysis, begin by graphing the data and looking for and correcting sources of bias.

13.5.2 Inputting data ▮▮▮

We have already looked at the data (Table 13.1) and the data file

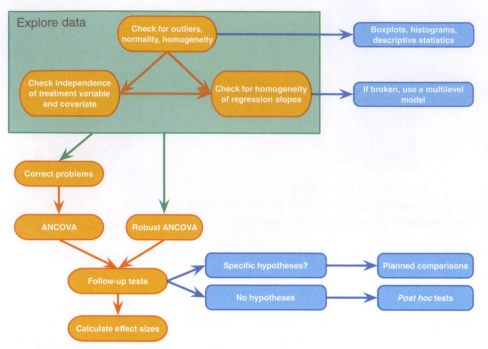

Figure 13.4 General procedure for analysis of covariance

ANOVA

Love of puppies (0–7)

	Sum of Squares	df	Mean Square	F	Sig.
Between Groups	12.769	2	6.385	1.979	.158
Within Groups	87.097	27	3.226		
Total	99.867	29			

Output 13.3

SELF TEST

Fit a model to test whether love of puppies (our covariate) is independent of the dose of puppy therapy (our independent variable).

(**Puppy Love.sav**). To remind you, the data file is set out like Table 13.1 and contains three columns: a coding variable called **Dose** (1 = control, 2 = 15 minutes, 3 = 30 minutes), a variable called **Happiness** containing the scores for the person's happiness, and a variable called **Puppy_love** containing the scores for love of puppies from 0 to 7. The 30 rows correspond to each person's scores on these three variables.

13.5.3 Testing the independence of the treatment variable and covariate

In Section 13.4.1, I mentioned that if the covariate and group means (independent variable) are independent then the interpretation of ANCOVA models is a lot more straightforward. In this case, the covariate is love of puppies, so we'd want

to check that the mean level of love of puppies is roughly equal across the three puppy therapy groups by fitting a linear model with **Puppy_love** as the outcome and **Dose** as the predictor.

Output 13.3 shows that the main effect of dose is not significant, $F(2, 27) = 1.98$, $p = 0.16$, which shows that the average level of love of puppies was roughly the same in the three puppy therapy groups. In other words, the means for love of puppies in Table 13.2 are not significantly different across the control, 15- and 30-minute groups. This result is good news for using love of puppies as a covariate in the model.

13.5.4 The main analysis

Most of the *General Linear Model* (GLM) procedures in SPSS Statistics contain the facility to include one or more covariates. For designs that don't involve repeated measures it is easiest to include covariates by selecting *Analyze* ▶ *General Linear Model* ▶ Univariate... to activate the dialog box in Figure 13.5. Drag the variable **Happiness** into the box labeled *Dependent Variable* (or click), drag **Dose** into the box labeled *Fixed Factor(s)* and drag **Puppy_love** into the box labeled *Covariate(s)*.

13.5.5 Contrasts

There are various dialog boxes that can be accessed from the main dialog box. If a covariate is selected, the *post hoc* tests are disabled because the tests that we used in the previous chapter are not designed for models that include covariates. However, comparisons can be done by clicking Contrasts... to access the *Contrasts* dialog box in Figure 13.6. You cannot enter codes to specify user-defined contrasts (but see SPSS Tip 13.1); instead you can select one of the standard contrasts that we met in Table 12.6. In this example, there was a control condition (coded as the first group), so a sensible set of contrasts would be simple contrasts comparing each experimental

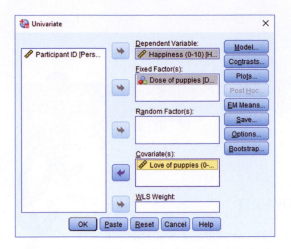

Figure 13.5 Main dialog box for GLM univariate

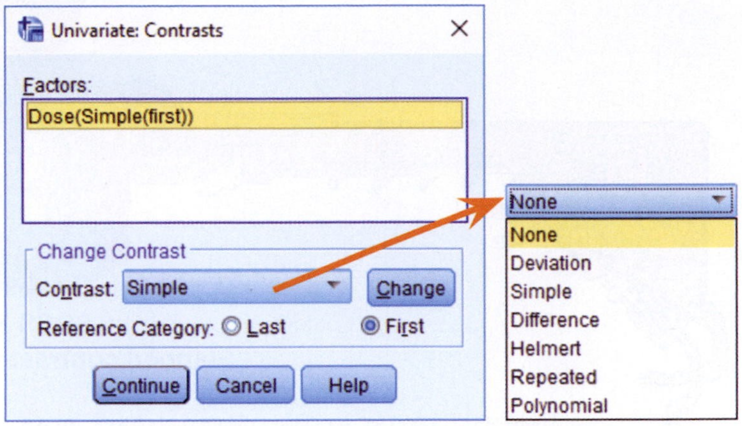

Figure 13.6 Options for standard contrasts in GLM univariate

group to the control (this results in the same contrasts as dummy coding). Click the drop-down list (None) and select a type of contrast (in this case *Simple*) from this list. For simple contrasts you need to specify the reference category (i.e., the category against which all other groups are compared). By default the last category is used, which for our data is the 30-minute group. We need to change the reference category to be the control group, which is the first category (assuming that you coded control as 1). We make this change by selecting ⦿ First. Having selected a contrast, click Change to register the selection. Figure 13.6 shows the completed dialog box. Click Continue to return to the main dialog box.

13.5.6 Other options ▮▮▮

You can get a limited range of *post hoc* tests by clicking EM Means... to access the *Estimated Marginal Means* dialog box (see Figure 13.7). To specify *post hoc* tests, drag the independent variable (in this case **Dose**) from the box labeled *Estimated Marginal Means: Factor(s) and Factor Interactions* to the box labeled *Display Means for* (or click ↦). Once a variable has been transferred, you'll be able to select ☑ Compare main effects to activate the drop-down list (LSD(none)) of *post hoc*

tests. The default is to perform a Tukey LSD *post hoc* test which makes no adjustment for multiple tests (and which I don't recommend). The other options are a Bonferroni *post hoc* test (recommended) and a **Šidák correction**, which is like the Bonferroni correction but is less conservative and so should be selected if you are concerned about the loss of power associated with Bonferroni. For this example we'll use the Šidák correction just for variety (we have used Bonferroni in

previous examples). As well as producing *post hoc* tests for the **Dose** variable, the options we've selected will create a table of estimated marginal means for this variable: these are the group means adjusted for the effect of the covariate. Click Continue Clicking Options... opens a dialog box containing the options described in Jane Superbrain Box 13.3. The most useful are (in my opinion) descriptive statistics, parameter estimates, residual plot and HC4 robust standard errors (see Figure 13.7).

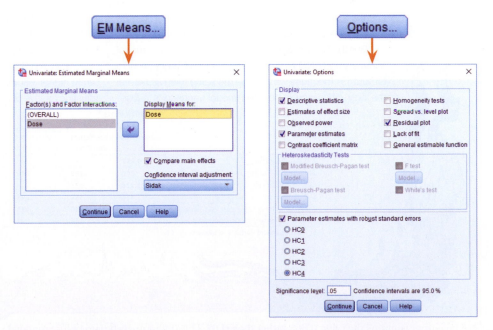

Figure 13.7 *Estimated marginal means* and *Options* dialog boxes for GLM univariate

SPSS Tip 13.1
Planned contrasts for ANCOVA ▐▐▐▐

There is no option for specifying planned contrasts like we used in the previous chapter (see Section 12.6.2). However, these contrasts can be done if we fit the model using the regression menu. Imagine you chose some planned contrasts as in Chapter 12, in which the first contrast compared the control group to all doses of puppy therapy, and the second contrast then compared the 30- and 15-minute groups (see Section 12.4). We saw in Sections 12.4 and 12.6.2 that we specify these contrasts with codes. For the first contrast we discovered that an appropriate set of codes was −2 for the control group and then 1 for both the 30- and 15-minute groups. For the second contrast the codes were 0 for the control group, −1 for the 15-minute group and 1 for the 30-minute group (see Table 12.4). To do these contrasts when a covariate is included in the model, enter these values as two dummy variables. In other words, add a column called **Dummy1** in which every person in the control group has a value of −2 and all other participants have a value of 1. Add a second column called **Dummy2**, in which everyone in the control group has the value 0, everyone in the 15-minute group has the value −1 and those in the 30-minute group have a value of 1. The file **Puppy Love Contrast.sav** includes these dummy variables.

Run the analysis as described in Section 13.3. The model summary and ANOVA table for the model will be identical to Output 13.1 (because we've done the same thing as before; the only difference is how the model variance is subse-

quently broken down with the contrasts). The b-values for the dummy variables will be different than before because we've specified different contrasts. Output 13.4 shows the model parameters. The first dummy variable compares the control group with the 15- and 30-minute groups. As such, it compares the adjusted mean of the control group (2.93) with the average of the adjusted means for the 15- and 30-minute groups ((4.71+5.15)/2 = 4.93). The b-value for the first dummy variable should reflect the difference between these values: 4.93 − 2.93 = 2. We discovered in a rather complex and boring bit of Section 12.4.2 that this value gets divided by the number of groups within the contrast (i.e., 3) and so will be 2/3 = 0.67 (as in Output 13.4).[4] The associated t-statistic is significant ($p = 0.010$), indicating that the control group was significantly different from the combined adjusted mean of the puppy therapy groups.

The second dummy variable compares the 15- and 30-minute groups, and so the b-value should reflect the difference between the adjusted means of these groups: 5.15 − 4.71 = 0.44. In Section 12.4.2 we discovered that this value gets divided by the number of groups within the contrast (i.e., 2) and so will be 0.44/2 = 0.22 (as in Output 13.4). The associated t-statistic is not significant ($p = 0.593$), indicating that the 30-minute group did not produce significantly higher happiness than the 15-minute group after adjusting for love of puppies.

Coefficients[a]

Model		Unstandardized Coefficients B	Std. Error	Standardized Coefficients Beta	t	Sig.
1	(Constant)	3.657	.634		5.764	.000
	Love of puppies (0–7)	.260	.193	.246	1.345	.189
2	(Constant)	3.126	.625		5.002	.000
	Love of puppies (0–7)	.416	.187	.395	2.227	.035
	Dummy 1: puppies vs. control	.668	.240	.478	2.785	.010
	Dummy 2: 15 mins vs. 30 mins	.220	.406	.094	.541	.593

a. Dependent Variable: Happiness (0–10)

Output 13.4

---

4 The output contains the value 0.668 rather than 0.67. This difference is because we've rounded values to 2 decimal places in our caculations whereas SPSS retains all decimal places in its calculations.

Jane Superbrain 13.3
Options for ANCOVA ▮▮▮▮

The remaining options in this dialog box are as follows:

- *Descriptive statistics*: This option produces a table of means and standard deviations for each group.
- *Estimates of effect size*: This option produces the value of partial eta squared (partial η^2) – see Section 13.10.
- *Observed power*: This option provides an estimate of the probability that the statistical test could detect the difference between the observed group means (see Section 2.9.7). This measure is pointless because if the F-test is significant then the probability that the effect was detected will, of course, be high. Likewise, if group differences were small, the observed power would be low. Do power calculations before the experiment is conducted, not after (see Section 2.9.8).
- *Parameter estimates*: This option produces a table of model parameters (b-values) and their tests of significance for the variables in the model (see Section 13.6.2).
- *Contrast coefficient matrix*: This option produces matrices of the coding values used for any contrasts in the analysis, which is useful for checking which groups are being compared in which contrast.
- *Homogeneity tests*: This option produces Levene's test of the homogeneity of variance assumption (see Section 9.3). You'll have seen by now that I think this test needs to be taken with a pinch of salt.
- *Spread vs. level plot*: This option produces a chart that plots the mean of each group of a factor (x-axis) against the standard deviation of that group (y-axis). This plot is

useful to check that there is no relationship between the mean and standard deviation. If a relationship exists then the data may need to be stabilized using a logarithmic transformation (see Chapter 6).

- *Residual plot*: This option produces a matrix scatterplot of all combinations of pairs of the following variables: observed values of the outcome, predicted values from the model, standardized residuals from the model. These plots can be used to assess the assumption of homoscedasticity. In particular, the plot of the standardized residuals against the predicted values from the model can be interpreted in a similar way to the *zpred vs. zresid* plot that we have discussed before.
- *Heteroskedasticity tests*: There are four tests for heteroscedasticity that you can select (two variants of the Breusch-Pagan test, White's test and an F-test). For the same reasons that I don't recommend Levene's test, I also don't recommend these (that is, because they are significance tests your decisions based on them will be confounded by your sample size).
- *Parameter estimates with robust standard errors*: This produces one of 5 methods (HC0 to HC4) to estimate standard errors (and, therefore, confidence intervals) for the model parameters that are robust to heteroscedasticity. These methods are described clearly in Hayes and Cai (2007). In short, HC3 has been shown to outperform HC0 to HC2 (Long & Ervin, 2000) but HC4 outperforms HC3 in some circumstances (Cribari-Neto, 2004). Basically choose HC3 or HC4.

Oditi's Lantern
ANCOVA

'I, Oditi, have discovered that covariates give us greater control. I like control, especially controlling people's minds and making them worship me, erm, I mean controlling people's minds for the benevolent purpose of helping them to seek truth and personal enlightenment. As long as they are personally enlightened to worship me. In any case, stare into my lantern to discover more about using covariates and ANCOVA.'

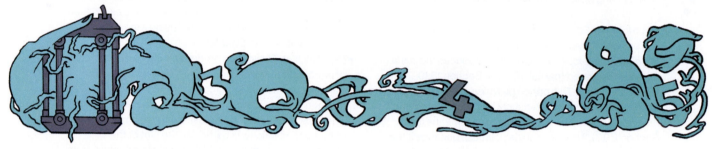

Tests of Between-Subjects Effects

Dependent Variable: Happiness (0–10)

Source	Type III Sum of Squares	df	Mean Square	F	Sig.
Corrected Model	16.844[a]	2	8.422	2.416	.108
Intercept	535.184	1	535.184	153.522	.000
Dose	16.844	2	8.422	2.416	.108
Error	94.123	27	3.486		
Total	683.000	30			
Corrected Total	110.967	29			

a. R Squared = .152 (Adjusted R Squared = .089)

Output 13.5

SELF TEST

Fit the model without the covariate to see whether the three groups differ in their levels of happiness.

13.5.7 Bootstrapping and plots ▮▮▮

There are other options available from the main dialog box. For example, if you have several independent variables you can plot them against each other (which is useful for interpreting interaction effects – see Section 14.7). There's also the `Bootstrap...` button, which you can use to activate bootstrapping. Selecting this option will bootstrap confidence intervals around the estimated marginal means, parameter estimates (*b*-values) and *post hoc* tests, but not the main *F*-statistic. Select the options described in Section 6.12.3 and click `OK` in the main dialog box to run the analysis.

13.6 Interpreting ANCOVA ▮▮▮

13.6.1 What happens when the covariate is excluded? ▮▮▮

Output 13.5 shows (for illustrative purposes) the ANOVA table for these data when the covariate is not included. It is clear from the significance value, which is greater than 0.05, that puppy therapy seems to have no significant effect on happiness. Note that the total amount of variation in happiness (SS_T) was 110.97 (*Corrected Total*), of which the therapy condition accounted for 16.84 units (SS_M), while 94.12 were unexplained (SS_R).

13.6.2 The main analysis ▮▮▮

The format of the ANOVA table in Output 13.6 is largely the same as without the covariate, except that there is an additional row of information about the covariate (**Puppy_love**). Looking first at the significance values, the covariate significantly predicts the dependent variable ($p = 0.035$, which is less than 0.05). Therefore, the person's happiness is significantly influenced by their love of puppies. What's more interesting is that when the effect of love of puppies is removed, the effect of puppy therapy is significant ($p = 0.027$, which is less than 0.05). The amount of variation accounted for by puppy therapy has increased to 25.19 units and the unexplained variance (SS_R) has been reduced to 79.05 units. Notice that SS_T has not changed; all that has changed is how that total variation is partitioned.[5]

This example illustrates how covariates can help us to exert stricter experimental control by taking account of confounding variables to give us a 'purer' measure of effect of the experimental manipulation. Looking back at the group means from Table 13.1, you might think that the significant F-statistic reflects a difference between the control group and the two experimental groups – because the 15- and 30-minute groups have very similar means (4.88 and 4.85) whereas the control group mean is much lower at 3.22. However, we can't use these group means to interpret the effect because they have not been adjusted for the effect of the covariate. These original means tell us nothing about the group differences reflected by the significant F. Output 13.7 gives the adjusted values of the group means (which we calculated in Section 13.3), and we use these values for

Tests of Between-Subjects Effects

Dependent Variable: Happiness (0–10)

Source	Type III Sum of Squares	df	Mean Square	F	Sig.
Corrected Model	31.920[a]	3	10.640	3.500	.030
Intercept	76.069	1	76.069	25.020	.000
Puppy_love	15.076	1	15.076	4.959	.035
Dose	25.185	2	12.593	4.142	.027
Error	79.047	26	3.040		
Total	683.000	30			
Corrected Total	110.967	29			

a. R Squared = .288 (Adjusted R Squared = .205)

Output 13.6

Estimates

Dependent Variable: Happiness (0–10)

Dose of puppies	Mean	Std. Error	95% Confidence Interval Lower Bound	95% Confidence Interval Upper Bound	Bootstrap for Mean[gn] Bias	Bootstrap for Mean[gn] Std. Error	BCa 95% Confidence Interval Lower	BCa 95% Confidence Interval Upper
Control	2.926[a]	.596	1.701	4.152	.030	.446	2.111	4.125
15 mins	4.712[a]	.621	3.436	5.988	.033[go]	.392[go]	3.988[go]	5.620[go]
30 mins	5.151[a]	.503	4.118	6.184	.041	.651	3.923	6.771

a. Covariates appearing in the model are evaluated at the following values: Love of puppies (0–7) = 2.73.

gn. Unless otherwise noted, bootstrap results are based on 1000 bootstrap samples

go. Based on 999 samples

Output 13.7

interpretation (this is why we selected *Display Means for* in Section 13.5.6). From these adjusted means you can see that happiness increased across the three doses.

Output 13.8 shows the parameter estimates selected in the *Options* dialog box and their bootstrapped confidence intervals and p-values (bottom table). These estimates result from **Dose** being coded using two dummy coding variables. The dummy variables are coded with the last category (the category coded with the highest value in the data editor, in this case the 30-minute group)

How do I interpret ANCOVA?

as the reference category. This reference category (labeled Dose=3 in the output) is coded with a 0 for both dummy variables (see Section 12.2 for a reminder of how dummy coding works). Dose=2, therefore, represents the difference between the group coded as 2 (15 minutes) and the reference category (30 minutes), and Dose=1 represents the difference between the group coded as 1 (control) and the reference category (30 minutes). The b-values represent the differences between the adjusted means in Output 13.7 and the significances of the t-tests tell us whether these adjusted group means differ significantly. The b for Dose=1 in Output 13.8 is the difference between the adjusted means for the control group and the 30-minute

---

5 I am often asked what the *Corrected Model* represents in this table. It is the fit of the model overall (i.e., the model containing the intercept, **Puppy_love** and **Dose**). Note that the SS of 31.92, *df* of 3, *F* of 3.5 and *p* of 0.03 are identical to the values in Output 13.1 (model 2), which tested the overall fit of this model when we ran the analysis as a regression.

Parameter Estimates

Dependent Variable: Happiness (0–10)

Parameter	B	Std. Error	t	Sig.	95% Confidence Interval	
					Lower Bound	Upper Bound
Intercept	4.014	.611	6.568	.000	2.758	5.270
Puppy_love	.416	.187	2.227	.035	.032	.800
[Dose=1]	-2.225	.803	-2.771	.010	-3.875	-.575
[Dose=2]	-.439	.811	-.541	.593	-2.107	1.228
[Dose=3]	0[a]	.	.	.	.	.

a. This parameter is set to zero because it is redundant.

Bootstrap for Parameter Estimates

Dependent Variable: Happiness (0–10)

Parameter	B	Bootstrap[a]			BCa 95% Confidence Interval	
		Bias	Std. Error	Sig. (2–tailed)	Lower	Upper
Intercept	4.014	.091[b]	.843[b]	.003[b]	1.969[b]	5.949[b]
Puppy_love	.416	-.029[b]	.202[b]	.052[b]	-.023[b]	.698[b]
[Dose=1]	-2.225	-.011[b]	.760[b]	.021[b]	-3.753[b]	-.823[b]
[Dose=2]	-.439	-.008[b]	.745[b]	.558[b]	-1.937[b]	.935[b]
[Dose=3]	0	0[b]	0[b]	.	.[b]	.[b]

a. Unless otherwise noted, bootstrap results are based on 1000 bootstrap samples

b. Based on 999 samples

Output 13.8

Parameter Estimates with Robust Standard Errors

Dependent Variable: Happiness (0–10)

Parameter	B	Robust Std. Error[a]	t	Sig.	95% Confidence Interval	
					Lower Bound	Upper Bound
Intercept	4.014	.805	4.989	.000	2.360	5.668
Puppy_love	.416	.190	2.187	.038	.025	.807
[Dose=1]	-2.225	.690	-3.226	.003	-3.642	-.807
[Dose=2]	-.439	.695	-.632	.533	-1.868	.990
[Dose=3]	0[b]	.	.	.	.	.

a. HC4 method

b. This parameter is set to zero because it is redundant.

Output 13.9

group, $2.926 - 5.151 = -2.225$, and the b for Dose=2 is the difference between the adjusted means for the 15-minute group and the 30-minute group, $4.712 - 5.151 = -0.439$.

The degrees of freedom for the t-test of the b-values are $N - k - 1$ (see Section 9.2.5), in which N is the total sample size (in this case 30) and k is the number of predictors (in this case 3, the two dummy variables and the covariate). For these data, $df = 30 - 3 - 1 = 26$. Based on the bootstrapped significance and confidence intervals

(remember you'll get different values than me because of how bootstrapping works), we could conclude that the 30-minute group differs significantly from the control group, $p = 0.021$ (Dose=1 in the table), but not from the 15-minute group, $p = 0.558$, (Dose=2 in the table).

The final thing to note is the value of b for the covariate (0.416), which is the same as in Output 13.2 (when we ran the analysis through the regression menu). This value tells us that if love of puppies increases by one unit, then the person's

happiness should increase by just under half a unit (although there is nothing to suggest a causal link between the two); because the coefficient is positive we know that as love of puppies increases so does happiness. A negative coefficient would mean the opposite: as one increases, the other decreases.

Output 13.9 repeats the parameter estimates from Output 13.8 but with standard errors, p-values and confidence intervals robust to heteroscedasticity (the HC4 estimates that we asked for). We can interpret the effects for **Dose** in the same way as for the regular and Bootstrap p-values and confidence intervals. For the effect of puppy love, the HC4 robust confidence interval and p-value supports the conclusion from the non-robust model: the p-value is 0.038, which is less than 0.05, and the confidence interval does not contain zero (0.025, 0.807). However, the bootstrap confidence interval (Output 13.8) contradicts this conclusion because it contains zero (-0.023, 0.698) and has a $p = 0.052$ (again, we're reminded of how daft it is to have a threshold that yields such opposing conclusions from such small differences in a value).

13.6.3 Contrasts

Output 13.10 shows the result of the contrast analysis specified in Figure 13.6 and compares level 2 (15 minutes) against level 1 (control) as a first comparison, and level 3 (30 minutes) against level 1 (control) as a second comparison. The group differences are displayed: a difference value, standard error, significance value and 95% confidence interval. These results show that both the 15-minute group (contrast 1, $p = 0.045$) and 30-minute group (contrast 2, $p = 0.010$) had significantly different happiness compared to the control group (note that contrast 2 is identical to the parameter for Dose=1 in the previous section).

Output 13.11 shows the results of the Šidák corrected *post hoc* comparisons that were requested in Section 13.5.6. The bottom table shows the bootstrapped significance and confidence intervals for these tests and because these will be robust we'll interpret this table (again, remember your values will differ because of how bootstrapping works). There is a significant difference between the control group and both the 15- ($p = 0.003$) and 30-minute ($p = 0.021$) groups. The 30- and 15-minute groups did not significantly differ ($p = 0.558$). It is interesting that the significant difference between the 15-minute and control groups when bootstrapped ($p = 0.003$) is not present for the normal *post hoc* tests ($p = 0.130$). This anomaly could reflect properties of the data that have biased the non-robust version of the *post hoc* test.

13.6.4 Interpreting the covariate ▌▐▐▐

I've already mentioned that the parameter estimates (Output 13.8) tell us how to interpret the covariate: the sign of the *b*-value tells us the direction of the relationship between the covariate and outcome variable. For these data the *b*-value was positive, indicating that as the love of puppies increases, so does the participant's happiness. Another way to discover the same thing is to draw a scatterplot of the covariate against the outcome.

Figure 13.8 confirms that the effect of the covariate is that as love of puppies increases, so does the participant's happiness (as shown by the slope of the line).

13.7 Testing the assumption of homogeneity of regression slopes ▌▐▐▐

Remember that the assumption of homogeneity of regression slopes means that the relationship between the covariate and outcome variable (in this

Contrast Results (K Matrix)

Dose of puppies Simple Contrast[a]		Dependent Variable Happiness (0–10)
Level 2 vs. Level 1	Contrast Estimate	1.786
	Hypothesized Value	0
	Difference (Estimate – Hypothesized)	1.786
	Std. Error	.849
	Sig.	.045
	95% Confidence Interval for Difference — Lower Bound	.040
	Upper Bound	3.532
Level 3 vs. Level 1	Contrast Estimate	2.225
	Hypothesized Value	0
	Difference (Estimate – Hypothesized)	2.225
	Std. Error	.803
	Sig.	.010
	95% Confidence Interval for Difference — Lower Bound	.575
	Upper Bound	3.875

a. Reference category = 1

Output 13.10

SELF TEST

Produce a scatterplot of love of puppies (horizontal axis) against happiness (vertical axis).

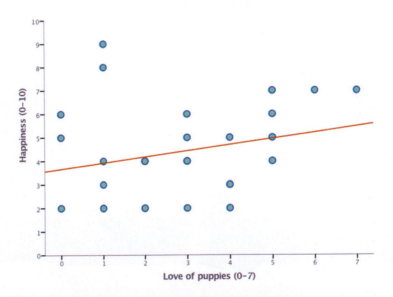

Figure 13.8 Scatterplot of happiness against love of puppies

case **Puppy_love** and **Happiness**) should be similar at different levels of the predictor variable (in this case in the three **Dose** groups). Figure 13.3 shows that the relationship between **Puppy_love** and **Happiness** looks comparable in the 15-minute and control groups, but seems different in the 30-minute group.

To test the assumption of homogeneity of regression slopes we need to refit the model but customize it to include the interaction between the covariate and categorical

Labcoat Leni's Real Research 13.1
Space invaders ▮▮▮▮

Muris, P., et al. (2008). *Child Psychiatry and Human Development, 39*(4), 469–480.

Anxious people tend to interpret ambiguous information in a negative way. For example, being highly anxious myself, if I overheard a student saying 'Andy Field's lectures are really *different*,' I would assume that 'different' meant rubbish, but it could also mean 'refreshing' or 'innovative'. Muris, Huijding, Mayer, and Hameetman (2008) addressed how these interpretational biases develop in children. Children imagined that they were astronauts who had discovered a new planet. They were given scenarios about their time on the planet (e.g., 'On the street, you encounter a space-man. He has a toy handgun and he fires at you …') and the child had to decide whether a positive ('You laugh: it is a water pistol and the weather is fine anyway') or negative ('Oops, this hurts! The pistol produces a red beam which burns your skin!') outcome occurred. After each response the child was told whether their choice was correct. Half of the children were *always* told that the negative interpretation was correct, and the reminder were told that the positive interpretation was correct.

Over 30 scenarios children were trained to interpret their experiences on the planet as negative or positive. Muris et al. then measured interpretational biases in everyday life to see whether the training had created a bias to interpret things negatively. In doing so, they could ascertain whether children might learn interpretational biases through feedback (e.g., from parents).

The data from this study are in the file **Muris et al (2008).sav**. The independent variable is **Training** (positive or negative) and the outcome is the child's interpretational bias score (**Interpretational_Bias**)—a high score reflects a tendency to interpret situations negatively. It is important to adjust for the **Age** and **Gender** of the child and also their natural anxiety level (which they measured with a standard questionnaire of child anxiety called the **SCARED**) because these things affect interpretational biases also. Labcoat Leni wants you to fit a model to see whether **Training** significantly affected children's **Interpretational_Bias** using **Age**, **Gender** and **SCARED** as covariates. What can you conclude? Answers are on the companion website (or look at pages 475–476 in the original article).

predictor. Access the main dialog box as before and place the variables in the same boxes as before (the finished dialog box should look like Figure 13.5). To customize the model, click [Model...] to access the dialog box in Figure 13.9 and select ◉ Custom. The variables specified in the main dialog box are listed on the

left-hand side. We need a model that includes the interaction between the covariate and grouping variable. To test this interaction term it's important to also include the main effects otherwise variance in the outcome (happiness) may be attributed to the interaction term that would otherwise be attributed to the main

effects. To begin with, then, select **Dose** and **Puppy_love** (you can select both simultaneously by holding down *Ctrl* or *Cmd* on a Mac), change the drop-down menu to [Main effects ▾], and click [➡] to transfer the main effects of **Dose** and **Puppy_love** to the box labeled *Model*. Next specify the interaction term by selecting **Dose** and

Puppy_love simultaneously (as just described), change the drop-down menu to Interaction ▼ and click ➡ to transfer the interaction of **Dose** and **Puppy_love** to the box labeled *Model*. The finished dialog box should look like Figure 13.9. Click Continue to return to the main dialog box and OK to run the analysis.

Output 13.11 shows the main summary table for the model including the interaction term. The effects of the dose of puppy therapy and love of puppies are still significant, but so is the covariate by outcome interaction (**Dose × Puppy_love**), implying that the assumption of homogeneity of regression slopes is not realistic ($p = 0.028$). Although this finding is not surprising given the pattern of relationships shown in Figure 13.3, it raises concerns about the main analysis.

13.8 Robust ANCOVA ▮▮▮▮

We have already looked at robust confidence intervals and *p*-values for the model parameters that were computed using bootstrapping and heteroscedasticity robust standard errors (Section 13.6.2). In addition, the companion website contains a syntax file (**robustANCOVA. sps**) for running a robust variant of ANCOVA (*ancboot*) that works on trimmed means and is described by Wilcox (2017). We need the *Essentials for R* plugin and *WRS2* package installed (Section 4.13). This test is limited to the situation where the independent variable (the categorical predictor) has two categories and there is one covariate. But it does enable you to ignore assumptions and get on with your life. Because this syntax only works when you have two groups, I have provided a data file called **PuppiesTwoGroup.sav**, which contains the example data for this chapter but excluding the 15-minute condition, so it compares the control (no puppies) with the 30-minute group (**Dose**), and has the scores for the love of puppies covariate too (**Puppy_love**). The syntax to run the robust test is as follows:

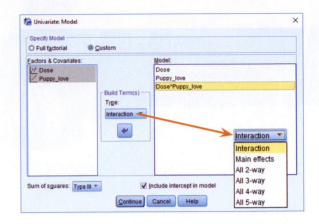

Figure 13.9 *Model* dialog box for GLM univariate

Pairwise Comparisons

Dependent Variable: Happiness (0–10)

(I) Dose of puppies	(J) Dose of puppies	Mean Difference (I–J)	Std. Error	Sig.[b]	95% Confidence Interval for Difference[b] Lower Bound	Upper Bound
Control	15 mins	−1.786	.849	.130	−3.953	.381
	30 mins	−2.225*	.803	.030	−4.273	−.177
15 mins	Control	1.786	.849	.130	−.381	3.953
	30 mins	−.439	.811	.932	−2.509	1.631
30 mins	Control	2.225*	.803	.030	.177	4.273
	15 mins	.439	.811	.932	−1.631	2.509

Based on estimated marginal means
*. The mean difference is significant at the .05 level.
b. Adjustment for multiple comparisons: Sidak.

Bootstrap for Pairwise Comparisons

Dependent Variable: Happiness (0–10)

(I) Dose of puppies	(J) Dose of puppies	Mean Difference (I–J)	Bootstrap[a] Bias	Std. Error	Sig. (2–tailed)	BCa 95% Confidence Interval Lower	Upper
Control	15 mins	−1.786	−.003[b]	.535[b]	.003[b]	−2.778[b]	−.765[b]
	30 mins	−2.225	−.011	.760	.021	−3.752	−.832
15 mins	Control	1.786	.003[b]	.535[b]	.003[b]	.663[b]	2.879[b]
	30 mins	−.439	−.008[b]	.745[b]	.558[b]	−1.937[b]	.935[b]
30 mins	Control	2.225	.011	.760	.021	.686	3.923
	15 mins	.439	.008[b]	.745[b]	.558[b]	−.938[b]	1.945[b]

a. Unless otherwise noted, bootstrap results are based on 1000 bootstrap samples
b. Based on 999 samples

Output 13.11

Tests of Between–Subjects Effects

Dependent Variable: Happiness (0–10)

Source	Type III Sum of Squares	df	Mean Square	F	Sig.
Corrected Model	52.346[a]	5	10.469	4.286	.006
Intercept	53.542	1	53.542	21.921	.000
Dose	36.558	2	18.279	7.484	.003
Puppy_love	17.182	1	17.182	7.035	.014
Dose * Puppy_love	20.427	2	10.213	4.181	.028
Error	58.621	24	2.443		
Total	683.000	30			
Corrected Total	110.967	29			

a. R Squared = .472 (Adjusted R Squared = .362)

Output 13.12

Cramming Sam's Tips
Covariates

- When the linear model is used to compare several means adjusted for the effect of one or more other variables (called *covariates*) it can be referred to as analysis of covariance (ANCOVA).
- Before the analysis check that the covariate(s) are independent of any independent variables by seeing whether those independent variables predict the covariate (i.e., the covariate should not differ across groups).
- In the table labeled *Tests of Between-Subjects Effects*, assuming you're using an alpha of 0.05, look to see if the value in the column labeled *Sig.* is below 0.05 for both the covariate and the independent variable. If it is for the covariate then this variable has a significant relationship

to the outcome variable; if it is for the independent variable then the means (adjusted for the effect of the covariate) are significantly different across categories of this variable.
- If you have generated specific hypotheses before the experiment use planned contrasts; if not, use *post hoc* tests.
- For parameters and *post hoc* tests, look at the columns labeled *Sig.* to discover if your comparisons are significant (they will be if the significance value is less than 0.05). Use bootstrapping to get robust versions of these tests.
- In addition to the assumptions in Chapter 6, test for *homogeneity of regression slopes* by customizing the model to look at the independent variable × covariate interaction.

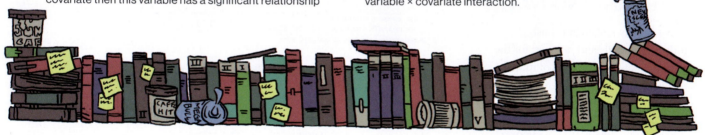

	n1	n2	diff	lower CI	upper CI	statistic	p-value
Puppy_love = 2	22	13	-1.0873	-3.1547	0.9801	-1.6952	0.098
Puppy_love = 3	27	15	-1.0719	-2.8097	0.6659	-1.9881	0.058
Puppy_love = 5	30	22	-0.6508	-2.4220	1.1204	-1.1843	0.250
Puppy_love = 6	23	21	-0.9846	-3.3281	1.3589	-1.3542	0.207
Puppy_love = 8	12	13	-1.5278	-4.3223	1.2667	-1.7622	0.119

Output 13.13

```
BEGIN PROGRAM R.
library(WRS2)
mySPSSdata = spssdata.
GetDataFromSPSS(factorMode =
"labels")
ancboot(Happiness ~ Dose + Puppy_
love, data = mySPSSdata, tr =
0.2, nboot = 1000)
END PROGRAM.
```

Select and run these five lines of syntax (see SPSS Tip 10.3). As Output 13.13 shows, the test works by identifying values of the covariate for which the relationship between the covariate and outcome are comparable in the two groups. In this example it identifies five values of **Puppy_love** (2, 3, 5, 6, and 8) for which the relationship between love of puppies and happiness is

comparable. At each of these design points, we're told the number of cases for the two groups (*n1* and *n2*) that have a value of the covariate (**Puppy_love**) close to these design points (not exactly *x*, but close to it). Based on these two samples, trimmed means (20% by default) are computed and the difference between them tested. This difference is stored in the column *Diff* along with the boundaries of the associated 95% bootstrap confidence interval (corrected to control for doing five tests) in the next two columns. The test statistic comparing the difference is in the column *statistic*, with its *p*-value in the final column. Output 13.12 shows no significant differences between trimmed means for any of the design points (all *p*-values are greater than 0.05).

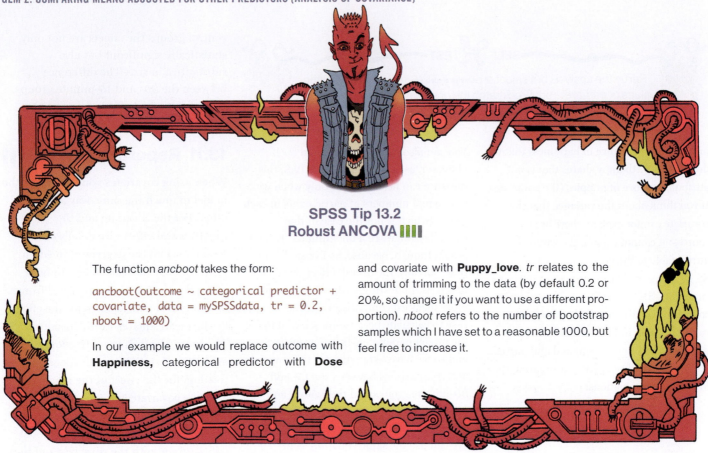

SPSS Tip 13.2
Robust ANCOVA ▮▮▮▮

The function *ancboot* takes the form:

```
ancboot(outcome ~ categorical predictor +
covariate, data = mySPSSdata, tr = 0.2,
nboot = 1000)
```

In our example we would replace outcome with **Happiness,** categorical predictor with **Dose** and covariate with **Puppy_love**. *tr* relates to the amount of trimming to the data (by default 0.2 or 20%, so change it if you want to use a different proportion). *nboot* refers to the number of bootstrap samples which I have set to a reasonable 1000, but feel free to increase it.

13.9 Bayesian analysis with covariates ▮▮▮▮

Because the model we have fitted is a linear model with a categorical predictor and a continuous predictor, you can use what you learned in Section 9.13 to run a Bayesian regression. You would need to manually create dummy variables (as in the file **Puppy Love Dummy.sav**) and drag these to the box labeled *Factor(s)* and drag **Puppy_Love** to the box labeled *Covariate(s)* see Figure 13.10. You would interpret in the same way as the model we fitted in Section 9.13.

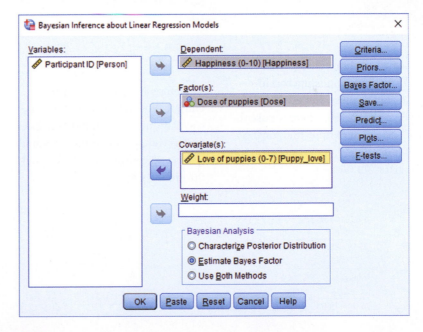

Figure 13.10

13.10 Calculating the effect size ▮▮▮▮

In the previous chapter we used eta squared, η^2, as an effect size measure when comparing means (Section 12.10). When we include a covariate too we have more than one effect and we could calculate eta squared for each effect. We can also use an effect size measure called **partial eta squared (partial η^2)**. This differs from eta squared in that it looks not at the proportion of total variance that a variable explains, but at the proportion of variance that a variable explains that *is not explained by other variables in the analysis*. Let's look at this with our example. Suppose we want to know the effect size of the dose of puppy therapy. Partial eta squared is the

Rerun the analysis but select ☑ Estimates of effect size in Figure 13.7. Do the values of partial eta squared match the ones we have just calculated?

proportion of variance in happiness that the dose of puppy therapy shares that is not attributed to love of puppies (the covariate). If you think about the variance that the covariate cannot explain, there are two sources: it cannot explain the variance attributable to the dose of puppy therapy, $SS_{puppy\ therapy}$, and it cannot explain the error variability, SS_R. Therefore, we use these two sources of variance instead of the total variability, SS_T, in the calculation. The difference between eta squared and partial eta squared is illustrated by comparing the following two equations:

$$\eta^2 = \frac{SS_{Effect}}{SS_{Total}} \quad (13.6)$$

$$Partial\ \eta^2 = \frac{SS_{Effect}}{SS_{Effect} + SS_{Residual}} \quad (13.7)$$

SPSS Statistics will produce partial eta squared for us (see Jane Superbrain Box 13.3), but to illustrate its calculation look at equation (13.8), where we use the sums of squares in Output 13.6 for the effect of dose (25.19), the covariate (15.08) and the error (79.05):

$$Partial\ \eta^2_{Dose} = \frac{SS_{Dose}}{SS_{Dose} + SS_{Residual}}$$
$$= \frac{25.19}{25.19 + 79.05} = \frac{25.19}{104.24} = 0.24$$
$$Partial\ \eta^2_{Puppy\ love} = \frac{SS_{Puppy\ love}}{SS_{Puppy\ love} + SS_{Residual}} \quad (13.8)$$
$$= \frac{15.08}{15.08 + 79.05} = \frac{15.08}{94.13} = 0.16$$

These values show that **Dose** explained a bigger proportion of the variance not attributable to other variables than **Puppy_love**.

You can also use omega squared (ω^2). However, as we saw in Section 12.8, this measure can be calculated only when we have equal numbers of participants in each group (which is not the case in this example). So, we're a bit stumped! Not all is lost, though, because, as I've said many times already, the overall effect size is not nearly as interesting as the effect size for more focussed comparisons. These are easy to calculate because we selected to see the model parameters (see Output 13.8) and so we have t-statistics for the covariate and comparisons between the 15- and 30-minute groups and the control and 30-minute group. These t-statistics have 26 degrees of freedom (see Section 13.6.1). We can use the same equation as in Section 10.9.5:[6]

$$r_{Contrast} = \sqrt{\frac{t^2}{t^2 + df}} \quad (13.9)$$

Therefore, we get (using t from Output 13.8) values of 0.40 for the covariate, and 0.48 and 0.11 respectively for the comparison of the 30-minute group and control, and the 15- and 30-minute groups:

$$r_{Covariate} = \sqrt{\frac{2.23^2}{2.23^2 + 26}} = \sqrt{\frac{4.97}{30.97}} = 0.40$$
$$r_{30\ mins\ vs.\ control} = \sqrt{\frac{(-2.77)^2}{(-2.77)^2 + 26}} = \sqrt{\frac{7.67}{33.67}} = 0.48$$
$$r_{30\ vs.\ 15\ mins} = \sqrt{\frac{(-0.54)^2}{(-0.54)^2 + 26}} = \sqrt{\frac{0.29}{26.29}} = 0.11$$

$$(13.10)$$

For the effect of the covariate and the difference between the 30-minute and control groups the effects are not only statistically significant but also substantive in size. The difference between the 30- and 15-minute groups was a fairly small effect.

13.11 Reporting results ▮▮▮▮

When using covariates you can report the model in much the same way as any other. For the covariate and the experimental effect give details of the F-statistic and the degrees of freedom from which it was calculated. In both cases, the F-statistic was derived from dividing the mean squares for the effect by the mean squares for the residual. Therefore, the degrees of freedom used to assess the F-statistic are the degrees of freedom for the effect of the model (df_M = 1 for the covariate and 2 for the experimental effect) and the degrees of freedom for the residuals of the model (df_R = 26 for both the covariate and the experimental effect)—see Output 13.6. The correct way to report the main findings would be:

✓ The covariate, love of puppies, was significantly related to the participant's happiness, $F(1, 26) = 4.96$, $p = 0.035$, $r = 0.40$. There was also a significant effect of puppy therapy on levels of happiness after controlling for the effect of love of puppies, $F(2, 26) = 4.14$, $p = 0.027$, partial $\eta^2 = 0.24$.

We can also report some contrasts (see Output 13.8):

✓ Planned contrasts revealed that having 30 minutes of puppy therapy significantly increased happiness compared to having a control, $t(26) = -2.77$, $p = 0.01$, $r = 0.48$, but not compared to having 15 minutes, $t(26) = -0.54$, $p = 0.59$, $r = 0.11$.

6 Strictly speaking, we should use a slightly more elaborate procedure when groups are unequal. It's a bit beyond the scope of this book, but Rosnow, Rosenthal, and Rubin (2000) give a very clear account.

13.12 Brian and Jane's Story ||||

The encounter in Blow Your Speakers had been beyond weird. Jane felt terrible. This Brian guy was so nice to her, and she'd just told him where to go—again! It had been easy to dismiss Brian at first, he'd seemed like a loser, a waste of her time. But there was more to him than that: he'd been working hard to learn statistics, and he'd made impressive progress. She liked how awkward he was around her, and how he always defaulted to talking stats. It was endearing. It could derail her research, though, and he could never know about that. She was a monster, and if he found out the truth it would be another let-down. Best to keep her distance.

The phone rang. It was her brother, Jake. She loved and admired Jake like no one else. Until he left home, he'd been her sanity in the madhouse that they grew up in. Their parents, both highly successful academics, were at home only long enough to pile the pressure on them both to succeed. Jane reacted by spending her youth in books, in a futile pursuit of their attention. Every set of straight As was met with 'these are just a step towards the exams that really matter, you'll need to up your game'. She was tired of trying to impress them. Jake was her opposite—he'd realized early on that he could never win. He let the pressure roll off him, and left home as soon as he could. But he always looked out for Jane.

'Mum is in hospital,' he said as the blood drained from Jane's legs.

'I don't care,' she replied, but she did. She also wanted to see Brian, because he was the closest thing she had to a friend in this town.

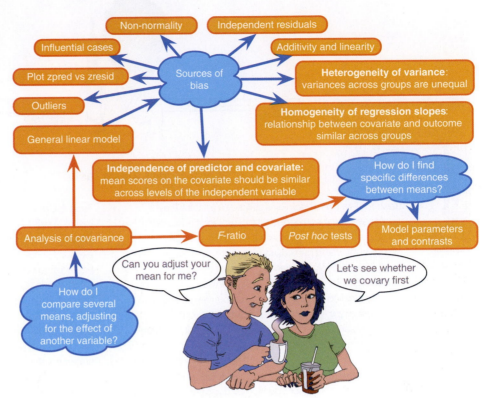

Figure 13.11   What Brian learnt from this chapter

13.13 What next? ||||

At the age of 13 I met my heroes, Iron Maiden, and very nice they were too. I've met them a couple of times since (not because they're my best buddies or anything exciting like that, but over the years the fan club put on various events where you were allowed to stand next to them and gibber like a fool while they humored you politely). You'll notice that the photo at the start of this chapter is signed by Dave Murray. This wasn't possible because I had my own darkroom installed backstage at the Hammersmith Odeon in which I could quickly process photographs or because I had access to time travel (sadly), but because I took the photo with me when I met him in 2000. I recounted the tale of how terrified I had been about meeting him in 1986. If he thought I was some strange stalker he certainly didn't let on. Uncharacteristic of most people who've sold millions of albums, they're lovely blokes.

Anyway, having seen Iron Maiden in their glory, I was inspired. They still inspire me: I still rate them as the best live band I've ever seen (and I've seen them over 35 times, so I ought to know). Although I had briefly been deflected from my destiny by the shock of grammar school, I was back on track. I *had* to form a band. There was just one issue: no one else played a musical instrument. The solution was easy: through several months of covert subliminal persuasion I convinced my two best friends (both called Mark, oddly enough) that they wanted nothing more than to start learning the drums and bass guitar. A power trio was in the making.

13.14 Key terms that I've discovered

Adjusted mean	Covariate	Partial eta squared (partial η^2)
Analysis of covariance (ANCOVA)	Homogeneity of regression slopes	Šidák correction

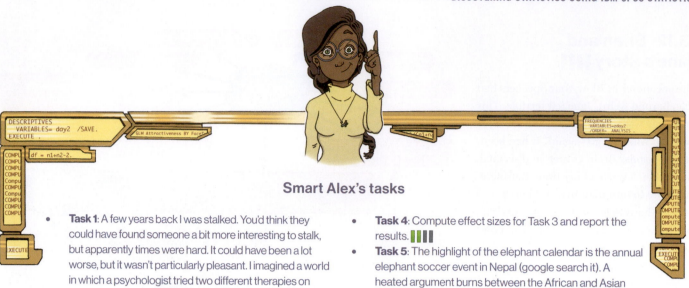

Smart Alex's tasks

- **Task 1**: A few years back I was stalked. You'd think they could have found someone a bit more interesting to stalk, but apparently times were hard. It could have been a lot worse, but it wasn't particularly pleasant. I imagined a world in which a psychologist tried two different therapies on different groups of stalkers (25 stalkers in each group—this variable is called **Group**). To the first group he gave cruel-to-be-kind therapy (every time the stalkers followed him around or sent him a letter, the psychologist attacked them with a cattle prod). The second therapy was psychodyshamic therapy, in which stalkers were hypnotized and regressed into their childhood to discuss their penis (or lack of penis), their father's penis, their dog's penis, the seventh penis of a seventh penis and any other penis that sprang to mind. The psychologist measured the number of hours stalking in one week both before (**stalk1**) and after (**stalk2**) treatment (**Stalker.sav**). Analyze the effect of therapy on stalking behavior after therapy, covarying for the amount of stalking behavior before therapy. ▌▌▌▌

- **Task 2**: Compute effect sizes for Task 1 and report the results. ▌▌▌▌

- **Task 3**: A marketing manager tested the benefit of soft drinks for curing hangovers. He took 15 people and got them drunk. The next morning as they awoke, dehydrated and feeling as though they'd licked a camel's sandy feet clean with their tongue, he gave five of them water to drink, five of them Lucozade (a very nice glucose-based UK drink) and the remaining five a leading brand of cola (this variable is called **drink**). He measured how well they felt (on a scale from 0 = I feel like death to 10 = I feel really full of beans and healthy) two hours later (this variable is called **well**). He measured how **drunk** the person got the night before on a scale of 0 = as sober as a nun to 10 = flapping about like a haddock out of water on the floor in a puddle of their own vomit (**HangoverCure.sav**). Fit a model to see whether people felt better after different drinks when covarying for how drunk they were the night before. ▌▌▌▌

- **Task 4**: Compute effect sizes for Task 3 and report the results. ▌▌▌▌

- **Task 5**: The highlight of the elephant calendar is the annual elephant soccer event in Nepal (google search it). A heated argument burns between the African and Asian elephants. In 2010, the president of the Asian Elephant Football Association, an elephant named Boji, claimed that Asian elephants were more talented than their African counterparts. The head of the African Elephant Soccer Association, an elephant called Tunc, issued a press statement that read 'I make it a matter of personal pride never to take seriously any remark made by something that looks like an enormous scrotum'. I was called in to settle things. I collected data from the two types of elephants (**elephant**) over a season and recorded how many goals each elephant scored (**goals**) and how many years of experience the elephant had (**experience**). Analyze the effect of the type of elephant on goal scoring, covarying for the amount of football experience the elephant has (**Elephant Football.sav**). ▌▌▌▌

- **Task 6**: In Chapter 4 (Task 6) we looked at data from people who had been forced to marry goats and dogs and measured their life satisfaction and also how much they like animals (**Goat or Dog.sav**). Fit a model predicting life satisfaction from the type of animal to which a person was married and their animal liking score (covariate). ▌▌▌▌

- **Task 7**: Compare your results for Task 6 to those for the corresponding task in Chapter 11. What differences do you notice and why? ▌▌▌▌

- **Task 8**: In Chapter 10 we compared the number of mischievous acts (**mischief2**) in people who had invisibility cloaks to those without (**cloak**). Imagine we also had information about the baseline number of mischievous acts in these participants (**mischief1**). Fit a model to see whether people with invisibility cloaks get up to more mischief than those without when factoring in their baseline level of mischief (**Invisibility Baseline.sav**). ▌▌▌▌

Answers & additional resources are available on the book's website at
https://edge.sagepub.com/field5e

GLM 3: FACTORIAL DESIGNS

14

14.1 What will this chapter tell me?

After persuading my two friends (Mark and Mark) to learn the bass and drums, I took the rather odd decision to *stop* playing the guitar. I didn't stop, as such, but I focussed on singing instead. In retrospect, this was a bad decision because I am *not* a good singer. Mind you, I'm not a good guitarist either. The upshot was that a classmate, Malcolm, ended up as our guitarist. I can't recall how or why we ended up in this configuration, but we called ourselves Andromeda, we learnt several Queen and Iron Maiden songs and we were truly awful. I have some recordings somewhere to prove just what a cacophony of tuneless drivel we produced, but the chances of them appearing on the companion website are slim at best. Suffice it to say, you'd be hard pushed to recognize *which* Iron Maiden and Queen songs we were trying to play. The fact that we were only 14 or 15 at the time cannot even begin to mitigate the depths of ineptitude to which we sank. On the plus side, we garnered a reputation for being too loud in school assembly and we did a successful tour of London's East End (well, our friends' houses). It's common for bands to tire of cover versions and to get lofty ambitions to write their own tunes. I wrote one called 'Escape From Inside' about the film *The Fly* that contained the rhyming couplet 'I am a fly, I want to die'—the great lyricists of the time quaked in their boots at the young new talent on the scene. The only thing we did that resembled the activities of a 'proper' band was to split up due to 'musical differences': Malcolm wanted to write 15-part symphonies about a boy's journey to worship electricity pylons and discover a mythical beast called the cuteasaurus, whereas I wanted to write songs about flies and dying (preferably both). When we could not agree on a musical direction the split became inevitable. Had I had the power of statistics in my hands back then, rather than split up we could have tested empirically the best musical direction for the band. Suppose Malcolm and I had each written a 15-part symphony and a 3-minute song about a fly. We could have played these songs to some very lucky people and measured their screams of agony. From these data, we could have ascertained the best musical direction to gain popularity. We have two variables that predict screams: whether Malcolm or I wrote the song (songwriter), and whether the song was a 15-part symphony or a song about a fly (song type). This design is called a factorial design, and this chapter looks at how the linear model extends to incorporate multiple categorical predictors.

Figure 14.1 Andromeda coming to a living room near you in 1988 (I'm the one wearing the *Anthrax* T-shirt) M

14.2 Factorial designs ▮▮▮

In the previous two chapters we have used the linear model to test for differences between group means when those groups have belonged to a single predictor variable (in experimental designs one independent variable has been manipulated). This chapter extends the linear model to situations in which there are two categorical predictors (independent variables).

Predictor variables (independent variables) often get lonely and want to have friends. Scientists are obliging individuals and often put a second (or third) independent variable into their designs to keep the others company. When an experiment has two or more independent variables it is known as a *factorial design* (because, as we have seen, independent variables are sometimes referred to as *factors*). There are several types of factorial design:

- **Independent factorial design**: There are several independent variables or predictors and each has been measured using different entities (between groups). We discuss this design in this chapter.
- **Repeated-measures (related) factorial design**: Several independent variables or predictors have been measured, but the same entities have been used in all conditions (see Chapter 15).
- **Mixed design**: Several independent variables or predictors have been measured; some have been measured with different entities, whereas others used the same entities (see Chapter 16).

As you might imagine, analyzing factorial designs can get quite complicated. Fortunately, we can offset this complexity by realizing that we can still fit a linear model to the design, which at least grounds everything in a model with which (I hope) by now are familiar. Remembering that for weird historical reasons (see Misconception Mutt 12.1) people often label this form of the linear model as 'ANOVA', people often refer to the linear model with two or more categorical predictors that represent experimental independent variables as **factorial ANOVA**. There are also a family of more specific labels that reflect the experimental design that is being analyzed (see Jane Superbrain Box 14.1). However, I think our lives would be simpler if we abandoned these labels because they deflect attention from the fact that the underlying model is the same.

14.3 Independent factorial designs and the linear model ▮▮▮

Throughout this chapter we'll use an example of an experimental design with two independent variables (a two-way independent design—Jane Superbrain Box 14.1). The study tested the prediction that subjective perceptions of physical attractiveness become inaccurate after drinking alcohol (the well-known **beer-goggles effect**). The example is based on real research by Chen, Wang, Yang, and Chen (2014) who looked at whether the beer-goggles effect was influenced by the attractiveness of the faces being rated. The logic is that alcohol consumption has been shown to reduce accuracy in symmetry judgements, and symmetric faces have been shown to be rated as more attractive. If the beer-goggles effect is driven by alcohol impairing symmetry judgements

then you'd expect a stronger effect for unattractive (asymmetric) faces (because alcohol will affect the perception of asymmetry) than attractive (symmetric) ones. The data we'll analyze are fictional, but the results mimic the findings of this research paper.

An anthropologist was interested in the effects of facial attractiveness on the beer-goggles effect. She randomly selected 48 participants. Participants were randomly subdivided into three groups of 16: (1) a placebo group drank 500 ml of alcohol-free beer; (2) a low-dose group drank 500 ml of average strength beer (4% ABV); and (3) a high-dose group drank 500 ml of strong beer (7% ABV). Within each group, half ($n = 8$) rated the attractiveness of 50 photos of unattractive faces on a scale from 0 (pass me a paper bag) to 10 (pass me their phone number) and the remaining half rated 50 photos of attractive faces.[1] The outcome for each participant was their median rating across the 50 photos (Table 14.1 and **Goggles.sav**).

To keep things simple, imagine for now that we have only two levels of the alcohol variable (placebo and high dose). As such, we have two predictor variables, each with two levels. We've seen many times that the general linear model takes the following general form:

$$Y_i = b_0 + b_1 X_{1i} + b_2 X_{2i} + \cdots + b_n X_{ni} + \varepsilon_i \tag{14.1}$$

When we first looked at using the linear model to compare means (see equation (12.2)) we used an example of the effect of three doses of puppy therapy (none, 15 and 30 minutes) on happiness. We saw that the linear model became:

$$\text{Happiness}_i = b_0 + b_1 \text{Long}_i + b_2 \text{Short}_i + \varepsilon_i \tag{14.2}$$

in which the **Long** and **Short** predictor variables were dummy variables that

---

1 These photographs were from a larger pool of 500 that had been pre-rated by a different sample. The 50 photos with the highest and lowest ratings were used.

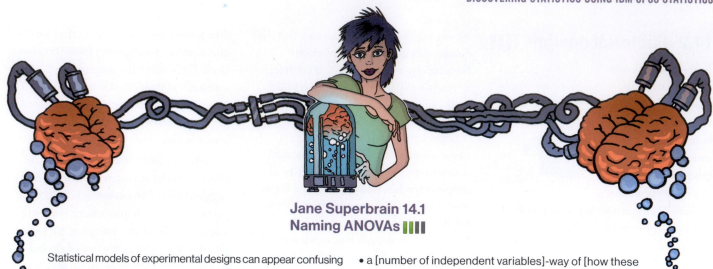

Jane Superbrain 14.1
Naming ANOVAs ▮▮▮▮

Statistical models of experimental designs can appear confusing because there appears to be an endless list of them. It can create the false impression that the models are completely distinct, rather than the truth that they are variations on a common model (the linear model). To add to the confusion, scientists refer to these models as 'ANOVA' rather than the linear model, because the *F*-statistic that tests the fit of the model (Section 9.2.4) partitions variance, and so is called 'analysis of variance'.

The names people use have two things in common: (1) they involve some quantity of independent (predictor) variables; and (2) they label whether these variables were measured using the same or different entities. If the same entities are tested multiple times the term *repeated measures* is applied, and if different entities take part in the treatment conditions the term *independent* is used. With two or more independent variables, it's possible that some were measured with the same entities and others with different entities; this is referred to as a *mixed* design. In general, people name models that compare means as:

- a [number of independent variables]-way of [how these variables were measured] ANOVA.

By remembering this you can decipher the name of any ANOVA model. Look at these examples and work out how many independent variables were used and how they were measured:

- One-way independent ANOVA
- Two-way repeated-measures ANOVA
- Two-way mixed ANOVA
- Three-way independent ANOVA

The answers you should get are:

- One independent variable measured using different entities.
- Two independent variables both measured using the same entities.
- Two independent variables: one measured using different entities and the other measured using the same entities.
- Three independent variables all of which are measured using different entities.

coded to which of the groups a participant belongs using values of 0 and 1. In our current example, we also have two variables that represent categories: **FaceType** (unattractive or attractive) and **Alcohol** (placebo and high dose). Just like we did for puppy therapy, we can code participant's category membership on these variables with zeros and ones; for example, we might code type of face as unattractive = 0, attractive = 1,

and alcohol group as 0 = placebo, 1 = high dose. We can copy the puppy therapy model (equation (14.2)) but replace the predictors with our two independent variables:

$$\text{Attractiveness}_i = b_0 + b_1 \text{FaceType}_i + b_2 \text{Alcohol}_i + \varepsilon_i$$

(14.3)

However, this model does not consider the interaction between type of face and

alcohol. To include this term, we extend the model to become:

$$\begin{aligned}\text{Attractiveness}_i &= b_0 + b_1 A_i + b_2 B_i + b_3 AB_i + \varepsilon_i \\ &= b_0 + b_1 \text{FaceType}_i + b_2 \text{Alcohol}_i \\ &\quad + b_3 \text{Interaction}_i + \varepsilon_i\end{aligned}$$

(14.4)

You might wonder how we code the interaction term, but we saw how to do this in Section 11.3. The interaction term represents the combined effect of **Alcohol** and **FaceType**, and it is obtained by multiplying the variables involved. This multiplication is the reason why interaction terms are written as type of face × alcohol. Table 14.2 shows the resulting predictor variables for the model (the group means for the combinations of type of face and alcohol are included because they'll come in useful in due course). Note that the interaction variable is the type of face dummy variable multiplied by the alcohol dummy variable. For example, someone receiving a high dose of alcohol and rating unattractive faces would have a value of 0 for the type of face variable, 1 for the alcohol variable and 0 for the interaction variable.

To see what the b-values in equation (14.4) represent we can insert values of our predictors and see what happens. Let's start with participants in the placebo group rating unattractive faces. In this case, the values of type of face, alcohol and the interaction are all 0. The predicted value of the outcome, as we have seen in Chapter 12, will be the group mean (3.500). Our model becomes:

$$\text{Attractiveness}_i = b_0 + b_1\text{FaceType}_i + b_2\text{Alcohol}_i$$
$$+ b_3\text{Interaction}_i + \varepsilon_i$$
$$\bar{X}_{\text{Unattractive, Placebo}} = b_0 + (b_1 \times 0) + (b_2 \times 0) + (b_3 \times 0)$$
$$b_0 = \bar{X}_{\text{Unattractive, Placebo}}$$
$$= 3.500 \qquad (14.5)$$

and you can see that the constant b_0 represents the mean of the group for which all variables are coded as 0. As such it's the mean value of the baseline category (in this case participants in the placebo group rating unattractive faces).

Now, let's see what happens when we look at participants in the placebo group rating attractive faces. The outcome is the mean rating of attractive faces after a placebo drink, the type of face variable is 1 and the alcohol and interaction variables are 0. Remember that b_0 is the mean rating of unattractive faces after a placebo drink. The model becomes:

Table 14.1   Data for the beer-goggles effect

Alcohol	Placebo		Low dose		High dose	
FaceType	Attractive	Unattractive	Attractive	Unattractive	Attractive	Unattractive
	6	2	7	3	5	5
	7	4	6	5	6	6
	6	3	8	7	7	8
	7	3	7	5	5	6
	6	4	6	4	7	7
	5	6	7	4	6	8
	8	5	6	5	5	7
	6	1	5	6	8	6
Total	51	28	52	39	49	53
Mean	6.375	3.500	6.500	4.875	6.125	6.625
Variance	0.839	2.571	0.857	1.554	1.268	1.125

Grand mean = 5.667
Grand variance = 2.525

Table 14.2   Coding scheme for factorial ANOVA

Type of faces rated	Alcohol	Dummy (FaceType)	Dummy (Alcohol)	Interaction	Mean
Unattractive	Placebo	0	0	0	3.500
Unattractive	High dose	0	1	0	6.625
Attractive	Placebo	1	0	0	6.375
Attractive	High dose	1	1	1	6.125

$$\bar{X}_{\text{Attractive, Placebo}} = b_0 + (b_1 \times 1) + (b_2 \times 0) + (b_3 \times 0)$$
$$= b_0 + b_1$$
$$= \bar{X}_{\text{Unattractive, Placebo}} + b_1$$
$$b_1 = \bar{X}_{\text{Attractive, Placebo}} - \bar{X}_{\text{Unattractive, Placebo}}$$
$$= 6.375 - 3.500$$
$$= 2.875 \qquad (14.6)$$

which shows that b_1 represents the difference between ratings of unattractive and attractive faces in people who drank a placebo. More generally we can say it's the effect of type of face for the baseline category of alcohol (the category coded with 0, in this case the placebo).

Let's look at people who had a high dose of alcohol and rated unattractive faces. The outcome is the mean rating of unattractive faces after a high dose of alcohol, the type of face variable is 0, the alcohol variable is 1 and the interaction variable is 0. We can

replace b_0 with the mean rating of unattractive faces after a placebo drink. The model becomes:

$$\bar{X}_{\text{Unattractive, High dose}} = b_0 + (b_1 \times 0) + (b_2 \times 1) + (b_3 \times 0)$$
$$= b_0 + b_2$$
$$= \bar{X}_{\text{Unattractive, Placebo}} + b_2$$
$$b_2 = \bar{X}_{\text{Unattractive, High dose}} \qquad (14.7)$$
$$- \bar{X}_{\text{Unattractive, Placebo}}$$
$$= 6.625 - 3.500$$
$$= 3.125$$

which shows that b_2 in the model represents the difference between ratings of unattractive faces after a high dose of alcohol compared to a placebo drink. Put more generally, it's the effect of alcohol in the baseline category of type of face (i.e., the category coded with a 0, in this case unattractive).

Finally, we can look at ratings of attractive faces after a high dose of alcohol. The

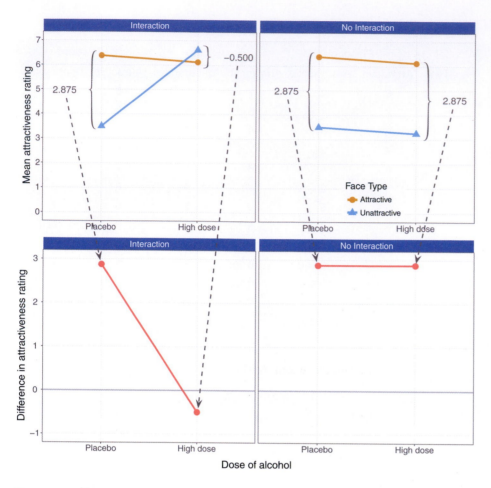

Figure 14.2 Breaking down what an interaction represents

drink to the effect of type of face after a high dose of alcohol.[2]

This explanation makes more sense if you think in terms of an interaction graph. Figure 14.2 (top left) shows the interaction graph for these data. The difference between ratings of unattractive and attractive faces in the two placebo groups is the distance between the lines on the graph for the placebo group (the difference between group means, which is 2.875). If we look at the same difference for the high-dose group, we find that the difference between ratings of unattractive and attractive faces is −0.500. If we plotted these two 'difference' values as a new graph we'd get a line connecting 2.875 to −0.500 (see Figure 14.2, bottom left). The slope of this line reflects the difference between the effect of type of face after a placebo compared to after a high dose of alcohol. We know that b-values represent gradients of lines and b_3 is the gradient of the line that connects the difference scores (this is (−0.500) − 2.875 = −3.375).

The right-hand side of Figure 14.2 illustrates what happens if there isn't an interaction effect by showing the same data as on the left except that the mean rating of unattractive pictures after a high dose of alcohol has been changed to 3.25. If we calculate the difference between ratings of unattractive and attractive faces after a placebo drink we get the same as before: 2.875. However, if we calculate the difference between ratings of unattractive and attractive faces after a high dose of alcohol we now also get 2.875. If we again plot these differences on a new graph, we find a completely flat line. So, when there's no interaction, the line connecting the effect of type of face after a placebo drink and after a high dose is flat and the resulting b_3 would be 0 (remember that a zero gradient means a flat line). If we calculate this difference, this is what we get: 2.875 − 2.875 = 0.

predicted outcome is the mean rating of attractive faces by people after a high dose of alcohol, and the type of face, alcohol and interaction variables are all 1. We can replace b_0, b_1, and b_2, with what we now know they represent from equations (14.5)–(14.7). The model becomes:

which is scary, but we'll break it down. It shows that b_3 compares the difference between ratings of unattractive and attractive faces in the placebo group to the same difference in the high-dose group. Put more generally, it compares the effect of type of face after a placebo

$$\bar{X}_{\text{Attractive, High dose}} = b_0 + (b_1 \times 1) + (b_2 \times 1) + (b_3 \times 1)$$
$$= b_0 + b_1 + b_2 + b_3$$
$$= \bar{X}_{\text{Unattractive, Placebo}} + (\bar{X}_{\text{Attractive, Placebo}} - \bar{X}_{\text{Unattractive, Placebo}})$$
$$+ (\bar{X}_{\text{Unattractive, High dose}} - \bar{X}_{\text{Unattractive, Placebo}}) + b_3$$
$$= \bar{X}_{\text{Attractive, Placebo}} + \bar{X}_{\text{Unattractive, High dose}} - \bar{X}_{\text{Unattractive, Placebo}} + b_3$$
$$b_3 = \bar{X}_{\text{Unattractive, Placebo}} - \bar{X}_{\text{Attractive, Placebo}} + \bar{X}_{\text{Attractive, High dose}}$$
$$- \bar{X}_{\text{Unattractive, High dose}}$$
$$= 3.500 - 6.375 + 6.125 - 6.625$$
$$= -3.375 \tag{14.8}$$

2 If you rearrange the terms in the equation you'll see that you can also phrase the interaction the opposite way around: it represents the effect of alcohol on ratings of attractiveness for attractive faces compared to unattractive ones.

Output 14.1 shows the resulting table of coefficients from the self-test. Note that the *b*-value for the type of face matches equation (14.6), the *b* for the alcohol group matches equation (14.7) and the *b* for the interaction matches equation (14.8). All of which I hope convinces you that we can use a linear model to analyze designs that incorporate multiple categorical predictors.

14.3.1 Behind the scenes of factorial designs ▮▮▮

Now that we have a good conceptual understanding of factorial designs as an extension of the linear model, we will turn our attention to the specific calculations that go on behind the scenes. The reason for doing this is that it should help you to understand what the output of the analysis means. Calculating the *F*-statistic with two categorical predictors is very similar to when we had only one: we still find the total sum of squared errors (SS_T) and break this variance down into variance that can be explained by the model/experiment (SS_M) and variance that cannot be explained (SS_R). The main difference is that with factorial designs, the variance explained by the model/experiment is made up of not one predictor (experimental manipulation) but two. Therefore, the model sum of squares gets further subdivided into variance explained by the first predictor/independent variable (SS_A), variance explained by the second predictor/independent variable (SS_B) and variance explained by the interaction of these two predictors ($SS_{A \times B}$)—see Figure 14.3.

14.3.2 Total sums of squares (SS_T) ▮▮▮

We start off in the same way as we did for a one-way ANOVA. That is, we calculate how much variability there is between scores when we ignore the experimental

SELF TEST

The file **GogglesRegression.sav** contains the dummy variables used in this example. Just to prove that this works, use this file to fit a linear model predicting attractiveness ratings from **FaceType**, **Alcohol** and the interaction variable.

Coefficients[a]

Model		Unstandardized Coefficients B	Std. Error	Standardized Coefficients Beta	t	Sig.
1	(Constant)	3.500	.426		8.219	.000
	Attractivenss of facial stimuli	2.875	.602	.851	4.774	.000
	Alcohol consumption	3.125	.602	.925	5.189	.000
	Interaction	−3.375	.852	−.866	−3.963	.000

a. Dependent Variable: Median attractiveness rating

Output 14.1

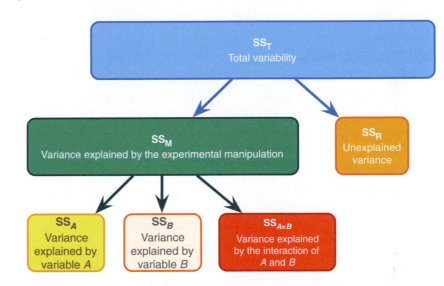

Figure 14.3 Breaking down the variance in a two-way factorial design

condition from which they came. Remember from one-way ANOVA (equation (12.9)) that SS_T is calculated using the following equation:

$$SS_T = \sum_{i=1}^{N}\left(x_i - \bar{x}_{grand}\right)^2 \qquad (14.9)$$
$$= s_{grand}^2 (N-1)$$

The grand variance is the variance of all scores when we ignore the group to which they belong. So we treat the data as one big group, calculate the variance of all scores, and find it's 2.525 as in Table 14.1 (try this on your calculator if

you don't trust me). We used 48 scores to generate this value, and so *N* is 48. As such the total sum of squares is 118.675 (equation 14.10). The degrees of freedom for this SS will be *N* – 1 or 47.

$$SS_T = s_{grand}^2 (N-1)$$
$$= 2.525(48-1) \qquad (14.10)$$
$$= 118.675$$

14.3.3 The model sum of squares, SS_M ▮▮▮

The next step is to work out the model sum of squares, which is then further broken down into the variance attributable

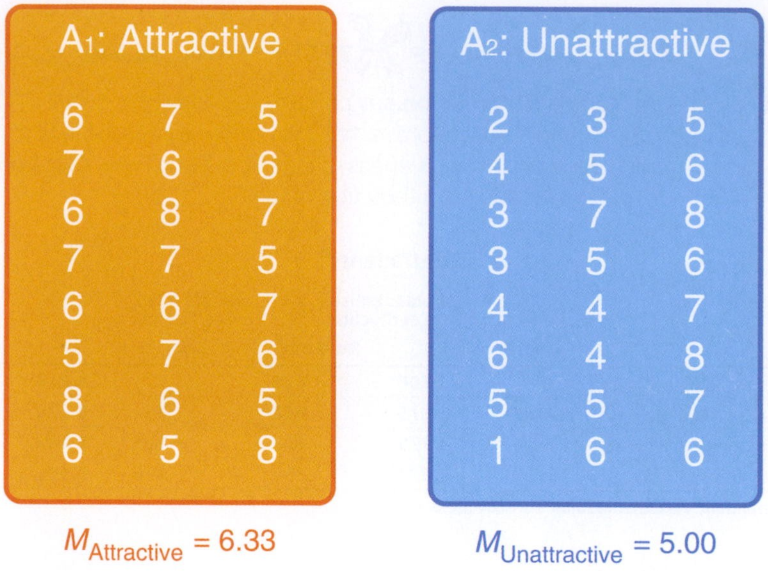

A₁: Attractive

6	7	5
7	6	6
6	8	7
7	7	5
6	6	7
5	7	6
8	6	5
6	5	8

$M_{\text{Attractive}} = 6.33$

A₂: Unattractive

2	3	5
4	5	6
3	7	8
3	5	6
4	4	7
6	4	8
5	5	7
1	6	6

$M_{\text{Unattractive}} = 5.00$

Figure 14.4 The main effect of type of face

to the first independent variable (SS_A), the variance attributable to the second independent variable (SS_B) and variance attributable to the interaction of these two variables ($\text{SS}_{A \times B}$). The model sum of squares (as we've seen a few times) is the difference between what the model predicts and the overall mean of the outcome variable. We've also seen that with predictors that represent group membership 'what the model predicts' is the group means. Therefore, we work out the model sum of squares by looking at the difference between each group mean and the overall mean (see Section 12.2.3). If we combine all levels of the two independent variables (three alcohol groups that rated unattractive faces and another three that rated attractive faces) we have six groups. We can apply the equation for the model sum of squares that we've used before (equation (12.11)):

$$\text{SS}_M = \sum_{g=1}^{k} n_g \left(\bar{x}_g - \bar{x}_{\text{grand}} \right)^2 \qquad (14.11)$$

The grand mean is the mean of all scores and is 5.667 (Table 14.1) and n is the number of scores in each group (8 in this case). Therefore, the model sum of squares is 61.17:

$$
\begin{aligned}
\text{SS}_M &= 8(6.375-5.667)^2 + 8(3.500-5.667)^2 + 8(6.500-5.667)^2 \\
&\quad + 8(4.875-5.667)^2 + 8(6.125-5.667)^2 + 8(6.625-5.667)^2 \\
&= 8(0.708)^2 + 8(-2.167)^2 + 8(0.833)^2 + 8(-0.792)^2 \\
&\quad + 8(0.458)^2 + 8(0.958)^2 \\
&= 4.010 + 37.567 + 5.551 + 5.018 + 1.678 + 7.342 \\
&= 61.17
\end{aligned}
\qquad (14.12)
$$

The degrees of freedom are the number of groups, k, minus 1; we had six groups and so $df = 5$. At this stage we know that the model (our experimental manipulations) can explain 61.17 units of variance out of the total of 118.675 units. The next stage is to break down this model sum of squares to separate the variance in attractiveness ratings explained by each of our independent variables.

14.3.4 The main effect of type of face, SS_A

To work out the variance accounted for by the first predictor/independent variable (in this case, type of face) we group attractiveness ratings according to which type of face was being rated. So, basically we ignore the dose of alcohol that was drunk, and we place all the ratings of unattractive faces into one group and all of the ratings of attractive ones into another. The data will look like Figure 14.4 (note

that the first box contains the three attractive columns from Table 14.1 and the second box contains the unattractive columns).

We then apply the same equation for the model sum of squares that we just used (compare equation (14.11)):

$$\text{SS}_A = \sum_{g=1}^{k} n_g \left(\bar{x}_g - \bar{x}_{\text{grand}} \right)^2 \qquad (14.13)$$

The grand mean is the mean of all scores (5.667, as above) and n is the number of scores in each group (i.e., the number of participants who rated unattractive and attractive faces; 24 in both cases). The means of the two groups have been calculated in Figure 14.4. The resulting model sum of squares for the main effect of the type of face is 21.32:

$$
\begin{aligned}
\text{SS}_{\text{FaceType}} &= 24(6.33-5.667)^2 + 24(5.00-5.667)^2 \\
&= 24(0.666)^2 + 24(-0.667)^2 \\
&= 10.645 + 10.677 \\
&= 21.32
\end{aligned}
\qquad (14.14)
$$

The degrees of freedom for this SS are the number of groups used, k, minus 1. We used two groups (unattractive and attractive) and so $df = 1$. To sum up, the main effect of type of face compares the mean of all ratings of unattractive faces to the corresponding mean for attractive faces (regardless of how much alcohol was consumed).

14.3.5 The main effect of alcohol, SS_B

To work out the variance accounted for by the second independent variable (dose of alcohol) we group the attractiveness ratings according to the dose of alcohol. In other words, we ignore the type of face the participant was rating, and we place all of the scores after the placebo drink in one group, the ratings after a low dose in another group and the ratings after a high dose in a third group. The data will look like

Figure 14.5. We apply the same equation as for the main effect of type of face:

$$SS_B = \sum_{g=1}^{k} n_g \left(\bar{x}_g - \bar{x}_{grand} \right)^2 \qquad (14.15)$$

The grand mean is the mean of all scores (5.667 as before), n is the number of scores in each group (i.e., the number of scores in each alcohol condition, in this case 16), and the group means are given in Figure 14.5. The resulting sum of squares is 16.53:

$$
\begin{aligned}
SS_{Alcohol} &= 16\left(4.938 - 5.667\right)^2 + 16\left(5.688 - 5.667\right)^2 \\
&\quad + 16\left(6.375 - 5.667\right)^2 \\
&= 16\left(-0.729\right)^2 + 16\left(0.021\right)^2 + 16\left(0.708\right)^2 \\
&= 8.503 + 0.007 + 8.020 \\
&= 16.53 \qquad\qquad (14.16)
\end{aligned}
$$

The degrees of freedom are the number of groups used minus 1 (see Section 12.2.3). We used three groups and so $df = 2$. To sum up, the main effect of alcohol compares the means of the placebo, low dose and high dose (regardless of whether the ratings were of unattractive or attractive faces).

14.3.6 The interaction effect, $SS_{A \times B}$ ▮▮▮▮

The final stage is to calculate how much variance is explained by the interaction of the two variables. The simplest way to do this is to remember that the SS_M is made up of three components (SS_A, SS_B and $SS_{A \times B}$). Therefore, given that we know SS_A and SS_B, we can calculate the interaction term by subtraction:

$$SS_{A \times B} = SS_M - SS_A - SS_B \qquad (14.17)$$

For these data, the value is 23.32:

$$
\begin{aligned}
SS_{A \times B} &= SS_M - SS_A - SS_B \\
&= 61.17 - 21.32 - 16.53 \\
&= 23.32 \qquad\qquad (14.18)
\end{aligned}
$$

The degrees of freedom can be calculated in the same way, but are also the product of the degrees of freedom for the main effects—either method works:

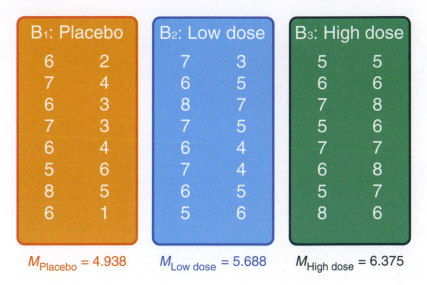

$$M_{Placebo} = 4.938 \qquad M_{Low\ dose} = 5.688 \qquad M_{High\ dose} = 6.375$$

Figure 14.5 The main effect of alcohol

$$
\begin{array}{ll}
df_{A \times B} = df_M - df_A - df_B & df_{A \times B} = df_A \times df_B \\
\quad\quad = 5 - 1 - 2 = 2 & \quad\quad = 1 \times 2 = 2
\end{array} \qquad (14.19)
$$

14.3.7 The residual sum of squares, SS_R ▮▮▮▮

The residual sum of squares is calculated in the same way as in Section 12.2.4. As ever, it represents errors in prediction from the model, but in experimental designs this also reflects individual differences in performance or variance that can't be explained by factors that were systematically manipulated. The value is calculated by taking the squared error between each data point and its corresponding group mean. An alternative way to express this is (see equation (12.14)):

$$
SS_R = \sum_{g=1}^{k} s_g^2 \left(n_g - 1 \right)
$$

$$
= s_{group\,1}^2 \left(n_1 - 1 \right) + s_{group\,2}^2 \left(n_2 - 1 \right) + \cdots + s_{group\,n}^2 \left(n_n - 1 \right) \qquad (14.20)
$$

We use the individual variances of each group from Table 14.1 and multiply them by one less than the number of people within the group (n), in this case $n = 8$. The resulting residual sum of squares is 57.50:

$$
\begin{aligned}
SS_R &= s_{group\,1}^2 \left(n_1 - 1 \right) + s_{group\,2}^2 \left(n_2 - 1 \right) + \cdots + s_{group\,6}^2 \left(n_6 - 1 \right) \\
&= 0.839(8-1) + 2.571(8-1) + 0.857(8-1) + 1.554(8-1) \\
&\quad + 1.268(8-1) + 1.125(8-1) \\
&= (0.839 \times 7) + (2.571 \times 7) + (0.857 \times 7) + (1.554 \times 7) \\
&\quad + (1.268 \times 7) + (1.125 \times 7) \\
&= 5.873 + 17.997 + 5.999 + 10.878 + 8.876 + 7.875 \\
&= 57.50 \qquad\qquad (14.21)
\end{aligned}
$$

The degrees of freedom for each group will be one less than the number of scores per group (i.e., 7). We add the degrees of freedom for each group to get a total of $6 \times 7 = 42$.

14.3.8 The *F*-statistics ▮▮▮▮

Each effect in a factorial design has its own *F*-statistic. In a two-way design this means we compute *F* for the two main effects and the interaction. To calculate these we first calculate the mean squares for each effect by taking the sum of squares and dividing by the respective degrees of freedom (think back to Section 12.2.5). We also need the mean squares for the residual term. So, for this example we'd have four mean squares:

$$
\begin{aligned}
MS_A &= \frac{SS_A}{df_A} = \frac{21.32}{1} = 21.32 \\[4pt]
MS_B &= \frac{SS_B}{df_B} = \frac{16.53}{2} = 8.27 \\[4pt]
MS_{A \times B} &= \frac{SS_{A \times B}}{df_{A \times B}} = \frac{23.32}{2} = 11.66 \\[4pt]
MS_R &= \frac{SS_R}{df_R} = \frac{57.50}{42} = 1.37
\end{aligned} \qquad (14.22)
$$

Misconception Mutt 14.1
F-statistics and Type I errors

The Misconception Mutt was very tired. His owner had been walking him early before lectures and he needed some sleep. This morning's game of fetch was killing him, and yet his little legs seemed incapable of not speeding his body off towards the stick that his owner seemed to lose so carelessly, persistently and forcefully. As he sprinted, his mind was wandering to *F*-statistics in factorial designs. He thought they were a neat way to control Type I errors. As he picked up the stick, it winked at him. Was it the sleep deprivation? He wasn't sure but then its bark turned a ginger hue, and he noticed fur shooting out from it. Soon legs began to poke out, and a tail. Before he knew it he was carrying a very heavy ginger cat in his mouth. This was all he needed.

'Did you notice how the *F*-statistics in factorial designs are computed?' the cat said.

Determined not to be outwitted by a cat, the mutt replied, 'Yes, the mean squares for the effect divided by the residual mean squares.'

'Exactly,' the cat grinned. 'They all use the residual mean squares. This means that they are not independent.'

The cat waited as though expecting applause. The mutt was confused, so the cat continued. 'The *F*-statistics are not independent, which means that they do not control the Type I error rate. In the example from your lecture there are three *F*s. The probability that at least one of them is a Type I error is not 0.05, but is 0.143 because the *F*s are not independent. If you had even more predictors and interactions the probability of at least one Type 1 error would be even greater.'

'What can be done?' the mutt asked.

'You could use a more stringent criterion for accepting an *F* as significant. With three effects you might use 0.05/3 = 0.017, which is effectively a Bonferroni correction. Or simply ignore any effects that don't test your substantive hypothesis. In this case interpret the interaction and ignore the main effects (even if they are significant).'

With that the legs, fur and grin retracted and the mutt dropped the stick at his owner's feet.

The *F*-statistic for each effect is then calculated, as we've seen before, by dividing its mean squares by the residual mean squares:

$$F_A = \frac{\mathrm{MS}_A}{\mathrm{MS}_R} = \frac{21.32}{1.37} = 15.56$$

$$F_B = \frac{\mathrm{MS}_B}{\mathrm{MS}_R} = \frac{8.27}{1.37} = 6.04 \qquad (14.23)$$

$$F_{A\times B} = \frac{\mathrm{MS}_{A\times B}}{\mathrm{MS}_R} = \frac{11.66}{1.37} = 8.51$$

IBM SPSS Statistics will compute an exact *p*-value for each of these *F*-statistics to tell us the probability (in the long run) of getting an *F* at least as big as the one we have if there were no effect in the population. One important issue here is that these *F*s do not control the familywise Type I error (Misconception Mutt 14.1).

To sum up, when you have a factorial design, the computations behind the linear model are basically the same as when you have only one categorical predictor, except that the model sum of squares is partitioned into three parts: the effect of each of the independent variables and the effect of how these variables interact.

14.4 Model assumptions in factorial designs ▮▮▮▮

When using the linear model to analyze a factorial design the sources of potential bias (and counteractive measures) discussed in Chapter 6 apply. If you have violated the assumption of homogeneity of variance there are corrections based on the Welch procedure that was described a couple of chapters back. However, this is quite technical, SPSS Statistics doesn't do it, and if you have anything more complicated than a 2 × 2 design then, really, it would be less painful to cover your body in paper cuts and then bathe in chilli sauce (see Algina & Olejnik, 1984). One practical solution is to bootstrap the *post hoc* tests so that these will be robust. You can also ask for confidence intervals and *p*-values for parameter estimates that are robust to heteroscedasticity (refer back to Section 13.5.6). This won't help for the *F*-statistics. There is a robust test based on trimmed means that can be done using the R extension (see Section 14.8).

14.5 Factorial designs using SPSS Statistics ▮▮▮

14.5.1 General procedure for factorial designs ▮▮▮

Using the linear model to test differences between means on several predictors/independent variables requires the same steps as for one predictor/independent variable, so refer back to Figure 12.4 as a guide.

14.5.2 Entering the data and accessing the main dialog box ▮▮▮

To enter the data we need two different coding variables to represent type of face and alcohol consumption. Create a variable called **FaceType** in the data editor. We have had a lot of experience with coding variables, so you should be able to define value labels to represent the two types of faces: I've used the codes unattractive = 0 and attractive = 1. Having created the variable, enter a code of 0 or 1 in the **FaceType** column indicating to which group the participant was assigned. Create a second variable called **Alcohol** and assign value labels of placebo = 0, low dose = 1 and high dose = 2. In the data editor, enter 0, 1 or 2 into the alcohol column to represent the amount of alcohol consumed by the participant. Remember that if you turn on the *value labels* option you will see text in the data editor rather than the numerical codes. The coding scheme I have suggested is summarized in Table 14.3. Once you have created the two coding variables, create a third called **Attractiveness** and use the *labels* option to give it the fuller name of *Median attractiveness rating*. Enter the scores in Table 14.1 into this column, taking care to ensure that each score is associated with the correct combination of type of face and alcohol.

Table 14.3 Coding two independent variables

FaceType	Alcohol	The participant . . .
0	0	Rated unattractive faces after a placebo drink
0	1	Rated unattractive faces after a low dose of alcohol
0	2	Rated unattractive faces after a high dose of alcohol
1	0	Rated attractive faces after a placebo drink
1	1	Rated attractive faces after a low dose of alcohol
1	2	Rated attractive faces after a high dose of alcohol

SELF ✂ TEST

Use the Chart Builder to plot an error bar graph of the attractiveness ratings with alcohol consumption on the *x*-axis and different colored lines to represent whether the faces being rated were unattractive or attractive.

Oliver Twisted
Please, Sir, Can I ... customize my model?

'My friend told me that there are different types of sums of squares,' complains Oliver with an air of impressive authority. 'Why haven't you told us about them? Is it because you have a microbe for a brain?' No, Oliver, it's because everyone but you will find this very tedious. If you want to find out more about what the [Model...] button does, and the different types of sums of squares that can be used, then the companion website will tell you.

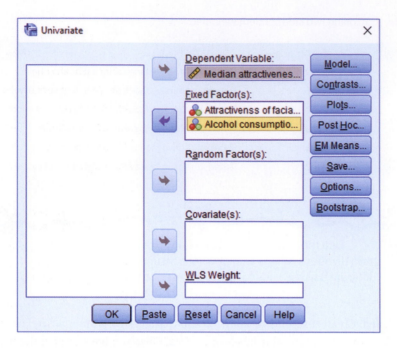

Figure 14.6 Main dialog box for independent factorial designs

To fit a linear model for independent factorial designs select *Analyze* ▶ *General Linear Model* ▶ 🔲 Univariate... to access the dialog box in Figure 14.6. Drag the outcome variable **Attractiveness** from the variables list on the left-hand side and to the space labeled *Dependent Variable* (or click 🔄). Drag **Alcohol** and **FaceType** (to select them simultaneously hold down *Ctrl* or *Cmd* on a Mac, while clicking each one) from the variables list to the *Fixed Factor(s)* box (or click 🔄). There are various other spaces that are available for conducting more complex analyses such as factorial ANCOVA, which extends the model at the beginning of this chapter to include a covariate (as in the previous chapter).

14.5.3 Graphing interactions ▮▮▮

Click ⌗ Plots... to access the dialog box in Figure 14.7, which allows you to specify line graphs of your data. These graphs can be useful for interpreting interaction effects, but they do tend to be scaled so as to create the impression

of massive differences between means, so, seriously, plot your data properly *before* the main analysis. If you want to use this plotting function to create an **interaction graph** to show the combined effect of type of face and alcohol consumption, drag **Alcohol** from the variables list to the space labeled *Horizontal Axis* (or click 🔄) and drag **FaceType** to the space labeled *Separate Lines*. Alternatively, you could plot the variables the other way round – it doesn't matter, so use your discretion as to which way produces the graph that helps make sense of the interaction. When you have moved the two variables to the appropriate boxes, click ⌗ Add to add the plot to the list at the bottom of the box. You can choose either a bar or line chart (I have chosen a line chart), and you should ask for error bars displaying confidence intervals. Using the default scaling, you can end up with a chart showing apparently massive group differences that turn out to be tiny when you inspect the *y*-axis. To avoid this crushing disappointment it is

generally (although not always) a good idea to select *Y axis starts at 0* to scale the *y*-axis from zero. In Figure 14.7 I have also specified graphs for the two main effects by dragging them (in turn) to the space labeled *Horizontal Axis* and clicking ⌗ Add. Click ⌗ Continue to return to the main dialog box.

14.5.4 Contrasts ▮▮▮

We saw in Chapter 12 that contrasts help us to break down main effects and tell us where the differences between groups lie. With one independent variable we could enter codes to define the contrasts we want. However, with two independent variables no such facility exists (although it can be done using syntax—see Oliver Twisted). Instead we are restricted to several standard contrasts that were described in Table 12.6.

Standard contrasts will be fine for this example. The type of face had only two levels, so we don't need contrasts for that main effect because it compares only two means. However, the effect of alcohol had three levels: placebo, low dose and high dose. We could select a simple contrast and use the first category as a reference. Doing so would compare the low-dose group to the placebo group, and then compare the high-dose group to the placebo group. As such, the alcohol groups would get compared to the placebo group. We could also select a *repeated* contrast, which would compare the low-dose group to the placebo group, and then the high-dose group to the low-dose group (i.e., it moves through the groups comparing each group to the one before). Again, this might be useful. We could also do a **Helmert contrast**, which compares each category against all subsequent categories. In this case it would compare the placebo group to the remaining categories (i.e., all of the

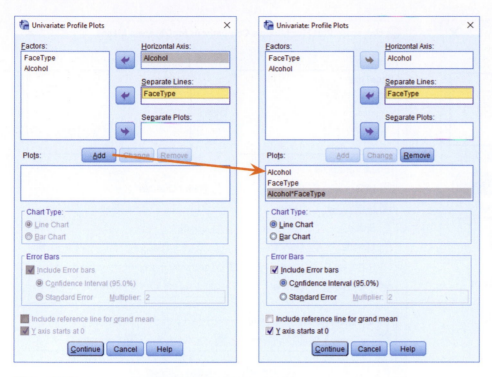

Figure 14.7 Defining plots for independent factorial designs

To get contrasts for the main effect of alcohol click Contrasts... in the main dialog box. We've used the *Contrasts* dialog box before (Section 13.5.5), so refer back to that section to help you select a Helmert contrast for the **Alcohol** variable (Figure 14.8). Click Continue to return to the main dialog box.

14.5.5 *Post hoc* tests ▮▮▮

Click Post Hoc... to access *post hoc* tests (Figure 14.9). The variable **FaceType** has only two levels and so we don't need *post hoc* tests (because any significant effects must reflect the difference between unattractive and attractive faces). However, there were three levels of **Alcohol** (placebo, low dose and high dose) so if we didn't have prior hypotheses to test we might want *post hoc* tests (remember that normally you would conduct contrasts *or post hoc* tests, not both). Drag the variable **Alcohol** from the box labeled *Factors* to the box labeled *Post Hoc Tests for:*. For my recommendations for *post hoc* procedures see Section 12.5 or select the ones in Figure 14.9. Click Continue to return to the main dialog box.

groups that had some alcohol) and then would move on to the low-dose group and compare it to the high. Any of these contrasts would be fine, but they give us contrasts only for the main effects. In reality, most of the time we want contrasts for our interaction term, and

they can be obtained only through syntax (it looks like you might have to look at Oliver Twisted after all!).

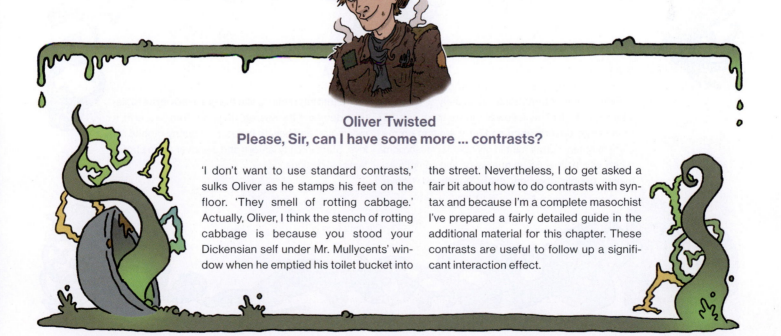

Oliver Twisted
Please, Sir, can I have some more ... contrasts?

'I don't want to use standard contrasts,' sulks Oliver as he stamps his feet on the floor. 'They smell of rotting cabbage.' Actually, Oliver, I think the stench of rotting cabbage is because you stood your Dickensian self under Mr. Mullycents' window when he emptied his toilet bucket into the street. Nevertheless, I do get asked a fair bit about how to do contrasts with syntax and because I'm a complete masochist I've prepared a fairly detailed guide in the additional material for this chapter. These contrasts are useful to follow up a significant interaction effect.

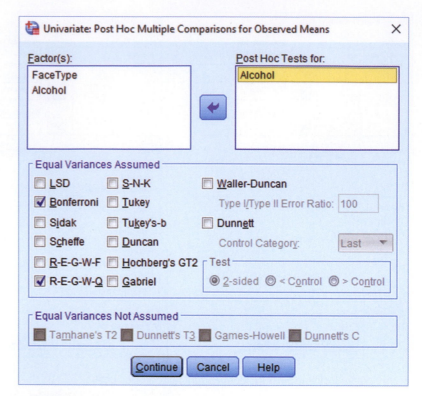

Figure 14.8 Defining contrasts for independent factorial designs

Figure 14.9 Dialog box for *post hoc* tests

Oditi's Lantern
Factorial ANOVA

'I, Oditi, enjoy interactions immensely. I want to interact with all of my followers, invite them around to my large desert ranch and let them sup on my tasty mint tea. I grow mint in my special mushroom patch, which gives it a unique flavor, and sometimes makes people obey my every command. I have learnt that interactions like these are powerful tools to understand the secrets of global domina … erm, I mean "life" and how to breed cute bunny rabbits of love. Stare into my lantern and discover more about factorial designs.'

14.5.6 Bootstrapping and other options ▮▮▮

Click [EM Means...] and [Options...] to activate the same dialog boxes that we saw in the previous chapter (Figure 13.7; the options are explained in Jane Superbrain Box 13.3). The main thing is to get estimated marginal means by transferring all effects into the box labeled *Display Means for* (Figure 14.10). Some people select *Homogeneity tests* to produce Levene's test (Section 6.11.2), but I'm not a fan. You can select ☑ *Estimates of effect size* to obtain partial eta squared (see Section 13.10).

The main dialog box contains the [Bootstrap...] button, which can be used to specify bootstrapped confidence intervals for the estimated marginal means, descriptive statistics and *post hoc* tests, but not the main *F*-statistic. This option is mainly useful if you plan to look at *post hoc* tests, which we will, so select the options described in Section 6.12.3. Once these options have been selected click [Continue] to return to the main dialog box, then click [OK].

14.6 Output from factorial designs ▮▮▮

If you're a Levene's test kind of person (which I'm not) and selected that option then Output 14.2 will appear in the viewer (see Jane Superbrain Box 6.6). With eight participants in each group this test will be horrifically underpowered so the fact that the result is non-significant ($p = 0.625$) could mean that the variance in

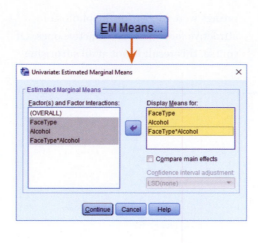
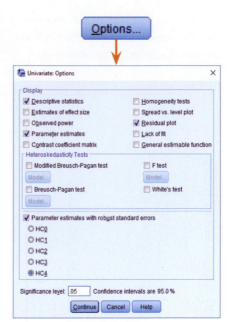

Figure 14.10 *Estimated marginal means* and *Options* dialog boxes for GLM univariate

attractiveness ratings is roughly equal across the combinations of type of face and alcohol or it could be that we don't have enough power to detect differences in the variance across groups.

14.6.1 The main effect of type of face ▮▮▮

Output 14.3 tells us whether any of the predictors/independent variables had a significant effect on attractiveness ratings. Note that the sums of squares, mean squares and *F*-statistics match (to within rounding error) the values we calculated in Sections 14.3.2–14.3.8. The main effect of type of face is significant because the *p* associated with the *F*-statistic is given as 0.000, which is less than 0.05. This effect means that overall *when we ignore how much alcohol had been drunk* the type of face being rated significantly affected attractiveness ratings. We can visualize this effect by plotting the average attractiveness rating at each level of type of face (ignoring the dose of alcohol completely).

Levene's Test of Equality of Error Variances[a]

Dependent Variable: Median attractiveness rating

F	df1	df2	Sig.
.702	5	42	.625

Tests the null hypothesis that the error variance of the dependent variable is equal across groups.

a. Design: Intercept + FaceType + Alcohol + FaceType * Alcohol

Output 14.2

Tests of Between-Subjects Effects

Dependent Variable: Median attractiveness rating

Source	Type III Sum of Squares	df	Mean Square	F	Sig.
Corrected Model	61.167[a]	5	12.233	8.936	.000
Intercept	1541.333	1	1541.333	1125.843	.000
FaceType	21.333	1	21.333	15.583	.000
Alcohol	16.542	2	8.271	6.041	.005
FaceType * Alcohol	23.292	2	11.646	8.507	.001
Error	57.500	42	1.369		
Total	1660.000	48			
Corrected Total	118.667	47			

a. R Squared = .515 (Adjusted R Squared = .458)

Output 14.3

Figure 14.11 plots the means and shows that the significant main effect reflects the fact that average attractiveness ratings were higher for the photos of attractive faces than unattractive ones. Of course, this result is not at all surprising because the attractive faces were pre-selected to be more attractive than the unattractive faces. This result is a useful manipulation check though: our participants, other things being equal, found the attractive faces more attractive than the unattractive ones.

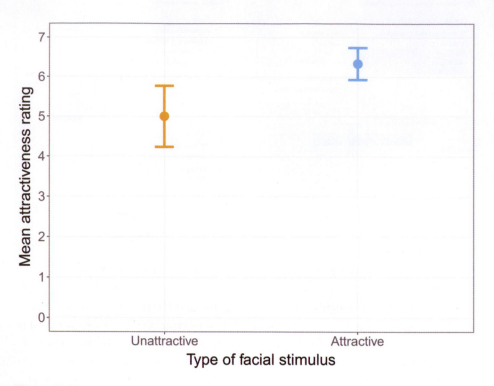

Figure 14.11   Graph to show the main effect of type of face on attractiveness ratings

14.6.2   The main effect of alcohol

Output 14.3 also shows a significant main effect of alcohol on attractiveness ratings (because the significance value of 0.005 is less than 0.05). This result means that *when we ignore whether the partici pant rated unattractive or attractive faces* the amount of alcohol influenced their attractiveness ratings. The best way to understand this effect is to plot the average attractiveness rating at each level of alcohol (ignoring the type of face completely)—we calculated these means in Section 14.3.5. Figure 14.12 shows such a graph, the mean attractiveness ratings increase quite linearly as more alcohol is drunk. This significant main effect is *likely* to reflect this trend. Looking at the error bars (95% confidence intervals), there is a lot of overlap between the placebo and low-dose groups (implying that these groups have similar average ratings) but the overlap between the placebo and high-dose groups is less, and quite possibly within what you'd expect from a significant difference (see Section 2.9.9). The confidence intervals for the low- and high-dose groups also overlap a lot, suggesting that these groups do not differ. Therefore, we might speculate based on the confidence intervals that this main effect reflects a difference between the placebo and high-dose groups but that no other groups differ. It could also reflect the linear increase in ratings as the dose of alcohol increases.

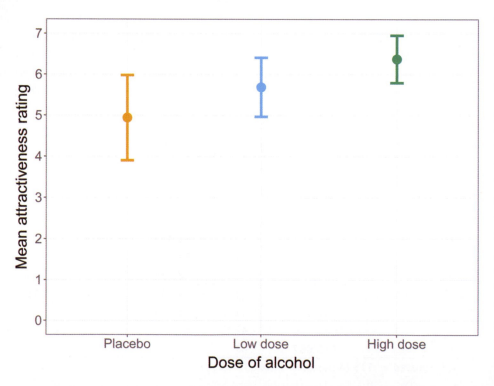

Figure 14.12   Graph showing the main effect of alcohol on attractiveness ratings

14.6.3 The interaction effect ▮▮▮▮

Finally, Output 14.3 tells us about the interaction between the effect of type of face and the effect of alcohol. The *F*-statistic is highly significant (because the observed *p*-value of 0.001 is less than 0.05). This effect means that the effect of alcohol on attractiveness ratings was different when rating unattractive faces compared to when rating attractive faces. The output will include the (probably badly scaled) plot that we asked for in Figure 14.7. Figure 14.13 is a nicer version of this graph that shows the estimated marginal means from Output 14.4. We can use this graph to get a handle on the interaction effect. Focus first on the blue line, which is flat and shows very little difference in average attractiveness ratings across the alcohol conditions. This line shows that *when rating attractive faces, alcohol has very little effect*. Now look at the orange line, which slopes upwards, showing that when rating unattractive faces ratings increase with the dose of alcohol. This line shows that *when rating unattractive faces, alcohol has an effect*. The significant interaction reflects the differing effect of alcohol when rating attractive and unattractive faces; that is, that alcohol has an effect on ratings of unattractive faces but not for attractive ones. This example illustrates an important point about interaction effects. We concluded earlier that alcohol significantly affected attractiveness ratings (the **Alcohol** main effect) but the interaction effect qualifies this conclusion by showing that this is true only when rating unattractive faces (ratings of attractive faces seem unaffected by alcohol). The take-home message is that *you should not interpret a main effect in the presence of a significant interaction involving that main effect*.

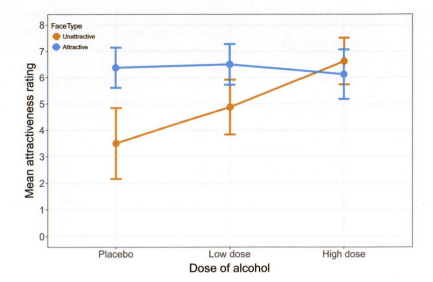

Figure 14.13 Graph of the interaction of type of face and alcohol consumption on attractiveness ratings

3. Attractivenss of facial stimuli * Alcohol consumption

Dependent Variable: Median attractiveness rating

Attractivenss of facial stimuli	Alcohol consumption	Mean	Std. Error	95% Confidence Interval Lower Bound	95% Confidence Interval Upper Bound	Bias	Std. Error	BCa 95% Confidence Interval Lower	BCa 95% Confidence Interval Upper
Unattractive	Placebo	3.500	.414	2.665	4.335	−.018	.570	2.429	4.571
	Low dose	4.875	.414	4.040	5.710	.006	.433	4.000	5.800
	High dose	6.625	.414	5.790	7.460	.002	.363	6.000	7.333
Attractive	Placebo	6.375	.414	5.540	7.210	−.002	.312	5.826	7.000
	Low dose	6.500	.414	5.665	7.335	.015	.316	5.833	7.200
	High dose	6.125	.414	5.290	6.960	.007	.404	5.375	7.000

a. Unless otherwise noted, bootstrap results are based on 1000 bootstrap samples

Output 14.4

Contrast Results (K Matrix)

Alcohol consumption Helmert Contrast			Dependent Variable Median attractiveness rating
Level 1 vs. Later	Contrast Estimate		−1.094
	Hypothesized Value		0
	Difference (Estimate − Hypothesized)		−1.094
	Std. Error		.358
	Sig.		.004
	95% Confidence Interval for Difference	Lower Bound	−1.817
		Upper Bound	−.371
Level 2 vs. Level 3	Contrast Estimate		−.688
	Hypothesized Value		0
	Difference (Estimate − Hypothesized)		−.688
	Std. Error		.414
	Sig.		.104
	95% Confidence Interval for Difference	Lower Bound	−1.522
		Upper Bound	.147

Output 14.5

14.6.4 Contrasts ▌▌▌▌

Output 14.5 shows the Helmert contrasts on the effect of alcohol. Ordinarily we wouldn't look at these because the interaction involving alcohol was significant, but for the sake of explanation I will. The top of the table shows the contrast for *Level 1 vs. Later*, which in this case means the placebo group compared to the two alcohol groups. This contrast tests whether the mean of the placebo group (4.938) is different from the mean of the low- and high-dose groups combined ((5.688 + 6.375)/2 = 6.032). This is a difference of –1.094 (4.938 – 6.032), which both the *Contrast Estimate* and the *Difference* in the table tell us. The value of *Sig.*

(0.004) tells us that this difference is significant because it is smaller than the criterion of 0.05. The confidence interval for this difference also doesn't cross zero so, assuming this sample is one of the 95 out of 100 that produce a confidence interval containing the true value of the difference, the real difference between the placebo and alcohol groups is not zero (between –1.817 and –0.371, to be precise). We could conclude that the effect of alcohol is that any amount of alcohol increases the attractiveness ratings of pictures compared to when a placebo was drunk. However, we need to look at the remaining contrast to qualify this statement.

The bottom of the table shows the contrast for *Level 2 vs. Level 3*, which in this case

means the low-dose group compared to the high. This contrast compares the mean of the low-dose group (5.688) to the mean of the high (6.375). This is a difference of –0.687 (5.688 – 6.375),[3] which both the *Contrast Estimate* and the *Difference* in the table tell us. This difference is not significant (because *Sig.* is 0.104, which is greater than 0.05). The confidence interval for this difference also crosses zero so, assuming this sample is one of the 95 out of 100 that produced confidence intervals that contain the true value of the difference, the real difference is between –1.522 and 0.0147 and could be zero. This contrast tells us that having high dose of alcohol doesn't significantly affect attractiveness ratings compared to having a low dose.

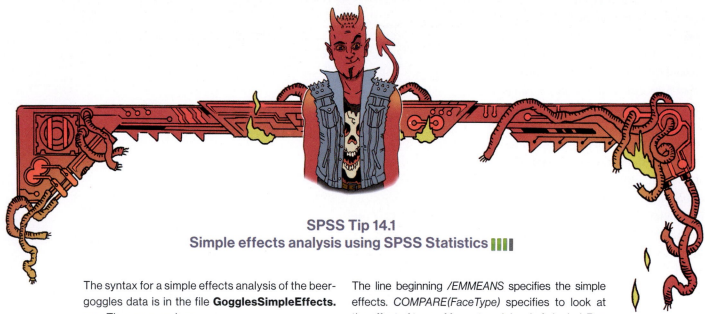

SPSS Tip 14.1
Simple effects analysis using SPSS Statistics ▌▌▌▌

The syntax for a simple effects analysis of the beer-goggles data is in the file **GogglesSimpleEffects.sps**. The commands are:

```
GLM Attractiveness by FaceType Alcohol

/EMMEANS = TABLES(FaceType*Alcohol)
COMPARE(FaceType).
```

The first line specifies the model using the *GLM* command, followed by the outcome variable (**Attractiveness**), the *BY* command and then a list of predictor/independent variables (**FaceType** and **Alcohol**).

The line beginning /EMMEANS specifies the simple effects. COMPARE(FaceType) specifies to look at the effect of type of face at each level of alcohol. Run the syntax (make sure you have **Goggles.sav** loaded) and you'll get the same output as in the chapter, but with an extra table containing the simple effects (Output 14.6). There was a significant difference in ratings of unattractive and attractive faces in the placebo group, $p < 0.001$, and in the low-dose group, $p = 0.008$, but not in the high-dose group, $p = 0.398$. These results confirm our speculation based on Figure 14.13 (see main text).

---

3 Because I've rounded the means to 3 decimal places, the value here differs from Output 14.5 very slightly.

Univariate Tests

Dependent Variable: Median attractiveness rating

Alcohol consumption		Sum of Squares	df	Mean Square	F	Sig.
Placebo	Contrast	33.063	1	33.063	24.150	.000
	Error	57.500	42	1.369		
Low dose	Contrast	10.563	1	10.563	7.715	.008
	Error	57.500	42	1.369		
High dose	Contrast	1.000	1	1.000	.730	.398
	Error	57.500	42	1.369		

Each F tests the simple effects of Attractivenss of facial stimuli within each level combination of the other effects shown. These tests are based on the linearly independent pairwise comparisons among the estimated marginal means.

Output 14.6

Oliver Twisted
Please, Sir, can I have some more ... simple effects?

'I want to impress my friends by doing a simple effects analysis by hand,' boasts Oliver. You don't really need to know how simple effects analyses are calculated to run them, Oliver, but seeing as you asked, it is explained on the companion website.

Multiple Comparisons

Dependent Variable: Median attractiveness rating

	(I) Alcohol consumption	(J) Alcohol consumption	Mean Difference (I–J)	Std. Error	Sig.	95% Confidence Interval Lower Bound	95% Confidence Interval Upper Bound
Bonferroni	Placebo	Low dose	−.75	.414	.231	−1.78	.28
		High dose	−1.44*	.414	.004	−2.47	−.41
	Low dose	Placebo	.75	.414	.231	−.28	1.78
		High dose	−.69	.414	.312	−1.72	.34
	High dose	Placebo	1.44*	.414	.004	.41	2.47
		Low dose	.69	.414	.312	−.34	1.72

Based on observed means.
The error term is Mean Square(Error) = 1.392.

*. The mean difference is significant at the .05 level.

Bootstrap for Multiple Comparisons

Dependent Variable: Median attractiveness rating

	(I) Alcohol consumption	(J) Alcohol consumption	Mean Difference (I–J)	Bootstrap[a] Bias	Bootstrap[a] Std. Error	BCa 95% Confidence Interval Lower	BCa 95% Confidence Interval Upper
Bonferroni	Placebo	Low dose	−.75	−.02	.59	−1.94	.40
		High dose	−1.44	−.01	.57	−2.54	−.41
	Low dose	Placebo	.75	.02	.59	−.40	1.94
		High dose	−.69	.01	.42	−1.49	.14
	High dose	Placebo	1.44	.01	.57	.31	2.60
		Low dose	.69	−.01	.42	−.12	1.45

a. Unless otherwise noted, bootstrap results are based on 1000 bootstrap samples

Output 14.7

Median attractiveness rating

	Alcohol consumption	N	Subset 1	Subset 2
Ryan–Einot–Gabriel–Welsch Range[a]	Placebo	16	4.94	
	Low dose	16	5.69	5.69
	High dose	16		6.38
	Sig.		.077	.104

Means for groups in homogeneous subsets are displayed.
Based on observed means.
The error term is Mean Square(Error) = 1.369.

a. Alpha = .05.

Output 14.8

14.6.5 Simple effects analysis ▮▮▮

A particularly effective way to break down interactions is **simple effects analysis**, which looks at the effect of one independent variable at individual levels of the other independent variable. For example, we could do a simple effects analysis looking at the effect of type of face at each level of alcohol. This would mean taking the average attractiveness rating of unattractive faces and comparing it to that for attractive faces after a placebo drink, then making the same comparison after a low dose of alcohol, and then finally for a high dose. Another way of looking at this

is to say we would look at the distance between the blue and orange dots in Figure 14.13 at each dose of alcohol: based on the graph, we might expect to find a difference in ratings after a placebo and possibly the low dose (the blue and orange dots are very far apart) but not after a high dose (the dots are in practically the same location).

An alternative is to quantify the effect of alcohol (the pattern of means across the placebo, low dose and high dose) separately for unattractive and attractive faces. This analysis would look at whether the means portrayed by the blue line in Figure 14.13 differ, and then separately do the same for the orange line. Simple effects analyses can't be run through dialog boxes—you need to use syntax (see SPSS Tip 14.1).

14.6.6 Post hoc analysis ▮▮▮

The Bonferroni *post hoc* tests (Output 14.7) break down the main effect of alcohol and can be interpreted as if **Alcohol** were the only predictor in the model (i.e., the reported effects for alcohol are collapsed with regard to the type of face). The tests show (both by the significance and whether the bootstrap confidence intervals cross zero) that when participants had a high dose of alcohol their ratings of faces were significantly higher than those who had a placebo drink ($p = 0.004$) but not than those who had a low dose of alcohol ($p = 0.312$), and that ratings were not significantly different between those who had a low dose of alcohol and those in the placebo group ($p = 0.231$).

The REGWQ test (Output 14.8) confirms that the means of the placebo and low-dose conditions were equivalent ($p = 0.077$), as were the means of the low- and high-dose groups ($p = 0.104$). I note again that we wouldn't normally interpret *post hoc* tests when there is a significant interaction involving that main effect (as there is here).

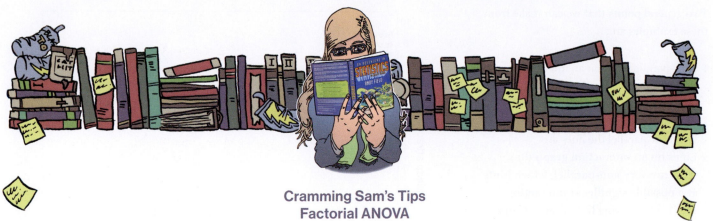

Cramming Sam's Tips
Factorial ANOVA

- Two-way independent designs compare several means when there are two independent variables and different entities have been used in all experimental conditions. For example, if you wanted to know whether different teaching methods worked better for different topics, you could take students from four courses (Psychology, Geography, Management, and Statistics) and assign them to either lecture-based or book-based teaching. The two variables are topic and method of teaching. The outcome might be the end-of-year mark (as a percentage).
- In the table labeled *Tests of Between-Subjects Effects*, look at the column labeled *Sig.* for all main effects and

- interactions; if the value is less than 0.05 then the effect is significant using the conventional criterion.
- To interpret a significant interaction, plot an interaction graph and conduct simple effects analysis.
- You don't need to interpret main effects if an interaction effect involving that variable is significant.
- If significant main effects are not qualified by an interaction then consult *post hoc* tests to see which groups differ: significance is shown by values smaller than 0.05 in the columns labeled *Sig.*, and bootstrap confidence intervals that do not contain zero.
- Test the same assumptions as for any linear model (see Chapter 6).

14.7 Interpreting interaction graphs

We can sum up the findings of the study as follows. Alcohol has an effect on ratings of the attractiveness of pictures of faces, but only when those pictures depict unattractive faces. When the faces are attractive, alcohol has no significant effect on judgements of attraction. This pattern of results was evident from the interaction graph (Figure 14.13).

Let's have a look at other examples of interaction graphs to practice interpreting them. Imagine we got the profile of results

in Figure 14.14. Do you think the interaction effect would still have been significant?

This profile of data probably would also give rise to a significant interaction term because the ratings of attractive and unattractive faces are different in the placebo and high-dose group, but relatively similar in the low-dose group. Visually, the blue line shows a different pattern than the orange one. This situation reflects a world in which the beer-goggles effect kicks in at low doses of alcohol, but after a higher dose reality kicks back in again. Theoretically more difficult to explain, but nevertheless this profile of

results reflects an interaction because the difference between ratings of attractive and unattractive faces varies depending on how much alcohol has been drunk.

Let's try another example. Is there a significant interaction in Figure 14.15?

The data in Figure 14.15 are unlikely to reflect a significant interaction because the effect of alcohol is the same for unattractive and attractive photos: ratings of attractiveness increase as the dose of alcohol increases, and this is true for both attractive and unattractive faces. Visually, the blue line shows the same pattern as the orange one.

Two general points that we can make from these examples are:

1 Non-parallel lines on an interaction graph indicate some degree of interaction, but how strong and whether the interaction is significant depends on how non-parallel the lines are.

2 Lines on an interaction graph that cross are very non-parallel, which hints at a possible significant interaction. However, crossing lines don't *always* reflect a significant interaction.

Sometimes people use bar charts to show interactions, so let's look at those in Figure 14.16. Panels (a) and (b) display the data from the example used in this chapter (have a go at plotting them for yourself). The data are presented in two different ways: panel (a) shows the data when levels of alcohol are placed along the *x*-axis and different-colored bars show the means for unattractive and attractive faces, and panel (b) shows the opposite scenario where type of face is plotted on the *x*-axis and different colors distinguish the dose of alcohol. Both graphs show the significant interaction effect from the example. You're looking for the differences in heights of colored bars to vary at different points along the *x*-axis. For example, in panel (a) you'd look at the difference between the light and dark blue bars for placebo, and then look at, say, the high dose and ask, 'Is the difference between the bars the same as for the placebo?' In this case the difference in height between the dark- and light-blue bars for the placebo is bigger than the same difference for the high-dose group, hence an interaction. Panel (b) shows the same thing, but plotted the other way around. Again look at the pattern of responses. First look at the unattractive faces and see that the pattern is that attractiveness ratings increase as the dose of alcohol goes up (the bars increase in height). Now look at the attractive faces. Is the pattern the same? No, the bars are all

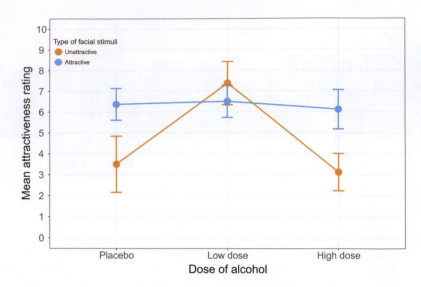

Figure 14.14 Another interaction graph

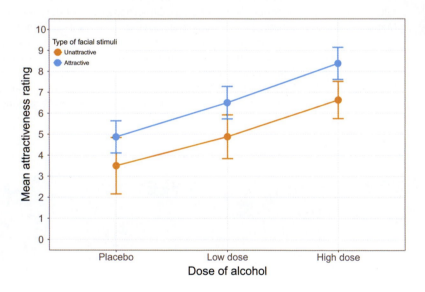

Figure 14.15 A 'lack of' interaction graph

SELF TEST

What about panels (c) and (d): do you think there is an interaction?

the same height, showing that ratings don't change as a function of alcohol. The interaction effect is shown up by the fact that for the attractive faces the bars follow a different pattern than the unattractive faces.

Panels (c) and (d) display the same data in two different ways, albeit different data than the example in the chapter. In the placebo group in panel (c) the dark bar is a little bit bigger than the light one; moving on to the low-dose group, the dark bar is

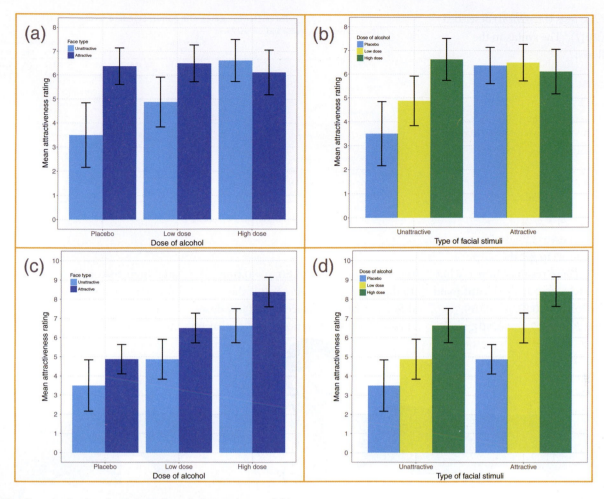

Figure 14.16   Bar charts showing interactions between two variables

also a little bit taller than the light bar; and finally, for the high-dose data the dark bar is again higher than the light one. In all conditions the same pattern is shown—the dark-blue bar is a similar amount taller than the light-blue one (i.e., the amount by which attractive faces are rated more attractive than unattractive faces is similar in all three alcohol groups)—therefore, there is no interaction. Panel (d) shows a similar result. For unattractive faces the pattern is that attractiveness ratings increase as more alcohol is drunk (the bars increase in height), and for the attractive faces the same pattern emerges, with ratings increasing as more alcohol is consumed. This again indicates no interaction: the change in attractiveness ratings due to alcohol is similar for unattractive and attractive faces.

14.8 Robust models of factorial designs ▐▐▐▐

If you requested robust standard errors in Figure 14.10 then you'll find a table of parameter estimates for the model in the output with confidence intervals and *p*-values robust to heteroscedasticity (Output 14.9). These are a little hard to unpick. SPSS Statistics automatically uses the last category as the reference category. The second parameter estimate for face type, because it had only two categories, gives us a robust estimate of the effect on ratings of attractive faces compared to unattractive ones. For the main effect of alcohol the first estimate *[Alcohol = 0]* compares the placebo to the high dose and *[Alcohol = 1]* compares the low dose to the high dose. Both parameters are not significant. For the interaction term, the

first parameter, *[FaceType = 0]*[Alcohol = 0]* suggests that the difference between ratings of attractive and unattractive faces in the placebo was significantly different to the same difference in the high dose group (*p* < 0.001). The second parameter, *[FaceType = 0]*[Alcohol = 1]* suggests that the difference between ratings of attractive and unattractive faces in the low dose group was significantly different to the same difference in the high dose group (*p* = 0.009). These conclusions are based on robust confidence intervals and *p*-values. It's possible to run a robust test for two- and three-way factorial designs using R (make sure the *Essentials for R* plugin and *WRS2* package are installed—see Section 4.13). The companion website contains the syntax file **robust2way.sps** for running a robust test of trimmed means for two independent variables/predictors (*t2way*) and associated robust *post hoc* tests (*mcp2atm*) described by Wilcox (2017).[4] The syntax in the file is as follows:

```
BEGIN PROGRAM R.
library(WRS2)
mySPSSdata = spssdata.
GetDataFromSPSS(factorMode = "labels")
t2way(Attractiveness ~ FaceType*
Alcohol, data = mySPSSdata, tr = 0.2)
mcp2atm(Attractiveness ~ FaceType*
Alcohol, data = mySPSSdata, tr = 0.2)
END PROGRAM.
```

Select and run these six lines of syntax (see SPSS Tip 14.2) to get the output in Output 14.10. These results confirm what the linear model showed: significant main effects of alcohol, $F_t = 10.31$, $p = 0.019$, and face type, $F_t = 14.57$, $p = 0.001$, and

Parameter Estimates with Robust Standard Errors

Dependent Variable: Median attractiveness rating

Parameter	B	Robust Std. Error[a]	t	Sig.	95% Confidence Interval Lower Bound	95% Confidence Interval Upper Bound
Intercept	6.125	.398	15.386	.000	5.322	6.928
[FaceType=0]	.500	.547	.914	.366	-.604	1.604
[FaceType=1]	0[b]	.	.	.	.	.
[Alcohol=0]	.250	.513	.487	.629	-.786	1.286
[Alcohol=1]	.375	.515	.728	.471	-.665	1.415
[Alcohol=2]	0[b]	.	.	.	.	.
[FaceType=0] * [Alcohol=0]	-3.375	.852	-3.963	.000	-5.094	-1.656
[FaceType=0] * [Alcohol=1]	-2.125	.775	-2.742	.009	-3.689	-.561
[FaceType=0] * [Alcohol=2]	0[b]	.	.	.	.	.
[FaceType=1] * [Alcohol=0]	0[b]	.	.	.	.	.
[FaceType=1] * [Alcohol=1]	0[b]	.	.	.	.	.
[FaceType=1] * [Alcohol=2]	0[b]	.	.	.	.	.

a. HC4 method

b. This parameter is set to zero because it is redundant.

Output 14.9

the interaction, $F_t = 16.60$, $p = 0.003$. In the *post hoc* tests, *Alcohol1* is the difference between placebo and low dose, *Alcohol2* is the difference between placebo and high dose, and *Alcohol3* is the difference between low and high dose. We can see from the *p*-values and confidence intervals that attractiveness ratings were

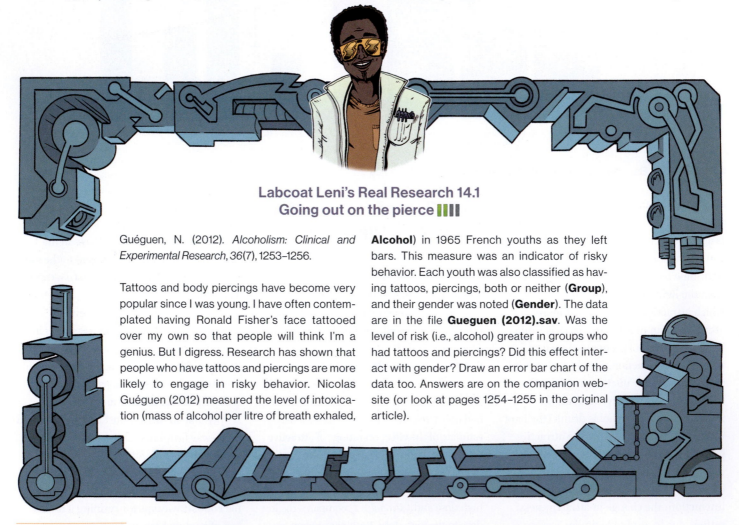

Labcoat Leni's Real Research 14.1
Going out on the pierce ▮▮▮▮

Guéguen, N. (2012). *Alcoholism: Clinical and Experimental Research*, 36(7), 1253–1256.

Tattoos and body piercings have become very popular since I was young. I have often contemplated having Ronald Fisher's face tattooed over my own so that people will think I'm a genius. But I digress. Research has shown that people who have tattoos and piercings are more likely to engage in risky behavior. Nicolas Guéguen (2012) measured the level of intoxication (mass of alcohol per litre of breath exhaled, **Alcohol**) in 1965 French youths as they left bars. This measure was an indicator of risky behavior. Each youth was also classified as having tattoos, piercings, both or neither (**Group**), and their gender was noted (**Gender**). The data are in the file **Gueguen (2012).sav**. Was the level of risk (i.e., alcohol) greater in groups who had tattoos and piercings? Did this effect interact with gender? Draw an error bar chart of the data too. Answers are on the companion website (or look at pages 1254–1255 in the original article).

---

4 You'll also find the file robust3way.sps, which includes the function *t3way* that performs the same test for designs with three independent variables and predictors, but you'll need to edit this file to include the variable names from your data.

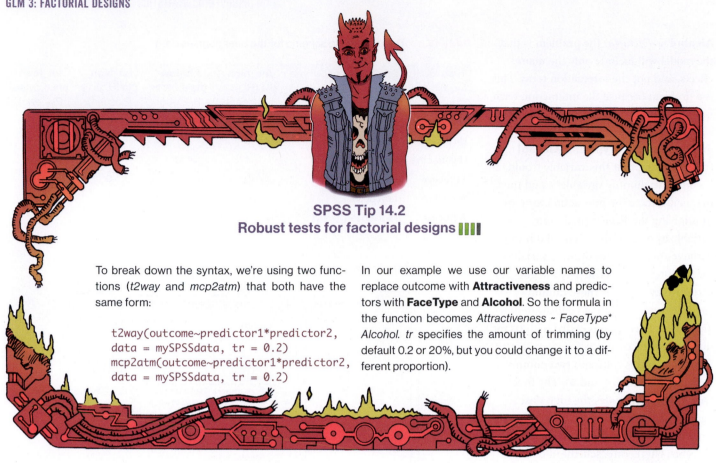

SPSS Tip 14.2
Robust tests for factorial designs ▮▮▮▮

To break down the syntax, we're using two functions (*t2way* and *mcp2atm*) that both have the same form:

```
t2way(outcome~predictor1*predictor2,
data = mySPSSdata, tr = 0.2)
mcp2atm(outcome~predictor1*predictor2,
data = mySPSSdata, tr = 0.2)
```

In our example we use our variable names to replace outcome with **Attractiveness** and predictors with **FaceType** and **Alcohol**. So the formula in the function becomes *Attractiveness ~ FaceType* Alcohol. tr* specifies the amount of trimming (by default 0.2 or 20%, but you could change it to a different proportion).

```
t2way(formula = Attractiveness ~ FaceType * Alcohol,
data = mySPSSdata
  tr = 0.2)
                       value  p.value
FaceType              14.5730   0.001
Alcohol               10.3117   0.019
FaceType:Alcohol      16.6038   0.003

Call:
mcp2atm(formula = Attractiveness ~ FaceType * Alcohol,
data = mySPSSdata,
  tr = 0.2)
                     psihat  ci.lower  ci.upper  p-value
FaceType1          -3.83333  -5.90979  -1.75688  0.00088
Alcohol1           -1.50000  -3.63960   0.63960  0.07956
Alcohol2           -2.83333  -5.17113  -0.49554  0.00543
Alcohol3           -1.33333  -3.38426   0.71759  0.10569
FaceType1:Alcohol1 -1.16667  -3.30627   0.97294  0.16314
FaceType1:Alcohol2 -3.50000  -5.83780  -1.16220  0.00110
FaceType1:Alcohol3 -2.33333  -4.38426  -0.28241  0.00802
```

Output 14.10

significantly different between the placebo and high-dose groups, $\hat{\psi} = -2.83$, $p = 0.005$, but not between the placebo and low-dose groups, $\hat{\psi} = -1.5$, $p = 0.080$ or between the low- and high-dose groups,

$\hat{\psi} = -1.33$, $p = 0.106$. Most interesting are the interaction terms, which look at the difference between ratings of attractive and unattractive faces across the group comparisons we've just looked at. These

effects tell us that the difference in ratings between attractive and unattractive faces in the high-dose group was significantly different from the corresponding difference in the placebo group, $\hat{\psi} = -3.5$, $p = 0.001$, and the low-dose group, $\hat{\psi} = -2.33$, $p = 0.008$. However, the difference in ratings between attractive and unattractive faces in the low-dose group was not significantly different from the corresponding difference in the placebo group, $\hat{\psi} = -1.17$, $p = 0.163$. In a nutshell, the interaction seems to be driven by effects of a high dose of alcohol compared to the other conditions.

14.9 Bayesian models of factorial designs ▮▮▮▮

You can't easily compute Bayes factors (Section 3.8.4) for factorial designs. Although we're using a linear model, so we could follow a similar process to that in Section 9.13 and specify **FaceType** and

Alcohol as *Factor(s)* the problem is that the model will include only the main effects, and not the interaction term. This is a problem because the interaction term is typically where our interests lie. The way around this would be to manually dummy code **FaceType** and **Alcohol** (remembering that this variable would become two dummy variables), and then manually create the interaction term by multiplying the **FaceType** dummy variable by each of the **Alcohol** dummy variables to create two dummy variables for the interaction term. The file **goggles_dummy.sav** contains these dummy variables and the coding is illustrated in Table 14.4. **FaceType** is coded with 0s for unattractive faces and 1s for attractive faces. **Alcohol** is split into two dummy variables (columns 4 and 5). The first (**Alc_high_pla**) codes the high dose against placebo and the second (**Alc_low_pla**) codes the low dose against placebo. This coding scheme should be familiar from Chapter 12. The final two columns code the interaction term. The first (**Int_high_pla**) are the codes from columns 3 and 4 multiplied (i.e., **FaceType × Alc_high_pla**) and represent the difference between ratings for unattractive faces compared to attractive ones in the placebo group relative to the same difference in the high dose group. The second (**Int_low_pla**) are the codes from columns 3 and 5 multiplied (i.e., **FaceType × Alc_low_pla**) and represent the difference between ratings for unattractive faces compared to attractive ones in the placebo group relative to the same difference in the low dose group.

To get Bayes factors and estimates with the default reference priors follow the process in Section 9.13 and drag all five dummy variables (**FaceType, Alc_high_pla, Alc_low_pla, Int_high_pla, Int_low_pla**) to the box labeled *Covariate(s)*. (If you drag them to box labeled *Factor(s)* then SPSS Statistics will recode them, so by using the *Covariate(s)* box the output will match the

Table 14.4 Dummy coding scheme for the beer goggles data

Type of faces rated	Alcohol	FaceType	**Alc_high_pla** (High vs Placebo)	**Alc_low_pla** (Low vs Placebo)	**Int_high_pla** (High vs Placebo)	**Int_low_pla** (Low vs Placebo)
Unattractive	Placebo	0	0	0	0	0
Unattractive	Low dose	0	0	1	0	0
Unattractive	High dose	0	1	0	0	0
Attractive	Placebo	1	0	0	0	0
Attractive	Low dose	1	0	1	0	1
Attractive	High dose	1	1	0	1	0

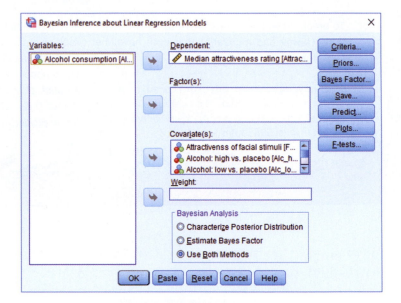

Figure 14.17 Dialog box for Bayesian analysis of factorial designs

dummy coding that we used in the data editor.) Select ⊙ Use Both Methods and leave the default options alone.

Output 14.11 shows the Bayes factor for the model including all of the predictors (all 5 dummy variables) against the null model (which I assume includes only the intercept). This Bayes factor is 1195.598. This is very strong evidence for the model. It means that the probability of the data given the model including all five predictors is about 1196 greater than the probability of the data given the model with only the intercept. We should shift our belief in the model (relative to the null model) by a factor of 1196!

The Bayesian estimate of the *b*s for the dummy variables can be found in the columns labeled *Posterior Mode* and

Posterior Mean. The 95% credible intervals for the model parameters contain the population value with a probability of 0.95 (95%). For the difference between ratings for unattractive faces compared to attractive ones in the placebo group relative to the same difference in the high dose group (i.e., the first interaction term), the relative difference has a 95% probability of being between about −5 and −1.7. For the difference between ratings for unattractive faces compared to attractive ones in the placebo group relative to the same difference in the low dose group (i.e., the second interaction term), the relative difference has a 95% probability of being between about −2.9 and 0.4. These credible intervals assume that the effect

Bayes Factor Model Summary[a,b]

Bayes Factor[c]	R	R Square	Adjusted R Square	Std. Error of the Estimate
1195.598	.718	.515	.458	1.17

a. Method: JZS

b. Model: (Intercept), Attractivenss of facial stimuli, Alcohol: high vs. placebo, Alcohol: low vs. placebo, Interaction: high vs placebo, att vs unatt, Interaction: low vs placebo, att vs unatt

c. Bayes factor: Testing model versus null model (Intercept).

Bayesian Estimates of Coefficients[a,b,c]

Parameter	Posterior Mode	Posterior Mean	Variance	95% Credible Interval Lower Bound	95% Credible Interval Upper Bound
(Intercept)	3.500	3.500	.180	2.665	4.335
Attractivenss of facial stimuli	2.875	2.875	.359	1.694	4.056
Alcohol: high vs. placebo	3.125	3.125	.359	1.944	4.306
Alcohol: low vs. placebo	1.375	1.375	.359	.194	2.556
Interaction: high vs placebo, att vs unatt	−3.375	−3.375	.719	−5.045	−1.705
Interaction: low vs placebo, att vs unatt	−1.250	−1.250	.719	−2.920	.420

a. Dependent Variable: Median attractiveness rating

b. Model: (Intercept), Attractivenss of facial stimuli, Alcohol: high vs. placebo, Alcohol: low vs. placebo, Interaction: high vs placebo, att vs unatt, Interaction: low vs placebo, att vs unatt

c. Assume standard reference priors.

Output 14.11

exists but in the later case it suggests that the effect could plausibly be positive, negative and very small indeed.

14.10 Calculating effect sizes ▌▌▌▌

SPSS can produce partial eta squared, η^2 (Section 13.8). However, we've seen before that omega squared (ω^2) is less biased. The calculation of omega squared becomes somewhat more cumbersome in factorial designs ('somewhat' being an understatement). Howell (2012), as ever, does a wonderful job of explaining it all (and has a nice table summarizing the various components for a variety of situations). Condensing all of this information, I'll say only that we first compute a variance component for each of the effects (the two main effects and the interaction term) and the error, and then use these to calculate effect sizes for each. If we call the first main effect A, the second main effect B and the interaction effect $A \times B$, then the variance

components for each of these are calculated from the mean squares of each effect and the sample sizes on which they're based:

$$\hat{\sigma}_{\alpha}^2 = \frac{(a-1)(MS_A - MS_R)}{nab}$$

$$\hat{\sigma}_{\beta}^2 = \frac{(b-1)(MS_B - MS_R)}{nab} \qquad (14.24)$$

$$\hat{\sigma}_{\alpha\beta}^2 = \frac{(a-1)(b-1)(MS_{A\times B} - MS_R)}{nab}$$

In these equations, a is the number of levels of the first independent variable, b is the number of levels of the second independent variable, and n is the number of people per condition.

Let's calculate these for our data. Output 14.3 contains the mean squares for each effect and for the error term. The first predictor, alcohol, had three levels (hence $a = 3$) and a mean squares of 8.271. Type of face had two levels (hence $b = 2$) and a mean squares of 21.333. The interaction had a mean squares of 11.646. The number of people in each group was 8 and the residual mean squares were 1.369. Therefore, our equations become:

$$\hat{\sigma}_{\alpha}^2 = \frac{(3-1)(8.271-1.369)}{8 \times 3 \times 2} = 0.288$$

$$\hat{\sigma}_{\beta}^2 = \frac{(2-1)(21.333-1.369)}{8 \times 3 \times 2} = 0.416 \qquad (14.25)$$

$$\hat{\sigma}_{\alpha\beta}^2 = \frac{(3-1)(2-1)(11.646-1.369)}{8 \times 3 \times 2} = 0.428$$

We estimate the total variability by adding the estimates in equation (14.25) to the residual mean squares:

$$\hat{\sigma}_{total}^2 = \hat{\sigma}_{\alpha}^2 + \hat{\sigma}_{\beta}^2 + \hat{\sigma}_{\alpha\beta}^2 + MS_R$$
$$= 0.288 + 0.416 + 0.428 + 1.369 \qquad (14.26)$$
$$= 2.501$$

The effect size is the variance estimate for the effect in which you're interested divided by the total variance estimate from equation (14.26):

$$\omega_{effect}^2 = \frac{\hat{\sigma}_{effect}^2}{\hat{\sigma}_{total}^2} \qquad (14.27)$$

For the main effect of alcohol we get 0.115:

$$\omega_{alcohol}^2 = \frac{\hat{\sigma}_{alcohol}^2}{\hat{\sigma}_{total}^2} = \frac{0.288}{2.501} = 0.115 \qquad (14.28)$$

For the main effect of type of face we get 0.166:

$$\omega_{type\,of\,face}^2 = \frac{\hat{\sigma}_{type\,of\,face}^2}{\hat{\sigma}_{total}^2} = \frac{0.416}{2.501} = 0.166 \qquad (14.29)$$

For the interaction of type of face and alcohol we get 0.171:

$$\omega_{alcohol\,\,type\,of\,face}^2 = \frac{\hat{\sigma}_{alcohol\,\,type\,of\,face}^2}{\hat{\sigma}_{total}^2} = \frac{0.428}{2.501} = 0.171 \qquad (14.30)$$

To make these values comparable to r we can take the square root, which gives us effect sizes of 0.34 for alcohol and 0.41 for both type of face and the interaction term. It's also possible to calculate effect sizes for our simple effects analysis (Section 14.6.5). These effects have 1 degree of freedom for the model (which means they're comparing only two things), and in these situations F can be converted to r using the following equation (which just uses the F-statistic and the residual degrees of freedom):[5]

---

5 If your F compares more than two things then a different equation is needed (see Rosenthal et al. (2000:44), but I think effect sizes for situations in which only two things are being compared are most useful because they have a clear interpretation.

$$r = \sqrt{\frac{F(1, df_R)}{F(1, df_R) + df_R}} \quad (14.31)$$

Looking at Output 14.6, we got Fs of 24.150, 7.715 and 0.730 for the effects of type of face at placebo, low dose and high dose, respectively. For each of these, the degrees of freedom were 1 for the model and 42 for the residual. Therefore, we get the following effect sizes:

$$r_{\text{Type of face (placebo)}} = \sqrt{\frac{24.15}{24.15 + 42}} = 0.604$$

$$r_{\text{Type of face (low dose)}} = \sqrt{\frac{7.715}{7.715 + 42}} = 0.394 \quad (14.32)$$

$$r_{\text{Type of face (high dose)}} = \sqrt{\frac{0.730}{0.730 + 42}} = 0.131$$

The effect of type of face is large at both placebo and low dose, but becomes small at a high dose of alcohol.

14.11 Reporting the results of factorial designs ||||

We report the details of the F-statistic and the degrees of freedom for each effect. For the effects of alcohol and the alcohol × type of face interaction, the model degrees of freedom were 2 ($df_M = 2$), but for the effect of type of face the degrees of freedom were 1 ($df_M = 1$). For all effects,

the degrees of freedom for the residuals were 42 ($df_R = 42$). We can, therefore, report the three effects as follows:

✓ There was a significant main effect of the amount of alcohol consumed on ratings of the attractiveness of faces, $F(2, 42) = 6.04$, $p = 0.005$, $\omega^2 = 0.12$. Bonferroni *post hoc* tests revealed that the attractiveness ratings were significantly higher after a high dose than after a placebo drink ($p = 0.004$). The attractiveness ratings were not significantly different after a low dose compared to a placebo ($p = 0.231$) or a high dose compared to a low dose ($p = 0.312$).

✓ Attractive faces were rated significantly higher than unattractive faces, $F(1, 42) = 15.58$, $p < 0.001$, $\omega^2 = 0.17$.

Labcoat Leni's Real Research 14.2
Don't forget your toothbrush? ||||

Davey, G. C. L., et al. (2003). *Journal of Behavior Therapy & Experimental Psychiatry, 34*, 141–160.

Many of us have experienced that feeling after we have left the house of wondering whether we remembered to lock the door, close the window or remove the bodies from the fridge in case the police turn up. However, some people with obsessive compulsive disorder (OCD) check things so excessively that they might, for example, take hours to leave the house. One theory is that this checking behavior is caused by the mood you are in (positive or negative) interacting with the rules you use to decide when to stop a task (do you continue until you feel like stopping or until you have done the task as best as you can?). Davey, Startup, Zara, MacDonald, and Field (2003) tested this hypothesis by getting people to think of as many things as they could that they should

check before going on holiday (**Checks**) after putting them into a negative, positive or neutral mood (**Mood**). Within each mood group, half of the participants were instructed to generate as many items as they could, whereas the remainder were asked to generate items for as long as they felt like continuing the task (**Stop_Rule**).

Plot an error bar chart and then conduct the appropriate analysis to test Davey et al.'s hypotheses that (1) people in negative moods who use an 'as many as can' stop rule would generate more items than those using a 'feel like continuing' stop rule; (2) people in a positive mood would generate more items when using a 'feel like continuing' stop rule compared to an 'as many as can' stop rule; (3) in neutral moods, the stop rule used won't have an effect (**Davey(2003). sav**). Answers are on the companion website (or look at pages 148–149 in the original article).

✓ There was a significant interaction between the amount of alcohol consumed and the type of face of the person rated on attractiveness, $F(2, 42) = 8.51$, $p = 0.001$, $\omega^2 = 0.17$. This effect indicates that ratings of unattractive and attractive faces were affected differently by alcohol. Simple effects analysis revealed that ratings of attractive faces were significantly higher than unattractive faces in the placebo group, $F(1, 42) = 24.15$, $p < 0.001$, and in the low-dose group, $F(1, 42) = 7.72$, $p = 0.008$, but not in the high dose group, $F(1, 42) = 0.73$, $p = 0.398$.

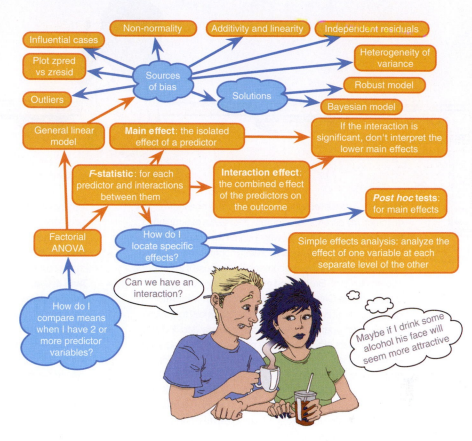

Figure 14.18 What Brian learnt from this chapter

14.12 Brian and Jane's Story ▮▮▮▮

After walking the streets alone with her thoughts, Jane arrived at Brian's apartment. It was going to seem weird that she was there: she barely knew his name, they'd never exchanged numbers, let alone addresses, but these things were easy enough to find out if you knew how to hack. She felt anxious. At the third attempt she mustered the courage to press the buzzer. No answer. She buzzed again. And again.

'Hi!' said the puzzled voice through the intercom.

'It's Jane. I need to talk.'

A minute later Brian opened the door, linked arms with her and walked. She had expected to be invited in, but this was cool too. They walked the neighborhood in the moonlight. She needed to talk, but she didn't: their lack of intimacy muted her. Brian filled the silence, as he always did, with his latest statistical knowledge. When he ran out of things to impress her with he ended, as he usually did, with a

question. She had no clever put-down this time, no enthusiasm for flirtatious banter, and no one else to talk to. She told him about her mum, about her ambivalence towards her parents, about her upbringing, her brother. It was unfiltered and unguarded and made her feel vulnerable. Brian listened and told her to go to her mother.

14.13 What next? ▮▮▮▮

No sooner had I started my first band than it disintegrated. I went with drummer Mark to sing in a band called the Outlanders, who

were much better musically but were not, if the truth be told, metal enough for me. They also sacked me after a very short period of time for not being able to sing like Bono (an insult at the time, but in retrospect . . .). So, that was two failed bands in very quick succession. You'd have thought that the message that perhaps singing wasn't the thing for me might have started to sink in. It hadn't and didn't for quite some time (still hasn't entirely). I needed a new plan, and one was hatched one evening while walking along a cliff top in Cornwall with drummer Mark. Fortunately, it wasn't a plan that involved throwing myself off into the sea . . .

14.14 Key terms that I've discovered

Beer-goggles effect

Factorial ANOVA

Independent factorial design

Interaction graph

Mixed design

Related factorial design

Simple effects analysis

Smart Alex's tasks

- **Task 1**: I've wondered whether musical taste changes as you get older: my parents, for example, after years of listening to relatively cool music when I was a kid, hit their mid-forties and developed a worrying obsession with country and western. This possibility worries me immensely, because if the future is listening to Garth Brooks and thinking 'oh boy, did I underestimate Garth's immense talent when I was in my twenties', then it is bleak indeed. To test the idea I took two groups (**age**): young people (which I arbitrarily decided was under 40 years of age) and older people (above 40 years of age). I split each of these groups of 45 into three smaller groups of 15 and assigned them to listen to Fugazi, ABBA or Barf Grooks[6] (**music**). Each person rated the music (**liking**) on a scale ranging from +100 (this is sick) through 0 (indifference) to −100 (I'm going to *be* sick). Fit a model to test my idea (**Fugazi.sav**).

- **Task 2**: Compute omega squared for the effects in Task 1 and report the results of the analysis.

- **Task 3**: In Chapter 5 we used some data that related to male and female arousal levels when watching *The Notebook* or a documentary about notebooks (**Notebook.sav**). Fit a model to test whether men and women differ in their reactions to different types of films.

- **Task 4**: Compute omega squared for the effects in Task 3 and report the results of the analysis.

- **Task 5**: In Chapter 4 we used some data that related to learning in men and women when either reinforcement or punishment was used in teaching (**Method Of Teaching.sav**). Analyze these data to see whether men and women's learning differs according to the teaching method used.

- **Task 6**: At the start of this chapter I described a way of empirically researching whether I wrote better songs than my old bandmate Malcolm, and whether this depended on the type of song (a symphony or song about flies). The outcome variable was the number of screams elicited by audience members during the songs. Draw an error bar graph (lines) and analyze these data (**Escape From Inside.sav**).

- **Task 7**: Compute omega squared for the effects in Task 6 and report the results of the analysis.

- **Task 8**: Using SPSS Tip 14.1, change the syntax in **GogglesSimpleEffects.sps** to look at the effect of alcohol at different levels of type of face.

- **Task 9**: There are reports of increases in injuries related to playing Nintendo Wii (http://ow.ly/ceWPj). These injuries were attributed mainly to muscle and tendon strains. A researcher hypothesized that a stretching warm-up before playing Wii would help lower injuries, and that athletes would be less susceptible to injuries because their regular activity makes them more flexible. She took 60 athletes and 60 non-athletes (**athlete**); half of them played Wii and half watched others playing as a control (**wii**), and within these groups half did a 5-minute stretch routine before playing/watching whereas the other half did not (**stretch**). The outcome was a pain score out of 10 (where 0 is no pain, and 10 is severe pain) after playing for 4 hours (**injury**). Fit a model to test whether athletes are less prone to injury, and whether the prevention program worked (**Wii.sav**).

- **Task 10**: A researcher was interested in what factors contributed to injuries resulting from game console use. She tested 40 participants who were randomly assigned to either an active or static game played on either a Wii or Xbox Kinect. At the end of the session their physical condition was evaluated on an injury severity scale. The data are in the file **Wii vs Xbox Injuries.sav** which contains the variables **Game** (0 = static, 1 = active), **Console** (0 = Wii, 1 = Xbox), and **InjurySeverity** (a score ranging from 0 (no injury) to 20 (severe injury)). Fit a model to see whether injury severity is significantly predicted from the type of game, the type of console and their interaction.

Answers & additional resources are available on the book's website at
https://edge.sagepub.com/field5e

---

6 A less well-known country musician not to be confused with anyone who has a similar name and produces music that makes you want to barf.

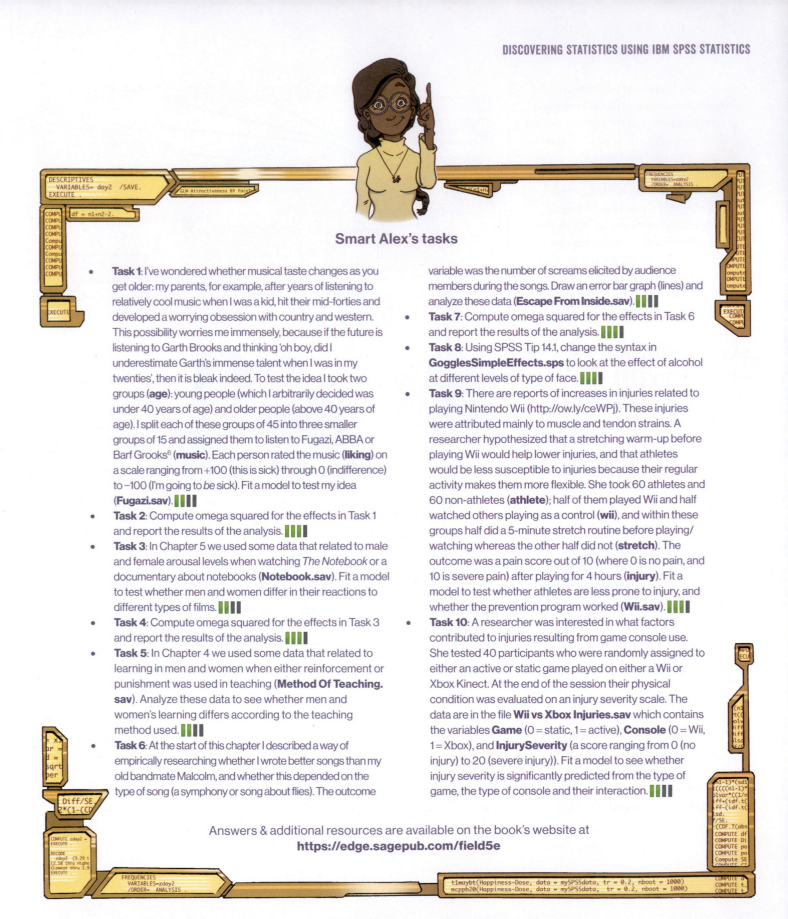

GLM 4: REPEATED-MEASURES DESIGNS

15

15.1 What will this chapter tell me?

At the age of 15, I was on holiday with my friend Mark (the drummer) in Cornwall. I had a pretty decent mullet by this stage (nowadays I wish I had enough hair to grow a mullet) and a respectable collection of heavy metal T-shirts from going to various gigs. We were walking along the clifftops one evening at dusk, reminiscing about Andromeda. We decided that the only thing we hadn't enjoyed about that band was Malcolm and that maybe we should re-form it with a different guitarist.[1] As I wondered who we could get to play guitar, Mark pointed out the blindingly obvious: I played guitar. So, when we got home Scansion was born.[2] As the singer, guitarist and songwriter I set about writing some songs. Lyrically speaking, flies were so 'last year', so I moved on to exploring existential angst (I have never really stopped). Lots of songs about the futility of existence, social leprosy, incapacitating fear of death. That sort of thing. Obviously we failed totally in our musical ambitions, otherwise I wouldn't be writing this book, but we did get reviewed in the music magazine Kerrang! They called us 'twee' in a live review, which is not really what you want to read as metal band, but were slightly more flattering about one of our demos (only slightly though).

---

1   Although this is what we thought at the time, in retrospect I feel bad because Malcolm was never anything other than a very nice, positive, enthusiastic guy. To be honest, at that age (and, some would argue, beyond) I could be a bit of a dolt.

2   Scansion is a term for the rhythm of poetry. We got the name by searching through a dictionary until we found a word that we liked. Originally we didn't think it was 'metal' enough, and we decided that any self-respecting heavy metal band needed to have a big spiky 'X' in their name. So, for the first couple of years we spelt it 'Scanxion'. Like I said, I could be a bit of a dolt back then.

Figure 15.1   Scansion in the early days; I used to stare a lot (L-R: me, Mark and Mark)

Our career high point was playing at the famous Marquee Club in London (which, like most of the venues of my youth, has closed now, but not because of us playing there).[3] This was the biggest gig of our career, and we needed to play like we never had before. As it turned out, we did: I ran on stage and fell over, de-tuning my guitar beyond recognition and breaking the zip on my trousers in the process. I spent the whole gig out of tune and spread-eagle to prevent my trousers falling down. Like I said, I'd never played like *that* before. We used to get quite obsessed with comparing how we played at different gigs. I didn't know about statistics then (happy days), but if I had I would have realized that we could rate ourselves and compare the mean ratings for different gigs; because we would always be the ones rating the gigs, this would be a repeated-measures design. That's what this chapter is about; hopefully it won't make our trousers fall down.

15.2 Introduction to repeated-measures designs ▍▍▍▍

So far in this book, when comparing means, we've concentrated on situations in which different entities contribute to different means. I have tended to focus on examples with different *people* taking part in different experimental conditions, but it could be different plants, companies, plots of land, viral strains, goats or even duck-billed platypuses (or whatever the plural is). I've completely ignored situations in

SELF ✂ TEST
What is a repeated-measures design? (Clue: it is described in Chapter 1.)

Figure 15.2 The Confusion machine has created different ways to refer to repeated-measures designs

which the same people (plants, goats, hamsters, seven-eyed green galactic leaders from space or whatever) contribute to the different means. I've put it off long enough, and now I'm going to take you through the models we fit to repeated-measures data. 'Repeated measures' is a term used when the same entities participate in all conditions of an experiment or provide data at multiple time points. For example, you might test the effects of alcohol on enjoyment of a party. Some people can drink a lot of alcohol without really feeling the consequences, whereas others, like myself, have only to sniff a pint of lager and they start flapping around on the floor waving their arms and legs around shouting 'Look at me, I'm Andy, King of the lost world of the Haddocks.' To control for these individual differences in tolerance to alcohol you can test the same people in all conditions of the experiment: participants could be given a questionnaire assessing their enjoyment of the party after they had consumed 1 pint, 2 pints, 3 pints and 4 pints of lager. There are lots of different ways to refer to this sort of design (Figure 15.2).

15.3 A grubby example ▍▍▍▍

I'm a Celebrity, Get Me Out of Here! is a TV show in which celebrities (more like ex-celebrities), in a pitiful attempt to salvage their careers (or to have careers in the first place), go and live in the jungle in Australia for a few weeks. During the show, they are subjected to various humiliating and degrading tasks to win food for their campmates. The tasks frequently involve creepy-crawlies; for example, a contestant might be locked in a coffin full of rats, forced to put their head in a bowl of large spiders or have eels and cockroaches poured onto them. It's cruel,

---

3 In its day the Marquee Club started the careers of people like Jimi Hendrix, The Who, Iron Maiden and Led Zeppelin. Google search it if you need to procrastinate.

Table 15.1 Data for the bushtucker example

Celebrity	Stick Insect	Kangaroo Testicle	Fish Eye	Witchetty Grub	Mean	s^2
1	8	7	1	6	5.50	9.67
2	9	5	2	5	5.25	8.25
3	6	2	3	8	4.75	7.58
4	5	3	1	9	4.50	11.67
5	8	4	5	8	6.25	4.25
6	7	5	6	7	6.25	0.92
7	10	2	7	2	5.25	15.58
8	12	6	8	1	6.75	20.92
Mean	**8.13**	**4.25**	**4.13**	**5.75**		

voyeuristic, gratuitous, car-crash TV, and I love it. As a vegetarian, for me the bushtucker trial in which the celebrities eat various animal-related things is particularly noxious. Some examples are eating live stick insects or witchetty grubs, and chewing on fish eyes, kangaroo testicles and (surprisingly elastic) kangaroo penises. Seeing a fish eye exploding in someone's mouth is a mental scar that's hard to shake off.

I've often wondered (perhaps a little too much) which of the bushtucker foods is the most revolting. Imagine that I answered this question by getting eight celebrities and forcing them to eat four different animals (the aforementioned stick insect, kangaroo testicle, fish eye and witchetty grub) in counterbalanced order. On each occasion I measured the time it took the celebrity to retch, in seconds. This is a repeated-measures design because every celebrity eats every food. The predictor/independent variable is the type of food eaten and the outcome/dependent variable is the time taken to retch. Table 15.1 shows the data: the eight rows show the different celebrities, and the columns indicate their time to retch after eating each animal. Because they'll

come in handy later, the mean time to retch (and variance) for each celebrity and the mean time to retch for each animal are shown. The overall variance in retching times will, in part, be caused by the fact that the foods differ in their palatability (the manipulation) and, in part, by differences in the celebrities' constitutions (individual differences).

15.4 Repeated-measures and the linear model ▮▮▮▮

Up until now we have considered everything as a variation of the general linear model. Repeated-measures designs can also be conceptualized in this way. First, imagine we wanted to conceptualize the experiment I've just described as a linear model in which different people ate the different animals. To keep things simple, let's imagine there were just two animals eaten: the stick insect and testicle. Thinking back to Chapters 10 and 12, we could write the model as:

$$Y_i = b_0 + b_1 X_{1i} + \varepsilon_i$$
$$\text{Retch}_i = b_0 + b_1 \text{Food}_i + \varepsilon_i \tag{15.1}$$

We predict retching time from the food that was eaten, plus the average retching time when food is equal to zero (b_0).

This model doesn't account for the fact that the same people took part in all conditions. First off, in an independent design we have one observation for the outcome from each person, so we predict the outcome for the individual (Y_i) based on the value of the predictor of that person (X_i), but with repeated measures the person has several values of the predictor so we predict the outcome from both the individual (i) and the specific value of the predictor that's of interest (g).[4] So we might write a simple model as:

$$\text{Retch}_{gi} = b_0 + b_1 \text{Food}_{gi} + \varepsilon_{gi} \tag{15.2}$$

which acknowledges that we predict the retch time (Y_{gi}) for food g within person i from the specific food eaten (X_{gi}), and the error in prediction, ε_{gi}, to which both the individual and the food eaten contribute. All that's changed is the subscripts in the model, which acknowledge that levels of the treatment condition (g) occur within individuals (i).

Equation (15.2) is a very simple model, and we might, for example, want to factor in the possibility that individuals will vary in their constitution. We can do this by adding a variance term to the intercept. Remember that the intercept represents the time to wretch when the predictor is 0, so if we allow this parameter to vary across individuals, we're effectively modeling the possibility that different people will have different natural retching latencies. This is known as a random intercept model, and we'll look at these in detail in Chapter 21. This model is written as:

$$\text{Retch}_{gi} = b_{0i} + b_1 X_{gi} + \varepsilon_{gi}$$
$$b_{0i} = b_0 + u_{0i} \tag{15.3}$$

All that's happened is the intercept has had an i added to the subscript to reflect

---

4 When using this model for repeated-measures designs people typically use subscripts i to represent different conditions and j to represent individuals, so my subscripts are a bit weird, but I think it's less confusing than referring to an earlier model (equation (15.2)) in which i represents individuals but then having i represent something else in equation (15.3).

that it is specific to an individual, and underneath we define the intercept as being made up of the group-level intercept (b_0) plus the deviation of the individual's intercept from the group-level intercept (u_{0i}). Put simply, u_{0i} reflects individual differences in retching. The first line of equation (15.3) becomes a model for an individual, and the group-level effects are incorporated into the second line.

We might also want to factor in the possibility that the effect of different foods varies across individuals. We can do this by adding a variance term to the slope. Remember that the slope represents the effect that the different foods have on retching. By allowing this parameter to vary across individuals, we model the possibility that the effect of the food on retching will be different in different participants. This is known as a random slope model (again, see Chapter 21 for a fuller discussion). We write this model as.

$$Retch_{gi} = b_{0i} + b_{1i}X_{gi} + \varepsilon_{gi}$$
$$b_{0i} = b_0 + u_{0i}$$
$$b_{1i} = b_1 + u_{1i}$$

(15.4)

Compare it to equation (15.3): the main change is that the slope (b_1) has had an i added to the subscript to reflect that it is specific to an individual, and underneath we define it as being made up of the group-level slope (b_1) plus the deviation of the individual's slope from the group-level slope (u_{1i}). As before, u_{1i} reflects individual differences in the effect of food on retching. As in equation (15.3), the top of equation (15.4) is a model for the individual, and the group-level effects are incorporated into the second and third lines.

We are getting ahead of ourselves a bit, but I just want to show that repeated measures can be incorporated into the general linear model. As we'll see in Chapter 21, these models get considerably more complex, but the take-home point is simply that we're again dealing with a linear model.

15.5 The ANOVA approach to repeated-measures designs

Although you can conceptualize repeated measures as a linear model, there are other ways too. The way that people typically handle repeated measures in IBM SPSS Statistics is to use a **repeated-measures ANOVA** approach. Very crudely, this is a linear model like I have just described, but with some very restrictive constraints. In a nutshell, we have seen that the standard linear model assumes that residuals are independent (not related to each other— see Section 12.3), but as you can see from the models above, this assumption isn't true for repeated-measures designs: the residuals are affected by both between-participant factors (which should be independent) and within-participant factors (which won't be independent). There are (broadly) two solutions. One is to model this within-participant variability, which is what the models in the previous section do. The other is to apply additional assumptions that allow a simpler, less flexible model to be fit. The later approach is the one that is historically popular.

15.5.1 The assumption of sphericity

The assumption that permits us to use a simpler model to analyze repeated-measures data is **sphericity**, which, trust me, is a pain in the butt to pronounce when you're giving statistics lectures at 9 a.m. on a Monday. Put crudely, sphericity is about assuming that the relationship between scores in pairs of treatment conditions is similar (i.e., the level of dependence between means is roughly equal).

The assumption of sphericity (denoted by ε and sometimes referred to as *circularity*) can be likened to the assumption of

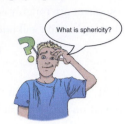

homogeneity of variance in between-group designs. It is a form of **compound symmetry**, which holds true when both the variances across conditions are equal (this is the same as the homogeneity of variance assumption in between-group designs) and the covariances between pairs of conditions are equal. So, we assume that the variation within conditions is similar and that no two conditions are any more dependent than any other two. Sphericity is a more general, less restrictive form of compound symmetry and refers to the equality of variances of the *differences* between treatment levels (Misconception Mutt 15.1).

The best way to explain what this means is by checking the assumption of sphericity by hand, which incidentally only a complete lunatic would do. Table 15.2 shows data from an experiment with three

Table 15.2 Hypothetical data to illustrate the calculation of the variance of the differences between conditions

Condition A	Condition B	Condition C	A – B	A – C	B – C
10	12	8	−2	2	4
15	15	12	0	3	3
25	30	20	−5	5	10
35	30	28	5	7	2
30	27	20	3	10	7
		Variance:	15.7	10.3	10.7

conditions and five people who have contributed scores to each condition. That's the first three columns. The second three show the differences between pairs of scores in all combinations of the treatment levels. I have calculated the variance of these differences in the bottom row. Sphericity is met when these variances are roughly equal. For these data, sphericity will hold when:

$$\text{variance}_{A-B} \approx \text{variance}_{A-C} \approx \text{variance}_{B-C} \qquad (5.5)$$

Table 15.2 shows that there is some deviation from sphericity in these data because the variance of the differences

between conditions A and B (15.7) is greater than the variance of the differences between both A and C (10.3) and B and C (10.7). However, these data have *local circularity* (or local sphericity) because two of the variances of differences are very similar, which means that sphericity can be assumed for any multiple comparisons involving these conditions (for a discussion of local circularity, see Rouanet & Lépine, 1970). For the data in Table 15.2 the biggest difference in variances is between A and B (15.7) and A and C (10.3), but how do we know whether this difference is large enough to be a problem?

15.5.2 Assessing the severity of departures from sphericity ▌▌▌

Mauchly's test assesses the hypothesis that the variances of the differences between conditions are equal. If Mauchly's test statistic is significant (i.e., has a probability value less than 0.05) it implies that there are significant differences between the variances of differences and, therefore, sphericity is not met. If Mauchly's test statistic is non-significant (i.e., $p > 0.05$) then the implication is that the variances of differences are roughly equal and sphericity is met. However, like any

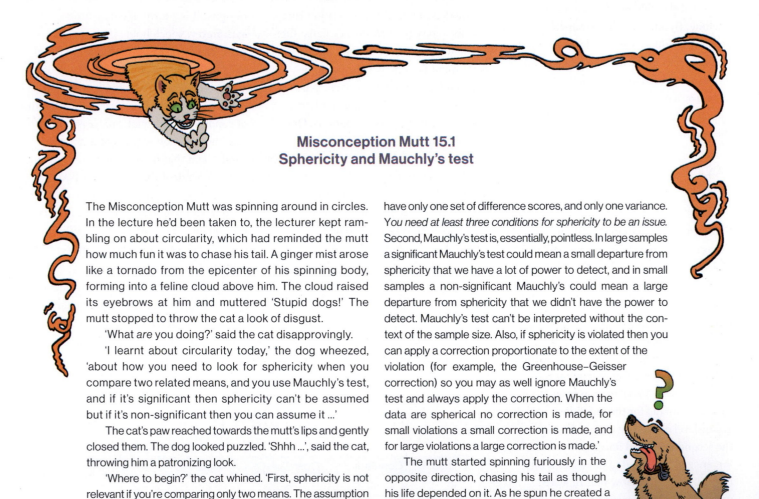

Misconception Mutt 15.1
Sphericity and Mauchly's test

The Misconception Mutt was spinning around in circles. In the lecture he'd been taken to, the lecturer kept rambling on about circularity, which had reminded the mutt how much fun it was to chase his tail. A ginger mist arose like a tornado from the epicenter of his spinning body, forming into a feline cloud above him. The cloud raised its eyebrows at him and muttered 'Stupid dogs!' The mutt stopped to throw the cat a look of disgust.

'What *are* you doing?' said the cat disapprovingly.

'I learnt about circularity today,' the dog wheezed, 'about how you need to look for sphericity when you compare two related means, and you use Mauchly's test, and if it's significant then sphericity can't be assumed but if it's non-significant then you can assume it ...'

The cat's paw reached towards the mutt's lips and gently closed them. The dog looked puzzled. 'Shhh ...', said the cat, throwing him a patronizing look.

'Where to begin?' the cat whined. 'First, sphericity is not relevant if you're comparing only two means. The assumption is that the variances of difference scores between pairs of treatment levels are equal, and with only two conditions you

have only one set of difference scores, and only one variance. *You need at least three conditions for sphericity to be an issue.* Second, Mauchly's test is, essentially, pointless. In large samples a significant Mauchly's test could mean a small departure from sphericity that we have a lot of power to detect, and in small samples a non-significant Mauchly's could mean a large departure from sphericity that we didn't have the power to detect. Mauchly's test can't be interpreted without the context of the sample size. Also, if sphericity is violated then you can apply a correction proportionate to the extent of the violation (for example, the Greenhouse–Geisser correction) so you may as well ignore Mauchly's test and always apply the correction. When the data are spherical no correction is made, for small violations a small correction is made, and for large violations a large correction is made.'

The mutt started spinning furiously in the opposite direction, chasing his tail as though his life depended on it. As he spun he created a reverse vortex that sucked the cat back into the ether. 'That'll learn him,' he thought to himself.

Oliver Twisted: Please, Sir, can I have some more ... sphericity?

'Balls!' says Oliver. 'Balls are spherical, and I like them. Maybe I'd like sphericity too if only you could explain it to me in more detail.' Be careful what you wish for, Oliver. In my youth I wrote an article called 'A bluffer's guide to sphericity', which I used to cite in this book, roughly on this page. Occasionally people ask me for it, so I thought I might as well reproduce it on the companion website.

Jane Superbrain 15.1
Sphericity and *post hoc* tests ▌▌▌▌

The violation of sphericity has implications for multiple comparisons. There is a more detailed summary of these online (see Oliver Twisted), but here are a few take-home messages. Boik (1981) recommends against using the *F*-statistic for repeated-measures contrasts because even very small departures from sphericity produce large biases. Maxwell (1980) compared the power and alpha levels for five *post hoc* tests: three variants of the Tukey procedure, the Bonferroni method, and a multivariate approach (the Roy–Bose simultaneous confidence interval). The multivariate approach was always 'too conservative for practical use' (p. 277), and this was most extreme when *n* (the number of participants) is small relative to *k* (the number of conditions). All variants of Tukey's test inflated the Type I error rate unacceptably with increasing departures from sphericity. The Bonferroni method was extremely robust (although *slightly* conservative) and controlled Type I error rates well. In terms of power (the Type II error rate), Tukey's wholly significant difference was the most powerful under non-sphericity in tiny samples ($n = 8$), but this advantage was severely reduced in even slightly larger samples ($n = 15$). Keselman and Keselman (1988) extended Maxwell's work to unbalanced designs and concluded that Bonferroni was more powerful than the multivariate approach as the number of repeated treatment levels increases. A simple take-home message is that the Bonferroni method has much to recommend it.

significance test, Mauchly's test depends upon sample size and is probably best ignored (Misconception Mutt 15.1). Instead we can estimate the degree of sphericity using the **Greenhouse–Geisser estimate,** $\hat{\varepsilon}$ (Greenhouse & Geisser, 1959) or the **Huynh–Feldt estimate** (Huynh & Feldt, 1976), $\tilde{\varepsilon}$. The Greenhouse–Geisser estimate varies between $1/(k-1)$ (where k is the number of repeated-measures conditions) and 1. For example, in a situation in which there are five conditions the lower limit of $\hat{\varepsilon}$ will be $1/(5-1)$ or 0.25 (known as the **lower-bound estimate** of sphericity). You can live a long, happy life oblivious to the precise calculation of these estimates (interested readers should consult Girden, 1992) so we won't go there. We need know only that these

estimates are used to correct for departures from sphericity and so are considerably more use than Mauchly's test (Misconception Mutt 15.1).

15.5.3 What's the effect of violating the assumption of sphericity? ▮▮▮

Rouanet and Lépine (1970) provided a detailed account of the validity of the F-statistic under violations of sphericity (see also Mendoza, Toothaker, & Crain, 1976). I summarized (Field, 1998) their findings in a very obscure newsletter that no one can ever access (see Oliver Twisted). The bottom line is that sphericity creates a loss of power and an F-statistic that doesn't have the distribution that it's supposed to have

(i.e., an F-distribution). Lack of sphericity also causes some amusing complications for *post hoc* tests. If you don't want to worry about what these complications are then when sphericity is violated, the Bonferroni method is the most robust in terms of power and control of the Type I error rate. When sphericity is definitely not violated, Tukey's test can be used (Jane Superbrain Box 15.1).

15.5.4 What do you do if you violate sphericity? ▮▮▮

You might think that if your data violate the sphericity assumption then you need to have a nervous breakdown or book in to see a

What do I do if sphericity is violated?

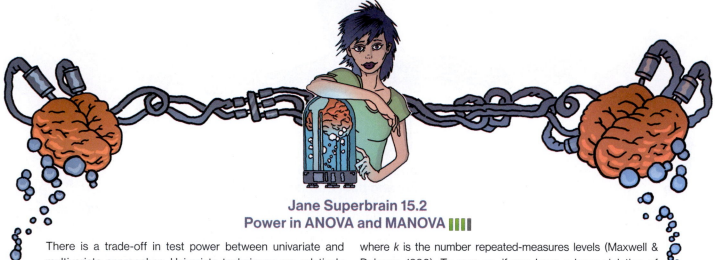

Jane Superbrain 15.2
Power in ANOVA and MANOVA ▮▮▮

There is a trade-off in test power between univariate and multivariate approaches. Univariate techniques are relatively powerless to detect small reliable changes between highly correlated conditions when other less correlated conditions are also present (Davidson, 1972). As the degree of violation of compound symmetry increases, the power of multivariate tests also increases, whereas for univariate tests it decreases (Mendoza, Toothaker, & Nicewander, 1974). However, multivariate approaches should probably not be used if n is less than $k + 10$,

where k is the number repeated-measures levels (Maxwell & Delaney, 1990). To sum up, if you have a large violation of sphericity ($\varepsilon < 0.7$) and your sample size is greater than $k + 10$ then multivariate procedures are more powerful, but with small sample sizes or when sphericity holds ($\varepsilon > 0.7$) use the univariate approach (Stevens, 2002). It is also worth noting that the power of MANOVA varies as a function of the correlations between dependent variables (see Jane Superbrain Box 17.1) and so the relationship between treatment conditions must be considered.

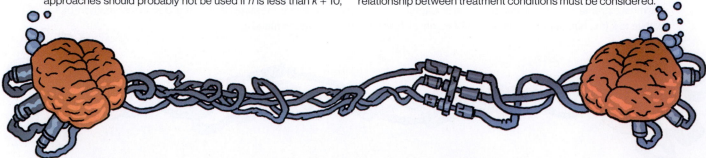

counsellor or something. You *can* do that, but a less costly option (emotionally and financially) is to adjust the degrees of freedom of any *F*-statistics affected. You can estimate sphericity in various ways (see above), resulting in a value that is 1 when your data are spherical and less than 1 when they are not. You multiply the degrees of freedom for an affected *F* by this estimate. The result is that when you have sphericity the degrees of freedom don't change (because you multiply them by 1) but when you don't the degrees of freedom get smaller (because you multiply them by a value less than 1). The greater the violation of sphericity, the smaller the estimate gets, and the smaller the degrees of freedom become. Smaller degrees of freedom make the *p*-value associated with the *F*-statistic less significant. By adjusting the degrees of freedom by the extent to which the data are not spherical, we also make the *F*-statistic more conservative. In doing so, the Type I error rate is controlled.

As I mentioned above, the degrees of freedom are adjusted using either the Greenhouse–Geisser or Huynh–Feldt estimates of sphericity. When the Greenhouse–Geisser estimate is greater than 0.75 the correction is too conservative (Huynh & Feldt, 1976), and this can also be true when the sphericity estimate is as high as 0.90 (Collier, Bakery, Mandeville, & Hages, 1967). However, the Huynh–Feldt estimate tends to overestimate sphericity (Maxwell & Delaney, 1990). Many authors recommend that when estimates of sphericity are greater than 0.75 the Huynh–Feldt estimate should be used, but when the Greenhouse–Geisser estimate of sphericity is less than 0.75 or nothing is known about sphericity the Greenhouse–Geisser correction should be used (Barcikowski & Robey, 1984; Girden, 1992; Huynh & Feldt, 1976). Stevens (2002) suggests taking an average of the two estimates and adjusting the *df* by this average. We will see how these values are used in due course.

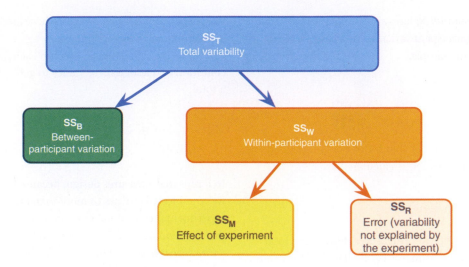

Figure 15.3 Partitioning variance for repeated-measures designs

Another option when you have data that violate sphericity is to fit the kind of model described in Section 15.4, which is known as a multilevel model (see Chapter 21 for more detail). A third option is to use multivariate test statistics (MANOVA), because they do not assume sphericity (see O'Brien & Kaiser, 1985). SPSS Statistics produces multivariate test statistics (see also Chapter 17). However, there may be trade-offs in power between these univariate and multivariate tests (see Jane Superbrain Box 15.2).

Grand mean = 5.56
Grand variance = 8.19

Figure 15.4 Treating the data as a single group

15.6 The *F*-statistic for repeated-measures designs ▮▮▮▮

In a repeated-measures design the effect of the experiment (the independent variable) is shown up in the within-participant variance (rather than in the between-group variance). Remember that in independent designs (Section 12.2) the within-participant variance is the residual sum of squares (SS_R); it is the variance created by individual differences in performance. When we carry out our experimental manipulation on the same entities, the within-participant variance will be made up of not just individual differences in performance but also the effect

of our manipulation. Therefore, the main difference with a repeated-measures design is that we look for the experimental effect (the model sum of squares) within the individual rather than within the group. Figure 15.3 illustrates how the variance is partitioned in a repeated-measures design. The important thing to note is that the types of variances are the same as in independent designs: we have a total sum of squares (SS_T), a model sum of squares (SS_M) and a residual sum of squares (SS_R). The *only* difference is from where those sums of squares come: in repeated-measures designs the model and residual sums of squares are both part of the within-participant variance. Let's have a look at an example.

15.6.1 The total sum of squares, SS$_T$ ▋▋▋▋

Remember that for one-way independent designs SS$_T$ is calculated as:

$$SS_T = s_{\text{grand}}^2 (N-1) \tag{15.6}$$

In repeated-measures designs the total sum of squares is calculated in the same way. The grand variance in the equation is the variance of all scores when we ignore the group to which they belong. So, if we treated the data as one big group (as in Figure 15.4), the variance of these scores is 8.19 (try this on your calculator), and this is the grand variance. We used 32 scores to generate this value, so N is 32. Our sum of squares becomes:

$$\begin{aligned} SS_T &= s_{\text{grand}}^2 (N-1) \\ &= 8.19(32-1) \\ &= 253.89 \end{aligned} \tag{15.7}$$

The degrees of freedom for this sum of squares, as for an independent design, are $N-1$ or 31.

15.6.2 The within-participant sum of squares, SS$_W$ ▋▋▋▋

The crucial difference from an independent design is that in a repeated-measures design there is a within-participant variance component, which represents individual differences within participants. In independent designs these individual differences were quantified with the residual sum of squares (SS$_R$) using the following equations:

$$\begin{aligned} SS_R &= \sum_{g=1}^{k} \sum_{i=1}^{n} (x_{ig} - \bar{x}_g)^2 \\ &= \sum_{g=1}^{k} s_g^2 (n_g - 1) \end{aligned} \tag{15.8}$$

In independent designs, because there are different participants within each condition, we calculated SS$_R$ within each condition and added these values to get a total:

$$\begin{aligned} SS_R &= s_{\text{group}1}^2 (n_1 - 1) + s_{\text{group}2}^2 (n_2 - 1) \\ &\quad + s_{\text{group}3}^2 (n_3 - 1) + \ldots + s_{\text{group}n}^2 (n_n - 1) \end{aligned} \tag{15.9}$$

In a repeated-measures design, because we've subjected entities to more than one experimental condition, we're interested in the variation not within a condition but *within an entity*. Therefore, we use the same equation but adapt it to look within participants rather than groups. If we call this sum of squares SS$_W$ (for within-participant SS), we adapt equation (15.9) to give:

$$\begin{aligned} SS_W &= s_{\text{entity}1}^2 (n_1 - 1) + s_{\text{entity}2}^2 (n_2 - 1) \\ &\quad + s_{\text{entity}3}^2 (n_3 - 1) + \ldots + s_{\text{entity}n}^2 (n_n - 1) \end{aligned} \tag{15.10}$$

This equation translates as looking at the variation in each individual's scores and then adding these variances for all the entities in the study. The ns represent the number of scores within the person (i.e., the number of experimental conditions or in this case the number of foods). All the variances we need are in Table 15.1, so we can calculate SS$_W$ as:

$$\begin{aligned} SS_W &= s_{\text{celebrity}1}^2 (n_1 - 1) + s_{\text{celebrity}2}^2 (n_2 - 1) \\ &\quad + \ldots + s_{\text{celebrity}n}^2 (n_n - 1) \\ &= 9.67(4-1) + 8.25(4-1) + 7.58(4-1) \\ &\quad + 11.67(4-1) + 4.25(4-1) \\ &\quad + 0.92(4-1) + 15.58(4-1) + 20.92(4-1) \\ &= 29 + 24.75 + 22.75 + 35 + 12.75 + 2.75 \\ &\quad + 46.75 + 62.75 \\ &= 236.50 \end{aligned} \tag{15.11}$$

The degrees of freedom for each entity are $n-1$ (i.e., the number of conditions minus 1). To get the total degrees of freedom we add the *df*s for all participants. With eight participants (celebrities) and four conditions (i.e., $n = 4$), there are 3 degrees of freedom for each celebrity and $8 \times 3 = 24$ degrees of freedom in total.

15.6.3 The model sum of squares, SS$_M$ ▋▋▋▋

So far, we know that the total amount of variation within the retching scores is 253.58 units. We also know that 236.50 of those units are explained by the variance created by individuals' (celebrities') performances under different conditions. Some of this variation is the result of our experimental manipulation and some of this variation is due to unmeasured factors. The next step is to work out how much variance is explained by our manipulation (different foods) and how much is not.

In independent designs, we worked out how much variation could be explained by our experiment (the model SS) by looking at the means for each group and comparing these to the overall mean. We measured the variance resulting from the differences between group means and the overall mean (see equation (12.11)). We do the same thing in a repeated-measures design:

$$SS_M = \sum_{g=1}^{k} n_g (\bar{x}_g - \bar{x}_{\text{grand}})^2 \tag{15.12}$$

Using the means from the bushtucker data (see Table 15.1), we calculate SS$_M$ as:

$$\begin{aligned} SS_M &= 8(8.13 - 5.56)^2 + 8(4.25 - 5.56)^2 \\ &\quad + 8(4.13 - 5.56)^2 + 8(5.75 - 5.56)^2 \\ &= 8(2.57)^2 + 8(-1.31)^2 + 8(-1.44)^2 + 8(0.196)^2 \\ &= 83.13 \end{aligned} \tag{15.13}$$

With independent designs the model degrees of freedom are the number of conditions (k) minus 1. The same is true here: there were four conditions (foods), so the degrees of freedom will be 3.

15.6.4 The residual sum of squares, SS$_R$ ▋▋▋▋

We now know that there are 253.58 units of variation to be explained in our data, and that the variation across our conditions accounts for 236.50 units. Of these 236.50 units, our experimental manipulation can explain 83.13 units. The final sum of squares is the residual sum of squares (SS_R), which tells us how much of the variation cannot be explained by the model. This value is the amount of variation caused by extraneous factors outside experimental control. Knowing SS_W and SS_M already, the simplest way to calculate SS_R is to subtract SS_M from SS_W:

$$\begin{aligned} SS_R &= SS_W - SS_M \\ &= 236.50 - 83.13 \\ &= 153.37 \end{aligned} \quad (15.14)$$

The degrees of freedom are calculated in a similar way:

$$\begin{aligned} df_R &= df_W - df_M \\ &= 24 - 3 \\ &= 21 \end{aligned} \quad (15.15)$$

15.6.5 The mean squares ▮▮▮▮

SS_M tells us how much variation the model (e.g., the experimental manipulation) explains and SS_R tells us how much variation is due to extraneous factors. Both these values are totals and depend on how many scores have contributed to them, so to make them comparable we convert to the mean (or average) sum of squares (MS) by dividing by the degrees of freedom:

$$\begin{aligned} MS_M &= \frac{SS_M}{df_M} = \frac{83.13}{3} = 27.71 \\ MS_R &= \frac{SS_R}{df_R} = \frac{153.37}{21} = 7.30 \end{aligned} \quad (15.16)$$

MS_M represents the average variation explained by the model (e.g., the average systematic variation), whereas MS_R is a gauge of the average variation explained by extraneous variables (the average unsystematic variation).

15.6.6 The F-statistic ▮▮▮▮

The F-statistic is the ratio of the variation explained by the model and the variation explained by unsystematic factors. As for independent designs, it is the model mean squares divided by the residual mean squares:

$$F = \frac{MS_M}{MS_R} \quad (15.17)$$

So, as with independent designs, the F-statistic is the ratio of systematic to unsystematic variation: it is the experimental effect on performance relative to the effect of unmeasured factors.

For the bushtucker data, the F-statistic is 3.79:

$$F = \frac{MS_M}{MS_R} = \frac{27.71}{7.30} = 3.79 \quad (15.18)$$

This value is greater than 1, which indicates that the experimental manipulation had some effect above and beyond the effect of unmeasured factors. This value can be compared against a critical value based on its degrees of freedom (which are df_M and df_R, which are 3 and 21 in this case), but more generally, it's possible to compute an exact p-value that is the probability of getting an F at least as big as the one we have observed if the null hypothesis is true.

15.6.7 The between-participant sum of squares ▮▮▮▮

We've sort of forgotten about the between-participant variation in Figure 15.3 because we didn't need it to calculate the F-statistic. I will briefly mention what it represents. The easiest way to calculate this term is by subtraction:

$$\begin{aligned} SS_T &= SS_B + SS_W \\ SS_B &= SS_T - SS_W \end{aligned} \quad (15.19)$$

We have already calculated SS_W and SS_T, so by replacing the values of these terms, we get:

$$\begin{aligned} SS_B &= SS_T - SS_W \\ &= 253.89 - 236.50 \\ &= 17.39 \end{aligned} \quad (15.20)$$

SS_B represents individual differences between cases. In this example different celebrities will have different tolerances for eating these foods. This variation is illustrated by the different means for the celebrities in Table 15.1. For example, celebrity 4 ($M = 4.50$) was, on average, more than 2 seconds quicker to retch than participant 8 ($M = 6.75$). Celebrity 8 had, on average, a stronger constitution than celebrity 4. The between-participant sum of squares reflects these differences between individuals. In this case only 17.39 units of variation in retching latencies is down to individual differences between celebrities.

15.7 Assumptions in repeated-measures designs ▮▮▮▮

Repeated-measures designs still use the linear model so all of the sources of potential bias (and counteractive measures) discussed in Chapter 6 apply. Using the ANOVA approach, the assumption of independence is replaced by assumptions about the relationships between differences scores (sphericity). In the multilevel approach, sphericity isn't required, and you have a lot more flexibility to model different types of assumptions about residuals (see Section 21.4.2).

If assumptions are not met, there is a robust variant of a one-way repeated-measures ANOVA that we'll cover in Section 15.10—and you have the option of Friedman's ANOVA (see Chapter 7) although I'd use that as a last resort. The Bootstrap... button is noticeable by its absence in repeated-measures designs, which is a sad story. If you have a factorial design using repeated measures then you're stuffed.

15.8 One-way repeated-measures designs using SPSS ▮▮▮

The general procedure for a one-way repeated-measures design (using ANOVA) is much the same as any other linear model, so, remind yourself of the general procedure in Chapter 9. Figure 15.5 shows a simpler overview that highlights some of the specific issues when using repeated measures.

15.8.1 The main analysis ▮▮▮

The data for the bushtucker example (**Bushtucker.sav**) can be entered into the data editor in the same format as Table 15.1 (although you don't need to include the columns labeled *Mean* or *s²*). If you're entering the data by hand, create a variable called **stick** and use the labels dialog box to give this variable a full title of 'Stick Insect'. In the next column, create a variable called **testicle**, with a full title of 'Kangaroo Testicle', and so on for variables called **eye** ('Fish Eye') and **witchetty** ('Witchetty Grub').

To compare means from a one-way repeated-measures design select *Analyze* ▶ *General Linear Model* ▶ Repeated Measures…. We use the *Define Factor(s)* dialog box to name our within-subject (repeated-measures) variable, which in this example is the type of animal eaten, so replace the word *factor1* with the word *Animal* (Figure 15.6).[5] Next, we specify how many levels the variable had (i.e., how many conditions participants took part in). There were four different animals eaten by each celebrity, so type '4' into the box labeled *Number of Levels*. Click Add to register this variable in the list of repeated-measures variables, where it appears as *Animal(4)* as in Figure 15.6. If your design has several repeated-measures variables you can add more factors to the list (see the next example). When you have finished creating repeated-measures factors, click Define to go to the main dialog box.

The main dialog box (Figure 15.7) has a space labeled *Within-Subjects Variables* that contains a list of four question marks followed by a number. These question marks are placeholders for the variables representing the four levels of the independent variable and need to be replaced with the variables corresponding to each level. The order of levels is not important for this example, so we can select all four variables in the data editor (click the stick insect variable then click the witchetty grub variable while holding the *Shift* key) and drag them to the box labeled

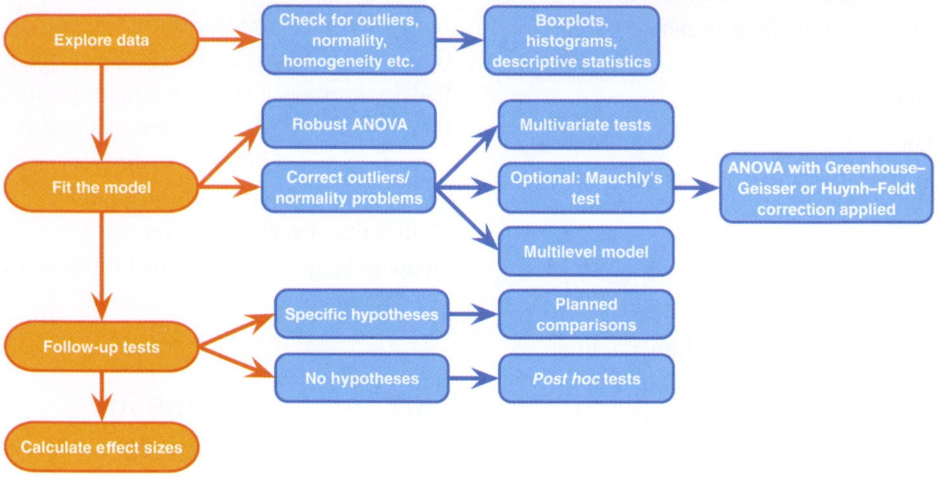

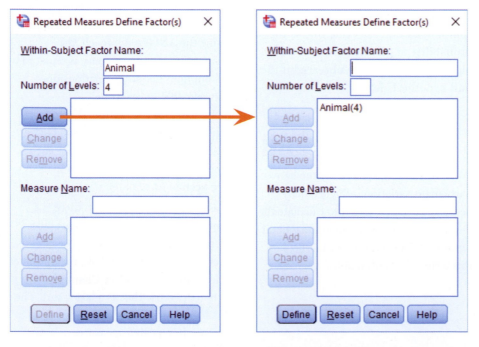

Figure 15.5 The process for analyzing repeated-measures designs

Figure 15.6 The *Define Factor(s)* dialog box for repeated-measures ANOVA

---

5 The name cannot have spaces in it.

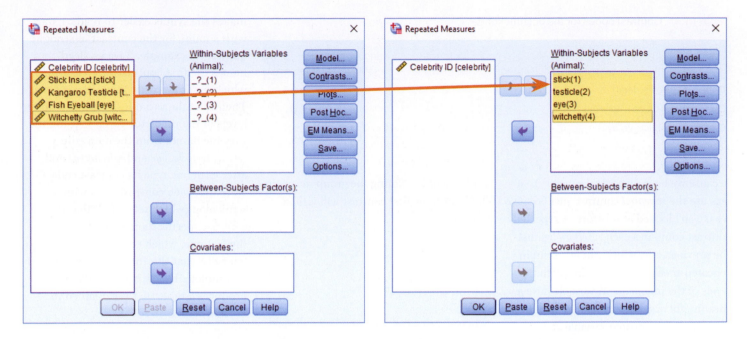

Figure 15.7 The main dialog box for repeated-measures designs (before and after completion)

Within-Subjects Variables (or click 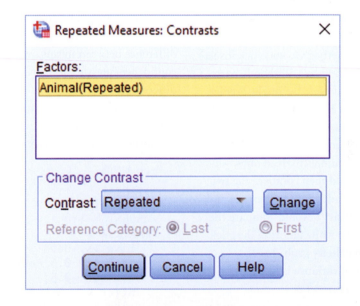). The finished dialog box is also shown in Figure 15.7.

15.8.2 Defining contrasts for repeated measures ▮▮▮

Click Contrasts... to access the dialog box in Figure 15.8, which is used to specify one of the many standard contrasts that we have come across before (see Section 12.4.4). The default is a polynomial contrast, but you can change it by selecting a variable in the box labeled *Factors*, clicking Polynomial ▾ , selecting a contrast from the list, and clicking Change . If you choose to conduct a simple contrast then you can specify whether to compare groups against the first or last category. The first category would be the one entered as (1) in the main dialog box and, for these data, the last category would be the one entered as (4). Therefore, the order in which you enter variables in the main dialog box is important for the contrasts you choose. There is no obvious contrast for this example (the simple contrast is not useful

Figure 15.8 Repeated-measures contrasts

SELF TEST

Devise some contrast codes for the contrasts described in the text.

Table 15.3 Contrast codes for the bushtucker data

	Stick insect	Witchetty grub	Kangaroo testicle	Fish eye
Contrast 1 (live vs. dead)	0.5	0.5	−0.5	−0.5
Contrast 2 (stick vs. witchetty)	1	−1	0	0
Contrast 3 (testicle vs. eye)	0	0	1	−1

because we have no control category), so let's use the *repeated* contrast, just because we haven't looked at it before. A repeated contrast compares each category against the previous one, which can be useful in repeated-measures designs in which the levels of the independent variable have a meaningful order—for example, if you've measured the outcome variable at successive points in time or administered increasing doses of a drug. When you have selected this contrast, click Continue .

15.8.3 Custom contrasts ▐▐▐▐

In Section 12.4 we looked at planned contrasts. Two of our animals are eaten alive (the witchetty grub and stick insect) whereas the others are dead body parts (the testicle and eye). We might predict that eating live animals is more disgusting than eating body parts and test this with a set of planned contrasts. Contrast 1 would compare live animals to dead ones, contrast 2 would then compare the two live animals (and ignore the testicle and eye), and contrast 3 would compare the eye and testicle while ignoring the stick insect and witchetty grub. The partitioning of variance is like that depicted in Figure 12.8.

The resulting contrast codes are in Table 15.3. If you follow the 'rules' in Section 12.4 you'll end up with codes of 2, 2, −2, −2 for contrast 1. I have divided these values by the number of groups (4) to get codes of 0.5, 0.5, −0.5, −0.5. I've done this so that the value for the contrast is the actual difference between means of the live and dead animals (rather than a

multiple of it). Noting the group means in Table 15.1, my first contrast will yield a value of 2.75:

$$\frac{\bar{X}_{\text{Stick}} + \bar{X}_{\text{Witchetty}}}{2} - \frac{\bar{X}_{\text{Testicle}} + \bar{X}_{\text{Eye}}}{2}$$
$$= \frac{8.13 + 5.75}{2} - \frac{4.25 + 4.13}{2} = 2.75 \qquad (15.21)$$

whereas using codes of 2 and −2 will yield a value 4 times this value (11). It doesn't affect the significance value at all, so in this sense you can use what codes you like, but it is handy to have the contrast value equal the actual difference between means because we can more easily interpret the confidence interval for the contrast (it will be the confidence interval for the difference between means, rather than the confidence interval for four times the difference).

To operationalize these contrasts you have to use the following syntax (**BushtuckerContrast.sps**):

```
GLM stick testicle eye
witchetty
/WSFACTOR=Animal 4 Polynomial
/WSDESIGN=Animal
/MMATRIX =
  'Live vs. dead' stick 0.5
  witchetty 0.5 testicle
  -0.5 eye -0.5;
  'Stick vs. witchetty'
  stick 1 witchetty -1
  testicle 0 eye 0;
  'Testicle vs. eye' stick 0
  witchetty 0 testicle 1 eye -1.
```

The first three lines specify a basic repeated-measures model. The contrasts are defined within the */MMATRIX* subcommand. I specify each contrast on a separate line (not essential, but it

makes the syntax easier to read). I start with a name for the contrast in straight quotes, for example, 'Live vs. dead' names the first contrast *Live vs. dead*, which describes what the contrast tests. Then I list the variables that make up the levels of the predictor variable using the variable names from the data editor (*stick*, *testicle*, *eye* and *witchetty*) and after each one type its contrast code. Note that each contrast ends with a semicolon (which tells SPSS that the contrast specification is finished) except the last one, which ends with a period to tell SPSS that the entire *GLM* command is complete. Executing this syntax produces a table that is explained in Section 15.9.4.

15.8.4 *Post hoc* tests and additional options ▐▐▐▐

Lack of sphericity creates entertaining complications for *post hoc* tests (see Jane Superbrain Box 15.1). When sphericity is definitely not violated, Tukey's test can be used, but if sphericity can't be assumed then the Games–Howell procedure is preferable. Because of these sphericity-related complications the standard *post hoc* tests for independent designs are not available for repeated-measures variables (the *post hoc* test dialog box does not list repeated-measured factors). The good news is that you can do some *post hoc* procedures by clicking EM Means... . To specify *post hoc* tests, drag the repeated-measures variable (in this case **Animal**) from the box labeled *Estimated Marginal Means: Factor(s) and Factor Interactions* to the box labeled *Display Means for* (or click ➡). Then select ☑ **Compare main effects** to activate the LSD(none) ▾ drop-down menu (Figure 15.9). The default is to have no adjustment for multiple tests (Tukey LSD), which I don't recommend, but you can choose a Bonferroni or Šidák correction (recommended for the reasons already mentioned). I've selected Bonferroni. Click Continue

Click [Options...] if you'd like to see things like descriptive statistics, a transformation matrix (which provides the coding values for any contrast selected in the *Contrasts* dialog box in Figure 15.8), and you can print the hypothesis, error and residual sum of squares and cross-products matrices (see Chapter 17)—Figure 15.9. If you have a between-group factor as well (mixed designs—see the next chapter) and you're a Levene's test kind of person then there is an option for homogeneity of variance tests. You can also change the level of significance at which to test any *post hoc* tests (you can change from the 0.05 level if you want to do a manual correction for multiple tests). Click [Continue] to return to the main dialog box, and [OK] to run the analysis.

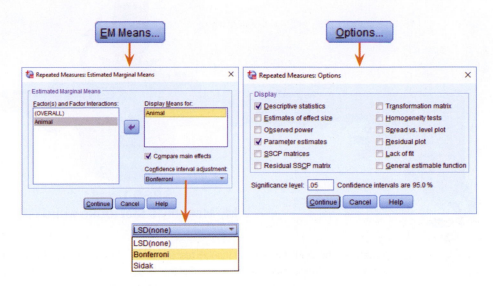

Figure 15.9 The *Options* dialog box

15.9 Output for one-way repeated-measures designs ▮▮▮▮

15.9.1 Descriptive statistics ▮▮▮▮

Output 15.1 shows two tables: the first (left) tells us the variables that represent each level of the predictor variable, which is a useful check that we entered the variables in the order that we intended to. The second table (right) shows that, on average, the time taken to retch was longest after eating the stick insect, and quickest after eating a testicle or eyeball. These means are useful for interpreting the main analysis.

15.9.2 Correcting for sphericity ▮▮▮▮

Output 15.2 shows Mauchly's test for the bushtucker data. The significance value (0.047) is less than the critical value of 0.05, which implies that the assumption of sphericity has been violated, but I suggested you ignore this test and

Within-Subjects Factors

Measure: MEASURE_1

Animal	Dependent Variable
1	stick
2	testicle
3	eye
4	witchetty

Descriptive Statistics

	Mean	Std. Deviation	N
Stick Insect	8.13	2.232	8
Kangaroo Testicle	4.25	1.832	8
Fish Eyeball	4.13	2.748	8
Witchetty Grub	5.75	2.915	8

Output 15.1

Mauchly's Test of Sphericity[a]

Measure: MEASURE_1

Within Subjects Effect	Mauchly's W	Approx. Chi-Square	df	Sig.	Greenhouse-Geisser	Huynh-Feldt	Lower-bound
					Epsilon[b]		
Animal	.136	11.406	5	.047	.533	.666	.333

Tests the null hypothesis that the error covariance matrix of the orthonormalized transformed dependent variables is proportional to an identity matrix.

a. Design: Intercept
   Within Subjects Design: Animal

b. May be used to adjust the degrees of freedom for the averaged tests of significance. Corrected tests are displayed in the Tests of Within-Subjects Effects table.

Output 15.2

routinely apply a correction for whatever deviation from sphericity is present in the data (Misconception Mutt 15.1). The more informative part of the table contains the

Greenhouse–Geisser ($\hat{\varepsilon} = 0.533$) and Huynh–Feldt ($\hat{\varepsilon} = 0.666$) estimates of sphericity.[6] If the data are perfectly spherical then these estimates will be 1.

---

6 The lowest possible value of the Greenhouse–Geisser estimate is $1/(k-1)$, which with four conditions is $1/(4-1) = 0.33$. This is given as the lower-bound estimate in Output 15.2.

SPSS Tip 15.1
My Mauchly's test looks weird ▮▮▮

Sometimes the significance for Mauchly's test shows a dot and no significance value, as in Output 15.3. Naturally, you fear that SPSS has gone crazy and is going to break into your bedroom at night and tattoo the equation for the Greenhouse–Geisser estimate on your face. Fear not, the reason for the dot is that you need at least three conditions for sphericity to be an issue (see

Misconception Mutt 15.1). Therefore, if the repeated-measures variable has only two levels then sphericity is met, the estimates of sphericity are 1 (perfect sphericity), the chi-square is zero and has no degrees of freedom and so a *p*-value can't be computed. It would be a lot easier if SPSS just printed in big letters 'Hooray! Hooray! Sphericity has gone away!' We can dream.

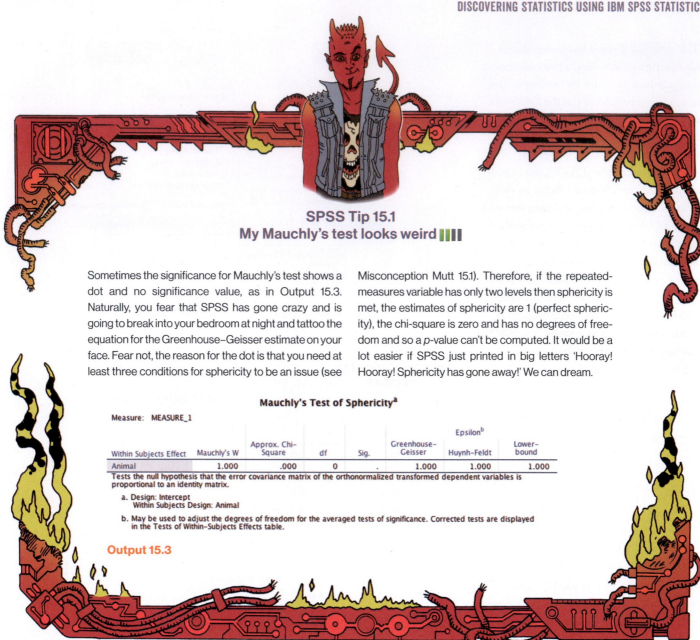

Mauchly's Test of Sphericity[a]

Measure: MEASURE_1

Within Subjects Effect	Mauchly's W	Approx. Chi-Square	df	Sig.	Epsilon[b] Greenhouse–Geisser	Huynh–Feldt	Lower–bound
Animal	1.000	.000	0	.	1.000	1.000	1.000

Tests the null hypothesis that the error covariance matrix of the orthonormalized transformed dependent variables is proportional to an identity matrix.

a. Design: Intercept
   Within Subjects Design: Animal

b. May be used to adjust the degrees of freedom for the averaged tests of significance. Corrected tests are displayed in the Tests of Within–Subjects Effects table.

Output 15.3

Therefore, both estimates indicate a departure from sphericity, so we may as well correct for it regardless of what Mauchly's test says. Jane Superbrain Box 15.3 explains how these estimates are used to correct the degrees of freedom for the *F*-statistic.

15.9.3 The *F*-statistic ▮▮▮

Output 15.4 shows the summary information for the *F*-statistic that tests

whether we can significantly predict retching times from the group means (i.e., are the means significantly different?). Note that the values of the sum of squares for the repeated-measures effect of **Animal**, the model sum of squares (SS_M), the residual sum of squares (SS_R), the mean squares and *F*-statistic are the same as we calculated in Sections 15.6.3–15.6.6. The *p*-value associated with the *F*-statistic is 0.026, which is significant because it is less than the criterion value of 0.05. This result

implies that there was a significant difference in the capacity of the four animals to induce retching when eaten. Remember though, the *F* does not tell us which animals differed from which.

Although the result seems plausible, we have learnt that departures from sphericity make the *F*-test inaccurate, and Output 15.2 shows that these data are not spherical. As well as showing the *F*-statistic and associated degrees of freedom when sphericity is assumed,

Tests of Within-Subjects Effects

Measure: MEASURE_1

Source		Type III Sum of Squares	df	Mean Square	F	Sig.
Animal	Sphericity Assumed	83.125	3	27.708	3.794	.026
	Greenhouse–Geisser	83.125	1.599	52.001	3.794	.063
	Huynh–Feldt	83.125	1.997	41.619	3.794	.048
	Lower-bound	83.125	1.000	83.125	3.794	.092
Error(Animal)	Sphericity Assumed	153.375	21	7.304		
	Greenhouse–Geisser	153.375	11.190	13.707		
	Huynh–Feldt	153.375	13.981	10.970		
	Lower-bound	153.375	7.000	21.911		

Output 15.4

Output 15.4 shows the results adjusted using the three estimates of sphericity in Output 15.2 (Greenhouse–Geisser, Huynh–Feldt, and the lower-bound value). These estimates are used to correct the degrees of freedom, which has the effect of increasing p (Jane Superbrain Box 15.3).

The adjustments result in the observed F being non-significant when using the Greenhouse–Geisser correction (because $p > 0.05$) but significant using the Huynh–Feldt correction (because the probability value of 0.048 is just below the criterion value of 0.05). I noted earlier that the

Greenhouse–Geisser correction is probably too strict and that the Huynh–Feldt correction is probably not strict enough, and we see this here because one of them takes the significance value above the conventional 0.05 threshold whereas the other doesn't (see Jane Superbrain Box 15.4). This leaves us with the puzzling dilemma of whether to accept this F-statistic as significant.

A recommendation mentioned earlier is to use the Greenhouse–Geisser estimate unless it's greater than 0.75 (which it's not in this case). Some people also suggest taking an average of the two estimates (Stevens, 2002). In practical terms, rather than averaging the estimates, correcting the degrees of freedom manually and trying with an abacus or two to generate exact p-values, we could average the two

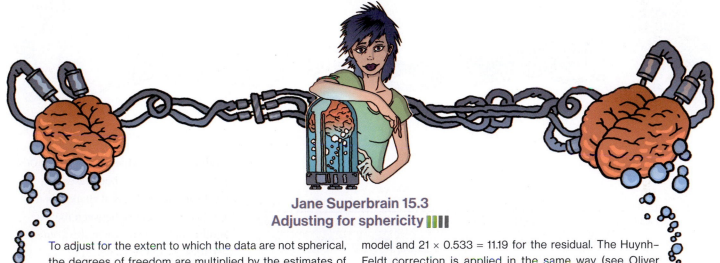

Jane Superbrain 15.3
Adjusting for sphericity ▮▮▮

To adjust for the extent to which the data are not spherical, the degrees of freedom are multiplied by the estimates of sphericity in Output 15.2. For example, the Greenhouse–Geisser estimate of sphericity was 0.533 (Output 15.2). The original degrees of freedom for the model sum of squares were 3 and for the residual sum of squares they were 21. These values are adjusted by multiplying by the estimate of sphericity (0.533) resulting in 3 × 0.533 = 1.599 for the

model and 21 × 0.533 = 11.19 for the residual. The Huynh–Feldt correction is applied in the same way (see Oliver Twisted on sphericity). The effect of reducing the degrees of freedom is to change the shape of the F-distribution that is used to obtain p, so the F-statistic is unchanged, but its p is based on a distribution with 1.599 and 11.19 degrees of freedom instead of 3 and 21. This increases the p-value associated with F.

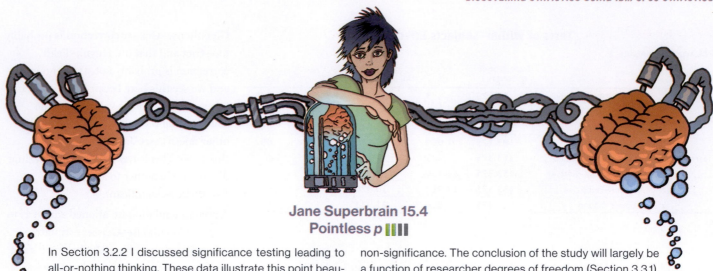

Jane Superbrain 15.4
Pointless p ▌▌▌▌

In Section 3.2.2 I discussed significance testing leading to all-or-nothing thinking. These data illustrate this point beautifully: the two sphericity corrections lead to significance values just above (0.063) or just below (0.048) the 0.05 criterion. These significance values differ by only 0.015 but lead to completely opposite conclusions. The decision about 'significance' becomes rather arbitrary: choose one correction for a result that is 'significant' but choose another for non-significance. The conclusion of the study will largely be a function of researcher degrees of freedom (Section 3.3.1). The means and the effect size are unaffected by sphericity corrections, and so whether the p falls slightly above or slightly below 0.05 side-tracks us from the more important question of how big the effect was. We might be well advised to look at an effect size to see whether the effect is substantive, regardless of its significance.

Multivariate Tests[a]

Effect		Value	F	Hypothesis df	Error df	Sig.
Animal	Pillai's Trace	.942	26.955[b]	3.000	5.000	.002
	Wilks' Lambda	.058	26.955[b]	3.000	5.000	.002
	Hotelling's Trace	16.173	26.955[b]	3.000	5.000	.002
	Roy's Largest Root	16.173	26.955[b]	3.000	5.000	.002

a. Design: Intercept
   Within Subjects Design: Animal

b. Exact statistic

Output 15.5

SELF ✂ TEST
What does contrast 3 (*Level 3 vs. Level 4*) compare?

p-values instead, which yields $p = (0.063 + 0.048)/2 = 0.056$. In both cases the conclusion would be that there was no significant difference between means. Another option is to use multivariate test statistics (MANOVA), which do not assume sphericity (see O'Brien & Kaiser, 1985). Output 15.5 shows the multivariate test statistics (details of these test statistics can be found in Section 17.4), all of which are significant (because p is 0.002, which is less than the criterion value of 0.05). Based on multivariate tests, we'd conclude that there are significant differences between the times taken to retch after eating different animals. It's easy to see how the decision rule applied to p-values can lead to results that don't replicate, conclusions that have been influenced by researcher degrees of freedom, and a lot of noise in the scientific literature.

15.9.4 Contrasts ▮▮▮

Output 15.6 shows the transformation matrix requested in the options. Thinking back to contrast coding (Chapter 12), a code of 0 means that the group is not included in a contrast. Therefore, contrast 1 (labeled *Level 1 vs. Level 2*) ignores the fish eyeball and witchetty grub. Also remember that groups with a negative weight are compared to groups with a positive weight. For this first contrast, this means that the stick insect is compared against the kangaroo testicle. Using the same logic, contrast 2 (labeled *Level 2 vs. Level 3*) ignores the stick insect and witchetty grub and compares the kangaroo testicle with the fish eye.

Contrast 3 compares the fish eyeball with the witchetty grub. This pattern is consistent with a repeated contrast in that all groups except the first are compared to the preceding category.

Output 15.7 lists each contrast and its *F*-statistic, which compares the two chunks of variation within the contrast. We can conclude that celebrities took significantly longer to retch after eating the stick insect compared to the kangaroo testicle, *p* = 0.002 (*Level 1 vs. Level 2*), but that the time to retch was roughly the same after eating the kangaroo testicle and the fish eyeball, *p* = 0.920 (*Level 2 vs. Level 3*) and after eating a fish eyeball compared to eating a witchetty grub, *p* = 0.402 (*Level 3 vs. Level 4*).

It's worth remembering that, by some criteria, our main effect of the type of animal eaten was not significant, and if this is the case then we really shouldn't look at these contrasts. However, given the multivariate tests, there is some justification in looking at these contrasts.

For those of you brave enough to attempt the custom contrasts in Section 15.8.3,

Output 15.8 shows the tables of contrasts. The top table is probably most useful because it includes the confidence interval for the contrast, but the bottom table reports the test statistics and degrees of freedom (note that the *p*-values are identical in the two tables). We can conclude from these tables that retching times were significantly longer after eating live animals than dead ones, *F*(1, 7) = 18.41, *p* = 0.004, but there was no significant difference between the stick insect and witchetty grub, *F*(1, 7) = 1.76, *p* = 0.227 or between the fish eye and kangaroo testicle, *F*(1, 7) = 0.011, *p* = 0.920. The confidence intervals tell us (assuming this sample is

one of the 95% that produce intervals containing the population value) that the difference in retching times after eating live animals compared to body parts is likely to fall between 1.24 and 4.27 seconds, the difference between stick insect and witchetty grub lies between –1.86 and 6.61

Animal[a]

Measure: MEASURE_1

Dependent Variable	Level 1 vs. Level 2	Level 2 vs. Level 3	Level 3 vs. Level 4
Stick Insect	1	0	0
Kangaroo Testicle	-1	1	0
Fish Eyeball	0	-1	1
Witchetty Grub	0	0	-1

a. The contrasts for the within subjects factors are: Animal: Repeated contrast

Output 15.6

Tests of Within-Subjects Contrasts

Measure: MEASURE_1

Source	Animal	Type III Sum of Squares	df	Mean Square	F	Sig.
Animal	Level 1 vs. Level 2	120.125	1	120.125	22.803	.002
	Level 2 vs. Level 3	.125	1	.125	.011	.920
	Level 3 vs. Level 4	21.125	1	21.125	.796	.402
Error(Animal)	Level 1 vs. Level 2	36.875	7	5.268		
	Level 2 vs. Level 3	80.875	7	11.554		
	Level 3 vs. Level 4	185.875	7	26.554		

Output 15.7

Contrast Results (K Matrix)

Contrast[a]		Live vs. dead	Stick vs. witchetty	Testicle vs. eye
L1	Contrast Estimate	2.750	2.375	.125
	Hypothesized Value	0	0	0
	Difference (Estimate – Hypothesized)	2.750	2.375	.125
	Std. Error	.641	1.792	1.202
	Sig.	.004	.227	.920
	95% Confidence Interval for Difference — Lower Bound	1.235	-1.863	-2.717
	Upper Bound	4.265	6.613	2.967

a. Estimable Function for Intercept

Univariate Test Results

Source	Transformed Variable	Sum of Squares	df	Mean Square	F	Sig.
Contrast	Live vs. dead	60.500	1	60.500	18.413	.004
	Stick vs. witchetty	45.125	1	45.125	1.756	.227
	Testicle vs. eye	.125	1	.125	.011	.920
Error	Live vs. dead	23.000	7	3.286		
	Stick vs. witchetty	179.875	7	25.696		
	Testicle vs. eye	80.875	7	11.554		

Output 15.8

seconds, and the difference between kangaroo testicle and fish eye lies between −2.72 and 2.97 seconds.

15.9.5 Post hoc tests

If you selected *post hoc* tests for the repeated-measures variable (see Section 15.8.3), then Output 15.9 is produced. The table should have a familiar feel from other *post hoc* tests we've looked at: it shows the difference between group means and their confidence intervals, the standard error, and the significance value. Based on the significance values and the means (in Output 15.1) we can conclude that the time to retch was significantly longer after eating a stick insect compared to a kangaroo testicle

Pairwise Comparisons

Measure: MEASURE_1

(I) Animal	(J) Animal	Mean Difference (I–J)	Std. Error	Sig.[b]	95% Confidence Interval for Difference[b] Lower Bound	Upper Bound
1	2	3.875*	.811	.012	.925	6.825
	3	4.000*	.732	.006	1.339	6.661
	4	2.375	1.792	1.000	−4.141	8.891
2	1	−3.875*	.811	.012	−6.825	−.925
	3	.125	1.202	1.000	−4.244	4.494
	4	−1.500	1.336	1.000	−6.359	3.359
3	1	−4.000*	.732	.006	−6.661	−1.339
	2	−.125	1.202	1.000	−4.494	4.244
	4	−1.625	1.822	1.000	−8.249	4.999
4	1	−2.375	1.792	1.000	−8.891	4.141
	2	1.500	1.336	1.000	−3.359	6.359
	3	1.625	1.822	1.000	−4.999	8.249

Based on estimated marginal means

*. The mean difference is significant at the .05 level.

b. Adjustment for multiple comparisons: Bonferroni.

Output 15.9

Cramming Sam's Tips
One-way repeated-measures designs

- One-way repeated-measures designs compares several means, when those means come from the same entities; for example, if you measured people's statistical ability each month over a year-long course.
- When you have three or more repeated-measures conditions there is an additional assumption: *sphericity*.
- You can test for sphericity using *Mauchly's test*, but it is better to always adjust for the departure from sphericity in the data.

- The table labeled *Tests of Within-Subjects Effects* shows the main *F*-statistic. Other things being equal, always read the row labeled *Greenhouse–Geisser* (or *Huynh–Feldt*, but you'll have to read this chapter to find out the relative merits of the two procedures). If the value in the column labeled *Sig.* is less than 0.05 then the means of the conditions are significantly different.
- For contrasts and *post hoc* tests, again look to the columns labeled *Sig.* to discover if your comparisons are significant (i.e., the value is less than 0.05).

(p = 0.012) and a fish eye (p = 0.006), but not compared to a witchetty grub (p = 1). The time to retch after eating a kangaroo testicle was not significantly different compared to after eating a fish eye or witchetty grub (both ps = 1). Finally, the time to retch was not significantly different after eating a fish eyeball compared to a witchetty grub (p = 1). Again, it's worth noting that we wouldn't interpret these effects if we decide that the main effect of the type of animal eaten wasn't significant.

15.10 Robust tests of one-way repeated-measures designs ▮▮▮▮

There is a robust test of several dependent means with *post hoc* tests that can be run using the syntax file **rmanova.sps**. As with other robust tests in this book, this syntax requires the *Essentials for R* plugin and *WRS2* package (Section 4.13). The syntax uses the *rmanova* and *rmmcp* functions described by Wilcox (2017). These tests assume neither normality nor homogeneity of variance, so you can ignore these assumptions and plough ahead. The syntax is a little more complicated than in previous chapters because the data need to be restructured, so I have written a bunch of stuff to try to make that process as painless as possible. There's a bunch of stuff that looks to see whether certain packages are installed and if they aren't it installs them, after which the core syntax is (see SPSS Tip 15.2 for an explanation):

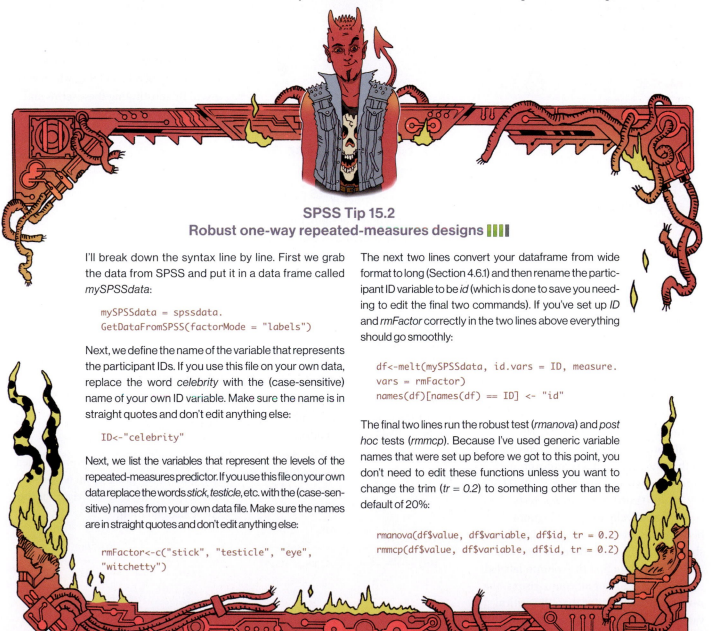

SPSS Tip 15.2
Robust one-way repeated-measures designs ▮▮▮▮

I'll break down the syntax line by line. First we grab the data from SPSS and put it in a data frame called *mySPSSdata*:

```
mySPSSdata = spssdata.
GetDataFromSPSS(factorMode = "labels")
```

Next, we define the name of the variable that represents the participant IDs. If you use this file on your own data, replace the word *celebrity* with the (case-sensitive) name of your own ID variable. Make sure the name is in straight quotes and don't edit anything else:

```
ID<-"celebrity"
```

Next, we list the variables that represent the levels of the repeated-measures predictor. If you use this file on your own data replace the words *stick*, *testicle*, etc. with the (case-sensitive) names from your own data file. Make sure the names are in straight quotes and don't edit anything else:

```
rmFactor<-c("stick", "testicle", "eye", "witchetty")
```

The next two lines convert your dataframe from wide format to long (Section 4.6.1) and then rename the participant ID variable to be *id* (which is done to save you needing to edit the final two commands). If you've set up *ID* and *rmFactor* correctly in the two lines above everything should go smoothly:

```
df<-melt(mySPSSdata, id.vars = ID, measure.
vars = rmFactor)
names(df)[names(df) == ID] <- "id"
```

The final two lines run the robust test (*rmanova*) and *post hoc* tests (*rmmcp*). Because I've used generic variable names that were set up before we got to this point, you don't need to edit these functions unless you want to change the trim (*tr = 0.2*) to something other than the default of 20%:

```
rmanova(df$value, df$variable, df$id, tr = 0.2)
rmmcp(df$value, df$variable, df$id, tr = 0.2)
```

```
Test statistic: 2.7528

Degrees of Freedom 1: 2.31

Degrees of Freedom 2: 11.55

p-value: 0.1002


Call:

rmmcp(y = df$value, groups = df$variable, blocks = df$id, tr = 0.2)
                    psihat  ci.lower ci.upper  p.value   p.crit    sig
stick vs. testicle   3.66667  -0.48300  7.81633  0.01360  0.01020  FALSE

stick vs. eye        4.00000  -0.35728  8.35728  0.01172  0.00851  FALSE

stick vs. witchetty  2.00000  -8.09920 12.09920  0.44148  0.01690  FALSE

testicle vs. eye     0.00000  -5.38802  5.38802  1.00000  0.05000  FALSE

testicle vs. witchetty -1.83333  -9.23266  5.56599  0.34371  0.01270  FALSE

eye vs. witchetty   -2.00000 -12.54827  8.54827  0.46001  0.02500  FALSE
```

Output 15.10

```
BEGIN PROGRAM R.
mySPSSdata = spssdata.
GetDataFromSPSS(factorMode =
"labels")
ID<-"celebrity"
rmFactor<-c("stick",
"testicle", "eye",
"witchetty")
df<-melt(mySPSSdata,
id.vars = ID, measure.
vars = rmFactor)
names(df)[names(df) == ID] <- "id"
rmanova(df$value,
df$variable, df$id, tr = 0.2)
rmmcp(df$value, df$variable,
df$id, tr = 0.2)
END PROGRAM.
```

Running this syntax (see SPSS Tip 15.2) should produce Output 15.10, which tells us that there was not a significant difference between means, $F_t (2.31, 11.55) = 2.75$, $p = 0.100$. The *post hoc* tests (which technically we should ignore because the overall test wasn't significant) show no significant difference between any groups (note that the column labeled *sig* shows FALSE for all comparisons, the *p*-values are all greater than the critical values (*p.crit*) and the confidence intervals all contain zero).

15.11 Effect sizes for one-way repeated-measures designs ▮▮▮▮

As with independent designs, the best measure of the overall effect size is omega squared (ω^2). However, just to make life even more complicated, the equations we've previously used for omega squared can't be used for repeated-measures designs: if you do use them the effect size will be overestimated. For the sake of simplicity some people do use the same equation, but I'm going to hit simplicity in the face with Stingy the particularly poison-ridden jellyfish, and embrace complexity like I do my children.

In repeated-measures designs, the equation for omega squared is (hang onto your hat):

$$\omega^2 = \frac{\left[\dfrac{k-1}{nk}\left(MS_M - MS_R\right)\right]}{MS_R + \dfrac{MS_B - MS_R}{k} + \left[\dfrac{k-1}{nk}\left(MS_M - MS_R\right)\right]} \tag{15.22}$$

I know what you're thinking and, no, I'm not having a laugh. Far from it. The

equation isn't too bad if you break it down. First, there are some mean squares that we've come across before: the mean square for the model (MS_M) and the residual mean square (MS_R), both of which can be obtained from Output 15.4. There's also k, the number of conditions, which for these data would be 4 (there were four animals), and n, the number of people who took part (in this case 8 celebrities).

The main problem is MS_B. At the beginning of Section 15.3 (Figure 15.3) I mentioned that the total variation is broken down into a within-participant variation and a between-participant variation (SS_B), which we can calculate using equation (15.19). SPSS doesn't give us SS_W, but we know that it's made up of SS_M and SS_R, which we are given. By substituting these terms and rearranging the equation we get:

$$SS_T = SS_B + SS_M + SS_R \tag{15.23}$$
$$SS_B = SS_T - SS_M - SS_R$$

SPSS, which is clearly trying to hinder us at every step, also doesn't give us SS_T, and I'm afraid (unless I've missed something in the output) you're going to have to calculate it by hand (see Section 15.6.1). For these data it's 17.39 (equation (15.20)). The next step is to convert this value to a mean squares by dividing by the degrees of freedom, which in this case are the number of people in the sample minus 1:

$$MS_B = \frac{SS_B}{df_B} = \frac{SS_B}{N-1} = \frac{17.38}{8-1} = 2.48 \tag{15.24}$$

Having done all this and probably died of boredom in the process, we now resurrect our corpses with renewed vigor for the effect size equation:

$$\omega^2 = \frac{\left[\dfrac{4-1}{8\times4}\left(27.71 - 7.30\right)\right]}{7.30 + \dfrac{2.48 - 7.30}{4} + \left[\dfrac{4-1}{8\times4}\left(27.71 - 7.30\right)\right]}$$
$$= \frac{1.91}{8.01}$$
$$= 0.24 \tag{15.25}$$

I hope you agree it was worth the effort. I've mentioned at various other points that it's more useful to have effect size measures for focussed comparisons anyway (rather than an overall F-statistic), and so an easier approach is to calculate effect sizes for the contrasts in Output 15.7. We can use the familiar equation to convert the F-values (because they all have 1 degree of freedom for the model) to r:

$$r = \sqrt{\frac{F(1, df_R)}{F(1, df_R) + df_R}} \qquad (15.26)$$

For the three contrasts we get the following values:

$$r_{\text{stick insect vs. kangaroo testicle}} = \sqrt{\frac{22.80}{22.80 + 7}} = 0.87$$

$$r_{\text{kangaroo testicle vs. fish eyeball}} = \sqrt{\frac{0.01}{0.01 + 7}} = 0.04$$

$$r_{\text{fish eyeball vs. witchetty grub}} = \sqrt{\frac{0.80}{0.80 + 7}} = 0.32$$

$$\qquad (15.27)$$

The difference between the stick insect and the testicle was large, between the fish eye and witchetty grub reasonable, but between the testicle and eyeball the effect was tiny.

15.12 Reporting one-way repeated-measures designs ▌▋▊▉

Reporting repeated-measures designs is much the same as independent designs, except that we need to pay attention to reporting the corrected degrees of freedom. The degrees of freedom used to assess the F-statistic are the degrees of freedom for the effect of the model ($df_M = 1.60$) and the degrees of freedom for the residuals of the model ($df_R = 11.19$). Therefore, we could report the main finding as:

✓ The Greenhouse–Geisser estimate of the departure from sphericity was $\varepsilon = 0.53$. The time to retch was not significantly affected by the type of animal eaten, $F(1.60, 11.19) = 3.79$, $p = 0.063$, $\omega^2 = 0.24$.

For Huynh–Feldt corrected values:

✓ The Huynh–Feldt estimate of the departure from sphericity was $\varepsilon = 0.67$. The time to retch was significantly affected by the type of animal eaten, $F(2, 13.98) = 3.79$, $p = 0.048$, $\omega^2 = 0.24$.

We could report multivariate tests. There are four different test statistics, but in most situations you should report Pillai's trace, V (see Chapter 17) and the associated F and its degrees of freedom (all from Output 15.6):

✓ The Greenhouse–Geisser estimate of sphericity showed a substantial deviation ($\varepsilon = 0.53$), therefore multivariate tests are reported. The time to retch was significantly affected by the type of animal eaten, $V = 0.94$, $F(3, 5) = 26.96$, $p = 0.002$, $\omega^2 = 0.24$.

Finally, robust tests:

✓ The Greenhouse–Geisser estimate of sphericity showed a substantial deviation ($\varepsilon = 0.53$). Robust tests of 20% trimmed means implemented with the WRS2 package in R (Mair et al., 2015) showed that the time to retch was not significantly affected by the type of animal eaten, $F_t(2.31, 11.55) = 2.75$, $p = 0.100$, $\omega^2 = 0.24$.

15.13 A boozy example: a factorial repeated-measures design ▌▋▊▉

We have seen that between-group designs can be extended to incorporate multiple predictor/independent variables. The same is true for repeated-measures designs. There is evidence that attitudes towards stimuli can be changed using positive and negative imagery (e.g., Hofmann, De Hawer, Perugini, Baegens, & Crombez, 2010; Stuart, Shrump, & Engel, 1987). As part of an initiative to stop binge drinking in teenagers, the government funded scientists to look at whether negative imagery could be used to make teenagers' attitudes towards alcohol more negative. The scientists compared the effects of negative imagery against positive and neutral imagery for different types of drinks. Table 15.4 illustrates the experimental design and contains the data for this example (each row represents a single participant).

Participants viewed a total of nine videos over three sessions. In one session, they saw three videos: (1) a brand of beer (Strange Brew) presented alongside negative imagery (a bunch of inanimate dead bodies in a trendy bar with the slogan 'Strange Brew: who needs a liver?'); (2) a brand of wine (Liquid Fire) presented within positive imagery (a bunch of sexy hipster types in a trendy bar with the slogan 'Liquid Fire: your life would be so much better if you were a sexy hipster type'); and (3) a brand of water (Backwater) presented with neutral imagery (some completely average people in a trendy bar accompanied by the slogan 'Backwater: it will make no difference to your life one way or another'). In a second session (a week later), the participants saw the same three brands, but this time Strange Brew was accompanied by the positive imagery, Liquid Fire by the neutral image and Backwater by the negative. In a third session, the participants saw Strange Brew accompanied by the neutral image, Liquid Fire by the negative image and Backwater by the positive. After each advert participants rated the drinks from –100 (dislike very much) through 0 (neutral) to 100 (like very much). The order of adverts was randomized, as was the order in which people participated in the

Table 15.4   Data from **Attitude.sav**

Drink	Beer			Wine			Water		
Image	+ve	−ve	Neut	+ve	−ve	Neut	+ve	−ve	Neut
Male	1	6	5	38	−5	4	10	−14	−2
	43	30	8	20	−12	4	9	−10	−13
	15	15	12	20	−15	6	6	−16	1
	40	30	19	28	−4	0	20	−10	2
	8	12	8	11	−2	6	27	5	−5
	17	17	15	17	−6	6	9	−6	−13
	30	21	21	15	−2	16	19	−20	3
	34	23	28	27	−7	7	12	−12	2
	34	20	26	24	−10	12	12	−9	4
	26	27	27	23	−15	14	21	−6	0
Female	1	−19	−10	28	−13	13	33	−2	9
	7	−18	6	26	−16	19	23	−17	5
	22	−8	4	34	−23	14	21	−19	0
	30	−6	3	32	−22	21	17	−11	4
	40	−6	0	24	−9	19	15	−10	2
	15	−9	4	29	−18	7	13	−17	8
	20	−17	9	30	−17	12	16	−4	10
	9	−12	−5	24	−15	18	17	−4	8
	14	−11	7	34	−14	20	19	−1	12
	15	−6	13	23	−15	15	29	−1	10

single participant. If a person participates in all conditions (in this case each person sees all types of drink presented with all types of imagery) then each condition is represented by a column. So, we need to create nine variables in the data editor with the names and value labels in Table 15.5.

Select *Analyze* ▶ *General Linear Model* ▶ *Repeated Measures…* to access the dialog boxes for a repeated-measures model. First we define the repeated-measures predictors. In this case there are two: **Drink** (beer, wine or water) and **Imagery** (positive, negative and neutral). Replace the word *factor1* with the word 'Drink', type '3' into the box labeled *Number of Levels*, and click Add. This variable appears in the list of variables as *Drink(3)*, which means we have defined a predictor called Drink that has three levels. We repeat this process for the second predictor by typing 'Imagery' into the space labeled *Within-Subject Factor Name*, typing '3' into the space labeled *Number of Levels*, and clicking Add. The variable

three sessions. This design is quite complex. There are two predictor/independent variables: the type of drink (beer, wine or water) and the type of imagery used (positive, negative or neutral). These two variables completely cross over, producing nine experimental conditions.

15.14 Factorial repeated-measures designs using SPSS Statistics ▮▮▮

When entering the data in Table 15.4 remember that each row represents a

Table 15.5   Variable names and labels for the Attitude.sav data

Variable name	Variable label
beerpos	Beer + sexy hipsters
beerneg	Beer + dead bodies
beerneut	Beer + average people
winepos	Wine + sexy hipsters
wineneg	Wine + dead bodies
wineneut	Wine + average people
waterpos	Water + sexy hipsters
waterneg	Water + dead bodies
waterneut	Water + average people

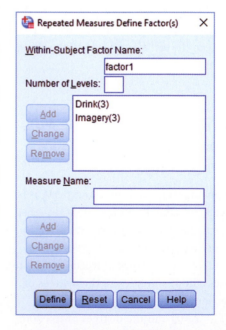

Figure 15.10   The *Define Factor(s)* dialog box for factorial repeated-measures designs

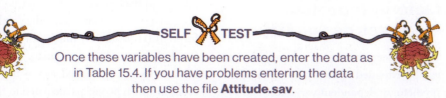

SELF TEST

Once these variables have been created, enter the data as in Table 15.4. If you have problems entering the data then use the file **Attitude.sav**.

will appear in the list as *Imagery(3)* (Figure 15.10). Click Define to go to the main dialog box.

The main dialog box is the same as for designs with one predictor variable, except that there are nine question marks (Figure 15.11). At the top of the *Within-Subjects Variables* box, SPSS lists the variables that we defined: **Drink** and **Imagery**. Underneath is a series of question marks followed by bracketed numbers. The numbers in brackets represent the levels of the predictor variables (independent variables):

- _?_(1,1) ⇒ variable representing 1st level of drink and 1st level of imagery
- _?_(1,2) ⇒ variable representing 1st level of drink and 2nd level of imagery
- _?_(1,3) ⇒ variable representing 1st level of drink and 3rd level of imagery
- _?_(2,1) ⇒ variable representing 2nd level of drink and 1st level of imagery
- _?_(2,2) ⇒ variable representing 2nd level of drink and 2nd level of imagery
- _?_(2,3) ⇒ variable representing 2nd level of drink and 3rd level of imagery
- _?_(3,1) ⇒ variable representing 3rd level of drink and 1st level of imagery
- _?_(3,2) ⇒ variable representing 3rd level of drink and 2nd level of imagery
- _?_(3,3) ⇒ variable representing 3rd level of drink and 3rd level of imagery

Because we have two predictors there are two numbers in the brackets: the first refers to levels of the first predictor listed above the box (in this case **Drink**) and the second refers to levels of the second predictor listed above the box (in this case **Imagery**). We need to replace these question marks with variable names from the data editor (which are listed on the left-hand side of the dialog box). At this stage we need to think about which conditions to assign to which level of each variable. For example, if we entered **beerpos** into the list first, then SPSS would treat beer as the first level of

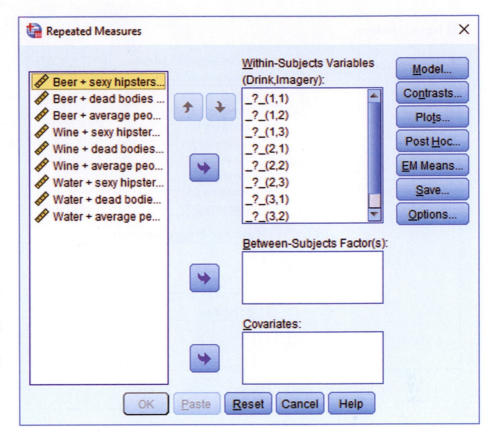

Figure 15.11 The main dialog box for factorial repeated-measures designs before completion

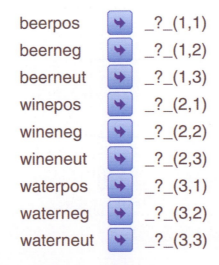

Figure 15.12 Variable allocations for the attitude data

Drink and positive imagery as the first level of the **Imagery** variable. However, if we entered **wineneg** into the list first, we're specifying wine as the first level of

Drink, and negative imagery as the first level of **Imagery**. For this reason, we need to think about what contrasts we might want *before* assigning variables in this dialog box.

The first variable, **Drink**, had three conditions, two of which involved alcoholic drinks. In a sense, the water condition acts as a control to whether the effects of imagery are specific to alcohol. Therefore, we might want to compare the beer and wine condition with the water condition. This comparison could be done by either specifying a simple contrast (see Table 12.6) in which the beer and wine conditions are compared to the water or using a difference contrast in which both alcohol conditions are compared to the water condition before being compared to each other. Either way the water condition should be

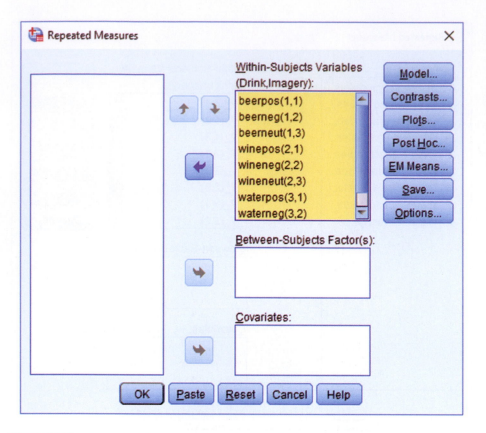

Figure 15.13 The main dialog box for factorial repeated-measures designs after completion

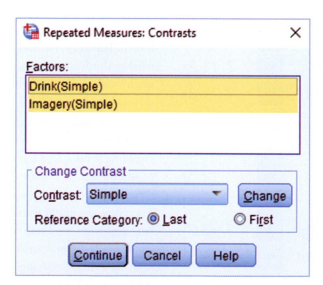

Figure 15.14 The *Contrasts* dialog box for factorial repeated-measures designs

entered as either the first or last level of **Drink**. The imagery variable also has a control category that was not expected to change attitudes: neutral imagery. This category might be a sensible reference category in a simple contrast[7] so, again, we'd want to enter it as either the first or last level.

Based on this discussion, it makes sense to have water as level 3 of **Drink** and neutral as level 3 of **Imagery**. The remaining levels can be decided arbitrarily. I have chosen beer as level 1 and wine as level 2 of **Drink**, and positive as level 1 and negative as level 2 of **Imagery**. I would, therefore, assign the variables as in Figure 15.12. Coincidentally, this is the order in which variables are listed in the data editor. (It's not a coincidence: I thought ahead about what contrasts I would do and entered the variables in the appropriate order.) When these variables have been assigned, the dialog box looks like Figure 15.13.

15.14.1 Contrasts

Click [Contrasts...] to access the dialog box in Figure 15.14. In the previous section I described why it might be interesting to use the water and neutral conditions as baseline categories for the drink and imagery factors, respectively. You should know how to assign contrasts by now (if not, read Sections 13.5.5 and 15.8.2), so select a simple contrast for both **Drink** and **Imagery**. We assigned variables such that the control category was the last one; therefore, we can leave the default reference category of ⊙ **Last**. Click [Continue] to return to the main dialog box.

---

7 We expect positive imagery to improve attitudes, whereas negative imagery should make attitudes more negative. Therefore, it does not make sense to do a Helmert or difference contrast for this factor because the effects of the two experimental conditions will cancel each other out.

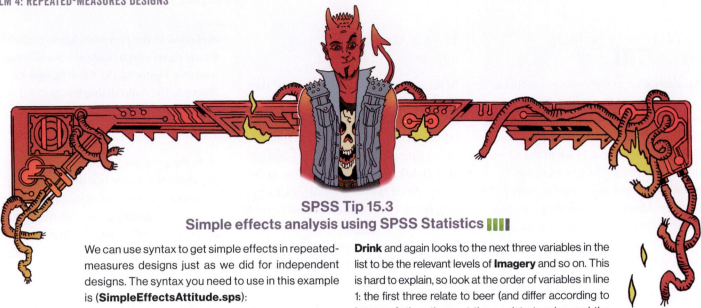

SPSS Tip 15.3
Simple effects analysis using SPSS Statistics ▐▐▐

We can use syntax to get simple effects in repeated-measures designs just as we did for independent designs. The syntax you need to use in this example is (**SimpleEffectsAttitude.sps**):

```
GLM beerpos beerneg beerneut winepos wineneg
wineneut waterposwaterneg waterneut
/WSFACTOR=Drink 3 Imagery 3
/EMMEANS = TABLES(Drink*Imagery)
COMPARE(Imagery).
```

The first line specifies the variables in the data editor that relate to the levels of our repeated-measures variables. The */WSFACTORS* command defines the two repeated-measures variables. The order of variables in the first line is important. Because we've defined *Drink 3 Imagery 3* in line 2, SPSS starts at level 1 of **Drink**, and then, because we've specified three levels of **Imagery**, it uses the first three variables listed as the levels of **Imagery** at level 1 of **Drink**. It then moves onto level 2 of

Drink and again looks to the next three variables in the list to be the relevant levels of **Imagery** and so on. This is hard to explain, so look at the order of variables in line 1: the first three relate to beer (and differ according to **Imagery**), then the next three relate to wine and the three levels of **Imagery**, and so on.[8]

The */EMMEANS* command specifies the simple effects. *TABLES(Drink*Imagery)* requests a table of means for the interaction of **Drink** and **Imagery**, and *COMPARE(Imagery)* give us the simple effect of **Imagery** at each level of **Drink**. If we wanted to look at the effect of **Drink** at each level of **Imagery**, then we'd use *COMPARE(Drink)* instead.

Run the syntax (make sure you have **Attitude. sav** loaded into the data editor) to get Output 15.11, which contains multivariate tests of the effect of **Imagery** at each level of **Drink**. Looking at the significance values, there were significant effects of **Imagery** at all levels of **Drink**.

Multivariate Tests

Drink		Value	F	Hypothesis df	Error df	Sig.
1	Pillai's trace	.593	13.122[a]	2.000	18.000	.000
	Wilks' lambda	.407	13.122[a]	2.000	18.000	.000
	Hotelling's trace	1.458	13.122[a]	2.000	18.000	.000
	Roy's largest root	1.458	13.122[a]	2.000	18.000	.000
2	Pillai's trace	.923	107.305[a]	2.000	18.000	.000
	Wilks' lambda	.077	107.305[a]	2.000	18.000	.000
	Hotelling's trace	11.923	107.305[a]	2.000	18.000	.000
	Roy's largest root	11.923	107.305[a]	2.000	18.000	.000
3	Pillai's trace	.939	138.795[a]	2.000	18.000	.000
	Wilks' lambda	.061	138.795[a]	2.000	18.000	.000
	Hotelling's trace	15.422	138.795[a]	2.000	18.000	.000
	Roy's largest root	15.422	138.795[a]	2.000	18.000	.000

Each F tests the multivariate simple effects of Imagery within each level combination of the other effects shown. These tests are based on the linearly independent pairwise comparisons among the estimated marginal means.

Output 15.11

8 It would also work to write the first two lines as:
```
GLMbeerpos winepos waterpos beerneg wineneg waterneg beerneut wineneut waterneut
/WSFACTORS Imagery 3 Drink 3
```

15.14.2 Simple effects analysis ▊▊▊

An alternative to the contrasts available here is to do a simple effects analysis (see Section 14.6.5) to look at the effect of one predictor at individual levels of another. For example, we could look at the effect of drink for positive imagery, then for negative imagery and then for neutral imagery. Alternatively, we could analyze the effect of imagery separately for beer, wine and water. To this analysis we need to use syntax (SPSS Tip 15.3).

15.14.3 Graphing interactions ▊▊▊

In the previous example we ignored the *Plots* dialog box, but with two predictors it is a convenient way to plot the interaction (although better to plot these graphs before fitting the model). Click `Plots...`, drag **Drink** from the variables list to the space labeled *Horizontal Axis* (or click ➡), drag **Imagery** to the space labeled *Separate Lines*, and click `Add`. Choose a *Line Chart*, ask for error bars that display 95% confidence intervals, and for reasons

explained in the previous chapter, select *Y axis starts at 0* to scale the *y*-axis from zero (see Figure 15.15). Click `Continue` to return to the main dialog box.

15.14.4 Other options ▊▊▊

As in the previous example, *post hoc* tests are disabled because this design has only repeated-measures variables. To get some we can click `EM Means...`, drag (or click ➡) all variables in the box labeled *Factor(s) and Factor Interactions* to the box labeled *Display Means for*, select ☑ Compare main effects and choose a correction from the drop-down menu (I chose Bonferroni). I've also asked for ☑ Descriptive statistics using `Options...` in the main dialog box (Figure 15.16).

15.15 Interpreting factorial repeated-measures designs ▊▊▊

Output 15.12 shows two tables. The first lists the variables from the data editor and the level of each predictor that they represent and is a useful way to verify that you entered the variables in the correct order for the comparisons that you want. The second table contains the means and standard deviations across the nine conditions. The names in this table are the variable labels that you entered in the data editor (refer to Table 15.5). The descriptives tell us that the variability among scores was greatest when beer was used as a product (compare the standard deviations of the beer variables against the others). Also, when dead bodies were used as imagery the ratings given to the products were negative (as expected) for wine and water but not for beer (for some reason negative imagery didn't have the expected effect when beer was used as a stimulus). Output 15.13 shows the sphericity estimates for each of the three effects in

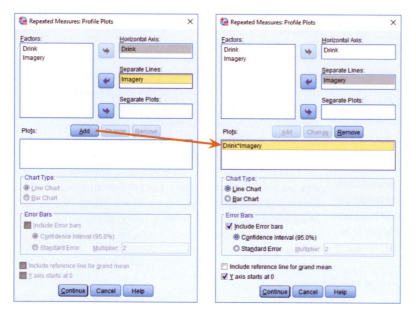

Figure 15.15 Defining profile plots in repeated-measures designs

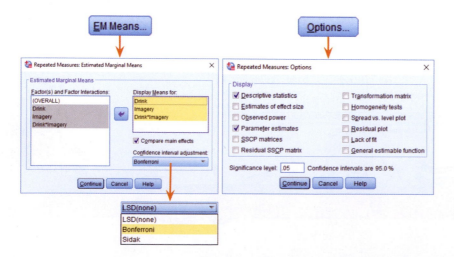

Figure 15.16

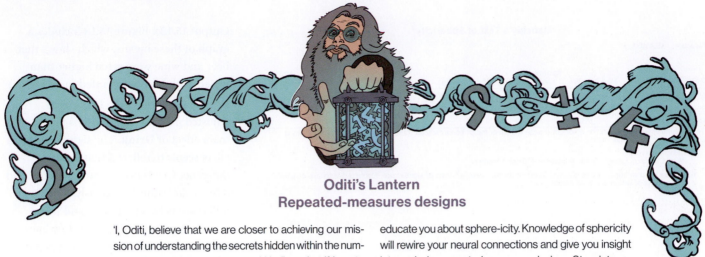

Oditi's Lantern
Repeated-measures designs

'I, Oditi, believe that we are closer to achieving our mission of understanding the secrets hidden within the numbers. The Earth is a sphere, and I believe that if I am to dominate, erm, I mean understand the Earth then I must educate you about sphere-icity. Knowledge of sphericity will rewire your neural connections and give you insight into analyzing repeated-measures designs. Stare into my lantern and feel your brain burn, but in a nice way.'

Within–Subjects Factors

Measure: MEASURE_1

Drink	Imagery	Dependent Variable
1	1	beerpos
	2	beerneg
	3	beerneut
2	1	winepos
	2	wineneg
	3	wineneut
3	1	waterpos
	2	waterneg
	3	waterneut

Descriptive Statistics

	Mean	Std. Deviation	N
Beer + sexy hipsters	21.05	13.008	20
Beer + dead bodies	4.45	17.304	20
Beer + average people	10.00	10.296	20
Wine + sexy hipsters	25.35	6.738	20
Wine + dead bodies	-12.00	6.181	20
Wine + average people	11.65	6.243	20
Water + sexy hipsters	17.40	7.074	20
Water + dead bodies	-9.20	6.802	20
Water + average people	2.35	6.839	20

Output 15.12

focusing on the information that we plan to use. For example, if, like me, you want to routinely report Greenhouse–Geisser corrected values then we can edit the table to show only these values (SPSS Tip 15.4). The significance values tell us that there is a significant main effect of the type of drink used as a stimulus, a significant main effect of the type of imagery used and a significant interaction between these two variables. I will examine each of these effects in turn.

the model (two main effects and one interaction). All three effects have estimates less than 1, indicating some deviation from sphericity, so we may as well correct for these.[9]

Output 15.14 shows the *F*-statistics (with corrections) and is split into sections that refer to each effect in the model and the associated error term. The table is quite mind-blowing, but we can stay calm by

15.15.1 The main effect of drink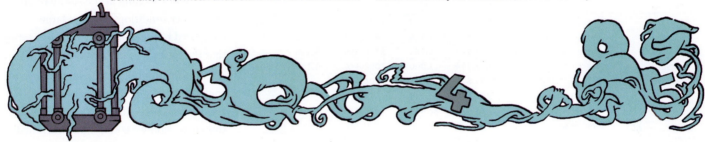

The type of drink used was significant, which tells us that if we ignore the type of imagery that was used, participants rated

---

9 I've tried to dissuade you form looking at Mauchly's test, but if you use it then you'd conclude that both the main effects of **Drink** and **Imagery** have violated this assumption (the *p*s are less than 0.05), but it is met for the interaction (because *p* > 0.05). However, compare the Greenhouse–Geisser estimates for **Imagery** (0.747) and the interaction (0.798). They're pretty similar, suggesting that the departure from sphericity is about the same. Despite this similarity, the *p*-values for Mauchly's test encourage you to conclude that sphericity is met for the interaction but not for the effect of imagery.

Mauchly's Test of Sphericity[a]

Measure: MEASURE_1

Within Subjects Effect	Mauchly's W	Approx. Chi-Square	df	Sig.	Epsilon[b]		
					Greenhouse-Geisser	Huynh-Feldt	Lower-bound
Drink	.267	23.753	2	.000	.577	.591	.500
Imagery	.662	7.422	2	.024	.747	.797	.500
Drink * Imagery	.595	9.041	9	.436	.798	.979	.250

Tests the null hypothesis that the error covariance matrix of the orthonormalized transformed dependent variables is proportional to an identity matrix.

a. Design: Intercept
   Within Subjects Design: Drink + Imagery + Drink * Imagery

b. May be used to adjust the degrees of freedom for the averaged tests of significance. Corrected tests are displayed in the Tests of Within-Subjects Effects table.

Output 15.13

Tests of Within-Subjects Effects

Measure: MEASURE_1

Source		Type III Sum of Squares	df	Mean Square	F	Sig.
Drink	Sphericity Assumed	2092.344	2	1046.172	5.106	.011
	Greenhouse-Geisser	2092.344	1.154	1812.764	5.106	.030
	Huynh-Feldt	2092.344	1.181	1770.939	5.106	.029
	Lower-bound	2092.344	1.000	2092.344	5.106	.036
Error(Drink)	Sphericity Assumed	7785.878	38	204.892		
	Greenhouse-Geisser	7785.878	21.930	355.028		
	Huynh-Feldt	7785.878	22.448	346.836		
	Lower-bound	7785.878	19.000	409.783		
Imagery	Sphericity Assumed	21628.678	2	10814.339	122.565	.000
	Greenhouse-Geisser	21628.678	1.495	14468.490	122.565	.000
	Huynh-Feldt	21628.678	1.594	13571.496	122.565	.000
	Lower-bound	21628.678	1.000	21628.678	122.565	.000
Error(Imagery)	Sphericity Assumed	3352.878	38	88.234		
	Greenhouse-Geisser	3352.878	28.403	118.048		
	Huynh-Feldt	3352.878	30.280	110.729		
	Lower-bound	3352.878	19.000	176.467		
Drink * Imagery	Sphericity Assumed	2624.422	4	656.106	17.155	.000
	Greenhouse-Geisser	2624.422	3.194	821.778	17.155	.000
	Huynh-Feldt	2624.422	3.914	670.462	17.155	.000
	Lower-bound	2624.422	1.000	2624.422	17.155	.001
Error(Drink*Imagery)	Sphericity Assumed	2906.689	76	38.246		
	Greenhouse-Geisser	2906.689	60.678	47.903		
	Huynh-Feldt	2906.689	74.373	39.083		
	Lower-bound	2906.689	19.000	152.984		

Output 15.14

some drinks significantly differently than others. In Section 15.14.4 we requested estimated marginal means for the effects in the model, and the means and standard errors for the main effect of **Drink** are shown in Figure 15.18.[10] The levels of

Drink are labeled 1, 2 and 3, so we must think back to the order in which we assigned variables to know which row of the table relates to which drink. We entered the beer condition first and the water condition last (as shown in

Output 15.12). Figure 15.18 includes a graph of these means, which shows that beer and wine were rated higher than water (with beer being rated most highly). Output 15.16 shows the Bonferroni adjusted pairwise comparisons for the main effect of **Drink**. The significant main effect seems to reflect a significant difference ($p = 0.001$) between levels 2 and 3 (wine and water). Curiously, the difference between the beer and water conditions is larger than that for wine and water, yet this effect is non-significant ($p = 0.066$). This inconsistency can be explained by looking at the standard error in the beer condition, which is large compared to the wine condition, indicating that the mean for beer is very noisy.

This finding highlights the importance of controlling the error rate by using a Bonferroni correction. Had we not used this correction we could have concluded erroneously that beer was rated significantly more highly than water.

15.15.2 The main effect of imagery ▌▌▌▌

The main effect of the type of imagery also had a significant influence on participants' ratings of the drinks (Output 15.14). This effect tells us that if we ignore the type of drink that was used, participants' ratings of those drinks were different according to the type of imagery that was used. Figure 15.19 shows the means that we requested in Section 15.14.4. The levels of imagery are labeled 1, 2 and 3, so we need to again think back to how we assigned variables. We assigned the positive condition to the first level and the neutral condition to the last. Figure 15.19 includes a graph of these means (and their confidence intervals) and

---

10 These means are obtained by taking the average of the means in Output 15.3 for a given condition. For example, the mean for the beer condition (ignoring type of imagery) is:

$$\bar{X}_{\text{Beer}} = \frac{\bar{X}_{\text{Beer, sexy hipsters}} + \bar{X}_{\text{Beer, dead bodies}} + \bar{X}_{\text{Beer, average people}}}{3} = \frac{21.05 + 4.45 + 10.00}{3} = 11.83$$

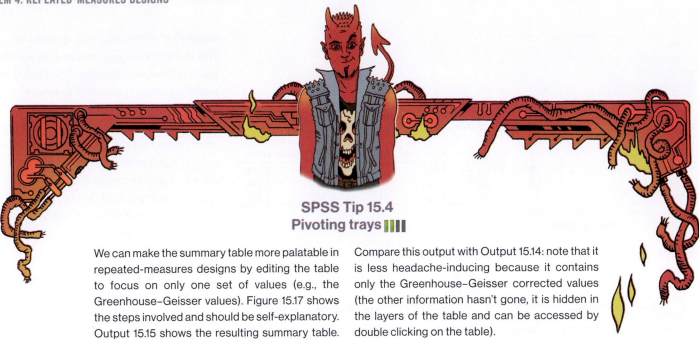

SPSS Tip 15.4
Pivoting trays ▮▮▮▮

We can make the summary table more palatable in repeated-measures designs by editing the table to focus on only one set of values (e.g., the Greenhouse–Geisser values). Figure 15.17 shows the steps involved and should be self-explanatory. Output 15.15 shows the resulting summary table.

Compare this output with Output 15.14: note that it is less headache-inducing because it contains only the Greenhouse–Geisser corrected values (the other information hasn't gone, it is hidden in the layers of the table and can be accessed by double clicking on the table).

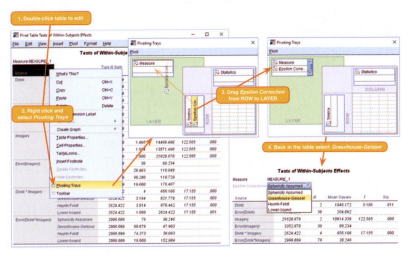

Figure 15.17 Using the *Pivot Trays* option to 'hide' parts of a summary table

Tests of Within-Subjects Effects

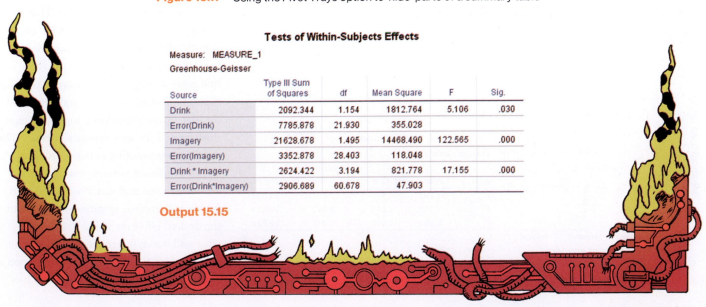

Measure: MEASURE_1
Greenhouse-Geisser

Source	Type III Sum of Squares	df	Mean Square	F	Sig.
Drink	2092.344	1.154	1812.764	5.106	.030
Error(Drink)	7785.878	21.930	355.028		
Imagery	21628.678	1.495	14468.490	122.565	.000
Error(Imagery)	3352.878	28.403	118.048		
Drink * Imagery	2624.422	3.194	821.778	17.155	.000
Error(Drink*Imagery)	2906.689	60.678	47.903		

Output 15.15

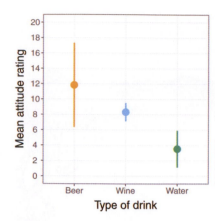

Figure 15.18 Output and graph of the main effect of **Drink**

Estimates

Measure: MEASURE_1

Drink	Mean	Std. Error	95% Confidence Interval	
			Lower Bound	Upper Bound
1	11.833	2.621	6.348	17.319
2	8.333	.574	7.131	9.535
3	3.517	1.147	1.116	5.918

SELF TEST

Try rerunning these *post hoc* tests but select the uncorrected values (LSD) in the options dialog box (see Section 13.8.5). You should find that the difference between beer and water is now significant (*p* = 0.02).

Pairwise Comparisons

Measure: MEASURE_1

(I) Drink	(J) Drink	Mean Difference (I–J)	Std. Error	Sig.[b]	95% Confidence Interval for Difference[b]	
					Lower Bound	Upper Bound
1	2	3.500	2.849	.703	–3.980	10.980
	3	8.317	3.335	.066	–.438	17.072
2	1	–3.500	2.849	.703	–10.980	3.980
	3	4.817*	1.116	.001	1.886	7.747
3	1	–8.317	3.335	.066	–17.072	.438
	2	–4.817*	1.116	.001	–7.747	–1.886

Based on estimated marginal means

*. The mean difference is significant at the .05 level.

b. Adjustment for multiple comparisons: Bonferroni.

Output 15.16

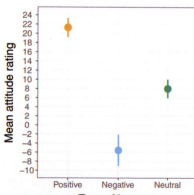

Figure 15.19 Output and graph of the main effect of **Imagery**

Estimates

Measure: MEASURE_1

Imagery	Mean	Std. Error	95% Confidence Interval	
			Lower Bound	Upper Bound
1	21.267	.977	19.222	23.312
2	–5.583	1.653	–9.043	–2.124
3	8.000	.969	5.972	10.028

shows that positive imagery resulted in very positive ratings (compared to neutral imagery) and negative imagery resulted in negative ratings (compared to neutral imagery). Output 15.17 shows the Bonferroni adjusted pairwise comparisons, which show that the significant main effect reflects significant differences (all *p*s < 0.001) between levels 1 and 2 (positive and negative), levels 1 and 3 (positive and neutral) and levels 2 and 3 (negative and neutral).

15.15.3 The interaction effect (drink × imagery)

The type of imagery interacted significantly with the type of drink used as a stimulus to affect ratings (Output 15.14). This effect tells us that the type of imagery used had a different effect depending on which type of drink was being rated. We can use the means that we requested in Section 15.14.4 to unpick this interaction. This table is shown in Output 15.18 and is essentially the same as the initial descriptive statistics in Output 15.12, except that the standard errors are displayed rather than the standard deviations.

Figure 15.20 displays the means from Output 15.18. The graph shows that the pattern of response across drinks was similar when positive and neutral imagery were used (blue and grey lines). That is, ratings were positive for beer, they were slightly higher for wine and they were lower for water. The fact that the (blue) line representing positive imagery is higher than the neutral (grey) line indicates that positive imagery produced higher ratings than neutral imagery across all drinks. The red line (representing negative imagery) shows a different pattern: ratings were lowest for wine and water but quite high for beer. Therefore, negative imagery had the desired effect on attitudes towards wine and water, but much less impact on ratings of beer. Therefore, the interaction is likely to reflect the fact that imagery has the expected effect for wine and water (that is, ratings are highest for positive imagery, lowest for negative imagery and neutral falls somewhere in between) but not for beer (where ratings after negative information do not seem to be particularly negative). To verify the interpretation of the interaction effect, we can look at the contrasts that we requested in Section 15.14.1.

15.15.4 Contrasts for the main effects ‖‖‖

In Section 15.14.1 we requested simple contrasts for both the **Drink** (water was used as the control category) and **Imagery** variables (neutral imagery was used as the control category). Output 15.19 shows these contrasts. The table is split into main effects and interactions, and within each are the contrasts. If you are confused as to which level is which, Output 15.12 lists them for you. For the main effect of drink, the first contrast shows a significant difference between level 1 (beer) and level 3 (water), $F(1, 19) = 6.22$, $p = 0.022$, which contradicts the equivalent *post hoc* test (see Output 15.16).

The next contrast shows a significant difference between level 2 (wine) and level 3 (water), $F(1, 19) = 18.61$, $p < 0.001$. For the imagery main effect, level 1 (positive) is significantly different than level 3 (neutral), $F(1, 19) = 142.19$, $p < 0.001$, and level 2 (negative imagery) is significantly different than level 3 (neutral), $F(1, 19) = 47.07$, $p < 0.001$.

15.15.5 Contrasts for the interaction effect ‖‖‖

The contrasts for the main effects tell us only what we already knew (although note the increased statistical power with these tests shown by the higher significance values). The contrasts for the interaction term are more interesting. To help us interpret these contrasts, Figure 15.21 breaks the interaction graph in Figure 15.20 into the four contrasts.

Pairwise Comparisons

Measure: MEASURE_1

(I) Imagery	(J) Imagery	Mean Difference (I–J)	Std. Error	Sig.[b]	95% Confidence Interval for Difference[b] Lower Bound	Upper Bound
1	2	26.850*	1.915	.000	21.824	31.876
	3	13.267*	1.113	.000	10.346	16.187
2	1	-26.850*	1.915	.000	-31.876	-21.824
	3	-13.583*	1.980	.000	-18.781	-8.386
3	1	-13.267*	1.113	.000	-16.187	-10.346
	2	13.583*	1.980	.000	8.386	18.781

Based on estimated marginal means
*. The mean difference is significant at the .05 level.
b. Adjustment for multiple comparisons: Bonferroni.

Output 15.17

3. Drink * Imagery

Measure: MEASURE_1

Drink	Imagery	Mean	Std. Error	95% Confidence Interval Lower Bound	Upper Bound
1	1	21.050	2.909	14.962	27.138
	2	4.450	3.869	-3.648	12.548
	3	10.000	2.302	5.181	14.819
2	1	25.350	1.507	22.197	28.503
	2	-12.000	1.382	-14.893	-9.107
	3	11.650	1.396	8.728	14.572
3	1	17.400	1.582	14.089	20.711
	2	-9.200	1.521	-12.384	-6.016
	3	2.350	1.529	-.851	5.551

Output 15.18

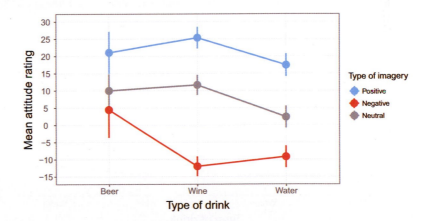

Figure 15.20 Interaction graph for **Attitude.sav**

SELF TEST
Why do you think that this contradiction has occurred?

Tests of Within-Subjects Contrasts

Measure: MEASURE_1

Source	Drink	Imagery	Type III Sum of Squares	df	Mean Square	F	Sig.
Drink	Level 1 vs. Level 3		1383.339	1	1383.339	6.218	.022
	Level 2 vs. Level 3		464.006	1	464.006	18.613	.000
Error(Drink)	Level 1 vs. Level 3		4226.772	19	222.462		
	Level 2 vs. Level 3		473.661	19	24.930		
Imagery		Level 1 vs. Level 3	3520.089	1	3520.089	142.194	.000
		Level 2 vs. Level 3	3690.139	1	3690.139	47.070	.000
Error(Imagery)		Level 1 vs. Level 3	470.356	19	24.756		
		Level 2 vs. Level 3	1489.528	19	78.396		
Drink * Imagery	Level 1 vs. Level 3	Level 1 vs. Level 3	320.000	1	320.000	1.576	.225
		Level 2 vs. Level 3	720.000	1	720.000	6.752	.018
	Level 2 vs. Level 3	Level 1 vs. Level 3	36.450	1	36.450	.235	.633
		Level 2 vs. Level 3	2928.200	1	2928.200	26.906	.000
Error(Drink*Imagery)	Level 1 vs. Level 3	Level 1 vs. Level 3	3858.000	19	203.053		
		Level 2 vs. Level 3	2026.000	19	106.632		
	Level 2 vs. Level 3	Level 1 vs. Level 3	2946.550	19	155.082		
		Level 2 vs. Level 3	2067.800	19	108.832		

Output 15.19

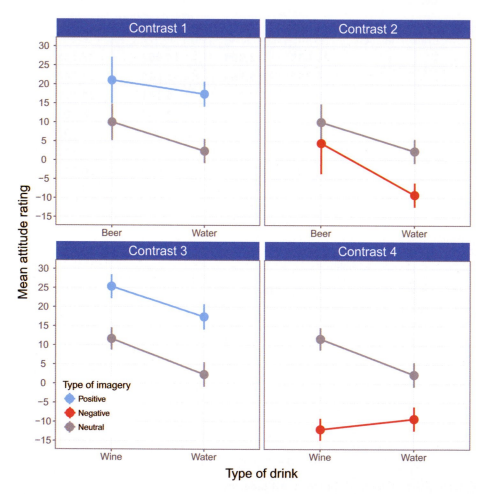

Figure 15.21 Graphs (generated using R, incidentally) illustrating the four contrasts in the attitude analysis

The first contrast for the interaction looks at level 1 of **Drink** (beer) compared to level 3 (water), when positive imagery (level 1) is used compared to neutral (level 3). This contrast is non-significant, $p = 0.225$. This result tells us that the higher ratings when positive imagery is used (compared to neutral imagery) are equivalent for beer and water. Figure 15.21 (top left) shows this contrast: the non-significance means that the distance between the lines in the beer condition is the same as the distance between the lines in the water condition. We could conclude that the improvement of ratings due to positive imagery compared to neutral is not affected by whether people are evaluating beer or water.

The second contrast for the interaction term looks at level 1 of **Drink** (beer) compared to level 3 (water), when negative imagery (level 2) is used compared to neutral (level 3). This contrast is significant, $F(1, 19) = 6.75$, $p = 0.018$. Figure 15.21 (top right) shows the contrast. The significance means that the distance between the red and grey line in the beer condition is significantly smaller than the distance between the red and grey line in the water condition. When beer is being rated the ratings are similar regardless of the type of imagery, but for water ratings are much lower with negative imagery than neutral imagery.

The third contrast looks at level 2 of **Drink** (wine) compared to level 3 (water), when positive imagery (level 1) is used compared to neutral (level 3). This contrast is non-significant, $p = 0.633$, indicating that the higher ratings when positive imagery is used (compared to neutral imagery) are similar for wine and water. Figure 15.21 (bottom left) shows this contrast. The non-significance implies that the distance between the grey and blue lines in the wine condition is similar to the distance between the lines in the water condition.

The final contrast for the interaction term looks at level 2 of **Drink** (wine) compared to level 3 (water), when negative imagery (level 2) is used compared to neutral (level 3). This contrast is significant, $F(1, 19) = 26.91$, $p < 0.001$. Figure 15.21 (bottom right) shows this contrast. The significance implies that the distance between the red and grey lines in the wine condition is significantly larger than the

Cramming Sam's Tips
Factorial repeated-measures designs

- Two-way repeated-measures designs compare means when there are two predictor/independent variables, and the same entities have been used in all conditions.
- You can test the assumption of *sphericity* when you have three or more repeated-measures conditions with *Mauchly's test*, but a better approach is to routinely interpret *F*-statistics that have been corrected for the amount by which the data are not spherical.
- The table labeled *Tests of Within-Subjects Effects* shows the *F*-statistics and their *p*-values. In a two-way design you will have a main effect of each variable and the interaction

between them. For *each* effect, read the row labeled *Greenhouse–Geisser* (you can also look at *Huynh–Feldt*, but you'll have to read this chapter to find out the relative merits of the two procedures). If the value in the column labeled *Sig.* is less than 0.05 then the effect is significant.
- Break down the main effects and interactions using contrasts. These contrasts appear in the table labeled *Tests of Within-Subjects Contrasts*. If the values in the column labeled *Sig.* are less than 0.05 the contrast is significant.

distance between the lines in the water condition. In short, the lower ratings due to negative imagery (compared to neutral) are significantly greater for wine than for water.

These contrasts tell us nothing about the differences between the beer and wine conditions (or the positive and negative conditions), and different contrasts would have to be run to find out more. However, they do tell us that, relative to the neutral condition, positive imagery increased liking for the products regardless of the product, whereas negative imagery affected ratings of wine but not so much beer. These differences were not predicted.

Interpreting interaction terms is complex, and even some well-respected researchers struggle with them, so don't feel disheartened if you find them hard. Try to be thorough

and break each effect down using contrasts and graphs, and you will get there.

15.16 Effect sizes for factorial repeated-measures designs

Calculating omega squared for a one-way repeated-measures design was hair-raising enough, and as I keep saying, effect sizes are more useful when they describe a focussed effect, so calculate effect sizes for your contrasts in factorial designs (and main effects that compare only two groups). You'll thank me for averting a nervous breakdown. Output 15.19 shows the values for the contrasts we requested, all of which have 1 degree of freedom for

the model (i.e., they represent a focussed and interpretable comparison) and have 19 residual degrees of freedom. We can convert these *F*-statistics to *r* using a formula we've come across before (equation (15.26)). For the two contrasts for the drink variable (Output 15.19), we get the following values:

$$r_{\text{beer vs. water}} = \sqrt{\frac{6.22}{6.22 + 19}} = 0.50$$

$$r_{\text{wine vs. water}} = \sqrt{\frac{18.61}{18.61 + 19}} = 0.70$$

(15.28)

For the two contrasts for the imagery variable (Output 15.19), we get:

$$r_{\text{positive vs. neutral}} = \sqrt{\frac{142.19}{142.19 + 19}} = 0.94$$

$$r_{\text{negative vs. neutral}} = \sqrt{\frac{47.07}{47.07 + 19}} = 0.84$$

(15.29)

Labcoat Leni's Real Research 15.1
Are splattered cadavers distracting? ▮▮▮▮

Perham, N., & Sykora, M. (2012). *Applied Cognitive Psychology, 26*(4), 550–555.

In Chapter 10, I used the example of whether listening to my favorite music would interefer with people's ability to write an essay. It turns out that Nick Perham has tested this hypothesis (sort of). He was interested in the effects of liked and disliked music (compared to quiet) on people's ability to remember things. Twenty-five participants remembered lists of eight letters. Perham and Sykora (2012) manipulated the background noise while each list was presented: silence (the control), liked music or disliked music. They used music that they believed most participants would like (a popular song called 'From Paris to Berlin'

by Infernal) and dislike (Repulsion's 'Acid Bath', 'Eaten Alive' and 'Splattered Cadavers'—in other words, the sort of thing I listen to, although I don't actually have any stuff by Repulsion). Participants recalled each list of eight letters, and the authors calculated the probability of correctly recalling a letter in each position in the list. There are two variables: position in the list (which letter in the sequence is being recalled, from 1 to 8) and sound playing when the list is presented (quiet, liked, disliked). Fit a model to see whether recall is affected by the type of sound played while learning the sequences (**Perham & Sykora (2012).sav**). Answers are on the companion website (or look at page 552 in the original article).

For the interaction term, we had four contrasts, but we can convert them to r because they all have 1 degree of freedom for the model (Output 15.19). You should get the following values:

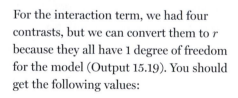

$$r_{\text{beer vs. water, positive vs. neutral}} = \sqrt{\frac{1.58}{1.58 + 19}} = 0.28$$

$$r_{\text{beer vs. water, negative vs. neutral}} = \sqrt{\frac{6.75}{6.75 + 19}} = 0.51$$

$$r_{\text{wine vs. water, positive vs. neutral}} = \sqrt{\frac{0.24}{0.24 + 19}} = 0.11$$

$$r_{\text{wine vs. water, negative vs. neutral}} = \sqrt{\frac{26.91}{26.91 + 19}} = 0.77$$

(15.30)

The two effects that were significant (beer vs. water, negative vs. neutral and wine vs.

water, negative vs. neutral) yield large effect sizes. The two effects that were not significant yielded a medium effect size (beer vs. water, positive vs. neutral) and a small effect size (wine vs. water, positive vs. neutral).

15.17 Reporting the results from factorial repeated-measures designs ▮▮▮▮

We've got three effects to report and we need to report corrected degrees of freedom for each, and these effects might have different degrees of

freedom. We can then report the three effects as follows:

- ✓ Unless otherwise stated $p < 0.001$. For the main effect of drink the Greenhouse–Geisser estimate of the departure from sphericity was $\varepsilon = 0.58$. This main effect was a significant, $F_{(1.15, 21.93)} = 5.11$, $p = 0.030$. Contrasts revealed that ratings of beer, $F_{(1, 19)} = 6.22$, $p = 0.022$, $r = 0.50$, and wine, $F_{(1, 19)} = 18.61$, $r = 0.70$, were significantly higher than water.
- ✓ For the main effect of imagery the Greenhouse–Geisser estimate of the departure from sphericity was $\varepsilon = 0.75$. The type of imagery also had a significant effect on ratings of the

drinks, $F(1.50, 28.40) = 122.57$. Contrasts revealed that ratings after positive imagery were significantly higher than after neutral imagery, $F(1, 19) = 142.19$, $r = 0.94$. Conversely, ratings after negative imagery were significantly lower than after neutral imagery, $F(1, 19) = 47.07$, $r = 0.84$.

✓ For the interaction the Greenhouse–Geisser estimate of the departure from sphericity was $\varepsilon = 0.80$. There was a significant interaction effect between the type of drink and the type of imagery used, $F(3.19, 60.68) = 17.16$. To break down this interaction, contrasts compared all drink types to their baseline (water) and all imagery types to their baseline (neutral imagery). These contrasts revealed significant interactions when comparing negative imagery to neutral imagery both for beer compared to water, $F(1, 19) = 6.75$, $p = 0.018$, $r = 0.51$, and wine compared to water, $F(1, 19) = 26.91$, $r = 0.77$. The interaction graph shows that these effects reflect that negative imagery (compared to neutral) lowered scores significantly more for water than for beer, and lowered scores significantly more for wine than for water. The remaining contrasts revealed no significant interaction when comparing positive imagery to neutral imagery both for beer compared to water, $F(1, 19) = 1.58$, $p = 0.225$, $r = 0.28$, and wine compared to water, $F(1, 19) = 0.24$, $p = 0.633$, $r = 0.11$. However, these contrasts did yield small to medium effect sizes.

15.18 Brian and Jane's Story ▮▮▮▮

The visit to the hospital had shaken Jane. Brian had offered to go with her, but the last thing she needed was a lecture from her sick mother on how distracting guys were to her studies. Seeing her mother so vulnerable and her father so lost without her had been strange. Their emotional wall was so perfectly formed that it had never occurred to her that her parents

might be deeply connected. Her dad was in pieces mentally, and her mum physically. They'd always seemed superhuman, like their intellect could defeat anything. She'd spent so many years trying to be what she thought they were, but seeing them bought home how little she knew them. She had become obsessed with intellect for the sake of intellect. But seeing her frail, frightened, mum made her feel helpless. All the knowledge she had, and she didn't know how to help her mum. Maybe instead of trying any means to become cleverer, she should put her brainpower to some use in the world.

She was thinking this as she climbed the steps to the basement door, the taste of formaldehyde fresh on her lips. Was this what she became when she spent too much time alone? She locked the door. It felt like it might be for the last time.

It had been a long day. Her legs took her to Brian's apartment. She went in, but was

too exhausted for a deep conversation. 'Talk stats,' she said, and relaxed into the sofa as he recited what he'd recently read on repeated-measures designs.

15.19 What next? ▮▮▮▮

By the age of 16 I had started my first 'serious' band. We stayed together for about 7 years (with the same line-up, and we're still friends now) before Mark (drummer) moved to Oxford, I moved to Brighton to do my PhD, and rehearsing became a mammoth feat of organization. We had a track on a CD, some radio play and transformed from a thrash metal band to a blend of Fugazi, Nirvana and metal. I never split my trousers during a gig again (although I did once split my head open). Why didn't we make it? Well, Mark is an astonishingly good drummer so it wasn't his fault, the other Mark was an extremely good bassist too, and so all

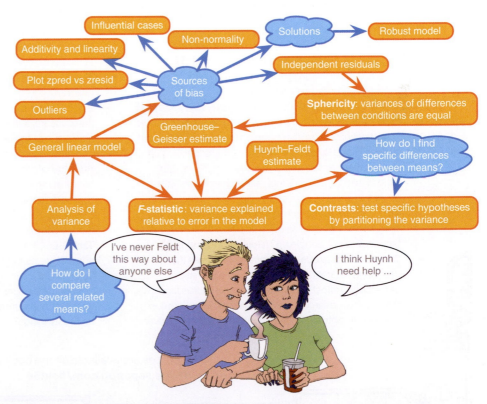

Figure 15.22 What Brian learnt from this chapter

logic points towards the weak link being me. This fact was especially unfortunate given that I had three roles in the band (guitar, singing, songs)—my poor bandmates never stood a chance.☺ I stopped playing music for quite a few years after we split. I still wrote songs (for personal consumption) but the three of us were such close friends that I couldn't bear the thought of playing with other people. At least not for a few years . . .

15.20 Key terms that I've discovered

Compound symmetry

Greenhouse–Geisser estimate

Huynh–Feldt estimate

Lower-bound estimate

Mauchly's test

Repeated-measures ANOVA

Sphericity

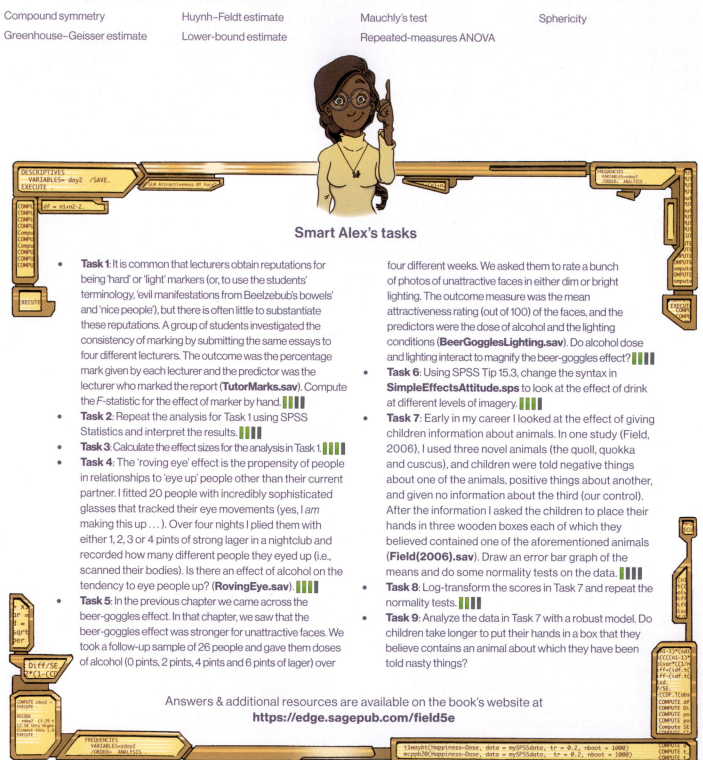

Smart Alex's tasks

- **Task 1**: It is common that lecturers obtain reputations for being 'hard' or 'light' markers (or, to use the students' terminology, 'evil manifestations from Beelzebub's bowels' and 'nice people'), but there is often little to substantiate these reputations. A group of students investigated the consistency of marking by submitting the same essays to four different lecturers. The outcome was the percentage mark given by each lecturer and the predictor was the lecturer who marked the report (**TutorMarks.sav**). Compute the *F*-statistic for the effect of marker by hand. ▮▮▮

- **Task 2**: Repeat the analysis for Task 1 using SPSS Statistics and interpret the results. ▮▮▮

- **Task 3**: Calculate the effect sizes for the analysis in Task 1. ▮▮▮

- **Task 4**: The 'roving eye' effect is the propensity of people in relationships to 'eye up' people other than their current partner. I fitted 20 people with incredibly sophisticated glasses that tracked their eye movements (yes, I *am* making this up . . .). Over four nights I plied them with either 1, 2, 3 or 4 pints of strong lager in a nightclub and recorded how many different people they eyed up (i.e., scanned their bodies). Is there an effect of alcohol on the tendency to eye people up? (**RovingEye.sav**). ▮▮▮

- **Task 5**: In the previous chapter we came across the beer-goggles effect. In that chapter, we saw that the beer-goggles effect was stronger for unattractive faces. We took a follow-up sample of 26 people and gave them doses of alcohol (0 pints, 2 pints, 4 pints and 6 pints of lager) over four different weeks. We asked them to rate a bunch of photos of unattractive faces in either dim or bright lighting. The outcome measure was the mean attractiveness rating (out of 100) of the faces, and the predictors were the dose of alcohol and the lighting conditions (**BeerGogglesLighting.sav**). Do alcohol dose and lighting interact to magnify the beer-goggles effect? ▮▮▮

- **Task 6**: Using SPSS Tip 15.3, change the syntax in **SimpleEffectsAttitude.sps** to look at the effect of drink at different levels of imagery. ▮▮▮

- **Task 7**: Early in my career I looked at the effect of giving children information about animals. In one study (Field, 2006), I used three novel animals (the quoll, quokka and cuscus), and children were told negative things about one of the animals, positive things about another, and given no information about the third (our control). After the information I asked the children to place their hands in three wooden boxes each of which they believed contained one of the aforementioned animals (**Field(2006).sav**). Draw an error bar graph of the means and do some normality tests on the data. ▮▮▮

- **Task 8**: Log-transform the scores in Task 7 and repeat the normality tests. ▮▮▮

- **Task 9**: Analyze the data in Task 7 with a robust model. Do children take longer to put their hands in a box that they believe contains an animal about which they have been told nasty things?

Answers & additional resources are available on the book's website at
https://edge.sagepub.com/field5e

GLM 5: MIXED DESIGNS

16

16.1 What will this chapter tell me?

Most teenagers have anxiety and depression, but I probably had more than my fair share. The all-boys' grammar school that I attended had feasted on my social skills like a parasitic leech. By the time I left school I was a terrified husk of a person. If I had my bandmates with me I had no real problem with playing guitar and 'singing' at people, but speaking to them was another matter entirely. In the band I felt at ease, in the real world I did not. Your 18th birthday is a time of great joy for most people. In the UK, it is a time to cast aside the shackles of childhood and embrace the exciting new world of adult life. You can drink alcohol and vote. Happy days. Your birthday cake might symbolize this happy transition by reflecting one of your great passions. Mine was decorated with a picture of a long-haired person who looked somewhat like me, slitting his wrists. That pretty much sums up the 18-year-old me.

Still, you can't lock yourself in your bedroom with your Iron Maiden albums for ever, and eventually I tried to integrate with society. Between the ages of 16 and 18 this pretty much involved getting drunk. I quickly discovered that getting drunk made it much easier to speak to people, and getting *really* drunk made you unconscious and then the problem of speaking to people went away entirely. This situation was exacerbated by the sudden presence of girls in my social circle. I hadn't seen a girl since Clair Sparks, which was about 7 years earlier. Girls were particularly problematic for the teenage me because not only were you expected to talk to them, but what you said had to be really impressive because then they might become your 'girlfriend', and having one of those meant that you might not be the social outcast that you imagined. The main problem with talking to girls (other than the incapacitating anxiety) was that in 1990 girls didn't like to talk about Iron Maiden—they probably still don't. Speed dating[1] didn't exist back then, but if it had I would have run a mile from it:
it would have been a sick and twisted manifestation of hell on earth for me.

Figure 16.1   My 18th birthday cake

1   In case speed dating goes out of fashion and no one knows what I'm going on about, the idea is that people wanting to find a romantic partner turn up at a venue where there will be lots of other people who also want a romantic partner. One-half of the group sit individually at small tables and the remainder choose a table, get 3 minutes to impress the other person at the table with their tales of heteroscedastic data, then a bell rings and they get up and move to the next table. Having worked around all the tables, the end of the evening is spent either locating the person who made you go gooey inside or avoiding the hideous mutant who was going on about heterosomethingorother.

I would have found it too chilling to contemplate entering a highly pressured social situation where I *had* to think of something witty and amusing to say or face 3 minutes in awkward silence contemplating my eternal loneliness. I feel much the same about giving talks: there's only so much disappointment in people's eyes that I can take. Anyway, perhaps as self-help or something, this chapter is about speed dating—oh, and mixed designs, but if I mention that you'll move swiftly on to the next chapter when the bell rings.

16.2 Mixed designs ▮▮▮▮

If you thought that the previous chapter was bad, I'm about to throw an added complication into the mix. This chapter looks at situations where we combine repeated-measures and independent designs. As if this wasn't bad enough, I'm going to use this as an excuse to show you a design with three independent variables (at this point you should imagine me leaning back in my chair, cross-eyed, dribbling and laughing maniacally). When a design includes some independent variables that were measured using different entities and others that used repeated measures it is called a **mixed design**. It should be obvious that a mixed design requires at least two independent variables, but you can have more complex scenarios too (e.g., two independent measures and one repeated measure, one independent measure and two repeated measures or even two of each). Because by adding independent variables we're simply adding predictors to the linear model, you can have virtually any number of independent variables if your sample size is big enough. However, as we shall see, interaction terms are very difficult to interpret with even three independent variables, so anything more than that really is the road to madness and best avoided.[2] We're still essentially using the linear model, with which you should be really familiar, so I'm going to dodge the theory (i.e., it's too complicated for me to understand) and assume that you can extend what we've learnt up to now to the more complex situation of having three independent variables. Like in the last chapter, because there are repeated measures involved, people typically use an 'ANOVA-style' model. That is, rather than using a flexible multilevel model to incorporate scores within individuals, they use a simpler, less flexible model that requires certain assumptions (sphericity). Therefore, the exact variety of linear model we'll use is sometimes known as **mixed ANOVA**.

16.3 Assumptions in mixed designs ▮▮▮▮

If you have read any of the previous chapters on comparing means you will be sick of me writing that we're using the linear model and so all the sources of potential bias (and counteractive measures) discussed in Chapter 6 apply. But, there you go, I've just written it again. Of course, because mixed designs include both repeated measures and independent measures you have the double whammy of concerning yourself with both homogeneity of variance *and* sphericity. It's enough to make you guzzle the ink from the octopus of inescapable despair. But don't: we can apply the Greenhouse–Geisser correction and forget about sphericity.

The various other woes in Chapter 6 are more troublesome. As we saw in the previous chapter, the Bootstrap... button is absent for repeated-measures designs.

'What about non-parametric tests?' you might ask. You wouldn't be alone: if I had £1 (or $1, €1 or whatever currency you fancy) for every time someone asked me what the non-parametric equivalent of mixed ANOVA was, then I'd have a shiny new drum kit. The short answer is, there isn't one. There is a robust model for mixed designs involving two independent variables that can be done using R (Wilcox, 2017), but it is not currently implemented in the *WRS2* package, which makes it tricky for me to demonstrate using the R extension in IBM SPSS Statistics. So, the options are: (1) learn R; (2) try out some of the stuff in Chapter 6 to shoehorn your data into a reasonable shape; or (3) stick an oxygen tank on your back and start swimming in the sea looking for that octopus . . .

16.4 A speed-dating example ▮▮▮▮

Lots of magazines go on about relationships (or perhaps it's just my wife's copies of *Marie Claire*, which I don't read— honestly). The big topic seems to be how to get a relationship in the first place, and within that lots of discussion of the relative importance of looks, personality, and dating strategies (whether you should 'treat them mean to keep them keen' and all that stuff). Scientists have looked at these issues too. For example, the top three most highly rated attributes of a partner in teenagers are reliability, honesty and kindness (Ha Overbeek, & Engels, 2010). Beyond that, in the same study boys tended to rate attractiveness slightly higher than girls, and girls rate a sense of humor more highly than boys (although both are ranked in the top 10 by both sexes). With regard to dating strategies, Dai, Dong, & Jair, (2014) suggest that if someone is committed to pursuing a relationship with a person who plays hard to get, they will find that person more desirable but less likeable.

---

2 Fans of irony will enjoy the four-way ANOVAs that I conducted in Field and Davey (1999) and Field and Moore (2005).

Table 16.1   Data from **LooksOrPersonality.sav**

Charisma	High Charisma			Some Charisma			Dullard		
Looks	Attractive	Average	Unattractive	Attractive	Average	Unattractive	Attractive	Average	Unattractive
Strategy									
Hard to get	86	84	67	88	69	50	97	48	47
	91	83	53	83	74	48	86	50	46
	89	88	48	99	70	48	90	45	48
	89	69	58	86	77	40	87	47	53
	80	81	57	88	71	50	82	50	45
	80	84	51	96	63	42	92	48	43
	89	85	61	87	79	44	86	50	45
	100	94	56	86	71	54	84	54	47
	90	74	54	92	71	58	78	38	45
	89	86	63	80	73	49	91	48	39
Normal	89	91	93	88	65	54	55	48	52
	84	90	85	95	70	60	50	44	45
	99	100	89	80	79	53	51	48	44
	86	89	83	86	74	58	52	48	47
	89	87	80	83	74	43	58	50	48
	80	81	79	86	59	47	51	47	40
	82	92	85	81	66	47	50	45	47
	97	69	87	95	72	51	45	48	46
	95	92	90	98	64	53	54	53	45
	95	93	96	79	66	46	52	39	47

Imagine a scientist designed a study to look at the interplay between looks, personality and dating strategies on evaluations of a date. She set up a speed-dating night with nine tables at which there sat a 'date'. All the dates were stooges selected to vary in their attractiveness (attractive, average, unattractive), their charisma (high charisma, average charisma, writes statistics books), and also the strategy they were told to employ during the conversation (normal or playing hard to get). The dates were trained before the study to act charismatically to varying degrees, and also how to act in a way that made them seem unobtainable (hard to get) or not. As such, across the nine dates/ stooges there were three attractive people, one of whom acted charismatically, one who acted normally (average) and another who acted like a dullard, and likewise for the three average-looking dates and the three unattractive dates. Therefore, each participant attending a speed-dating night would be exposed to all combinations of attractiveness and charisma (these are repeated measures).[3] Upon arrival participants were randomly assigned a blue or red sticker. For the participants with the red sticker the stooges played hard to get (unobtainable) and for those with a blue sticker they acted normally. Over the course of a few nights 20 people attended, spent 5 minutes with each of the nine 'dates' and then rated how much they'd like to have a proper date with the person as a percentage (100% = 'I'd pay a large sum of money for their phone number', 0% = 'I'd pay a large sum of money for a plane ticket to get me as far away from them as possible').

To be clear, each participant rated nine different people who varied in their attractiveness and charisma. These are two repeated-measures variables: **Looks** (with three levels because the 'date' could be attractive, average or unattractive) and **Charisma** (with three levels because the person could act with high charisma, with some charisma or like a dullard). In addition the 'date' employed a 'hard to get' strategy for half of the participants and acted normally for the rest, so we can include **Strategy** as a between-group variable. The data are in Table 16.1.

3   There was a set of nine male stooges and nine females so that those attending could meet 'dates' of whichever sex interested them.

16.5 Mixed designs using SPSS Statistics ▌▌▌▌

The general procedure for mixed designs is the same as for any other linear model (see Chapter 9). Figure 16.2 shows a simpler overview that highlights some of the specific issues when using a mixed design.

16.5.1 Entering data ▌▌▌▌

We enter these data in the same way as the previous chapter. Remember that each row in the data editor represents a single participant and levels of repeated-measures variables are placed in columns. In this experiment there are nine experimental conditions and so the data need to be entered in nine columns (the format is identical to Table 16.1). You will also need to create a coding variable to enter values for the dating strategy employed by the 'dates'.

16.5.2 Fitting the model ▌▌▌▌

Select *Analyze* ▸ *General Linear Model* ▸ GLM REP *Repeated Measures…* to access the dialog box in Figure 16.4. As in the previous chapter, we first name our repeated-measures variables and specify how many levels they have. We have two repeated measures: **Looks** (attractive, average and unattractive) and **Charisma** (high charisma, some charisma and dullard). In the *Define Factor(s)* dialog box replace the word *factor1* with the word 'Looks' and type '3' into the box labeled *Number of Levels*. Click [Add] to register this variable in the list of repeated-measures variables (it appears as *Looks(3)*). Next, type 'Charisma' into the box labeled *Within-Subject Factor Name* and type '3' into the space labeled *Number of Levels*. Click [Add] and *Charisma(3)* will appear in the list (see Figure 16.4). Click [Define] to go to the main dialog box.

The main dialog box in Figure 16.5 looks the same as in the previous chapter. At the top of the *Within-Subjects Variables* box the two variables we just defined (**Looks** and **Charisma**) are listed and underneath is a series of question marks followed by numbers in parentheses. The numbers in parentheses represent the levels of the independent variables—see the previous chapter for a more detailed explanation. We have two repeated-measures independent variables and so there are two numbers in the brackets. The first number refers to levels of the first variable listed above the box (in this case **Looks**), and the second refers to levels of the second variable listed above the box (in this case **Charisma**). As with the other repeated-measures designs we've come across, we

must assign variables to the question marks. Before we do this assignment, we need to think about contrasts.

Variable name	Variable label
att_high	Attractive and highly charismatic
av_high	Average and highly charismatic
ug_high	Ugly and highly charismatic
att_some	Attractive and some charisma
av_some	Average and some charisma
ug_some	Ugly and some charisma
att_none	Attractive and a dullard
av_none	Average and a dullard
ug_none	Ugly and a dullard

Figure 16.3   Variable names and labels

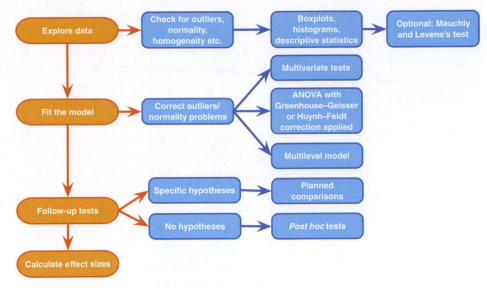

Figure 16.2   The process for analyzing mixed designs

In the data editor create nine variables with the names and variable labels given in Figure 16.3. Create a variable **Strategy** with value labels 0 = normal, 1 = hard to get.

Enter the data as in Table 16.1. If you have problems then use the file **LooksOrPersonality.sav.**

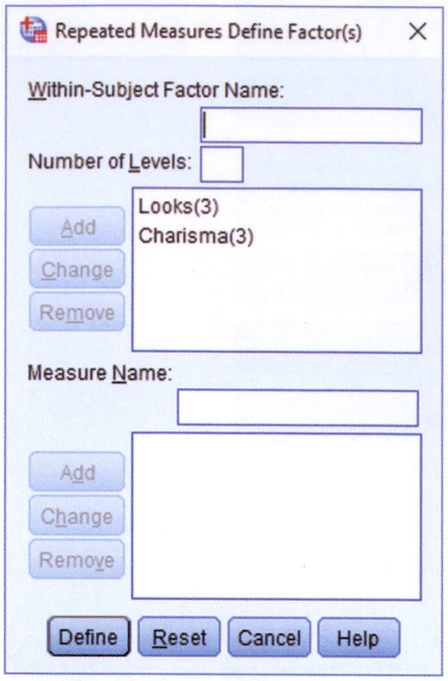

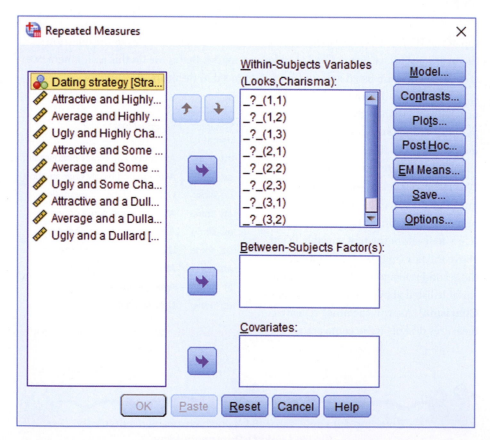

Figure 16.4 The dialog box for defining repeated measures

The first variable, **Looks**, had three conditions: attractive, average and unattractive. It makes sense to compare the attractive and unattractive conditions to the average because the average person represents the norm (although it wouldn't be wrong to, for example, compare attractive and average to unattractive). This comparison could be done using a simple contrast (see Table 12.6) if we assign 'average' to the first or last category. **Charisma** also has a category that represents the norm: some charisma. Again, we could use this as a control against which to compare our two extremes (high charisma and acting like a dullard). As with **Looks**, we could use a simple contrast to compare everything against 'some charisma' if we assign this category to either the first or last level.

Based on the proposed contrasts, it makes sense to have 'average' as level 3 of **Looks**

and 'some charisma' level 3 of **Charisma**. The remaining levels can be assigned arbitrarily. I assigned 'attractive' to level 1 and 'unattractive' to level 2 of **Looks**, and for **Charisma** I assigned 'high charisma' to level 1 and 'none' to level 2. These decisions mean that the variables should be entered as in Figure 16.6. I've deliberately made the order different from how the variables are listed in the data editor to mess with your head. It makes me feel better about the dating ineptitude that the teenage me endured.

So far the procedure has been similar to other factorial repeated-measures designs. However, we have a mixed design, so we also need to specify any between-group variables as well. We do this by dragging **Strategy** to the box labeled *Between-Subjects Factors* (or click). The completed dialog box is shown in Figure 16.7.

16.5.3 Other options

As we saw in the previous chapter, you can only enter custom contrast codes using syntax, but we're going to use the built-in contrasts (see Table 12.6) anyway. Click on Contrasts... to activate the dialog box in Figure 16.8. In the previous section I described why it might be interesting to use the 'average' attractiveness and 'some charisma' as baseline categories for the **Looks** and **Charisma** variables, respectively. We used the *Contrasts*

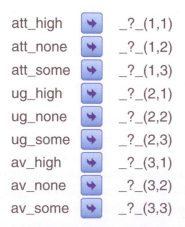

Figure 16.6 Variable allocations for the speed-dating data

Figure 16.5 The main dialog box for mixed designs before completion

dialog box in Sections 13.5.5 and 15.8.2 and so you should know how to select a simple contrast for both **Looks** and **Charisma**. In both cases, we specified the variables such that the control category was the last one; therefore, we can leave the reference category as ⦿ Last. **Strategy** has only two levels (hard to get and normal) so we don't need to specify contrasts for this variable, nor do we need to select *post hoc* tests.[4] Click Continue to return to the main dialog box.

We can plot a graph of the looks × charisma × strategy interaction effect by clicking Plots... to access the dialog box in Figure 16.9. Drag **Looks** to the slot labeled *Horizontal Axis*, **Charisma** to the slot labeled *Separate Line*, and **Strategy** to the slot labeled *Separate Plots*. Click Add to register this plot. Specifying the graph in this way plots the interaction graph for **Looks** and **Charisma**, but produces separate versions for those whose dates used a 'hard to get' strategy and those whose dates did not. Choose a *Line Chart* and *Include Error bars* showing 95% confidence intervals. Finally, as we've done before select *Y axis starts at 0* to scale the graph at zero. You can also use this dialog box to plot graphs of the main effects and the various two-way interactions.

As far as other options are concerned, select the same ones that were chosen for the example in the previous chapter (see Section 15.14.4): it is worth selecting estimated marginal means for all effects (because these values will help you to understand any significant effects). If you must, select ☑ Homogeneity tests.

16.6 Output for mixed factorial designs ▌▌▌▌

Output 16.1 contains a table listing the repeated-measures variables from the data editor and the level of each independent

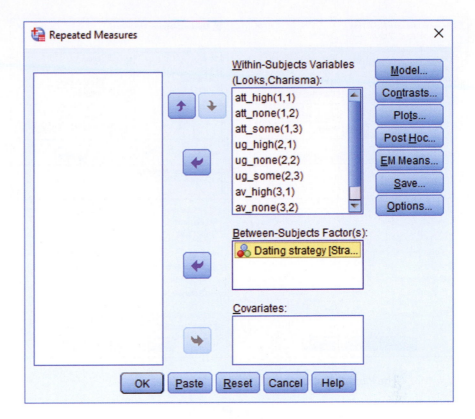

Figure 16.7 The main dialog box for mixed designs after completion

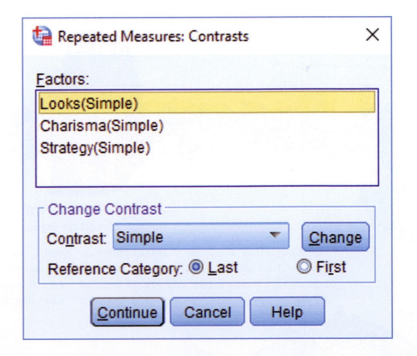

Figure 16.8

---

4 If, for your own data, you want *post hoc* tests click Post Hoc... to activate the *post hoc* test dialog box, which can be used as explained in Section 12.6.3.

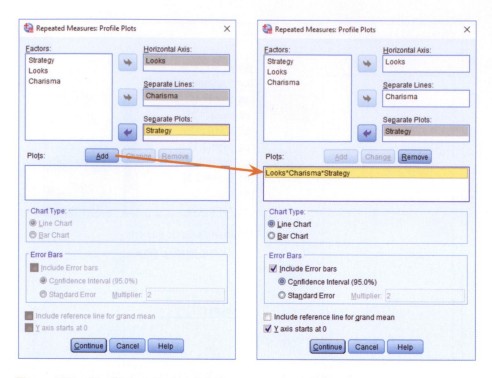

Figure 16.9 The *Plots* dialog box for a three-way mixed design

Output 16.2 shows information about sphericity. Based on what you have already learnt, what would you conclude from this information?

variable that they represent. A second table contains descriptive statistics (mean and standard deviation) for each of the nine repeated measures conditions split according to whether participants sat with dates who played hard to get or not.

Output 16.2 shows information about sphericity for each of the three repeated-measures effects in the model. Although I advised correcting for sphericity by default, the estimates show very little deviation from sphericity (the Huynh–Feldt estimates are all 1, which equates to spherical data). Given that the Huynh–Feldt estimates show no deviation form sphericity it's reasonable not to correct for it, but it's also the case that correcting (e.g., by Greenhouse–Geisser) will have little impact (these estimates too are all close to 1) so we may as well do it. If you have more enthusiasm for Mauchly's test than I do you might note that all the values in the column labeled *Sig.* are above 0.05, indicating no significant departures from sphericity.

Output 16.3 contains the *F*-statistics. The version of this table that you will see will

Oditi's Lantern
Mixed designs

'I, Oditi, may be low on attractiveness but I am high on charisma. I'm high on a few other things too. Look deep into my charming eyes and you will find that you want to join the cult of undiscovered numerical truths. Our next lesson is mixed designs, so stare into my lantern and immerse yourself ever deeper in the cult. You will awaken with a strange love of three-way interactions, and a desire to do only as I say.'

look a lot more hideous, but I have used the instructions in SPSS Tip 15.4 to hide the values that I don't want to see. The table is split into sections for each of the effects in the model and their associated error terms. The interactions between the between-groups variable of strategy and the repeated-measures effects are included in this table also. Working down from the top of the table, we find significant effects (the value in the column *Sig.* is less than 0.05) of **Looks**, the **Looks** × **Strategy** interaction, **Charisma**, the **Charisma** × **Strategy** interaction, the **Looks** × **Charisma** interaction and the **Looks** × **Charisma** × **Strategy** interaction. Everything, basically. You wouldn't normally be interested in main effects when there are significant interactions, but for completeness we'll interpret each effect in turn, starting with the main effect of **Strategy**.

16.6.1 The main effect of strategy ▌▌▌

Before looking at the main effect of strategy, some people (but not me) would use Levene's test to check the assumption of homogeneity of variance (see Section 6.11.2).

Output 16.4 shows Levene's test for whether the variances were equivalent in the hard to get and normal conditions across all nine combined levels of the repeated-measures variables; because all significance values are greater than 0.05, variances are homogeneous for all levels of the repeated-measures variables.

The main effect of strategy is listed separately from the repeated-measures effects in Output 16.5. It had a non-significant effect on ratings of dates because the significance of 0.946 is greater than the standard cut-off of 0.05. This effect tells us that if we ignore all other variables, ratings were equivalent regardless of whether the date adopted a hard to get persona or not. If you requested *Estimated Marginal Means* in the options (I'll assume you did from now

Within-Subjects Factors

Measure: MEASURE_1

Looks	Charisma	Dependent Variable
1	1	att_high
	2	att_none
	3	att_some
2	1	ug_high
	2	ug_none
	3	ug_some
3	1	av_high
	2	av_none
	3	av_some

Descriptive Statistics

	Dating strategy	Mean	Std. Deviation	N
Attractive and Highly Charismatic	Normal	89.60	6.637	10
	Hard to get	88.30	5.697	10
	Total	88.95	6.057	20
Attractive and a Dullard	Normal	51.80	3.458	10
	Hard to get	87.30	5.438	10
	Total	69.55	18.743	20
Attractive and Some Charisma	Normal	87.10	6.806	10
	Hard to get	88.50	5.740	10
	Total	87.80	6.170	20
Ugly and Highly Charismatic	Normal	86.70	5.438	10
	Hard to get	56.80	5.731	10
	Total	71.75	16.274	20
Ugly and a Dullard	Normal	46.10	3.071	10
	Hard to get	45.80	3.584	10
	Total	45.95	3.252	20
Ugly and Some Charisma	Normal	51.20	5.453	10
	Hard to get	48.30	5.376	10
	Total	49.75	5.476	20
Average and Highly Charismatic	Normal	88.40	8.329	10
	Hard to get	82.80	7.005	10
	Total	85.60	8.022	20
Average and a Dullard	Normal	47.00	3.742	10
	Hard to get	47.80	4.185	10
	Total	47.40	3.885	20
Average and Some Charisma	Normal	68.90	5.953	10
	Hard to get	71.80	4.417	10
	Total	70.35	5.314	20

Output 16.1

Mauchly's Test of Sphericity[a]

Measure: MEASURE_1

Within Subjects Effect	Mauchly's W	Approx. Chi-Square	df	Sig.	Epsilon[b] Greenhouse-Geisser	Huynh-Feldt	Lower-bound
Looks	.960	.690	2	.708	.962	1.000	.500
Charisma	.929	1.246	2	.536	.934	1.000	.500
Looks * Charisma	.613	8.025	9	.534	.799	1.000	.250

Tests the null hypothesis that the error covariance matrix of the orthonormalized transformed dependent variables is proportional to an identity matrix.

a. Design: Intercept + Strategy
Within Subjects Design: Looks + Charisma + Looks * Charisma

b. May be used to adjust the degrees of freedom for the averaged tests of significance. Corrected tests are displayed in the Tests of Within-Subjects Effects table.

Output 16.2

Tests of Within-Subjects Effects

Measure: MEASURE_1
Greenhouse-Geisser

Source	Type III Sum of Squares	df	Mean Square	F	Sig.
Looks	20779.633	1.923	10803.275	423.733	.000
Looks * Strategy	3944.100	1.923	2050.527	80.427	.000
Error(Looks)	882.711	34.622	25.496		
Charisma	23233.600	1.868	12437.761	328.250	.000
Charisma * Strategy	4420.133	1.868	2366.252	62.449	.000
Error(Charisma)	1274.044	33.624	37.891		
Looks * Charisma	4055.267	3.197	1268.295	36.633	.000
Looks * Charisma * Strategy	2669.667	3.197	834.945	24.116	.000
Error(Looks*Charisma)	1992.622	57.554	34.622		

Output 16.3

SELF TEST
What is the difference between a main effect and an interaction?

SELF TEST
Based on Output 16.4, was the assumption of homogeneity of variance met?

on) you will get the table in Figure 16.10. I've also included a plot of these means. It is clear from this graph that, overall, ratings of dates playing hard to get were equivalent to dates who were not.

16.6.2 The main effect of looks ▋▋▋▋

Output 16.3 showed a significant main effect of looks, $F(1.92, 34.62) = 423.73$, $p < 0.001$, which means that if we ignore all other variables, ratings of attractive, average and unattractive dates differed. Figure 16.11

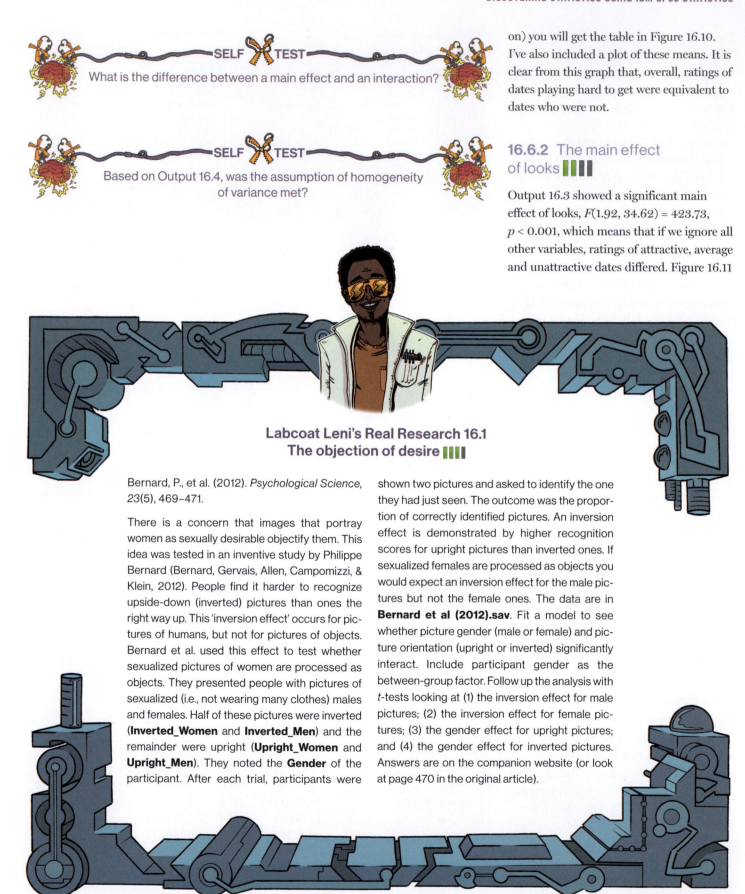

Labcoat Leni's Real Research 16.1
The objection of desire ▋▋▋▋

Bernard, P., et al. (2012). *Psychological Science,* *23*(5), 469–471.

There is a concern that images that portray women as sexually desirable objectify them. This idea was tested in an inventive study by Philippe Bernard (Bernard, Gervais, Allen, Campomizzi, & Klein, 2012). People find it harder to recognize upside-down (inverted) pictures than ones the right way up. This 'inversion effect' occurs for pictures of humans, but not for pictures of objects. Bernard et al. used this effect to test whether sexualized pictures of women are processed as objects. They presented people with pictures of sexualized (i.e., not wearing many clothes) males and females. Half of these pictures were inverted (**Inverted_Women** and **Inverted_Men**) and the remainder were upright (**Upright_Women** and **Upright_Men**). They noted the **Gender** of the participant. After each trial, participants were

shown two pictures and asked to identify the one they had just seen. The outcome was the proportion of correctly identified pictures. An inversion effect is demonstrated by higher recognition scores for upright pictures than inverted ones. If sexualized females are processed as objects you would expect an inversion effect for the male pictures but not the female ones. The data are in **Bernard et al (2012).sav**. Fit a model to see whether picture gender (male or female) and picture orientation (upright or inverted) significantly interact. Include participant gender as the between-group factor. Follow up the analysis with *t*-tests looking at (1) the inversion effect for male pictures; (2) the inversion effect for female pictures; (3) the gender effect for upright pictures; and (4) the gender effect for inverted pictures. Answers are on the companion website (or look at page 470 in the original article).

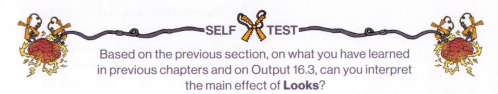

SELF TEST

Based on the previous section, on what you have learned in previous chapters and on Output 16.3, can you interpret the main effect of **Looks**?

shows the *Estimated Marginal Means* and a plot of them. The levels of **Looks** are labeled as 1, 2 and 3, and it's down to you to remember how you entered the variables (or refer to Output 16.1). If you assigned variables as I did then level 1 is attractive, level 2 is unattractive and level 3 is average. From this table and plot you can see that as attractiveness falls, the mean rating falls too. This main effect seems to reflect that the raters were more likely to express a greater interest in going out with attractive people than with average or unattractive people. However, contrasts will help us to understand exactly what's going on.

Output 16.6 shows the contrasts that we requested. For the time being, just look at the row labeled *Looks*. Remember that we did a simple contrast, and so we get a contrast comparing level 1 to level 3, and then comparing level 2 to level 3; because of the order in which we entered the variables, these contrasts represent attractive compared to average (level 1 vs. level 3) and unattractive compared to average (level 2 vs. level 3). The values of F for each contrast, and their related significance values, tell us that the main effect of **Looks** represented the fact that attractive dates were rated significantly higher than average dates, $F_{(1, 18)} = 226.99$, $p < 0.001$, and average dates were rated significantly higher than unattractive ones, $F_{(1, 18)} = 160.07$, $p < 0.001$.

16.6.3 The main effect of charisma

In Output 16.3 there was a significant main effect of charisma, $F_{(1.87, 33.62)} = 328.25$, $p < 0.001$, which tells us that if we

ignore all other variables, ratings for highly charismatic, a bit charismatic and dull dates differed. Figure 16.12 shows the estimated marginal means from the

output together with a plot. Again, the levels of **Charisma** are labeled as 1, 2 and 3. If you followed what I did then level 1 is high charisma, level 2 is dullard and level 3 is some charisma. This main effect seems to reflect that as charisma declines, the mean rating of the date falls too: raters expressed a greater interest in going out with charismatic people than average people or dullards.

Levene's Test of Equality of Error Variances[a]

	F	df1	df2	Sig.
Attractive and Highly Charismatic	1.131	1	18	.302
Attractive and a Dullard	1.949	1	18	.180
Attractive and Some Charisma	.599	1	18	.449
Ugly and Highly Charismatic	.005	1	18	.945
Ugly and a Dullard	.082	1	18	.778
Ugly and Some Charisma	.124	1	18	.729
Average and Highly Charismatic	.102	1	18	.753
Average and a Dullard	.004	1	18	.950
Average and Some Charisma	1.763	1	18	.201

Tests the null hypothesis that the error variance of the dependent variable is equal across groups.

a. Design: Intercept + Strategy
Within Subjects Design: Looks + Charisma + Looks * Charisma

Output 16.4

Tests of Between-Subjects Effects

Measure: MEASURE_1
Transformed Variable: Average

Source	Type III Sum of Squares	df	Mean Square	F	Sig.
Intercept	94027.756	1	94027.756	20036.900	.000
Strategy	.022	1	.022	.005	.946
Error	84.469	18	4.693		

Output 16.5

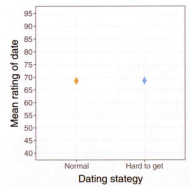

1. Dating strategy

Measure: MEASURE_1

			95% Confidence Interval	
Dating strategy	Mean	Std. Error	Lower Bound	Upper Bound
Normal	68.533	.685	67.094	69.973
Hard to get	68.600	.685	67.161	70.039

Figure 16.10 Means and graph of the main effect of **Strategy**

525

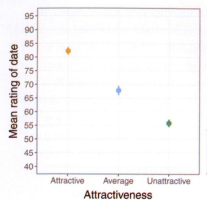

Figure 16.11 Means and graph of the main effect of **Looks**

2. Looks

Measure: MEASURE_1

Looks	Mean	Std. Error	95% Confidence Interval Lower Bound	95% Confidence Interval Upper Bound
1	82.100	.652	80.729	83.471
2	55.817	.651	54.449	57.184
3	67.783	.820	66.061	69.505

Tests of Within-Subjects Contrasts

Measure: MEASURE_1

Source	Looks	Charisma	Type III Sum of Squares	df	Mean Square	F	Sig.
Looks	Level 1 vs. Level 3		4099.339	1	4099.339	226.986	.000
	Level 2 vs. Level 3		2864.022	1	2864.022	160.067	.000
Looks * Strategy	Level 1 vs. Level 3		781.250	1	781.250	43.259	.000
	Level 2 vs. Level 3		540.800	1	540.800	30.225	.000
Error(Looks)	Level 1 vs. Level 3		325.078	18	18.060		
	Level 2 vs. Level 3		322.067	18	17.893		
Charisma		Level 1 vs. Level 3	3276.800	1	3276.800	109.937	.000
		Level 2 vs. Level 3	4500.000	1	4500.000	227.941	.000
Charisma * Strategy		Level 1 vs. Level 3	810.689	1	810.689	27.199	.000
		Level 2 vs. Level 3	665.089	1	665.089	33.689	.000
Error(Charisma)		Level 1 vs. Level 3	536.511	18	29.806		
		Level 2 vs. Level 3	355.356	18	19.742		
Looks * Charisma	Level 1 vs. Level 3	Level 1 vs. Level 3	3976.200	1	3976.200	21.944	.000
		Level 2 vs. Level 3	441.800	1	441.800	4.091	.058
	Level 2 vs. Level 3	Level 1 vs. Level 3	911.250	1	911.250	6.231	.022
		Level 2 vs. Level 3	7334.450	1	7334.450	88.598	.000
Looks * Charisma * Strategy	Level 1 vs. Level 3	Level 1 vs. Level 3	168.200	1	168.200	.928	.348
		Level 2 vs. Level 3	6552.200	1	6552.200	60.669	.000
	Level 2 vs. Level 3	Level 1 vs. Level 3	1711.250	1	1711.250	11.701	.003
		Level 2 vs. Level 3	110.450	1	110.450	1.334	.263
Error(Looks*Charisma)	Level 1 vs. Level 3	Level 1 vs. Level 3	3261.600	18	181.200		
		Level 2 vs. Level 3	1944.000	18	108.000		
	Level 2 vs. Level 3	Level 1 vs. Level 3	2632.500	18	146.250		
		Level 2 vs. Level 3	1490.100	18	82.783		

Output 16.6

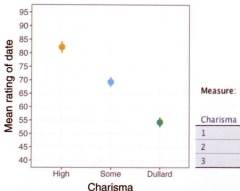

Figure 16.12 Means and graph of the main effect of **Charisma**

3. Charisma

Measure: MEASURE_1

Charisma	Mean	Std. Error	95% Confidence Interval Lower Bound	95% Confidence Interval Upper Bound
1	82.100	1.010	79.978	84.222
2	54.300	.573	53.096	55.504
3	69.300	.732	67.763	70.837

We requested simple contrasts (the row labeled *Charisma* in Output 16.6), and because of the order that we entered variables these contrasts represent high charisma compared to some charisma (level 1 vs. level 3) and no charisma compared to some charisma (level 2 vs. level 3). These contrasts tell us that the main effect of **Charisma** is that highly charismatic dates were rated significantly higher than dates with some charisma, $F(1, 18) = 109.94$, $p < 0.001$, and dates with some charisma were rated significantly higher than dullards, $F(1, 18) = 227.94$, $p < 0.001$.

16.6.4 The interaction between strategy and looks

Strategy significantly interacted with the attractiveness of the date, $F(1.92, 34.62) = 80.43$, $p < 0.001$ (Output 16.3). This effect tells us that the profile of ratings across dates of different attractiveness was different depending on whether or not they played hard to get. The estimated marginal means and interaction graph (you can obtain a version of this using the dialog box in Figure 16.9) are in Figure 16.13. The graph shows that for average-looking dates it doesn't make a difference whether they played hard to get (the orange and blue dots are in a similar location). For attractive dates, ratings were higher when the date played hard to get (blue dot) than when they didn't (orange dot), and for unattractive dates the opposite was true—ratings were lower when dates played hard to get. In short, playing hard to get only has an effect at the extremes of attractiveness. Another way to look at this is the slope of the lines: when dates played hard to get the slope (blue line) is steeper than when they didn't (orange line), implying that attractiveness has a greater impact on ratings when dates play hard to get. This interaction can be clarified using the contrasts in Output 16.6.

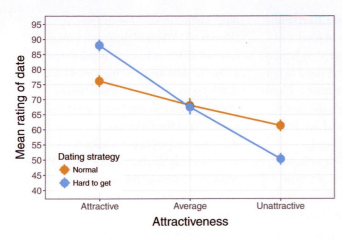

4. Dating strategy * Looks

Measure: MEASURE_1

Dating strategy	Looks	Mean	Std. Error	95% Confidence Interval	
				Lower Bound	Upper Bound
Normal	1	76.167	.923	74.228	78.105
	2	61.333	.921	59.399	63.267
	3	68.100	1.159	65.665	70.535
Hard to get	1	88.033	.923	86.095	89.972
	2	50.300	.921	48.366	52.234
	3	67.467	1.159	65.031	69.902

Figure 16.13   Means and graph of the **Strategy** × **Looks** interaction

The first contrast for the interaction term looks at level 1 of **Looks** (attractive) compared to level 3 (average), comparing playing hard to get to normal. This contrast is highly significant, $F(1, 18) = 43.26$, $p < 0.001$, suggesting that the increased interest in attractive dates compared to average-looking dates found when dates played hard to get is significantly more than when they acted normally. So, in Figure 16.13 the slope of the blue line (hard to get) between attractive dates and average dates is steeper than the comparable orange line (normal). The preferences for attractive dates, compared to average-looking dates, are greater when they play hard to get than when they don't.

The second contrast, which compares playing hard to get to normal at level 2 of looks (unattractive) relative to level 3 (average) is also highly significant, $F(1, 18) = 30.23$, $p < 0.001$. This contrast tells us that the decreased interest in unattractive dates compared to average-looking dates found when dates played hard to get is significantly more than when they did not. In Figure 16.13 the slope of the blue line between the unattractive and average dates is steeper than the corresponding orange line. The preferences for average-looking dates, compared to unattractive dates, are greater when they play hard to get than when they don't.

16.6.5 The interaction between strategy and charisma ▮▮▮▮

Output 16.3 showed that strategy significantly interacted with how charismatic the date was, $F(1.87, 33.62) = 62.45$, $p < 0.001$. This effect means that the profile of ratings across dates of different levels of charisma was influenced by the dating strategy employed. The estimated marginal means and plot in Figure 16.14 show almost the reverse pattern to the **Strategy** × **Looks** interaction. For dates with normal amounts of charisma the dating strategy they adopted had little impact (the blue and orange dots

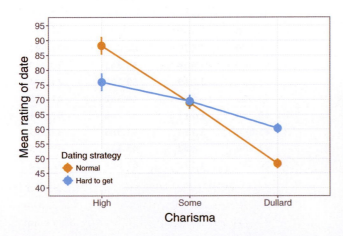

5. Dating strategy * Charisma

Measure: MEASURE_1

Dating strategy	Charisma	Mean	Std. Error	95% Confidence Interval	
				Lower Bound	Upper Bound
Normal	1	88.233	1.428	85.233	91.234
	2	48.300	.810	46.598	50.002
	3	69.067	1.035	66.893	71.240
Hard to get	1	75.967	1.428	72.966	78.967
	2	60.300	.810	58.598	62.002
	3	69.533	1.035	67.360	71.707

Figure 16.14   Means and graph of the **Strategy** × **Charisma** interaction

coincide). However, highly charismatic dates were rated higher when they acted normally compared to when they played hard to get (the blue dot is below the orange). Conversely, dull dates were rated higher when they played hard to get compared to when they acted normally (the blue dot is above the orange). Although interest in dating decreases as charisma decreases, this decrease is less pronounced when the dates play hard to get.

We can break this interaction down using the contrasts in Output 16.6. The first one, which looks at level 1 of **Charisma** (high charisma) compared to level 3 (some charisma), for playing hard to get relative to normal, is highly significant, $F(1, 18) = 27.20$, $p < 0.001$. This result tells us that the increased interest in highly charismatic dates compared to averagely charismatic dates found when dates acted normally is significantly more than when they played hard to get. In Figure 16.14 the slope of the orange line (hard to get) between high charisma and some charisma is steeper than the corresponding blue line (normal). The preferences for very charismatic dates, compared to averagely charismatic dates, are smaller when dates play hard to get.

The second contrast for the **Charisma × Strategy** interaction looks at level 2 of **Charisma** (dullard) compared to level 3 (some charisma), for playing hard to get relative to normal. This contrast is highly significant, $F(1, 18) = 33.69$, $p < 0.001$, suggesting that the decreased interest in dull dates compared to averagely charismatic dates found is significantly less when dates play hard to get than when they act normally. In Figure 16.14 the slope of the orange line (normal) between some charisma and dullard is steeper than the corresponding blue line (hard to get): the preferences for dates with some charisma over dullards is greater when dates act normally than when they play hard to get.

16.6.6 The interaction between looks and charisma

There was a significant **Looks × Charisma** interaction, $F(3.20, 57.55) = 36.63$, $p < 0.001$ (Output 16.3). This effect tells us that the profile of ratings across dates of different levels of charisma was different for attractive, average and unattractive dates. We can unpick this interaction using the estimated marginal means, a plot (use the dialog box in Figure 16.9 to get a similar

one), and contrasts. The graph (Figure 16.15) shows the mean ratings of dates of different levels of attractiveness when the date also had high levels of charisma (orange line), some charisma (blue line) and no charisma (green line). Look first at the difference between attractive and average-looking dates. The interest in highly charismatic dates doesn't change (the orange line is more or less flat between these two points), but for dates with some charisma or no charisma interest levels decline (the blue and green lines slope down). If you have lots of charisma you can get away with being average-looking and people will still want to date you. Now look at the difference between average-looking and unattractive dates. A different pattern is observed: for dates with no charisma there is little difference between unattractive and average-looking people (the green line is flat) but for those with any charisma, there is a decline in interest if you're unattractive (the orange and blue lines slope down). If you're a dullard you need to be really attractive before people want to date you, and if you're unattractive then having 'some' charisma won't help you much.

The contrasts in Output 16.6 help to pick apart this interaction. The first contrast for the **Looks × Charisma** interaction

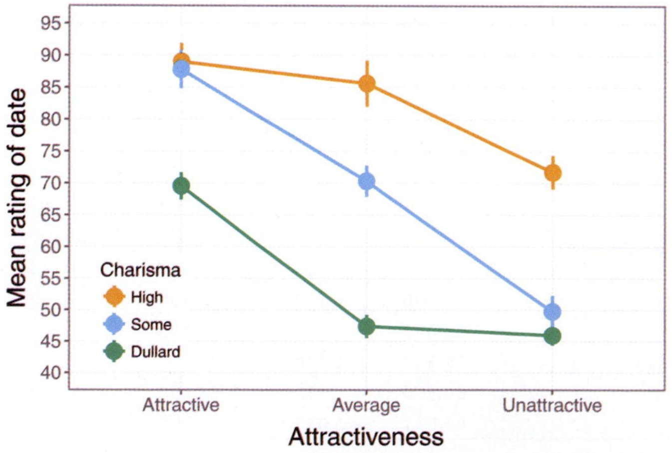

6. Looks * Charisma

Measure: MEASURE_1

Looks	Charisma	Mean	Std. Error	95% Confidence Interval Lower Bound	Upper Bound
1	1	88.950	1.383	86.045	91.855
	2	69.550	1.019	67.409	71.691
	3	87.800	1.408	84.842	90.758
2	1	71.750	1.249	69.126	74.374
	2	45.950	.746	44.382	47.518
	3	49.750	1.211	47.206	52.294
3	1	85.600	1.721	81.985	89.215
	2	47.400	.888	45.535	49.265
	3	70.350	1.172	67.888	72.812

Figure 16.15 Means and graph of the **Looks × Charisma** interaction

investigates level 1 of **Looks** (attractive) compared to level 3 (average-looking), for level 1 of **Charisma** (high charisma) relative to level 3 (some charisma). This is like asking 'Is the difference between high charisma and some charisma the same for attractive people and average-looking people?' The best way to understand this contrast is to focus on the relevant bit of the interaction graph in Figure 16.15, which I have reproduced in Figure 16.16 (top left). Interest (as indicated by high ratings) in attractive dates was the same regardless of whether they had high or some charisma; however, for average-looking dates, there was more interest when that person had high charisma rather than some. The contrast is highly significant, $F(1, 18) = 21.94$, $p < 0.001$, and tells us that as attractiveness is reduced there is a significantly greater decline in interest when charisma is average compared to when it is high.

The second contrast asks the question 'Is the difference between no charisma and some charisma the same for attractive people and average-looking people? It explores level 1 of **Looks** (attractive) compared to Level 3 (average-looking), for level 2 of **Charisma** (dullard) relative to level 3 (some charisma). We can again focus on the relevant part of the interaction graph (Figure 16.15) which is reproduced in Figure 16.16 (top right). This graph shows that interest in attractive dates was higher when they had some charisma (blue) than when they were a dullard (green); the same is also true for average-looking dates. The two lines are fairly parallel, which is reflected in the non-significant contrast, $F(1, 18) = 4.09$, $p = 0.058$. It seems that as the attractiveness of dates is reduced there is a decline in interest both when charisma is average and when the date is dull.

The third contrast investigates level 2 of **Looks** (unattractive) relative to level 3

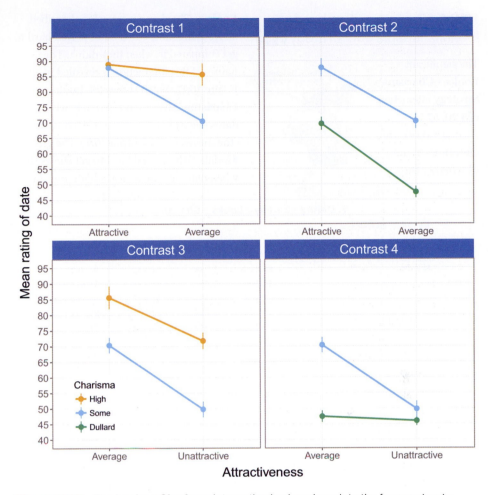

Figure 16.16   The **Looks** × **Charisma** interaction broken down into the four contrasts

(average-looking), comparing level 1 of **Charisma** (high charisma) to level 3 (some charisma). This contrast asks 'Is the difference between high charisma and some charisma the same for unattractive people and average-looking people?' The relevant part of the interaction graph is shown in Figure 16.16 (bottom left). Interest in dating decreases from average-looking dates to unattractive ones in dates with both high and some charisma; however, this fall is significantly greater in the low-charisma dates (the blue line is slightly steeper than the orange), $F(1, 18) = 6.23$, $p = 0.022$. As dates' attractiveness is reduced there is a significantly greater decline in interest when dates have some charisma compared to when they have a lot.

The final contrast addresses the question 'Is the difference between no charisma and some charisma the same for unattractive people and average-looking people?' It compares level 2 of **Looks** (unattractive) to level 3 (average-looking), in level 2 of **Charisma** (dullard) relative to level 3 (some charisma). The relevant part of the interaction graph is shown in Figure 16.16 (bottom right). For average-looking dates, ratings were higher when they had some charisma than when they were a dullard, but for unattractive dates the ratings were roughly the same regardless of the level of charisma. This contrast is highly significant, $F(1, 18) = 88.60$, $p < 0.001$.

16.6.7 The interaction between looks, charisma and strategy ▮▮▮▮

The significant **Looks × Charisma × Strategy** interaction, $F(3.20, 57.55) = 24.12, p < 0.001$ (in Output 16.3), tells us whether the

How do I interpret a three-way interaction?

Looks × Charisma interaction described above is the same when dates played hard to get compared to when they didn't. The nature of this interaction is revealed in Figure 16.17, which shows the **Looks × Charisma** interaction separately when dates played hard to get and acted normally (the means are in Output 16.7). The graph for dates who played hard to get shows that when dates are attractive, a high interest

was expressed regardless of charisma levels (the orange, blue and green lines meet). At the opposite end of the attractiveness scale, when a date is unattractive, regardless of charisma, very little interest is expressed (ratings are all low). If the date plays hard to get, the only time charisma makes a difference is if the date is average-looking, in which case high charisma (orange) boosts interest, being a dullard (green) reduces interest, and having 'some' charisma leaves things somewhere in between. The take-home message is that playing hard to get only works if you're averagely attractive: if you're highly charismatic it will boost your appeal, but it's a disastrous strategy if you're dull.[5]

The picture when the date doesn't play hard to get (acts normally) is different. If someone has high levels of charisma then what they look like won't affect interest in them (the orange line is relatively flat). At the other extreme, if the date is a dullard, then they will have very little interest expressed in them regardless of how attractive they are (the green line is relatively flat). The only time attractiveness makes a difference is when someone has an average amount of charisma (the blue line), in which case being attractive boosts interest, and being unattractive reduces it. If you don't play hard to get and you are averagely attractive then you can influence others' interest you with your charisma.

Again, we can use contrasts to further break this interaction down (Output 16.6). These contrasts are similar to those for the **Looks × Charisma** interaction, but they now take into account the effect of dating strategy as well. The first contrast for the **Looks × Charisma × Strategy** interaction explores level 1 of **Looks** (attractive) relative to level 3 (average-looking), when level 1 of **Charisma** (high charisma) is compared to level 3 (some charisma), when dates played hard to get relative to when they didn't, $F(1, 18) = 0.93, p = 0.348$.

7. Dating strategy * Looks * Charisma

Measure: MEASURE_1

Dating strategy	Looks	Charisma	Mean	Std. Error	95% Confidence Interval	
					Lower Bound	Upper Bound
Normal	1	1	89.600	1.956	85.491	93.709
		2	51.800	1.441	48.773	54.827
		3	87.100	1.991	82.917	91.283
	2	1	86.700	1.767	82.989	90.411
		2	46.100	1.055	43.883	48.317
		3	51.200	1.712	47.603	54.797
	3	1	88.400	2.434	83.287	93.513
		2	47.000	1.255	44.363	49.637
		3	68.900	1.657	65.418	72.382
Hard to get	1	1	88.300	1.956	84.191	92.409
		2	87.300	1.441	84.273	90.327
		3	88.500	1.991	84.317	92.683
	2	1	56.800	1.767	53.089	60.511
		2	45.800	1.055	43.583	48.017
		3	48.300	1.712	44.703	51.897
	3	1	82.800	2.434	77.687	87.913
		2	47.800	1.255	45.163	50.437
		3	71.800	1.657	68.318	75.282

Output 16.7

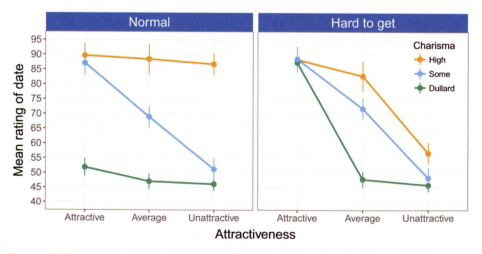

Figure 16.17 Graphs showing the **Looks × Charisma** interaction for different dating strategies. Lines represent high charisma (orange), some charisma (blue) and no charisma (green)

---

5 These data are made up, so please don't base your life decisions on this example 😊

The relevant parts of Figure 16.17 are shown in the first column of Figure 16.18. It seems that interest in dating (as indicated by high ratings) attractive dates was the same regardless of whether they had high or average charisma (the blue and orange dots are in a similar place). However, for average-looking dates, there was more interest when that person had high charisma rather than some charisma (the blue dot is lower than the orange dot). The non-significance of this contrast indicates that this pattern of results is very similar when dates played hard to get and when they didn't.

The second contrast explores level 1 of **Looks** (attractive) relative to level 3 (average-looking), when level 2 of **Charisma** (dullard) is compared to level 3 (some charisma), when dates played hard to get relative to when they didn't. The relevant means are shown in the second column of Figure 16.18. The contrast is significant, $F(1, 18) = 60.67$, $p < 0.001$, which reflects the fact that the pattern of means is different when dates played hard to get compared to when they didn't. First, if we look at average-looking dates, more interest was expressed when the date has some charisma than when they have none, and this is true whether or not dates played hard to get (the distance between the blue and green lines is about the same in the two dating strategy groups). So, the difference created by playing hard to get doesn't appear to be here. Now look at attractive dates. When dates played hard to get (bottom) the interest in the date is high regardless of their charisma (the lines meet). However, when dates acted normally (top) interest in dating an attractive person is much lower if they are a dullard (the green dot is much lower than the blue).

Another way to look at it is that for dates with some charisma, the reduction in interest as attractiveness goes down is about the same regardless of whether they played hard to get (the blue lines have the same slope). However, for dates who are dullards, the decrease in interest if these dates are average-looking rather than attractive is much more dramatic if they play hard to get (the green line is steeper in the hard to get group).

The third contrast was also significant, $F(1, 18) = 11.70$, $p = 0.003$. This contrast compares level 2 of **Looks** (unattractive) to level 3 (average-looking), in level 1 of **Charisma** (high charisma) relative to level 3 (some charisma), when dates played hard to get relative to when they didn't. The third column of Figure 16.18 shows the relevant means. First, let's look at when dates played hard to get (bottom). As attractiveness goes down, so does interest when the date has high charisma

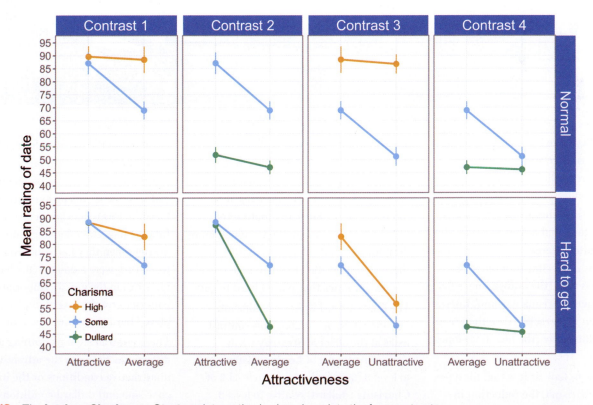

Figure 16.18 The **Looks** × **Charisma** × **Strategy** interaction broken down into the four contrasts

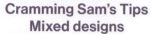

Cramming Sam's Tips
Mixed designs

- Mixed designs compare several means when there are two or more independent variables, and at least one of them has been measured using the same entities and at least one other has been measured using different entities.

- Correct for deviations from sphericity for the repeated-measures variable(s) by routinely interpreting the Greenhouse–Geisser corrected effects. (Some people do this only if Mauchly's test is significant, but this approach is problematic because the results of the test depend on the sample size.)

- The table labeled *Tests of Within-Subjects Effects* shows the *F*-statistic(s) for any repeated-measures variables and all of the interaction effects. For *each* effect, read the row labeled *Greenhouse–Geisser* or *Huynh–Feldt* (read the

previous chapter to find out the relative merits of the two procedures). If the value in the *Sig.* column is less than 0.05 then the means are significantly different.

- The table labeled *Tests of Between-Subjects Effects* shows the *F*-statistic(s) for any between-group variables. If the value in the *Sig.* column is less than 0.05 then the means of the groups are significantly different.

- Break down the main effects and interaction terms using contrasts. These contrasts appear in the table labeled *Tests of Within-Subjects Contrasts*; again look to the columns labeled *Sig.* to discover if your comparisons are significant (they are if the significance value is less than 0.05).

- Look at the means—or, better still, draw graphs—to help you interpret the contrasts.

and when they have some charisma (the slopes of the orange and blue lines are similar). So, regardless of charisma, there is a similar reduction in interest as attractiveness declines. Now let's look at when the dates acted normally (top). The picture is quite different: when charisma is high, there is no decline in interest as attractiveness falls (the orange line is flat); however, when charisma is 'some', interest is lower in an unattractive date than in an average-looking date (the blue line slopes down).

Another way to look at it is that for dates with some charisma, the reduction in interest as attractiveness goes down is

about the same regardless of whether dates play hard to get (the blue lines have similar slopes). However, for dates who have high charisma, the decrease in interest if these dates are unattractive rather than average-looking is much more dramatic when dates played hard to get than when they didn't (the orange line is steeper for dates that played hard to get).

The final contrast was not significant, $F(1, 18) = 1.33$, $p = 0.263$. This contrast looks at the effect of **Strategy** when comparing level 2 of **Looks** (unattractive) to level 3 (average-looking), in level 2 of **Charisma** (dullard) relative to level 3 (some charisma). The relevant means are

displayed in the fourth column of Figure 16.18. Interest in unattractive dates was the same regardless of whether they had some charisma or were a dullard (the blue and green dots are in the same place). Interest in average-looking dates was greater when they had some charisma than when they were a dullard (the blue dot is higher than the green). Importantly, this pattern of results is very similar when dates played hard to get and when they didn't.

These contrasts tell us nothing about the differences between the attractive and unattractive conditions or the high-charisma and dullard conditions, because these were never compared. We could rerun the analysis and specify our contrasts differently to get these effects. What should be clear from this chapter is that when more than two independent variables are used and you're comparing means it yields complex interaction effects that require a great deal of concentration to interpret. Imagine just how much your brain would throb if interpreting a four-way interaction. If faced with this particularly unpleasant scenario, my best advice is to take a systematic approach to interpretation, and plotting graphs is a useful way to proceed. It is also advisable to think carefully about the most useful contrasts to use to answer the questions that your data were collected to test.

16.7 Calculating effect sizes ▮▮▮

I keep emphasizing that effect sizes are more useful when they summarize a focussed effect. This also gives me a useful excuse to circumvent the complexities of omega squared in mixed designs (trust me, you don't want to go there). A straightforward approach is to calculate effect sizes for your contrasts. Output 16.6 shows the values for several contrasts, all of which have 1 degree of freedom for the model (i.e., they represent a focussed and interpretable comparison) and have 18 residual degrees of freedom. We can convert these F-ratios to r using equation (14.31) from Chapter 14. First, let's deal with the main effect of Strategy because it compares only two groups:

$$r_{strategy} = \sqrt{\frac{0.005}{0.005 + 18}} = 0.02 \quad (16.1)$$

For the two contrasts we did for the **Looks** variable (Output 16.6), we get:

$$r_{attractive\,vs.\,average} = \sqrt{\frac{226.99}{226.99 + 18}} = 0.96$$

$$r_{unattractive\,vs.\,average} = \sqrt{\frac{160.07}{160.07 + 18}} = 0.95 \quad (16.2)$$

For the two contrasts we did for the **Charisma** variable (Output 16.6), we get:

$$n_{high\,vs.\,some} = \sqrt{\frac{109.94}{109.94 + 18}} = 0.93$$

$$r_{dullard\,vs.\,some} = \sqrt{\frac{227.94}{227.94 + 18}} = 0.96 \quad (16.3)$$

For the **Looks × Strategy** interaction, we get:

$$r_{attractive\,vs.\,average,\,normal\,vs.\,hard\,to\,get} = \sqrt{\frac{43.26}{43.26 + 18}} = 0.84$$

$$r_{unattractive\,vs.\,average,\,normal\,vs.\,hard\,to\,get} = \sqrt{\frac{30.23}{30.23 + 18}} = 0.79 \quad (16.4)$$

For the **Charisma × Strategy** interaction, the two contrasts give us:

$$n_{high\,vs.\,some,\,normal\,vs.\,hard\,to\,get} = \sqrt{\frac{27.20}{27.20 + 18}} = 0.78$$

$$r_{dullard\,vs.\,some,\,normal\,vs.\,hard\,to\,get} = \sqrt{\frac{33.69}{33.69 + 18}} = 0.81 \quad (16.5)$$

Moving on to the **Looks × Charisma** interaction, we get the following four contrasts:

$$r_{attractive\,vs.\,average,\,high\,vs.\,some} = \sqrt{\frac{21.94}{21.94 + 18}} = 0.74$$

$$r_{attractive\,vs.\,average,\,dullard\,vs.\,some} = \sqrt{\frac{4.09}{4.09 + 18}} = 0.43$$

$$r_{unattractive\,vs.\,average,\,high\,vs.\,some} = \sqrt{\frac{6.23}{6.23 + 18}} = 0.51$$

$$r_{unattractive\,vs.\,average,\,dullard\,vs.\,some} = \sqrt{\frac{88.60}{88.60 + 18}} = 0.91 \quad (16.6)$$

Finally, for the **Looks × Charisma × Strategy** interaction we have:

$$r_{attractive\,vs.\,average,\,high\,vs.\,some,\,normal\,vs.\,hard\,to\,get} = \sqrt{\frac{0.93}{0.93 + 18}} = 0.22$$

$$r_{attractive\,vs.\,average,\,dullard\,vs.\,some,\,normal\,vs.\,hard\,to\,get} = \sqrt{\frac{60.67}{60.67 + 18}} = 0.88$$

$$r_{unattractive\,vs.\,average,\,high\,vs.\,some,\,normal\,vs.\,hard\,to\,get} = \sqrt{\frac{11.70}{11.70 + 18}} = 0.63$$

$$r_{unattractive\,vs.\,average,\,dullard\,vs.\,some,\,normal\,vs.\,hard\,to\,get} = \sqrt{\frac{1.33}{1.33 + 18}} = 0.26 \quad (16.7)$$

16.8 Reporting the results of mixed designs ▮▮▮

As you've probably gathered, when you have more than two independent variables there's a hell of a lot of information to report. I've mentioned a few times that when interaction effects are significant there's no point in interpreting main effects, so you can save space by not reporting them; however, some journals will expect you to report them anyway. In any case, certainly reserve the most detail for the effects that are central to your main hypothesis.

Assuming we want to report all our effects, we could do it something like this (though not as a list!):

✓ All effects are reported as significant at $p < 0.001$ unless otherwise stated. There was a significant main effect of the attractiveness of the date on interest expressed by participants, $F(1.92, 34.62) = 423.73$. Contrasts revealed that attractive dates were significantly more desirable than average-looking ones, $F(1, 18) = 226.99$, $r = 0.96$, and unattractive dates were significantly less desirable than average-looking ones, $F(1, 18) = 160.07$, $r = 0.95$.

✓ There was also a significant main effect of the amount of charisma the date exhibited on the interest expressed in dating them, $F(1.87, 33.62) = 328.25$. Contrasts revealed that dates displaying high charisma were significantly more desirable than dates displaying some charisma, $F(1, 18) = 109.94$, $r = 0.93$, and dull dates were significantly less desirable than dates exhibiting some charisma, $F(1, 18) = 227.94$, $r = 0.96$.

✓ There was no significant effect of strategy, indicating that ratings of dates who played hard to get was similar to those who acted normally, $F(1, 18) = 0.005$, $p = 0.946$, $r = 0.02$.

✓ There was a significant interaction effect between the attractiveness of the date and the dating strategy of the date, $F(1.92, 34.62) = 80.43$. This effect indicates that

Labcoat Leni's Real Research 16.2
Keep the faith(ful)? ▮▮▮▮

Schützwohl, A., (2008). *Personality and Individual Differences*, *44*, 633–644.

People can be jealous when they think that their partner is being unfaithful. An evolutionary view suggests that men and women have evolved distinctive types of jealousy: specifically, a woman's sexual infidelity deprives her mate of a reproductive opportunity and could burden him with years investing in a child that is not his. Conversely, a man's sexual infidelity does not burden his mate with unrelated children, but may divert his resources from his mate's progeny. This diversion of resources is signalled by emotional attachment to another female. Consequently, men's jealousy mechanism should have evolved to prevent a mate's *sexual* infidelity, whereas in women it has evolved to prevent emotional infidelity. If this is the case, women should be 'on the look-out' for emotional infidelity, whereas men should be watching out for sexual infidelity.

Whether or not you buy into this theory, it can be tested. Achim Schützwohl exposed men and women to sentences on a computer screen (Schützwohl, 2008). At each trial, participants saw a target sentence that was emotionally neutral (e.g., 'The gas station is at the other side of the street'). However, before each of these targets, a distractor sentence was presented that could also be affectively neutral or could

indicate sexual infidelity (e.g., 'Your partner suddenly has difficulty becoming sexually aroused when he and you want to have sex') or emotional infidelity (e.g., 'Your partner doesn't say "I love you" to you anymore'). Schützwohl reasoned that if these distractor sentences grabbed a person's attention then (1) they would remember them, and (2) they would not remember the target sentence that came afterwards (because their attentional resources were focussed on the distractor). These effects should show up only in people currently in a relationship. The outcome was the number of sentences that a participant could remember (out of 6), and the predictors were whether the person had a partner or not (**Relationship**), whether the trial used a neutral distractor, an emotional infidelity distractor or a sexual infidelity distractor, and whether the sentence was a distractor or the target following a distractor. Schützwohl analyzed the men's and women's data separately. The predictions are that women should remember more emotional infidelity sentences (distractors) but fewer of the targets that followed those sentences (target). For men, the same effect should be found but for sexual infidelity sentences (**Schützwohl(2008).sav**). Labcoat Leni wants you to fit two models (one for men and the other for women) to test these hypotheses. Answers are on the companion website (or look at pages 638–642 in the original article).

the desirability of dates of different levels of attractiveness differed according to whether they played hard to get. Contrasts compared each level of attractiveness to average-looking, across dating strategies. These contrasts revealed significant interactions when comparing ratings of dates who played hard to get and those who acted normally when comparing attractive to average-looking dates, $F_{(1, 18)} = 43.26$, $r = 0.84$, and to unattractive dates compared to average dates, $F_{(1, 18)} = 30.23$, $r = 0.79$. The interaction graph shows that although interest decreased as attractiveness decreased regardless of date's strategy, this decrease was more pronounced when they played hard to get, suggesting that when charisma is ignored, attractiveness had a greater impact on ratings of the date when they played hard to get than when they acted normally.

✓ There was a significant interaction effect between the level of charisma of the date and the strategy of the date, $F_{(1.87, 33.62)} = 62.45$, indicating that the desirability of dates of different levels of charisma differed according to whether they played hard to get. Contrasts compared each level of charisma to the middle category of 'some charisma' across dating strategies. These contrasts revealed significant interactions when comparing ratings of dates when they played hard to get relative to when they did not, in dates with some charisma compared to those with high charisma, $F_{(1, 18)} = 27.20$, $r = 0.78$, and dullards, $F_{(1, 18)} = 33.69$, $r = 0.81$. The interaction graph reveals that interest decreased as charisma decreased, but this decrease was less pronounced when the date played hard to get, suggesting charisma influenced dating preferences more when the dates acted normally than when they played hard to get.

✓ There was a significant charisma × attractiveness interaction, $F_{(3.20, 57.55)} = 36.63$, indicating that the desirability of dates of different levels of charisma differed according to their attractiveness.

Contrasts compared each level of charisma to the middle category of 'some charisma' across each level of attractiveness compared to the category of average attractiveness. The first contrast revealed a significant interaction when comparing attractive dates to average-looking dates when the date had high charisma compared to some charisma, $F_{(1, 18)} = 21.94$, $r = 0.74$, and tells us that as attractiveness was reduced there was a greater decline in interest when charisma was low compared to when charisma was high. The second contrast, which compared attractive dates to average-looking dates when the date was a dullard compared to when they had some charisma, was not significant, $F_{(1, 18)} = 4.09$, $p = 0.058$, $r = 0.43$. This result suggests that as attractiveness was reduced there was a decline in interest both when charisma was average and when the date had no charisma at all. The third contrast, which compared unattractive dates to average-looking dates when they had high charisma compared to average charisma, was significant, $F_{(1, 18)} = 6.23$, $p = 0.022$, $r = 0.51$. This contrast implies that as attractiveness of the date was reduced there was a greater decline in interest when charisma was average compared to when it was high. The final contrast compared unattractive dates to average-looking dates for dullards compared to when they had some charisma. This contrast was highly significant, $F_{(1, 18)} = 88.60$, $r = 0.91$, and suggests that as attractiveness was reduced the decline in interest in dates with a bit of charisma was significantly greater than for dullards.

✓ Finally, the looks × charisma × strategy interaction was significant $F_{(3.20, 57.55)} = 24.12$. This indicates that the looks × charisma interaction described previously was moderated by whether the date played hard to get. Contrasts were used to break down this interaction; these contrasts compared scores at each level of charisma relative to the middle category of 'some charisma' across each level of attractiveness relative to the category of average attractiveness when dates played hard to get compared to when they did not. The first contrast revealed a non-significant effect of playing hard to get when comparing attractive dates to average-looking dates when the dates had high charisma compared to some charisma, $F_{(1, 18)} = 0.93$, $p = 0.348$, $r = 0.22$. This effect suggests that, regardless of whether the date played hard to get, as the date's attractiveness was reduced there was a greater decline in interest when charisma was average compared to high. The second contrast investigated the effect of playing hard to get when comparing attractive dates to average-looking dates for a dullard compared to when they had average charisma, $F_{(1, 18)} = 60.67$, $r = 0.88$. This finding indicates that for dates with average charisma, the reduction in interest as attractiveness went down was unaffected by whether the date played hard to get, but for dullard dates, the decrease in interest if these dates were average-looking rather than attractive was much more dramatic when they played hard to get. The third contrast looked at the effect of playing hard to get when comparing unattractive dates to average-looking dates when they had high charisma compared to average charisma, $F_{(1, 18)} = 11.70$, $p = 0.003$, $r = 0.63$, and tells us that for dates with average charisma, the reduction in interest as attractiveness went down was unaffected by playing hard to get, but for dates who had high charisma, the decrease in interest if they were unattractive rather than average-looking was much more dramatic when they played hard to get. The final contrast looked at the effect of playing hard to get when comparing unattractive dates to average-looking dates for dullards compared to when they had average charisma, $F(1, 18) = 1.33$, $p = 0.263$, $r = 0.26$. This effect suggests that regardless of whether dates played hard to get, as physical attractiveness was reduced the decline in interest in dates with average charisma was significantly greater than for dullards.

16.9 Brian and Jane's Story ▌▌▌▌

Jane's mother was slowly regaining strength. Jane visited regularly, mainly because Brian insisted. She had seen a lot of Brian since her mother's collapse. He was stable, reassuring and emotional in a way that she found difficult to relate to. Their meetings had developed a predictable pattern: Jane would demand that he talked to her about anything he wanted to that wasn't family or emotional, he'd nervously recite whatever he'd been learning about as a safe emotion-free topic, and once his soothing words of statistics had calmed her, she'd slowly give up a little more of her past to him. His advice never wavered: no matter what, her parents were her parents and she should be there for them. His view irritated her a little; he didn't know what she'd endured. This ritual had gone on for weeks. It was only now that it occurred to her to ask him about his own family. He smiled when she asked, but with tears in his eyes.

'My dad's awesome,' he said. 'He brought me up. He worked 50+ hours a week and still always had time for me. He was there getting me ready for school every morning, and there to pick me up. He never missed anything. *Anything*. The guy never slept for cramming in work while I was asleep so he could be there when I was awake. I don't know how he did it. He gave up his life to give me a happy childhood.'

'And your mum?'

'She died,' he said after a long pause. He choked. 'I was 10. It was a hit and run.'

Jane wasn't equipped for this kind of revelation. 'That must have been hard,' she said.

'The weird thing is I don't remember anything.' The levee in his eyes broke. 'I know that I had 10 years with my mum—I've seen photos, but they feel like someone else's life. It's like she was wiped from my mind when she died. I remember the look of desolation in my dad's eyes when he told me, and everything since, but I don't remember anything before that day. It kills me.'

Jane did something she'd never done in her adult life. She put her arm around another person and hugged him. She quickly felt awkward, not knowing whether it was weirder to keep holding him or to let go.

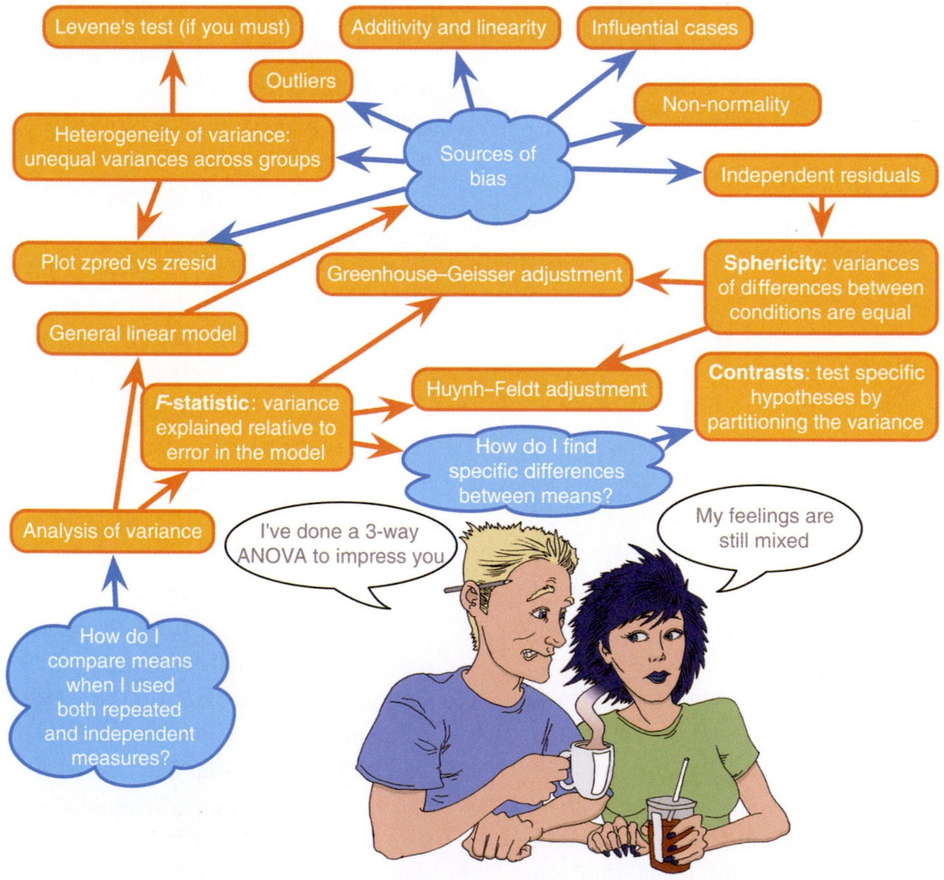

16.10 What next? ▌▌▌▌

We've discovered in this chapter that if you are dull then it doesn't matter how attractive you are you won't get a date. Unless you play hard to get and are *really* attractive. This is why as a 16–18-year-old my life was so complicated, because I wouldn't know how to do hard to get, I wasn't attractive anyway, and where on earth do you discover your hidden charisma? Before you get out your tiny violins, I had one small bit of dating fortune, which was that some girls from Essex find alcoholics appealing. The girl (Nicola) I was particularly keen on at 16 was, as it turned out, keen on me too. I refused to believe this for quite some time. All our friends were getting so bored with us declaring our undying love for each other to them but not actually speaking to each

Figure 16.19 What Brian learnt from this chapter

other that they held an intervention. At a party one evening all of Nicola's friends had spent hours convincing me to ask her on a date, guaranteeing me that she would say 'yes'. I psyched myself up, I was going to do it, I was actually going to ask a girl out on a date. My whole life had been leading up to this moment, I thought to myself, and I must not do anything to ruin it. By the time she arrived she had to step over my paralytic corpse to get into the house. My nerves had got the better of me, and nerves made me drink alcohol. Later that evening, once I'd returned to semi-consciousness, my friend Paul Spreckley (see Figure 11.1) physically carried Nicola from another room, put her next to me and said something to the effect of 'Andy, I'm going to sit here until you ask her out.' He had a long wait, but eventually, miraculously, the words came out of my mouth. What happened next is the topic for another book, not about statistics.

16.11 Key terms that I've discovered

Mixed ANOVA

Mixed design

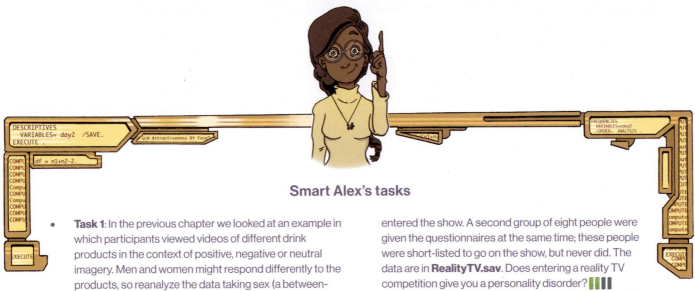

Smart Alex's tasks

- **Task 1**: In the previous chapter we looked at an example in which participants viewed videos of different drink products in the context of positive, negative or neutral imagery. Men and women might respond differently to the products, so reanalyze the data taking sex (a between-group variable) into account. The data are in the file **MixedAttitude.sav**.
- **Task 2**: Text messaging and Twitter encourage communication using abbreviated forms of words (if u no wat I mean). A researcher wanted to see the effect this had on children's understanding of grammar. One group of 25 children was encouraged to send text messages on their mobile phones over a six-month period. A second group of 25 was forbidden from sending text messages for the same period (to ensure adherence, this group were given armbands that administered painful shocks in the presence of a phone signal). The outcome was a score on a grammatical test (as a percentage) that was measured both before and after the experiment. The data are in the file **TextMessages.sav**. Does using text messages affect grammar?
- **Task 3**: A researcher hypothesized that reality TV show contestants start off with personality disorders that are exacerbated by being forced to spend time with people as attention-seeking as them (see Chapter 1). To test this hypothesis, she gave eight contestants a questionnaire measuring personality disorders before and after they entered the show. A second group of eight people were given the questionnaires at the same time; these people were short-listed to go on the show, but never did. The data are in **RealityTV.sav**. Does entering a reality TV competition give you a personality disorder?
- **Task 4**: Angry Birds is a video game in which you fire birds at pigs. Some daft people think this sort of thing makes people more violent. A (fabricated) study was set up in which people played Angry Birds and a control game (Tetris) over a two-year period (one year per game). They were put in a pen of pigs for a day before the study, and after 1 month, 6 months and 12 months. Their violent acts towards the pigs were counted. Does playing Angry Birds make people more violent to pigs compared to a control game? (**Angry Pigs.sav**)
- **Task 5**: A different study was conducted with the same design as in Task 4. The only difference was that the participant's violent acts in real life were monitored before the study, and after 1 month, 6 months and 12 months. Does playing Angry Birds make people more violent in general compared to a control game? (**Angry Real.sav**)
- **Task 6**: My wife believes that she has received fewer friend requests from random men on Facebook since she changed her profile picture to a photo of us both. Imagine we took 40 women who had profiles on a social networking website; 17 of them had a relationship status of 'single' and the remaining 23 had their status as 'in a relationship'

(relationship_status). We asked these women to set their profile picture to a photo of them on their own (alone) and to count how many friend request they got from men over 3 weeks, then to switch it to a photo of them with a man (couple) and record their friend requests from random men over 3 weeks. Fit a model to see if friend requests are affected by relationship status and type of profile picture (ProfilePicture.sav). ▮▮▮▮

- **Task 7**: Labcoat Leni's Real Research 5.2 described a study by Johns et al. (2012) in which they reasoned

that if red was a proxy signal to indicate sexual proceptivity then men should find red female genitalia more attractive than other colors. They also recorded the men's sexual experience (Partners) as 'some' or 'very little'. Fit a model to test whether attractiveness was affected by genitalia color (PalePink, LightPink, DarkPink, Red) and sexual experience (Johns et al. (2012). sav). Look at page 3 of Johns et al. to see how to report the results ▮▮▮▮

Answers & additional resources are available on the book's website at
https://edge.sagepub.com/field5e

MULTIVARIATE ANALYSIS OF VARIANCE (MANOVA)

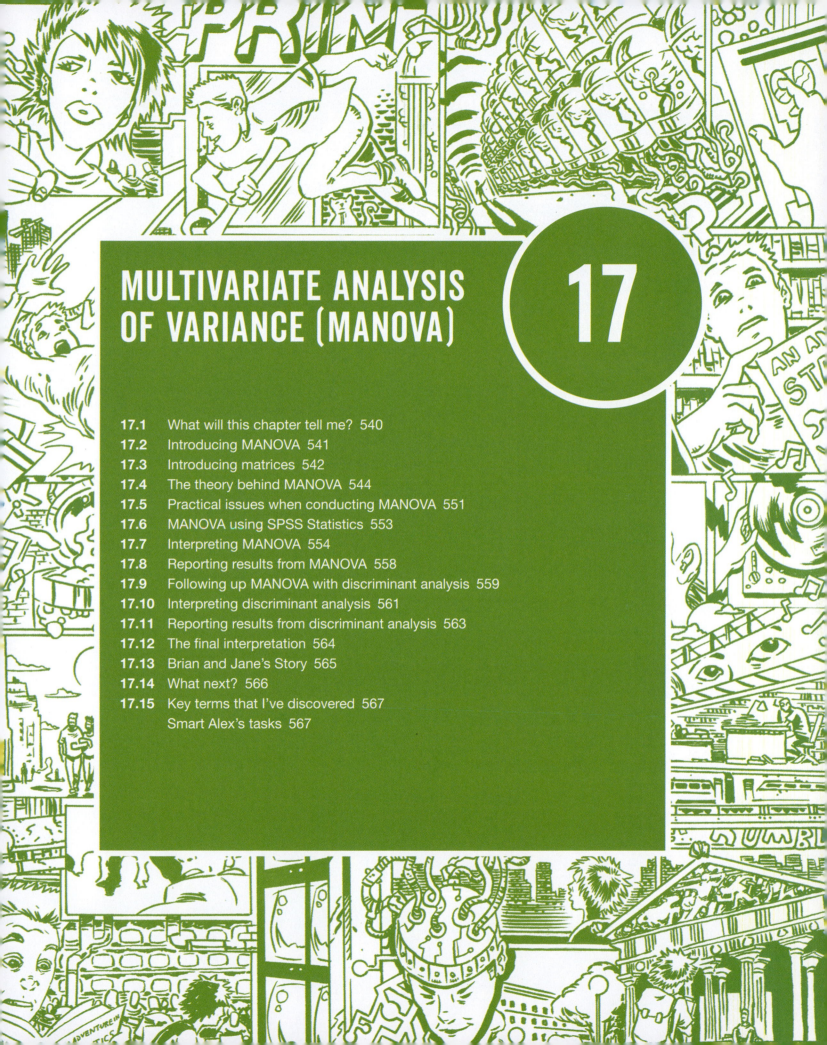

17

17.1 What will this chapter tell me?

Having had what little confidence I had squeezed out of me by my first forays into dating and my band's unqualified failure to have an impact on the musical world, as I reached adulthood I decided that I could either kill myself or get a cat. I'd wanted to do both for years but when I was introduced to a little 4-week-old bundle of gingerness the choice was made. Fuzzy (as I named him) was born on 8 April 1996 and was my right-hand feline for over 20 years. Like the Cheshire Cat in Lewis Carroll's *Alice's Adventures in Wonderland*[1] he used to vanish and reappear seemingly at will: I'd go to get clothes from my wardrobe and spot a ginger face peering out at me, I'd put my pants in the laundry basket and he'd look up at me from a pile of smelly socks, I'd go to have a bath and he'd be sitting in it, and I'd shut the bedroom door yet wake up to find him asleep next to me. His best vanishing act was when I moved house one time. He'd been locked up in his travel basket (which he hated) during the move, so once we were in our new house I thought I'd let him out as soon as possible. I found a quiet room, checked the doors and windows to make sure he couldn't escape, opened the basket, gave him a cuddle and left him to get to know his new territory. When I returned five minutes later, he was gone. The door had been shut, the windows closed and the walls were solid (I checked). He had literally vanished into thin air and he didn't even leave behind his smile.

Before his dramatic disappearance, Fuzzy had stopped my suicidal tendencies, and we saw in Chapter 12 that there is a belief that having a pet is good for your mental health. If you wanted to test this you could compare people with pets against those without to see if they had better mental health. However, the term *mental health* covers a wide range of concepts including (to name a few) anxiety, depression, general distress and psychosis. As such, we might have several outcome measures, and the linear model we've looked at so far only deals with one. That is, until now, where we discover that it can mutate into MANOVA. Yes, it's as scary as it sounds.

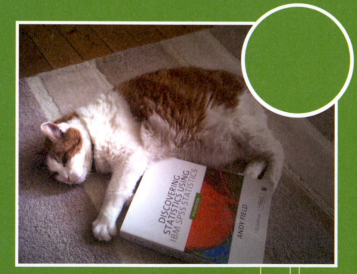

1 This is one of my favorite books from my childhood. For those who haven't read it, the Cheshire Cat is a big fat cat mainly remembered for vanishing and reappearing out of nowhere; on one occasion it vanished leaving only its smile behind.

Figure 17.1 Fuzzy doing some light reading

17.2 Introducing MANOVA ▌▌▌▌

Over Chapters 10–16, we have seen how the general linear model can be used to detect group differences on a single outcome. However, there may be circumstances in which we are interested in several outcomes, and in these cases we use **multivariate analysis of variance** (or **MANOVA**). The principles of the linear model extend to MANOVA in that we can use MANOVA when there is one independent/predictor variable or several, we can look at interactions between outcome variables, and we can do contrasts to see which groups differ. When we have only one outcome variable the model is known as **univariate** (meaning 'one variable'), but when we include several outcome variables simultaneously the model is **multivariate** (meaning 'many variables'). There is a lengthy theory section explaining the workings of MANOVA, but for those of you who value the little time you have on Earth, accept that we're extending the linear model again and go straight to the sections on applying and interpreting MANOVA. This process leads us to another statistical tool known as *discriminant function analysis* (or *discriminant analysis* for short).

If we have scores on several outcome variables we could simply fit separate linear models (*F*-statistics) to each outcome (it is not unusual for researchers to do this). However, we learnt in Section 2.9.7 that when we carry out multiple tests on the same data the Type I errors mount up. For this reason, we shouldn't really fit separate linear models to each outcome variable. Also, if separate models are fitted to each outcome, then any relationship between these outcome variables is ignored—we lose this important information. By including all outcome variables in the model MANOVA factors in the relationship between them. Related to this point, separate models can tell us only whether groups differ along a single dimension, whereas MANOVA has the power to detect whether groups differ along a combination of dimensions.

For example, we might be able to distinguish people who are married, living together or single by their happiness. 'Happiness' is a complex construct, so we might want to measure their happiness with work, socially, sexually and within themselves (self-esteem). It might not be possible to distinguish people who are married, living together or single by only one aspect of happiness (which is what a univariate model tests), but these groups might be distinguished by *a combination* of their happiness across all four domains (which is what a MANOVA tests). In this sense MANOVA has greater potential power to detect an effect (see Jane Superbrain Box 17.1).

17.2.1 Choosing outcomes ▌▌▌▌

MANOVA is probably looking like a pretty good way to measure hundreds of outcome variables and then sling them into an analysis without getting accused of *p*-hacking. Not so. It is a bad idea to lump outcome measures together in a MANOVA unless you have a good theoretical or empirical basis for doing so. The adage of 'garbage in, garbage out' applies here. Where there is a good theoretical basis for including some, but not all, of your outcome measures, then fit separate models: one for the outcomes being tested on a heuristic basis and one for the theoretically meaningful outcomes. The point here is not to include lots of outcome variables in a MANOVA just because you measured them.

17.2.2 An intrusive example ▌▌▌▌

Obsessive compulsive disorder (OCD) is a mental health problem characterized by intrusive images or thoughts that the sufferer finds abhorrent—in my case the thought of someone fitting non-robust models, but more commonly thoughts with themes of aggression, sexuality or disease/contamination (Julien, O'Connor, & Aardema, 2007). These thoughts lead the sufferer to engage in activities to neutralize the unpleasantness of these thoughts (these activities can be mental, such as doing a MANOVA in my head to make me feel better about the non-robust models in the world or physical, such as touching the floor 23 times so that you won't murder your statistics lecturer). A clinical psychologist was interested in the effects of cognitive behavior therapy (CBT) on OCD. She compared people with OCD after sessions of CBT or behavior therapy (BT) with a group who were awaiting treatment (a no treatment condition, NT).[2] Most psychopathologies have both behavioral and cognitive elements to them. For example, for someone with OCD who has an obsession with germs and contamination, the disorder might manifest itself in the number of times they both wash their hands (behavior) and *think about* washing their hands (cognition). To gauge the success of therapy, it is not enough to look only at behavioral outcomes (such as whether obsessive behaviors are reduced); we need to look at whether cognitions are changed too. Hence, the clinical

---

2 A note for non-psychologists: behavior therapy assumes that if you stop the maladaptive behaviors the disorder will go away, whereas cognitive therapy assumes that treating the maladaptive cognitions will stop the disorder. CBT does a bit of both.

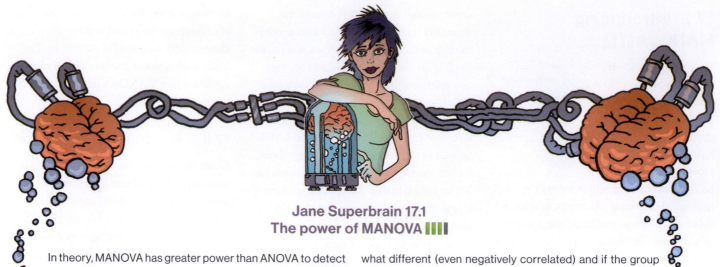

Jane Superbrain 17.1
The power of MANOVA ▌▌▌▌

In theory, MANOVA has greater power than ANOVA to detect effects because it takes account of the correlations between outcome variables (Huberty & Morris, 1989). However, the issue is complicated (when isn't it?). The evidence is contradictory, with some studies showing *diminishing* power as the correlation between outcome variables increases, whereas others show that power with high correlations between outcome variables is generally higher than for moderate correlations (Stevens, 1980). Work by Cole, Maxwell, Arvey, & Salas (1994) suggests that if you are expecting to find a large effect, MANOVA will have greater power if the measures are some-

what different (even negatively correlated) and if the group differences are in the same direction for each measure. If you have two outcome variables, one of which exhibits a large group difference and one of which exhibits a small or no group difference, then power will be increased if these variables are highly correlated. Although Cole et al.'s work is limited to the case where two groups are being compared, the take-home message is that if you are interested in how powerful the MANOVA is likely to be you should consider not only the correlation between outcome variables but also the size and pattern of group differences that you expect to get.

psychologist measured two outcomes: the occurrence of obsession-related behaviors (**Actions**) and the occurrence of obsession-related cognitions (**Thoughts**) on a single day. The data are in Table 17.1 (and **OCD.sav**). Participants belonged to group 1 (CBT), group 2 (BT) or group 3 (NT), and within these groups all participants had both actions and thoughts measured.

17.3 Introducing matrices ▌▌▌▌

The theory of MANOVA requires knowing a bit about matrix algebra, which is way

beyond the scope of this book. I intend to give a flavor of the conceptual basis of MANOVA, using matrices, without getting into the actual algebra. Those wanting more detail can read Bray and Maxwell (1985). We can't avoid everything to do with matrices, though, so we'll have a brief introduction to some key concepts.

Despite what Hollywood would have you believe, a **matrix** does not enable you to jump acrobatically through the air, Ninja style, as time seemingly slows to a point where you can gracefully contort to avoid high-velocity objects. I have worked with matrices many times, and I have never (to my knowledge) stopped time, and would certainly end up in a pool of my own

innards if I ever tried to dodge a bullet. The mundane reality is that a matrix is a grid of numbers arranged in columns and rows. In fact, throughout this book you have been using a glorified matrix: the data editor, which is often just numbers arranged in columns and rows (i.e. a matrix). A matrix can have many columns and rows, and we specify its dimensions using numbers. A 2 × 3 matrix is a matrix with two rows and three columns, and a 5 × 4 matrix is one with five rows and four columns (Figure 17.2).

A matrix could represent wide data (Section 4.6.1) in which case each row contains the data from a single participant and each column has scores on a particular variable. So, the 5 × 4 matrix would represent five participants tested on four measures: the first participant scored 3 on the first variable and 20 on the fourth variable. The values within a matrix are *components* or *elements* and the rows and columns are *vectors*.

A **square matrix** has an equal number of columns and rows (Figure 17.3). When using square matrices we sometimes use the diagonal components (i.e., the values that lie on the diagonal line from the top left component to the bottom right component) and the off-diagonal ones (the values that do not lie on the diagonal). In Figure 17.3 (right) the diagonal components are 3, 21, 9 and 11 (the highlighted values) and the off-diagonal components are the other values. An **identity matrix** is a square matrix in which the diagonal elements are 1 and the off-diagonal elements are 0 (Figure 17.3 (left)). Hopefully, the concept of a matrix is less scary than you thought it might be: it is not some magical mathematical entity, merely a way of representing data—just like a spreadsheet.

When we have a single outcome variable, we are interested in computing an *F*-statistic that represents how much variance can be explained by the fact that certain scores appear in certain groups (which in experimental research represents our manipulation) relative to the error in prediction in the model. We basically want to do the same, but with the complication of having several outcome measures. To achieve this aim we need a multivariate analogue of the sums of squares that we used for univariate models (Chapters 9 and 12); these are the sum of squares due to the model/grouping variable (the model sum of squares, SS_M), the error in prediction from the model (the residual sum of squares, SS_R) and of course the total amount of variation in the outcome(s) that needs to be explained (SS_T). It turns out that matrices are a good way to operationalize multivariate versions

Table 17.1 Data from **OCD.sav**

Group:	DV 1: Actions			DV 2: Thoughts		
	CBT (1)	BT (2)	NT (3)	CBT (1)	BT (2)	NT (3)
	5	4	4	14	14	13
	5	4	5	11	15	15
	4	1	5	16	13	14
	4	1	4	13	14	14
	5	4	6	12	15	13
	3	6	4	14	19	20
	7	5	7	12	13	13
	6	5	4	15	18	16
	6	2	6	16	14	14
	4	5	5	11	17	18
$\overline{X}$	4.90	3.70	5.00	13.40	15.20	15.00
s	1.20	1.77	1.05	1.90	2.10	2.36
s^2	1.43	3.12	1.11	3.60	4.40	5.56

$$\overline{X}_{grand(Actions)} = 4.53 \qquad \overline{X}_{grand(Thoughts)} = 14.53$$
$$S^2_{grand(Actions)} = 2.1195 \qquad S^2_{grand(Thoughts)} = 4.8782$$

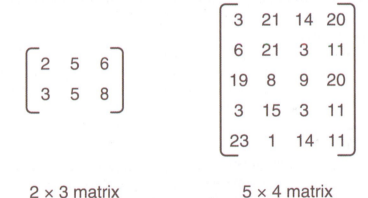

$$\begin{bmatrix} 2 & 5 & 6 \\ 3 & 5 & 8 \end{bmatrix} \qquad \begin{bmatrix} 3 & 21 & 14 & 20 \\ 6 & 21 & 3 & 11 \\ 19 & 8 & 9 & 20 \\ 3 & 15 & 3 & 11 \\ 23 & 1 & 14 & 11 \end{bmatrix}$$

2 × 3 matrix   5 × 4 matrix

Figure 17.2   Some examples of matrices

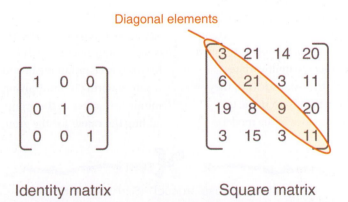

Diagonal elements

$$\begin{bmatrix} 1 & 0 & 0 \\ 0 & 1 & 0 \\ 0 & 0 & 1 \end{bmatrix} \qquad \begin{bmatrix} 3 & 21 & 14 & 20 \\ 6 & 21 & 3 & 11 \\ 19 & 8 & 9 & 20 \\ 3 & 15 & 3 & 11 \end{bmatrix}$$

Identity matrix   Square matrix

Figure 17.3   Both matrices are square, and the one on the left is also an identity matrix

of these sums of squares (they're called **sum of squares and cross-products (SSCP) matrices**).

The matrix that represents the systematic variance (or the model sum of squares for all variables) is denoted by the letter H and is called the **hypothesis sum of squares and cross-products matrix** (or **hypothesis SSCP**). The matrix that represents the unsystematic variance (the residual sums of squares for all variables) is denoted by the letter E and is called the **error sum of squares and cross-products matrix** (or **error SSCP**). Finally, the matrix that represents the total amount of variance present for each outcome variable (the total sums of squares for each outcome) is denoted by T and is called the **total sum of squares and cross-products matrix** (or **total SSCP**). It should be obvious why these matrices are referred to as sum of squares matrices, but why is there a reference to cross-products in their name?

Cross-products represent a total value for the combined error between two variables (in some sense they represent an unstandardized estimate of the total correlation between two variables). As such, whereas the sum of squares of a variable is the total squared difference between the observed values and the mean value, the cross-product is the total combined error *between* two variables. I mentioned earlier that MANOVA had the power to account for correlation between outcome variables, and it does this by using these cross-products. Later, I will show how these SSCP matrices are used in the same way as the simple sums of squares (SS_M, SS_R and SS_T) in univariate linear models to derive test statistics that are multivariate equivalents of the F-statistic (i.e., they represent the ratio of systematic to unsystematic variance in the model).

17.4 The theory behind MANOVA ▮▮▮

To begin with let's calculate univariate F-statistics on each of the two outcome variables in the OCD example (see Table 17.1). I will draw heavily on the assumption that you have read Chapter 12.

17.4.1 Univariate F for outcome 1 (actions) ▮▮▮

There are three sums of squares that need to be calculated. First, we assess how much variability there is to be explained within the outcome (SS_T), and then break this variability down into that explained by the model (SS_M) and the error in prediction (SS_R). Referring to the equations in Chapter 12, we calculate each of these values as follows.

The total sum of squares, $SS_{T(Actions)}$, is obtained by calculating the difference between each of the 30 scores and the mean of those scores, then squaring these differences and adding them. Alternatively, you calculate the variance for the action scores (regardless of which group the score falls into) and multiply this value by the number of scores minus 1:

$$SS_T = s_{grand}^2(N-1)$$
$$= 2.1195(30-1)$$
$$= 2.1195 \times 29$$
$$= 61.47$$
(17.1)

The degrees of freedom will be $N - 1 = 29$. The model sum of squares, $SS_{M(Actions)}$, is calculated by taking the difference between each group mean and the grand mean, squaring it, multiplying by the number of scores in the group and then adding the values for the groups together:

$$SS_M = 10(4.90-4.53)^2 + 10(3.70-4.53)^2 + 10(5.00-4.53)^2$$
$$= 10(0.37)^2 + 10(-0.83)^2 + 10(0.47)^2$$
$$= 1.37 + 6.89 + 2.21$$
$$= 10.47$$
(17.2)

The degrees of freedom will be $k - 1 = 2$. The residual sum of squares, $SS_{R(Actions)}$, value is calculated by taking the difference between each score and the mean of the group from which it came. These differences are then squared and added together. Alternatively multiply each group variance by the number of scores minus 1 and then add the results:

$$SS_R = s_{CBT}^2(n_{CBT}-1) + s_{BT}^2(n_{BT}-1) + s_{NT}^2(n_{NT}-1)$$
$$= 1.433(10-1) + 3.122(10-1) + 1.111(10-1)$$
$$= (1.433 \times 9) + (3.122 \times 9) + (1.111 \times 9)$$
$$= 12.9 + 28.1 + 10$$
$$= 51$$
(17.3)

The degrees of freedom will be the sample size of each group minus 1 (e.g., 9) multiplied by the number of groups, $3 \times 9 = 27$. The next step is to calculate the average sums of squares (the mean square) by dividing the sums of squares by their degrees of freedom:

$$MS_M = \frac{SS_M}{df_M} = \frac{10.47}{2} = 5.235$$
$$MS_R = \frac{SS_R}{df_R} = \frac{51}{27} = 1.889$$
(17.4)

The F-statistic is the mean squares for the model divided by the mean squares for the error in the model:

$$F = \frac{MS_M}{MS_R} = \frac{5.235}{1.889} = 2.771$$
(17.5)

17.4.2 Univariate F for outcome 2 (thoughts) ▮▮▮

The three sums of squares for thoughts are calculated in the same way as for actions (the degrees of freedom for each is the same as above). For the total sum of squares, $SS_{T(Thoughts)}$, we have:

SELF TEST

What is a cross-product? (See Chapter 8.)

$$SS_T = s^2_{grand}(n-1)$$
$$= 4.878(30-1)$$
$$= 4.878 \times 29 \quad (17.6)$$
$$= 141.46$$

The model sum of squares, $SS_{M(Thoughts)}$, is:

$$SS_M = 10(13.40-14.53)^2 + 10(15.2-14.53)^2 + 10(15-14.53)^2$$
$$= 10(-1.13)^2 + 10(0.67)^2 + 10(0.47)^2 \quad (17.7)$$
$$= 12.77 + 4.49 + 2.21$$
$$= 19.47$$

The residual sum of squares, $SS_{R(Thoughts)}$, is:

$$SS_R = s^2_{CBT}(n_{CBT}-1) + s^2_{BT}(n_{BT}-1) + s^2_{NT}(n_{NT}-1)$$
$$= 3.6(10-1) + 4.4(10-1) + 5.56(10-1)$$
$$= (3.6 \times 9) + (4.4 \times 9) + (5.56 \times 9) \quad (17.8)$$
$$= 32.4 + 39.6 + 50$$
$$= 122$$

The mean sums of squares are the sums of squares divided by the degrees of freedom:

$$MS_M = \frac{SS_M}{df_M} = \frac{19.47}{2} = 9.735$$
$$\quad (17.9)$$
$$MS_R = \frac{SS_R}{df_R} = \frac{122}{27} = 4.519$$

The F-statistic is the mean squares for the model divided by the residual mean squares:

$$F = \frac{MS_M}{MS_R} = \frac{9.735}{4.519} = 2.154 \quad (17.10)$$

17.4.3 The relationship between outcomes: cross-products ▮▮▮▮

OK, we have the sums of squares associated with each outcome variable, now let's look at the relationship between them. If we want a measure of the relationship that is comparable to a sum of squares then we need something that quantifies the *total* relationship (because sums of squares are totals). We saw in Chapter 8 that the cross-product does this job. There are three relevant cross-products that correspond to the three sums of squares that we just calculated: the total cross-product, the cross-product for the model, and a

residual cross-product. Let's look at the total cross-product (CP_T) first.

The cross-product is the difference between the scores and the mean for one variable multiplied by the difference between the scores and the mean for another variable (Chapter 8). In the case of the total cross-product, the mean of interest is the grand mean for each

outcome variable (see Table 17.2). We can apply the cross-product equation described in Chapter 8 to the two outcome variables:

$$CP_T = \sum_{i=1}^{n} \left(x_{i(Actions)} - \bar{X}_{grand(Actions)} \right) \quad (17.11)$$
$$\left(x_{i(Thoughts)} - \bar{X}_{grand(Thoughts)} \right)$$

Table 17.2 Calculation of the total cross-product

Group	Actions	Thoughts	Actions – $\bar{X}_{grand(Actions)}$ (D_1)	Actions – $\bar{X}_{grand(Actions)}$ (D_2)	$D_1 \times D_2$
CBT	5	14	0.47	−0.53	−0.25
	5	11	0.47	−3.53	−1.66
	4	16	−0.53	1.47	−0.78
	4	13	−0.53	−1.53	0.81
	5	12	0.47	−2.53	−1.19
	3	14	−1.53	−0.53	0.81
	7	12	2.47	−2.53	−6.25
	6	15	1.47	0.47	0.69
	6	16	1.47	1.47	2.16
	4	11	−0.53	−3.53	1.87
BT	4	14	−0.53	−0.53	0.28
	4	15	−0.53	0.47	−0.25
	1	13	−3.53	−1.53	5.40
	1	14	−3.53	−0.53	1.87
	4	15	−0.53	0.47	−0.25
	6	19	1.47	4.47	6.57
	5	13	0.47	−1.53	−0.72
	5	18	0.47	3.47	1.63
	2	14	−2.53	−0.53	1.34
	5	17	0.47	2.47	1.16
NT	4	13	−0.53	−1.53	0.81
	5	15	0.47	0.47	0.22
	5	14	0.47	−0.53	−0.25
	4	14	−0.53	−0.53	0.28
	6	13	1.47	−1.53	−2.25
	4	20	−0.53	5.47	−2.90
	7	13	2.47	−1.53	−3.78
	4	16	−0.53	1.47	−0.78
	6	14	1.47	−0.53	−0.78
	5	18	0.47	3.47	1.63
	4.53	**14.53**			

$$CP_T = \sum D_1 \times D_2 = 5.47$$

Table 17.3 Calculating the model cross-product

	$\overline{X}_{group}$ Actions	$\overline{X}_{group} - \overline{X}_{grand}$ (D_1)	$\overline{X}_{group}$ Thoughts	$\overline{X}_{group} - \overline{X}_{grand}$ (D_2)	$D_1 \times D_2$	$N(D_1 \times D_2)$
CBT	4.9	0.37	13.4	−1.13	−0.418	−4.18
BT	3.7	−0.83	15.2	0.67	−0.556	−5.56
NT	5.0	0.47	15.0	0.47	0.221	2.21
	4.53		**14.53**			$\text{CP}_M = \sum N(D_1 \times D_2) = -7.53$

For each outcome variable you take each score and subtract from it the grand mean for that variable. This leaves you with two values per participant (one for each outcome variable) that are multiplied together to get the cross-product for each participant. The total can then be found by adding the cross-products of all participants. Table 17.2 illustrates this process. The total cross-product is a gauge of the overall relationship between the two variables. We can decompose this total. First, we can quantify how the relationship between the outcome variables is influenced by our experimental manipulation using the model cross-product (CP_M):

$$\text{CP}_M = \sum_{g=1}^{k} n \left[\begin{array}{c} \left(\overline{x}_{g(\text{Actions})} - \overline{X}_{\text{grand(Actions)}} \right) \\ \left(\overline{x}_{g(\text{Thoughts})} - \overline{X}_{\text{grand(Thoughts)}} \right) \end{array} \right] \quad (17.12)$$

The CP_M is calculated in a similar way to the model sum of squares. First, the difference between each group mean and the grand mean is calculated for each outcome variable. The cross-product is calculated by multiplying the differences found for each group. Each product is then multiplied by the number of scores within the group (as was done with the sum of squares). This principle is illustrated in Table 17.3. Finally, we can quantify how the relationship between the two outcome variables is influenced by individual differences/unmeasured variables using the residual cross-product (CP_R):

$$\text{CP}_R = \sum_{i=1}^{n} \left(x_{i(\text{Actions})} - \overline{X}_{\text{group(Actions)}} \right) \quad (17.13)$$
$$\left(x_{i(\text{Thoughts})} - \overline{X}_{\text{group(Thoughts)}} \right)$$

The CP_R is calculated in a similar way to the total cross-product, except that the group means are used rather than the grand mean. So, to calculate each of the difference scores, we take each score and subtract from it the mean of the group to which it belongs (see Table 17.4). The residual cross-product can also be calculated by subtracting the model cross-product from the total cross-product:

$$\begin{aligned}\text{CP}_R &= \text{CP}_T - \text{CP}_M \\ &= 5.47 - (-7.53) \\ &= 13 \end{aligned} \quad (17.14)$$

Each of the cross-products tells us something important about the relationship between the two outcome variables. Although I have used a simple scenario to keep the maths relatively simple, these principles can be easily extended to more complex scenarios. For example, if we have measured three outcome variables then the cross-products between pairs of outcome variables are calculated (as they were in this example). As the complexity of the situation increases, so does the amount of calculation that needs to be done. At times such as these the benefit of software like IBM SPSS Statistics becomes apparent!

17.4.4 The total SSCP matrix (T)

SSCP matrices are square. With two outcome variables (as in this example) the SSCP matrices will be 2×2 matrices, with three outcome variables they'd be 3×3

matrices, and so on. The total SSCP matrix, T, contains the total sums of squares for each outcome variable and the total cross-product between the outcome variables. You can think of the first column and first row as representing one outcome variable and the second column and row as representing the second outcome variable (Figure 17.4). We calculated the values in the matrix earlier in this section, and if we place them in the appropriate cells of the matrix we get:

$$T = \begin{pmatrix} 61.47 & 5.47 \\ 5.47 & 141.47 \end{pmatrix} \quad (17.15)$$

From the values in the matrix (and what they represent) it should be clear that the total SSCP represents both the total amount of variation that exists within the outcome variables and the total co-dependence that exists between them. Also note that the off-diagonal components are the same (they are both the total cross-product), because this value is equally important for both outcome variables.

17.4.5 The residual SSCP matrix (E)

The residual (or error) sum of squares and cross-products matrix, E, contains the residual sums of squares for each outcome variable and the residual cross-product between the two outcome variables. This SSCP matrix is like the total SSCP, except that the information relates to the error in the model (Figure 17.5). Placing the values that we calculated earlier in this section in the appropriate cells of the matrix, we get:

$$E = \begin{pmatrix} 51 & 13 \\ 13 & 122 \end{pmatrix} \quad (17.16)$$

The residual SSCP represents both the unsystematic variation that exists for each outcome variable and the co-dependence between the outcome variables that is due to unmeasured factors. As with the total SSCP, the off-diagonal elements are the same (they are both the residual cross-product).

17.4.6 The model SSCP matrix (H)

The model (or hypothesis) sum of squares and cross-product matrix, H, contains the model sums of squares for each outcome variable and the model cross-product between the two outcome variables (Figure 17.6). Placing the values that we calculated earlier in this section in the appropriate cells of the matrix, we get:

$$H = \begin{pmatrix} 10.47 & -7.53 \\ -7.53 & 19.47 \end{pmatrix} \quad (17.17)$$

The model SSCP represents both the systematic variation that exists for each outcome variable and the co-dependence between the outcome variables that is due to the model (i.e., which in experimental research is the experimental manipulation). Matrices are additive, which means that you can add (or subtract) two matrices together by adding (or subtracting) corresponding components. When we calculated univariate models we saw that the total sum of squares was the sum of the model sum of squares and the residual sum of squares (i.e., $SS_T = SS_M + SS_R$). The same is true in MANOVA, except that we add the SSCPs rather than single values:

$$T = H + E$$
$$= \begin{pmatrix} 10.47 & -7.53 \\ -7.53 & 19.47 \end{pmatrix} + \begin{pmatrix} 51 & 13 \\ 13 & 122 \end{pmatrix}$$
$$= \begin{pmatrix} 10.47+51 & -7.53+13 \\ -7.53+13 & 19.47+122 \end{pmatrix} \quad (17.18)$$
$$= \begin{pmatrix} 61.47 & 5.47 \\ 5.47 & 141.47 \end{pmatrix}$$

The demonstration that these matrices add up to should (hopefully) reinforce the idea that the MANOVA calculations are conceptually the same as for univariate models—the difference is that matrices are used rather than single values.

Table 17.4   Calculation of CP_R

Group	Actions	Actions – X̄ grand (Actions) (D_1)	Thoughts	Actions – X̄ grand (Thoughts) (D_2)	D_1 × D_2
CBT	5	0.10	14	0.60	0.06
	5	0.10	11	-2.40	-0.24
	4	-0.90	16	2.60	-2.34
	4	-0.90	13	-0.40	0.36
	5	0.10	12	-1.40	-0.14
	3	-1.90	14	0.60	-1.14
	7	2.10	12	-1.40	-2.94
	6	1.10	15	1.60	1.76
	6	1.10	16	2.60	2.86
	4	-0.90	11	-2.40	2.16
	4.9		**13.4**		**Σ = 0.40**
BT	4	0.30	14	-1.20	-0.36
	4	0.30	15	-0.20	-0.06
	1	-2.70	13	-2.20	5.94
	1	-2.70	14	-1.20	3.24
	4	0.30	15	-0.20	-0.06
	6	2.30	19	3.80	8.74
	5	1.30	13	-2.20	-2.86
	5	1.30	18	2.80	3.64
	2	-1.70	14	-1.20	2.04
	5	1.30	17	1.80	2.34
	3.7		**15.2**		**Σ = 22.60**
NT	4	-1.00	13	-2.00	2.00
	5	0.00	15	0	0.00
	5	0.00	14	-1.00	0.00
	4	-1.00	14	-1.00	1.00
	6	1.00	13	-2.00	-2.00
	4	-1.00	20	5.00	-5.00
	7	2.00	13	-2.00	-4.00
	4	-1.00	16	1.00	-1.00
	6	1.00	14	-1.00	-1.00
	5	0.00	18	3.00	0.00
	5		15		**Σ = -10.00**

$$CP_R = \sum D_1 \times D_2 = 13$$

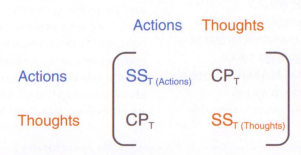

Figure 17.4   The components of the total SSCP

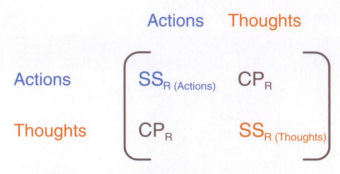

Figure 17.5   The components of the residual SSCP

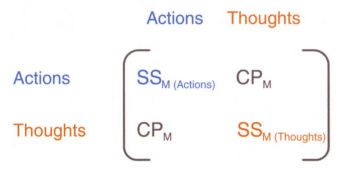

Figure 17.6   The components of the model SSCP

17.4.7 HE^{-1}: an analogue of F ▮▮▮▮

The univariate F is the ratio of systematic variance to unsystematic variance (i.e., it is a function of SS_M divided by SS_R).[3] The conceptual equivalent would therefore be to divide the matrix H by the matrix E. The matrix equivalent to division is to multiply by what's known as the inverse of a matrix. So, an analogue of F would be to divide H by E or in matrix terms to multiply H by the inverse of E (denoted by E^{-1}). The resulting matrix is called **HE^{-1}** and is a multivariate analogue of the univariate F (with respect to what it represents conceptually).

Calculating the inverse of a matrix is tricky, and there is no need for you to know how to do it to get a good conceptual grasp of MANOVA. If you're interested, Stevens (2002) and Namboodiri (1984) provide accessible

accounts of how to derive an inverse matrix, and having read them, you could look at Oliver Twisted. Everyone else can trust me that we end up with:

$$E^{-1} = \begin{pmatrix} 0.0202 & -0.0021 \\ -0.0021 & 0.0084 \end{pmatrix}$$
$$HE^{-1} = \begin{pmatrix} 0.2273 & -0.0852 \\ -0.1930 & 0.1794 \end{pmatrix} \tag{17.19}$$

Remember that HE^{-1} represents the ratio of systematic variance in the model to the unsystematic variance in the model, and so the resulting matrix is conceptually the same as the univariate F-statistic. There is another problem, though. In univariate models we get a single value for F, but, as equation (17.19) shows, when we divide matrices we end up with several values (four in this case). In fact, HE^{-1} will always contain p^2 values, where p is the number of outcome variables. The

problem is how to convert this matrix of values into a meaningful single value for which we can compute a p-value. This is the point at which we abandon any hope of understanding the maths and talk conceptually instead.

17.4.8   Discriminant function variates ▮▮▮▮

The problem of having several values with which to assess statistical significance can be simplified by converting the outcome variables into underlying dimensions or factors (this process will be discussed in Chapter 18). Most of this book has looked at how we can use a linear model to predict an outcome variable. Linear models are made up of a combination of predictor variables, each of which makes a unique contribution to the model. We can do a similar thing here, except that we want to do the opposite: predict an independent variable from a set of outcome variables. It is possible to calculate underlying linear dimensions of the outcome variables known as *variates* (or sometimes called *components*). In this context, we want to use these linear variates to predict to which group a person belongs (i.e., whether they were given CBT, BT or no treatment). Because we are using them to discriminate groups of people/cases these variates are called *discriminant functions* or **discriminant function variates**.

That's the theory in simplistic terms, but how do we discover these discriminant functions? Without going into too much detail, we use a mathematical procedure of maximization, such that the first discriminant function (V_1) is the linear combination of outcome variables that maximizes the differences between groups. It follows from this that the ratio of systematic to unsystematic variance

---

3 We use the mean squares, not the sum of squares, but the means squares are directly proportional to the sums of squares because they are the sums of squares divided by the degrees of freedom.

(SS_M/SS_R) will be maximized for this first variate, but subsequent variates will have smaller values of this ratio. Remember that this ratio is an analogue of what the F-statistic represents in univariate models, and so in effect we obtain the maximum possible value of the F-statistic when we look at the first discriminant function. This variate can be described in terms of a linear model equation because it is a linear combination of the outcome variables:

$$y_i = b_0 + b_1 X_{1i} + b_2 X_{2i}$$
$$V_{1i} = b_0 + b_1 \text{Outcome} 1_{1i} + b_2 \text{Outcome} 2_{2i} \quad (17.20)$$
$$= b_0 + b_1 \text{Actions}_i + b_2 \text{Thoughts}_i$$

Equation (17.20) shows the equation for two predictors and then extends this to show how a comparable form of this equation can describe discriminant functions. The b-values in the equation are weights that tell us something about the contribution of each outcome variable to the variate in question. In the linear models we have looked at in this book the values of b are estimated using the method of least squares. The values of b for the discriminant functions are obtained from the *eigenvectors* (see Jane Superbrain Box 9.3) of HE^{-1}. We can ignore b_0 as well because it serves only to locate the variate in geometric space, which isn't necessary when we're using the variate to discriminate groups.

In a situation in which there are only two outcome variables and two groups to predict, there will be only one variate. This makes the scenario very simple: by looking at the discriminant function of the outcome variables, rather than looking at the outcome variables themselves, we can obtain a single value of SS_M/SS_R for the discriminant function, and then assess this value for significance. However, in more complex cases where there are more than two outcome variables or three or more categories to predict (as is the case in our example) there will be more than one variate. The number of variates obtained will be the smaller of p (the number of

outcome variables) and $k-1$ (where k is the number of categories/groups to be predicted). In our example, both p and $k-1$ are 2, so we will find two variates. I mentioned earlier that the b-values that describe the variates are obtained by calculating the eigenvectors of the matrix HE^{-1}, and in fact, there will be two eigenvectors derived from this matrix: one with the b-values for the first variate, and one with the b-values of the second variate. Conceptually speaking, eigenvectors are the vectors associated with a given matrix that are unchanged by transformation of that matrix to a diagonal matrix (refer to Jane Superbrain Box 9.3 for a visual explanation of eigenvectors and eigenvalues). In an identity matrix the off-diagonal elements are zero (Figure 17.3) and by changing HE^{-1} into an identity matrix we eliminate all of the off-diagonal elements (thus reducing the number of values to be considered for significance testing). Therefore, by calculating the eigenvectors and eigenvalues, we still end up with values that represent the ratio of systematic to unsystematic variance (because they are unchanged by the transformation), but there are considerably fewer of them.

The calculation of eigenvectors is extremely complex (insane students can consider reading Namboodiri, 1984), so trust me that for the matrix HE^{-1} the eigenvectors obtained are:

$$\text{eigenvector}_1 = \begin{pmatrix} 0.603 \\ -0.335 \end{pmatrix}$$
$$\text{eigenvector}_2 = \begin{pmatrix} 0.425 \\ 0.339 \end{pmatrix} \quad (17.21)$$

Replacing these values into the two equations for the variates (equation (17.20)), and bearing in mind we can ignore b_0, we obtain the models described in the following equation:

$$V_{1i} = 0.603 \text{Actions}_i - 0.335 \text{Thoughts}_i \quad (17.22)$$
$$V_{2i} = 0.425 \text{Actions}_i + 0.339 \text{Thoughts}_i$$

It is possible to use the equations for each variate to calculate a score for each person

on the variate. For example, the first participant in the CBT group carried out 5 obsessive actions, and had 14 obsessive thoughts. Therefore, this participant's score on variate 1 would be –1.675:

$$V_1 = (0.603 \times 5) - (0.335 \times 14) = -1.675 \quad (17.23)$$

Their score for variate 2 would be 6.87:

$$V_2 = (0.425 \times 5) + (0.339 \times 14) = 6.871 \quad (17.24)$$

If we calculated these variate scores for each participant and then calculated the SSCP matrices (e.g., H, E, T and HE^{-1}) that we used previously, we would find that all of them have cross-products of zero. The reason for this is that the variates extracted from the data are orthogonal, which means that they are uncorrelated. In short, the variates extracted are independent dimensions constructed from a linear combination of the outcome variables that were measured.

This data reduction has a very useful property in that if we look at the matrix HE^{-1} calculated from the variate scores (rather than the outcome variables) we find that the off-diagonal elements (the cross-products) are zero. The diagonal elements of this matrix represent the ratio of the systematic variance to the unsystematic variance (i.e., SS_M/SS_R) for each of the underlying variates. So, in this example, this means that instead of having four values representing the ratio of systematic to unsystematic variance, we now have only two. This reduction may not seem a lot. However, in general if we have p outcome variables, then ordinarily we would end up with p^2 values representing the ratio of systematic to unsystematic variance; by using discriminant functions, we reduce this number back to p. For example, with four outcome variables we would end up with four values rather than 16.

For the data in our example, the matrix HE^{-1} calculated from the variate scores is:

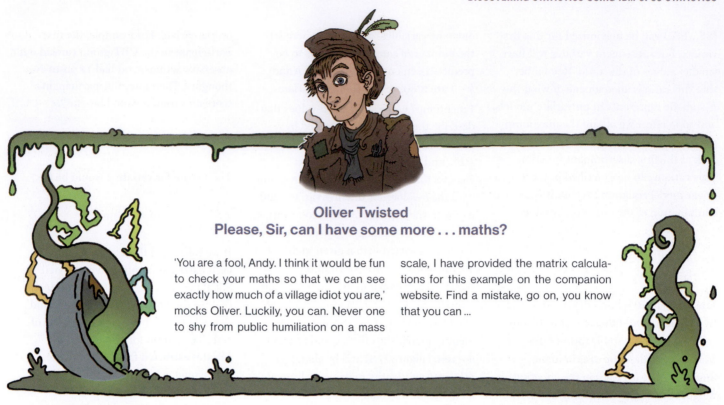

Oliver Twisted
Please, Sir, can I have some more . . . maths?

'You are a fool, Andy. I think it would be fun to check your maths so that we can see exactly how much of a village idiot you are,' mocks Oliver. Luckily, you can. Never one to shy from public humiliation on a mass scale, I have provided the matrix calculations for this example on the companion website. Find a mistake, go on, you know that you can ...

$$HE_{\text{variate}}^{-1} = \begin{pmatrix} 0.335 & 0 \\ 0 & 0.073 \end{pmatrix} \qquad (17.25)$$

It is evident from this matrix that we have two values to consider when assessing the significance of the group differences. It may seem like a complex procedure to reduce the data in this way: however, it transpires that the values along the diagonal of the matrix for the variates (namely 0.335 and 0.073) are the *eigenvalues* of the original HE^{-1} matrix. Therefore, these values can be calculated directly from the original scores without first forming the eigenvectors. If you have lost all sense of rationality and want to know how these eigenvalues are calculated, then see Oliver Twisted. These eigenvalues are conceptually equivalent to the univariate F-statistic and so the final step is to assess how large these values are compared to what we would expect if there were no effect in the population. There are four common ways to do this.

17.4.9 The Pillai–Bartlett trace (V) ▮▮▮▮

The **Pillai–Bartlett trace** (also known as Pillai's trace) is given by

$$V = \sum_{i=1}^{s} \frac{\lambda_i}{1 + \lambda_i} \qquad (17.26)$$

in which λ represents the eigenvalues for each of the discriminant variates and s represents the number of variates. Pillai's trace is the sum of the proportion of explained variance on the discriminant functions. It is similar to R^2 (the ratio of SS_M/SS_T). For our data, Pillai's trace turns out to be 0.319, which can be transformed to a value that has an approximate F-distribution:

$$V = \frac{0.335}{1 + 0.335} + \frac{0.073}{1 + 0.073} = 0.319 \qquad (17.27)$$

17.4.10 Hotelling's T^2 ▮▮▮▮

The **Hotelling–Lawley trace** (also known as Hotelling's T^2; Figure 17.7) is the sum of the eigenvalues for each variate:

$$T = \sum_{i=1}^{s} \lambda_i \qquad (17.28)$$

so for these data its value is 0.335 + 0.073 = 0.408. This test statistic is the

Figure 17.7   Harold Hotelling enjoying my favorite activity of drinking tea

sum of SS_M/SS_R for each of the variates and so it compares directly to the univariate F-statistic.

17.4.11 Wilks's lambda (Λ)

Wilks's lambda is the product of the *unexplained* variance on each of the variates:

$$\Lambda = \prod_{i=1}^{s} \frac{1}{1+\lambda_i} \qquad (17.29)$$

(the ∏ symbol is like the summation symbol (Σ) that we have encountered already except that it means *multiply* rather than add up). Wilks's lambda represents the ratio of error variance to total variance (SS_R/SS_T) for each variate. Large eigenvalues (which in themselves represent a large experimental effect) lead to small values of Wilks's lambda: hence statistical significance is found when Wilks's lambda is small. In this example Wilks's lambda is 0.698:

$$\Lambda = \frac{1}{1+0.335} \times \frac{1}{1+0.073} = 0.698 \qquad (17.30)$$

17.4.12 Roy's largest root

Roy's largest root makes me think of some bearded statistician with a garden spade digging up an enormous parsnip (or similar root vegetable). It isn't a parsnip but, as the name suggests, it is the eigenvalue (or 'root') for the first variate:

$$\Theta = \lambda_{largest} \qquad (17.31)$$

So, in a sense it is the same as the Hotelling–Lawley trace but for the first variate only. Roy's largest root represents the proportion of explained variance to unexplained variance (SS_M/SS_R) for the first discriminant function.[4] For the data in this example, the value of Roy's largest root is 0.335 (the eigenvalue for the first variate). This value is conceptually the same as the univariate *F*-statistic. It should be apparent, from what we have learnt about the maximizing properties of these discriminant variates, that Roy's root represents the maximum possible between-group difference given the data collected. Therefore, this statistic should in many cases be the most powerful.

17.5 Practical issues when conducting MANOVA

17.5.1 Assumptions and how to check them

MANOVA has similar assumptions to all the models in this book (see Chapter 6) but extended to the multivariate case:

- **Independence**: Residuals should be statistically independent.
- **Random sampling**: Data should be randomly sampled from the population of interest and measured at an interval level.
- **Multivariate normality**: In univariate models we assume that our residuals are normally distributed. In the case of MANOVA, we assume that the residuals have multivariate normality.
- **Homogeneity of covariance matrices**: In univariate models, it is assumed that the variances in each group are roughly equal (homogeneity of variance). In MANOVA we assume that this is true for each outcome variable, but also that the correlation between any two outcome variables is the same in all groups. This assumption is examined by testing whether the population **variance–covariance matrices** of the different groups in the analysis are equal.[5]

We can correct for bias in the usual ways; however, the assumption of multivariate normality cannot be tested using SPSS Statistics and so the only practical solution is to check the assumption of univariate normality of residuals for each outcome variable in turn (see Chapter 6). This solution is practical (because it is easy to implement) and useful (because univariate normality is a necessary condition for multivariate normality), but it does not *guarantee* multivariate normality.

The effect of violating the assumption of equality of covariance matrices is unclear, except that Hotelling's T^2 is robust in the two-group situation when sample sizes are equal (Hakstian, Roed, & Lind, 1979). The assumption can be tested using **Box's test**, which should be non-significant if the matrices are similar. Box's test is notoriously susceptible to deviations from multivariate normality and so can be non-significant not because the matrices are similar, but because the assumption of multivariate normality is not tenable. Also, as with any significance test, in large samples Box's test could be significant even when covariance matrices are relatively similar.

If sample sizes are equal then people tend to disregard Box's test, because (1) it is unstable, and (2) in this situation we can assume that Hotelling's and Pillai's statistics are robust (see Section 17.5.3). However, if group sizes are different, then robustness cannot be assumed. The more outcome variables you have measured, and the greater the differences in sample sizes, the more distorted the probability values become. Tabachnick and Fidell (2012) suggest that if the larger samples produce greater variances and covariances then the probability values will be conservative (and so significant findings can be trusted). However, if it is the smaller samples that produce the larger variances and covariances then the probability values will be liberal and so significant

---

4 This statistic is sometimes characterized as $\lambda_{largest}/(1+\lambda_{largest})$, but this is not the statistic reported by SPSS Statistics.

5 For those of you who read about SSCP matrices, if you think about the relationship between sums of squares and variance, and cross-products and correlations, it should be clear that a variance–covariance matrix is basically a standardized form of an SSCP matrix.

differences should be treated with caution (although non-significant effects can be trusted). Therefore, the variance–covariance matrices for samples should be inspected to assess whether the printed probabilities for the multivariate test statistics are likely to be conservative or liberal. In the event that you cannot trust the printed probabilities, there is little you can do except equalize the samples by randomly deleting cases in the larger groups (although with this loss of information comes a loss of power, and of course your results will have been influenced by the process of deletion, so you should do a sensitivity analysis by comparing results when you delete different sets of random cases).

17.5.2 What to do when assumptions are violated ▮▮▮▮

SPSS Statistics doesn't offer a non-parametric version of MANOVA; however, some ideas have been put forward based on ranked data. There are some techniques that can be beneficial when multivariate normality or homogeneity of covariance matrices cannot be assumed (Zwick, 1985). In addition, there are robust methods for straightforward designs with multiple outcome variables, such as the Munzel–Brunner method, which can be implemented in the software R (Wilcox, 2017); although it can't be done using the *WRS2* package, Field et al. (2012) have a step-by-step guide if you fancy getting into R. Although you will see a `Bootstrap...` button in the dialog box for MANOVA it does not bootstrap the main tests and is, ultimately, disappointing.

17.5.3 Choosing a test statistic ▮▮▮▮

Only when there is one underlying variate will the four test statistics necessarily be the same, which raises the question of which one is 'best'. As ever,

when addressing this question we really need to know which has the most power, the least error and the greatest robustness to violations of test assumptions.

Which test statistic should I use?

Research investigating power (Olson, 1974, 1976, 1979; Stevens, 1980) suggests that: (1) for small and moderate sample sizes the four statistics differ little; (2) if group differences are concentrated on the first variate Roy's statistic should have the most power (because it takes account of only that first variate) followed by Hotelling's trace, Wilks's lambda and Pillai's trace; (3) when groups differ along more than one variate, this power order is reversed (i.e., Pillai's trace is most powerful and Roy's root is least); (4) unless sample sizes are large it's probably wise to use fewer than 10 outcome variables.

In terms of robustness, all four test statistics are relatively robust to violations of multivariate normality (although Roy's root is affected by platykurtic distributions—see Olson, 1976). Roy's root is not robust when the homogeneity of covariance matrix assumption is untenable (Stevens, 1979). Bray and Maxwell (1985) conclude that when sample sizes are equal the Pillai–Bartlett trace is the most robust to violations of assumptions, but when sample sizes are unequal this statistic is affected by violations of the assumption of equal covariance matrices. As a rule, with unequal group sizes, check the homogeneity of covariance matrices; if they seem homogeneous and if the assumption of multivariate normality is tenable, then assume that Pillai's trace is accurate.

17.5.4 Follow-up analysis ▮▮▮▮

The traditional approach is to follow a significant MANOVA with separate

univariate models (ANOVA) on each of the outcome variables. You might think that this approach is daft, given that I wrote earlier that multiple univariate models would inflate the Type I error rate. I'd agree that it's daft, but some people argue that the univariate *F*-statistics are 'protected' by the initial MANOVA (Bock, 1975). The logic is that the overall multivariate test protects against inflated Type I error rates because if that initial test is non-significant (i.e., the null hypothesis is true) then the subsequent univariate *F*-statistics are ignored (because any significant *F* must be a Type I error because the null hypothesis is true). This notion of protection is dubious because a significant MANOVA usually reflects a significant difference for some, but not all, of the outcome variables. This argument of 'protection' applies only to the outcome variables for which group differences genuinely exist (see Bray & Maxwell, 1985, pp. 40–41), not for all outcome variables that you've included in the model. Despite this limitation, people tend to interpret univariate *F*s on *all* outcome variables. Therefore, if you do use univariate *F*s then you ought to apply a Bonferroni correction (Harris, 1975).

The bigger problem I have with univariate *F*s is that they don't relate to what the multivariate tests look at. Remember that the multivariate test statistic quantifies the extent to which groups can be differentiated by a *linear combination of the outcome variables*. Subsequent univariate *F*s look at the outcome variables as independent entities, not as a linear combination. Therefore, they make no sense as a follow-up strategy. An alternative that *is* consistent with the multivariate test statistics is discriminant analysis, which finds the linear combination(s) of the outcome variables that best *separates* (or discriminates) the groups. The major advantage of this approach over multiple univariate *F*s is that it reduces the outcome variables to a set of underlying dimensions thought to reflect substantive theoretical dimensions. As such, it is true to the ethos of MANOVA.

17.6 MANOVA using SPSS Statistics ▮▮▮▮

In the remainder of this chapter we will use the OCD data (Section 17.2.2) to illustrate the application and interpretation of MANOVA. Either load the data in the file **OCD.sav** or enter the data manually. If you enter the data manually you need three columns: one is a coding variable for the **Group** variable (I used the codes CBT = 1, BT = 2, NT = 3), and in the remaining two columns enter the scores for each outcome variable. Figure 17.8 overviews the analysis procedure: basically, explore the data as you normally would, run the MANOVA, then follow up this analysis with a discriminant function analysis. Some of you will want to look at univariate ANOVAs so I've included that in the diagram, but personally I'd avoid them.

17.6.1 The main analysis ▮▮▮▮

Select *Analyze* ▶ *General Linear Model* ▶ [GLM] **Multivariate...** to access the dialog box in Figure 17.9, which is very similar to the dialog box for factorial designs (Chapter 14) except that the box labeled *Dependent Variables* has room for several variables. Drag the two outcome variables (i.e., **Actions** and **Thoughts**) to the *Dependent Variables* box (or click ➡). Drag (or click ➡) **Group** to the *Fixed Factor(s)* box. There is also a box for covariates. For this analysis there are none, but you can apply the principles of ANCOVA to the multivariate case and conduct multivariate analysis of covariance (MANCOVA). The buttons in this dialog box are pretty much the same as we have seen in the past few chapters.

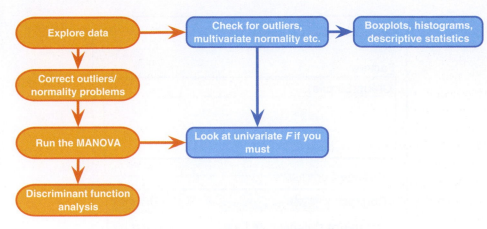

Figure 17.8 Overview of the general procedure for MANOVA

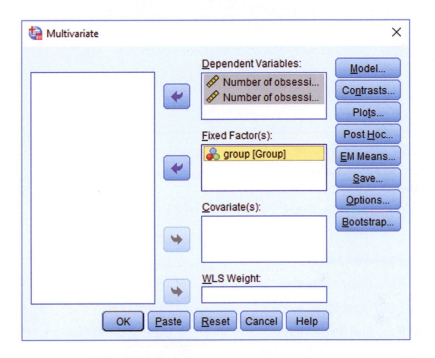

Figure 17.9 Main dialog box for MANOVA

17.6.2 Multiple comparisons in MANOVA ▮▮▮▮

The default, but silly, way to follow up a MANOVA is to look at individual univariate *F*s for each outcome variable. You have the same options as we've seen for other univariate models (see Chapter 12). Click [Contrasts...] to open the dialog box (Figure 17.10) for specifying standard contrasts for the categorical predictors in the model. Refer back to Table 12.6, which describes what the standard contrasts

compare. For this example it makes sense to use a *simple* contrast that compares each of the therapy groups to the no treatment control group. The no treatment control was coded as the last category (it had the largest code in the data editor), so select **Group**, change the contrast to *Simple* and select ◉ **Last** (see Figure 17.10).

Instead of a contrast, we could carry out *post hoc* tests that compare each group to all others by clicking [Post Hoc...] to access a dialog box that is the same as for

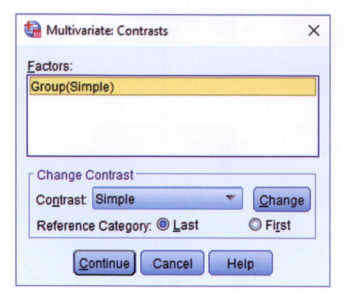

Figure 17.10 Contrasts for independent variable(s) in MANOVA

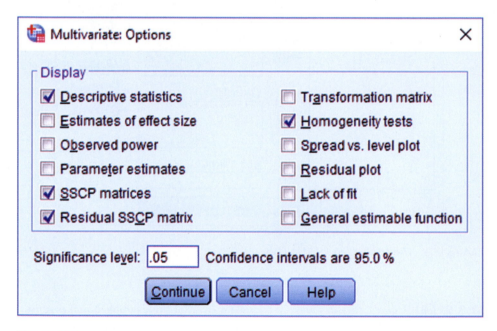

Figure 17.11 Additional options in MANOVA

- *SSCP matrices*: If this option is selected, SPSS will produce the model, error and total SSCP matrices, which can be useful for understanding the computation of the MANOVA. If you skipped the theory section you might be happy not to select this option and live in blissful ignorance!
- *Residual SSCP matrix*: If this option is selected, SPSS produces the error SSCP matrix, the error variance–covariance matrix and the error correlation matrix. **Bartlett's test of sphericity** examines whether the variance–covariance matrix is proportional to an identity matrix (i.e., the covariances are zero and the variances—the values along the diagonal—are roughly equal).

The remaining options are the same as for factorial designs and were described in Chapter 14.

17.7 Interpreting MANOVA ▌▌▌▌

17.7.1 Preliminary analysis and testing assumptions ▌▌▌▌

Output 17.1 contains the overall and group means and standard deviations for each outcome variable in turn. These values correspond to those in Table 17.1. It is clear from the means that participants had many more obsession-related thoughts than behaviors.

Output 17.2 shows Box's test of the assumption of equality of covariance matrices (see Section 17.5.1). This statistic should be *non-significant*, which it is, $p = 0.18$ (which is greater than 0.05): hence, the covariance matrices are roughly equal as assumed. Bartlett's test of sphericity tests whether the assumption of sphericity has been met and is useful only in univariate repeated-measures designs because MANOVA does not require this assumption. Basically I included it just to point out that you should ignore it. That's just the kind of maverick I am.

factorial designs (see Figure 14.9). To choose a test, see the discussion in Section 12.5. For the purposes of this example, I suggest selecting two of my usual recommendations: REGWQ and Games–Howell. Once you have selected *post hoc* tests return to the main dialog box.

17.6.3 Additional options ▌▌▌▌

Click [Options…] to access the dialog box in Figure 17.11, which the same as the one we used for factorial designs (see Section 14.5.6). There are a few options that are worth mentioning and haven't been discussed before.

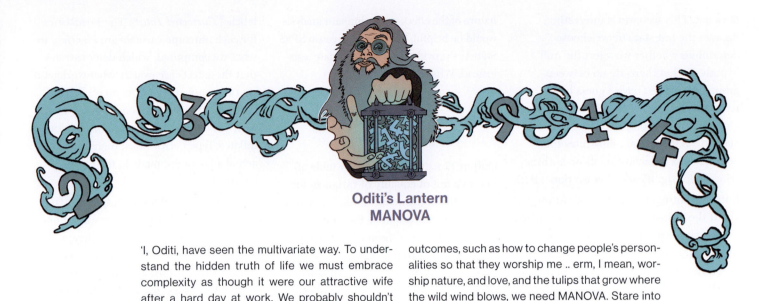

Oditi's Lantern
MANOVA

'I, Oditi, have seen the multivariate way. To understand the hidden truth of life we must embrace complexity as though it were our attractive wife after a hard day at work. We probably shouldn't have sex with it, though. To understand complex outcomes, such as how to change people's personalities so that they worship me .. erm, I mean, worship nature, and love, and the tulips that grow where the wild wind blows, we need MANOVA. Stare into my lantern and discover all about it.'

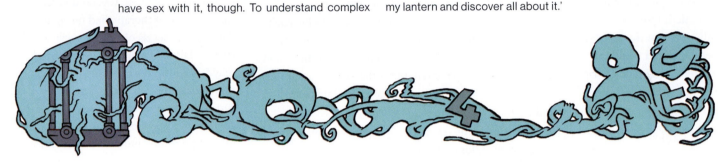

17.7.2 MANOVA test statistics ▮▮▮▮

Output 17.3 shows the test statistics for the intercept of the model (see, I told you, even MANOVA can be characterized as a linear model, although how this is done is beyond the scope of my brain) and for the group variable. The group effect tells us whether the therapies had different effects on the OCD clients. The four multivariate test statistics and their values correspond to those calculated in Sections 17.4.9–17.4.12. These values are transformed into an *F-statistic* (the degrees of freedom vary for each test statistic) and the *p*-value associated with this *F* is in the column labeled *Sig.* For these data, Pillai's trace ($p = 0.049$), Wilks's lambda ($p = 0.050$) and Roy's largest root ($p = 0.020$) all reach the criterion for significance (they are below 0.05), but Hotelling's trace ($p = 0.051$)

Descriptive Statistics

	group	Mean	Std. Deviation	N
Number of obsession–related behaviours	CBT	4.90	1.197	10
	BT	3.70	1.767	10
	No Treatment Control	5.00	1.054	10
	Total	4.53	1.456	30
Number of obsession–related thoughts	CBT	13.40	1.897	10
	BT	15.20	2.098	10
	No Treatment Control	15.00	2.357	10
	Total	14.53	2.209	30

Output 17.1

Box's Test of Equality of Covariance Matrices

Box's M	9.959
F	1.482
df1	6
df2	18168.923
Sig.	.180

Tests the null hypothesis that the observed covariance matrices of the dependent variables are equal across groups.

a. Design: Intercept + Group

Bartlett's Test of Sphericity[a]

Likelihood Ratio	.042
Approx. Chi-Square	5.511
df	2
Sig.	.064

Tests the null hypothesis that the residual covariance matrix is proportional to an identity matrix.

a. Design: Intercept + Group

Output 17.2

does not. This scenario is interesting, because the test statistic we choose determines whether we reject the null hypothesis that there are no between-group differences (have a guess how researcher degrees of freedom might have a negative role to play here—see Section 3.3.2). It again illustrates the pointlessness of having an all-or-nothing criterion for significance (see Section 3.2.2).

Enough ranting. Given what we know about the robustness of Pillai's trace when sample sizes are equal, we might trust the result of that test statistic, which indicates a significant difference. This example also highlights the additional power associated with Roy's root (note how this statistic is more significant than all others) when the test assumptions have been met and when the group differences are focussed on one variate (which, as we will see later, they are).

We should probably conclude that the type of therapy employed had a significant effect on OCD. The nature of this effect is not clear from the multivariate test statistic: it tells us nothing about which groups differed from which or about whether the effect of therapy was on the obsession-related thoughts, the obsession-related behaviors or a combination of both. To determine the

nature of the effect, a discriminant analysis would be helpful, but for some reason SPSS Statistics provides us with univariate tests instead. What a bilge rat.

17.7.3 Univariate test statistics ▮▮▮▮

Output 17.4 shows a summary table of Levene's test of equality of variances for each of the outcome variables. These tests are the same as would be found if univariate tests had been conducted on each outcome variable. Levene's test should be non-significant for all outcome variables if the assumption of homogeneity of variance has been met (but see Section 6.11.2). If you buy into Levene's test being useful (ho hum), Output 17.4 shows that the assumption has been met, which strengthens the case for assuming that the multivariate test statistics are robust.

The second table shows the univariate Fs for each outcome variable. In Sections 17.4.1 and 17.4.2 we computed various values for both actions and thoughts that are in this table: the model sum of squares (in the row labeled $Group$), the residual sum of squares (in the row labeled $Error$) and total sums of squares (in the row

labeled $Corrected\ Total$). The F-statistics for each outcome variable are *identical* to what we computed, which demonstrates that these tests are exactly what we'd get if we ran one-way ANOVAs on each outcome variable independently. As such, MANOVA offers *hypothetical* protection of inflated Type I error rates: there is no actual adjustment made to the values obtained.

Based on the p-values in Output 17.4 (in the column labeled $Sig.$) there was a non-significant difference between therapy groups in terms of both obsession-related thoughts ($p = 0.136$) and obsession-related behaviors ($p = 0.080$). Based on univariate tests, then, we should conclude that the type of therapy had no significant effect on the levels of OCD experienced by clients. Those of you who are still awake may have noticed something odd: the multivariate test statistics led us to conclude that therapy had had a significant impact on OCD, yet the univariate results indicate that therapy has not been successful.

The reason for the anomaly is that the multivariate test takes account of the correlation between outcome variables and looks at whether groups can be distinguished by a *linear combination of the outcome variables*. This suggests that it is not thoughts or actions in themselves that distinguish the therapy groups, but some combination of them. The discriminant function analysis will provide more insight into this conclusion.

17.7.4 SSCP matrices ▮▮▮▮

Outputs 17.5 and 17.6 will be produced if you selected the two options to display SSCP matrices (Section 17.6.3). Output 17.5 displays the model SSCP (H), which is labeled *Hypothesis Group* (I have shaded this matrix blue) and the error SSCP (E) which is labeled *Error* (shaded yellow). The values in the model and error

Multivariate Tests[a]

Effect		Value	F	Hypothesis df	Error df	Sig.
Intercept	Pillai's Trace	.983	745.230[b]	2.000	26.000	.000
	Wilks' Lambda	.017	745.230[b]	2.000	26.000	.000
	Hotelling's Trace	57.325	745.230[b]	2.000	26.000	.000
	Roy's Largest Root	57.325	745.230[b]	2.000	26.000	.000
Group	Pillai's Trace	.318	2.557	4.000	54.000	.049
	Wilks' Lambda	.699	2.555[b]	4.000	52.000	.050
	Hotelling's Trace	.407	2.546	4.000	50.000	.051
	Roy's Largest Root	.335	4.520[c]	2.000	27.000	.020

a. Design: Intercept + Group

b. Exact statistic

c. The statistic is an upper bound on F that yields a lower bound on the significance level.

Output 17.3

MULTIVARIATE ANALYSIS OF VARIANCE (MANOVA)

SELF TEST

Why might the univariate tests be non-significant when
the multivariate tests were significant?

(i.e., group variances are the same), and whether the off-diagonal elements are approximately zero (i.e., the outcome variables are not correlated). For these data, the variances are quite different (1.89 compared to 4.52) and the covariances slightly different from zero (0.48),

matrices correspond to the values we calculated in Sections 17.4.6 and 17.4.5, respectively. These matrices provide insight into the pattern of the data, looking at the values of the cross-products to indicate the relationship between outcome variables. In this example, the sums of squares for the error SSCP matrix are substantially bigger than in the model (or group) SSCP matrix, whereas the absolute values of the cross-products are fairly similar. This pattern suggests that if the MANOVA is significant then it might be the relationship between outcome variables that is important rather than the individual outcome variables themselves. Output 17.6 shows the residual SSCP matrix again, but this time it includes the variance–covariance matrix and the correlation matrix. These matrices are related. If you think back to Chapter 8, you might remember that the covariance is the average cross-product. Likewise, the variance is the average sum of squares. Hence, the variance–covariance matrix is the average form of the SSCP matrix. Similarly, we saw in Chapter 8 that the correlation was a standardized version of the covariance and so the correlation matrix represents the standardized form of the variance–covariance matrix. As with the SSCP matrix, these other matrices are useful for assessing the extent of the error in the model. The variance–covariance matrix is especially useful because Bartlett's test of sphericity is based on it. Bartlett's test examines whether this matrix is proportional to an identity matrix (see earlier). Therefore, Bartlett's test tests whether the diagonal elements of the variance–covariance matrix are equal

Levene's Test of Equality of Error Variances[a]

	F	df1	df2	Sig.
Number of obsession–related behaviours	1.828	2	27	.180
Number of obsession–related thoughts	.076	2	27	.927

Tests the null hypothesis that the error variance of the dependent variable is equal across groups.

a. Design: Intercept + Group

Tests of Between-Subjects Effects

Source	Dependent Variable	Type III Sum of Squares	df	Mean Square	F	Sig.
Corrected Model	Number of obsession–related behaviours	10.467[a]	2	5.233	2.771	.080
	Number of obsession–related thoughts	19.467[b]	2	9.733	2.154	.136
Intercept	Number of obsession–related behaviours	616.533	1	616.533	326.400	.000
	Number of obsession–related thoughts	6336.533	1	6336.533	1402.348	.000
Group	Number of obsession–related behaviours	10.467	2	5.233	2.771	.080
	Number of obsession–related thoughts	19.467	2	9.733	2.154	.136
Error	Number of obsession–related behaviours	51.000	27	1.889		
	Number of obsession–related thoughts	122.000	27	4.519		
Total	Number of obsession–related behaviours	678.000	30			
	Number of obsession–related thoughts	6478.000	30			
Corrected Total	Number of obsession–related behaviours	61.467	29			
	Number of obsession–related thoughts	141.467	29			

a. R Squared = .170 (Adjusted R Squared = .109)
b. R Squared = .138 (Adjusted R Squared = .074)

Output 17.4

Between–Subjects SSCP Matrix

			Number of obsession–related behaviours	Number of obsession–related thoughts
Hypothesis	Intercept	Number of obsession–related behaviours	616.533	1976.533
		Number of obsession–related thoughts	1976.533	6336.533
	Group	Number of obsession–related behaviours	10.467	-7.533
		Number of obsession–related thoughts	-7.533	19.467
Error		Number of obsession–related behaviours	51.000	13.000
		Number of obsession–related thoughts	13.000	122.000

Based on Type III Sum of Squares

Output 17.5

Residual SSCP Matrix

		Number of obsession-related behaviours	Number of obsession-related thoughts
Sum-of-Squares and Cross-Products	Number of obsession-related behaviours	51.000	13.000
	Number of obsession-related thoughts	13.000	122.000
Covariance	Number of obsession-related behaviours	1.889	.481
	Number of obsession-related thoughts	.481	4.519
Correlation	Number of obsession-related behaviours	1.000	.165
	Number of obsession-related thoughts	.165	1.000

Based on Type III Sum of Squares

Output 17.6

SELF ✂ TEST

Based on what you have learnt in previous chapters, interpret the table of contrasts in your output.

so Bartlett's test has come out as nearly significant (see Output 17.2). Although this discussion is irrelevant to the multivariate tests, I hope that by expanding upon them here you can relate these ideas back to the issues of sphericity raised in Chapter 15, and see more clearly how this assumption is tested.

17.7.5 Contrasts ▮▮▮▮

The univariate ANOVAs were non-significant, so we should not interpret the contrasts that we requested. However, just for practice, try the self-test.

17.8 Reporting results from MANOVA ▮▮▮▮

Reporting a MANOVA is much like reporting a univariate comparison of group means. As you can see in Output 17.3, the multivariate tests are converted into approximate Fs, and people often just report these Fs in the usual way. Personally, I think the multivariate test statistic should be quoted as well. There are four different multivariate tests reported in Output 17.3;

Cramming Sam's Tips
MANOVA

- MANOVA is used to test the difference between groups across several outcome variables/outcomes simultaneously.
- Box's test looks at the assumption of equal covariance matrices. This test can be ignored when sample sizes are equal because when they are, some MANOVA test statistics are robust to violations of this assumption. If group sizes differ this test should be inspected. If the value of *Sig.* is less than 0.001 then the results of the analysis should not be trusted (see Section 17.7.1).
- The table labeled *Multivariate Tests* gives us four test statistics (Pillai's trace, Wilks's lambda, Hotelling's trace

and Roy's largest root). I recommend using Pillai's trace. If the value of *Sig.* for this statistic is less than 0.05 then the groups differ significantly with respect to a linear combination of the outcome variables.
- Univariate *F*-statistics can be used to follow up the MANOVA (a different *F*-statistic for each outcome variable). The results of these are listed in the table entitled *Tests of Between-Subjects Effects*. These *F*-statistics can in turn be followed up using contrasts. Personally I recommend *discriminant function analysis* over this approach.

I'll report each one in turn (note that the degrees of freedom and value of F change), but in reality you would just report one of the four:

✓ Using Pillai's trace, there was a significant effect of therapy on the number of obsessive thoughts and behaviors, $V = 0.32$, $F_{(4, 54)} = 2.56$, $p = 0.049$.

✓ Using Wilks's statistic, there was a significant effect of therapy on the number of obsessive thoughts and behaviors, $\Lambda = 0.70$, $F_{(4, 52)} = 2.56$, $p = 0.05$.

✓ Using Hotelling's trace statistic, there was not a significant effect of therapy on the number of obsessive thoughts and behaviors, $T = 0.41$, $F_{(4, 50)} = 2.55$, $p = 0.051$.

✓ Using Roy's largest root, there was a significant effect of therapy on the number of obsessive thoughts and behaviors, $\Theta = 0.35$, $F_{(2, 27)} = 4.52$, $p = 0.02$.

We can also report the follow-up univariate Fs in the usual way (see Outputs 17.3 and 17.4):

✓ Using Pillai's trace, there was a significant effect of therapy on the number of obsessive thoughts and behaviors, $V = 0.32$, $F_{(4, 54)} = 2.56$, $p = 0.049$. However, separate univariate tests on the outcome variables revealed non-significant treatment effects on obsessive thoughts, $F_{(2, 27)} = 2.15$, $p = 0.136$, and behaviors, $F_{(2, 27)} = 2.77$, $p = 0.08$.

17.9 Following up MANOVA with discriminant analysis ▌▌▌▌

I mentioned earlier on that a significant MANOVA could be followed up using discriminant analysis (sometimes called discriminant function analysis). In my opinion this method is the best way to follow up a significant MANOVA because a MANOVA looks at whether groups differ along *a linear combination* of outcome variables, and discriminant analysis (unlike univariate Fs) breaks down the linear combination in more detail. In discriminant analysis we look to see how we can best separate (or discriminate) a set of groups using several predictors (it is a little like logistic regression but where there are several groups rather than two).[6] In MANOVA we predicted a set of outcome measures from a grouping variable, whereas in discriminant function analysis we do the opposite by predicting a grouping variable from a set of outcome measures. The core underlying principles of these tests are the same: remember from the theory of MANOVA that it works by identifying linear variates that best differentiate the groups and these 'linear variates' are the 'functions' in discriminant function analysis.

Select *Analyze* ▶ *Classify* ▶ 🟥 **Discriminant**... to access the dialog box in Figure 17.12. Drag **Group** to the box labeled *Grouping Variable* (or click ➡), then click Define Range... to activate a dialog box in which you specify the value of the highest and lowest coding values (1 and 3 in this case). Once you have specified the codes used for the grouping variable, drag **Actions** and **Thoughts** to the box labeled *Independents* (or click ➡). There are two options that determine how predictors are entered into the model. Because in MANOVA the outcome variables are analyzed simultaneously, we want to select the default option of ⊙ Enter independents together.

Click Statistics... to activate the dialog box in Figure 17.13. This dialog box allows us to request group means, univariate ANOVAs (*F*s) and *Box's test of equality of covariance matrices*, all of which have already been provided in the MANOVA output (so don't ask for them again). Furthermore, we can ask for the within-group correlation and covariance matrices, which are the same as the residual correlation and covariance matrices seen in Output 17.6. There is also an option to display a *Separate-groups covariance* matrix, which can be useful for gaining insight into the relationships between outcome variables for each group (this matrix is something that the MANOVA procedure doesn't display, and I recommend selecting it). Finally, we can ask for a total covariance matrix, which displays covariances and variances of the outcome variables overall. Another useful option is to select *Unstandardized* function coefficients to produce the unstandardized *b*s for each variate (see equation (17.22)). Click Continue to return to the main dialog box.

Click Classify... to access the dialog box in Figure 17.14, in which you select how prior probabilities are determined. If your group sizes are equal then leave the default setting alone; however, if you have an unbalanced design then it is beneficial to base prior probabilities on the observed group sizes. The default option for basing the analysis on the within-group covariance matrix is fine (because this is the matrix upon which the MANOVA is based). You should also request a combined-groups plot, which will plot the variate scores for each participant grouped according to the therapy they were given. The separate-groups plots show the same thing but using different graphs for each of the groups; when the number of groups is small it is better to select a combined plot because it is easier to interpret. The

---

6 I could just as easily describe discriminant analysis rather than logistic regression in Chapter 20 because they are different ways of achieving the same end result. However, logistic regression has far fewer restrictive assumptions and is generally more robust, which is why I have limited the coverage of discriminant analysis to this chapter.

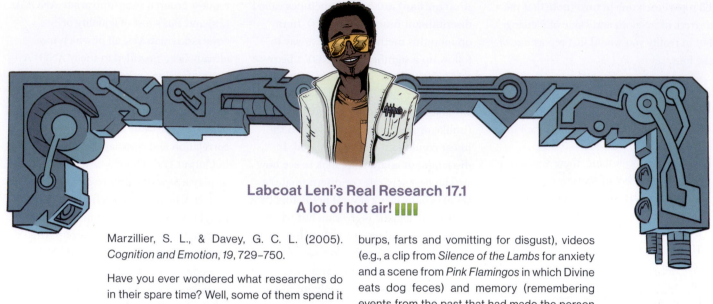

Labcoat Leni's Real Research 17.1
A lot of hot air! ||||

Marzillier, S. L., & Davey, G. C. L. (2005). *Cognition and Emotion, 19*, 729–750.

Have you ever wondered what researchers do in their spare time? Well, some of them spend it tracking down the sounds of people burping and farting! Anxious people are, typically, easily disgusted. Throughout this book I have talked about how you cannot infer causality from relationships between variables. This has been a bit of a conundrum for anxiety researchers: does anxiety cause feelings of disgust or does a low threshold for being disgusted cause anxiety? Two colleagues at Sussex addressed this by inducing feelings of anxiety, disgust or a neutral mood. They looked at the effect of these induced moods on feelings of anxiety, sadness, happiness, anger, disgust and contempt. To induce these moods, they used three different types of manipulation: vignettes (e.g., 'You're swimming in a dark lake and something brushes your leg' for anxiety, and 'You go into a public toilet and find it has not been flushed. The bowl of the toilet is full of diarrhea' for disgust), music (e.g., some scary music for anxiety, and a tape of

burps, farts and vomitting for disgust), videos (e.g., a clip from *Silence of the Lambs* for anxiety and a scene from *Pink Flamingos* in which Divine eats dog feces) and memory (remembering events from the past that had made the person anxious, disgusted or neutral).

Different people underwent anxious, disgust and neutral mood inductions. Within these groups, the induction was done using either vignettes and music, videos or memory recall and music for different people. The outcome variables were the change (from before to after the induction) in six moods: anxiety, sadness, happiness, anger, disgust and contempt. Draw an error bar graph of the changes in moods in the different conditions, then conduct a 3 (**Mood**: anxiety, disgust, neutral) × 3 (**Induction**: vignettes + music, videos, memory recall + music) MANOVA on these data (**Marzillier and Davey (2005).sav**). Whatever you do, don't imagine what their fart tape sounded like while you do the analysis. Answers are engraved on a turd on the companion website (or look at page 738 of the original article).

remaining options are of little interest when using discriminant analysis to follow up MANOVA except for the *Summary table*, which provides an overall gauge of how well the discriminant variates classify the actual participants. Click Continue to return to the main dialog box.

Click Save... to access the dialog box in Figure 17.15. There are three options available, two of which relate to the predicted group memberships and probabilities of group memberships from the model. The final option is to provide the **discriminant scores**. These are the

scores for each person, on each variate, obtained from equation (17.22). These scores can be useful because the variates that the analysis identifies may represent underlying social or psychological constructs. If these constructs are identifiable, then it is useful for interpretation to know what a participant scores on each dimension.

17.10 Interpreting discriminant analysis ▐▐▐▐

Output 17.7 shows the covariance matrices for separate groups (selected in Figure 17.13). These matrices are made up of the variances of each outcome variable for each group (in fact these values are shown in Table 17.1). The covariances are obtained by taking the cross-products between the outcome variables for each group (shown in Table 17.4 as 0.40, 22.6 and –10) and dividing each by 9, the degrees of freedom, $N - 1$ (where N is the number of observations). The values in this table give us some idea of how the relationship between outcome variables changes from group to group. For example, in the CBT group behaviors and thoughts have virtually no relationship because the covariance is almost zero. In the BT group thoughts and actions are positively related, so as the number of behaviors decreases, so does the number of thoughts. In the NT condition there is a negative relationship, so if the number of thoughts increases then the number of behaviors decreases. It is important to note that these matrices don't tell us about the substantive importance of the relationships because they are unstandardized (see Chapter 18), but they give a basic indication.

Output 17.8 shows the initial statistics from the discriminant analysis. The first table contains the eigenvalues for each variate—note that the values correspond to the values of the diagonal elements of the matrix HE^{-1} in equation (17.19). These eigenvalues are converted into percentage of variance accounted for; the first variate accounts for 82.2% of variance, whereas the second accounts for only 17.8%. This table also shows the canonical correlation, which we can square to use as an effect size (just like R^2 in a standard linear model).

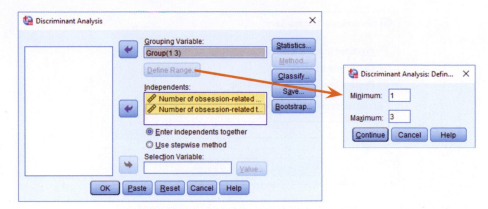

Figure 17.12 Main dialog box for discriminant analysis

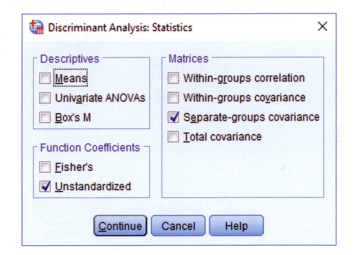

Figure 17.13 Statistics options for discriminant analysis

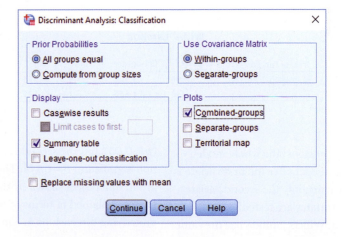

Figure 17.14 Discriminant analysis classification options

The second table in Output 17.8 shows the significance tests of both variates (*1 through 2* in the table), and the significance after the first variate has been removed (*2* in the table). So, effectively we test the model as a whole, and then peel away variates one at a time to see whether what's left is significant. With two variates

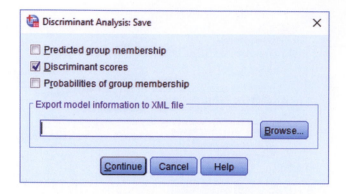

Figure 17.15 The save new variables dialog box in discriminant analysis

Covariance Matrices

group		Number of obsession–related behaviours	Number of obsession–related thoughts
CBT	Number of obsession–related behaviours	1.433	.044
	Number of obsession–related thoughts	.044	3.600
BT	Number of obsession–related behaviours	3.122	2.511
	Number of obsession–related thoughts	2.511	4.400
No Treatment Control	Number of obsession–related behaviours	1.111	-1.111
	Number of obsession–related thoughts	-1.111	5.556

Output 17.7

Eigenvalues

Function	Eigenvalue	% of Variance	Cumulative %	Canonical Correlation
1	.335ª	82.2	82.2	.501
2	.073ª	17.8	100.0	.260

a. First 2 canonical discriminant functions were used in the analysis.

Wilks' Lambda

Test of Function(s)	Wilks' Lambda	Chi-square	df	Sig.
1 through 2	.699	9.508	4	.050
2	.932	1.856	1	.173

Output 17.8

we get only two steps: the whole model, and then the model after the first variate is removed (leaving only the second variate). When both variates are tested in combination, Wilks's lambda has the same value (0.699), degrees of freedom (4) and significance value (0.05) as in the MANOVA (see Output 17.3). The important point to note from this table is that the two variates significantly discriminate the groups in combination

($p = 0.05$), but the second variate alone is non-significant, $p = 0.173$. Therefore, the group differences shown by the MANOVA can be explained in terms of *two* underlying dimensions in combination.

The tables in Output 17.9 are the most important for interpretation. The first shows the standardized discriminant function coefficients for the two variates. These values are standardized versions of the values in the eigenvectors calculated in

Section 17.4.8. Recall that if the variates can be expressed in terms of a linear regression equation (see equation (17.20)), the standardized discriminant function coefficients are equivalent to the standardized *b*-values in a linear model. The structure matrix shows the same information, but in a slightly different form. The values in this matrix are the canonical variate correlation coefficients. These values indicate the substantive nature of the variates. Bargman (1970) argues that when some outcome variables have high canonical variate correlations while others have low ones, the ones with high correlations contribute most to group separation. As such they represent the relative contribution of each outcome variable to group separation (see Bray & Maxwell, 1985, pp. 42–45). Hence, the coefficients in these tables tell us the relative contribution of each variable to the variates.

If we look at variate 1 first, thoughts and behaviors have the opposite effect (behavior has a positive relationship with this variate, whereas thoughts have a negative relationship). Given that these values (in both tables) can vary between 1 and –1, we can also see that both relationships are strong (although behaviors have a slightly larger contribution to the first variate). The first variate, then, could be seen as one that differentiates thoughts and behaviors (it affects thoughts and behaviors in the opposite way). Both thoughts and behaviors have a strong positive relationship with the second variate. This tells us that this variate represents something that affects thoughts and behaviors in a similar way. Remembering that ultimately these variates are used to differentiate groups, we could say that the first variate differentiates groups by some factor that affects thoughts and behaviors differently, whereas the second variate differentiates groups on some dimension that affects thoughts and behaviors in the same way.

Output 17.10 tells us first the canonical discriminant function coefficients, which are the unstandardized versions of the standardized coefficients described above. These values are the values of b in equation (17.20), and you'll notice that these values correspond to the values in the eigenvectors derived in Section 17.4.8 and used in equation (17.22). The values are less useful than the standardized versions, but demonstrate from where the standardized versions come.

The centroids are the mean variate scores for each group. For interpretation look at the sign of the centroid (positive or negative). We can also use a combined-groups plot (selected using the dialog box in Figure 17.14). This graph plots the variate scores for each person, grouped according to the experimental condition to which that person belonged. In addition, the group centroids from Output 17.10 are shown as red squares. The graph (Figure 17.16) and the tabulated values of the centroids (Output 17.10) tell us that variate 1 discriminates the BT group (look at the big squares labeled with the group initials) from the CBT (look at the horizontal distance between these centroids). The second variate differentiates the no treatment group from the two interventions (look at the vertical distances), but this difference is not as dramatic as for the first variate. Remember that the variates significantly discriminate the groups in combination (i.e., when both are considered).

17.11 Reporting results from discriminant analysis ||||

The guiding principle in presenting data is to give the readers enough information to be able to judge for themselves what

Standardized Canonical Discriminant Function Coefficients

	Function	
	1	2
Number of obsession-related behaviours	.829	.584
Number of obsession-related thoughts	-.713	.721

Structure Matrix

	Function	
	1	2
Number of obsession-related behaviours	.711*	.703
Number of obsession-related thoughts	-.576	.817*

Pooled within-groups correlations between discriminating variables and standardized canonical discriminant functions
Variables ordered by absolute size of correlation within function.

*. Largest absolute correlation between each variable and any discriminant function

Output 17.9

Canonical Discriminant Function Coefficients

	Function	
	1	2
Number of obsession-related behaviours	.603	.425
Number of obsession-related thoughts	-.335	.339
(Constant)	2.139	-6.857

Unstandardized coefficients

Functions at Group Centroids

	Function	
group	1	2
CBT	.601	-.229
BT	-.726	-.128
No Treatment Control	.125	.357

Unstandardized canonical discriminant functions evaluated at group means

Output 17.10

the data mean. Personally, I would suggest reporting percentage of variance explained (which gives the reader the same information as the eigenvalue, but in a more palatable form) and the squared canonical correlation for each variate (this is the appropriate effect size measure for discriminant analysis). I would also report the chi-square significance tests of the variates. These values can be found in Output 17.8 (but remember to square the canonical correlation). It is probably also useful to quote the values in the structure matrix in Output 17.9 (which will tell the reader about how the outcome variables relate to the underlying variates). Finally, although I won't reproduce it below, you could consider including a (well-edited) copy of the combined-groups centroid plot (Figure 17.16), which will help readers to determine how the variates contribute to distinguishing your groups. We could, therefore, write something like this:

✓ The MANOVA was followed up with discriminant analysis, which revealed two discriminant functions. The first explained 82.2% of the variance, canonical $R^2 = 0.25$, whereas the second explained only 17.8%, canonical $R^2 = 0.07$. In combination these discriminant functions significantly differentiated the treatment groups, $\Lambda = 0.70$, $\chi^2(4) = 9.51$, $p = 0.05$, but removing the first function indicated that the second function did not significantly differentiate the treatment groups, $\Lambda = 0.93$, $\chi^2(1) = 1.86$, $p = 0.173$. The correlations between outcomes and the discriminant functions revealed that obsessive behaviors loaded highly onto both functions ($r = 0.71$ for the first function and $r = 0.70$ for the second); obsessive thoughts loaded more highly on the second function ($r = 0.82$) than the first function ($r = -0.58$). The discriminant function plot showed that the first function discriminated the BT

Canonical Discriminant Functions

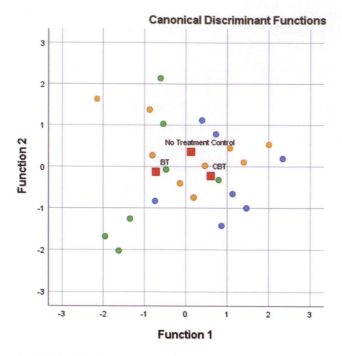

Figure 17.16   Combined-groups plot

group from the CBT group, and the second function differentiated the no treatment group from the two interventions.

17.12 The final interpretation ▊▊▊▊

How can we bring all of this information together to answer our research question: can therapy improve OCD, and, if so, which therapy is best? The MANOVA tells us that therapy can have a significant effect on OCD symptoms, but the non-significant univariate *F*s suggested that this improvement is not simply in terms of either thoughts or behaviors. The discriminant analysis suggests that the group separation can be best explained in terms of one underlying dimension.

Cramming Sam's Tips
Discriminant function analysis

- Discriminant function analysis can be used after MANOVA to see how the outcome variables discriminate the groups.
- Discriminant function analysis identifies variates (combinations of the outcome variables). To find out how many variates are significant look at the tables labeled *Wilks's Lambda*: if the value of *Sig.* is less than 0.05 then the variate is significantly discriminating the groups.
- Once the significant variates have been identified, use the table labeled *Canonical Discriminant Function Coefficients*

to find out how the outcome variables contribute to the variates. High scores indicate that an outcome variable is important for a variate, and variables with positive and negative coefficients are contributing to the variate in opposite ways.
- Finally, to find out which groups are discriminated by a variate look at the table labeled *Functions at Group Centroids:* for a given variate, groups with values opposite in sign are being discriminated by that variate.

The dimension is likely to be OCD itself (which we can realistically assume is made up of both thoughts and behaviors). Therapy doesn't necessarily change behaviors or thoughts *per se*, it influences the underlying dimension of OCD. So, the answer to the first question seems to be: yes, therapy can influence OCD in general.

The next question is more complex: which therapy is best? Figure 17.17 shows that for actions, BT reduces the number of obsessive behaviors, whereas CBT and NT do not. For thoughts, CBT reduces the number of obsessive thoughts, whereas BT and NT do not (check the pattern of the bars). Figure 17.18 shows the relationships between thoughts and actions across the groups. In the BT group there is a positive relationship between thoughts and actions, so the more obsessive thoughts a person has, the more obsessive behaviors they carry out. In the CBT group there is no relationship at all (thoughts and actions vary quite independently). In the no treatment group there is a negative (and non-significant, incidentally) relationship between thoughts and actions. What we have discovered from the discriminant analysis is that BT and CBT can be differentiated from the control group based on variate 2, a variate that has a similar effect on both thoughts and behaviors. We could say then that BT and CBT are both better than a no treatment group at changing obsessive thoughts and behaviors. We also discovered that BT and CBT could be distinguished by variate 1, a variate that had the opposite effects on thoughts and behaviors. Combining this information with that in Figure 17.17, we could conclude that BT is better at changing behaviors and CBT is better at changing thoughts. So, the NT group can be distinguished from the CBT and BT groups using a variable that affects both thoughts and behaviors. Also, the CBT and BT groups can be distinguished by a variate that has opposite effects on thoughts and behaviors. So, some therapy

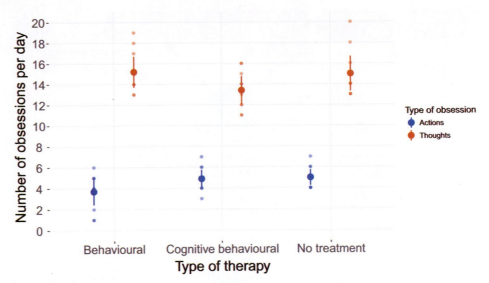

Figure 17.17 Graph showing the means and 95% confidence intervals between the outcome variables in each therapy group

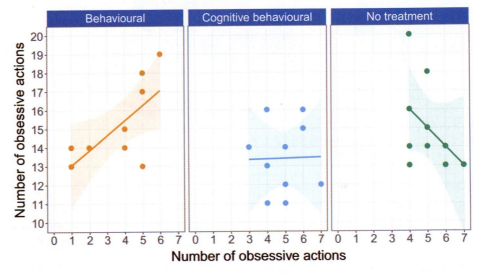

Figure 17.18 Graphs showing the relationships between the outcome variables across each therapy group

is better than none, but the choice of CBT or BT depends on whether you think it's more important to target thoughts (CBT) or behaviors (BT).

17.13 Brian and Jane's Story ▍▍▍▍

Brian was confused. It had been five days since he'd opened up to Jane and he'd heard nothing since. He rarely told people about his mother, because it made them

feel awkward. They acted differently. It was hard to say exactly how they acted, but they were sort of safe with him, as though scared of saying or doing something that might trigger him. It was easier not to talk about her. It was amazing how many friends you could have who never asked about your parents. After the recent weeks of talking to Jane about her family it seemed wrong not to tell her when she asked. It must've freaked her, though, because her phone was dead, and

Figure 17.19 What Brian learnt from this chapter

look as if to say 'Have you lost something?' (see Figure 17.20). Yep, freaked out by the whole moving experience, he had done the only sensible thing and hidden up the chimney! Cats, you gotta love 'em!

At some point everything ends (even my chapters), and for a few editions of this book I have been aware that the day would come when I had to edit my story about Fuzzy to be in the past tense. That day has come. Fuzzy took on a whole other existence as 'the cat in the stat' (book). I started writing this book when he was a kitten, and he sat by my side, on my lap, on the desk for 20 years. He saw me through the end of my PhD and four editions of this book (and a few others as well). I used to joke that he was immortal, but that was only to make myself feel better about the fact that he wasn't. He was just a soppy old ginger cat like lots of other soppy old ginger cats, except he was my soppy old ginger cat, and I loved him because while people came and went from my life, he was the constant that saw me through the day: the feline friend I could always rely on for a cuddle. Even a year on, it's weird not having him here while I type. So now, despite being widely known as 'a cat person' I don't have a cat because Fuzzy is irreplaceable. Ironically, I *do* have a crazy cocker spaniel (see Chapter 12) who *is* on the floor next to me. I like to think that when my back is turned, a ginger cloud emerges from the ether and gives him a hard time about stats. But not too hard a time.

this was the longest, since Jane's mother got ill, that she hadn't contacted him.

Jane was exhausted. She emerged from the lab into the dark. She'd hit a wall and needed to sleep. Her battery had long died on her phone and she had no idea what time it was. She felt unsteady on her feet. A girl walking past asked if she was OK. She said Jane looked terrible. Jane asked her for the time. It was 9:30 pm. As an afterthought she asked what day it was. 'Friday,' the girl replied, giving her the kind of look you'd give someone who didn't know what day it was. 'Holy shit!' Jane thought. She'd been working five days. How was that possible? She panicked, realizing that she had vanished from Brian for so long without explanation. Why hadn't it occurred to her to call him? She guessed because there were things she didn't want

him to know yet. Jane caught the train into the city. She stood in front of Brian's apartment a little scared to press the buzzer. Brian recoiled when he saw her. 'You look exhausted,' he said. She lay on his couch. 'Talk to me,' she demanded, and she drifted into sleep to the sound of his calming voice.

17.14 What next? ▮▮▮

At the beginning of this chapter we discovered that pets can be therapeutic. I left the whereabouts of Fuzzy a mystery. After frantically searching the house, I went back to the room he had vanished from to check again whether there was a hole that he could have wriggled through. As I scuttled around on my hands and knees tapping the walls, a little ginger (and sooty) face popped out from the fireplace with a

Figure 17.20 Fuzzy hiding up a fireplace

17.15 Key terms that I've discovered

Bartlett's test of sphericity

Box's test

Discriminant analysis

Discriminant function variates

Discriminant scores

Error SSCP (*E*)

HE^{-1}

Homogeneity of covariance matrices

Hotelling–Lawley trace (T^2)

Hypothesis SSCP (*H*)

Identity matrix

Matrix

Multivariate

Multivariate analysis of variance (MANOVA)

Multivariate normality

Pillai–Bartlett trace (*V*)

Roy's largest root

Square matrix

Sum of squares and cross-products matrix (SSCP)

Total SSCP (*T*)

Univariate

Variance–covariance matrix

Wilks's lambda (Λ)

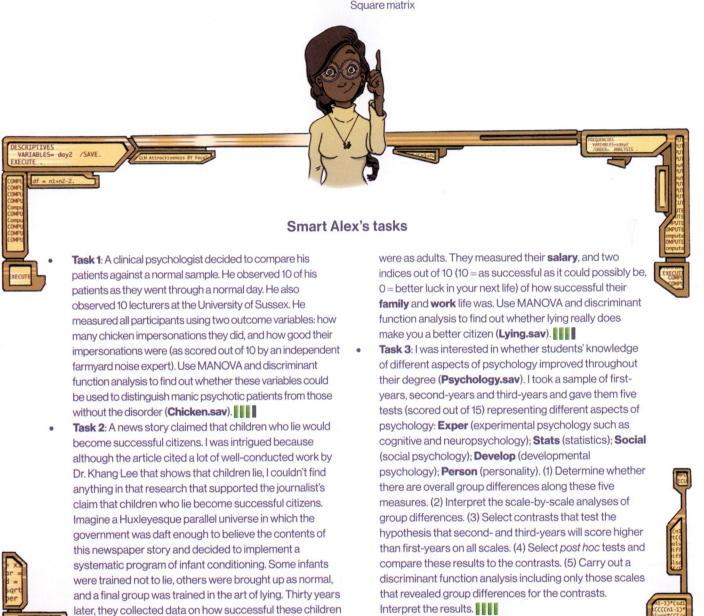

Smart Alex's tasks

- **Task 1**: A clinical psychologist decided to compare his patients against a normal sample. He observed 10 of his patients as they went through a normal day. He also observed 10 lecturers at the University of Sussex. He measured all participants using two outcome variables: how many chicken impersonations they did, and how good their impersonations were (as scored out of 10 by an independent farmyard noise expert). Use MANOVA and discriminant function analysis to find out whether these variables could be used to distinguish manic psychotic patients from those without the disorder (**Chicken.sav**). ▊▊▊▊

- **Task 2**: A news story claimed that children who lie would become successful citizens. I was intrigued because although the article cited a lot of well-conducted work by Dr. Khang Lee that shows that children lie, I couldn't find anything in that research that supported the journalist's claim that children who lie become successful citizens. Imagine a Huxleyesque parallel universe in which the government was daft enough to believe the contents of this newspaper story and decided to implement a systematic program of infant conditioning. Some infants were trained not to lie, others were brought up as normal, and a final group was trained in the art of lying. Thirty years later, they collected data on how successful these children were as adults. They measured their **salary**, and two indices out of 10 (10 = as successful as it could possibly be, 0 = better luck in your next life) of how successful their **family** and **work** life was. Use MANOVA and discriminant function analysis to find out whether lying really does make you a better citizen (**Lying.sav**). ▊▊▊▊

- **Task 3**: I was interested in whether students' knowledge of different aspects of psychology improved throughout their degree (**Psychology.sav**). I took a sample of first-years, second-years and third-years and gave them five tests (scored out of 15) representing different aspects of psychology: **Exper** (experimental psychology such as cognitive and neuropsychology); **Stats** (statistics); **Social** (social psychology); **Develop** (developmental psychology); **Person** (personality). (1) Determine whether there are overall group differences along these five measures. (2) Interpret the scale-by-scale analyses of group differences. (3) Select contrasts that test the hypothesis that second- and third-years will score higher than first-years on all scales. (4) Select *post hoc* tests and compare these results to the contrasts. (5) Carry out a discriminant function analysis including only those scales that revealed group differences for the contrasts. Interpret the results. ▊▊▊▊

Answers & additional resources are available on the book's website at
https://edge.sagepub.com/field5e

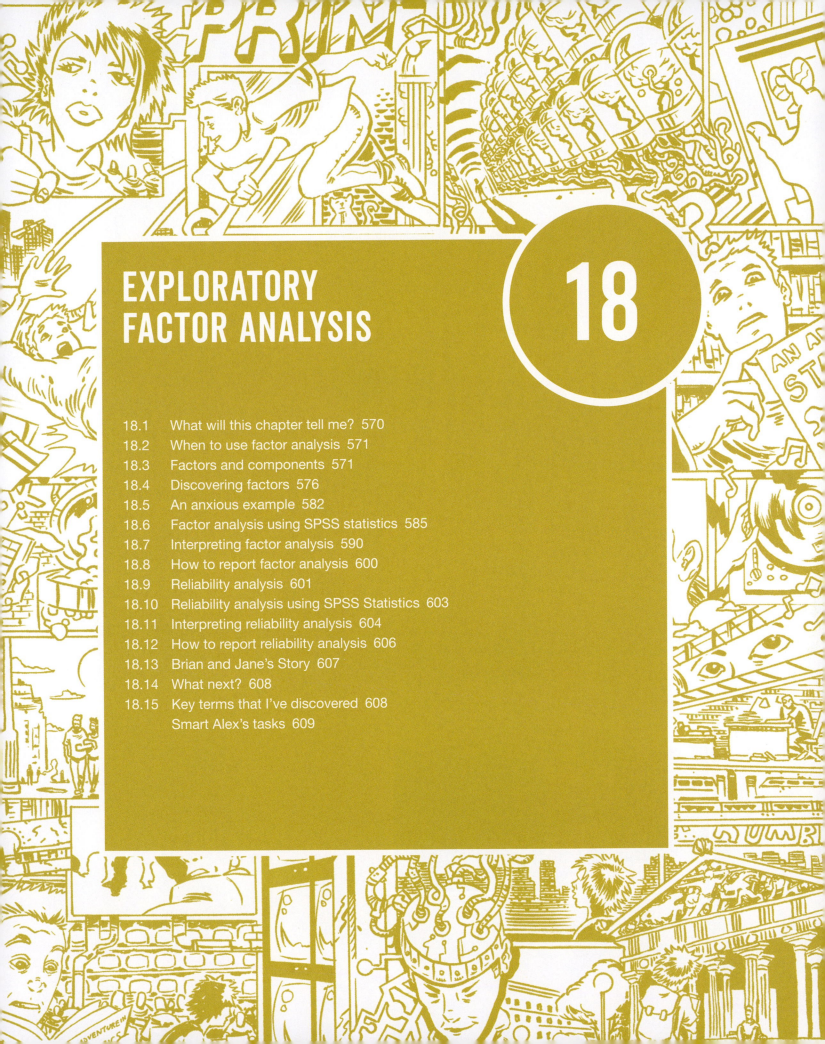

EXPLORATORY FACTOR ANALYSIS

18

18.1 What will this chapter tell me?

Having failed to become a rock star, I went to university and eventually ended up doing a PhD (in Psychology) at the University of Sussex. Like many postgraduates, I taught to survive. I was allocated to second-year undergraduate statistics. I was very shy at the time, and I didn't have a clue about statistics, so standing in front of a room full of strangers and talking to them about ANOVA was about as appealing as dislocating my knees and running a marathon. I obsessively prepared for my first session so that it would go well; I created handouts, I invented examples, I rehearsed what I would say. I went in terrified but knowing that if preparation was any predictor of success then I would be OK. About half way through one of the students rose majestically from her chair. An aura of bright white light surrounded her and she appeared to me as though walking through dry ice. I guessed that she had been chosen by her peers to impart a message of gratitude for the hours of preparation I had done and the skill with which I was unclouding their brains of statistical mysteries. She stopped inches away from me. She looked into my eyes, and mine raced around the floor looking for the reassurance of my shoelaces. 'No one in this room has a rabbit[1] clue what you're going on about,' she spat before storming out. Scales have not been invented yet to measure how much I wished I'd run the dislocated-knees marathon that morning. To this day I have intrusive thoughts about students in my lectures walking zombie-like towards the front of the lecture theater chanting 'No one knows what you're going on about' before devouring my brain in a rabid feeding frenzy.

The aftermath of this trauma was that I threw myself into trying to be the best teacher in the universe. I wrote detailed handouts and started using wacky examples. Based on these I was signed up by a publisher to write a book. This book. At the age of 23 I didn't realize that this was academic suicide (really, textbooks take a long time to write and they are not at all valued compared to research articles). I also didn't realize the emotional pain I was about to inflict on myself. I soon discovered that writing a statistics book was like doing a factor analysis: in factor analysis

Figure 18.1   In my office during my PhD, probably preparing some teaching—I had quite long hair back then because it hadn't started falling out at that point

---

1   She didn't say 'rabbit', but she did say a word that describes what rabbits do a lot; it begins with an 'f' and my publishers think that it will offend you.

we take a lot of information (variables) and a computer effortlessly reduces this mass of confusion into a simple message (fewer variables). A computer does this in a few seconds. Similarly, my younger self took a mass of information about statistics that I didn't understand and filtered it down into a simple message that I *could* understand: I became a living, breathing factor analysis . . . except that, unlike a computer, it took me two years and some considerable effort.

18.2 When to use factor analysis ▮▮▮▮

In science we often need to measure something that cannot be accessed directly (a so-called **latent variable**). For example, management researchers might be interested in measuring 'burnout', which is when someone who has been working very hard on a project (a book, for example) for a prolonged period of time suddenly finds himself devoid of motivation, inspiration, and wants to repeatedly head-butt his computer, screaming 'Please, SAGE, unlock the door, let me out of the basement, I need to feel the soft warmth of sunlight on my skin'. You can't measure burnout directly: it has many facets. However, you can measure different aspects of burnout: you could get some idea of motivation, stress levels, whether the person has any new ideas and so on. Having done this, it would be helpful to

know whether these facets reflect a single variable. Put another way, are these observable measures driven by the same underlying variable?

This chapter explores **factor analysis** and **principal component analysis (PCA)**— techniques for identifying clusters of variables. These techniques have three main uses: (1) to understand the structure of a set of variables (e.g., Spearman and Thurstone used factor analysis to try to understand the structure of the latent variable 'intelligence'); (2) to construct a questionnaire to measure an underlying variable (e.g., you might design a questionnaire to measure burnout); and (3) to reduce a data set to a more manageable size while retaining as much of the original information as possible (e.g., factor analysis can be used to solve the problem of multicollinearity that we discovered in Chapter 9 by combining variables that are collinear).

There are numerous examples of the use of factor analysis in science. Most readers will be familiar with the extroversion–introversion and neuroticism traits measured by Eysenck (1953). Most other personality questionnaires are also based on factor analysis—notably Cattell's (1966a) 16 personality factors questionnaire—and these inventories are frequently used for recruiting purposes in industry (and even by some religious groups). Economists, for example, might also use factor analysis to see whether productivity, profits and workforce can be reduced to an underlying dimension of company growth, and Jeremy Miles told me of a biochemist who used it to analyze urine samples.

Both factor analysis and PCA aim to reduce a set of variables into a smaller set of dimensions (called 'factors' in factor analysis and 'components' in PCA). To non-statisticians, like me, the differences between a component and a factor are difficult to conceptualize (they are both linear models), and the differences are hidden away in the maths behind the techniques.[2] However, there are important differences, which I'll discuss in due course. Most of the practical issues are the same regardless of whether you do factor analysis or PCA, so once the theory is over you can apply any advice I give to either factor analysis or PCA.

18.3 Factors and components ▮▮▮▮

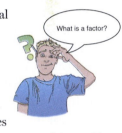

If we measure several variables or ask someone several questions about themselves, the correlation between each pair of variables (or questions) can be arranged in a table (just like the output from a correlation analysis as seen in Chapter 8). This table is sometimes called an *R*-matrix, just to scare you. The diagonal elements of an *R*-matrix are all ones because each variable will correlate perfectly with itself. The off-diagonal elements are the correlation coefficients between pairs of variables or questions.[3] Factor analysis attempts to achieve parsimony by explaining the maximum amount of *common variance* in a correlation matrix using the smallest

What is a factor?

---

2 Principal component analysis is not the same as factor analysis. This doesn't stop idiots like me from discussing them as though they are. I tend to focus on the similarities between the techniques, which will reduce some statisticians (and psychologists) to tears. I'm banking on these people not needing to read this book, so I'll take my chances because I think it's easier for you if I give you a general sense of what the procedures do and not obsess too much about their differences. Once you have got the basics under your belt, feel free to obsess about their differences and complain to your friends about how awful the book by that imbecile Field is ...

3 This matrix is called an *R*-matrix or *R*, because it contains correlation coefficients and *r* usually denotes Pearson's correlation (see Chapter 8)—the *r* turns into a capital letter when it denotes a matrix.

number of explanatory constructs. These 'explanatory constructs' are known as latent variables (or *factors*) and they represent clusters of variables that correlate highly with each other. PCA differs in that it tries to explain the maximum amount of *total variance* (not just common variance) in a correlation matrix by transforming the original variables into linear components.

Imagine that we wanted to measure different aspects of what might make a person popular. We could administer several measures that we believe tap different aspects of popularity. So, we might measure a person's social skills (**Social skills**), their selfishness (**Selfish**), how interesting others find them (**Interest**), the proportion of time they spend talking about the other person during a conversation (**Talk (other)**), the proportion of time they spend talking about themselves (**Talk (self)**), and their propensity to lie (**Liar**). We calculate the correlation coefficients for each pair of variables and create an *R*-matrix. Figure 18.2 shows this matrix. There appear to be two clusters of interrelating variables. First, the amount that someone talks about the other person during a conversation correlates highly with both the level of social skills and how interesting the other finds that person; social skills also correlate well with how interesting others perceive a person to be. The relationships between these three variables indicate that the

better your social skills, the more interesting and talkative you are likely to be. Second, the amount that people talk about themselves within a conversation correlates well with how selfish they are and how much they lie. Being selfish also correlates highly with the degree to which a person tells lies. In short, selfish people are likely to lie and talk about themselves.

Factor analysis and PCA both aim to reduce this *R*-matrix into a smaller set of dimensions. In factor analysis these dimensions or factors, are estimated from the data and are believed to reflect constructs that can't be measured directly. In this example, there appear to be two clusters that fit the bill. The first 'factor' seems to relate to general sociability, whereas the second 'factor' seems to relate to the way in which a person treats others socially (we might call it 'inconsideration'). It might, therefore, be assumed that popularity depends not only on your ability to socialize, but also on whether you are inconsiderate towards others. PCA, in contrast, transforms the data into a set of linear components; it does not estimate unmeasured variables, it just transforms measured ones. Strictly speaking, then, we shouldn't interpret components as unmeasured variables. Despite these differences, both techniques look for variables that correlate highly with a group of other variables, but do not correlate with variables outside of that group.

18.3.1 Graphical representation ▮▮▮

Factors and components can be visualized as the axis of a graph along which we plot variables. The coordinates of variables along each axis represent the strength of relationship between that variable and each factor. In an ideal world a variable will have a large coordinate for one of the axes, and small coordinates for any others. This scenario indicates that this particular variable is related to only one factor. Variables that have large coordinates on the same axis are assumed to measure different aspects of some common underlying dimension. The coordinate of a variable along a classification axis is known as a **factor loading** (or *component loading*). The factor loading can be thought of as the Pearson correlation between a factor and a variable (see Jane Superbrain Box 18.1). From what we know about interpreting correlation coefficients (see Section 8.4.2) it should be clear that if we square the factor loading for a variable we get a measure of its substantive importance to a factor.

Figure 18.3 shows such a plot for the popularity data (in which there were only two factors). Notice that for both factors, the axis line ranges from −1 to +1, which are the outer limits of a correlation coefficient. The triangles represent the three variables that have high factor loadings (i.e., a strong relationship) with factor 1 (sociability: horizontal axis) but have a low correlation with factor 2 (inconsideration: vertical axis). Conversely, the circles represent variables that have high factor loadings with consideration but low loadings with sociability. This plot shows what we found in the *R*-matrix: selfishness, the amount a person talks about themselves and their propensity to lie contribute to a factor which could be called inconsideration of others, and how much a person takes an interest in other

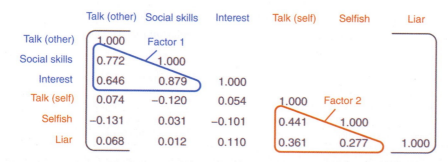

Figure 18.2 An *R*-matrix

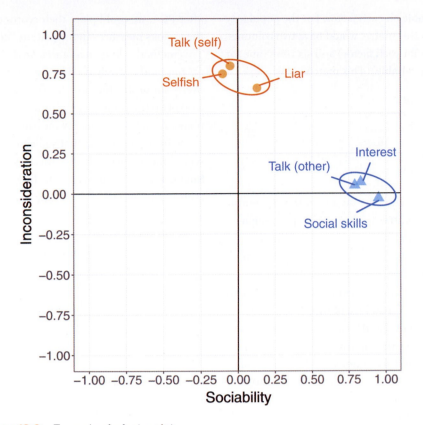

Figure 18.3 Example of a factor plot

What is the equation of a straight line/linear model?

people, how interesting they are and their level of social skills contribute to a second factor, sociability. Of course, if a third factor existed within these data it could be represented by a third axis (creating a 3-D graph). If more than three factors exist in a data set, then they cannot all be represented by a 2-D drawing.

18.3.2 Mathematical representation ▌▌▌▌

The axes in Figure 18.3, which represent factors, are straight lines and any straight line can be described mathematically by a familiar equation.

The following equation reminds us of the equation describing a linear model:

$$Y_i = b_1 X_{1i} + b_2 X_{2i} + \cdots + b_n X_{ni} \tag{18.1}$$

A component in PCA can be described in the same way:

$$\text{Component}_i = b_1 \text{Variable}_{1i} + b_2 \text{Variable}_{2i} + \cdots + b_n \text{Variable}_{ni}$$

You'll notice that there is no intercept in the equation because the lines intersect at zero (hence the intercept is zero), and there is also no error term because we are simply transforming the variables. The bs in the equation represent the loadings. Sticking with our example of popularity, we found that there were two components: general sociability and inconsideration. We can, therefore, construct an equation that

describes each factor in terms of the variables that have been measured:

$$Y_i = b_1 X_{1i} + b_2 X_{2i} + \cdots + b_n X_{ni}$$
$$\text{Sociability}_i = b_1 \text{Talk(other)}_i + b_2 \text{Social skills}_i$$
$$+ b_3 \text{Interest}_i + b_4 \text{Talk(self)}_i + b_5 \text{Selfish}_i$$
$$+ b_6 \text{Liar}_i$$
$$\text{Inconsideration}_i = b_1 \text{Talk(other)}_i + b_2 \text{Social skills}_i$$
$$+ b_3 \text{Interest}_i + b_4 \text{Talk(self)}_i + b_5 \text{Selfish}_i$$
$$+ b_6 \text{Liar}_i \tag{18.2}$$

First, notice that the equations are identical in form: they both include all the variables that were measured. However, the values of b in the two equations will be different (depending on the relative importance of each variable to the component). In fact, we can replace each value of b with the coordinate of that variable on the graph in Figure 18.3 (i.e., replace the values of b with the factor loadings). The resulting equations are:

$$Y_i = b_1 X_{1i} + b_2 X_{2i} + \cdots + b_n X_{ni}$$
$$\text{Sociability}_i = 0.87 \text{Talk(other)}_i + 0.96 \text{Social skills}_i$$
$$+ 0.92 \text{Interest}_i + 0.00 \text{Talk(self)}_i$$
$$- 0.10 \text{Selfish}_i + 0.09 \text{Liar}_i$$
$$\text{Inconsideration}_i = 0.01 \text{Talk(other)}_i - 0.03 \text{Social skills}_i$$
$$+ 0.04 \text{Interest}_i + 0.82 \text{Talk(self)}_i$$
$$+ 0.75 \text{Selfish}_i + 0.70 \text{Liar}_i \tag{18.3}$$

For the **Sociability** component, the values of b are high for **Talk (other)**, **Social skills** and **Interest**. For the remaining variables (**Talk (self)**, **Selfish** and **Liar**) the values of b are very low (close to 0). This tells us that three of the variables are very important for that component (the ones with high values of b) and three are relatively unimportant (the ones with low values of b). The way the three variables clustered on the factor plot confirms this interpretation (Figure 18.3). The factor plot and these equations represent the same thing: the factor loadings in the plot are the b-values in these equations. For the second factor, **Inconsideration**, the opposite pattern can be seen: **Talk (self)**, **Selfish** and **Liar** have high b-values, but the values for the remaining three variables are close to 0. Ideally, variables would have very high b-values for one component and very low b-values for all other components.

The factors in factor analysis are not represented in quite the same way as components. A factor is defined as follows:

$$x = \mu + \Lambda\xi + \delta$$
$$\text{Variables} = \text{Variable Means}$$
$$+ (\text{Loadings} \times \text{Common Factor}) \quad (18.4)$$
$$+ \text{Unique Factor}$$

The Greek letters represent matrices or vectors containing numbers. If we put the Greek letters through Andy's magical translation machine then we can stop worrying about what the matrices contain and focus on what they represent. In factor analysis, scores on the measured variables are predicted from the means of those variables plus a person's scores on the **common factors** (i.e., factors that explain the correlations between variables) multiplied by their factor loadings, plus scores on any **unique factors** within the data (factors that cannot explain the correlations between variables).

In a sense, the factor analysis model flips PCA on its head: in PCA we predict components from the measured variables, but in factor analysis we predict the measured variables from the underlying factors. Psychologists are usually interested in factors, because they're interested in how the stuff going on inside people's heads (the latent variables) affects how they answer the questions (the measured variables). The other big difference is that, unlike PCA, factor analysis contains an error term (δ is made up of both scores on unique factors and measurement error). The fact that PCA assumes that there is no measurement error upsets a lot of people who use factor analysis.

Both factor analysis and PCA are linear models in which loadings are used as weights. In both cases, these loadings can be expressed as a matrix in which the columns represent each factor and the rows represent the loadings of each variable on each factor. For the popularity data this matrix would have two columns (one for each factor) and six rows (one for each variable). This matrix, Λ, is:

$$\Lambda = \begin{pmatrix} 0.87 & 0.01 \\ 0.96 & -0.03 \\ 0.92 & 0.04 \\ 0.00 & 0.82 \\ -0.10 & 0.75 \\ 0.09 & 0.70 \end{pmatrix} \quad (18.5)$$

and is called the **factor matrix** or **component matrix** (if doing principal component analysis)—see Jane Superbrain Box 18.1 to find out about the different forms of this matrix. Try relating the elements to the loadings in equation (18.3) to give you an idea of what this matrix represents (in the case of PCA). For example, the top row represents the first variable, **Talk (other)**, which had a loading of 0.87 for the first factor (**Sociability**) and a loading of 0.01 for the second factor (**Inconsideration**).

The major assumption in factor analysis (but not PCA) is that these algebraic factors represent real-world dimensions, the nature of which must be *guessed at* by inspecting which variables have high loads on the same factor. So, psychologists might believe that factors represent dimensions of the psyche, education researchers might believe they represent abilities, and sociologists might believe they represent races or social classes. However, it is an extremely contentious point: some believe that the dimensions derived from factor analysis are real only in the statistical sense—and are real-world fictions.

18.3.3 Factor scores

Having discovered which factors exist, and estimated the equation that describes them, it should be possible to estimate a person's score on a factor, based on their scores for the constituent variables; these are known as **factor scores** (or *component scores* in PCA). For example, if we wanted to derive a sociability score for someone after PCA, we could place their scores on the various measures into equation (18.3). This method is known as a *weighted average* and is rarely used because it is overly simplistic, but it is the easiest way to explain the principle. For example, imagine our six personality measures range from 1 to 10 and that someone scored the following: **Talk** (**other**) = 4, **Social skills** = 9, **Interest** = 8, **Talk** (**self**) = 6, **Selfish** = 8, and **Liar** = 6. We could plug these values into equation (18.3) to get a score for this person's sociability and inconsideration:

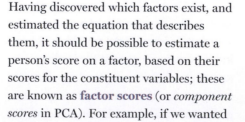

$$\text{Sociability}_i = 0.87\,\text{Talk (other)}_i + 0.96\,\text{Social skills}_i$$
$$+ 0.92\,\text{Interest}_i + 0.00\,\text{Talk (self)}_i$$
$$- 0.10\,\text{Selfish}_i + 0.09\,\text{Liar}_i$$
$$\text{Sociability}_i = (0.87 \times 4) + (0.96 \times 9)$$
$$+ (0.92 \times 8) + (0.00 \times 6) - (0.10 \times 8)$$
$$+ (0.09 \times 6)$$
$$= 19.22$$
$$\text{Inconsideration}_i = 0.01\,\text{Talk (other)}_i - 0.03\,\text{Social skills}_i$$
$$+ 0.04\,\text{Interest}_i + 0.82\,\text{Talk (self)}_i$$
$$+ 0.75\,\text{Selfish}_i + 0.70\,\text{Liar}_i$$
$$\text{Inconsideration}_i = (0.01 \times 4) - (0.03 \times 9) + (0.04 \times 8)$$
$$+ (0.82 \times 6) + (0.75 \times 8) + (0.70 \times 6)$$
$$= 15.21$$

$$(18.6)$$

The resulting scores of 19.22 and 15.21 reflect the degree to which this person is sociable and their inconsideration to others, respectively. This person scores higher on sociability than inconsideration. However, the scales of measurement used will influence the resulting scores, and if different variables use different measurement scales, then factor scores for different factors cannot be compared. As such, this method of calculating factor scores is poor, and more sophisticated methods are usually used. There are several techniques for calculating factor scores that use factor score coefficients as weights rather than the factor loadings. Factor score coefficients can be calculated in several ways. The simplest way is the regression method, in which the factor loadings are adjusted to take account of the initial correlations between variables; in doing

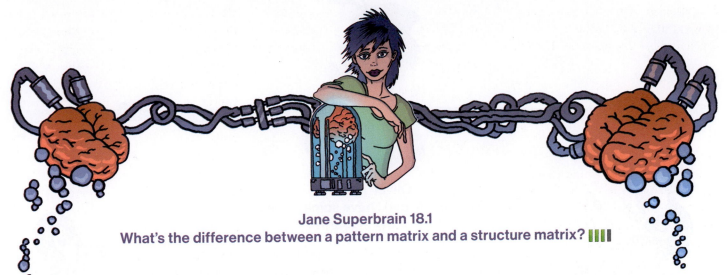

Jane Superbrain 18.1
What's the difference between a pattern matrix and a structure matrix? ▮▮▮▮

So far I've been a bit vague about factor loadings. Sometimes I've said that loadings can be thought of as the correlation between a variable and a given factor, then at other times I've described them as b-values (like in regression). Broadly speaking, both correlation coefficients and b-values represent the relationship between a variable and linear model, so my vagueness might not be the evidence of buffoonery that it seems. The take-home message is that factor loadings tell us about the relative contribution that a variable makes to a factor. If you understand that much, you'll be OK.

However, the factor loadings in a given model can be both correlation coefficients and b-values. In a few sections' time we'll discover that the interpretation of factor analysis is helped greatly by a technique known as *rotation*. There are two types: orthogonal and oblique rotation (see Section 18.4.6). When orthogonal rotation is used, underlying factors are assumed to be independent, and the factor loading *is* the correlation between the factor and the variable, but it is also the b-value. Put another way, the values of the correlation coefficients are the same as the values of the b-values. However, oblique rotation is used when the underlying factors are assumed to be related to each other, resulting in correlated factors. In these situations, the resulting correlations between variables and factors will differ from the corresponding b-values. In this case, there are, in effect, two different sets of factor loadings: the correlation coefficients between each variable and factor (contained in the factor *structure matrix*) and the b-values for each variable on each factor (contained in the factor *pattern matrix*). These coefficients can have quite different interpretations (see Graham, Guthrie, & Thompson 2003).

so, differences in units of measurement and variable variances are stabilized.

To obtain the matrix of factor score coefficients (B) we multiply the matrix of factor loadings by the inverse (R^{-1}) of the original correlation or R-matrix (this is the same process that is used to estimate the bs in ordinary regression). You might remember from the previous chapter that multiplying by the inverse of a matrix is like division (see Section 17.4.8), so by multiplying the matrix of factor loadings by the inverse of the correlation matrix we are, conceptually speaking, dividing the factor loadings by the correlation coefficients. As such, the resulting factor score matrix represents the relationship between each variable and each factor, adjusting for the original relationships between pairs of variables. This matrix represents a purer measure of the *unique* relationship between variables and factors.

Using the regression technique, the resulting factor scores have a mean of 0 and a variance equal to the squared multiple correlation between the estimated factor scores and the true factor values. The downside is that the scores can correlate not only with factors other than the one on which they are based, but also with other factor *scores* from a different orthogonal factor. To overcome this problem two adjustments have been proposed: the Bartlett method and the **Anderson–Rubin method**. The Bartlett method produces scores that are unbiased

Oliver Twisted
Please, Sir, can I have some more . . .
matrix algebra?

'*The Matrix*—that was a good film,' enthuses Oliver. I want to dress in black and glide through the air as though time has stood still. Maybe the matrix of factor scores is as cool as the film.' I think you might be disappointed, Oliver, but we'll give it a shot. The matrix calculations of factor score coefficients for this example are detailed on the companion website. Be afraid, be very afraid ...

and that correlate only with their own factor. The mean and standard deviation of the scores is the same as for the regression method. However, factor scores can still correlate with each other. The Anderson–Rubin method is a modification of the Bartlett method that produces factor scores that are uncorrelated and standardized (they have a mean of 0 and a standard deviation of 1). Tabachnick and Fidell (2012) conclude that the Anderson–Rubin method is best when uncorrelated scores are required but that the regression method is preferred in other circumstances simply because it is most easily understood. Although it isn't important that you understand the maths behind any of the methods, it is important that you understand what the factor scores represent, namely, a composite score for each individual on a particular factor.

There are several uses of factor scores. First, if the purpose of the factor analysis is to reduce a large set of data into a smaller subset of measurement variables, then the factor scores tell us an individual's score on this subset of measures. Any further analysis can be carried out on the factor scores rather than the original data. For example, we could carry out a *t*-test to see whether extroverts are significantly more sociable than introverts using the factor scores for *sociability*. A second use is in overcoming collinearity problems in linear models. If we have identified sources of multicollinearity in a linear model (see Section 9.9.3) then a solution is to reduce collinear predictors to a subset of uncorrelated factors using PCA and enter the component scores as predictors instead of the raw variable scores. By using uncorrelated component scores (e.g., by using the Anderson–Rubin method—see above) as predictors we can be confident that there will be no correlation between predictors—hence, no multicollinearity.

18.4 Discovering factors ▮▮▮▮

By now, you should have some grasp of what factors and components are. We will now delve into how to find or estimate these mythical beasts.

18.4.1 Choosing a method ▮▮▮▮

There are several methods for unearthing factors and the one you chose will depend on what you hope to do with the analysis (for a review, see Tinsley & Tinsley, 1987). There are two things to consider: whether you want to generalize the findings from your sample to a population, and whether you are exploring your data or testing a specific hypothesis. This chapter describes techniques for exploring data using factor analysis. Testing hypotheses about the structures of latent variables and their relationships to each other is a big topic and needs a different software package (IBM SPSS AMOS, R or MPlus to name a few), so I'm not going to cover it. For those interested in hypothesis testing techniques (known as **confirmatory factor analysis**) I recommend Brown (2015).

Assuming we want to explore, we need to consider whether we want to apply our findings to the sample collected (descriptive method) or to generalize our findings to a population (inferential methods). Factor analysis was originally developed to explore data to generate future hypotheses, and it was assumed that the technique would be applied to the entire population of interest. In other words, certain techniques assume that the sample used *is* the population and results cannot be extrapolated beyond that sample. Principal component analysis is an example, as are principal factors analysis (*principal axis factoring*) and image covariance analysis (*image factoring*). Of these, principal component analysis and principal factors analysis are the preferred methods and usually result in similar solutions (see Section 18.4.3). If you use one of these methods then you should restrict your conclusions to the sample collected. If you want to generalize the results then you would need to cross-validate the factor structure in a different sample.

A different approach assumes that participants are randomly selected but that the variables measured constitute the population of variables in which we're interested. By assuming this, it is possible to generalize from the sample to a larger population, but with the caveat that any findings hold true only for the set of variables measured (because we've assumed this set constitutes the entire population of variables). Techniques in this category include the *maximum-likelihood method* (see Harman, 1976) and Kaiser's **alpha factoring** (Kaiser & Caffrey, 1965). The choice of method depends largely on what generalizations, if any, you want to make from your data.

18.4.2 Communality ▊▊▊

The idea of what variance is and how it is calculated should, by now, be an old friend with whom you enjoy tea and biscuits (if not, see Chapter 2). The total variance for

a variable in the *R*-matrix will have two components: some of it will be shared with other variables or measures (**common variance**) and some of it will be specific to that measure (**unique variance**). We tend to use the term *unique variance* to refer to variance that can be reliably attributed to only one measure. However, there is also variance that is specific to one measure but not reliably so, known as error or **random variance**. The proportion of common variance present in a variable is known as the **communality**. As such, a variable that has no unique variance (or random variance) would have a communality of 1; a variable that shares none of its variance with any other variable would have a communality of 0.

Factor analysis tries to find common underlying dimensions within the data and so is primarily concerned with the common variance. In short, we want to find out how much of the variance in our data is common. This aim presents us with a logical impasse: to do the factor analysis we need to know the proportion of common variance in the data, but the only way to find out the extent of the common variance is by carrying out a factor analysis! There are two solutions. The first is to assume that all variance is common by assuming that the communality of every variable is 1. By making this assumption we transpose our original data into constituent linear components. This procedure is PCA. Remember that I said earlier that PCA assumes no measurement error? Well, by setting the communalities to 1, we assume that all variance is common and there is no random variance at all.

The second solution is used in factor analysis and is to estimate the amount of common variance by estimating communality values for each variable. There are various methods of estimating communalities, but the most widely used (including alpha factoring) is to use the squared multiple correlation (SMC) of

each variable with all others. So, for the popularity data, imagine you fit a linear model with one measure (**Selfish**) as the outcome and the other five measures as predictors: the resulting multiple R^2 (see Section 9.2.4) would act as an estimate of the communality for the variable **Selfish**. These estimates allow the factor analysis to be done. Once the underlying factors have been extracted, new communalities can be calculated that represent the multiple correlation between each variable and the factors extracted. Therefore, the communality is a measure of the proportion of variance explained by the extracted factors.

18.4.3 Factor analysis or PCA? ▊▊▊

Should I use factor analysis or PCA?

I have just explained that there are two approaches to locating underlying dimensions of a data set: factor analysis and principal component analysis. These techniques differ in the communality estimates that are used. As I have hinted before, factor analysis derives a mathematical model from which factors are estimated, whereas principal component analysis decomposes the original data into a set of linear variates (see Dunteman, 1989, Chapter 8, for more detail on the differences between the procedures). As such, only factor analysis can estimate the underlying factors and it relies on various assumptions for these estimates to be accurate. PCA is concerned only with establishing which linear components exist within the data and how a particular variable might contribute to a given component.

Based on an extensive literature review, Guadagnoli and Velicer (1988) concluded that the solutions generated from PCA differ little from those derived from factor-analytic techniques. In reality, with 30 or more variables and communalities greater than 0.7 for all variables, different solutions are unlikely; however, with fewer than 20 variables and any low communalities (less than 0.4) differences can occur (Stevens, 2002).

The flip side of this argument is eloquently described by Cliff (1987), who observed that proponents of factor analysis 'insist that components analysis is at best a common factor analysis with some error added and at worst an unrecognizable hodgepodge of things from which nothing can be determined' (p. 349). Indeed, feeling is strong on this issue, with some arguing that when PCA is used it should not be described as a factor analysis (oops!) and that you should not impute substantive meaning to the resulting components. Ultimately, as I hope to have made clear, they are doing slightly different things.

18.4.4 Theory behind PCA ▮▮▮▮

The theory behind factor analysis is, frankly, a bit of an arse; an arse tattooed with matrix algebra. No one wants to look at matrix algebra when they're admiring an arse, so we'll look at the squeezable buttocks of PCA instead. Principal component analysis works in a very similar way to MANOVA and discriminant function analysis (see Chapter 17). In MANOVA, various sum of squares and cross-products matrices were calculated that contained information about the relationships between dependent variables. I mentioned before that these SSCP matrices can be converted to variance–covariance matrices, which represent the same information, but in averaged form (i.e., taking account of the number of observations). I also pointed out that by dividing each element by the relevant standard deviation the variance–covariance

matrices become standardized. The result is a correlation matrix. In PCA we usually deal with correlation matrices (though it is possible to analyze a variance–covariance matrix too) and my point is that this matrix represents the same information as an SSCP matrix in MANOVA.

In MANOVA, because we were comparing groups we ended up looking at the variates or components of the SSCP matrix that represented the ratio of the model variance to the error variance. These variates were linear dimensions that separated the groups tested, and we saw that the dependent variables mapped onto these underlying components. In short, we looked at whether the groups could be separated by some linear combination of the dependent variables. These variates were found by calculating the eigenvectors of the SSCP. The number of variates obtained was the smaller of p (the number of dependent variables) and $k - 1$ (where k is the number of groups).

In PCA we do much the same thing, but using the overall correlation matrix (because we're not interested in comparing groups of scores). So, and I'm simplifying things a little, we take a correlation matrix and calculate the variates. There are no groups of observations, and so the number of variates calculated will always equal the number of variables measured (p). The variates are described, as for MANOVA, by the eigenvectors associated with the correlation matrix. The elements of the eigenvectors are the weights of each variable on the variate. These values are the loadings described earlier (i.e., the b-values in equation (17.22)). The largest eigenvalue associated with each of the eigenvectors provides a single indicator of the substantive importance of each component. The basic idea is that we retain components with relatively large eigenvalues and ignore those with relatively small eigenvalues.

Factor analysis works differently, but there are similarities. Rather than using the correlation matrix, factor analysis starts by estimating the communalities between variables using the SMC (as described earlier). It then replaces the diagonal of the correlation matrix (the 1s) with these estimates. Then the eigenvectors and associated eigenvalues of this matrix are computed. Again, these eigenvalues tell us about the substantive importance of the factors, and, based on them, a decision is made about how many factors to retain. Loadings and communalities are then estimated using only the retained factors.

18.4.5 Factor extraction: eigenvalues and the scree plot ▮▮▮▮

How many factors should I extract?

In both PCA and factor analysis, not all factors are retained. The process of deciding how many factors to keep is called **extraction**. I mentioned above that eigenvalues associated with a variate indicate the substantive importance of that factor. Therefore, it is logical to retain only factors with large eigenvalues. This section looks at how we determine whether an eigenvalue is large enough to represent a meaningful factor.

Cattell (1966b) suggested plotting each eigenvalue (Y-axis) against the factor with which it is associated (X-axis). This graph is known as a **scree plot** (because it looks like a rock face with a pile of debris or scree, at the bottom). I mentioned earlier that it is possible to obtain as many factors as there are variables and that each has an associated eigenvalue. By graphing the eigenvalues, the relative importance of each factor becomes apparent. Typically there will be a few factors with quite high eigenvalues, and many factors with relatively low eigenvalues, and so this graph has a very characteristic shape: there is a sharp descent in the curve followed by a tailing off (see Figure 18.4). The point of inflexion is where the slope of the line changes dramatically, and Cattell (1966b) suggested using this point as the cut-off for retaining factors. In Figure 18.4, imagine drawing two straight lines (the red dashed lines): one summarizing the vertical part of the plot and the other summarizing the horizontal part. The point of inflexion is the data point at which these two lines meet. You retain only factors to the left of the point of inflexion (and do not include the factor at the point of inflexion itself),[4] so in both examples in Figure 18.4 we would extract two factors because the point of inflexion occurs at the third data point (factor). With a sample of more than 200 participants, the scree plot provides a fairly reliable criterion for factor selection (Stevens, 2002).

An alternative to the scree plot is to use the eigenvalues, because these represent the amount of variation explained by a factor. You set a criterion value that represents a substantial amount of variation and retain factors with eigenvalues above this criterion. There are two common criteria: **Kaiser's criterion** (Kaiser, 1960, 1970) is to retain factors with eigenvalues greater than 1 (followed by normal varimax rotation),[5] or a more liberal value of 0.7 (Jolliffe, 1972, 1986). The difference between how many factors are retained using these two methods can be dramatic. Generally speaking, Kaiser's criterion overestimates the number of factors to retain (see Jane Superbrain Box 18.2), which means that

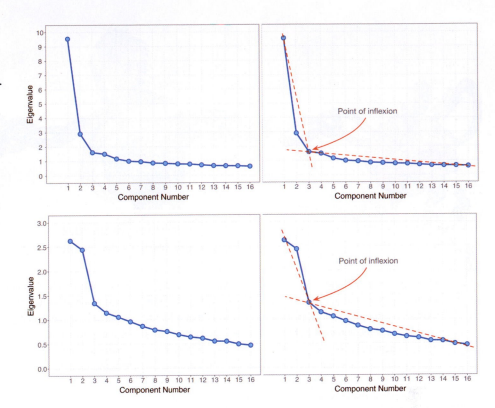

Figure 18.4   Examples of scree plots for data that probably have two underlying factors

Joliffe's criterion *really* overestimates. There is evidence that Kaiser's criterion is accurate when the number of variables is less than 30 and the resulting communalities (after extraction) are all greater than 0.7; it can also be accurate when the sample size exceeds 250 and the average communality is greater than or equal to 0.6. In any other circumstances, if the sample size is greater than 200 use a scree plot (see Stevens, 2002, for more detail). By default, IBM SPSS Statistics uses Kaiser's criterion to extract factors, so if the scree plot throws up a different number of factors to extract then you may need to rerun the analysis specifying the number of factors you want to retain.

As is often the case in statistics, the three criteria often provide different answers. In these situations consider the communalities of the factors. Remember that communalities represent the common variance: if the values are 1 then all common variance is accounted for, and if the values are 0 then no common variance is accounted for. In both PCA and factor analysis we determine how many factors/components to extract and then re-estimate the communalities. The factors we retain will not explain all the variance in the data (because we have discarded some information), and so the communalities after extraction will always be less than 1. The factors retained do not map perfectly

---

4   If you read Cattell's original paper he advised including the factor at the point of inflexion as well because it represents an error factor or 'garbage can', as he put it. However, Thurstone argued that it is better to retain too few than too many factors and in practice the 'garbage can' factor is rarely retained.

5   A colleague of Kaiser, Chester Harris, referred to this procedure as a 'little Jiffy'.

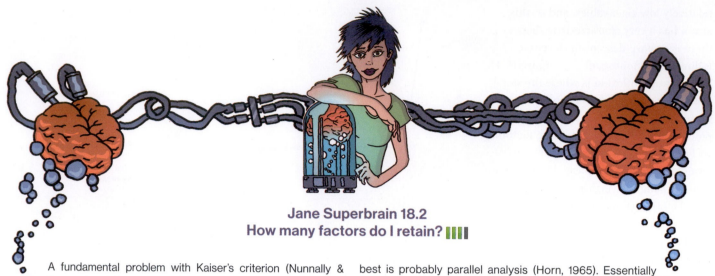

Jane Superbrain 18.2
How many factors do I retain? IIII

A fundamental problem with Kaiser's criterion (Nunnally & Bernstein, 1994) is that an eigenvalue of 1 means different things in different analyses: with 100 variables it means that a factor explains 1% of the variance, but with 10 variables it means that a factor explains 10% of the variance. These two situations are very different and a single rule that covers both is inappropriate. An eigenvalue of 1 also means only that the factor explains as much variance as a variable, which rather defeats the original intention of the analysis to reduce variables down to 'more substantive' underlying factors. Consequently, Kaiser's criterion often over-estimates the number of factors. Jolliffe's criterion is even worse (a factor explains less variance than a variable).

There are other ways to determine how many factors to retain, but they are not easy to do using SPSS Statistics. The best is probably parallel analysis (Horn, 1965). Essentially each eigenvalue (which represents the size of the factor) is compared against an eigenvalue for the corresponding factor in many randomly generated data sets that have the same characteristics as the data being analyzed. In doing so, each eigenvalue is compared to an eigenvalue from a data set that has no underlying factors. This is a bit like asking whether our observed factor is bigger than a non-existing factor. Factors that are bigger than their 'random' counterparts are retained. Of parallel analysis, the scree plot and Kaiser's criterion, Kaiser's criterion is, in general, worst and parallel analysis best (Zwick & Velicer, 1986). If you want to do parallel analysis then SPSS syntax is available (O'Connor, 2000) from https://people.ok.ubc.ca/brioconn/nfactors/nfactors.html.

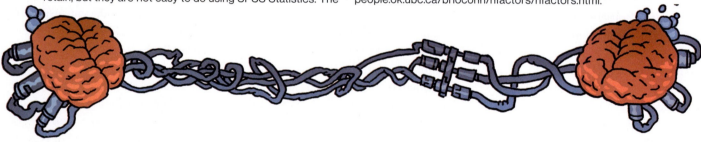

onto the original variables—they merely reflect the common variance in the data. Because communalities represent a loss of information they are important statistics. The closer the communalities are to 1, the better our factors are at explaining the original data. It is logical that the greater the number of factors retained, the greater the communalities will be (because less information is discarded); therefore, the communalities are good indices of whether too few factors have been retained. In fact, with generalized least squares factor analysis and maximum-likelihood factor analysis you can get a statistical measure of the goodness of fit of the factor solution (see the next chapter for more on goodness-of-fit tests). This basically measures the proportion of variance that the factor solution explains (so can be thought of as comparing communalities before and after extraction). As a final word of advice, your decision on how many factors to extract will depend also on why you're doing the analysis; for example, if you're trying to overcome multicollinearity problems in regression, then it might be better to extract too many factors than too few.

18.4.6 Factor rotation IIII

Should I rotate factors?

Once factors have been extracted, it is possible to calculate the degree to which variables load onto these factors (i.e., calculate the loadings for each variable on each factor). Generally, you will find that most variables have high loadings on the most important factor and small loadings on all other factors. This characteristic makes interpretation difficult, and so a technique called factor **rotation** is used to discriminate factors. If we visualize our factors as an axis along which variables can be plotted, then factor rotation effectively rotates these axes such that variables are loaded maximally to only one factor. Let's return to our example where we had measures of popularity that produced two factors (sociability and inconsideration). Earlier we looked at having just three measures loading highly onto each factor. Imagine now that we'd measured 20 variables, and it turned out that 10 seemed to reflect sociability and the other 10 inconsideration. Figure 18.5 shows two scenarios. As with the factor plot in Figure 18.3, the full lines represent the factors, and by looking at the coordinates it should be clear that the blue circles have high loadings for inconsideration (they are a long way up this axis) and small-to-medium loadings for sociability (they are not very far along this axis). Conversely, the orange circles have high loadings for sociability and small to medium loadings for inconsideration. Factor rotation amounts to rotating the axes (the red dashed lines) to try to ensure that both clusters of variables are intersected by the factor to which they relate most. After rotation, the loadings of the variables are maximized on one factor (the factor that intersects the cluster) and minimized on the remaining factor(s). If an axis passes through a cluster of variables, then these variables will have a loading close to zero on the opposite axis. If this idea is confusing, then look at Figure 18.5 and think about the values of the coordinates before and after rotation (this is best achieved by turning the book when you look at the rotated axes).

There are two flavors of rotation. **Orthogonal rotation** is shown in Figure 18.5 (left). We saw in Chapter 12 that the term *orthogonal* means unrelated, and in this context it means that we rotate factors while keeping them independent or uncorrelated. Before rotation, all

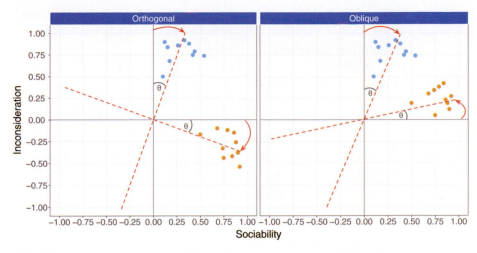

Figure 18.5 Schematic representations of factor rotation. The left-hand graph displays orthogonal rotation whereas the right-hand graph displays oblique rotation (see text for more details). θ is the angle through which the axes are rotated

factors are independent (i.e., they have a correlation of zero), and orthogonal rotation ensures that the factors remain this way. That is why in Figure 18.5 the axes remain perpendicular as they rotate. **Oblique rotation** allows factors to correlate, hence, the axes of Figure 18.5 (right) do not remain perpendicular: they can rotate by different amounts and in different directions if need be.

SPSS Statistics implements three methods of orthogonal rotation (**varimax**, **quartimax** and **equamax**) and two methods of oblique rotation (**direct oblimin** and **promax**). Quartimax rotation attempts to maximize the spread of factor loadings for a variable across all factors. Therefore, interpreting variables becomes easier but often results in lots of variables loading highly on a single factor. Varimax is the opposite in that it attempts to maximize the dispersion of loadings within factors. Therefore, it tries to load a smaller number of variables highly onto each factor, resulting in more interpretable clusters of factors. Equamax is a hybrid of the other two approaches and is reported to behave fairly erratically (see Tabachnick & Fidell, 2012). If you use orthogonal rotation then you should

probably select varimax because it is a good general approach that simplifies the interpretation of factors.

The case with oblique rotations is more complex because correlation between factors is permitted. Direct oblimin determines the degree to which factors are allowed to correlate by the value of a constant called delta. The default value in SPSS Statistics is 0, and this ensures that high correlation between factors is not allowed (this is known as *direct quartimin rotation*). If you choose to set delta to a value greater than 0 (up to 0.8), then you can expect highly correlated factors; if you set delta less than 0 (down to −0.8) you can expect less correlated factors. The default setting of zero is sensible for most analyses, and I don't recommend changing it unless you know what you are doing (see Pedhazur & Schmelkin, 1991, 620). Promax is a faster procedure designed for very large data sets. If you use oblique rotation then, other things being equal, use direct oblimin.

The choice of orthogonal or oblique rotation depends on: (1) whether there is a good theoretical reason to suppose that the factors should correlate or be independent; and (2) how the variables cluster on the factors before rotation.

On the first point, it is probably rare that you would measure a set of related variables and expect their underlying dimensions to be completely independent. For example, we wouldn't expect sociability to be completely independent of inconsideration (presumably inconsiderate people might not like other people much, which might make them antisocial). Therefore, on theoretical grounds, we would choose oblique rotation. There are strong grounds to believe that orthogonal rotations are a complete nonsense for naturalistic data, and certainly for any data involving humans (can you think of any psychological construct that is not in any way correlated with some other psychological construct?). As such, some argue that orthogonal rotations should never be used.

On the second point, Figure 18.5 demonstrates how the positioning of clusters will determine how successful a rotation will be (note the position of the orange circles). If an orthogonal rotation was carried out on the right-hand diagram it would be considerably less successful in maximizing loadings than the oblique rotation that is displayed.

A compromise is to run the analysis using both types of rotation. Pedhazur and Schmelkin (1991) suggest that if the oblique rotation demonstrates a negligible correlation between the extracted factors then it is reasonable to use the orthogonally rotated solution. If the oblique rotation reveals a correlated factor structure, then the orthogonally rotated solution should be discarded. We can check the relationships between factors using the **factor transformation matrix**, which is used to convert the unrotated factor loadings into the rotated ones. Values in this matrix represent the angle through which the axes have been rotated or the degree to which factors have been rotated.

18.4.7 Interpreting the factor structure ▮▮▮

Once a factor structure has been found, it needs to be interpreted. Earlier I wrote that the loadings were a gauge of the substantive importance of a given variable to a given factor. Therefore, it makes sense that we use these values to place variables with factors. Every variable will have a loading on every factor, so what we're looking for is variables that load highly on a given factor. Once we've identified these variables, we look for a theme within them.

What do I mean by 'load highly'? It is possible to assess the statistical significance of a loading (after all, it is simply a correlation coefficient or b-value), but, as with all significance tests, the p-value depends on the sample size. For example, based on Stevens (2002), for a sample size of 100 the loading should be greater than 0.512, but for 300 it should be greater than 0.298, and for 1000 greater than 0.162. Therefore, the significance of a loading gives little indication of the substantive importance of a variable to a factor because it depends on sample size (e.g., in very large samples, even small loadings will be 'significant'). Instead, we can gauge importance by squaring the loading to give an estimate of the amount of variance in a factor accounted for by a variable (like R^2). Stevens (2002) recommends interpreting factor loadings with an absolute value greater than 0.4 (the factor explains around 16% of the variance in the variable). Some researchers opt for the lower criterion of 0.3.

18.5 An anxious example ▮▮▮

Factor analysis is used frequently to develop questionnaires. I have noticed that a lot of students become very stressed about SPSS Statistics. Imagine that I wanted to design a questionnaire to measure a trait that I termed 'SPSS anxiety'. I devised a questionnaire to measure various aspects of students' anxiety towards learning SPSS, the SAQ (Figure 18.6). I generated questions based on interviews with anxious and non-anxious students and came up with 23 possible questions to include. Each question was a statement followed by a five-point Likert scale: 'strongly disagree', 'disagree', 'neither agree nor disagree', 'agree' and 'strongly agree' (SD, D, N, A and SA, respectively). What's more, I wanted to know whether anxiety about SPSS could be broken down into specific forms of anxiety. In other words, what latent variables contribute to anxiety about SPSS?[6]

With a little help from a few lecturer friends I collected 2571 completed questionnaires. Load the data (**SAQ.sav**) into SPSS Statistics. Note that each question (variable) is represented by a different column: there are 23 variables labeled **Question_01** to **Question_23** and each has a label indicating the question. By labeling my variables I can be very clear about what each variable represents (this is the value of giving your variables full titles rather than just using restrictive column headings).

18.5.1 General procedure ▮▮▮

Figure 18.7 shows the general procedure for conducting factor analysis or PCA. First screen the data. Then, once you

---

6 Such is my 'talent' (cough) for seamlessly blurring fact and fiction that I have had some people think that this is an actual bit of research that I've done. It's not, I made the example and data up.

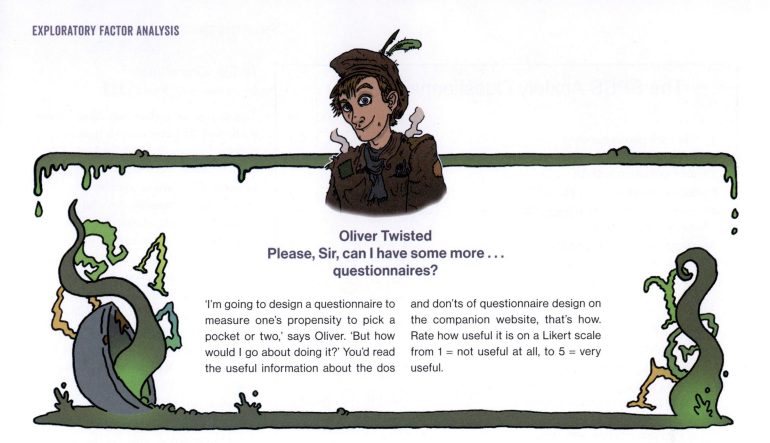

Oliver Twisted
Please, Sir, can I have some more . . .
questionnaires?

'I'm going to design a questionnaire to measure one's propensity to pick a pocket or two,' says Oliver. 'But how would I go about doing it?' You'd read the useful information about the dos and don'ts of questionnaire design on the companion website, that's how. Rate how useful it is on a Likert scale from 1 = not useful at all, to 5 = very useful.

embark on the main analysis, consider how many factors to retain and what rotation to use. If you are using the analysis to look at the factor structure of a questionnaire, then follow up with a reliability analysis (see Section 18.9).

18.5.2 Sample size ▊▊▊

Correlation coefficients fluctuate from sample to sample, much more so in small samples than in large. Therefore, the reliability of factor analysis depends on sample size. Many 'rules of thumb' exist for the ratio of cases to variables; a common one is to have at least 10–15 participants per variable. Although I've heard this rule bandied about on numerous occasions, its empirical basis is unclear (although Nunnally, 1978, did recommend having 10 times as many participants as variables). Based on real data, Arrindell and van der Ende (1985) concluded that the cases-to-variables ratio made little difference to the stability of factor solutions.

What does matter is the overall sample size, factor loadings and communalities. Test parameters tend to be stable regardless of the cases-to-variables ratio (Kass & Tinsley, 1979), which is why Comrey and Lee (1992) class 100 as a poor sample size, 300 as good and 1000 as excellent. With respect to factor loadings, Guadagnoli and Velicer (1988) found that if a factor has four or more loadings greater than 0.6 then it is reliable regardless of sample size; factors with 10 or more loadings greater than 0.40 are reliable if the sample size is greater than 150; and factors with a few low loadings should not be interpreted unless the sample size is 300 or more. With respect to communalities, MacCallum, Widaman, Zhang, & Hong (1999) showed that as communalities become lower the importance of sample size increases. With all communalities above 0.6, relatively small samples (less than 100) may be perfectly adequate. With communalities in the 0.5 range, samples between 100 and 200 can be good enough provided there are relatively

few factors each with only a small number of indicator variables. In the worst scenario of low communalities (well below 0.5) and a larger number of underlying factors they recommend samples above 500.

What's clear from this work is that a sample of 300 or more will probably provide a stable factor solution, but that a wise researcher will measure enough variables to measure adequately all the factors that theoretically they would expect to find.

There are measures of sampling adequacy such as the **Kaiser–Meyer–Olkin (KMO) measure of sampling adequacy** (Kaiser, 1970; Kaiser & Rice, 1974).[7] The KMO can be calculated for individual and multiple variables and represents the ratio of the squared correlation between variables to the squared partial correlation between variables. The KMO statistic varies between 0 and 1. A value of 0 indicates that the sum of partial correlations is large relative to the sum of correlations, indicating diffusion in the pattern of correlations (hence, factor analysis is likely to be inappropriate).

---

7 There are different versions of the KMO. SPSS Statistics implements the one in Kaiser and Rice (1974).

The SPSS Anxiety Questionnaire (SAQ)

	SD	D	N	A	SA
1. Statistics makes me cry	○	○	○	○	○
2. My friends will think I'm stupid for not being able to cope with SPSS	○	○	○	○	○
3. Standard deviations excite me	○	○	○	○	○
4. I dream that Pearson is attacking me with correlation coefficients	○	○	○	○	○
5. I don't understand statistics	○	○	○	○	○
6. I have little experience of computers	○	○	○	○	○
7. All computers hate me	○	○	○	○	○
8. I have never been good at mathematics	○	○	○	○	○
9. My friends are better at statistics than me	○	○	○	○	○
10. Computers are useful only for playing games	○	○	○	○	○
11. I did badly at mathematics at school	○	○	○	○	○
12. People try to tell you that SPSS makes statistics easier to understand but it doesn't	○	○	○	○	○
13. I worry that I will cause irreparable damage because of my incompetence with computers	○	○	○	○	○
14. Computers have minds of their own and deliberately go wrong whenever I use them	○	○	○	○	○
15. Computers are out to get me	○	○	○	○	○
16. I weep openly at the mention of central tendency	○	○	○	○	○
17. I slip into a coma whenever I see an equation	○	○	○	○	○
18. SPSS always crashes when I try to use it	○	○	○	○	○
19. Everybody looks at me when I use SPSS	○	○	○	○	○
20. I can't sleep for thoughts of eigenvectors	○	○	○	○	○
21. I wake up under my duvet thinking that I am trapped under a normal distribution	○	○	○	○	○
22. My friends are better at SPSS than I am	○	○	○	○	○
23. If I am good at statistics people will think I am a nerd	○	○	○	○	○

Figure 18.6 The SPSS anxiety questionnaire (SAQ)

18.5.3 Correlations between variables ▮▮▮▮

The 'garbage in, garbage out' adage applies particularly to factor analysis because a factor solution will usually be found to a set of variables, but will have little meaning if the variables put into the analysis are not sensible. A useful first step is to look at the correlations between variables. There are essentially two potential problems: (1) correlations that are not high enough; and (2) correlations that are too high. In both cases the remedy is to remove variables from the analysis. The correlations between variables can be checked using the *correlate* procedure (see Chapter 12) to create a correlation matrix of all variables. This matrix can also be created as part of the factor analysis. We will look at each problem in turn.

If our test questions measure the same underlying dimension (or dimensions) then we would expect them to correlate with each other (because they are measuring the same thing). Even if questions measure different aspects of the same things (e.g., we could measure overall anxiety in terms of sub-components such as worry, intrusive thoughts and physiological arousal), there should still be moderate correlations between the variables relating to these sub-traits. We wouldn't expect to see variables that have small correlations with each other. We can visually scan the correlation matrix and look for correlations below about 0.3 (you could use the *p*-values, but this approach isn't helpful because very small correlations will be significant in large samples and factor analysis typically employs large samples). If any variables have lots of correlations below 0.3 then consider excluding them. Of course this approach is subjective, but analyzing data is a skill, and there's more to it than following a recipe book!

If you want an objective test of whether correlations (overall) are too small then you

A value close to 1 indicates that patterns of correlations are relatively compact and so factor analysis should yield distinct and reliable factors. Kaiser and Rice (1974) provided appealing guidelines, especially if you like the letter M:

- Marvellous: values in the 0.90s
- Meritorious: values in the 0.80s
- Middling: values in the 0.70s
- Mediocre: values in the 0.60s
- Miserable: values in the 0.50s
- Merde: values below 0.50. (They used the word 'unacceptable', but I don't like the fact that it doesn't start with the letter 'M' so I have changed it.)

To sum up, values smaller than 0.5 should lead you either to collect more data or to rethink which variables to include.

can test whether the correlation matrix resembles an identity matrix (see Section 17.3). This would mean that the off-diagonal components would be zero—in other words, the correlations between variables are all zero. This is a pretty extreme scenario. *Bartlett's test* (Section 17.6.3) tells us whether our correlation matrix is significantly different from an identity matrix. If it is significant then it means that the correlations between variables are (overall) significantly different from zero. The trouble is that because significance depends on sample size (see Section 2.9.10) and in factor analysis sample sizes are very large, Bartlett's test will nearly always be significant: even when the correlations between variables are very small indeed. As such, it's not a useful test (although in the unlikely event that it is non-significant you certainly have a big problem).

The opposite problem is when variables correlate too highly. Although mild multicollinearity is not a problem for factor analysis, it is important to avoid extreme multicollinearity (i.e., variables that are very highly correlated) and **singularity** (variables that are perfectly correlated). As with linear models, multicollinearity causes problems in factor analysis because it becomes impossible to determine the unique contribution to a factor of the variables that are highly correlated. Multicollinearity does not cause a problem for principal component analysis.

Multicollinearity can be detected by looking at the determinant of the R-matrix, denoted $|R|$ (see Jane Superbrain Box 18.3). One heuristic is that the determinant of the R-matrix should be greater than 0.00001.

To avoid or correct for multicollinearity you could look through the correlation matrix for variables that correlate very highly ($r > 0.8$) and consider eliminating one of the variables (or more depending on the extent of the problem) before

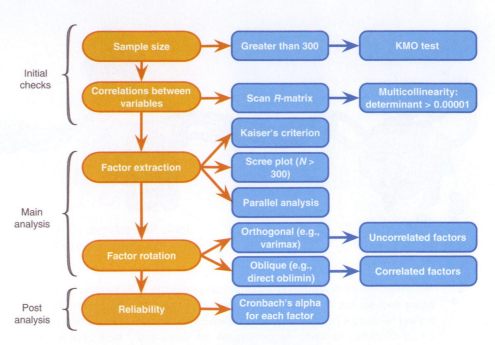

Figure 18.7   General procedure for factor analysis and PCA

proceeding. The problem with a heuristic such as this is that the effect of two variables correlating with $r = 0.9$ might be less than the effect of, say, three variables that all correlate with $r = 0.6$. In other words, eliminating such highly correlating variables might not be getting at the cause of the multicollinearity (Rockwell, 1975). It may take trial and error to work out which variables are creating the problem.

18.5.4 The distribution of data

The assumption of normality is important if you wish to generalize the results of your analysis beyond the sample collected or do significance tests, but otherwise it's not. You can do factor analysis on non-continuous data; for example, if you had dichotomous variables, it's possible (using syntax) to do the factor analysis direct from the correlation matrix, but you should construct the correlation matrix from tetrachoric correlation coefficients. The only hassle is computing the correlations, but there's lots of help to be found at the website http://www.john-uebersax.com/stat/tetra.htm.

18.6 Factor analysis using SPSS statistics ▮▮▮

Access the main dialog box (Figure 18.9) by selecting *Analyze* ▸ *Dimension Reduction* ▸ 👤 *Factor…*. Drag the variables you want to include in the analysis (or select them and click ➡) to the box labeled *Variables*. Remember to exclude any variables that were identified as problematic during the data screening. Click Descriptives… to access the dialog box in Figure 18.10. Checking ☑ Univariate descriptives produces means and standard deviations for each variable. Most of the other options relate to the correlation matrix of variables (the R-matrix described earlier): selecting ☑ Coefficients produces it, ☑ Significance levels includes the significance value of each correlation within it, and its ☑ Determinant is useful for testing for multicollinearity or singularity (see Section 18.5.3). Checking ☑ KMO and Bartlett's test of sphericity produces the Kaiser–Meyer–Olkin (see Section 18.5.2) measure of sampling adequacy and Bartlett's test (see Section 18.5.3). We have already seen the various

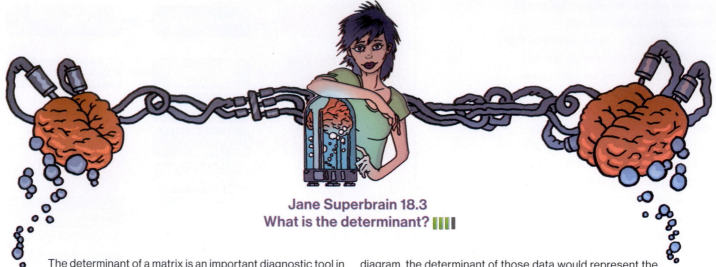

Jane Superbrain 18.3
What is the determinant? ▮▮▮▮

The determinant of a matrix is an important diagnostic tool in factor analysis, but the question of what it is is not easy to answer because it has a mathematical definition and I'm not a mathematician. However, we can bypass the maths and think about the determinant conceptually. The way that I think of the determinant is as describing the 'area' of the data. In Jane Superbrain Box 9.3 we saw the diagram reproduced in Figure 18.8. At the time I used these to describe eigenvectors and eigenvalues (which describe the shape of the data). The determinant is related to eigenvalues and eigenvectors, but instead of describing the height and width of the data it describes the overall area. So, in the left-hand diagram, the determinant of those data would represent the area inside the blue dashed ellipse. These variables have a low correlation so the determinant (area) is big; the maximum value is 1. In the right-hand diagram, the variables are perfectly correlated or singular, and the ellipse (blue dashed line) has been squashed down to basically a straight line. In other words, the opposite sides of the ellipse will meet with no distance between them at all. Put another way, the area or determinant, is 0. Therefore, the determinant tells us whether the correlation matrix is singular (determinant is 0) or if all variables are completely unrelated (determinant is 1) or somewhere inbetween.

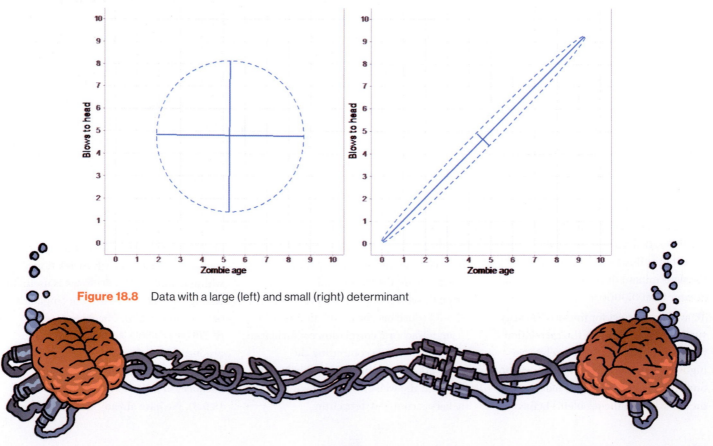

Figure 18.8   Data with a large (left) and small (right) determinant

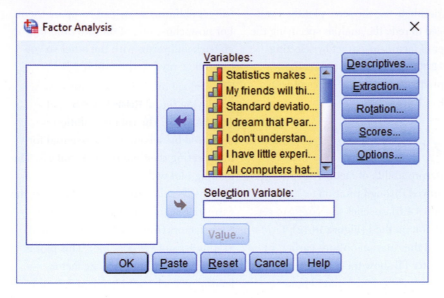

Figure 18.9 Main dialog box for factor analysis

Figure 18.10 Descriptives in factor analysis

criteria for adequacy, but with a sample of 2571 we shouldn't have cause to worry.

Selecting ☑ Reproduced produces a correlation matrix based on the model (rather than the real data). Differences between the matrix based on the model and the matrix based on the observed data constitute the residuals of the model. These residuals appear in the lower table of the reproduced matrix and we want relatively few of these values to be greater than 0.05. To save us scanning this matrix, a summary of how many residuals lie above 0.05 is produced. Checking ☑ Anti-image produces an anti-image matrix of covariances and correlations. This matrix contains measures of sampling adequacy for each variable along the diagonal and the negatives of the partial correlations/covariances on the off-diagonals. The diagonal elements, like the KMO measure, should all be greater than 0.5 at a bare minimum if the sample is adequate for a given pair of variables. If any pair of variables has a value less than this, consider dropping one of them from the analysis. The off-diagonal elements should all be very small (close to zero) in

a good model. Click Continue to return to the main dialog box.

18.6.1 Factor extraction using SPSS Statistics ▌▌▌

Click Extraction... to set the method of factor extraction (see Section 18.4.1). We will use *principal axis factoring* (Principal axis factoring ▼) as in Figure 18.11. In the *Analyze* box we can choose between analyzing the *Correlation matrix*

or *Covariance matrix* (SPSS Tip 18.1). We can choose to display the ☑ Unrotated factor solution and a ☑ Scree plot. The scree plot is a useful way to establish how many factors to retain, and the unrotated factor solution is useful in assessing how much rotation improves interpretation of the factor solution. If the rotated solution is little better than the unrotated one then it is possible that an inappropriate (or suboptimal) rotation method has been used.

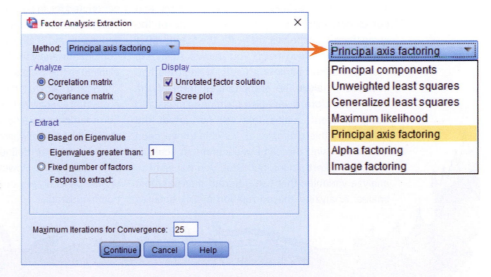

Figure 18.11 Dialog box for factor extraction

The *Extract* box provides options pertaining to the retention of factors. You have the choice of either extracting factors ⦿ Bas<u>e</u>d on Eigenvalue greater than a user-specified value (by default Kaiser's criterion of 1, but you can change this value) or retaining a ⦿ Fixed <u>n</u>umber of factors. It is probably best to run a primary analysis ⦿ Bas<u>e</u>d on Eigenvalue greater than 1, select a scree plot and compare the results. If looking at the scree plot and the eigenvalues over 1 lead you to retain the same number of factors then happy days. If the two criteria give different results then examine the communalities and decide for yourself which of the two criteria to believe.

If you decide to use the scree plot then you may need to redo the analysis specifying the number of factors to extract by selecting ⦿ Fixed <u>n</u>umber of factors and typing the appropriate number in the space provided (e.g., 4). Click Continue .

18.6.2 Rotation ▮▮▮▮

The interpretability of factors can be improved through rotation (Section 18.4.6), so, click Rotation... to set the rotation method (Figure 18.12). I discussed the rotation options earlier. In this chapter I'll show the output for both an orthogonal (varimax) and oblique

(direct oblimin) so we can compare them. For now, chose one of the two (and maybe go back and rerun with the other so you can follow my interpretation).

The dialog box also has options for displaying the ☑ Rotated solution and a *Loading plot*. The rotated solution is displayed by default and is essential for interpreting the final rotated analysis. The loading plot will provide a graphical display of each variable plotted against the extracted factors up to a maximum of three factors (four- or five-dimensional graphs are not yet possible). This plot is like Figure 18.3 and uses the factor loading of each variable for each factor.

SPSS Tip 18.1
Correlation or covariance matrix? ▮▮▮▮

This far into the book I hope you're happy with the idea that the variance–covariance matrix and correlation matrix are different versions of the same thing. Despite this, generally the results differ depending on which matrix you analyze. Analyzing the correlation matrix is a useful default method because it takes the standardized form of the matrix; therefore, if variables have been measured using different scales this will not affect the solution. In this example, all variables have been measured using the same measurement scale (a five-point Likert scale), but often you will want to analyze variables that use different measurement scales. Analyzing the correlation matrix ensures

that differences in measurement scales are accounted for. In addition, even variables measured using the same scale can have very different variances and this creates problems for principal component analysis. Using the correlation matrix eliminates this problem.

Having said that, there are statistical reasons for preferring to analyze the covariance matrix: correlation coefficients are not sensitive to variations in the dispersion of data, whereas the covariance is and so it produces better-defined factor structures (Tinsley & Tinsley, 1987). However, the covariance matrix should be analyzed only when your variables are commensurable.

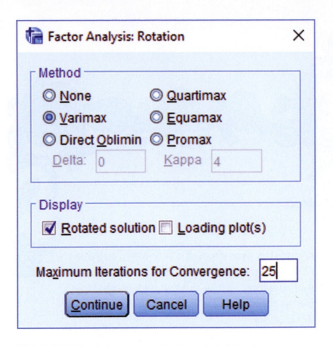

Figure 18.12 *Factor Analysis: Rotation* dialog box

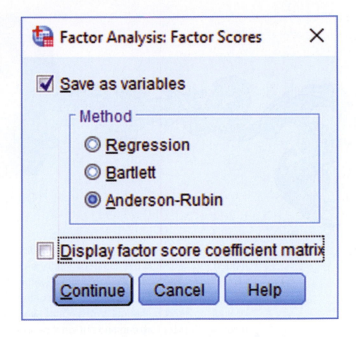

Figure 18.13 *Factor Analysis: Factor Scores* dialog box

With two factors these plots are interpretable: you hope to see one group of variables clustered close to the *X*-axis and a different group clustered around the *Y*-axis. If all variables are clustered between the axes, then the rotation has been relatively unsuccessful in maximizing the loading of a variable onto a single factor. With three factors these plots will strain even the most dedicated visual system, so unless you have only two factors I would probably avoid them.

A final option is to set the *Maximum Iterations for Convergence* (see SPSS Tip 20.1), which specifies the number of times that the computer will search for an optimal solution. In most circumstances the default of 25 is adequate; however, if you get an error message about convergence then increase this value. Click Continue.

18.6.3 Scores ▌▌▌

Chose a method to compute factor scores (Figure 18.13) by clicking Scores... . This option will save factor scores (see Section 18.3.3) for each case in the data editor.

A new column is created for each factor extracted and the factor score for each case is placed within that column. These scores can be used for further analysis or to identify groups of participants who score high on particular factors. If you want to ensure that factor scores are uncorrelated then select
⦿ Anderson-Rubin; if correlations between factor scores are acceptable then choose
⦿ Regression. As a final option, you can produce the factor score coefficient matrix, which realistically, we don't need to see. Click Continue.

18.6.4 Options ▌▌▌

Click Options... to set the remaining options (Figure 18.14). Missing data are a problem for factor analysis just like

most other procedures, and the options reflect the usual choices that were explained in SPSS Tip 6.1. If data are missing at random, and there's not too much missing data, there really is no

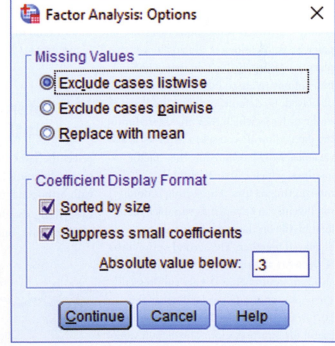

Figure 18.14 *Factor Analysis: Options* dialog box

589

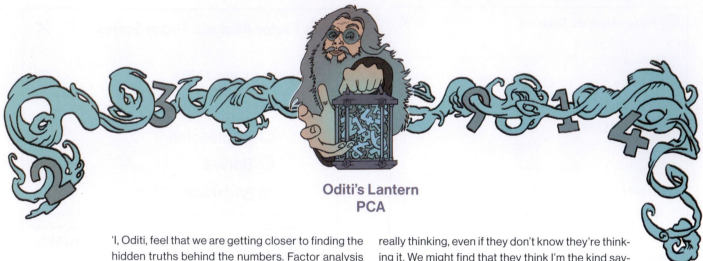

Oditi's Lantern
PCA

'I, Oditi, feel that we are getting closer to finding the hidden truths behind the numbers. Factor analysis allows us to estimate variables "hidden" within the data. This technique is the very essence of the cult of undiscovered numerical truths. Once we have mastered this tool we can find out what people are really thinking, even if they don't know they're thinking it. We might find that they think I'm the kind saviour of cute furry gerbils, but that underneath they know the truth ... stare into my lantern to discover factor analysis.'

reason (other than that I don't cover it in this book) not to use methods more sophisticated than ⦿ R̲eplace with mean (Enders, 2010). If you exclude cases pairwise your estimates can go all over the place, so definitely don't do that. The final two options relate to how coefficients are displayed. By default, variables are listed in the order in which they are entered into the data editor. If you select ☑ S̲orted by size variables will be ordered by their factor loadings, which can be useful for interpreting factors. The sorting is done intelligently in as much as variables that load highly on the same factor are displayed together. The second option is to ☑ S̲uppress small coefficients (by default absolute values below 0.1 are suppressed). This option hides factor loadings within ±0.1 in the output. Again, this option is useful for interpretation. The default value is probably sensible, but on the first run-through I recommend changing it to 0.3

to make interpretation simpler. We know that a loading of 0.4 is substantial, but so I don't throw out the baby with the bath water I tend to set the value at 0.3 as in Figure 18.14. Click Continue .

18.7 Interpreting factor analysis ▮▮▮▮

Select the same options as I have in the screen diagrams and run a factor analysis with orthogonal rotation.

Everything should go smoothly with the example data, but with your own data

you might be unlucky enough to get an error message about a 'non-positive definite matrix' (see SPSS Tip 18.2). A 'non-positive definite matrix' sounds a bit like a collection of depressed numbers that lack certainty about their lives. In some respects it is.

To make the larger outputs vaguely legible, each variable is referred to by its name in the data editor (e.g., Question_12) rather than the variable label (the question itself), which is what you will see in *your* output. When using my output refer to Figure 18.6 if you need to know the question content.

SELF ✗ TEST

Having done this, select the *Direct O̲blimin* option in Figure 18.12 and repeat the analysis. You should obtain two outputs identical in all respects except that one used an orthogonal rotation and the other an oblique.

SPSS Tip 18.2
Error messages about a 'non-positive definite matrix' ▮▮▮

Factor analysis works by looking at your correlation matrix. This matrix has to be 'positive definite' for the analysis to work. This term means lots of horrible things mathematically (e.g., the eigenvalues and determinant of the matrix are positive), but in more basic terms, factors are like lines floating in space, and eigenvalues measure the length of those lines. If your eigenvalue is negative then it means that the length of your line/factor is negative too. It's a bit like me asking you how tall you are, and you responding 'I'm minus 175 cm tall'. That would be nonsense. If a factor has negative length, that's nonsense too. When SPSS decomposes the correlation matrix to look for factors, if it comes across a negative eigenvalue it starts thinking 'Oh dear, I've entered some weird parallel universe where the usual rules of maths no longer apply and things can have negative lengths, and this probably means that time runs backwards, my mum is my dad, my sister is a dog, my head is a fish, and my toe is a frog called Gerald.'

It does the sensible thing and decides not to proceed. Things like the KMO test and the determinant rely on a positive definite matrix too: they can't be computed without one.

The most likely reason for having a non-positive definite R-matrix is that you have too many variables and too few cases of data, which makes the correlation matrix a bit unstable. It could also be that you have too many highly correlated items in your matrix (singularity, for example, tends to mess things up). In any case, it means that your data are bad, naughty data, and not to be trusted; if you let them loose then you have only yourself to blame for the consequences.

Other than cry, there's not that much you can do to rectify the situation. You could try to limit your items or selectively remove items (especially highly correlated ones) to see if that helps. Collecting more data can help too. There are some mathematical fudges you can do, but they're not as tasty as vanilla fudge and they are hard to implement.

18.7.1 Preliminary analysis ▮▮▮

The first body of output concerns data screening, assumption testing and sampling adequacy. You'll find several large tables (or matrices) that tell us interesting things about our data. If you selected

☑ Univariate descriptives (Figure 18.10) then you'll get a table of descriptive statistics for each variable (the mean, standard deviation

and number of cases). I haven't reproduced it here because I'm confident that you can interpret it by now. The table is a useful way to determine the extent of missing data.

Output 18.1 shows the R-matrix (i.e., the correlation matrix):[8] the top half contains the Pearson correlation coefficient between all pairs of questions and the bottom half contains the one-tailed p-values of these

coefficients. Note correlations less than 0.3 and greater than 0.9. Variables with very few correlations above 0.3 might not 'fit' with the pool of items, and variables with correlations greater than 0.9 might be collinear. You can also check the determinant of the correlation matrix and, if necessary, eliminate variables that you think are causing the problem. The determinant is listed at the bottom of the

---

8 To save space only columns for the first and last five questions in the questionnaire are included here.

Correlation Matrix[a]

		Question_01	Question_02	Question_03	Question_04	Question_05	Question_19	Question_20	Question_21	Question_22	Question_23
Correlation	Question_01	1.000	-.099	-.337	.436	.402	-.189	.214	.329	-.104	-.004
	Question_02	-.099	1.000	.318	-.112	-.119	.203	-.202	-.205	.231	.100
	Question_03	-.337	.318	1.000	-.380	-.310	.342	-.325	-.417	.204	.150
	Question_04	.436	-.112	-.380	1.000	.401	-.186	.243	.410	-.098	-.034
	Question_05	.402	-.119	-.310	.401	1.000	-.165	.200	.335	-.133	-.042
	Question_06	.217	-.074	-.227	.278	.257	-.167	.101	.272	-.165	-.069
	Question_07	.305	-.159	-.382	.409	.339	-.269	.221	.483	-.168	-.070
	Question_08	.331	-.050	-.259	.349	.269	-.159	.175	.296	-.079	-.050
	Question_09	-.092	.315	.300	-.125	-.096	.249	-.159	-.136	.257	.171
	Question_10	.214	-.084	-.193	.216	.258	-.127	.084	.193	-.131	-.062
	Question_11	.357	-.144	-.351	.369	.298	-.200	.255	.346	-.162	-.086
	Question_12	.345	-.195	-.410	.442	.347	-.267	.298	.441	-.167	-.046
	Question_13	.355	-.143	-.318	.344	.302	-.227	.204	.374	-.195	-.053
	Question_14	.338	-.165	-.371	.351	.315	-.254	.226	.399	-.170	-.048
	Question_15	.246	-.165	-.312	.334	.261	-.210	.206	.300	-.168	-.062
	Question_16	.499	-.168	-.419	.416	.395	-.267	.265	.421	-.156	-.082
	Question_17	.371	-.087	-.327	.383	.310	-.163	.205	.363	-.126	-.092
	Question_18	.347	-.164	-.375	.382	.322	-.257	.235	.430	-.160	-.080
	Question_19	-.189	.203	.342	-.186	-.165	1.000	-.249	-.275	.234	.122
	Question_20	.214	-.202	-.325	.243	.200	-.249	1.000	.468	-.100	-.035
	Question_21	.329	-.205	-.417	.410	.335	-.275	.468	1.000	-.129	-.068
	Question_22	-.104	.231	.204	-.098	-.133	.234	-.100	-.129	1.000	.230
	Question_23	-.004	.100	.150	-.034	-.042	.122	-.035	-.068	.230	1.000
Sig. (1-tailed)	Question_01		.000	.000	.000	.000	.000	.000	.000	.000	.410
	Question_02	.000		.000	.000	.000	.000	.000	.000	.000	.000
	Question_03	.000	.000		.000	.000	.000	.000	.000	.000	.000
	Question_04	.000	.000	.000		.000	.000	.000	.000	.000	.043
	Question_05	.000	.000	.000	.000		.000	.000	.000	.000	.017
	Question_06	.000	.000	.000	.000	.000	.000	.000	.000	.000	.000
	Question_07	.000	.000	.000	.000	.000	.000	.000	.000	.000	.000
	Question_08	.000	.006	.000	.000	.000	.000	.000	.000	.000	.005
	Question_09	.000	.000	.000	.000	.000	.000	.000	.000	.000	.000
	Question_10	.000	.000	.000	.000	.000	.000	.000	.000	.000	.001
	Question_11	.000	.000	.000	.000	.000	.000	.000	.000	.000	.000
	Question_12	.000	.000	.000	.000	.000	.000	.000	.000	.000	.009
	Question_13	.000	.000	.000	.000	.000	.000	.000	.000	.000	.004
	Question_14	.000	.000	.000	.000	.000	.000	.000	.000	.000	.007
	Question_15	.000	.000	.000	.000	.000	.000	.000	.000	.000	.001
	Question_16	.000	.000	.000	.000	.000	.000	.000	.000	.000	.000
	Question_17	.000	.000	.000	.000	.000	.000	.000	.000	.000	.000
	Question_18	.000	.000	.000	.000	.000	.000	.000	.000	.000	.000
	Question_19	.000	.000	.000	.000	.000		.000	.000	.000	.000
	Question_20	.000	.000	.000	.000	.000	.000		.000	.000	.039
	Question_21	.000	.000	.000	.000	.000	.000	.000		.000	.000
	Question_22	.000	.000	.000	.000	.000	.000	.000	.000		.000
	Question_23	.410	.000	.000	.043	.017	.000	.039	.000	.000	

a. Determinant = .001

Output 18.1

KMO and Bartlett's Test

Kaiser–Meyer–Olkin Measure of Sampling Adequacy.		.930
Bartlett's Test of Sphericity	Approx. Chi-Square	19334.492
	df	253
	Sig.	.000

Output 18.2

matrix (blink and you'll miss it). For these data its value is 0.001, which is greater than the necessary value of 0.00001 (see Section 18.5.3).[9] All questions in the SAQ correlate reasonably well with all others and none of the correlation coefficients are excessively large; therefore, we won't eliminate any questions at this stage.

Output 18.2 shows the Kaiser–Meyer–Olkin measure of sampling adequacy and Bartlett's test of sphericity. The KMO statistic is 0.93, which is well above the minimum criterion of 0.5 and falls into the range of 'marvellous' (see Section 18.5.2), so we might take comfort that the sample size is probably adequate for factor analysis. The KMO values for individual variables are produced on the diagonal of the anti-image correlation matrix in Output 18.3 (I have highlighted these cells).[10] We should check that the diagonal elements of the anti-image matrix are

above the bare minimum of 0.5 (and preferably higher). For these data all values are well above 0.5, which is good news. If you find variables with values below 0.5 then consider excluding them from the analysis (or run the analysis with and without that variable and note the difference). Removal of a variable affects the KMO statistics, so if you remove a variable be sure to re-examine the new anti-image correlation matrix. As for the rest of the anti-image correlation matrix, the off-diagonal elements represent the partial correlations between variables. We want these correlations to be very small (the smaller, the better). So, as a final check you can look through to see that the off-diagonal elements are small (they should be for these data).

Bartlett's measure (Output 18.2) tests the null hypothesis that the original correlation matrix is an identity matrix. We want this test to be *significant* (see Section 18.5.3). Given the large sample sizes usually used in factor analysis, this test will almost certainly be significant, and it is ($p < 0.001$). A non-significant test would certainly indicate a massive problem, but this significant value only really tells us that we don't have a massive problem, which is nice to know, I suppose.

18.7.2 Factor extraction

The first part of the factor extraction process is to determine the linear components within the variables—the eigenvectors (see Section 18.4.5). There are as many components (eigenvectors) in the *R*-matrix as there are variables, but most will be unimportant. To determine which vectors to retain we apply criteria based on the magnitude of the associated eigenvalues. By default, Kaiser's criterion of retaining factors with eigenvalues greater than 1 is used (see Figure 18.11).

9 Actually the determinant of this matrix is 0.0005271. I have no idea why SPSS reports this value as 0.001.

10 In your output the anti-image correlation appears with the covariance matrices. These matrices provide similar information (remember the relationship between covariance and correlation) but the anti-image correlation matrix is the most informative.

Anti-image Matrices

Anti-image Correlation

	Question_01	Question_02	Question_03	Question_04	Question_05	Question_19	Question_20	Question_21	Question_22	Question_23
Question_01	.930	−.020	.053	−.167	−.156	.012	−.016	.006	.001	−.059
Question_02	−.020	.875	−.157	−.041	.010	−.029	.059	.041	−.121	−.002
Question_03	.053	−.157	.951	.084	.037	−.121	.078	.070	−.007	−.076
Question_04	−.167	−.041	.084	.955	−.134	−.034	−.004	−.086	−.033	−.017
Question_05	−.156	.010	.037	−.134	.960	−.018	−.011	−.046	.035	−.005
Question_06	.020	−.053	−.042	−.007	−.035	−.015	.051	.039	.040	.018
Question_07	.023	.016	.072	−.087	−.044	.068	.048	−.208	.013	−.008
Question_08	−.049	−.033	−.007	−.075	−.027	.047	.021	−.020	−.023	.002
Question_09	−.016	−.193	−.142	.030	−.020	−.111	.038	−.031	−.126	−.092
Question_10	−.012	−.012	−.016	.006	−.093	−.009	.043	.017	.019	.015
Question_11	−.041	.038	.064	−.022	−3.269E-5	−.006	−.082	−.005	.034	.010
Question_12	−.007	.031	.087	−.154	−.058	.040	−.065	−.079	.018	−.028
Question_13	−.085	−.008	−.032	.023	.004	.009	.018	−.033	.052	−.030
Question_14	−.040	.023	.069	−.004	−.026	.044	.001	−.063	.029	−.026
Question_15	.089	.037	.008	−.062	.014	.009	−.037	.035	.025	−.024
Question_16	−.264	−.011	.081	−.036	−.096	.047	−.005	−.085	−.003	.023
Question_17	−.047	−.029	.035	−.035	−.018	−.047	.015	−.041	.010	.055
Question_18	−.023	.018	.039	−.025	.002	.030	−.003	−.072	−.024	.023
Question_19	.012	−.029	−.121	−.034	−.018	.941	.091	.031	−.115	−.038
Question_20	−.016	.059	.078	−.004	−.011	.091	.889	−.323	−.011	−.028
Question_21	.006	.041	.070	−.086	−.046	.031	−.323	.929	−.024	.013
Question_22	.001	−.121	−.007	−.033	.035	−.115	−.011	−.024	.878	−.176
Question_23	−.059	−.002	−.076	−.017	−.005	−.038	−.028	.013	−.176	.766

Output 18.3

Output 18.4 lists the eigenvalues associated with each factor before extraction, after extraction and after rotation. Before extraction, 23 factors are identified (there should be as many eigenvectors as there are variables and so there will be as many factors as variables— see Section 18.4.5). The eigenvalues associated with each factor represent the variance explained by that particular factor and the output contains this information: the eigenvalue is translated into the percentage of variance explained (e.g., factor 1 explains 31.696% of total variance). The first few factors explain relatively large amounts of variance (especially factor 1) whereas subsequent factors explain only small amounts. All factors with eigenvalues greater than 1 are then extracted, leaving us with four factors. The eigenvalues associated with these factors (and the percentage of variance explained) are displayed under the heading of *Extraction Sums of Squared Loadings*. In the part of the table labeled *Rotation Sums of Squared Loadings*, the eigenvalues of the factors after rotation are displayed. Rotation has the effect of optimizing the factor structure, and one consequence for these data is that the relative importance of the four factors is equalized a bit. Before rotation, factor 1

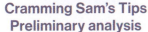

Cramming Sam's Tips
Preliminary analysis

- Scan the correlation matrix for variables that have very small correlations with most other variables or correlate very highly ($r = 0.9$) with one or more other variables.
- In factor analysis, check that the determinant of this matrix is bigger than 0.00001; if it is then multicollinearity isn't a problem. You don't need to worry about this for principal component analysis.
- In the table labeled *KMO and Bartlett's Test* the KMO statistic should be greater than 0.5 as a bare minimum; if it isn't, collect more data. You should check the KMO statistic for individual variables by looking at the diagonal of the anti-image matrix. These values should also be above 0.5 (this is useful for identifying problematic variables if the overall KMO is unsatisfactory).
- Bartlett's test of sphericity will usually be significant (the value of *Sig.* will be less than 0.05), if it's not, you've got a disaster on your hands.

Total Variance Explained

Factor	Initial Eigenvalues			Extraction Sums of Squared Loadings			Rotation Sums of Squared Loadings		
	Total	% of Variance	Cumulative %	Total	% of Variance	Cumulative %	Total	% of Variance	Cumulative %
1	7.290	31.696	31.696	6.744	29.323	29.323	3.033	13.188	13.188
2	1.739	7.560	39.256	1.128	4.902	34.225	2.855	12.415	25.603
3	1.317	5.725	44.981	.814	3.539	37.764	1.986	8.636	34.238
4	1.227	5.336	50.317	.624	2.713	40.477	1.435	6.239	40.477
5	.988	4.295	54.612						
6	.895	3.893	58.504						
7	.806	3.502	62.007						
8	.783	3.404	65.410						
9	.751	3.265	68.676						
10	.717	3.117	71.793						
11	.684	2.972	74.765						
12	.670	2.911	77.676						
13	.612	2.661	80.337						
14	.578	2.512	82.849						
15	.549	2.388	85.236						
16	.523	2.275	87.511						
17	.508	2.210	89.721						
18	.456	1.982	91.704						
19	.424	1.843	93.546						
20	.408	1.773	95.319						
21	.379	1.650	96.969						
22	.364	1.583	98.552						
23	.333	1.448	100.000						

Extraction Method: Principal Axis Factoring.

Output 18.4

Communalities

	Initial	Extraction
Question_01	.373	.373
Question_02	.188	.260
Question_03	.398	.472
Question_04	.385	.419
Question_05	.291	.299
Question_06	.427	.594
Question_07	.470	.489
Question_08	.490	.646
Question_09	.220	.339
Question_10	.197	.197
Question_11	.530	.629
Question_12	.424	.453
Question_13	.451	.474
Question_14	.393	.425
Question_15	.344	.322
Question_16	.463	.458
Question_17	.494	.575
Question_18	.492	.544
Question_19	.209	.245
Question_20	.270	.266
Question_21	.454	.468
Question_22	.167	.247
Question_23	.086	.116

Extraction Method: Principal Axis Factoring.

Factor Matrix[a]

	Factor			
	1	2	3	4
Question_18	.684			
Question_07	.663			
Question_16	.653			
Question_13	.650			
Question_11	.646	.313		
Question_12	.643			
Question_21	.633			
Question_17	.632	.359		
Question_14	.628			
Question_04	.607			
Question_03	-.605			
Question_15	.559			
Question_01	.557			
Question_06	.552		.489	
Question_08	.546	.483		
Question_05	.522			
Question_20	.407			
Question_10	.404			
Question_19	-.397			
Question_09		.460		
Question_02		.372		
Question_22				
Question_23				

Extraction Method: Principal Axis Factoring.

a. 4 factors extracted. 11 iterations required.

Output 18.5

accounted for considerably more variance than the remaining three (29.32% compared to 4.90%, 3.54% and 2.71%), but after rotation it accounts for only 13.19% of variance (compared to 12.42%, 8.64% and 6.24%).

Output 18.5 (left) shows the table of communalities before and after extraction. Remember that the communality is the proportion of common variance within a variable (see Section 18.4.2). Factor analysis starts by estimating the variance

that is common; therefore, before extraction the communalities are a kind of best guess. Once factors have been extracted, we can better estimate how much variance is common. The communalities in the column labeled *Extraction* reflect this common variance. So, for example, we can say that 37.3% of the variance associated with question 1 is common or shared, variance. Another way to look at these communalities is in terms of the proportion of variance explained by the underlying factors. Remember that after extraction we have discarded some factors (in this case we've retained only four), so the communalities after extraction represent the amount of variance in each variable that can be explained by the retained factors. Output 18.5 (right) shows the factor matrix before rotation. This matrix contains the loadings of each variable onto each factor. Because we requested that loadings less than 0.3 be suppressed (see Figure 18.14) there are blank spaces for many of the loadings. This matrix is not particularly important for interpretation, but it is interesting to note that before rotation most variables load highly onto the first factor (that is why this factor accounts for most of the variance in Output 18.4). Because you should never let a computer make important decisions for you, we need to think about the number of factors to extract (Section 18.4.5). By Kaiser's criterion we should extract four factors (which is what has been done). This criterion is accurate when there are less than 30 variables and communalities after extraction are greater than 0.7 or when the sample size exceeds 250 and the average communality is greater than 0.6. For these data, no communalities exceed 0.7 (Output 18.5), and the average communality is quite low: adding up the communalities and dividing by how many there are gives us 9.31/23 = 0.405. Both criteria suggest Kaiser's rule might be inappropriate for these data. Using Jolliffe's criterion (retain factors with eigenvalues greater than 0.7) we'd end up with 10 factors (see Output 18.4) most of which equate to small portions of variance, so I think that would be a silly idea. The scree plot (Output 18.6) is a little difficult to interpret because there are points of inflexion at both 3 and 5 factors, meaning that we could justify retaining either two or four factors.

So how many factors *should* we extract? The recommendations for Kaiser's criterion are for much smaller samples than we have. Given our huge sample and that there is some consistency between Kaiser's criterion and the scree plot, it is reasonable to extract four factors; however, you could rerun the analysis and ask for only two factors (see Figure 18.11) and compare the results.

Output 18.7 shows an edited version of the reproduced correlation matrix. The top half of this matrix (labeled *Reproduced Correlations*) contains the correlation coefficients between the questions based on the factor model. The diagonal of this matrix contains the communalities after extraction for each variable (you can check the values against Output 18.5). The correlations in the reproduced matrix differ from those in the *R*-matrix because they stem from the model rather than the observed data. If the model were a perfect fit to the data then we would expect the reproduced correlation coefficients to be the same as the original correlation coefficients. Therefore, to assess the fit of the model we can look at the differences between the observed correlations and the correlations based on the model. For example, if we take the correlation between questions 1 and 2, the correlation based on the observed data is −0.099 (Output 18.1), and based on the model is −0.112 (Output 18.7). The difference is 0.013:

$$\text{residual} = r_{\text{observed}} - r_{\text{from model}}$$
$$\text{residual}_{Q_1 Q_2} = (-0.099) - (-0.112) \qquad (18.7)$$
$$= 0.013$$

This value is the same as the one in the lower half of the reproduced matrix (labeled *Residual*) for questions 1 and 2

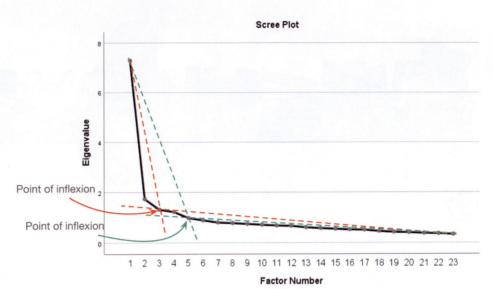

Scree Plot

Point of inflexion

Point of inflexion

Output 18.6

(highlighted in blue). More generally, the lower half of the reproduced matrix contains the differences between the observed correlation coefficients and the ones predicted from the model for all pairs of variables. For a good model these values will all be small: ideally, we want most values to be less than 0.05. Rather than scan this huge matrix, the footnote to the matrix states how many residuals have an

Reproduced Correlations

		Question_01	Question_02	Question_03	Question_04	Question_05	Question_19	Question_20	Question_21	Question_22	Question_23
Reproduced Correlation	Question_01	.373	-.112	-.338	.393	.328	-.191	.266	.398	-.072	-.013
	Question_02	-.112	.260	.295	-.129	-.119	.237	-.192	-.201	.227	.146
	Question_03	-.338	.295	.472	-.367	-.316	.328	-.336	-.431	.242	.133
	Question_04	.393	-.129	-.367	.419	.353	-.214	.282	.429	-.092	-.021
	Question_05	.328	-.119	-.316	.353	.299	-.190	.237	.364	-.091	-.025
	Question_06	.221	-.078	-.218	.269	.249	-.167	.078	.259	-.175	-.072
	Question_07	.349	-.154	-.363	.393	.344	-.243	.230	.408	-.173	-.066
	Question_08	.345	-.044	-.258	.345	.277	-.129	.172	.283	-.086	-.055
	Question_09	-.071	.290	.295	-.092	-.092	.255	-.174	-.174	.272	.178
	Question_10	.191	-.096	-.210	.218	.194	-.149	.116	.223	-.130	-.061
	Question_11	.362	-.131	-.345	.375	.311	-.210	.213	.339	-.178	-.110
	Question_12	.374	-.189	-.407	.412	.356	-.265	.291	.447	-.158	-.057
	Question_13	.329	-.143	-.341	.371	.325	-.231	.202	.375	-.182	-.078
	Question_14	.342	-.155	-.359	.381	.333	-.238	.237	.400	-.160	-.061
	Question_15	.289	-.160	-.327	.319	.277	-.223	.204	.331	-.180	-.091
	Question_16	.401	-.193	-.426	.430	.364	-.267	.315	.457	-.152	-.063
	Question_17	.379	-.089	-.321	.393	.324	-.181	.212	.351	-.123	-.066
	Question_18	.355	-.155	-.369	.402	.354	-.249	.230	.419	-.179	-.066
	Question_19	-.191	.237	.328	-.214	-.190	.245	-.218	-.271	.211	.124
	Question_20	.266	-.192	-.336	.282	.237	-.218	.266	.329	-.122	-.059
	Question_21	.398	-.201	-.431	.429	.364	-.271	.329	.468	-.142	-.051
	Question_22	-.072	.227	.242	-.092	-.091	.211	-.122	-.142	.247	.163
	Question_23	-.013	.146	.133	-.021	-.025	.124	-.059	-.051	.163	.116
Residual[b]	Question_01		.013	.001	.042	.074	.002	-.052	-.069	-.032	.009
	Question_02	.013		.023	.017	-.001	-.034	-.010	-.004	.004	-.046
	Question_03	.001	.023		-.014	.006	.014	.011	.014	-.039	.017
	Question_04	.042	.017	-.014		.048	.028	-.039	-.018	-.006	-.013
	Question_05	.074	-.001	.006	.048		.025	-.037	-.030	-.041	-.017
	Question_06	-.004	.004	-.009	.009	.009	.000	.022	.013	.010	.003
	Question_07	-.044	-.006	-.019	.016	-.005	-.026	-.009	.075	.005	-.004
	Question_08	-.014	-.005	.000	.004	-.009	-.030	.003	.013	.006	.005
	Question_09	-.022	.024	.005	-.033	-.003	-.005	.015	.038	-.015	-.007
	Question_10	.023	.012	.017	-.003	.064	.022	-.032	-.030	-.001	-.001
	Question_11	-.005	-.013	-.006	-.007	-.013	.011	.042	.007	.016	.023
	Question_12	-.028	-.006	-.003	.030	-.009	-.001	.007	-.007	-.009	.011
	Question_13	.025	-2.146E-5	.023	-.026	-.024	.004	.002	-.001	-.014	.025
	Question_14	-.004	-.009	-.012	-.030	-.017	-.016	-.011	-.001	-.009	.012
	Question_15	-.044	-.005	.015	.015	-.016	.013	.002	-.031	.012	.029
	Question_16	.098	.025	.007	-.014	.030	-3.481E-5	-.050	-.036	-.003	-.019
	Question_17	-.009	.002	-.006	-.010	-.014	.018	-.007	.012	-.003	-.026
	Question_18	-.008	-.009	-.006	-.020	-.032	-.007	.005	.011	.019	-.014
	Question_19	.002	-.034	.014	.028	.025		-.031	-.004	.023	-.002
	Question_20	-.052	-.010	.011	-.039	-.037	-.031		.139	.022	.024
	Question_21	-.069	-.004	.014	-.018	-.030	-.004	.139		.013	-.017
	Question_22	-.032	.004	-.039	-.006	-.041	.023	.022	.013		.067
	Question_23	.009	-.046	.017	-.013	-.017	-.002	.024	-.017	.067	

Extraction Method: Principal Axis Factoring.
b. Residuals are computed between observed and reproduced correlations. There are 12 (4.0%) nonredundant residuals with absolute values greater than 0.05.

Output 18.7

Cramming Sam's Tips
Factor extraction

- To decide how many factors to extract, look at the table labeled *Communalities* and the column labeled *Extraction*. If these values are all 0.7 or above and you have less than 30 variables then the default (Kaiser's criterion) for extracting factors is fine. Likewise, if your sample size exceeds 250 and the average of the communalities is 0.6 or greater.

Alternatively, with 200 or more participants the scree plot can be used.
- Check the bottom of the table labeled *Reproduced Correlations* for the percentage of 'nonredundant residuals with absolute values greater than 0.05'. This percentage should be less than 50% and the smaller it is, the better.

absolute value greater than 0.05. For these data there are only 12 residuals (4%)[11] that are greater than 0.05. There are no hard-and-fast rules about what proportion of residuals should be below 0.05; however, if more than 50% are greater than 0.05 you probably have grounds for concern; 4% (as we have here) is certainly nothing to worry about.

18.7.3 Orthogonal rotation (varimax)

The first analysis I asked you to run was using an orthogonal rotation, but I also asked you to rerun the analysis using oblique rotation. The results of both analyses will be presented to highlight the

differences. This comparison will also be a useful way to show the circumstances in which one type of rotation might be preferable to another.

Output 18.8 shows the rotated factor matrix (called the rotated component matrix in PCA), which is a matrix of the factor loadings for each variable on each factor. This matrix contains the same information as the factor matrix in Output 18.5, but calculated *after* rotation. Factor loadings less than 0.3 have not been displayed because we asked for these to be suppressed and the variables are listed in the order of size of their factor loadings because we asked for the output to be sorted by size (Figure 18.14). For all other parts of the output I suppressed the

variable labels (to save space) but at this point my outputs show the variable labels because it makes interpretation easier.

Before rotation (Output 18.5) most variables loaded highly on the first factor and the remaining factors didn't really get a look-in. The rotation of the factor structure has clarified things considerably: there are four factors, and most variables load very highly on only one factor.[12] In cases where a variable loads highly on more than one factor the loading is typically higher for one factor than another. For example, 'SPSS always crashes when I try to use it' loads highly on both factor 1 and 2, but the loading for factor 2 (0.612) is greater than for factor 1 (0.366), so it makes sense to think of it as making a

---

11 There are 253 unique correlation coefficients in the table and 12 residuals greater than 0.05, which is $(12/253) \times 100 = 4.74\%$.
   Weirdly, SPSS seems to round down to the nearest whole percentage value.
12 The suppression of loadings less than 0.3 and ordering variables by their loading size makes this pattern easy to see.

Rotated Factor Matrix[a]

	Factor			
	1	2	3	4
I wake up under my duvet thinking that I am trapped under a normal distribution	.594			
I weep openly at the mention of central tendency	.543			
I dream that Pearson is attacking me with correlation coefficients	.527			
People try to tell you that SPSS makes statistics easier to understand but it doesn't	.510	.398		
Standard deviations excite me	-.505			.399
Statistics makes me cry	.504			
I can't sleep for thoughts of eigenvectors	.465			
I don't understand statistics	.436			
I have little experience of computers		.753		
SPSS always crashes when I try to use it	.366	.612		
I worry that I will cause irreparable damage because of my incompetence with computers		.564		
All computers hate me	.364	.559		
Computers have minds of their own and deliberately go wrong whenever I use them	.388	.485		
Computers are useful only for playing games		.380		
Computers are out to get me		.377		
I have never been good at mathematics			.759	
I did badly at mathematics at school			.688	
I slip into a coma whenever I see an equation			.641	
My friends are better at statistics than me				.559
My friends are better at SPSS than I am				.465
My friends will think I'm stupid for not being able to cope with SPSS				.464
Everybody looks at me when I use SPSS				.375
If I'm good at statistics my friends will think I'm a nerd				.329

Extraction Method: Principal Axis Factoring.
Rotation Method: Varimax with Kaiser Normalization.[a]

a. Rotation converged in 7 iterations.

Output 18.8

bigger contribution to factor 2 than to factor 1. Remember that every variable has a loading on every factor, it just appears as though they don't in Output 18.8 because we asked that they not be printed if they were lower than 0.3. It's not the case that a variable loads on one factor but not on another (although people often use that turn of phrase); variables load on all factors, but to interpret the factors we assign variables to a factor based on them having a greater loading on that factor than on the others. If a variable has similar-sized loadings across two or more factors this could be because the factors reflect related constructs (and if you expect factors to correlate this can be fine) or that it is not a good item for distinguishing those constructs.

The next step is to look at the content of questions that load highly on the same factor to try to identify common themes. If the mathematical factors represent some real-world construct then common themes among highly loading questions can help us identify what that construct might be.

The questions that load highly on factor 1 seem to relate to different aspects of statistics; therefore, we might label this factor *fear of statistics*. The questions that load highly on factor 2 seem to relate to using computers or SPSS. Therefore, we might label this factor *fear of computers*. The three questions that load highly on factor 3 relate to mathematics, and we might label this factor *fear of mathematics*. Finally, the questions that load highly on factor 4 contain some component of social evaluation and might reflect *peer evaluation*. This analysis seems to reveal that the questionnaire is composed of four subscales: fear of statistics, fear of computers, fear of maths and fear of negative peer evaluation. There are two possibilities here. The first is that the SAQ failed to measure what it set out to (namely, SPSS anxiety) but instead measures related constructs. The second is that these four constructs are sub-components of SPSS anxiety; however, the factor analysis does not indicate which of these possibilities is true.

18.7.4 Oblique rotation (direct oblimin) ▮▮▮▮

With oblique rotation the factor matrix is split into two: the *pattern matrix* and the *structure matrix* (see Jane Superbrain Box 18.1). For orthogonal rotation these matrices are the same. The pattern matrix contains the factor loadings and is comparable to the factor matrix that we interpreted for the orthogonal rotation. The structure matrix adjusts for the relationship between factors; it is a product of the pattern matrix and the matrix containing the correlation coefficients between factors). Most researchers interpret the pattern matrix, because it is usually simpler, but there are situations in which values in the pattern matrix are suppressed because of relationships between the factors. Therefore, the structure matrix is a useful double-check and Graham et al. (2003) recommend reporting both (with some useful examples of why this can be important).

The same four factors as for the orthogonal rotation seem to have emerged from the pattern matrix in Output 18.9. Factor 1 seems to represent fear of statistics, factor 2 represents fear of peer evaluation, factor 3 represents fear of computers and factor 4 represents fear of mathematics. The structure matrix (Output 18.10) differs in that shared variance is not ignored. The picture becomes more complicated because, with the exception of factor 2, several variables load highly on more than one factor. This has occurred because of the relationship between factors 1 and 3 and factors 3 and 4. This example highlights why the pattern matrix is preferable for interpretative reasons: because it contains information about the *unique* contribution of a variable to a factor.

Output 18.11 is a matrix containing the correlation coefficients between factors. As predicted from the structure matrix, factor 2 has small relationships with the

Pattern Matrix[a]

	Factor			
	1	2	3	4
I wake up under my duvet thinking that I am trapped under a normal distribution	.536			
I can't sleep for thoughts of eigenvectors	.470			
I weep openly at the mention of central tendency	.449			
I dream that Pearson is attacking me with correlation coefficients	.441			
Standard deviations excite me	-.435	.324		
Statistics makes me cry	.432			
People try to tell you that SPSS makes statistics easier to understand but it doesn't	.412		.358	
I don't understand statistics	.357			
My friends are better at statistics than me		.559		
My friends are better at SPSS than I am		.465		
My friends will think I'm stupid for not being able to cope with SPSS		.453		
If I'm good at statistics my friends will think I'm a nerd		.345		
Everybody looks at me when I use SPSS		.336		
I have little experience of computers			.862	
SPSS always crashes when I try to use it			.635	
All computers hate me			.562	
I worry that I will cause irreparable damage because of my incompetence with computers			.558	
Computers have minds of their own and deliberately go wrong whenever I use them			.473	
Computers are useful only for playing games			.386	
Computers are out to get me			.318	
I have never been good at mathematics				-.851
I did badly at mathematics at school				-.734
I slip into a coma whenever I see an equation				-.675

Extraction Method: Principal Axis Factoring.
Rotation Method: Oblimin with Kaiser Normalization. [a]

   a. Rotation converged in 17 iterations.

Output 18.9

Structure Matrix

	Factor			
	1	2	3	4
I wake up under my duvet thinking that I am trapped under a normal distribution	.657		.475	-.391
I weep openly at the mention of central tendency	.621		.493	-.469
Standard deviations excite me	-.596	.486	-.409	.369
People try to tell you that SPSS makes statistics easier to understand but it doesn't	.593		.564	-.366
I dream that Pearson is attacking me with correlation coefficients	.586		.472	-.458
Statistics makes me cry	.552		.407	-.449
I can't sleep for thoughts of eigenvectors	.496			
I don't understand statistics	.492		.422	-.374
My friends are better at statistics than me		.572		
My friends will think I'm stupid for not being able to cope with SPSS		.486		
My friends are better at SPSS than I am		.484		
Everybody looks at me when I use SPSS	-.360	.425		
If I'm good at statistics my friends will think I'm a nerd		.328		
I have little experience of computers			.746	-.341
SPSS always crashes when I try to use it	.486		.720	-.407
All computers hate me	.479		.676	-.415
I worry that I will cause irreparable damage because of my incompetence with computers	.414		.673	-.457
Computers have minds of their own and deliberately go wrong whenever I use them	.489		.613	-.390
Computers are out to get me	.384		.510	-.428
Computers are useful only for playing games			.437	
I have never been good at mathematics	.314		.353	-.798
I did badly at mathematics at school	.369		.478	-.783
I slip into a coma whenever I see an equation	.404		.476	-.750

Extraction Method: Principal Axis Factoring.
Rotation Method: Oblimin with Kaiser Normalization.

Output 18.10

other factors, but all other factors have largish correlations. In other words, the latent constructs represented by the factors are related. If the constructs were independent, oblique rotation should produce an identical solution to an orthogonal rotation and the factor correlation matrix would be an identity matrix (i.e., all factors would have correlation coefficients of 0). Therefore, this matrix is useful for assessing how reasonable it is to assume independence between factors; for these data it appears that we cannot assume independence and so the obliquely rotated solution is a more reasonable representation of reality.

On a theoretical level the dependence between factors does not cause concern; we might expect a strong relationship between fear of maths, fear of statistics and fear of computers. Generally, the less mathematically and technically minded people struggle with statistics. However, we would not necessarily expect these constructs to correlate strongly with fear of peer evaluation (because this construct is more socially based) and this is the factor that least correlates with all others—on a theoretical level, things have turned out well.

18.7.5 Factor scores

Having reached a suitable solution and rotated it we can look at the factor scores. SPSS will display the component score matrix B (see Section 18.3.3) from which the factor scores are calculated. I haven't reproduced this table here because I can't think of a reason why most people would want to look at it. We asked for scores to be calculated based on the Anderson–Rubin method, and these scores will be in the data editor in columns labeled *FAC1_1*, *FAC2_1*, *FAC3_1* and *FAC4_1* for each factor, respectively. If you asked for factor scores in the subsequent obliquely rotated solution then these will appear in the data editor in four columns labeled *FAC2_1* and so on.

Output 18.12 shows the factor scores for the first 10 participants. Participant 9 scored highly on factors 1–3 and so this person is very anxious about statistics, computing and maths, but less so about peer evaluation (factor 4). Factor scores can be used in this way to assess the

Cramming Sam's Tips
Interpretation

- If you've conduced orthogonal rotation then look at the table labeled *Rotated Factor Matrix*.
  For each variable, note the factor/component for which the variable has the highest loading (above about 0.3–0.4 when you ignore the plus or minus sign). Try to make sense of

what the factors represent by looking for common themes in the items that load highly on the same factor.
- If you've conducted oblique rotation then do the same as above but for the *Pattern Matrix*. Double-check what you find by doing the same for the *Structure Matrix*.

Factor Correlation Matrix

Factor	1	2	3	4
1	1.000	-.296	.483	-.429
2	-.296	1.000	-.302	.186
3	.483	-.302	1.000	-.532
4	-.429	.186	-.532	1.000

Extraction Method: Principal Axis Factoring.
Rotation Method: Oblimin with Kaiser Normalization.

Output 18.11

Use the *case summaries* command (Section 9.11.6) to list the factor scores for these data (given that there are over 2500 cases, restrict the output to the first 10).

relative fear of one person compared to another or we could add the scores up to obtain a single score for each participant (which we might assume represents SPSS anxiety as a whole). We can also use factor scores in regression when groups of predictors correlate so highly that there is multicollinearity. However, normally people do not use factor scores themselves but instead sum scores on items that they have decided load onto the same factor (for example, create a score for statistics anxiety by adding up a person's scores on items 1, 3, 4, 5, 12, 16, 20, and 21).

Case Summaries[a]

		A-R factor score 1 for analysis 1	A-R factor score 2 for analysis 1	A-R factor score 3 for analysis 1	A-R factor score 4 for analysis 1
1		-1.12974	.05090	-1.58646	-.55242
2		-.04484	-.47739	-.22126	.64055
3		.15620	-.72240	.08299	-.90901
4		.79370	.61178	-.79341	-.31779
5		-.98251	.66284	-.35819	.54788
6		-.59551	2.13562	-.53156	-.52313
7		-1.33140	-.19415	.08213	.87306
8		-.91760	-.20011	-.02149	.96984
9		1.70800	1.45700	3.03959	.65963
10		-.37637	-.77093	.06181	1.58454
Total	N	10	10	10	10

a. Limited to first 10 cases.

Output 18.12

18.7.6 Summary ▊▊▊

To sum up, the analyses revealed four underlying subscales in our questionnaire that may or may not, relate to genuine sub-components of SPSS anxiety. It also seems as though an obliquely rotated solution was preferred due to the relationships between factors. The use of factor analysis is purely exploratory; it should be used only to guide future hypotheses or to inform researchers about patterns within data sets. A great many decisions are left to the researcher using factor analysis and so researcher degrees of freedom (Section 3.3.2) come into play a lot: try to make informed, impartial decisions and resist the lure of decisions that give you the outcomes you want to get.

18.8 How to report factor analysis ▊▊▊

When reporting factor analysis, provide readers with enough information to make an informed opinion about what you've done. We need to be clear about our criteria for extracting factors and the method of rotation used. We should produce a table of the rotated factor loadings of all items and flag (in bold) values above a criterion level (I would personally choose 0.40, but see Section 18.4.7). Report the percentage of variance that each factor explains and possibly the eigenvalue too. Table 18.1 shows an example of such a table for the SAQ data (oblique rotation); note that I have reported the sample size in the title.

A table of factor loadings and a description of the analysis are a bare minimum. You should provide some information on sample size adequacy. You could consider depositing the table of correlations from which someone could reproduce your analysis (should they want to) on an open science repository such as the Open Science Framework (https://osf.io/). For example:

✓ A principal axis factor analysis (FA) was conducted on the 23 items with oblique rotation (direct oblimin). The Kaiser–Meyer–Olkin measure verified the sampling adequacy for the analysis, KMO = 0.93 ('marvellous' according to Kaiser & Rice, 1974), and all KMO values for individual items were greater than 0.77, which is well above the acceptable limit of 0.5 (Kaiser & Rice, 1974). An initial analysis was run to obtain eigenvalues for each factor in the data. Four factors had eigenvalues over Kaiser's criterion of 1 and in

Table 18.1 Summary of exploratory factor analysis results for the SPSS anxiety questionnaire (N = 2571)

Item	Rotated Factor Loadings			
	Fear of Statistics	Peer Evaluation	Fear of Computers	Fear of Maths
I wake up under my duvet thinking that I am trapped under a normal distribution	**0.54**	−0.04	0.17	−0.06
I can't sleep for thoughts of eigenvectors	**0.47**	−0.14	−0.08	−0.05
I weep openly at the mention of central tendency	**0.45**	−0.05	0.17	−0.18
I dream that Pearson is attacking me with correlation coefficients	**0.44**	0.08	0.18	−0.19
Standard deviations excite me	**−0.44**	0.32	−0.05	0.10
Statistics makes me cry	**0.43**	0.10	0.11	−0.23
People try to tell you that SPSS makes statistics easier to understand but it doesn't	**0.41**	−0.04	0.36	0.01
I don't understand statistics	0.36	0.05	0.20	−0.13
My friends are better at statistics than me	−0.09	**0.56**	−0.02	−0.11
My friends are better at SPSS than I am	0.07	**0.47**	−0.11	0.04
My friends will think I'm stupid for not being able to cope with SPSS	−0.18	**0.45**	0.04	−0.05
If I'm good at statistics my friends will think I'm a nerd	0.10	0.35	0.00	0.07
Everybody looks at me when I use SPSS	−0.22	0.34	**−0.08**	0.01
I have little experience of computers	−0.22	−0.01	**0.86**	0.03
SPSS always crashes when I try to use it	0.18	−0.01	**0.64**	0.01
All computers hate me	0.19	−0.02	**0.56**	−0.03
I worry that I will cause irreparable damage because of my incompetence with computers	0.08	−0.04	**0.56**	−0.12
Computers have minds of their own and deliberately go wrong whenever I use them	0.24	−0.02	**0.47**	−0.03
Computers are useful only for playing games	0.00	−0.06	0.39	−0.06
Computers are out to get me	0.11	−0.13	0.32	−0.19
I have never been good at mathematics	0.01	0.05	−0.09	**−0.85**
I did badly at mathematics at school	−0.01	−0.11	0.06	**−0.73**
I slip into a coma whenever I see an equation	0.08	0.02	0.09	**−0.68**
Eigenvalues	7.29	1.74	1.32	1.23
% of variance	31.70	7.56	5.73	5.34
α	0.82	0.57	0.82	0.82

Note: Factor loadings over 0.40 appear in bold.

combination explained 50.32% of the variance. The scree plot was ambiguous and showed inflexions that would justify retaining both two and four factors. We retained four factors because of the large sample size and the convergence of the scree plot and Kaiser's criterion on this value. Table 18.1 shows the factor loadings after rotation. The items that cluster on the same factor suggest that factor 1 represents a fear of statistics, factor 2 represents peer evaluation concerns, factor 3 a fear of computers and factor 4 a fear of maths.

18.9 Reliability analysis ▌▌▌

18.9.1 Measures of reliability ▌▌▌

If you're using factor analysis to validate a questionnaire, it is useful to check the reliability of your scale.

Reliability means that a measure (or in this case questionnaire) should consistently reflect the construct that it is measuring. One way to think of this is that, other things being equal, a person should get the same score on a questionnaire if they

complete it at two different points in time (we have already discovered that this is called *test–retest reliability*). So, someone who is terrified of SPSS and who scores highly on our SAQ should score similarly highly if we tested them a month later (assuming they hadn't completed some kind of SPSS-anxiety therapy in that month). Another way to look at reliability is to say that two

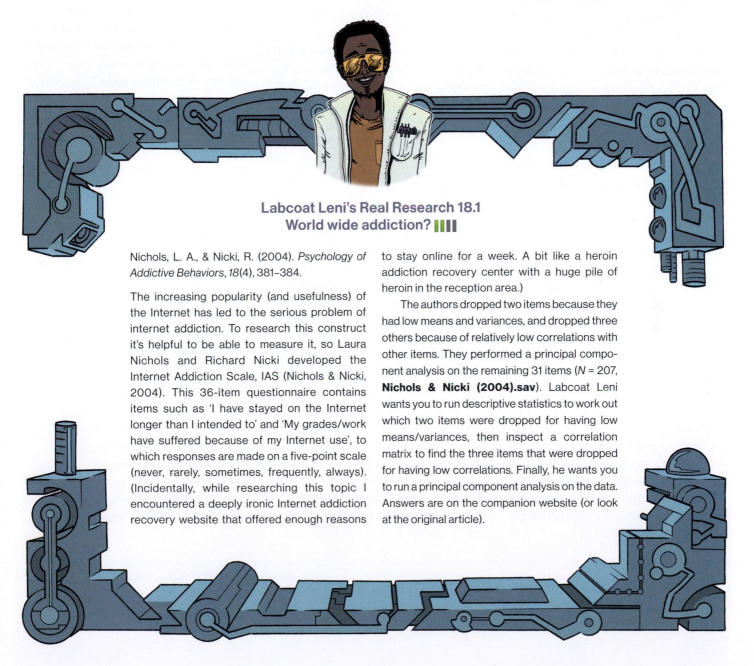

Labcoat Leni's Real Research 18.1
World wide addiction? ▌▌▌

Nichols, L. A., & Nicki, R. (2004). *Psychology of Addictive Behaviors*, *18*(4), 381–384.

The increasing popularity (and usefulness) of the Internet has led to the serious problem of internet addiction. To research this construct it's helpful to be able to measure it, so Laura Nichols and Richard Nicki developed the Internet Addiction Scale, IAS (Nichols & Nicki, 2004). This 36-item questionnaire contains items such as 'I have stayed on the Internet longer than I intended to' and 'My grades/work have suffered because of my Internet use', to which responses are made on a five-point scale (never, rarely, sometimes, frequently, always). (Incidentally, while researching this topic I encountered a deeply ironic Internet addiction recovery website that offered enough reasons

to stay online for a week. A bit like a heroin addiction recovery center with a huge pile of heroin in the reception area.)

The authors dropped two items because they had low means and variances, and dropped three others because of relatively low correlations with other items. They performed a principal component analysis on the remaining 31 items (*N* = 207, **Nichols & Nicki (2004).sav**). Labcoat Leni wants you to run descriptive statistics to work out which two items were dropped for having low means/variances, then inspect a correlation matrix to find the three items that were dropped for having low correlations. Finally, he wants you to run a principal component analysis on the data. Answers are on the companion website (or look at the original article).

people who are the same in terms of the construct being measured should get the same score. So, if we took two people who were equally SPSS-anxious, they should get more or less identical scores on the SAQ. Likewise, if we took two people who loved SPSS, they should get equally low scores. The SAQ wouldn't be an accurate measure of SPSS anxiety if we took someone who loved SPSS and someone who was terrified of it and they got the same score! In statistical terms, the usual way to look at reliability is based on the idea that individual items (or sets of items) should produce results consistent with the overall questionnaire. So, if we take someone scared of SPSS, then their overall score on the SAQ will be high; if the SAQ is reliable then if we randomly select some items from it the person's score on those items should also be high.

The simplest way to do this in practice is to use **split-half reliability**. This method splits the scale set into two randomly selected sets of items. A score for each participant is calculated on each half of the scale. If a scale is reliable a person's score on one half of the scale should be the same as (or similar to) their score on the other half. Across several participants, scores from the two halves of the questionnaire should correlate very highly. The correlation between the two halves is the statistic computed in the split-half method, with large correlations being a sign of reliability. The problem with this method is that there are several ways in which a set of data can be randomly split into two and so the results could be a product of the way in which the data were split. To overcome this problem, Cronbach (1951) came up with a measure that is loosely equivalent to creating two sets of

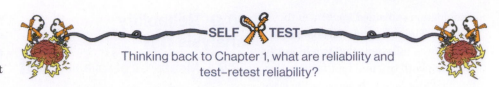
SELF TEST

Thinking back to Chapter 1, what are reliability and test–retest reliability?

items in every way possible and computing the correlation coefficient for each split. The average of these values is equivalent to **Cronbach's alpha,** α, which is the most common measure of scale reliability:[13]

$$\alpha = \frac{N^2 \overline{\text{cov}}}{\sum s_{\text{item}}^2 + \sum \text{cov}_{\text{item}}} \qquad (18.8)$$

This equation is less complicated than it looks. For each item on our scale we can calculate two things: the variance within the item, and the covariance between a particular item and any other item on the scale. Put another way, we can construct a variance–covariance matrix of all items. In this matrix the diagonal elements will be the variance within a particular item, and the off-diagonal elements will be covariances between pairs of items. The top half of the equation is the number of items (N) squared multiplied by the average covariance between items (the average of the off-diagonal elements in the variance–covariance matrix). The bottom half is the sum of all the item variances and item covariances (i.e., the sum of everything in the variance–covariance matrix).

There is a standardized version of the coefficient too, which essentially uses the same equation except that correlations are used rather than covariances, and the bottom half of the equation uses the sum of the elements in the correlation matrix of items (including the 1s that appear on the diagonal of that matrix). The normal alpha is appropriate when items on a scale are summed to produce a single score for that

scale (the standardized alpha is not appropriate in these cases). The standardized alpha is useful, though, when items on a scale are standardized before being summed.

18.9.2 Interpreting Cronbach's α: some cautionary tales

You'll often see in books or journal articles or be told by people, that a value of 0.7 to 0.8 is an acceptable value for Cronbach's α and that values substantially lower indicate an unreliable scale. Kline (1999) notes that although the generally accepted value of 0.8 is appropriate for cognitive tests such as intelligence tests, for ability tests a cut-off point of 0.7 is more suitable. He goes on to say that when dealing with psychological constructs values below even 0.7 can, realistically, be expected because of the diversity of the constructs being measured. Some even suggest that in the early stages of research values as low as 0.5 will suffice (Nunnally, 1978). However, there are many reasons not to use these general guidelines, not least of which is that they distract you from thinking about what the value means within the context of the research you're doing (Pedhazur & Schmelkin, 1991).

We'll now look at some issues in interpreting alpha, which have been discussed particularly well by Cortina (1993) and Pedhazur and Schmelkin (1991). First, the value of α depends on the number of items on the scale. You'll notice that the top half of the equation for α

---

13 Although this is the easiest way to conceptualize Cronbach's α, whether it is exactly equal to the average of all possible split-half reliabilities depends on exactly how you calculate the split-half reliability (see the glossary for computational details). If you use the Spearman–Brown formula, which takes no account of item standard deviations, then Cronbach's α will be equal to the average split-half reliability only when the item standard deviations are equal; otherwise α will be smaller than the average. However, if you use a formula for split-half reliability that does account for item standard deviations (such as Flanagan, 1937; Rulon, 1939) then α will always equal the average split-half reliability (see Cortina, 1993).

includes the number of items squared. Therefore, as the number of items on the scale increases, α will increase. Therefore, it's possible to get a large value of α because you have a lot of items on the scale, and not because your scale is reliable. For example, Cortina (1993) reports data from two scales, both of which have $\alpha = 0.8$. The first scale has only three items, and the average correlation between items was a respectable 0.57; however, the second scale had 10 items with an average correlation between these items of a less respectable 0.28. Clearly the internal consistency of these scales differs, but according to Cronbach's α they are both equally reliable.

Second, people tend to think that alpha measures 'unidimensionality' or the extent to which the scale measures one underlying factor or construct. This is true when there is one factor underlying the data (see Cortina, 1993), but Grayson (2004) demonstrates that data sets with the same α can have very different factor structures. He showed that $\alpha = 0.8$ can be achieved in a scale with one underlying factor, with two moderately correlated factors and with two uncorrelated factors. Cortina (1993) has also shown that with more than 12 items, and fairly high correlations between items ($r > 0.5$), α can reach values around and above 0.7 (0.65 to 0.84). These results show that α should not be used as a measure of 'unidimensionality'. Indeed, Cronbach (1951) suggested that if several factors exist then the formula should be applied separately to items relating to different factors. In other words, if your questionnaire has subscales, α should be applied separately to these subscales.

The final warning is about items that have a reverse phrasing. For example, in the SAQ there is one item (question 3) that was phrased the opposite way

around to all other items. The item was 'standard deviations excite me'. Compare this to any other item and you'll see it requires the opposite response.

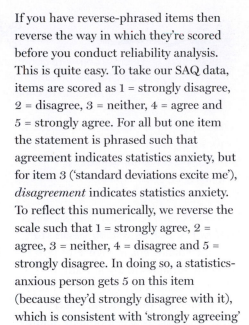

Eek! My alpha is negative, have I done something wrong?

For example, item 1 is 'statistics make me cry'. If you don't like statistics then you'll strongly agree with this statement and so will get a score of 5 on our scale. For item 3, if you hate statistics then standard deviations are unlikely to excite you so you'll strongly disagree and get a score of 1 on the scale. These reverse-phrased items are important for reducing response bias because participants need to pay attention to the questions. For factor analysis, this reverse phrasing doesn't matter; all that happens is you get a negative factor loading for any reversed items (in fact, you'll see that item 3 has a negative factor loading in Output 18.9). However, these reverse-scored items will affect alpha. To see why, think about the equation for Cronbach's α. The top half incorporates the *average* covariance between items. If an item is reverse-phrased then it will have a negative relationship with other items, hence the covariances between this item and other items will be negative. The average covariance is the sum of covariances divided by the number of covariances, and by including a bunch of negative values we reduce the sum of covariances, and hence we also reduce Cronbach's α, because the top half of the equation gets smaller. In extreme cases, it is even possible to get a negative value for Cronbach's α, simply because the magnitude of negative covariances is bigger than the magnitude of positive ones. A negative Cronbach's α doesn't make much sense, but it does happen, and if it does, ask yourself whether you included any reverse-phrased items.

If you have reverse-phrased items then reverse the way in which they're scored before you conduct reliability analysis. This is quite easy. To take our SAQ data, items are scored as 1 = strongly disagree, 2 = disagree, 3 = neither, 4 = agree and 5 = strongly agree. For all but one item the statement is phrased such that agreement indicates statistics anxiety, but for item 3 ('standard deviations excite me'), *disagreement* indicates statistics anxiety. To reflect this numerically, we reverse the scale such that 1 = strongly agree, 2 = agree, 3 = neither, 4 = disagree and 5 = strongly disagree. In doing so, a statistics-anxious person gets 5 on this item (because they'd strongly disagree with it), which is consistent with 'strongly agreeing' to an item like 'statistics makes me cry'.

To reverse the scoring find the maximum value of your response scale (in this case 5) and add 1 to it (6 in this case). For each person, subtract from this value the score they got. For example, someone who scored 5 originally now scores 6 − 5 = 1, and someone who scored 1 originally now gets 6 − 1 = 5. Someone in the middle of the scale with a score of 3 will still get 6 − 3 = 3. We can get SPSS to do it for all participants simultaneously.

18.10 Reliability analysis using SPSS Statistics ▮▮▮▮

Let's test the reliability of the SAQ (**SAQ. sav**). You should have reverse-scored item 3 (see above), but if you can't be bothered, load up the file **SAQ (Item 3 Reversed). sav**. Remember that I said we should conduct reliability analysis on individual subscales. Using the results from the oblique rotation (Output 18.9) we have these four subscales:

Subscale 1 (*Fear of statistics*): items 1, 3, 4, 5, 12, 16, 20, 21

Subscale 2 (*Peer evaluation*): items 2, 9, 19, 22, 23

SELF TEST

Use the *compute* command to reverse-score item 3 (see Chapter 6; remember that you are changing the variable to 6 minus its original value)

Subscale 3 (*Fear of computers*): items 6, 7, 10, 13, 14, 15, 18

Subscale 4 (*Fear of mathematics*): items 8, 11, 17

To conduct each reliability analysis select *Analyze* ▶ *Scale* ▶ Reliability Analysis... to access the dialog box in Figure 18.15. Drag any variables that you want to analyze to the box labeled *Items* (or click 🡆). Remember that you can select several items at the same time by holding down *Ctrl* (*Cmd* on a Mac) while you select the variables. To begin with, let's select the items from the fear of statistics subscale: items 1, 3, 4, 5, 12, 16, 20 and 21 (Figure 18.15).

There are several reliability statistics available. The default option is Cronbach's α, which is fine for our purpose, but you can change the method (e.g., to the split-half method) by clicking on Alpha to reveal a drop-down list of possibilities. Also, it's a good idea to type the name of the scale (in this case 'Fear of Statistics') into the box labeled *Scale label* because doing so adds a header to the output with whatever you type in this box: typing a

sensible name here will make your output easier to follow.

Click Statistics... to access the dialog box in Figure 18.16. One of most useful options for questionnaire reliability is *Scale if item deleted*, which tells us what the value of α would be if each item were deleted. If our questionnaire is reliable then we would not expect any one item to greatly affect the overall reliability. In other words, no item should cause a substantial change in α if it is removed. If an item's removal changes alpha substantially then you should review your items (and that probably means going back to square one with the factor analysis).

The inter-item correlations and covariances (and summaries) provide us with correlation coefficients and averages for items on our scale. We should already have these values from our factor analysis, so there is little point in selecting these options. Options like the *F-test*, *Friedman chi-square* (if your data are ranked), *Cochran chi-square* (if your data are dichotomous), and *Hotelling's T-square* compare the central tendency of different items on the

questionnaire. These tests might be useful to check that items have similar distributional properties (i.e., the same average value), but given the large sample sizes you ought to be using for factor analysis, they will inevitably produce significant results even when only small differences exist between the questionnaire items.

You can also request an **intraclass correlation coefficient** (**ICC**). The correlation coefficients that we encountered earlier in this book measure the relation between variables that measure different things. For example, the correlation between listening to Deathspell Omega and Satanism involves two classes of measures: the type of music a person likes and their religious beliefs. Intraclass correlations measure the relationship between two variables that measure the same thing (i.e., variables within the same class). Two common uses are in comparing paired data (such as twins) on the same measure, and assessing the consistency between judges' ratings of a set of objects (hence the reason why it is found in the reliability statistics). If you'd like to know more, see Section 21.2.1.

Use the simple set of options in Figure 18.16 to run a basic reliability analysis. Click Continue to return to the main dialog box and OK to run the analysis.

18.11 Interpreting reliability analysis ▌▌▌▌

Output 18.13 shows the results for the fear of statistics subscale. The value of Cronbach's α is presented in a small table and indicates the overall reliability of the scale. Bearing in mind what we've already noted about effects from the number of items, and how daft it is to apply general rules, we're looking for values in the region of about 0.7–0.8. In this case α is 0.821, which is certainly in the region indicated by Kline (1999), and probably indicates good reliability.

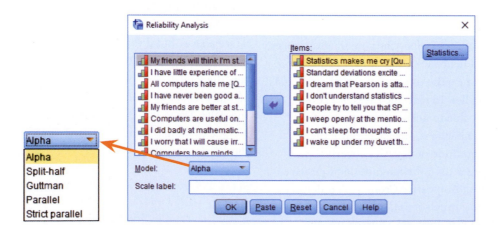

Figure 18.15 Main dialog box for reliability analysis

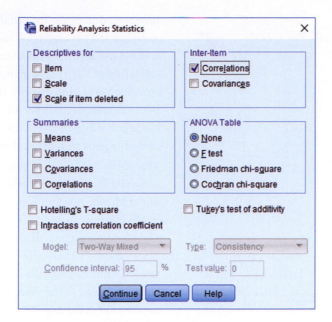

Figure 18.16   Statistics for reliability analysis

In the table labeled *Item-Total Statistics* the column labeled *Corrected Item-Total Correlation* shows the correlations between each item and the total score from the questionnaire. In a reliable scale all items should correlate well with the total. So, we're looking for items that don't correlate well with the overall score from the scale: if any of these values are less than about 0.3 then we've got problems, because it means that a particular item does not correlate very well with the scale overall. Items with low correlations may have to be dropped. For these data, all data have item–total correlations above 0.3, which is encouraging.

The values in the column labeled *Cronbach's Alpha if Item Deleted* are the values of the overall α if that item isn't included in the calculation. As such, they reflect the change in Cronbach's α that would be seen if an item were deleted. The overall α is 0.821, and so all values in this column should be around that same value. We're mainly looking for values of alpha greater than the overall α. If you think about it, if the deletion of an item increases Cronbach's α then this means that the deletion of that item improves reliability. Therefore, any items that have values of α in the column labeled *Cronbach's Alpha if Item Deleted* greater

than the overall α may need to be deleted from the scale to improve its reliability. None of the items here would increase alpha if they were deleted, which is good news. It's worth noting that if items do need to be removed at this stage then you should rerun your factor analysis as well to make sure that the deletion of the item has not affected the factor structure.

To illustrate the importance of reverse-scoring items before running reliability analysis, Output 18.14 shows the reliability analysis for the fear of statistics subscale, but done on the original data (i.e., without item 3 being reverse-scored). Note that (1) the overall α is considerably lower (0.605 rather than 0.821); (2) this item has a negative item–total correlation (which is a good way to spot if you have a potential reverse-scored item in the data that haven't been reverse-scored); and (3) the α if item deleted is 0.800. That is, if this item were deleted then the reliability goes up from about 0.6 to about 0.8. This example, I hope, illustrates that failing to reverse-score items that have been phrased oppositely to other items on the scale will mess up your reliability analysis.

Let's now look at our subscale of peer evaluation (Output 18.15). The overall α is 0.570, which is nothing to bake a cake for; it is quite low and, although in keeping with what Kline says we should expect for this kind of social science data, it is well below the statistics subscale and (as we shall see) the other two. The scale has five items, compared to seven, eight and three on the other scales, so its reliability relative to the other scales is not going to be dramatically affected by the number of items. The values in the column labeled *Corrected Item-Total Correlation* are all around 0.3, and smaller for item 23. These results again indicate questionable internal consistency and identify item 23 as a potential problem. The values in the column labeled *Cronbach's Alpha if Item Deleted* indicate that none of the items

Item–Total Statistics

	Scale Mean if Item Deleted	Scale Variance if Item Deleted	Corrected Item–Total Correlation	Squared Multiple Correlation	Cronbach's Alpha if Item Deleted
Statistics makes me cry	21.76	21.442	.536	.343	.802
Standard deviations excite me	20.72	19.825	.549	.309	.800
I dream that Pearson is attacking me with correlation coefficients	21.35	20.410	.575	.355	.796
I don't understand statistics	21.41	20.942	.494	.272	.807
People try to tell you that SPSS makes statistics easier to understand but it doesn't	20.97	20.639	.572	.337	.796
I weep openly at the mention of central tendency	21.25	20.451	.597	.389	.793
I can't sleep for thoughts of eigenvectors	20.51	21.176	.419	.244	.818
I wake up under my duvet thinking that I am trapped under a normal distribution	20.96	19.939	.606	.399	.791

Reliability Statistics

Cronbach's Alpha	Cronbach's Alpha Based on Standardized Items	N of Items
.821	.823	8

Output 18.13

Item-Total Statistics

	Scale Mean if Item Deleted	Scale Variance if Item Deleted	Corrected Item-Total Correlation	Squared Multiple Correlation	Cronbach's Alpha if Item Deleted
Statistics makes me cry	20.93	12.125	.505	.343	.521
Standard deviations excite me	20.72	19.825	-.549	.309	.800
I dream that Pearson is attacking me with correlation coefficients	20.52	11.447	.526	.355	.505
I don't understand statistics	20.58	11.714	.466	.272	.523
People try to tell you that SPSS makes statistics easier to understand but it doesn't	20.14	11.739	.501	.337	.515
I weep openly at the mention of central tendency	20.42	11.584	.529	.389	.507
I can't sleep for thoughts of eigenvectors	19.68	12.107	.353	.244	.558
I wake up under my duvet thinking that I am trapped under a normal distribution	20.13	11.189	.541	.399	.497

Reliability Statistics

Cronbach's Alpha	Cronbach's Alpha Based on Standardized Items	N of Items
.605	.641	8

Output 18.14

Item-Total Statistics

	Scale Mean if Item Deleted	Scale Variance if Item Deleted	Corrected Item-Total Correlation	Squared Multiple Correlation	Cronbach's Alpha if Item Deleted
My friends will think I'm stupid for not being able to cope with SPSS	11.46	8.119	.339	.134	.515
My friends are better at statistics than me	10.24	6.395	.391	.167	.476
Everybody looks at me when I use SPSS	10.79	7.381	.316	.106	.522
My friends are better at SPSS than I am	10.20	7.282	.378	.144	.487
If I'm good at statistics my friends will think I'm a nerd	9.65	7.988	.239	.069	.563

Reliability Statistics

Cronbach's Alpha	Cronbach's Alpha Based on Standardized Items	N of Items
.570	.572	5

Output 18.15

Item-Total Statistics

	Scale Mean if Item Deleted	Scale Variance if Item Deleted	Corrected Item-Total Correlation	Squared Multiple Correlation	Cronbach's Alpha if Item Deleted
I have little experience of computers	15.87	17.614	.619	.398	.791
All computers hate me	15.17	17.737	.619	.395	.790
Computers are useful only for playing games	15.81	20.736	.400	.167	.824
I worry that I will cause irreparable damage because of my incompetence with computers	15.64	18.809	.607	.384	.794
Computers have minds of their own and deliberately go wrong whenever I use them	15.22	18.719	.577	.350	.798
Computers are out to get me	15.33	19.322	.491	.250	.812
SPSS always crashes when I try to use it	15.52	17.832	.647	.447	.786

Reliability Statistics

Cronbach's Alpha	Cronbach's Alpha Based on Standardized Items	N of Items
.823	.821	7

Output 18.16

SELF TEST

Run reliability analysis on the other three subscales.

here would increase the reliability if they were deleted because all values in this column are less than the overall reliability of 0.570. The items on this subscale cover quite diverse themes of peer evaluation, and this might explain the relative lack of consistency. We probably need to rethink this subscale.

Moving on to the fear of computers subscale, Output 18.16 shows an overall α of 0.823, which is pretty good. The values in the column labeled *Corrected Item-Total Correlation* are again all above 0.3, which is also good. The values in the column labeled *Cronbach's Alpha if Item Deleted* show that none of the items would increase the reliability if they were deleted. This indicates that all items are positively contributing to the overall reliability.

Finally, the fear of maths subscale (Output 18.17) shows an overall reliability of 0.819, which indicates good reliability. The values in the column labeled *Corrected Item-Total Correlation* are all above 0.3, which is good, and the values in the column labeled *Cronbach's Alpha if Item Deleted* indicate that none of the items here would increase the reliability if they were deleted because all values in this column are less than the overall reliability value.

18.12 How to report reliability analysis ▮▮▮▮

Report the reliabilities in the text using the symbol α and remember that if you follow APA practice (which I'm not doing), because Cronbach's α can't be larger than 1, then drop the zero before the decimal:

✓ The fear of computers, fear of statistics and fear of maths subscales of the SAQ all had high reliabilities, all Cronbach's α = 0.82. However, the fear of negative peer evaluation subscale had relatively low reliability, Cronbach's α = 0.57.

However, the most common way to report reliability analysis when it follows a factor

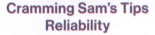

Cramming Sam's Tips
Reliability

- Reliability analysis is used to measure the consistency of a measure.
- Remember to reverse-score any items that were reverse-phrased on the original questionnaire before you run reliability analysis.
- Run separate reliability analyses for all subscales of your questionnaire.
- Cronbach's α indicates the overall reliability of a questionnaire, and values around 0.8 are good (or 0.7 for ability tests and the like).

- The *Cronbach's Alpha if Item Deleted* column tells you whether removing an item will improve the overall reliability: values greater than the overall reliability indicate that removing that item will improve the overall reliability of the scale. Look for items that dramatically increase the value of α and remove them.
- If you remove items, rerun the factor analysis to check that the factor structure still holds.

Item–Total Statistics

	Scale Mean if Item Deleted	Scale Variance if Item Deleted	Corrected Item–Total Correlation	Squared Multiple Correlation	Cronbach's Alpha if Item Deleted
I have never been good at mathematics	4.72	2.470	.684	.470	.740
I did badly at mathematics at school	4.70	2.453	.682	.467	.742
I slip into a coma whenever I see an equation	4.49	2.504	.652	.425	.772

Reliability Statistics

Cronbach's Alpha	Cronbach's Alpha Based on Standardized Items	N of Items
.819	.819	3

Output 18.17

analysis is to report the values of Cronbach's α as part of the table of factor loadings. For example, notice that in the last row of Table 18.1 I quote the value of Cronbach's α for each subscale in turn.

18.13 Brian and Jane's Story

Brian put his duvet over Jane, wedged a pillow under her head, turned the light out and went to his bedroom. He dug out a sleeping bag to replace the duvet that was keeping Jane warm on the sofa. He found it hard to sleep. Everything in his head was a mess. Just when he thought he'd scared Jane off, she turns up looking and sounding spent. With no explanation for the past five days, she passes out on his sofa. She was an enigma, that was for sure. A brilliant, dazzling, slightly odd enigma. Brian's head told him to stay away, but it was too late: he was strapped in for whatever the ride was and however dangerous it may be.

Brian woke too soon. He felt disorientated. Why was he in a sleeping bag? Where was his duvet? As the previous night came back to him, he raced clumsily into the

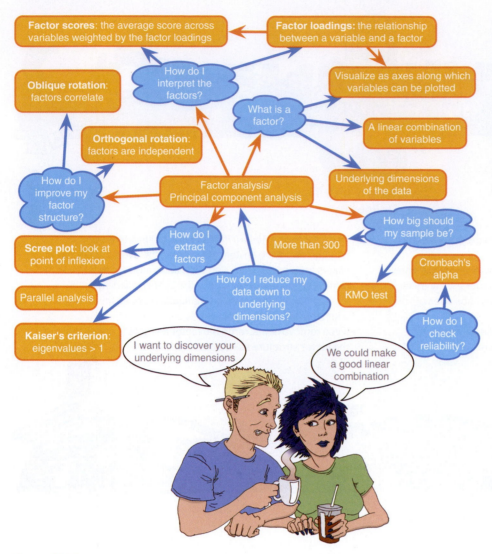

Figure 18.17   What Brian learnt from this chapter

lounge. Neatly folded on the sofa was his duvet, with the pillow on top. Jane had gone. He went to the bathroom. The shower was wet, his towel neatly folded on the floor. It smelt of his coconut shampoo. Back in the lounge, he noticed cereals spilled on the breakfast bar, but as he got closer he realized they were carefully arranged into a word. 'Wait,' it said, with a Rice Krispie smile.

He took a shower and emerged to find Jane at the breakfast bar with two coffees and bagels. She looked down at the towel covering his waist and, with a knowing smile, said, 'Put some clothes on, breakfast is getting cold.'

18.14 What next?

At the age of 23 I took it upon myself to become a living homage to the digestive system. I furiously devoured articles and books on statistics (some of them I even understood), I mentally chewed over them, I broke them down with the stomach acid of my intellect, I stripped them of their goodness and nutrients, I compacted them down, and after about two years I forced the smelly brown remnants of those intellectual meals out of me in the form of a book. I was mentally exhausted at the end of it; 'It's a good job I'll never have to do that again,' I thought.

18.15 Key terms that I've discovered

Alpha factoring
Anderson–Rubin method
Common factor
Common variance
Communality
Component matrix
Confirmatory factor analysis
Cronbach's α
Direct oblimin

Extraction
Equamax
Factor analysis
Factor loading
Factor matrix
Factor scores
Factor transformation matrix, Λ
Intraclass correlation coefficient (ICC)
Kaiser's criterion

Latent variable
Kaiser–Meyer–Olkin (KMO) measure of sampling adequacy
Oblique rotation
Orthogonal rotation
Pattern matrix
Principal component analysis (PCA)
Promax
Quartimax

Random variance
Rotation
Scree plot
Singularity
Split-half reliability
Structure matrix
Unique factor
Unique variance
Varimax

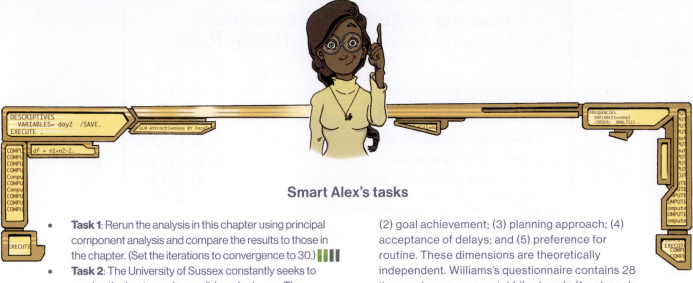

Smart Alex's tasks

- **Task 1**: Rerun the analysis in this chapter using principal component analysis and compare the results to those in the chapter. (Set the iterations to convergence to 30.) ▌▌▌▌

- **Task 2**: The University of Sussex constantly seeks to employ the best people possible as lecturers. They wanted to revise the 'Teaching of Statistics for Scientific Experiments' (TOSSE) questionnaire, which is based on Bland's theory that says that good research methods lecturers should have: (1) a profound love of statistics; (2) an enthusiasm for experimental design; (3) a love of teaching; and (4) a complete absence of normal interpersonal skills. These characteristics should be related (i.e., correlated). The University revised this questionnaire to become the 'Teaching of Statistics for Scientific Experiments—Revised' (TOSSE-R; Figure 18.18). They gave this questionnaire to 239 research methods lecturers to see if it supported Bland's theory. Conduct a factor analysis (with appropriate rotation) and interpret the factor structure (**TOSSE-R.sav**). ▌▌▌▌

- **Task 3**: Dr Sian Williams (University of Brighton) devised a questionnaire to measure organizational ability. She predicted five factors to do with organizational ability: (1) preference for organization; (2) goal achievement; (3) planning approach; (4) acceptance of delays; and (5) preference for routine. These dimensions are theoretically independent. Williams's questionnaire contains 28 items using a seven-point Likert scale (1 = strongly disagree, 4 = neither, 7 = strongly agree). She gave it to 239 people. Run a principal component analysis on the data in **Williams.sav**. ▌▌▌▌

- **Task 4**: Zibarras, Port, & Woods (2008) looked at the relationship between personality and creativity. They used the Hogan Development Survey (HDS), which measures 11 dysfunctional dispositions of employed adults: being **volatile**, **mistrustful**, **cautious**, **detached**, **passive_aggressive**, **arrogant**, **manipulative**, **dramatic**, **eccentric**, **perfectionist**, and **dependent**. Zibarras et al. wanted to reduce these 11 traits down and, based on parallel analysis, found that they could be reduced to three components. They ran a principal component analysis with varimax rotation. Repeat this analysis (**Zibarras et al. (2008).sav**) to see which personality dimensions clustered together (see page 210 of the original paper). ▌▌▌▌

Teaching of Statistics for Scientific Experiments — Revised (TOSSE-R)

		SD	D	N	A	SA
1.	I once woke up in a vegetable patch hugging a turnip that I'd mistakenly dug up thinking it was Roy's largest root	○	○	○	○	○
2.	If I had a big gun I'd shoot all the students I have to teach	○	○	○	○	○
3.	I memorize probability values for the *F*-distribution	○	○	○	○	○
4.	I worship at the shrine of Pearson	○	○	○	○	○
5.	I still live with my mother and have little personal hygiene	○	○	○	○	○
6.	Teaching others makes me want to swallow a large bottle of bleach because the pain of my burning oesophagus would be light relief in comparison	○	○	○	○	○
7.	Helping others to understand sums of squares is a great feeling	○	○	○	○	○
8.	I like control conditions	○	○	○	○	○
9.	I calculate 3 ANOVAs in my head before getting out of bed	○	○	○	○	○
10.	I could spend all day explaining statistics to people	○	○	○	○	○
11.	I like it when I've helped people to understand factor rotation	○	○	○	○	○
12.	People fall asleep as soon as I open my mouth to speak	○	○	○	○	○
13.	Designing experiments is fun	○	○	○	○	○
14.	I'd rather think about appropriate dependent variables than go to the pub	○	○	○	○	○
15.	I soil my pants with excitement at the mere mention of factor analysis	○	○	○	○	○
16.	Thinking about whether to use repeated- or independent-measures thrills me	○	○	○	○	○
17.	I enjoy sitting in the park contemplating whether to use participant observation in my next experiment	○	○	○	○	○
18.	Standing in front of 300 people in no way makes me lose control of my bowels	○	○	○	○	○
19.	I like to help students	○	○	○	○	○
20.	Passing on knowledge is the greatest gift you can bestow an individual	○	○	○	○	○
21.	Thinking about Bonferroni corrections gives me a tingly feeling in my groin	○	○	○	○	○
22.	I quiver with excitement when thinking about designing my next experiment	○	○	○	○	○
22.	I often spend my spare time talking to the pigeons ... and even they die of boredom	○	○	○	○	○
23.	I tried to build myself a time machine so that I could go back to the 1930s and follow Fisher around on my hands and knees licking the floor on which he'd just trodden	○	○	○	○	○
25.	I love teaching	○	○	○	○	○
26.	I spend lots of time helping students	○	○	○	○	○
27.	I love teaching because students have to pretend to like me or they'll get bad marks	○	○	○	○	○
28.	My cat is my only friend	○	○	○	○	○

Figure 18.18

Answers & additional resources are available on the book's website at
https://edge.sagepub.com/field5e

CATEGORICAL OUTCOMES: CHI-SQUARE AND LOGLINEAR ANALYSIS

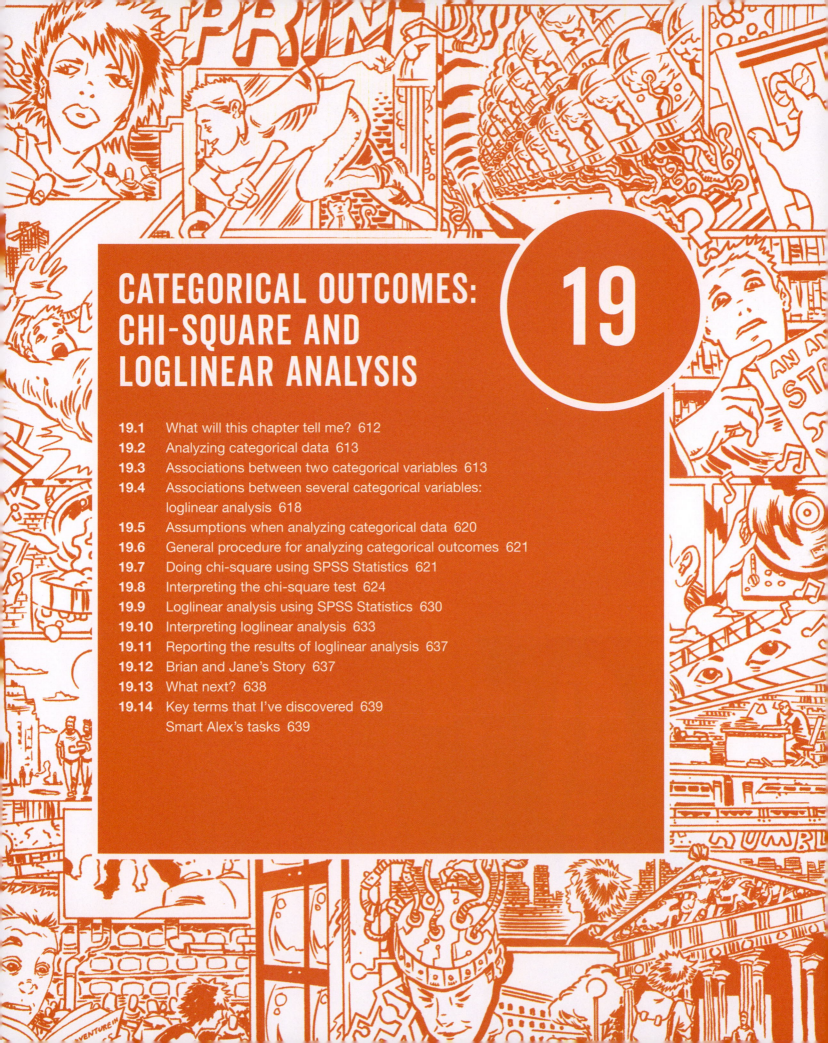

19

19.1 What will this chapter tell me?

We discovered in the previous chapter that I wrote a book. This book. There are a lot of good things about writing books. For one thing, your parents are impressed. They're not *that* impressed, because they think that a successful book sells as many copies as *Harry Potter* and that people should queue outside bookshops for the latest enthralling instalment of *Discovering Statistics* .... My parents are, consequently, quite baffled about how this book is perceived as successful, yet I don't get invited to dinner by the Queen. Nevertheless, given that my family don't really understand what I do, books are tangible proof that I do *something*. The size of this book and the fact it has equations in it is a bonus because it makes me look cleverer than I actually am. But not as clever as 'the clever one'.☺ The downside to writing books is the immeasurable mental anguish. In England we don't talk about our emotions, because we fear that if they get out into the open, civilization will collapse. I'm going to buck my national trend and reveal that the writing process for the second edition of this book was so stressful that I came within one of Fuzzy's whiskers of a total meltdown. It took me two years to recover, just in time to start thinking about the third edition.[1] The pain is worth it when people tell me that they found the book vaguely useful. Of course, the publishers focus less on warm fuzzy feelings of altruism and more on sales figures and comparisons with other books. They collate data on sales figures for this book and its competitors in different 'markets' (you are not a person, you are a 'consumer', and you don't live in a country, you live in a 'market') and they gibber and twitch at their consoles creating pink frequency distributions (with 3-D effects) of these values. The data they get are frequency data (the number of books sold in a certain market or discipline). If they wanted to compare sales of this book to its competitors, in different countries, they would need to read this chapter because it's all about analyzing data for which we know only the frequency with which events occur. They won't read this chapter, but they should . . .

Figure 19.1   Midway through writing the second edition of this book, things became a little strange

1   Writing this fifth edition has been an insane amount of work, I started it with burnout from writing an entirely new book (*An Adventure in Statistics*), and I had a tight deadline. I've not had a lot of sleep in the past 5 months. Despite this I've enjoyed this update the most of the five editions, which (and you should never interpret correlation as cause) I attribute to it being the first edition that I've written since having children (see Chapter 3).

19.2 Analyzing categorical data ▮▮▮▮

So far we have looked at fitting linear models with categorical predictor variables, but always predicting a continuous outcome variable. Sometimes we want to predict categorical outcome variables. In other words, we want to predict into which category an entity falls. For example, we might want to predict whether someone is pregnant or not, for which political party a person will vote, whether a tumor is benign or malignant, whether a sports team will win, lose or draw. In these examples, an entity can fall into only one category, for example a woman is either pregnant or not and a team can't win *and* lose the same match. The next two chapters deal with statistical models for categorical outcomes. We'll begin with modeling associations between categorical variables, then look at predicting categorical outcomes from categorical predictors, then in the next chapter we'll move on to look at predicting categorical outcomes from both categorical and continuous predictor variables.

19.3 Associations between two categorical variables ▮▮▮▮

We will begin by looking at the simplest situation of quantifying the relationship between two categorical variables. With categorical variables we can't use the mean or any similar statistic because the mean of a categorical variable is meaningless: the numeric values you attach to different categories are arbitrary, and the mean of those numeric values will depend on how many members each category has. Therefore, when we've measured only categorical variables, we analyze the number of things that fall into each combination of categories (i.e., the frequencies).

19.3.1 A furry example ▮▮▮▮

A researcher was interested in whether animals could be trained to line-dance. He took 200 cats and tried to train them to line-dance by giving them either food or affection as a reward for dance-like behavior. At the end of the week he counted how many animals could line-dance and how many could not. There are two categorical variables here: **training** (the animal was trained using either food or affection, not both) and **dance** (the animal either learnt to line-dance or it did not). By combining categories, we end up with four different categories. All we then need to do is to count how many cats fall into each category. Table 19.1 shows a contingency table (Section 3.7.3) of these data.

19.3.2 Pearson's chi-square test ▮▮▮▮

To see whether there's a relationship between two categorical variables (i.e., does the number of cats that line-dance relate to the type of training used?) we can use Pearson's **chi-square test** (Fisher, 1922; Pearson, 1900). This statistic is based on the simple idea of comparing the frequencies you observe in certain categories to the frequencies you might expect to get in those categories by chance. We saw in Chapter 2 (equation (2.11)), that if we want to calculate the fit (or total error) of a model we add up the squared differences between the observed values of the outcome and the predicted values that come from the model:

$$\text{total error} = \sum_{i=1}^{n}\left(\text{observed}_i - \text{model}_i\right)^2 \tag{19.1}$$

This equation was the basis of our sums of squares in the linear model. We use essentially the same equation when variables are categorical. There is a slight variation in that we divide by the model scores as well, which is much the same process as dividing the sum of squares by the degrees of freedom to get the mean squares: basically, it standardizes the deviation for each observation. If we add these standardized deviations together the resulting statistic is Pearson's chi-square (χ^2) given by:

$$\chi^2 = \sum \frac{\left(\text{observed}_{ij} - \text{model}_{ij}\right)^2}{\text{model}_{ij}} \tag{19.2}$$

in which i represents the rows in the contingency table and j represents the columns.

The observed data are the frequencies in Table 19.1, but what is the model? When we predict a continuous outcome from categorical predictors (e.g., the linear model) the model we use is group means, but we can't work with means when we have a categorical outcome variable (see above) so we work with frequencies instead. We use 'expected frequencies'. A simple way to estimate the expected frequencies would be to say 'We've got 200 cats in total, and four categories, so the expected value is 200/4 = 50'. This approach would be fine if, for example, we had the same number of cats that had affection as a reward as we did cats that had food as a reward, but we didn't: 38 got food and 162 got affection as a reward. Likewise, there are not equal

Table 19.1 Contingency table showing how many cats will line-dance after being trained with different rewards

		Training		
		Food as Reward	Affection as Reward	Total
Could They Dance?	Yes	28	48	76
	No	10	114	124
	Total	38	162	200

numbers that could and couldn't dance. To adjust for these inequalities, we calculate expected frequencies for each cell in the table using the column and row totals for that cell. By doing so we factor in the total number of observations that *could have* contributed to that cell. The following equation, in which n is the total number of observations (in this case 200), shows this process:

$$\text{model}_{ij} = E_{ij} = \frac{\text{row total}_i \times \text{column total}_j}{n} \quad (19.3)$$

We can calculate these expected frequencies for the four cells within our table as follows (where row total and column total are abbreviated to RT and CT, respectively):

$$
\begin{aligned}
\text{model}_{\text{Food, Yes}} &= \frac{\text{RT}_{\text{Yes}} \times \text{CT}_{\text{Food}}}{n} \\
&= \frac{76 \times 38}{200} = 14.44 \\
\text{model}_{\text{Food, No}} &= \frac{\text{RT}_{\text{No}} \times \text{CT}_{\text{Food}}}{n} \\
&= \frac{124 \times 38}{200} = 23.56 \\
\text{model}_{\text{Affection, Yes}} &= \frac{\text{RT}_{\text{Yes}} \times \text{CT}_{\text{Affection}}}{n} \\
&= \frac{76 \times 162}{200} = 61.56 \\
\text{model}_{\text{Affection, No}} &= \frac{\text{RT}_{\text{No}} \times \text{CT}_{\text{Affection}}}{n} \\
&= \frac{124 \times 162}{200} = 100.44
\end{aligned}
\quad (19.4)
$$

These are the model values that we put into equation (19.2).

We now have the model values and the observed values are in Table 19.1. All we need to do is take each value in each cell of Table 19.1, subtract from it the corresponding model value, square the result, and then divide by the corresponding model value. Once we've done this for each cell we add them up:

$$
\begin{aligned}
\chi^2 &= \frac{(28-14.44)^2}{14.44} + \frac{(10-23.56)^2}{23.56} + \\
&\quad \frac{(48-61.56)^2}{61.56} + \frac{(114-100.44)^2}{100.44} \\
&= \frac{13.56^2}{14.44} + \frac{(-13.56)^2}{23.56} + \frac{(-13.56)^2}{61.56} + \frac{13.56^2}{100.44} \\
&= 12.73 + 7.80 + 2.99 + 1.83 \\
&= 25.35
\end{aligned}
\quad (19.5)
$$

This statistic has a distribution with known properties called the **chi-square distribution**, which has a shape determined by the degrees of freedom which are $(r-1)(c-1)$, in which r is the number of rows and c is the number of columns. Another way to think of it is the number of levels of each variable minus one multiplied together. In this example we get $df = (2-1)(2-1) = 1$.

In the days when people did these things by hand, they would get their pet diplodocus to find a critical value for the chi-square distribution with (for these data) $df = 1$. If the value of the observed chi-square statistic was bigger than this critical value they would conclude that there was a significant relationship between the two variables, rub some sticks together to make a fire and invite their friends over to celebrate. For those of you still living in caves, critical values are in the Appendix; for $df = 1$ the critical values are 3.84 ($p = 0.05$) and 6.63 ($p = 0.01$), and because the observed chi-square is bigger than these values it is significant at $p < 0.01$. For the rest of us, we can get a computer to compute the precise probability of obtaining a chi-square statistic at least as big as (in this case) 25.35 if there were no association between the variables in the population.

19.3.3 Fisher's exact test ▮▮▮▮

The chi-square statistic has a sampling distribution that is only *approximately* a chi-square distribution. The larger the sample is, the better this approximation becomes, and in large samples the approximation is good enough to not worry about the fact that it is an approximation. In small samples, the approximation is not good enough, making significance tests of the chi-square statistic inaccurate. This is why you'll often read about the chi-square test needing expected frequencies in each cell to be greater than 5 (see Section 19.5).

When the expected frequencies are greater than 5, the sampling distribution is probably close enough to a chi-square distribution for us not to worry. However, when the expected frequencies are too low, it probably means that the sampling distribution of the test statistic is too deviant from a chi-square distribution to be accurate.

Fisher came up with a solution to this problem called **Fisher's exact test** (Fisher, 1922). It's not a test as such, it's a way to compute the exact probability of the chi-square statistic in small samples. This procedure is normally used on 2×2 contingency tables (i.e., two variables each with two options) and with small samples. It *can* be used on larger contingency tables and with large samples, but there's no point because it was designed to overcome the problem of small samples, and in larger contingency tables it becomes computationally intensive and your computer might have a meltdown.

19.3.4 The likelihood ratio ▮▮▮▮

An alternative to Pearson's chi-square is the likelihood ratio statistic, which is based on maximum-likelihood theory. The general idea behind this theory is that you collect some data and create a model for which the probability of obtaining the observed set of data is maximized, then you compare this model to the probability of obtaining those data under the null hypothesis. The resulting statistic is based on comparing observed frequencies with those predicted by the model. The computation is:

$$L\chi^2 = 2 \sum \text{observed}_{ij} \ln\left(\frac{\text{observed}_{ij}}{\text{model}_{ij}}\right) \quad (19.6)$$

in which i and j are the rows and columns of the contingency table and ln is the natural logarithm (a standard mathematical function that we came across in Chapter 6). Using the same model and observed values as in the previous section, we get a likelihood ratio of 24.94:

$$
\begin{aligned}
L\chi^2 &= 2 \left[\begin{array}{l} 28 \times \ln\left(\dfrac{28}{14.44}\right) + 10 \times \ln\left(\dfrac{10}{23.56}\right) + 48 \times \\[4pt] \ln\left(\dfrac{48}{61.56}\right) + 114 \times \ln\left(\dfrac{114}{100.44}\right) \end{array} \right] \\[6pt]
&= 2 \left[\begin{array}{l} 28 \times 0.662 + 10 \times -0.857 + \\ 48 \times -0.249 + 114 \times 0.127 \end{array} \right] \\[6pt]
&= 2 \left[18.54 - 8.57 - 11.94 + 14.44 \right] \\
&= 24.94
\end{aligned}
\tag{19.7}
$$

As with Pearson's chi-square, this statistic has a chi-square distribution with the same degrees of freedom (in this case 1). We would use the same critical values as before and again conclude that the test statistic is significant because the observed value of 24.94 is bigger than the critical values of 3.84 ($p = 0.05$) and 6.63 ($p = 0.01$). A computer would give us a precise p-value. For large samples the likelihood ratio will be roughly the same as Pearson's chi-square, but is preferred when samples are small.

19.3.5 Yates's correction ▮▮▮▯

When you have a 2 × 2 contingency table (i.e., two categorical variables each with two categories) then Pearson's chi-square tends to produce significance values that are too small (it tends to make a Type I error). Yates suggested a correction to the Pearson formula (usually referred to as **Yates's continuity correction**). The basic idea is that when you calculate the deviation from the model (observed$_{ij}$ − model$_{ij}$ in equation (19.2)) you subtract 0.5 from the absolute value of this deviation before you square it. Put simply, you calculate the deviation, ignore whether it is positive or negative, subtract 0.5 from it and then square it. With Yates's correction applied Pearson's equation becomes:

$$
\chi^2 = \sum \frac{\left(\left| \text{observe } d_{ij} - \text{model}_{ij} \right| - 0.5 \right)^2}{\text{model}_{ij}}
\tag{19.8}
$$

For the data in our example this translates into a value of 23.52:

$$
\begin{aligned}
\chi^2 &= \frac{(13.56 - 0.5)^2}{14.44} + \frac{(13.56 - 0.5)^2}{23.56} + \\[6pt]
&\quad \frac{(13.56 - 0.5)^2}{61.56} + \frac{(13.56 - 0.5)^2}{100.44} \\[6pt]
&= 11.81 + 7.24 + 2.77 + 1.70 \\
&= 23.52
\end{aligned}
\tag{19.9}
$$

Note that the correction lowers the value of the chi-square statistic and, therefore, makes it less significant. There is a fair bit of evidence that this adjustment overcorrects and produces chi-square values that are too small. Howell (2012) provides an excellent discussion, if you're interested; all I will say is that although the correction is worth knowing about, it's probably best ignored.

19.3.6 Other measures of association ▮▮▮▯

There are measures of the strength of association that modify the chi-square statistic to take account of sample size and degrees of freedom and try to restrict the range of the test statistic from 0 to 1 (to make them similar to the correlation coefficient described in Chapter 8). Three related measures are:

- *Phi*: This statistic is accurate for 2 × 2 contingency tables. However, for tables with more than two dimensions the value of phi may not lie between 0 and 1 because the chi-square value can exceed the sample size. Therefore, Pearson suggested the use of the contingency coefficient.
- *Contingency coefficient*: This coefficient ensures a value between 0 and 1 but, unfortunately, it seldom

reaches its upper limit of 1 and for this reason Cramér devised an alternative denoted by V.

- *Cramér's V*: When both variables have only two categories, phi and Cramér's V are identical. However, when variables have more than two categories Cramér's statistic can attain its maximum of 1 (unlike the other two) and so it is the most useful.

19.3.7 The chi-square test as a linear model ▮▮▮▮

As with all the models in this book, the chi-square test can be conceptualized as a general linear model. The general linear model is expressed as:

$$
Y_i = b_0 + b_1 X_{1i} + b_2 X_{2i} + \cdots + b_n X_{ni} + \varepsilon_i
\tag{19.10}
$$

which is an equation we have seen several times throughout this book (e.g., Chapter 9). We've also seen that this linear model is perfectly capable of accommodating categorical predictor variables. For example, in Chapter 12 (equation (12.2)) when we wanted to compare the means of puppy therapy groups we included these groups using categorical dummy variables:

$$
\text{Happiness}_i = b_0 + b_1 \text{Long}_i + b_2 \text{Short}_i + \varepsilon_i
\tag{19.11}
$$

With a categorical outcome variable we can use essentially the same model. Let's see how. In our dancing cat example we have two categorical variables: training (food or affection) and dance (yes they did dance or no they didn't dance). Both variables have two categories and we can represent each one with a single dummy variable (see Section 11.5.1) in which one category is coded

Table 19.2 Coding scheme for dancing cats

Training	Dance	Dummy (Training)	Dummy (Dance)	Interaction	Frequency
Food	No	0	0	0	10
Food	Yes	0	1	0	28
Affection	No	1	0	0	114
Affection	Yes	1	1	1	48

as 0 and the other as 1. Let's code training as 0 for food and 1 for affection, and dancing as 1 for yes and 0 for no (see Table 19.2).

This situation is like the factorial design that we looked at in Section 14.3: in that example, we also had two variables as predictors and the general linear model became:

$$\text{Outcome}_i = b_0 + b_1 A_i + b_2 B_i + b_3 AB_i + \varepsilon_i \quad (19.12)$$

in which A represents the first variable, B represents the second and AB represents the interaction between the two variables (think back to equation (14.4)). Therefore, we can construct a linear model using the dummy variables in Table 19.2 that is like the one we used for factorial designs:

$$\text{outcome}_i = (\text{model}) + \text{error}_i$$
$$\text{outcome}_{ij} = \begin{pmatrix} b_0 + b_1 \text{Training}_i \\ + b_2 \text{Dance}_j \\ + b_3 \text{Interaction}_{ij} \end{pmatrix} + \varepsilon_{ij} \quad (19.13)$$

The interaction term will be the training variable multiplied by the dance variable (look at Section 11.3.2, and if it doesn't make sense look at Section 14.3 because the coding is the same as this example). However, because the *outcome variable* is categorical, to make this model linear we have to use log values. The model becomes:[2]

$$\ln(O_i) = \ln(\text{model}) + \ln(\varepsilon_i)$$
$$\ln(O_{ij}) = \begin{pmatrix} b_0 + b_1 \text{Training}_i \\ + b_2 \text{Dance}_j \\ + b_3 \text{Interaction}_{ij} \end{pmatrix} + \ln(\varepsilon_{ij}) \quad (19.14)$$

The training and dance variables and the interaction can take the values 0 and 1, depending on which combination of categories we're looking at (Table 19.2). Therefore, to work out what the b-values represent in this model we can do the same as we did for other linear models and look at what happens when we replace training and dance with different values of 0 and 1.

To begin with, let's see what happens when we look at when training and dance are both zero. This situation represents the category of cats that got food reward and didn't line-dance. When we've used the linear model before the outcomes were taken from the observed data we used the group means (see, for example, Sections 10.4 and 12.2). With a categorical outcome we use the observed frequencies (rather than observed means). In Table 19.1 there were 10 cats that had food for a reward and didn't line-dance. If we use this as the observed outcome then the model can be written as:

$$\ln(O_{ij}) = b_0 + b_1 \text{Training}_i \\ + b_2 \text{Dance}_j + b_3 \text{Interaction}_{ij} \quad (19.15)$$

if we ignore the error term for the time being. For cats that had food reward and didn't dance, the training and dance variables and the interaction will all be 0 and the equation reduces to:

$$\ln(O_{\text{Food, No}}) = b_0 + (b_1 \times 0) + (b_2 \times 0) + (b_3 \times 0)$$
$$= b_0$$
$$\ln(10) = b_0 \quad (19.16)$$
$$b_0 = 2.303$$

Therefore, b_0 in the model represents the log of the observed value when all categories are zero: it's the log of the observed value of the base category (in this case cats that got food and didn't dance).

Now, let's see what happens when we look at cats that had affection as a reward and didn't dance. In this case, the training variable is 1 and the dance variable and the interaction are still 0. Also, our outcome now changes to be the observed value for cats that received affection and didn't dance (from Table 19.1 the value is 114). The equation becomes:

$$\ln(O_{\text{Affection, No}}) = b_0 + (b_1 \times 1) + (b_2 \times 0) \\ + (b_3 \times 0) \quad (19.17)$$
$$= b_0 + b_1$$
$$b_1 = \ln(O_{\text{Affection, No}}) - b_0$$

Remembering that b_0 is the expected value for cats that had food and didn't dance, we get:

$$b_1 = \ln(O_{\text{Affection, No}}) - \ln(O_{\text{Food, No}})$$
$$= \ln(114) - \ln(10)$$
$$= 4.736 - 2.303 \quad (19.18)$$
$$= 2.433$$

The important thing is that b_1 is the difference between the log of the observed frequency for cats that received affection and didn't dance, and the log of the observed values for cats that received food and didn't dance. Put another way, within the group of cats that didn't dance it represents the difference between those trained using food and those trained using affection.

When we look at cats that had food as a reward and danced the training variable is 0, the dance variable is 1 and the interaction is 0. Our outcome is the observed frequency for cats that received food and danced (from Table 19.1 the value is 28). The equation becomes:

$$\ln(O_{\text{Food, Yes}}) = b_0 + (b_1 \times 0) + (b_2 \times 1) + (b_3 \times 0)$$
$$= b_0 + b_2 \quad (19.19)$$
$$b_2 = \ln(O_{\text{Food, Yes}}) - b_0$$

Let's replace b_0 with what we know it represents (the expected value for cats that had food and didn't dance):

$$b_2 = \ln(O_{\text{Food, Yes}}) - \ln(O_{\text{Food, No}})$$
$$= \ln(28) - \ln(10)$$
$$= 3.332 - 2.303 \quad (19.20)$$
$$= 1.029$$

Therefore, b_2 is the difference between the log of the observed frequency for cats that received food and danced, and the log of the observed frequency for cats that received food and didn't dance. Put another way, within the group of cats that received food as a reward it represents the difference between cats that didn't dance and those that did.

Finally, let's look at cats that had affection and danced. The training and dance

---

2 The convention is to denote b_0 as θ and the b-values as λ, but I think these notational changes serve only to confuse people so I'm sticking with b because I want to emphasize the similarities to the linear model.

variables are both 1 and the interaction (which is the value of training multiplied by the value of dance) is also 1. We can replace b_0, b_1 and b_2, with what we now know they represent. The outcome is the log of the observed frequency for cats that received affection and danced (this value is 48—see Table 19.1). Therefore, the equation becomes:

$$\ln\left(O_{A,Y}\right) = b_0 + (b_1 \times 1) + (b_2 \times 1) + (b_3 \times 1)$$
$$= b_0 + b_1 + b_2 + b_3$$
$$= \ln\left(O_{F,N}\right) + \left(\ln\left(O_{A,N}\right) - \ln\left(O_{F,N}\right)\right) +$$
$$\left(\ln\left(O_{F,Y}\right) - \ln\left(O_{F,N}\right)\right) + b_3$$
$$= \ln\left(O_{A,N}\right) + \ln\left(O_{F,Y}\right) -$$
$$\ln\left(O_{F,N}\right) + b_3 \tag{19.21}$$
$$b_3 = \ln\left(O_{A,Y}\right) - \ln\left(O_{F,Y}\right) +$$
$$\ln\left(O_{F,N}\right) - \ln\left(O_{A,N}\right)$$
$$= \ln(48) - \ln(28) + \ln(10) - \ln(114)$$
$$= -1.895$$

(I've used the shorthand of A for affection, F for food, Y for yes, and N for no), which shows that b_3 compares the difference between affection and food when the cats didn't dance to the difference between affection and food when the cats *did* dance. Put another way, it compares the effect of training when cats didn't dance to the effect of training when they did dance. Putting all these b-values together, we get the following model:

$$\ln\left(O_{ij}\right) = 2.303 + 2.433 \, \text{Training}_i + 1.029 \, \text{Dance}_j$$
$$- 1.895 \, \text{Interaction}_{ij} + \ln\left(\varepsilon_{ij}\right) \tag{19.22}$$

The important thing to take from this is that everything is the same as in factorial designs except that we dealt with log-transformed values (compare this section to Section 14.3 to see just how similar everything is). In case you don't believe that the chi-square test works as a general linear model, load **Cat Regression.sav**, which contains the two variables **Dance** (0 = no, 1 = yes) and **Training** (0 = food, 1 = affection) and the interaction (**Interaction**). There is a variable called **Observed** that contains the observed frequencies in Table 19.1 for each combination of **Dance** and **Training** and a variable called **LnObserved**, which is the natural logarithm of these observed frequencies (remember that throughout this section we've dealt with the log observed values).

Output 19.1 shows the resulting coefficients. Note that the constant, b_0, is 2.303 as calculated above, the b-value for type of training, b_1, is 2.434 and for dance, b_2, is 1.030, both of which are within rounding error of what was calculated above. Finally, the coefficient for the interaction, b_3, is −1.895 as predicted. One weird thing is that the standard errors are

all zero: there is *no* error whatsoever in this model. This lack of error is because the various combinations of coding variables completely explain the observed values. This is known as a **saturated model** and I will return to it later, so make a mental note of it.

This is all very well, but the heading of this section did rather imply that I would show you how the chi-square test can be conceptualized as a linear model. Here goes. The chi-square test looks at whether two variables are independent, therefore, it has no interest in their combined effect (the interaction), only their main effect. So, we take the interaction out of the saturated model and the model becomes:

$$\ln\left(\text{model}_{ij}\right) = b_0 + b_1 \, \text{Training}_i + b_2 \, \text{Dance}_j \tag{19.23}$$

With this new model, we cannot predict the observed values perfectly (like the saturated model) because we've lost information (namely, the interaction term). Therefore, the outcome from the model changes, and the b-values too. We saw earlier that the chi-square test is based on 'expected frequencies'. Our outcome becomes these expected values as in the following equation:

$$\ln\left(E_{ij}\right) = b_0 + b_1 \, \text{Training}_i + b_2 \, \text{Dance}_j \tag{19.24}$$

We already computed the expected values for this example in equation (19.4). We can recalculate the beta values based on these expected values. For cats that had food reward and didn't dance, the training and dance variables will be 0 and the equation becomes:

$$\ln\left(E_{\text{Food, No}}\right) = b_0 + (b_1 \times 0) + (b_2 \times 0)$$
$$= b_0 \tag{19.25}$$
$$b_0 = \ln(23.56)$$
$$= 3.16$$

Therefore, b_0 represents the log of the expected value when all categories are zero. When we look at cats that had affection as a reward and didn't dance, the training variable is 1 and the dance

SELF TEST

Fit a linear model with **LnObserved** as the outcome, and **Training**, **Dance** and **Interaction** as the three predictors.

Coefficients[a]

Model		Unstandardized Coefficients B	Std. Error	Standardized Coefficients Beta	t	Sig.
1	(Constant)	2.303	.000		72046662.8	.000
	Type of Training	2.434	.000	1.385	73011512.2	.000
	Did they dance?	1.030	.000	.725	27654265.1	.000
	Interaction	−1.895	.000	−1.174	−46106003	.000

a. Dependent Variable: LN (Observed Frequencies)

Output 19.1

variable is still 0. Also, our outcome now changes to be the expected value for cats that received affection and didn't dance:

$$\ln\left(E_{\text{Affection, No}}\right) = b_0 + (b_1 \times 1) + (b_2 \times 0)$$
$$= b_0 + b_1$$
$$b_1 = \ln\left(E_{\text{Affection, No}}\right) - b_0$$
$$= \ln\left(E_{\text{Affection, No}}\right) - \ln\left(E_{\text{Food, No}}\right)$$
$$= \ln(100.44) - \ln(23.56)$$
$$= 1.45 \qquad (19.26)$$

The important thing is that b_1 is the difference between the log of the expected frequency for cats that received affection and didn't dance and the log of the expected values for cats that received food and didn't dance. This value is the same as the column marginal, that is, the difference between the total number of cats getting affection and the total number of cats getting food: $\ln(162) - \ln(38) = 1.45$. Put simply, it represents the main effect of the type of training.

When we look at cats that had food as a reward and danced, the training variable is 0 and the dance variable is 1. Our outcome now changes to be the expected frequency for cats that received food and danced:

$$\ln\left(E_{\text{Food, Yes}}\right) = b_0 + (b_1 \times 0) + (b_2 \times 1)$$
$$= b_0 + b_2$$
$$b_2 = \ln\left(E_{\text{Food, Yes}}\right) - b_0$$
$$= \ln\left(E_{\text{Food, Yes}}\right) - \ln\left(E_{\text{Food, No}}\right) \quad (19.27)$$
$$= \ln(14.44) - \ln(23.56)$$
$$= -0.49$$

Therefore, b_2 is the difference between the log of the expected frequencies for cats that received food and did or didn't dance. In fact, the value is the same as the row marginal, that is the difference between the total number of cats that did and didn't dance: $\ln(76) - \ln(124) = -0.49$. In simpler terms, it is the main effect of whether the cat danced.

We can double-check all of this by looking at the final cell (cats that had affection and danced):

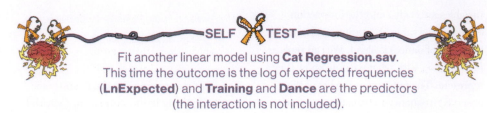

SELF TEST
Fit another linear model using **Cat Regression.sav**.
This time the outcome is the log of expected frequencies (**LnExpected**) and **Training** and **Dance** are the predictors (the interaction is not included).

$$\ln\left(E_{\text{Affection, Yes}}\right) = b_0 + (b_1 \times 1) + (b_2 \times 1)$$
$$= b_0 + b_1 + b_2$$
$$\ln(61.56) = 3.16 + 1.45 - 0.49 \qquad (19.28)$$
$$4.12 = 4.12$$

If we put the b-values into the model then the final chi-square is:

$$\ln(O_i) = \ln(\text{model}) + \ln(\varepsilon_i)$$
$$\ln(O_i) = 3.16 + 1.45\,\text{Training} - 0.49\,\text{Dance} + \ln(\varepsilon_i) \qquad (19.29)$$

We can rearrange this equation to get residuals (the error term):

$$\ln(\varepsilon_i) = \ln(O_i) - \ln(\text{model}) \qquad (19.30)$$

The model is the expected frequencies that were calculated for the chi-square test, so the residuals are the differences between the observed and expected frequencies. This section demonstrates how chi-square can be thought of as a linear model in which the beta values tell us something about the relative differences in frequencies across categories of our two variables. The take-home point is simply that even with categorical outcomes you're using the same model that you have been learning about throughout this book.

19.4 Associations between several categorical variables: loglinear analysis

Often we want to analyze more complex contingency tables in which there are three or more variables. For example, suppose we took the example we've just used but also collected data from a sample of

70 dogs? We might want to compare the behavior in dogs to that in cats. We would have three variables: **Animal** (dog or cat), **Training** (food as reward or affection as reward) and **Dance** (did they dance or not?). This couldn't be analyzed with the Pearson chi-square and instead has to be analyzed with a technique called **loglinear analysis**.

In the previous section, after nearly reducing my brain to even more of a rotting vegetable than it already is trying to explain how categorical data analysis is just another form of regression, I fitted an ordinary regression to prove that I wasn't talking gibberish. At the time I rather glibly said 'Oh, by the way, there's no error in the model, that's odd, isn't it?' and sort of passed this off by telling you that it was a 'saturated' model and not to worry too much about it because I'd explain it all later just as soon as I'd worked out what the hell was going on. That was a good avoidance tactic while it lasted, but I now have to explain what I was going on about.

To begin with, I hope you're now happy with the idea that categorical data can be expressed in the form of a linear model if we use log values (this, incidentally, is why the technique we're discussing is called *log*linear analysis). From what you hopefully already know about linear models generally, you should also be cosily tucked up in bed with the idea that we can extend any linear model to include any number of predictors and any resulting interaction terms between predictors. If we can represent a simple two-variable categorical analysis as a linear model, then it shouldn't amaze you to discover that if we have more than two variables the model simply extends to include new categorical predictors and their interactions with existing predictors. Each predictor will have a parameter (b). This is all you really need to know. By thinking in terms of a linear model, it becomes conceptually very easy to understand how the chi-square model expands to incorporate new predictor variables. For example, if we have three predictors (A, B and C) in a linear model we end up with three two-way interactions (AB, AC, BC) and one three-way interaction (ABC):

$$\text{outcome}_{ijk} = b_0 + b_1 A_i + b_2 B_j + b_3 C_k \\ + b_4 AB_{ij} + b_5 AC_{ik} + b_6 BC_{jk} \quad (19.31) \\ + b_7 ABC_{ijk} + \varepsilon_{ij}$$

If the outcome is categorical we get an identical model, but with an outcome expressed as logs:

$$\ln\left(O_{ijk}\right) = b_0 + b_1 A_i + b_2 B_j + b_3 C_k + b_4 AB_{ij} + \\ b_5 AC_{ik} + b_6 BC_{jk} + b_7 ABC_{ijk} + \ln\left(\varepsilon_{ij}\right) \\ (19.32)$$

The calculation of b-values and expected values from the model becomes considerably more cumbersome and confusing than with just one predictor and one outcome, but that's why we invented computers—so that we don't have to worry about it. Just imagine that a puppy called Tobin collects numbers into a pile and then circles them while chasing his tail. As the numbers get sucked into the vortex that his spinning creates, a value eventually flies out like a leaf floating on the wind and lands on the grass beside him. Loglinear analysis works on these principles. Sort of.

As we saw in the two-variable case, when our outcome is categorical and we include all the available terms (main effects and interactions) we get no error: our predictors perfectly predict the outcome (the expected values). The model is saturated. If we start with this model, we get no error. The job of loglinear analysis is to try to fit a simpler model without any substantial loss of predictive power. Therefore, loglinear analysis typically works on a principle of backward elimination (yes, the same kind of backward elimination as in Section 9.9.1). We begin with the saturated model, remove a predictor from the model, re-estimate the model and use it to predict our outcome (calculate expected frequencies, just like the chi-square test) and see how well it fits the data (i.e., are the expected frequencies close to the observed frequencies?). If the fit of the new model is not very different from the more complex model, then we abandon the complex model in favor of the new, simpler, one. Put another way, we assume the term we removed was not having a significant impact on the ability of our model to predict the observed outcome.

We don't remove terms randomly, we do it hierarchically. So, we start with the saturated model, remove the highest-order interaction, and assess the effect that this has. If removing the highest-order interaction term has no substantial impact on the model then we get rid of it and move on to remove the next highest-order interactions. If removing these interactions has no effect then we remove them and so on down to the main effects. We carry on until we find an effect that *does* affect the fit of the model when it is removed.

To put this in more concrete terms, at the beginning of this section I asked you to imagine we'd extended our training and line-dancing example to incorporate a sample of dogs. So, we now have three variables: **Animal** (dog or cat), **Training** (food or affection) and **Dance** (did they dance or not?). This model has three main effects:

- **Animal**
- **Training**
- **Dance**

three interactions involving two variables:

- **Animal** × **Training**
- **Animal** × **Dance**
- **Training** × **Dance**

and one interaction involving all three variables:

- **Animal** × **Training** × **Dance**

When I talk about backward elimination I mean that loglinear analysis starts by including all these effects. The highest-order interaction (in this case the three-way interaction of **Animal** × **Training** × **Dance**) is removed. A new model is estimated without this interaction, and expected frequencies are computed from this new model. These expected frequencies (or model frequencies) are compared to the observed frequencies using the standard equation for the likelihood ratio statistic (see Section 19.3.4). If the new model significantly changes the likelihood ratio statistic, then removing the interaction term has a significant effect on the fit of the model, which tells us that this effect is statistically important. If this is the case then we stop here and conclude that we have a significant three-way interaction. We won't test any other effects because all lower-order effects are subsumed within higher-order effects. If, however, removing the three-way interaction doesn't significantly affect the fit of the model then we move on to lower-order interactions. Therefore, we look at the **Animal** × **Training**, **Animal** × **Dance** and **Training** × **Dance** interactions in turn and construct models in which these terms are not present. For each model expected frequencies are computed and compared to the observed frequencies using a likelihood ratio statistic.[3] Again, if any one of these

---

3 It's worth mentioning that for every model, the computation of expected values differs, and as the designs get more complex, the computation gets increasingly tedious and incomprehensible (at least to me); however, you don't need to know the calculations to get a feel for what is going on.

models results in a significant change in the likelihood ratio then the term is retained and we won't move on to look at any main effects involved in that interaction (so if the **Animal** × **Training** interaction is significant the computer won't look at the main effects of **Animal** or **Training**). However, if the likelihood ratio is unchanged then the offending interaction term is removed and the computer moves on to look at main effects.

I mentioned that the likelihood ratio statistic (Section 19.3.4) is used to assess each model. This equation (equation (19.6)) can be adapted to fit any model: the observed values are the same throughout, and the model frequencies are the expected frequencies from the model being tested. For the saturated model, this statistic will always be 0 (because the observed and model frequencies are the same, so the ratio of observed to model frequencies will be 1, and ln(1) = 0), but in other situations it will provide a measure of how well the model fits the observed frequencies. To test whether a new model has changed the likelihood ratio, we take the likelihood ratio for a model and subtract from it the likelihood statistic for the previous model (provided the models are hierarchically structured):

$$L\chi^2_{Change} = L\chi^2_{Current\ Model} - L\chi^2_{Previous\ Model} \quad (19.33)$$

I've tried to give you a flavor of how loglinear analysis works without getting bogged down in the nitty-gritty of the calculations. The curious among you might want to know exactly how everything is calculated, and to these people I have two things to say: 'I don't know' and 'I know a really good place where you can buy a straitjacket'. Tabachnick and Fidell (2012) has a wonderfully detailed and lucid chapter on the subject, which puts my feeble attempt to shame.

19.5 Assumptions when analyzing categorical data

The chi-square test does not rely on the assumptions discussed in Chapter 6 (e.g., categorical data cannot have a normal sampling distribution because they aren't continuous). However, it has two important assumptions relating to (1) independence and (2) expected frequencies.

19.5.1 Independence

The general linear model makes an assumption about the independence of residuals, and the chi-square test, being a linear model of sorts, is no exception. For the chi-square test to be meaningful each person, item or entity must contribute to only one cell of the contingency table. Therefore, you cannot use a chi-square test on a repeated-measures design (e.g., if we had trained some cats with food to see if they would dance and then trained the same cats with affection to see if they would dance we couldn't analyze the resulting data with Pearson's chi-square test). If you find yourself in this situation you need to get yourself a good book about generalized linear mixed models (GLMMs) because you effectively need to fit a variant of a multilevel model (Chapter 21) for categorical outcomes.

19.5.2 Expected frequencies

With 2 × 2 contingency tables (i.e., two categorical variables both with two categories) no expected values should be below 5. In larger tables, and when looking at associations between three or more categorical variables (loglinear analysis), the rule is that all expected counts should be greater than 1 and no more than 20% of expected counts should be less than 5. Howell (2012) gives a nice explanation of why violating this assumption creates problems. If this

assumption is broken the result is a radical reduction in test power—so dramatic in fact that it may not be worth bothering with the analysis at all.

In terms of remedies, if you're looking at associations between only two variables then consider using Fisher's exact test (Section 19.3.3). With three or more variables (i.e., loglinear analysis) your options are to: (1) collapse the data across one of the variables (preferably the one you least expect to have an effect); (2) collapse levels of one of the variables; (3) collect more data; or (4) accept the loss of power. If you want to collapse data across one of the variables then:

1 The highest-order interaction should be non-significant.
2 At least one of the lower-order interaction terms involving the variable to be deleted should be non-significant.

Let's think about our loglinear example in which we're looking at the relationship between training (food vs. affection), whether the animal danced (yes vs. no), and the species of animal (cats vs. dogs). Say we wanted to delete the animal variable; then for this to be valid, the **Animal** × **Training** × **Dance** variable should be non-significant, and either the **Animal** × **Training** or the **Animal** × **Dance** interaction should also be non-significant.

You can also collapse categories within a variable. So, if you had a variable of 'season' relating to spring, summer, autumn and winter, and you had very few observations in winter, you could consider reducing the variable to three categories: spring, summer, autumn/winter perhaps. However, you should combine only categories for which it makes theoretical sense.

Finally, some people overcome the problem by simply adding a constant to all cells of the table, but there really is no point in doing this because it doesn't address the issue of power.

19.5.3 More doom and gloom ▮▮▮▮

Finally, although it's not an assumption, it seems fitting to mention in a section in which a gloomy and foreboding tone is being used that proportionately small differences in cell frequencies can result in statistically significant associations between variables if the sample is large enough (although it might need to be very large indeed). Therefore, we must look at row and column percentages to interpret the significant effects that we get. These percentages will reflect the patterns of data far better than the frequencies themselves (because these frequencies will be dependent on the sample sizes in different categories).

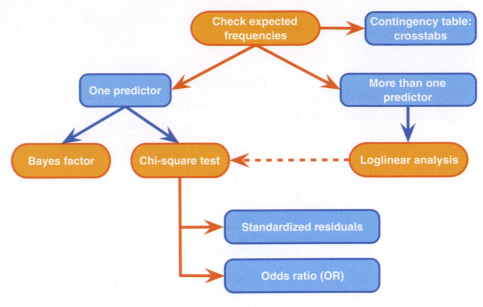

Figure 19.2   The general process for fitting models in which both predictors and the outcome are categorical

19.6 General procedure for analyzing categorical outcomes ▮▮▮▮

Figure 19.2 shows a general procedure for analyzing data when you want to fit models that have both an outcome and predictor(s) that are categorical. Essentially you first look at a contingency table and check the expected frequencies. If you have one predictor then head straight to a chi-square test or Bayes factor, but if you have more than one predictor first do a loglinear analysis (Section 19.9) and then follow up any significant effects with one or more chi-square tests. After a chi-square test it's useful to inspect the standardized residuals and compute an odds ratio, which is an effect size quantifying the relationship between variables.

19.7 Doing chi-square using SPSS Statistics ▮▮▮▮

To begin with, let's imagine we're looking at the data from only cats. We want to input data about whether the 200 cats danced and what type of training they had. There are two ways to do this.

19.7.1 Entering raw scores ▮▮▮▮

The first way is to do what we normally do: enter each cat's data as a row of the data. You would create two coding variables (**Training** and **Dance**) and, in keeping with Table 19.2, **Training** could be coded 0 to represent a food reward and 1 to

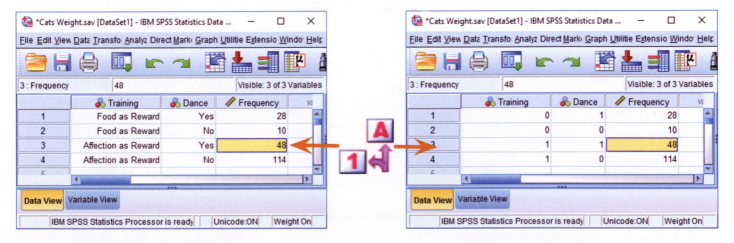

Figure 19.3   Data entry using weighted cases

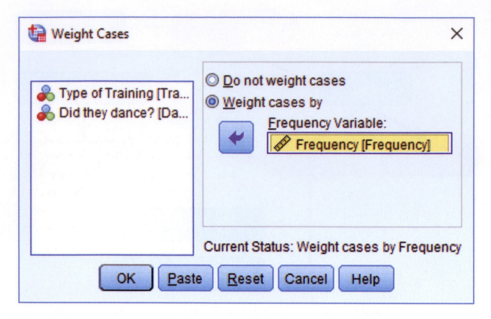

Figure 19.4 The dialog box for the *weight cases* command

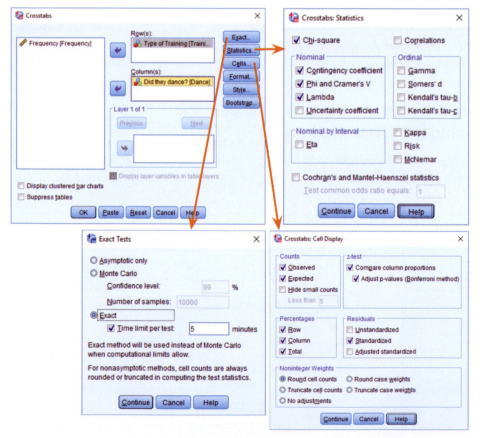

Figure 19.5 Dialog boxes for the *crosstabs* command

represent affection, and **Dance** could be coded 1 to represent an animal that danced and 0 to represent one that did not. For each animal, you put the appropriate numeric code into each column. For example, a cat that was trained with food that did not dance would have 0 in the training column and 0 in the dance column as in **Cats.sav**. Note that there are 200 cats and, therefore, 200 rows of data.

19.7.2 Entering frequencies and weighting cases

The second way to enter data is to create the same coding variables as before, but have a third variable that represents the number of animals that fell into each combination of categories. We could call this variable **Frequency**. Instead of having 200 rows, each one representing a different animal, we have one row representing each combination of categories and in the **Frequency** variable we enter the number of animals that fell into this category combination. Figure 19.3 shows the data set up in this way: the first row represents cats that had food as a reward and then danced and the value in **Frequency** tells us that there were 28 cats that had food as a reward and then danced. Extending this principle, we can see that when affection was used as a reward 114 cats did not dance. Data entered in this way are in the file **Cats Weight.sav**.

This method of data entry saves a lot of time, but if you use this method to enter data you must tell the computer that the variable **Frequency** represents the number of cases that fell into a particular combination of categories. To do this, select *Data* ▶ Weight Cases... to access the dialog box in Figure 19.4, select ⦿ **Weight cases by** and drag the variable in which the number of cases is specified (in this case **Frequency**) to the box labeled *Frequency Variable* (or click ➡). IBM SPSS Statistics will now weight each category combination by the number in the column labeled **Frequency**. For example, the computer will pretend that there are 28 rows of data that have the category combination 0, 1 (representing cats trained with food and that danced).

19.7.3 Specifying a chi-square test ▮▮▮▮

The first step in Figure 19.2 is to create a contingency table using the *crosstabs* command, check the expected frequencies and then do the chi-square test. We can do these steps simultaneously. *Select*

Analyze ▶ *Descriptive Statistics* ▶ 🔢 *Crosstabs…* to access the dialog box in Figure 19.5 (the variable **Frequency** is shown in the diagram because I ran the analysis on the **Cats Weight.sav** data). First, drag one of the variables of interest (I chose **Training**) to the box labeled *Row(s)* (or select it and click ➡). Next, drag the other variable of interest (**Dance**) to the box labeled *Column(s)* (or click ➡). It is also possible to select a layer variable (i.e., to split the rows of the table into further categories). If you had a third categorical variable (as we will later in this chapter) you could split the

contingency table by this variable (so layers of the table represent different categories of this third variable).

Click Statistics… to specify various statistical tests (see SPSS Tip 19.1). Select the chi-square test, the contingency coefficient, phi and lambda. Click Cells… in the main dialog box to specify what values are displayed in the crosstabulation table. Select *Expected* because we use these to check the assumptions about the expected frequencies (Section 19.5). It is also useful to have a look at the row, column and total percentages because these values are usually more easily interpreted than the

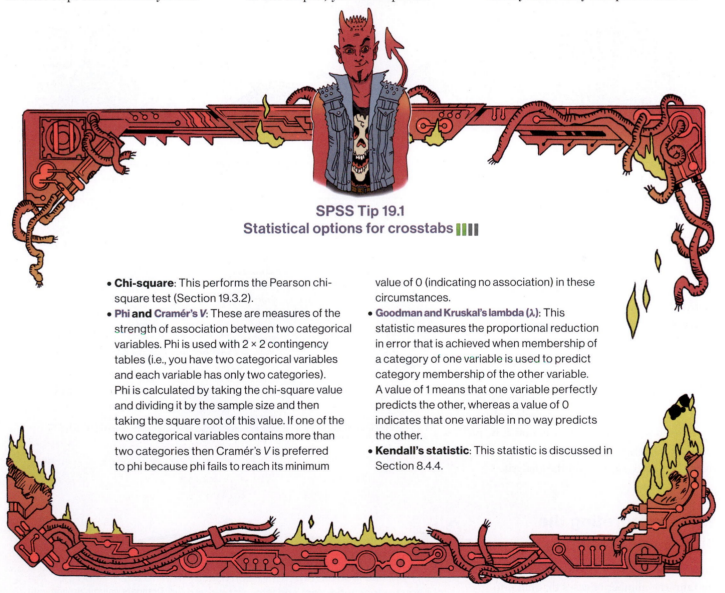

SPSS Tip 19.1
Statistical options for crosstabs ▮▮▮▮

- **Chi-square**: This performs the Pearson chi-square test (Section 19.3.2).
- **Phi** and **Cramér's V**: These are measures of the strength of association between two categorical variables. Phi is used with 2 × 2 contingency tables (i.e., you have two categorical variables and each variable has only two categories). Phi is calculated by taking the chi-square value and dividing it by the sample size and then taking the square root of this value. If one of the two categorical variables contains more than two categories then Cramér's *V* is preferred to phi because phi fails to reach its minimum

value of 0 (indicating no association) in these circumstances.
- **Goodman and Kruskal's lambda (λ)**: This statistic measures the proportional reduction in error that is achieved when membership of a category of one variable is used to predict category membership of the other variable. A value of 1 means that one variable perfectly predicts the other, whereas a value of 0 indicates that one variable in no way predicts the other.
- **Kendall's statistic**: This statistic is discussed in Section 8.4.4.

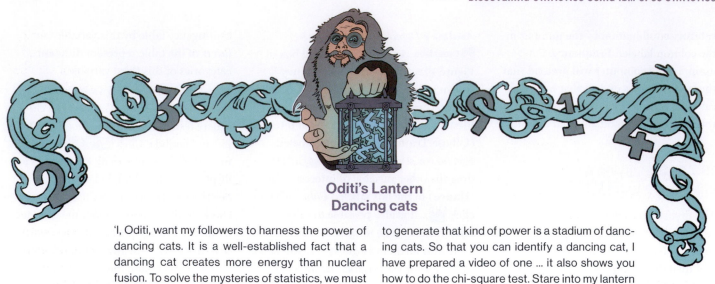

Oditi's Lantern
Dancing cats

'I, Oditi, want my followers to harness the power of dancing cats. It is a well-established fact that a dancing cat creates more energy than nuclear fusion. To solve the mysteries of statistics, we must power thousands of computers, and the only way to generate that kind of power is a stadium of dancing cats. So that you can identify a dancing cat, I have prepared a video of one ... it also shows you how to do the chi-square test. Stare into my lantern and be amazed.'

raw frequencies. There are two options that are useful for breaking down a significant effect (should we get one): (1) a *z*-test that compares cell counts across columns of the contingency table (☑ Compare column proportions), and it's a good idea to

☑ Adjust p-values (Bonferroni method) because there will be multiple tests; and (2)

☑ Standardized residuals. Click Exact... in the main dialog box to compute Fisher's exact test (Section 19.3.3) if your sample is small or if your expected frequencies are too low (see Section 19.5). Even though we don't need it for these data I have selected the *Exact* test option to show you how it is used. Click Continue to return to the main dialog box and OK to run the analysis.

Type of Training * Did they dance? Crosstabulation

			Did they dance? No	Did they dance? Yes	Total
Type of Training	Food as Reward	Count	10a	28b	38
		Expected Count	23.6	14.4	38.0
		% within Type of Training	26.3%	73.7%	100.0%
		% within Did they dance?	8.1%	36.8%	19.0%
		% of Total	5.0%	14.0%	19.0%
		Standardized Residual	-2.8	3.6	
	Affection as Reward	Count	114a	48b	162
		Expected Count	100.4	61.6	162.0
		% within Type of Training	70.4%	29.6%	100.0%
		% within Did they dance?	91.9%	63.2%	81.0%
		% of Total	57.0%	24.0%	81.0%
		Standardized Residual	1.4	-1.7	
Total		Count	124	76	200
		Expected Count	124.0	76.0	200.0
		% within Type of Training	62.0%	38.0%	100.0%
		% within Did they dance?	100.0%	100.0%	100.0%
		% of Total	62.0%	38.0%	100.0%

Each subscript letter denotes a subset of Did they dance? categories whose column proportions do not differ significantly from each other at the .05 level.

Output 19.2

19.8 Interpreting the chi-square test ▮▮▮▮

The contingency table (Output 19.2) contains the number of cases that fall into each combination of categories. We can

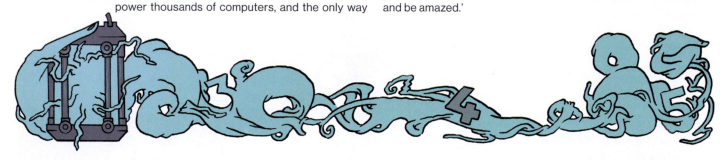

SELF ✂ TEST

Using the **Cats Weight.sav** data, change the frequency of cats that had food as reward and didn't dance from 10 to 28. Redo the chi-square test and select and interpret *z*-tests (☑ Compare column proportions). Is there anything about the results that seems strange?

see that in total 76 cats danced (38% of the total) and of these 28 were trained using food (36.8% of the total that danced) and 48 were trained with affection (63.2% of the total that danced). Further, 124 cats didn't dance at all (62% of the total) and of those that didn't dance, 10 were trained using food as a reward (8.1% of the total that didn't dance) and a massive 114 were trained using affection (91.9% of the total that didn't dance). The numbers of cats can be read from the rows labeled *Count* and the percentages are read from the rows labeled *% within Did they dance?* We can also look at the percentages within the training categories by looking at the rows labeled *% within Type of Training*. This tells us, for example, that of those trained with food as a reward, 73.7% danced and 26.3% did not. Similarly, for those trained with affection, only 29.6% danced compared to 70.4% that didn't. In summary, when food was used as a reward most cats would dance, but when affection was used most cats refused to dance.

First, let's check the expected frequencies (Section 19.5). We have a 2 × 2 table, so all expected frequencies need to be greater than 5. Looking at the expected counts in the contingency table (which incidentally are the same as we calculated earlier), we see that the smallest expected count is 14.4 (for cats that were trained with food and did dance). This value exceeds 5 and so the assumption has been met. If you found an expected count lower than 5, calculate a Bayes factor instead.

If you're wondering why the counts have subscript letters it's because we selected *Compare column proportions* in Figure 19.5. These subscripts tell us the results of the *z*-test that we asked for: columns with different subscripts have significantly different column proportions. It's not immediately obvious what's being tested here, and to be honest with you it took me a while to fathom it out because I could interpret the SPSS Statistics help files in

different ways (perhaps it's just me). I got there in the end (I think). We need to look within rows of the table. So, for *Food as Reward* the columns have different subscripts (the count of 10 has a subscript letter *a* and the count of 28 has a subscript letter *b*), which means that proportions within the column variable (i.e., *Did they dance?*) are significantly different. The *z*-test compares the *proportion* of the total frequency of the first column that falls into the first row against the *proportion* of the total frequency of the second column that falls into the first row. The different subscripts tell us that these proportions are significantly different. So, the proportion of cats that danced after food (36.8%) was significantly more than the proportion that *didn't* dance after food (8.1%). The test compares the proportions and *not the counts themselves*, so it is not the case that the count of 28 is different from the count of 10 (in this example). The self-test uses an example to illustrate this point.

Moving on to the row labeled *Affection as Reward*, the count of 114 has a subscript letter *a* and the count of 48 has a subscript letter *b*; as before, the fact they have different letters tells us that the column proportions are significantly different: in other words, 91.9% is significantly different from 63.2%. The proportion of cats that danced after affection was significantly less than the proportion that didn't dance after affection.

Output 19.3 shows the chi-square statistic and its significance value. The value of the chi-square statistic is given in the table (together with the degrees of freedom), as is the significance value. The value of the chi-square statistic is 25.356, which is within rounding error of what we calculated in Section 19.3.2, and this value is highly significant ($p < 0.001$), indicating that the type of training was significantly associated with whether an animal danced.

The table also includes other statistics that you requested in Figure 19.5. *Continuity*

Chi-Square Tests

	Value	df	Asymptotic Significance (2-sided)	Exact Sig. (2-sided)	Exact Sig. (1-sided)	Point Probability
Pearson Chi-Square	25.356[a]	1	.000	.000	.000	
Continuity Correction[b]	23.520	1	.000			
Likelihood Ratio	24.932	1	.000	.000	.000	
Fisher's Exact Test				.000	.000	
Linear-by-Linear Association	25.229[c]	1	.000	.000	.000	.000
N of Valid Cases	200					

a. 0 cells (.0%) have expected count less than 5. The minimum expected count is 14.44.

b. Computed only for a 2x2 table

c. The standardized statistic is –5.023.

Output 19.3

Symmetric Measures

		Value	Approximate Significance	Exact Significance
Nominal by Nominal	Phi	–.356	.000	.000
	Cramer's V	.356	.000	.000
	Contingency Coefficient	.335	.000	.000
N of Valid Cases		200		

Output 19.4

Correction is Yates's continuity corrected chi-square (Section 19.3.5), which matches the value we calculated earlier (23.52). This test is probably best ignored. The *Likelihood Ratio* (Section 19.3.4), which we'd prefer to the chi-square test if the sample were small, is within rounding error of the value we calculated (24.93) and is also highly significant ($p < 0.001$). There are several footnotes. The first is a summary of expected counts in case you forgot to check this yourself. We're told that there were no expected frequencies less than 5 so the chi-square statistic should be accurate.

Output 19.4 contains the measures of association discussed in Section 19.3.6 (if requested). Cramér's statistic is 0.36 out of a possible maximum value of 1, which represents a medium association between the type of training and whether the cats danced or not (think of it like a correlation coefficient). This value is highly significant ($p < 0.001$), indicating that a value of the test statistic that is at least this big if the null hypothesis were true is unlikely to have happened if there were no association in the population. These results confirm what the chi-square test already told us but also give us an estimate of the effect size.

19.8.1 Using standardized residuals

In a 2×2 contingency table like the one we have in this example, the nature of a significant association can be clear from just the cell percentages or counts. In larger contingency tables this may not be the case and you need a finer-grained investigation of the contingency table. You can think of a significant chi-square test in much the same way as a significant interaction in a linear model: it is an effect that needs to be broken down further. We have already looked at the z-tests in the contingency table, but we can also use the standardized residual.

Like in any linear model, the residual is the error between what the model predicts (the expected frequency) and the observed data (the observed frequency):

$$residual_{ij} = observed_{ij} - model_{ij} \qquad (19.34)$$

in which i and j represent the two variables (i.e., the rows and columns in the contingency table). This residual is the same conceptually as every other residual or deviation in this book (compare this equation to, for example, equation (2.11)). To standardize this equation, we divide by the square root of the expected frequency:

$$standardized\ residual = \frac{observed_{ij} - model_{ij}}{\sqrt{model_{ij}}} \qquad (19.35)$$

Does this equation look familiar? Well, it's basically part of equation (19.2). The only difference is that rather than looking at squared deviations, we're looking at the pure deviations. Remember that the rationale for squaring deviations is to make them positive so that they don't cancel out when we add them to get the chi-square statistic. If we're not planning to add up the deviations/residuals then we don't need to square them; in fact the direction of the value (plus or minus) is useful information about whether the model over- or underestimates. There are two important things about these standardized residuals:

1 Given that the chi-square statistic is the sum of these standardized residuals (sort of), if we want to decompose what contributes to the overall association that the chi-square statistic measures, then looking at the individual standardized residuals is a good idea because they have a direct relationship with the test statistic.

2 These standardized residuals behave like any other (see Section 9.3): each one is

a z-score. This is very useful because by looking at a standardized residual we can assess its significance (see Section 1.8.6): if the value lies outside of ± 1.96 then it is significant at $p < 0.05$, if it lies outside ± 2.58 then it is significant at $p < 0.01$, and if it lies outside ± 3.29 then it is significant at $p < 0.001$.

Because we selected in Figure 19.5 the standardized residuals are in Output 19.2. There are four residuals: one for each combination of the type of training and whether the cats danced. When food was used as a reward the standardized residual was significant[4] for both those that danced ($z = 3.6$) and those that didn't dance ($z = -2.8$). The plus or minus sign tells us something about the direction of the effect, as do the counts and expected counts within the cells. We can interpret these standardized residuals as follows: when food was used as a reward significantly more cats than expected danced, and significantly fewer cats than expected did not dance. When affection was used as a reward the standardized residual was not significant[5] for both those that danced ($z = -1.7$) and those that didn't dance ($z = 1.4$). This tells us that when affection was used a reward as many cats as expected danced and did not dance. In a nutshell, the cells for when food was used as a reward both significantly contribute to the overall chi-square statistic: the association between the type of reward and dancing is mainly driven by when food is a reward.

19.8.2 Summary

The highly significant result indicates that there is an association between the type of training and whether the cat

How do I interpret a chi-square test?

---

4 Because both values are larger than 1.96 (when you ignore the minus sign).

5 Because both values are smaller than 1.96 (when you ignore the minus sign).

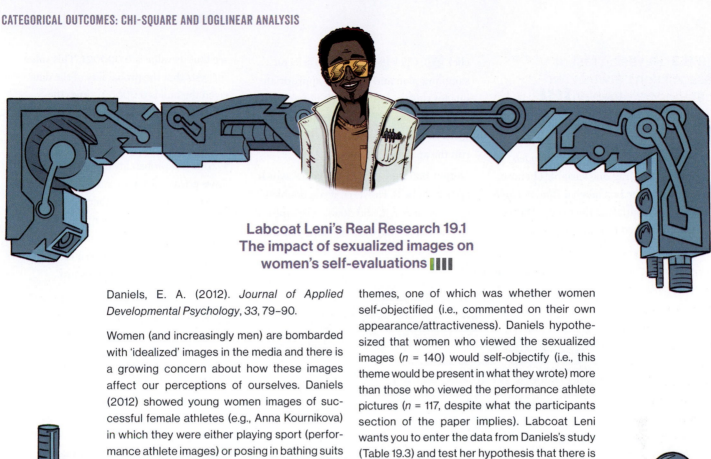

Labcoat Leni's Real Research 19.1
The impact of sexualized images on
women's self-evaluations ||||

Daniels, E. A. (2012). *Journal of Applied Developmental Psychology, 33*, 79–90.

Women (and increasingly men) are bombarded with 'idealized' images in the media and there is a growing concern about how these images affect our perceptions of ourselves. Daniels (2012) showed young women images of successful female athletes (e.g., Anna Kournikova) in which they were either playing sport (performance athlete images) or posing in bathing suits (sexualized images). Participants completed a short writing exercise after viewing these images. Each participant saw only one type of image, but several examples. Daniels then coded these written exercises and identified themes, one of which was whether women self-objectified (i.e., commented on their own appearance/attractiveness). Daniels hypothesized that women who viewed the sexualized images (*n* = 140) would self-objectify (i.e., this theme would be present in what they wrote) more than those who viewed the performance athlete pictures (*n* = 117, despite what the participants section of the paper implies). Labcoat Leni wants you to enter the data from Daniels's study (Table 19.3) and test her hypothesis that there is a significant association between the type of image viewed, and whether or not the women self-objectified (**Daniels (2012).sav**). The answers are on the companion website or on page 85 of Daniels's paper.

danced or not. In other words, the pattern of responses (i.e., the proportion of cats that danced compared to the proportion that did not) in the two training conditions is significantly different. We saw from the earlier *z*-tests that of the cats trained with food a significantly greater proportion danced, and conversely of those trained with affection a significantly greater proportion didn't dance. From the standardized residuals we know that when food is used as a reward, more cats danced (and fewer did not dance)

Table 19.3 Some data from Daniels (2012)

	Theme Present	Theme Absent	Total
Performance athletes	20	97	117
Sexualized athletes	56	84	140

than expected: about 74% of cats learn to dance and 26% do not. When affection is used, the opposite is true (about 70% refuse to dance and 30% do dance), which is consistent with expected frequencies. We can conclude that the type of training used

significantly influences the cats: they will dance for food but not for love. Having lived with a lovely cat for many years, this supports my cynical view that they will do nothing unless there is a bowl of cat-food waiting for them at the end of it!

627

19.8.3 Bayesian test of association between two categorical variables

To compute a Bayes factor for a contingency table select *Analyze* ▶ *Bayesian Statistics* ▶ *Loglinear Models* (Figure 19.6). Drag one variable (I chose **Training**) to the box labeled *Row variable* (or click ➡), and drag the other (**Dance**) to the box labeled *Column variable* (or

click ➡). Click Bayes Factor... to select a sampling plan and the most appropriate one for this example is *Multinomial Model* with ⦿ Row Sum as *Fixed Margins* (SPSS Tip 19.1). In the main dialog box click OK to run the analysis.

Output 19.5 shows the Bayes factor, which appears to be 0. However, if you double-click the table to edit it, and double-click the cell containing the Bayes factor value you will

see that its value is 0.000021. This value suggests that the probability of the data given the null is 0.000021 times the probability of the data given the alternative hypothesis. We can flip the interpretation by dividing 1 by this value: 1/0.000021 = 47,619. The probability of the data is 47,000 times greater given the alternative hypothesis than given the null. In other words, we should strongly change our prior beliefs towards the alternative hypothesis: there is extremely strong evidence that dancing is associated with the type of reward.

19.8.4 Calculating an effect size

Cramér's *V* is an adequate effect size (it is constrained to fall between 0 and 1 and is, therefore, easily interpretable), but a more common and useful measure of effect size for categorical data is the odds ratio. Odds ratios are most interpretable in 2×2 contingency tables and are not as useful for larger contingency tables. I've said many times already that effect sizes are most useful to summarize a focussed comparison, and a 2×2 contingency table is the categorical data equivalent of a focussed comparison.

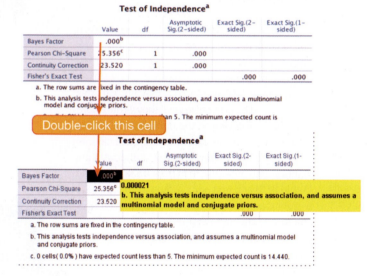

Output 19.5

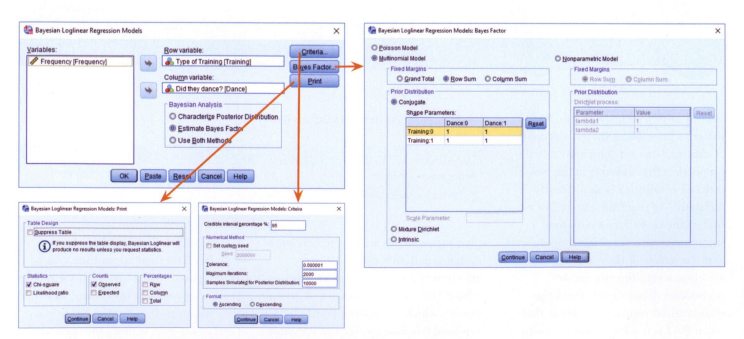

Figure 19.6   Dialog box for a Bayesian test of independence of two categorical variables

SPSS Tip 19.1
Sampling plans for Bayes factors ▮▮▮▮

The *BayesFactor* package allows you to select one of four sampling plans. I'll look at them with respect to our cat example.

Poisson Model: This option assumes that observations occur as a Poisson process in which the total sample size is not fixed. In effect, we assume that the four cells of the contingency table are independent Poisson random variables. It's like assuming the four cells represent different groups of cats who materialized randomly in our study. This model would be unrealistic because the researcher determined which cats had affection and which had reward – these were not pre-existing groups.

Multinomial Model (Grand Total): This option assumes that total *N* is fixed, and observations are assigned to cells with fixed probability. This is more

realistic than the Poisson sampling plan because our researcher chose to study 200 cats (so the total *N* is fixed). However, it is not realistic to assume that cells are assigned with fixed probability because the researcher determined gave more cats food than affection (which will affect the expected frequencies).

Multinomial Model (Row Sum or Column Sum): This option assumes that either the row or column totals are fixed. This sampling plan is the most realistic for our data because the experimenter determined how many cats were given affection and how many were given food (i.e., the row totals are fixed) and then observed how many danced or not (the column totals are not fixed). Therefore, we should set the sampling plan to be independent multinomial with fixed rows (because we specified **Training** as the rows).

The odds ratio is simple enough to calculate. Using our example, we'd first calculate the odds that a cat danced given that they had food as a reward, which is the number of cats that were given food and danced, divided by the number of cats given food that didn't dance:

$$odds_{dancing\,after\,food} = \frac{number\,that\,had\,food\,and\,danced}{number\,that\,had\,food\,but\,didn't\,dance}$$

$$= \frac{28}{10} = 2.8 \qquad (19.36)$$

Next we calculate the odds that a cat danced given that they had affection as a reward, which is the number of cats that were given

affection and danced, divided by the number of cats given affection that didn't dance:

$$odds_{dancing\,after\,affection} = \frac{number\,that\,had\,affection\,and\,danced}{number\,that\,had\,affection\,but\,didn't\,dance}$$

$$= \frac{48}{114} = 0.421 \qquad (19.37)$$

The odds ratio is the odds of dancing after food divided by the odds of dancing after affection:

$$odds\,ratio = \frac{odds_{dancing\,after\,food}}{odds_{dancing\,after\,affection}}$$

$$= \frac{2.8}{0.421} = 6.65 \qquad (19.38)$$

It tells us that if a cat was trained with food the odds of their dancing were 6.65 times higher than if they had been trained with affection. The odds ratio is an extremely elegant and easily understood metric for expressing this type of effect.

19.8.5 Reporting the results of a chi-square test ▮▮▮▮

When reporting Pearson's chi-square we report the value of the test statistic with its associated degrees of freedom and the significance value. The test statistic, as we've seen, is denoted by χ^2. The output

Cramming Sam's Tips
Associations between two categorical variables

- To test the relationship between two categorical variables use *Pearson's chi-square test* or the *likelihood ratio statistic*.
- Look at the table labeled *Chi-Square Tests*; if the *Exact Sig.* value is less than 0.05 for the row labeled *Pearson Chi-Square* then there is a significant relationship between your two variables.
- Check underneath this table to make sure that no expected frequencies are less than 5.
- Look at the contingency table to work out what the relationship between the variables is: look out for significant standardized residuals (values outside of ±1.96), and columns that have different letters as subscripts (this indicates a significant difference).

- Calculate the *odds ratio*.
- The Bayes factor reported by SPSS Statistics tells you the probability of the data under the null hypothesis relative to the alternative. Divide 1 by this value to see the probability of the data under the alternative hypothesis relative to the null. Values greater than 1 indicate that your belief should change towards the alternative hypothesis, with values greater than 3 starting to indicate a change in beliefs that has substance.
- Report the χ^2 statistic, the degrees of freedom, the significance value and odds ratio. Also report the contingency table.

tells us that the value of χ^2 was 25.36, that the degrees of freedom on which this was based were 1, and that it was significant at $p < 0.001$ (too small to report the exact p-value). It's also useful to reproduce the contingency table and my vote would go to quoting the odds ratio and Bayes factor too. As such, we could report:

✓ There was a significant association between the type of training and whether cats would dance $\chi^2(1) = 25.36$, $p < 0.001$. The Bayes factor strongly supported the alternative hypothesis, $BF_{10} = 47619$. The odds ratio showed that the odds of cats dancing were 6.65 times higher if they were trained with food than if trained with affection.

19.9 Loglinear analysis using SPSS Statistics ▮▮▮▮

19.9.1 Initial considerations ▮▮▮▮

Data are entered for loglinear analysis in the same way as for the chi-square test (see Sections 19.7.1 and 19.7.2). Let's extend the previous example to include dogs as well as cats. The data are in the file **Cats and Dogs.sav**; open this file. There are three variables (**Animal**, **Training** and **Dance**) that each have codes to represent the different categories within these

variables. The process for fitting the model is outlined in Figure 19.2—it's the same as for the chi-square test. First, let's check the expected frequencies in the contingency table (Section 19.5.2) using the *crosstabs* command.

The contingency table (Output 19.6) contains the number of cases that fall into each combination of categories. The top half of this table is the same as Output 19.2 because the data are the same (we've just added some dogs), so look back in this chapter for a summary of what this tells

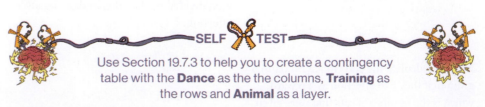

SELF TEST

Use Section 19.7.3 to help you to create a contingency table with the **Dance** as the the columns, **Training** as the rows and **Animal** as a layer.

Labcoat Leni's Real Research 19.2
Is the black American happy? ▌▌▌▌

Beckham, A. S. (1929). *Journal of Abnormal and Social Psychology, 24*, 186–190.

During my psychology degree I spent a lot of time reading about the civil rights movement in the USA. Instead of reading psychology, I read about Malcolm X and Martin Luther King Jr. For this reason I find Beckham's 1929 study of black Americans a fascinating historical piece of research. Beckham was a black American who founded the psychology laboratory at Howard University, Washington, DC, and his wife Ruth was the first black woman ever to be awarded a PhD (also in psychology) at the University of Minnesota. To put some context on the study, it was published 36 years before the Jim Crow laws were finally overthrown by the Civil Rights Act of 1964, and in a time when black Americans were segregated, openly discriminated against and victims of the most abominable violations of civil liberties and human rights (I recommend James Baldwin's superb *The Fire Next Time* for an insight into the times). The language of the study and the data from it are an uncom-

fortable reminder of the era in which it was conducted.

Beckham sought to measure the psychological state of 3443 black Americans with three questions. He asked them to answer yes or no to whether they thought black Americans were happy, whether they personally were happy as a black American, and whether black Americans *should* be happy. Beckham did no formal statistical analysis of his data (Fisher's article containing the popularized version of the chi-square test was published only 7 years earlier in a statistics journal that would not have been read by psychologists). I love this study, because it demonstrates that you do not need elaborate methods to answer important and far-reaching questions; with just three questions, Beckham told the world an enormous amount about very real and important psychological and sociological phenomena.

The frequency data (number of yes and no responses within each employment category) from this study are in the file **Beckham(1929). sav**. Labcoat Leni wants you to carry out three chi-square tests (one for each question that was asked). What conclusions can you draw?

us. For the dogs, 49 danced (70% of the total) and of these 20 were trained using food (40.8% of the total that danced) and 29 were trained with affection (59.2% of the total that danced). Twenty-one dogs didn't dance at all (30% of the total) and of those that didn't dance, 14 were trained using food as a reward (66.7% of the total

that didn't dance) and 7 were trained using affection (33.3% of the total that didn't dance). In summary, dogs (70%) seem more willing to dance than cats (38%), and they're not too worried what training method is used: about half of those that danced were trained with affection and about half with food.

For loglinear analysis there should be no expected counts less than 1, and no more than 20% less than 5 (Section 19.5.2). The smallest expected count in the contingency table is 10.2 (for dogs that were trained with food but didn't dance), which exceeds 5, and so the assumption has been met.

Type of Training * Did they dance? * Animal Crosstabulation

Animal					Did they dance? No	Yes	Total
Cat	Type of Training	Food as Reward		Count	10a	28b	38
				Expected Count	23.6	14.4	38.0
				% within Type of Training	26.3%	73.7%	100.0%
				% within Did they dance?	8.1%	36.8%	19.0%
				% of Total	5.0%	14.0%	19.0%
				Standardized Residual	-2.8	3.6	
		Affection as Reward		Count	114a	48b	162
				Expected Count	100.4	61.6	162.0
				% within Type of Training	70.4%	29.6%	100.0%
				% within Did they dance?	91.9%	63.2%	81.0%
				% of Total	57.0%	24.0%	81.0%
				Standardized Residual	1.4	-1.7	
	Total			Count	124	76	200
				Expected Count	124.0	76.0	200.0
				% within Type of Training	62.0%	38.0%	100.0%
				% within Did they dance?	100.0%	100.0%	100.0%
				% of Total	62.0%	38.0%	100.0%
Dog	Type of Training	Food as Reward		Count	14a	20b	34
				Expected Count	10.2	23.8	34.0
				% within Type of Training	41.2%	58.8%	100.0%
				% within Did they dance?	66.7%	40.8%	48.6%
				% of Total	20.0%	28.6%	48.6%
				Standardized Residual	1.2	-.8	
		Affection as Reward		Count	7a	29b	36
				Expected Count	10.8	25.2	36.0
				% within Type of Training	19.4%	80.6%	100.0%
				% within Did they dance?	33.3%	59.2%	51.4%
				% of Total	10.0%	41.4%	51.4%
				Standardized Residual	-1.2	.8	
	Total			Count	21	49	70
				Expected Count	21.0	49.0	70.0
				% within Type of Training	30.0%	70.0%	100.0%
				% within Did they dance?	100.0%	100.0%	100.0%
				% of Total	30.0%	70.0%	100.0%

Each subscript letter denotes a subset of Did they dance? categories whose column proportions do not differ significantly from each other at the .05 level.

Output 19.6

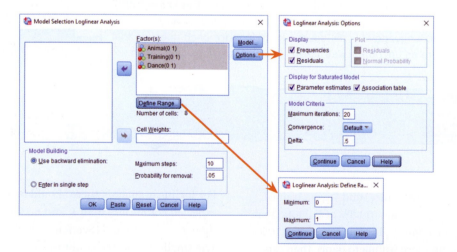

Figure 19.7   Main dialog box for loglinear analysis

19.9.2 The main analysis ▮▮▮▮

Having established that the assumptions have been met we select *Analyze* ▶ *Loglinear* ▶ *Model Selection…* to access the dialog box in Figure 19.7. Drag any variables that you want to include in the analysis to the box labeled *Factor(s)* (or click ⬆).[6] We have to specify the codes that we've used to define our categorical variables by selecting one or more variables in the *Factor(s)* box and clicking `Define Range…` to activate a dialog box in which you type the value of the minimum and maximum code that you used for the selected variables. Because all three variables in this example have the same codes (they all have two categories and were coded with 0 and 1) we can select all three, click `Define Range…`, type '0' in the *Minimum* box and '1' in the *Maximum* box, then click `Continue`.

The default options in the main box are fine. The default method is backward elimination (as I've described elsewhere). Alternatively, you can select *Enter in a single step*, which is a non-hierarchical method (in which all effects are entered and evaluated, like forced entry in the linear model). In loglinear analysis the combined effects take precedence over lower-order effects and so there is little to recommend non-hierarchical methods. Click `Model…` to open a dialog box like those we saw in ANCOVA (e.g., Figure 13.9). By default the saturated model is fitted and unless you have a very good reason for not fitting it, leave well alone. Click `Options…` in the main dialog box to open the dialog box in Figure 19.7. The default options are fine but you can select ☑ *Parameter estimates* to produce a table of parameter estimates for each effect (a *z*-score and associated confidence interval), and ☑ *Association table* to produce chi-square statistics for all of the effects in the

---

6 Remember that you can select several at the same time by holding down the *Ctrl* key or *Cmd* on a Mac.

model. This table can be useful in some situations, but as I've said before, if the higher-order interactions are significant then we shouldn't really be interested in the lower-order effects because they're confounded with the higher-order effects. Click **Continue** to return to the main dialog box and click **OK** to run the analysis.

19.10 Interpreting loglinear analysis ▮▮▮▮

Output 19.7 contains three tables. The first tells us that we have 270 cases (remember that we had 200 cats and 70 dogs and this is a useful check that no cats or dogs have been lost—they do tend to wander off). To begin the saturated model is fitted (all terms are in the model, including the highest-order interaction, in this case the **Animal** × **Training** × **Dance** interaction). The second table gives us the observed and expected counts for each of the combinations of categories in our model. These values should be the same as the original contingency table, except that each cell has 0.5 added to it (this value is the default and is fine, but if you want to change it you can do so by changing *Delta* in Figure 19.7). The final table contains two goodness-of-fit statistics: Pearson's chi-square and the likelihood ratio. These statistics are testing the hypothesis that the frequencies predicted by the model (the expected frequencies) are significantly different from the observed frequencies in the data. If our model is a good fit of the data then the observed and expected frequencies should be very similar (i.e., not significantly different). A significant result means that our model predictions are significantly different from our data (i.e., the model is a poor fit). In large samples these statistics should give the same results but the likelihood ratio statistic is preferred in small samples. Both statistics are 0 and yield a probability

Data Information

		N
Cases	Valid	270
	Out of Range[a]	0
	Missing	0
	Weighted Valid	270
Categories	Animal	2
	Type of Training	2
	Did they dance?	2

a. Cases rejected because of out of range factor values.

Cell Counts and Residuals

Animal	Type of Training	Did they dance?	Observed Count[a]	Observed %	Expected Count	Expected %	Residuals	Std. Residuals
Cat	Food as Reward	No	10.500	3.9%	10.500	3.9%	.000	.000
		Yes	28.500	10.6%	28.500	10.6%	.000	.000
	Affection as Reward	No	114.500	42.4%	114.500	42.4%	.000	.000
		Yes	48.500	18.0%	48.500	18.0%	.000	.000
Dog	Food as Reward	No	14.500	5.4%	14.500	5.4%	.000	.000
		Yes	20.500	7.6%	20.500	7.6%	.000	.000
	Affection as Reward	No	7.500	2.8%	7.500	2.8%	.000	.000
		Yes	29.500	10.9%	29.500	10.9%	.000	.000

a. For saturated models, .500 has been added to all observed cells.

Goodness-of-Fit Tests

	Chi-Square	df	Sig.
Likelihood Ratio	.000	0	.
Pearson	.000	0	.

Output 19.7

K-Way and Higher-Order Effects

	K	df	Likelihood Ratio Chi-Square	Likelihood Ratio Sig.	Pearson Chi-Square	Pearson Sig.	Number of Iterations
K-way and Higher Order Effects[a]	1	7	200.163	.000	253.556	.000	0
	2	4	72.267	.000	67.174	.000	2
	3	1	20.305	.000	20.778	.000	4
K-way Effects[b]	1	3	127.896	.000	186.382	.000	0
	2	3	51.962	.000	46.396	.000	0
	3	1	20.305	.000	20.778	.000	0

a. Tests that k-way and higher order effects are zero.

b. Tests that k-way effects are zero.

Output 19.8

value, *p*, of '.', which is a rather confusing way of saying that the probability cannot be computed. The reason why is that at this stage the model predicts the data *perfectly* (I explained why in Section 19.4). The next question is what bits of the model we can remove without significantly affecting the fit.

Output 19.8 tells us about the effects of removing parts of the model. The part of the table labeled *K-Way and Higher-Order Effects* has rows showing likelihood ratio

and Pearson chi-square statistics when *K* = 1, 2 and 3 (as we go down the rows of the table). The first row (*K* = 1) tells us whether removing the one-way effects (i.e., the main effects of **Animal**, **Training** and **Dance**) and any higher-order effects will significantly affect the fit of the model. There are lots of higher-order effects here—there are the two-way interactions and the three-way interaction—and so this is basically testing whether if we remove everything from the model there will be a

Partial Associations

Effect	df	Partial Chi–Square	Sig.	Number of Iterations
Animal*Training	1	13.760	.000	2
Animal*Dance	1	13.748	.000	2
Training*Dance	1	8.611	.003	2
Animal	1	65.268	.000	2
Training	1	61.145	.000	2
Dance	1	1.483	.223	2

Output 19.9

Parameter Estimates

Effect	Parameter	Estimate	Std. Error	Z	Sig.	95% Confidence Interval Lower Bound	95% Confidence Interval Upper Bound
Animal*Training*Dance	1	−.360	.083	−4.320	.000	−.523	−.197
Animal*Training	1	−.402	.083	−4.823	.000	−.565	−.239
Animal*Dance	1	.197	.083	2.364	.018	.034	.360
Training*Dance	1	−.104	.083	−1.251	.211	−.268	.059
Animal	1	.404	.083	4.843	.000	.240	.567
Training	1	−.328	.083	−3.937	.000	−.492	−.165
Dance	1	−.232	.083	−2.782	.005	−.395	−.069

Output 19.10

significant effect on the fit of the model. This effect is highly significant. If this test was non-significant (if the values of *Sig.* were above 0.05) then this would tell you that removing everything from your model would not affect the fit (overall the combined effect of your variables and interactions is not significant). The next row (*K* = 2) tells us whether removing the two-way interactions (i.e., the **Animal × Training, Animal × Dance** and **Training × Dance** interactions) and any higher-order effects (i.e. the three-way interaction) will affect the model. This test is also highly significant, indicating that if we removed the two-way interactions and the three-way interaction then this would have a significant detrimental effect on the model. The final row (*K* = 3) tests whether removing the three-way effect *and* higher-order effects will significantly affect the fit of the model. The three-way interaction *is* the highest order effect that we have, so this test evaluates removal of three-way interaction (i.e., the **Animal ×Training ×**

Dance interaction). Both chi-square and likelihood ratio tests agree that removing this interaction will significantly affect the fit of the model (because the probability value is less than 0.05).

The bottom of the table (*K-way Effects*) expresses the same thing but without including the higher-order effects. The first row (*K* = 1), tests whether removing the main effects (the one-way effects of **Animal**, **Training** and **Dance**) has a significant detrimental effect on the model, and it does (because the *p*-value is less than 0.05). The second row (*K* = 2) tests whether removing the two-way interactions (**Animal × Training, Animal × Dance** and **Training × Dance**) has a significant detrimental effect on the model, and again it does ($p < 0.001$). This finding tells us that one or more of these two-way interactions is a significant predictor. The final row (*K* = 3) tests whether removing the three-way interaction (**Animal × Training × Dance**) has a detrimental effect on the

model. It does ($p < 0.001$), suggesting that this interaction is a significant predictor of the data. The results in this row are identical to the final row of the top half of the table (the *K-way and Higher-Order Effects*) because it is the highest-order effect and so in the top part of the table there were no higher-order effects to include.

In a nutshell, Output 19.8 tells us that the three-way interaction is significant: removing it from the model has a significant effect on how well the model fits the data. We also know that removing all two-way interactions has a significant effect on the model, but remember that loglinear analysis should be done hierarchically and so these two-way interactions aren't of interest to us because the three-way interaction is significant (we'd look only at these effects if the three-way interaction were non-significant).

Output 19.9 shows the *Partial Associations* table (which you might have selected by ticking ☑ Association table in Figure 13.7). This table breaks down the model into specific components. For example, the previous output told us that removing all of the two-way interactions significantly affects the model fit, but we don't know which of the two-way interactions specifically make a difference; this table tells us. The Pearson chi-square tests are significant for all three interactions (*Sig.* is less than 0.05). Likewise, the previous output told us that removing the main effects significantly affected the fit of the model, and Output 19.9 shows specifically that the main effects of **Animal** and **Training** are both significant ($p < 0.001$), but the main effect of **Dance** is not ($p = 0.223$). We should ignore all of these effects, though, because they are confounded with the higher-order interaction of **Animal × Training × Dance**.

Output 19.10 shows the ☑ Parameter estimates (selected in Figure 19.7) of the model.

It tests each effect in the model with a z-score, and also gives us confidence intervals. If you ignore the plus or minus sign, the bigger the z, the more significant the effect is; therefore, the value of z gives us a useful comparison between effects. The main effect of **Animal** is the most important effect in the model ($z = 4.84$) followed by the **Animal** $\times$ **Training** interaction ($z = -4.82$) and then the **Animal** $\times$ **Training** $\times$ **Dance** interaction ($z = -4.32$) and so on. However, it's worth reiterating that in this case we don't need to concern ourselves with anything other than the three-way interaction.

Output 19.11 deals with the backward elimination. We begin with the highest-order effect (in this case the **Animal** $\times$ **Training** $\times$ **Dance** interaction); we remove it from the model, see what effect this has, and if it doesn't have a significant effect then we move on to the next highest effects (in this case the two-way interactions). However, we've already seen that removing the three-way interaction will have a significant effect, and this is confirmed at this stage by the table labeled *Step Summary*. Therefore, the analysis stops here: the three-way interaction is not removed and this final model is evaluated using the likelihood ratio statistic. We're looking for a non-significant test statistic, which indicates that the expected values generated by the model are not significantly different from the observed data (put another way, the model is a good fit to the data). In this case the model is a perfect fit to the data.[7]

Step Summary

Step[a]		Effects	Chi-Square[c]	df	Sig.	Number of Iterations
0	Generating Class[b]	Animal*Training*Dance	.000	0	.	
	Deleted Effect 1	Animal*Training*Dance	20.305	1	.000	4
1	Generating Class[b]	Animal*Training*Dance	.000	0	.	

a. At each step, the effect with the largest significance level for the Likelihood Ratio Change is deleted, provided the significance level is larger than .050.

b. Statistics are displayed for the best model at each step after step 0.

c. For 'Deleted Effect', this is the change in the Chi-Square after the effect is deleted from the model.

Cell Counts and Residuals

Animal	Type of Training	Did they dance?	Observed Count	Observed %	Expected Count	Expected %	Residuals	Std. Residuals
Cat	Food as Reward	No	10.000	3.7%	10.000	3.7%	.000	.000
		Yes	28.000	10.4%	28.000	10.4%	.000	.000
	Affection as Reward	No	114.000	42.2%	114.000	42.2%	.000	.000
		Yes	48.000	17.8%	48.000	17.8%	.000	.000
Dog	Food as Reward	No	14.000	5.2%	14.000	5.2%	.000	.000
		Yes	20.000	7.4%	20.000	7.4%	.000	.000
	Affection as Reward	No	7.000	2.6%	7.000	2.6%	.000	.000
		Yes	29.000	10.7%	29.000	10.7%	.000	.000

Goodness-of-Fit Tests

	Chi-Square	df	Sig.
Likelihood Ratio	.000	0	.
Pearson	.000	0	.

Output 19.11

> I don't need a loglinear analysis to tell me that cats are vastly superior to dogs

SELF TEST

Use the *split file* command (see Section 6.10.4) to run a chi-square test on **Dance** and **Training** for dogs and cats.

Chi-Square Tests

	Value	df	Asymptotic Significance (2-sided)	Exact Sig. (2-sided)	Exact Sig. (1-sided)
Pearson Chi-Square	3.932[a]	1	.047		
Continuity Correction[b]	2.966	1	.085		
Likelihood Ratio	3.984	1	.046		
Fisher's Exact Test				.068	.042
Linear-by-Linear Association	3.876	1	.049		
N of Valid Cases	70				

a. 0 cells (0.0%) have expected count less than 5. The minimum expected count is 10.20.

b. Computed only for a 2x2 table

Output 19.12

---

7 The fact that the analysis has stopped here is unhelpful because I can't show you how it would proceed in the event of a non-significant three-way interaction. However, it does keep things simple, and if you're interested in exploring loglinear analysis further, there is a task at the end of the chapter that shows you what happens when the highest-order interaction is not significant.

19.10.1 Following up loglinear analysis ▮▮▮

An alternative way to interpret a three-way interaction is to conduct chi-square analysis at different levels of one of your variables. For example, to interpret our **Animal** × **Training** × **Dance** interaction, we could perform a chi-square test on **Training** and **Dance**, but do this separately for dogs and cats (in fact the analysis for cats will be the same as the example we used for chi-square). You can then compare the results in the different animals.

The results and interpretation for cats are in Output 19.3 and for dogs in Output 19.12. For dogs there is still a significant relationship between the types of training and whether they danced, but it is weaker (the chi-square is 3.93, compared to 25.2 for cats).[8] This finding seems to suggest that dogs are more likely to dance if given affection than if given food, the opposite of cats.

19.10.2 Interpreting the interaction ▮▮▮

Let's piece this together to make sense of the three-way interaction. Let's plot the frequencies across all the different categories. Figure 19.8 plots the percentage of the total (these values can be found in Output 19.6 in the rows labeled *% of total*). Look at food first: the pattern for dogs and cats is almost identical in that the percentage of yes responses is slightly higher than the percentage of no responses for both animals (the blue dot is a similar distance above the orange one). Compare this with affection, where for cats the percentage of no responses was much higher than yes responses (the orange circle is a long way above the blue), but for dogs the opposite is true—the percentage of yes responses was much higher than no responses (the blue circle is a long way above the orange). So cats are sensible

creatures that do stupid stuff only when there's something in it for them (i.e., food), whereas dogs are just daft. ☺

19.10.3 Effect sizes in loglinear analysis ▮▮▮

As with Pearson's chi-square, let's look at odds ratios again. Odds ratios are easiest to understand for 2 × 2 contingency tables and so if you have significant higher-order interactions or your variables have more than two categories, it is worth trying to break these effects down into logical 2 × 2 tables and calculating odds ratios that reflect the nature of the interaction. In this example we could calculate odds ratios for dogs and cats separately. We have the odds ratio for cats already (Section 19.8.3), and for dogs we would get 0.35:

$$\text{odds}_{\text{dancing after food}} = \frac{\text{number that had food and danced}}{\text{number that had food but didn't dance}}$$

$$= \frac{20}{14} = 1.43 \qquad (19.39)$$

$$\text{odds}_{\text{dancing after affection}} = \frac{\text{number that had affection and danced}}{\text{number that had affection but didn't dance}}$$

$$= \frac{29}{7} = 4.14 \qquad (19.40)$$

$$\text{odds ratio} = \frac{\text{odds}_{\text{dancing after food}}}{\text{odds}_{\text{dancing after affection}}}$$

$$= \frac{1.43}{4.14} = 0.35 \qquad (19.41)$$

This tells us that if a dog was trained with food the odds of their dancing were 0.35 times the odds if they were rewarded with affection (i.e., they were less likely to dance). Another way to say this is that the odds of their dancing were 1/0.35 = 2.90 times lower if they were trained with food instead of affection. Compare this to cats where the odds of dancing were 6.65 higher if they were trained with food rather than affection. As you can see, comparing the odds ratios for dogs and cats is an

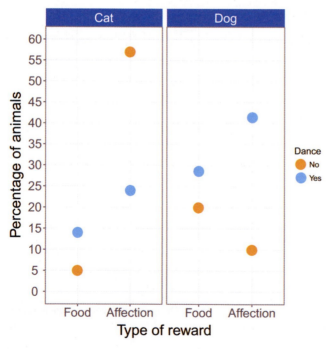

Figure 19.8 Percentage of different animals who danced or not after being trained with affection or food

---

8 The chi-square statistic depends on the sample size, so really you need to calculate effect sizes and compare them to make this kind of statement (unless you had equal numbers of dogs and cats).

elegant way to present the three-way interaction term in the model.

19.11 Reporting the results of loglinear analysis

For loglinear analysis report the likelihood ratio statistic for the final model, usually denoted just by χ^2. For any terms that are significant you should report the chi-square change or you could consider reporting the z-score for the effect and its associated confidence interval. If you break down any higher-order interactions in subsequent analyses then you need to report the relevant chi-square statistics (and odds ratios). For this example we could report:

✓ The three-way loglinear analysis produced a final model that retained all effects. The likelihood ratio of this model was $\chi^2(0) = 0$, $p = 1$. This indicated that the highest-order interaction (the animal × training × dance interaction) was significant, $\chi^2(1) = 20.31$, $p < 0.001$. To break down this effect, separate chi-square tests on the training and dance variables were performed separately for dogs and cats. For cats, there was a significant association between the type of training and whether they would dance, $\chi^2(1) = 25.36$, $p < 0.001$; this was true in dogs also, $\chi^2(1) = 3.93$, $p = 0.047$. Odds ratios indicated that the odds of dancing were 6.65 higher after food than affection in cats, but only 0.35 in dogs (i.e., in dogs, the odds of dancing were 2.90 times lower if trained with food compared to affection). The analysis seems to reveal a fundamental difference between dogs and cats: cats are more likely to dance for food than affection, whereas dogs are more likely to dance for affection than food.

19.12 Brian and Jane's Story

Jane looked at the discolored brass of the jar in front of her. It had a timeless class. She brushed dust from the glass and stared at the brain within. What to do? She'd decided to stop her crazy experiment weeks ago. It had been easy at first: the time she'd spent with Brian made her feel more human than she ever had; she hadn't given the corridors of the Pleiades building a thought. But like an addict, she returned there as soon as life got tough. Maybe she'd been working too

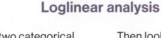

Cramming Sam's Tips
Loglinear analysis

- Test the relationship between more than two categorical variables with *loglinear analysis*.
- Loglinear analysis is hierarchical: the initial model contains all main effects and interactions. Starting with the highest-order interaction, terms are removed to see whether their removal significantly affects the fit of the model. If it does then this term is not removed and all lower-order effects are ignored.
- Look at the table labeled *K-Way and Higher-Order Effects* to see which effects have been retained in the final model.

- Then look at the table labeled *Partial Associations* to see the individual significance of the retained effects (look at the column labeled *Sig.*—values less than 0.05 indicate significance).
- Look at the *Goodness-of-Fit Tests* for the final model: if this model is a good fit of the data then this statistic should be non-significant (*Sig.* should be bigger than 0.05).
- Look at the contingency table to interpret any significant effects (percentage of total for cells is the best thing to look at).

hard. Weeks of little sleep and long stretches of work were taking their toll on her resolve. She'd hit a wall with what she was working on, and no matter how much she chiselled away at it with her mind, it was defeating her. She didn't like defeat. Like a ghost she had been floating in and out of Brian's life. When she ran out of energy to work she would go to him and he would take care of her before she disappeared again. He rolled with it. He didn't ask questions, but he listened and supported. It was so important to her to find a way, any way, to break the impasse and finish what she'd started. 'Just one more,' she thought, 'one more to give my mind the kick it needs.'

She looked deeper into the jar, trying to find a way into the soul of whoever used to possess this mind. She felt a tingle of excitement at her own depravity, and then disgust as she thought of Brian, waiting

for her somewhere, oblivious to her current dilemma, and her past.

'Just one more,' she thought.

She had no choice, did she? She had to finish her work. This was the only way.

'Just one more.'

She thought of Brian. Her hands trembled. Her chest felt tight as she tried to suppress her emotions.

'Just one more,' she whispered, 'one more . . . one more . . . one more . . . NO MORE!'

Jane punched the lever that retuned the jar to its cubby hole and ran. She didn't stop until she reached Brian's.

19.13 What next? ▮▮▮▮

When I wrote the first edition of this book I had no idea what journey I was starting. My main ambition was to write a statistics book that I would enjoy

reading. I didn't expect anyone else to enjoy reading it, but it turns out some people do, which is a nice feeling. One of the weird side effects of writing statistics books is that everyone assumes that I'm a statistician. I'm not, which means that I constantly disappoint people by not being able to answer their statistics questions. This book is pretty much my total knowledge about statistics. It's probably more than the sum. The multinomial logistic regression section in the next chapter, for example, was the result of reading a lot of stuff about multinomial logistic regression, which I have now forgotten (because I never used it). Should I ever need to do a multinomial logistic regression I will read the chapter in this book and be amazed at how good I am at pretending to know what I'm talking about. If you knew how often I look stuff up in my own book, you probably wouldn't have bought it.

Over the editions of this book I've had 'fans' set up appreciation societies, had videos on YouTube dedicated to me, had hundreds of people send me pictures of their dogs, cats, lizards, children, horses, birds posing with my book, been befriended by the manager of a black metal band, been invited to an autopsy (seriously ...), had people randomly turn up at my office from all over the world to say 'hi', had strangers ask for selfies with me at conferences, and all because I did the most uncool thing you can think of: wrote stats books. It bemuses me, but it is profoundly heart-warming to have strangers from all walks of life want to spend their valuable time saying hi to you. In its own tiny, unimportant (to most people) microcosm, it's a bit like being the rock star that the younger me always wanted to be, except writing about stats is nowhere near as much fun as playing music. It was only a matter of time before the musical itch needed to be scratched once again.

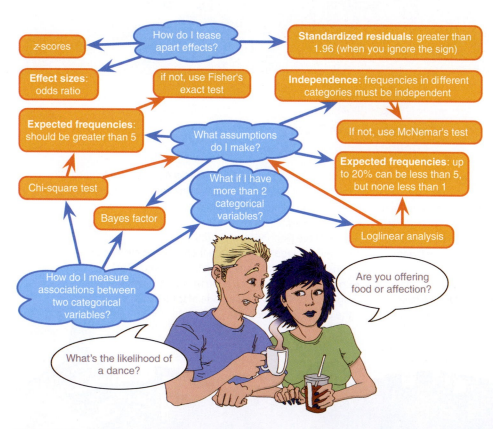

Figure 19.9 What Brian learnt from this chapter

19.14 Key terms that I've discovered

Chi-square distribution

Chi-square test

Cramér's V

Fisher's exact test

Goodman and Kruskal's λ

Loglinear analysis

Phi

Saturated model

Yates's continuity correction

Smart Alex's tasks

- **Task 1**: Research suggests that people who can switch off from work (**Detachment**) during off-hours are more satisfied with life and have fewer symptoms of psychological strain (Sonnentag, 2012). Factors at work, such as time pressure, affect your ability to detach when away from work. A study of 1709 employees measured their time pressure (**Time_Pressure**) at work (no time pressure, low, medium, high and very high time pressure). Data generated to approximate Figure 1 in Sonnentag (2012) are in the file **Sonnentag (2012).sav**. Carry out a chi-square test to see if time pressure is associated with the ability to detach from work. ▮▮▮

- **Task 2**: Labcoat Leni's Real Research 19.1 describes a study (Daniels, 2012) that looked at the impact of sexualized images of athletes compared to performance pictures on women's perceptions of the athletes and of themselves. Women looked at different types of pictures (**Picture**) and then did a writing task. Daniels identified whether certain themes were present or absent in each written piece (**Theme_Present**). We looked at the self-evaluation theme, but Daniels identified others: commenting on the athlete's body/appearance (**Athletes_Body**), indicating admiration or jealousy for the athlete (**Admiration**), indicating that the athlete was a role model or motivating (**Role_Model**), and their own physical activity (**Self_Physical_Activity**). Test whether the type of picture viewed was associated with commenting on the athlete's body/appearance (**Daniels (2012).sav**). ▮▮▮

- **Task 3**: Using the data in Task 2, see whether the type of picture viewed was associated with indicating admiration or jealousy for the athlete. ▮▮▮

- **Task 4**: Using the data in Task 2, see whether the type of picture viewed was associated with indicating that the athlete was a role model or motivating. ▮▮▮

- **Task 5**: Using the data in Task 2, see whether the type of picture viewed was associated with the participant commenting on their own physical activity. ▮▮▮

- **Task 6**: I wrote much of the third edition of this book in the Netherlands (I have a soft spot for it). The Dutch travel by bike much more than the English. I noticed that many more Dutch people cycle while steering with only one hand. I pointed this out to one of my friends, Birgit Mayer, and she said that I was a crazy English fool and that Dutch people did not cycle one-handed. Several weeks of me pointing at one-handed cyclists and her pointing at two-handed cyclists ensued. To put it to the test I counted the number of Dutch and English cyclists who ride with one or two hands on the handlebars (**Handlebars.sav**). Can you work out which one of us is correct? ▮▮▮

- **Task 7**: Compute and interpret the odds ratio for Task 6. ▮▮▮

- **Task 8**: Certain editors at SAGE like to think they're great at football (soccer). To see whether they are better than Sussex lecturers and postgraduates we invited employees of SAGE to join in our football matches. Every person played in one match. Over many matches, we counted the number of players that scored goals. Is there a significant relationship between scoring goals and whether you work for SAGE or Sussex? (**Sage Editors Can't Play Football.sav**). ▮▮▮

- **Task 9**: Compute and interpret the odds ratio for Task 8. ▮▮▮

- **Task 10**: I was interested in whether horoscopes are tosh. I recruited 2201 people, made a note of their star sign (this variable, obviously, has 12 categories: Capricorn, Aquarius, Pisces, Aries, Taurus, Gemini, Cancer, Leo, Virgo, Libra, Scorpio and Sagittarius) and whether they believed in horoscopes (this variable has two categories: believer or unbeliever). I sent them an identical horoscope about events in the next month, which read:

August is an exciting month for you. You will make friends with a tramp in the first week and cook him a cheese omelette. Curiosity is your greatest virtue, and in the second week, you'll discover knowledge of a subject that you previously thought was boring. Statistics perhaps. You might purchase a book around this time that guides you towards this knowledge. Your new wisdom leads to a change in career around the third week, when you ditch your current job and become an accountant. By the final week you find yourself free from the constraints of having friends, your boy/girlfriend has left you for a Russian ballet dancer with a glass eye, and you now spend your weekends doing loglinear analysis by hand with a pigeon called Hephzibah for company.

At the end of August I interviewed these people and I classified the horoscope as having come true or not, based on how closely their lives had matched the fictitious horoscope. Conduct a loglinear analysis to see whether there is a relationship between the person's star sign, whether they believe in horoscopes and whether the horoscope came true (**Horoscope.sav**). ▌▌▌▌

- **Task 11**: On my statistics module students have weekly SPSS classes in a computer laboratory. I've noticed that many students are studying Facebook more than the very interesting statistics assignments that I have set them. I wanted to see the impact that this behavior had on their exam performance. I collected data from all 260 students on my module. I classified their **Attendance** as being either more or less than 50% of their lab classes, and I classified them as someone who looked at **Facebook** during their lab class or someone who never did. After the exam, I noted whether they passed or failed (**Exam**). Do a loglinear analysis to see if there is an association between studying Facebook and failing your exam (**Facebook.sav**). ▌▌▌▌

Answers & additional resources are available on the book's website at
https://edge.sagepub.com/field5e

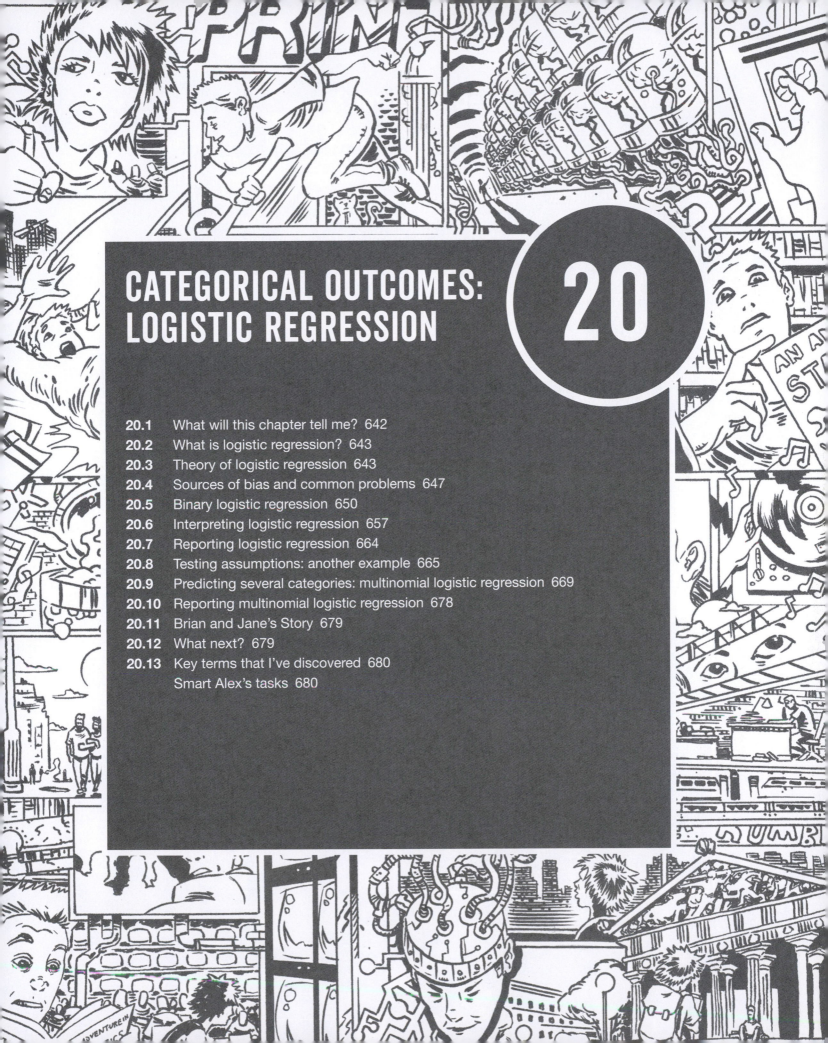

CATEGORICAL OUTCOMES: LOGISTIC REGRESSION

20

20.1 What will this chapter tell me?

Over the last few chapters we've seen how my childhood dreams of rock stardom crumbled as I became a statistical model in human form. I can scarcely imagine a more dramatic failure to achieve one's ambitions. I felt like I had become numbers without a soul, and I needed salvation before the transformation became complete. It was time to unlock the latent rock star once more, and get some therapy in the process. So, the hollow husk of a 29-year-old Andy decided to learn the drums (feel free to insert your own joke about it being the perfect instrument for a failed musician, but they're much harder to play than people think). A couple of years later I had a call from an old friend of mine, Doug. Back in my Scansion days, Doug had played in another local band, so we went back a long way. The conversation went a bit like this:

'Remember the last time I saw you we talked about you coming and having a jam with us?' asked Doug.

I had absolutely no recollection whatsoever of him saying this, so I responded 'Yes'.

'How about it then?' he said.

'OK,' I said, 'you arrange it and I'll bring my guitar'.

'No, you whelk,' he said, 'we need a drummer. Learn some of the songs on the CD I gave you last year.'

I'd played his band's CD and I liked it, but their songs were ridiculously fast and there was no way on earth that I could play them. 'Sure, no problem,' I lied.

I spent the next two weeks trying to become a much better drummer than I was. It'd be nice to report that at the rehearsal I astounded them with my brilliance, but I didn't. I did, however, nearly have a heart attack and herniate everything in my body. Still, we had another rehearsal, and then another, and we're still having them.[1] It's curious that I started off playing guitar (which I can still play, incidentally) and then took up drums, because there are always assumptions about the personalities of musicians in rock bands: the singers are egocentric, guitarists are the cool ones, bassists are laid back and introverted, and drummers are alleged to be wild hedonists, high on the Asperger's

Figure 20.1   Having a therapy session

---

1 Although not with Doug in the band as it turns out, and not very frequently these days because of, you know, having children, jobs, being old and boring. That sort of thing.

spectrum (enjoying counting *does* help) or both. There's definitely more Asperger's about me than hedonism. These assumptions are speculative, though. If we wanted to test what personality characteristics predict the instrument you choose to play, we'd have a categorical outcome (type of instrument) with several categories (drums, guitar, bass, singing, keyboard, tuba, etc.) and continuous predictors (neuroticism, extroversion, etc.). We've looked at how we can quantify associations between purely categorical variables, but if we have continuous predictors too then surely there's no model on earth that can handle that kind of complexity so we should just go to the pub and have a good time instead? Actually, we can do **logistic regression**—bugger!

20.2 What is logistic regression? ▮▮▮▮

In the previous chapter we looked models of the relationships between categorical variables, and we'll now extend this discussion to logistic regression—a model for predicting categorical outcomes from categorical and continuous predictors. In its simplest form, this means predicting which of two categories a person is likely to belong to, given their scores on predictors. For example, in medical research logistic regression is used to generate models from which predictions can be made about the likelihood that a tumor is cancerous or benign. Based on existing data, the logistic regression model is used to establish variables that predict the malignancy of a tumor. These variables can then be measured for a new patient and their values placed in the

logistic regression model to obtain a probability of malignancy. If the probability of the tumor being malignant is low then the doctor may decide not to carry out expensive and painful surgery that is likely to be unnecessary. When we are trying to predict membership of only two categories the model is known as **binary logistic regression**, but when we want to predict membership of more than two categories we use **multinomial** (or polychotomous) logistic regression.

20.3 Theory of logistic regression ▮▮▮▮

I won't dwell on the maths behind logistic regression (I am living proof that you don't need to know it). Instead I'll draw a few parallels to the linear model to give you the conceptual gist of what's going on. To keep things simple I'll explain binary logistic regression, but most of the principles extend to predicting membership of more than two categories. We've seen many times that the linear model can be expressed as:

$$Y_i = b_0 + b_1 X_{1i} + \varepsilon_i \qquad (20.1)$$

when there is one predictor variable and as:

$$Y_i = b_0 + b_1 X_{1i} + b_2 X_{2i} + \cdots + b_n X_{ni} + \varepsilon_i \qquad (20.2)$$

when there are two or more. Remember that b_0 is the value of the outcome when the predictors are zero (the intercept), the bs quantify the relationship between each predictor and outcome, X is the value of each predictor variable and ε is the error in prediction (the residual).

One of the assumptions of the linear model is that the relationship between the predictors and outcome is linear

Can I use a linear model with a categorical outcome?

(Section 9.4.1). When the outcome variable is categorical, this assumption is violated (Berry, 1993). One way around this problem is to transform the data using the logarithmic transformation (see Berry & Feldman, 1985), which is a way of expressing a non-linear relationship in a linear way. Logistic regression uses this transformation to express the linear model equation in logarithmic terms (called the *logit*). In doing so it allows us to predict categorical outcomes using the standard linear model that we've discussed throughout this book. Righteous.

In logistic regression, instead of predicting the value of a variable Y from a predictor variable X_1 or several predictor variables (Xs), we predict the *probability* of Y occurring, $P(Y)$, from known (log-transformed) values of X_1 (or Xs). Let's look at the equation—equations are always fun. Here is the logistic regression model with one predictor:

$$P(Y) = \frac{1}{1 + e^{-(b_0 + b_1 X_{1i})}} \qquad (20.3)$$

$P(Y)$ is the probability of Y occurring, e is the base of natural logarithms, and the linear model that you've seen countless times already (equation (20.1)) is cosily tucked up inside the parentheses. When there are several predictors the model becomes:

$$P(Y) = \frac{1}{1 + e^{-(b_0 + b_1 X_{1i} + b_2 X_{2i} + \cdots + b_n X_{ni})}} \qquad (20.4)$$

Note that all that changes is that instead of the parentheses containing the linear model for one predictor, they have invited some friends around for tea and the parentheses party now contains the full linear model (equation (20.2)). The model can be presented in other ways, but the version I have chosen expresses the outcome as the probability of Y occurring (i.e., the probability that a case belongs in a certain category). The resulting value from the model will, therefore, lie between 0 and 1. A value close to 0 means that Y is very unlikely to have occurred, and a value close to 1 means that Y is very likely to have occurred. Just like the linear model, each predictor variable has its own parameter (b), which is estimated from the sample data. In linear models these parameters are (typically) estimated using the method of least squares (Section 2.6), whereas in logistic regression **maximum-likelihood estimation** is used, which selects coefficients that make the observed values most likely to have occurred. Essentially, the chosen estimates of the bs will be ones that, when values of the predictor variables are placed in it, result in values of Y closest to the observed values.

20.3.1 Assessing the model: the log-likelihood statistic ▮▮▮

The logistic regression model predicts the probability of an event occurring for a given person (we denote this as $P(Y_i)$, the probability that Y occurs for the ith person), based on observations of whether the event *did* occur for that person (we could denote this as Y_i, the observed outcome for the ith person). For a given person, the *observed* Y will be either 0 (the outcome didn't occur) or 1 (the outcome did occur), but the *predicted* Y, $P(Y)$, will be a value between 0 (there is no chance that the outcome will occur) and 1 (the outcome will certainly occur). When we assessed the fit of the linear model we compared the observed and predicted values of the outcome (if you remember, we use R^2, which is the squared Pearson correlation between the observed values of the outcome and the values predicted by the model). We do much the same in logistic regression using the **log-likelihood**:

$$\text{log-likelihood} = \sum_{i=1}^{N} \left[Y_i \ln\left(P(Y_i)\right) + (1 - Y_i)\ln\left(1 - P(Y_i)\right) \right]$$

(20.5)

The log-likelihood is based on summing the probabilities associated with the predicted, $P(Y_i)$, and actual, Y_i, outcomes. It is analogous to the residual sum of squares in the sense that it is an indicator of how much unexplained information there is after the model has been fitted. It follows, therefore, that large values of the log-likelihood statistic indicate poorly fitting statistical models, because the larger the value of the log-likelihood, the more unexplained observations there are.

20.3.2 Assessing the model: the deviance statistic ▮▮▮

The **deviance** is closely related to the log-likelihood: it's given by

$$\text{deviance} = -2 \times \text{log-likelihood} \qquad (20.6)$$

The deviance is often referred to as **−2LL** because of the way it is calculated. It's rather convenient to (almost) always use the deviance rather than the log-likelihood because it has a chi-square distribution (see Chapter 19 and the Appendix), which makes it easy to calculate the significance of it.

One important use of the log-likelihood and deviance is to compare models. For example, it's useful to compare a logistic regression model against a baseline state—usually the model when only the intercept is included (i.e., no predictors). In the standard linear model, the baseline model we use is the mean of all scores (i.e., we predict the outcome from the intercept). With a categorical outcome it doesn't make sense to use the overall mean (because the outcome is whether an event happened or not), so instead we use the frequency with

which the outcome occurred. If the outcome occurs 107 times, and doesn't occur 72 times, then our best guess of the outcome will be that it occurs (because it occurs more times than it doesn't). As such, like the linear model, our baseline model is the model that gives us the best prediction when we know nothing other than the values of the outcome: in logistic regression this will be the outcome category that occurs most often, which is the same as predicting the outcome from the intercept. If we add one or more predictors to the model, we can compute the improvement of the model as:

$$\chi^2 = \left(-2LL_{\text{baseline}}\right) - \left(-2LL_{\text{new}}\right)$$
$$= 2LL_{\text{new}} - 2LL_{\text{baseline}} \qquad (20.7)$$
$$df = k_{\text{new}} - k_{\text{baseline}}$$

We merely take the new model deviance and subtract from it the deviance for the baseline model (the model when only the constant is included). This difference is known as a likelihood ratio[2] and has a chi-square distribution with degrees of freedom equal to the number of parameters, k, in the new model minus the number of parameters in the baseline model. The number of parameters in the baseline model will always be 1 (the constant is the only parameter); any subsequent model will have degrees of freedom equal to the number of predictors plus 1 (i.e., the number of predictors plus one parameter representing the constant).

If we build up models hierarchically (i.e., adding one predictor at a time) we can also use equation (20.7) to compare these models. For example, if you have a model (we'll call it the 'old' model) and you add a predictor (the 'new' model), you can see whether the new model has improved the fit using equation (20.7) in which the baseline model is the 'old'

---

2 You might wonder why it is called a 'ratio' when a 'ratio' usually means something is divided by something else, and we're not dividing anything here: we're subtracting. The reason is that if you subtract logs of numbers, it's the same as dividing the numbers. For example, $10/5 = 2$ and (try it on your calculator) $\log(10) - \log(5) = \log(2)$.

model. The degrees of freedom will be the difference between the degrees of freedom of the two models.

20.3.3 Assessing the model: R and R^2 ▮▮▮

When we discussed the linear model we saw that the multiple correlation coefficient R and the corresponding R^2 are useful measures  of how well the model fits the data. The likelihood ratio is similar in that it is based on the level of correspondence between predicted and observed values of the outcome. However, there is a more literal analogue of the multiple correlation in logistic regression known as the R-statistic. It is the partial correlation between the outcome variable and each of the predictor variables and it varies between –1 and +1. A positive value indicates that as the predictor variable increases, so does the likelihood of the event occurring. A negative value implies that as the predictor value increases, the likelihood of the outcome occurring decreases. If a predictor variable has a small value of R then it contributes only a small amount to the model.

To compute R use the following equation:

$$R = \sqrt{\frac{z^2 - 2df}{-2LL_{\text{baseline}}}} \tag{20.8}$$

in which the $-2LL$ is the deviance for the original model, the Wald statistic (z) is calculated as described in Section 20.3.4, and the degrees of freedom can be read from the SPSS output for the variables in the equation. We'll see in the next section that the Wald statistic can be inaccurate under certain circumstances and, because of this, R is by no means an accurate measure. For this reason treat the value of

R with caution, and it is invalid to square it and interpret it as you would in a linear model with a continuous outcome.

There is some controversy over what *would* make a good analogue to R^2 in logistic regression. **Hosmer and Lemeshow's** (1989) measure, R_{L}^2, is calculated by dividing the model chi-square, which represents the change from the baseline (based on the log-likelihood) by the baseline $-2LL$ (the deviance of the model before any predictors were entered):

$$R_{\text{L}}^2 = \frac{\chi^2_{\text{model}}}{-2LL_{\text{baseline}}} \tag{20.9}$$

Given what the model chi-square represents (see equation (20.7)), another way to express this statistic is:

$$R_{\text{L}}^2 = \frac{\left(-2LL_{\text{baseline}}\right) - \left(-2LL_{\text{new}}\right)}{-2LL_{\text{baseline}}} \tag{20.10}$$

R_{L}^2 is the proportional reduction in the absolute value of the log-likelihood measure. It is a measure of how much the badness of fit improves as a result of the inclusion of the predictor variables. It can vary between 0 (indicating that the predictors are useless at predicting the outcome variable) and 1 (indicating that the model predicts the outcome variable perfectly).

IBM SPSS Statistics uses **Cox and Snell's** (1989) measure, R_{CS}^2, which is based on the deviance of the model ($-2LL_{\text{new}}$), the deviance of the original model ($-2LL_{\text{baseline}}$), and the sample size, n:

$$R_{\text{CS}}^2 = 1 - \exp\left(\frac{-2LL_{\text{new}} - (-2LL_{\text{baseline}})}{n}\right) \tag{20.11}$$

This statistic never reaches its theoretical maximum of 1, therefore, Nagelkerke (1991) suggested the following amendment (**Nagelkerke's** R_{N}^2):

$$R_{\text{N}}^2 = \frac{R_{\text{CS}}^2}{1 - \exp\left(-\frac{-2LL_{\text{baseline}}}{n}\right)} \tag{20.12}$$

Although all these measures differ in their computation (and the answers you get), conceptually they are somewhat the same. They are somewhat similar to the R^2 in linear models with continuous outcomes in that they provide a gauge of the substantive significance of the model.

20.3.4 Assessing the contribution of predictors: the Wald statistic ▮▮▮

In addition to assessing the fit of the model overall we want to know the individual contribution of predictors. In the linear model, we used the estimated regression coefficients (b) and their standard errors to compute a t-statistic. In logistic regression there is an analogous statistic—the z-statistic—which follows the normal distribution. Like the t-statistic in the linear model, the z-statistic tells us whether the b-value for that predictor is significantly different from zero. If the coefficient is significantly different from zero then we assume that the predictor is making a significant contribution to the prediction of the outcome (Y).

The following equation shows how the z-statistic is calculated, and it's basically identical to the t-statistic in the linear model (see equation (9.14)): it is the b-value divided by its associated standard error:

$$z = \frac{b}{SE_b} \tag{20.13}$$

The z-statistic was developed by Abraham Wald (Figure 20.2), and is known as the **Wald statistic**. SPSS Statistics reports the Wald statistic as z^2, which transforms it so that it has a chi-square distribution. The z statistic should be used a little cautiously because, when the b-value is large, the standard error tends to become inflated, resulting in the z-statistic being underestimated (see Menard, 1995). The inflation of the standard error increases the probability of rejecting a predictor as being

Figure 20.2 Abraham Wald writing 'I must not devise test statistics prone to having inflated standard errors' on the blackboard 100 times

significant when, in reality, it is making a significant contribution to the model (i.e., a Type II error). When assessing whether predictors significantly predict the outcome, it is probably more accurate to enter predictors hierarchically and examine the change in likelihood ratio statistics.

20.3.5 The odds ratio: exp(*B*)

The odds ratio is crucial to the *interpretation* of logistic regression (Section 3.7.3). The odds ratio is the exponential of B (i.e., e^B or $\exp(B)$) and it is an indicator of the change in odds resulting from a unit change in the predictor. As such, it is like the b-value but easier to understand because it doesn't require a logarithmic transformation. When the predictor variable is categorical the odds ratio is easier to explain, so imagine we had a simple example in which we were trying to predict whether someone subscribes to Spotify from whether they listen to pop or metal music. As we saw in Section 3.7.3, the odds of an event occurring are defined as the probability of an event occurring divided by the probability of that event not occurring (see equation (20.14)) and should not be confused with the more colloquial usage of the word to refer to probability. So,

for example, the odds of subscribing to Spotify are the probability of subscribing to Spotify divided by the probability of not subscribing to Spotify:

$$odds = \frac{P(\text{event})}{P(\text{no event})}$$

$$P(\text{event } Y) = \frac{1}{1 + e^{-(b_0 + b_1 X_{1i})}}$$

$$P(\text{no event } Y) = 1 - P(\text{event } Y)$$

(20.14)

To calculate the change in odds that results from a unit change in the predictor, we must first calculate the odds of subscribing to Spotify given that someone is a pop fan. We then calculate the odds of subscribing to Spotify given that someone is a metal fan. Finally, we calculate the proportionate change in these two odds.

To calculate the first set of odds, we use equation (20.3) to calculate the probability of subscribing to Spotify given that someone is a pop fan. If we had more than one predictor we would use equation (20.4). There are three unknown quantities in this equation: the coefficient of the constant (b_0), the coefficient for the predictor (b_1) and the value of the predictor itself (X). We'll know the value of X from how we coded the type of music variable (e.g., we might have used 0 = pop fan and 1 = metal fan). The values of b_1 and b_0 will be

estimated for us. We can calculate the odds as in equation (20.14).

Next, we calculate the same thing after the predictor variable has changed by one unit. In this case, because the predictor variable is dichotomous, we need to calculate the odds of subscribing to Spotify, given that someone is a metal fan. So, the value of X is now 1 (rather than 0). We now know the odds before and after a unit change in the predictor variable. It is a simple matter to calculate the proportionate change in odds by dividing the odds after a unit change in the predictor by the odds before that change:

$$\text{odds ratio} = \frac{\text{odds after a unit change in the predictor}}{\text{original odds}}$$

(20.15)

This proportionate change in odds is the odds ratio, and we interpret it in terms of the change in odds: if the value is greater than 1 then it indicates that as the predictor increases, the odds of the outcome occurring increase. Conversely, a value less than 1 indicates that as the predictor increases, the odds of the outcome occurring decrease. We'll see how this works with a real example shortly.

20.3.6 Model building and parsimony

When you have more than one predictor, you can choose between the same methods to build your model as described for the linear model (Section 9.9.1). Forced entry and hierarchical methods are preferred, but if you are undeterred by the criticisms of stepwise methods then you can choose a forward or backward stepwise method. Stepwise methods work in the same way as we've seen for the linear model, except that different statistics are used to determine whether predictors are entered or removed

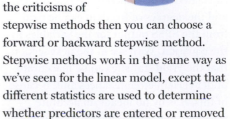

Which method should I use to enter variables?

from the model. The forward method enters predictors based on their score statistic, then assesses removal based on one of three statistics: the likelihood ratio statistic described in Section 19.3.4 (*Forward: LR*), an arithmetically less intense version of the likelihood ratio statistic called the conditional statistic (*Forward: Conditional*) or significance values of the Wald statistic (*Forward: Wald*) above a removal criterion (0.1 by default). The likelihood ratio method is the best removal criterion because the Wald statistic can be unreliable (see Section 20.3.4). Backward stepwise methods begin with all predictors in the model and remove them if their removal is not detrimental to the fit of the model (assessed using the same three methods as the forward approach).

As we have seen before, stepwise methods are best avoided for theory testing but can be used when no previous research exists on which to base hypotheses, and in situations where causality is not of interest and you wish only to find a model to fit your data (Agresti & Finlay, 1986; Menard, 1995). If you use a stepwise method then the backward method is preferable because forward methods are more likely to exclude predictors involved in **suppressor effects**.

As with the linear model, it is best to use hierarchical methods and to build models in a systematic and theory-driven way. Although we haven't yet discussed this for linear models, when building a model we should strive for **parsimony**. In a scientific context, parsimony refers to the idea that simpler explanations of a phenomenon are preferable to complex ones. The statistical implication of using a parsimony heuristic is that models be kept as simple as possible. In other words, do not include predictors unless they have explanatory benefit. To implement this strategy we first need to fit the model that includes all potential predictors, and then systematically remove any that don't seem

to contribute to the model. This is a bit like a backward stepwise method, except that the decision-making process is in the researcher's hands: they make informed decisions about what predictors should be removed. It's also worth bearing in mind that if you have interaction terms in your model then for an interaction term to be valid *you must retain the main effects involved in the interaction term as well* (even if they don't appear to contribute much).

20.4 Sources of bias and common problems ▮▮▮▮

20.4.1 Assumptions ▮▮▮▮

Logistic regression, like any linear model, is open to the sources of bias discussed in Chapter 6 and Section 9.3. In the context of logistic regression, it's worth noting a couple of points about the assumptions of linearity and independence:

- *Linearity*: In the linear model we assume that the outcome has linear relationships with the predictors. In logistic regression the outcome is categorical and so this assumption is violated, so we use the log (or *logit*) of the data. The assumption of linearity in logistic regression, therefore, assumes that there is a linear relationship between any continuous predictors and *the logit of the outcome variable*. This assumption can be tested by looking at whether the interaction term between the predictor and its log transformation is significant (Hosmer & Lemeshow, 1989). We will go through an example in Section 20.8.1.
- *Independence of errors*: In logistic regression, violating this assumption produces overdispersion, which we'll discuss in Section 20.4.4.

Logistic regression also has some unique problems. These are not sources of bias so much as things that can go wrong. Logistic regression parameters are estimated by an

iterative process (SPSS Tip 20.1). Sometimes, instead of pouncing on the correct solution quickly, you'll notice nothing happening: the computer begins to move infinitely slowly or appears to have just got fed up with you asking it to do stuff and has gone on strike. If it can't find a correct solution, then sometimes it actually does give up, quietly offering you a result which is completely incorrect. Usually this is revealed by implausibly large standard errors. Two situations can provoke this situation, both of which are related to the ratio of cases to variables: incomplete information and complete separation.

20.4.2 Incomplete information from the predictors ▮▮▮▮

Imagine you're trying to predict lung cancer from smoking and whether or not you eat tomatoes (which are believed to reduce the risk of cancer). You collect data from people who do and don't smoke, and from people who do and don't eat tomatoes; however, this isn't sufficient unless you collect data from all combinations of smoking and tomato eating. Imagine you ended up with the data in Table 20.1. Observing only the first three possibilities does not prepare you for the outcome of the fourth. You have no way of knowing whether this last person will have cancer or not based on the other data you've collected. Therefore, unless you've collected data from all combinations of your variables it'll be tricky to estimate the model. You can check for incomplete information using a contingency table before you fit the model (I describe how to do this in Chapter 19). While you're checking these tables, look at the expected frequencies in each cell of the table to make sure that they are greater than 1 and no more than 20% are less than 5 (see Section 19.5). This is because the goodness-of-fit tests in logistic regression make this assumption.

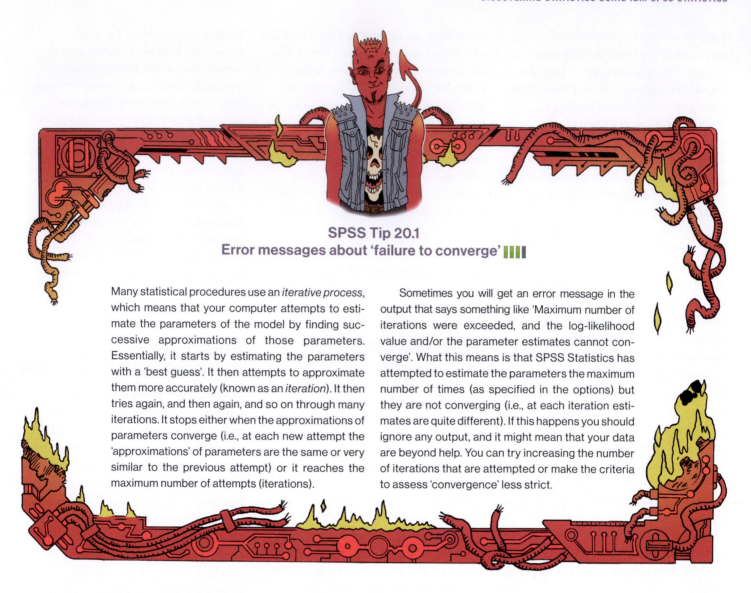

SPSS Tip 20.1
Error messages about 'failure to converge' ▮▮▮▮

Many statistical procedures use an *iterative process*, which means that your computer attempts to estimate the parameters of the model by finding successive approximations of those parameters. Essentially, it starts by estimating the parameters with a 'best guess'. It then attempts to approximate them more accurately (known as an *iteration*). It then tries again, and then again, and so on through many iterations. It stops either when the approximations of parameters converge (i.e., at each new attempt the 'approximations' of parameters are the same or very similar to the previous attempt) or it reaches the maximum number of attempts (iterations).

Sometimes you will get an error message in the output that says something like 'Maximum number of iterations were exceeded, and the log-likelihood value and/or the parameter estimates cannot converge'. What this means is that SPSS Statistics has attempted to estimate the parameters the maximum number of times (as specified in the options) but they are not converging (i.e., at each iteration estimates are quite different). If this happens you should ignore any output, and it might mean that your data are beyond help. You can try increasing the number of iterations that are attempted or make the criteria to assess 'convergence' less strict.

This point applies not only to categorical predictors, but also to continuous ones. Suppose that you wanted to investigate factors related to human happiness. These might include age, sex, sexual orientation, religious beliefs, levels of anxiety and even whether a person is right-handed. You interview 1000 people, record their characteristics, and whether they are happy ('yes' or 'no'). Although a sample of 1000 seems quite large, is it likely to include many 80-year-old, highly anxious, Buddhist, left-handed lesbians? Probably not. If you found one or two such people and they were happy, should you conclude that everyone else in the same category is happy? It would, obviously, be better to

have several more people in this category to confirm that this combination of characteristics is associated with happiness. One solution is to collect more data.

As a general point, whenever samples are broken down into categories and one or more combinations are empty it creates problems. These will probably be signalled by coefficients that have unreasonably large standard errors. Conscientious researchers produce and check multi-way crosstabulations of all categorical independent variables. Lazy but cautious ones don't bother with crosstabulations, but look carefully at the standard errors. Those who don't bother with either should expect trouble.

20.4.3 Complete separation ▮▮▮▮

A second situation in which logistic regression collapses might surprise you: it's when the outcome variable can be perfectly predicted by one variable or a combination of variables. This situation is known as **complete separation**. Imagine you placed a pressure pad under your door mat and connected it to your security system so that you could detect burglars when they creep in at night. However, because your teenage children (which you would have if you're old enough and rich enough to have security systems and pressure pads) and their friends are often coming home in the middle of the night, when they tread on the pad you want it to work out the probability that the person is a burglar and not one of your teenagers. Therefore, you could measure the weight of some burglars and some teenagers and use logistic regression to predict the outcome (teenager or burglar) from the weight. The graph (Figure 20.3) would show a line of triangles at zero (the data points for the teenagers you weighed) and a line of triangles at 1 (the data points for burglars you weighed). Note that these lines of triangles overlap (some teenagers are as heavy as burglars). In logistic regression, we predict the probability of the outcome given a value of the predictor. In this case, at low weights the fitted probability follows the bottom line of the plot, and at high weights it follows the top line. At intermediate values it tries to follow the probability as it changes.

Imagine that we had the same pressure pad, but our teenage children had left home to go to university. We're now interested in distinguishing burglars from our pet cat based on weight. Again, we can weigh some cats and weigh some burglars. This time the graph (Figure 20.4) still has a row of triangles at zero (the cats) and a row at 1 (the burglars) but this time the rows of triangles do not overlap: there is no burglar who weighs the same as a cat—obviously there were no cat burglars in the sample (groan). This situation is known as complete (or sometimes perfect) separation: the outcome (cats and burglars) can be perfectly predicted from weight (anything less than 15 kg is a cat, anything more than 40 kg is a burglar). If we try to calculate the probabilities of the outcome given a certain weight then we run into trouble. When the weight is low, the probability is 0, and when the weight is high, the probability is 1, but what happens in between? We have no data in between 15 and 40 kg on which to base these probabilities. The figure shows two possible probability curves that we could fit to these data, one steeper than the other.

Table 20.1 Incomplete contingency table

Do you smoke?	Do you eat tomatoes?	Do you have cancer?
Yes	No	Yes
Yes	Yes	Yes
No	No	Yes
No	Yes	??????

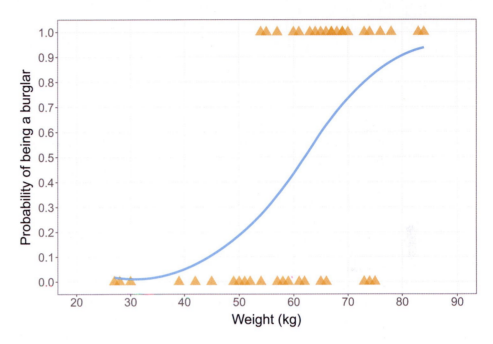

Figure 20.3 An example of the relationship between weight and a dichotomous outcome variable (being a burglar or not). Note that the weights of burglars ($y = 1$) and non-burglars ($y = 0$) overlap

Either one of these curves is valid based on the data we have available. The lack of data means that the computer will be uncertain about how steep it should make the intermediate slope and it will try to bring the center as close to vertical as possible, but its estimates veer unsteadily towards infinity (hence large standard errors). Complete separation often arises when too many variables are fitted to too few cases. Often the only satisfactory solution is to collect more data, but sometimes a neat answer is found by using a simpler model.

20.4.4 Overdispersion ▍▍▍▍

Logistic regression is not only used to predict a two category outcome (coded 0 and 1), it can also be used to, for example, predict proportions or outcomes with several categories (see section 20.9). In these latter cases you can get **overdispersion**. I'm a psychologist, not a statistician, and most of what I've read on overdispersion doesn't make an awful lot of sense to me. From what I can gather, it is when the observed variance is bigger than expected from the logistic regression model. This can happen for two reasons. The first is correlated observations (i.e., when the assumption of independence is broken) and the second is due to variability in success probabilities. For example, imagine our outcome was whether a puppy in a litter survived or died. Genetic factors mean that within a

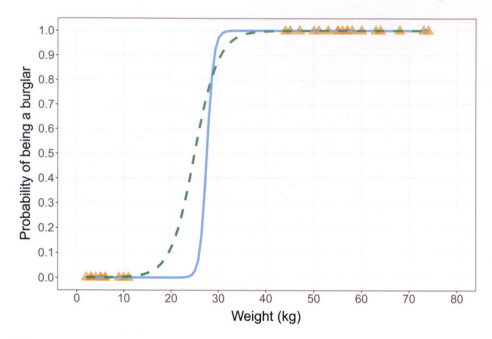

Figure 20.4 An example of *complete separation*. Note that the weights of the burglars (*y* = 1) and non-burglars (*y* = 0) do not overlap

too, and whether you rescale using this statistic or the Pearson chi-square statistic depends on which one is bigger. The bigger statistic will have the bigger dispersion parameter (because their degrees of freedom are the same), and will make the bigger correction; therefore, correct by the bigger of the two.

20.5 Binary logistic regression ▍▍▍▍

20.5.1 An example that will make you feel eel ▍▍▍▍

A hobby of mine is unearthing bizarre academic papers (really, if you find any, please email them to me)—it's amazing what you find. I like finding research that makes me laugh, and a research paper by Lo, Wong, Leung, Law, & Yip (2004) made me laugh *a lot*. They describe a case of a 50-year-old man who reported to the emergency department of a hospital with abdominal pain. A physical examination revealed peritonitis, so they X-rayed the man's abdomen. The X-ray revealed the shadow of an eel. The authors don't directly quote the man's response to this news, but I like to imagine it was something to the effect of 'Oh, that! Erm, yes, well, I didn't think it was terribly relevant to my abdominal pain so I didn't mention it, but I did insert an eel into my anus earlier today. Do you think that's the problem?' He probably didn't say that, but he did admit that he inserted the eel to 'relieve constipation'.

I have a lively imagination, and I can't help thinking about the poor eel. There it was, minding its own business swimming about, thinking to itself 'Today seems like a nice day, there are no eel-eating sharks about,

Can an eel cure constipation?

given litter the chances of success (living) depend on the litter from which the puppy came. As such success probabilities vary across litters (Halekoh & Højsgaard, 2007), this example of dead puppies— despite making my spaniel cower in the corner with his paws over his eyes—shows how variability in success probabilities can create correlation between observations (the survival rates of puppies from the same litter are not independent).

Overdispersion tends to limit standard errors, which creates two problems. Firstly, test statistics of regression parameters are computed by dividing by the standard error (see equation (20.13)), so if the standard error is too small then the test statistic will be too big and falsely deemed significant. Secondly, confidence intervals are computed from standard errors, so if the standard error is too small then the confidence interval will be too narrow and make us overconfident about the likely relationship between predictors and the outcome in the population. In short, overdispersion doesn't affect the model parameters (*b*-values) themselves but

biases our conclusions about their significance and population value.

SPSS Statistics produces a chi-square goodness-of-fit statistic, and overdispersion is present if the ratio of this statistic to its degrees of freedom is greater than 1 (this ratio is called the *dispersion parameter*, ϕ). Overdispersion is likely to be problematic if the dispersion parameter approaches or is greater than 2. (Incidentally, *under*dispersion is shown by values less than 1, but this problem is much less common than *over*dispersion.) There is also the *deviance* goodness-of-fit statistic, and the dispersion parameter can be based on this statistic instead (again by dividing by the degrees of freedom). When the chi-square and deviance statistics are very discrepant, overdispersion is likely.

The effects of overdispersion can be reduced by using the dispersion parameter to rescale the standard errors and confidence intervals. For example, the standard errors are multiplied by $\sqrt{\phi}$ to make them bigger (as a function of how big the overdispersion is). You can base these corrections on the deviance statistic

Cramming Sam's Tips
Issues in logistic regression

- In logistic regression, we assume the same things as the linear model.
- The linearity assumption is that each predictor has a linear relationship with the log of the outcome variable.
- If we created a table that combined all possible values of all variables then we should ideally have some data in every cell of this table. If you don't then watch out for big standard errors.

- If the outcome variable can be predicted perfectly from one predictor variable (or a combination of predictor variables) then we have *complete separation*. This problem creates large standard errors too.
- *Overdispersion* is where the variance is larger than expected from the model. This can be caused by violating the assumption of independence. This problem makes the standard errors too small.

the sun is out, the water is nice, what could possibly go wrong?' The next thing it knows, it's being shoved up a man's anus. 'I didn't see that coming,' thinks the eel. It finds itself in a tight, dark tunnel, there's no light, there's a distinct lack of water compared to its usual habitat, and it's scared. Its day has gone *very* wrong. It considers its fate and, noticing that the walls of the prison cell are fairly soft, it does what any self-respecting eel would do: it decides, 'Bugger this,[3] I'll *eat* my way out of here'. Unfortunately the eel didn't make it, but it went out with a

fight (there's a fairly unpleasant photograph in the article of the eel biting the splenic flexure). Lo et al. conclude that 'Insertion of a live animal into the rectum causing rectal perforation has never been reported. This may be related to a bizarre healthcare belief, inadvertent sexual behavior or criminal assault. However, the true reason may never be known.' Quite.

This is a really grim[4] and bizarre[5] tale. I'm no medic, but if constipation is a failure to empty the bowel, inserting more stuff up there seems, at best, a counterintuitive

remedy. But upon reflection I wondered if I was being harsh on the man—maybe an eel up the anus really can cure constipation. To test this hypothesis, we could do a randomized controlled trial of eel therapy. Our outcome variable would be 'constipated' vs. 'not constipated', which is a dichotomous variable that we're trying to predict. The main predictor variable would be the intervention condition (eel up the anus vs. waiting list/ no treatment), but we might also factor in how many days the patient had been

---

3 Literally.

4 As it happens, it isn't an isolated grim tale. Through this article I found myself hurtling down a wormhole of rectal insertion that involved a very large stone (Sachdev, 1967), a test tube (Hughes, Marice, & Gathright, 1976), a baseball (McDonald & Rosenthal, 1977), an aerosol deodorant can, hose pipe, iron bar, broomstick, penknife, marijuana, banknotes, blue plastic tumbler, vibrator and primus stove (Clarke, Buccimazza, Anderson, & Thompson, 2005), and a toy pirate ship, with or without pirates I'm not sure (Bemelman & Hammacher, 2005). I encourage you to send me bizarre research, but if it involves objects in the rectum then probably don't, unless someone has managed to put Buckingham Palace up there.

5 Possibly not as bizarre as the study I subsequently found of a 14-year-old boy who reported to hospital unable to urinate (Vezhaventhan & Jeyaraman, 2007). A small fish was discovered in his bladder that had swum up his penis while he was having a wee while cleaning a fish tank. Yes, of course it did.

constipated before treatment. This scenario is perfect for logistic regression (but not for eels).

Some statistics lecturers don't share my unbridled joy at discussing eel-created rectal perforations with students, so in the data file (**Eel.sav**) I have used general variable names and descriptions:

- *Outcome* (dependent variable): **Cured** (cured or not cured).
- *Predictor* (independent variable): **Intervention** (intervention or no treatment).
- *Predictor* (independent variable): **Duration** (the number of days before treatment that the patient had the problem).

In doing so, your tutor can adapt the example to something more palatable if they wish, but you will secretly know that it's all about having eels up your bum.

20.5.2 Building a model ▮▮▮▮

In Section 20.3.6 we discussed the idea of building models based on the principle of parsimony. In this example, we have three potential predictors: **Intervention**, **Duration** and the **Intervention ×️ Duration** interaction. The most complex model we can fit would include all these variables, but the parsimony principle would suggest building up to this model in steps, looking at what predictors didn't improve the model and going back to a simpler model that doesn't include them. The key effect of interest is whether the intervention has an effect, so the first model would have only **Intervention** as a predictor. Figure 20.5 shows how we then build this model up by adding in the other main effect of **Duration** (model 2) and then the interaction term (model 3). Our job is to determine which of these models best fits the data, while adhering to the general idea of parsimony. If adding the interaction term doesn't improve the model then we roll back to model 2 as our final model, and if **Duration** doesn't add anything we roll back to the first model. Remember, though, that if you want to look at an interaction *you must include any main effects involved in that interaction in the model even if the main effects weren't significant*. In this example, if we want to assess the contribution of the **Duration ×️**

Intervention interaction, we must also include **Intervention** and **Duration** in the model involving the interaction.

20.5.3 Logistic regression: the general procedure ▮▮▮▮

Figure 20.6 shows the general process of conducting logistic regression. First, we run an initial hierarchical analysis to fit competing models and decide which one is best (i.e., the three models identified in Section 20.5.2). Having done this, we refit the model that we chose but save diagnostic statistics and inspect them to look for signs of bias (outliers and influential cases). We then check for linearity of the logit (in fact, it's a good idea to do this first, but it's a little complicated so I want to deal with it later in the chapter), and check for multicollinearity.

20.5.4 Data entry ▮▮▮▮

The data should be entered as for the linear model: for this example they are arranged in three columns (one for each variable). Look at **Eel.sav** in the data editor, and note that both categorical variables are coding variables (Section 4.6.5) in which numbers specify categories. In general, to ease interpretation, code the outcome variable as 1 (event occurred) and 0 (event did not occur); in this example, 1 represents being cured and 0 represents not being cured. For the intervention a similar coding has been used (1 = intervention, 0 = no treatment).

20.5.5 Building the models using SPSS Statistics ▮▮▮▮

To build the three models in Figure 20.5 select *Analyze* ▶ *Regression* ▶ **Binary Logistic**... to access the dialog box in Figure 20.7. The models need to be specified in blocks, with each block adding a new variable to the model. To specify the first model, drag the outcome variable

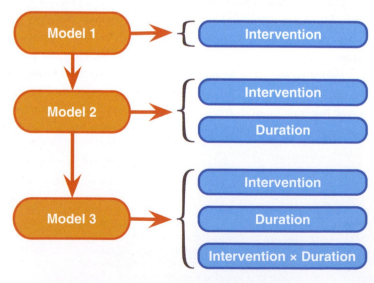

Figure 20.5 Building models based on the principle of parsimony

(**Cured**) to the *Dependent* box (or select it and click ➡). There is a box labeled *Covariates* for specifying the predictor variables. Model 1 has one predictor variable (the main effect of **Intervention**), so drag this variable to the *Covariates* box (or click ➡). Make sure that the *Method* is set to Enter ▾ . That's model 1 sorted. Click Next , which clears the *Covariates* box and re-labels it to indicate which block of the model you are in. To add a new predictor in a second block, model 2 adds the main effect of **Duration**, so drag this variable to the *Covariates* box (or click ➡). Because **Intervention** is already forced into the model in block 1, this second block creates a model that includes both **Intervention** and **Duration** (i.e., model 2). Again, make sure that the method is set to Enter ▾ . That's model 2 done.

To specify the final model we need to add in the **Duration × Intervention** interaction. Click Next to create a new block (block 3), then select **Duration** and **Intervention** simultaneously by holding down the *Ctrl* (Windows) or *Cmd* (Mac OS) key when selecting them and click >a*b> . This should add the interaction term. Model 3 is complete. To move between models use the Previous and Next buttons.

20.5.6 Method of regression ▮▮▮▮

For each of the models that we specified you can select a method (see Section 20.5.5) of variable entry by clicking Enter ▾ and selecting a method in the drop-down list. We are doing this analysis hierarchically so we want to use the Enter ▾ method in each block (i.e., you don't need to change anything)—just bear in mind that other methods exist. If you want to try out stepwise then do the self-test, which has a detailed explanation and interpretation on the companion website.

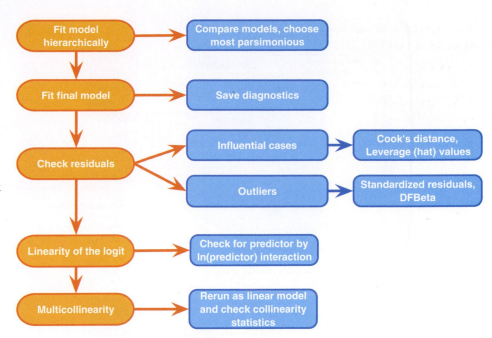

Figure 20.6   The process of fitting a logistic regression model

20.5.7 Categorical predictors ▮▮▮▮

Click Categorical... in the main dialog box to activate the dialog box in Figure 20.8, which you use to specify categorical predictors. Drag any categorical predictor variables in your model (in this example we have only one, **Intervention**) to the *Categorical Covariates* box (or click ➡). There are different ways to code categorical predictors, and we have discussed some of them (Sections 11.5.1 and 12.4). By default, *Indicator* coding is used, which is standard dummy variable coding where you choose either the first or last category as the baseline category. To use a different type of coding click Indicator ▾ and select from simple contrasts, difference contrasts, Helmert contrasts, repeated contrasts, polynomial contrasts and deviation contrasts in the drop-down list (see Table 12.6). Let's use standard dummy coding (*indicator*) for this example. We need to decide whether to use the ⦿ First or ⦿ Last category as a baseline. In this

example, it doesn't make any difference because we have only two categories, but if you had a categorical predictor with more than two categories then you should either use the highest number to code your control category in the data editor and select ⦿ Last or use the lowest number to code your control category and select ⦿ First. In our data, I coded 'cured' as 1 and 'not cured' (our control category) as 0; therefore, select the contrast, click ⦿ First and then Change so that the completed dialog box looks like Figure 20.8.

20.5.8 Comparing the models ▮▮▮▮

Before we look at some of the other options, let's compare our models with just the basic settings (which is all we need to assess their fit). Having selected the options that I have already described, click OK in the main dialog box. Output 20.1 tells us both how we coded our outcome variable (it reminds us that 0 = not cured, and 1 = cured)[6] and how it has coded the categorical predictors

---

6   These values are the same as the data editor, so this table might seem pointless; however, had we used codes other than 0 and 1 (e.g., 1 = not cured, 2 = cured) then these codes change to zeros and ones and this table informs you of which category is represented by 0 and which by 1, which is important for interpretation.

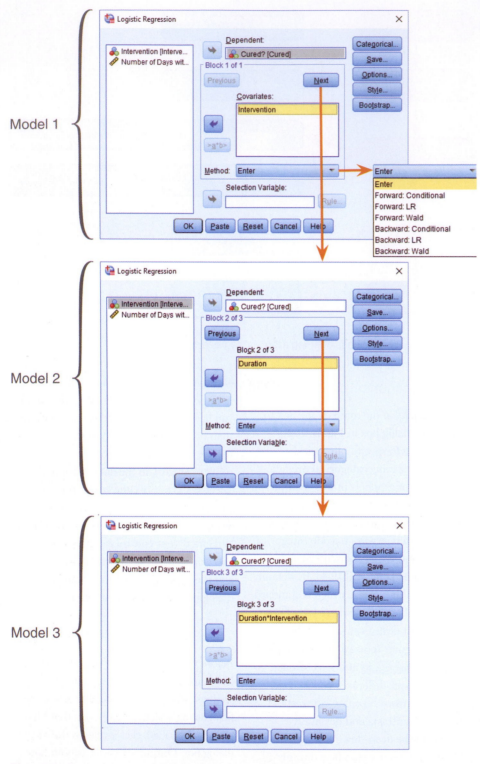

Model 1

Model 2

Model 3

Figure 20.7 Specifying models using the *Logistic Regression* dialog box

category the codes would have been −0.5 (**Intervention** = no treatment) and 0.5 (**Intervention** = treatment) and if ⦿ **Last** was selected the signs of the codes would be reversed. The parameter codes are important for calculating the probability of the outcome variable ($P(Y)$), but we will come to that later.

Output 20.2 shows the overall model summary statistics for each of the three models. The table labeled *Omnibus Tests of Model Coefficients* includes the chi-square statistic (which is related to −2LL) for the model overall (*Model*) and the change since the previous model (*Block*). Model 1 yields a chi-square of 9.926, which is highly significant, $p = 0.002$. We can compare the models using equation (20.7). For model 1, the −2LL would have been compared to that obtained from a model that included only the intercept, so we're comparing a model that includes **Intervention** against a model that has no predictors. The chi-square tells us that the model has improved significantly by adding **Intervention** as a predictor. In model 2, we added the effect of **Duration**, and this model is a significant fit of the data because the *Model* chi-square in the table labeled *Omnibus Tests of Model Coefficients* is significant, $\chi^2(2) = 9.93$, $p = 0.007$. However, we're not interested in the model overall because the previous model was also a significant fit; we're interested in the *improvement* of model 2 over model 1, and this information is given by the chi-square for *Block*. The *Block* chi-square tells us about the change in chi-square in this block: it is the *change* in the chi-square resulting from adding **Duration** to the model. The value is obtained by taking the difference between the model chi-square for the two models (in this case 9.928 − 9.926 = 0.002). This change is very non-significant, $\chi^2(1) = 0.002$, $p = 0.964$, indicating that adding **Duration** to the model had virtually no effect on the fit (the chi-square has hardly changed).

(the parameter codings for **Intervention**). We chose indicator coding and so the coding is the same as the values in the data editor (0 = no treatment, 1 = treatment). If

deviation coding had been chosen then the coding would have been −1 (treatment) and 1 (no treatment). With a *simple* contrast, if ⦿ **First** was selected as the reference

In model 3, we added the **Intervention ×
Duration** interaction. Again, this model
is a significant fit of the data because the
Model chi-square is significant, $\chi^2(3) =$
9.99, $p = 0.019$. However, as with model
2, we're interested in the *improvement* of
model 3 over the previous one (the chi-
square for *Block*). As before, the value is
obtained by taking the difference
between the model chi-square for the two
models (in this case $9.989 - 9.928 =$
0.061). This change is very non-
significant,
$\chi^2(1) = 0.061$, $p = 0.805$, indicating that
adding the interaction term has had
virtually no effect on the fit.

We could, if we wanted, look at the
difference between models 1 and 3 as well
using equation (20.7). The result is:

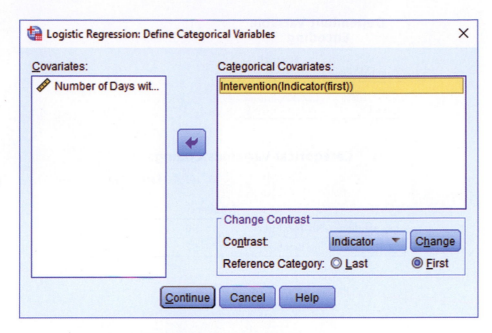

Figure 20.8   Defining categorical variables in logistic regression

**Oditi's Lantern
Logistic regression**

'I, Oditi, believe that my loyal brethren will find it difficult to master the secrets within the data if their
bowels are creaking at the seams because the
curse of constipation has afflicted them. You could
do the magic dance of the turtle head and hope that
it brings you relief, but it is my belief that to remove
an intestinal log we need log-istic regression. Stare
into my lantern and feel immediate relief.'

Dependent Variable Encoding

Original Value	Internal Value
Not Cured	0
Cured	1

Categorical Variables Codings

		Frequency	Parameter coding (1)
Intervention	No Treatment	56	.000
	Intervention	57	1.000

Output 20.1

Omnibus Tests of Model Coefficients

Model 1

		Chi-square	df	Sig.
Step 1	Step	9.926	1	.002
	Block	9.926	1	.002
	Model	9.926	1	.002

Omnibus Tests of Model Coefficients

Model 2

		Chi-square	df	Sig.
Step 1	Step	.002	1	.964
	Block	.002	1	.964
	Model	9.928	2	.007

Omnibus Tests of Model Coefficients

Model 3

		Chi-square	df	Sig.
Step 1	Step	.061	1	.805
	Block	.061	1	.805
	Model	9.989	3	.019

Output 20.2

$$\chi^2 = \chi^2_{model3} - \chi^2_{model1} = 9.989 - 9.926 = 0.063$$
$$df = df_{model3} - df_{model1} = 3 - 1 = 2 \qquad (20.16)$$

We could compare this against the critical values for the chi-square distribution with 2 degrees of freedom, but we don't need to because 0.063 is so small it barely warrants turning the pages to confirm what we already know: **Duration** and the **Duration × Intervention** interaction add nothing to the model. Based on this comparison, we would choose model 1.

20.5.9 Refitting the model

Comparing the models shows that **Duration** and the **Duration × Intervention** interaction add little and that we should proceed with model 1. We set up the main dialog box as we did before for model 1, with only **Intervention** as a predictor (the top dialog box in Figure 20.7). Set the same options as before, but we'll now also get some more detailed information about the model.

20.5.10 Obtaining residuals

To save residuals (see Section 9.2.3) click Save... in the main dialog box to access the dialog box in Figure 20.9. Most of the options are similar to what we encountered for the linear model (Section 9.10.4). The *predicted probabilities* and *predicted group memberships* are unique to logistic regression. The predicted probabilities are the probabilities of Y occurring (derived from equation (20.3)), given the values of each predictor for a given case. The predicted group membership tells us to which of the two outcome categories a participant is most likely to belong, based on the model. The group memberships are based on the predicted probabilities, and I will explain these values in due course. As a bare minimum select the options in Figure 20.9. Note that these variables won't save if you activate bootstrapping (see below).

20.5.11 Further options

Finally, click Options... in the main dialog box to obtain the dialog box in Figure 20.10. For the most part, the default settings are fine. I mentioned in Section 20.5.6 that when a stepwise method is used there are default criteria for selecting and removing predictors from the model. These default settings are displayed under *Probability for Stepwise*. The probability thresholds can be changed, but there is really no need. Another default is to arrive at a model after a maximum of 20 iterations (SPSS Tip 20.1). Unless you have a very complex model, 20 iterations will be enough. Linear models (like logistic regression) usually contain a constant (the

b_0 in equation (20.4)) and by default it is included, but you can deselect this option and force the model to pass through the origin (i.e., Y is 0 when X is 0). Normally we don't want to do this.

A useful way to assess the fit of the model to the observed data is a histogram of the actual and predicted values of the outcome variable (select ☑ Classification plots). It is possible to do a ☑ Casewise listing of residuals for any cases for which the standardized residual is greater than 2 standard deviations (this default value is sensible, but change it if you like) or for all cases. I recommend a more thorough examination of residuals, but this option can be useful for a quick inspection. Select ☑ CI for exp(B): to produce a 95% confidence interval (see Section 2.8) for the odds ratio (you can change it, but 95% is what is conventionally reported). The ☑ Hosmer-Lemeshow goodness-of-fit statistic is used to assess how well the chosen model fits the data. The remaining options are fairly unimportant: you can choose to display all statistics and graphs at each stage of an analysis (the default) or only after the final model has been fitted. Finally, you can display a correlation matrix of parameter estimates for the terms in the model (*Correlation of estimates*)—the practical function of doing this is lost on most of us mere mortals. You can display coefficients and log-likelihood values at each iteration of the parameter estimation process (☑ Iteration history), which is useful because it's the only way you can display the initial −2LL, and we need this value if we want to compute R. When you have selected the options I've just described, click OK and watch the output spew forth.

20.5.12 Bootstrapping ▮▮▮

If you use forced entry then you can bootstrap your model by clicking Bootstrap... in the main dialog box and selecting appropriate options. This function doesn't

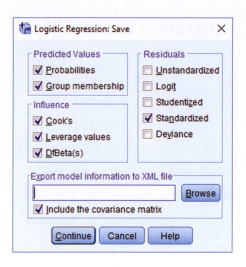

Figure 20.9 Dialog box for obtaining residuals for logistic regression

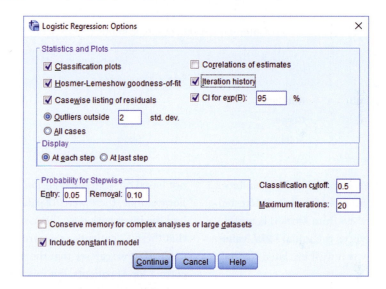

Figure 20.10 Dialog box for logistic regression options

work with stepwise methods, so the button will be inactive unless you choose Enter ▾ . It's also worth remembering that if you activate bootstrapping then any residuals that you have asked to be saved won't be saved. This is annoying because it means that to bootstrap the model parameters we have to run the analysis again, but deselecting the options in Figure 20.9 (I would also deselect the options in Figure 20.10 just to keep things simple). Let's do this, select the usual bootstrapping options (Section 6.12.3), and rerun the model by clicking OK .

20.5.13 Listing residuals ▮▮▮

As we saw for the linear model, residuals are saved in columns in the data editor. We can list them in the output viewer using the *Analyze ▸ Reports ▸* 🔢 Case Summaries... dialog box (see Section 9.11.6).

20.6 Interpreting logistic regression ▮▮▮

20.6.1 Block 0 ▮▮▮

The output is split into two blocks: the model before (block 0) and after (block 1)

Iteration History[a,b,c]

Iteration		−2 Log likelihood	Coefficients Constant
Step 0	1	154.084	.301
	2	154.084	.303
	3	154.084	.303

a. Constant is included in the model.

b. Initial −2 Log Likelihood: 154.084

c. Estimation terminated at iteration number 3 because parameter estimates changed by less than .001.

Output 20.3

Table 20.2 Crosstabulation of intervention with outcome status (cured or not)

		Intervention or Not	
		No Treatment	Intervention
Cured? (Cured)	**Not Cured**	32	16
	Cured	24	41
	Total	56	57

Intervention is included. As such, block 1 is the main bit in which we're interested, but Output 20.3 from block 0 is useful because it tells us the initial −2LL value (154.084), which we'll use later, so don't forget it.[7]

20.6.2 Model summary

With **Intervention** included in the model a patient is classified as being cured or not based on whether they had an intervention or not. To illustrate this principle, look at the crosstabulation for the variables **Intervention** and **Cured** in Table 20.2.[8] The model applies this classification table to decide whether a patient was cured or

not based on what intervention they had. For example, there were 57 patients who had the intervention, so the model predicts that these 57 patients were cured. From Table 20.2 we can see that the model is correct for 41 patients, but misclassifies 16 patients as 'cured' who were not. Similarly, the model predicts that all the 56 patients who received no treatment were not cured and in doing so it correctly classifies 32 patients but misclassifies 24 patients as 'not cured' who were. The classification table in Output 20.4 shows this pattern of predicted values and observed values, with the diagonal values being the correctly classified cases and the off-diagonal elements being misclassified cases. This

output also tells us that the model correctly classifies 66.7% of not cured cases and 63.1% of cured cases and the overall accuracy is 64.6%[9] (the weighted average of these two values).

Output 20.4 also shows summary statistics for the model,[10] but we should also look at the table that we already inspected in Output 20.2 that showed the chi-square statistic for the model (remember it was 9.926 and highly significant, $p = 0.002$). This chi-square statistic is derived from equation (20.7) and is the difference between the current −2LL (which for this model is 144.158) and the baseline −2LL (i.e., the value before **Intervention** was added, which is reported as 154.084 in Output 20.3): 154.084 − 144.158 = 9.926. Output 20.4 also tells us the values of Cox and Snell's and Nagelkerke's R^2, but we will discuss these a little later.

Output 20.5 tells us the estimates for the b-values, confidence intervals, p-values and odds ratio for the predictors in the model (namely **Intervention** and the constant). We can replace the b-values in equation (20.3) to establish the probability that a case falls into a certain category. In linear models of continuous outcomes the value of b is the change in the outcome resulting from a unit change in the predictor variable. The interpretation is very similar in logistic regression: it is the change in the *logit* of the outcome variable associated with a one-unit change in the predictor variable. The logit of the outcome is the natural logarithm of the odds of Y occurring.

The output also tells us the Wald statistic (equation (20.13)),[11] from which a p-value can be computed. If the coefficient is significantly different from zero then we

---

7 If you can't see this output it'll be because you didn't select ☑ Iteration history in Figure 20.10.

8 The dialog box to produce this table can be obtained by selecting *Analyze* ▶ *Descriptive Statistics* ▶ ⊞ C̲rosstabs....

9 If you go back and look at the classification tables for models 2 and 3, you'll notice they are identical to the one reported for this model, which means that adding **Duration** and the interaction term did not lead to even a single person being more accurately classified than when we include only **Intervention** as a predictor.

10 If you use bootstrapping you'll notice a load of guff below this table. Ignore it.

11 As we have seen, the Wald statistic is b divided by its standard error (1.229/0.40 = 3.0725); however, SPSS quotes the Wald statistic squared, $3.0725^2 = 9.44$ as reported (within rounding error) in the table.

can assume that the predictor is making a significant contribution to the prediction of the outcome (Y). For these data, the Wald statistic indicates that having the intervention (or not) is a significant predictor of whether the patient was cured because the p-value is 0.002, which is less than the conventional threshold of 0.05.

If you followed my advice and reran the model with bootstrapping activated you'll get the table labeled *Bootstrap for Variables in the Equation* in Output 20.5. This table reports the b values again, but estimates the standard error using bootstrap resampling (Section 6.12.3). The change in the standard error results in a different p-value for the b (it is 0.004 instead of 0.002), but it is still significant. The bootstrap confidence interval for the b-values tells us that the population value of b falls between 0.399 and 2.223 (assuming this sample is one of the 95% for which the confidence interval contains the population value). This interval doesn't include zero, so we might conclude that there is a genuine positive relationship between having the intervention (or not) and being cured (or not). The bootstrap confidence intervals will differ slightly every time you run the analysis, but they are nevertheless robust to violations of the underlying assumptions of the test.

20.6.3 Analogues of R ▮▮▮▯

In Section 20.3.3 we saw that we could calculate an analogue of R using equation (20.8). For these data, z^2 (the Wald statistic in the output) and its df are in Output 20.5 (9.447 and 1, respectively), and the baseline $-2LL$ was 154.084 (Output 20.3). Therefore, R is 0.22:

$$R = \sqrt{\frac{9.447 - (2 \times 1)}{154.084}}$$
$$= 0.22$$

(20.17)

Hosmer and Lemeshow's measure (R_L^2) from Section 20.3.3 is 0.06:

Model Summary

Step	-2 Log likelihood	Cox & Snell R Square	Nagelkerke R Square
1	144.158[a]	.084	.113

a. Estimation terminated at iteration number 3 because parameter estimates changed by less than .001.

Classification Table[a]

			Predicted		
			Cured?		Percentage Correct
Observed			Not Cured	Cured	
Step 1	Cured?	Not Cured	32	16	66.7
		Cured	24	41	63.1
	Overall Percentage				64.6

a. The cut value is .500

Output 20.4

Variables in the Equation

		B	S.E.	Wald	df	Sig.	Exp(B)	95% C.I.for EXP(B) Lower	Upper
Step 1[a]	Intervention(1)	1.229	.400	9.447	1	.002	3.417	1.561	7.480
	Constant	-.288	.270	1.135	1	.287	.750		

a. Variable(s) entered on step 1: Intervention.

Bootstrap for Variables in the Equation

		Bootstrap[a]					
		B	Bias	Std. Error	Sig. (2-tailed)	BCa 95% Confidence Interval Lower	Upper
Step 1	Intervention(1)	1.229	.034	.421	.004	.399	2.223
	Constant	-.288	-.015	.280	.293	-.804	.154

a. Unless otherwise noted, bootstrap results are based on 1000 bootstrap samples

Output 20.5

$$R_L^2 = \frac{(-2LL_{\text{baseline}}) - (-2LL_{\text{new}})}{-2LL_{\text{baseline}}}$$
$$= \frac{154.084 - 144.158}{154.084}$$
$$= 0.06$$

(20.18)

This is the same as dividing the model chi-square after **Intervention** has been entered into the model (9.93) by the baseline $-2LL$ (before any variables were entered). The resulting value of 0.06 is different than the one we would get by squaring R above ($R^2 = 0.22^2 = 0.05$). Two other measures of R^2, which were described in Section 20.3.3, are in Output 20.4: Cox and Snell's measure (0.084) and Nagelkerke's adjusted value (0.113). All of these R^2 values differ, but they give us an approximate effect size measure for the model.

20.6.4 The odds ratio ▮▮▮▮

Output 20.5 also gives us the odds ratio (*Exp(B)* in the output), which was described in Section 20.3.5. The reason why the odds ratio is labeled *Exp(B)* is that it is literally the exponential of the b for the predictor, in this case $e^{1.229} = 3.42$. However, most people are more familiar with the term 'odds ratio'. In the options (see Section 20.5.11), we requested a confidence interval for the odds ratio, and it can also be found in the output. Assuming the current sample is one of the 95% for which the confidence interval contains the true value, the population value of the odds ratio lies between 1.56 and 7.48. However, our sample could be one of the 5% that produce a confidence interval that 'misses' the population value. The important thing is

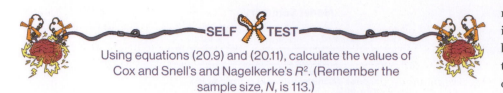

SELF TEST

Using equations (20.9) and (20.11), calculate the values of Cox and Snell's and Nagelkerke's R^2. (Remember the sample size, N, is 113.)

mean that as the predictor variable increases, so do the odds of (in this case) being cured, but values less than 1 mean that as the predictor variable increases, the odds of being cured *decrease*. If the confidence interval contains 1 then the population value might be one that suggests that the intervention improves the probability of cure, but equally it might be

that the interval doesn't contain 1 (both values are greater than 1). The value of 1 is important because it is the threshold at which the direction of the effect changes. Think about what the odds ratio represents (Section 20.3.5): values greater than 1

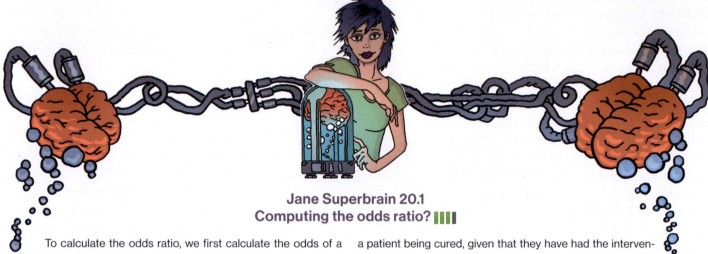

Jane Superbrain 20.1
Computing the odds ratio?

To calculate the odds ratio, we first calculate the odds of a patient being cured given that they *didn't* have the intervention, using equation (20.3). The parameter coding at the beginning of the output told us that patients who did not have the intervention were coded with a 0, so we use this value as X. The value of b_1 has been estimated for us as 1.229 (see Output 20.5), and the coefficient for the constant can be taken from the same table and is −0.288. We can calculate the odds as:

$$P(\text{cured}) = \frac{1}{1+e^{-(b_0+b_1X_{1i})}}$$
$$= \frac{1}{1+e^{-[-0.288+(1.229\times0)]}} = 0.428$$
$$P(\text{not cured}) = 1 - P(\text{cured})$$
$$= 1 - 0.428$$
$$= 0.572$$
$$\text{odds} = \frac{0.428}{0.572}$$
$$= 0.748$$

(20.19)

Now, we calculate the same thing *after the predictor variable has changed by one unit*. In this case, because the predictor variable is dichotomous it means calculating the odds of

a patient being cured, given that they have had the intervention. The value of the intervention variable, X, is now 1 (rather than 0). The resulting calculations are:

$$P(\text{cured}) = \frac{1}{1+e^{-(b_0+b_1X_{1i})}}$$
$$= \frac{1}{1+e^{-[-0.288+(1.229\times1)]}} = 0.719$$
$$P(\text{not cured}) = 1 - P(\text{cured})$$
$$= 1 - 0.719$$
$$= 0.281$$
$$\text{odds} = \frac{0.719}{0.281} = 2.559$$

(20.20)

Now that we know the odds before and after a unit change in the predictor variable, it is a simple matter to calculate the odds ratio as in equation (20.15). The result is 3.42:

$$\text{odds ratio} = \frac{\text{odds after a unit change in the predictor}}{\text{original odds}}$$
$$= \frac{2.559}{0.748}$$
$$= 3.42$$

(20.21)

a value that suggests that the intervention decreases the probability of being cured. For our confidence interval, the fact that both limits are above 1 suggests that the direction of the relationship that we have observed is true in the population (i.e., it's likely that having an intervention compared to not increases the odds of being cured). If the lower limit had been below 1 then it would tell us that there is a chance that in the population the direction of the relationship is the opposite to what we have observed. This would mean that it's ambiguous whether the intervention increases the odds of being cured.

20.6.5 Classification plots

Output 20.6 displays the classification plot, which is a histogram of the predicted probabilities of a patient being cured. If the model perfectly fits the data, then this histogram should show all the cases for which the event has occurred on the right-hand side, and all the cases for which the event hasn't occurred on the left-hand side. In this example, all the patients who were cured should appear on the right and all those who were not cured should appear on the left. Because the only predictor is dichotomous there are just two columns of cases on the plot. If the predictor is a continuous variable, the cases will be spread across many columns. As a rule of thumb, the more the cases cluster at each end of the graph, the better; such a plot would show that when the outcome did occur (i.e., the patient was cured) the predicted probability of the event occurring is also high (i.e., close to 1). Likewise, at the other end of the plot it would show that when the event didn't occur (i.e., the patient still had a problem) the predicted probability of the event occurring is also low (i.e., close to 0). This situation represents a model that correctly predicts the observed outcome data. If lots of points cluster in the center of the plot then it shows that for many cases the

model is predicting a probability of 0.5; in other words, there is close to a 50:50 chance that these cases are predicted correctly by the model—you could predict these cases as accurately as the model by tossing a coin. In Output 20.6 cured cases are predicted relatively well by the model (the probabilities are not that close to 0.5), but for not cured cases the model is less good (the probability of classification is only slightly lower than 0.5). Also, a good model will ensure that few cases are misclassified; for these data there are a few Ns (not cured) appearing on the cured side, but more worryingly there are quite a few Cs (cured) appearing on the N side.

20.6.6 Listing predicted probabilities

In Section 20.5.10 we saved residuals and predicted probabilities. The predicted probabilities and predicted group memberships are saved as variables in the

data editor with the names **PRE_1** and **PGR_1**, respectively. These probabilities can be listed using the *Analyze* ▶ *Reports* ▶ Case Summaries... dialog box (see Section 9.11.6).

Output 20.7 shows a selection of the predicted probabilities (because the only predictor in the model was a dichotomous variable, there will be only two different probability values). I have also listed the predictor variables to clarify from where the predicted probabilities come. The only predictor in the final model was having the intervention, which could have a value of either 1 (had the intervention) or 0 (no intervention). If these two values are placed into equation (20.3) with the respective regression coefficients, then the two probability values in Output 20.7 are derived. In fact, we calculated these values in Jane Superbrain Box 20.1: the calculated probabilities (P(cured) in these equations) correspond to the values in **PRE_1**. These values tell us that when a patient is not

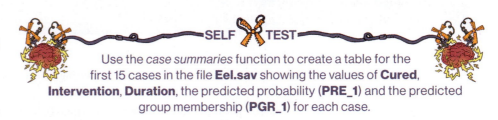

SELF TEST

Use the *case summaries* function to create a table for the first 15 cases in the file **Eel.sav** showing the values of **Cured**, **Intervention**, **Duration**, the predicted probability (**PRE_1**) and the predicted group membership (**PGR_1**) for each case.

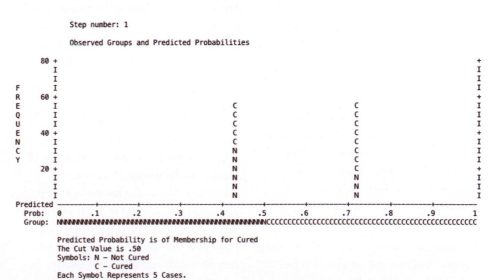

Output 20.6

Cramming Sam's Tips
Model fit

- Build your model systematically and choose the most parsimonious model as the final one.
- The overall fit of the model is shown by −2LL and its associated chi-square statistic. If the significance of the chi-square statistic is less than 0.05, then the model is a significant fit to the data.
- Check the table labeled *Variables in the Equation* to see the regression parameters for any predictors you have in the model.

- For each variable in the model, look at the Wald statistic and its significance (which again should be below 0.05). Use the odds ratio, *Exp*(*B*), for interpretation. If the value is greater than 1 then as the predictor increases, the odds of the outcome occurring increase. Conversely, a value less than 1 indicates that as the predictor increases, the odds of the outcome occurring decrease. For the aforementioned interpretation to be reliable the confidence interval of *Exp*(*B*) should not cross 1.

treated (**Intervention** = 0, no treatment), there is a probability of 0.428 that they will be cured—basically, about 43% of people get better without any treatment. However, if the patient does have the intervention (**Intervention** = 1, yes), there is a probability of 0.719 that they will get better—about 72% of people treated get better. When you consider that a probability of 0 indicates no chance of getting better, and a probability of 1 indicates that the patient will definitely get better, the values obtained suggest that having the intervention increases your chances of getting better (although the probability of recovery without the intervention is still not bad).

Assuming we are content that the model is accurate and that the intervention has some substantive significance, then we could conclude that our intervention

(which, to remind you, was putting an eel up the anus) is the best predictor of getting better (not being constipated). Furthermore, including the duration of the constipation pre-intervention and its interaction with the intervention did not improve how well we could predict whether a person got better.

20.6.7 Interpreting residuals

Fitting a model without checking how well it fits the data is like buying a new pair of trousers without trying them on: they might look fine on the hanger but get them home and you find you're Jenny or Johnny Tight-pants. The trousers do their job (they cover your legs and keep you warm) but they have little real-life value (because they cut off the blood circulation to your legs, which then need to be amputated).

Likewise, a model does its job regardless of the data, but the real-life value of the model may be limited. So, our conclusions so far are fine in themselves, but to be sure that the model is a good one, it is important to examine the residuals.

We saw in Chapter 9 that the main purpose of examining residuals is to (1) isolate points for which the model fits poorly, and (2) isolate points that exert an undue influence on the model. To assess the former we examine the residuals, especially the studentized residuals, standardized residuals and deviance statistics. To assess the latter we use influence statistics such as Cook's distance, DFBeta and leverage statistics. These statistics were explained in Section 9.3 and their interpretation in logistic regression is the same; therefore,

Table 20.3 summarizes the main statistics that you should look at and what to look for, but for more detail consult Chapter 9. Remember that these residuals are saved in the data editor. If you inspect them you'll see that Cook's distance, leverage, standardized residuals and DFBeta values are pretty good: all cases have DFBetas less than 1, and leverage statistics (**LEV_1**) are very close to the calculated expected value of 0.018. There are also no unusually high values of Cook's distance (**COO_1**) which, all in all, means that there are no influential cases. The standardized residuals all have values of less than ±2 and so there seems to be very little here to concern us.

The residuals in this model are slightly unusual because they are based on a single predictor that is categorical. Consequently there isn't a lot of variability in the values of the residuals. Also, remember that if substantial outliers or influential cases had been isolated, you would not be justified in eliminating these cases to make the model fit better. Instead these cases should be inspected closely to try to isolate a good reason why they were unusual. It might simply be an error in inputting data or it could be that the case was one which had a special reason for being unusual: for example, there were other medical complications that might contribute to the constipation and that were noted during the patient's assessment. In such a case, you may have good reason to exclude the case and duly note the reasons why.

Case Summaries[a]

	Cured?	Intervention	Number of Days with Problem before Treatment	Predicted probability	Predicted group
1	Not Cured	No Treatment	7	.42857	Not Cured
2	Not Cured	No Treatment	7	.42857	Not Cured
3	Not Cured	No Treatment	6	.42857	Not Cured
4	Cured	No Treatment	8	.42857	Not Cured
5	Cured	Intervention	7	.71930	Cured
6	Cured	No Treatment	6	.42857	Not Cured
7	Not Cured	Intervention	7	.71930	Cured
8	Cured	Intervention	7	.71930	Cured
9	Cured	No Treatment	8	.42857	Not Cured
10	Not Cured	No Treatment	7	.42857	Not Cured
11	Cured	Intervention	7	.71930	Cured
12	Cured	No Treatment	7	.42857	Not Cured
13	Cured	No Treatment	5	.42857	Not Cured
14	Not Cured	Intervention	9	.71930	Cured
15	Not Cured	No Treatment	6	.42857	Not Cured
Total N	15	15	15	15	15

a. Limited to first 15 cases.

Output 20.7

Oliver Twisted
Please, Sir, can I have some more … diagnostics?

'What about the trees?' protests eco-warrior Oliver. 'These SPSS outputs take up so much room, why don't you put them on the website instead?' It's a valid point so I have produced a table of the diagnostic statistics for this example, but it's on the companion website.

Table 20.3 Summary of residual statistics saved by SPSS

Label	Name	Comment
PRE_1	Predicted value	
PGR_1	Predicted group	
COO_1	Cook's distance	Should be less than 1
LEV_1	Leverage	Lies between 0 (no influence) and 1 (complete influence). The expected leverage is $(k +1)/N$, where k is the number of predictors and N is the sample size. In this example it would be 2/113 = 0.018
SRE_1	Studentized residual	Only 5% should lie outside ±1.96, and about 1% should lie outside ±2.58. Cases above 3 are cause for concern, and cases close to 3 warrant inspection
ZRE_1	Standardized residual	
DEV_1	Deviance	
DFB0_1	DFBeta for the constant	Should be less than 1
DFB1_1	DFBeta for the first predictor (**Intervention**)	

20.6.8 Calculating the effect size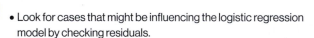

The best effect size to use in the context of logistic regression is the odds ratio, which we looked at in Section 20.6.4 (but see Jane Superbrain Box 20.2).

20.7 Reporting logistic regression

I'd report logistic regression much the same as any linear model (see Section 9.13). I'd be inclined to tabulate the results, unless it's a very simple model.

As a bare minimum, report the b-values (and their standard errors and significance value), the odds ratio (and its confidence interval) and some general statistics about the model (such as the R^2 and goodness-of-fit statistics). I'd also include the constant so that readers of your work can construct the full regression model if they need to. You might also consider reporting the variables that were not significant predictors, because this can be as valuable as knowing about which predictors were significant.

For the example in this chapter we might produce something like Table 20.4. Hopefully you can work out from where the values came by looking back through the chapter so far. I've rounded off to 2 decimal places throughout. If you use APA style then for the R^2 and p-values there should be no zero before the decimal point (because these values cannot exceed 1). I have reported the bootstrap confidence intervals for b.

Cramming Sam's Tips
Diagnostic statistics

- Look for cases that might be influencing the logistic regression model by checking residuals.
- Look at standardized residuals and check that no more than 5% of cases have absolute values above 2, and that no more than about 1% have absolute values above 2.5. Any case with a value above about 3 could be an outlier.

- Look in the data editor for the values of Cook's distance: any value above 1 indicates a case that might be influencing the model.
- Calculate the average leverage (the number of predictors plus 1, divided by the sample size) and then look for values greater than twice or three times this average value.
- Look for absolute values of DFBeta greater than 1.

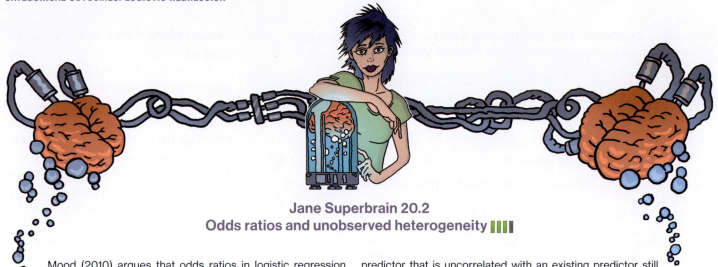

Jane Superbrain 20.2
Odds ratios and unobserved heterogeneity ▊▊▊▊

Mood (2010) argues that odds ratios in logistic regression reflect not just the effect of a predictor but unobserved heterogeneity in the model. This observation is demonstrated by model parameters in logistic regression changing when an uncorrelated predictor is added. Imagine a linear model in which Y is predicted from X_1. Remember from our discussion of linear models that b-values represent the effect of a predictor adjusted for their relationship to other predictors in the model. If we add the predictor X_2, then the b-value for X_1 will be adjusted for its relationship with X_2: its value will change as a function of whether X_2 is in the model. If X_1 and X_2 are uncorrelated (i.e. independent), then the 'adjustment' will be zero, and so the b-value for X_1 will be unaffected by whether X_2 is also in the model. This doesn't happen in logistic regression: adding a predictor that is uncorrelated with an existing predictor still changes the b-value for the existing predictor. This is because estimates of residual variance are affected by what predictors are in the model. Consequently, Mood argues that: (1) it is problematic to use odds ratios (or log odds ratios) as effect sizes measures because they reflect unobserved heterogeneity as well as the size of effect; (2) odds ratios (and log odds ratios) should not be compared across models with different predictors because unobserved heterogeneity will vary across these models; and (3) where models use the same predictors, it is still problematic to compare odds ratios (and log odds ratios) across different samples, across groups within the same sample, and over time because unobserved heterogeneity will vary across samples, groups and time points.

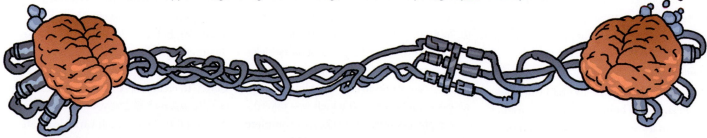

20.8 Testing assumptions: another example ▊▊▊▊

I am English, and a very important part of being English is believing that we can win sports events despite the crushing weight of historical  evidence to the contrary. English people are genetically programd to fail in high-pressure environments; that's a fact,[12] but as each new tournament arrives we are programd by the media to believe that somehow England will be victorious. With every defeat, we lose a little bit of our soul. My writing of each edition of this book has coincided with a soccer-related national failure. In 1998 when I wrote the first edition, England was knocked out of the World Cup by losing a penalty shootout. In 2004 (second edition), we were knocked out of the European Championship in another penalty shootout. We didn't even manage to qualify for the 2008 European Championship (third edition); not a penalty shootout this time, just playing like cretins. In 2012 (fourth edition), we lost yet another penalty shootout that sent us home from the European

Why do the England football team always miss penalties?

---

12 The rugby World Cup winning side of 2003 was the exception that proves the rule. Oh, and I think we won the Ashes in 2005, but we lost them the previous 18 years in a row, so that victory is hardly anything to feel smug about. Also, I don't like cricket (and not in the 10cc sense).

Table 20.4 Coefficients of the model predicting whether a patient was cured (95% BCa bootstrap confidence intervals based on 1000 samples in brackets)

	b	95% CI for Odds Ratio		
		Lower	Odds Ratio	Upper
Included				
Constant	−0.29 [−0.77, 0.20]			
Intervention	1.23* 0.42, 2.06]	1.56	3.42	7.48

Note. R^2 = 0.06 (Hosmer–Lemeshow), 0.08 (Cox–Snell), 0.11 (Nagelkerke). Model $\chi^2(1)$ = 9.93, *p* = 0.002. **p* < 0.01.

Championship. And last year (2016), just before this edition, we lost to a group of semi-professional Icelandic people.[13] What is *wrong* with English footballers?[14]

If I were the England soccer team coach, I'd take each one of the overpaid prima donnas and give them a kick in the testicles. A really hard kick. The kind of kick that lets them know that I'm the kind of guy that'll give you a good hard kick in your testicles if you miss a penalty. It wouldn't be as much fun, but I suppose, if push came to shove, I could also use science to find out which factors predict whether a player will score a penalty. Then I'd kick them in the testicles. One way or another, their testicles are getting kicked.

Those of you who hate football can read this example as being factors that predict success in a free throw in basketball or netball, a penalty in hockey, a penalty kick in rugby or a field goal in American football. This research question is perfect for logistic regression because our outcome variable is a dichotomy: a penalty can be either scored or missed. Imagine that past research (Hoddle et al., 1998; Eriksson et al., 2004; Hodgson et al., 2012) had shown that there are two factors that reliably predict whether a penalty kick will be missed or scored. The first factor is whether the player taking the kick is a worrier (measured using a

scale such as the Penn State Worry Questionnaire, PSWQ). The second factor is the player's past success rate at scoring (so whether the player has a good track record of scoring penalty kicks). It is well accepted that anxiety has detrimental effects on the performance of a variety of tasks, and so it was also predicted that state anxiety might be able to account for some of the unexplained variance in penalty success.

This example is a classic case of building on a well-established model, because two predictors are already known and we want to test the effect of a new one. So, 75 football players were selected at random and before taking a penalty kick in a competition they were given a state anxiety questionnaire to complete (to assess anxiety before the kick was taken). These players were also asked to complete the PSWQ to give a measure of how much they worried about things generally, and their past success rate was obtained from a database. Finally, a note was made of whether the penalty was scored or missed. The file **Penalty.sav**

contains four variables, each in a separate column:

- **Scored**: This variable is our outcome and it is coded such that 0 = penalty missed and 1 = penalty scored.
- **PSWQ**: The first predictor is a measure of the degree to which a player worries.
- **Previous**: This variable is the percentage of penalties scored by a player in their career. It represents previous success at scoring penalties.
- **Anxious**: This variable is a measure of state anxiety before taking the penalty. It's our third predictor and it is a variable that has not previously been used to predict penalty success.

20.8.1 Testing for linearity of the logit

In this example we have three continuous variables, therefore we have to check that each one is linearly related to the log of the outcome variable (**Scored**). I mentioned earlier in this chapter that to test this assumption we need to run the logistic regression but include predictors that are the interaction between each predictor and the log of itself (Hosmer & Lemeshow, 1989). To create these interaction terms, we use *Transform* ▶ Compute Variable... (Section 6.12.4). For each variable create a new variable that is the log of the original variable. For example, for **PSWQ**, create a new variable called **LnPSWQ** by entering this name into the box labeled *Target Variable*, click Type & Label... and give the variable a name such as *Ln(PSWQ)*. In the

SELF TEST
Conduct a hierarchical logistic regression analysis on these data. Enter **Previous** and **PSWQ** in the first block and **Anxious** in the second (forced entry). There is a full guide on how to do the analysis and its interpretation on the companion website.

---

13 Which, honestly, I found hilarious.
14 More accurately, what is wrong with *male* English footballers? The women's team does us proud year after year.

list labeled *Function group* click *Arithmetic* and then in the box labeled *Functions and Special Variables* click *Ln* (this is the natural log transformation) and transfer it to the command area by clicking ⬆. The command will appear in the command area as 'LN(?)' and the question mark should be replaced with a variable name (which can be typed manually or transferred from the variables list). Replace the question mark with the variable **PSWQ** by either selecting the variable in the list and clicking ➡ or typing 'PSWQ' where the question mark is. Click OK to create the variable.

To test the assumption we redo the analysis, but force all variables in a single block (i.e., we don't need to do it hierarchically). We also need to put in three new interaction terms of each predictor and their logs. Select *Analyze* ▸ *Regression* ▸ 🔳 **Binary Logistic**…, then in the main dialog box drag **Scored** to the *Dependent* box (or click ➡). Specify the main effects by clicking on **PSWQ**, **Anxious** and **Previous** while holding down the *Ctrl* key (or *Cmd* on a Mac) and dragging them to the *Covariates* box (or click ➡). To input the interactions, click the two variables in the interaction while holding down the *Ctrl* key (or *Cmd* on a Mac): for example, click **PSWQ** then, while holding down *Ctrl*, click **Ln(PSWQ))** and click >a*b> to move them to the *Covariates* box. This action specifies the **PSWQ** × **Ln(PSWQ)** interaction. Specify the **Anxious** × **Ln(Anxious)** and **Previous** × **Ln(Previous)** interactions in the same way. The completed dialog box is in Figure 20.11.

Output 20.8 shows the part of the output that tests the assumption. We're interested only in whether the

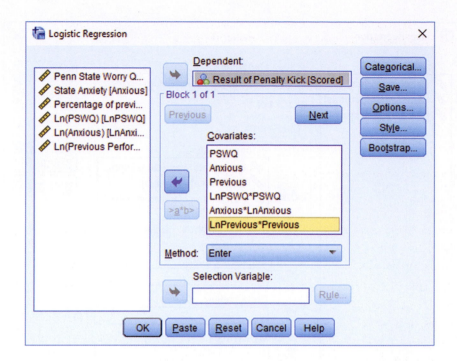

Figure 20.11   Dialog box for testing the assumption of linearity in logistic regression

Variables in the Equation

		B	S.E.	Wald	df	Sig.	Exp(B)	95% C.I.for EXP(B) Lower	Upper
Step 1ᵃ	Penn State Worry Questionnaire	−.422	1.102	.147	1	.702	.656	.076	5.690
	State Anxiety	−2.650	2.784	.906	1	.341	.071	.000	16.564
	Percentage of previous penalties scored	1.669	1.473	1.285	1	.257	5.309	.296	95.202
	Ln(PSWQ) by PSWQ	.044	.297	.022	1	.883	1.045	.584	1.869
	State Anxiety by Ln(Anxious)	.682	.650	1.102	1	.294	1.978	.553	7.069
	Ln(Previous Performance) by Percentage of previous penalties scored	−.319	.315	1.025	1	.311	.727	.392	1.348
	Constant	−3.874	14.924	.067	1	.795	.021		

a. Variable(s) entered on step 1: Penn State Worry Questionnaire, State Anxiety, Percentage of previous penalties scored, Ln(PSWQ) * Penn State Worry Questionnaire , State Anxiety * Ln(Anxious) , Ln(Previous Performance) * Percentage of previous penalties scored .

Output 20.8

interaction terms are significant. Any interaction that is significant indicates that the main effect has violated the assumption of linearity of the logit. All three interactions have significance values greater than 0.05, indicating that the assumption of linearity of the logit has been met for **PSWQ**, **Anxious** and **Previous**.

20.8.2 Testing for multicollinearity ▌▌▌▌

In Section 9.9.3 we saw how multicollinearity can affect the parameters of a linear model. Logistic regression is just as prone to the biasing effect of collinearity, so we need to test for it. Unfortunately, SPSS does not produce collinearity diagnostics in logistic regression (which creates the illusion that multicollinearity doesn't matter). However, you can obtain statistics such as the tolerance and VIF by running a linear regression using the same outcome and predictors.

SELF TEST
Try creating two new variables that are the natural logs of **Anxious** and **Previous**.

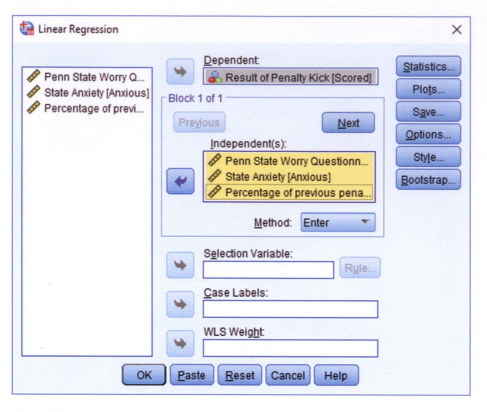

Figure 20.12 *Linear Regression* dialog box for the penalty data

For the penalty example, select *Analyze ▸ Regression ▸* Linear.... The completed dialog box is shown in Figure 20.12. It is unnecessary to specify lots of options (we are using this technique only to obtain tests of collinearity), but it is essential that you click Statistics... and then select ☑ Collinearity diagnostics and switch off all the default options. Click Continue to return to the main dialog box, and click OK to run the analysis.

The first table in Output 20.9 shows tolerance values of 0.015 for **Previous**, 0.014 for **Anxious** and 0.575 for **PSWQ**. To recap, tolerance values less than 0.1 (Menard, 1995) and VIF values greater than 10 (Myers, 1990) indicate a problem (Chapter 9). In these data the VIF values are around 70 for both **Anxious** and **Previous**, indicating an issue of collinearity between the predictor variables. We can investigate this issue further by examining the table labeled *Collinearity Diagnostics* (Section

9.9.3). For these data the final dimension has a condition index of 80.5, which is massive compared to the other dimensions. Although there are no hard-and-fast rules about how much larger a condition index needs to be to indicate collinearity problems, this case clearly shows that a problem exists. The variance proportions tell us the proportion of the variance of each predictor's *b*-value that is attributed to each eigenvalue. These proportions can be converted to percentages by multiplying them by 100. So, for example, for **PSWQ** 95% of the variance of the *b*-value is associated with eigenvalue number 3, 4% is associated with eigenvalue number 2 and 1% is

associated with eigenvalue number 1. In terms of collinearity, we are looking for predictors that have high proportions on the same *small* eigenvalue, because this would indicate that the variances of their *b*-value are dependent. So we are interested mainly in the bottom few rows of the table (which represent small eigenvalues). In this example, 99% of the variance in the regression coefficients of both **Anxiety** and **Previous** is associated with eigenvalue number 4 (the smallest eigenvalue), which clearly indicates dependency between these variables. The result of this analysis is clear-cut: there is collinearity between state anxiety and previous experience of taking penalties, and this dependency results in the model becoming biased.

If you have identified collinearity then, unfortunately, there's not much that you can do about it. One obvious solution is to omit one of the variables (so, for example, we might stick with a model that ignores state anxiety). The problem with this should be obvious: there is no way of knowing which variable to omit. The resulting theoretical conclusions are meaningless because, statistically speaking, any of the collinear variables could be omitted. There are no statistical grounds for omitting one variable over another. Even if a predictor is removed, Bowerman and O'Connell (1990) recommend that another equally important predictor that does not have such strong multicollinearity replace it. They also suggest collecting more data to see whether the multicollinearity can be lessened. Another possibility when there are several predictors involved in the multicollinearity is to run a PCA on these predictors and to use the resulting

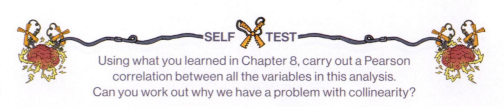

SELF TEST

Using what you learned in Chapter 8, carry out a Pearson correlation between all the variables in this analysis. Can you work out why we have a problem with collinearity?

component scores as a predictor (see Chapter 18). The safest (although unsatisfactory) remedy is to acknowledge the unreliability of the model. So, if we were to report the analysis of which factors predict penalty success, we might acknowledge that previous experience significantly predicted penalty success in the first model, but propose that this experience might affect penalty taking by increasing state anxiety. This statement would be highly speculative because the correlation between **Anxious** and **Previous** tells us nothing of the direction of causality, but it would acknowledge the inexplicable link between the two predictors.

Coefficients[a]

| Model | | Collinearity Statistics | |
		Tolerance	VIF
1	Penn State Worry Questionnaire	.575	1.740
	State Anxiety	.014	70.028
	Percentage of previous penalties scored	.015	68.777

a. Dependent Variable: Result of Penalty Kick

Collinearity Diagnostics[a]

Model	Dimension	Eigenvalue	Condition Index	(Constant)	Penn State Worry Questionnaire	State Anxiety	Percentage of previous penalties scored
1	1	3.436	1.000	.00	.01	.00	.00
	2	.490	2.647	.00	.04	.00	.00
	3	.073	6.875	.00	.95	.01	.00
	4	.001	80.491	1.00	.00	.99	.99

a. Dependent Variable: Result of Penalty Kick

Output 20.9 Collinearity diagnostics for the penalty data

20.9 Predicting several categories: multinomial logistic regression ▌▌▌

If you want to predict membership of more than two categories, the logistic regression model extends to *multinomial logistic regression*. The outcome categories can have a meaningful order (ordinal categories). Multinomial logistic regression works in much the same way as the binary case, so there's no need for any additional equations to explain what is going on (hooray!). The model breaks the outcome variable into a series of comparisons between two categories (that's why no extra equations are necessary). For example, if you have three outcome categories (A, B and C), then the model will consist of two comparisons that depend on how you specify the model: you can compare categories against the first outcome category (e.g., A vs. B and A vs. C), the last category (e.g., A vs. C and B vs. C) or a custom category, for example category B (e.g., B vs. A and

B vs. C). Therefore, in practice, you have to select a baseline outcome category. The important parts of the model and output are much the same for binary logistic regression.

Let's look at an example. Research on how men and women evaluate chat-up lines (Bale, Morrison, & Caryl, 2006; Cooper, O'Donnell, Caryl, Morrison, & Bale, 2007) has looked at how the content (e.g., whether the chat-up line is funny, has sexual content or reveals desirable personality characteristics) affects how favorably the chat-up line is viewed. The take-home message from this work is that men and women like different things: men prefer chat-up lines with a high sexual content, and women prefer chat-up lines that are funny and show good moral fiber.

Imagine that we wanted to assess how *successful* these chat-up lines were. We recorded the chat-up lines used by 348 men and 672 women in a nightclub. Our outcome was whether the chat-up line resulted in one of the following

three events: the person got no response or the recipient walked away, the person obtained the recipient's phone number or the person left the nightclub with the recipient. This is an ordinal outcome (the three outcome categories map onto increasing levels of 'success'). Afterwards, the chat-up lines used in each case were rated by a panel of judges for how funny they were (0 = not funny at all, 10 = the funniest thing that I have ever heard), sexuality (0 = no sexual content at all, 10 = very sexually direct) and whether the chat-up line reflected good moral values (0 = does not reflect good characteristics, 10 = very indicative of good characteristics). For example, 'I may not be Fred Flintstone, but I bet I could make your bed rock' would score high on sexual content, low on good characteristics and medium on humor; 'I've been looking all over for you, the woman of my dreams' would score high on good characteristics, low on sexual content and low on humor (but high on cheese, had it been measured). Based on the research, we predict that the success of different types of chat-up line will interact with the biological sex of the recipient.

Labcoat Leni's Real Research 20.1
Mandatory suicide? ▌▌▌

Lacourse, E., Claes & Villeneuve (2001). *Journal of Youth & Adolescence, 30*, 321–332.

My favorite kind of music is heavy metal. One thing that is mildly irritating about liking heavy music is that everyone assumes that you're either a moron or a miserable or aggressive bastard. When not listening (and often while listening) to heavy metal, I research clinical psychology in youths. Therefore, I was literally beside myself with excitement when I stumbled on a paper that combined these two interests: Lacourse, Claes, & Villeneuve(2001) carried out a study to see whether a love of heavy metal could predict suicide risk. Fabulous stuff!

Eric Lacourse and his colleagues used questionnaires to measure: suicide risk (yes or no), marital status of parents (together or divorced/separated), the extent to which the person's mother and father were neglectful, self-estrangement/powerlessness (adolescents who have negative self-perceptions, are bored with life, etc.), social isolation (feelings of a lack of support), normlessness (beliefs that socially disapproved behaviors can be used to achieve certain goals), meaninglessness (doubting that school is relevant to gaining employment) and drug use. In addition, the authors measured liking of heavy metal; they included the sub-genres of classic (Black Sabbath, Iron Maiden), thrash metal (Slayer, Metallica), death/black metal (Obituary, Burzum) and gothic (Marilyn Manson). As well as liking they measured behavioral manifestations of worshipping these bands (e.g., hanging posters, hanging out with other metal fans) and what the authors termed 'vicarious music listening' (whether music was used when angry or to bring out aggressive moods). They used logistic regression to predict suicide risk from these variables for males and females separately.

The data for the female sample are in the file **Lacourse et al. (2001) Females.sav**. Labcoat Leni wants you to carry out a logistic regression predicting **Suicide_Risk** from all the predictors (forced entry). (To make your results easier to compare to the published results, enter the predictors in the same order as Table 3 in the paper: **Age, Marital_Status, Mother_Negligence, Father_Negligence, Self_Estrangement, Isolation, Normlessness, Meaninglessness, Drug_Use, Metal, Worshipping, Vicarious**). Does listening to heavy metal predict girls' suicide? If not, what does? Answers are on the companion website (or look at Table 3 in the original paper).

The data are in the file **Chat-Up Lines. sav**. There is one outcome variable (**Success**) with three categories (no response, phone number, go home with recipient) and four predictors: funniness of the chat-up line (**Funny**), sexual content of the chat-up line (**Sex**), degree to which the chat-up line reflects good characteristics/moral fiber (**Moral**) and the biological sex of the person being chatted up (**Recipient_Sex**).

20.9.1 Multinomial logistic regression using SPSS Statistics ▐▐▐▐

To do a multinomial logistic regression select *Analyze* ▸ *Regression* ▸ ▐ *Multinomial Logistic...* to access the main dialog box (Figure 20.13). Our outcome variable is **Success** so drag this variable to the box labeled *Dependent* (or select it and click ▾). Next, we need to set the baseline category for the outcome variable. By default the last category will be used.

It makes most sense to use the first category as a baseline because it represents failure (the chat-up line did not have the desired effect and resulted in no response or the recipient walking off) whereas the other two categories represent some form of success (getting a phone number or leaving the club together). To change the reference category to be the first category, click Reference Category... , select ⦿ First Category and then click Continue to return to the main dialog box (Figure 20.13).

Next we specify the predictor variables. We have one categorical predictor variable, which is **Recipient_Sex**, so drag (or select and click ▾) this variable to the box labeled *Factor(s)*. Finally, we have three continuous predictors or covariates (**Funny**, **Sex** and **Moral**). Select all of these variables simultaneously by holding down the *Ctrl* key (*Cmd* on a Mac) as you click each one. Drag all three to the box labeled *Covariate(s)* (or click ▾). This is all we need to do to get a model into which these predictors are forced. However, our hypotheses involve interaction terms (between chat-up line content and the recipient's sex), and to get these we need to customize the model.

20.9.2 Customizing the model using SPSS Statistics ▐▐▐▐

Unlike binary logistic regression, with multinomial logistic regression we can't specify interactions between predictor

SELF ❊ TEST

Think about the three categories that we have as an outcome variable. Which of these categories do you think makes most sense as a baseline category?

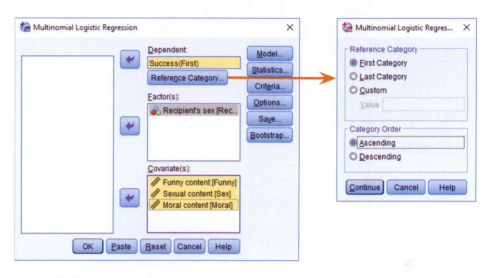

Figure 20.13 Main dialog box for multinomial logistic regression

variables in the main dialog box. Instead we specify a 'custom model' by clicking Model... to open the dialog box in Figure 20.14. You'll see that, by default, only the main effects are included. In this example, the main effects are not particularly interesting: based on past research we don't expect funny chat-up lines to be successful overall, we expect them to be more successful in female recipients than in male ones. This prediction implies a significant *interaction* between **Recipient_Sex** and **Funny**. Similarly, chat-up lines with a high sexual content should not be successful overall, only when the recipient is male. This means that we might not expect the **Sex** main effect to be significant, but we do expect the **Recipient_Sex** × **Sex** interaction to be significant.

To add these interaction terms select ⦿ Custom/Stepwise. There are two main ways to specify terms: we can force them in (by moving them to the box labeled *Forced Entry Terms*) or we can put them into the

model using a stepwise procedure (by moving them to the box labeled *Stepwise Terms*). If we look at interaction terms, we must force the main effects into the model otherwise we allow the interaction term to explain variance that might otherwise be attributed to the main effect (in other words, we're not really looking at the interaction any more). Select the variables in the box labeled *Factors & Covariates* by clicking on them while holding down *Ctrl* (*Cmd* on a Mac) or by selecting the first variable and then clicking on the last variable while holding down *Shift*. There is a drop-down list that determines whether you transfer these effects as main effects or interactions. We want to transfer them as main effects so set this box to Main effects ▾ and click ▾.

We specify interactions in much the same way: select two or more variables, set the drop-down box to Interaction ▾ and click ▾. If, for example, we selected **Funny** and **Sex**, then this process specifies the **Funny** × **Sex** interaction. We can specify multiple

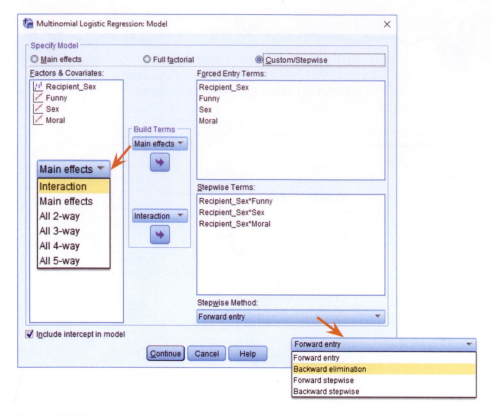

Figure 20.14   Specifying a custom model

to carry out the stepwise analysis. There is a drop-down list of methods under the heading *Stepwise Method*. I've described these methods elsewhere. Select forward entry for this analysis. Click Continue to return to the main dialog box.

20.9.3 Statistics ▮▮▮▮

Click Statistics... to access the dialog box in Figure 20.15, in which you can request:

- *Pseudo R-square*: This option produces the Cox and Snell and Nagelkerke R^2 statistics, which can be used as effect sizes.
- *Step summary*: This option produces a table that summarizes the predictors entered or removed at each step. We should select it because we have a stepwise component to the model.
- *Model fitting information*: This option produces a table that compares the model (or models in a stepwise analysis) to the baseline (the model with only the intercept in it). This table can be useful to compare whether the model has improved (from the baseline) because of entering the predictors that you have.
- *Information Criteria*: This option produces Akaike's information criterion (AIC) and Schwarz's Bayesian information criterion (BIC), which are both useful for comparing models (see Section 21.4.1). Select this option if you're using stepwise methods or if you want to compare different models containing different combinations of predictors.
- *Cell probabilities*: This option produces a table of the observed and expected frequencies, which is basically the same as the classification table produced in binary logistic regression and is probably worth inspecting.
- *Classification table*: This option produces a contingency table of observed versus predicted responses for all combinations of predictor variables. I wouldn't select this option, unless you're running a relatively small analysis (i.e., a small number of predictors made up of a

interactions at once. For example, if we selected **Funny**, **Sex** and **Recipient_Sex** and then set the drop-down box to All 2-way ▾ , it transfers *all* of the interactions involving two variables (i.e., **Funny × Sex, Recipient_Sex × Funny** and **Recipient_Sex × Sex**). You get the general idea. We could also select ⦿ Full factorial which would automatically enter all main effects (**Funny, Sex, Moral, Recipient_Sex**), all interactions with two variables (**Funny × Sex, Recipient_Sex × Funny, Funny × Moral, Recipient_Sex × Sex, Sex × Moral, Recipient_Sex × Moral**), all interactions with three variables (**Funny × Recipient_Sex × Sex, Funny × Sex × Moral, Recipient_Sex × Moral × Sex, Funny × Recipient_Sex × Moral**) and the interaction of all four variables (**Funny × Recipient_Sex × Sex × Moral**).

For our example we want to specify interactions between the ratings of the chat-up lines and **Recipient_Sex** only (we're not interested in any interactions

involving three variables or all four variables). We can either force these interaction terms into the model by putting them in the box labeled *Forced Entry Terms* or we can put them into the model using a stepwise procedure (by moving them into the box labeled *Stepwise Terms*). We're going to do the latter, so interactions will be entered into the model only if they are significant predictors of the success of the chat-up line. Let's first enter the **Recipient_Sex × Funny** interaction. Click **Recipient_Sex** and **Funny** in the *Factors & Covariates* box while holding down the *Ctrl* key (*Cmd* on a Mac). Next to the box labeled *Stepwise Terms* change the drop-down menu to Interaction ▾ and then click ➡. You should now see **Recipient_Sex × Funny** listed in the *Stepwise Terms* box. Specify the **Recipient_Sex × Sex** and **Recipient_Sex × Moral** interactions in the same way. Once the three interaction terms have been specified we can decide how we want

small number of possible values). In this example, we have three covariates with 11 possible values and one predictor (**Recipient_Sex**) with two possible values. Tabulating all combinations of these variables will create a very big table indeed.

- *Goodness-of-fit*: This option is important because it produces Pearson and likelihood ratio chi-square statistics for the model.

- *Monotonicity measures*: This option is worth selecting only if your outcome variable has two outcomes (which in our case it doesn't). It will produce measures of monotonic association such as the concordance index, which measures the probability that, using a previous example, a person who scored a penalty kick is classified by the model as having scored and can range from 0.5 (guessing) to 1 (perfect prediction).

- *Estimates*: This option produces the *b*-values, test statistics and confidence intervals for predictors in the model and is essential.

- *Likelihood ratio tests*: The model overall is tested using likelihood ratio statistics, but this option will compute the same test for individual effects in the model. (It tells us the same as the significance values for individual predictors.)

- *Asymptotic correlations* and *Asymptotic covariances*: This option produces a table of correlations (or covariances) between the betas in the model.

Set the options as in Figure 20.15 and click Continue to return to the main dialog box.

20.9.4 Other options

Click Criteria... to access the dialog box in Figure 20.16 (right). Logistic regression works through an iterative process (SPSS Tip 20.1). The options available here relate to this process. For example, by default, 100 attempts (iterations) are made to fit the model and the threshold for how similar parameter estimates have to be to 'converge' can be made more or less strict

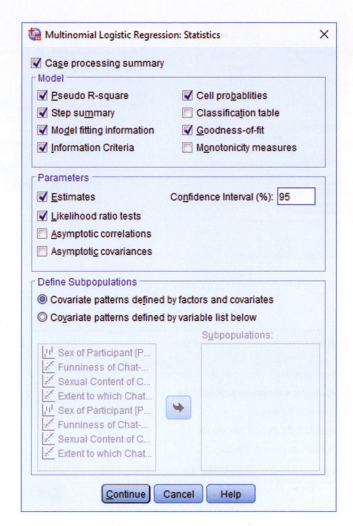

Figure 20.15 Statistics options for multinomial logistic regression

(the default is 0.0000001). Leave these options alone unless when you run the analysis you get an error message saying something about 'failing to converge', in which case you could try increasing the *Maximum iterations* (to 150 or 200), the *Parameter convergence* (to 0.00001) or *Log-likelihood convergence* (to greater than 0). However, bear in mind that a failure to converge can reflect messy data, and forcing the model to converge does not necessarily mean that parameters are accurate or stable across samples.

Click Options... in the main dialog box to access the dialog box in Figure 20.16 (left). The *Scale* option here can be quite useful; I mentioned in Section 20.4.4 that

overdispersion can be a problem in logistic regression with several outcome categories because it reduces the standard errors that are used to test the significance and construct the confidence intervals of the parameter estimates for individual predictors. I also mentioned that this problem could be counteracted by rescaling the standard errors. Should you be in a situation where you need to do this (i.e., you have run the analysis and found evidence of overdispersion) then return to this dialog box and use the drop-down list to select to correct the standard errors by the dispersion parameter based on either the Deviance ▼ or Pearson ▼ statistic. Select whichever of these two statistics was

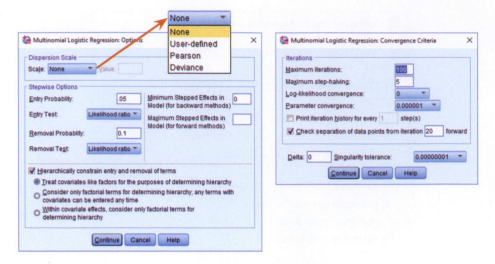

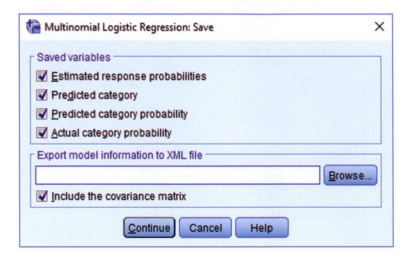

Figure 20.16 Criteria and options for multinomial logistic regression

Figure 20.17 Save options for multinomial logistic regression

Once we have ignored the warnings, like all the best researchers do, the first part of the output tells us about the model overall (Output 20.10). First, because we requested a stepwise analysis for our interaction terms, we get a table summarizing the steps in the analysis. You can see here that after the main effects were entered (model 0), the **Recipient_Sex × Funny** interaction term was entered (model 1) followed by the **Recipient_Sex × Sex** interaction (model 2). The chi-square statistics for steps 1 and 2 are highly significant, indicating that these interactions significantly improve the model's ability to predict the outcome of a chat-up line (also, these terms wouldn't have entered the model had they not been significant). The AIC gets smaller as these terms are added to the model, indicating that the fit of the model is improving (the BIC changes less, but shows a broadly similar pattern). Underneath the step summary, the statistics for the final model replicate the model-fitting criteria from the last line of the step summary table.

Remember that the log-likelihood is a measure of how much unexplained variability there is in the outcome and the *change* in the log-likelihood indicates how much new variance has been explained by a model relative to an earlier model. The decrease in the log-likelihood from the baseline model (1149.53) to the final model (871.00) is assessed with a chi-square statistic that is the difference between the two (1149.53 − 871 = 278.53). This change is significant, meaning that the final model is a better fit than the original model (it accounts for more variability in the outcome).

Output 20.12 relates to the fit of the model. The Pearson and deviance statistics test the same thing, which is whether the predicted

biggest in the original analysis (because this will produce the bigger correction). Finally, click [Save...] in the main dialog box to save predicted probabilities and predicted group membership (Figure 20.17) as we did for binary logistic regression (they are labeled *Estimated response probabilities* and *Predicted category*, respectively).

preparation, weeks entering data, years reading chapters of stupid statistics textbooks, and sleepless nights with equations chipping at your brain with little pickaxes, to see at the start of your analysis: 'Warning! Warning! Abandon ship! Flee for your life! Bad data alert! Bad data alert!' Welcome to the world of data analysis.

20.9.5 Interpreting multinomial logistic regression ▮▮▮▮

The output begins with a warning (SPSS Tip 20.2). It's always nice after months of

SELF TEST
What does the log-likelihood measure?

values from the model differ significantly from the observed values. If these statistics are not significant then the model is a good fit. Here we have contrasting results: the deviance statistic says that the model is a good fit to the data ($p = 0.45$, which is much higher than 0.05), but the Pearson test indicates the opposite: predicted values are significantly different from the observed values ($p < 0.001$). Oh dear.

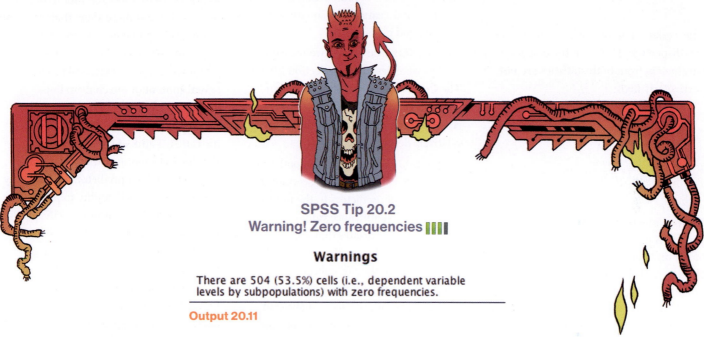

Step Summary

Model	Action	Effect(s)	Model Fitting Criteria			Effect Selection Tests		
			AIC	BIC	-2 Log Likelihood	Chi-Square[a]	df	Sig.
0	Entered	Intercept, Recipient's sex, Funny content, Sexual content, Moral content	937.572	986.848	917.572	.		
1	Entered	Recipient's sex * Funny content	908.451	967.582	884.451	33.121	2	.000
2	Entered	Recipient's sex * Sexual content	899.002	967.987	871.002	13.450	2	.001

Stepwise Method: Forward Entry

a. The chi-square for entry is based on the likelihood ratio test.

Model Fitting Information

Model	Model Fitting Criteria			Likelihood Ratio Tests		
	AIC	BIC	-2 Log Likelihood	Chi-Square	df	Sig.
Intercept Only	1153.526	1163.382	1149.526			
Final	899.002	967.987	871.002	278.525	12	.000

Output 20.10

SPSS Tip 20.2
Warning! Zero frequencies ▌▌▌▌

Warnings

There are 504 (53.5%) cells (i.e., dependent variable levels by subpopulations) with zero frequencies.

Output 20.11

Sometimes in logistic regression you get a warning about zero frequencies. This relates to the problem that I discussed in Section 20.4.2 of 80-year-old, highly anxious, Buddhist left-handed lesbians (well, incomplete information). Imagine we had looked only at the recipient's sex as a predictor of chat-up line success. We have three outcome categories and two categories for the recipient's sex. There are six possible combinations of these two variables, and ideally we want a large number of observations in each of these combinations. However, we didn't just look at recipient's sex, we had three continuous predictor variables (**Funny**, **Sex** and **Moral**) with 11 possible values, **Recipient_Sex** with two possible outcomes and an outcome variable with 3 outcome categories. By including the three covariates, the number of combinations of values for these variables has escalated considerably. This error message tells us that there are some combinations of

these variables for which there are no observations. For example, you can imagine it might be difficult to find a chat-up line that was rated a top score of 10/10 on funny, moral fiber, *and* sexual content (because lines that reflect good moral fiber are unlikely to contain highly sexualized content). If you consider that you'd then also need to find such a rarity of a chat-up line used on both a male and a female recipient, you can see how you'd get zero frequencies. In fact 53.5% of our possible combinations of variables had no data despite 1020 cases!

Whenever you have covariates it is inevitable that you will have empty cells, so you will get this kind of error message. To some extent, given its inevitability, we can just ignore it. However, it is worth reiterating that empty cells create problems, and that when you get a warning like this you should look for coefficients that have unreasonably large standard errors and if you find them be wary of them.

Goodness-of-Fit

	Chi-Square	df	Sig.
Pearson	886.616	614	.000
Deviance	617.481	614	.453

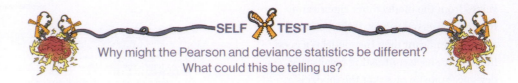

SELF TEST

Why might the Pearson and deviance statistics be different?
What could this be telling us?

Pseudo R-Square

Cox and Snell	.239
Nagelkerke	.277
McFadden	.138

Output 20.12

One explanation for this contradiction is overdispersion. However, the dispersion parameters from both statistics are not particularly high:

$$\phi_{Pearson} = \frac{\chi^2_{Pearson}}{df} = \frac{886.62}{614} = 1.44$$

$$\phi_{deviance} = \frac{\chi^2_{deviance}}{df} = \frac{617.48}{614} = 1.01$$
(20.22)

The one based on the deviance statistic is close to the ideal value of 1 and the value based on Pearson is greater than 1, but not close to 2. Based on these statistics, there's not a lot to suggest that the data are overdispersed.[15] Another possibility is that the Pearson statistic can be very inflated

by low expected frequencies, which we would have because there were so many empty cells (as indicated by our warning). One thing that is certain is that conflicting deviance and Pearson chi-square statistics are not good news.

Output 20.12 also shows two measures of R^2 that were described in Section 20.3.3. The Cox and Snell measure (0.24) and Nagelkerke's adjusted value (0.28) are reasonably similar values and represent relatively decent-sized effects.

Output 20.13 shows the likelihood ratio tests, which can be used to ascertain the significance of predictors to the model. Note that no significance values are produced for covariates that are involved in higher-order interactions (hence the blank spaces in the *Sig.* column for the effects of **Funny** and **Sex**). This table tells us that recipient's sex had a significant main effect on success rates of chat-up

lines, $\chi^2(2) = 18.54$, $p < 0.001$, as did whether the chat-up lines showed evidence of moral fiber, $\chi^2(2) = 6.32$, $p = 0.042$. The interactions are most relevant to our hypotheses, and these show that (1) humor in the chat-up line interacted with recipient's sex to predict their reaction, $\chi^2(2) = 35.81$, $p < 0.001$; and (2) the sexual content of the chat-up line interacted with the recipient's sex in predicting their reaction, $\chi^2(2) = 13.45$, $p = 0.001$. Think of these likelihood statistics as looking at the overall effect: they tell us which predictors significantly improve the model's ability to predict the outcome category, but not specifically which categories it helps to predict.

The individual parameter estimates (Output 20.14) help us to break down the overall effects. The table is split into two halves because each parameter compares pairs of outcome categories. We specified the first category (*No response/walked away*) as the reference category so the top half of the table (labeled *Get Phone Number*) compares the *Get Phone Number* outcome to the *No response/walked away* outcome. Likewise, the bottom half compares the *Go Home with Person* category to the *No response/walked away* category.

Let's look at the top half of Output 20.14 first. We'll look at each effect in turn; because we are comparing two categories the interpretation is the same as for binary logistic regression (so if you don't understand my conclusions reread the start of this chapter):

Likelihood Ratio Tests

	Model Fitting Criteria			Likelihood Ratio Tests		
Effect	AIC of Reduced Model	BIC of Reduced Model	-2 Log Likelihood of Reduced Model	Chi-Square	df	Sig.
Intercept	899.002	967.987	871.002[a]	.000	0	.
Recipient's sex	913.540	972.671	889.540	18.538	2	.000
Funny content	899.002	967.987	871.002[a]	.000	0	.
Sexual content	899.002	967.987	871.002[a]	.000	0	.
Moral content	901.324	960.454	877.324	6.322	2	.042
Recipient's sex * Funny content	930.810	989.941	906.810	35.808	2	.000
Recipient's sex * Sexual content	908.451	967.582	884.451	13.450	2	.001

The chi-square statistic is the difference in -2 log-likelihoods between the final model and a reduced model. The reduced model is formed by omitting an effect from the final model. The null hypothesis is that all parameters of that effect are 0.

a. This reduced model is equivalent to the final model because omitting the effect does not increase the degrees of freedom.

Output 20.13

---

15 Incidentally, large dispersion parameters can occur for reasons other than overdispersion, for example omitted variables or interactions (in this example there were several interaction terms that we could have entered but chose not to), and predictors that violate the linearity of the logit assumption.

- **Recipient_Sex**: The sex of the person being chatted up significantly predicted whether they gave out their phone number or gave no response, $b = -1.65$, Wald $\chi^2(1) = 4.27$, $p = 0.039$. Remember that 0 = female and 1 = male, so this is the effect of females compared to males. The odds ratio tells us that as recipient's sex changes from female (0) to male (1) the change in the odds of giving out a phone number compared to not responding is 0.19. In other words, the odds of a man giving out his phone number compared to not responding are $1/0.19 = 5.26$ times more than for a woman.

- **Funny**: Whether the chat-up line was funny did not significantly predict whether you got a phone number or no response, $b = 0.14$, Wald $\chi^2(1) = 1.60$, $p = 0.206$. Although this predictor is not significant, the odds ratio is approximately the same as for the previous predictor (which was significant). So, the effect size is comparable, but the non-significance stems from a relatively higher standard error. (Note that this effect is superseded by the interaction with recipient's sex below.)

- **Sex**: The sexual content of the chat-up line significantly predicted whether you got a phone number or no response/walked away, $b = 0.28$, Wald $\chi^2(1) = 9.59$, $p = 0.002$. The odds ratio tells us that as the sexual content increased by a unit, the change in the odds of getting a phone number (rather than no response) is 1.32. In short, you're more likely to get a phone number than not if you use a chat-up line with high sexual content. (But this effect is superseded by the interaction with recipient's sex.)

- **Moral**: Whether the chat-up line showed good moral fiber significantly predicted whether you got a phone number or no response/walked away, $b = 0.13$, Wald $\chi^2(1) = 6.02$, $p = 0.014$. The odds ratio tells us that as chat-up lines show one more unit of moral fiber, the change in the odds of getting a phone number

Parameter Estimates

Success of chat-up line[a]		B	Std. Error	Wald	df	Sig.	Exp(B)	95% Confidence Interval for Exp(B) Lower Bound	Upper Bound
Get Phone Number	Intercept	-1.783	.670	7.087	1	.008			
	[Recipient's sex=0]	-1.646	.796	4.274	1	.039	.193	.040	.918
	[Recipient's sex=1]	0[b]	.	.	0	.	.	.	.
	Funny content	.139	.110	1.602	1	.206	1.150	.926	1.427
	Sexual content	.276	.089	9.589	1	.002	1.318	1.107	1.570
	Moral content	.132	.054	6.022	1	.014	1.141	1.027	1.268
	[Recipient's sex=0] * Funny content	.492	.140	12.374	1	.000	1.636	1.244	2.153
	[Recipient's sex=1] * Funny content	0[b]	.	.	0	.	.	.	.
	[Recipient's sex=0] * Sexual content	-.348	.106	10.824	1	.001	.706	.574	.869
	[Recipient's sex=1] * Sexual content	0[b]	.	.	0	.	.	.	.
Go Home with Person	Intercept	-4.286	.941	20.731	1	.000			
	[Recipient's sex=0]	-5.626	1.329	17.934	1	.000	.004	.000	.049
	[Recipient's sex=1]	0[b]	.	.	0	.	.	.	.
	Funny content	.318	.125	6.459	1	.011	1.375	1.076	1.758
	Sexual content	.417	.122	11.683	1	.001	1.518	1.195	1.928
	Moral content	.130	.084	2.423	1	.120	1.139	.967	1.341
	[Recipient's sex=0] * Funny content	1.172	.199	34.627	1	.000	3.230	2.186	4.773
	[Recipient's sex=1] * Funny content	0[b]	.	.	0	.	.	.	.
	[Recipient's sex=0] * Sexual content	-.477	.163	8.505	1	.004	.621	.451	.855
	[Recipient's sex=1] * Sexual content	0[b]	.	.	0	.	.	.	.

a. The reference category is: No response/Walk Off.
b. This parameter is set to zero because it is redundant.

Output 20.14

(rather than no response/walked away) is 1.14. You're more likely to get a phone number than not if you use a chat-up line that demonstrates good moral fiber.

- **Recipient_Sex × Funny**: The success of funny chat-up lines depended significantly on whether they were delivered to a man or a woman, $b = 0.49$, Wald $\chi^2(1) = 12.37$, $p < 0.001$. Bearing in mind how we interpreted the effect of recipient's sex above, the odds ratio tells us that as recipient's sex changes from female (0) to male (1) in combination with funniness increasing, the change in the odds of giving out a phone number compared to not responding was 1.64. In other words, as funniness increases, women become more likely to hand out their phone number than men. In line with past research, funny chat-up lines are more successful when used on women than men.

- **Recipient_Sex × Sex**: The success of chat-up lines with sexual content depended significantly on whether they were delivered to a man or a woman, $b = -0.35$, Wald $\chi^2(1) = 10.82$, $p = 0.001$. Bearing in mind how we interpreted the interaction above (note that b is negative here but positive above), the odds ratio tells us that as recipient's sex changes from female (0)

to male (1) in combination with the sexual content increasing, the change in the odds of giving out a phone number compared to not responding is 0.71. In other words, as sexual content increases, women become *less* likely than men to hand out their phone number. Consistent with past research, chat-up lines with higher sexual content are more successful when used on men than women.

We can interpret the bottom half of Output 20.14 in much the same way except that we're now comparing the *Go Home with Person* category to the *No response/walked away* category:

- **Recipient_Sex**: The sex of the person being chatted up significantly predicted whether he or she went home with the person or gave no response, $b = -5.63$, Wald $\chi^2(1) = 17.93$, $p < 0.001$. The odds ratio tells us that as recipient's sex changes from female (0) to male (1) the change in the odds of going home with the person compared to not responding is 0.004. The odds of a man going home with someone compared to not responding are $1/0.004 = 250$ times more likely than for a woman.

Table 20.5 How to report multinomial logistic regression

	b (SE)	Lower	Odds Ratio	Upper
		95% CI for Odds Ratio		
Phone Number vs. No Response				
Intercept	−1.78 (0.67)**			
Recipient's sex	−1.65 (0.80)*	0.04	0.19	0.92
Funny content	0.014 (0.11)	0.93	1.15	1.43
Sexual content	0.28 (0.09)**	1.11	1.32	1.57
Moral content	0.13 (0.05)*	1.03	1.14	1.27
Recipient's sex × Funny content	0.049 (0.14)***	1.24	1.64	2.15
Recipient's sex × Sexual content	−0.35 (0.11)*	0.57	0.71	0.87
Going Home vs. No Response				
Intercept	−4.29 (0.94)***			
Recipient's sex	−5.63 (1.33)***	0.00	0.00	0.05
Funny content	0.32 (0.13)*	1.08	1.38	1.76
Sexual content	0.42 (0.12)**	1.20	1.52	1.93
Moral content	0.13 (0.08)	0.97	1.14	1.34
Recipient's sex × Funny content	1.17 (0.20)***	2.19	3.23	4.77
Recipient's sex × Sexual content	−0.48 (0.16)**	0.45	0.62	0.86

Note. R^2 = 0.24 (Cox–Snell), 0.28 (Nagelkerke). Model $\chi^2(12)$ = 278.53, $p < 0.001$. *$p < .05$, **$p < 0.01$, ***$p < 0.001$.

- **Funny:** Whether the chat-up line was funny significantly predicted whether the recipient went home with the date or gave no response, $b = 0.32$, Wald $\chi^2(1) = 6.46$, $p = 0.011$. The odds ratio tells us that as chat-up lines are one unit funnier, the change in the odds of going home with the person (rather than no response) is 1.38. A person is more likely to go home with you than give no response if you use a chat-up line that is funny. (This effect is superseded by the interaction with recipient's sex below.)

- **Sex:** The sexual content of the chat-up line significantly predicted whether the recipient went home with the date or got a slap in the face, $b = 0.42$, Wald $\chi^2(1) = 11.68$, $p = 0.001$. The odds ratio tells us that as the sexual content increased by a unit, the change in the odds of going home with the person (rather than no response) is 1.52: you're more likely to have someone go home with you than not if you use a chat-up line with high sexual content. (This effect is superseded by the interaction with recipient's sex below.)

- **Moral:** Whether the chat-up line showed signs of good moral fiber did not significantly predict whether you went home with the date or got no response, $b = 0.13$, Wald $\chi^2(1) = 2.42$, $p = 0.120$. The recipient is not significantly more likely to go home with you if you use a chat-up line that demonstrates good moral fiber.

- **Recipient_Sex × Funny:** The success of funny chat-up lines depended on whether they were delivered to a man or a woman, $b = 1.17$, Wald $\chi^2(1) = 34.63$, $p < 0.001$. The odds ratio tells us that as recipient's sex changes from female (0) to male (1) in combination with funniness increasing, the change in the odds of going home with the person compared to not responding is 3.23. As funniness increases, women become more likely to go home with the person than men. Consistent with past research, funny chat-up lines are more successful when used on women than on men.

- **Recipient_Sex × Sex:** The success of chat-up lines with sexual content depended on whether they were delivered to a man or a woman, $b = −0.48$, Wald $\chi^2(1) = 8.51$, $p = 0.004$. The odds ratio tells us that as recipient's sex changes from female (0) to male (1) in combination with the sexual content increasing, the change in the odds of going home with the date compared to not responding is 0.62. As sexual content increases, women become *less* likely than men to go home with the person. Consistent with past research, chat-up lines with sexual content are more successful when used on men than on women.

20.10 Reporting multinomial logistic regression ▋▋▋▋

We can report the results using a table (see Table 20.5). Note that I have split the table by the outcome categories being compared, but otherwise it is the same as before. If you use APA format drop the leading zeros before the p-values (i.e., .001 not 0.001).

SELF TEST

Use what you learnt earlier in this chapter to check the assumptions of multicollinearity and linearity of the logit.

20.11 Brian and Jane's Story ▌▌▌▌

'Do you and your dad talk about your mum?' Jane asked. She usually avoided emotional conversations, but since he'd opened up to her Jane had felt guilty that she hadn't asked Brian anything more about it. She wondered how someone could grow up so balanced with such a shadow lurking in their past.

Brian looked surprised by the question, but he was open to it. 'When I was growing up . . . but not so much now,' he replied. 'I think it upsets him to talk about her.'

'But he used to . . . ,' Jane said, challenging Brian's assumption.

'He had this thing about me not remembering her. It really hurt him, so he'd talk about her. He'd pull out photos every few weeks from when she was alive and talk about what we'd done, when the photo was taken . . . he talk about their routines, about things we did together, anything to keep her alive in the house and to jog my memory. But nothing. He hated the fact I couldn't remember her, it made him so desolate. I wanted to bring him some peace, so eventually I lied.'

'Lied?'

'Yeah, you know . . . just little lies, like, I picked up on their routine: mum took me to school, he picked me up, so I'd tell him "I remember this one time on the way to school and mum and I saw this cat . . ."— just invent some cute little incident that he couldn't verify. His eyes sparkled every time. That was all I needed to keep going. I kept lying and eventually we stopped talking about her.'

'Do you miss her?'

'How can you miss what you don't remember? That's what you'd think, wouldn't you? But, yeah, I think about her every day. It's like, I know my dad so well, I look at myself and I see him in

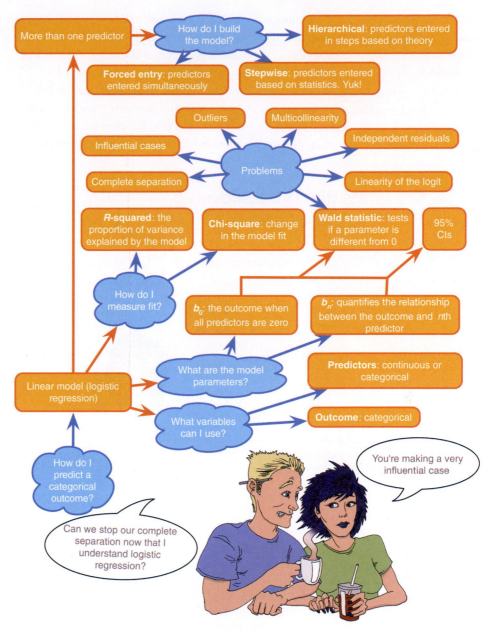

Figure 20.18 What Brian learnt from this chapter

what I look like and what I do, but then there's this bunch of stuff about me, that I guess comes from her, but I just don't know. I know who I am and yet half of me is missing ... But I'm grateful for what I have. My dad is awesome. Some people have horrible, destructive parents, that mess them up. All I've ever had is love.'

Jane wondered what that would feel like.

20.12 What next? ▌▌▌▌

At the age of 10 I thought I was going to be a rock star. Such was my conviction about this that even today (many years on) I'm still not entirely sure how I ended up *not* being a rock star (possible explanations are lack of talent, being so uncool that I still use the word 'cool', inability to write good songs, the list is depressingly long). Instead of the glitzy and fun life that I anticipated I am instead reduced to writing textbook chapters about things that I don't even remotely understand.

The other thing that I thought at the age of 10 was that I would marry Clair Sparks. Such was my conviction that even today I'm still not entirely sure …. Nah, I'm just kidding. However, as a boy I was (for some inexplicable reason) convinced that I would get married at the age of 28. It was something of a shock to me when I reached 28 and discovered that I didn't have a wife and probably wasn't in any fit state to have one. I'd spent far too much time dedicated to music, then to working ridiculously hard to try to convince everyone I was clever enough to be a scientist, and the two years dating someone who destroyed my very soul didn't help either. Time ticked on until in my mid-thirties I found myself alone, without a wife. 'I'd better get one of those,' I thought to myself.[16]

20.13 Key terms that I've discovered

−2LL	Deviance	Log-likelihood	Overdispersion
Binary logistic regression	Exp(B)	Maximum-likelihood estimation	Parsimony
Complete separation	Hosmer and Lemeshow's R^2_L	Multinomial logistic regression	Suppressor effects
Cox and Snell's R^2_{CS}	Logistic regression	Nagelkerke's R^2_N	Wald statistic

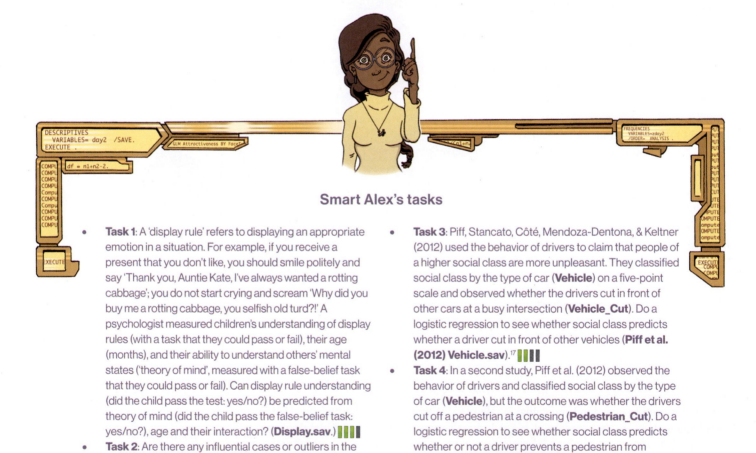

Smart Alex's tasks

- **Task 1**: A 'display rule' refers to displaying an appropriate emotion in a situation. For example, if you receive a present that you don't like, you should smile politely and say 'Thank you, Auntie Kate, I've always wanted a rotting cabbage'; you do not start crying and scream 'Why did you buy me a rotting cabbage, you selfish old turd?!' A psychologist measured children's understanding of display rules (with a task that they could pass or fail), their age (months), and their ability to understand others' mental states ('theory of mind', measured with a false-belief task that they could pass or fail). Can display rule understanding (did the child pass the test: yes/no?) be predicted from theory of mind (did the child pass the false-belief task: yes/no?), age and their interaction? (**Display.sav**.)
- **Task 2**: Are there any influential cases or outliers in the model for Task 1?

- **Task 3**: Piff, Stancato, Côté, Mendoza-Dentona, & Keltner (2012) used the behavior of drivers to claim that people of a higher social class are more unpleasant. They classified social class by the type of car (**Vehicle**) on a five-point scale and observed whether the drivers cut in front of other cars at a busy intersection (**Vehicle_Cut**). Do a logistic regression to see whether social class predicts whether a driver cut in front of other vehicles (**Piff et al. (2012) Vehicle.sav**).[17]
- **Task 4**: In a second study, Piff et al. (2012) observed the behavior of drivers and classified social class by the type of car (**Vehicle**), but the outcome was whether the drivers cut off a pedestrian at a crossing (**Pedestrian_Cut**). Do a logistic regression to see whether social class predicts whether or not a driver prevents a pedestrian from crossing (**Piff et al. (2012) Pedestrian.sav**).

16 Needless to say, I'm skipping over a fair few romantic events between the ages of 10 and my mid-thirties, some of them more pleasant than others.

17 I reconstructed the raw data from Figure 1 of the paper, so you will get basically the same values as reported there, but not the exact ones because they also controlled for the age and gender of drivers (and we don't have those variables).

- **Task 5**: Four hundred and sixty-seven lecturers completed questionnaire measures of **Burnout** (burnt out or not), **Perceived Control** (high score = low perceived control), **Coping Style** (high score = high ability to cope with stress), **Stress from Teaching** (high score = teaching creates a lot of stress for the person), **Stress from Research** (high score = research creates a lot of stress for the person) and **Stress from Providing Pastoral Care** (high score = providing pastoral care creates a lot of stress for the person). Cooper, Sloan, & Williams's (1988) model of stress indicates that perceived control and coping style are important predictors of burnout. The remaining predictors were measured to see the unique contribution of different aspects of a lecturer's work to their burnout. Conduct a logistic regression to see which factors predict burnout (**Burnout.sav**). ▌▌▌▌

- **Task 6**: An HIV researcher explored the factors that influenced condom use with a new partner (relationship less than 1 month old). The outcome measure was whether a condom was used (**Use**: condom used = 1, not used = 0). The predictor variables were mainly scales from the Condom Attitude Scale (CAS) by Sacco, Levine, Reed, & Thompson (1991): **Gender**; the degree to which the person views their relationship as 'safe' from sexually transmitted disease (**Safety**); the degree to which previous experience influences attitudes towards condom use (**Sexexp**); whether or not the couple used a condom in their previous encounter (**Previous**: 1 = condom used, 0 = not used, 2 = no previous encounter with this partner); the degree of self-control that a person has when it comes to condom use (**Selfcon**); the degree to which the person perceives a risk from unprotected sex (**Perceive**). Previous research (Sacco, Rickman, Thompson, Levine, & Reed, 1993) has shown that gender, relationship safety and perceived risk predict condom use. Verify these previous findings and test whether self-control, previous usage and sexual experience predict condom use (**Condom.sav**). ▌▌▌▌

- **Task 7**: How reliable is the model in Task 6? ▌▌▌▌

- **Task 8**: Using the final model from Task 6, what are the probabilities that participants 12, 53 and 75 will use a condom? ▌▌▌▌

- **Task 9**: A female who used a condom in her previous encounter scores 2 on all variables except perceived risk (for which she scores 6). Use the model in Task 6 to estimate the probability that she will use a condom in her next encounter. ▌▌▌▌

- **Task 10**: At the start of the chapter we looked at whether the type of instrument a person plays is connected to their personality. A musicologist measured **Extroversion** and **Agreeableness** in 200 singers and guitarists (**Instrument**). Use logistic regression to see which personality variables (ignore their interaction) predict which instrument a person plays (**Sing or Guitar.sav**). ▌▌▌▌

- **Task 11**: Which problem associated with logistic regression might we have in the analysis in Task 10? ▌▌▌▌

- **Task 12**: In a new study, the musicologist in Task 10 extended her previous one by collecting data from 430 musicians who played their voice (singers), guitar, bass or drums (**Instrument**). She measured the same personality variables but also their **Conscientiousness** (**Band Personality.sav**). Use multinomial logistic regression to see which of these three variables (ignore interactions) predict which instrument a person plays (use drums as the reference category). ▌▌▌

Answers & additional resources are available on the book's website at
https://edge.sagepub.com/field5e

21 MULTILEVEL LINEAR MODELS

21.1 What will this chapter tell me?

Years at an all-boys' school carefully nurturing a morbid fear of women and a love of heavy metal had made the world of relationships a tricky place for me to inhabit. I'd always dreamt that by my mid-thirties I would have a wife and a cute little child or two to remind me of the important things in life. But as I took my first furtive and depressing steps into middle age I found myself single and the closest thing to a child was a ginger cat and this book. However, something remarkable had happened since my teens: rock music had become popular again, and some women liked to talk about Iron Maiden. I needed to capitalize before the ephemeral guillotine of fashion spliced this opportunity from me. I met Zoë, who was not only happy to discuss Iron Maiden, but owned my favorite of their albums (*Piece of Mind*). She had no aversion to statistics or soccer, and happened to be the most lovely woman ever placed on the face of the earth. Result. 'I'd better marry her before she realizes I'm a balding geek with slight hoarding tendencies,' I thought. So that's what we did. A little later than anticipated, my dreams had come true: I started my late thirties with a wife and a cute little … book about a statistics package called R to which my wife contributed. I made a mental note to next time create a little human, not a big book.[1]

Marriage is a leap of faith into the unknown, a shared adventure full of challenges. A bit like this chapter really, because, when I started writing it, multilevel linear models were an 'unknown': I knew nothing about them. If you're reading this section then you probably don't know much about them either. So, we'll learn together—a shared adventure, and it sure will include some challenges …

---

1 A mental note that I heeded (see Chapter 3).

Figure 21.1 On the road to happiness

21.2 Hierarchical data ▮▮▮▮

So far we have treated data as though they are organized at a single level. The exception was repeated-measures designs, where I talked about observations being nested within participants. This 'nesting' creates a hierarchy in the data. This happens in situations other than repeated-measures designs. For example, when I'm not writing statistics books I research how anxiety develops in schoolchildren. When I run research in a school, I test children who have been assigned to different classes, and who are taught by different teachers. The classroom that a child is in could affect my results. Imagine I collect data in two different classrooms. Mr. Nervous, who teaches the first class, is very anxious. He tells children to be careful, that things that they do are dangerous, and that they might hurt themselves. Little Miss Daredevil,[2] who teaches the second, is carefree, tells children not to be scared of things and gives them the freedom to explore new situations.

My experiment involves telling children information about an animal in a big wooden box. Some children I focus on the positive aspects of the animal (e.g., soft fur) whereas others I focus on the negative (e.g., big teeth). I then ask the children to put their hand into the box and stroke the animal (which they can't see). I measure whether they will. Children taught by Mr. Nervous have grown up in an environment that reinforces caution, whereas children taught by Miss Daredevil have been encouraged to embrace new experiences. Therefore, irrespective of the information I have given the children, we might expect Mr. Nervous's children to be more reluctant to put their hand in the box than Miss Daredevil's children because of the

classroom experiences that they have had. Similarly, the information itself might have a different effect for the two classes (e.g., Mr. Nervous' children might be more sensitive to the negative information). The classroom is a **contextual variable**. Figure 21.2 illustrates this scenario: the child (or case) is at the bottom of the hierarchy (known as a *level 1* variable) and is nested (or grouped) within the class that they are in (the class is a level up from the child in the hierarchy and is said to be a *level 2* variable).

You can have more complex hierarchies than a two-level one. Sticking with our

example, an obvious third level is that classrooms are nested within schools. If I collected data not just in different classrooms but at different schools, then I'd have another level in the hierarchy. The logic is the same as before: children in the same school will be more like each other than like children in different schools because schools have their own teaching environments and reflect their social demographic (which differs from school to school). Figure 21.3 shows this three-level hierarchy: the child (level 1) is nested within the class (level 2) to which he or

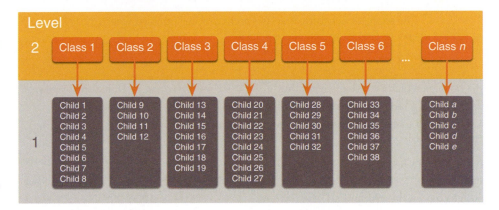

Figure 21.2   An example of a two-level hierarchical data structure. Children (level 1) are organized within classrooms (level 2)

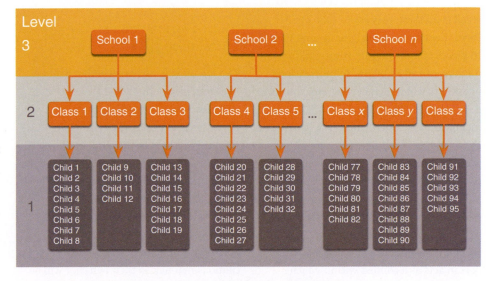

Figure 21.3   An example of a three-level hierarchical data structure

---

2 Those of you who don't spot the Mr. Men/Little Miss references here, check out http://www.mrmen.com. Mr. Nervous used to be called Mr. Jelly and was a pink jelly-shaped blob, which in my opinion was better than his current incarnation.

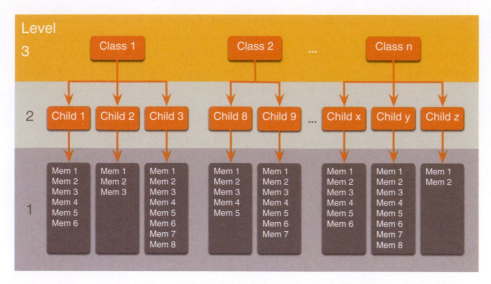

Figure 21.4 An example of a three-level hierarchical data structure, where the level 1 variable is a repeated measure (memories recalled)

she belongs which is nested within the school (level 3) to which the class belongs. There are two contextual variables: school and classroom.

As we saw in Chapter 15, hierarchical data structures apply to repeated-measures designs too. In these situations observations (level 1) are nested within entities (level 2). Let's take a memory example. Imagine that a week after I gave children information about my boxed animal I asked them to recall everything they could that I told them. Let's say that I originally gave them 15 pieces of information; some children might recall all 15 pieces of information, but others will remember less (maybe only 2 or 3 things). The bits of information or memories, are nested within the person and their recall depends on the person. The probability of a given memory being recalled depends on what other memories are available, and the recall of one memory may have knock-on effects for what other memories are recalled. Therefore, memories are not independent units. As such, the person acts as a context within which memories are recalled (Wright, 1998). Figure 21.4 illustrates this scenario: the child is the level 2 variable, and within each child there are memories (the level 1

variable). Of course we can also have levels of the hierarchy above the child, for example, the class from which they came could be a level 3 variable, and we could even include the school as a level 4 variable. Growth models (Section 21.7) are a widely used example of where observations (at different times) are a level 1 variable nested within some other level 2 variable (such as people or organizations).

21.2.1 The intraclass correlation

The reason why we care whether data are hierarchical (or not) is that the contextual variables in the hierarchy introduce dependency in observations. In plain English, residuals in the model will be correlated. To understand why, imagine that Charlotte and Emily are two children taught by Mr. Nervous, and Kiki and Jip are taught by Miss Daredevil. Charlotte and Emily's responses to the animal in the carrier have both been influenced by Mr. Nervous's cautious manner, so their behavior will be similar. Their scores are likely to be correlated or dependent (because of the contextual influence of Mr. Nervous). Likewise, Kiki and Jip's responses to the

animal in the box are likely to be correlated because they have both been influenced by Miss Daredevil's carefree manner. However, Charlotte and Emily's responses shouldn't correlate with Kiki and Jip's, because the former pair have not been influenced by Mr. Nervous and the latter have not been influenced by Miss Daredevil.

This similarity is a problem because the linear model assumes that errors are independent (Chapter 6), and when entities are sampled from similar contexts this independence is unlikely to be true. By thinking about contextual variables and factoring them into the model we can overcome this problem of non-independent observations.

We can use the intraclass correlation (which we came across as a measure of inter-rater reliability in Section 18.10) to estimate the dependency between scores. We'll skip the formalities of calculating the ICC (see Oliver Twisted if you're keen to know), and I'll try to give you a conceptual grasp of what it represents. In our two-level example of children within classes, the ICC represents the proportion of the total variability in the outcome that is attributable to the classes. It follows that if a class has had a big effect on the children within it then the variability within the class will be small (the children will behave similarly). As such, variability in the outcome within classes is minimized, and variability in the outcome between classes is maximized; therefore, the ICC is large. Conversely, if the class has little effect on the children then the outcome will vary a lot within classes, which will make differences between classes relatively small. Therefore, the ICC is small too. Thus, the ICC tells us that variability within levels of a contextual variable (in this case the class to which a child belongs) is small, but between levels of a contextual variable (comparing classes) is large. As such the ICC is a good gauge of the extent to which a contextual variable has an effect on the outcome.

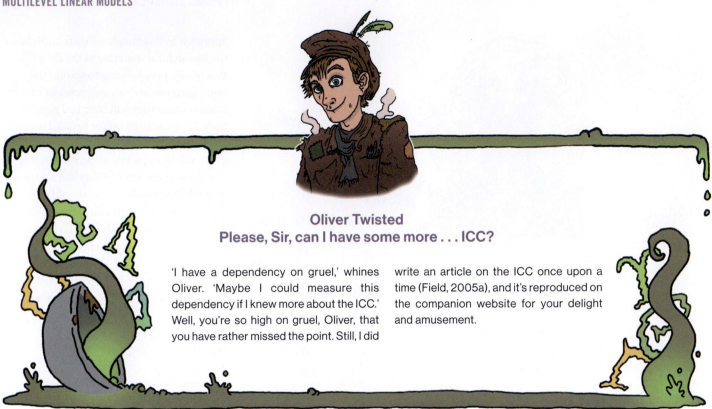

Oliver Twisted
Please, Sir, can I have some more . . . ICC?

'I have a dependency on gruel,' whines Oliver. 'Maybe I could measure this dependency if I knew more about the ICC.' Well, you're so high on gruel, Oliver, that you have rather missed the point. Still, I did write an article on the ICC once upon a time (Field, 2005a), and it's reproduced on the companion website for your delight and amusement.

21.2.2 Benefits of multilevel models ▌▌▌▌

To convince you that trawling through this chapter is going to reward you with statistical possibilities beyond your wildest dreams, here are just a few (slightly overstated) benefits of **multilevel linear models** (Figure 21.5):

- **Cast aside the assumption of homogeneity of regression slopes**: When we use analysis of covariance we assume that the relationship between the covariate and outcome is the same across the different groups that make up our predictor variable (Chapter 13). In multilevel models we can explicitly model this variability in regression slopes and forget about this assumption.
- **Say 'bye bye' to the assumption of independence**: Linear models assume independent errors (Chapter 6). If errors are dependent, little lizards climb out of your mattress while you're asleep and eat you. Multilevel models allow you to model dependencies between residuals.

- **Laugh in the face of missing data**: I've spent a lot of this book extolling the virtues of balanced designs and not having missing data. Linear models do strange things when data are missing or when (experimental) designs are not balanced. Missing data are a particular problem within clinical trials and other longitudinal designs in which you might want to collect follow-up data months or years after an intervention (or other baseline) took place and patients/participants might be difficult to track down. There are ways to correct for and impute missing data, but these techniques are quite complicated (Enders, 2011; Yang, Li, & Shoptaw, 2008), and often people simply delete the case if a single time point is missing. Multilevel models do not require complete data sets and so when data are missing for one time point they do not need to be imputed, nor does the whole case need to be deleted. Instead parameters can be estimated successfully with the available data, which offers a relatively easy solution to dealing with missing data. It is important to stress that no statistical procedure can overcome data that are missing. Good methods, designs and research execution should be used to minimize missing values, and reasons for missing values should always be explored.

I think you'll agree that multilevel models are pretty funky. 'Is there anything they can't do?' I hear you cry. Well, I've never had one make me tea.

21.3 Theory of multilevel linear models ▌▌▌▌

The underlying theory of multilevel models is far too complicated for my little peanut of a brain to comprehend. Fortunately, the advent of computers and software makes it possible for feeble-minded individuals such as myself to take advantage of this tool without needing to know the maths. Better still, this means I can get away with not explaining the maths (and really, I'm not kidding, I don't understand any of it). What I will do is try to give you a flavor of what multilevel models are and what they do by describing

Figure 21.5 Thanks to the Confusion machine, there are many ways to refer to a multilevel model

the key concepts within the framework of linear models that has permeated this whole book.

21.3.1 A surgical example ▌▌▌

In the USA, there was a 1600% increase in cosmetic surgical and non-surgical treatments between 1992 and 2002, and in the UK, 65,000 people underwent privately and publicly funded operations in 2004 (Kellett, Clark, & McGill, 2008). There are two main reasons to have cosmetic surgery: (1) to help a physical problem (i.e., breast reduction surgery to relieve backache, skin grafts after accidents); and (2) to change your external appearance when there is no underlying physical pathology. Related to this second reason, some have suggested that in the future cosmetic surgery might be performed as a psychological intervention to improve self-esteem (Cook, Rosser, & Salman, 2006; Kellett et al., 2008). Our first example looks at the effects of cosmetic surgery on quality of life. The variables in the data file (**Cosmetic Surgery.sav**) are:

- **Post_QoL**: This is the outcome variable and it measures quality of life after cosmetic surgery.

- **Base_QoL**: We need to adjust our outcome for quality of life before the surgery.
- **Surgery**: This dummy variable specifies whether the person has undergone cosmetic surgery (1) or is on the waiting list (0), which acts as our control group.
- **Clinic**: This variable specifies which of 10 clinics the person attended to have their surgery.
- **Age**: This variable tells us the person's age in years.
- **BDI**: People volunteering for cosmetic surgery (especially when the surgery is purely for vanity) have different personality profiles than the general public (Cook, Rosser, Toone, James, & Salmon, 2006). In particular, these people might have low self-esteem or be depressed. This variable measures natural levels of depression using the Beck Depression Inventory (BDI).
- **Reason**: This dummy variable specifies whether the person had or is waiting to have surgery purely to change their appearance (0) or because of a physical reason (1).
- **Sex**: This variable specifies whether the person was a man (1) or a woman (0).

When conducting hierarchical models we work up from a very simple model to more complicated models, and we will take that

approach in this chapter. Figure 21.6 shows the hierarchical structure of the data. Essentially, people being treated in the same surgeries are not independent of each other because they will have had surgery from the same surgeon (or team). Surgeons will vary in how good they are, and quality of life will to some extent depend on how well the surgery went (if they did a nice neat job then quality of life should be higher than if they left you with unpleasant scars). Therefore, people within clinics will be more similar to each other than people in different clinics. As such, the person undergoing surgery is the level 1 variable, and the clinic attended is a level 2 variable.

21.3.2 Fixed and random coefficients ▌▌▌

The concepts of effects and variables should be very familiar to you by now. Throughout the book we have viewed these concepts a bit simplistically by ignoring whether an effect is fixed or random. The terms 'fixed' and 'random' are confusing because they are used differently in a variety of contexts. For example, an effect in an experiment is said to be a **fixed effect** if all possible treatment conditions in which a researcher is interested are present in the experiment, but it is a **random effect** if the experiment contains only a random sample of possible treatment conditions. This distinction is important because fixed effects can be generalized only to the situations in your experiment, whereas random effects can be generalized beyond the treatment conditions in the experiment (provided that the treatment conditions are representative). For example, in our puppy therapy example from Chapter 12, the effect is fixed if we say that we are interested only in the three conditions that we had (no puppies, 15 minutes and 30 minutes) and we can generalize our findings only to the situation of 0, 15 and 30 minutes of puppy therapy. However, if we were to say that the three doses were only a sample of possible doses (we could have tried an hour of puppy therapy), then it is a random effect and we can generalize beyond just 0, 15 and 30 minutes of therapy. The predictor variables in the examples up to now have been treated as fixed effects and the vast majority of academic research that you read will treat predictor variables as fixed effects.

People also talk about **fixed variables** and **random variables**. A fixed variable is one that is not supposed to change over time (e.g., for most people their biological sex is a fixed variable—it never changes), whereas a random one varies over time (e.g., your weight is likely to fluctuate over time).

In the context of multilevel models we make a distinction between **fixed coefficients** and **random coefficients**. In the linear models we have used we have assumed that the b-values (parameters) are fixed. For example, we have seen numerous times (e.g., Chapter 9) that the general linear model with one predictor is expressed as:

$$Y_i = b_0 + b_1 X_{1i} + \varepsilon_i \tag{21.1}$$

The outcome (Y), predictor (X) and error (ε) all vary as a function of i, which normally represents a particular case of data. In other words, it represents the level 1 variable. If, for example, we wanted to predict Sam's score, we could replace the is with her name:

$$Y_{\text{Sam}} = b_0 + b_1 X_{1,\text{Sam}} + \varepsilon_{\text{Sam}} \tag{21.2}$$

To review: when we fit this linear model we assume that the bs are fixed and we estimate them from the data. In doing so, we assume that the model holds true across the entire sample and that for every case of data we can predict a score using the same b-values for the predictor and intercept. These parameters can also be

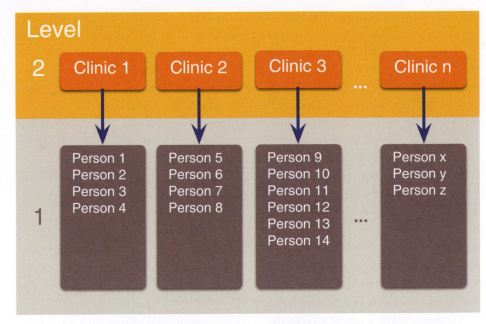

Figure 21.6 Diagram to show the hierarchical structure of the cosmetic surgery data set. People are clustered within clinics. For each person several variables were measured: surgery, BDI, age, sex, reason and pre-surgery quality of life

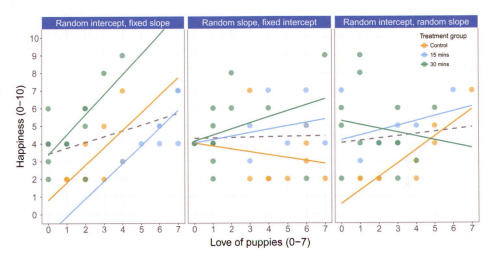

Figure 21.7 Data sets showing an overall model (dashed line) and the models for separate contexts within the data (i.e., groups of cases)

conceptualized as being 'random',[3] which means that they can vary from case to case. In other words, their values are not fixed. Up until now we have thought of linear models as having **fixed intercepts** and **fixed slopes**, but the idea that parameters can vary opens up three possibilities shown in Figure 21.7. This figure uses the example from Chapter 12 and shows the relationship between a person's love of puppies and their happiness separately for the three groups in the study (groups that had 0, 15 and 30 minutes of puppy therapy).

---

3 'Random' isn't an intuitive term for us non-statisticians because it implies that values are plucked out of thin air (randomly selected). However, this is not the case; they are carefully estimated just as fixed parameters are.

The first and simplest possibility is to model intercepts that vary across contexts (or groups)—a **random intercepts** model. For our puppy data this is like assuming that the relationship between love of puppies and happiness is the same in the 0-, 15- and 30-minute groups (i.e., the slope is the same), but that level of happiness when love of puppies is zero differs across the groups (i.e., the intercepts are different). This model is shown in the left-hand panel of Figure 21.7, in which the same relationship exists between love of puppies and happiness in the three groups (the slopes are the same) but level of happiness when love of puppies is zero differs across groups (the intercepts vary, which can be seen by the lines starting at different points on the happiness scale).

A second possibility is that slopes vary across contexts—that is, we assume **random slopes**. For our puppy data this is like assuming that the relationship between love of puppies and happiness differs in the 0-, 15- and 30-minute groups (i.e., the slope is different), but that levels of happiness are the same across the groups when love of puppies is zero (i.e., the intercepts are the same). This is what happens when we violate the assumption of homogeneity of regression slopes in ANCOVA. This situation is shown in the middle panel of Figure 21.7, in which the models within the different contexts (colors) converge on a single intercept but have different slopes.

It would be unusual to assume random slopes without also assuming random intercepts because variability in the relationship (slopes) would normally create variability in the level of the outcome variable (intercepts). Therefore, if you assume that slopes are random you would normally also assume that intercepts are random, which is our final

scenario (shown in the right-hand panel of Figure 21.7). In this situation the models within the different contexts (colors) have different slopes but are also located in different geometric space and so have different intercepts. For our puppy data this would be to assume that the relationship between love of puppies and happiness differs across the 0-, 15- and 30-minute groups, and that levels of happiness when love of puppies is zero also differ across the groups.

21.4 The multilevel model ▌▌▌▌

Having seen conceptually what a random intercept, random slope and a random intercept and slope model look like, let's look at how we represent the models. Let's use our cosmetic surgery example and imagine that we first wanted to predict someone's quality of life (QoL) after cosmetic surgery. We can represent this as the following linear model:

$$\text{QoL After Surgery}_i = b_0 + b_1 \text{Surgery}_i + \varepsilon_i \qquad (21.3)$$

In this example, we have a contextual variable of the clinic in which the cosmetic surgery was conducted. We might expect the effect of surgery on QoL to vary as a function of which clinic the surgery was conducted at because clinics use different surgeons, may have different care plans and so on. This is a level 2 variable. We could allow the model that represents the effect of surgery on QoL to vary across the different contexts (clinics) by allowing the intercepts, slopes or both to vary across clinics.

To begin with, let's include a random intercept for QoL. All we do is add a component to the intercept that measures the variability in intercepts, u_{0j}. Therefore,

the intercept changes from b_0 to become $b_0 + u_{0j}$. This term estimates the intercept of the overall model fitted to the data, b_0, and the variability of intercepts around that overall model, u_{0j}. The overall model becomes:[4]

$$Y_{ij} = \left(b_0 + u_{0j}\right) + b_1 X_{ij} + \varepsilon_{ij} \qquad (21.4)$$

The js in the equation reflect levels of the variable over which the intercept varies (in this case the clinic)—the level 2 variable. A common way to write this model is to define the random intercept separately so that the model looks like an ordinary linear model equation, except that the intercept has changed from a fixed, b_0, to a random one, b_{0j}, which is defined in a separate equation:

$$Y_{ij} = b_{0j} + b_1 X_{ij} + \varepsilon_{ij} \qquad (21.5)$$
$$b_{0j} = b_0 + u_{0j}$$

Therefore, if we want to know the estimated intercept for clinic 7, we replace the j with 'clinic 7' in the second equation:

$$b_{0,\text{clinic}7} = b_0 + u_{0,\text{clinic}7} \qquad (21.6)$$

If we want to include random slopes for the effect of surgery on QoL, then we add a component to the slope of the overall model that measures the variability in slopes, u_{1j}. Therefore, the gradient changes from b_1 to $b_1 + u_{1j}$. This term estimates the slope of the overall model fitted to the data, b_1, and the variability of slopes in different contexts around that overall model, u_{1j}. The overall model becomes (compare to the random intercept model above):

$$Y_{ij} = b_0 + \left(b_1 + u_{1j}\right) X_{ij} + \varepsilon_{ij} \qquad (21.7)$$

Again it's common to define the random slope in a separate equation. You end up

---

4 Some people use gamma (γ), not b, to represent the parameters, but I prefer b because it makes the link to the other linear models that we have used in this book clearer.

with an equation that looks like a standard linear model except that the slope has changed from a fixed, b_1, to a random one, b_{1j}, which is defined underneath:

$$Y_{ij} = b_{0i} + b_{1j}X_{ij} + \varepsilon_{ij}$$
$$b_{1j} = b_1 + u_{1j}$$
(21.8)

If we want to model a situation with random slopes *and* intercepts, then we combine the two models above. We still estimate the intercept and slope of the overall model (b_0 and b_1) but we also include the two terms that estimate the variability in intercepts, u_{0j}, and slopes, u_{1j}. The overall model becomes (compare to the two models above):

$$Y_{ij} = (b_0 + u_{0j}) + (b_1 + u_{1j})X_{ij} + \varepsilon_{ij}$$
(21.9)

We can link this more directly to a linear model if we take some of these extra terms out into separate equations. We begin with a standard linear model that replaces the fixed intercept and slope (b_0 and b_1) with their random counterparts (b_{0j} and b_{1j}), both of which are defined underneath:

$$Y_{ij} = b_{0j} + b_{1j}X_{ij} + \varepsilon_{ij}$$
$$b_{0j} = b_0 + u_{0j}$$
$$b_{1j} = b_1 + u_{1j}$$
(21.10)

The take-home point is that we're basically doing a posh linear model.

Now imagine we wanted to add in another predictor, for example quality of life before surgery. Knowing what we do about multiple regression, we shouldn't be invading the personal space of the idea that we can add this variable in with an associated beta:

$$\text{QoL After Surgery}_i = b_0 + b_1\text{Surgery}_i$$
$$+ b_2\text{QoL Before Surgery}_i + \varepsilon_i$$
(21.11)

This is a review of ideas from earlier in the book. Remember that the i represents the level 1 variable, in this case the people we tested. Therefore, we can predict a given person's QoL after surgery by replacing the i with their name:

$$\text{QoL After Surgery}_{\text{Sam}} = b_0 + b_1\text{Surgery}_{\text{Sam}}$$
$$+ b_2\text{QoL Before Surgery}_{\text{Sam}} + \varepsilon_{\text{Sam}}$$
(21.12)

Now, if we want to allow the intercept of the effect of surgery on QoL after surgery to vary across contexts then we simply replace b_0 with b_{0j}. If we want to allow the slope of the effect of surgery on QoL after surgery to vary across contexts then we replace b_1 with b_{1j}. So, even with a random intercept and slope, our model stays much the same:

$$\text{QoL After Surgery}_{ij} = b_{0j} + b_{1j}\text{Surgery}_{ij}$$
$$+ b_2\text{QoL Before Surgery}_{ij} + \varepsilon_{ij}$$
$$b_{0j} = b_0 + u_{0j}$$
$$b_{1j} = b_1 + u_{1j}$$
(21.13)

Remember that the j in the equation relates to the level 2 contextual variable (clinic in this case). So, if we wanted to predict someone's score we wouldn't just do it from their name, but also from the clinic they attended. Imagine our guinea pig Sam had her surgery done at clinic 7, then we could replace the is and js as follows:

$$\text{QoL After Surgery}_{\text{Sam,clinic 7}} = b_{0,\text{clinic 7}} + b_{1,\text{clinic 7}}\text{Surgery}_{\text{Sam,clinic 7}}$$
$$+ b_2\text{QoL Before Surgery}_{\text{Sam,clinic 7}}$$
$$+ \varepsilon_{\text{Sam,clinic 7}}$$
(21.14)

I want to sum up by reiterating that all we're doing in a multilevel model is a fancy linear model in which we allow either the intercepts or slopes or both, to vary across different contexts. All that changes is that for every parameter that we allow to be random, we get an estimate of the variability of that parameter as well as the parameter itself. So, there isn't anything terribly complicated; we can add new predictors to the model and for each one decide whether its regression parameter is fixed or random.

21.4.1 Assessing fit and comparing multilevel models ▮▮▮▮

As in logistic regression (Chapter 20) the overall fit of a multilevel model is tested using a chi-square likelihood ratio

test (see Section 19.3.4); SPSS Statistics reports the deviance, which is minus twice the log-likelihood, $-2LL$ (see Section 20.3.1). Essentially, the smaller the value of the log-likelihood, the better. SPSS Statistics also produces four adjusted versions of the log-likelihood value, which can be interpreted in the same way as the log-likelihood:

- *Akaike's information criterion* (**AIC**): This statistic is a goodness-of-fit measure that is corrected for model complexity. That means that it takes into account how many parameters have been estimated.
- *Hurvich and Tsai's criterion* (**AICC**): This version of the AIC is designed for small samples.
- *Bozdogan's criterion* (**CAIC**): This version of the AIC corrects not only for model complexity but also for sample size.
- *Schwarz's Bayesian criterion* (**BIC**): This statistic is comparable to the AIC, but it is slightly more conservative (it corrects more harshly for the number of parameters being estimated). It should be used when sample sizes are large and the number of parameters is small.

The AIC and BIC are the most commonly used. It's not meaningful to talk about any of their values being large or small *per se*, but their values can all be compared across models that are built up hierarchically. In all cases smaller values mean better-fitting models. For example, it is recommended to build up multilevel models starting with a 'basic' model in which all parameters are fixed and then adding in random coefficients as appropriate and exploring confounding variables (Raudenbush & Bryk, 2002; Twisk, 2006). You compare the fit of the model as you make parameters random or as you add in predictor variables. Models can be compared using the AIC and BIC values or by subtracting the log-likelihood of the new model from the value for the old:

$$\chi^2_{\text{change}} = (-2LL_{\text{old}}) - (-2LL_{\text{new}})$$
$$df_{\text{change}} = k_{\text{old}} - k_{\text{new}}$$
(21.15)

in which k is the number of parameters in the respective model. This equation is basically the same as equations (19.33) and (20.7). There are two caveats to this equation: (1) it works only if full maximum-likelihood estimation is used (and not restricted maximal likelihood—see SPSS Tip 21.2); and (2) the new model must contain all the effects of the older model.

21.4.2 Types of covariance structures ▮▮▮▮

If you have random effects or repeated measures in your multilevel model then you can fit a *covariance structure* to each. The covariance structure specifies the form of the variance–covariance matrix (a matrix in which the diagonal elements are variances and the off-diagonal elements are covariances). There are various forms that this matrix could take. Most of the time we'll be taking an educated guess so it is useful to run the model with different covariance structures and use the goodness-of-fit indices (the AIC, AICC, CAIC and BIC) to see whether changing the covariance structure improves the fit of the model. If the fit is improved then the covariance structure you've selected is likely to be a good choice.

The covariance structure is important because it is used as a starting point to estimate the model parameters. You will get different results depending on which covariance structure you choose. If you specify a covariance structure that is too simple then you are more likely to make a Type I error (finding a parameter is significant when it is not), but if you specify one that is too complex then you run the risk of a Type II error (finding parameters to be non-significant when in reality they are significant). SPSS Statistics has 17 covariance structures that you can use. We will look at four of the most common. In each case I use a representation of the variance–covariance matrix to illustrate, for which you could

imagine that the rows and columns represent four different clinics in our cosmetic surgery data:

$$\begin{pmatrix} 1 & 0 & 0 & 0 \\ 0 & 1 & 0 & 0 \\ 0 & 0 & 1 & 0 \\ 0 & 0 & 0 & 1 \end{pmatrix}$$

Variance Components: This covariance structure assumes that all random effects are independent (hence, the covariances in the matrix are 0). Variances of random effects are assumed to be the same (hence why they are 1 in the matrix) and sum to the variance of the outcome variable. This is the default covariance structure for random effects, and it is sometimes called the independence model.

$$\begin{pmatrix} \sigma_1^2 & 0 & 0 & 0 \\ 0 & \sigma_2^2 & 0 & 0 \\ 0 & 0 & \sigma_3^2 & 0 \\ 0 & 0 & 0 & \sigma_4^2 \end{pmatrix}$$

Diagonal: This variance structure is like variance components, except that variances are assumed to be heterogeneous (this is why the diagonal of the matrix is made up of different variance terms). This structure again assumes that variances are independent and, therefore, that all of the covariances are 0. This is the default covariance structure for repeated measures.

$$\begin{pmatrix} 1 & \rho & \rho^2 & \rho^3 \\ \rho & 1 & \rho & \rho^2 \\ \rho^2 & \rho & 1 & \rho \\ \rho^3 & \rho^2 & \rho & 1 \end{pmatrix}$$

AR(1): This stands for first-order autoregressive structure. In layman's terms, this means that the relationship between variances changes in a systematic way. If you imagine the rows and columns of the matrix to be points in time, then it assumes that the correlation between repeated measurements is highest at adjacent time points. So, in the first column, the correlation between time points 1 and 2 is ρ; let's assume that this value is 0.3. As we move to time point 3, the correlation between time point 1 and 3 is ρ^2

or 0.09. In other words, it has decreased: scores at time point 1 are more related to scores at time 2 than they are to scores at time 3. At time 4, the correlation goes down to ρ^3 or 0.027. So, the correlations between time points next to each other are assumed to be ρ, scores two intervals apart are assumed to have correlations of ρ^2, and scores three intervals apart are assumed to have correlations of ρ^3. So the correlation between scores gets smaller over time. Variances are assumed to be homogeneous, but there is a version of this covariance structure where variance can be heterogeneous. This structure is often used for repeated-measures data (especially when measurements are taken over time, such as in growth models).

$$\begin{pmatrix} \sigma_1^2 & \sigma_{21} & \sigma_{31} & \sigma_{41} \\ \sigma_{21} & \sigma_2^2 & \sigma_{32} & \sigma_{42} \\ \sigma_{31} & \sigma_{32} & \sigma_3^2 & \sigma_{43} \\ \sigma_{41} & \sigma_{42} & \sigma_{43} & \sigma_4^2 \end{pmatrix}$$

Unstructured: This covariance structure is completely general and is, therefore, the default option for random effects. Covariances are assumed to be completely unpredictable: they do not conform to a systematic pattern.

21.5 Some practical issues ▮▮▮▮

21.5.1 Assumptions ▮▮▮▮

Multilevel linear models are an extension of the linear model, so the usual assumptions apply (see Chapter 6). There is a caveat, which is that a multilevel model can sometimes solve the assumptions of independence and independent errors because we can factor in the correlations between cases caused by higher-level variables.

Cramming Sam's Tips
Multilevel models

- Multilevel models should be used to analyze data that have a hierarchical structure. For example, you might measure depression after psychotherapy. In your sample, patients will see different therapists within different clinics. This is a three-level hierarchy with depression scores from patients (level 1), nested within therapists (level 2), who are themselves nested within clinics (level 3).
- Hierarchical models are just linear models in which you can allow parameters to vary (this is called a random effect). In the standard linear model, parameters generally are a fixed value estimated from the sample (a fixed effect).
- If we estimate a linear model within each context (the therapist or clinic, to use the example above) rather than the

sample as a whole, then we can assume that the intercepts of these models vary (a random intercepts model) or that the slopes of these models differ (a random slopes model) or that both vary.
- We can compare different models by looking at the difference in the value of −2LL. Usually we would do this when we have changed only one parameter (added one new thing to the model).
- For any model we have to assume a covariance structure. For random intercepts models the default of variance components is fine, but when slopes are random an unstructured covariance structure is often assumed. When data are measured over time an autoregressive structure (AR(1)) is often assumed.

As such, if a lack of independence is being caused by a level 2 or level 3 variable then a multilevel model should make this problem go away (although not always). Also in repeated-measures designs we need not restrict the covariance structure to being spherical (i.e., we don't need to assume sphericity) because we can model less restrictive structures.

There are two additional assumptions in multilevel models that relate to the random coefficients. These coefficients are assumed to be normally distributed around the overall model. So, in a random

intercepts model the intercepts in the different contexts are assumed to be normally distributed around the overall model. Similarly, in a random slopes model, the slopes of the models in different contexts are assumed to be normally distributed.

Finally, it's worth mentioning that multicollinearity can be a particular problem in multilevel models if you have interactions that cross levels in the data hierarchy (cross-level interactions). However, centering predictors (Section 21.5.4) can help matters enormously (Kreft & de Leeuw, 1998).

21.5.2 Robust multilevel models ▌▌▌▌

Although we don't use these methods within this chapter, the main dialog box for specifying a multilevel model (e.g., Figure 21.11) has a `Bootstrap...` button, which can be used to access a bootstrap dialog box (Section 6.12.3). Use this dialog box to obtain robust confidence intervals of the model parameters (e.g., to get robust versions of the information in Output 21.5). Be warned that the analysis may take some time to run, and that, for complex models especially, bootstrap confidence intervals cannot always be computed.

21.5.3 Sample size and power ▮▮▮▯

As you might well imagine, the situation with power and sample size is very complex indeed. One complexity is that we are trying to make decisions about our power to detect both fixed and random effects coefficients. Kreft and de Leeuw (1998) do a tremendous job of making sense of things for us. Essentially, the message is the more data, the better. As more levels are introduced into the model, more parameters need to be estimated and the larger the sample sizes need to be. Kreft and de Leeuw conclude that if you are looking for cross-level interactions then you should aim to have more than 20 contexts (groups) in the higher-level variable, and that group sizes 'should not be too small'. They conclude that there are so many factors involved in multilevel analysis that it is impossible to produce any meaningful rules of thumb.

Twisk (2006) agrees that the number of contexts relative to individuals within those contexts is important. He also points out that standard sample size and power calculations can be used but then 'corrected' for the multilevel component of the analysis (by factoring, among other things, the intraclass correlation). However, he discusses two corrections that yield very different sample sizes. He recommends using sample size calculations with caution.

Having said all of that, there are tools available for power calculations if you use Windows (but no Mac OS versions that I can find). Probably the most flexible is *MLPowSIm*. There's also Optimal Design and, for two-level hierarchies, PinT.[5]

21.5.4 Centering predictors ▮▮▮▮

Centering is the process of transforming a variable into deviations around a fixed point (Section 11.3.3). One  such fixed point is the mean of a variable (**grand mean centering**). This form of centering is used in multilevel models too, but sometimes **group mean centering** is used instead. Group mean centering occurs when for a given variable we take each score and subtract from it the mean of the scores (for that variable) *within a given group*. For multilevel models, it is usually only level 1 predictors that are centered (in our cosmetic surgery example this would be predictors such as age, BDI and pre-surgery quality of life). If group mean centering is used then a level 1 variable is typically centered on means of a level 2 variable (in our cosmetic surgery data this would mean that, for example, the age of a person would be centered around the mean of age for the clinic at which the person had their surgery).

In multilevel models centering can be a useful way to combat multicollinearity between predictor variables. It's also helpful when predictors do not have a meaningful zero point. Multilevel models with centered predictors tend to be more stable, and estimates from these models can be treated as more or less independent of each other, which might be desirable. However, as with standard linear models (Section 11.3.3), centering affects the model. I'll try to sum up some excellent reviews (Kreft, 1995; Kreft & de Leeuw, 1998; Enders & Tofighi, 2007). Essentially, if you fit a multilevel model using the raw score predictors and then fit the same model but with grand mean centered predictors then the resulting models are equivalent. They will fit the data

equally well, have the same predicted values, and the residuals will be the same. The parameters themselves (the *b*s) will, of course, be different, but there will be a direct relationship between the parameters from the two models. Therefore, grand mean centering doesn't change the multilevel model, but it would change your interpretation of the parameters (you can't interpret them as though they are raw scores). Group mean centering is more complicated because the raw score model is not equivalent to the centered model in either the fixed part or the random part. One exception is when only the intercept is random (which arguably is an unusual situation), and the group means are reintroduced into the model as level 2 variables (Kreft & de Leeuw, 1998).

People learning statistics often worry about there being a 'best' way to do things, but the 'best' method usually depends on what you're trying to do. Centering is a good example. Although some people make a decision about whether to use group or grand mean centering based on some statistical criterion, there is no statistically correct choice between not centering, grand mean centering and group mean centering (Kreft et al., 1995). Enders and Tofighi (2007) make four recommendations when analyzing data with a two-level hierarchy: (1) group mean centering should be used if the primary interest is in an association between variables measured at level 1 (i.e., the aforementioned relationship between surgery and quality of life after surgery); (2) grand mean centering is appropriate when the primary interest is in the level 2 variable but you want to adjust for the level 1 covariate (i.e., you want to look at the effect of clinic on quality of life after surgery while adjusting for the type of surgery); (3) both types of centering can be used to look at the differential influence of a variable at level 1 and 2 (i.e., is the effect of surgery on quality of life different at the clinic level

---

5 Because URLs change, I suggest the search terms 'MLPowSIm', 'optimal design software' and 'PinT multilevel'.

Oliver Twisted

Please, Sir, can I have some more . . . group mean centering?

'Centering was so much fun when we did it in Chapter 11. It was like being gently tickled to sleep in a warm bath of octopuses. I want some more!' gurgles Oliver as he splashes around in his bath. Fair enough, Oliver, if you want to know how to do group mean centering using SPSS, then the material on the companion website will tell you. Oh, and be careful with that toy pirate ship …

compared to the client level?); and (4) group mean centering is preferable for examining cross-level interactions (e.g., the interactive effect of clinic and surgery on quality of life). If group mean centering is used then the group means should be reintroduced as a level 2 variable unless you want to look at the effect of your 'group' or level 2 variable uncorrected for the mean effect of the centered level 1 predictor, such as when fitting a model when time is your main explanatory variable (Kreft & de Leeuw, 1998).

21.6 Multilevel modeling using SPSS Statistics

Most people who do multilevel modeling tend to use specialist software such as MLwiN, HLM and R. In books that compare the various packages SPSS tends to fare relatively badly (Twisk, 2006; Tabachnick & Fidell, 2012). Apart from anything else, SPSS Statistics has a completely indecipherable interface for

multilevel models (and I'm not the only one to say this).

Figure 21.8 shows a very stripped-down version of how we proceed with the analysis. After initial checks of the data, it is useful to build up models starting with a 'basic' model in which all parameters are fixed and then add random coefficients as appropriate before exploring confounding variables (as I mentioned in Section 21.4.1).

21.6.1 Entering the data

Data entry depends a bit on the study design: when the same variables are measured at several points in time or you have several scores nested within an entity, you need to use the long format, but otherwise you can use the wide

format (Section 4.6.1). The surgery example contains a single outcome score from each person (not several outcomes scores nested within each person) so we use the wide format shown in Figure 21.9. Each row represents a case of data (in this case a person who had surgery). Their scores on the various variables are entered in different columns. So, for example, the first person was 31 years old, had a BDI score of 12, was in the waiting list control group at clinic 1, female and waiting for surgery to change her appearance.

21.6.2 Ignoring the data structure

Let's ground the example in something familiar to us: the linear model. For the

SELF TEST

Conduct a linear model (one-way ANOVA) using **Surgery** as the predictor and **Post_QoL** as the outcome.

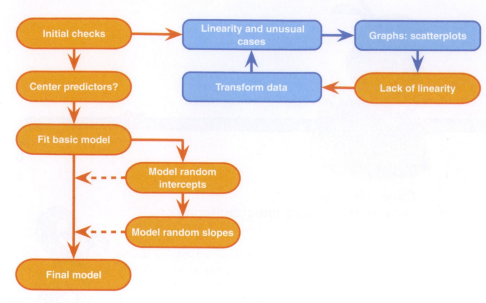

Figure 21.8 The basic process of fitting a multilevel model

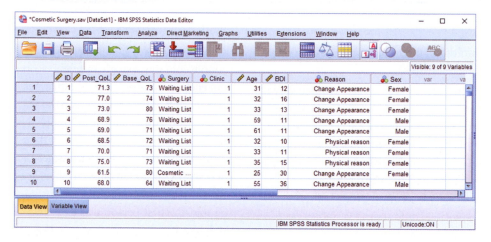

Figure 21.9 Data layout for multilevel modeling when outcome scores are not nested within cases

ANOVA

Quality of Life After Cosmetic Surgery

	Sum of Squares	df	Mean Square	F	Sig.
Between Groups	28.620	1	28.620	.330	.566
Within Groups	23747.883	274	86.671		
Total	23776.504	275			

Output 21.1

time being imagine that we're interested only in the effect that surgery has on post-operative quality of life. We could fit the linear model described by equation (21.3). In reality we wouldn't do this—I'm using it to show you that multilevel models are not big and scary, they are simply extensions of what we have done before. If you did the self-test you'll get Output 21.1, which shows a non-significant effect of surgery on quality of life, $F(1, 274) = 0.33$, $p = 0.566$.

To run the same analysis through the multilevel model dialog boxes select *Analyze* ▶ *Mixed Models* ▶ 🔲 Linear… to activate the dialog box in Figure 21.10. This dialog box is for specifying the hierarchical structure of the data, but because, for now, we are ignoring this structure, click Continue to move to the dialog box in Figure 21.11. First specify the outcome variable, which is quality of life after surgery, by dragging **Post_QoL** to the box labeled *Dependent Variable* (or select it and click ➡). Next, specify the predictor, which is whether the person had surgery, by dragging **Surgery** to the box labeled *Covariate(s)* or click ➡ (SPSS Tip 21.1).

In the dialog box in Figure 21.11 we use Fixed… to specify fixed effects in the model, and Random… to specify—yes, you've guessed it—random effects. To begin with we are going to treat our effects as fixed, so click Fixed… to bring up the dialog box in Figure 21.12. We have only one variable specified as a predictor, and we want this to be treated as a fixed effect, so select it in the list labeled *Factors and Covariates* and click Add to transfer it to the *Model*. Click Continue to return to the main dialog box.

Click Estimation… to open the dialog box in Figure 21.13 (left), in which you can change the settings for the estimation process. For example, if you don't get a solution then you could increase the number of iterations (SPSS Tip 20.1). The defaults can be left alone, but you should decide whether to use the maximum-likelihood or restricted maximum-likelihood estimation method. There are pros and cons to both (see SPSS Tip 21.2), but, because we want to compare models as we build them up, we will select ◉ Maximum Likelihood (ML). Click Continue to return to the main dialog box.

Click Statistics... to open the dialog box in Figure 21.13 (right). There are two useful options in this dialog box. The first is ☑ Parameter estimates, which will give us *b*-values for each effect and their significance. The second is ☑ Tests for covariance parameters, which will give us a significance test of each of the covariance estimates in the model (i.e., the values of *u* in equations (21.5), (21.8) and (21.10)). These estimates tell us about the variability of intercepts or slopes across our contextual variable and so significance-testing them can be useful (we can then say that there was significant or non-significant, variability in intercepts or slopes). Select these two options and click Continue to return to the main dialog box and OK to fit the model.

Output 21.2 shows the *F*-statistic for the effect of surgery on quality of life: compare this output with Output 21.1. There's basically no difference: we get a non-significant effect of surgery with an *F* of 0.33, and a *p* of 0.565.[6] The point is that if we ignore the hierarchical structure of the data then what we are left with is something very familiar: a linear model. The numbers are the same—we just reached them via different menus.

21.6.3 Ignoring the data structure: covariates ▮▮▮▮

OK, so there's no significant effect of cosmetic surgery on quality of life, but we

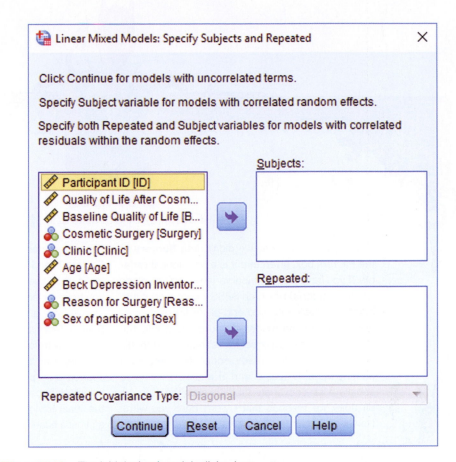

Figure 21.10 The initial mixed models dialog box

didn't factor in quality of life *before* surgery. Let's do that now. Our model is now described by equation (21.11). This model is an example of an ANCOVA, which we could run through the univariate GLM menu. As in the previous section, we'll run the analysis both ways to illustrate that we're doing the same thing when we run a hierarchical model.

Output 21.3 shows the results of the self-test. With baseline quality of life included we find a significant effect of surgery on

quality of life, $F(1, 273) = 4.04$, $p = 0.045$. Baseline quality of life also significantly predicted quality of life after surgery, $F(1, 273) = 214.89$, $p < 0.001$.

Let's fit the model again through the *Analyze* ▸ *Mixed Models* ▸ Linear... menu. As before, ignore the first dialog box because, for now, we are ignoring the hierarchical structure of our data. We can leave the main dialog box set up as it was before, except that we add baseline quality of life as another predictor (Figure 21.14).

SELF TEST

Fit a linear model (a one-way ANCOVA) using **Surgery** as the predictor, **Post_QoL** as the outcome and **Base_QoL** as the covariate.

6 The slight difference is because here we're using maximum likelihood methods to estimate the parameters of the model whereas the original linear model is estimated using ordinary least squares.

SPSS Tip 21.1
Factor(s) or Covariate(s) ▐▐▐▐

You might wonder why we didn't drag **Surgery** to the *Factors* box, given that it is a categorical variable. If you have a categorical variable and you place it in the *Factor(s)* box then SPSS will convert it into dummy variables for you and place these dummy variables into the model. If you place it into the *Covariate(s)* box it will treat it as a linear trend. In this example, we have already coded **Surgery** as a dummy variable (i.e., 0 and 1) and so it's fine to specify it as a covariate (and it makes the output a bit tidier for reasons that I won't bore you with). However, if your categorical variable had more than two categories you should certainly drag it into *Factor(s)* (and if you try out the end-of-chapter tasks you'll see that we do this). The exception is if you have ordered categories. In the second example we have a variable representing different time points. Technically, this variable is a categorical variable with four levels (each representing a point in time), but because it represents four equally spaced time points we would again treat it as a covariate because by doing so we'd be looking at the linear trend of time (rather than each time point against a baseline).

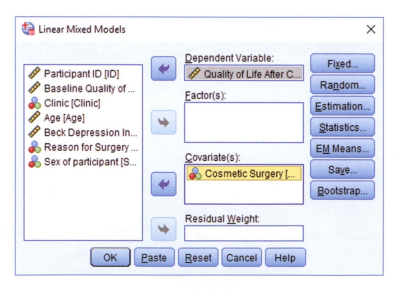

Figure 21.11 The main mixed models dialog box

To do this, drag **Base_QoL** to the box labeled *Covariate(s)* (or click ➡). Again, we want this new variable to go into the model as a fixed effect, so click Fixed..., select **Base_QoL** in the list labeled *Factors and Covariates* and click Add to transfer it to the *Model* (Figure 21.15). Click Continue to return to the main dialog box and OK to fit the model.

Output 21.4 shows the main statistics for the model: compare it with Output 21.3. The results are pretty similar to when we ran the analysis as ANCOVA:[7] we get a significant effect of surgery with an *F* of 4.08, *p* = 0.044, and a significant effect of baseline quality of life with an *F* of 217.25,

---

7 Again, the slight differences in values are because of using maximum likelihood estimation instead of ordinary least squares.

$p < 0.001$. We can also see that the *b*-value for surgery is –1.70.

The purpose of this exercise was to convince you that a multilevel model is just an extension of the linear model, and we've spent most of this book learning about them. If you think about multilevel models as an extension of something that you already (I hope) understand then they become less overwhelming (I hope again). Now let's look at how we factor in the hierarchical structure of the data.

21.6.4 Including random intercepts in the model

We have seen that when we factor in the pre-surgery quality of life scores, which themselves significantly predict post-surgery quality of life scores, surgery seems to negatively affect quality of life. However, we have ignored the fact that our data have a hierarchical structure. Essentially we have violated the independence assumption, because scores from people who had their surgery at the same clinic are likely to be related to each other (and certainly more related than with people at different clinics). Violating the assumption of independence can have some quite drastic consequences (see Section 12.3), but rather than panic and gibber about our *F*-statistic being inaccurate, we can model this covariation within clinics by including the hierarchical data structure in our model.

To begin with, we will include the hierarchy in the most basic way by assuming that intercepts vary across clinics. This model is described as follows:

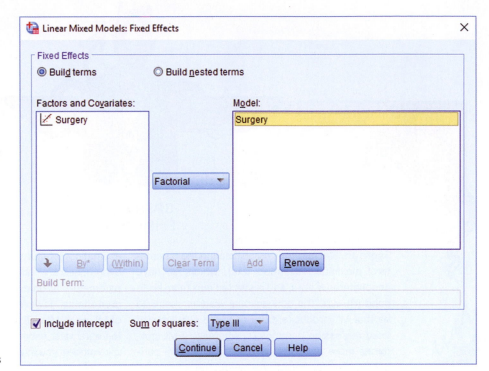

Figure 21.12 The dialog box for specifying fixed effects in mixed models

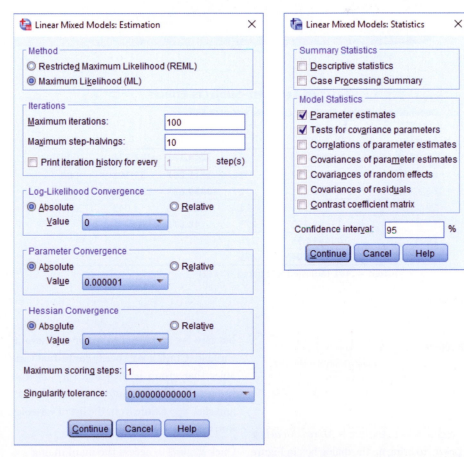

Figure 21.13 The *Estimation* and *Statistics* options for mixed models

Type III Tests of Fixed Effects[a]

Source	Numerator df	Denominator df	F	Sig.
Intercept	1	276	6049.727	.000
Surgery	1	276	.333	.565

a. Dependent Variable: Quality of Life After Cosmetic Surgery.

Output 21.2

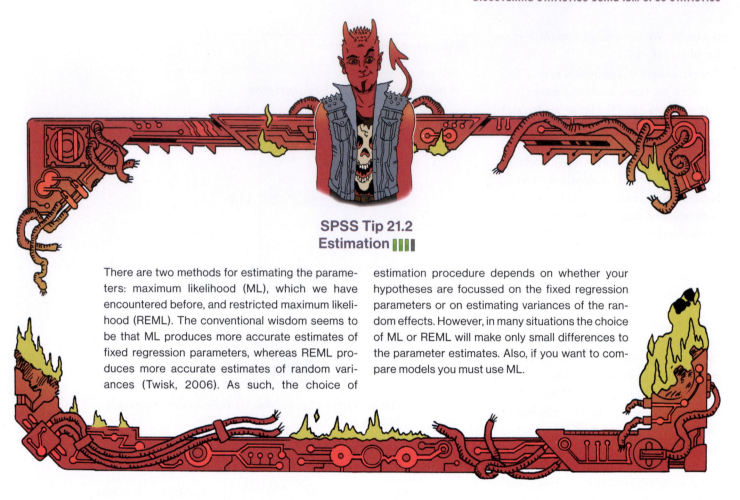

SPSS Tip 21.2
Estimation ||||

There are two methods for estimating the parameters: maximum likelihood (ML), which we have encountered before, and restricted maximum likelihood (REML). The conventional wisdom seems to be that ML produces more accurate estimates of fixed regression parameters, whereas REML produces more accurate estimates of random variances (Twisk, 2006). As such, the choice of estimation procedure depends on whether your hypotheses are focussed on the fixed regression parameters or on estimating variances of the random effects. However, in many situations the choice of ML or REML will make only small differences to the parameter estimates. Also, if you want to compare models you must use ML.

Tests of Between–Subjects Effects

Dependent Variable: Quality of Life After Cosmetic Surgery

Source	Type III Sum of Squares	df	Mean Square	F	Sig.
Corrected Model	10488.253[a]	2	5244.127	107.738	.000
Intercept	1713.257	1	1713.257	35.198	.000
Base_QoL	10459.633	1	10459.633	214.888	.000
Surgery	196.816	1	196.816	4.043	.045
Error	13288.250	273	48.675		
Total	1004494.53	276			
Corrected Total	23776.504	275			

a. R Squared = .441 (Adjusted R Squared = .437)

Output 21.3

$$QoL\,After\,Surgery_{ij} = b_{0j} + b_1 Surgery_{ij}$$
$$+ b_2 QoL\,Before\,Surgery_{ij} + \varepsilon_{ij}$$
$$b_{0j} = b_0 + u_{0j}$$

(21.16)

We again select *Analyze* ▶ *Mixed Models* ▶ Linear... to bring up the dialog box in Figure 21.10. Previously we ignored this dialog box, but now we will use it to specify our level 2 variable (**Clinic**). We specify contextual variables that group participants (or subjects) using the box labeled *Subjects*. Drag (or select and click) **Clinic** from the list of variables into this box as in Figure 21.16.

Click Continue to access the main dialog box. We don't need to change this from the previous model (Figure 21.14) because all we are going to do is change the intercept from being fixed to random. We also don't need to re-specify our fixed effects, so there is no need to click Fixed... unless you want to check that the dialog box still looks like Figure 21.15. However, we do need to specify a random effect for the first time, so click Random... to access the dialog box in Figure 21.17. First, we specify our contextual variable by selecting it (**Clinic**) from the section labeled *Subjects* (which will contain any variables that we specified in Figure 21.16) and dragging it (or clicking) to the area labeled *Combinations*. To specify that the intercept is random select ☑ Include intercept (Figure 21.17). There is a drop-down list to specify the type of covariance (Variance Components ▼), but for a random intercept model this default option is fine. Click Continue to return to the main dialog box and then [Add] to fit the model.

Output 21.5 shows the edited output. We can test whether allowing the intercept to vary across clinics has made a difference to the model using the change in *–2LL* (equation (21.15)). In the current model *–2LL* is 1837.49 based on five parameters (Output 21.5) and in the previous model (Output 21.4) *–2LL* was 1852.54, based on four parameters. Therefore, the change in chi-square is 15.05 with 1 degree of freedom:

$$\chi^2_{change} = 1852.54 - 1837.49 = 15.05$$
$$df_{change} = 5 - 4 = 1$$

(21.17)

The critical values for the chi-square statistic with 1 degree of freedom (in the Appendix) are 3.84($p < 0.05$) and 6.63 ($p < 0.01$), and because the change in chi-square is bigger than these values it is highly significant. Put another way, modeling the variability in intercepts significantly improvesthe fit of our model. We can conclude, then, that the intercepts for the relationship between surgeryand quality of life (when controlling for baseline quality of life) vary significantly across the different clinics.

An alternative way to see whether intercepts vary across clinics is to test the variance estimate for the intercept (9.24) using the Wald statistic (Output 21.5), which is a standard *z*-score ($z = 1.69$, $p = 0.09$). This test contradicts our conclusion from the change in *–2LL*; however, the change in *–2LL* is a much more reliable way to assess the significance of changes to the model. In general, be cautious in interpreting the Wald statistic because, for random parameters especially (like we have here), it can be quite unpredictable (for fixed effects it should be OK).

By allowing the intercept to vary we get a different *b*-value for the effect of surgery,

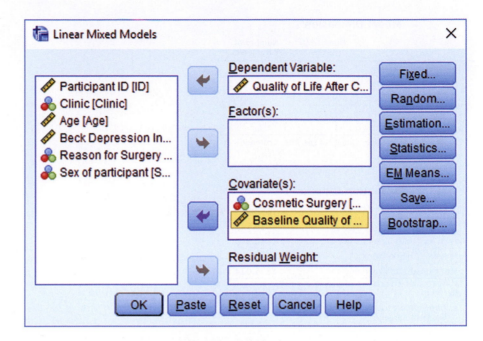

Figure 21.14   The main mixed models dialog box

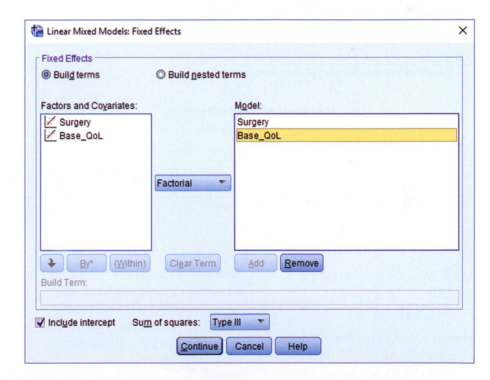

Figure 21.15   The dialog box for specifying fixed effects in mixed models

which is –0.31 compared to –1.70 when the intercept was fixed (Output 21.4). In other words, by allowing the intercepts to vary over clinics, the effect of surgery has decreased dramatically. In fact, it is not significant any more, $F(1, 275.63) = 0.14$, $p = 0.709$. This shows how, had we ignored the hierarchical structure in our data, we would have reached very different conclusions than what we have found here.

Model Dimension[a]

		Number of Levels	Number of Parameters
Fixed Effects	Intercept	1	1
	Surgery	1	1
	Base_QoL	1	1
Residual			1
Total		3	4	← df for –2LL

a. Dependent Variable: Quality of Life After Cosmetic Surgery.

Information Criteria[a]

–2 Log Likelihood	1852.543	← –2LL
Akaike's Information Criterion (AIC)	1860.543
Hurvich and Tsai's Criterion (AICC)	1860.690
Bozdogan's Criterion (CAIC)	1879.024
Schwarz's Bayesian Criterion (BIC)	1875.024

The information criteria are displayed in smaller-is-better form.

a. Dependent Variable: Quality of Life After Cosmetic Surgery.

Type III Tests of Fixed Effects[a]

Source	Numerator df	Denominator df	F	Sig.
Intercept	1	276	39.379	.000
Surgery	1	276	4.088	.044
Base_QoL	1	276	217.249	.000

a. Dependent Variable: Quality of Life After Cosmetic Surgery.

Estimates of Fixed Effects[a]

Parameter	Estimate	Std. Error	df	t	Sig.	95% Confidence Interval Lower Bound	Upper Bound
Intercept	18.147025	2.891820	276	6.275	.000	12.454198	23.839851
Surgery	–1.697233	.839442	276	–2.022	.044	–3.349756	–.044710
Base_QoL	.665036	.045120	276	14.739	.000	.576213	.753858

a. Dependent Variable: Quality of Life After Cosmetic Surgery.

Output 21.4

21.6.5 Including random intercepts and slopes in a model ▮▮▮▮

Including a random intercept is important for this model (it changes the log-likelihood significantly). Let's now look at whether adding a random slope also improves our ability to predict quality of life after surgery. This model is described by equation (21.13). All we are doing is adding another random term to the model, so the only changes we need to make are in the dialog box accessed by clicking on [Random…]. (If you are starting from scratch then follow the instructions for setting up the dialog box in the previous section.) Select the predictor (**Surgery**) from the list of *Factors and Covariates* and add it to the *Model* by clicking [Add] (Figure 21.18). Click [Continue] to return to the main dialog box and then click [OK] to run the analysis.

To see whether including the variance in slopes (estimated as 29.63 in Output 21.6) helps the model, we can again look at the change in –2LL. In the current model –2LL is 1816 (Output 21.6), based on six parameters, and in the previous model (Output 21.5) it was 1837.49, based on five parameters. Therefore, the change in chi-square is 21.49 with 1 degree of freedom:

$$\chi^2_{change} = 1837.49 - 1816 = 21.49$$
$$df_{change} = 6 - 5 = 1$$

(21.18)

Comparing this value to the same critical values (3.84 and 6.63) as before for the chi-square distribution with $df = 1$ shows that this change is highly significant (because 21.49 is much larger than the critical values). Put another way, the fit of our model significantly improves when the variance of slopes is included: there is significant variability in slopes.

Given that there is significant variability in slopes, we should estimate the degree to which the slopes and intercepts correlate (or covary). By selecting [Variance Components ▾] in the previous analysis, we assumed that the covariances between the intercepts and slopes were zero. Therefore, we estimated only the variance of slopes, which was useful because it allowed us to look at this effect in isolation. To include the *covariance* between random slopes and random intercepts click [Variance Components ▾] in Figure 21.18 to access the drop-down list, and select [Unstructured ▾]. By changing to [Unstructured ▾], we remove the assumption that the covariances between slopes and intercepts are zero, and estimate this covariance. Basically, we add a term to the model that estimates the covariance between random slopes and intercepts. Refit the previous model, but change [Variance Components ▾] to [Unstructured ▾] in Figure 21.18. We assess the degree to which adding the covariance between slopes and intercepts has made a difference to the model using the change in –2LL (equation (21.15)). In our current model –2LL is 1798.62 (Output 21.7), based on a total of seven parameters, and in the previous model (Output 21.6) it was 1816, based on six parameters. Therefore, the change in –2LL is 17.38 with 1 degree of freedom:

$$\chi^2_{Change} = 1816 - 1798.62 = 17.38 \qquad (21.19)$$
$$df_{Change} = 7 - 6 = 1$$

This change is highly significant at $p < 0.01$ because 17.38 is bigger than the critical value of 6.63 for the chi-square statistic with 1 degree of freedom (see the Appendix). The fit of our model is significantly improved when the covariance term is included. The variance estimates for the intercept (37.60) and slopes (−36.68 and 38.41), and their associated significance based on the Wald test, confirm this because all three estimates are close to significance (although I reiterate my earlier point that the Wald statistic should be interpreted with caution).

Notice that the random part of the slopes now has two values (−36.68 and 38.41). The reason is because we changed from a covariance structure of Variance Components ▼, which assumes that slopes and intercepts are uncorrelated, to Unstructured ▼, which makes no such assumption, and estimates the covariance too. The first of the two values (−36.68) is the covariance between slopes and intercepts, and the second (38.41) is the variance of the random slopes. Let's look at the covariance first.

Remember that the covariance (Chapter 8) is an unstandardized measure of the relationship between variables. It's like a correlation. Therefore, the covariance term tells us whether there is a relationship between the random slope and the random intercept within the model. The size of this value is not terribly important because it is unstandardized (so we can't compare the size of covariances measured across different models containing different variables), but its direction is. In this model the covariance is negative (−36.68), indicating a negative relationship between the intercepts and the slopes. Remember that we are looking at the effect of surgery on quality of life in 10 different clinics, which means that, across these clinics, as the intercept for the relationship between surgery and quality of life increases, the

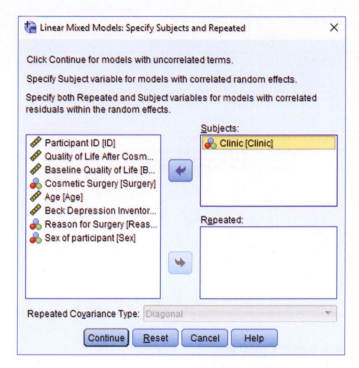

Figure 21.16 Specifying a level 2 variable in a hierarchical linear model

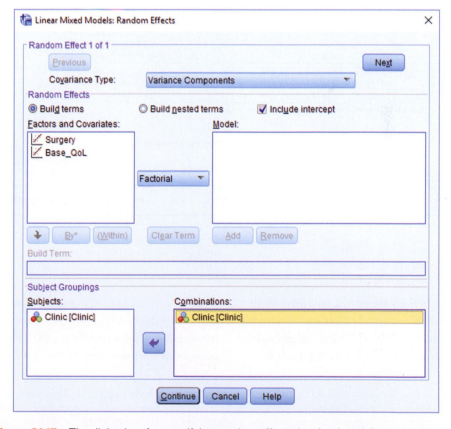

Figure 21.17 The dialog box for specifying random effects in mixed models

Model Dimension[a]

		Number of Levels	Covariance Structure	Number of Parameters	Subject Variables
Fixed Effects	Intercept	1		1	
	Surgery	1		1	
	Base_QoL	1		1	
Random Effects	Intercept[b]	1	Variance Components	1	Clinic
Residual				1	
Total		4		5 ←	*df* for −2LL

a. Dependent Variable: Quality of Life After Cosmetic Surgery.

b. As of version 11.5, the syntax rules for the RANDOM subcommand have changed. Your command syntax may yield results that differ from those produced by prior versions. If you are using version 11 syntax, please consult the current syntax reference guide for more information.

Information Criteria[a]

−2 Log Likelihood	1837.490 ← −2LL
Akaike's Information Criterion (AIC)	1847.490
Hurvich and Tsai's Criterion (AICC)	1847.712
Bozdogan's Criterion (CAIC)	1870.592
Schwarz's Bayesian Criterion (BIC)	1865.592

The information criteria are displayed in smaller-is-better form.

a. Dependent Variable: Quality of Life After Cosmetic Surgery.

Type III Tests of Fixed Effects[a]

Source	Numerator df	Denominator df	F	Sig.
Intercept	1	163.879	73.305	.000
Surgery	1	275.631	.139	.709
Base_QoL	1	245.020	83.159	.000

a. Dependent Variable: Quality of Life After Cosmetic Surgery.

bs

Estimates of Fixed Effects[a]

Parameter	Estimate	Std. Error	df	t	Sig.	95% Confidence Interval Lower Bound	95% Confidence Interval Upper Bound
Intercept	29.563601	3.452958	163.879	8.562	.000	22.745578	36.381624
Surgery	−.312999	.838551	275.631	−.373	.709	−1.963776	1.337779
Base_QoL	.478630	.052486	245.020	9.119	.000	.375248	.582012

a. Dependent Variable: Quality of Life After Cosmetic Surgery.

$Var(\varepsilon_{ij})$

Estimates of Covariance Parameters[a]

Parameter		Estimate	Std. Error	Wald Z	Sig.	95% Confidence Interval Lower Bound	95% Confidence Interval Upper Bound
Residual		42.497179	3.703949	11.473	.000	35.823786	50.413718
Intercept [subject = Clinic]	Variance	9.237126	5.461678	1.691	.091	2.898965	29.432742

a. Dependent Variable: Quality of Life After Cosmetic Surgery.

$Var(u_{0j})$

Output 21.5

value of the slope decreases. Figure 21.19 shows this pattern by plotting the observed values of quality of life after surgery against those predicted by our model. Each line

represents a different clinic and it's clear that the 10 clinics differ: those with low intercepts (low values on the *y*-axis) have quite steep positive slopes. As the intercept

increases (as we go from the line that crosses the *y*-axis at the lowest point up to the line that hits the *y*-axis at the highest point) the slopes of the lines tend to get flatter (the slope decreases). The negative covariance between slope and intercept reflects this relationship. Had it been positive it would mean the opposite: as intercepts increase, the slopes increase also.

Moving on to the variance of slopes (38.41), this tells us how much the slopes vary around a single slope fitted to the entire data set (i.e., ignoring the clinic from which the data came). This value confirms what our chi-square test showed us: that the slopes across clinics are significantly different. This is also evident from Figure 21.19.

To sum up, we can conclude that the intercepts and slopes for the relationship between surgery and quality of life (when controlling for baseline quality of life) vary significantly across the different clinics. Allowing the intercept and slopes to vary results in a new *b*-value for the effect of surgery, which is −0.65 compared to −0.31 when the slopes were fixed (Output 21.5). By allowing the intercepts to vary over clinics, the effect of surgery has increased, although it is still not significant, $F(1, 9.518) = 0.10$, $p = 0.762$.

21.6.6 Adding an interaction to the model ▮▮▮▮

We can build up the model by adding in another variable. In the data we recorded the reason for the person having cosmetic surgery: was it to resolve a physical problem or purely for vanity? Let's add this predictor to the model and also look at whether it interacts with surgery in predicting quality of life.[8] Our model will expand to incorporate these new terms, and

---

8 In reality, because we would use the change in −2LL to see whether effects are significant, we would build this new model up a term at a time. Therefore, we would first include only **Reason** in the model, then in a separate model we would add the interaction. By doing so we can calculate the change in −2LL for each effect. To save space I'm going to put both into the model in a single step.

each term will have a *b*-value (which we select to be fixed). This model is described as follows (note that all that has changed is that there are two new predictors):

$$\text{QoL After Surgery}_{ij} = b_{0j} + b_{1j}\text{Surgery}_{ij} + b_2\text{QoL Before Surgery}_{ij}$$
$$+ b_3\text{Reason}_{ij} + b_4\left(\text{Reason} \times \text{Surgery}\right)_{ij} + \varepsilon_{ij}$$
$$b_{0j} = b_0 + u_{0j}$$
$$b_{1j} = b_1 + u_{1j} \qquad\qquad (21.20)$$

This model requires only minor changes to the dialog boxes that we have already set up. First, select *Analyze* ▶ *Mixed Models* ▶ Linear... to access the initial dialog box, which should be set up as for the previous analysis (it should look like Figure 21.16). Click Continue to access the main dialog box. Assuming you're continuing the previous model then this dialog box will already be set up (it should look like Figure 21.14). We have two new covariates to add to the model: the effect of the reason for the surgery (**Reason**) and the interaction of **Reason** and **Surgery**. At this stage we add **Reason** as a covariate, so drag this variable to the box labeled *Covariate(s)* (or click ⬇).[9] The completed dialog box is in Figure 21.20.

To add the new predictors to the model as fixed effects, click Fixed... to bring up the dialog box in Figure 21.21. To specify the main effect of **Reason**, select this variable in the list labeled *Factors and Covariates* and click Add to transfer it to *Model*. To specify the interaction term, first click Factorial ▾ and change it to Interaction ▾. Next, select **Surgery** from *Factors and Covariates* and, while holding down the *Ctrl* (*Cmd* on a Mac) key, select **Reason**. With both variables selected, click Add to transfer them to *Model* as an interaction effect. The completed dialog box should look like Figure 21.21. Click Continue to return to the main dialog box. We don't need to specify any new random coefficients, so leave the dialog box accessed through Random... as it is in Figure 21.18, and we can leave the other

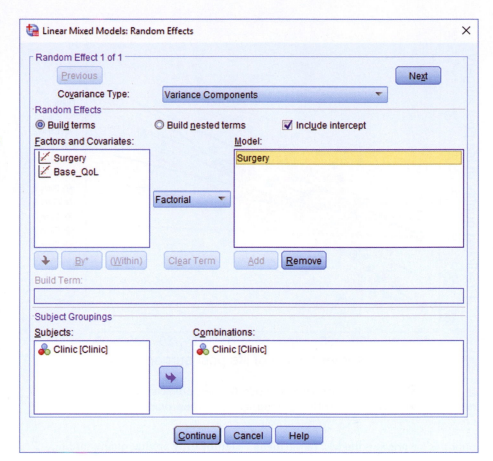

Figure 21.18 The dialog box for specifying random effects in mixed models

Information Criteria[a]

–2 Log Likelihood	1816.001 ← –2LL
Akaike's Information Criterion (AIC)	1828.001
Hurvich and Tsai's Criterion (AICC)	1828.314
Bozdogan's Criterion (CAIC)	1855.724
Schwarz's Bayesian Criterion (BIC)	1849.724

The information criteria are displayed in smaller-is-better form.

a. Dependent Variable: Quality of Life After Cosmetic Surgery.

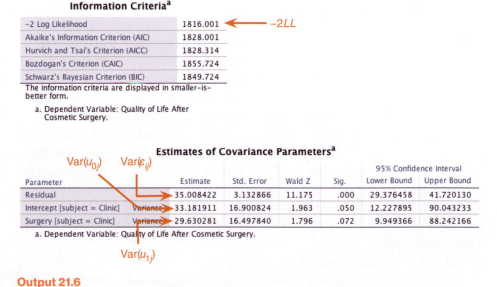

Estimates of Covariance Parameters[a]

						95% Confidence Interval	
Parameter		Estimate	Std. Error	Wald Z	Sig.	Lower Bound	Upper Bound
Residual		35.008422	3.132866	11.175	.000	29.376458	41.720130
Intercept [subject = Clinic]	Variance	33.181911	16.900824	1.963	.050	12.227895	90.043233
Surgery [subject = Clinic]	Variance	29.630281	16.497840	1.796	.072	9.949366	88.242166

a. Dependent Variable: Quality of Life After Cosmetic Surgery.

$\text{Var}(u_{0j})$ $\text{Var}(\varepsilon_{ij})$ $\text{Var}(u_{1j})$

Output 21.6

---

9 As with **Surgery**, I've dragged **Reason** to the *Covariate(s)* box because it is already dummy-coded (SPSS Tip 21.1).

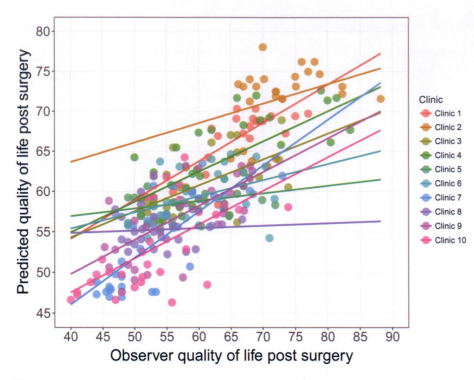

Figure 21.19 Predicted values from the model (surgery predicting quality of life after controlling for baseline quality of life) plotted against the observed values

options as for previous models. In the main dialog box, click OK to fit the model.

Output 21.8 is similar to the previous output, except that we have two new fixed effects. To assess whether these fixed effects have improved the fit of the model we can use the change in the log-likelihood statistics again:

$$\chi^2_{change} = 1798.62 - 1789.05 = 9.57$$
$$df_{change} = 9 - 7 = 2$$ (21.21)

The change is 9.57 based on 2 degrees of freedom, which is greater than the critical value for the chi-square statistic (with 2 *df*) in the Appendix, which is 5.99 ($p < 0.05$, $df = 2$). Adding the reason for surgery and the interaction between surgery and the reason improves the fit of the model significantly.

We can look at the effects individually in the table of fixed effects. Quality of life before surgery significantly predicted quality of life after surgery, $F(1, 268.92)$ = 33.65, $p < 0.001$, surgery still did not

significantly predict quality of life, $F(1, 15.86)$ = 2.17, $p = 0.161$, but the reason for surgery, $F(1, 259.89)$ = 9.67, $p = 0.002$, and the interaction of surgery and the reason for surgery, $F(1, 217.09)$ = 6.28, $p = 0.013$, both did significantly predict quality of life. The table of estimates of fixed effects tells us much the same thing, except it also gives us the *b*-values and their confidence intervals.

Broadly speaking our conclusions about our random parameters stay much the same as in the previous model. The values of the variance for the intercept (30.06) and the slope (29.35) are lower than in the previous model and the associated *p*-values are slightly larger, but have not changed dramatically. Also the covariance

between the slopes and intercepts is still negative (−28.08).

The interaction term is the most interesting effect, because this tells us the effect of the reason for surgery taking account of whether the person has had the surgery. To break down this interaction we could rerun the analysis separately for the two 'reason groups'. Obviously we would remove the interaction term and the main effect of **Reason** from this model (because we are analyzing the physical reason group separately from the group that wanted to change their appearance). As such, you need to fit the model in the previous section, but first split the file by **Reason**.

Output 21.9 shows the parameter estimates from these analyses. It shows that for those operated on only to change their appearance, surgery had a *negative* (and close to significant) relationship with quality of life after surgery, $b = -4.31$, $t(7.72) = -1.92$, $p = 0.09$. The negative *b* shows that in these people, quality of life was lower after surgery compared to the control group. However, for those who had surgery to solve a physical problem surgery had a *positive* (and very nonsignificant) relationship with quality of life, $b = 1.20$, $t(7.61) = 0.58$, $p = 0.58$. The positive *b* indicates that people who had surgery scored higher on quality of life than those on the waiting list (even if not significantly so). The interaction effect, therefore, reflects the difference in slopes for surgery as a predictor of quality of life in those who had surgery for physical problems (slight positive slope) and those who had surgery purely for vanity (a negative slope).

To sum up, quality of life after surgery, after controlling for quality of life before surgery, was lower for those who had

SELF TEST
Split the file by **Reason** and then run a multilevel model predicting **Post_QoL** with a random intercept, and random slopes for **Surgery**, and including **Base_QoL** and **Surgery** as predictors.

Model Dimension[a]

		Number of Levels	Covariance Structure	Number of Parameters	Subject Variables
Fixed Effects	Intercept	1		1	
	Surgery	1		1	
	Base_QoL	1		1	
Random Effects	Intercept + Surgery[b]	2	Unstructured	3	Clinic
Residual				1	
Total		5		7	← *df* for −2LL

a. Dependent Variable: Quality of Life After Cosmetic Surgery.

b. As of version 11.5, the syntax rules for the RANDOM subcommand have changed. Your command syntax may yield results that differ from those produced by prior versions. If you are using version 11 syntax, please consult the current syntax reference guide for more information.

Information Criteria[a]

−2 Log Likelihood	1798.624 ← −2LL
Akaike's Information Criterion (AIC)	1812.624
Hurvich and Tsai's Criterion (AICC)	1813.042
Bozdogan's Criterion (CAIC)	1844.967
Schwarz's Bayesian Criterion (BIC)	1837.967

The information criteria are displayed in smaller-is-better form.

a. Dependent Variable: Quality of Life After Cosmetic Surgery.

surgery to change their appearance than those who had surgery for a physical reason. This makes sense because for those having surgery to correct a physical problem, the surgery has probably bought relief and their quality of life improved. However, those having surgery for vanity might well discover that having a different appearance wasn't at the root of their unhappiness, so their quality of life is lower.

21.7 Growth models ▮▮▮▮

Growth models are widely used in many areas of science, including psychology, medicine, physics, chemistry and economics. In a growth model the aim is to look at the rate of change of a variable over time: for example, we could look at white blood cell counts, attitudes, radioactive decay or profits. In all cases we're trying to see which model best describes the change over time.

Type III Tests of Fixed Effects[a]

Source	Numerator df	Denominator df	F	Sig.
Intercept	1	84.954	107.284	.000
Surgery	1	9.518	.097	.762
Base_QoL	1	265.933	33.984	.000

a. Dependent Variable: Quality of Life After Cosmetic Surgery.

Estimates of Fixed Effects[a]

bs

Parameter	Estimate	Std. Error	df	t	Sig.	95% Confidence Interval Lower Bound	95% Confidence Interval Upper Bound
Intercept	40.102525	3.871729	84.954	10.358	.000	32.404430	47.800620
Surgery	−.654530	2.099413	9.518	−.312	.762	−5.364643	4.055583
Base_QoL	.310218	.053214	265.933	5.830	.000	.205443	.414993

a. Dependent Variable: Quality of Life After Cosmetic Surgery.

$Var(u_{0j})$
variance of intercepts

$Var(\varepsilon_{ij})$
variance of residuals

Estimates of Covariance Parameters[a]

Parameter		Estimate	Std. Error	Wald Z	Sig.	95% Confidence Interval Lower Bound	95% Confidence Interval Upper Bound
Residual		34.955705	3.116670	11.216	.000	29.351106	41.630504
Intercept + Surgery [subject = Clinic]	UN (1,1)	37.609439	18.726052	2.008	.045	14.173482	99.796926
	UN (2,1)	−36.680707	18.763953	−1.955	.051	−73.457378	.095965
	UN (2,2)	38.408857	20.209811	1.901	.057	13.694612	107.724142

a. Dependent Variable: Quality of Life After Cosmetic Surgery.

$Var(u_{1j})$
variance of slopes

$Cov(u_{0j}, u_{1j})$
covariance of slopes and intercepts

Output 21.7

21.7.1 Growth curves (polynomials) ▮▮▮▮

Figure 21.22 shows three examples of **growth curves**: three **polynomials** representing a linear trend (the orange line) otherwise

What is a growth curve?

known as a first-order polynomial, a quadratic trend (the blue line) otherwise known as a second-order polynomial, and a cubic trend (the green line) otherwise known as a third-order polynomial. Notice that the linear trend is a straight line, but as the polynomials increase they get more and more curved, indicating more rapid growth over time. Also, as polynomials increase, the change in the curve is quite dramatic (so dramatic that I had to adjust the scale of the *y*-axis on each graph to fit all three on the same diagram). This

observation highlights the fact that any growth curve higher than a quadratic (or possibly cubic) trend is very unrealistic in real data. By fitting a growth model we can see which trend best describes the

growth of an outcome variable over time (although no one will believe that a significant fifth-order polynomial is telling us anything meaningful about the real world!).

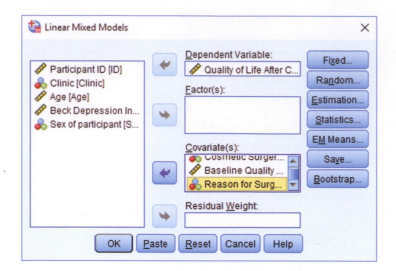

Figure 21.20   The main mixed models dialog box

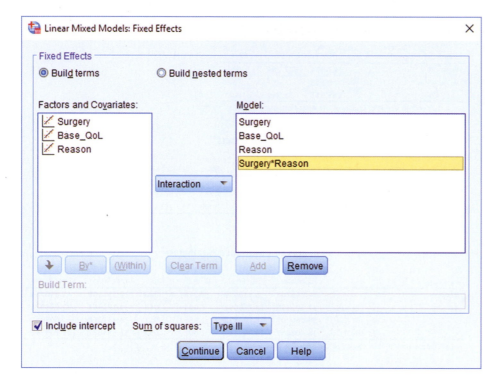

Figure 21.21   Specifying a fixed effect interaction in mixed models

The growth curves that I have just described should seem familiar: they are the same as the trends that we described for ordered means in Section 12.4.5. What we're discussing now is really no different. There are two important things to remember when fitting growth curves: (1) you can fit polynomials up to one less than the number of time points that you have; and (2) a polynomial is defined by a simple power function.

On the first point, this means that with three time points you can fit a linear and quadratic growth curve (or a first- and second-order polynomial), but you cannot fit any higher-order growth curves. Similarly, if you have six time points you can fit up to a fifth-order polynomial. This is the same basic idea as having one less contrast than the number of groups when comparing ordered means (see Section 12.4).

On the second point, we define growth curves manually in multilevel models: there is not a convenient option that does it for us. However, it is quite easy to do. If *time* is our predictor variable, then a linear trend is tested by including this variable alone. A quadratic or second-order polynomial is tested by including a predictor that is $time^2$, a cubic or third-order polynomial is tested by including a predictor that is $time^3$, and so on. Any polynomial is tested by including a variable that is the predictor to the power of the order of polynomial that you want to test: for a fifth-order polynomial we need a predictor of $time^5$ and for an nth-order polynomial we would include $time^n$ as a predictor. Hopefully you get the general idea.

21.7.2 A honeymoon example

I once saw a brilliant talk given by Professor Daniel Kahneman, who won the 2002 Nobel Prize for Economics. In this talk Kahneman assimilated research on life satisfaction (he explored questions such as whether people are happier if they are richer). There was one graph in this talk that particularly grabbed my attention. It showed that, leading up to marriage, people reported greater life satisfaction, but by about two years after marriage this life satisfaction decreased back to its baseline level. This graph perfectly illustrated what people talk about as the 'honeymoon period': a new relationship/ marriage is great at first (no matter how ill-suited you may be), but after six months or so the cracks start to appear and everything turns to shit. Kahneman argued that people adapt to marriage; it does not make them happier in the long run (Kahneman & Krueger, 2006).[10]

This talk got me thinking about whether we could apply this argument to any new relationship. Therefore, in a completely fictitious parallel world in which I concern myself with people's life satisfaction I organized a massive speed-dating event (see Chapter 16). At the start of the night I measured everyone's life satisfaction (**Satisfaction_Baseline**) on a 10-point scale (0 = completely dissatisfied, 10 = completely satisfied) and recorded their biological sex (**Sex**). After the event I noted who had found dates. If they ended up in a relationship with the person they met on the speed-dating night then I kept in touch with these people over the next 18 months of that relationship. I obtained measures of their life satisfaction at 6 months (**Satisfaction_6_Months**), 12 months (**Satisfaction_12_Months**) and 18 months (**Satisfaction_18_Months**) after they entered the relationship. Then I got bored and stopped harassing them. None of the people measured were in the same relationship (i.e., I measured life satisfaction only from one of the people in the couple).[11] Also, as is often the case with longitudinal data, I didn't have scores for all people at all time points because not everyone was available at the follow-up sessions. One of the benefits of a multilevel approach is that these missing data do not pose a problem. The data are in the file **Honeymoon Period.sav**.

Figure 21.23 shows the data. Each dot is a data point and the line shows the average life satisfaction over time. Basically, from baseline, life satisfaction rises slightly at time 1 (6 months) but then starts to decrease over the next 12 months. There are two things to note about the data.

Model Dimension[a]

		Number of Levels	Covariance Structure	Number of Parameters	Subject Variables
Fixed Effects	Intercept	1		1	
	Surgery	1		1	
	Base_QoL	1		1	
	Reason	1		1	
	Surgery * Reason	1		1	
Random Effects	Intercept + Surgery[b]	2	Unstructured	3	Clinic
Residual				1	
Total		7		9	

← df for −2LL

a. Dependent Variable: Quality of Life After Cosmetic Surgery.

b. As of version 11.5, the syntax rules for the RANDOM subcommand have changed. Your command syntax may yield results that differ from those produced by prior versions. If you are using version 11 syntax, please consult the current syntax reference guide for more information.

Information Criteria[a]

−2 Log Likelihood	1789.045
Akaike's Information Criterion (AIC)	1807.045
Hurvich and Tsai's Criterion (AICC)	1807.722
Bozdogan's Criterion (CAIC)	1848.629
Schwarz's Bayesian Criterion (BIC)	1839.629

← −2LL

The information criteria are displayed in smaller-is-better form.

a. Dependent Variable: Quality of Life After Cosmetic Surgery.

Type III Tests of Fixed Effects[a]

Source	Numerator df	Denominator df	F	Sig.
Intercept	1	108.853	122.593	.000
Surgery	1	15.863	2.167	.161
Base_QoL	1	268.920	33.647	.000
Reason	1	259.894	9.667	.002
Surgery * Reason	1	217.087	6.278	.013

a. Dependent Variable: Quality of Life After Cosmetic Surgery.

Estimates of Fixed Effects[a]

bs

Parameter	Estimate	Std. Error	df	t	Sig.	95% Confidence Interval Lower Bound	Upper Bound
Intercept	42.517820	3.840055	108.853	11.072	.000	34.906839	50.128800
Surgery	−3.187677	2.165484	15.863	−1.472	.161	−7.781510	1.406157
Base_QoL	.305356	.052642	268.920	5.801	.000	.201713	.408999
Reason	−3.515148	1.130552	259.894	−3.109	.002	−5.741357	−1.288939
Surgery * Reason	4.221288	1.684798	217.087	2.506	.013	.900633	7.541944

a. Dependent Variable: Quality of Life After Cosmetic Surgery.

$Var(u_{0j})$
variance of intercepts

$Var(\varepsilon_{ij})$
variance of residuals

Estimates of Covariance Parameters[a]

Parameter		Estimate	Std. Error	Wald Z	Sig.	95% Confidence Interval Lower Bound	Upper Bound
Residual		33.859719	3.024395	11.196	.000	28.421886	40.337948
Intercept + Surgery [subject = Clinic]	UN (1,1)	30.056340	15.444593	1.946	.052	10.978478	82.286775
	UN (2,1)	−28.083657	15.195713	−1.848	.065	−57.866706	1.699393
	UN (2,2)	29.349323	16.404492	1.789	.074	9.813593	87.774453

a. Dependent Variable: Quality of Life After Cosmetic Surgery.

$Var(u_{1j})$
variance of slopes

$Cov(u_{0j}, u_{1j})$
covariance of slopes and intercepts

Output 21.8

---

10 The romantics among you might be relieved to know that others have used the same data to argue the complete opposite: that married people are happier than non-married people in the long term (Easterlin, 2003).

11 However, I could have measured both people in the couple because, using a multilevel model, I could have treated people as being nested within 'couples' to take account of the dependency in their data.

Surgery for cosmetic reasons

Estimates of Fixed Effects[a,b]

Parameter	Estimate	Std. Error	df	t	Sig.	95% Confidence Interval Lower Bound	Upper Bound
Intercept	41.786055	5.487873	77.331	7.614	.000	30.859052	52.713059
Surgery	-4.307014	2.239912	7.719	-1.923	.092	-9.505157	.891130
Base_QoL	.338492	.079035	88.619	4.283	.000	.181440	.495543

a. Reason for Surgery = Change Appearance
b. Dependent Variable: Quality of Life After Cosmetic Surgery.

Surgery for physical reasons

Estimates of Fixed Effects[a,b]

Parameter	Estimate	Std. Error	df	t	Sig.	95% Confidence Interval Lower Bound	Upper Bound
Intercept	38.020790	4.666154	93.558	8.148	.000	28.755460	47.286119
Surgery	1.196550	2.081999	7.614	.575	.582	-3.647282	6.040382
Base_QoL	.317710	.068883	172.816	4.612	.000	.181749	.453670

a. Reason for Surgery = Physical reason
b. Dependent Variable: Quality of Life After Cosmetic Surgery.

Output 21.9

First, time 0 is before the people enter into their new relationship, yet already there is a lot of variability in their responses (reflecting the fact that people will vary in their satisfaction due to other reasons such as finances, personality and so on). This suggests that intercepts for life satisfaction differ across people. Second, there is also a lot of variability in life satisfaction after the relationship has started (time 1) and at all subsequent time points, which suggests that the slope of the relationship between time and life satisfaction might vary across people also. If we think of the time points as a level 1 variable that is nested within people (a level 2 variable) then we can easily model this variability in intercepts and slopes within people. We have a

Cramming Sam's Tips
Multilevel models output

- The *Information Criteria* table can be used to assess the overall fit of the model. The value of −2LL can be tested for significance with df equal to the number of parameters being estimated. It is mainly used, though, to compare models that are the same in all but one parameter by testing the difference in −2LL in the two models against df = 1 (if only one parameter has been changed). The AIC, AICC, CAIC and BIC can also be compared across models (but not tested for significance).
- The table of *Type III Tests of Fixed Effects* tells you whether your predictors significantly predict the outcome: look in the column labeled *Sig.* If the value is less than 0.05 then the effect is significant.

- The table of *Estimates of Fixed Effects* gives us the b-values for each effect and its confidence interval. The direction of these coefficients tells us whether the relationship between each predictor and the outcome is positive or negative.
- The table labeled *Estimates of Covariance Parameters* tells us about random effects in the model. These values can tell us how much intercepts and slopes varied over our level 1 variable. The significance of these estimates should be treated cautiously. The exact labeling of these effects depends on which covariance structure you selected for the analysis.

Oditi's Lantern
Multilevel models

'I, Oditi, believe that you know that experimental manipulations happen within ~~cults~~ contexts; and people within ~~cults~~ contexts become more similar to each other than to people outside such ~~cults~~ contexts. To eliminate this dependency we must make everyone join our cult, erm, no, seriously, I mean we must factor the dependency by using a multilevel model. Stare into my lantern one last time and you will become worthy to call yourself one of my cult of undiscovered numerical truths.'

situation like Figure 21.4 (except with two levels instead of three).

21.7.3 Restructuring the data ▌▌▌▌

The first problem with having data measured over time is that to do a multilevel model the data need to be in a different format than we are used to. Figure 21.24 shows how we would normally set up the data editor for a repeated-measures design: each row represents a person, and the repeated-measures variable of time is represented by four different columns. If we were going to run an ordinary repeated-measures ANOVA this data layout would be fine; however, for a multilevel model we need the variable **Time** to be represented by a single column (i.e., the long format). We could enter all of the data again, but that would be tedious, so it's lucky that the *restructure* command will do it for us: it's also tedious,

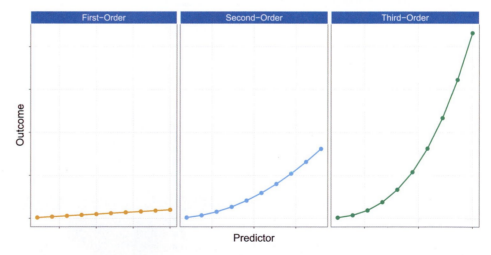

Figure 21.22 Illustration of a first-order (linear orange), second-order (quadratic, blue) and third-order (cubic, green) polynomial

SELF TEST

Use Oliver Twisted's guide to restructure the data file. Save the restructured file as **Honeymoon Period Restructured.sav**.

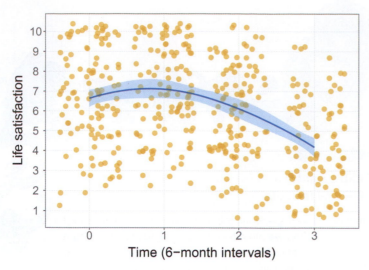

Figure 21.23 Life satisfaction over time (a jitter has been used to avoid points overlapping)

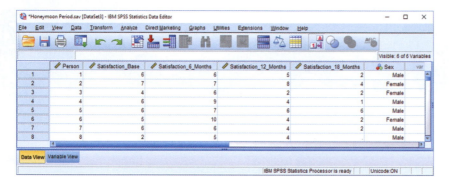

Figure 21.24 The data editor for a normal repeated-measures data set

but not as tedious as retyping data. The *restructure* command enables you to take a data set and create a new one that is organized differently (see Oliver Twisted). The restructured data are shown in Figure 21.25; compare the restructured data with the old data file in Figure 21.24. Notice that each person is now represented by four rows (one for each time point) and that variables such as sex that are invariant over the time have the same value within each person. The outcome variable (**Life_Satisfaction**), which is not invariant over time, is different at the four time points (the four rows for each person). The time points have values from 1 to 4. However, it's useful to anchor this variable at 0 (Section 21.5.4) because our initial life satisfaction was measured before the new relationship. Therefore, an intercept of 0 is meaningful for these data: it is the value of life satisfaction when not in a relationship. By anchoring the scores on a baseline value of 0 we can interpret the intercept more intuitively. The easiest way to change the values is using the *compute* command to recompute **Time** to be **Time** minus 1. This will change the values from 1–4 to 0–3. If you can't be bothered with all of this, use **Honeymoon Period Restructured.sav**.

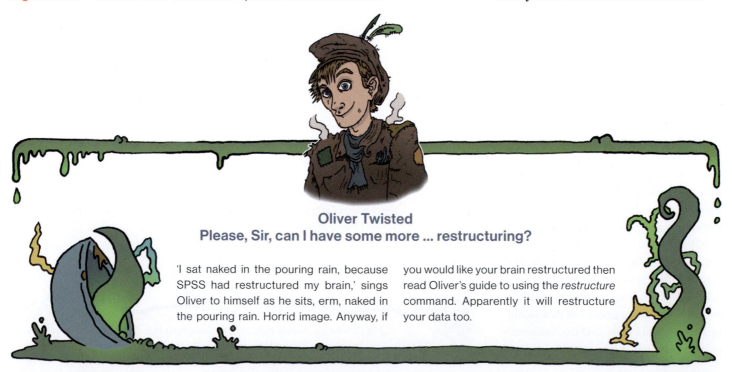

Oliver Twisted
Please, Sir, can I have some more ... restructuring?

'I sat naked in the pouring rain, because SPSS had restructured my brain,' sings Oliver to himself as he sits, erm, naked in the pouring rain. Horrid image. Anyway, if you would like your brain restructured then read Oliver's guide to using the *restructure* command. Apparently it will restructure your data too.

21.7.4 Growth models using SPSS Statistics ▮▮▮▮

We can set up this model in a very similar way to the previous example. First, select *Analyze* ▸ *Mixed Models* ▸ 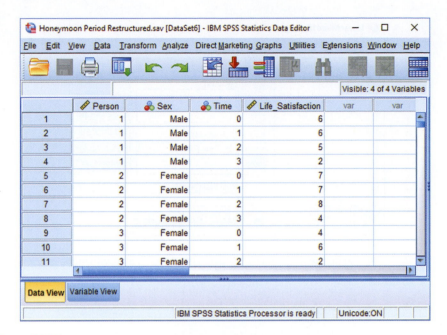 *Linear...* and in the initial dialog box set up the level 2 variable. In this example, life satisfaction at multiple time points is nested within people. Therefore, the level 2 variable is the person so drag **Person** to the box labeled *Subjects* (or click ➡) as in Figure 21.26. Click Continue to access the main dialog box, in which we set up our predictors and outcome. The outcome was life satisfaction, so drag **Life_Satisfaction** to the box labeled *Dependent Variable* (or click ➡). Our predictor or growth variable, was **Time**, so drag this variable to the box labeled *Covariate(s)* or click ➡, as in Figure 21.27.[12]

To add the potential growth curves as fixed effects to our model, click Fixed... to bring up the dialog box in Figure 21.28. With four time points we can fit up to a third-order polynomial (Section 21.7.1). As in the previous example, we want to build the model up step by step, so we'd start with just the linear effect (**Time**), then run a new model with the linear and quadratic (**Time^2**) polynomials to see if the quadratic trend improves the model. Finally, run a third model with the linear, quadratic and cubic (**Time^3**) polynomial in, and see if the cubic trend adds to the model. So, basically, we add polynomials one at a time and assess the change in −2LL. To specify the linear polynomial click **Time** and then Add to add it into the model. Click Continue to return to the main dialog box.

Figure 21.25 Data entry for a multilevel model in which scores are nested within cases

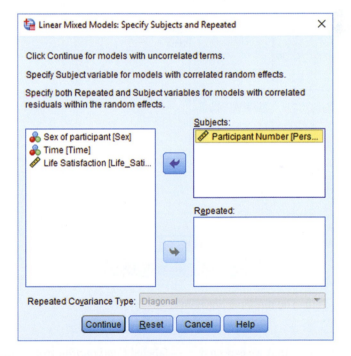

Figure 21.26 Setting up the level 2 variable in a growth model

SELF TEST

Use the *compute* command to transform **Time** into **Time** minus 1.

---

12 I have dragged the **Time** to the *Covariate(s)* box because I want to treat it as a linear trend and not as a categorical variable (see SPSS Tip 21.1).

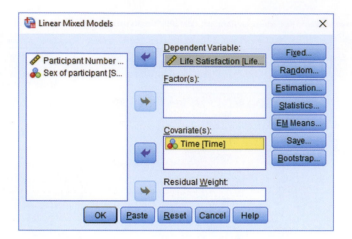

Figure 21.27 Setting up the outcome variable and predictor in a growth model

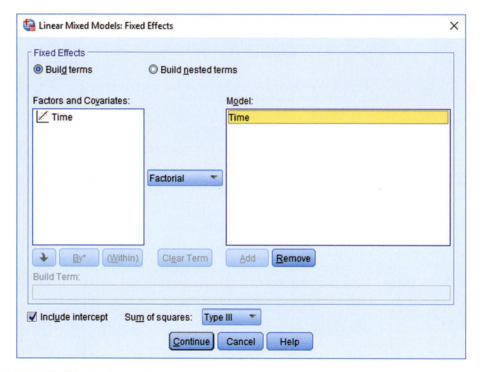

Figure 21.28 Setting up the linear polynomial

I mentioned earlier on that we expected the relationship between time and life satisfaction to have both a random intercept and a random slope. We define these parameters by clicking on Random... to access the dialog box in Figure 21.29. First, we specify our contextual variable by dragging **Person** from the section labeled *Subjects* (which will list any variables specified in Figure 21.26) to the area

labeled *Combinations* (or click ➡). To specify that the intercept is random select ☑ Include intercept, and to specify random slopes for the effect of **Time**, click this variable in the *Factors and Covariates* list and then click Add to include it in *Model*. Finally, we need to specify the covariance structure. By default, the covariance structure is set to be Variance Components ▾. However, when we have

repeated measures over time it can be useful to specify a covariance structure that assumes that scores become less correlated over time (Section 21.4.2). Therefore, let's choose an autoregressive covariance structure, AR(1), and let's also assume that variances will be heterogeneous. Therefore, select AR(1): Heterogeneous ▾ from the drop-down list (Figure 21.29). Click Continue to return to the main dialog box. Click Estimation... and select ⦿ Maximum Likelihood (ML), and then click Statistics... and select ☑ Parameter estimates and ☑ Tests for covariance parameters (see Figure 21.13). Click Continue to return to the main dialog box. To fit the model, click OK.

Output 21.10 shows that the linear trend was significant, $F(1, 106.72) = 134.26$, $p < 0.001$. For evaluating the improvement in the model when we add in new polynomials, we need to note the value of $-2LL$, which is 1862.63, and the degrees of freedom, which are 6 (look at the row labeled *Total* in the column labeled *Number of Parameters*, in the table called *Model Dimension*).

Now, let's add the quadratic trend. To do this follow the instructions to run this analysis again until you reach the point where you click Fixed... . The linear polynomial should already be specified from the last model and the dialog box will look like Figure 21.28. To add the higher-order polynomials select ⦿ Build nested terms. For the quadratic or second-order polynomial we need to define **Time**^{2} (**Time** multiplied by itself). Select **Time** in the *Factors and Covariates* list and ⬇ will become active; click this button and **Time** will appear in the space labeled *Build Term*. Next, click By* to add a multiplication symbol to the term, then select **Time** again and click ⬇. The *Build Term* bar should now read *Time*Time* (or, put another way, **Time**^{2}). This term is the second-order polynomial; click Add to put it into the model (it will appear in the space labeled *Model*).

Click [Continue] to return to the main dialog box and click [OK] to fit the model.

Output 21.11 includes the quadratic polynomial. To see whether it has improved the model we use the change in −2*LL*. For the model containing only the linear term −2*LL* was 1862.63, with 6 parameters being estimated (Output 21.10), and with the quadratic term included it is 1802.03, with 7 parameters being estimated (Output 21.11). The difference is 60.60 with 1 degree of freedom

$$\chi^2_{change} = 1862.63 - 1802.03 = 60.60$$
$$df_{change} = 7 - 6 = 1$$

(21.22)

which is greater than the critical values for the chi-square statistic for *df* = 1 in the Appendix (3.84, *p* < 0.05 and 6.63, *p* < 0.01). Adding the quadratic term significantly improves the fit of the model.

Finally, let's add the cubic trend, which is defined as **Time**³ (or *Time*Time*Time*). Return to the dialog box for fixed effects: the linear and quadratic polynomials should already be specified and the dialog box will look like Figure 21.30. Make sure ⦿ **Build nested terms** is selected, then select **Time**, click [↓], click [By*], select **Time** again, click [↓], click [By*] again, select **Time** for a third time, click [↓] and finally click [Add]. This process adds the third-order polynomial (or *Time*Time*Time*) to the model as in Figure 21.31.[13] Click [Continue] to return to the main dialog box and [OK] to fit the model.

Output 21.12 includes the cubic polynomial. To see whether it has improved the model we use the change in −2*LL*. For the model containing the linear and quadratic term −2*LL* was 1802.03, with 7 parameters being estimated (Output 21.11), and with the cubic term included it is 1798.86, with 8 parameters

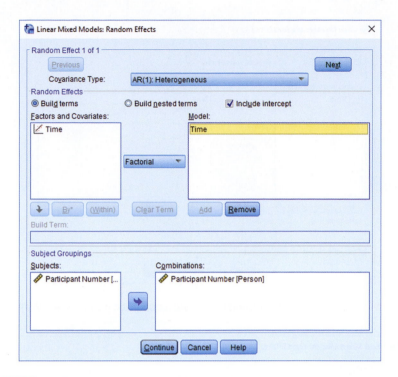

Figure 21.29 Defining a random intercept and random slopes in a growth model

Model Dimension[a]

		Number of Levels	Covariance Structure	Number of Parameters	Subject Variables
Fixed Effects	Intercept	1		1	
	Time	1		1	
Random Effects	Intercept + Time	2	Heterogeneous First-Order Autoregressive	3	Person
Residual				1	
Total		4		6	

a. Dependent Variable: Life Satisfaction.

Information Criteria[a]

−2 Log Likelihood	1862.626
Akaike's Information Criterion (AIC)	1874.626
Hurvich and Tsai's Criterion (AICC)	1874.821
Bozdogan's Criterion (CAIC)	1905.119
Schwarz's Bayesian Criterion (BIC)	1899.119

The information criteria are displayed in smaller-is-better form.

a. Dependent Variable: Life Satisfaction.

Type III Tests of Fixed Effects[a]

Source	Numerator df	Denominator df	F	Sig.
Intercept	1	113.653	1137.088	.000
Time	1	106.715	134.264	.000

a. Dependent Variable: Life Satisfaction.

Output 21.10

---

13 Should you ever want even high-order polynomials (notwithstanding my remark about them having little real-world relevance) then you can extrapolate from what I have told you about the other polynomials; for example, for a fourth-order polynomial you go through the whole process again, but this time creating **Time**⁴ (or *Time*Time*Time*Time*).

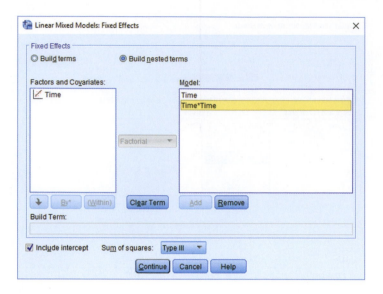

Figure 21.30 Specifying a linear trend (*Time*) and a quadratic trend (*Time*Time*)

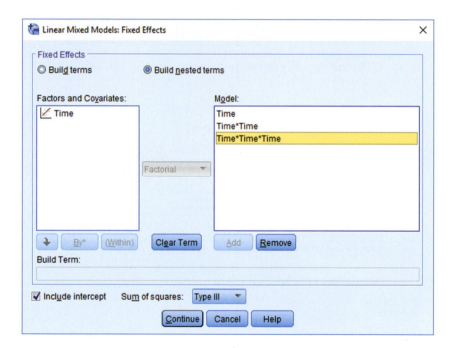

Figure 21.31 Specifying linear (*Time*), quadratic (*Time*Time*) and cubic (*Time*Time*Time*) trends

Information Criteria[a]

–2 Log Likelihood	1802.026
Akaike's Information Criterion (AIC)	1816.026
Hurvich and Tsai's Criterion (AICC)	1816.287
Bozdogan's Criterion (CAIC)	1851.602
Schwarz's Bayesian Criterion (BIC)	1844.602

The information criteria are displayed in smaller–is–better form.

a. Dependent Variable: Life Satisfaction.

Output 21.11

trends both significantly describe the pattern of the data over time. These results confirm what we already know from comparing the fit of successive models. The trend in the data is best described by a second-order polynomial or a quadratic trend. This trend reflects the initial increase in life satisfaction 6 months after finding a new partner but a subsequent reduction in life satisfaction at 12 and 18 months after the start of the relationship (Figure 21.23). The parameter estimates tell us much the same thing. It's worth remembering that this quadratic trend is only an *approximation*: if it were completely accurate then we would predict from the model that couples who had been together for 10 years would have negative life satisfaction, which is impossible given the scale we used to measure it.

The final part of the output tells us about the random parameters in the model. First, the variance of the random intercepts was $\text{Var}(u_{0j}) = 3.87$. This suggests that we were correct to assume that life satisfaction at baseline varied significantly across people. Also, the variance of the people's slopes varied significantly, $\text{Var}(u_{1j}) = 0.24$. This suggests also that the change in life satisfaction over time varied significantly across people too. Finally, the covariance between the slopes and intercepts (-0.37) suggests that as intercepts increased, the slope decreased. (Ideally, these terms would have been added individually so that we could calculate the chi-square statistic for the change in $-2LL$ for each of them.)

being estimated (Output 21.12). The difference is 3.17 with 1 degree of freedom:

$$\chi^2_{\text{Change}} = 1802.03 - 1798.86 = 3.17$$
$$df_{\text{Change}} = 8 - 7 = 1$$
(21.23)

which is less than the critical value of 3.84, $p < 0.05$ (see above). Adding the cubic term does not significantly improve the fit of the model.

In the interests of parsimony, we should interpret the model that contained the quadratic term (because adding the cubic term did not improve the fit of the model). Output 21.13 shows the model with the linear and quadratic trends included. The tables of fixed effects and parameter estimates tell us that the linear, $F(1, 273.22) = 13.26, p < 0.001$, and quadratic, $F(1, 226.86) = 72.07, p < 0.001$,

Information Criteria[a]

–2 Log Likelihood	1798.857
Akaike's Information Criterion (AIC)	1814.857
Hurvich and Tsai's Criterion (AICC)	1815.193
Bozdogan's Criterion (CAIC)	1855.515
Schwarz's Bayesian Criterion (BIC)	1847.515

The information criteria are displayed in smaller-is–better form.

a. Dependent Variable: Life Satisfaction.

Output 21.12

Type III Tests of Fixed Effects[a]

Source	Numerator df	Denominator df	F	Sig.
Intercept	1	133.626	912.307	.000
Time	1	273.219	13.261	.000
Time * Time	1	226.857	72.069	.000

a. Dependent Variable: Life Satisfaction.

Estimates of Fixed Effects[a]

Parameter	Estimate	Std. Error	df	t	Sig.	95% Confidence Interval	
						Lower Bound	Upper Bound
Intercept	6.684546	.221310	133.626	30.204	.000	6.246822	7.122270
Time	.754482	.207185	273.219	3.642	.000	.346601	1.162364
Time * Time	–.562231	.066228	226.857	–8.489	.000	–.692731	–.431731

a. Dependent Variable: Life Satisfaction.

Estimates of Covariance Parameters[a]

Parameter		Estimate	Std. Error	Wald Z	Sig.	95% Confidence Interval	
						Lower Bound	Upper Bound
Residual		1.855235	.181149	10.241	.000	1.532095	2.246530
Intercept + Time [subject = Person]	Var: Intercept	3.867628	.699590	5.528	.000	2.713165	5.513318
	Var: Time	.242175	.097544	2.483	.013	.109972	.533308
	ARH1 rho	–.373673	.153978	–2.427	.015	–.631228	–.041891

a. Dependent Variable: Life Satisfaction.

Output 21.13

21.7.5 Further analysis ▮▮▮▮

I've kept this growth curve simple to give you the basic tools. In the example I allowed only the linear term to have a random intercept and slopes, but given that we discovered that a second-order polynomial described the change in responses, we could redo the analysis and allow random intercepts and slopes for the second-order polynomial also. To do these we would just have to specify these terms in Figure 21.29 in much the same way as we set them up as fixed effects in Figure 21.30. If we were to do this it would make sense to add the random components one at a time and test whether they have a significant impact on the model by comparing the log-likelihood values or other fit indices. Also, the polynomials I have described are not the only ones that can be used. You could test for a logarithmic trend over time or even an exponential one (Long, 2012).

21.8 How to report a multilevel model ▮▮▮▮

Multilevel models take on so many forms that giving standard advice is not straightforward. If you have built up your model from one with only fixed parameters to one with a random intercept, and then random slope, it is advisable to report all stages of this process (or at the very least report the fixed-effects-only model and the final model). For any model you need to say something about the random effects.

For the final model of the cosmetic surgery example you could write something like:

✓ The relationship between surgery and quality of life showed significant variance in intercepts across participants, $Var(u_{0j}) = 30.06$, $x^2(1) = 15.05$, $p < 0.01$. In addition, the slopes varied across participants, $Var(u_{1j}) = 29.35$, $x^2(1) = 21.49$, $p < 0.01$, and the slopes and intercepts negatively and significantly covaried, $Cov(u_{0j}, u_{1j}) = -28.08$, $x^2(1) = 17.38$, $p < 0.01$.

For the model itself, you have two choices. The first is to report the results with the Fs and degrees of freedom for the fixed effects, and then report the parameters for the random effects in the text as well. The second is to produce a table of parameters as you would for a linear model. For example, you might report the cosmetic surgery example as follows:

✓ Quality of life before surgery significantly predicted quality of life after surgery, $F(1, 268.92) = 33.65$, $p < 0.001$, surgery did not significantly predict quality of life,

$F(1, 15.86) = 2.17$, $p = 0.161$, but the reason for surgery, $F(1, 259.89) = 9.67$, $p = 0.002$, and the interaction of surgery and the reason for surgery, $F(1, 217.09) = 6.28$, $p = 0.013$, both significantly predicted quality of life. This interaction was broken down by conducting separate multilevel models on the 'physical reason' and 'attractiveness reason'. The models specified were the same as the main model but excluded the main effect and interaction term involving the reason for surgery. These analyses showed that for those operated on only to change their appearance, surgery had a negative relationship to quality of life that was close to significance, $b = -4.31$, $t(7.72) = -1.92$, $p = 0.09$: quality of life was lower after surgery compared to the control group. However, for those who had surgery to solve a physical problem, surgery did not significantly predict quality of life, $b = 1.20$, $t(7.61) = 0.58$, $p = 0.58$. The interaction effect, therefore, reflects the difference in slopes for surgery as a predictor of quality of life in those who had surgery for physical problems (slight positive slope) and those who had surgery purely for vanity (a negative slope).

Cramming Sam's Tips
Growth models

- Growth models are multilevel models in which changes in an outcome over time are modeled using potential growth patterns.
- These growth patterns can be linear, quadratic, cubic, logarithmic, exponential or anything you like really.
- The hierarchy in the data is that time points are nested within people (or other entities). As such, it's a way of analyzing repeated-measures data that have a hierarchical structure.
- The *Information Criteria* table can be used to assess the overall fit of the model. The value of $-2LL$ can be tested for significance with df equal to the number of parameters being estimated. It is mainly used, though, to compare models that are the same in all but one parameter by testing the difference in $-2LL$ in the two models against $df = 1$ (if only one parameter has been changed). The AIC, AICC,

CAIC and BIC can also be compared across models (but not tested for significance).
- The table of *Type III Tests of Fixed Effects* tells you whether the growth functions in the model significantly predict the outcome: look in the column labeled *Sig*. If the value is less than 0.05 then the effect is significant.
- The table labeled *Estimates of Covariance Parameters* tells us about random effects in the model. These values can tell us how much intercepts and slopes varied over our level 1 variable. The significance of these estimates should be treated cautiously. The exact labeling of these effects depends on which covariance structure you selected for the analysis.
- An autoregressive covariance structure, AR(1), is often assumed in time course data such as that in growth models.

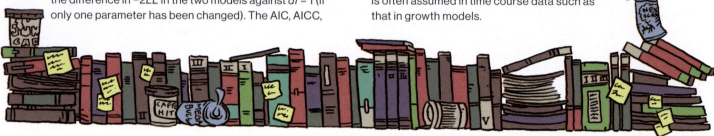

Table 21.1   Summary of the surgery multilevel model

	b	SE b	95% CI
Baseline QoL	0.31	0.05	0.20, 0.41
Surgery	−3.19	2.17	−7.78, 1.41
Reason	−3.51	1.13	−5.74, −1.29
Surgery × Reason	4.22	1.68	0.90, 7.54

Alternatively, you could present parameter information as in Table 21.1.

21.9 A message from the octopus of inescapable despair ▌▌▌

When I started writing this chapter I didn't know anything about multilevel models, but by its completion I felt a tiny bit smug that I had them nailed. However, I don't, and if you now feel like you understand multilevel models too then you're wrong. You're not wrong because you're daft, but because multilevel modeling is very complicated and this chapter barely scratches the surface of what there is to

know. Multilevel models often fail to converge, with no apology or explanation, and trying to fathom out what's happening can feel like hammering nails into your head.

21.10 Brian and Jane's Story ▌▌▌

Jane was giddy with excitement. Theoretically the device would work, but she wouldn't know for sure until she tried it. It had been easy enough to steal a portable TMS device, but getting it to focus its magnetic pulse into a precise location had beaten her at first. Then last week it came to her, in the shower of all places, and she'd spent another three weeks away from the world, beating away at

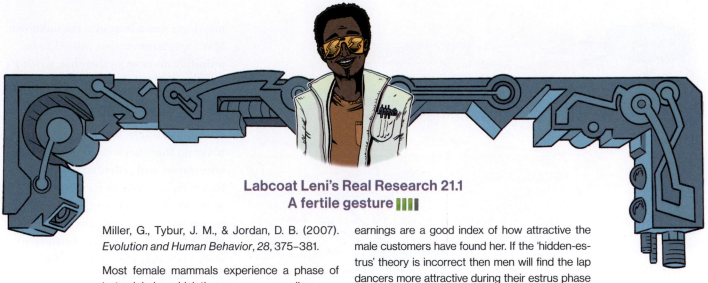

Labcoat Leni's Real Research 21.1
A fertile gesture ▮▮▮▮

Miller, G., Tybur, J. M., & Jordan, D. B. (2007). *Evolution and Human Behavior, 28*, 375–381.

Most female mammals experience a phase of 'estrus' during which they are more sexually receptive, proceptive, selective and attractive. The evolutionary benefit to this phase is believed to be to attract mates of superior genetic stock. Some people have argued that this important phase became uniquely lost or hidden in human females. Geoffrey Miller and his colleagues reasoned that if the 'hidden-estrus' theory is incorrect then men should find women most attractive during the fertile phase of their menstrual cycle compared to the pre-fertile (menstrual) and post-fertile (luteal) phase.

To measure how attractive men found women in an ecologically valid way, they collected data from women working at lap-dancing clubs (**Miller et al. (2007).sav.**). These women maximize their tips from male visitors by attracting more dances. In effect the men 'try out' several dancers before choosing a dancer for a prolonged dance. For each dance the male pays a 'tip', therefore the greater the number of men choosing a particular woman, the more her earnings will be. As such, each dancer's earnings are a good index of how attractive the male customers have found her. If the 'hidden-estrus' theory is incorrect then men will find the lap dancers more attractive during their estrus phase (i.e., they will earn more money during this phase).

The researchers collected data from several dancers (**ID**), who provided data for multiple lap-dancing shifts (so for each person there are several rows of data). They measured what phase of the menstrual cycle the women were in at a given shift (**Cyclephase**), and whether they were using hormonal contraceptives (**Contraceptive**), because this would affect their cycle. The outcome was their earnings on a given shift in dollars (**Tips**). The data are unbalanced: the women differed in the number of shifts for which they provided data (the range was from 9 to 29 shifts).

Labcoat Leni wants you to fit a multilevel model to see whether **Tips** can be predicted from **Cyclephase**, **Contraceptive** and their interaction. Is the 'hidden-estrus' hypothesis supported? Answers are on the companion website (or see page 378 in the original article).

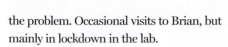

the problem. Occasional visits to Brian, but mainly in lockdown in the lab.

Looking at the machine, Jane felt proud, perhaps for the first time. She'd solved this problem on her own, with no visits to the Pleiades building. She'd taken her intellect for granted, played it down even, but this was a real achievement.

She was desperate to test the device, but she needed to sleep, and to double-check everything with a fresh head. The field test would have to wait. She went home, showered, ate, and messaged Brian. 'Sorry, been busy, I'll come over soon.' She curled up under her duvet and felt utterly content.

21.11 What next? ▮▮▮▮

This brings my life story up to date. I left out some of the more colorful bits, but only because I couldn't find even an extremely tenuous way to link them to statistics. We saw that over my life I managed to completely fail to achieve

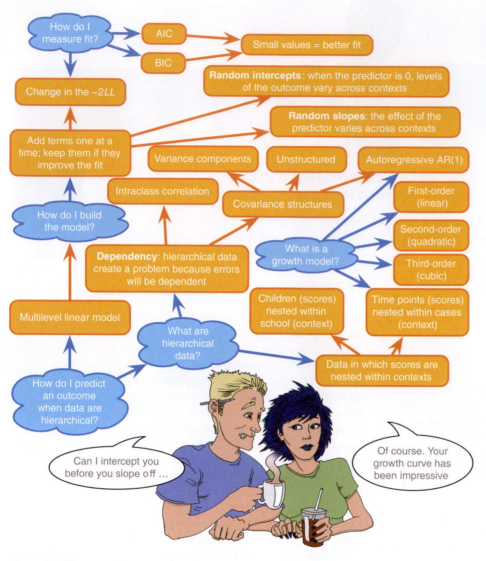

Figure 21.32 What Brian learnt from this chapter

marriage, was a leap into the unknown. Marriage, however, has proved to be infinitely more enjoyable than writing about multilevel models. I think marriage is a useful metaphor for learning about statistics: if you think about both things logically you might never do them because they are full of uncertainty and potential scariness. However, you have to go with your heart, knowing that jumping in will enrich you. Admittedly the kind of enrichment that marriage bestows is more obviously pleasant than knowing about autoregressive covariance structures, but statistics does give you enormous power to negotiate the scientific world (not just as practicing scientists, but as normal people evaluating the, often misleading, media reports of scientific findings).

My wife and I think a lot about what makes a marriage work, and we think it comes down to reciprocal effort to enrich the other person's life. There is a parallel to this book: you and I have entered into a statistical relationship of sorts. For my part, I've put as much effort as I can into trying to pass on what I know about statistics, and if you have reciprocated that effort in reading the book and working through the examples, then hopefully our time together has enriched you. In return, your reactions to this book, more often than not, enrich me.

any of my childhood dreams. It's OK, I have other ambitions now (on a slightly smaller scale than 'rock star') and I'm looking forward to failing to achieve them too. I did at least manage to marry my lovely wife. Writing this chapter, like

21.12 Key terms that I've discovered

AIC	Diagonal	Group mean centering	Random slope
AICC	Fixed coefficient	Growth curve	Random variable
AR(1)	Fixed effect	Multilevel linear model	Unstructured
BIC	Fixed intercept	Polynomial	Variance components
CAIC	Fixed slope	Random coefficient	
Centering	Fixed variable	Random effect	
Contextual variable	Grand mean centering	Random intercept	

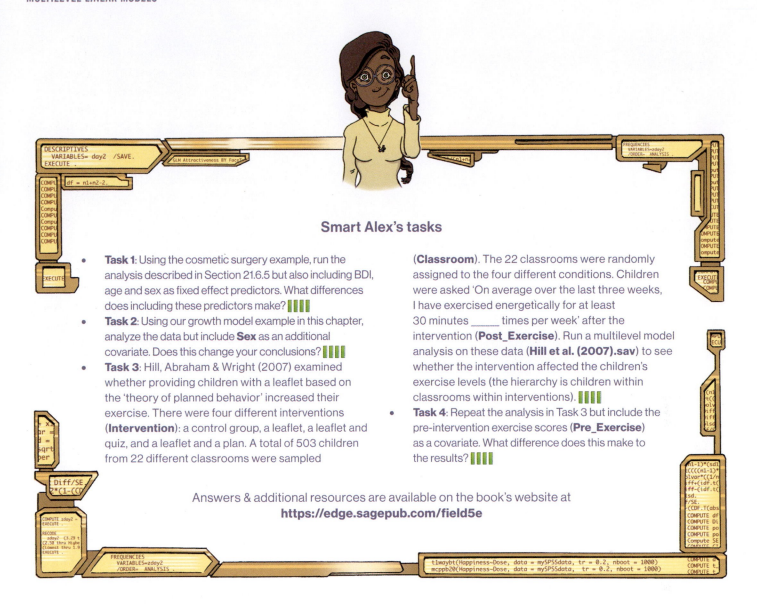

Smart Alex's tasks

- **Task 1**: Using the cosmetic surgery example, run the analysis described in Section 21.6.5 but also including BDI, age and sex as fixed effect predictors. What differences does including these predictors make?

- **Task 2**: Using our growth model example in this chapter, analyze the data but include **Sex** as an additional covariate. Does this change your conclusions?

- **Task 3**: Hill, Abraham & Wright (2007) examined whether providing children with a leaflet based on the 'theory of planned behavior' increased their exercise. There were four different interventions (**Intervention**): a control group, a leaflet, a leaflet and quiz, and a leaflet and a plan. A total of 503 children from 22 different classrooms were sampled (**Classroom**). The 22 classrooms were randomly assigned to the four different conditions. Children were asked 'On average over the last three weeks, I have exercised energetically for at least 30 minutes _____ times per week' after the intervention (**Post_Exercise**). Run a multilevel model analysis on these data (**Hill et al. (2007).sav**) to see whether the intervention affected the children's exercise levels (the hierarchy is children within classrooms within interventions).

- **Task 4**: Repeat the analysis in Task 3 but include the pre-intervention exercise scores (**Pre_Exercise**) as a covariate. What difference does this make to the results?

Answers & additional resources are available on the book's website at
https://edge.sagepub.com/field5e

EPILOGUE

Here's some questions that the writer sent
Can an observer be a participant?
Have I seen too much?
Does it count if it doesn't touch?
If the view is all I can ascertain,
Pure understanding is out of range

(Fugazi, 2001)

Brian woke with a spasm. A sharp pain had woken him up, but he couldn't move. His arms and legs were tied to the bed. His head felt heavy, like it was clamped. As his eyes focussed he saw Jane walking away with a syringe in her hand. He realized that the top of his head was in a metal device of some description. He sensed it humming slightly, vibrating his skin. He followed the wires from his head to a metal box by the side of the bed. It looked like a failed DIY electronics project. He panicked. There had been rumors about Jane around campus, that she was a psychopath, but Brian put it down to her quirkiness, and other girls envying her genius. As he considered his predicament, he felt like a fool, and terrified. She placed the syringe on the breakfast bar and turned to face him.

His throat congealed. He was sweating. He remembered reading that hostages should humanize themselves to their captors. 'Please,' he rasped, 'think what this will do to my dad.'

Jane looked confused. She walked calmly over, stroked his hand and said, 'Let yourself go, it won't hurt.'

Brian was sure that brain extraction would hurt a great deal. Whatever Jane had injected was taking effect, though, and he felt drowsy and weak. She turned to the machine, flicked some switches, and the thing on Brian's head hummed more intensely, pulsing periodically. Jane faded from view.

—

Brian's eyes opened. He couldn't speak, and he was cocooned. His eyes darted around, everything was blurred. He could make out contrast, but nothing tangible. His gaze fixed on a familiar configuration of blurred shapes. They made him feel content and safe. A voice was quietly singing near him. 'Lavender's blue, dilly dilly, lavender's green. When you are king, dilly dilly, I will be queen. A penny for your thoughts my dear, a penny for your thoughts my dear, IOU for your love, IOU for your love, for your love,' she sang, and Brian fell asleep.

—

Brian's eyes opened. He was standing, but his legs felt like he was too heavy for them. Two big hands grabbed his own from behind. In front of him was a woman in her early thirties crouching down, smiling, her arms open. She had long brown hair, big brown eyes. Seeing her made him smile, but he didn't know why. She was talking to him, beckoning with her hands. He couldn't understand her, but her face and voice excited him. He bounced a little, trying out his legs. A deeper voice spoke as the giant hands that propped him up let go. Brian had one intense urge: to go to the woman. He put one foot forward and his weight shifted awkwardly onto it. He felt himself falling and pulled his other leg forward to regain his balance. Moving felt exciting. He moved the first leg again, then the other, then again, and again, and faster and faster until he tumbled into the arms of the woman. The two voices cheered and the woman looked him in the eyes. It felt as though she was pouring her soul into him. He'd never felt so happy.

—

Brian's eyes opened. He was in a children's playground. A woman embraced him. His head was nestled in her brown hair. It smelt of coconut. He loved the smell. His wrist hurt and he was crying. The woman kissed his wrist. 'It'll be better soon,' she said. It was his mum. Her words made him feel safe. A man came over. He looked concerned. Brian recognized him. It was his dad. He looked young.

'Is he OK?' he asked the woman.

'Yeah, he just had a little fall, nothing major,' she replied.

Brian's mum locked eyes with his dad for perhaps just a second and they exchanged reassuring smiles. Brian studied them. Each of them had given him the same look a thousand times. 'Pouring their souls into each other,' he thought, but he wasn't sure why.

———

Brian had dismissed the idea of your life flashing before your eyes before you die, but maybe he'd been wrong. His head was pulsing from the flood of thoughts and images. Christmas mornings opening presents, sitting on his bed listening to storybooks, building Lego, walking through woods, sitting in cafés, scooting in the park, eating beans on toast, splashing in the bath, and the sound of that lavender song at bedtime when he couldn't sleep. From the excitement of him scoring a goal for his team at sports day, to the mundane putting on of shoes. Brian remembered a thousand things and more, all with one thing in common: the brown-haired woman. His mum. He saw her tears as she left him at nursery, her worry as he climbed in the playground, her patience as he asked questions and pushed boundaries, her empathy when he was upset, her rage when he was wronged, and her joy when he was happy. But these were not the photographic images that his dad showed him, these were like bite-sized chunks of reality sprouting in his consciousness. He could hear her, touch her, smell her, feel her. As his memories engulfed him, he found it harder to breathe. Was this the end? He tensed.

———

Brian's eyes opened. He saw school gates. His mum knelt to face him like she did every morning. Brian couldn't look her in the eye. He felt guilty. At breakfast he had told her that it was embarrassing having his mum take him to school. He was 10, he could go on his own. When he'd seen how hurt she'd looked, he wanted to take the words back. He'd be at high school next term, and he'd get the bus. He didn't need to say anything, he just needed to be patient.

'About this morning,' she began. 'Maybe you're right, maybe you are old enough to walk here on your own. It's hard, though, . . . as a parent. ... Children cannot help but take their parents for granted because they've never known a world without them, but parents had a life before their children, so they know what they have to lose. It can make you . . . overprotective. I'm proud of you for having the courage to tell me. If you don't want me to walk you to school, that's fine.'

Brian hugged her tightly. He didn't want to let go. He secretly hoped that she'd change her mind because he didn't want this to be the last time that she walked him to school, but it was, although not because of what he'd said.

———

Brian's eyes opened. He was in his apartment. Jane smiled at him. He convulsed violently as reality sank in, his juddering body eventually settling into a rhythm of uninhibited sobbing. Simultaneously he felt as euphoric and as lonely as he could ever remember. He wished that another human being wasn't here to witness his vulnerability.

Jane switched off the machine, removed the cap from Brian's twitching head, and slowly untied the binding on each of his limbs. Brian curled up into a ball as if trying to hide from Jane. She got on the bed and lay next to him, put her arm around him and pulled herself into his back. It didn't feel awkward at all. She held him until his breathing returned to normal and he was still. Brian turned to face Jane. His eyes were red. He looked spent.

'I remember her,' he said.

'You're welcome,' Jane replied.

APPENDIX

A.1 Table of the standard normal distribution

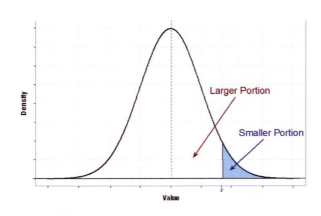

z	Larger Portion	Smaller Portion	y
.00	.50000	.50000	.3989
.01	.50399	.49601	.3989
.02	.50798	.49202	.3989
.03	.51197	.48803	.3988
.04	.51595	.48405	.3986
.05	.51994	.48006	.3984
.06	.52392	.47608	.3982
.07	.52790	.47210	.3980
.08	.53188	.46812	.3977
.09	.53586	.46414	.3973
.10	.53983	.46017	.3970
.11	.54380	.45620	.3965
.12	.54776	.45224	.3961
.13	.55172	.44828	.3956
.14	.55567	.44433	.3951
.15	.55962	.44038	.3945
.16	.56356	.43644	.3939
.17	.56749	.43251	.3932

z	Larger Portion	Smaller Portion	y
.18	.57142	.42858	.3925
.19	.57535	.42465	.3918
.20	.57926	.42074	.3910
.21	.58317	.41683	.3902
.22	.58706	.41294	.3894
.23	.59095	.40905	.3885
.24	.59483	.40517	.3876
.25	.59871	.40129	.3867
.26	.60257	.39743	.3857
.27	.60642	.39358	.3847
.28	.61026	.38974	.3836
.29	.61409	.38591	.3825
.30	.61791	.38209	.3814
.31	.62172	.37828	.3802
.32	.62552	.37448	.3790
.33	.62930	.37070	.3778
.34	.63307	.36693	.3765
.35	.63683	.36317	.3752
.36	.64058	.35942	.3739
.37	.64431	.35569	.3725
.38	.64803	.35197	.3712
.39	.65173	.34827	.3697
.40	.65542	.34458	.3683
.41	.65910	.34090	.3668
.42	.66276	.33724	.3653
.43	.66640	.33360	.3637
.44	.67003	.32997	.3621
.45	.67364	.32636	.3605
.46	.67724	.32276	.3589
.47	.68082	.31918	.3572
.48	.68439	.31561	.3555
.49	.68793	.31207	.3538
.50	.69146	.30854	.3521
.51	.69497	.30503	.3503

z	Larger Portion	Smaller Portion	y	z	Larger Portion	Smaller Portion	y
.52	.69847	.30153	.3485	.96	.83147	.16853	.2516
.53	.70194	.29806	.3467	.97	.83398	.16602	.2492
.54	.70540	.29460	.3448	.98	.83646	.16354	.2468
.55	.70884	.29116	.3429	.99	.83891	.16109	.2444
.56	.71226	.28774	.3410	1.00	.84134	.15866	.2420
.57	.71566	.28434	.3391	1.01	.84375	.15625	.2396
.58	.71904	.28096	.3372	1.02	.84614	.15386	.2371
.59	.72240	.27760	.3352	1.03	.84849	.15151	.2347
.60	.72575	.27425	.3332	1.04	.85083	.14917	.2323
.61	.72907	.27093	.3312	1.05	.85314	.14686	.2299
.62	.73237	.26763	.3292	1.06	.85543	.14457	.2275
.63	.73565	.26435	.3271	1.07	.85769	.14231	.2251
.64	.73891	.26109	.3251	1.08	.85993	.14007	.2227
.65	.74215	.25785	.3230	1.09	.86214	.13786	.2203
.66	.74537	.25463	.3209	1.10	.86433	.13567	.2179
.67	.74857	.25143	.3187	1.11	.86650	.13350	.2155
.68	.75175	.24825	.3166	1.12	.86864	.13136	.2131
.69	.75490	.24510	.3144	1.13	.87076	.12924	.2107
.70	.75804	.24196	.3123	1.14	.87286	.12714	.2083
.71	.76115	.23885	.3101	1.15	.87493	.12507	.2059
.72	.76424	.23576	.3079	1.16	.87698	.12302	.2036
.73	.76730	.23270	.3056	1.17	.87900	.12100	.2012
.74	.77035	.22965	.3034	1.18	.88100	.11900	.1989
.75	.77337	.22663	.3011	1.19	.88298	.11702	.1965
.76	.77637	.22363	.2989	1.20	.88493	.11507	.1942
.77	.77935	.22065	.2966	1.21	.88686	.11314	.1919
.78	.78230	.21770	.2943	1.22	.88877	.11123	.1895
.79	.78524	.21476	.2920	1.23	.89065	.10935	.1872
.80	.78814	.21186	.2897	1.24	.89251	.10749	.1849
.81	.79103	.20897	.2874	1.25	.89435	.10565	.1826
.82	.79389	.20611	.2850	1.26	.89617	.10383	.1804
.83	.79673	.20327	.2827	1.27	.89796	.10204	.1781
.84	.79955	.20045	.2803	1.28	.89973	.10027	.1758
.85	.80234	.19766	.2780	1.29	.90147	.09853	.1736
.86	.80511	.19489	.2756	1.30	.90320	.09680	.1714
.87	.80785	.19215	.2732	1.31	.90490	.09510	.1691
.88	.81057	.18943	.2709	1.32	.90658	.09342	.1669
.89	.81327	.18673	.2685	1.33	.90824	.09176	.1647
.90	.81594	.18406	.2661	1.34	.90988	.09012	.1626
.91	.81859	.18141	.2637	1.35	.91149	.08851	.1604
.92	.82121	.17879	.2613	1.36	.91309	.08691	.1582
.93	.82381	.17619	.2589	1.37	.91466	.08534	.1561
.94	.82639	.17361	.2565	1.38	.91621	.08379	.1539
.95	.82894	.17106	.2541	1.39	.91774	.08226	.1518

z	Larger Portion	Smaller Portion	y	z	Larger Portion	Smaller Portion	y
1.40	.91924	.08076	.1497	1.84	.96712	.03288	.0734
1.41	.92073	.07927	.1476	1.85	.96784	.03216	.0721
1.42	.92220	.07780	.1456	1.86	.96856	.03144	.0707
1.43	.92364	.07636	.1435	1.87	.96926	.03074	.0694
1.44	.92507	.07493	.1415	1.88	.96995	.03005	.0681
1.45	.92647	.07353	.1394	1.89	.97062	.02938	.0669
1.46	.92785	.07215	.1374	1.90	.97128	.02872	.0656
1.47	.92922	.07078	.1354	1.91	.97193	.02807	.0644
1.48	.93056	.06944	.1334	1.92	.97257	.02743	.0632
1.49	.93189	.06811	.1315	1.93	.97320	.02680	.0620
1.50	.93319	.06681	.1295	1.94	.97381	.02619	.0608
1.51	.93448	.06552	.1276	1.95	.97441	.02559	.0596
1.52	.93574	.06426	.1257	1.96	.97500	.02500	.0584
1.53	.93699	.06301	.1238	1.97	.97558	.02442	.0573
1.54	.93822	.06178	.1219	1.98	.97615	.02385	.0562
1.55	.93943	.06057	.1200	1.99	.97670	.02330	.0551
1.56	.94062	.05938	.1182	2.00	.97725	.02275	.0540
1.57	.94179	.05821	.1163	2.01	.97778	.02222	.0529
1.58	.94295	.05705	.1145	2.02	.97831	.02169	.0519
1.59	.94408	.05592	.1127	2.03	.97882	.02118	.0508
1.60	.94520	.05480	.1109	2.04	.97932	.02068	.0498
1.61	.94630	.05370	.1092	2.05	.97982	.02018	.0488
1.62	.94738	.05262	.1074	2.06	.98030	.01970	.0478
1.63	.94845	.05155	.1057	2.07	.98077	.01923	.0468
1.64	.94950	.05050	.1040	2.08	.98124	.01876	.0459
1.65	.95053	.04947	.1023	2.09	.98169	.01831	.0449
1.66	.95154	.04846	.1006	2.10	.98214	.01786	.0440
1.67	.95254	.04746	.0989	2.11	.98257	.01743	.0431
1.68	.95352	.04648	.0973	2.12	.98300	.01700	.0422
1.69	.95449	.04551	.0957	2.13	.98341	.01659	.0413
1.70	.95543	.04457	.0940	2.14	.98382	.01618	.0404
1.71	.95637	.04363	.0925	2.15	.98422	.01578	.0396
1.72	.95728	.04272	.0909	2.16	.98461	.01539	.0387
1.73	.95818	.04182	.0893	2.17	.98500	.01500	.0379
1.74	.95907	.04093	.0878	2.18	.98537	.01463	.0371
1.75	.95994	.04006	.0863	2.19	.98574	.01426	.0363
1.76	.96080	.03920	.0848	2.20	.98610	.01390	.0355
1.77	.96164	.03836	.0833	2.21	.98645	.01355	.0347
1.78	.96246	.03754	.0818	2.22	.98679	.01321	.0339
1.79	.96327	.03673	.0804	2.23	.98713	.01287	.0332
1.80	.96407	.03593	.0790	2.24	.98745	.01255	0325
1.81	.96485	.03515	.0775	2.25	.98778	.01222	.0317
1.82	.96562	.03438	.0761	2.26	.98809	.01191	.0310
1.83	.96638	.03362	.0748	2.27	.98840	.01160	.0303

z	Larger Portion	Smaller Portion	y	z	Larger Portion	Smaller Portion	y
2.28	.98870	.01130	.0297	2.68	.99632	.00368	.0110
2.29	.98899	.01101	.0290	2.69	.99643	.00357	.0107
2.30	.98928	.01072	.0283	2.70	.99653	.00347	.0104
2.31	.98956	.01044	.0277	2.71	.99664	.00336	.0101
2.32	.98983	.01017	.0270	2.72	.99674	.00326	.0099
2.33	.99010	.00990	.0264	2.73	.99683	.00317	.0096
2.34	.99036	.00964	.0258	2.74	.99693	.00307	.0093
2.35	.99061	.00939	.0252	2.75	.99702	.00298	.0091
2.36	.99086	.00914	.0246	2.76	.99711	.00289	.0088
2.37	.99111	.00889	.0241	2.77	.99720	.00280	.0086
2.38	.99134	.00866	.0235	2.78	.99728	.00272	.0084
2.39	.99158	.00842	.0229	2.79	.99736	.00264	.0081
2.40	.99180	.00820	.0224	2.80	.99744	.00256	.0079
2.41	.99202	.00798	.0219	2.81	.99752	.00248	.0077
2.42	.99224	.00776	.0213	2.82	.99760	.00240	.0075
2.43	.99245	.00755	.0208	2.83	.99767	.00233	.0073
2.44	.99266	.00734	.0203	2.84	.99774	.00226	.0071
2.45	.99286	.00714	.0198	2.85	.99781	.00219	.0069
2.46	.99305	.00695	.0194	2.86	.99788	.00212	.0067
2.47	.99324	.00676	.0189	2.87	.99795	.00205	.0065
2.48	.99343	.00657	.0184	2.88	.99801	.00199	.0063
2.49	.99361	.00639	.0180	2.89	.99807	.00193	.0061
2.50	.99379	.00621	.0175	2.90	.99813	.00187	.0060
2.51	.99396	.00604	.0171	2.91	.99819	.00181	.0058
2.52	.99413	.00587	.0167	2.92	.99825	.00175	.0056
2.53	.99430	.00570	.0163	2.93	.99831	.00169	.0055
2.54	.99446	.00554	.0158	2.94	.99836	.00164	.0053
2.55	.99461	.00539	.0154	2.95	.99841	.00159	.0051
2.56	.99477	.00523	.0151	2.96	.99846	.00154	.0050
2.57	.99492	.00508	.0147	2.97	.99851	.00149	.0048
2.58	.99506	.00494	.0143	2.98	.99856	.00144	.0047
2.59	.99520	.00480	.0139	2.99	.99861	.00139	.0046
2.60	.99534	.00466	.0136	3.00	.99865	.00135	.0044
2.61	.99547	.00453	.0132	...	...	...	...
2.62	.99560	.00440	.0129	3.25	.99942	.00058	.0020
2.63	.99573	.00427	.0126	...	...	...	...
2.64	.99585	.00415	.0122	3.50	.99977	.00023	.0009
2.65	.99598	.00402	.0119	...	...	...	...
2.66	.99609	.00391	.0116	4.00	.99997	.00003	.0001
2.67	.99621	.00379	.0113				

Values computed by the author using IBM SPSS Statistics

A.2 Critical values of the *t*-distribution

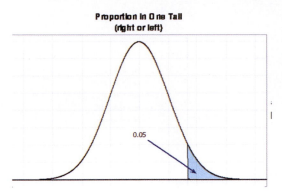

Proportion in One Tail (right or left)

0.05

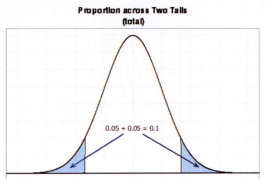

Proportion across Two Tails (total)

0.05 + 0.05 = 0.1

	Proportion across One Tails				
	0.10	0.05	0.025	0.01	0.005
	Proportion across Two Tails				
df	0.20	0.10	0.05	0.02	0.01
1	3.078	6.314	12.706	31.821	63.657
2	1.886	2.920	4.303	6.965	9.925
3	1.638	2.353	3.182	4.541	5.841
4	1.533	2.132	2.776	3.747	4.604
5	1.476	2.015	2.571	3.365	4.032
6	1.440	1.943	2.447	3.143	3.707
7	1.415	1.895	2.365	2.998	3.499
8	1.397	1.860	2.306	2.896	3.355
9	1.383	1.833	2.262	2.821	3.250
10	1.372	1.812	2.228	2.764	3.169
11	1.363	1.796	2.201	2.718	3.106
12	1.356	1.782	2.179	2.681	3.055
13	1.350	1.771	2.160	2.650	3.012
14	1.345	1.761	2.145	2.624	2.977
15	1.341	1.753	2.131	2.602	2.947
16	1.337	1.746	2.120	2.583	2.921
17	1.333	1.740	2.110	2.567	2.898
18	1.330	1.734	2.101	2.552	2.878
19	1.328	1.729	2.093	2.539	2.861
20	1.325	1.725	2.086	2.528	2.845
21	1.323	1.721	2.080	2.518	2.831
22	1.321	1.717	2.074	2.508	2.819
23	1.319	1.714	2.069	2.500	2.807
24	1.318	1.711	2.064	2.492	2.797
25	1.316	1.708	2.060	2.485	2.787
26	1.315	1.706	2.056	2.479	2.779
27	1.314	1.703	2.052	2.473	2.771
28	1.313	1.701	2.048	2.467	2.763
29	1.311	1.699	2.045	2.462	2.756
30	1.310	1.697	2.042	2.457	2.750
35	1.306	1.690	2.030	2.438	2.724
40	1.303	1.684	2.021	2.423	2.704
45	1.301	1.679	2.014	2.412	2.690
50	1.299	1.676	2.009	2.403	2.678
55	1.297	1.673	2.004	2.396	2.668
60	1.296	1.671	2.000	2.390	2.660
70	1.294	1.667	1.994	2.381	2.648
80	1.292	1.664	1.990	2.374	2.639
90	1.291	1.662	1.987	2.368	2.632
100	1.290	1.660	1.984	2.364	2.626
∞	1.282	1.645	1.960	2.326	2.576

Values computed by the author using R

A.3 Critical values of the *F*-distribution

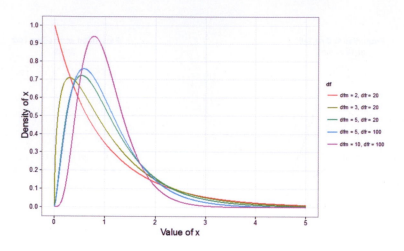

df_R						df_M								
	1	2	3	4	5	6	7	8	9	10	15	25	50	100
1	161.45	199.50	215.71	224.58	230.16	233.99	236.77	238.88	240.54	241.88	245.95	249.26	251.77	253.04
2	18.51	19.00	19.16	19.25	19.30	19.33	19.35	19.37	19.38	19.40	19.43	19.46	19.48	19.49
3	10.13	9.55	9.28	9.12	9.01	8.94	8.89	8.85	8.81	8.79	8.70	8.63	8.58	8.55
4	7.71	6.94	6.59	6.39	6.26	6.16	6.09	6.04	6.00	5.96	5.86	5.77	5.70	5.66
5	6.61	5.79	5.41	5.19	5.05	4.95	4.88	4.82	4.77	4.74	4.62	4.52	4.44	4.41
6	5.99	5.14	4.76	4.53	4.39	4.28	4.21	4.15	4.10	4.06	3.94	3.83	3.75	3.71
7	5.59	4.74	4.35	4.12	3.97	3.87	3.79	3.73	3.68	3.64	3.51	3.40	3.32	3.27
8	5.32	4.46	4.07	3.84	3.69	3.58	3.50	3.44	3.39	3.35	3.22	3.11	3.02	2.97
9	5.12	4.26	3.86	3.63	3.48	3.37	3.29	3.23	3.18	3.14	3.01	2.89	2.80	2.76
10	4.96	4.10	3.71	3.48	3.33	3.22	3.14	3.07	3.02	2.98	2.85	2.73	2.64	2.59
12	4.75	3.89	3.49	3.26	3.11	3.00	2.91	2.85	2.80	2.75	2.62	2.50	2.40	2.35
14	4.60	3.74	3.34	3.11	2.96	2.85	2.76	2.70	2.65	2.60	2.46	2.34	2.24	2.19
16	4.49	3.63	3.24	3.01	2.85	2.74	2.66	2.59	2.54	2.49	2.35	2.23	2.12	2.07
18	4.41	3.55	3.16	2.93	2.77	2.66	2.58	2.51	2.46	2.41	2.27	2.14	2.04	1.98
20	4.35	3.49	3.10	2.87	2.71	2.60	2.51	2.45	2.39	2.35	2.20	2.07	1.97	1.91
22	4.30	3.44	3.05	2.82	2.66	2.55	2.46	2.40	2.34	2.30	2.15	2.02	1.91	1.85
24	4.26	3.40	3.01	2.78	2.62	2.51	2.42	2.36	2.30	2.25	2.11	1.97	1.86	1.80
26	4.23	3.37	2.98	2.74	2.59	2.47	2.39	2.32	2.27	2.22	2.07	1.94	1.82	1.76
28	4.20	3.34	2.95	2.71	2.56	2.45	2.36	2.29	2.24	2.19	2.04	1.91	1.79	1.73
30	4.17	3.32	2.92	2.69	2.53	2.42	2.33	2.27	2.21	2.16	2.01	1.88	1.76	1.70
33	4.14	3.28	2.89	2.66	2.50	2.39	2.30	2.23	2.18	2.13	1.98	1.84	1.72	1.66
35	4.12	3.27	2.87	2.64	2.49	2.37	2.29	2.22	2.16	2.11	1.96	1.82	1.70	1.63
40	4.08	3.23	2.84	2.61	2.45	2.34	2.25	2.18	2.12	2.08	1.92	1.78	1.66	1.59
45	4.06	3.20	2.81	2.58	2.42	2.31	2.22	2.15	2.10	2.05	1.89	1.75	1.63	1.55
50	4.03	3.18	2.79	2.56	2.40	2.29	2.20	2.13	2.07	2.03	1.87	1.73	1.60	1.52
55	4.02	3.16	2.77	2.54	2.38	2.27	2.18	2.11	2.06	2.01	1.85	1.71	1.58	1.50

df_R						df_M									
	1	2	3	4	5	6	7	8	9	10	15	25	50	100
60	4.00	3.15	2.76	2.53	2.37	2.25	2.17	2.10	2.04	1.99	1.84	1.69	1.56	1.48
65	3.99	3.14	2.75	2.51	2.36	2.24	2.15	2.08	2.03	1.98	1.82	1.68	1.54	1.46
70	3.98	3.13	2.74	2.50	2.35	2.23	2.14	2.07	2.02	1.97	1.81	1.66	1.53	1.45
75	3.97	3.12	2.73	2.49	2.34	2.22	2.13	2.06	2.01	1.96	1.80	1.65	1.52	1.44
80	3.96	3.11	2.72	2.49	2.33	2.21	2.13	2.06	2.00	1.95	1.79	1.64	1.51	1.43
85	3.95	3.10	2.71	2.48	2.32	2.21	2.12	2.05	1.99	1.94	1.79	1.64	1.50	1.42
90	3.95	3.10	2.71	2.47	2.32	2.20	2.11	2.04	1.99	1.94	1.78	1.63	1.49	1.41
95	3.94	3.09	2.70	2.47	2.31	2.20	2.11	2.04	1.98	1.93	1.77	1.62	1.48	1.40
100	3.94	3.09	2.70	2.46	2.31	2.19	2.10	2.03	1.97	1.93	1.77	1.62	1.48	1.39

df_R \ df_M	1	2	3	4	5	6	7	8	9	10	15	25	50	100
1	4052.18	4999.50	5403.35	5624.58	5763.65	5858.99	5928.36	5981.07	6022.47	6055.85	6157.28	6239.83	6302.52	6334.11
2	98.50	99.00	99.17	99.25	99.30	99.33	99.36	99.37	99.39	99.40	99.43	99.46	99.48	99.49
3	34.12	30.82	29.46	28.71	28.24	27.91	27.67	27.49	27.35	27.23	26.87	26.58	26.35	26.24
4	21.20	18.00	16.69	15.98	15.52	15.21	14.98	14.80	14.66	14.55	14.20	13.91	13.69	13.58
5	16.26	13.27	12.06	11.39	10.97	10.67	10.46	10.29	10.16	10.05	9.72	9.45	9.24	9.13
6	13.75	10.92	9.78	9.15	8.75	8.47	8.26	8.10	7.98	7.87	7.56	7.30	7.09	6.99
7	12.25	9.55	8.45	7.85	7.46	7.19	6.99	6.84	6.72	6.62	6.31	6.06	5.86	5.75
8	11.26	8.65	7.59	7.01	6.63	6.37	6.18	6.03	5.91	5.81	5.52	5.26	5.07	4.96
9	10.56	8.02	6.99	6.42	6.06	5.80	5.61	5.47	5.35	5.26	4.96	4.71	4.52	4.41
10	10.04	7.56	6.55	5.99	5.64	5.39	5.20	5.06	4.94	4.85	4.56	4.31	4.12	4.01
12	9.33	6.93	5.95	5.41	5.06	4.82	4.64	4.50	4.39	4.30	4.01	3.76	3.57	3.47
14	8.86	6.51	5.56	5.04	4.69	4.46	4.28	4.14	4.03	3.94	3.66	3.41	3.22	3.11
16	8.53	6.23	5.29	4.77	4.44	4.20	4.03	3.89	3.78	3.69	3.41	3.16	2.97	2.86
18	8.29	6.01	5.09	4.58	4.25	4.01	3.84	3.71	3.60	3.51	3.23	2.98	2.78	2.68
20	8.10	5.85	4.94	4.43	4.10	3.87	3.70	3.56	3.46	3.37	3.09	2.84	2.64	2.54
22	7.95	5.72	4.82	4.31	3.99	3.76	3.59	3.45	3.35	3.26	2.98	2.73	2.53	2.42
24	7.82	5.61	4.72	4.22	3.90	3.67	3.50	3.36	3.26	3.17	2.89	2.64	2.44	2.33
26	7.72	5.53	4.64	4.14	3.82	3.59	3.42	3.29	3.18	3.09	2.81	2.57	2.36	2.25
28	7.64	5.45	4.57	4.07	3.75	3.53	3.36	3.23	3.12	3.03	2.75	2.51	2.30	2.19
30	7.56	5.39	4.51	4.02	3.70	3.47	3.30	3.17	3.07	2.98	2.70	2.45	2.25	2.13
33	7.47	5.31	4.44	3.95	3.63	3.41	3.24	3.11	3.00	2.91	2.63	2.39	2.18	2.06
35	7.42	5.27	4.40	3.91	3.59	3.37	3.20	3.07	2.96	2.88	2.60	2.35	2.14	2.02
40	7.31	5.18	4.31	3.83	3.51	3.29	3.12	2.99	2.89	2.80	2.52	2.27	2.06	1.94
45	7.23	5.11	4.25	3.77	3.45	3.23	3.07	2.94	2.83	2.74	2.46	2.21	2.00	1.88
50	7.17	5.06	4.20	3.72	3.41	3.19	3.02	2.89	2.78	2.70	2.42	2.17	1.95	1.82
55	7.12	5.01	4.16	3.68	3.37	3.15	2.98	2.85	2.75	2.66	2.38	2.13	1.91	1.78
60	7.08	4.98	4.13	3.65	3.34	3.12	2.95	2.82	2.72	2.63	2.35	2.10	1.88	1.75
65	7.04	4.95	4.10	3.62	3.31	3.09	2.93	2.80	2.69	2.61	2.33	2.07	1.85	1.72
70	7.01	4.92	4.07	3.60	3.29	3.07	2.91	2.78	2.67	2.59	2.31	2.05	1.83	1.70
75	6.99	4.90	4.05	3.58	3.27	3.05	2.89	2.76	2.65	2.57	2.29	2.03	1.81	1.67
80	6.96	4.88	4.04	3.56	3.26	3.04	2.87	2.74	2.64	2.55	2.27	2.01	1.79	1.65
85	6.94	4.86	4.02	3.55	3.24	3.02	2.86	2.73	2.62	2.54	2.26	2.00	1.77	1.64
90	6.93	4.85	4.01	3.53	3.23	3.01	2.84	2.72	2.61	2.52	2.24	1.99	1.76	1.62
95	6.91	4.84	3.99	3.52	3.22	3.00	2.83	2.70	2.60	2.51	2.23	1.98	1.75	1.61
100	6.90	4.82	3.98	3.51	3.21	2.99	2.82	2.69	2.59	2.50	2.22	1.97	1.74	1.60

df_M

Values computed by the author using R

A.4 Critical values of the chi-square distribution

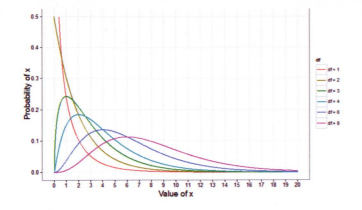

df	p = 0.05	p = 0.01
19	30.14	36.19
20	31.41	37.57
21	32.67	38.93
22	33.92	40.29
23	35.17	41.64
24	36.42	42.98
25	37.65	44.31
26	38.89	45.64
27	40.11	46.96
28	41.34	48.28
29	42.56	49.59
30	43.77	50.89
35	49.80	57.34
40	55.76	63.69
45	61.66	69.96
50	67.50	76.15
60	79.08	88.38
70	90.53	100.43
80	101.88	112.33
90	113.15	124.12
100	124.34	135.81
200	233.99	249.45
300	341.40	359.91
400	447.63	468.72
500	553.13	576.49
600	658.09	683.52
700	762.66	789.97
800	866.91	895.98
900	970.90	1001.63
1000	1074.68	1106.97

df	p = 0.05	p = 0.01
1	3.84	6.63
2	5.99	9.21
3	7.81	11.34
4	9.49	13.28
5	11.07	15.09
6	12.59	16.81
7	14.07	18.48
8	15.51	20.09
9	16.92	21.67
10	18.31	23.21
11	19.68	24.72
12	21.03	26.22
13	22.36	27.69
14	23.68	29.14
15	25.00	30.58
16	26.30	32.00
17	27.59	33.41
18	28.87	34.81

Values computed by the author using R

GLOSSARY

0: how much clue Sage have about the amount of effort I put into writing this book.

–2LL: the *log-likelihood* multiplied by minus 2. This version of the likelihood is used in *logistic regression*.

α-level: the probability of making a *Type I error* (usually this value is 0.05).

A life: what you don't have when writing statistics textbooks.

Adjusted mean: in the context of *analysis of covariance* this is the value of the group mean adjusted for the effect of the *covariate*.

Adjusted predicted value: a measure of the influence of a particular case of data. It is the predicted value of a case from a model estimated without that case included in the data. The value is calculated by re-estimating the model without the case in question, then using this new model to predict the value of the excluded case. If a case does not exert a large influence over the model then its predicted value should be similar regardless of whether the model was estimated including or excluding that case. The difference between the predicted value of a case from the model when that case was included and the predicted value from the model when it was excluded is the *DFFit*.

Adjusted R^2: a measure of the loss of predictive power or *shrinkage* in regression. The adjusted R^2 tells us how much variance in the outcome would be accounted for if the model had been derived from the population from which the sample was taken.

AIC (Akaike's information criterion): a *goodness-of-fit* measure that is corrected for model complexity. That just means that it takes account of how many parameters have been estimated. It is not intrinsically interpretable, but can be compared in different models to see how changing the model affects the fit. A small value represents a better fit to the data.

AICC (Hurvich and Tsai's criterion): a *goodness-of-fit* measure that is similar to *AIC* but is designed for small samples. It is not intrinsically interpretable, but can be compared in different models to see how changing the model affects the fit. A small value represents a better fit to the data.

Alpha factoring: a method of *factor analysis*.

Alternative hypothesis: the prediction that there will be an effect (i.e., that your experimental manipulation will have some effect or that certain variables will relate to each other).

Analysis of covariance: a statistical procedure that uses the *F-statistic* to test the overall fit of a linear model, adjusting for the effect that one or more *covariates* have on the *outcome variable*. In experimental research this linear model tends to be defined in terms of group means and the resulting ANOVA is therefore an overall test of whether group means differ after the variance in the outcome variable explained by any *covariates* has been removed.

Analysis of variance: a statistical procedure that uses the *F-statistic* to test the overall fit of a linear model. In experimental research this linear model tends to be defined in terms of group means, and the resulting ANOVA is therefore an overall test of whether group means differ.

ANCOVA: acronym for *analysis of covariance*.

Anderson–Rubin method: a way of calculating *factor scores* which produces scores that are uncorrelated and *standardized* with a mean of 0 and a standard deviation of 1.

ANOVA: acronym for *analysis of variance*.

AR(1): this stands for first-order autoregressive structure. It is a covariance structure used in *multilevel linear models* in which the relationship between scores changes in a systematic way. It is assumed that the correlation between scores gets smaller over time and that variances are assumed to be homogeneous. This structure is often used for repeated-measures data (especially when measurements are taken over time such as in growth models).

Autocorrelation: when the *residuals* of two observations in a regression model are correlated.

b_i: unstandardized regression coefficient. Indicates the strength of relationship between a given predictor, i, of many and an outcome in the units of measurement of the predictor. It is the change in the outcome associated with a unit change in the predictor.

β_i: standardized regression coefficient. Indicates the strength of relationship between a given predictor, i, of many and an outcome in a *standardized* form. It is the change in the outcome (in standard deviations) associated with a one standard deviation change in the predictor.

β-level: the probability of making a *Type II error* (Cohen, 1992, suggests a maximum value of 0.2).

Bar chart: a graph in which a summary statistic (usually the mean) is plotted on the y-axis against a categorical variable on the x-axis (this categorical variable could represent, for example, groups of people, different times or different experimental conditions). The value of the mean for each category is shown by a bar. Different-colored bars may be used to represent levels of a second categorical variable.

Bartlett's test of sphericity: unsurprisingly, this is a test of the assumption of *sphericity*. This test examines whether a *variance–covariance matrix* is proportional to an *identity matrix*. Therefore, it effectively tests whether the diagonal elements of the variance–covariance matrix are equal (i.e., group variances are the same), and whether the off-diagonal elements are approximately zero (i.e., the *dependent variables* are not *correlated*). Jeremy Miles, who does a lot of multivariate stuff, claims he's never ever seen a matrix that reached non-significance using this test and, come to think of it, I've never seen one either (although I do less multivariate stuff), so you've got to wonder about its practical utility.

Bayes factor: the ratio of the probability of the observed data given the *alternative hypothesis* to the probability of the observed data given the *null hypothesis* although SPSS statistics tends to express it the other way around. Put another way, it is the *likelihood* of the alternative hypothesis relative to the null. A Bayes factor of 3, for example, means that the observed data are 3 times more likely under the alternative hypothesis than under the null hypothesis. A Bayes factor less than 1 supports the null hypothesis by suggesting that the probability of the data given the null is higher than the probability of the data given the alternative hypothesis. Conversely, a Bayes factor greater than 1 suggests that the observed data are more likely given the alternative hypothesis than the null. Values between 1 and 3 are considered evidence for the alternative hypothesis that is 'barely worth mentioning', values between 3 and 10 are considered 'substantial evidence' ('having substance' rather than 'very strong') for the alternative hypothesis, and values greater than 10 are strong evidence for the alternative hypothesis.

Bayesian statistics: a branch of statistics in which hypotheses are tested or model parameters are estimated using methods based on *Bayes' theorem*.

Bayes' theorem: a mathematical description of the relationship between the *conditional probability* of events A and B, $p(A|B)$, their reverse conditional probability, $p(B|A)$, and individual probabilities of the events, $p(A)$ and $p(B)$. The theorem states that

$$p(A|B) = \frac{p(B|A)p(A)}{p(B)}$$

Beer-goggles effect: the phenomenon that people of the opposite sex (or the same, depending on your sexual orientation) appear much more attractive after a few alcoholic drinks.

Between-groups design: another name for *independent design*.

Between-subjects design: another name for *independent design*.

BIC (Schwarz's Bayesian information criterion): a *goodness-of-fit* statistic comparable to the *AIC*, although it is slightly more conservative (it corrects more harshly for the number of parameters being estimated). It should be used when sample sizes are large and the number of parameters is small. It is not intrinsically interpretable, but can be compared in different models to see how changing the model affects the fit. A small value represents a better fit to the data.

Bimodal: a description of a distribution of observations that has two *modes*.

Binary logistic regression: *logistic regression* in which the outcome variable has exactly two categories.

Binary variable: a *categorical variable* that has only two mutually exclusive categories (e.g., being dead or alive).

Biserial correlation: a standardized measure of the strength of relationship between two variables when one of the two variables is *dichotomous*. The biserial correlation coefficient is used when one variable is a continuous dichotomy (e.g., has an underlying continuum between the categories).

Bivariate correlation: a correlation between two variables.

Blockwise regression: another name for *hierarchical regression*.

Bonferroni correction: a correction applied to the α-*level* to control the overall *Type I error rate* when multiple significance tests are carried out. Each test conducted should use a criterion of significance of the α-*level* (normally 0.05) divided by the number of tests conducted. This is a simple but effective correction, but tends to be too strict when lots of tests are performed.

Bootstrap: a technique from which the sampling distribution of a statistic is estimated by taking repeated samples (with replacement) from the data set (in effect, treating the data as a population from which smaller samples are taken). The statistic of interest (e.g., the *mean* or b coefficient) is calculated for each sample, from which the sampling distribution of the statistic is estimated. The standard error of the statistic is estimated as the standard deviation of the sampling distribution created from the bootstrap samples. From this, confidence intervals and significance tests can be computed.

Boredom effect: refers to the possibility that performance in tasks may be influenced (the assumption is a negative influence) by boredom or lack of concentration if there are many tasks or the task goes on for a long period of time. In short, what you are experiencing reading this glossary is a boredom effect.

Boxplot (a.k.a. box–whisker diagram): a graphical representation of some important characteristics of a set of observations. At the center of the plot is the *median*, which is surrounded by a box, the top and bottom of which are the limits within which the middle 50% of observations fall (the *interquartile range*). Sticking out of the top and bottom of the box are two whiskers which extend to the highest and lowest extreme scores, respectively.

Box's test: a test of the assumption of *homogeneity of covariance matrices*. This test should be non-significant if the matrices are roughly the same. Box's test is very susceptible to deviations from *multivariate normality* and so may be non-significant not because the *variance–covariance matrices* are similar across groups, but because the assumption of multivariate normality is not tenable. Hence, it is vital to have some idea of whether the data meet the multivariate normality assumption (which is extremely difficult) before interpreting the result of Box's test.

Box–whisker plot: see *boxplot*.

Brown–Forsythe *F*: a version of the *F-statistic* designed to be accurate when the assumption of *homogeneity of variance* has been violated.

CAIC (Bozdogan's criterion): a *goodness-of-fit* measure similar to the *AIC*, but correcting for model complexity and sample size. It is not intrinsically interpretable, but can be compared in different models to see how changing the model affects the fit. A small value represents a better fit to the data.

Categorical variable: any variable made up of categories of objects/entities. The university you attend is a good example of a categorical variable: students who attend the University of Sussex are not also enrolled at Harvard or UV Amsterdam, therefore, students fall into distinct categories.

Catterplot: when your data has fried your mind so much that every graph you look at transforms before your eyes into cute kittens that beckon you to 'come play with a ball of string'. Under no circumstances follow them.

Central limit theorem: this theorem states that when samples are large (above about 30) the *sampling distribution* will take the shape of a *normal distribution* regardless of the shape of the population from which the sample was drawn. For small samples the *t*-distribution better approximates the shape of the sampling distribution. We also know from this theorem that the *standard deviation* of the sampling distribution (i.e., the *standard error* of the sample *mean*) will be equal to the standard deviation of the sample (*s*) divided by the square root of the sample size (*N*).

Central tendency: a generic term describing the center of a *frequency distribution* of observations as measured by the *mean*, *mode* and *median*.

Centering: the process of transforming a variable into deviations around a fixed point. This fixed point can be any value that is chosen, but typically a mean is used. To center a variable the mean is subtracted from each score. See *grand mean centering*, *group mean centering*.

Chartjunk: superfluous material that distracts from the data being displayed on a graph.

Chi-square distribution: a *probability distribution* of the sum of squares of several normally distributed variables. It tends to be used to test hypotheses about categorical data, and to test the fit of models to the observed data.

Chi-square test: although this term can apply to any *test statistic* having a *chi-square distribution*, it generally refers to Pearson's chi-square test of the independence of two categorical variables. Essentially it tests whether two categorical variables forming a *contingency table* are associated.

Cochran's *Q*: this test is an extension of *McNemar's test* and is basically a *Friedman's ANOVA* for *dichotomous* data. So imagine you asked 10 people whether they'd like to shoot Justin Timberlake, David Beckham and Simon Cowell and they could answer only 'yes' or 'no'. If we coded responses as 0 (no) and 1 (yes) we could do Cochran's test on these data.

Coefficient of determination: the proportion of variance in one variable explained by a second variable. It is *Pearson's correlation coefficient* squared.

Cohen's *d*: an *effect size* that expresses the difference between two means in standard deviation units. In general it can be estimated using:

$$\hat{d} = \frac{\bar{X}_1 - \bar{X}_2}{s}$$

Common factor: a *factor* that affects all measured *variables* and, therefore, explains the *correlations* between those variables.

Common variance: variance shared by two or more variables.

Communality: the proportion of a variable's variance that is *common variance*. This term is used primarily in *factor analysis*. A variable that has no *unique variance* (or *random variance*) would have a communality of 1, whereas a variable that shares none of its variance with any other variable would have a communality of 0.

Complete separation: a situation in *logistic regression* when the outcome variable can be perfectly predicted by one predictor or a combination of predictors! Suffice it to say this situation makes your computer have the equivalent of a nervous breakdown: it'll start gibbering, weeping and saying it doesn't know what to do.

Component matrix: general term for the *structure matrix* in *principal components analysis*.

Compound symmetry: a condition that holds true when both the variances across conditions are equal (this is the same as the *homogeneity of variance* assumption) and the *covariances* between pairs of conditions are also equal.

Concurrent validity: a form of *criterion validity* where there is evidence that scores from an instrument correspond to concurrently recorded external measures conceptually related to the measured construct.

Confidence interval: for a given statistic calculated for a sample of observations (e.g., the mean), the confidence interval is a range of values around that statistic that are believed to contain, in a certain proportion of samples (e.g., 95%), the true value of that statistic (i.e., the population parameter). What that also means is that for the other proportion of samples (e.g., 5%), the confidence interval won't contain that true value. The trouble is, you don't know which category your particular sample falls into.

Confirmatory factor analysis (CFA): a version of *factor analysis* in which specific hypotheses about structure and relations between the *latent variables* that underlie the data are tested.

Confounding variable: a variable (that we may or may not have measured) other than the *predictor variables* in which we're interested that potentially affects an *outcome variable*.

Contaminated normal distribution: see *mixed normal distribution*.

Content validity: evidence that the content of a test corresponds to the content of the construct it was designed to cover.

Contingency table: a table representing the cross-classification of two or more *categorical variables*. The levels of each variable are arranged in a grid, and the number of observations falling into each category is noted in the cells of the table. For example, if we took the categorical variables of glossary (with two categories: whether an author was made to write a glossary or not), and mental state (with three categories: normal, sobbing uncontrollably and utterly psychotic), we could construct a table as below. This instantly tells us that 127 authors who were made to write a glossary ended up as utterly psychotic, compared to only 2 who did not write a glossary.

| | | Glossary | | |
		Author made to write glossary	No glossary	Total
Mental state	Normal	5	423	428
	Sobbing uncontrollably	23	46	69
	Utterly psychotic	127	2	129
	Total	155	471	626

Continuous variable: a variable that can be measured to any level of precision. (Time is a continuous variable, because there is in principle no limit on how finely it could be measured.)

Cook's distance: a measure of the overall influence of a case on a model. Cook and Weisberg (1982) have suggested that values greater than 1 may be cause for concern.

Correlation coefficient: a measure of the strength of association or relationship between two variables. See *Pearson's correlation coefficient*, *Spearman's correlation coefficient*, *Kendall's tau*.

Correlational research: a form of research in which you observe what naturally goes on in the world without directly interfering with it. This term implies that data will be analyzed so as to look at relationships between naturally occurring variables rather than making statements about cause and effect. Compare with *cross-sectional research*, *longitudinal research* and *experimental research*.

Counterbalancing: a process of systematically varying the order in which experimental conditions are conducted. In the simplest case of there being two conditions (A and B), counterbalancing simply implies that half of the participants complete condition A followed by condition B, whereas the remainder do condition B followed by condition A. The aim is to remove systematic bias caused by *practice effects* or *boredom effects*.

Covariance: a measure of the 'average' relationship between two variables. It is the average *cross-product deviation* (i.e., the cross-product divided by one less than the number of observations).

Covariance ratio (CVR): a measure of whether a case influences the variance of the parameters in a *regression model*. When this ratio is close to 1 the case has very little influence on the variances of the model parameters. Belsey et al. (1980) recommend the following: if the CVR of a case is greater than $1 + [3(k + 1)/n]$ then deleting that case will damage the precision of some of the model's parameters, but if it is less than $1 - [3(k + 1)/n]$ then deleting the case will improve the precision of some of the model's parameters (k is the number of predictors and n is the sample size).

Covariate: a variable that has a relationship with (in terms of *covariance*) or has the potential to be related to, the *outcome variable* we've measured.

Cox and Snell's R^2_{CS}: a version of the *coefficient of determination* for logistic regression. It is based on the log-likelihood of a model (LL_{new}), the log-likelihood of the original model ($LL_{baseline}$) and the sample size, n. However, it is notorious for not reaching its maximum value of 1 (see *Nagelkerke's R^2_N*).

Cramér's V: a measure of the strength of association between two *categorical variables* used when one of these variables has more than two categories. It is a variant of *phi* used because when one or both of the categorical variables contain more than two categories, phi fails to reach its minimum value of 0 (indicating no association).

Credible interval: in Bayesian statistics, a credible interval is an interval within which a certain percentage of the *posterior distribution* falls (usually 95%). It can be used to express the limits within which a parameter falls with a fixed probability. For example, if we estimated the average length of a romantic relationship to be 6 years with a 95% credible interval of 1 to 11 years, then this would mean that 95% of the *posterior distribution* for the length of romantic relationships falls between 1 and 11 years. A plausible estimate of the length of romantic relationships would, therefore, be 1 to 11 years.

Criterion validity: evidence that scores from an instrument correspond with (*concurrent validity*) or predict (*predictive validity*) external measures conceptually related to the measured construct.

Cronbach's α: a measure of the reliability of a scale defined by

$$a = \frac{N^2 \overline{Cov}}{\sum s^2_{item} + \sum Cov_{item}}$$

in which the top half of the equation is simply the number of items (N) squared multiplied by the average covariance between items (the average of the off-diagonal elements in the *variance–covariance matrix*). The bottom half is the sum of all the elements in the *variance–covariance matrix*.

Cross-product deviations: a measure of the 'total' relationship between two variables. It is the deviation of one variable from its mean multiplied by the other variable's deviation from its mean.

Cross-sectional research: a form of research in which you observe what naturally goes on in the world without directly interfering with it by measuring several variables at a single time point. In psychology, this term usually implies that data come from people at different age points, with different people representing each age point. See also *correlational research*, *longitudinal research*.

Cross-validation: assessing the accuracy of a model across different samples. This is an important step in *generalization*. In a *regression model* there are two main methods of cross-validation: *adjusted R²* or data splitting, in which the data are split randomly into two halves, and a regression model is estimated for each half and then compared.

Crying: what you feel like doing after writing statistics textbooks.

Cubic trend: if you connected the means in ordered conditions with a line then a cubic trend is shown by two changes in the direction of this line. You must have at least four ordered conditions.

Currency variable: a variable containing values of money.

Data view: there are two ways to view the contents of the *data editor* window. The data view shows you a spreadsheet and can be used for entering raw data. See also *variable view*.

Date variable: a variable made up of dates. The data can take forms such as dd-mmm-yyyy (e.g., 21-Jun-1973), dd-mmm-yy (e.g., 21-Jun-73), mm/dd/yy (e.g., 06/21/73), dd.mm.yyyy (e.g., 21.06.1973).

Degrees of freedom: an impossible thing to define in a few pages, let alone a few lines. Essentially it is the number of 'entities' that are free to vary when estimating some kind of statistical parameter. In a more practical sense, it has a bearing on significance tests for many commonly used *test statistics* (such as the *F-statistic*, *t-Statistic*, *chi-square test*) and determines the exact form of the *probability distribution* for these *test statistics*. The explanation involving soccer players in Chapter 2 is far more interesting…

Deleted residual: a measure of the influence of a particular case of data. It is the difference between the *adjusted predicted value* for a case and the original observed value for that case.

Density plot: similar to a *histogram* except that rather than having a summary bar representing the frequency of scores, it shows each individual score as a dot. They can be useful for looking at the shape of a distribution of scores.

Dependent *t*-test: see *paired-samples t-test*.

Dependent variable: another name for *outcome variable*. This name is usually associated with experimental methodology (which is the only time it really makes sense) and is used because it is the variable that is not manipulated by the experimenter and so its value depends on the variables that have been manipulated. To be honest, I just use the term *outcome variable* all the time – it makes more sense (to me) and is less confusing.

Deviance: the difference between the observed value of a variable and the value of that variable predicted by a statistical model.

Deviation contrast: a non-orthogonal *planned contrast* that compares the mean of each group (except for the first or last, depending on how the contrast is specified) to the overall mean.

DFBeta: a measure of the influence of a case on the values of b_i in a *regression model*. If we estimated a regression parameter b_i and then deleted a particular case and re-estimated the same regression parameter b_i, then the difference between these two estimates would be the DFBeta for the case that was deleted. By looking at the values of the DFBetas, it is possible to identify cases that have a large influence on the parameters of the regression model; however, the size of DFBeta will depend on the units of measurement of the regression parameter.

DFFit: a measure of the influence of a case. It is the difference between the *adjusted predicted value* and the original predicted value of a particular case. If a case is not influential then its DFFit should be zero – hence, we expect non-influential cases to have small DFFit values. However, we have the problem that this statistic depends on the units of measurement of the outcome, and so a DFFit of 0.5 will be very small if the outcome ranges from 1 to 100, but very large if the outcome varies from 0 to 1.

Diagonal: a covariance structure used in *multilevel linear models*. In this variance structure variances are assumed to be heterogeneous and all of the covariances are 0.

Dichotomous: description of a variable that consists of only two categories (e.g., biological sex is a dichotomous variable because it consists of only two categories: male and female).

Difference contrast: a non-orthogonal *planned contrast* that compares the mean of each condition (except the first) to the overall mean of all previous conditions combined.

Direct effect: the effect of a *predictor variable* on an *outcome variable* when a *mediator* is present in the model (cf. *indirect effect*).

Direct oblimin: a method of *oblique rotation*.

Discrete variable: a variable that can only take on certain values (usually whole numbers) on the scale.

Discriminant analysis: see *discriminant function analysis*.

Discriminant function analysis: identifies and describes the *discriminant function variates* of a set of variables and is useful as a follow-up test to *MANOVA* as a means of seeing how these variates allow groups of cases to be discriminated.

Discriminant function variate: a linear combination of variables created such that the differences between group means on the transformed variable are maximized. It takes the general form: $Variate_{1i} = b_0 + b_1 X_{1i} + b_2 X_{2i} + … + b_n X_{ni}$.

Discriminant score: a score for an individual case on a particular *discriminant function variate* obtained by substituting that case's scores on the measured variables into the equation that defines the variate in question.

Dummy variables: a way of recoding a categorical variable with more than two categories into a series of variables all of which are *dichotomous* and can take on values of only 0 or 1. There are seven basic steps to create such variables: (1) count the number of groups you want to recode and subtract 1; (2) create as many new variables as the value you calculated in step 1 (these are your dummy variables); (3) choose one of your groups as a baseline (i.e., a group against which all other groups should be compared, such as a control group); (4) assign that baseline group values of 0 for all of your dummy variables; (5) for your first dummy variable, assign the value 1 to the first group that you want to compare against the baseline group (assign all other groups 0 for this variable); (6) for the second dummy variable assign the value 1 to the second group that you want to compare against the baseline group (assign all other groups 0 for this variable); (7) repeat this process until you run out of dummy variables.

Durbin–Watson test: a test for serial correlations between errors in *regression models*. Specifically, it tests whether adjacent residuals are correlated, which is useful in assessing the assumption of *independent errors*. The test statistic can vary between 0 and 4, with a value of 2 meaning that the residuals are uncorrelated. A value greater than 2 indicates a negative correlation between adjacent residuals, whereas a value below 2 indicates a positive correlation. The size of the Durbin–Watson statistic depends upon the number of predictors in the model and the number of observations. For accuracy, look up the exact acceptable values in Durbin and Watson's (1951) original paper. As a very conservative rule of thumb, values less than 1 or greater than 3 are definitely cause for concern; however, values closer to 2 may still be problematic, depending on the sample and model.

Ecological validity: evidence that the results of a study, experiment or test can be applied, and allow inferences, to real-world conditions.

Eel: long, snakelike, scaleless fish that lacks pelvic fins. From the order Anguilliformes or Apodes, eels should probably not be inserted into your anus to cure constipation (or for any other reason).

Effect size: an objective and (usually) standardized measure of the magnitude of an observed effect. Measures include Cohen's *d*, Glass's *g* and Pearson's correlations coefficient, *r*.

Empirical probability: the empirical probability is the *probability* of an *event* based on the observation of many trials. For example, if you define the collective as all men, then the empirical probability of infidelity in men will be the proportion of men who have been unfaithful while in a relationship. The probability applies to the collective and not to the individual events. You can talk about there being a 0.1 probability of men being unfaithful, but the individual men were either faithful or not, so their *individual* probability of infidelity was either 0 (they were faithful) or 1 (they were unfaithful).

Equamax: a method of *orthogonal rotation* that is a hybrid of *quartimax* and *varimax*. It is reported to behave fairly erratically (see Tabachnick & Fidell, 2012) and so is probably best avoided.

Error bar chart: a graphical representation of the mean of a set of observations that includes the 95% confidence interval of the mean. The mean is usually represented as a circle, square or rectangle at the value of the mean (or a bar extending to the value of the mean). The confidence interval is represented by a line protruding from the mean (upwards, downwards or both) to a short horizontal line representing the limits of the confidence interval. Error bars can be drawn using the standard error or standard deviation instead of the 95% confidence interval.

Error SSCP (*E*): the error sum of squares and cross-products matrix. This is a *sum of squares and cross-products matrix* for the error in a predictive *linear model* fitted to *multivariate* data. It represents the *unsystematic variance* and is the multivariate equivalent of the *residual sum of squares*.

Eta squared (η^2): an *effect size* measure that is the ratio of the *model sum of squares* to the *total sum of squares*. So, in essence, the *coefficient of determination* by another name. It doesn't have an awful lot going for it: not only is it biased, but it typically measures the overall effect of an ANOVA and effect sizes are more easily interpreted when they reflect specific comparisons (e.g., the difference between two means).

Exp(*B*): the label that SPSS applies to the *odds ratio*. It is an indicator of the change in *odds* resulting from a unit change in the predictor in *logistic regression*. If the value is greater than 1 then it indicates that as the predictor increases, the odds of the outcome occurring increase. Conversely, a value less than 1 indicates that as the predictor increases, the odds of the outcome occurring decrease.

Experimental hypothesis: synonym for *alternative hypothesis*.

Experimental research: a form of research in which one or more variables are systematically manipulated to see their effect (alone or in combination) on an *outcome variable*. This term implies that data will be able to be used to make statements about cause and effect. Compare with *cross-sectional research* and *correlational research*.

Experimentwise error rate: the probability of making a *Type I error* in an experiment involving one or more statistical comparisons when the null hypothesis is true in each case.

Extraction: a term used for the process of deciding whether a *factor* in *factor analysis* is statistically important enough to 'extract' from the data and interpret. The decision is based on the magnitude of the eigenvalue associated with the factor. See *Kaiser's criterion*, *scree plot*.

F_{max}: see *Hartley's F_{max}*.

F-statistic: a test statistic with a known *probability distribution* (the *F*-distribution). It is the ratio of the average variability in the data that a given model can explain to the average variability unexplained by that same model. It is used to test the overall fit of the model in *simple regression* and *multiple regression*, and to test for overall differences between group means in experiments.

Factor: another name for an *independent variable* or *predictor* that's typically used when describing experimental designs. However, to add to the confusion, it is also used synonymously with *latent variable* in factor analysis.

Factor analysis: a *multivariate* technique for identifying whether the correlations between a set of observed variables stem from their relationship to one or more *latent variables* in the data, each of which takes the form of a *linear model*.

Factor matrix: general term for the *structure matrix* in *factor analysis*.

Factor loading: the *regression coefficient* of a variable for the *linear model* that describes a *latent variable* or *factor* in *factor analysis*.

Factor score: a single score from an individual entity representing their performance on some *latent variable*. The score can be crudely conceptualized as follows: take an entity's score on each of the variables that make up the factor and multiply it by the corresponding *factor loading* for the variable, then add these values up (or average them).

Factor transformation matrix, Λ: a matrix used in *factor analysis*. It can be thought of as containing the angles through which factors are rotated in factor *rotation*.

Factorial ANOVA: an analysis of variance involving two or more *independent variables* or *predictors*.

Falsification: the act of disproving a hypothesis or theory.

Familywise error rate: the probability of making a *Type I error* in any family of tests when the null hypothesis is true in each case. The 'family of tests' can be loosely defined as a set of tests conducted on the same data set and addressing the same empirical question.

Fisher's exact test: Fisher's exact test (Fisher, 1922) is not so much a test as a way of computing the exact probability of a statistic. It was designed originally to overcome the problem that with small samples the sampling distribution of the chi-square statistic deviates substantially from a chi-square distribution. It should be used with small samples.

Fit: how sexually attractive you find a statistical test. Alternatively, it's the degree to which a statistical model is an accurate representation of some observed data. (Incidentally, it's just plain *wrong* to find statistical tests sexually attractive.)

Fixed coefficient: a coefficient or model parameter that is fixed; that is, it cannot vary over situations or contexts (cf. *random coefficient*).

Fixed effect: an effect in an experiment is said to be a fixed effect if all possible treatment conditions that a researcher is interested in are present in the experiment. Fixed effects can be generalized only to the situations in the experiment. For example, the effect is fixed if we say that we are interested only in the conditions that we had in our experiment (e.g., placebo, low dose and high dose) and we can generalize our findings only to the situation of a placebo, low dose and high dose.

Fixed intercept: a term used in *multilevel linear modeling* to denote when the intercept in the model is fixed. That is, it is not free to vary across different groups or contexts (cf. *random intercept*).

Fixed slope: a term used in *multilevel linear modeling* to denote when the slope of the model is fixed. That is, it is not free to vary across different groups or contexts (cf. *random slope*).

Fixed variable: a fixed variable is one that is not supposed to change over time (e.g., for most people their gender is a fixed variable – it never changes).

Frequency distribution: a graph plotting values of observations on the horizontal axis, and the frequency with which each value occurs in the data set on the vertical axis (a.k.a. *histogram*).

Friedman's ANOVA: a non-parametric test of whether more than two related groups differ. It is the non-parametric version of one-way *repeated-measures ANOVA*.

General linear model: a term to represent the fact that the *linear model* can encompass a range of different research designs such as multiple outcome variables (a.k.a. *MANOVA*), comparing means of categorical predictors (a.k.a. *t-test*, ANOVA), and including both categorical and continuous predictors (a.k.a. ANCOVA).

Generalization: the ability of a statistical model to say something beyond the set of observations that spawned it. If a model generalizes it is assumed that predictions from that model can be applied not just to the sample on which it is based, but to a wider population from which the sample came.

Glossary: a collection of grossly inaccurate definitions (written late at night when you really ought to be asleep) of things that you thought you understood until some evil book publisher forced you to try to define them.

Goodman and Kruskal's λ: measures the proportional reduction in error that is achieved when membership of a category of one variable is used to predict category membership of the other variable. A value of 1 means that one variable perfectly predicts the other, whereas a value of 0 indicates that one variable in no way predicts the other.

Goodness of fit: an index of how well a model fits the data from which it was generated. It's usually based on how well the data predicted by the model correspond to the data that were actually collected.

Grand mean: the *mean* of an entire set of observations.

Grand mean centering: grand mean *centering* means the transformation of a variable by taking each score and subtracting the mean of all scores (for that variable) from it (cf. *group mean centering*).

Grand variance: the *variance* within an entire set of observations.

Greenhouse–Geisser estimate: an estimate of the departure from *sphericity*. The maximum value is 1 (the data completely meet the assumption of sphericity) and the minimum is the *lower bound*. Values below 1 indicate departures from sphericity and are used to correct the *degrees of freedom* associated with the corresponding *F-statistics* by multiplying them by the value of the estimate. Some say the Greenhouse–Geisser correction is too conservative (strict) and recommend the *Huynh–Feldt correction* instead.

Group mean centering: group mean *centering* means the transformation of a variable by taking each score and subtracting from it the mean of the scores (for that variable) for the group to which that score belongs (cf. *grand mean centering*).

Growth curve: a curve that summarizes the change in some outcome over time. See *polynomial*.

HARKing: the practice in research articles of presenting a hypothesis that was made *after* data were collected as though it were made *before* data collection.

Harmonic mean: a weighted version of the *mean* that takes account of the relationship between variance and sample size. It is calculated by summing the reciprocal of all observations, then dividing by the number of observations. The reciprocal of the end product is the harmonic mean:

$$H = \frac{1}{\frac{1}{n} \sum_{i=1}^{n} \frac{1}{x_i}}$$

Hartley's F_{max}: also known as the *variance ratio*, is the ratio of the variances between the group with the biggest variance and the group with the smallest variance. This ratio is compared to critical values in a table published by Hartley as a test of *homogeneity of variance*. Some general rules are that with sample sizes (n) of 10 per group, an F_{max} less than 10 is more or less always going to be non-significant, with 15–20 per group the ratio needs to be less than about 5, and with samples of 30–60 the ratio should be below about 2 or 3.

Hat values: another name for *leverage*.

HE^{-1}: this is a matrix that is functionally equivalent to the *hypothesis SSCP* divided by the *error SSCP* in *MANOVA*. Conceptually it represents the ratio of *systematic* to *unsystematic variance*, so is a *multivariate* analogue of the *F-statistic*.

Helmert contrast: a non-orthogonal *planned contrast* that compares the mean of each condition (except the last) to the overall mean all subsequent conditions combined.

Heterogeneity of variance: the opposite of *homogeneity of variance*. This term means that the variance of one variable varies (i.e., is different) across levels of another variable.

Heteroscedasticity: the opposite of *homoscedasticity*. This occurs when the residuals at each level of the predictor variables(s) have unequal variances. Put another way, at each point along any predictor variable, the spread of residuals is different.

Hierarchical regression: a method of *multiple regression* in which the order in which predictors are entered into the regression model is determined by the researcher based on previous research: variables already known to be predictors are entered first, new variables are entered subsequently.

Histogram: a *frequency distribution*.

Homogeneity of covariance matrices: an assumption of some *multivariate* tests such as *MANOVA*. It is an extension of the *homogeneity of variance* assumption in *univariate* analyses. However, as well as assuming that *variances* for each *dependent variable* are the same across groups, it also assumes that relationships (*covariances*) between these dependent variables are roughly equal. It is tested by comparing the population *variance–covariance matrices* of the different groups in the analysis.

Homogeneity of regression slopes: an assumption of *analysis of covariance*. This is the assumption that the relationship between the *covariate* and *outcome variable* is constant across different treatment levels. So, if we had three treatment conditions, if there's a positive relationship between the covariate and the outcome in one group, we assume that there is a similar-sized positive relationship between the covariate and outcome in the other two groups too.

Homogeneity of variance: the assumption that the variance of one variable is stable (i.e., relatively similar) at all levels of another variable.

Homoscedasticity: an assumption in regression analysis that the residuals at each level of the predictor variable(s) have similar variances. Put another way, at each point along any predictor variable, the spread of residuals should be fairly constant.

Hosmer and Lemeshow's R_L^2: a version of the *coefficient of determination* for logistic regression. It is a fairly literal translation in that it is the *−2LL* for the model divided by the original *−2LL*, in other words, it's the ratio of what the model can explain compared to what there was to explain in the first place.

Hotelling–Lawley trace (T^2): a *test statistic* in *MANOVA*. It is the sum of the eigenvalues for each *discriminant function variate* of the data and so is conceptually the same as the *F-statistic* in *ANOVA*: it is the sum of the ratio of *systematic* and *unsystematic variance* (SS_M/SS_R) for each of the variates.

Huynh–Feldt estimate: an estimate of the departure from *sphericity*. The maximum value is 1 (the data completely meet the assumption of sphericity). Values below this indicate departures from sphericity and are used to correct the *degrees of freedom* associated with the corresponding *F-statistics* by multiplying them by the value of the estimate. It is less conservative than the *Greenhouse–Geisser estimate*, but some say it is too liberal.

Hypothesis: a proposed explanation for a fairly narrow phenomenon or set of observations. It is not a guess, but an informed, theory-driven attempt to explain what has been observed. A hypothesis cannot be tested directly but must first be operationalized as predictions about variables that can be measured (see *experimental hypothesis* and *null hypothesis*).

Hypothesis SSCP (H): the hypothesis sum of squares and cross-products matrix. This is a *sum of squares* and *cross-products matrix* for a predictive *linear model* fitted to *multivariate* data. It represents the *systematic variance* and is the multivariate equivalent of the *model sum of squares*.

Identity matrix: a square matrix (i.e., having the same number of rows and columns) in which the diagonal elements are equal to 1, and the off-diagonal elements are equal to 0. The following are all examples:

$$\begin{pmatrix} 1 & 0 \\ 0 & 1 \end{pmatrix} \quad \begin{pmatrix} 1 & 0 & 0 \\ 0 & 1 & 0 \\ 0 & 0 & 1 \end{pmatrix} \quad \begin{pmatrix} 1 & 0 & 0 & 0 \\ 0 & 1 & 0 & 0 \\ 0 & 0 & 1 & 0 \\ 0 & 0 & 0 & 1 \end{pmatrix}$$

Independence: the assumption that one data point does not influence another. When data come from people, it basically means that the behavior of one person does not influence the behavior of another.

Independent ANOVA: *analysis of variance* conducted on any design in which all *independent variables* or *predictors* have been manipulated using different participants (i.e., all data come from different entities).

Independent design: an experimental design in which different treatment conditions utilize different organisms (e.g., in psychology, this would mean using different people in different treatment conditions) and so the resulting data are independent (a.k.a. between-groups or between-subjects designs).

Independent errors: for any two observations in regression the *residuals* should be uncorrelated (or independent).

Independent factorial design: an experimental design incorporating two or more *predictors* (or *independent variables*) all of which have been manipulated using different participants (or whatever entities are being tested).

Independent *t*-test: a test using the *t-statistic* that establishes whether two means collected from independent samples differ significantly.

Independent variable: another name for a *predictor variable*. This name is usually associated with experimental methodology (which is the only time it makes sense) and is used because it is the variable that is manipulated by the experimenter and so its value does not depend on any other variables (just on the experimenter). I just use the term *predictor variable* all the time because the meaning of the term is not constrained to a particular methodology.

Index of mediation: a standardized measure of an *indirect effect*. In a mediation model, it is the *indirect effect* multiplied by the ratio of the standard deviation of the *predictor variable* to the standard deviation of the *outcome variable*.

Indirect effect: the effect of a *predictor variable* on an *outcome variable* through a *mediator* (cf. *direct effect*).

Informative prior distribution: in Bayesian statistics an informative *prior distribution* is a distribution representing your beliefs in a model parameter where the distribution narrows those beliefs to some degree. For example, a prior distribution that is normal with a peak at 5 and range from 2 to 8 would narrow your beliefs in a parameter such that you most strongly believe that its value will be 5, and you think it is impossible for the parameter to be less than 2 or greater than 8. As such, this distribution constrains your prior beliefs. Informative priors can vary from weakly informative (you are prepared to believe a wide range of values) to strongly informative (your beliefs are very constrained) (cf. *uninformative prior distribution*).

Interaction effect: the combined effect of two or more *predictor variables* on an *outcome variable*. It can used to gauge *moderation*.

Interaction graph: a graph showing the means of two or more *independent variables* in which means of one variable are shown at different levels of the other variable. Unusually the means are connected with lines or are displayed as bars. These graphs are used to help understand *interaction effects*.

Interquartile range: the limits within which the middle 50% of an ordered set of observations fall. It is the difference between the value of the *upper quartile* and *lower quartile*.

Interval variable: data measured on a scale along the whole of which intervals are equal. For example, people's ratings of this book on Amazon.com can range from 1 to 5; for these data to be interval it should be true that the increase in appreciation for this book represented by a change from 3 to 4 along the scale should be the same as the change in appreciation represented by a change from 1 to 2 or 4 to 5.

Intraclass correlation (ICC): a *correlation coefficient* that assesses the consistency between measures of the same class, that is, measures of the same thing (cf. *Pearson's correlation coefficient*, which measures the relationship between variables of a different class). Two common uses are in comparing paired data (such as twins) on the same measure, and assessing the consistency between judges' ratings of a set of objects. The calculation of these correlations depends on whether there is a measure of consistency (in which the order of scores from a source is considered but not the actual value around which the scores are anchored) or absolute agreement (in which both the order of scores and the relative values are considered), and whether the scores represent averages of many measures or just a single measure is required. This measure is also used in *multilevel linear models* to measure the dependency in data within the same context.

Jonckheere–Terpstra test: this statistic tests for an ordered pattern of medians across independent groups. Essentially it does the same thing as the *Kruskal–Wallis test* (i.e., test for a difference between the medians of the groups) but it incorporates information about whether the order of the groups is meaningful. As such, you should use this test when you expect the groups you're comparing to produce a meaningful order of medians.

Journal: in the context of academia a journal is a collection of articles on a broadly related theme, written by scientists, that report new data, new theoretical ideas or reviews/critiques of existing theories and data. Their main function is to induce learned helplessness in scientists through a complex process of self-esteem regulation using excessively harsh or complimentary peer feedback that has seemingly no obvious correlation with the actual quality of the work submitted.

Kaiser–Meyer–Olkin (KMO) measure of sampling adequacy: the KMO can be calculated for individual and multiple variables and represents the ratio of the squared correlation between variables to the squared *partial correlation* between variables. It varies between 0 and 1: a value of 0 means that the sum of partial correlations is large relative to the sum of correlations, indicating diffusion in the pattern of correlations (hence, *factor analysis* is likely to be inappropriate); a value close to 1 indicates that patterns of correlations are relatively compact and so factor analysis should yield distinct and reliable factors. Values between 0.5 and 0.7 are mediocre, values between 0.7 and 0.8 are good, values between 0.8 and 0.9 are great and values above 0.9 are superb (see Kaiser & Rice, 1974).

Kaiser's criterion: a method of *extraction* in *factor analysis* based on the idea of retaining factors with associated eigenvalues greater than 1. This method appears to be accurate when the number of variables in the analysis is less than 30 and the resulting *communalities* (after *extraction*) are all greater than 0.7 or when the sample size exceeds 250 and the average communality is greater than or equal to 0.6.

Kendall's tau: a non-parametric correlation coefficient similar to *Spearman's correlation coefficient*, but should be used in preference for a small data set with a large number of tied ranks.

Kendall's W: this is much the same as *Friedman's ANOVA* but is used specifically for looking at the agreement between raters. So, if, for example, we asked 10 different women to rate the attractiveness of Justin Timberlake, David Beckham and Brad Pitt we could use this test to look at the extent to which they agree. Kendall's *W* ranges from 0 (no agreement between judges) to 1 (complete agreement between judges).

Kolmogorov–Smirnov test: a test of whether a distribution of scores is significantly different from a *normal distribution*. A significant value indicates a deviation from normality, but this test is notoriously affected by large samples in which small deviations from normality yield significant results.

Kolmogorov–Smirnov Z: not to be confused with the *Kolmogorov-Smirnov test* that tests whether a sample comes from a normally distributed population. This tests whether two groups have been drawn from the same population (regardless of what that population may be). It does much the same as the *Mann–Whitney test* and *Wilcoxon rank-sum test*! This test tends to have better power than the Mann–Whitney test when sample sizes are less than about 25 per group.

Kruskal–Wallis test: a non-parametric test of whether more than two independent groups differ. It is the non-parametric version of one-way *independent ANOVA*.

Kurtosis: this measures the degree to which scores cluster in the tails of a frequency distribution. Kurtosis is calculated such that no kurtosis yields a value of 3. To make the measure more intuitive, SPSS Statistics (and some other packages) subtract 3 from the value so that no kurtosis is expressed as 0 and positive and negative kurtosis take on positive and negative values, respectively. A distribution with positive kurtosis (*leptokurtic*, kurtosis > 0) has too many scores in the tails and is too peaked, whereas a distribution with negative kurtosis (*platykurtic*, kurtosis < 0) has too few scores in the tails and is quite flat.

Latent variable: a variable that cannot be directly measured, but is assumed to be related to several variables that can be measured.

Leptokurtic: see *kurtosis*.

Levels of measurement: the relationship between what is being measured and the numbers obtained on a scale.

Levene's test: tests the hypothesis that the variances in different groups are equal (i.e., the difference between the variances is zero). It basically does a one-way ANOVA on the *deviations* (i.e., the absolute value of the difference between each score and the mean of its group). A significant result indicates that the variances are significantly different – therefore, the assumption of *homogeneity of variances* has been violated. When sample sizes are large, small differences in group variances can produce a significant Levene's test. I do not recommend using this test – instead interpret statistics that have been adjusted for the degree of heterogeneity in variances.

Leverage: leverage statistics (or hat values) gauge the influence of the observed value of the outcome variable over the predicted values. The average leverage value is $(k+1)/n$, in which k is the number of predictors in the model and n is the number of participants. Leverage values can lie between 0 (the case has no influence whatsoever) and 1 (the case has complete influence over prediction). If no cases exert undue influence over the model then we would expect all of the leverage values to be close to the average value. Hoaglin and Welsch (1978) recommend investigating cases with values greater than twice the average $(2(k+1)/n)$ and Stevens (2002) recommends using three times the average $(3(k+1)/n)$ as a cut-off point for identifying cases having undue influence.

Likelihood: the probability of obtaining a set of observations given the parameters of a model fitted to those observations. When using *Bayes' theorem* to test a hypothesis, the likelihood is the probability that the observed data could be produced given the hypothesis or model being considered, $p(\text{data}|\text{model})$. It is the inverse conditional probability of the *posterior probability*. See also *marginal likelihood*.

Linear model: a statistical model that is based upon an equation of the form $Y = BX + E$, in which Y is a vector containing scores from an outcome variable, B represents the *b*-values, X the predictor variables and E the error terms associated with each predictor. The equation can represent a solitary predictor variable (B, X and E are vectors) as in *simple regression* or multiple predictors (B, X and E are matrices) as in *multiple regression*. The key is the form of the model, which is linear (e.g., with a single predictor the equation is that of a straight line).

Line chart: a graph in which a summary statistic (usually the mean) is plotted on the *y*-axis against a categorical variable on the *x*-axis (this categorical variable could represent, for example, groups of people, different times or different experimental conditions). The value of the mean for each category is shown by a symbol, and means across categories are connected by a line. Different-colored lines may be used to represent levels of a second categorical variable.

Logistic regression: a version of *multiple regression* in which the outcome is a *categorical variable*. If the categorical variable has exactly two categories the analysis is called *binary logistic regression*, and when the outcome has more than two categories it is called *multinomial logistic regression*.

Log-likelihood: a measure of error or unexplained variation, in categorical models. It is based on summing the probabilities associated with the predicted and actual outcomes and is analogous to the *residual sum of squares* in multiple regression in that it is an indicator of how much unexplained information there is after the model has been fitted. Large values of the log-likelihood statistic indicate poorly fitting statistical models, because the larger value of the log-likelihood, the more unexplained observations there are. The log-likelihood is the logarithm of the *likelihood*.

Loglinear analysis: a procedure used as an extension of the *chi-square test* to analyze situations in which we have more than two *categorical variables* and we want to test for relationships between these variables. Essentially, a *linear model* is fitted to the data that predicts expected frequencies (i.e., the number of cases expected in a given category). In this respect it is much the same as *analysis of variance* but for entirely categorical data.

Long format data: data that are arranged such that scores on an outcome variable appear in a single column and rows represent a combination of the attributes of those scores – the entity from which the scores came, when the score was recorded, etc. In long format data, scores from a single entity can appear over multiple rows where each row represents a combination of the attributes of the score – for example, levels of an independent variable or time point at which the score was recorded (cf. *wide format data*).

Longitudinal research: a form of research in which you observe what naturally goes on in the world without directly interfering with it, by measuring several variables at multiple time points. See also *correlational research*, *cross-sectional research*.

Lower-bound estimate: the name given to the lowest possible value of the *Greenhouse–Geisser estimate* of *sphericity*. Its value is $1/(k-1)$, in which k is the number of treatment conditions.

Lower quartile: the value that cuts off the lowest 25% of the data. If the data are ordered and then divided into two halves at the median, then the lower quartile is the median of the lower half of the scores.

M-estimator: a robust measure of location. One example is the median. In some cases it is a measure of location computed after outliers have been removed; unlike a *trimmed mean*, the amount of trimming used to remove outliers is determined empirically.

Mahalanobis distances: these measure the influence of a case by examining the distance of cases from the mean(s) of the predictor variable(s). One needs to look for the cases with the highest values. It is not easy to establish a cut-off point at which to worry, although Barnett and Lewis (1978) have produced a table of critical values dependent on the number of predictors and the sample size. From their work it is clear that even with large samples ($N = 500$) and five predictors, values above 25 are cause for concern. In smaller samples ($N = 100$) and with fewer predictors (namely three) values greater than 15 are problematic, and in very small samples ($N = 30$) with only two predictors values greater

than 11 should be examined. However, for more specific advice, refer to Barnett and Lewis's (1978) table.

Main effect: the unique effect of a *predictor variable* (or *independent variable*) on an *outcome variable*. The term is usually used in the context of *ANOVA*.

Mann–Whitney test: a *non-parametric test* that looks for differences between two independent samples. That is, it tests whether the populations from which two samples are drawn have the same location. It is functionally the same as *Wilcoxon's rank-sum test*, and both tests are non-parametric equivalents of the *independent t-test*.

MANOVA: acronym for *multivariate analysis of variance*.

Marginal likelihood (evidence): when using *Bayes' theorem* to test a hypothesis, the marginal likelihood (sometimes called evidence) is the probability of the observed data, $p(\text{data})$. See also *likelihood*.

Matrix: a collection of numbers arranged in columns and rows. The values within a matrix are typically referred to as *components* or *elements*.

Mauchly's test: a test of the assumption of *sphericity*. If this test is significant then the assumption of *sphericity* has not been met and an appropriate correction must be applied to the *degrees of freedom* of the *F-statistic* in *repeated-measures ANOVA*. The test works by comparing the *variance–covariance matrix* of the data to an *identity matrix*; if the variance–covariance matrix is a scalar multiple of an *identity matrix* then sphericity is met.

Maximum-likelihood estimation: a way of estimating statistical parameters by choosing the parameters that make the data most likely to have happened. Imagine for a set of parameters that we calculated the probability (or likelihood) of getting the observed data; if this probability was high then these particular parameters yield a good fit of the data, but conversely if the probability was low, these parameters are a bad fit to our data. Maximum-likelihood estimation chooses the parameters that maximize the probability.

McNemar's test: this tests differences between two related groups (see *Wilcoxon signed-rank test* and *sign test*), when *nominal data* have been used. It's typically used when we're looking for changes in people's scores and it compares the proportion of people who changed their response in one direction (i.e., scores increased) to those who changed in the opposite direction (scores decreased). So, this test needs to be used when we've got two related dichotomous variables.

Mean: a simple statistical model of the center of a distribution of scores. A hypothetical estimate of the 'typical' score.

Mean squares: a measure of average variability. For every *sum of squares* (which measure the total variability) it is possible to create mean squares by dividing by the number of things used to calculate the sum of squares (or some function of it).

Measurement error: the discrepancy between the numbers used to represent the thing that we're measuring and the actual value of the thing we're measuring (i.e., the value we would get if we could measure it directly).

Median: the middle score of a set of ordered observations. When there is an even number of observations the median is the average of the two scores that fall either side of what would be the middle value.

Median test: a non-parametric test of whether samples are drawn from a population with the same median. So, in effect, it does the same thing as the *Kruskal–Wallis test*. It works on the basis of producing a contingency table that is split for each group into the number of scores that fall above and below the observed median of the entire data set. If the groups are from the same population then these frequencies would be expected to be the same in all conditions (about 50% above and about 50% below).

Mediation: perfect mediation occurs when the relationship between a *predictor variable* and an *outcome variable* can be completely explained by their relationships with a third variable. For example, taking a dog to work reduces work stress. This relationship is mediated by positive mood if (1) having a dog at work increases positive mood; (2) positive mood reduces work stress; and (3) the relationship between having a dog at work and work stress is reduced to zero (or at least weakened) when positive mood is included in the model.

Mediator: a variable that reduces the size and/or direction of the relationship between a *predictor variable* and an *outcome variable* (ideally to zero) and is associated statistically with both.

Meta-analyses: this is a statistical procedure for assimilating research findings. It is based on the simple idea that we can take effect sizes from individual studies that research the same question, quantify the observed effect in a standard way (using *effect sizes*) and then combine these effects to get a more accurate idea of the true effect in the population.

Method of least squares: a method of estimating parameters (such as the *mean* or a regression coefficient) that is based on minimizing the *sum of squared errors*. The parameter estimate will be the value, out of all of those possible, which has the smallest *sum of squared errors*.

Mixed ANOVA: *analysis of variance* used for a *mixed design*.

Mixed design: an experimental design incorporating two or more *predictors* (or *independent variables*) at least one of which has been manipulated using different participants (or whatever entities are being tested) and at least one of which has been manipulated using the same participants (or entities). Also known as a split-plot design because Fisher developed ANOVA for analyzing agricultural data involving 'plots' of land containing crops.

Mixed normal distribution: a normal-looking distribution that is contaminated by a small proportion of scores from a different distribution. These distributions are not normal and have too many scores in the tails (i.e., at the extremes). The effect of these heavy tails is to inflate the estimate of the population variance. This, in turn, makes significance tests lack power.

Mode: the most frequently occurring score in a set of data.

Model sum of squares: a measure of the total amount of variability for which a model can account. It is the difference between the *total sum of squares* and the *residual sum of squares*.

Moderation: moderation occurs when the relationship between two variables changes as a function of a third variable. For example, the relationship between watching horror films (predictor) and feeling scared at bedtime (outcome) might increase as a function of how vivid an imagination a person has (moderator).

Moderator: a variable that changes the size and/or direction of the relationship between two other variables.

Monte Carlo method: a term applied to the process of using data simulations to solve statistical problems. Its name comes from the use of Monte Carlo roulette tables to generate 'random' numbers in the pre-computer age. Karl Pearson, for example, purchased copies of *Le Monaco*, a weekly Paris periodical that published data from the Monte Carlo casinos' roulette wheels. He used these data as pseudo-random numbers in his statistical research.

Moses extreme reactions: a non-parametric test that compares the variability of scores in two groups, so it's a bit like a non-parametric *Levene's test*.

Multicollinearity: a situation in which two or more variables are very closely linearly related.

Multilevel linear model (MLM): a linear model (just like regression, ANCOVA, ANOVA, etc.) in which the hierarchical structure of the data is explicitly considered. In this analysis regression parameters can be fixed (as in regression and ANOVA) but also random (i.e., free to vary across different contexts at a higher level of the hierarchy). This means that for each regression parameter there is a fixed component but also an estimate of how much the parameter varies across contexts (see *fixed coefficient, random coefficient*).

Multimodal: description of a distribution of observations that has more than two *modes*.

Multinomial logistic regression: *logistic regression* in which the outcome variable has more than two categories.

Multiple R: the multiple correlation coefficient. It is the correlation between the observed values of an outcome and the values of the outcome predicted by a *multiple regression* model.

Multiple regression: an extension of *simple regression* in which an outcome is predicted by a linear combination of two or more predictor variables. The form of the model is:

$$Y_i = \left(b_0 + b_1 X_{1i} + b_2 X_{2i} + \cdots + b_n X_{ni}\right) + \varepsilon_i$$

in which the outcome is denoted by Y, and each predictor is denoted by X. Each predictor has a regression coefficient b associated with it, and b_0 is the value of the outcome when all predictors are zero.

Multivariate: means 'many variables' and is usually used when referring to analyses in which there is more than one *outcome variable* (*MANOVA, principal component analysis*, etc.).

Multivariate analyses of variance: a family of tests that extend the basic *analysis of variance* to situations in which more than one *outcome variable* has been measured.

Multivariate normality: an extension of a normal distribution to multiple variables. It is a *probability distribution* of a set of variables $v' = [v_1, v_2 ... v_n]$ given by:

$$f(v') = 2\pi^{-n/2} |\Sigma|^{-1/2} \exp \left\{ -\frac{1}{2} (v - \mu)' \Sigma^{-1} (v - \mu) \right\}$$

in which μ is the vector of means of the variables, and Σ is the *variance–covariance* matrix. If that made any sense to you then you're cleverer than I am.

Nagelkerke's R_N^2: a version of the *coefficient of determination* for logistic regression. It is a variation on *Cox and Snell's R_{CS}^2* which overcomes the problem that this statistic has of not being able to reach its maximum value.

Negative skew: see *skew*.

Nominal variable: where numbers merely represent names. For example, the numbers on sports players shirts: a player with the number 1 on her back is not necessarily worse than a player with a 2 on her back. The numbers have no meaning other than denoting the type of player (full back, center forward, etc.).

Noniles: a type of *quantile*; they are values that split the data into nine equal parts. They are comonly used in educational research.

Non-parametric tests: a family of statistical procedures that do not rely on the restrictive assumptions of parametric tests. In particular, they do not assume that the sampling distribution is normally distributed.

Normal distribution: a *probability distribution* of a random variable that is known to have certain properties. It is perfectly symmetrical (has a *skew* of 0), and has a *kurtosis* of 0.

Null hypothesis: the reverse of the *experimental hypothesis*, it states that your prediction is wrong and the predicted effect doesn't exist.

Numeric variables: variables involving numbers.

Oblique rotation: a method of *rotation* in *factor analysis* that allows the underlying factors to be correlated.

Odds: the probability of an event occurring divided by the probability of that event not occurring.

Odds ratio: the ratio of the *odds* of an event occurring in one group compared to another. So, for example, if the odds of dying after writing a glossary are 4, and the odds of dying after not writing a glossary are

0.25, then the odds ratio is $4/0.25 = 16$. This means that the *odds* of dying if you write a glossary are 16 times higher than if you don't. An odds ratio of 1 would indicate that the *odds* of a particular outcome are equal in both groups.

Omega squared: an *effect size* measure associated with ANOVA that is less biased than *eta squared*. It is a (sometimes hideous) function of the *model sum of squares* and the *residual sum of squares* and isn't actually much use because it measures the overall effect of the ANOVA and so can't be interpreted in a meaningful way. In all other respects it's great, though.

One-tailed test: a test of a directional hypothesis. For example, the hypothesis 'the longer I write this glossary, the more I want to place my editor's genitals in a starved crocodile's mouth' requires a one-tailed test because I've stated the direction of the relationship. I would generally advise against using them because of the temptation to interpret interesting effects in the opposite direction to that predicted. See also *two-tailed test*.

Open science: a movement to make the process, data and outcomes of scientific research freely available to everyone.

Ordinal variable: data that tell us not only that things have occurred, but also the order in which they occurred. These data tell us nothing about the differences between values. For example, gold, silver and bronze medals are ordinal: they tell us that the gold medallist was better than the silver medallist, but they don't tell us how much better (was gold a lot better than silver or were gold and silver very closely competed?).

Ordinary least squares (OLS): a method of *regression* in which the parameters of the model are estimated using the *method of least squares*.

Orthogonal: means perpendicular (at right angles) to something. It tends to be equated to *independence* in statistics because of the connotation that perpendicular *linear models* in geometric space are completely independent (one is not influenced by the other).

Orthogonal rotation: a method of *rotation* in *factor analysis* that keeps the underlying factors independent (i.e., not correlated).

Outcome variable: a variable whose values we are trying to predict from one or more *predictor variables*.

Outlier: an observation or observations very different from most others. Outliers bias statistics (e.g., the mean) and their standard errors and confidence intervals.

Overdispersion: when the observed variance is bigger than expected from the logistic regression model. Like leprosy, you don't want it.

p-curve: a curve summarizing the frequency distribution of *p*-values you'd expect to see in published research. On a graph that shows the value of the *p*-value on the horizontal axis against the frequency (or proportion) on the vertical axis, the p-curve is the line reflecting how frequently (or proportionately) each value of *p* should occur for a given *effect size*.

p-hacking: research practices that lead to selective reporting of significant *p*-values. Some examples of *p*-hacking are: (1) trying multiple analyses and reporting only the one that yields significant results; (2) stopping collecting data at a point other than when the predetermined sample size is reached; (3) deciding whether to include data based on the effect they have on the *p*-value; (4) including (or excluding) variables in an analysis based on how they affect the *p*-value; (5) measuring multiple *outcome* or *predictor variables* but reporting only those for which the effects are significant; (6) merging groups of variables or scores to yield significant results; and (7) transforming or otherwise manipulating, scores to yield significant *p*-values.

Paired-samples *t*-test: a test using the *t-statistic* that establishes whether two means collected from the same sample (or related observations) differ significantly.

Pairwise comparisons: comparisons of pairs of means.

Parameter: a very difficult thing to describe. When you fit a statistical model to your data, that model will consist of *variables* and parameters: variables are measured constructs that vary across entities in the sample, whereas parameters describe the relations between those variables in the population. In other words, they are constants believed to represent some fundamental truth about the measured variables. We use sample data to estimate the likely value of parameters because we don't have direct access to the population. Of course, it's not quite as simple as that.

Parametric test: a test that requires data from one of the large catalogue of distributions that statisticians have described. Normally this term is used for parametric tests based on the *normal distribution*, which require four basic assumptions that must be met for the test to be accurate: a normally distributed sampling distribution (see *normal distribution*), *homogeneity of variance*, *interval* or *ratio data*, and *independence*.

Parsimony: in a scientific context, parsiomony refers to the idea that simpler explanations of a phenomenon are preferable to complex ones. This idea relates to Ockham's (or Occam's if you prefer) razor, which is a phrase referring to the principle of 'shaving' away unnecessary assumptions or explanations to produce less complex theories. In statistical terms, parsimony tends to refer to a general heuristic that models be kept as simple as possible – in other words, not including variables that don't have real explanatory benefit.

Part correlation: another name for a *semi-partial correlation*.

Partial correlation: a measure of the relationship between two variables while the effect that one or more additional variables has on both.

Partial eta squared (partial η²): a version of *eta squared* that is the proportion of variance that a variable explains when excluding other variables in the analysis. Eta squared is the proportion of total variance explained by a variable, whereas partial eta squared is the proportion of variance that a variable explains that is not explained by other variables.

Partial out: to partial out the effect of a variable is to remove the variance that the variable shares with other variables in the analysis before looking at their relationships (see *partial correlation*).

Pattern matrix: a matrix in *factor analysis* containing the *regression coefficients* for each variable on each *factor* in the data. See also *structure matrix*.

Pearson's correlation coefficient: Pearson's product-moment correlation coefficient, to give it its full name, is a *standardized* measure of the strength of relationship between two variables. It can take any value from –1 (as one variable changes, the other changes in the opposite direction by the same amount), through 0 (as one variable changes the other doesn't change at all), to +1 (as one variable changes, the other changes in the same direction by the same amount).

Peer Reviewers' Openness Initiative: an initiative to get scientists to commit to the principles of *open science* when they act as expert reviewers for journals. Signing up is a pledge to review submissions only if the data, stimuli, materials, analysis scripts and so on are made publically available (unless there is a good reason, such as a legal requirement, not to).

Percentiles: a type of *quantile*; they are values that split the data into 100 equal parts.

Perfect collinearity: exists when at least one predictor in a *regression model* is a perfect linear combination of the others (the simplest example being two predictors that are perfectly correlated – they have a correlation coefficient of 1).

Phi: a measure of the strength of association between two *categorical variables*. Phi is used with 2–2 *contingency tables* (tables which have two categorical variables and each variable has only two categories). Phi is a variant of the *chi-square test*, χ²:

$$\phi = \sqrt{\frac{\chi^2}{N}}$$

in which *N* is the total number of observations.

Pillai–Bartlett trace (*V*): a *test statistic* in *MANOVA*. It is the sum of the proportion of explained variance on the *discriminant function variates* of the data. As such, it is similar to the ratio of SS_M/SS_T.

Pilot fish (*Naucrates ductor*): a carnivorous fish in the Carangidae family known for congregating around larger more impressive beings (e.g., sharks) and feeding parasitically from their bodies. A bit like Courtney Love.

Planned comparisons: another name for *planned contrasts*.

Planned contrasts: a set of comparisons between group means that are constructed before any data are collected. These are theory-led comparisons and are based on the idea of partitioning the variance created by the overall effect of group differences into gradually smaller portions of variance. These tests have more power than *post hoc tests*.

Platykurtic: see *kurtosis*.

Point-biserial correlation: a standardized measure of the strength of relationship between two variables when one of the two variables is *dichotomous*. The point-biserial correlation coefficient is used when the dichotomy is a discrete or true, dichotomy (i.e., one for which there is no underlying continuum between the categories). An example of this is pregnancy: you can be either pregnant or not, there is no in between.

Polychotomous logistic regression: another name for *multinomial logistic regression*.

Polynomial: a posh name for a *growth curve* or trend over time. If *time* is our predictor variable, then any polynomial is tested by including a variable that is the predictor to the power of the order of polynomial that we want to test: a linear trend is tested by *time* alone, a quadratic or second-order polynomial is tested by including a predictor that is *time*^{2}, for a fifth-order polynomial we need a predictor of *time*^{5} and for an *n*th-order polynomial we would have to include *time*^{n} as a predictor.

Polynomial contrast: a contrast that tests for trends in the data. In its most basic form it looks for a linear trend (i.e., that the group means increase proportionately).

Population: in statistical terms this usually refers to the collection of units (be they people, plankton, plants, cities, suicidal authors, etc.) to which we want to generalize a set of findings or a statistical model.

Positive skew: see *skew*.

Post hoc tests: a set of comparisons between group means that were not thought of before data were collected. Typically these tests involve comparing the means of all combinations of pairs of groups. To compensate for the number of tests conducted, each test uses a strict criterion for significance. As such, they tend to have less power than *planned contrasts*. They are usually used for exploratory work for which no firm hypotheses were available on which to base planned contrasts.

Posterior distribution: a distribution of *posterior probabilities*. This distribution should contain our subjective beliefs about a parameter or hypothesis after considering the data. The posterior distribution can be used to ascertain a value of the posterior probability (usually by examining some measure of where the peak of the distribution lies or a *credible interval*).

Posterior odds: the ratio of *posterior probability* for one hypothesis to another. In Bayesian hypothesis testing the posterior odds are the ratio of the probability of the alternative hypothesis given the data, *p*(alternative|data), to the probability of the null hypothesis given the data, *p*(null|data).

Posterior probability: when using *Bayes' theorem* to test a hypothesis, the posterior probability is our belief in a hypothesis or model *after* we have considered the data, *p*(model|data). This is the value that we are usually interested in knowing. It is the inverse conditional probability of the *likelihood*.

Power: the ability of a test to detect an effect of a particular size (a value of 0.8 is a good level to aim for).

P-P plot: short for 'probability–probability plot'. A graph plotting the cumulative probability of a variable against the cumulative probability of a particular distribution (often a normal distribution). Like a *Q-Q plot*, if values fall on the diagonal of the plot then the variable shares the same distribution as the one specified. Deviations from the diagonal show deviations from the distribution of interest.

Practice effect: refers to the possibility that participants' performance in a task may be influenced (positively or negatively) if they repeat the task because of familiarity with the experimental situation and/or the measures being used.

Predictive validity: a form of *criterion validity* where there is evidence that scores from an instrument predict external measures (recorded at a different point in time) conceptually related to the measured construct.

Predicted value: the value of an outcome variable based on specific values of the predictor variable or variables being placed into a statistical model.

Predictor variable: a variable that is used to try to predict values of another variable known as an *outcome variable*.

Pre-registration: a term referring to the practice of making all aspects of your research process (rationale, hypotheses, design, data processing strategy, data analysis strategy) publically available before data collection begins. This can be done in a *registered report* in an academic journal or more informally (e.g., on a public website such as the Open Science Framework). The aim is to encourage adherence to an agreed research protocol, thus discouraging threats to the validity of scientific results such as *researcher degrees of freedom*.

Principal component analysis (PCA): a *multivariate* technique for identifying the linear components of a set of variables.

Prior distribution: a distribution of *prior probabilities*. This distribution should contain our subjective beliefs about a parameter or hypothesis before or prior to, considering the data. The prior distribution can be an *informative prior* or an *uninformative prior*.

Prior odds: the ratio of the probability of one hypothesis/model to a second. In Bayesian hypothesis testing, the prior odds are the probability of the *alternative hypothesis*, *p*(alternative), divided by the probability of the *null hypothesis*, *p*(null). The prior odds should reflect your belief in the alternative hypothesis relative to the null *before* you look at the data.

Prior probability: when using *Bayes' theorem* to test a hypothesis, the prior probability is our belief in a hypothesis or model before or prior to, considering the data, *p*(model). See also *posterior probability*, *likelihood*, *marginal likelihood*.

Probability density function (PDF): the function that describes the probability of a random variable taking a certain value. It is the mathematical function that describes the *probability distribution*.

Probability distribution: a curve describing an idealized *frequency distribution* of a particular variable from which it is possible to ascertain the probability with which specific values of that variable will occur. For categorical variables it is simply a formula yielding the probability with which each category occurs.

Promax: a method of *oblique rotation* that is computationally faster than *direct oblimin* and so useful for large data sets.

Publication bias: the fact that articles published in scientific journals tend to over-represent positive findings. This can be because (1) non-significant findings are less likely to be published; (2) scientists don't submit their non-significant results to journals; (3) scientists selectively report their results to focus on significant findings and exclude non-significant ones; and (4) scientists capitalize on *researcher degrees of freedom* to shed their results in the most favourable light possible.

Q-Q plot: short for 'quantile–quantile plot'. A graph plotting the *quantiles* of a variable against the quantiles of a particular distribution (often a normal distribution). Like a *P-P plot*, if values fall on the diagonal of the plot then the variable shares the same distribution as the one specified. Deviations from the diagonal show deviations from the distribution of interest.

Quadratic trend: if the means in ordered conditions are connected with a line then a quadratic trend is shown by one change in the direction of this line (e.g., the line is curved in one place); the line is, therefore, U-shaped. There must be at least three ordered conditions.

Qualitative methods: extrapolating evidence for a theory from what people say or write (contrast with *quantitative methods*).

Quantiles: values that split a data set into equal portions. *Quartiles*, for example, are a special case of quantiles that split the data into four equal parts. Similarly, *percentiles* are points that split the data into 100 equal parts and *noniles* are points that split the data into nine equal parts (you get the general idea).

Quantitative methods: inferring evidence for a theory through measurement of variables that produce numeric outcomes (cf. *qualitative methods*).

Quartic trend: if the means in ordered conditions are connected with a line then a quartic trend is shown by three changes in the direction of this line. There must be at least five ordered conditions.

Quartiles: a generic term for the three values that cut an ordered data set into four equal parts. The three quartiles are known as the first or *lower quartile*, the second quartile (or *median*) and the third or *upper quartile*.

Quartimax: a method of *orthogonal rotation*. It attempts to maximize the spread of factor loadings for a variable across all *factors*. This often results in lots of variables loading highly on a single *factor*.

Random coefficient: a coefficient or model parameter that is free to vary over situations or contexts (cf. *fixed coefficient*).

Random effect: an effect is said to be random if the experiment contains only a sample of possible treatment conditions. Random effects can be generalized beyond the treatment conditions in the experiment. For example, the effect is random if we say that the conditions in our experiment (e.g., placebo, low dose and high dose) are only a sample of possible conditions (perhaps we could have tried a very high dose). We can generalize this random effect beyond just placebos, low doses and high doses.

Random intercept: a term used in *multilevel linear modeling* to denote when the intercept in the model is free to vary across different groups or contexts (cf. *fixed intercept*).

Random slope: a term used in *multilevel linear modeling* to denote when the slope of the model is free to vary across different groups or contexts (cf. *fixed slope*).

Random variable: a random variable is one that varies over time (e.g., your weight is likely to fluctuate over time).

Random variance: variance that is unique to a particular variable but not reliably so.

Randomization: the process of doing things in an unsystematic or random way. In the context of experimental research the word usually applies to the random assignment of participants to different treatment conditions.

Range: the range of scores is the value of the smallest score subtracted from the highest score. It is a measure of the dispersion of a set of scores. See also *variance*, *standard deviation* and *interquartile range*.

Ranking: the process of transforming raw scores into numbers that represent their position in an ordered list of those scores. The raw scores are ordered from lowest to highest and the lowest score is assigned a rank of 1, the next highest score is assigned a rank of 2, and so on.

Ratio variable: an *interval variable* but with the additional property that ratios are meaningful. For example, people's ratings of this book on Amazon.com can range from 1 to 5; for these data to be ratio not only must they have the properties of *interval variables*, but in addition a rating of 4 should genuinely represent someone who enjoyed this book twice as much as someone who rated it as 2. Likewise, someone who rated it as 1 should be half as impressed as someone who rated it as 2.

Registered report: an article in a journal usually outlining an intended research process (rationale, hypotheses, design, data processing strategy, data analysis strategy). The report is reviewed by relevant expert scientists, ensuring that authors get useful feedback before data collection. If the protocol is accepted by the journal editor it typically comes with a guarantee to publish the findings no matter what they are, thus reducing *publication bias* and discouraging *researcher degrees of freedom* aimed at achieving significant results.

Regression coefficient: see b_i and β_i.

Regression model: see *multiple regression* and *simple regression*.

Regression line: a line on a scatterplot representing the *regression model* of the relationship between the two variables plotted.

Related design: another name for a *repeated-measures design*.

Related factorial design: an experimental design incorporating two or more *predictors* (or *independent variables*) all of which have been manipulated using the same participants (or whatever entities are being tested).

Reliability: the ability of a measure to produce consistent results when the same entities are measured under different conditions.

Repeated contrast: a non-orthogonal *planned contrast* that compares the mean in each condition (except the first) to the mean of the preceding condition.

Repeated-measures ANOVA: an *analysis of variance* conducted on any design in which the *independent variable* (*predictor*) or *variables* (*predictors*) have all been measured using the same participants in all conditions.

Repeated-measures design: an experimental design in which different treatment conditions utilize the same organisms (i.e., in psychology, this would mean the same people take part in all experimental conditions) and so the resulting data are related (a.k.a. related design or within-subject design).

Researcher degrees of freedom: the analytic decisions a researcher makes that potentially influence the results of the analysis. Some examples are: when to stop data collection, which control variables to include in the statistical model, and whether to exclude cases from the analysis.

Residual: the difference between the value a model predicts and the value observed in the data on which the model is based. Basically, an error. When the residual is calculated for each observation in a data set the resulting collection is referred to as the *residuals*.

Residuals: see *Residual*.

Residual sum of squares: a measure of the variability that cannot be explained by the model fitted to the data. It is the total squared *deviance* between the observations, and the value of those observations predicted by whatever model is fitted to the data.

Reverse Helmert contrast: another name for a *difference contrast*.

Robust test: a term applied to a family of procedures to estimate statistics that are reliable even when the normal assumptions of the statistic are not met.

Rotation: a process in *factor analysis* for improving the interpretability of factors. In essence, an attempt is made to transform the *factors* that emerge from the analysis in such a way as to maximize *factor loadings* that are already large, and minimize factor loadings that are already small. There are two general approaches: *orthogonal rotation* and *oblique rotation*.

Roy's largest root: a *test statistic* in *MANOVA*. It is the eigenvalue for the first *discriminant function variate* of a set of observations. So, it is the same as the *Hotelling–Lawley trace*, but for the first variate only. It represents the proportion of explained variance to unexplained variance (SS_M/SS_R) for the first discriminant function.

Sample: a smaller (but hopefully representative) collection of units from a *population* used to determine truths about that population (e.g., how a given population behaves in certain conditions).

Sampling distribution: the *probability distribution* of a statistic. We can think of this as follows: if we take a *sample* from a *population* and calculate some statistic (e.g., the *mean*), the value of this statistic will depend somewhat on the sample we took. As such the statistic will vary slightly from sample to sample. If, hypothetically, we took lots and lots of samples from the population and calculated the statistic of interest we could create a frequency distribution of the values we get. The resulting distribution is what the sampling distribution represents: the distribution of possible values of a given statistic that we could expect to get from a given population.

Sampling variation: the extent to which a statistic (the mean, median, t, F, etc.) varies in samples taken from the same population.

Saturated model: a model that perfectly fits the data and, therefore, has no error. It contains all possible *main effects* and *interactions* between variables.

Scatterplot: a graph that plots values of one variable against the corresponding values of another variable (and the corresponding values of a third variable can also be included on a 3-D scatterplot).

Scree plot: a graph plotting each *factor* in a *factor analysis* (X-axis) against its associated eigenvalue (Y-axis). It shows the relative importance of each factor. This graph has a very characteristic shape (there is a sharp descent in the curve followed by a tailing off), and the point of inflexion of this curve is often used as a means of *extraction*. With a sample of more than 200 participants, this provides a fairly reliable criterion for *extraction* (Stevens, 2002).

Second quartile: another name for the *median*.

Semi-partial correlation: a measure of the relationship between two variables while adjusting for the effect that one or more additional variables have on one of those variables. If we call our variables x and y, it gives us a measure of the variance in y that x alone shares.

Shapiro–Wilk test: a test of whether a distribution of scores is significantly different from a *normal distribution*. A significant value indicates a deviation from normality, but this test is notoriously affected by large samples in which small deviations from normality yield significant results.

Shrinkage: the loss of predictive power of a regression model if the model has been derived from the population from which the sample was taken, rather than the sample itself.

Šidák correction: a slightly less conservative variant of a *Bonferroni correction*.

Sign test: tests whether two related samples are different. It does the same thing as the *Wilcoxon signed-rank test*. Differences between the conditions are calculated and the sign of this difference (positive or negative) is analyzed because it indicates the direction of differences. The magnitude of change is completely ignored (unlike in Wilcoxon's test, where the rank tells us something about the relative magnitude of change), and for this reason it lacks *power*. However, its computational simplicity makes it a nice party trick if ever anyone drunkenly accosts you needing some data quickly analyzed without the aid of a computer – doing a sign test in your head really impresses people. Actually it doesn't, they just think you're a sad gimboid.

Simple contrast: a non-orthogonal *planned contrast* that compares the mean in each condition to the mean of either the first or last condition, depending on how the contrast is specified.

Simple effects analysis: this analysis looks at the effect of one *independent variable* (categorical *predictor variable*) at individual levels of another *independent variable*.

Simple regression: a *linear model* in which one variable or outcome is predicted from a single predictor variable. The model takes the form:

$$Y_i = (b_0 + b_1 X_{1i}) + \varepsilon_i$$

in which Y is the outcome variable, X is the predictor, b_1 is the regression coefficient associated with the predictor and b_0 is the value of the outcome when the predictor is zero.

Simple slopes analysis: an analysis that looks at the relationship (i.e., the *simple regression*) between a *predictor variable* and an *outcome variable* at low, mean and high levels of a third (*moderator*) variable.

Singularity: a term used to describe variables that are perfectly correlated (i.e., the *correlation coefficient* is 1 or –1).

Skew: a measure of the symmetry of a *frequency distribution*. Symmetrical distributions have a skew of 0. When the frequent scores are clustered at the lower end of the distribution and the tail points towards the higher or more positive scores, the value of skew is positive. Conversely, when the frequent scores are clustered at the higher end of the distribution and the tail points towards the lower or more negative scores, the value of skew is negative.

Smartreader: a free piece of software that can be downloaded from the IBM SPSS website and enables people who do not have *SPSS Statistics* installed to open and view SPSS output files.

Sobell test: a significance test of *mediation*. It tests whether the relationship between a *predictor variable* and an *outcome variable* is significantly reduced when a mediator is included in the model. It tests the *indirect effect* of the predictor on the outcome.

Spearman's correlation coefficient: a standardized measure of the strength of relationship between two variables that does not rely on the assumptions of a *parametric test*. It is *Pearson's correlation coefficient* performed on data that have been converted into ranked scores.

Sphericity: a less restrictive form of *compound symmetry* which assumes that the variances of the differences between data taken from the same participant (or other entity being tested) are equal. This assumption is most commonly found in *repeated-measures ANOVA* but applies only where there are more than two points of data from the same participant. See also *Greenhouse–Geisser correction*, *Huynh–Feldt correction*.

Split-half reliability: a measure of *reliability* obtained by splitting items on a measure into two halves (in some random fashion) and obtaining a score from each half of the scale. The correlation between the two scores, corrected to take account of the fact the correlations are based on only half of the items, is used as a measure of reliability. There are two popular ways to do this. Spearman (1910) and Brown (1910) developed a formula that takes no account of the standard deviation of items:

$$r_{sh} = \frac{2r_{12}}{1 + r_{12}}$$

in which r_{12} is the correlation between the two halves of the scale. Flanagan (1937) and Rulon (1939), however, proposed a measure that does account for item variance:

$$r_{sh} = \frac{4r_{12} \times s_1 \times s_2}{s_T^2}$$

in which, s_1 and s_2 are the standard deviations of each half of the scale, and S_T^2 is the variance of the whole test. See Cortina (1993) for more details.

Square matrix: a *matrix* that has an equal number of columns and rows.

Standard deviation: an estimate of the average variability (spread) of a set of data measured in the same units of measurement as the original data. It is the square root of the *variance*.

Standard error: the standard deviation of the *sampling distribution* of a statistic. For a given statistic (e.g., the *mean*) it tells us how much variability there is in this statistic across *samples* from the same *population*. Large values, therefore, indicate that a statistic from a given sample may not be an accurate reflection of the population from which the sample came.

Standard error of differences: if we were to take several pairs of samples from a population and calculate their means, then we could also calculate the difference between their means. If we plotted these differences between sample means as a *frequency distribution*, we would have the *sampling distribution* of differences. The standard deviation of this sampling distribution is the standard error of differences. As such it is a measure of the variability of differences between sample means.

Standard error of the mean (SE): the *standard error* associated with the mean. Did you really need a glossary entry to work that out?

Standardization: the process of converting a variable into a standard unit of measurement. The unit of measurement typically used is *standard deviation* units (see also *z-scores*). Standardization allows us to compare data when different units of measurement have been used (we could compare weight measured in kilograms to height measured in inches).

Standardized: see *standardization*.

Standardized DFBeta: a *standardized* version of *DFBeta*. These standardized values are easier to use than DFBeta because universal cut-off points can be applied. Stevens (2002) suggests looking at cases with absolute values greater than 2.

Standardized DFFit: a *standardized* version of *DFFit*.

Standardized residuals: the *residuals* of a model expressed in standard deviation units. Standardized residuals with an absolute value greater than 3.29 (actually, we usually just use 3) are cause for concern because in an average sample a value this high is unlikely to happen by chance; if more than 1% of our observations have standardized residuals with an absolute value greater than 2.58 (we usually just say 2.5) there is evidence that the level of error within our model is unacceptable (the model is a fairly poor fit to the sample data); and if more than 5% of observations have standardized residuals with an absolute value greater than 1.96 (or 2 for convenience) then there is also evidence that the model is a poor representation of the actual data.

Stepwise regression: a method of *multiple regression* in which variables are entered into the model based on a statistical criterion (the *semi-partial correlation* with the *outcome variable*). Once a new variable is entered into the model, all variables in the model are assessed to see whether they should be removed.

String variables: variables involving words (i.e., letter strings). Such variables could include responses to open-ended questions such as 'How much do you like writing glossary entries?'; the response might be 'About as much as I like placing my ballbag on hot coals'.

Structure matrix: a matrix in *factor analysis* containing the *correlation coefficients* for each variable on each *factor* in the data. When *orthogonal rotation* is used this is the same as the *pattern matrix*, but when oblique rotation is used these matrices are different.

Studentized deleted residual: a measure of the influence of a particular case of data. This is a standardized version of the *deleted residual*.

Studentized residuals: a variation on *standardized residuals*. Studentized residuals are the *unstandardized residual* divided by an estimate of its standard deviation that varies point by point. These residuals have the same properties as the *standardized residuals* but usually provide a more precise estimate of the error variance of a specific case.

Sum of squared errors: another name for the *sum of squares*.

Sum of squares (SS): an estimate of total variability (spread) of a set of observations around a parameter (such as the *mean*). First the *deviance* for each score is calculated, and then this value is squared. The SS is the sum of these squared deviances.

Sum of squares and cross-products matrix (SSCP matrix): a *square matrix* in which the diagonal elements represent the *sum of squares* for a particular variable, and the off-diagonal elements represent the *cross-products* between pairs of variables. The SSCP matrix is basically the same as the *variance–covariance matrix*, except that the SSCP matrix expresses variability and between-variable relationships as total values, whereas the variance–covariance matrix expresses them as average values.

Suppressor effect: a situation where a predictor has a significant effect but only when another variable is held constant.

Syntax: predefined written commands that instruct SPSS Statistics what you would like it to do (writing 'bugger off and leave me alone' doesn't seem to work …).

Systematic variation: variation due to some genuine effect (be that the effect of an experimenter doing something to all of the participants in one sample but not in other samples or natural variation between sets of variables). We can think of this as variation that can be explained by the model that we've fitted to the data.

t-statistic: a *test statistic* with a known *probability distribution* (the *t*-distribution). In the context of the *linear model* it is used to test whether a *b*-value is significantly different from zero; in the context of experimental work this *b*-value represents the difference between two means and so *t* is a test of whether the difference between those means is significantly different from zero. See also *paired-samples t-test* and *independent t-test*.

Tertium quid: the possibility that an apparent relationship between two variables is actually caused by the effect of a third variable on them both (often called the *third-variable problem*).

Test of excess success (TES): a procedure designed for identifying sets of results within academic articles that are 'too good to be true'. For an article reporting multiple scientific studies examining the same effect, the test computes (based on the size of effect being measured and sample size of the studies) the probability that you would get significant results for all of the studies. If this probability is low it is highly unlikely that the researcher would get these results and the results appear 'too good to be true', implying *p-hacking* (Francis, 2013). It is noteworthy that the TES is not universally accepted as testing what it sets out to test (e.g., Morey, 2013).

Test–retest reliability: the ability of a measure to produce consistent results when the same entities are tested at two different points in time.

Test statistic: a statistic for which we know how frequently different values occur. The observed value of such a statistic is typically used to test *hypotheses*.

Theory: although it can be defined more formally, a theory is a hypothesized general principle or set of principles that explain known findings about a topic and from which new hypotheses can be generated. Theories have typically been well-substantiated by repeated testing.

Tolerance: tolerance statistics measure *multicollinearity* and are simply the reciprocal of the *variance inflation factor* (1/VIF). Values below 0.1 indicate serious problems, although Menard (1995) suggests that values below 0.2 are worthy of concern.

Total SSCP (*T*): the total sum of squares and cross-products matrix. This is a *sum of squares and cross-products matrix* for an entire set of observations. It is the *multivariate* equivalent of the *total sum of squares*.

Total sum of squares: a measure of the total variability within a set of observation. It is the total squared *deviance* between each observation and the overall mean of all observations.

Transformation: the process of applying a mathematical function to all observations in a data set, usually to correct some distributional abnormality such as *skew* or *kurtosis*.

Trimmed mean: a statistic used in many *robust tests*. It is a mean calculated using trimmed data. For example, a 20% trimmed mean is a mean calculated after the top and bottom 20% of ordered scores have been removed. Imagine we had 20 scores representing the annual income of students (in thousands), rounded to the nearest thousand: 0, 1, 2, 2, 3, 3, 3, 3, 3, 4, 4, 4, 4, 4, 4, 4, 4, 4, 4, 40. The mean income is 5 (£5000), which is biased by an outlier. A 10% trimmed mean will remove 10% of scores from the top and bottom of ordered scores before the mean is calculated. With 20 scores, removing 10% of scores involves removing the top and bottom two scores. This gives us: 2, 2, 3, 3, 3, 3, 3, 4, 4, 4, 4, 4, 4, 4, 4, 4, the mean of which is 3.44. The mean depends on a symmetrical distribution to be accurate, but a trimmed mean produces accurate results even when the distribution is not symmetrical. There are more complex examples of robust methods such as the *bootstrap*.

Two-tailed test: a test of a non-directional hypothesis. For example, the hypothesis 'writing this glossary has some effect on what I want to do with my editor's genitals' requires a two-tailed test because it doesn't suggest the direction of the relationship. See also *one-tailed test*.

Type I error: occurs when we believe that there is a genuine effect in our population, when in fact there isn't.

Type II error: occurs when we believe that there is no effect in the population, when in fact there is.

Uninformative prior distribution: in Bayesian statistics an uninformative *prior distribution* is a distribution representing your beliefs in a model parameter where the distribution assigns equal probability to all values of the model/parameter. For example, a prior distribution that is uniform across all potential values of a parameter suggests that you are prepared to believe that the parameter can take on any value with equal probability. As such, this distribution does not constrain your prior beliefs (cf. *informative prior*).

Unique factor: a *factor* that affects only one of many measured *variables* and, therefore, cannot explain the *correlations* between those variables.

Unique variance: variance that is specific to a particular variable (i.e., is not shared with other variables). We tend to use the term 'unique variance' to refer to variance that can be reliably attributed to only one measure, otherwise it is called *random variance*.

Univariate: means 'one variable' and is usually used to refer to situations in which only one *outcome variable* has been measured (*ANOVA*, *t-tests*, *Mann–Whitney tests*, etc.).

Unstructured: a covariance structure used in *multilevel linear modeling*. This covariance structure is completely general. Covariances are assumed to be completely unpredictable: they do not conform to a systematic pattern.

Unstandardized residuals: the *residuals* of a model expressed in the units in which the original outcome variable was measured.

Unsystematic variation: this is variation that isn't due to the effect in which we're interested (so could be due to natural differences between people in different samples such as differences in intelligence or motivation). We can think of this as variation that can't be explained by whatever model we've fitted to the data.

Upper quartile: the value that cuts off the highest 25% of ordered scores. If the scores are ordered and then divided into two halves at the median, then the upper quartile is the median of the top half of the scores.

Validity: evidence that a study allows correct inferences about the question it was aimed to answer or that a test measures what it set out to measure conceptually. See also *content validity*, *criterion validity*.

Variables: anything that can be measured and can differ across entities or across time.

Variable view: there are two ways to view the contents of the *data editor* window. The variable view allows you to define properties of the variables for which you wish to enter data. See also *data view*.

Variance: an estimate of average variability (spread) of a set of data. It is the sum of squares divided by the number of values on which the sum of squares is based minus 1.

Variance components: a covariance structure used in *multilevel linear modeling*. This covariance structure is very simple and assumes that all random effects are independent and that the variances of random effects are assumed to be the same and sum to the variance of the outcome variable.

Variance–covariance matrix: a square matrix (i.e., same number of columns and rows) representing the variables measured. The diagonals represent the *variances* within each variable, whereas the off-diagonals represent the *covariances* between pairs of variables.

Variance inflation factor (VIF): a measure of *multicollinearity*. The VIF indicates whether a predictor has a strong linear relationship with the other predictor(s). Myers (1990) suggests that a value of 10 is a good value at which to worry. Bowerman and O'Connell (1990) suggest that if the average VIF is greater than 1, then multicollinearity may be biasing the regression model.

Variance ratio: see *Hartley's F_{max}*.

Variance sum law: states that the variance of a difference between two independent variables is equal to the sum of their variances.

Varimax: a method of *orthogonal rotation*. It attempts to maximize the dispersion of *factor loadings* within *factors*. Therefore, it tries to load a smaller number of variables highly onto each factor, resulting in more interpretable clusters of factors.

VIF: see *variance inflation factor*.

Wald statistic: a *test statistic* with a known *probability distribution* (a *normal distribution* or a *chi-square distribution* when squared) that is used to test whether the *b* coefficient for a predictor in a *logistic regression* model is significantly different from zero. It is analogous to the *t-statistic* in a *regression model* in that it is simply the *b* coefficient divided by its standard error. The Wald statistic is inaccurate when the regression coefficient (*b*) is large, because the standard error tends to become inflated, resulting in the Wald statistic being underestimated.

Wald–Wolfowitz runs: another variant on the *Mann–Whitney test*. Scores are rank-ordered as in the Mann–Whitney test, but rather than analyzing the ranks, this test looks for 'runs' of scores from the same group within the ranked order. Now, if there's no difference between groups then obviously ranks from the two groups should be randomly interspersed. However, if the groups are different then one should see more ranks from one group at the lower end, and more ranks from the other group at the higher end. By looking for clusters of scores in this way the test can determine if the groups differ.

Weight: a number by which something (usually a variable in statistics) is multiplied. The weight assigned to a variable determines the influence that variable has within a mathematical equation: large weights give the variable a lot of influence.

Weighted least squares: a method of *regression* in which the parameters of the model are estimated using the *method of least squares* but observations are weighted by some other variable. Often they are weighted by the inverse of their *variance* to combat *heteroscedasticity*.

Welch's *F*: a version of the *F-statistic* designed to be accurate when the assumption of *homogeneity of variance* has been violated. Not to be confused with the squelch test which is where you shake your head around after writing statistics books to see if you still have a brain.

Wide format data: data that are arranged such that scores from a single entity appear in a single row and levels of independent or predictor variables are arranged over different columns. As such, in designs with multiple measurements of an outcome variable within a case the *outcome variable* scores will be contained in multiple columns each representing a level of an independent variable or a time point at which the score was observed. Columns can also represent attributes of the score or entity that are fixed over the duration of data collection, such as participant sex, employment status etc. (cf. *long format data*).

Wilcoxon's rank-sum test: a *non-parametric test* that looks for differences between two independent samples. That is, it tests whether the populations from which two samples are drawn have the same location. It is functionally the same as the *Mann–Whitney test*, and both tests are non-parametric equivalents of the *independent t-test*.

Wilcoxon signed-rank test: a *non-parametric test* that looks for differences between two related samples. It is the non-parametric equivalent of the *related t-test*.

Wilks's lambda (Λ)**:** a *test statistic* in *MANOVA*. It is the product of the unexplained variance on each of the *discriminant function variates*, so it represents the ratio of error variance to total variance (SS_R/SS_T) for each variate.

Within-subject design: another name for a *repeated-measures design*.

Writer's block: something I suffered from a lot while writing this edition. It's when you can't think of any decent examples and so end up talking about sperm the whole time. Seriously, look at this book, it's all sperm this, sperm that, quail sperm, human sperm. Frankly, I'm amazed donkey sperm didn't get in there somewhere. Oh, it just did.

Yates's continuity correction: an adjustment made to the *chi-square test* when the *contingency table* is 2 rows by 2 columns (i.e., there are two categorical variables both of which consist of only two categories). In large samples the adjustment makes little difference and is slightly dubious anyway (see Howell, 2012).

***z*-score:** the value of an observation expressed in standard deviation units. It is calculated by taking the observation, subtracting from it the mean of all observations, and dividing the result by the standard deviation of all observations. By converting a distribution of observations into *z*-scores a new distribution is created that has a mean of 0 and a standard deviation of 1.

REFERENCES

Agresti, A., & Finlay, B. (1986). *Statistical methods for the social sciences* (2nd ed.). San Francisco: Dellen.

Aiken, L. S., & West, S. G. (1991). *Multiple regression: Testing and interpreting interactions*. Newbury Park, CA: Sage.

Algina, J., & Olejnik, S. F. (1984). Implementing the Welch-James procedure with factorial designs. *Educational and Psychological Measurement, 44*, 39–48.

American Pyschological Association (2010). *Publication manual of the American Psychological Association* (6th ed.). Washington, DC: APA Books.

Anderson, C. A., & Bushman, B. J. (2001). Effects of violent video games on aggressive behavior, aggressive cognition, aggressive affect, physiological arousal, and prosocial behavior: A meta-analytic review of the scientific literature. *Psychological Science, 12*(5), 353–359.

Arrindell, W. A., & van der Ende, J. (1985). An empirical test of the utility of the observer-to-variables ratio in factor and components analysis. *Applied Psychological Measurement, 9*, 165–178.

Baguley, T. (2004). Understanding statistical power in the context of applied research. *Applied Ergonomics, 35*(2), 73–80.

Bale, C., Morrison, R., & Caryl, P. G. (2006). Chat-up lines as male sexual displays. *Personality and Individual Differences, 40*(4), 655–664.

Barcikowski, R. S., & Robey, R. R. (1984). Decisions in single group repeated measures analysis: Statistical tests and three computer packages. *American Statistician, 38*(2), 148–150.

Bargman, R. E. (1970). Interpretation and use of a generalized discriminant function. In R. C. Bose et al. (Eds.), *Essays in probability and statistics*. Chapel Hill: University of North Carolina Press.

Barnard, G. A. (1963). Ronald Aylmer Fisher, 1890–1962: Fisher's contributions to mathematical statistics. *Journal of the Royal Statistical Society, Series A, 126*, 162–166.

Barnett, V., & Lewis, T. (1978). *Outliers in statistical data*. New York: Wiley.

Baron, R. M., & Kenny, D. A. (1986). The moderator–mediator variable distinction in social psychological research – conceptual, strategic, and statistical considerations. *Journal of Personality and Social Psychology, 51*(6), 1173–1182.

Beckham, A. S. (1929). Is the Negro happy? A psychological analysis. *Journal of Abnormal and Social Psychology, 24*, 186–190.

Belia, S., Fidler, F., Williams, J., & Cumming, G. (2005). Researchers misunderstand confidence intervals and standard error bars. *Psychological Methods, 10*(4), 389–396.

Belsey, D. A., Kuh, E., & Welsch, R. (1980). *Regression diagnostics: Identifying influential data and sources of collinearity*. New York: Wiley.

Bemelman, M., & Hammacher, E. R. (2005). Rectal impalement by pirate ship: A case report. *Injury Extra, 36*, 508–510.

Berger, J. O. (2003). Could Fisher, Jeffreys and Neyman have agreed on testing? *Statistical Science, 18*(1), 1–12.

Bernard, P., Gervais, S. J., Allen, J., Campomizzi, S., & Klein, O. (2012). Integrating sexual objectification with object versus person recognition: The sexualized-body-inversion hypothesis. *Psychological Science, 23*(5), 469–471.

Berry, W. D. (1993). *Understanding regression assumptions*. Sage University Paper Series on Quantitative Applications in the Social Sciences, 07–092. Newbury Park, CA: Sage.

Berry, W. D., & Feldman, S. (1985). *Multiple regression in practice*. Sage University Paper Series on Quantitative Applications in the Social Sciences, 07–050. Beverly Hills, CA: Sage.

Bishop, D. V. M., & Thompson, P. A. (2016). Problems in using *p*-curve analysis and text-mining to detect rate of *p*-hacking and evidential value. *PeerJ, 4*, e1715.

Board, B. J., & Fritzon, K. (2005). Disordered personalities at work. *Psychology, Crime & Law, 11*(1), 17–32.

Bock, R. D. (1975). *Multivariate statistical methods in behavioral research*. New York: McGraw-Hill.

Boik, R. J. (1981). A priori tests in repeated measures designs: Effects of nonsphericity. *Psychometrika, 46*(3), 241–255.

Bowerman, B. L., & O'Connell, R. T. (1990). *Linear statistical models: An applied approach* (2nd ed.). Belmont, CA: Duxbury.

Bray, J. H., & Maxwell, S. E. (1985). *Multivariate analysis of variance*. Sage University Paper Series on Quantitative Applications in the Social Sciences, 07–054. Newbury Park, CA: Sage.

Brown, M. B., & Forsythe, A. B. (1974). The small sample behaviour of some statistics which test the equality of several means. *Technometrics, 16*, 129–132.

Brown, T. A. (2015). *Confirmatory factor analysis for applied research* (2nd ed.). New York: Guilford.

Brown, W. (1910). Some experimental results in the correlation of mental abilities. *British Journal of Psychology, 3*, 296–322.

Bruns, S. B., & Ioannidis, J. P. A. (2016). *p*-curve and *p*-hacking in observational research. *PLoS One, 11*(2), e0149144.

Budescu, D. V. (1982). The power of the F test in normal populations with heterogeneous variances. *Educational and Psychological Measurement, 42*, 609–616.

Budescu, D. V., & Appelbaum, M. I. (1981). Variance stabilizing transformations and the power of the F test. *Journal of Educational Statistics, 6*(1), 55–74.

Carter, S. P., Greenberg, K., & Walker, M. (2016). *The impact of computer usage on academic performance: Evidence from a randomized trial at the United States Military Academy*. SEII Discussion Paper #2016.02, School Effectiveness & Inequality Initiative, MIT Department of Economics, Cambride, MA. Retrieved from http://seii.mit.edu/wp-content/uploads/2016/05/SEII-Discussion-Paper-2016.02-Payne-Carter-Greenberg-and-Walker-2.pdf

Cattell, R. B. (1966a). *The scientific analysis of personality*. Chicago: Aldine.

Cattell, R. B. (1966b). The scree test for the number of factors. *Multivariate Behavioral Research, 1*, 245–276.

Çetinkaya, H., & Domjan, M. (2006). Sexual fetishism in a quail (*Coturnix japonica*) model system: Test of reproductive success. *Journal of Comparative Psychology, 120*(4), 427–432.

Chambers, C. D., Dienes, Z., McIntosh, R. D., Rotshtein, P., & Willmes, K. (2015). Registered reports: Realigning incentives in scientific publishing. *Cortex, 66*, A1–A2.

Chamorro-Premuzic, T., Furnham, A., Christopher, A. N., Garwood, J., & Martin, N. (2008). Birds of a feather: Students' preferences for lecturers' personalities as predicted by their own personality and learning approaches. *Personality and Individual Differences, 44*, 965–976.

Chen, P. Y., & Popovich, P. M. (2002). *Correlation: Parametric and nonparametric measures*. Thousand Oaks, CA: Sage.

Chen, X., Wang, X. Y., Yang, D., & Chen, Y. G. (2014). The moderating effect of stimulus attractiveness on the effect of alcohol consumption on attractiveness ratings. *Alcohol and Alcoholism, 49*(5), 515–519.

Chen, X. Z., Luo, Y., Zhang, J. J., Jiang, K., Pendry, J. B., & Zhang, S. A. (2011). Macroscopic invisibility cloaking of visible light. *Nature Communications, 2*, 176.

Clarke, D. L., Buccimazza, I., Anderson, F. A., & Thomson, S. R. (2005). Colorectal foreign bodies. *Colorectal Disease, 7*(1), 98–103.

Claxton, A., O'Rourke, N., Smith, J. Z., & DeLongis, A. (2012). Personality traits and marital satisfaction within enduring relationships: An intra-couple discrepancy approach. *Journal of Social and Personal Relationships, 29*(3), 375–396.

Cliff, N. (1987). *Analyzing multivariate data*. New York: Harcourt Brace Jovanovich.

Cohen, J. (1968). Multiple regression as a general data-analytic system. *Psychological Bulletin, 70*(6), 426–443.

Cohen, J. (1988). *Statistical power analysis for the behavioral sciences* (2nd ed.). New York: Academic Press.

Cohen, J. (1990). Things I have learned (so far). *American Psychologist, 45*(12), 1304–1312.

Cohen, J. (1992). A power primer. *Psychological Bulletin, 112*(1), 155–159.

Cohen, J. (1994). The earth is round (*p* < .05). *American Psychologist, 49*(12), 997–1003.

Coldwell, J., Pike, A., & Dunn, J. (2006). Household chaos – links with parenting and child behaviour. *Journal of Child Psychology and Psychiatry, 47*(11), 1116–1122.

Cole, D. A., Maxwell, S. E., Arvey, R., & Salas, E. (1994). How the power of MANOVA can both increase and decrease as a function of the intercorrelations among the dependent variables. *Psychological Bulletin, 115*(3), 465–474.

Collier, R. O., Baker, F. B., Mandeville, G. K., & Hayes, T. F. (1967). Estimates of test size for several test procedures based on conventional variance ratios in the repeated measures design. *Psychometrika, 32*(2), 339–352.

Comrey, A. L., & Lee, H. B. (1992). *A first course in factor analysis* (2nd ed.). Hillsdale, NJ: Erlbaum.

Cook, R. D., & Weisberg, S. (1982). *Residuals and influence in regression*. New York: Chapman & Hall.

Cook, S. A., Rosser, R., & Salmon, P. (2006). Is cosmetic surgery an effective psychotherapeutic intervention? A systematic review of the evidence. *Journal of Plastic, Reconstructive & Aesthetic Surgery, 59*, 1133–1151.

Cook, S. A., Rosser, R., Toone, H., James, M. I., & Salmon, P. (2006). The psychological and social characteristics of patients referred for NHS cosmetic surgery: Quantifying clinical need. *Journal of Plastic, Reconstructive & Aesthetic Surgery, 59*, 54–64.

Cooper, C. L., Sloan, S. J., & Williams, S. (1988). *Occupational Stress Indicator Management Guide*. Windsor: NFER-Nelson.

Cooper, H. M. (2010). *Research synthesis and meta-analysis: A step-by-step approach* (4th ed.). Thousand Oaks, CA: Sage.

Cooper, M., O'Donnell, D., Caryl, P. G., Morrison, R., & Bale, C. (2007). Chat-up lines as male displays: Effects of content, sex, and personality. *Personality and Individual Differences, 43*(5), 1075–1085.

Cortina, J. M. (1993). What is coefficient alpha? An examination of theory and applications. *Journal of Applied Psychology, 78*, 98–104.

Coursol, A., & Wagner, E. E. (1986). Effect of positive findings on submission and acceptance rates: A note on meta-analysis bias. *Professional Psychology, 17*, 136–137.

Cox, D. R., & Snell, D. J. (1989). *The analysis of binary data* (2nd ed.). London: Chapman & Hall.

Cribari-Neto, F. (2004). Asymptotic inference under heteroskedasticity of unknown form. *Computational Statistics & Data Analysis, 45*(2), 215–233. doi:10.1016/s0167-9473(02)00366-3

Cronbach, L. J. (1951). Coefficient alpha and the internal structure of tests. *Psychometrika, 16*, 297–334.

Cronbach, L. J. (1957). The two disciplines of scientific psychology. *American Psychologist, 12*, 671–684.

Cumming, G. (2012). *Understanding the new statistics: Effect sizes, confidence intervals, and meta-analysis*. New York: Routledge.

Cumming, G., & Finch, S. (2005). Inference by eye – confidence intervals and how to read pictures of data. *American Psychologist, 60*(2), 170–180.

Dai, X. C., Dong, P., & Jia, J. S. (2014). When does playing hard to get increase romantic attraction? *Journal of Experimental Psychology: General, 143*(2), 521–526.

Daniels, E. A. (2012). Sexy versus strong: What girls and women think of female athletes. *Journal of Applied Developmental Psychology, 33*, 79–90.

Davey, G. C. L., Startup, H. M., Zara, A., MacDonald, C. B., & Field, A. P. (2003). Perseveration of checking thoughts and mood-as-input hypothesis. *Journal of Behavior Therapy & Experimental Psychiatry, 34*, 141–160.

Davidson, M. L. (1972). Univariate versus multivariate tests in repeated-measures experiments. *Psychological Bulletin, 77*, 446–452.

Davies, P., Surridge, J., Hole, L., & Munro-Davies, L. (2007). Superhero-related injuries in paediatrics: A case series. *Archives of Disease in Childhood, 92*(3), 242–243.

REFERENCES

De Groot, A. D. (1956/2014). The meaning of 'significance' for different types of research [translated and annotated by Eric-Jan Wagenmakers, Denny Borsboom, Josine Verhagen, Rogier Kievit, Marjan Bakker, Angelique Cramer, Dora Matzke, Don Mellenbergh, and Han L. J. van der Maas]. *Acta Psychologica, 148,* 188–194.

DeCarlo, L. T. (1997). On the meaning and use of kurtosis. *Psychological Methods, 2*(3), 292–307.

DeCoster, J., Gallucci, M., & Iselin, A.-M. R. (2011). Best practices for using median splits, artificial categorization, and their continuous alternatives. *Journal of Experimental Psychopathology, 2*(2), 197–209.

DeCoster, J., Iselin, A.-M. R., & Gallucci, M. (2009). A conceptual and empirical examination of justifications for dichotomization. *Psychological Methods, 14*(4), 349–366.

Di Falco, A., Ploschner, M., & Krauss, T. F. (2010). Flexible metamaterials at visible wavelengths. *New Journal of Physics, 12,* 113006.

Dickersin, K., Min, Y.-I., & Meinert, C. L. (1992). Factors influencing publication of research results: Follow-up of applications submitted to two institutional review boards. *Journal of the American Medical Association, 267,* 374–378.

Dienes, Z. (2011). Bayesian versus orthodox statistics: Which side are you on? *Perspectives on Psychological Science, 6*(3), 274–290.

Domjan, M., Blesbois, E., & Williams, J. (1998). The adaptive significance of sexual conditioning: Pavlovian control of sperm release. *Psychological Science, 9*(5), 411–415.

Donaldson, T. S. (1968). Robustness of the *F*-test to errors of both kinds and the correlation between the numerator and denominator of the *F*-ratio. *Journal of the American Statistical Association, 63,* 660–676.

Dunlap, W. P., Cortina, J. M., Vaslow, J. B., & Burke, M. J. (1996). Meta-analysis of experiments with matched groups or repeated measures designs. *Psychological Methods, 1*(2), 170–177.

Dunteman, G. E. (1989). *Principal components analysis.* Sage University Paper Series on Quantitative Applications in the Social Sciences, 07–069. Newbury Park, CA: Sage.

Durbin, J., & Watson, G. S. (1951). Testing for serial correlation in least squares regression, II. *Biometrika, 30,* 159–178.

Easterlin, R. A. (2003). Explaining happiness. *Proceedings of the National Academy of Sciences, 100*(19), 11176–11183.

Efron, B., & Tibshirani, R. (1993). *An introduction to the bootstrap.* New York: Chapman & Hall.

Enders, C. K. (2010). *Applied missing data analysis.* New York: Guilford.

Enders, C. K. (2011). Analyzing longitudinal data with missing values. *Rehabilitation Psychology, 56*(4), 267–288.

Enders, C. K., & Tofighi, D. (2007). Centering predictor variables in cross-sectional multilevel models: A new look at an old issue. *Psychological Methods, 12*(2), 121–138.

Eriksson, S.-G., Beckham, D., & Vassell, D. (2004). Why are the English so shit at penalties? A review. *Journal of Sporting Ineptitude, 31,* 231–1072.

Erlebacher, A. (1977). Design and analysis of experiments contrasting the within- and between-subjects manipulations of the independent variable. *Psychological Bulletin, 84,* 212–219.

Eysenck, H. J. (1953). *The structure of human personality.* New York: Wiley.

Famous People (2015). Prasanta Chandra Mahalanobis biography. Retrieved from www.thefamouspeople.com/profiles/prasanta-chandra-mahalanobis-6572.php

Fanelli, D. (2009). How many scientists fabricate and falsify research? A systematic review and meta-analysis of survey data. *PLoS One, 4*(5), e5738.

Fanelli, D. (2010a). Do pressures to publish increase scientists' bias? An empirical support from US states data. *PLoS One, 5*(4), e10271.

Fanelli, D. (2010b). 'Positive' results increase down the hierarchy of the sciences. *PLoS One, 5*(3), e10068.

Fanelli, D. (2012). Negative results are disappearing from most disciplines and countries. *Scientometrics, 90*(3), 891–904.

Feng, J., Spence, I., & Pratt, J. (2007). Playing an action video game reduces gender differences in spatial cognition. *Psychological Science, 18*(10), 850–855.

Feng, L., Gwee, X., Kua, E. H., & Ng, T. P. (2010). Cognitive function and tea consumption in community dwelling older Chinese in Singapore. *Journal of Nutrition Health & Aging, 14*(6), 433–438.

Fesmire, F. M. (1988). Termination of intractable hiccups with digital rectal massage. *Annals of Emergency Medicine, 17*(8), 872.

Field, A. P. (1998). A bluffer's guide to sphericity. *Newsletter of the Mathematical, Statistical and Computing Section of the British Psychological Society, 6*(1), 13–22.

Field, A. P. (2000). *Discovering statistics using SPSS for Windows: Advanced techniques for the beginner.* London: Sage.

Field, A. P. (2001). Meta-analysis of correlation coefficients: A Monte Carlo comparison of fixed- and random-effects methods. *Psychological Methods, 6*(2), 161–180.

Field, A. P. (2003). Can meta-analysis be trusted? *Psychologist, 16*(12), 642–645.

Field, A. P. (2005a). Intraclass correlation. In B. Everitt & D. C. Howell (Eds.), *Encyclopedia of statistics in behavioral science* (Vol. 2, pp. 948–954). New York: Wiley.

Field, A. P. (2005b). Is the meta-analysis of correlation coefficients accurate when population correlations vary? *Psychological Methods, 10*(4), 444–467.

Field, A. P. (2005c). Meta-analysis. In J. Miles & P. Gilbert (Eds.), *A handbook of research methods in clinical and health psychology* (pp. 295–308). Oxford: Oxford University Press.

Field, A. P. (2005d). Sir Ronald Aylmer Fisher. In B. S. Everitt & D. C. Howell (Eds.), *Encyclopedia of statistics in behavioral science* (Vol. 2, pp. 658–659). Chichester: Wiley.

Field, A. P. (2006). The behavioral inhibition system and the verbal information pathway to children's fears. *Journal of Abnormal Psychology, 115*(4), 742–752.

Field, A. P. (2010). Teaching statistics. In D. Upton & A. Trapp (Eds.), *Teaching psychology in higher education* (pp. 134–163). Chichester: Wiley-Blackwell.

Field, A. P. (2012). Meta-analysis in clinical psychology research. In J. S. Comer & P. C. Kendall (Eds.), *The Oxford handbook of research strategies for clinical psychology.* Oxford: Oxford University Press.

Field, A. P. (2016). *An adventure in statistics: The reality enigma.* London: Sage.

Field, A. P., & Davey, G. C. L. (1999). Reevaluating evaluative conditioning: A nonassociative explanation of conditioning effects in the visual evaluative conditioning paradigm. *Journal of Experimental Psychology: Animal Behavior Processes, 25*(2), 211–224.

Field, A. P., & Gillett, R. (2010). How to do a meta-analysis. *British Journal of Mathematical & Statistical Psychology, 63*, 665–694.

Field, A. P., & Hole, G. J. (2003). *How to design and report experiments*. London: Sage.

Field, A. P., Miles, J. N. V., & Field, Z. C. (2012). *Discovering statistics using R: And sex and drugs and rock 'n' roll*. London: Sage.

Field, A. P., & Moore, A. C. (2005). Dissociating the effects of attention and contingency awareness on evaluative conditioning effects in the visual paradigm. *Cognition and Emotion, 19*(2), 217–243.

Field, A. P., & Wilcox, R. R. (in press). Robust statistical methods: a primer for clinical psychology and experimental psychopathology researchers. *Behaviour Research and Therapy*. doi: 10.1016/j.brat.2017.05.013

Fienberg, S. E., Stigler, S. M., & Tanur, J. M. (2007). The William Kruskal legacy: 1919–2005. *Statistical Science, 22*(2), 255–261.

Fisher, R. A. (1921). On the probable error of a coefficient of correlation deduced from a small sample. *Metron, 1*, 3–32.

Fisher, R. A. (1922). On the interpretation of chi square from contingency tables, and the calculation of P. *Journal of the Royal Statistical Society, 85*, 87–94.

Fisher, R. A. (1925). *Statistical methods for research workers*. Edinburgh: Oliver & Boyd.

Fisher, R. A. (1925/1991). *Statistical methods, experimental design, and scientific inference* (reprint). Oxford: Oxford University Press.

Fisher, R. A. (1956). *Statistical methods and scientific inference*. New York: Hafner.

Flanagan, J. C. (1937). A proposed procedure for increasing the efficiency of objective tests. *Journal of Educational Psychology, 28*, 17–21.

Francis, G. (2013). Replication, statistical consistency, and publication bias. *Journal of Mathematical Psychology, 57*(5), 153–169.

Francis, G. (2014a). The frequency of excess success for articles in psychological science. *Psychonomic Bulletin & Review, 21*(5), 1180–1187.

Francis, G. (2014b). Too much success for recent groundbreaking epigenetic experiments. *Genetics, 198*(2), 449–451. doi:10.1534/genetics.114.163998

Francis, G., Tanzman, J., & Matthews, W. J. (2014). Excess success for psychology articles in the journal *Science*. *PLoS One, 9*(12), e114255.

Friedman, M. (1937). The use of ranks to avoid the assumption of normality implicit in the analysis of variance. *Journal of the American Statistical Association, 32*, 675–701.

Gallup, G. G. J., Burch, R. L., Zappieri, M. L., Parvez, R., Stockwell, M., & Davis, J. A. (2003). The human penis as a semen displacement device. *Evolution and Human Behavior, 24*, 277–289.

Games, P. A. (1983). Curvilinear transformations of the dependent variable. *Psychological Bulletin, 93*(2), 382–387.

Games, P. A. (1984). Data transformations, power, and skew: A rebuttal to Levine and Dunlap. *Psychological Bulletin, 95*(2), 345–347.

Games, P. A., & Lucas, P. A. (1966). Power of the analysis of variance of independent groups on non-normal and normally transformed data. *Educational and Psychological Measurement, 26*, 311–327.

Gelman, A., & Hill, J. (2007). *Data analysis using regression and multilevel/hierarchical models*. Cambridge: Cambridge University Press.

Gelman, A., & Weakliem, D. (2009). Of beauty, sex and power: Too little attention has been paid to the statistical challenges in estimating small effects. *American Scientist, 97*, 310–316.

Girden, E. R. (1992). *ANOVA: Repeated measures*. Sage University Paper Series on Quantitative Applications in the Social Sciences, 07–084. Newbury Park, CA: Sage.

Glass, G. V. (1966). Testing homogeneity of variances. *American Educational Research Journal, 3*(3), 187–190.

Glass, G. V., Peckham, P. D., & Sanders, J. R. (1972). Consequences of failure to meet assumptions underlying the fixed effects analyses of variance and covariance. *Review of Educational Research, 42*(3), 237–288.

Gönen, M., Johnson, W. O., Lu, Y. G., & Westfall, P. H. (2005). The Bayesian two-sample t test. *American Statistician, 59*(3), 252–257. doi:10.1198/000313005x55233

Graham, J. M., Guthrie, A. C., & Thompson, B. (2003). Consequences of not interpreting structure coefficients in published CFA research: A reminder. *Structural Equation Modeling, 10*(1), 142–153.

Grayson, D. (2004). Some myths and legends in quantitative psychology. *Understanding Statistics, 3*(1), 101–134.

Green, C. S., & Bavelier, D. (2007). Action-video-game experience alters the spatial resolution of vision. *Psychological Science, 18*(1), 88–94.

Green, C. S., Pouget, A., & Bavelier, D. (2010). Improved probabilistic inference as a general learning mechanism with action video games. *Current Biology, 20*(17), 1573–1579.

Greenhouse, S. W., & Geisser, S. (1959). On methods in the analysis of profile data. *Psychometrika, 24*, 95–112.

Greenland, S., Senn, S. J., Rothman, K. J., Carlin, J. B., Poole, C., Goodman, S. N., & Altman, D. G. (2016). Statistical tests, *P* values, confidence intervals, and power: A guide to misinterpretations. *European Journal of Epidemiology, 31*, 337–350.

Greenwald, A. G. (1975). Consequences of prejudice against null hypothesis. *Psychological Bulletin, 82*(1), 1–19.

Guadagnoli, E., & Velicer, W. F. (1988). Relation of sample size to the stability of component patterns. *Psychological Bulletin, 103*(2), 265–275.

Guéguen, N. (2012). Tattoos, piercings, and alcohol consumption. *Alcoholism: Clinical and Experimental Research, 36*(7), 1253–1256.

Ha, T., Overbeek, G., & Engels, R. C. M. E. (2010). Effects of attractiveness and social status on dating desire in heterosexual adolescents: An experimental study. *Archives of Sexual Behavior, 39*(5), 1063–1071.

Hakstian, A. R., Roed, J. C., & Lind, J. C. (1979). Two-sample T2 procedure and the assumption of homogeneous covariance matrices. *Psychological Bulletin, 86*, 1255–1263.

Halekoh, U., & Højsgaard, S. (2007). Overdispersion. Retrieved from http://gbi.agrsci.dk/statistics/courses/phd07/material/Day7/overdispersion-handout.pdf (accessed 10/01/08).

Hall, J., & Sammons, P. M. (2014). Mediation, moderation, & interaction: Definitions, discrimination & (some) means of testing. In T. Teo (Ed.), *Handbook of quantitative methods for educational research* (pp. 267–286). Rotterdam: Sense.

Hamilton, B. L. (1977). An empirical investigation of effects of heterogeneous regression slopes in analysis of covariance. *Educational and Psychological Measurement, 37*(3), 701–712.

Hardy, M. A. (1993). *Regression with dummy variables*. Sage University Paper Series on Quantitative Applications in the Social Sciences, 07–093. Newbury Park, CA: Sage.

REFERENCES

Harman, B. H. (1976). *Modern factor analysis* (3rd ed., revised). Chicago: University of Chicago Press.

Harris, R. J. (1975). *A primer of multivariate statistics*. New York: Academic Press.

Hartgerink, C. H. J., van Aert, R. C. M., Nuijten, M. B., Wicherts, J. M., & van Assen, M. (2016). Distributions of p-values smaller than .05 in psychology: what is going on? *PeerJ, 4*, e1935.

Hayes, A. F. (2017). *Introduction to Mediation, Moderation, and Conditional Process Analysis: A Regression-Based Approach* (2nd ed.). New York: The Guilford Press.

Hayes, A. F., & Cai, L. (2007). Using heteroskedasticity-consistent standard error estimators in OLS regression: An introduction and software implementation. *Behavior Research Methods, 39*(4), 709–722.

Hayes, A. F., & Matthes, J. (2009). Computational procedures for probing interactions in OLS and logistic regression: SPSS and SAS implementations. *Behavior Research Methods, 41*, 924–936.

Head, M. L., Holman, L., Lanfear, R., Kahn, A. T., & Jennions, M. D. (2015). The extent and consequences of p-hacking in science. *PLoS Biology, 13*(3), e1002106.

Hedges, L. V. (1984). Estimation of effect size under non-random sampling: The effects of censoring studies yielding statistically insignificant mean differences. *Journal of Educational Statistics, 9*, 61–85.

Hedges, L. V. (1992). Meta-analysis. *Journal of Educational Statistics, 17*(4), 279–296.

Hill, C., Abraham, C., & Wright, D. B. (2007). Can theory-based messages in combination with cognitive prompts promote exercise in classroom settings? *Social Science & Medicine, 65*, 1049–1058.

Hoaglin, D., & Welsch, R. (1978). The hat matrix in regression and ANOVA. *American Statistician, 32*, 17–22.

Hoagwood, K. E., Acri, M., Morrissey, M., & Peth-Pierce, R. (2017). Animal-assisted therapies for youth with or at risk for mental health problems: A systematic review. *Applied Developmental Science, 21*(1), 1–13.

Hoddle, G., Batty, D., & Ince, P. (1998). How not to take penalties in important soccer matches. *Journal of Cretinous Behaviour, 1*, 1–2.

Hodgson, R., Cole, A., & Young, A. (2012). The name of the game: Why can't people called Ashley score from a penalty kick? *Sporting Weakness Review, 24*(6), 574–581.

Hoffmann, F., Musolf, K., & Penn, D. J. (2012). Spectrographic analyses reveal signals of individuality and kinship in the ultrasonic courtship vocalizations of wild house mice. *Physiology & Behavior, 105*, 766–771.

Hofmann, W., De Houwer, J., Perugini, M., Baeyens, F., & Crombez, G. (2010). Evaluative conditioning in humans: A meta-analysis. *Psychological Bulletin, 136*(3), 390–421.

Hollingsworth, H. H. (1980). An analytical investigation of the effects of heterogeneous regression slopes in analysis of covariance. *Educational and Psychological Measurement, 40*(3), 611–618.

Horn, J. L. (1965). A rationale and test for the number of factors in factor analysis. *Psychometrika, 30*, 179–185.

Hosmer, D. W., & Lemeshow, S. (1989). *Applied logistic regression*. New York: Wiley.

Howell, D. C. (2012). *Statistical methods for psychology* (8th ed.). Belmont, CA: Wadsworth.

Huberty, C. J., & Morris, J. D. (1989). Multivariate analysis versus multiple univariate analysis. *Psychological Bulletin, 105*(2), 302–308.

Hughes, J. P., Marice, H. P., & Gathright, J. B. (1976). Method of removing a hollow object from the rectum. *Diseases of the Colon & Rectum, 19*(1), 44–45.

Hume, D. (1739–40/1965). *A treatise of human nature* (Ed. L. A. Selby-Bigge). Oxford: Clarendon Press.

Hunter, J. E., & Schmidt, F. L. (2004). *Methods of meta-analysis: Correcting error and bias in research findings* (2nd ed.). Newbury Park, CA: Sage.

Hutcheson, G., & Sofroniou, N. (1999). *The multivariate social scientist*. London: Sage.

Huynh, H., & Feldt, L. S. (1976). Estimation of the Box correction for degrees of freedom from sample data in randomised block and split-plot designs. *Journal of Educational Statistics, 1*(1), 69–82.

Jeffreys, H. (1961). *Theory of probability* (3rd ed.). Oxford: Oxford University Press.

Johns, S. E., Hargrave, L. A., & Newton-Fisher, N. E. (2012). Red is not a proxy signal for female genitalia in humans. *PLoS One, 7*(4), e34669.

Johnson, P. O., & Neyman, J. (1936). Tests of certain linear hypotheses and their applications to some educational problems. *Statistical Research Memoirs, 1*, 57–93.

Jolliffe, I. T. (1972). Discarding variables in a principal component analysis, I: Artificial data. *Applied Statistics, 21*, 160–173.

Jolliffe, I. T. (1986). *Principal component analysis*. New York: Springer.

Jonckheere, A. R. (1954). A distribution-free *k*-sample test against ordered alternatives. *Biometrika, 41*, 133–145.

Judd, C. M., & Kenny, D. A. (1981). Process analysis: Estimating mediation in evaluation research. *Evaluation Research, 5*, 602–619.

Julien, D., O'Connor, K. P., & Aardema, F. (2007). Intrusive thoughts, obsessions, and appraisals in obsessive-compulsive disorder: A critical review. *Clinical Psychology Review, 27*(3), 366–383.

Kahneman, D., & Krueger, A. B. (2006). Developments in the measurement of subjective well-being. *Journal of Economic Perspectives, 20*(1), 3–24.

Kaiser, H. F. (1960). The application of electronic computers to factor analysis. *Educational and Psychological Measurement, 20*, 141–151.

Kaiser, H. F. (1970). A second-generation little jiffy. *Psychometrika, 35*, 401–415.

Kaiser, H. F., & Caffrey, J. (1965). Alpha factor analysis. *Psychometrika, 30*(1), 1–14. doi: 10.1007/bf02289743

Kaiser, H. F., & Rice, J. (1974). Little jiffy, mark 4. *Educational and Psychological Measurement, 34*(1), 111–117.

Kanazawa, S. (2007). Beautiful parents have more daughters: A further implication of the generalized Trivers-Willard hypothesis. *Journal of Theoretical Biology, 244*, 133–140.

Kass, R. A., & Tinsley, H. E. A. (1979). Factor analysis. *Journal of Leisure Research, 11*, 120–138.

Kellett, S., Clarke, S., & McGill, P. (2008). Outcomes from psychological assessment regarding recommendations for cosmetic surgery. *Journal of Plastic, Reconstructive & Aesthetic Surgery, 61*, 512–517.

Kerr, N. L. (1998). HARKing: Hypothesizing after the results are known. *Personality and Social Psychology Review, 2*(3), 196–217.

Keselman, H. J., & Keselman, J. C. (1988). Repeated measures multiple comparison procedures: Effects of violating multisample sphericity in unbalanced designs. *Journal of Educational Statistics, 13*(3), 215–226.

Kimmel, H. D. (1957). Three criteria for the use of one-tailed tests. *Psychological Bulletin, 54*(4), 351–353.

Kirk, R. E. (1996). Practical significance: A concept whose time has come. *Educational and Psychological Measurement, 56*(5), 746–759.

Kline, P. (1999). *The handbook of psychological testing* (2nd ed.). London: Routledge.

Klockars, A. J., & Sax, G. (1986). *Multiple comparisons.* Sage University Paper Series on Quantitative Applications in the Social Sciences, 07–061. Newbury Park, CA: Sage.

Koot, V. C. M., Peeters, P. H. M., Granath, F., Grobbee, D. E., & Nyren, O. (2003). Total and cause specific mortality among Swedish women with cosmetic breast implants: Prospective study. *British Medical Journal, 326*(7388), 527–528.

Kreft, I. G. G., & de Leeuw, J. (1998). *Introducing multilevel modeling.* London: Sage.

Kreft, I. G. G., de Leeuw, J., & Aiken, L. S. (1995). The effect of different forms of centering in hierarchical linear models. *Multivariate Behavioral Research, 30*, 1–21.

Kruschke, J. K. (2010a). Bayesian data analysis. *Wiley Interdisciplinary Reviews: Cognitive Science, 1*(5), 658–676.

Kruschke, J. K. (2010b). What to believe: Bayesian methods for data analysis. *Trends in Cognitive Sciences, 14*(7), 293–300.

Kruschke, J. K. (2013). Bayesian estimation supersedes the t test. *Journal of Experimental Psychology: General, 142*(2), 573–603.

Kruschke, J. K. (2014). *Doing Bayesian data analysis: A tutorial with R, JAGS and STAN* (2nd ed.). Burlington, MA: Academic Press.

Kruskal, W. H., & Wallis, W. A. (1952). Use of ranks in one-criterion variance analysis. *Journal of the American Statistical Association, 47*, 583–621.

Lacourse, E., Claes, M., & Villeneuve, M. (2001). Heavy metal music and adolescent suicidal risk. *Journal of Youth and Adolescence, 30*(3), 321–332.

Lakens, D. (2015). What p-hacking really looks like: A comment on Masicampo and LaLande (2012). *Quarterly Journal of Experimental Psychology, 68*(4), 829–832.

Lakens, D., Hilgard, J., & Staaks, J. (2016). On the reproducibility of meta-analyses: Six practical recommendations. *BMC Psychology, 4*(1), 24.

Lambert, N. M., Negash, S., Stillman, T. F., Olmstead, S. B., & Fincham, F. D. (2012). A love that doesn't last: Pornography consumption and weakened commitment to one's romantic partner. *Journal of Social and Clinical Psychology, 31*(4), 410–438.

Leggett, N. C., Thomas, N. A., Loetscher, T., & Nicholls, M. E. R. (2013). The life of p: 'Just significant' results are on the rise. *Quarterly Journal of Experimental Psychology, 66*(12), 2303–2309.

Lehmann, E. L. (1993). The Fisher, Neyman-Pearson theories of testing hypotheses: One theory or two? *Journal of the American Statistical Association, 88*, 1242–1249.

Lenth, R. V. (2001). Some practical guidelines for effective sample size determination. *American Statistician, 55*(3), 187–193.

Levene, H. (1960). Robust tests for equality of variances. In I. Olkin, S. G. Ghurye, W. Hoeffding, W. G. Madow, & H. B. Mann (Eds.), *Contributions to probability and statistics: Essays in honor of Harold Hotelling* (pp. 278–292). Stanford, CA: Stanford University Press.

Levine, D. W., & Dunlap, W. P. (1982). Power of the F test with skewed data: Should one transform or not? *Psychological Bulletin, 92*(1), 272–280.

Levine, D. W., & Dunlap, W. P. (1983). Data transformation, power, and skew: A rejoinder to Games. *Psychological Bulletin, 93*(3), 596–599.

Levy, K. N., Johnson, B. N., Clouthier, T. L., Scala, J. W., & Temes, C. M. (2015). An attachment theoretical framework for personality disorders. *Canadian Psychology/Psychologie Canadienne, 56*(2), 197–207.

Liang, F., Paulo, R., Molina, G., Clyde, M. A., & Berger, J. O. (2008). Mixtures of g priors for Bayesian variable selection. *Journal of the American Statistical Association, 103*(481), 410–423.

Lo, S. F., Wong, S. H., Leung, L. S., Law, I. C., & Yip, A. W. C. (2004). Traumatic rectal perforation by an eel. *Surgery, 135*(1), 110–111.

Loftus, G. R., & Masson, M. E. J. (1994). Using confidence intervals in within-subject designs. *Psychonomic Bulletin and Review, 1*(4), 476–490.

Lombardi, C. M., & Hurlbert, S. H. (2009). Misprescription and misuse of one-tailed tests. *Austral Ecology, 34*(4), 447–468.

Long, J. D. (2012). *Longitudinal data analysis for the behavioral sciences using R.* Thousand Oaks, CA: Sage.

Long, J. S., & Ervin, L. H. (2000). Using heteroscedasticity consistent standard errors in the linear regression model. *American Statistician, 54*(3), 217–224. doi:10.2307/2685594

Lord, F. M. (1967). A paradox in the interpretation of group comparisons. *Psychological Bulletin, 68*(5), 304–305.

Lord, F. M. (1969). Statistical adjustments when comparing preexisting groups. *Psychological Bulletin, 72*(5), 336–337.

Lumley, T., Diehr, P., Emerson, S., & Chen, L. (2002). The importance of the normality assumption in large public health data sets. *Annual Review of Public Health, 23*, 151–169.

Lunney, G. H. (1970). Using analysis of variance with a dichotomous dependent variable: An empirical study. *Journal of Educational Measurement, 7*(4), 263–269.

MacCallum, R. C., Widaman, K. F., Zhang, S., & Hong, S. (1999). Sample size in factor analysis. *Psychological Methods, 4*(1), 84–99.

MacCallum, R. C., Zhang, S., Preacher, K. J., & Rucker, D. D. (2002). On the practice of dichotomization of quantitative variables. *Psychological Methods, 7*(1), 19–40.

MacKinnon, D. P. (2008). *Introduction to statistical mediation analysis.* Mahwah, NJ: Erlbaum.

Mair, P., Schoenbrodt, F., & Wilcox, R. R. (2015). WRS2: Wilcox robust estimation and testing. R package version (Version 0.4–0). Retrieved from http://cran.r-project.org/package=WRS2

Mann, H. B., & Whitney, D. R. (1947). On a test of whether one of two random variables is stochastically larger than the other. *Annals of Mathematical Statistics, 18*, 50–60.

Marzillier, S. L., & Davey, G. C. L. (2005). Anxiety and disgust: Evidence for a unidirectional relationship. *Cognition and Emotion, 19*(5), 729–750.

Masicampo, E. J., & Lalande, D. R. (2012). A peculiar prevalence of p values just below .05. *Quarterly Journal of Experimental Psychology*, *65*(11), 2271–2279.

Massar, K., Buunk, A. P., & Rempt, S. (2012). Age differences in women's tendency to gossip are mediated by their mate value. *Personality and Individual Differences*, *52*, 106–109.

Mather, K. (1951). R. A. Fisher's *Statistical Methods for Research Workers*: An appreciation. *Journal of the American Statistical Association*, *46*, 51–54.

Matthews, R. C., Domjan, M., Ramsey, M., & Crews, D. (2007). Learning effects on sperm competition and reproductive fitness. *Psychological Science*, *18*(9), 758–762.

Maxwell, S. E. (1980). Pairwise multiple comparisons in repeated measures designs. *Journal of Educational Statistics*, *5*(3), 269–287.

Maxwell, S. E., & Delaney, H. D. (1990). *Designing experiments and analyzing data*. Belmont, CA: Wadsworth.

McDonald, P. T., & Rosenthal, D. (1977). An unusual foreign body in the rectum – a baseball: Report of a case. *Diseases of the Colon & Rectum*, *20*(1), 56–57.

McElreath, R. (2016). *Statistical rethinking: A Bayesian course with examples in R and Stan*. Boca Raton, FL: Chapman & Hall/CRC Press.

McGrath, R. E., & Meyer, G. J. (2006). When effect sizes disagree: The case of r and d. *Psychological Methods*, *11*(4), 386–401.

McKiernan, E. C., Bourne, P. E., Brown, C. T., Buck, S., Kenall, A., Lin, J., …, & Yarkoni, T. (2016). How open science helps researchers succeed. *eLife*, *5*, e16800.

McNulty, J. K., Neff, L. A., & Karney, B. R. (2008). Beyond initial attraction: Physical attractiveness in newlywed marriage. *Journal of Family Psychology*, *22*(1), 135–143.

Meehl, P. E. (1978). Theoretical risks and tabular asterisks: Sir Karl, Sir Ronald, and the slow progress of soft psychology. *Journal of Consulting and Clinical Psychology*, *46*, 806–834.

Menard, S. (1995). *Applied logistic regression analysis*. Sage University Paper Series on Quantitative Applications in the Social Sciences, 07–106. Thousand Oaks, CA: Sage.

Mendoza, J. L., Toothaker, L. E., & Crain, B. R. (1976). Necessary and sufficient conditions for F ratios in the $L \times J \times K$ factorial design with two repeated factors. *Journal of the American Statistical Association*, *71*, 992–993.

Mendoza, J. L., Toothaker, L. E., & Nicewander, W. A. (1974). A Monte Carlo comparison of the univariate and multivariate methods for the groups by trials repeated measures design. *Multivariate Behavioural Research*, *9*, 165–177.

Meston, C. M., & Frohlich, P. F. (2003). Love at first fright: Partner salience moderates roller-coaster-induced excitation transfer. *Archives of Sexual Behavior*, *32*(6), 537–544.

Miles, J. N. V., & Shevlin, M. (2001). *Applying regression and correlation: A guide for students and researchers*. London: Sage.

Mill, J. S. (1865). *A system of logic: Ratiocinative and inductive*. London: Longmans, Green.

Miller, G., Tybur, J. M., & Jordan, B. D. (2007). Ovulatory cycle effects on tip earnings by lap dancers: Economic evidence for human estrus? *Evolution and Human Behavior*, *28*, 375–381.

Miller, G. A., & Chapman, J. P. (2001). Misunderstanding analysis of covariance. *Journal of Abnormal Psychology*, *110*(1), 40–48.

Mishra, J., Zinni, M., Bavelier, D., & Hillyard, S. A. (2011). Neural basis of superior performance of action videogame players in an attention-demanding task. *Journal of Neuroscience*, *31*(3), 992–998.

Mood, C. (2010). Logistic regression: Why we cannot do what we think we can do, and what we can do about it. *European Sociological Review*, *26*(1), 67–82.

Morewedge, C. K., Huh, Y. E., & Vosgerau, J. (2010). Thought for food: Imagined consumption reduces actual consumption. *Science*, *330*(6010), 1530–1533.

Morey, R. D. (2013). The consistency test does not – and cannot – deliver what is advertised: A comment on Francis (2013). *Journal of Mathematical Psychology*, *57*(5), 180–183.

Morey, R. D., Chambers, C. D., Etchells, P. J., Harris, C. R., Hoekstra, R., Lakens, D., …, & Zwaan, R. A. (2016). The Peer Reviewers' Openness Initiative: Incentivizing open research practices through peer review. *Royal Society Open Science*, *3*(1).

Muris, P., Huijding, J., Mayer, B., & Hameetman, M. (2008). A space odyssey: Experimental manipulation of threat perception and anxiety-related interpretation bias in children. *Child Psychiatry and Human Development*, *39*(4), 469–480.

Myers, R. (1990). *Classical and modern regression with applications* (2nd ed.). Boston: Duxbury.

Nagelkerke, N. J. D. (1991). A note on a general definition of the coefficient of determination. *Biometrika*, *78*, 691–692.

Namboodiri, K. (1984). *Matrix algebra: An introduction*. Sage University Paper Series on Quantitative Applications in the Social Sciences, 07–38. Beverly Hills, CA: Sage.

Neyman, J., & Pearson, E. S. (1933). On the problem of the most efficient tests of statistical hypotheses. *Philosophical Transactions of the Royal Society of London, Series A*, *231*, 289–337.

Nichols, L. A., & Nicki, R. (2004). Development of a psychometrically sound internet addiction scale: A preliminary step. *Psychology of Addictive Behaviors*, *18*(4), 381–384.

Nosek, B. A., Alter, G., Banks, G. C., Borsboom, D., Bowman, S. D., Breckler, S. J., …, & Yarkoni, T. (2015). Promoting an open research culture. *Science*, *348*(6242), 1422–1425.

Nosek, B. A., & Lakens, D. (2014). Registered reports: A method to increase the credibility of published results. *Social Psychology*, *45*(3), 137–141.

Nosek, B. A., Spies, J. R., & Motyl, M. (2012). Scientific utopia: II. Restructuring incentives and practices to promote truth over publishability. *Perspectives on Psychological Science*, *7*(6), 615–631.

Nunnally, J. C. (1978). *Psychometric theory*. New York: McGraw-Hill.

Nunnally, J. C., & Bernstein, I. H. (1994). *Psychometric theory* (3rd ed.). New York: McGraw-Hill.

O'Brien, M. G., & Kaiser, M. K. (1985). MANOVA method for analyzing repeated measures designs: An extensive primer. *Psychological Bulletin*, *97*(2), 316–333.

O'Connor, B. P. (2000). SPSS and SAS programs for determining the number of components using parallel analysis and Velicer's MAP test. *Behavior Research Methods, Instrumentation, and Computers*, *32*, 396–402.

Ofcom (2008). Media literacy audit: Report on children's media literacy. Retrieved from https://www.ofcom.org.uk/__data/assets/pdf_file/0021/55182/ml_childrens08.pdf

Olson, C. L. (1974). Comparative robustness of six tests in multivariate analysis of variance. *Journal of the American Statistical Association, 69*, 894–908.

Olson, C. L. (1976). On choosing a test statistic in multivariate analysis of variance. *Psychological Bulletin, 83*, 579–586.

Olson, C. L. (1979). Practical considerations in choosing a MANOVA test statistic: A rejoinder to Stevens. *Psychological Bulletin, 86*, 1350–1352.

Ong, E. Y. L., Ang, R. P., Ho, J. C. M., Lim, J. C. Y., Goh, D. H., Lee, C. S., & Chua, A. Y. K. (2011). Narcissism, extraversion and adolescents' self-presentation on Facebook. *Personality and Individual Differences, 50*(2), 180–185.

Oxoby, R. J. (2008). On the efficiency of AC/DC: Bon Scott versus Brian Johnson. *Economic Enquiry, 47*(3), 598–602.

Pearson, E. S., & Hartley, H. O. (1954). *Biometrika tables for statisticians, volume I*. New York: Cambridge University Press.

Pearson, K. (1894). Science and Monte Carlo. *The Fortnightly Review, 55*, 183–193.

Pearson, K. (1900). On the criterion that a given system of deviations from the probable in the case of a correlated system of variables is such that it can be reasonably supposed to have arisen from random sampling. *Philosophical Magazine, 50*(5), 157–175.

Pedhazur, E., & Schmelkin, L. (1991). *Measurement, design and analysis: An integrated approach*. Hillsdale, NJ: Erlbaum.

Peirce, C. S. (1878). Deduction, induction, and hypothesis. *Popular Science Monthly, 13*, 470–482.

Perham, N., & Sykora, M. (2012). Disliked music can be better for performance than liked music. *Applied Cognitive Psychology, 26*(4), 550–555.

Piff, P. K., Stancato, D. M., Côté, S., Mendoza-Dentona, R., & Keltner, D. (2012). Higher social class predicts increased unethical behavior. *Proceedings of the National Academy of Sciences, 109*(11), 4086–4091.

Plackett, R. L. (1983). Karl Pearson and the chi-squared test. *International Statistical Review, 51*(1), 59–72.

Preacher, K. J., & Hayes, A. F. (2004). SPSS and SAS procedures for estimating indirect effects in simple mediation models. *Behavior Research Methods Instruments & Computers, 36*(4), 717–731.

Preacher, K. J., & Hayes, A. F. (2008a). Asymptotic and resampling strategies for assessing and comparing indirect effects in multiple mediator models. *Behavior Research Methods, 40*(3), 879–891.

Preacher, K. J., & Hayes, A. F. (2008b). Contemporary approaches to assessing mediation in communication research. In A. F. Hayes, M. D. Slater, & L. B. Snyder (Eds.), *The SAGE sourcebook of advanced data analysis methods for communication research* (pp. 13–54). Thousand Oaks, CA: Sage.

Preacher, K. J., & Kelley, K. (2011). Effect size measures for mediation models: Quantitative strategies for communicating indirect effects. *Psychological Methods, 16*(2), 93–115.

R Core Team (2016). *R: A language and environment for statistical computing*. Vienna: R Foundation for Statistical Computing. Retrieved from http://www.r-project.org/

Ratcliff, R. (1993). Methods for dealing with reaction-time outliers. *Psychological Bulletin, 114*(3), 510–532.

Raudenbush, S. W., & Bryk, A. S. (2002). *Hierarchical linear models* (2nd ed.). Thousand Oaks, CA: Sage.

Rauscher, F. H., Shaw, G. L., & Ky, K. N. (1993). Music and spatial task performance. *Nature, 365*(6447), 611.

Rockwell, R. C. (1975). Assessment of multicollinearity: The Haitovsky test of the determinant. *Sociological Methods and Research, 3*(4), 308–320.

Rogosa, D. (1981). On the relationship between the Johnson-Neyman region of significance and statistical tests of parallel within group regressions. *Educational and Psychological Measurement, 41*(1), 73–84.

Rosenthal, R. (1991). *Meta-analytic procedures for social research* (2nd ed.). Newbury Park, CA: Sage.

Rosenthal, R., Rosnow, R. L., & Rubin, D. B. (2000). *Contrasts and effect sizes in behavioural research: A correlational approach*. Cambridge: Cambridge University Press.

Rosnow, R. L., Rosenthal, R., & Rubin, D. B. (2000). Contrasts and correlations in effect-size estimation. *Psychological Science, 11*, 446–453.

Rouanet, H., & Lépine, D. (1970). Comparison between treatments in a repeated-measurement design: ANOVA and multivariate methods. *British Journal of Mathematical and Statistical Psychology, 23*, 147–163.

Rouder, J. N., & Morey, R. D. (2012). Default Bayes factors for model selection in regression. *Multivariate Behavioral Research, 47*(6), 877–903.

Rouder, J. N., Speckman, P. L., Sun, D., Morey, R. D., & Iverson, G. (2009). Bayesian t tests for accepting and rejecting the null hypothesis. *Psychonomic Bulletin & Review, 16*(2), 225–237.

Rowe, R., Costello, E. J., Angold, A., Copeland, W. E., & Maughan, B. (2010). Developmental pathways in oppositional defiant disorder and conduct disorder. *Journal of Abnormal Psychology, 119*(4), 726–738.

Rozeboom, W. W. (1960). The fallacy of the null-hypothesis significance test. *Psychological Bulletin, 57*(5), 416–428.

Rulon, P. J. (1939). A simplified procedure for determining the reliability of a test by split-halves. *Harvard Educational Review, 9*, 99–103.

Ruxton, G. D., & Neuhaeuser, M. (2010). When should we use one-tailed hypothesis testing? *Methods in Ecology and Evolution, 1*(2), 114–117.

Sacco, W. P., Levine, B., Reed, D., & Thompson, K. (1991). Attitudes about condom use as an AIDS-relevant behavior: Their factor structure and relation to condom use. *Psychological Assessment: A Journal of Consulting and Clinical Psychology, 3*(2), 265–272.

Sacco, W. P., Rickman, R. L., Thompson, K., Levine, B., & Reed, D. L. (1993). Gender differences in AIDS-relevant condom attitudes and condom use. *AIDS Education and Prevention, 5*(4), 311–326.

Sachdev, Y. V. (1967). An unusual foreign body in the rectum. *Diseases of the Colon & Rectum, 10*(3), 220–221.

Salsburg, D. (2002). *The lady tasting tea: How statistics revolutionized science in the twentieth century*. New York: Owl Books.

Sana, F., Weston, T., & Cepeda, N. J. (2013). Laptop multitasking hinders classroom learning for both users and nearby peers. *Computers & Education, 62*, 24–31.

Savage, L. J. (1976). On re-reading R. A. Fisher. *Annals of Statistics, 4*, 441–500.

Scanlon, T. J., Luben, R. N., Scanlon, F. L., & Singleton, N. (1993). Is Friday the 13th bad for your health? *British Medical Journal, 307*, 1584–1586.

Scariano, S. M., & Davenport, J. M. (1987). The effects of violations of independence in the one-way ANOVA. *American Statistician*, *41*(2), 123–129.

Schützwohl, A. (2008). The disengagement of attentive resources from task-irrelevant cues to sexual and emotional infidelity. *Personality and Individual Differences*, *44*, 633–644.

Senn, S. (2006). Change from baseline and analysis of covariance revisited. *Statistics in Medicine*, *25*, 4334–4344.

Shee, J. C. (1964). Pargyline and the cheese reaction. *British Medical Journal*, *1*(539), 1441.

Silberzahn, R., & Uhlmann, E. L. (2015). Many hands make tight work. *Nature*, *526*(7572), 189–191.

Silberzahn, R., Uhlmann, E. L., Martin, D., Anselmi P., Aust, F., Awtrey, E. C., …, & Nosek, B. A. (2015). Many analysts, one dataset: Making transparent how variations in analytical choices affect results. Retrieved from http://osf.io/gvm2z

Simmons, J. P., Nelson, L. D., & Simonsohn, U. (2011). False-positive psychology: Undisclosed flexibility in data collection and analysis allows presenting anything as significant. *Psychological Science*, *22*(11), 1359–1366.

Simonsohn, U., Nelson, L. D., & Simmons, J. P. (2014). P-curve: A key to the file-drawer. *Journal of Experimental Psychology: General*, *143*(2), 534–547.

Sobel, M. E. (1982). Asymptotic intervals for indirect effects in structural equations models. In S. Leinhart (Ed.), *Sociological methodology 1982* (pp. 290–312). San Francisco: Jossey-Bass.

Sonnentag, S. (2012). Psychological detachment from work during leisure time: The benefits of mentally disengaging from work. *Current Directions in Psychological Science*, *21*(2), 114–118.

Spearman, C. (1910). Correlation calculated with faulty data. *British Journal of Psychology*, *3*, 271–295.

Stephens, R., Atkins, J., & Kingston, A. (2009). Swearing as a response to pain. *Neuroreport*, *20*(12), 1056–1060.

Stevens, J. P. (1979). Comment on Olson: Choosing a test statistic in multivariate analysis of variance. *Psychological Bulletin*, *86*, 355–360.

Stevens, J. P. (1980). Power of the multivariate analysis of variance tests. *Psychological Bulletin*, *88*, 728–737.

Stevens, J. P. (2002). *Applied multivariate statistics for the social sciences* (4th ed.). Hillsdale, NJ: Erlbaum.

Strahan, R. F. (1982). Assessing magnitude of effect from rank-order correlation coeffients. *Educational and Psychological Measurement*, *42*, 763–765.

Stuart, E. W., Shimp, T. A., & Engle, R. W. (1987). Classical conditioning of consumer attitudes: Four experiments in an advertising context. *Journal of Consumer Research*, *14*(3), 334–349.

Studenmund, A. H., & Cassidy, H. J. (1987). *Using econometrics: A practical guide*. Boston: Little Brown.

Student (1908). The probable error of a mean. *Biometrika*, *6*(1), 1–25.

Tabachnick, B. G., & Fidell, L. S. (2012). *Using multivariate statistics* (6th ed.). Boston: Allyn & Bacon.

Terpstra, T. J. (1952). The asymptotic normality and consistency of Kendall's test against trend, when ties are present in one ranking. *Indagationes Mathematicae*, *14*, 327–333.

Tinsley, H. E. A., & Tinsley, D. J. (1987). Uses of factor analysis in counseling psychology research. *Journal of Counseling Psychology*, *34*, 414–424.

Tomarken, A. J., & Serlin, R. C. (1986). Comparison of ANOVA alternatives under variance heterogeneity and specific noncentrality structures. *Psychological Bulletin*, *99*, 90–99.

Toothaker, L. E. (1993). *Multiple comparison procedures*. Sage University Paper Series on Quantitative Applications in the Social Sciences, 07–089. Newbury Park, CA: Sage.

Tufte, E. R. (2001). *The visual display of quantitative information* (2nd ed.). Cheshire, CT: Graphics Press.

Tuk, M. A., Trampe, D., & Warlop, L. (2011). Inhibitory spillover: Increased urination urgency facilitates impulse control in unrelated domains. *Psychological Science*, *22*(5), 627–633.

Tukey, J. W. (1960). A survey of sampling from contaminated normal distributions. In I. Olkin, S. G. Ghurye, W. Hoeffding, W. G. Madow, & H. B. Mann (Eds.), *Contributions to probability and statistics: Essays in honor of Harold Hotelling, Issue 2* (pp. 448–485). Stanford, CA: Stanford University Press.

Twenge, J. M. (2000). The age of anxiety? Birth cohort change in anxiety and neuroticism, 1952–1993. *Journal of Personality and Social Psychology*, *79*(6), 1007–1021.

Twisk, J. W. R. (2006). *Applied multilevel analysis: A practical guide*. Cambridge: Cambridge University Press.

Umpierre, S. A., Hill, J. A., & Anderson, D. J. (1985). Effect of Coke on sperm motility. *New England Journal of Medicine*, *313*(21), 1351.

Vandekerckhove, J., Guan, M., & Styrcula, S. A. (2013). The consistency test may be too weak to be useful: Its systematic application would not improve effect size estimation in meta-analyses. *Journal of Mathematical Psychology*, *57*(5), 170–173.

Vezhaventhan, G., & Jeyaraman, R. (2007). Unusual foreign body in urinary bladder: A case report. *Internet Journal of Urology*, *4*(2).

Wainer, H. (1972). A practical note on one-tailed tests. *American Psychologist*, *27*(8), 775–776.

Wainer, H. (1984). How to display data badly. *American Statistician*, *38*(2), 137–147.

Wasserstein, R. L. (Ed.) (2016). ASA statement on statistical significance and P-values. *American Statistician*, *70*(2), 131–133.

Wasserstein, R. L., & Lazar, N. A. (2016). The ASA's statement on p-values: Context, process, and purpose. *American Statistician*, *70*(2), 129–131.

Weaver, B., & Koopman, R. (2014). An SPSS macro to compute confidence intervals for Pearson's correlation. *Quantitative Methods for Psychology*, *10*(1), 29–39.

Weaver, B., & Wuensch, K. L. (2013). SPSS and SAS programs for comparing Pearson correlations and OLS regression coefficients. *Behavior Research Methods*, *45*(3), 880–895.

Welch, B. L. (1951). On the comparison of several mean values: An alternative approach. *Biometrika*, *38*, 330–336.

Wen, Z. L., & Fan, X. T. (2015). Monotonicity of effect sizes: Questioning kappa-squared as mediation effect size measure. *Psychological Methods*, *20*(2), 193–203.

Wilcox, R. R. (2010). *Fundamentals of modern statistical methods: Substantially improving power and accuracy* (2nd ed.). New York: Springer.

Wilcox, R. R. (2016). *Understanding and applying basic statistical methods using R*. Hoboken, NJ: John Wiley & Sons.

Wilcox, R. R. (2017). *Introduction to robust estimation and hypothesis testing* (4th ed.). Burlington, MA: Elsevier.

Wilcox, R. R., Carlson, M., Azen, S., & Clark, F. (2013). Avoid lost discoveries, because of violations of standard assumptions, by using modern robust statistical methods. *Journal of Clinical Epidemiology, 66*(3), 319–329.

Wilcoxon, F. (1945). Individual comparisons by ranking methods. *Biometrics, 1,* 80–83.

Wildt, A. R., & Ahtola, O. (1978). *Analysis of covariance.* Sage University Paper Series on Quantitative Applications in the Social Sciences, 07–012. Newbury Park, CA: Sage.

Wilkinson, L. (1999). Statistical methods in psychology journals: Guidelines and explanations. *American Psychologist, 54*(8), 594–604.

Wright, D. B. (1998). Modeling clustered data in autobiographical memory research: The multilevel approach. *Applied Cognitive Psychology, 12,* 339–357.

Wright, D. B. (2003). Making friends with your data: Improving how statistics are conducted and reported. *British Journal of Educational Psychology, 73,* 123–136.

Wright, D. B., London, K., & Field, A. P. (2011). Using bootstrap estimation and the plug-in principle for clinical psychology data. *Journal of Experimental Psychopathology, 2*(2), 252–270.

Wu, Y. W. B. (1984). The effects of heterogeneous regression slopes on the robustness of two test statistics in the analysis of covariance. *Educational and Psychological Measurement, 44*(3), 647–663.

Yang, X. W., Li, J. H., & Shoptaw, S. (2008). Imputation-based strategies for clinical trial longitudinal data with nonignorable missing values. *Statistics in Medicine, 27*(15), 2826–2849.

Yates, F. (1951). The influence of *Statistical methods for research workers* on the development of the science of statistics. *Journal of the American Statistical Association, 46,* 19–34.

Yuen, K. K. (1974). The two-sample trimmed t for unequal population variances. *Biometrika, 61*(1), 165–170.

Zabell, S. L. (1992). R. A. Fisher and fiducial argument. *Statistical Science, 7*(3), 369–387.

Zellner, A., & Siow, A. (1980). Posterior odds ratios for selected regression hypotheses. In J. M. Bernardo, M. H. DeGroot, D. V. Lindley, & A. F. M. Smith (Eds.), *Bayesian Statistics: proceedings of the first international meeting held in Valencia (Spain)* (pp. 585–603): University of Valencia.

Zhang, S., Schmader, T., & Hall, W. M. (2013). L'eggo my ego: Reducing the gender gap in math by unlinking the self from performance. *Self and Identity, 12*(4), 400–412.

Zibarras, L. D., Port, R. L., & Woods, S. A. (2008). Innovation and the 'dark side' of personality: Dysfunctional traits and their relation to self-reported innovative characteristics. *Journal of Creative Behavior, 42*(3), 201–215.

Ziliak, S. T., & McCloskey, D. N. (2008). *The cult of statistical significance: How the standard error costs us jobs, justice and lives.* Ann Arbor: University of Michigan.

Zimmerman, D. W. (2004). A note on preliminary tests of equality of variances. *British Journal of Mathematical & Statistical Psychology, 57,* 173–181.

Zwick, R. (1985). Nonparametric one-way multivariate analysis of variance: A computational approach based on the Pillai-Bartlett trace. *Psychological Bulletin, 97*(1), 148–152.

Zwick, W. R., & Velicer, W. F. (1986). Comparison of five rules for determining the number of components to retain. *Psychological Bulletin, 99*(3), 432–442.

INDEX

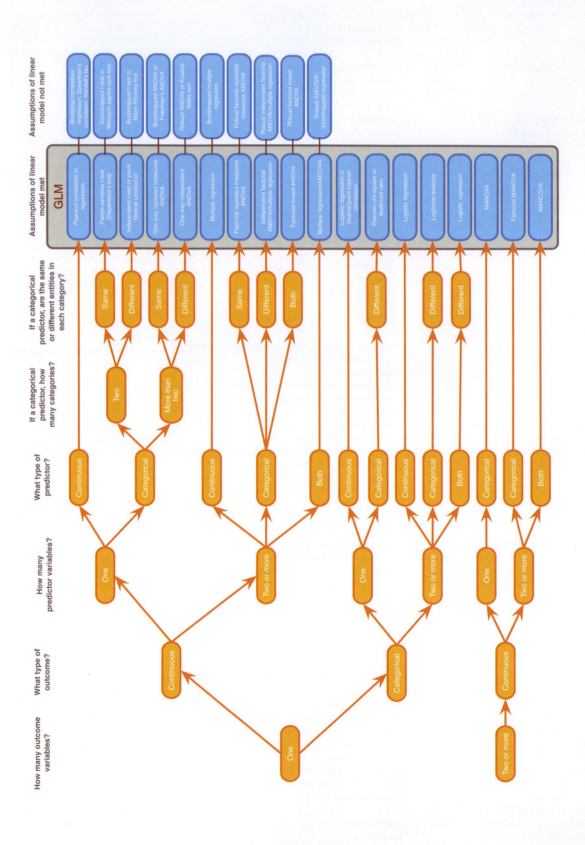